2025 IEEE Workshop on Wide Bandgap Power Devices and Applications in Asia (WiPDA Asia 2025)

Beijing, China
15-17 August 2025

Pages 1-519

IEEE Catalog Number: CFP25O09-POD
ISBN: 979-8-3315-1110-4

**Copyright © 2025 by the Institute of Electrical and Electronics Engineers, Inc.
All Rights Reserved**

Copyright and Reprint Permissions: Abstracting is permitted with credit to the source. Libraries are permitted to photocopy beyond the limit of U.S. copyright law for private use of patrons those articles in this volume that carry a code at the bottom of the first page, provided the per-copy fee indicated in the code is paid through Copyright Clearance Center, 222 Rosewood Drive, Danvers, MA 01923.

For other copying, reprint or republication permission, write to IEEE Copyrights Manager, IEEE Service Center, 445 Hoes Lane, Piscataway, NJ 08854. All rights reserved.

*** This is a print representation of what appears in the IEEE Digital Library. Some format issues inherent in the e-media version may also appear in this print version.

IEEE Catalog Number: CFP25O09-POD
ISBN (Print-On-Demand): 979-8-3315-1110-4
ISBN (Online):   979-8-3315-1109-8
ISSN: 2831-3704

Additional Copies of This Publication Are Available From:

Curran Associates, Inc
57 Morehouse Lane
Red Hook, NY 12571 USA
Phone: (845) 758-0400
Fax: (845) 758-2633
E-mail: curran@proceedings.com
Web: www.proceedings.com

2025 IEEE Workshop on Wide Bandgap Power Devices and Applications in Asia (WiPDA Asia 2025)

Beijing, China
15-17 August 2025

Pages 1-519

IEEE Catalog Number: CFP25O09-POD
ISBN: 979-8-3315-1110-4

Copyright © 2025 by the Institute of Electrical and Electronics Engineers, Inc.
All Rights Reserved

Copyright and Reprint Permissions: Abstracting is permitted with credit to the source. Libraries are permitted to photocopy beyond the limit of U.S. copyright law for private use of patrons those articles in this volume that carry a code at the bottom of the first page, provided the per-copy fee indicated in the code is paid through Copyright Clearance Center, 222 Rosewood Drive, Danvers, MA 01923.

For other copying, reprint or republication permission, write to IEEE Copyrights Manager, IEEE Service Center, 445 Hoes Lane, Piscataway, NJ 08854. All rights reserved.

*** This is a print representation of what appears in the IEEE Digital Library. Some format issues inherent in the e-media version may also appear in this print version.

IEEE Catalog Number: CFP25O09-POD
ISBN (Print-On-Demand): 979-8-3315-1110-4
ISBN (Online):   979-8-3315-1109-8
ISSN: 2831-3704

Additional Copies of This Publication Are Available From:

Curran Associates, Inc
57 Morehouse Lane
Red Hook, NY 12571 USA
Phone: (845) 758-0400
Fax: (845) 758-2633
E-mail: curran@proceedings.com
Web: www.proceedings.com

TABLE OF CONTENTS

Design of Short-Circuit and Overload Protection Chips for SiC MOSFETs 1
Jiahui Lv, Yuan Yang, Minmin Zhang

Repetitive Recovery Analysis and Island Width Optimization of 4H-SiC Floating Island Devices 6
Runze Xia, Hengyu Wang, Ce Wang, Kuang Sheng

High-Temperature Device Model for GaN-Based MIS-HEMTs 11
Hanlin Cao, Pingyu Cao, Sang Lam, Miao Cui

Dynamic Model of Non-Segmented SiC MOSFETs Considering Switching Behavior and
Temperature Effects 16
Wenzheng Dou, Xin Lan, Ning Zhao

Investigation of the Threshold Voltage Drift in Trench-Type 1200V SiC MOSFETs Under Bipolar
Dynamic Gate Stress 21
*Jiaying Cao, Hao Guan, Jiuyang Tang, Liudan Kong, Yifei Chang, Yuhan Duan, Qingchun
Zhang, Pan Liu*

Analysis of Planar and SBD-Embedded SiC MOSFET Under Avalanche Stress 26
*Hao Huang, Yuting Jin, Dong Hai, Dingyuan Bao, Taohui Zhang, Haoze Luo, Francesco
Iannuzzo, Wuhua Li*

Mechanism Analysis of Different Undamped Oscillations for GaN Devices 32
Qiang Hu, Jian Chen, Ziyang Wang, Hao Yue, Wensheng Song

Multi-Objective Optimization for Three-Level Power Module Based on Electro-Thermal Co-
Design 38
Runze Wang, Jianing Wang, Shaolin Yu, Xing Li, Xiahao Wang, Baolong Yan, Zhenchun Xia

A Novel Zero-Pressure Packaging Structure of High-Voltage SiC IGBT Device 43
Tang Xinling, Wangliang, Du Yujie, Wei Xiaoguang

Design and Characterization of a Novel Embedded SiC Power Module with Double-Sided RDL 47
Yifei Du, Min Chen, Xinnan Sun, Jie Li, Yucheng Wu, Han Zhou, Feng Jiang

Low-Profile Negative Coupled Inductor for AI Chip Vertical Power Delivery Voltage Regulator 52
Ruibo Cao, Xiangan You, Fengze Hou, Na Yan, Jian Song, Qidong Wang, Liqiang Cao

Shape Optimization of High-Stress-Buffering Bridge-Typed Spacers for Double-Sided Power
Modules 57
Gaojia Zhu, Ke Xu, Bingru Li, Youzheng Wang, Longnv Li, Yun-Hui Mei

A Study of Multi-Chip Carrier Structure for SiC PCB-Embedded Power Module 62
*Jia-Jun Lai, Xuelun Zhang, Hao Sun, Ye Zhu, Yunbin Pan, Mingfu Li, Ziliang Shi, Xiong Yang,
Xunjin Xu, Min Chen*

CNN-Based Rapid Co-Optimization of BV and RON,sp for 4H-SiC SJ MOSFET 67
Tiefu Wang, Haoyuan Cheng, Chi Zhang, Hengyu Wang, Kuang Sheng

Thermal Dissipation in GaN-On-Sapphire Power HEMTs: Synergy of Substrate Thinning and
High-Conductivity Thermal Interface Materials 71
*Chang Liu, Junsong Jiang, Kun Tan, Jie Lu, Suxia Guo, Cungang Hu, Zhaofu Zhang, Xi Tang,
Wenping Cao*

Analysis of Influencing Factors on the Thermal Performance of SiC Multi-Chip Parallel Packaging Structure 77
Lei Wang, Rui Jin, Xuebao Li, Peng Zhang, Chao Li

A Synchronously Implanted Termination for 1200V SiC Trench-Gate MOSFET 83
Zhengyun Zhu, Jian Luo, Kai Huang, Qian Wang, Mingyang Gao, Aoxue Xu, Liang Zhou

Research on a Crosstalk Parasitic Oscillation Suppression Method Based on Series Resonance Principle 87
Tiancong Shao, Pei Hu, Yuhan Sun, Kai Wang, Bin Qi, Fangwei Zhao, Zhe Zhang

Effect of Argon-Oxygen Ratios on Crystallization and Surface Morphology of Gallium Oxide Thin Films by Low-Pressure Chemical Vapor Deposition 91
Zhihao Yang, Jichao Hu, Jinhui Yu, Xiaodong Yang, Xiaomin He, Bo Peng

Design of an Active Gate Driver IC for E-Mode GaN HEMTs Based on Adaptive Phased Regulation 95
Wei Liu, Yuan Yang, Wenbin Xing

A Vertical Ultra-Thin Voltage Regulator Module (VRM) with Dual-Phase Integrated Inductor for High-Performance Computing Applications 100
Nianzheng Wang, Hang Kong, Laili Wang

Design of a GaN 1-MHz/48-V-To-1-V/320-A Series-Capacitor Buck Converter 104
Wendi Fan, Yenan Chen

GaN-Based Low-Voltage, High-Current DC-DC Converter for Aerospace Applications 109
Xiao Chen, Zhihao Zhang, Zhongjie Wang, Mingming Ji, Lingxiao Xue

A New Method for Estimating the DC-DC Converter Performance Under Low Voltage and High Current Application 114
Maohang Qiu, Hongbo Ma

A Totem Pole PFC Control Based on Dynamic Root Locus with Wide Output Voltage Range and Low THD 118
Hao Chu, Xiangan Xiao, Xuchen Sun, Jiajia Guan, Cai Chen, Yong Kang

A SiC/Si Hybrid Pulsed Power Supply for High-Current Laser Diode Driving 122
Yihui Tang, Xingye Chen, Yan Su, Helong Li, Zhiqing Yang, Xianbin Qi

Interleaved LLC Resonant Converter with Hybrid PSM/PFM Control for Wide Voltage Gain Applications 128
Jintong Dong, Yijie Wang, Xiaohui Xu, Kai Ji

A Reconfigurable Half-Bridge Buck Based on Dickson SC Converter with DCX-LLC for Wide Input Voltage Range Applications 133
Lichang Man, Yijie Wang, Shanshan Gao, Hongqi Ben, Dianguo Xu

Degradation Evaluation of Single-Event Effects Induced by Heavy-Ion Irradiation on E-Mode p-GaN Gate HEMT Devices 138
Yujie Cheng, Zhangzhe Yan, Jianjun Zhou, Denggui Wang, Zhuangzhuang Hu, Hongfei Wu, Haifeng Cheng

Development of a High-Power-Density SiC Module 142
Jiajun Yang, Xiaoshuang Hui, Puqi Ning

Minimizing Sensor Usage in TSEPs-Based Junction Temperature Estimation of SiC Power Transistor Through Time-Series Analysis 147
  Valentyna Afanasenko, Oleksandr Solomakha, Kevin Muñoz Barón, Ingmar Kallfass

Molecular Dynamics Simulations of AgCu Core-Shell Nanoparticle Sintered with Nanoflake 151
  Yifeng Chen, Xin Lan, Ziyang Zhang, Xin Li

The Impact of Power Module Layout on Stray Inductance 156
  Yuhui Kang, Puqi Ning, Xiaoshuang Hui, Jiajun Yang

Cu/Diamond Composite Electrodes for Low Thermal Resistance Packaging in High-Power IGCTs 160
  Tang Xinling, Wei Xiaoguang, Lin Zhongkang, Yang Guang, Yu Kefan, Wang Jingfei

1200V SiC MOSFETs Under Dynamic Reverse Bias Testing in High-Temperature and High-Frequency Environments 164
  Liudan Kong, Jiuyang Tang, Jiaying Cao, Yifei Chang, Hao Guan, Yuhan Duan, Qingchun Zhang, Pan Liu

Investigation of Electrical Property and Reliability in AlGaN/GaN MIS-HEMTs Fabricated by Using N2/O2 Composite Plasma Treatment 169
  Qingyuan Zuo, Huolin Huang, Yun Lei, Jiayu Zhang, Jianyu Zhao, Jianxun Dai

Investigation on Surge Current Failure in 1200V SiC MOSFETs with Varied Gate Biases in the Third Quadrant 173
  Chengyuan Zhou, Hengyu Wang, Yang Zou, Kuang Sheng

Study on SiC Power Module Cooling Technologies Suitable for High-Junction-Temperature and High-Power-Density Applications 177
  Baihan Liu, Jianwei Lv, Yipeng Liu, Yifan Zhang, Siqi Dai, Zexiang Zheng, Suhang Wei, Cai Chen, Yong Kang

An Enhanced Over-Current Fault Detection System for a 7.2kV Power Module Using Series-Connected SiC MOSFETs 181
  Chunyao Hou, An Lou, Yue Wu, Aozu Luan, Shuai Shao

Modeling of Static Current Sharing of Direct-Paralleled 1.7kV 800A SiC Power Modules 186
  Zhongjie Wang, Hui Liao, Lingxiao Xue, Xiangqian Zhang, Liwen Zou, Xiaonan Dong

A 500W 13.56 MHz Amplifier Realized by Two Combining Class-E Circuits 190
  Yanfei Ji, Mei Liang, Jiwen Chen, Pengyu Jia, Yihang Zhang

A Method to Enlarge the Soft-Switching Range for the Dual Active Bridge Series Resonant Converter 195
  Pengyu Jia, Yimei Xing, Kai Qiu, Mingjun Liu

Bidirectional Energy-Storage Converter Based on Partial Power Processing 201
  Zhihao Zhang, Xiao Chen, Fan Zhai, Zhongjie Wang, Lingxiao Xue

3kV/42kW Pump-Back Test for Medium Voltage DC Transformer with Series-Connected SiC MOSFETs 206
  Yujian Zong, Shuai Shao, Chaojun Wang, Wentao Cui, Junming Zhang

Thermal Network of Double-Sided Cooling Power Module Based on Cauer Model 211
  Jie Li, Min Chen, Xinnan Sun, Jun Huang, Yucheng Wu, Yifei Du, Han Zhou, Feng Jiang

Multi-Objective Automatic Design of Power Module Packaging Based on Artificial Neural Network and Deep Reinforcement Learning 215

Weina Mao, Jianing Wang, Shaolin Yu, Zhenchun Xia, Honghong Li, Xiang Pan, Xiahao Wang

A High-Performance Double-Sided Cooling SiC Power Module Packaging Design for EV Inverters 221

Haobin Chen, Haidong Yan, Maosheng Zhang, Ji Cheng, Yakun Zhang, Chaohui Liu

Optimal Design of High-Temperature SiC Power Module Based on Gene Algorithm 226

Zhenchun Xia, Jianing Wang, Shaolin Yu, Weina Mao, Xiahao Wang, Honghong Li, Runze Wang

Monolithically Integrated Over-Temperature Protection Circuit Based on GaN HEMTs 232

Pingyu Cao, Kepeng Zhao, Yihao Xu, Harm Van Zalinge, Ping Zhang, Miao Cui, Fei Xue

Lifetime Modeling of IGBT Devices Based on Micro-Defect Topography Inversion 236

Miaomiao Shangguan, Wei Lai, Hao Wang, Yunjie Wu, Yu Liu, Hang Zhao

A 6.78-MHz 2.3-KW Full-Bridge GaN Inverter with Bottom-Side Cooled Transistor GS-065-030-2-L for Wireless Power Transfer 241

Jianping Ning, Zhen Sun, Yao Wang, Shuang Zhao

Two-Step Turn-Off Delay Time Control for Efficiency Enhancement in Si/SiC Hybrid Switch-Based Inverters 246

Zijie Zheng, Jun Wang, Xuanting Song, Yongzhou Zou, Yuxing Dai, Kamal Al-Haddad

6.78 MHz 2.8 kW Resonant DC-DC Power Conversion Utilizing 650 V GaN HEMT-Based H-Bridge Inverter 250

Zhen Sun, Jianping Ning, Yao Wang, Yun Yang

A Passive Cancellation Circuit Incorporating the Motor Stator Windings for Reducing the Common-Mode Noise in Motor Drive Systems 255

Lihong Xie, Yuzhen Wu, Xinbo Ruan

A Voltage-Sharing Cascode Switch Structure Using SiC JFET and GaN HEMT 260

Yin Fang, Shan Jayamaha, Carl Ngai Man Ho

CLLLC Resonant Converter with Full-Bridge Half-Bridge Switching for Improved Voltage Gain Range and Light-Load Efficiency 264

Jinfeng Yu, Sinuo Liu, Yu Gu, Hui Liu

Analysis and Prediction Method of Gate Voltage Peak in Switching Transient of GaN E-HEMT 269

Yushan Liu, Yirui Hu, Xiao Li

An Active Gate Charge Controlled Driver for SiC MOSFET with Zero Turn-Off Loss 275

Yuze Zheng, Xu Cheng, Fan Zhang, Xiaolu Zhang, Yukun Niu, Xuhui Song

A Modulation Optimization Strategy for Phase Shift Full Bridge with Low Freewheeling Loss 280

Enyou Wu, Xuchen Sun, Jiajia Guan, Siyuan Feng, Tianxi Li, Cai Chen, Yong Kang

Optimized Trajectory Control for Soft Start-Up of Multiple-Mode Resonant Switched Capacitor Converter 284

Jingjing Qi, Kai Zhang, Haining Zhang

Datasheet-Driven Non-Segmented SPICE Model of SiC MOSFET with Improved Accuracy 290

Ziqi Jia, Yu Jiang, Yifan Hu, Hailong He, Chunping Niu, Yi Wu

Investigation and Improvement on Surge Robustness of Double-Trench SiC MOSFETs in Synchronous Rectification Mode 295
  Hexin Zhu, Xiaochuan Deng, Haohao Dai, Qian Huang, Xuan Li, Xu Li

A High-Output-Current Radiation Hardened Half Bridge DC-DC Converter for Space Applications 299
  Dawei Li, Xiang Zhou, Yeerzhati Nuerdebieke

The Thermal Simulation Study of DC-DC Converters Based on Phase Change Cooling 304
  Boyang Liu, Haoyu Zhang, Yu Han, Jialiang Chi, Yao Zhao, Zhiqiang Wang

A 2:1 Capacitive Isolated Resonant Switched-Capacitor DC-DC Converter 308
  Xinyu Zhang, Yu Fu, Yucheng Zhao, Shouxiang Li

An Auxiliary Power Supply Utilizing Planar Winding Transformer with High Insulation and Low Coupling Capacitance Considerations 314
  Yong Chen, Xuhui Song, Fan Zhang, Yuze Zheng, Xiaolu Zhang, Yukun Niu, Zheyuan Yu, Min Wu, Kaixiang Gong

Gate-Source Voltage Oscillation of Multi-Chip SiC Power Modules Considering Differential Gate Resistance 319
  Longnv Li, Chuyuan Liu, Lu Wang, Youzheng Wang, Gaojia Zhu, Yun-Hui Mei

Development and Electrical Characteristics Study of a 10kV/125A SiC MOSFET Module 324
  Yaodong Zhang, Yong Chen, Xu Cheng, Xiaotian Zhang

Study of Thermal Stress on Devices in Si/SiC Hybrid Half-Bridge Inverter 328
  Yulin Wang, Ruixiao Dong, Jun Wang, Yuxing Dai, Liuchen Chang

A 8:1 Multi-Resonant Switched-Capacitor Converter 334
  Yuxin Yan, Yu Fu, Yucheng Zhao, Shouxiang Li

Sustainability of Power Devices: A Perspective on Design for Recycling 339
  Jinpeng Cheng, Shuyu Liu, Hao Feng, Li Ran

Cryogenic Output Capacitance Loss of GaN HD-GIT 343
  Yudong Wang, Zilong Chen, Yukun Zhang, Chong Dou, Qian Cui, Yuqi Wei

Improved Control Strategy Based on BM Theory for Grid Forming Type VSG System 349
  Shuanglong Li, Yajing Zhang, Bin Liu, Baoying Huang, Siyu Pan, Hao Ma

Baseplate Temperature Gradient-Based Health Status Monitoring for Power Module Bonding Wires 355
  Yongxin Chen, Lei Xu, Kun Tan, Xi Tang, Cungang Hu, Wenping Cao

Study on Influencing Factors and Mechanisms of Single-Event Gate Rupture in SiC STP-MOSFETs 361
  Ying Yang, Zixuan Liu, Chen Wang, Xulong Wang

SiC Power Module with Staggered Terminals Layout Design to Reduce Parasitic Inductance 366
  Jiahang Wang Guangzhou, Xi Jiang Guangzhou, Runze Ouyang, Song Yuan, Yuanzhi Zhao, Qingrong Hu, Ying Wang, Xiaowu Gong

Novel Tri-Gate Multichannel Device for Improved Vth Controllability 371
  Quanbo He, Hengyu Wang, Florin Udrea

Linear-ESO Based Control for Wide Input Range Partial Power Regulated DCDC Converter 376
  Xikun Sang, Shanshan Gao, Yijie Wang, Dianguo Xu

A Multi-Objective Optimization Method Based on NSGA-II Algorithm for Electro-Thermal-Stress Collaborative Design of Intelligent Power Modules 382
Tao Xu, Lei Ming, Ningbo Li, Zihang Gu, Zhiwei Jiao, Zhen Xin

GaN-Based Partial Power DC-DC Converter with Four-Quadrant Operation Capability 388
Chao Liu, Zhe Zhang, Shunqing Wu, Zeqi Yang, Chuang Liu

UIS Ruggedness of Si/SiC Hybrid Switches 393
Chuanqi Zhang, Xuanting Song, Shiwei Liang, Yuxing Dai, Jun Wang, Kamal Al-Haddad

Research on Thermal Conductivity and Mechanical Properties Control of Epoxy Resin in Extreme Environments 398
Zhen Li, Liang Zou, Qingsong Liu, Zhiyun Han, Jinyang Bai

A 3L-ANPC SiC MOSFET Power Module with Low Thermal Coupling and Parasitic Inductance 404
Zedong Xue, Zhiyuan Qi, Zihao Chen, Hao Yuan, Qingwen Song, Yuming Zhang

Research on Transient Electric Field Calculation in Welded Devices Considering Dielectric Relaxation of Insulating Materials 410
Zihan Sang, Zhaocheng Liu, Hao Li, Xuebao Li, Ying Cao, Peng Shu

A Two-Stage Turn-On Gate Driver for SiC MOSFET with Short-Circuit Current Suppression 416
Yong Chen, Xiaolu Zhang, Xu Cheng, Yuze Zheng, Xuhui Song, Yukun Niu, Kaixiang Gong, Fan Zhang

A Zero-Sequence Injection Method to Reduce Electromagnetic Interference 420
Hui Liu, Dong Jiang, Junzhao Zhang

Development of a Novel Analytical Trapped Charge Model for Total Ionizing Dose Effects of SiC MOSFETs 424
Qingmao Hu, Xin Yang, Qingzhong Gui

Method for Estimating Power Loss of IGBT Module in Wind Power Converter Based on Measured Temperature Information 430
Ye Tian, Dawei Chen, Zhijie Zeng, Lixuan Zhu, Guojun Bao, Zhixiang Zou

A Method for Enhancing the Heat Dissipation Performance of Power Devices Through Graphene Coating 436
Xin Li, Jianing Wang, Shaolin Yu, Runze Wang, Xiahao Wang, Baolong Yan, Zhenchun Xia

Research on DC Characteristics of β-(Al0.22Ga0.78)2O3/β-Ga2O3 Modfet 440
Haitao Zhang, Xiaomin He

A Structure-Reconfigurable Electronic Transformer for Renewable Energy DC Distribution Syste 445
Yu Feng, Xianbin Qi, Zhiqing Yang, Jinxiao Wei, Peng Qin, Helong Li, Liu Fang, Lijian Ding

Pyrolysis Process Analysis of Polyimide at High Temperature Based on Molecular Dynamics Simulation 450
Yuteng Jiang, Minglei Xie, Wenzhi He, Zhi Wang, Bingxin Chen, Zhiyun Han, Sixiao Xin, Liang Zou

Analysis of SiC Output Capacitance Effects on Soft-Switching Characteristics in Three-Level Resonant Converters 456
Zhe Shao, Zhiyuan Wang, Binbin Li

On the First Demonstration and Analysis of the HTGB Induced Electrical Degradation of High Voltage SiC IGBT Devices 462

Tuanzhuang Wu, Jiaxing Wei, Junhou Cao, Hao Fu, Zhaoxiang Wei, Desheng Ding, Siyang Liu, Weifeng Sun, Xiaolei Yang, Song Bai

A New Si/SiC Hybrid Interleaved Three-Level ANPC Inverter with Cost and Performance Tradeoff 468

Ruixiao Dong, Jun Wang, Yulin Wang, Yuxing Dai, Liuchen Chang, Chao Zhang

Turn-Off Analysis and Modeling of Releasing Loss in Snubber Capacitor Self-Balancing Circuits for Series-Connected SiC MOSFETs Applied to High-Voltage Pulsed Power Systems 474

Jiaxuan Niu, Xu Cheng, Xu Yang, Yong Chen, Fan Zhang, Kexin Zhao

Optimization and Compensation of Leakage-Induced Deviation of CTTC Magnetic Integrated Structure in CLLC Resonant Converter 480

Liwen Jia, Bodong Li, Yahong Yang, Jianyu Lan, Jiarui Zhang, Kelin Chen, Feng Jiang, Min Chen

Sustained Oscillation Characterization of GaN HEMT at Cryogenic Temperature 485

Zilong Chen, Yuqi Wei, Yanjie He, Yukun Zhang, Chong Dou, Qian Cui

Design of an All-SiC On-Board Auxiliary Inverter for Urban Rail Vehicles 491

Xuefei Li, Yongang Chen, Zixiao Li, Shuiyuan He, Yuwen Qi, Lijun Diao

A Symmetrical Double-Sided Cooled SiC Power Module for Multi-Parallel Applications 495

Guolian Guan, Zhiqiang Zhao, Mingzhi Zhao, Tongyu Zhang, Laili Wang, Dewen Wang

A Transistor Clamp Circuit of On-State Voltage Drop for SiC MOSFET Temperature Monitoring 500

Yixiang Zhao, Hong Li, Xiaofei Hu, Kuang Zhang

A Dynamic Current Balancing Method for Multichip SiC MOSFET Modules with Separate Gate Drive Structures 506

Zicong Li, Zenan Shi, Yifei Luo, Xin Li

A Novel Analytical Physical Model of Gate-Drain Capacitance and Output Characteristics for SiC-MOSFET 512

Ze Tao, Zenan Shi, Yifei Luo, Xin Li

Reliability Analysis for 1200V SiC MOSFETs Under Repetitive Surge Current Operation with Negative Gate-Source Bias 516

Xinbin Zhan, Yanjing He, Jiankun Lai, Xi Jiang, Song Yuan, Hao Yuan, Qingwen Song, Xiaoyan Tang, Xiaowu Gong, Yuming Zhang

A Novel Modulation Method with Simultaneous Reduction of Common-Mode Voltage and Switching Losses for Three-Level SNPC Inverter 520

Shuangxi Zhu, Jiajia Guan, Cai Chen, Yong Kang

An Active Gate Driver of Voltage Overshoot Suppressing for SiC MOSFETs 524

Tingwen Hu, Wensheng Song, Jian Chen, Tao Tang, Hao Yue, Guoyou Liu

A High-Frequency LLC Resonant Converter Incorporating Matrix Transformer 528

Haochen Zhang, Yueshi Guan, Yijie Wang, Dianguo Xu

An Active Gate Driver Addressing GaN HEMT Gate-Source Voltage Overshoots for Both Turn-On and Turn-Off Periods 534

Lurenhang Wang, Yishun Yan, Shuaiqing Zhi, Mingcheng Ma, Xizhi Sun, Dianguo Xu

Application of Wide Bandgap High Frequency Inverter in 10 kW Magnetic Field-Coupled Undersea Wireless Power Transfer Systems.. 539
  Lei Yang, Jiahua Sun, Yuanfeng Wang

Minimizing Eddy Current Loss for Implanted Wireless Charger Through Phase Difference Optimization... 544
  Pengyu Chen, Siyi Yao, Xiyuan Lin, Hongjun Zheng, Congcong Zhang, Minfan Fu

An Accurate Leakage Inductance Model for Magnetic Integrated Planar Transformers in LLC Resonant Converters.. 550
  Zhili Mo, Xuetong Zhou, Yufei Tian, Li Zheng, Xinhong Cheng

Subharmonic Oscillations in High-Frequency Switched-Mode Power Amplifiers 556
  Wei Liu, Ming Liu

Novel Structure for High-Voltage Vertical β-Ga2O3 Schottky Barrier Diodes: A TCAD Study..................... 562
  Yan Liu, Meng-Qi Fan, Jia-Xiang Chen, Teng Jiao, Mao-Jin Yang, Xian-Hu Zha, Xiao-Ping Wang, Xiang-Jin Ding, Yu-Xi Wan, Dao-Hua Zhang

Efficiency-Oriented Adaptive Dead Time Control for the Dual Active Bridge Converter 566
  Shaoyan Jiang, Xichen Fu, Xingque Xu, Jiabin Ruan, Chuanwei Xiao

Simulation Study of Breakdown Characteristic of Vertical Diamond Schottky Barrier Diodes with Different Drift Layer Parameters and Termination Structures... 572
  Tianhe Mi, Peng Wang, Yan Liu, Teng Jiao, Xinchun Cui, Haolin Hu, Yuxi Wan, Daohua Zhang

Multi-Parameter Degradation Modeling Method for Power MOSFETs Integrating Semiconductor Physics Degradation Data.. 577
  Chenyi Wang, Cen Chen, Weixuan Kong, Haodong Wang, Zhenning Zhou

1.5KW HSC Converter Power Density-Efficiency Advancement: Enabled by Planar Transformer............... 582
  Pengfei Wang, Vickie Qu, Qianru Shi, Qingchang Liu, Ian Yj Chan, Minfan Fan

Control Strategy of Dual Phase Shifting Dual Active Bridge Converter Based on BM................................ 588
  Yiting Huo, Yajing Zhang, Bin Liu, Baoying Huang, Siyu Pan, Hao Ma

β-Type High-Gain Boost Converter with Diode-Capacitor Cell 594
  Peng Sun, Hong Li, Yidi Liang, Xu Shangguan, Mingbo Wei, Huizhu Zhuang

Soft-Switching Fixed-Frequency Control Strategy for Three-Level Buck-Boost Converter 600
  Fang Li, Yao Xue, Fangwei Zhao, Yajing Zhang, Jun Xu

High-Robust Power Integrated Synchronization Scheme for the Grid-Forming Inverter 606
  Wen Zou, Yuying He, Li Zhang, Dongsheng Yang

Charge Control for Critical Conduction Mode Soft-Switching Grid-Tied Inverters 609
  Zhengzi Lei, Zhongshu Zheng, Wenbo An, Li Zhang, Yuying He

Dynamic Analysis of Hybrid Photovoltaic-Battery System with Dual-Mode Grid-Forming Control............. 614
  Wenjie Ning, Yiyang Liao, Yaoyu Hu, Shaoze Zhou, Zheng Wei, Yitong Li

Optimization Design of High-Speed and High-Voltage Switching Transient Characteristics for GaN HEMTs Using Multi-Objective Particle Swarm Optimization 619
  Xiao Li, Zhuofan Xiong, Yushan Liu, Qiang Zhou

A Compact EMI Filter Design Method Based on Chaotic SVPWM in Motor Drive System 624
Yanjun Li, Hong Li, Zuoxing Wang, Aojie Liao, Mingxin Shi

A SiC MOSFET Accelerated Degradation Test Platform that Accounts for Turn-On and Turn-Off
Times 630
Bohang Lu, Cen Chen, Zicheng Wang, Xuanyu Lin

Bias Temperature Instability of SiC Trench MOSFET Under DC and AC Gate Stress.................................. 634
Kanghua Yu, Qian Wang, Yuwei Wang, Jun Wang

A Novel Approach to Estimate the Thermal Destruction Point of SiC MOSFETs Under Short
Circuit Conditions 638
Rony Thomas, Zhe Yu, Sebastian Fahlbusch

Investigation on Electrical Properties and Safe-Operating Area for Novel Split-Gate IGBTs 644
Xuanting Song, Jun Wang, Gaoqiang Deng, Zijie Zheng, Yongzhou Zou, Shiwei Liang, Yuxing
Dai, Kamal Al-Haddad

Influencing Factors and Mechanisms of Single-Event Burnout in STP SiC MOSFETs 649
Ying Yang, Xulong Wang, Chen Wang, Zixuan Liu

A Multi-Operation Characterization System for Quantifying Dynamic RON Degradation in p-GaN
Gate GaN HEMTs Under Real-World Switching Conditions... 655
Junbo Wang, Xiangdong Li, Xi Jiang, Haonan Jiang, Shuzhen You, Yue Hao, Jincheng Zhang

The Establishment of Quasi-3D TCAD Simulation for Bipolar Degradation Occurs in SiC
MOSFET Body Diode.. 659
Junhou Cao, Xinyu Zhou, Chenlu Wang, Lei Huang, Hao Fu, Zhaoxiang Wei, Tuanzhuang
Wu, Jiaxing Wei, Siyang Liu, Weifeng Sun

Study on AlxGa1-XN Graded Composite Barrier GaN HEMT 663
Ruihao Zhang, Fayu Wan

Chip Screening Strategy for SiC MOSFET Based on Simulated Annealing Algorithm................................. 667
Zhanshan Zhu, Chuangye Li, Helong Li, Zhiqiang Liu, Lijian Ding

Research on Power Loss Calculation and Optimization of FB-VSCC Active EMI Filter............................... 673
Daozhen He, Hong Li, Yuanheng He, Mingxin Shi

Analysis of Common-Mode EMI in Multi-Converter Systems with EMI Filters 678
Runquan Jiang, Peng Zhou, Guifeng Geng, Xuejun Pei

Research on Key Technology of Low Noise Transformer and Its Application on 110kV Transformer 684
Shouhui Han, Qingsong Liu, Zheng Liu, Liang Zou, Zhiyun Han

Study on Cross-Scale Collaborative Control of Composite Properties for Power Device Packaging............. 688
Xianfeng Li, Jian Wang, Hanwen Ren, Wei Wang, Chen Chen, Qingmin Li

Partial Discharge Denoising Technology for GIS Equipment Based on CEEMDAN 692
Zhongyue Liu, Pei Cao, Zeyu Li, Hanwen Ren

Improvement of Thermal Conductivity of Epoxy Composites Using the Synergistic Effect of Boron
Nitride Nanosheets and Boron Nitride Whiskers 696
Tiandong Zhang, Chenghai Wang, Xinle Zhang, Tangman Xue, Changhai Zhang, Qingguo
Chi

Study on Surface Discharge Characteristics of Silicone Gel in Salt Fog Environment 700
Feng Wang, Zhihui Li, Yateng Yany, Shanzhen Fan, Hanwen Ren, Jian Wang, Wei Wang,
Qingmin Li

Effect of Atomic-Level Roughness on Vertical β-Ga2O3 Schottky Barrier Diode and High-
Temperature Performance 704
Jiaxiang Chen, Xingye Zhang, Maojin Yang, Xinpeng Lin, Teng Jiao, Yan Liu, Xianhu Zha,
Haolin Hu, Xiangjin Ding, Yuxi Wan, Jun Ma, Mengyuan Hua, Dao-Hua Zhang

U-Shaped Sloped Field Plate Edge Termination Structure in β-Ga2O3 Vertical SBD 709
Yanzuo Li, Song Yuan, Xi Jiang, Yanjing He, An Xu, Guorui Mo, Yunxuan Zhao, Zhaoheng
Yan, Qifan Liu, Xinbin Zhan, Ying Wang, Linhai Zhong, Xiaowu Gong

Enhancement-Mode Vertical (001) β-Ga2O3 Power Transistor Enabled by Dual Ion Implantation 713
Anjing Luo, Gaofu Guo, Li Zhang, Tiwei Chen, Dengrui Zhao, Zhili Zou, Chunhong Zeng,
Huanyu Zhang, Baoshun Zhang, Zhucheng Li, Xiaodong Zhang, Zhongming Zeng

Partial Discharge Classification and Intelligent Maintenance Decision-Making Based on
Spikingformer and LLMs 717
Changdong Wang, Jingli Yang, Xunran Yin, Shuangyan Yin, Tianyu Gao, Yongqi Chang,
Huamin Jie, Zhou Shu, Zhenyu Zhao

Junction Temperature Fluctuation Suppression Strategy for SiC MOSFETs Based on Equivalent
Gate Resistance Control 721
Ruoyin Wang, Xiaoyong Zhu, Hong Zheng

Study of a High-Sensitivity Recessed-Anode GaN MIS Diode for Temperature Sensing 726
Yunxuan Zhao, Xi Jiang, Song Yuan, Zhaoheng Yan, Yanzuo Li, Chaofan Deng, Xiangdong Li,
Xiaowu Gong

Warping Model of SiC Power Module After Reflow Soldering 731
Chang Liu, Yingxin Cui, Yanao Guo, Shoulai Gong, Jisheng Han

A Novel Mechanical Stress Related Failure Mechanism of the State-Of-The-Art SiC Double Trench
MOSFET Under Surge Current Stress 737
Shikang Xu, Xuan Li, Hanqing Zhao, Yi Wen, Wensong Peng, Zekun Zhou, Xu Li, Xiaochuan
Deng, Bo Zhang

Impact of Elevated Humidity Conditions on the Thermal and Mechanical Reliability of Silicon
Carbide MOSFETs 741
Jiayu Zhang, Dong Xie, Zhiliang Xu, Zepeng Jiang, Xinglai Ge

Reliability Analysis and Implementation of Silver Sintering Connection in SiC High-Temperature
Package 747
Zizhen Cheng, Wenjie Xu, Fengtao Yang, Zhiqiang Zhao, Dewen Wang, Laili Wang

Impact of Kelvin Source Connection (KSC) Vs. Common Source Connection (CSC) on Power
Cycling Aging Characteristics of Silicon Carbide (SiC) MOSFETs 753
Huaihao Cheng, Yong Chen, Xu Cheng, Xingyu Pei, Jianbiao Li, Hongyuan Wu

SiC VDMOSFET Performance Prediction Method Based on Neural Network 757
Xiamin Hao, Feng He, Rui Jin

Oxidation-Free Sintered Copper Die Attachment for Power Electronics Packaging 762
Junyang Chen, Haiqiang Zhao, Zewei Zhang, Zhuo Pang, Tsung-Huan Sheng, Yi Chiu, Zhi-
Ying Huang, Yen-Liang Lin, Meiyu Wang

Advanced Packaging for High-Voltage SiC Module with Excellent Thermal, Mechanical, and Electrical Properties.. 767
  Meiyu Wang, Yiting Han, Peng Gao, Zhuo Pang, Yingkun Yang, Haidong Yan

A One-Inductor Two-Switch Three-Port DC-DC Converter for PV-Battery Systems..................................... 772
  Yidi Liang, Hong Li, Huizhu Zhuang, Peng Sun, Xu Shangguan

A Parameter Design Method for Bidirectional LLC-C Resonant Converter for Bidirectional On-Board EV Charger Application... 777
  Yiheng Zhang, Pengyu Jia, Mingjun Liu, Yimei Xing

A Phase Shift Control Strategy for the DCX to Realize a Wide Voltage Range by a Variable Mode Transformer 783
  Pengyu Jia, Mingjun Liu, Yimei Xing

A Multi-Port Converter-Based Equalization Architecture with Wide Voltage Gain for Long Series Battery Packs.. 788
  Haipeng Hu, Xianbin Qi, Helong Li, Mingzhu Fang, Peng Qin, Zhiqing Yang, Lijian Ding

Design of Planar Integrated Transformer for LLC Resonant Converters Based on Hybrid Magnetic Materials... 792
  Jianguang Yao, Xiaoyi Xu, Yanfang Mao, Zhujian Ou, Runyang Ji

Common-Mode Noise Analysis of Hall-Effect Sensor for SiC Power Converter 798
  Guifeng Geng, Peng Zhou, Runquan Jiang, Xuejun Pei

Research on Online Reliability Assessment Technology for IGBT Based on Driver-Side Measurement Fusion 802
  Yi Liu, Zhicheng Liu, Yunhui Mei

An Active Current Sharing Strategy Based on Master-Slave Cooperative Control for Paralleled SiC Power Modules... 808
  Baolong Yan, Jianing Wang, Shaolin Yu, Honghong Li, Runze Wang, Xin Li

Investigation on SiO2 Gate Formed Through Ultrahigh-Temperature NO Oxidation 813
  Yingfeng He, Given Shucheng Chang, Decai Liu, Tao Zhu, Zheyang Li, Rui Jin

The Method for Chip Sorting Based on SiC MOSFET Parameters for Parallel Current Sharing Under a Symmetrical Layout... 817
  Yi Li, Bowen Tian, Bin Zhao, Peng Sun, Zhibin Zhao, Xu Cheng, Yong Chen, Xingyu Pei, Jianbiao Li, Hongyuan Wu

Charaterization of a Nonlinear Conductivity Encapsulant for Electric Field Reduction in High-Voltage Power Module Packaging... 822
  Meiyu Wang, Peng Gao, Yiting Han, Haidong Yan

Configuration Selection for Degradation Trajectory Prediction of Power Modules Based LSTM Model 827
  Yichi Zhang, Yi Zhang, Jie Kong, Jiahong Liu, Bo Yao, Huai Wang

A Review and Analysis of Grid-Forming Technologies in Renewable Energy Power Systems...................... 833
  Zhicheng Liu, Dezheng Zhang, Yehan Fu, Yuying He, Li Zhang

Switching Oscillation Suppression Based on Embedded SiC Power Module with Low Parasitic Inductance for CLLC Resonant Converter 839
  Jiarui Zhang, Bodong Li, Xinnan Sun, Jiahui Wang, Liwen Jia, Kelin Chen, Feng Jiang, Min Chen

Modeling and Design of the Planar Magnetic Integration for Dual-Stage EMI Filters 844
Haiyan Liang, Yitao Liu, Zijian Lu

A Method to Decrease the Submodule Capacitor Voltage Fluctuations in Voltage-Source Modular
Multilevel Converter with Wide-Bandgap Power Devices 849
Qian Kang, Tiancong Shao, Trillion Zheng, Yuqing Geng, Yaqi Li, Zhitong Bai, Xiaofeng Yang

Junction Temperature Control and Thermal Stability Enhancement Method for Power Devices
Based on Vapor Phase-Change Principle 853
Ruya Song, Shuang Zhao, Jinxiao Wei, Waleed Alhosaini, Lijian Ding

A Self-Clamped L-Shaped Trench Gate SiC MOSFET with Improved Breakdown and Short-Circuit
Reliability 857
Xiaobo Cao, Jing Liu, Shaowei Zhang, Qian Zhang, Zhonggang Yin

Degradation of Planar-Gate SiC MOSFETs Under Repetitive Short-Circuit Stress in Different Gate
Bias 863
Yifan Wu, Chi Li, Jianwei Liu, Zedong Zheng

Low Roughness and Shape Reforming 4H-SiC Trench Process Optimized by CCP Etching with
High Temperature Annealing and Sacrificial Oxidation 867
*Qiongyang Zhuang, Xixi Luo, Yu Chen, Caixin Gu, Lei Song, Kaiju Liao, Qin Hu, Jiamin
Tian, Yidan Chen, Gang Chen, Jinliang He*

Investigation of L-FER ESD Protection Capability on E-Mode p-GaN HEMT 873
*Junye Wu, Yitian Gu, Chao Feng, Danfeng Mao, Yanlin Wu, Haolin Hu, Wei Zeng, David
Zhou, Yuxi Wan*

Study on 4H-SiC Trench Gate Dual-Mode Composite Transistors (T-DCT) Structure to Reduce
Ron and Enhance SCWT 877
Wenyu Xi, Cailin Wang, Lei Guan

Super-Junction IGBT with Adaptive Hole Channel Around Stepped Trench Gate for Low On-State
Voltage and Low Turn-Off Loss 881
Xuelei Zhou, Hengyu Wang, Yifan Wang, Kuang Sheng

Effect of Voltage Probes on the Characterisation of Switching Processes in Wide-Bandgap
Semiconductor Devices 885
Yishun Yan, Lurenhang Wang, Xuchong Cai, Mingcheng Ma, Yanchen Pan, Dianguo Xu

Performance Improvement of SiC n-LTT by Semi-Through Via Structure 890
Yulei Zhang, Xi Wang, Xuhui Pu, Jichao Hu, Hongbin Pu, Yuan Yang

Comparative Investigation of Gate Oxide Degradation in 1.2 kV Planar, Double-Trench, and
Asymmetric-Trench SiC MOSFETs 894
Dingkun Zhao, Xin Yang

Understanding the Role of Buffer Traps in GaN HEMTs: A Simulation Study on Dynamic Ron 898
*Haiyang Li, Mengqi Fan, Xinyue Dai, Xiaoping Wang, David Zhou, Danfeng Mao, Yan Wang,
Yuxi Wan*

A 3×1 Silicon Carbide Bidirectional Switch Power Module with Balanced Inductance During
Current Commutation 902
Zhiwei Jiao, Lei Ming, Yufeng Cao, Tao Xu, Zihang Gu, Zhen Xin

An Experimental Study on Single Pulse Avalanche Characteristics of Si/SiC Hybrid Switch 907
Hangzhi Liu, Yuming Zhou

Substrate Coupling Considerations for Monolithic Integration of High-Voltage Power Transistors with Low-Voltage Devices and Circuits in GaN-On-Si Technology 911
Rui Ray Yao, Miao Cui, Zhao Wang, Sang Lam, Stephen Taylor

Analysis and Compensation Method of Transient Unbalanced Current in Parallel Connection of SiC MOSFET 916
Yong Chen, Xu Cheng, Xingyu Pei, Jianbiao Li, Hongyuan Wu, Bin Zhao

Design and Fabrication of a SiC Trench MOSFET with Multi-Step P-Type Shielding and Multiple CSL Layers...920
Wei Chen, Fei Guo, Yangyang Wu, Kuan Wang, Zhijie Cheng, Jun Yuan, Rong Zhang, Guoqing Xin, Zhiqiang Wang

A Novel Integrated Power Module Package Method with SiC MOSFETs and Energy Absorber in Solid-State Circuit Breaker Application 924
Dongxin Jin, Jie Gong, Yuchen Wang, Cheng Luo, Xiaojun Dong, Guangyin Lei

Dry Oxidation and SiO$_2$ Deposition Strategies for High-Performance Gate Dielectrics in 3.3 kV SiC Power MOSFETs...929
Zijian Hu, Hongyi Xu, Na Ren, Kuang Sheng

Low Parasitic Repackaging and Integration of Multiple GaN HEMT Devices................................933
Yue Chen, Mingrui Zou, Dongjun Jiang, Senhao Liang, Jiakun Gong, Zheng Zeng

Impact of Electroluminescence Spectrum Sampling on SiC MOSFET Junction Temperature and Current Sensing 939
Yuting Jin, Hao Huang, Jingyang Hu, Shengjie Luo, Haoze Luo, Wuhua Li

Research on Threshold Voltage Instability of SiC MOSFETs at High Temperature 944
Xu Cheng, Yong Chen, Xingyu Pei, Jianbiao Li, Hongyuan Wu, Cong Chen

A High-Isolation X-Ray Power Supply with Multi-Transformer Series Configuration 949
Ziyang An, Ye Tian, Jie Ming, Yu Dou, Chushan Li

A High-Speed Dynamic Gate Driver with Low Oscillation for GaN HEMTs 954
Xuetong Zhou, Li Zheng, Xinhong Cheng, Lingyan Shen

Research on an Anti-Offset Wireless Power Transfer System with Auxiliary Resonant Circuit 959
Youzheng Wang, Shengxiu Xu, Shuyu Wang, Hongchen Liu, Longnv Li, Gaojia Zhu, Yunhui Mei

Research on Low Speed Power Boosting Technology for High Speed Maglev Trains 963
Zheyi Zheng, Fuao Chen, Xiaojun Zhang, Haoyun Wang, Yang Chen, Ruikun Mai

Current Overshoot and Oscillation Suppression in SiC MOSFETs Through Variable Gate Capacitance During Turn-On Transient...968
Xuchong Cai, Yishun Yan, Yanchen Pan, Mingcheng Ma, Binbo Xu, Dianguo Xu

An Active Clamped Resonant Ultra-High Frequency Quasi-Square Wave Gate Driver for SiC MOSFET 974
Zhiqing Liang, Zhixing He, Haoyi Sheng, Renfeng Guan, Zhenyuan Ou, Yang Liu, Zongjian Li, Jun Wang

48V-To-0.9V Voltage Regulator with GaN Devices and Integrated Magnetic Design 980
Zikang Li, Jingyang Tan, Yijie Wang, Dianguo Xu

A Bidirectional-Signal Transmission Method for Gate Drive Application Using Single Isolation Transformer 985
Junru Lin, Junming Zhang

A Dual-Voltage 650 V and 100 V GaN Integrated Platform Featuring High-Performance Monolithic Components 990
Yanlin Wu, Junye Wu, Zuoheng Jiang, Danfeng Mao, Keping Wu, Chao Feng, Jiawei Chen, David Zhou, Yuxi Wan

Exploring the Soft-Switching Benefits of TZCM Mode in Three-Level DC-DC Converters Using Wide Bandgap Power Devices 994
Zhigang Yao, Jingrui Liu, Sankun Yao, Bac-Bien Ngo, Ziheng Xiao, Yi Tang

Design and Current-Sharing Study of the TL-Boost Power Unit Based on Discrete Devices in Parallel 1000
Jianing Wang, Honghong Li, Shaolin Yu, Donglei Zhang, Baolong Yan, Zhenchun Xia, Weina Mao

Study on New Structures of EST to Inhibit Snapback Effect and Enhance MCC 1006
Wuhua Yang, Jia Liping, Guo Jiarui, Zhang Chao, Shen Sihao, Wang Cailin

A Transient Interaction Mechanism Analysis Method for the Grid-Forming Voltage Support Device Integrated into LCC-HVDC System 1011
Yanlin Song, Zhichang Yang, Hong Li

A Numerical Model of SiC MOSFET for Electro-Thermal Characteristics Based on TCAD 1017
Yujie Zhang, Yongle Huang, Yifei Luo

The Method to Evaluate the SOA of Drain-Source Voltage for SiC MOSFET 1022
Fengming Yang, Xin Li, Yifei Luo, Lin Liang, Yongle Huang

GaN-Based 1.5 MHz Synchronous Buck Converter with Partial Soft-Switching Control Scheme 1028
Zeqi Yang, Yuan Liu, Zhe Zhang

Design and Demonstration of a Novel SiC Trench MOSFET with Periodically Grounded P Shield Island Based on Secondary Epitaxy Process 1033
Yangyang Wu, Fei Guo, Kuan Wang, Wei Chen, Zhijie Cheng, Yuan Jun, Rong Zhang, Guoqing Xin, Zhiqiang Wang

Author Index

Technical Program Committee

Hao Bai	Northwestern Polytechnical University	China
Bo Chen	Tianjin University	China
Cai Chen	Huazhong University of Science and Technology	China
Cen Chen	Harbin Institute of Technology	China
Hui Chen	Hangzhou City University	China
Jian Chen	Southwest Jiaotong University	China
Min Chen	Zhejiang University	China
Wenjie Chen	Xi'an Jiaotong University	China
Yenan Chen	Zhejiang University	China
Bin Cui	Tsinghua University	China
Erping Deng	Hefei University of Technology	China
Xiaochuan Deng	University of ElectronicScience and Technology of China	China
Xiaofeng Ding	BeiHang University	China
Jiajie Fan	Fudan University	China
Shanshan Gao	Harbin Institute of Technology	China
Zhiyun Han	Shandong University	China
Junping He	Harbin Institute of Technology, Shen zhen	China
Yuying He	Hohai University	China
Fengze Hou	Institute of Microelectronics of Chinese Academy of Sciences	China
Borong Hu	University of Cambridge	United Kingdom (Great Britain)
Huolin Hunag	Dalian University of Technology	China
Pengyu Jia	North China University of Technology	China

Xi Jiang	Xidian University	China
Jiabao Kou	Wenzhou University	China
Wei Lai	Chongqing Uiversity	China
Binbin Li	Harbin Institute of Technology	China
Bodong Li	Zhejiang University	China
Helong Li	Hefei University of Technology	China
Hong Li	Zhejiang University	China
Shouxiang Li	Beijing Institute of Technology	China
Xiangdong Li	Xidian University	China
Xu Li	Southwest Jiaotong University	China
Xuebao Li	North China Electric Power University	China
Yi Li	Wuhan University	China
Yitong Li	Xi'an Jiaotong University	China
Shiwei Liang	Hunan University	China
Zhijuan Liao	China University of Ming and Technology	China
Zhiheng Lin	University of Alberta	Canada
Fei Liu	Nanjing University of Aeronautics and Astronautics	China
Yitao Liu	Shenzhen University	China
Gang Lyu	Beijing University of Aeronautics and Astronautics & Beihang University	China
Hongbo Ma	Southwest Jiaotong University	China
Yunhui Mei	Tiangong University	China
Hao Niu	China Electronic Product Reliability and EnvironmentalTesting Research Institute	China
Dimosthenis Peftitsis	Norwegian University of Science and Technology, NTNU & Pulsed Power Engineering Section, CERN	Norway

Xuejun Pei	Huazhong University of Science and Technology	China
Jingjing Qi	Tsinghua University	China
Xianbin Qi	Hefei University of Technology	China
Maohang Qiu	Southwest Jiaotong University	China
Hanwen Ren	North China Electric Power University	China
Na Ren	Zhejiang University	China
Zhenyu Shan	Beihang University	China
Shuai Shao	Zhejiang University	China
Tiancong Shao	Beijing Jiaotong University	China
Peng Sun	North China Electric Power University	China
Kun Tan	Anhui University	China
Xinling Tang	Beijing Huairou Laboratory	China
Qingxin Tian	Sichuan University	China
Changdong Wang	Harbin Institute of Technology	China
Hengyu Wang	Zhejiang University	China
Jian Wang	North China Electric Power University	China
Yijie Wang	Harbin Institute of Technology	China
Youzheng Wang	Tiangong University	China
Jiaxing Wei	Southeast University	China
Jinxiao Wei	Hefei University of Technology	China
Man Hoi Wong	Hong Kong University of Science and Technology	Hong Kong
Chao Wu	Shanghai Jiao Tong University	China
Dong Xie	Southwest Jiaotong University	China

Haidong Yan	Zhejiang University	China
Haidong Yan	Zhejiang University	China
Fengtao Yang	Xi'an Jiaotong University	China
Jingxi Yang	City University of Hong Kong	Hong Kong
Xin Yang	Hunan University	China
Yanyong Yang	University of Mining and Technology-Beijing	China
Yangbin Zeng	South China University of Technology	China
Zheng Zeng	Chongqing University	China
Li Zhang	Hohai University	China
Tiandong Zhang	Harbin University of Scienceand Technology	China
Yajing Zhang	Beijing Information Science and Technology University	China
Yichi Zhang	Aalborg University	Denmark
Yuhao Zhang	University of Hong Kong	Hong Kong
Zhaofu Zhang	Wuhan University	China
Shuang Zhao	Hefei University of Technology	China
Zhenyu Zhao	National University of Singapore	Singapore
Jialin Zheng	Tsinghua University	China
Peng Zhou	Huazhong University of Science and Technology	China
Wenpeng Zhou	The Hong Kong University of Science and Technology	Hong Kong
Ye Zhu	Zhejiang GEENER Microelectronics Co. Ltd.	China
Liang Zou	Shandong University	China

Additional Reviewers

Zemin Bu	Tiangong University	China
Ruibo Cao	Institute of Microelectronics, Chinese Academy of Sciences	China
Wentao Cui	Zhejiang University	China
Lunbo Deng	Southwest Jiaotong University	China
Yujie Du	Beijing Huairou Laboratory	China
Mingzhu Fang	Deep Space Exploration Lab	China
Zhi Gao	Zhejiang University	China
Hehe Gong	The University of Hong Kong	Hong Kong
Chao Gu	Fudan University	China
Tao Guo	Zhejiang University	China
Wei Han	The Hong Kong University of Science and Technology (Guangzhou)	China
Chunyao Hou	Zhejiang University	China
Liqian Huang	Harbin Institute of Technology	China
Yin Hui	Hefei University of Technology	China
Hongchuan Jia	Nanjing University of Information Science and Technology	China
Liwen Jia	Zhejiang University	China
Hang Kong	Xi'an Jiaotong University	China
Rafał Kopacz	Warsaw University of Technology	Poland
Jiajun Lai	Zhejiang Jingneng Microelectronics Co. Ltd.	China
Chaochao Li	Harbin Institute of Technology	China
Daohang Li	Tiangong University	China
Longnv Li	Tiangong University	China
Wei Li	Tianjin University of Technology	China
Wenyu Li	Fudan University	China
Haolin Liu	Shanghai Jiao Tong University	China
Jiahong Liu	Aalborg University	Denmark
An Lou	Zhejiang University	China
Aozu Luan	Zhejiang University	China
Zhichao Luo	South China University of Technology	China

Xiaoyong Ma	Tianjin University	China
Lichang Man	Harbin Institute of Technology	China
Zishun Peng	University of Hunan	China
Xikun Sang	Harbin Institute of Technology	China
Chaochao Shen	Nanjing University of Aeronautics and Astronautics	China
Jiapeng Shen	Hohai University	China
Lingyan Shen	Shanghai Institute of Microsystem and Information Technology	China
Wei Song	University of Electronic Science and Technology	China
Hao Sun	Zhejiang Geener	China
Song Tang	Nanjing Institute of Technology	China
Long Tao	Tianjin University of Technology	China
Yangyang Tao	Zhejiang Geener	China
Zhang Tongyu	Xi'an Jiaotong University	China
Jiangbin Wan	Zhejiang University	China
Jingfei Wang	Beijing Huairou Laboratory	China
Long Wang	Nanjing University of Information Science & Technology	China
Meiyu Wang	Nankai University	China
Peiyue Wang	Chongqing University of Posts and Telecommunications	China
Shuyu Wang	Harbin Institute of Technology	China
Yaohua Wang	Beijing Huairou Laboratory	China
Li Wangyun	Guilin University of Electronic Technology	China
Lingxuan Xiao	Nanjing University of Aeronautics and Astronautics	China
Lihong Xie	Nanjing University of Aeronautics and Astronautics and Astronautics	China
Guo Xinhua	Huaqiao University	China
Nengmou Xu	Southwest Jiaotong University	China
Zhiliang Xu	Southwest Jiaotong University	China
Yiran Yan	Nanjing University of Aeronautics and Astronautics	China
Xin Yang	The University of Hong Kong	Hong Kong
Zhiqing Yang	Hefei University of Technology	China
Lihai Yao	Hangzhou City University	China

Qianchen Yin	Hefei University of Technology	China
Hongyu Yu	Southern University of Science and Technology	China
Huiying Yu	Harbin Institute of Technology	China
Kefan Yu	Beijing Huairou Laboratory	China
Ziheng Yuan	Hefei University of Technology	China
Mo Yun	Sun Yat-sen University	China
Wenjun Zeng	Southwest Jiaotong University	China
Baifu Zhang	Taiyuan University of Technology	China
Man Zhang	Hefei University of Technology	China
Shibo Zhang	Nanjing University of Aeronautics and Astronautics	China
Xinsheng Zhang	Harbin Institute of Technology	China
Xuelun Zhang	Zhejiang Geener	China
Yaoyao Zhang	China University of Mining and Technology	China
Yixue Zhang	Harbin Institute of Technology	China
Wei Zhou	Southwestern Jiaotong University	China
Weiqiang Zhou	Zhejiang University	China
Yujian Zong	Zhejiang University	China
Yuhang Zou	Tsinghua University	China

Design of Short-Circuit and Overload Protection Chips for SiC MOSFETs

1st Jiahui Lv
Department of Automation and Information Engineering
Xi'an University of Technology
Xi'an, China
1240310004@stu.xaut.edu.cn

2nd Yuan Yang
Department of Automation and Information Engineering
Xi'an University of Technology
Xi'an, China
yangyuan@xaut.edu.cn

3rd Minmin Zhang
Department of Automation and Information Engineering
Xi'an University of Technology
Xi'an, China
2200320083@stu.xaut.edu.cn

Abstract—This paper proposes a fast short-circuit (SC) and overload (OL) protection integrated circuit (IC) for silicon carbide (SiC) metal-oxide-semiconductor field-effect transistors (MOSFETs). The method achieves short-circuit protection (SCP) by detecting the voltage drop across the inductance between the Kelvin source and power source of the SiC MOSFET, which is generated by the di/dt. OL protection is achieved by detecting the current and voltage of the SiC MOSFET, thereby accumulating energy for protection. The proposed protection IC is implemented using a 0.5 μm Bipolar-CMOS-DMOS (BCD) process, with a chip area of approximately 2 mm². The effectiveness of SC and OL protection is validated on a 1200V/80mΩ SiC MOSFET test platform. Experimental results show that when a SC fault occurs in the SiC MOSFET, the chip can detect the fault and safely turn off the device in approximately 150 ns. Similarly, when an OL fault occurs, the chip can quickly detect the fault and shut down the device.

Index Terms—SiC MOSFET, short-circuit, protection, energy accumulation.

I. INTRODUCTION

With the rapid development of power electronics technology, traditional silicon-based power devices are increasingly constrained by their material properties, making it difficult to meet the growing demands for enhanced system efficiency, higher power density, and faster switching frequencies [1]–[3]. Power electronic devices based on wide-bandgap semiconductors such as silicon carbide (SiC) and gallium nitride (GaN) have emerged as crucial technologies for overcoming these limitations due to their superior performance [4]. Silicon carbide metal-oxide-semiconductor field-effect transistors (SiC MOSFETs), in particular, are widely used in high-voltage, high-frequency, and high-power applications, owing to their high switching frequency, low switching losses, and excellent high-voltage tolerance [5]–[8]. However, their relatively low short-circuit (SC) withstand capability requires more stringent and reliable short-circuit protection (SCP) circuit designs [9], [10].

This work was supported in part by the National Natural Science Foundation of China under Grants 62174134, in part by the Shaanxi Innovation Capability Support Project under Grant 2021TD-25, and in part by the Xi'an key industrial chain key core technology research projects under Grant 23LLRH0044.

Currently, several methods are available for SC detection in SiC MOSFETs, including desaturation detection, di/dt detection, gate voltage detection, and current sensing using Rogowski coils.

To provide SC protection for SiC MOSFETs, an accelerated desaturation detection method is proposed, which combines the desaturation detection circuit with gate-emitter voltage V_{GE} sensing [11]. The feedback path of V_{GE} is employed to enhance the speed of fault detection. This method achieves a low fault detection delay and simplifies the circuit structure; however, its detection accuracy is susceptible to variations in the gate voltage.

To provide both SC and overload (OL) protection for SiC MOSFET, a dual protection scheme for SC and OL faults in gate drivers is presented [12]. This approach employs a novel parasitic inductance measurement method, based on the principle of di/dt, by incorporating an additional network to detect OL faults. The presented method is independent of temperature, supports a wide current range, offers high-speed response, and features low power consumption and low complexity. However, it imposes stringent requirements on the cutoff frequency of the filter formed by the integrator circuit, as any deviation from the required frequency can result in severe distortion of the measurement results. A di/dt SC detection method utilizing an RC network with a diode is introduced, which is applicable to various load and fault conditions [13]. This method addresses the issue of requiring different reference voltage thresholds for detecting the hard switching fault (HSF) and the fault under load (FUL). However, the inclusion of the diode complicates the integration of the detection and protection circuitry.

A fault detection method is proposed, which monitors the gate-source voltage and gate charge during the turn-on transient process [14]. This method eliminates the need for blanking time, with the reference gate-source voltage set to the Miller plateau voltage. However, this circuit is limited to detecting the HSF within 1 μs.

The other method for SC fault detection using an improved Rogowski coil is proposed [15]. This method optimizes the parameters of the Rogowski coil and employs a sample-filter-hold circuit for reliable integrator calibration. It preserves the inherent high-bandwidth advantage of conventional Rogowski coils while minimizing droop distortion, thereby improving the detection accuracy of SiC MOSFET SC faults. However, the design is relatively complex.

This paper proposes an SCP method based on $\mathrm{d}i/\mathrm{d}t$ detection and an OL protection method based on energy accumulation, which aims to develop a highly reliable, fast, accurate, and easily integrated protection scheme tailored for SiC MOSFETs.

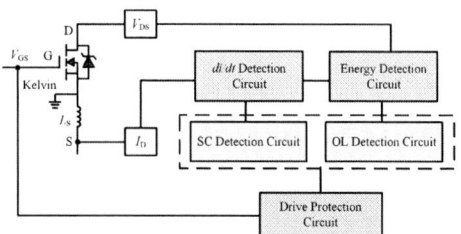

Fig. 1. The circuit block diagram of SC and OL protection

II. THE PRINCIPLE OF THE PROPOSED PROTECTION INTEGRATED CIRCUIT

The structural block diagram of the SC and OL protection methods is shown in Fig. 1. It primarily consists of a di/dt detection circuit, an energy detection circuit, a SC detection circuit, an OL detection circuit, and a drive protection circuit. The di/dt detection can trigger SCP, while energy detection can trigger OL protection. Regardless of whether a SC or OL fault occurs, the circuit will activate a fault signal. Once the fault signal is latched, it is sent to the driver circuit to protect and turn off the SiC MOSFET. To prevent excessive voltage overshoot during the fast turn-off process, which could risk device breakdown, a soft turn-off method is employed to ensure safe and reliable turn-off of the driver circuit.

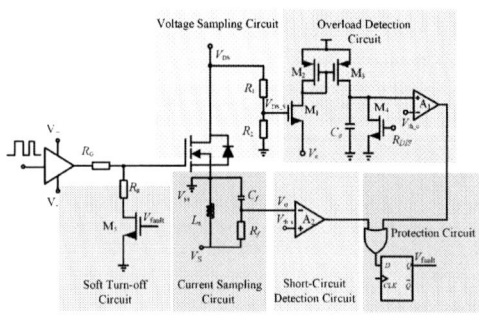

Fig. 2. The circuit schematic of the proposed SC and OL protection method

A. SC Protection Circuit

As shown in Fig. 2, the SCP circuit consists of a current sensing circuit and a SC detection circuit. The current detection unit is connected between the

module power source and the Kelvin source of the SiC MOSFET. There is a parasitic inductance L_S between the power source and the Kelvin source. When an SC fault occurs, the current rises rapidly to 8-10 times the rated load current. The large $\mathrm{d}i/\mathrm{d}t$ flowing through the parasitic inductance generates a negative induced voltage V_S. The transfer function from the current I_D to the induced voltage V_S exhibits a constant gain within a specific frequency range, allowing the current signal to be transmitted to the detection voltage without distortion. The transfer function $G_1(s)$ from I_D to V_S is expressed by

$$G_1(s) = \frac{-V_S(s)}{I_D(s)} = L_s \cdot s \tag{1}$$

The integration of V_S is performed using an integrator consisting of a passive RC filter, which aims to reconstruct the rapidly rising current I_D from the measured $\mathrm{d}I_D/\mathrm{d}t$ during an SC fault. The high-frequency transfer function $G_2(s)$ can be expressed by

$$G_2(s) = \frac{V_O(s)}{V_S(s)} = \frac{1}{R_f C_f \cdot s + 1} \tag{2}$$

where V_O, R_f, and C_f represent the output voltage, resistance, and capacitance of the RC filter, respectively. Therefore, the transfer function $G_3(s)$ is given by

$$G_3(s) = \frac{-V_O(s)}{I_D(s)} = \frac{L_s \cdot s}{R_f C_f \cdot s + 1} = \frac{L_s}{R_f C_f} \cdot \frac{R_f C_f \cdot s}{R_f C_f \cdot s + 1} \tag{3}$$

When the SC frequency component is at high frequencies, i.e., $R_f C_f \cdot s \gg 1$, the relationship between I_D and V_O can be expressed as:

$$V_O(s) = \frac{L_s}{R_f C_f} \cdot I_D \tag{4}$$

As shown in (4), the output voltage V_O of the current detection circuit is proportional to I_D, and thus V_O reflects the level of fault current. By comparing V_O with the preset threshold voltage $V_{\mathrm{th\_s}}$, when V_O reaches $V_{\mathrm{th\_s}}$, the comparator outputs a high level, indicating the detection of an SC fault. This approach effectively converts the induced voltage across the parasitic inductance into current for SC detection and helps mitigate false protection triggered by the large $\mathrm{d}I_D/\mathrm{d}t$ when the SiC MOSFET is turned on. When the device operates normally without an SC fault, V_O initially increases with $\mathrm{d}I_D/\mathrm{d}t$, and once I_D reaches the load current, $\mathrm{d}I_D/\mathrm{d}t$ becomes zero. V_O is then discharged through R_f and L_S. Throughout this entire process, the level of $\mathrm{d}I_D/\mathrm{d}t$ remains much smaller than that during an SC fault, so V_O does not reach $V_{\mathrm{th\_s}}$, and the comparator continuously outputs a low level.

B. OL Protection Circuit

As shown in Fig. 2, the OL protection circuit consists of a voltage sampling circuit and an OL detection circuit. When an OL fault occurs in the SiC MOSFET, a negative induced voltage is generated across the parasitic inductance L_s, similar to that in an SC fault. However, the variation in V_O is not as significant as that during an SC fault. In this case, the entire bus voltage is applied across the drain-source terminals of the device, leading to a large value of V_{DS}. By simultaneously monitoring the variations of V_O and V_{DS}, the OL fault can be detected, which is equivalent to detecting both the current and voltage under OL conditions. The voltage detection circuit is designed to sample the drain-source voltage V_{DS} of the SiC MOSFET. Since the range of V_{DS} is very large, a resistive voltage divider is used. The sampled voltage V_{DS_S} is applied to the gate of the OL detection transistor M_1, while the output signal V_O from the current detection unit is applied to the source of M_1. When an OL fault occurs, both V_{DS_S} and V_O increase rapidly. Since V_O is an inversely increasing voltage (i.e., a negative-going voltage), the gate voltage of M_1 increases rapidly, while the source voltage decreases sharply, causing M_1 to turn on and generate current. The expression for this current is given by

$$I_1 = K_n(V_{GS} - V_{TH})^2 = K_n(V_{DS\_S} - V_s - V_{TH})^2 \tag{5}$$

where $K_n = \frac{1}{2}\mu_n C_{ox} \frac{W}{L}$, μ_n is the carrier (electron) mobility of the n-channel MOSFET, C_{ox} is the gate oxide capacitance per unit area, W is the channel width of the transistor, and L is the channel length. The current is mirrored through the current mirror formed by M_2 and M_3. The width-to-length ratio of M_2 and M_3 is $1:n$, so the current in the M_3 branch is nI_1. This current is used to charge the capacitor C_e, and the voltage across the capacitor, V_e, can be expressed as

$$V_e = \frac{1}{C_e} \int nI_1 dt \approx \frac{n}{C_e} I_1 \cdot t = \frac{n}{C_e} K_n(V_{DS\_S} - V_s - V_{TH})^2 \cdot t \tag{6}$$

The voltage is compared with the preset OL threshold voltage V_{th_o}. When the voltage V_e across the capacitor accumulates to a value greater than V_{th_o}, the comparator flips and outputs a high level, indicating that an OL fault has been detected. M_4 simulates the heat dissipation of the SiC MOSFET during normal operation by providing a small current path to slowly discharge C_e. The voltage accumulated on C_e must be much greater than the voltage discharged by M_4 to trigger the fault, indicating that the rate of heat generation during an OL fault is significantly higher than the rate of heat dissipation. The OL detection voltage V_e is proportional to the current I_1 and time t, and the magnitude of I_1 is controlled by V_{DS_S} and V_S. V_{DS_S} reflects the magnitude of the drain-source voltage V_{DS}, while V_S reflects the magnitude of the fault current I_{DS}. Therefore, the OL detection voltage V_e can be considered to be determined by V_{DS}, I_{DS}, and t. The larger the values of V_{DS}, I_{DS}, and t, the greater the resulting V_e will be. Thus, the OL fault detection process can be equivalently viewed as

an energy accumulation process, determined by three factors: voltage, current, and time.

C. Soft Turn-OFF Circuit

As shown in Fig. 2, the soft turn-off protection circuit consists of both the fault detection circuit and the soft turn-off circuit. When a short-circuit or OL fault occurs in the SiC MOSFET, the signal V_{fault} switches from low to high. This signal is then input into a latch composed of a D flip-flop, which holds the fault condition until the arrival of the next turn-on signal of the SiC MOSFET, at which point the latch is reset. Once latched, the high-level fault signal is sent to the main driver circuit to initiate soft turn-off of the SiC MOSFET. The gate pull-down transistor M_5 is turned on, and a resistor with relatively large resistance, R_g, is inserted into the gate driving loop. At this point, the gate voltage is gradually discharged through M_5 at a slower rate, thereby reducing voltage overshoot during the turn-off process.

III. EXPERIMENTAL RESULTS

The test circuit diagram of the designed driver protection IC is shown in Fig. 3. V_{DC} represents the bus voltage, C_{DC} denotes the bus capacitance, L_{fault} is the OL fault inductance, and S_1 is the switch. During SCP testing, S_1 is turned off, and a short, thick copper bar is used to directly short the upper switch to simulate a SC fault. For OL fault testing, S_1 is turned on, and the OL fault inductance L_{fault} is connected into the circuit to simulate an OL. Fig. 4 shows that the high-voltage DC bus capacitance C_{DC} is composed of multiple 30 μF/1.1 kV film capacitors. The low-side driver uses the proposed protection integrated circuit chip. The low-side DUT and the high-side M_H are the 1200 V/80 mΩ SiC MOSFETs (SCT3080KR) from Rohm Semiconductor.

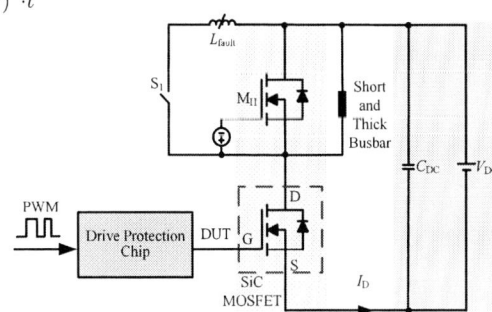

Fig. 3. Schematic diagram of SC and OL protection test circuit

A. SCP Test

The SCP functionality of the driver protection chip was tested under different bus voltages, with the bus voltage set to 200 V, 300 V, and 400 V, respectively. The test results are presented in Fig. 5. At a bus voltage of 200 V, the SC current peak reached 149 A, the voltage overshoot during shutdown was 500 V, and the total protection response time of the circuit was

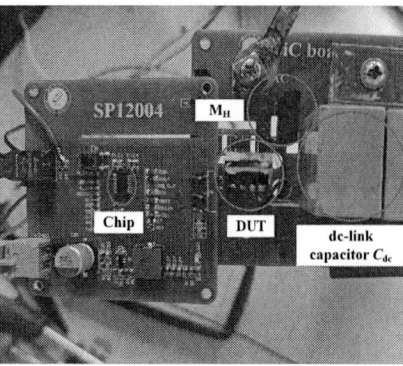

Fig. 4. Test for the proposed protection IC

136 ns. When the bus voltage was increased to 300 V, the SC current peak rose to 151 A, the voltage overshoot during shut-down reached 620 V, and the total protection response time was approximately 125 ns. At a bus voltage of 400 V, the SC current peak further increased to 155 A, the voltage overshoot during shutdown was 710 V, and the total protection response time was around 120 ns. As the bus voltage increased, the peak SC current rose, while the SCP response time decreased.

To evaluate the SCP functionality, tests were conducted with different reference voltages for the SC detection unit, with the DC bus voltage set to 400 V. The reference voltage was adjusted to 1 V and 1.5 V, respectively, and the results are shown in Fig. 6. When the reference voltage was set to 1 V, the peak SC current reached 170 A, the voltage overshoot during shutdown was 690 V, and the total protection response time was approximately 140 ns. When the reference voltage was set to 1.5 V, the peak SC current decreased to 155 A, the voltage overshoot during shutdown increased to 710 V, and the total protection response time was approximately 120 ns. A smaller reference voltage results in a longer detection time for the SC fault, leading to a higher peak SC current.

B. OL Protection Test

Other test conditions remain unchanged, and the current rise rate is varied by adjusting the bus voltage to test the OL fault. The test results are shown in Fig. 7. When the bus voltage is 350V, 380V, and 400V, faults are triggered at the 18th, 13th, and 11th pulses, respectively. The higher the bus voltage, the faster the current rise rate, leading to faster triggering of the protection function under the same threshold conditions.

Tests were conducted by setting different OL detection reference voltages V_{th_o}, with the bus voltage fixed at 400V. The test results are shown in Fig. 8. It can be observed that when V_{th_o} is 0.8V, the SiC MOSFET triggers an OL fault at the 9th pulse and then turns off. In contrast, when V_{th_o} is 1V, the OL fault is triggered at the 11th pulse. This is because, at a constant current

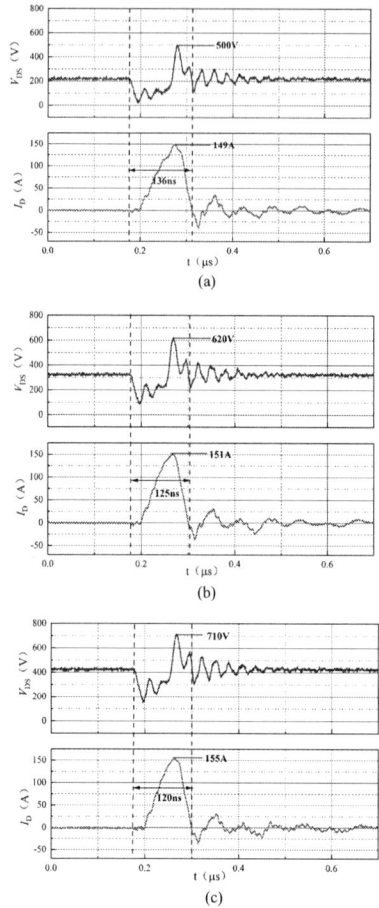

Fig. 5. SC protection test waveforms at different bus voltages (a) 200V (b) 300V (c) 400V

rise rate, a lower threshold leads to faster triggering of the OL fault, resulting in a smaller current value at the turn-off point.

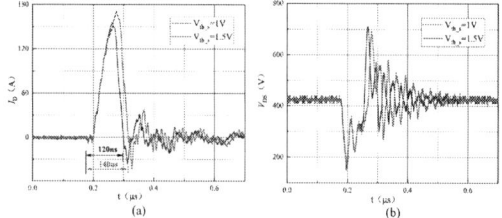

Fig. 6. SCP test waveforms at different reference voltages (a) SC current (b) drain-source voltage

IV. CONCLUSION

The fast SC and OL protection integrated circuit (IC) proposed in this paper was fabricated based on a 0.5μm BCD process. The effectiveness of the proposed method was validated using a SiC MOSFET test platform. Experimental results show that the driver protection IC can simultaneously detect and prevent

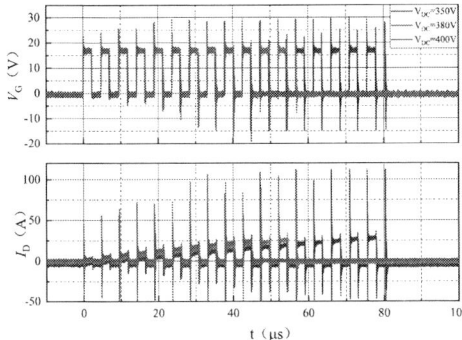

Fig. 7. OL protection test waveforms at different bus voltages

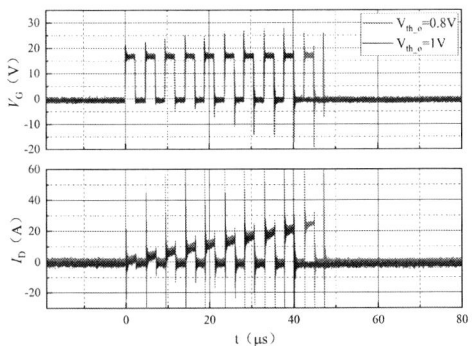

Fig. 8. OL protection test waveforms at different reference voltages

SC and OL faults. When a SC fault occurs in the SiC MOSFET, the chip is able to detect the fault within approximately 150 ns and safely turn off the device. Similarly, when an OL fault occurs in the SiC MOSFET, it can also quickly detect the fault and turn off the device.

REFERENCES

[1] F. Yang, L. Wang, H. Kong, M. Zhu, X. Liu, X. Lu, M. Qin, T. Zhang, Y. Gan, and L. Jia, "Compact-interleaved packaging method of power module with dynamic characterization of 4H SiC MOSFET and development of power electronic converter at extremely high junction temperature," *IEEE Transactions on Power Electronics*, vol. 38, no. 1, pp. 417–434, 2023.

[2] Y. Cai, H. Xu, P. Sun, J. Ke, E. Deng, Z. Zhao, X. Li, and Z. Chen, "Effect of threshold voltage hysteresis on switching characteristics of silicon carbide mosfets," *IEEE Transactions on Electron Devices*, vol. 68, no. 10, pp. 5014–5021, 2021.

[3] J. Wang and X. Jiang, "Review and analysis of SiC MOSFETs' ruggedness and reliability," *IET Power Electronics*, vol. 13, no. 3, pp. 445–455, 2020.

[4] K. Yao, H. Yano, H. Tadano, and N. Iwamuro, "Investigations of SiC MOSFET short-circuit failure mechanisms using electrical, thermal, and mechanical stress analyses," *IEEE Transactions on Electron Devices*, vol. 67, no. 10, pp. 4328–4334, 2020.

[5] A. B. Jørgensen, T. S. Aunsborg, S. Bęczkowski, C. Uhren-feldt, and S. Munk-Nielsen, "High-frequency resonant operation of an integrated medium-voltage SiC MOSFET power module," *IET Power Electronics*, vol. 13, no. 3, pp. 475–482, 2020.

[6] X. Yuan, I. Laird, and S. Walder, "Opportunities, challenges, and potential solutions in the application of fast-switching SiC power devices and converters," *IEEE Transactions on Power Electronics*, vol. 36, no. 4, pp. 3925–3945, 2020.

[7] S. Ji, M. Laitinen, X. Huang, J. Sun, W. Giewont, F. Wang, and L. M. Tolbert, "Short-circuit characterization and protection of 10-kV SiC MOSFET," *IEEE Transactions on Power Electronics*, vol. 34, no. 2, pp. 1755–1764, 2018.

[8] L. Zhang, X. Yuan, X. Wu, C. Shi, J. Zhang, and Y. Zhang, "Performance evaluation of high-power SiC MOSFET modules in comparison to Si IGBT modules," *IEEE Transactions on Power Electronics*, vol. 34, no. 2, pp. 1181–1196, 2018.

[9] C. D. Fuentes, S. Kouro, and S. Bernet, "Comparison of 1700-V SiC-MOSFET and Si-IGBT modules under identical test setup conditions," *IEEE Transactions on Industry Applications*, vol. 55, no. 6, pp. 7765–7775, 2019.

[10] S. Lee, K. Kim, M. Shim, and I. Nam, "A digital signal processing based detection circuit for short-circuit protection of SiC MOSFET," *IEEE Transactions on Power Electronics*, vol. 36, no. 12, pp. 13 379–13 382, 2021.

[11] T. Krone, C. Xu, and A. Mertens, "Fast and easily implementable detection circuits for short circuits of power semiconductors," *IEEE Transactions on Industry Applications*, vol. 53, no. 3, pp. 2871–2879, 2016.

[12] K. Sun, J. Wang, R. Burgos, and D. Boroyevich, "Design, analysis, and discussion of short circuit and overload gate-driver dual-protection scheme for 1.2-kV, 400-A SiC MOSFET modules," *IEEE Transactions on Power Electronics*, vol. 35, no. 3, pp. 3054–3068, 2019.

[13] J. Xue, Z. Xin, H. Wang, P. C. Loh, and F. Blaabjerg, "An improved di/dt-RCD detection for short-circuit protection of SiC MOSFET," *IEEE Transactions on Power Electronics*, vol. 36, no. 1, pp. 12–17, 2020.

[14] T. Horiguchi, S.-i. Kinouchi, Y. Nakayama, and H. Akagi, "A fast short-circuit protection method using gate charge characteristics of SiC MOSFETs," in *2015 IEEE Energy Conversion Congress and Exposition (ECCE)*. IEEE, 2015, pp. 4759–4764.

[15] Q. Xu, Y. Feng, P. Guo, N. Mo, B. Xu, Z. Qing, Y. Chen, and A. Luo, "Design of pcb rogowski coil current sensor with low droop distortion," *IEEE Transactions on Power Electronics*, vol. 38, no. 4, pp. 5513–5523, 2023.

Repetitive Recovery Analysis and Island Width Optimization of 4H-SiC Floating Island Devices

Runze Xia
College of Electrical Engineering
Zhejiang University
Hangzhou, China
22410181@zju.edu.cn

Hengyu Wang
College of Electrical Engineering
Zhejiang University
Hangzhou, China
wanghengyu@zju.edu.cn

Ce Wang
College of Electrical Engineering
Zhejiang University
Hangzhou, China
wangce@zju.edu.cn

Kuang Sheng
College of Electrical Engineering
Zhejiang University
Hangzhou, China
shengk@zju.edu.cn

Abstract— The 4H-SiC Floating Island (FI) device can break the 1-D SiC limit due to its ability to modulate the electric field in the drift region. In the previous research, we utilized the N+ buffer to overcome the forward recovery issue of the FI device. However, the FI device with N+ buffer can still suffer from repetitive forward recovery. In this work, we take the Floating Island JBS (FIJBS) as the research object. We will elucidate the cause of the repetitive recovery issue in the FI device and analyze the dynamic equilibrium mechanism of electron-hole pairs. We also simulate the impact of the floating island width on the device's static and dynamic characteristics, which is beneficial for selecting an appropriate island width. To further improve the repetitive recovery performance of the device, we propose a novel structure with width-optimized islands. Simulation results indicate that the recovered resistance of the new structure is significantly reduced even after repetitive switching tests compared with the conventional FJBS. Moreover, this novel structure only changes the island width without requiring complex manufacturing processes. Those results demonstrate the great potential of this new structure for high performance power device.

Keywords—Floating Island, repetitive forward recovery, dynamic equilibrium, floating island width, nonuniform width, SiC limit

I. INTRODUCTION

With the rapid development of power electronics technology, higher requirements are placed on power semiconductor devices, such as higher breakdown voltage (BV) and smaller specific ON-resistance ($R_{on,sp}$). However, due to the inherent limitations of silicon carbide, there is a theoretical trade-off between BV and $R_{on,sp}$, which is known as the 1-D SiC limit[1, 2]. This limits the development of devices with high blocking voltage and low on-resistance. Researchers have made some achievements by designing Super Junction (SJ)[3, 4] and Floating Island (FI)[5-7] to break through this limit. However, SJ structure requires complex processes and stricter doping control. Compared to SJ structure, the FI structure only needs multiple floating island implantation and epitaxy growth, making the process simpler[7-13]. Therefore, the floating island structure has a promising prospect.

Nevertheless, the FI device has a severe forward recovery problem. For instance, during the forward recovery process of the FI diode, there is a significant voltage overshoot[14-16]. Additionally, during the transition from Off-state to On-state in the FI MOSFETs, the drain-to-source voltage drops slowly,

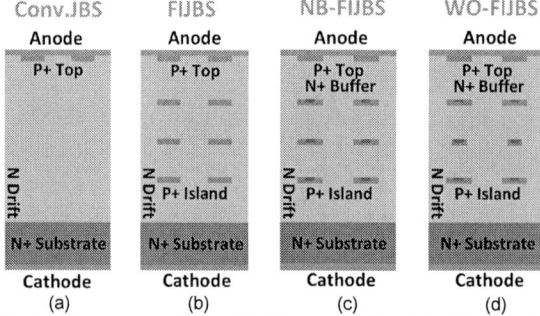

Fig. 1. The cross sections of (a) the conventional Junction Barrier Schottky Diode (JBS), (b) the Floating Island JBS (FIJBS), (c) the FIJBS with N+ Buffer (NB-FIJBS) and (d) the FIJBS with Width-Optimized Island (WO-FIJBS).

exhibiting hysteresis problem[17, 18]. Although P-Bus has been introduced to try to solve the turn-on issue, the actual performance still falls short of expectations, which affects the practical application of FI devices.

In prior research endeavors, we introduced an innovative design by integrating an N+ buffer into the FI structure[19]. The results revealed that this novel structure effectively mitigated the voltage overshoot and hysteresis issues. In this study, with the aim of enhancing performance during repetitive switching tests, we analyzed the reason for repetitive recovery issues and the influence mechanism of floating island width. Subsequently, we propose a structure featuring wider and Nonuniform-Width floating island (termed Width-Optimized island) to enhance dynamic performance. This work utilizes TCAD simulations to substantiate the efficacy of the proposed structure and to elucidate its underlying mechanism.

II. DESCRIPTION OF THE STRUCTURE

The analysis in this work will focus on the FIJBS. Taking 3.3kV FIJBS with three layers of floating islands as an example, the parameters of the device refer to the previous model[20]. The thickness of the drift region is 32μm and the doping concentration is 1.4e16 cm^{-3}. In order to clearly show the structures of different devices, the cross-sections of conventional Junction Barrier Schottky Diode (Conv. JBS), the Floating Island JBS (FIJBS), the FIJBS with N+ Buffer (NB-FIJBS) and the FIJBS with Width-Optimized Island (WO-FIJBS) are depicted in Fig. 1 (a), (b), (c) and (d), respectively.

979-8-3315-1110-4/25 $31.00 © 2025 IEEE

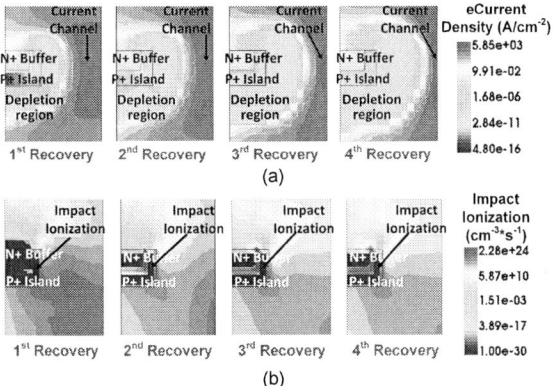

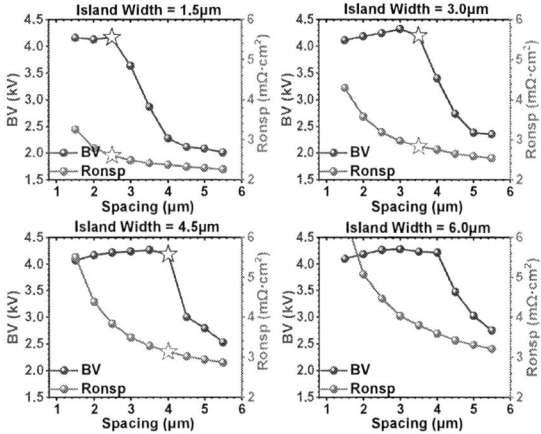

(a)

(b)

Fig. 2. Dynamic changes inside the FI device during the first four forward recoveries: (a) The distribution of depletion region and current channel around the floating island, (b) The distribution of impact ionization at the floating island.

Obviously, the key distinction of WO-FIJBS lies in its utilization of floating islands with nonuniform widths. Our findings demonstrate that this structure can further enhance the dynamic characteristics of FI devices, particularly their repetitive recovery capability. The underlying mechanisms and design rationale will be elaborated in following sections.

III. MECHANISM OF REPETITIVE RECOVERY ISSUE

The essence of the repetitive recovery issue of floating islands is the balance problem of internal electron-hole pairs. In previous studies, in order to solve the single recovery problem, we added the N+ buffer to the floating island. The principle is that during forward recovery, N+ buffer injects electrons into the drift region, thereby neutralizing the space charge within the drift region[19]. This process mitigates the potential barrier of the drift region and subsequently reduces the voltage drop. However, during repetitive switching tests, we observed that the turn-on voltage of the device continued to rise until it reached a stable value. This will greatly increase the conduction loss of the device.

Through simulation and theoretical analysis, we have discovered the cause of the repetitive recovery issue. This is because at the moment of turn-on, N+ buffer injects electrons outward, but at the moment of turn-off, the electrons in the drift region do not return to N+ buffer. This will cause the space charge inside N+ buffer to accumulate continuously after multiple recoveries, and the space charge in the P+ floating island will also accumulate. Due to the existence of the space charge inside the floating island, the potential barrier of the floating island will increase, causing a corresponding depletion region to form in the drift region around the island, hindering the flow of electrons at the spacing between islands as shown in the Fig. 2 (a). When the internal space charge of the floating island accumulates continuously and the internal electric field approaches the critical electric field, strong impact ionization will occur at the moment of turn-on, generating electron-hole pairs to replenish the lost carriers as shown in the Fig. 2 (b). Therefore, stability will be reached after multiple recoveries.

It is worth noting that if the doping concentrations of the P+ island and N+ buffer remain fixed, the critical electric field inside the floating island will not change, nor will the width of the space charge region or the potential barrier within the island. Consequently, the width of the space charge region in the surrounding drift region will also remain unchanged. This

Fig. 3. The simulated BV and static $R_{on,sp}$ of the FIJBS with different island widths and lateral spacings. Data points used for comparing dynamic characteristic in the Fig. 5 have been marked with pentacles.

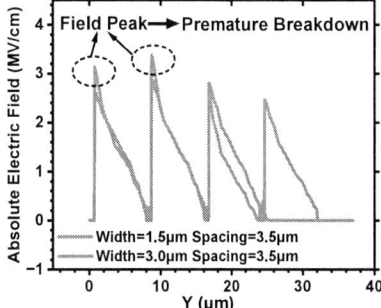

Fig. 4. The simulated electric field in the FIJBS with different island widths under the same spacing at their breakdown voltages. The electric field peak in the FIJBS with narrower island are circled.

means that increasing the spacing can widen the conductive path between the islands under dynamic equilibrium, thereby reducing the on-state voltage drop.

However, simply increasing the spacing would weaken the ability of the floating islands to regulate the electric field during blocking state, thereby reduce the breakdown voltage. In our previous studies, the width of the islands was usually unified to 1.5μm[19, 21, 22], which limited the increase in spacing. In this work, we perform simulation analysis on devices with different floating island widths and select the optimal device from different widths for lateral comparison.

IV. STATIC CHARACTERISTIC

To ensure the static characteristics of FI devices meet specifications, particularly breakdown voltage, this section conducts static simulations on FI devices with different island widths and analyzes their underlying mechanisms.

As shown in the Fig. 3, the static characteristics of FI devices with different floating island widths are compared. Obviously, increasing the width of the floating island enables the utilization of wider spacing. However, wider islands lead to an overall increase in $R_{on,sp}$ because their enlarged proportion within the cell, which constrict conduction paths. This adverse effect can be mitigated by optimizing the spacing design. The optimal spacing for each island width is marked with pentacles in the Fig. 3. These devices will be compared in the next section to evaluate their repetitive recovery

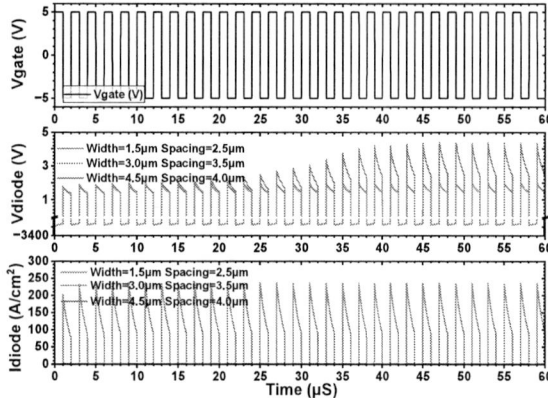

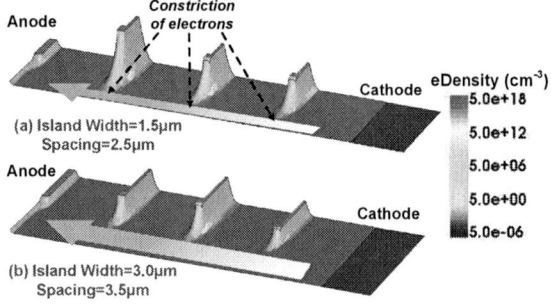

Fig. 5. Waveforms in repetitive switching tests, including the gate voltage of the MOSFET, the anode-to-cathode voltage and the current density of the FIJBS with different island widths and spacings.

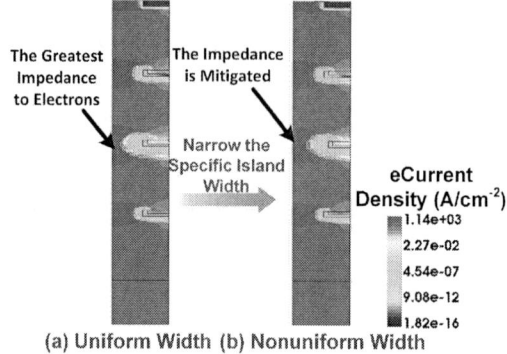

(a) Uniform Width (b) Nonuniform Width

Fig. 7. The distribution of electron current after multiple recoveries, for (a) the NB-FIJBS and (b) the WO-FIJBS. The widths of the floating islands in NB-FIJBS are the same, while the widths of the floating islands in WO-FIJBS are different.

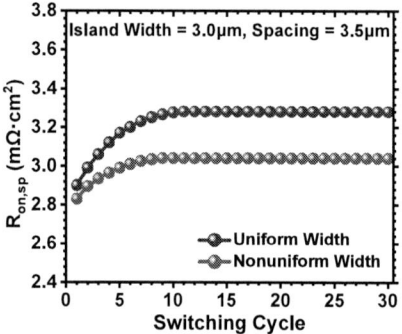

Fig. 6. The distribution of the conduction band energy (indicated by height) and the electron density (indicated by color) after multiple recoveries, for (a) the FIJBS with 1.5μm island width and 2.5μm spacing, (b) the FIJBS with 3μm island width and 3.5μm spacing. The electron flow and constriction of electrons are also illustrated.

Fig. 8. Recovered $R_{on,sp}$ of the NB-FIJBS and the WO-FIJBS during repetitive switching tests.

performance. It should be noted that we did not select device with 6μm island width, as their optimal spacing is comparable to that of the device with 4.5μm island width, while the latter exhibits clear advantages.

Here, we briefly analyze how increasing island width affects spacing optimization. To put it briefly, wider islands can enhance electric field regulation, which improves blocking capability. This allows the utilization for larger spacing. More precisely, wider floating islands reduce the lateral electric field gradient, which in turn decrease the vertical field gradient, thus suppressing local field peak at the floating island and prevent premature breakdown as shown in the Fig. 4.

V. REPETITIVE SWITCHING PERFORMANCE

As shown in the Fig. 5, the device with increased spacing demonstrates significantly improved dynamic performance. Its on-state voltage remains consistently low even after multiple switching cycles. In contrast, the device with 1.5μm island width exhibits notable on-state voltage increase, which would lead to higher energy loss and degraded system stability.

To better illustrate the advantages of FI device with wider islands, Fig. 6 shows the conduction band energy during forward recovery when the dynamic equilibrium is established. Evidently, the conduction path in narrow-spacing devices is severely constrained, leading to significantly increased on-resistance. In contrast, wide-spacing

configuration maintain wider current path, thereby enhancing conductivity.

Additionally, the simulation results indicate that the dynamic performance of the device with 4.5μm floating island width is nearly identical to those with 3μm width. However, considering practical fabrication process, we still recommend selecting devices with a 3μm floating island width.

VI. OPTIMIZATION OF FLOATING ISLAND WIDTH

In previous work, the width and spacing of floating islands in a device were usually set to uniform values. Although this can reduce the number of layouts and simplify the manufacturing process, it affects the optimal performance of the FI devices. We found that in repetitive switching tests, the depletion region around the islands of certain layers is significantly larger than that of other floating islands, contributing the greatest impedance to electrons as shown in Fig. 7. The possible reason for this phenomenon is that the N+ buffer of the lower islands inject relatively fewer electrons outward, while the N+ buffer of the upper islands can accept a certain amount of electrons during reverse recovery. Therefore, the middle islands require stronger impact ionization to provide more electrons. Based on this, we believe that it is possible to optimize certain layers of floating islands specifically.

In this study, we propose a simple method, namely reducing the width of the floating islands at the certain layers as shown in the Fig. 7 (b). Certainly, reducing the island width necessitates re-simulation to ensure the breakdown voltage meets specifications. The Fig. 8 shows that the dynamic

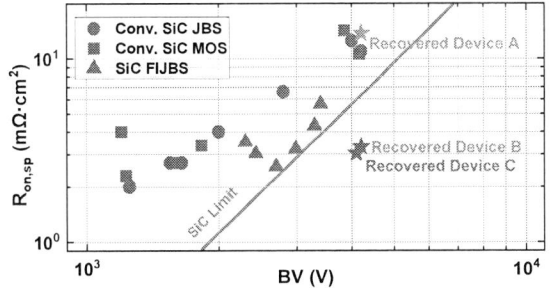

Fig. 9. The dynamic trade-off of the NB-FIJBS with different island width, the newly proposed WO-FIJBS, and the static trade-off of other SiC FIJBSs, the Conv SiC JBSs and the Conv SiC MOSs in literature.

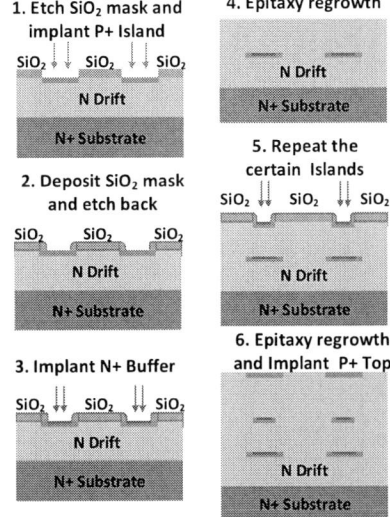

Fig. 10. The manufacture process of the WO-FIJBS.

characteristics of the optimized device are superior to those of devices with uniformly set floating islands. The dynamic trade-off of the recovered $R_{on,sp}$ and BV of this newly proposed WO-FIJBS, together with the dynamic trade-off of the conventional NB-FIJBS are plotted in Fig. 9. Device A represents the NB-FIJBS with 1.5μm island width and 2.5μm spacing; Device B represents the NB-FIJBS with 3μm island width and 3.5μm spacing; Device C is the newly proposed WO-FIJBS. The static trade-off of other SiC FIJBSs in literature [14, 15, 23-25], the Conv SiC JBSs, the 1-D SiC Limit [1, 2] are also shown. It is evident that the NB-FIJBS with narrow floating island exhibits an extremely high recovered $R_{on,sp}$, potentially even failing to function properly. In contrast, the NB-FIJBS with wide floating island successfully break the 1-D SiC limit even after repetitive recoveries. The newly proposed WO-FIJBS further reduces the recovered $R_{on,sp}$, demonstrating excellent application potential.

Besides, the obvious advantage of this structure lies in its straightforward fabrication process. As illustrated in the Fig. 10, the floating island layers can be fabricated simply by repeating the lithography pattern with different island widths.

VII. CONCLUSION

Dynamic issue is the limiting factor hindering practical applications of Floating Island devices. In order to investigate the underlying mechanisms and propose potential solutions,

this work focuses on three aspects: (1) analyzing the intrinsic mechanism of repetitive recovery issue of floating island devices; (2) examining the impact of floating island width and spacing on the static and dynamic characteristics of floating islands; and (3) proposing the idea for structural optimization of specific floating islands. In summary, the newly proposed structure enhances the dynamic performance of FI devices without requiring complex fabrication processes, demonstrating promising application potential. We hope that this work can provide a reference for subsequent research on SiC floating island devices.

ACKNOWLEDGMENT

This work is supported by the National Natural Science Foundation of China under Grant U23B20136.

REFERENCE

[1] H. Niwa, J. Suda, and T. Kimoto, "Impact Ionization Coefficients in 4H-SiC Toward Ultrahigh-Voltage Power Devices," *IEEE Transactions on Electron Devices,* vol. 62, no. 10, pp. 3326-3333, 2015.

[2] J. A. Cooper and D. T. Morisette, "Performance Limits of Vertical Unipolar Power Devices in GaN and 4H-SiC," *IEEE Electron Device Letters,* vol. PP, no. 99, pp. 1-1, 2020.

[3] X. Chen, "Semiconductor power devices with alternating conductivity type high-voltage breakdown regions," *U.S. Patent 5216275,* 1993.

[4] FUJIHIRA and Tatsuhiko, "Theory of Semiconductor Superjunction Devices," *Japanese journal of applied physics. Pt. 1, Regular papers & short notes,* vol. 36, no. 10, pp. 6254-6262, 1997.

[5] N. Cezac, F. Morancho, P. Rossel, H. Tranduc, and A. Payre-Lavigne, "A new generation of power unipolar devices: the concept of the FLoating islands MOS transistor (FLIMOST)," in *12th International Symposium on Power Semiconductor Devices & ICs.,* May 2000, pp. 69-72, doi: 10.1109/ISPSD.2000.856775.

[6] N. Cezac, P. Rossel, F. Morancho, H. Tranduc, A. Peyre-Lavigne, and I. Pages, "A new generation of power devices based on the concept of the "Floating Islands"," in *2000 22nd International Conference on Microelectronics. Proceedings (Cat. No.00TH8400),* 14-17 May 2000 2000, vol. 2, pp. 637-640 vol.2, doi: 10.1109/ICMEL.2000.838771.

[7] X. B. Chen, X. Wang, and J. K. O. Sin, "A novel high-voltage sustaining structure with buried oppositely doped regions," *IEEE Transactions on Electron Devices,* vol. 47, no. 6, pp. 1280-1285, Jun. 2000, doi: 10.1109/16.842974.

[8] W. Saitoh, I. Omura, K. i. Tokano, T. Ogura, and H. Ohashi, "Ultra low On-resistance SBD with P-buried floating layer," in *Proceedings of the 14th International Symposium on Power Semiconductor Devices and Ics,* 2002: IEEE, pp. 33-36.

[9] W. Saito, I. Omura, K. Tokano, T. Ogura, and H. Ohashi, "A novel low on-resistance Schottky-barrier diode with p-buried floating layer structure," *IEEE Transactions on Electron Devices,* vol. 51, no. 5, pp. 797-802, 2004, doi: 10.1109/ted.2004.826886.

[10] S. Alves, F. Morancho, J. Reynes, J. Margheritta, I. Deram, K. Isoird, and H. Tranduc, "Technological realization of low on-resistance FLYMOS/spl trade/ transistors dedicated to automotive applications," in *European Conference on Power Electronics and Applications,* Sept. 2005, pp. 1-10, doi: 10.1109/EPE.2005.219700.

[11] S. Alves, F. Morancho, J. M. Reynès, J. Margheritta, I. Deram, and K. Isoird, "Experimental validation of the "FLoating Island" concept: realization of low on-resistance FLYMOS™ transistors," *The European Physical Journal Applied Physics,* vol. 32, no. 1, pp. 7-13, 2005, doi: 10.1051/epjap:2005058.

[12] H. Takaya, K. Miyagi, K. Hamada, Y. Okura, N. K. Tokura, and A. Kuroyanagi, "Floating island and thick bottom oxide trench gate MOSFET (FITMOS) - a 60V ultra low on-resistance novel MOSFET with superior internal body diode," *Proceedings. ISPSD*

'05. The 17th International Symposium on Power Semiconductor Devices and ICs, 2005., pp. 43-46, 2005.

[13] S. Alves, F. Morancho, J. M. Reynes, J. Margheritta, I. Deram, K. Isoird, and B. Beydoun, "Experimental validation of the 'FLoating Islands' concept: 95 V breakdown voltage vertical FLIDiode," *IEE Proceedings - Circuits, Devices and Systems,* vol. 153, no. 1, 2006, doi: 10.1049/ip-cds:20050048.

[14] C. Ota, J. Nishio, K. Takao, T. Hatakeyama, T. Shinohe, K. Kojima, S. I. Nishizawa, and H. Ohashi, "Doping Concentration Optimization for Ultra-Low-Loss 4H-SiC Floating Junction Schottky Barrier Diode (Super-SBD)," *Materials Science Forum,* vol. 615-617, pp. 655-658, 2009, doi: 10.4028/www.scientific.net/MSF.615-617.655.

[15] A. Bolotnikov, P. A. Losee, R. Ghandi, S. Kennerly, R. Datta, and X. She, "SiC Charge-Balanced Devices Offering Breakthrough Performance Surpassing the 1-D Ron versus BV Limit," *Materials Science Forum,* vol. 963, pp. 655-659, 2019, doi: 10.4028/www.scientific.net/MSF.963.655.

[16] R. Ghandi, A. Bolotnikov, D. Lilienfeld, S. Kennerly, and R. Ravisekhar, "3kV SiC Charge-Balanced Diodes Breaking Unipolar Limit," in *2019 31st International Symposium on Power Semiconductor Devices and ICs (ISPSD),* 2019: IEEE, pp. 179-182.

[17] R. Ghandi, A. Bolotnikov, S. Kennerly, C. Hitchcock, P. Tang, and T. P. Chow, "4.5kV SiC Charge-Balanced MOSFETs with Ultra-Low On-Resistance," in *2020 32nd International Symposium on Power Semiconductor Devices and ICs (ISPSD),* 13-18 Sept. 2020 2020, pp. 126-129, doi: 10.1109/ISPSD46842.2020.9170171.

[18] J. Knoll, M. Shawky, S. H. Yen, I. Eshera, C. DiMarino, R. Ghandi, S. Kennerly, and C. Buttay, "Characterization of 4.5 kV Charge-Balanced SiC MOSFETs," in *2021 IEEE Applied Power Electronics Conference and Exposition (APEC),* 14-17 June 2021 2021, pp. 2217-2223, doi: 10.1109/APEC42165.2021.9487454.

[19] C. Wang, H. Wang, H. Cheng, and K. Sheng, "A Novel Solution to the Turn-On Recovery Problem of the Floating Island Device," *IEEE Transactions on Electron Devices,* vol. 70, no. 9, pp. 4596-4603, 2023, doi: 10.1109/TED.2023.3299898.

[20] C. Wang, H. Wang, N. Ren, Q. Guo, and K. Sheng, "Analytical Model and Optimization for SiC Floating Island Structure," *IEEE Transactions on Electron Devices,* vol. 68, no. 1, pp. 222-229, 2021, doi: 10.1109/TED.2020.3039433.

[21] C. Wang, H. Wang, Q. Que, M. Bai, Y. Wu, H. Cheng, J. Wan, J. Li, Y. Li, C. Zhang, H. Xu, N. Ren, Q. Guo, and K. Sheng, "2kV SiC Floating Island SBD with N+ Buffer," in *2024 36th International Symposium on Power Semiconductor Devices and ICs (ISPSD),* 2-6 June 2024 2024, pp. 1-4, doi: 10.1109/ISPSD59661.2024.10579606.

[22] C. Wang, H. Wang, and K. Sheng, "Floating Island Structure With Metal Bridge to Resolve the Turn-On Recovery Problem," *IEEE Transactions on Electron Devices,* pp. 1-4, 2024, doi: 10.1109/TED.2024.3438102.

[23] C. Ota, J. Nishio, T. Hatakeyama, T. Shinohe, K. Kojima, S. I. Nishizawa, and H. Ohashi, "Fabrication of 4H-SiC floating junction Schottky barrier diodes (super-SBDs) and their electrical properties," *Materials science forum,* vol. 527, pp. 1175-1178, 2006, doi: https://doi.org/10.4028/www.scientific.net/MSF.527-529.1175.

[24] C. Ota, J. Nishio, T. Hatakeyama, T. Shinohe, K. Kojima, S. I. Nishizawa, and H. Ohashi, "Simulation, Fabrication and Characterization of 4H-SiC Floating Junction Schottky Barrier Diodes (Super-SBDs)," *Materials Science Forum,* vol. 556-557, pp. 881-884, 2007, doi: 10.4028/www.scientific.net/MSF.556-557.881.

[25] H. Yuan, C. Wang, X. Tang, Q. Song, Y. He, Y. Zhang, Y. Zhang, L. Xiao, L. Wang, and Y. Wu, "Experimental Study of High Performance 4H-SiC Floating Junction JBS Diodes," *IEEE Access,* vol. 8, pp. 93039-93047, 2020, doi: 10.1109/access.2020.2994625.

High-Temperature Device Model for GaN-based MIS-HEMTs

Hanlin Cao[1,2], Pingyu Cao[1,2], Sang Lam[1,2], Miao Cui[1,2]*

[1]Department of Electrical and Electronic Engineering, Xi'an Jiaotong-Liverpool University, Suzhou, China

[2]School of Advanced Technology, Xi'an Jiaotong-Liverpool University, Suzhou, China

*Miao.Cui02@xjtlu.edu.cn

Abstract- This study develops a high-temperature model for GaN-based MIS-HEMTs, enabling accurate simulation from 25 °C to 200 °C. Using the Advanced Curtice Quadratic Model, key parameters like threshold voltage and transconductance were fitted to experimental data and adjusted with temperature. Simulated I-V characteristics closely matched measurements, showing clear thermal trends. Dynamic DCFL inverter simulations confirmed the model's reliability, with rise time increasing at higher temperatures. The model accounts for parasitic effects and supports both static and dynamic analysis, making it suitable for high-temperature applications in automotive, aerospace, and renewable energy.

Introduction

Gallium nitride (GaN) has attracted significant attention due to its excellent performance in high-temperature, high-power, and high-frequency applications. Compared to silicon (Si) and silicon carbide (SiC), GaN offers superior electron mobility, higher saturation velocity, and a stronger breakdown electric field, making it an ideal candidate for next-generation power devices [1–3].

The Metal-Insulator-Semiconductor HEMT (MIS-HEMT) structure introduces a gate insulator, which reduces leakage current, enhances breakdown voltage, and improves thermal stability. GaN MIS-HEMTs can reliably operate above 200 °C, surpassing the 150 °C limit of Si-based devices, making them suitable for harsh environments [4, 5].

Despite their advantages, existing simulation models for GaN devices lack precision at elevated temperatures. While the Curtice quadratic model has been adapted for MIS-HEMTs with increasing temperatures, it primarily addresses static behavior [6-8]. Parker *et al.* improved the FET model, which enhances consistency between small-and large-signal simulations [9]. However, these models lack dynamic performance modeling across temperatures. Moreover, models like ASM-GaN focus on dynamic characteristics by capturing parasitic capacitances

[10], although they did not consider temperature effects.

Accurate modeling is crucial for integrated circuit design, particularly for predicting dynamic performance across various geometries and thermal conditions [11]. To improve simulation fidelity, this study employs the Advanced Curtice Quadratic Model and enhances it by fitting experimental data, incorporating parasitic effects, and analyzing temperature-dependent parameters such as threshold voltage and transconductance.

This work presents a high-temperature GaN model in Advanced Design System (ADS), accurately simulating static and dynamic behavior from 25 °C to 200 °C. The model was validated through experimental calibration and inverter simulations, demonstrating good agreement with measured data. This work provides a reliable GaN model for high-temperature applications.

High-Temperature Model and Results

A. Static simulation

Fig.1 shows the schematic of E/D mode GaN MIS-HEMTs, a 20 nm Al_2O_3 layer was deposited to reduce gate leakage current. The AlGaN barrier of E-mode device was fully recessed to achieve a positive threshold voltage, while the D-mode device had an etch depth of 18 nm. The detailed fabrication process was described in [12].

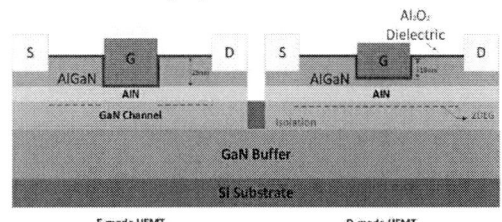

Fig. 1 Schematic of E/D mode GaN MIS-HEMTs.

The device model must be properly calibrated to simulate the electrical properties of AlGaN/GaN MIS-HEMTs accurately. The dataset utilized in this project is based on experimental data of enhancement-mode (E-mode) and depletion-mode (D-mode) GaN MIS-

HEMTs from 25 °C to 200 °C. The dataset comprises current-voltage curves, specifically illustrating the dependencies between drain-source current, drain-source voltage, and gate-source voltage.

For simulation in ADS, the Advanced Curtice Quadratic Model [7] was used to represent multi-channel heterojunction field-effect transistors (MIS-HEMTs). the drain current is described by the following equation:

$$I_{ds} = \beta_U \cdot \left(V_{gs} - V_{to}\right)^2 \times (1 + \lambda \cdot V_{ds})$$
$$\times \tanh(\alpha \cdot V_{ds}) \cdot A \quad (1)$$

where

$$\beta_U = \frac{\beta}{\left[1 + \left(V_{gs} - V_{to}\right) \cdot U_{crit}\right]} \quad (2)$$

V_{to} is the threshold voltage, β is the transconductance parameter, β_U is the degenerated transconductance parameter, λ is the channel length modulation, α is the saturation parameter, A defines the device width parameter, and U_{crit} refers to the critical field for mobility degradation.

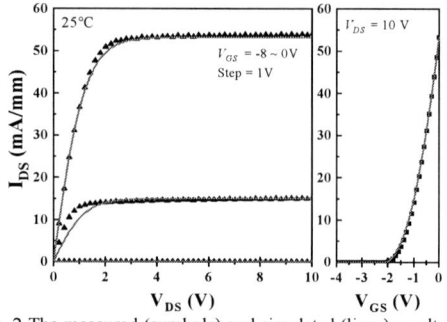

Fig. 2 The measured (symbols) and simulated (lines) results of D-mode GaN devices at 25 °C.

Fig. 2 presents a comparison between the measured and simulated I-V characteristics. Following parameter adjustment, the improved fitting accuracy demonstrates the effectiveness of this model in the static simulation.

The Advanced Curtice Quadratic Model incorporates detailed temperature dependent equations [8] that are used to analyze GaN MIS-HEMTs behavior under various ambient conditions. The temperature dependent equations can be expressed as:

$$V_{to}(T) = V_{to} + V_{totc} \times (T - T_{nom}) \quad (3)$$

$$\alpha(T) = \alpha \times (1.01)^{\alpha_{tc} \times (T - T_{nom})} \quad (4)$$

$$\beta(T) = \beta \times (1.01)^{\beta_{tc} \times (T - T_{nom})} \quad (5)$$

Where T represents the characterization temperature, T_{nom} denotes the temperature at which the model parameters were originally determined. The threshold

voltage V_{to} scales linearly with temperature. The scale factor is V_{totc}. $\alpha(T)$ is the high-temperature saturation parameter, α_{tc} is the α temperature coefficient, $\beta(T)$ is the high-temperature transconductance, β_{tc} is the β temperature coefficient.

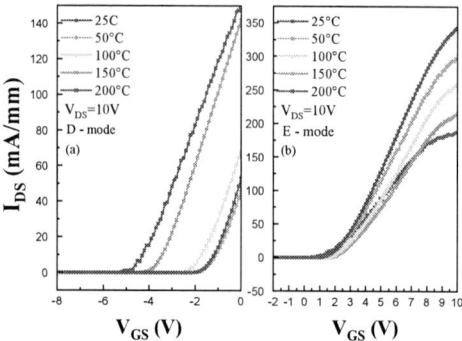

Fig. 3 The measured I_{DS}-V_{GS} curves of D-mode (a) and E-mode (b) devices with V_{DS}=10 V at various temperatures.

The measured transfer characteristics are shown in Fig. 3. Based on equation (3), the temperature dependent threshold voltage can be expressed as follows:

$$V_{to\_D}(T) = -1.72 - 0.018 \times (T - 25) \quad (6)$$
$$V_{to\_E}(T) = 1.30 + 0.005 \times (T - 25) \quad (7)$$

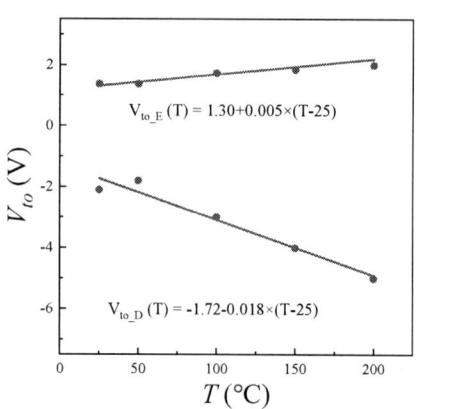

Fig. 4 The measured (symbols) and linear fit (lines) of threshold voltage of D-mode and E-mode.

In Fig. 4, the threshold voltage of D-mode device shows a negative shift trend with temperature, which is caused by the interface traps emission at elevated temperature. While the threshold voltage of E-mode exhibits a slightly positive shift, which might be caused by traps and lattice deformation of the GaN layer at elevated temperatures.

The parameters U_{crit} and β should be considered under different temperatures. The degradation coefficient, also referred to as the critical electric field for mobility degradation U_{crit}, is related to the drain-source current I_{ds} through equations (1) and (2). When devices operate in the saturation region, the term $(1 + \lambda \cdot V_{ds}) \times \tanh(\alpha \cdot V_{ds})$ can be approximated as 1. Consequently, U_{crit} can be determined by using two drain-source current values I_{ds} measured at different gate-source voltages, as described by the following equation:

$$\frac{I_{ds1}}{I_{ds2}} = \frac{1 + (V_{gs} - V_{to})_{\_2} U_{crit}}{1 + (V_{gs} - V_{to})_{\_1} U_{crit}} \left[\frac{(V_{gs} - V_{to})_{\_1}}{(V_{gs} - V_{to})_{\_2}} \right]^2 \quad (8)$$

The value of U_{crit} can be determined by utilizing the drain-source current I_{ds} measured under different temperatures. With the extracted U_{crit} and the laboratory-measured data, the transconductance parameter β for each condition can be expressed as:

$$\beta = \beta_U \times \left[1 + (V_{gs} - V_{to}) \times U_{crit} \right]$$
$$= \frac{I_{ds}}{\left[(V_{gs} - V_{to})^2 \times (1 + \lambda \cdot V_{ds}) \times \tanh(\alpha \cdot V_{ds}) \cdot A \right]}$$
$$\times \left[1 + (V_{gs} - V_{to}) \times U_{crit} \right] \quad (9)$$

λ is a negligibly small parameter and can be approximated as zero. As a result, the transconductance parameter β and the saturation parameter α can be extracted under various GaN HEMT operating conditions. This process was performed by implementing the I_{ds} model in MATLAB and applying a nonlinear least-squares fitting algorithm to determine the values of β and α accurately.

The calculated parameter values were validated in ADS, and the corresponding β and α were extracted at different temperatures to analyze their temperature-dependent behavior. By applying the temperature scaling equations for β and α, this approach facilitates the identification of the optimal fitting relationships for both $\beta - T$ and $\alpha - T$.

For D-mode devices, the two parameters as a function of temperature can be expressed as:

$$\alpha(T) = 0.87 \times (1.01)^{-0.72 \times (T-25)} \quad (10)$$

$$\beta(T) = \begin{cases} 2.34 \times 10^{-4} \times (1.01)^{4.76 \times (T-25)}, & (T < 100) \\ 7.4 \times 10^{-2} \times (1.01)^{-2.96 \times (T-25)}, & (T \geq 100) \end{cases}$$
$$(11)$$

Fig. 5 shows the curves of $\beta - T$ and $\alpha - T$ for D-mode GaN MIS-HEMTs. As shown in the figure, α exhibits an exponential decrease as the temperature increases, which is consistent with [8]. The value of β initially increases with temperature but starts to decrease once the temperature reaches 100 °C. This phenomenon is related to the temperature-dependent variation of D-mode I_{ds} in Fig. 6. When the temperature is below 100 °C, the threshold voltage dominates, and the negative shift of threshold voltage will lead to an increase in β. However, when the temperature exceeds 100 °C, mobility is the dominant factor for the degradation of β at high temperatures.

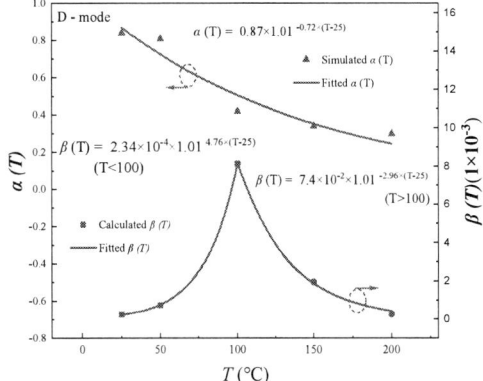

Fig. 5 The calculated (symbols) and fitted equation curves (lines) of α and β of D-mode devices at various temperatures.

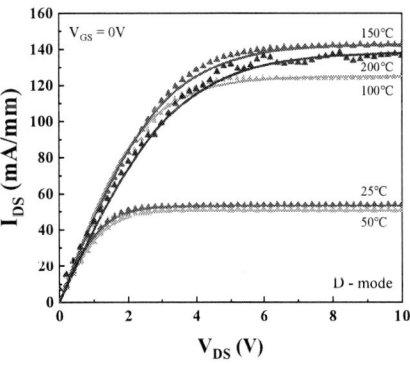

Fig. 6 The measured (symbols) and simulated (lines) I_{DS}-V_{DS} curves of D-mode GaN MIS-HEMTs.

Fig. 6 shows a comparison between the simulated and measured results of I_{DS}. It can be observed that the simulation accuracy remains consistently high across different temperatures. In Fig. 6, the saturation current of D-mode MIS-HEMT exhibits a non-monotonic with respect to temperature, reaching a peak around 150 °C before subsequently decreasing. This behavior is primarily driven by a negative shift in the threshold voltage (V_{to}), as described in Equation (6). At room temperature, electrons are trapped at the

Al_2O_3/AlGaN interface. As the temperature increases, trapped electrons are released, causing a negative shift of V_{to} with temperature. This shift enhances the saturation current in Equation (1).

Similarly, for E-mode GaN devices, α and β as a function of temperature can be expressed as:

$$\alpha(T) = 0.16 \times (1.01)^{0.14 \times (T-25)} \quad (12)$$
$$\beta(T) = 2.89 \times 10^{-3} \times (1.01)^{-1.45 \times (T-25)} \quad (13)$$

Fig. 7 shows the curves of $\beta - T$ and $\alpha - T$ for the E-mode GaN MIS-HEMTS. The temperature dependence of the parameters in the E-mode devices follows a similar trend in. α increases exponentially with temperature, while β decreases exponentially with temperature.

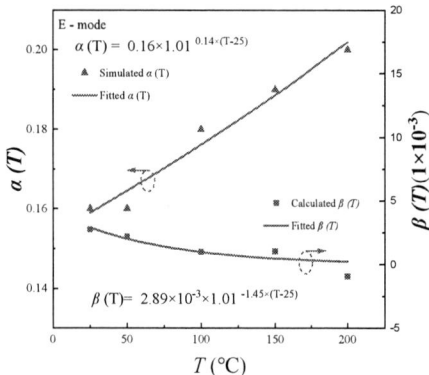

Fig. 7 The calculated (symbols) and fitted equation curves (lines) of α and β of E-mode devices at various temperatures.

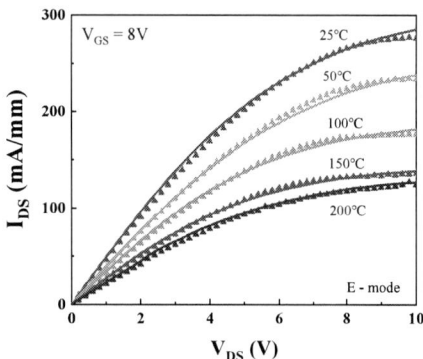

Fig. 8 The measured (symbols) and simulated (lines) I_{DS}-V_{DS} curves of E-mode MIS-HEMTs.

Fig. 8 presents the I_{DS}-V_{DS} characteristics of the E-mode MIS-HEMTs. As the temperature increases, a gradual decline in saturation current is observed. This behavior aligns with the anticipated effect of elevated temperatures, which leads to a degradation in the performance of the GaN HEMTs. Overall, the E-

mode simulation results demonstrate strong agreement with experimental data, indicating that the high-temperature models are effective.

B. Dynamic Simulation

Following the successful completion of static simulations for GaN MIS-HEMTs under various temperature conditions, dynamic simulation at high temperatures is investigated in this work. Previous studies on the performance of DCFL inverters at high temperatures have consistently shown that the inverter's rise time increases as the temperature rises [12]. Accordingly, this study will conduct dynamic simulations and calibration of the inverter based on the high-temperature model, utilizing the measured data. The GaN inverter consists of one D-mode device and one E-mode device. The dynamic output waveforms were measured from 25 °C to 200 °C in Fig. 9. The rise time exhibits an obvious increasing trend with temperature due to current degradation at high temperatures.

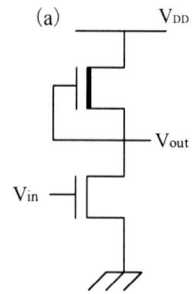

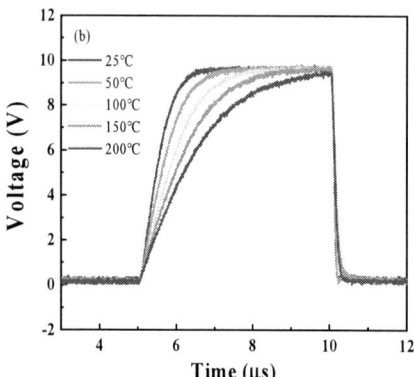

Fig. 9 (a) The circuit diagram of GaN DCFL inverter. (b) Dynamic output voltage at various temperatures, the frequency is 100 kHz.

In the dynamic simulation of the DCFL inverter, parasitic capacitance plays a crucial role, as it significantly influences the inverter's performance, particularly its rise time. In this work, the parasitic capacitances were considered and cited from [10] to simulate dynamic behavior at various temperatures.

Fig. 10 shows simulated results of the GaN inverter at various temperatures. The temperature-dependent trend is consistent with findings from previous studies [12]. The inverter's rise time obviously increases as temperature increases, while the fall time experiences a slight increase at elevated temperatures. This observation confirms that the preliminary simulation results align with the actual thermal behavior, thereby providing valuable reference data at high-temperature applications.

Regarding dynamic simulations, this study did not perform a quantitative analysis or validation of the simulation results. Instead, the performance of the thermal model was evaluated based on temperature-dependent trends. Therefore, future work should carefully consider temperature-dependent parasitic, and a comparative analysis between the simulation results and the actual performance of DCFL inverters will be evaluated to enhance the dynamic behavior at elevated temperatures.

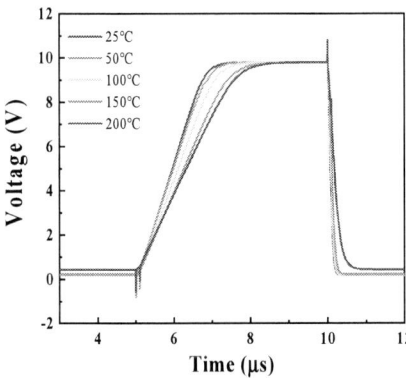

Fig. 10 The simulated results of the GaN inverter at different temperatures.

Conclusion

This study successfully developed a high-temperature device model for both D-mode and E-mode GaN MIS-HEMTs, utilizing the Advanced Curtice Quadratic Model to accurately simulate device behavior across a temperature range from 25 °C to 200 °C. Through calibration with experimental data, key parameters such as threshold voltage (V_{to}), transconductance (β), and saturation parameter (α) were calibrated with high temperatures, demonstrating a strong alignment between simulated and measured I-V characteristics.

Dynamic simulations of a DCFL inverter further validated the model's feasibility, showing an increase in rise time with temperature. The high-temperature model considers parasitic capacitance effects and aligns with experimental results.

This work provides a comprehensive framework for designing GaN-based circuits in automotive, aerospace, and renewable energy applications, where thermal reliability and efficiency are important. The high-temperature model contributes to further progress in GaN integrated circuits, particularly for next-generation high-temperature applications.

Acknowledgments

This work was supported by XJTLU Research Development Fund (RDF21-02-031, PGRS2206039)

References

[1] B. J. Baliga, "Gallium nitride devices for power electronic applications," *Semiconductor Science and Technology*, vol. 28, no. 8, pp. 1-10, 2013.

[2] S. Chowdhury, "Gallium nitride based power switches for next generation of power conversion," *physica status solidi (a)*, vol. 212, pp. 1066–1074, 2015.

[3] T. Imada, M. Kanamura and T. Kikkawa, "Enhancement-mode GaN MIS-HEMTs for power supplies," *The 2010 International Power Electronics Conference - ECCE ASIA -*, Sapporo, Japan, 2010, pp. 1027-1033.

[4] Luis Felipe de Oliveira Bergamim, Bertrand Parvais, Eddy Simoen, Maria Glória Caño de Andrade, "Analog performance of GaN/AlGaN high-electron-mobility transistors", *Solid-State Electronics*, vol.183, pp.108048, 2021.

[5] Z. Mutsafi, K. Shimanovich, V. Kairys, R. Shima-Edelstein, Y. Roizin and Y. Rosenwaks, "High-Temperature Sensitivity of a Depletion-Mode AlGaN/GaN MIS-HEMT," *IEEE Transactions on Electron Devices*, vol. 68, no. 11, pp. 5695-5700, Nov. 2021.

[6] W. R. Curtice and M. Ettenberg, "A Nonlinear GaAs FET Model for Use in the Design of Output Circuits for Power Amplifiers," *IEEE Transactions on Microwave Theory and Techniques*, vol. 33, no. 12, pp. 1383-1394, Dec. 1985.

[7] W. R. Curtice, "A MESFET Model for Use in the Design of GaAs Integrated Circuits," *IEEE Transactions on Microwave Theory and Techniques*, vol. 28, no. 5, pp. 448-456, May 1980.

[8] R. Sun, Y. C. Liang, Y.-C. Yeo, Y.-H. Wang, and C. Zhao, "Design of power integrated circuits in full AlGaN/GaN MIS-HEMT configuration for power conversion," *physica status solidi (a)*, vol. 213, no. 11, pp. 2937-2944, Nov. 2016

[9] A. E. Parker and D. J. Skellern, "An improved FET model for computer simulators," *IEEE Transactions on Computer-Aided Design of Integrated Circuits and Systems*, vol. 9, no. 5, pp. 551-553, May 1990.

[10] F. Li et al., "Investigation on the Dynamic Characteristics of Hydrogen Plasma Treated p-GaN HEMTs Circuit Using ASM-GaN Model," *IEEE Journal of the Electron Devices Society*, vol. 12, pp. 457-463, 2024.

[11] Z. Liu, X. Huang, F. C. Lee and Q. Li, "Package Parasitic Inductance Extraction and Simulation Model Development for the High-Voltage Cascode GaN HEMT," *IEEE Transactions on Power Electronics*, vol. 29, no. 4, pp. 1977-1985, April 2014.

[12] M. Cui, et al., "The Impact of Etch Depth of D-mode AlGaN/GaN MIS-HEMTs on DC and AC Characterizations of 10 V Input Direct-Coupled FET Logic (DCFL) Inverters", *2019 International Conference on IC Design and Technology (ICICDT)*, Suzhou, China, 2019, pp. 1-4.

Dynamic Model of Non-Segmented SiC MOSFETs Considering Switching Behavior and Temperature Effects

Wenzheng Dou
School of Nuclear Science, Energy and Power Engineering
Shandong University
Jinan, China
202420740@mail.sdu.edu.cn

Xin Lan
School of Nuclear Science, Energy and Power Engineering
Shandong University
Jinan, China
lanxin@sdu.edu.cn

Ning Zhao
School of Nuclear Science, Energy and Power Engineering
Shandong University
Jinan, China
202234583@mail.sdu.edu.cn

Abstract—In this work, a behavioral model of SiC MOSFETs is presented, with a focus on dynamic switching behavior. The model is constructed in Matlab/Simulink, and non-segmented equations are employed to describe the static output characteristics. Additionally, the body diode and parasitic capacitance are modeled separately. The validation results demonstrate the high accuracy in output characteristics, transfer characteristics, and body diode characteristics. The model can also accurately predict dynamic parameters, including switching loss, current transition characteristics, and switching waveforms in dynamic characteristics.

Keywords—*SiC MOSFET, behavioral model, temperature-dependent model, device modeling*

I. INTRODUCTION

With the continuous development of power semiconductor device technology, power device simulation technology has become an indispensable and important way to evaluate and optimize device performance. As a key tool in the design and development of power electronic systems, accurate device simulation models are important for the optimization of system performance [1].

Device modeling is defined as the predictive description of a device's current, voltage, and other parameters through a function that incorporates variables such as applied voltage, current, environmental regulation, and physical characteristics. Depending on the modeling approach, SiC MOSFET device models can be classified into five different types: semi-physical [2], physical, half-value, numerical, and behavioral [3]. The modeling approach for SiC MOSFETs primarily entails the utilization of physical, semi-physical, and behavioral models. Behavioral modeling involves the construction of an equivalent circuit model, thereby circumventing the necessity for an analysis of the device's internal physical structure. This approach entails the design of an equivalent circuit that reflects the device's operating characteristics based on its actual operating environment. The efficacy and practicality of this modeling method have led to its extensive application in the domain of power device modeling.

The SiC MOSFET behavioral model is comprised of two components: a static model and a dynamic model. Numerous findings have been obtained in the area of modeling till recently. Mukunoki et al. [4] employed piecewise functions to model the static characteristics of SiC MOSFETs in both linear and saturation regions. However, the ambiguous boundary between these two operational regions complicates the determination of critical thresholds, thereby affecting the accuracy of parameter extraction. Furthermore, this modeling approach typically requires multi-step parameter extraction procedures, significantly increasing the computational and experimental effort required. J. Wang et al.[5] has proposed a PSpice model utilized for 10kV SiC MOSFETs. This model constructs the channel current model and the body diode model from PSpice's proprietary model editor. It also accounts for the variation of static characteristics with temperature. In contrast, Zhou et al. proposed a compact behavioral model for SiC MOSFETs, emphasizing enhanced accuracy and convergence of the model, along with reduced simulation time [6]. It is achieved by leveraging additional degrees of freedom in the channel current model, leading to the elimination of drift resistance and rectifier junction capacitance mathematical models. A fast and accurate model proposed by Sochor et al. is characterized by its compactness, reduced cell count, and the capacity to execute tens of thousands of switching pulses within a brief timeframe [7]. This model is proposed for application in the virtual prototyping of complex power supply circuits. In addition, the temperature characteristics is considered for SiC MOSFET half-bridge power modules by Yang et al.[8] The methodology accurately models the drain-source current, body diode, and parasitic capacitance with temperature. Furthermore, it describes in detail the methodology for parameter extraction during the modeling process. Oggier et al. [9] conducted stage-segmented mathematical modeling of turn-on and turn-off transients in SiC MOSFETs, achieving comprehensive characterization of a 10-kV SiC MOSFET module encompassing both static and dynamic properties. This integrated approach enables the model to reliably predict device behavior across diverse operating conditions.

In this work, a behavioral model of a SiC MOSFET is developed, considering both the static and dynamic switching behavior. The model is constructed by Simulink and uses non-segmented equations to describe the output characteristics. Additionally, the body diode and parasitic capacitance are modeled. The model has been validated to have high accuracy in terms of the output characteristics, transfer characteristics, and body diode characteristics. Furthermore, the model can effectively predict the switching process and obtain the dynamic parameters, is noteworthy.

The National Natural Science Foundation of China(No. 52376118).

979-8-3315-1110-4/25 $31.00 © 2025 IEEE

II. MODEL ANALYSIS

A. I_{DS} model

To address convergence challenges inherent in piecewise modeling approaches, this study proposes a voltage-controlled current source-based non-segmented equation formulation for characterizing the static output characteristics of SiC MOSFETs, and the transfer characteristic model of the device is described based on the Angelov model, which is multiplied to obtain the model of I_{DS}:

$$I_{DS} = k \cdot \{1 + \tanh[a \cdot (V_{GS} + c) + b \cdot (V_{GS} + d)^2]\} \cdot p \cdot \frac{V_{DS}}{1 + q \cdot V_{DS}} \quad (1)$$

where k, a, b, c, and d are parameters related to the transfer characteristics; p and q are parameters of the output characteristics, related to the V_{GS}:

$$p = e_1 \cdot e^{f_1 V_{GS}} + g_1 \quad (2)$$

$$q = e_2 \cdot e^{f_2 V_{GS}} + g_2 \quad (3)$$

where $e1$, $f1$, $g1$, and $e2$, $f2$, $g2$ are the parameters of p and q, respectively. Since both output and transfer characteristics are affected by temperature, it is necessary to introduce a temperature variable as shown in Eq. (4).

$$(e, f, g)_{1,2} = s_1 \cdot (T - 25)^2 + s_2 \cdot (T - 25) + s_3 \quad (4)$$

These parameters were extracted at multiple temperatures and converted into temperature-dependent parameters using quadratic equations, thereby enabling the I_{DS} model to incorporate temperature effects. The introduction of second-order equations also reduces the possibility of model non-convergence.

B. Body diode model

A SiC MOSFET body diode behavior model considering the temperature effect is proposed based on the physical equations, and the forward characteristics can be described as follows:

$$I_{fwd} = \exp\left(\frac{V_{jun} - V_{itr}}{V_{ref}}\right) \quad (5)$$

Where, V_{jun} is the voltage across the PN junction; V_{itr} is the intrinsic voltage; V_{ref} is the reference voltage. These three variables can be expressed as:

$$V_{jun} = V_d - R_s I_{fwd} \quad (6)$$

$$R_s = R_{s(nom)} (1 + \alpha_{R_{s1}} T_j' + \alpha_{R_{s1}} T_j'^2) \quad (7)$$

$$V_{itr} = V_{itr(nom)} (1 + \alpha_{Vitr} T_j') \quad (8)$$

$$V_{ref} = V_{ref(nom)} (1 + \alpha_{Vref} T_j') \quad (9)$$

$$T_j' = T_j - T_{nom} \quad (10)$$

Where V_d is the voltage across the diode; Rs is the series resistance; α_{Rs1}, α_{Rs2}, α_{Vitr} and α_{Vref} are empirical coefficients; T_j' is the relative junction temperature; $V_{itr(nom)}$ is the intrinsic voltage at the reference temperature; $V_{ref(nom)}$ is the reference voltage at the reference temperature; and T_j is the junction temperature.

C. Parasitic capacitance model

(1) Gate-Drain Capacitor C_{GD}

A two-stage equation is proposed to describe C_{GD}:

$$C_{GD} = \begin{cases} M_1 \exp(M_2 \cdot V_{DS}) & 0 < V_{DS} \leq V_0 \\ M_3 V_{DS}^{M_4} & V_0 < V_{DS} < 1000 \end{cases} \quad (11)$$

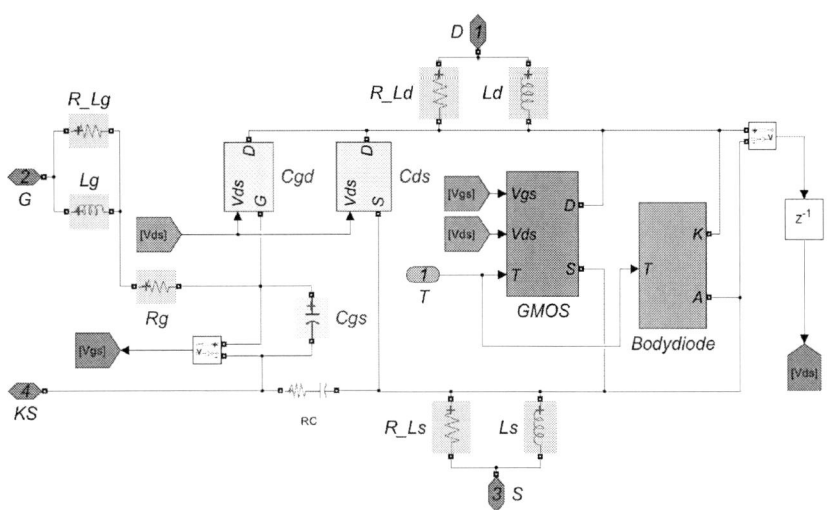

Fig. 1. SiC MOSFET model structure

Where, $M_1\sim M_4$ are the model parameters; V_0 is the voltage value when C_{GD} generates a sudden change.

(2) Drain-Source Capacitor C_{DS}

An exponential function is used to describe C_{DS}:

$$C_{DS} = M_5 + M_6 \exp(V_{DS}/N_1) + M_7 \exp(V_{DS}/N_2) \quad (12)$$

where $M_5\sim M_7$, $N_1\sim N_2$ are model parameters.

(3) Gate-Source Capacitor C_{GS}

Typically, C_{GS} is set to a constant value.

D. Simulink modeling

Figure 1 shows the structure of the SiC MOSFET model, and the main part consists of GMOS, C_{GS}, C_{GD}, C_{DS} and Body diode. In addition, structures such as parasitic inductors and parasitic resistors are also included.

III. MODEL VALIDATION

The accuracy of the proposed electrical model is verified by comparing the simulated results of the dynamic and static performances for the SiC MOSFET C3M0075120K together with those predicted by the official models of Wolfspeed.

A. Parameter Extraction for Static Models

The parameter extraction and validation of the static characteristics of the electrical model are based on the datasheet provided by Wolfspeed. The fixed parameters of the I_{DS} at 25°C are shown in Table 1, and the temperature-dependent characteristic parameters are shown in Table 2.

TABLE I.   FIXED PARAMETERS AT 25°C

Parameter	k	a	b	c	d
Value	4.356	3.293	-18.845	5.471	25.019

TABLE II.   TEMPERATURE CHARACTERIZATION PARAMETERS

Para meter	e_1	f_1	g_1	e_2	f_2	g_2
S_1	-1.81E-4	1.13E-6	2.39E-4	8.67E-3	-7.89E-6	-2.40E-6
S_2	0.036	-1.14E-3	-0.045	-1.36	2.34E-5	3.25E-4
S_3	-4.70	-0.11	2.62	118.06	-0.77	0.085

For the body diode, since Eq. (5) becomes an implicit function after the introduction of each parameter, V_{jun} can be expressed after pending into Eq. (5) and mathematically transforming it:

$$V_d = V_{ref} * \ln I_{fwd} + V_{itr} + Rs * I_{fwd} \quad (13)$$

The SiC MOSFET body diode parameters extracted from the datasheet are shown in Table 3.

TABLE III.   BODY DIODE PARAMETERS

Parameter	Value	Parameter	Value
$R_{s(nom)}$	0.016	α_{Vitr}	-0.0011
α_{Rs1}	-3.56E-4	$V_{ref(nom)}$	1.00
α_{Rs2}	1.01E-05	α_{Vref}	-6.39E-4
$V_{itr(nom)}$	2.56		

B. Static Model Validation

In this work, to verify the static characteristics of SiC MOSFETs, the simulated results of the model with Matlab/Simulink are compared with the characteristics in the datasheet. The simulated results of the output characteristics are compared with the datasheet for various fixed temperatures -40°C, 25°C and 175°C in Figure 2(a), (b) and (c), respectively.

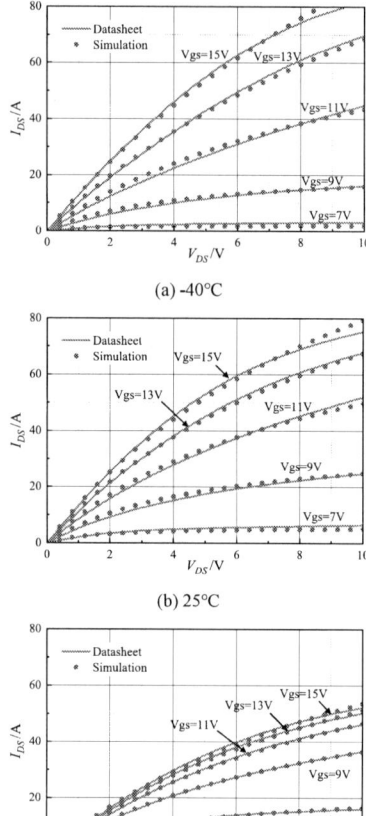

(a) -40°C

(b) 25°C

(c) 175°C

Fig. 2.   Output Characteristics Simulated vs. Datasheet

The simulation results of transfer characteristics under different temperature conditions are plotted against the datasheet as shown in Fig. 3.

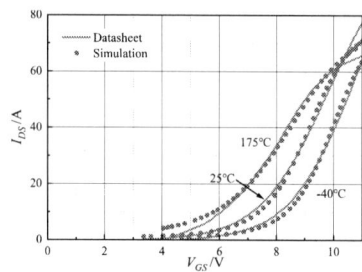

Fig. 3.   Transfer characteristics Simulation vs. Datasheet

The simulation results of body diode forward current at different temperatures are plotted against the datasheet as shown in Fig. 4.

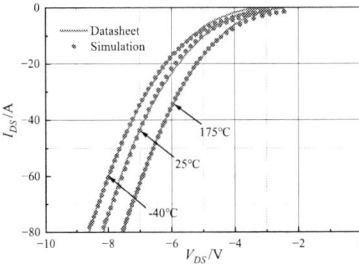

Fig. 4.   Body Diode Characteristics Simulation vs. Datasheet

The results show that the simulation model of SiC MOSFET proposed in this research has high accuracy in output characteristics, transfer characteristics and body diode characteristics, which fully verifies its ability to accurately characterize the static electrical behavior of SiC MOSFETs.

C. Parameter Extraction for Dynamic Models

The C_{GS} is taken from the data sheet as 1388.295pF. The C_{GD} and C_{DS} model parameters are given in tables 4 and 5.

TABLE IV. C_{GD} MODEL PARAMETERS

Parameter	V_0	M_1	M_2	M_3	M_4
Value	20	322.557	-0.182	39.531	-0.575

TABLE V. C_{GS} MODEL PARAMETERS

Parameter	M_5	M_6	N_1	M_7	N_2
Value	57.913	138.904	-146.703	599.046	-13.635

The simulation results of C_{GS}, C_{GD} and C_{DS} are shown in Fig. 5 in comparison with the datasheet.

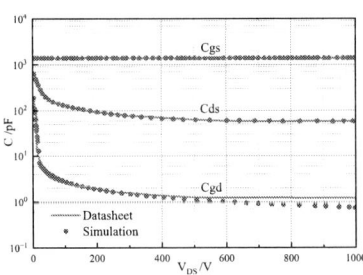

Fig. 5.   Parasitic capacitance Simulated vs. Datasheet

In order to better reflect the dynamic characteristics of SiC MOSFETs, the remaining parasitic parameters in the model are appropriately adjusted according to the datasheet, and the values of each parameter are shown in Table 6.

TABLE VI.   *SIMULATION MODEL PARASITIC PARAMETER VALUES*

Parameter	Value	Parameter	Value
R_Ld	5mΩ	R_Ls	100Ω
Ld	4.37nH	Ls	3.46nH
R_Lg	100Ω	Rg	3.32mΩ
Lg	11.36nH		

D. Dynamic Models Validation

In this section, the dynamic characteristics of the SiC MOSFET model are verified by the officially given device model combined with LTspice simulation, and the switching waveforms are compared as shown in Fig. 6, and the associated switching data are shown in Table 7.

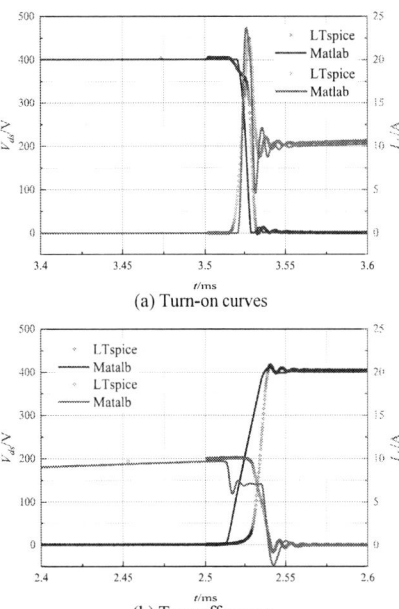

(a) Turn-on curves

(b) Turn-off curves

Fig. 6.   Output Characteristics Simulated vs. Datasheet

As demonstrated in Fig. 6(b), a decline in the current I_{DS} of the simulation model is evident during the shutdown phase. This is attributable to the discharge of the upper body diode junction capacitance, which diverts a proportion of the current from the load inductor. Consequently, the Simulink model exhibits a lower E_{off} value. Furthermore, the turn-on time and turn-off time between the two models demonstrate minimal deviations.

TABLE VII.   *SIMULATION MODEL PARASITIC PARAMETER VALUES*

Model	$t_{d(on)}$	t_r	$t_{d(off)}$	t_f	E_{on}	E_{off}
LTspice	22ns	6ns	29ns	8ns	52μJ	10μJ
Simulink	23ns	7ns	5.5ns	27ns	42μJ	32μJ

The accurate description of the dynamic electrical behavior in the switching process is verified by comparing the simulation results of the proposed model together with those of the official Spice model.

IV. CONCLUSION

In this work, a SiC MOSFET simulation model considering dynamic switching behavior is constructed based on Matlab/Simulink, and the non-segmented equations are used to describe the static output characteristics, and the body diode and parasitic capacitance are modelled separately. The model was validated to have high prediction accuracy in both the static and dynamic performances. The model can be used in modelling and simulation of the electrical behavior of SiC MOSFETs.

ACKNOWLEDGMENT

This work was supported by the National Natural Science Foundation of China(No. 52376118).

REFERENCES

[1] M. R. Nielsen et al., "High-Power Electronic Applications Enabled by Medium Voltage Silicon-Carbide Technology: An Overview," in *IEEE Transactions on Power Electronics.*, vol. 40, no. 1, pp. 987–1011, Jan. 2025.

[2] Q. Liu, P. Sun, G. Peng, and X. Ma, "A Semi-Physical Model of SiC MOSFETs for Improved Static Characteristic," in *2024 IEEE 7th International Electrical and Energy Conference (CIEEC)*, pp. 1842–1845, May. 2024.

[3] Y. Xu, C. N. M. Ho, A. Ghosh, and D. Muthumuni, "A Datasheet-Based Behavioral Model of SiC MOSFET for Power Loss Prediction in Electromagnetic Transient Simulation," in *2019 IEEE Applied Power Electronics Conference and Exposition (APEC)*, pp. 521–526., Mar. 2019.

[4] Y. Mukunoki, K. Konno, T. Matsuo, et al., "An Improved Compact Model for a Silicon-Carbide MOSFET and Its Application to Accurate Circuit Simulation," *IEEE Transactions on Power Electronics*, vol. 33, no. 11, pp. 9834–9842, Nov. 2018.

[5] J. Wang et al., "Characterization, Modeling, and Application of 10-kV SiC MOSFET," in *IEEE Transactions on Electron Devices*, vol. 55, no. 8, pp. 1798-1806, Aug. 2008.

[6] Z. Zhou, Q. Ge, L. Zhao and B. Yang, "A modified general model for sic power MOSFET in rail transportation application,"in *2017 20th International Conference on Electrical Machines and Systems (ICEMS)*, pp. 1–5., 2017.

[7] P. Sochor, A. Huerner and R. Elpelt, "A Fast and Accurate SiC MOSFET Compact Model for Virtual Prototyping of Power Electronic Circuits,"in *PCIM Europe 2019; International Exhibition and Conference for Power Electronics, Intelligent Motion, Renewable Energy and Energy Management*, pp. 1–8, 2019.

[8] P. Yang, W. Ming and J. Liang, "A Step-by-step Modelling Approach for SiC Half-bridge Modules Considering Temperature Characteristics,"in *2020 IEEE Energy Conversion Congress and Exposition (ECCE)*, pp. 2827–2834, 2020.

[9] G. G. Oggier, R. G. Jimenez, Y. Zhao, and J. C. Balda, "Modeling and Characterization of 10-kV SiC MOSFET Modules for Medium-Voltage Distribution Systems," in *2020 IEEE 11th International Symposium on Power Electronics for Distributed Generation Systems (PEDG)*, Sep. 2020, pp. 583–590.

Investigation of the Threshold Voltage Drift in Trench-Type 1200V SiC MOSFETs Under Bipolar Dynamic Gate Stress

1st Jiaying Cao
College of Intelligent Robotics and Advanced Manufacturing
Fudan University
Shanghai, China
23210860002@m.fudan.edu.cn

2nd Hao Guan
College of Intelligent Robotics and Advanced Manufacturing
Fudan University
Shanghai, China
23210860040@m.fudan.edu.cn

3rd Jiuyang Tang
College of Intelligent Robotics and Advanced Manufacturing
Fudan University
Shanghai, China
tangjy23@m.fudan.edu.cn

4th Liudan Kong
College of Intelligent Robotics and Advanced Manufacturing
Fudan University
Shanghai, China
23210860049@m.fudan.edu.cn

5th Yifei Chang
College of Intelligent Robotics and Advanced Manufacturing
Fudan University
Shanghai, China
22210860070@m.fudan.edu.cn

6th Yuhan Duan
College of Intelligent Robotics and Advanced Manufacturing
Fudan University
Shanghai, China
22110860002@m.fudan.edu.cn

Qingchun Zhang*
College of Intelligent Robotics and Advanced Manufacturing
Fudan University
Shanghai, China
Research Institute of Fudan University in Ningbo
Fudan University
Ningbo, China
Qingchun_Zhang@fudan.edu.cn

Pan Liu*
College of Intelligent Robotics and Advanced Manufacturing
Fudan University
Shanghai, China
Research Institute of Fudan University in Ningbo
Fudan University
Ningbo, China
panliu@fudan.edu.cn

Abstract—Since the threshold voltage (V_{th}) drift of Silicon Carbide (SiC) MOSFETs under dynamic gate stress (DGS) brought reliability concerns, this study investigated the threshold voltage drift of trench-type 1200 V SiC MOSFETs under the dynamic gate bias test. The V_{th} drift was monitored during stress lasting for over 6×10^{11} switching cycles under different case temperatures (25 °C, 75 °C, 125 °C, 175 °C) and different positive bias voltages (20 V, 21 V, 22 V, 24 V, 26 V). The influence of such a reliability test on the on-resistance and the breakdown voltage was also investigated. Results showed that the V_{th} drift under dynamic gate stress was positive. The drift decreased with increasing temperature but increased with higher positive gate bias. The on-resistance increased with the positive drift of the V_{th}, while the breakdown characteristics of the device remained unchanged. Besides, the mechanism of the V_{th} drift behavior was further studied, focusing on the trench-type electric field reconstruction under dynamic stress. The V_{th} drift under dynamic gate stress, followed by cross-section analysis and TCAD simulation, provides insights to improve the understanding of dynamic reliability tests of SiC MOSFETs in power applications.

Keywords—SiC MOSFET, dynamic gate bias (DGB), gate oxide reliability, bias temperature instability

I. INTRODUCTION

Silicon Carbide (SiC) MOSFETs have been widely adopted in commercial applications such as electric vehicles and wind power generators. The gate oxide reliability, as a critical factor affecting the stability and lifetime of device performance, has been attractive for researchers in recent years. It is known that the switching cycles of SiC MOSFETs could reach up to 10^{14} times throughout their entire application period. Defect states at the SiC/SiO$_2$ interface cause carrier trapping and detrapping processes, which in turn lead to long-term and short-term variations in the threshold voltage (V_{th}). Studies have revealed that under bipolar mode (positive and negative gate-source voltage) operation with a large number of switching cycles, both trench-type and planar-type SiC MOSFETs exhibit an additional shift in the V_{th} [1]. Such V_{th} drift under dynamic gate stress (DGS) is usually larger than that under static gate stress conditions [2]. The positive drift of V_{th} usually leads to an increase in the on-resistance, while the negative shift could cause spurious turning on. These phenomena seriously threaten the device's reliability. Therefore, investigating the V_{th} drift of SiC MOSFETs under dynamic gate stress is of great importance for improving the understanding of dynamic stress influence on device reliability, thus accelerating chip design/manufacturing processes.

Currently, the impact of experimental conditions (eg, frequency, temperature, etc.) is analyzed on the V_{th} drift of SiC MOSFETs with different structures. Previous work [3] pointed out that the V_{th} drift caused by switching events was independent of the operating frequency. The effect of temperature on V_{th} drift depends on the specific device type; some devices do not exhibit strong temperature dependence, while others show an increase or decrease in V_{th} drift with rising temperature [4-6]. Furthermore, studies on the relationship between dv/dt during turn-on and turn-off and the V_{th} drift of devices indicated that the influence of switching time on V_{th} drift also varied with device structure [7, 8].

To further understand why threshold voltage drift is greater under dynamic gate stress, the underlying physical mechanisms were investigated. Two primary mechanisms have been proposed to explain the V_{th} shift under dynamic gate stress conditions. One is based on the local electric field enhancement, and the other is related to the recombination-enhanced defect reactions (REDRs) [4, 9, 10]. The former points out that under dynamic stress conditions, the charge trapping and detrapping processes at defects lag behind the rapid changes in the gate voltage, leading to a local electric field enhancement in the channel region that exceeds the electric field generated by the static gate voltage. Such enhanced field excites the trapping behavior of deeper energy level traps, resulting in an increased V_{th} drift. The latter theory suggests that the non-radiative recombination of electrons and holes at defect sites releases energy sufficient to activate additional acceptor-like defects, thereby causing a positive shift in the V_{th}. Although preliminary theoretical frameworks have existed, the physical mechanism of the V_{th} drift under dynamic gate stress remains unclear, requiring further in-depth analysis and experimental verification.

Synthesizing existing research findings, it is concluded that experimental investigations often focus on the V_{th}, with limited comparative analysis of other key static parameters before and after stress. Furthermore, simulation-based experimental approaches are rarely explored. In this work, we focus on the influence of different case temperatures and positive gate biases on the V_{th} drift of trench-type devices. The variation in on-resistance and breakdown voltage before and after the experiments was monitored. Additionally, scanning electron microscopy (SEM) and focused ion beam (FIB) techniques are employed to image the cell structure. TCAD simulations are utilized to analyze the electric field variations under different experimental conditions, aiming to further explain the mechanism of V_{th} drift under dynamic gate stress.

II. DEVICE AND TEST INFORMATION

A. Device Information

Commercially available 1200 V trench-type devices (TO247-4 package) were selected as the devices under test (DUTs), with a V_{th} of 4.8 V and an on-resistance of 36 mΩ, as shown in Fig. 1(a). For subsequent simulation work, the cross-section of the device cell region was characterized by SEM, as shown in Fig. 1(b).

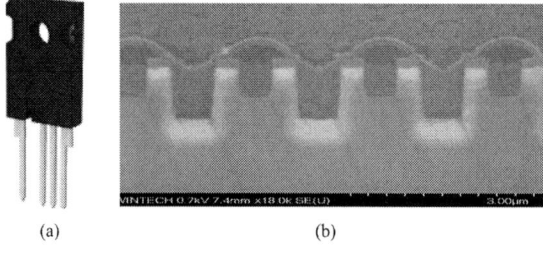

(a)   (b)

Fig. 1. Basic information of the DUT (a) TO247-4 package, (b) SEM results of the DUT cell part.

B. DGS Test Setup and Procedures

The test circuit is shown in Fig. 2. During the experiment, a voltage square wave was applied to the gate at a frequency of 500 kHz after the specified experimental temperature was reached and maintained for 15 minutes. As shown in Fig. 2, the source and drain of the device were shorted simultaneously. Fig. 3 illustrates the test procedure for the

DGS experiment. When the stress application time reached 20 minutes, the program monitored the V_{th} at 20-minute intervals, with the temperature maintained at the experimental temperature. The V_{th} was measured following the JEP183 standard, which involved applying a 20 V pre-pulse for 10 ms, followed by reading the V_{th} at a drain current (I_d) of 10 mA.

The DUTs are characterized for the V_{th}, on-resistance, and breakdown voltage using Keysight B1505A before the dynamic stress test. After the test, these static parameters were remeasured with a 24-hour recovery period at room temperature in between.

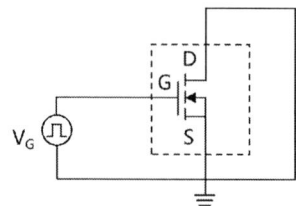

Fig. 2. Test circuit diagram for DGB experiment.

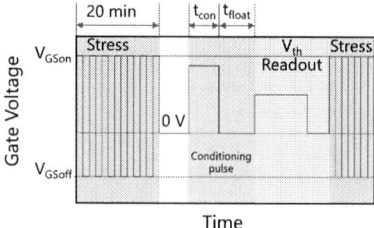

Fig. 3. Dynamic gate stress test setup.

To thoroughly investigate the impact of experimental conditions on the V_{th}, 12 devices were used for each experimental condition. To ensure consistency and comparability of experimental results, these 12 devices were selected from a large batch characterized by a Keysight B1505A, possessing similar initial V_{th} and on-resistance. Table I shows the experimental conditions used in this study to investigate the effects of temperature and positive bias on V_{th} drift and other static parameters. The duration of the temperature experiments was maintained for 672 hours, while that of the positive bias experiments was 336 hours.

TABLE I.   DEVICE PARAMETERS

Temperature (°C) @20 V/-8 V	Gate Bias (V) @25 °C
25	20/-8
75	21/-8
125	22/-8
175	24/-8
-	26/-8

III. DEGRADATION UNDER DYNAMIC GATE BIAS

In this section, the V_{th} drift under different temperatures and positive bias conditions was studied.

A. Degradation Under Different Temperatures

The case temperature of the devices was controlled at 25 °C, 75 °C, 125 °C, and 175 °C to study its impact on V_{th} drift. Fig. 4 shows the relationship between V_{th} drift and the number of switching cycles at different temperatures. The results showed that the V_{th} drift increased with the number of cycles. After approximately 3×10^9 cycles, an abrupt change in the slope of the drift curve was observed. The V_{th} drift decreased as the temperature increased. Furthermore, no saturation behavior was observed even after 10^{12} switching cycles across all tested temperatures, with the drift at room temperature exceeding 2 V.

The change in V_{th} after the devices were left to rest for 24 hours is shown in Fig. 5, revealing that the V_{th} exhibited minimal recovery, and the negative correlation with temperature remained. This suggests that the drift could be considered a permanent shift. The suppression of V_{th} drift with increasing temperature has been reported in many studies on trench-type devices [5, 8]. A possible reason for such a phenomenon is that higher temperatures enhance the capture and emission rates of carriers, leading to a reduction in the peak electric field at the gate oxide during the transition from negative to positive gate voltage, thereby resulting in less electron trapping and a weaker V_{th} drift.

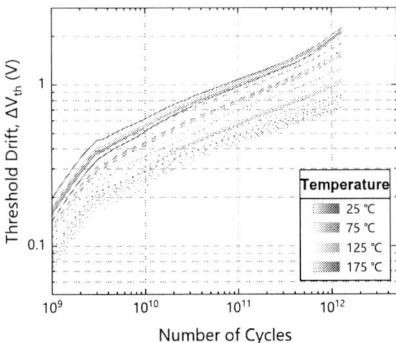

Fig. 4. Threshold voltage drift of trench-type devices versus the number of cycles at different case temperatures (25 °C, 75 °C, 125 °C, 175 °C).

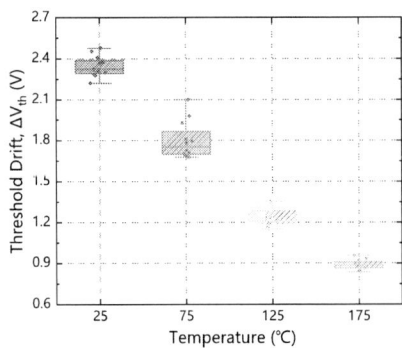

Fig. 5. Threshold voltage drift of trench-type devices at different case temperatures (25 °C, 75 °C, 125 °C, and 175 °C) after 24 hours of resting.

The impact of V_{th} drift is further reflected in the change of on-resistance, as shown in Fig. 6(a). A larger V_{th} drift led to a greater increase in on-resistance, therefore, the change in on-resistance also exhibited a negative correlation with temperature. When the temperature was 25 °C, the V_{th} drift exceeded 2 V, and the average increase in on-resistance was more than 6 mΩ. This could result in substantial conduction losses during device operation. The change in breakdown voltage was less than 2 V, as shown in Fig. 6(b), which

illustrated that the dynamic gate stress applied to the devices did not induce any notable degradation in their breakdown characteristics.

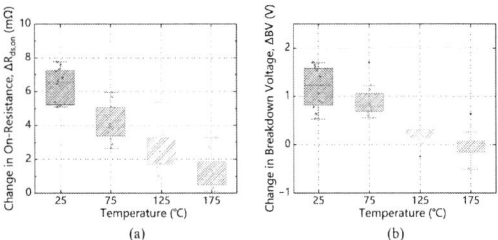

Fig. 6. Changes in static parameters measured after 24 hours of resting: (a) on-resistance, (b) breakdown voltage, under case temperatures of 25 °C, 75 °C, 125 °C, and 175 °C.

B. Degradation Under Different Biases

To investigate the influence of positive gate voltage on V_{th} drift, the positive voltage conditions applied to the devices were 20 V, 21 V, 22 V, 24 V, and 26 V, while the negative voltage was consistently maintained at -8 V. The result is shown in Fig. 7, which demonstrates that a larger positive bias led to a greater V_{th} drift and accelerated the rate of V_{th} drift. Simultaneously, results showed an abrupt change in the V_{th} drift rate, with a notable slowdown, after approximately 3×10^9 cycles across all positive bias conditions.

The drift after a 24-hour rest period, as shown in Fig. 8, further confirmed this. When the positive bias reached 26 V, the V_{th} drift remained above 2.4 V. This behavior could be attributed to a stronger electric field and electric field peaks near the interface under a larger positive voltage, which results in more electron trapping during each cycle. These trapped electrons occupy deeper energy levels, leading to a more severe V_{th} drift.

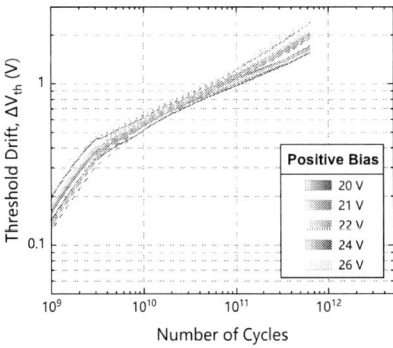

Fig. 7. Threshold voltage drift of trench-type devices versus the number of cycles at different positive bias voltages (20 V, 21 V, 22 V, 24 V, 26 V).

Changes in on-resistance and breakdown voltage were also observed under different positive voltage conditions. As shown in Fig. 9(a), a larger positive bias led to a greater increase in V_{th}, which in turn resulted in a larger increase in on-resistance in general. The breakdown characteristics of the devices are shown in Fig. 9(b). After undergoing dynamic gate stress tests under different positive bias voltages, the breakdown characteristics of the devices did not show pronounced changes

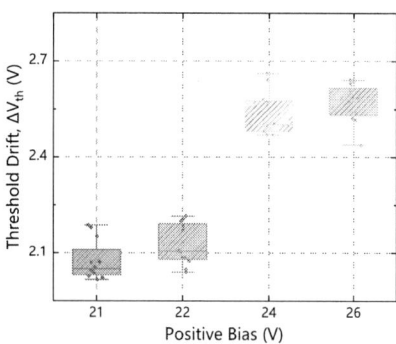

Fig. 8. Threshold voltage drift of trench-type devices at different positive bias voltages (21 V, 22 V, 24 V, 26 V) after 24 hours of resting.

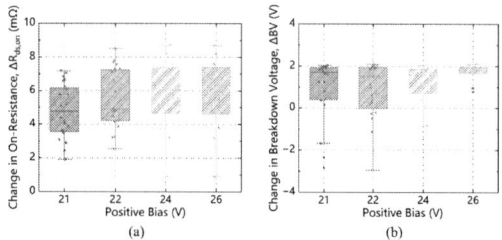

Fig. 9. Changes in static parameters measured after 24 hours of resting: (a) on-resistance, (b) breakdown voltage, under positive bias voltages of 21 V, 22 V, 24 V, and 26 V.

In summary, the dependence of the static characteristics of 1200 V devices on temperature and positive bias under DGS was explored in this part. The study revealed a positive shift in the V_{th} of the devices. Elevated temperatures mitigated the extent of this drift, while increased positive bias exacerbated it. The rise in V_{th} affected the on-state performance of the devices, while it hardly impacted the breakdown characteristics. Based on the current experimental findings, optimization of the device application conditions could be implemented to extend their operational lifetime.

IV. SIMULATION RESULTS

To further validate and interpret the experimental observations, TCAD simulations were conducted to investigate the underlying physical mechanisms. In this section, the variation of the electric field near the SiC/SiO$_2$ interface during a single switching cycle was simulated under different temperature and positive bias conditions by building a TCAD model with electron and hole traps introduced at the interface. The half-cell structure and the location for electric field extraction are shown in Fig. 10(a). The voltage waveform applied to the gate is depicted in Fig. 10(b). The electric field at the designated location was extracted at the five time points marked in Fig. 10 (b).

The electric field drives the capture and emission of electrons and holes by interface states. The simulated electric field variation over time under different temperature and positive bias conditions is shown in Fig. 11. During the transition of the gate voltage from negative to positive, an electric field spike exists near the interface within the channel, indicating that the charge trapping indeed caused local electric field enhancement. As shown in Fig. 11(a), the peak value of the electric field spike did not exhibit clear variation with

increasing temperature, while the duration of the spike decreased. This could be attributed to the fact that higher temperature accelerates the emission rate of holes, leading to a weakened local electric field enhancement effect, and consequently, reduced electron trapping and a smaller drift. When the positive voltage increased, the electric field spike became larger, resulting in more electron trapping and hole release within the same time, which led to a greater V_{th} drift and a faster decay of the local electric field. Thus, the simulation results further support the hypothesis that the V_{th} drift caused by dynamic gate stress may be due to local electric field enhancement resulting from charges within defects that cannot respond instantaneously to the changing gate voltage.

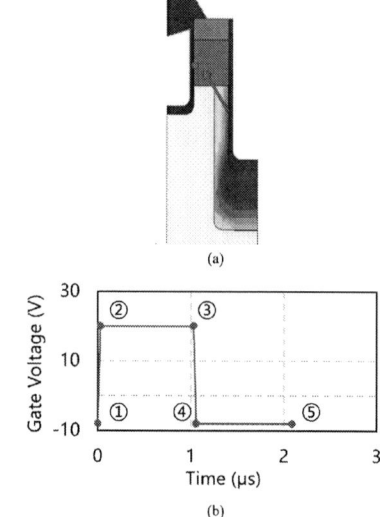

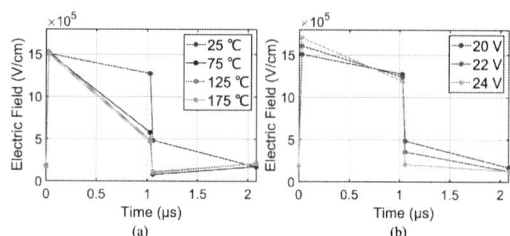

Fig. 10. (a) Schematic of the device cells for TCAD simulation, (b) Gate voltage waveform.

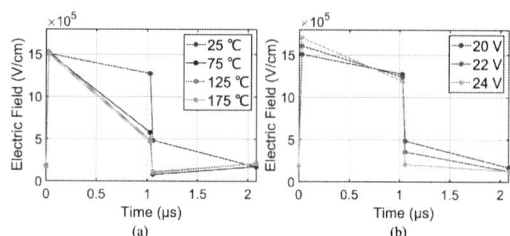

Fig. 11. (a) Electric field variation over time at different temperatures (25 °C, 75 °C, 125 °C, 175 °C). (b) Electric field variation over time under different positive biases (20 V, 22 V, 24 V).

The TCAD simulations in this paper primarily aim to investigate the variation of the electric field with the applied voltage waveform under different conditions, attempting to explore the influence mechanisms of temperature and positive bias on V_{th} drift. However, due to certain discrepancies between the interface trap parameters set in the current model and the actual conditions, the simulation results may deviate from the real behavior. Therefore, future research should further calibrate the interface defect settings to provide a more reliable basis for mechanism analysis.

V. CONCLUSION

In this work, the threshold voltage drift of trench-type 1200 V SiC MOSFETs under bipolar dynamic gate stress was investigated through experiments and simulation. After more than 6×10^{11} cycle experiments, the effects of case temperature and positive gate bias on device degradation were examined. Experimental results showed that V_{th} drift increased with higher positive bias and decreased with elevated temperatures, with negligible recovery after stress removal. Post-stress measurements revealed that there was a positive correlation between V_{th} drift and on-resistance, whereas breakdown voltage remained unaffected. TCAD simulations further confirmed that transient local electric field enhancements, caused by delayed charge trapping and detrapping, were the likely origin of the observed V_{th} drift behavior. These findings provide insight into the reliability mechanisms of SiC MOSFETs under dynamic switching conditions and highlight the need for optimized gate drive strategies to mitigate long-term degradation. Future work needs to involve detailed analysis of the key factors influencing V_{th} drift through microstructural characterization, failure analysis, and high-accuracy modeling and simulation.

ACKNOWLEDGMENT

The work is supported by Shanghai SiC Power Device Engineering and Technology Research Center (19DZ2253400), Zhejiang Provincial Science and Technology Program (2024C01247(SD2)), and industrial cooperation project (KCH2310169).

REFERENCES

[1] X. Zhong, H. Jiang, L. Tang, X. Qi, P. Jiang, and L. Ran, "Gate stress polarity dependence of AC bias temperature instability in silicon carbide MOSFETs," IEEE Transactions on Electron Devices, vol. 69, no. 6, pp. 3328–3333, 2022.

[2] H. Jiang, X. Zhong, G. Qiu, L. Tang, X. Qi, and L. Ran, "Dynamic gate stress induced threshold voltage drift of silicon carbide MOSFET," IEEE Electron Device Letters, vol. 41, no. 9, pp. 1284–1287, 2020.

[3] M. W. Feil et al., "On the frequency dependence of the gate switching instability in silicon carbide MOSFETs," in Materials Science Forum, 2023, vol. 1092: Trans Tech Publ, pp. 109–117.

[4] M. W. Feil et al., "Gate switching instability in silicon carbide MOSFETs—part I: experimental," IEEE Transactions on Electron Devices, vol. 71, no. 7, pp. 4210–4217, 2024.

[5] D. B. Habersat and A. J. Lelis, "AC-stress degradation and its anneal in SiC MOSFETs," IEEE Transactions on Electron Devices, vol. 69, no. 9, pp. 5068–5073, 2022.

[6] C. Chen et al., "Degradation dependency analysis and modeling of 1700 V planar-gate SiC MOSFETs under gate switching instability," IEEE Transactions on Power Electronics, vol. 40, no. 2, pp. 3566–3577, 2025.

[7] M. W. Feil et al., "Towards understanding the physics of gate switching instability in silicon carbide MOSFETs," in 2023 IEEE International Reliability Physics Symposium (IRPS), 26–30 March 2023, pp. 1–10.

[8] X. Zhong et al., "Bias temperature instability of silicon carbide power MOSFET under AC gate stresses," IEEE Transactions on Power Electronics, vol. 37, no. 2, pp. 1998–2008, 2022.

[9] H. Jiang et al., "A physical explanation of threshold voltage drift of SiC MOSFET induced by gate switching," IEEE Transactions on Power Electronics, vol. 37, no. 8, pp. 8830–8834, 2022.

[10] T. Aichinger, M. W. Feil, and P. Salmen, "Assessing, controlling and understanding parameter variations of SiC power MOSFETs in switching operation," Key Engineering Materials, vol. 947, pp. 69–75, 2023.

Analysis of Planar and SBD-Embedded SiC MOSFET Under Avalanche Stress

Hao Huang
College of Electrical Engineering
Zhejiang University
Hangzhou, China
12310083@zju.edu.cn

Yuting Jin
College of Electrical Engineering
Zhejiang University
Hangzhou, China
yutingjin@zju.edu.cn

Dong Hai
College of Electrical Engineering
Zhejiang University
Hangzhou, China
donghai@zju.edu.cn

Dingyuan Bao
College of Electrical Engineering
Zhejiang University
Hangzhou, China
22410017@zju.edu.cn

Taohui Zhang
College of Electrical Engineering
Zhejiang University
Hangzhou, China
22410136@zju.edu.cn

Haoze Luo
College of Electrical Engineering
Zhejiang University
Hangzhou, China
haozeluo@zju.edu.cn

Francesco Iannuzzo
DENERG: Dipartimento Energia
"Galileo Ferraris"
Politecnico di Torino
Turin, Italy
francesco.iannuzzo@polito.it

Wuhua Li
College of Electrical Engineering
Zhejiang University
Hangzhou, China
woohualee@zju.edu.cn

Abstract—**In this paper, Commercial Planar and SBD-Embedded MOSFETs are studied under Avalanche stress. After screening the devices with threshold voltage, unclamped inductive switching tests are carried out to evaluate avalanche reliability of the devices. SBD-Embedded MOSFETs can only endure 27.1% current of the counterpart. The experimental results are analyzed by TCAD simulation that avalanche failure mechanism in SBD-Embedded MOSFET is closely related to the Schottky barrier diode embedded region. Transient TCAD simulations uncover localized heating at the Schottky contact region during avalanche, exacerbated by current-path competition between the MOSFET channel, body diode, and SBD during turn-off. The results highlight a fundamental trade-off between static efficiency gains from SBD integration and dynamic ruggedness boundary for further application.**

Keywords— *Avalanche ruggedness, SBD-Embedded MOSFET, UIS Test*

I. INTRODUCTION

In recent years, silicon carbide (SiC) MOSFETs have been widely applied in the market, especially in converter applications demanding high switching speeds and extremely high junction - temperature tolerance [1]. In some high speed switching applications, particularly synchronous rectification and motor drives, body diodes endure harsh stress, with increasing switching loss e. Compared with the intrinsic body diode, The anti-parallel Schottky Barrier Diode (SBD) of SiC MOSFET, can not only prevent the bipolar degradation but also accelerate the reverse recovery process [2].

To enhance power density and reduce manufacturing costs, monolithic SBD Embedded MOSFETs have emerged [3]. This type of power device integrates the SBD chip into a single SiC MOSFET chip. The two share the conduction loop

and terminals, thus achieving a reduction in chip area and simplification of the manufacturing process [4]. With the application of this device, the ruggedness and reliability of its diode remains a problem in converter applications. In order to characterize the extremely high voltage and current stress of the diodes in practical condition, unclamped inductive switching (UIS) tests can be conducted on different commercial devices ,which helps to characterize the substrate defects of power devices under the depletion region [5]. In planar SiC MOSFET, the literature [6-8] indicates that under surge conditions, the body diode will endure significant voltage and current stresses, which is related to the parasitic transistor latching during the switching off phase, mainly resulting in the avalanche breakdown of the device. It is concluded that the failure mechanism of SBD Embedded MOS is the positive feedback between SBD leakage and temperature, leading to thermal runaway[9]. Heat conduction model was established to verify that increasing the Schottky barrier height of the SBD region to reduce the leakage current in the SBD region can improve the avalanche ruggedness of JMOS[10]. However, the transient thermoelectric coupling mechanism and the circuit behavior model were not explicitly combined.

This paper will focus on anti-parallel diode of planar and SBD embedded SiC MOSFET. To analyze the Third quadrant characteristic characteristics of the two devices, Unclamped Inductive Switching (UIS) tests are carried out. Based on the TCAD simulation, the current path shifted to the Schottky contact region, causing localized heating and potential thermal runaway.

II. EXPERIMENT SETUP

A. Devices under test

To compare the body diode between commercial Planar and SBD-Embedded MOSFET, the TW045N120C is elected for SBD Embedded type MOSFET and C3M0040120K for planar type MOSFET. SBD-Embedded MOS and Planar

---

This work was supported by the National Key Research and Development Program of China under Grant 2022YFE0138400, in part by Zhejiang Provincial Natural Science Foundation of China under Grant LR24E070001, and in part by the Joint Foundation for Basic Research on Railways under Grant U2368206.

MOS exhibit significant differences in structure and performance. Structurally, SBD-Embedded MOS builds upon the traditional MOSFET by integrating a Schottky Barrier Diode (SBD) near the source or drain, forming a composite structure. This design allows the SBD-Embedded MOS to effectively suppress leakage current during reverse conduction while maintaining low on-resistance. In contrast, Planar MOS features a conventional planar structure where the source, drain, and gate are all on the same plane. Its structure is relatively simple, and its manufacturing process is well-established. The cross-sectional figure of the device cells are as shown below :

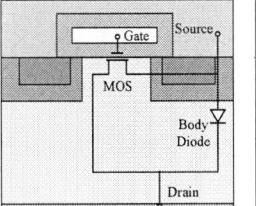 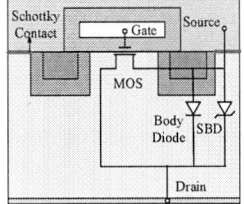

Fig. 1. Structures of SBD Embedded and Planar SiC MOSFET

Both devices share similar forward characteristics, which is more convincing for the third quadrant characteristics comparison. Considering the cell structure, SBD Embedded region, though more vulnerable than body diode region, could be suitable for applications at 1.2kV if designed carefully [4]. Regarding on-resistance, elected devices share similar characteristics. the typical on-resistance $R_{DS}(on)$ of the SBD Embedded type is 45mΩ, whereas the planar is 40mΩ. Static Electrical parameters are listed below:

TABLE I.   ELECTRICAL PARAMETER OF SELECTED DEVICES

	Planar	*SBD Embedded*
Drain source voltage	1.2kV	1.2kV
Drain source on-resistance	40mΩ	45mΩ
Threshold Voltage	3-5V(25°C)	3-5V(25°C)
Diode Forward Voltage	-4.26V	-1.35V

However, when comparing the reverse characteristics of the two, due to the integration of the Schottky Barrier Diode, the forward voltage drop of the diode in the SBD Embedded device is reduced by 30%. Before further Unclamped inductive switching test of the reverse diode, switching performance should be considered as well. 8 samples with threshold voltage parameters similar to the typical values were selected for testing to ensure the devices consistency , and the relevant parameters are shown below:

TABLE II.   THRESHOLD VOLTAGE OF SELECTED DEVICES

Device	No.	Threshold Voltage
SBD-Embedded	SBD1	3.02V
	SBD2	3.15V
	SBD3	3.12V
	SBD4	3.07V
Planar	Planar1	2.92V
	Planar2	3.02V
	Planar3	2.90V
	Planar4	2.95V

B. UIS test setup

UIS (Unclamped Inductive Switching) possesses the unique capability to detect latent defects by inducing a fully depleted condition during turn-off, even at low current densities. The corresponding process is shown below:

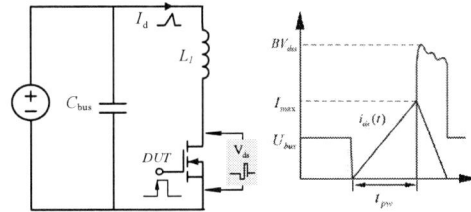

Fig. 2.   UIS Test Topology and Waveform

Initially, the device under test (DUT) is in the turn-off state, and the bus capacitor is charged. Since the power device is in the off state, the drain-source voltage (V_{ds}) is equal to the bus voltage (U_{in}), and no energy is stored in the inductor (L_l). Subsequently, the power device is subjected to a pulse with a specific duration (t_{pw}), which is calculated based on the load inductance and the anticipated maximum current (I_{max}). The selection of the pulse width (t_{pw}) is critical to ensure that the device is tested under conditions that accurately reflect its operational limits and stress tolerance:

$$t_{pw} = \frac{L_l I_{max}}{U_{bus}}$$

When the gate-source voltage (V_{gs}) is raised, the channel of the device is opened. At this point, the current from the capacitor flows through the inductor, with the drain current (I_{ds}) conducting from the drain to the source through the load inductor. During this process, energy is stored in the load inductor until the current reaches the target peak current.

$$\frac{di_{ds}}{dt} = \frac{U_{bus}}{L_l}$$

Subsequently, when the gate-source voltage (V_{gs}) is lowered, the channel of the device is closed. However, due to the energy conservation characteristic of the inductor itself, the drain current (I_{ds}) continues to flow out from the inductor. Therefore, when the device is in the off-state, the current flowing into the device causes the drain-source voltage (VDS) to rise sharply to an extremely high value(BV_{dss}) leading to high voltage stress of the diode. When avalanche breakdown of the diode occurs, the current from the inductor still flows through the diode even though the MOSFET is in the off-state. This energy release process can subject the MOSFET to high voltage and high current stresses, which may lead to device damage. Specifically, the energy stored in the inductor is entirely dissipated through the MOSFET, and if this energy exceeds the avalanche breakdown energy (EAS) of the MOSFET, it may cause the device to fail due to overheating.

$$EAS = \frac{1}{2} L_l I_{max}^2 \frac{BV_{dss}}{BV_{dss} - U_{bus}}$$

To evaluate the avalanche reliability of both devices, a non-inductive clamped switching experiment will be conducted. The non-inductive clamped switching test platform is shown in Fig.3 , with the circuit diagram depicted. During the test, the gate-source voltage (Vgs) for both devices ranges from -4V to +20V. The gate resistor is set to 5Ω to simulate the switching conditions in practical applications. The gate-source voltage is 100V, and the power load inductor has four taps, with the inductance values used in the experiment being 272μH. To test the failure boundaries of both devices, the current is increased from initial current set value to avalanche breakdown current in 2A increments.

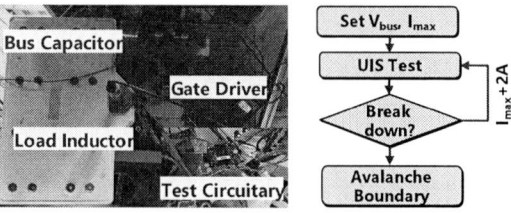

Fig. 3.   UIS Test Platform Setup

III. EXPERIMENT ANALYSIS AND MODELING

A. UIS Test Results

The UIS test was firstly performed on the SBD-Embedded device according to the above test setup.The following behavior was observed when the device underwent the test:

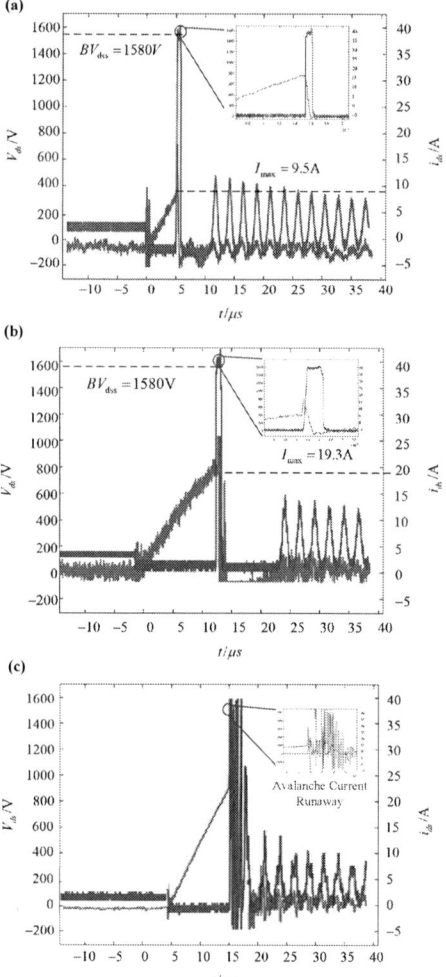

Fig. 4.   UIS Test Waveform of SBD-Embedded SiC MOSFET

(a) UIS test under 10A (b) Last test before Failure (c) Avalanche failure

Fig.4(a) depicts the voltage and current waveform of a certain selected SBD-Embedded MOSFET during a single UIS test failure event. With the current set at 10A, the pulse width (t_{pw}) is set to 5μs. Once the current reaches the set value, the gate signal is turned off, causing the drain-source voltage to surge dramatically, and the device immediately enters avalanche mode, with the voltage reaching a maximum of around 1600V. As the current continues to increase, it reaches 19.3A as shown in Fig.4(b). When the current further increases as is shown in Fig.4(c), oscillations in both current and voltage occur during turn-off unable to completely shut off the device.

In contrast, a planar gate device was also subjected to UIS testing. As is shown in Fig.5, with the similar voltage resistance in avalanche, the current reaches 71.2A, almost 4 times of the SBD embedded type. With slower reverse recovery speed, there is no oscillation before the avalanche breakdown in planar type though. The following behavior was observed when the elected planar type SiC MOSFETs underwent the UIS test:

979-8-3315-1110-4/25 $31.00 © 2025 IEEE

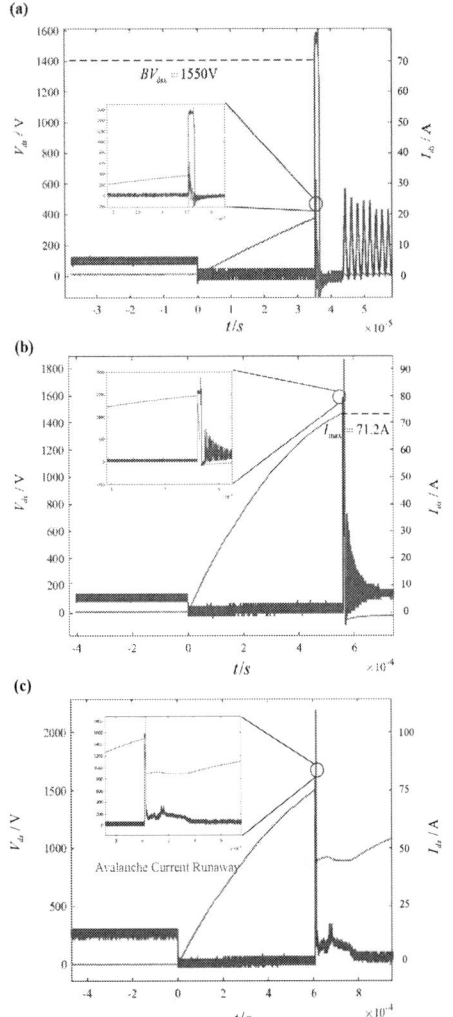

Fig. 5.   UIS Test Waveform of Planar SiC MOSFET

(a) UIS test under 20A (b) Last test before failure (c) Avalanche failure

Replicating the aforementioned experimental procedure, the selected set of eight devices was subjected to identical testing conditions in order to evaluate the consistency and reliability of the observed phenomena. Each device underwent the same sequence of gate switching operations, current measurements, and voltage stress conditions as previously described:

TABLE III.   AVALANCHE FAILURE OF SELECTED DEVICES

Device	No.	Avalanche Voltage	Current Tolerance	Failure Mode
SBD-Embedded	SBD1	1592V	19.3A	• g,d,s Short
	SBD2	1580V	18.7A	• Vds Oscillation Current Runaway
	SBD3	1570V	5A	
	SBD4	1597V	15.8A	• Accidental early avalanche failure
Planar	Planar1	1550V	71A	• g,d,s Short
	Planar2	1490V	67.2A	

	Planar3	1500V	72.2A	• Mainly current runaway failure
	Planar4	1530V	70.4A	

B. TCAD Simulation Analysis

To further investigate the behavioral differences between the two types of devices under UIS testing, cell models for both devices were established, with their main parameters and cell structures as follows：

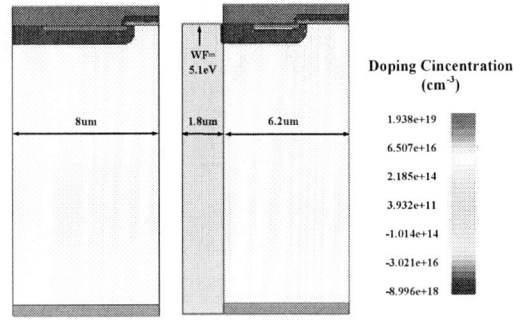

Fig. 6.   Cell Structure of Device Under Test

The cross-sectional view of the semiconductor device illustrates the doping concentration profile and electrical potential distribution across its structure. The left panel displays the doping levels, showing the heavily doped n+ source region at the top with a concentration of approximately $1.53 \times 10^{19} \text{cm}^{-3}$, followed by the p-type body region with moderate doping around $6.5 \times 10^{17} \text{cm}^{-3}$, and the lightly doped n-type substrate at the bottom near $-6.99 \times 10^{18} \text{cm}^{-3}$. The right panel shows the corresponding electrical potential under a gate work function of 5.2 eV , revealing a SBD interface that induces an inversion layer in the p-type body.

Based on the actual test conditions, a TCAD (Technology Computer-Aided Design) simulation was conducted to validate the Unclamped Inductive Switching (UIS) performance of the device under a critical stress condition. The simulation was specifically designed to replicate the experimental setup, where the device was subjected to a unclamped inductive switching event with a peak voltage of 100 V and an inductance value of 50 mH:

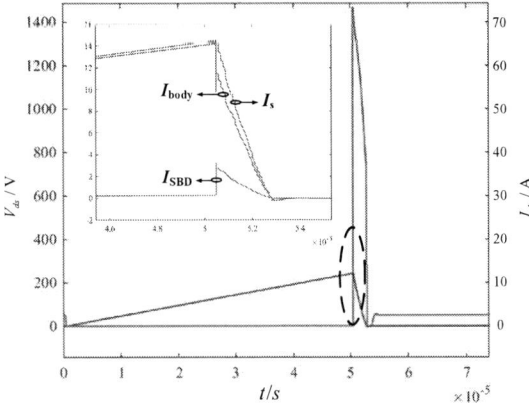

Fig. 7.   UIS Simulation of SBD Embedded SiC MOSFET

As shown in the experimental data, the Schottky Area current (I_{SBD}) drops to nearly zero when the gate is initially turned on. During the gate-on period, the leakage current in the Schottky Barrier Diode (SBD) region remains almost constant over time. Just before the gate is turned off, the source current (I_{body}) is measured to be approximately 14 A. The simulation results demonstrate a clear transition of the current path from the MOSFET channel to the Schottky diode after the gate is turned off, highlighting the role of the Schottky contact in current conduction during avalanche mode is to bypass the body diode current. To further analyze the thermal performance, the junction temperature of the cell simulation during the turn-off period is compared as follows::

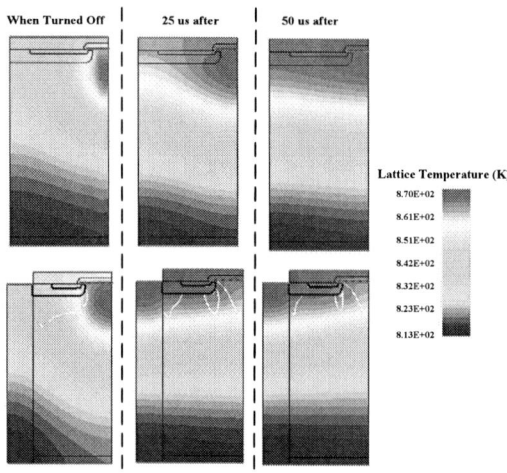

Fig. 8. Heat Distribution of SBD Embedded SiC MOSFET

Upon gate turned off, the SBD Embedded device enters the avalanche operation mode, and the drain current manifests as the avalanche current. At this point, the source current begins to decrease, while the Schottky current starts to increase after 25 us, coinciding with a sharp rise in device temperature. This indicates that the primary current path shifts to the Schottky contact region after 25 us and heat transfer horizontally beneath the aluminum after 50 us. Compared with planar SiC MOSFET, the hot spot is mainly on the SBD region rather than the channel and body diode in the process, which is further indicating that the SBD Embedded device is more likely breakdown and the V_{ds} oscillation may be caused by the current competition and thermal behavior between gate channel, bode diode, and SBD region in the turn off phase. During the turn-off transient of the device, a notable oscillation between the Schottky Barrier Diode (SBD) current and the body diode current is observed, especially when the SBD experiences avalanche breakdown. This phenomenon can be attributed to the dynamic competition between the two conduction paths, driven by differences in carrier response times, forward voltage characteristics, and thermal feedback mechanisms. When the drain-source voltage rises rapidly, the SBD may enter the breakdown region first due to its lower knee voltage, initiating significant current conduction. However, due to its negative temperature coefficient and potential local heating effects, the SBD's conduction state becomes unstable, leading to fluctuations in current distribution.Concurrently, parasitic inductances and capacitances in the circuit can couple with the nonlinear I-V characteristics of both the SBD and the body diode, forming an oscillatory loop. Additionally, non-uniformities in SBD material properties or structural defects may exacerbate localized electric field enhancements, further contributing to the instability. These combined factors result in periodic switching of the dominant current path between the SBD and the body diode, manifesting as current oscillations in the measured waveforms. Understanding and mitigating this behavior is critical for improving device reliability and suppressing electromagnetic interference (EMI) in high-speed and high-power applications.

IV. CONCLUSION

This paper compared the antiparallel diodes of commercial planar and SBD-Embedded SiC MOSFETs via UIS tests and TCAD simulations. The SBD Embedded MOSFETs, with lower forward voltage drops due to the Embedded Schottky Barrier Diode, showed inferior performance in UIS tests and had 27.1% current tolerance of planar type, with failure modes including current runaway, and early avalanche failure. TCAD simulations revealed that during turn off, the current path shifted to the Schottky contact region, causing localized heating and potential thermal runaway. The hot spot was mainly on the SBD region, making it more prone to breakdown and V_{ds} oscillation due to current and thermal coupling.

References

[1] A. Kumar et al., "Effect of capacitive current on reverse recovery of body diode of 10kV SiC MOSFETs and external 10kV SiC JBS diodes," 2017 IEEE 5th Workshop on Wide Bandgap Power Devices and Applications (WiPDA), Albuquerque, NM, USA, 2017, pp. 208-212.

[2] Y. Wu et al., "Evaluation of Bipolar Degradation in SiC MOSFETs for Converter Design," 2023 IEEE Energy Conversion Congress and Exposition (ECCE), Nashville, TN, USA, 2023, pp. 5359-5365.

[3] S. Hino et al., ""Demonstration of SiC-MOSFET embedding Schottky barrier diode for inactivation of parasitic body diode,"" 2016 European Conference on Silicon Carbide & Related Materials (ECSCRM), Halkidiki, Greece, 2016, pp. 1-1

[4] T. Ohashi et al., "Improved Clamping Capability of Parasitic Body Diode Utilizing New Equivalent Circuit Model of SBD-embedded SiC MOSFET," 2021 33rd International Symposium on Power Semiconductor Devices and ICs (ISPSD), Nagoya, Japan, 2021, pp. 79-82.

[5] K. Hasegawa et al., "Which is harder SOA test for SiC MOSFET to do Unclamped Inductive Switching (UIS) or Unloaded Short Circuit mode Switching (USCS)? Does UIS play a role of USCS?," 2020 32nd International Symposium on Power Semiconductor Devices and ICs (ISPSD), Vienna, Austria, 2020, pp. 62-65

[6] T. Luo, R. Luo, Z. Xiang, J. Zhuang, G. Zhang and J. Fan, ""Avalanche Ruggedness Evaluation on Planar and Trench SiC MOSFETs: An Experimental and TCAD Simulation Study,"" 2024 25th International Conference on Electronic Packaging Technology (ICEPT), Tianjin, China, 2024, pp. 1-5

[7] A. Fayyaz et al., "UIS failure mechanism of SiC power MOSFETs," 2016 IEEE 4th Workshop on Wide Bandgap Power Devices and Applications (WiPDA), Fayetteville, AR, USA, 2016, pp. 118-122.

[8] W. Saito, Z. Lou and S. -I. NIshizawa, ""Unclamped Inductive Switching Robustness of SiC Devices With Parallel-Connected Varistor,"" in IEEE Transactions on Electron Devices, vol. 69, no. 10, pp. 5671-5677

[9] J. Yu, Z. Li, Z. He, X. Jiang, C. Zhang and J. Wang, "Performance Comparison of Traditional and JBS Integrated SiC MOSFETs in Si/SiC Hybrid Switch," 2019 IEEE Energy Conversion Congress and Exposition (ECCE), Baltimore, MD, USA, 2019, pp. 1918-1921

[10] Y. Zhang et al., "Investigation and Modeling of the Avalanche Failure Mechanism of 1.2-kV 4H-SiC JMOS," in IEEE Transactions on Electron Devices, vol. 68, no. 12, pp. 6313-6320

Mechanism analysis of different undamped oscillations for GaN devices

Qiang Hu
School of Electrical Engineering
Southwest Jiaotong University
Chengdu, China
hq2019111758@my.swjtu.edu.cn

Jian Chen
School of Integrated Circuits Science
and Engineering
Southwest Jiaotong University
Chengdu, China
chenjian@swjtu.edu.cn

Ziyang Wang
School of Electrical Engineering
Southwest Jiaotong University
Chengdu, China
2022210727@my.swjtu.edu.cn

Hao Yue
School of Electrical Engineering
Southwest Jiaotong University
Chengdu, China
yuehao6866@my.swjtu.edu.cn

Wensheng Song
School of Integrated Circuits Science
and Engineering
Southwest Jiaotong University
Chengdu, China
songwsh@swjtu.edu.cn

Abstract—**Gallium nitride high electron mobility transistors (GaN HEMTs) are widely used due to their advantages of fast switching speed and low on-resistance. However, these may lead to undamped oscillations. Current research primarily focuses on the mechanism analysis and modeling of single-mode undamped oscillation, without considering transition between different undamped oscillations. There is a lack of theoretical research on systems that satisfy mechanism of different undamped oscillations. This article analyses the transition mechanism of undamped oscillations in the turn-OFF process. First, the generation mechanism of different oscillation is analysed. Then, a high-frequency equivalent model is established to study the influence of various parasitic parameters on oscillations. Finally, the accuracy of proposed theory is verified by experiment. The theory and modelling in this article comprehensively analyse the generation and transition mechanism of undamped oscillations and provide a systematic theoretical basis for studying undamped oscillations.**

Keywords—*Gallium nitride(GaN), high-electron-mobility transistors (HEMTs), undamped oscillation.*

I. INTRODUCTION

Silicon (Si)-based devices have reached their performance thresholds due to material limitations[1]. In this context, wide bandgap (WBG) semiconductors, including gallium nitride (GaN) and silicon carbide (SiC), have gained significant attentions. In particular, GaN high electron mobility transistors (HEMTs) exhibit high electric breakdown field, high electron mobility and low on-resistance, which provide advantages in applications requiring high frequency and high power density[2][3].

However, high switching speed and low on-impedance loop may induce undamped oscillations[4][5]. Undamped oscillations can lead to instantaneous shoot-through in the bridge-leg, excessive power loss or even more serious device breakdown. Therefore, it is essential to analyze undamped oscillatons.

A lot of research have been conducted on undamped oscillations in WBG devices. In [5] and [6], the generation mechanism of false triggering oscillations (FTO) is analyzed and the negative impedance model is developed. However, the common-source inductance (CSI) is neglected to simplify the model, which significantly affects gate-source voltage

oscillation and cannot be overlooked. In [7] and [8], FTO is investigated and a passive suppression method with the addition of ferrite beads in power loop is introduced to suppress oscillation. However, the drive resistance is not considered, that has a significant effect on the FTO and may affect the accuracy of the model. In [9] and [10], based on the Barkhausen criterion, it is concluded that the ratio between the CSI and the gate-drain capacitance is a critical factor in effectively preventing FTO. However, the influence of the drive resistance is ignored for easier computation. In addition to FTO, WBG devices may also generate other modes of undamped oscillations due to external conditions or inherent device characteristics. In [11], the generation mechanism of self-sustained oscillations (SSO) of the drain-source voltage is explored and the influence of parasitic parameters on system stability is analyzed.

The above analysis shows that the current studies mainly focus on single undamped oscillations. In practical engineering, the authors find that multiple undamped oscillations, such as FTO and SSO, occur at the same time, which is more complicated and harmful than a single oscillation. Therefore, this article emphasizes studying the generation and switching of different undamped oscillations. The main contributions of this work are as follows:

1) Analyzing the switching mechanism of two modes of undamped oscillations.

2) Analyzing the impact of different parameters on the generation and switching of different undamped oscillations.

The rest of this article is organized as follows: In Section II, the generation mechanism of SSO and FTO are analyzed. Based on the modes, Section III investigates the influence of parasitic parameters on the generation and switching of different undamped oscillations. To validate the theoretical analysis, Section IV presents simulation and experimental results. Finally, Section V concludes the article.

II. CIRCUIT MODELING

The half-bridge circuit is used to analyse the instability problem of the system, the schematic diagram is shown in Fig. 1. Both Q_1 and Q_2 are implemented using GaN HEMTs, where the top device Q_1 operates as a passive device, and the

This work was supported in part by the National Natural Science Foundation of China under Grant 52307224.

bottom device Q_2 operates as an active device, driven by a double-pulse signal.

A. The Generation Mechanism of undamped oscillations

The reverse conduction characteristics of GaN HEMTs differ significantly from those of Si or SiC MOSFETs. In MOSFETs, current flows through the intrinsic body diode. In contrast, GaN HEMTs lack a body diode due to their unique structure and rely on the conduction of the channel for freewheeling. When the drain-source voltage is negative, the channel current is controlled by the gate-drain voltage. During the turn-OFF period of Q_2, the gate-source voltage of Q_1 is 0, then the following equation is obtained

$$\begin{cases} v_{gd} > V_{gd,th} \\ v_{sd} > v_{gd} - V_{gd,th} \end{cases} \quad (1)$$

where $V_{th,gd}$ refers to the threshold voltage of the gate-drain voltage. From (1), it can be concluded that Q_1 remains in the saturation region if no driving voltage. In this case, the channel current is controlled by v_{gd1}, which can be expressed as

$$i_{ch1} = g_{m1} \cdot v_{gd1} \quad (2)$$

where g_{m1} represents reverse transconductance of the GaN HEMTs, and i_{ch1} represents the reverse channel current of Q_1. In this case, if the feedback network of the circuit becomes positive feedback due to the reverse conduction characteristics and parasitic parameters, Q_1 will occur SSO of drain-source voltage.

In addition to SSO, Q_2 may also occur other undamped oscillation during its turn-OFF period. At this point, Q_2 operates in the forward cutoff region. The gate-source voltage of Q_2 can be calculated as

$$v_{gs2} = -(L_{G2} + L_{S2}) \cdot \frac{di_{g2}}{dt} - L_{S2} \cdot \frac{di_d}{dt} - i_{g2} \cdot R_{G2} \quad (3)$$

where v_{gs2}, i_{g2}, and i_d are gate-source voltage, gate current, and drain current of Q_2.

From Fig. 1, it can be seen that the high dv/dt applied to the Miller capacitance generates a displacement current flowing into the gate node, causing an increase in di_{g2}/dt. Simultaneously, the di/dt acting on the CSI generates a negative voltage, both of which contribute to an increase in v_{gs2}, eventually exceeding the V_{th}. Once gate-source voltage of Q_2 exceeds the V_{th}, the channel of Q_2 will be forcefully turned on, which is called false turn-ON. The channel current is controlled by the gate-source voltage. At this point, since Q_2 is subjected to the bus voltage, it transitions from the cutoff region to the saturation region, where the following condition is satisfied

$$\begin{cases} v_{gs} > V_{gs,th} \\ v_{ds} > v_{gs} - V_{gs,th} \end{cases} \quad (4)$$

If the feedback network of the circuit becomes positive feedback, Q_2 will occur FTO of gate-source voltage.

In summary, the oscillation during turn-OFF period of Q_2 can be categorized into three modes:

(1) Q_1 occurs SSO of drain-source voltage, while Q_2 works in the cutoff region.

(2) Q_2 occurs FTO, while Q_1 continues to conduct channel due to reverse conduction characteristics.

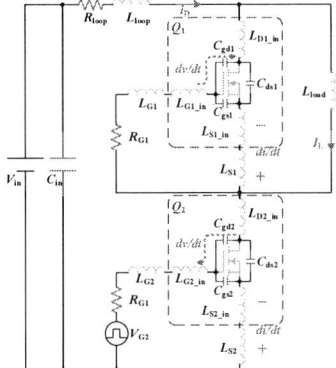

Fig. 1. The half-bridge circuit based on GaN HEMTs.

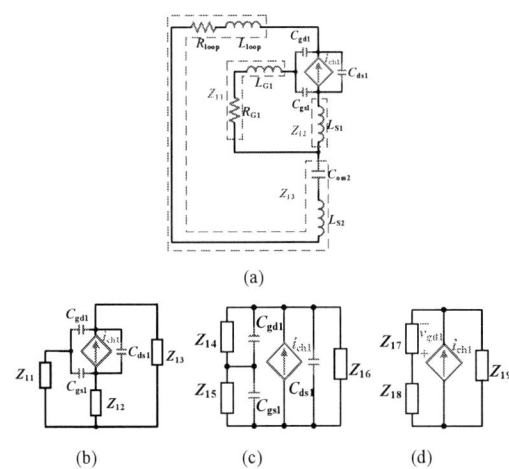

Fig. 2 Equivalent model of SSO. (a) Initial model of SSO. (b) Simplified model of SSO. (c) Further simplified model of SSO. (d) Final simplified model of SSO.

(3) Both Q_1 and Q_2 occur oscillations. In this scenario, the circuit works as a unified system that cannot sustain two independent oscillations simultaneously. As a result, one oscillation tends to be dominate, while the other manifests as a weaker part within overall oscillation. Since both Q_1 and Q_2 generate channel current, it is consistent with the case (2) and feedback of FTO cannot be interrupted. Furthermore, in the case of SSO, most of the energy is exchanged between the loop parasitic inductance and the output capacitance of Q_2. However, if Q_2 occurs false turn-ON, the energy transfer pathway becomes imbalanced, thereby destabilizing the feedback loop of SSO. Consequently, when both Q_1 and Q_2 occur oscillation, Q_2 is more likely to dominate, making FTO the primary oscillation.

B. Equivalent Circuit Setting and Stability Criterion

In this section, based on the analysis above, high-frequency equivalent models are established to analyze all oscillation mechanism during the turn-OFF period of Q_2. To simplify the equivalent circuit as shown in Fig. 1, the following assumptions are made.

1) In the high-frequency equivalent models, the bus capacitor C_{bus} and bus voltage V_{bus} are considered as

short-circuited, while the load inductance L_{load} is considered as open-circuited.

2) Since undamped oscillations occur during the turn-OFF period of Q_2, the V_G remains at a low level.

3) The existence of channel is a necessary condition for undamped oscillation to occur and is represented by a voltage-controlled current source.

When circuit occurs SSO, high-frequency equivalent model is shown in Fig. 2(a). Q_1 is equivalently represented as a current source controlled by the gate-drain voltage. At this time, Q_2 can be equivalently represented by its output capacitance C_{oss2}.

To simplify the equivalent circuit of SSO, as shown in Fig. 2(a). Z_{11}, Z_{12}, and Z_{13} is transformed into a triangle connection, as shown in Fig. 2(b) and (c):

$$\begin{cases} Z_{14} = \dfrac{Z_{11}Z_{12} + Z_{12}Z_{13} + Z_{11}Z_{13}}{Z_{12}} \\ Z_{15} = \dfrac{Z_{11}Z_{12} + Z_{12}Z_{13} + Z_{11}Z_{13}}{Z_{13}} \\ Z_{16} = \dfrac{Z_{11}Z_{12} + Z_{12}Z_{13} + Z_{11}Z_{13}}{Z_{11}} \end{cases} \quad (5)$$

Where $Z_{11} = R_{G1} + sL_{G1}$, $Z_{12} = sL_{S1}$, $Z_{13} = R_{loop} + s(L_{loop} + L_{S2}) + 1/(sC_{oss2})$. After further simplification, Fig. 2(d) can be obtained. Z_{17}, Z_{18}, and Z_{19} are obtained using the following equations:

$$\begin{cases} Z_{17} = Z_{14} // (1/(sC_{gd1})) \\ Z_{18} = Z_{15} // (1/(sC_{gs1})) \\ Z_{19} = Z_{16} // (1/(sC_{ds1})) \end{cases} \quad (6)$$

Fig. 3 illustrates the block diagram of the feedback system, where A is the forward transfer function, F is the feedback transfer function, and d represents the disturbance. A_1 can be calculated using the following equation:

$$A_1 = \frac{i_{ch1}(s)}{v_{gd1}(s)} = g_{m1} \quad (7)$$

The channel current is fed back to v_{gd1} through the circuit, and F_1 can be determined using the following equation:

$$F_1 = \frac{v_{gd1}(s)}{i_{ch1}(s)} = \frac{Z_{17}Z_{19}}{Z_{17} + Z_{18} + Z_{19}} \quad (8)$$

Based on Fig. 2(d), the closed-loop transfer function of the SSO can be expressed as

$$T_1(s) = \frac{A_1(s)}{1 - A_1(s)F_1(s)} = \frac{g_{m1}s(a_4s^4 + a_3s^3 + a_2s^2 + a_1s^1 + a_0)}{b_5s^5 + b_4s^4 + b_3s^3 + b_2s^2 + b_1s^1 + b_0} \quad (9)$$

where a_i (i = 0, 1, 2, 3, 4) and b_j (j = 0, 1, 2, 3, 4, 5) represent the coefficients of each term in the numerator and denominator of the closed-loop transfer function, respectively. The derivation shows that the closed-loop transfer function for analyzing the SSO is a fifth-order system.

When Q_2 occurs FTO, both Q_1 and Q_2 work in the saturation region. Gate disturbances in both Q_1 and Q_2 can theoretically impact the stability of the circuit. At this stage, the circuit operates as a multi-input system. The formation of the channel in Q_2 interrupts the energy transfer path of SSO, thereby impairing its positive feedback mechanism. As a result, when both Q_1 and Q_2 are subjected to disturbances, the

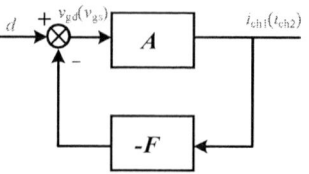

Fig. 3 Positive feedback block diagram.

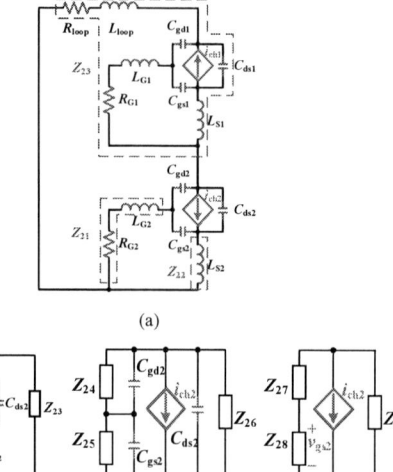

(a)

(b)   (c)   (d)

Fig. 4 Equivalent model of FTO. (a) Initial model of FTO. (b) Simplified model of FTO. (c) Further simplified model of FTO. (d) Final simplified model of FTO.

circuit occurs FTO dominated by Q_2. The modeling and experimental analysis in the subsequent sections of this article will further illustrate this phenomenon.

The equivalent model of the FTO is shown in Fig. 4(a), in which Q_1 is represented by an equivalent impedance Z_{23}, the star connection of Z_{21}, Z_{22}, and Z_{23} is transformed into a triangle connection of Z_{24}, Z_{25} and Z_{26}, as shown in Fig. 4(b) and (c), respectively. Subsequently, Z_{27}, Z_{28}, Z_{29} are obtained by further simplification, as shown in Fig. 4(d). Similarly, transfer function of the FTO can be derived as follows:

$$A_2(s) = \frac{i_{ch2}(s)}{v_{gs2}(s)} = g_{m2} \quad (10)$$

$$F_2 = \frac{v_{gs2}(s)}{i_{ch2}(s)} = \frac{Z_{27}Z_{29}}{Z_{27} + Z_{28} + Z_{29}} \quad (11)$$

$$\begin{aligned} T_2(s) &= \frac{A_2(s)}{1 - A_2(s)F_2(s)} \\ &= \frac{g_{m2}(a_7s^7 + a_6s^6 + a_5s^5 + a_4s^4 + a_3s^3 + a_2s^2 + a_1s^1 + a_0)}{b_7s^7 + b_6s^6 + b_5s^5 + b_4s^4 + b_3s^3 + b_2s^2 + b_1s^1 + b_0} \end{aligned} \quad (12)$$

where a_i (i = 0, 1, 2, 3, 4, 5, 6, 7) and b_j (j = 0, 1, 2, 3, 4, 5, 6, 7) represent the coefficients of each term in the numerator and denominator of the closed-loop transfer function, respectively. The FTO model involves two channels, making it a higher-order system with $T_2(s)$ as a seventh-order system

According to the Routh-Hurwitz criterion, the system is deemed unstable if any pole of the transfer function lies in the right half of the complex frequency domain.

979-8-3315-1110-4/25 $31.00 © 2025 IEEE

TABLE I. SPECIFICATIONS AND KEY CIRCUIT PARAMETER

Parameter	Value	Parameter	Value
g_{m1}	10S	R_{loop}	0.21Ω
g_{m2}	55S	C_{oss1}	483pF
L_{loop}	7.4nH	C_{oss2}	161pF
L_{G1}	4.47nH	C_{iss1}	274pF
L_{G2}	6.5nH	C_{iss2}	240pF
L_{S1}	0.7nH	C_{rss1}	60pF
L_{S2}	0.6nH	C_{rss2}	1.72pF

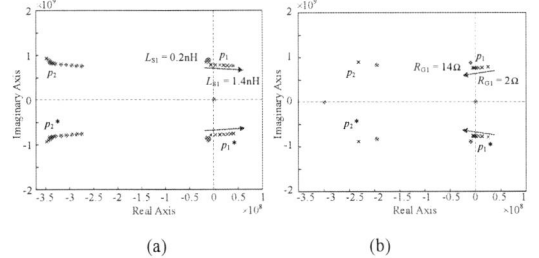

(a)   (b)

Fig. 5 The trajectory of the poles and zeros. (a) L_{S1} varies from 0.2 nH to 1.4 nH. (b) R_{G1} varies from 2 Ω to 14 Ω.

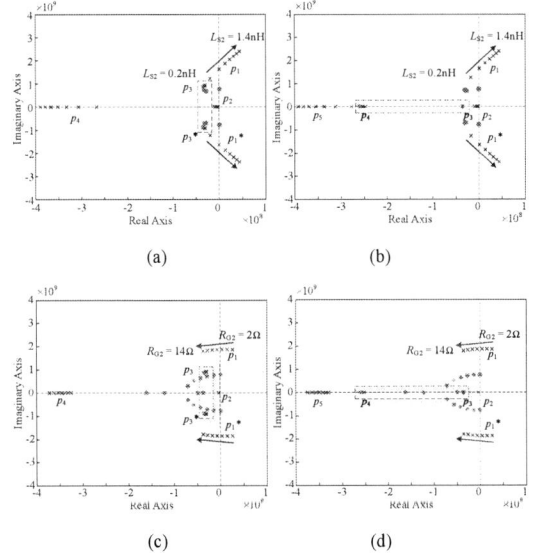

(a)   (b)

(c)   (d)

Fig. 6 The trajectory of the poles and zeros. (a) L_{S2} varies from 0.2 nH to 1.4 nH when R_{G1} is 3.2Ω. (b) L_{S2} varies from 0.2 nH to 1.4 nH when R_{G1} is 10Ω. (c) R_{G2} varies from 2 Ω to 14 Ω when R_{G1} is 3.2Ω. (d) R_{G2} varies from 2 Ω to 14 Ω when R_{G1} is 10Ω.

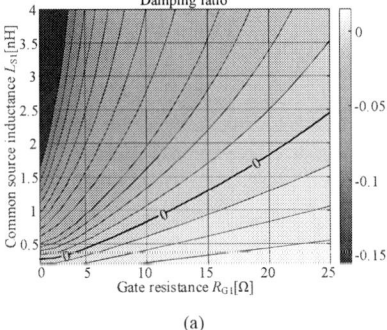

(a)

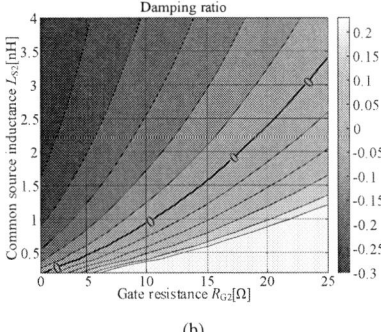

(b)

Fig. 7 The contour plot of parasitic parameters for undamped oscillations. (a) SSO. (b) FTO.

The parasitic inductances and high-frequency resistances are extracted using Ansys Q3D software. The high-frequency equivalent model is linearized around the DC operating point, therefore, the parasitic parameter values are selected at the DC operating point. Specifically, the parasitic capacitances C_{oss1}, C_{iss1}, and C_{rss1} are extracted at v_{ds1}=0, while C_{oss2}, C_{iss2}, and C_{rss2} are extracted at $v_{ds2} = V_{bus}$. The g_{m1} and g_{m2} are approximated by the slopes of the transfer characteristics near the threshold voltage. Key parameters are summarized in Table I.

Fig. 5(a) exhibits the influence of parasitic parameters on SSO, poles represent in blue and zeros in red. It can be observed that as L_{S1} increases, the dominant poles shift into the right half-plane of the complex frequency domain, causing the system to become unstable, based on Routh-Hurwitz criterion. Fig. 5(b) illustrates the trajectory of poles and zeros when R_{G1} varies from 2Ω to 14 Ω. The results show that as R_{G1} increases, the dominant poles shift into the left half-plane of the complex frequency domain.

Fig. 6 exhibits the influence of parasitic parameters on FTO. It is evident that increasing L_{S2} causes the system to become unstable. Fig. 6 (a) and (b) validate the effect of the CSI L_{S2} on the system's stability at $R_{G1} = 3.2$ Ω and 10Ω, respectively. It is evident that increasing L_{S2} causes the system to become unstable. Fig. 6(c) and (d) demonstrate the influence of R_{G2} on system stability under the same conditions. The results indicate that increasing R_{G2} can enhance system's stability.

It should be noted that variations in R_{G1} within the Q_1 driving loop primarily convert the dipole-forming conjugate poles p_3 and p_3^* into two real poles, while the dominant poles p_1 and p_1^* remains nearly unchanged, as shown in Fig. 6(a) and (b). This indicates that when both Q_1 and Q_2 work as oscillation sources, Q_2 occupies a dominant position, resulting in FTO being the primary oscillation, while the other appears as a weaker part within oscillation.To further investigate their influence on system stability, the locations of these poles can be analyzed, and their damping ratios computed.

III. IMPACT OF PARAMETERS ON OSCILLATIONS

Different modes of undamped oscillations fundamentally originate from the formation of distinct feedback systems, each involving specific core oscillatory devices and associated parasitic parameters. This section investigates the influence of parasitic parameters on different oscillations. The damping ratio ζ is employed as an indicator to evaluate the system's stability, which is defined as:

$$\zeta = \frac{\sigma}{|p_1|} = \frac{\sigma}{\sqrt{\sigma^2 + \omega^2}} \qquad (13)$$

where σ and ω is the real and imaginary parts of the complex poles p_1 and p_1^*, respectively. When $\zeta > 0$, the system is in a stable state, and the oscillation exhibits a decaying trend; When $\zeta = 0$, the system has zero damping, resulting in sustained oscillations with constant amplitude; when $\zeta < 0$, the system exhibits negative damping, leading to divergent oscillations. However, due to the limitations of the circuit and device, the divergent oscillations will eventually transition into sustained oscillations with constant amplitude.

Fig. 7 (a) illustrate the damping ratio when only Q_1 occurs SSO, with experimental conditions set to $V_{\text{bus}} = 100\text{V}$ and $I_L = 25\text{A}$. It should be noted that increasing L_{S1} and decreasing R_{G1} make the feedback system of SSO more unstable. Fig. 7 (b) illustrate the damping ratio when Q_2 occurs FTO. It should be noted that increasing L_{S2} and decreasing R_{G2} make the feedback system of FTO more unstable.

IV. EXPERIMENTAL VERIFICATION

As shown in Fig. 8, it is the experimental platform of a GaN-based half-bridge circuit with an inductance load. The switching devices are 650V e-mode GaN HEMT (GS66508T) produced by Infineon, and the driver chip is model SI8271AB with a driving voltage of 6V. A coaxial shunt resistor (SSDN-10) is used to measure the current, which has a high bandwidth (2000MHz), a resistance of 0.1Ω, and a parasitic inductance of 2nH.

1) Bus Voltage: Fig. 9 shows the voltage and current waveforms. When the bus voltage is 40V, the feedback system of FTO is stable, but the feedback system of SSO is unstable, thus, the circuit occurs SSO. As the bus voltage increases to 80V, the feedback system of FTO begins to be unstable and the channel of Q_2 interferes with the feedback loop of SSO, Consequently, the oscillation mode transitions from SSO to FTO.

2) Load Current: The magnitude of the load current directly affects the initial condition di/dt, so generation of FTO is constrained by load current. Fig. 10 shows the undamped oscillation at different current levels. When load current is 15A, Q_1 occurs SSO. Although the feedback system of FTO is alse unstable, Q_2 does no work in the saturation region, because v_{gs2} does not occur false turn-ON. So Q_2 does not occur FTO. When load current increases to 25A, false turn-ON is occurred and the oscillation mode transitions from SSO to FTO.

3) Common Source Inductance: The effect of CSI on the circuit system stability is relatively complex. If the CSI and the oscillation source are located within the same devices, it will have a significant impact. On the other hand, if the CSI and the oscillation source are located in the different devices, respectively, the CSI can be regarded as a part of the loop inductance. When both modes of feedback systems are in an unstable state, increasing L_{S2} causes the system to switch from SSO to FTO, as shown in Fig. 11(a) and (b). When only the feedback system of FTO is unstable, increasing L_{S2} causes the system to switch from damped oscillation to FTO, as shown in Fig. 11(c) and (d). When only the feedback system of SSO is unstable, increasing L_{S1} causes the system to switch from damped oscillation to SSO, as shown in Fig. 11(e) and (f).

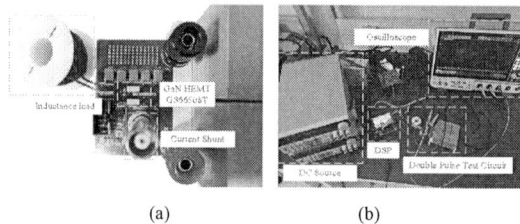

(a)   (b)

Fig. 8. The experimental platform. (a) Half-bridge circuit. (b) Top view of the experimental platform.

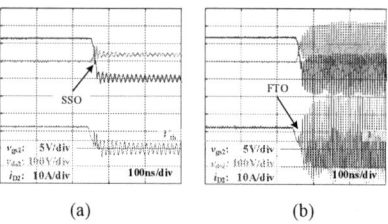

(a)   (b)

Fig. 9 Measured waveforms under different bus voltage where $R_{G1}=3.2\Omega$, $L_{S1}=0.7\text{nH}$, $R_{G2}=4.3\Omega$, $L_{S2}=0.6\text{nH}$, $I_L=25\text{A}$. (a) $V_{\text{bus}}=40\text{V}$. (b) $V_{\text{bus}}=80\text{V}$.

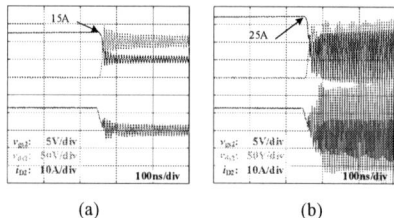

(a)   (b)

Fig. 10. Measured waveforms under different load current where $R_{G1}=3.2\Omega$, $L_{S1}=0.7\text{nH}$, $R_{G2}=4.3\Omega$, $L_{S2}=0.6\text{nH}$, $V_{\text{bus}}=100\text{V}$. (a) $I_L=15\text{A}$. (b)$I_L=25\text{A}$.

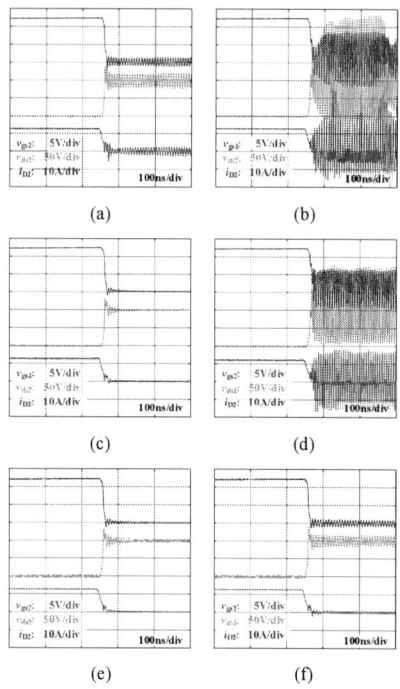

(a)   (b)

(c)   (d)

(e)   (f)

Fig. 11 Measured waveforms under different CSI where $V_{\text{bus}}=100\text{V}$, $I_L=25\text{A}$. (a) $R_{G1}=2\Omega$, $L_{S1}=0.5\text{nH}$, $R_{G2}=3.7\Omega$, $L_{S2}=0.4\text{nH}$. (b) $R_{G1}=2\Omega$, $L_{S1}=0.5\text{nH}$, $R_{G2}=3.7\Omega$, $L_{S2}=0.65\text{nH}$. (c) $R_{G2}=6.1\Omega$, $L_{S2}=0.5\text{nH}$. (d) $R_{G2}=6.1\Omega$, $L_{S2}=1\text{nH}$. (e) $R_{G1}=3.2\Omega$, $L_{S1}=0.3\text{nH}$. (f) $R_{G1}=3.2\Omega$, $L_{S1}=0.65\text{nH}$.

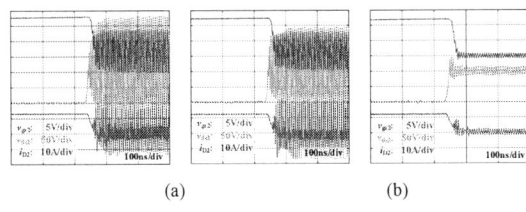

(a)　　　　(b)

Fig. 12. Measured waveforms under different gate resistance where V_{bus}=100V, I_L=25A, L_{S1}=0.5nH, L_{S2}=0.65nH. (a) R_{G1}=2Ω, R_{G2}=4.3Ω. (b) R_{G1}=19Ω, R_{G2}=4.3Ω. (C) R_{G1}=2Ω, R_{G2}=11Ω.

4) Gate Resistance: The gate resistance is one of the most significant factors influencing system stability. Increasing R_{G2} could suppress FTO, resulting in channel of Q_2 to be turned off, as shown in Fig. 12(a) and (c). At this point, the feedback system of SSO is satisfied, so the circuit generates SSO rather than damped oscillation. On the other hand, increasing R_{G1} does not affect FTO but merely reduces the amplitude of the v_{ds}, as shown in Fig. 12(a) and (b). It is because the oscillation source of FTO is Q_2, and the oscillation from Q_1 only adds a weak part to the FTO.

V. CONCLUSION

This article proposes a comprehensive study on generation and switching of different undamped oscillations. First, the mechanism of SSO and FTO are investigated. It is found that Q_1 occurs SSO when Q_2 is turned off and the feedback system SSO is unstable. Q_2 occurs FTO when Q_2 occurs false turn-ON and the feedback system of FTO is unstable. during the turn-OFF period of Q_2, when the circuit simultaneously satisfies the conditions for both FTO and SSO, the oscillation is dominated by FTO. Finally, the switching mechanisms of different undamped oscillations are verified through experiments. This article can provide systematical guidance for further research of undamped oscillations.

REFERENCES

[1] J. Chen, X. Du, Q. Luo, X. Zhang, P. Sun, and L. Zhou, "A review of switching oscillations of wide bandgap semiconductor devices," *IEEE Trans. Power Electron.*, vol. 35, no. 12, pp. 13182-13199, Dec. 2020.

[2] J. Chen, Q. Luo, J. Huang, Q. He, and X. Du, "A complete switching analytical model of low-voltage eGaN HEMTs and its application in loss analysis," *IEEE Trans. Ind. Electron.*, vol. 67, no. 2, pp. 1615-1625, Feb. 2020.

[3] T. Liu, T. T. Y. Wong, and Z. J. Shen, "A survey on switching oscillations in power converters," *IEEE J. Emerg. Sel. Topics Power Electron.*, vol. 8, no. 1, pp. 893-908, March 2020.

[4] P. Xue and F. Iannuzzo, "Self-sustained turn-off oscillation of cascode GaN HEMTs: occurrence mechanism, instability analysis, and oscillation suppression," *IEEE Trans. Power Electron.*, vol. 37, no. 5, pp. 5491-5500, May 2022.

[5] A. Lemmon, M. Mazzola, J. Gafford and C. Parker, "Stability considerations for silicon carbide field-effect transistors," *IEEE Trans. Power Electron.*, vol. 28, no. 10, pp. 4453-4459, Oct. 2013.

[6] A. Lemmon, M. Mazzola, J. Gafford and C. Parker, "Instability in half-bridge circuits switched with wide band-gap transistors," *IEEE Trans. Power Electron.*, vol. 29, no. 5, pp. 2380-2392, May 2014.

[7] F. Zhao, Y. Li, Y. Zheng, and N. Zhang, "Optimized design method of RC damper based on modified negative conductance model to suppress switching oscillations in synchronous rectifier circuits with GaN devices," *CPSS. Trans. Power. Electron. Appl.*, vol. 8, no. 3, pp. 314-324, Sept. 2023.

[8] F. Zhao, Y. Li, Z. Chen, S. Yang, and J. Chen, "Negative conductance modeling and stability analysis of high-frequency oscillation based on cascode GaN circuits," *IEEE Access*, vol. 8, pp. 114100-114111, 2020.

[9] K. Umetani, R. Matsumoto and E. Hiraki, "Prevention of oscillatory false triggering of GaN-FETs by balancing gate-drain capacitance and common-source inductance," *IEEE Trans. Ind. Appl.*, vol. 55, no. 1, pp. 610-619, Jan.-Feb. 2019.

[10] R. Matsumoto, K. Umetani, and E. Hiraki, "Optimization of the balance between the gate-drain capacitance and the common source inductance for preventing the oscillatory false triggering of fast switching GaN-FETs," in *Proc. IEEE Energy Convers. Congr. Expo.*, Cincinnati, OH, United states, 2017, pp. 405-412.

[11] K. Wang, X. Yang, L. Wang and P. Jain, "Instability analysis and oscillation suppression of enhancement-mode GaN devices in half-bridge circuits," *IEEE Trans. Power Electron.*, vol. 33, no. 2, pp. 1585-1596, Feb. 2018.

Multi-objective Optimization for Three-Level Power Module Based on Electro-Thermal Co-Design

Runze Wang
School of Electrical and Information Engineering
Anhui University of Science and Technology
Huainan, China
2023201831@aust.edu.cn

Jianing Wang
School of Electrical Engineering and Automation
Hefei University of Technology
Hefei, China
jianingwang@hfut.edu.cn

Shaolin Yu
The Institute of Energy, Hefei Comprehensive National Science Center(Anhui Energy Laboratory)
Hefei, China
yusl@ie.ah.cn

Xing Li
School of Electrical and Information Engineering
Anhui University of Science and Technology
Huainan, China
2023201813@aust.edu.cn

Xiahao Wang
School of Electrical Engineering and Automation
Hefei University of Technology
Hefei, China
2023170534@mail.hfut.edu.cn

Baolong Yan
School of Electrical and Information Engineering
Anhui University of Science and Technology
Huainan, China
2023201832@aust.edu.cn

Zhenchun Xia
School of Electrical Engineering and Automation
Hefei University of Technology
Hefei, China
2023170526@mail.hfut.edu.cn

Abstract—**Due to the characteristics of multi-chip, multi-commutation loops, and multiple operating modes in three-level power modules, there often exists a design conflict between the loop parasitic inductance and the junction temperature of the chips, making it challenging to simultaneously achieve low parasitic inductance for different loops and low junction temperature for chips under various operating modes. To address this issue, this paper proposes a multi-objective optimization design method for three-level power modules based on electro-thermal co-design, targeting loop parasitic inductance (L_{loop}), chip junction temperature (T_j), and thermal stress. By establishing models for multi-loop parasitic inductance and chip junction temperature under different operating modes, this method explores the influence of chip layout on L_{loop} for different loops, under various modes, and thermal stress through electro-thermal co-simulation. Furthermore, guided by the principle of trade-off design, a three-level power module with balanced electro-thermal performance is designed. Based on a self-developed packaging and interconnection platform, the designed power module is fabricated. Electrical-thermal characterization and reliability tests are conducted on the assembled module. The results demonstrate that the proposed design method significantly improves the comprehensive performance of the module.**

Keywords—Three-level power module, parasitic inductance, c hip junction temperature, power density, electrothermal coordina tion

I. Introduction

Insulated Gate Bipolar Transistor (IGBT) modules, as core components in modern power electronic systems, are often subjected to harsh natural environments in photovoltaic inverters. During continuous operation, IGBT devices endure extreme conditions such as high voltage, current stress, and thermal stress, making them prone to failure, which significantly impacts the service life of photovoltaic inverters [1-2]. IGBT modules come in various types, including different circuit structures and packaging forms. Among them, multi-level power modules, particularly three-level power modules, have been widely adopted in medium- and high-voltage applications due to their significant advantages in reducing switching losses, improving efficiency, and enhancing output waveform quality. However, with the increasing power density, the electro-thermal coupling effects within the modules have become increasingly pronounced, emerging as a major factor affecting their reliability and performance. The high operating temperature inside the chips is the root cause of device failures [3]. Electro-thermal coupling effects not only lead to localized overheating but may also trigger thermal stress concentration, material aging, and solder joint fatigue, thereby shortening the module's lifespan.

In fact, there has been extensive research on the packaging design of IGBT modules. References[3-4]optimized the chip layout to reduce the parasitic inductance of half-bridge loops. Reference [5] proposed a method of integrating shunt damping capacitors inside the module, significantly reducing loop parasitic inductance. Reference[6]introduced a simplified thermal analysis model to study the thermal coupling effects between module chips. Reference[7]presented a double-sided liquid-cooled module, which reduces the junction-to-case thermal resistance by 50%. Reference[8]reduced thermomechanical stress by replacing solid copper interconnects with sintered silver. However, the designs or analysis methods proposed in the aforementioned studies focus solely on single objectives, such as parasitic inductance, chip junction temperature, or thermal stress, neglecting the coupling between different objectives. While this approach may improve the performance of a single objective, it often deteriorates other objectives, thereby compromising the long-

This work was supported by Anhui Province Key Research and Development Program Project, Project Number:JZ2024AKKG0057.

979-8-3315-1110-4/25 $31.00 © 2025 IEEE

term reliability of the module. Additionally, although some studies, such as reference[9], have addressed the electro-thermal coupling characteristics of modules by proposing an electro-thermal model for IGBT modules based on SPICE subcircuit formulation, there has been no detailed report on the multi-objective optimization design for three-level power modules.

This paper proposes an electro-thermal co-design method to optimize the multi-objective characteristics of three-level power modules. By establishing models for multi-loop parasitic inductance and chip junction temperature under different operating modes, the method investigates the influence of chip layout on parasitic inductance across different loops, chip junction temperature under various modes, and thermal stress through electro-thermal co-simulation. The designed three-level power module is then fabricated using the proposed method. Furthermore, corresponding experimental tests are conducted to validate the effectiveness of the proposed approach.

II. THE ELECTRO-THERMAL SIMULATION

A. The specification of the designed power module

The optimized power module in this study is designed with a specification of 650V/300 A, and its internal circuit topology adopts a split-type NPC (Neutral Point Clamped) three-level circuit topology. The circuit topology is illustrated in Fig 1, where T_1, T_2, T_3, and T_4 each consist of four IGBT chips connected in parallel. D_5 and D_6 are clamping diodes, each comprising four FRD (Fast Recovery Diode) chips in parallel. D_1–D_4 are individual small-current freewheeling diodes, while D_7 and D_8 are large-current freewheeling diodes, each consisting of three FRD chips in parallel. The chip specifications used in the module of this article are shown in Table 1. The CAD model of the module is shown in Fig 2.

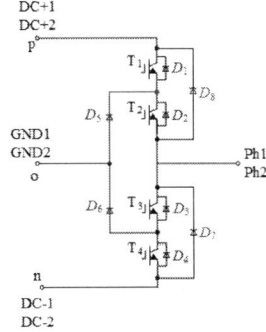

Fig 1. Equivalent circuit topology.

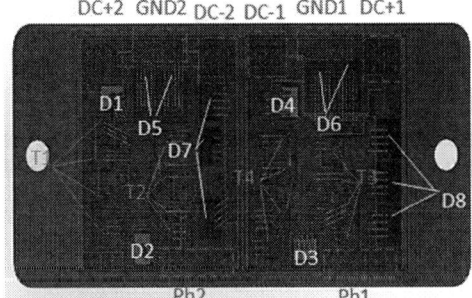

Fig 2. Module CAD model diagram.

Chip	Type	Specification
T1-T4	SMG065N60WSA1	650V/60A
D1-D4	MMK65A0F04	650V/100A
D5-D6	MMK65A0F04	650V/100A
D7-D8	MMKC575F04	1250V/75A

Table 1. Chip specification

B. Modeling of the parasitic inductance, the junction temperature and the power density

The parasitic inductance model of the power module is established based on the current loops during the commutation transients. For three-level power modules, different operating modes correspond to different commutation loops[10]. Taking the commutation process with the current direction being positive (flowing toward the load side) as an example, Figure 3(a) illustrates the commutation loop and the inductance distribution during the transition from the positive level to the zero level. Before commutation, the current flows from the positive terminal through transistors T_1 and T_2 to the load side. At this point, T_1 turns off, and the load current cannot change abruptly. The current then flows from the clamping diode D_5 and transistor T_2 to the load side. Therefore, the commutation loop includes the path from the positive terminal (p) to the collector (c) of T_1, the emitter (e) of T_1 to the cathode (k) of D_5, and the anode (a) of D_5 to the neutral point (o). The loop inductance is the sum of the inductances corresponding to these three paths, as shown in Equation (1). Here, L_{loop1} represents the parasitic inductance of the loop, while L_{pc}, L_{ek} and L_{ao} denote the equivalent parasitic inductances of the three segments, respectively.

$$L_{loop1} = L_{pc} + L_{ek} + L_{ao} \quad (1)$$

Further, the parasitic inductance of the loop is extracted by ANSYS Q3D. In Q3D, the CAD model of IGBT module is imported and different material properties are set. According to the above current path, the corresponding current excitation, namely Source and Sink, is set for each path segment. Fig 3 (b) shows the schematic diagram extracted by simulation in Q3D, in which the corresponding conductor segments of each part of the path are marked. The parasitic inductance of each conductor segment can be obtained through the excitation assignment :DC+ is Source and GND2 is Sink. In fact, the form of parasitic inductance matrix extracted by simulation includes self-inductance and mutual inductance of each conductor segment, as shown in formula (2). Take the path from the positive terminal p to the collector c of the T_1 tube as an example, where $L_{pc\text{-}pc}$ is the self-inductance corresponding to the path, $M_{pc\text{-}ck}$, and $M_{pc\text{-}ao}$ are the mutual inductance corresponding to the path and other paths.

$$\begin{bmatrix} L_{pc\text{-}pc} & M_{pc\text{-}ek} & M_{pc\text{-}ao} \\ M_{ek\text{-}pc} & L_{ek\text{-}ek} & L_{ek\text{-}ao} \\ M_{ao\text{-}pc} & M_{ao\text{-}ek} & L_{ao\text{-}ao} \end{bmatrix} \quad (2)$$

Finally, based on the circuit relationship of each path in series in the loop, the corresponding parasitic inductance of the loop can be calculated by the following formula (3).

$$\begin{aligned} L_{loop1} = &\, L_{pc\text{-}pc} + L_{ek\text{-}ek} + L_{ao\text{-}ao} \\ &- 2M_{pc\text{-}ek} - 2M_{pc\text{-}ao} - 2M_{ek\text{-}ao} \end{aligned} \quad (3)$$

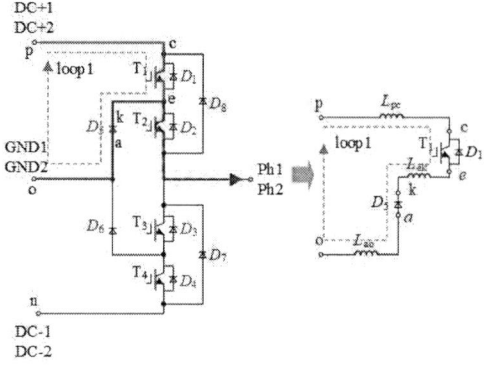

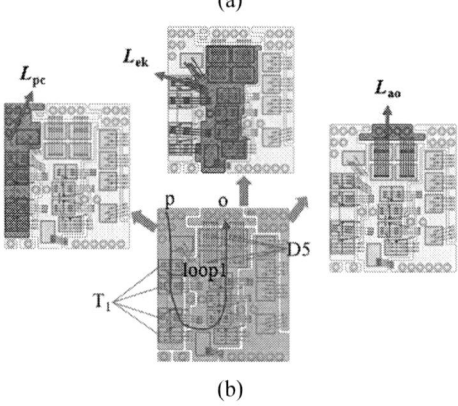

(a)

(b)

Fig 3. Converter loop and ANSYS simulation.

The junction temperature distribution of each chip in the power module is related to its power losses. The method for calculating losses in NPC three-level inverter circuits under Space Vector Pulse Width Modulation (SVPWM) has been provided in reference[10]. Based on this method and the parameters specified in the chip datasheets, the power losses of the chips under different operating modes can be calculated. Due to the symmetry of the upper and lower arms, Fig 4(a) and (b) present the calculated power losses of the chips in the upper arm under inverter mode with modulation indices (m) of 0.1 and 1, respectively. The loss distribution of each chip in the rectification mode is shown in Figure 5.

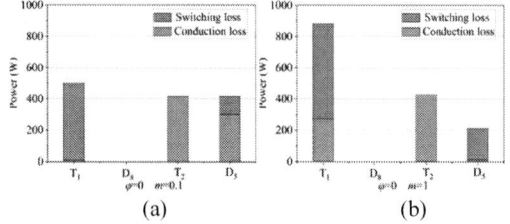

(a)

(b)

Fig 4. Chip loss in inverter mode.

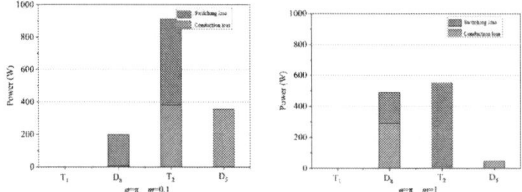

(a)   (b)

Fig 5. Chip loss in rectification mode.

Using COMSOL software, the temperature distribution of the module under different operating modes can be simulated. The CAD model of the module is imported into COMSOL, and the corresponding material properties are assigned. In the module, the chips are defined as heat sources, and their respective power losses are input. The heat sink baseplate of the module is set as the heat flux surface. The heat transfer coefficient is set to 2000 W/m², following conventional air-cooling standards. The solver in the study is configured for a steady-state analysis, and the steady-state temperature distribution contour of the module is obtained through simulation. The average junction temperature of each chip is selected as the observation point, and the temperature distribution in each mode is shown in Figure 6(a)-(d).

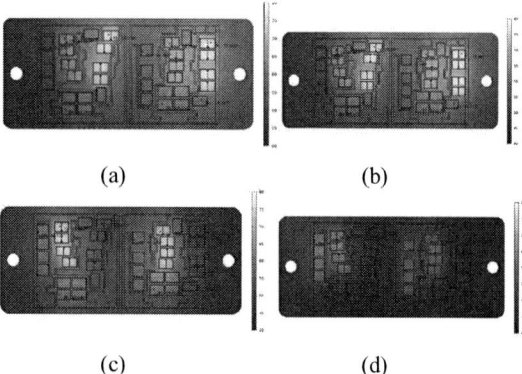

(a)   (b)

(c)   (d)

Fig 6. Cloud image of module surface temperature.

Furthermore, the thermal stress distribution under different losses can be obtained by COMSOL multi-physics simulation. The thermal stress distribution under different modes is shown in Figure 7(a)-(d).

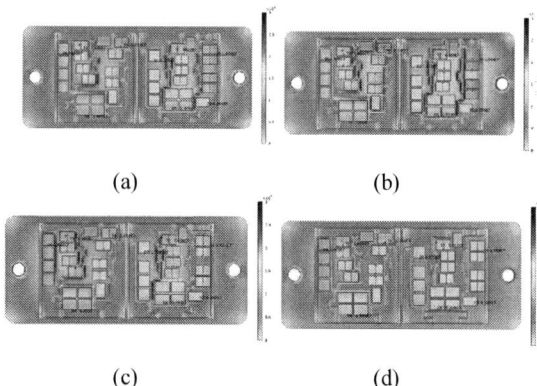

(a)   (b)

(c)   (d)

Fig 7. Cloud image of stress on the module surface.

C. Co-Design based on Electro-Thermal simulation

When the layout of the chips inside the module changes, the corresponding loop parasitic inductance, thermal distribution, and thermal stress distribution of the module also vary accordingly. In this study, the spacing d_1 between the chips is used as a design variable for the multi-objective optimization of the module. Taking the spacing d_1 of transistor T_1 as an example, Fig 8 illustrates the relationship between parasitic inductance and chip junction temperature for different values of d_1. It can be observed that as the chip spacing increases, the parasitic inductance of the module increases, while the junction temperature decreases. This is because the increase in chip spacing leads to a longer current path, resulting in higher self-inductance and mutual inductance, which ultimately increases the loop parasitic inductance. On the other hand, the increased chip spacing reduces the thermal coupling effect, thereby lowering the chip junction temperature.

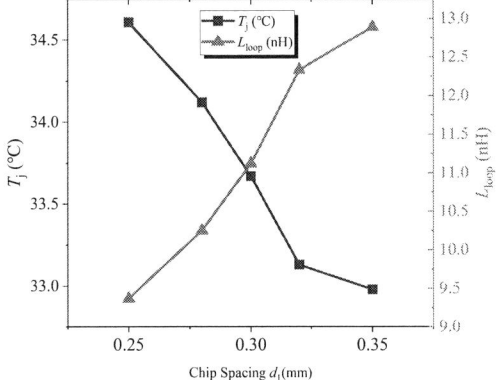

Fig 8. Temperature and inductance variation with chip spacing.

III. THE FABRICATION OF THE DESIGNED POWER MODULE

A. The material selection

The DBC (Direct Bonded Copper) substrate serves as the primary current-carrying component in the module. The substrate used in the designed module is standard alumina, with upper and lower copper layers of 0.3 mm thickness and an alumina ceramic layer of 0.38 mm thickness. The surface is treated with anti-oxidation technology. The pins are S-type integrated pins, with dimensions of 0.8 mm × 14.6 mm × 14 mm, featuring a gold-plated surface and tin-plated bottoms, which are directly soldered onto the DBC substrate. The bottom heat-dissipating copper baseplate measures 47 mm × 107 mm. Table 2 provides the specifications of the various materials used.

	IGBT	FRD	DBC	Pin	Copper Substrate
specification	650V/60A	650V/100A 1200V/75A	Cu 0.3mm Al₂O₃ 0.38mm	\	\
Size(mm)	4.2*4.2	5.4*3.8 4.5*6.3	37*40	0.8*14.6*14	47*107

Table 2. Material specifications.

B. The process of the module fabrication

First, solder paste is uniformly applied to the DBC substrate using a printing machine. Next, the chips are mounted onto the solder paste using a pick-and-place machine. After this step, the DBC substrate is placed into a reflow oven, where the reflow profile is adjusted to allow the solder to reach its melting point, completing the soldering process in the high-temperature zone. Subsequently, an ultrasonic wire bonder is used to achieve electrical interconnection between the chips and the substrate. Furthermore, a designed fixture is employed to secure the module as a whole, and the assembly is placed into the reflow oven for a second reflow soldering process to solder the pins and copper baseplate to the substrate. Finally, the housing is assembled, and the module is encapsulated and

cured. Fig 9 illustrates the overall flowchart of the module assembly process.

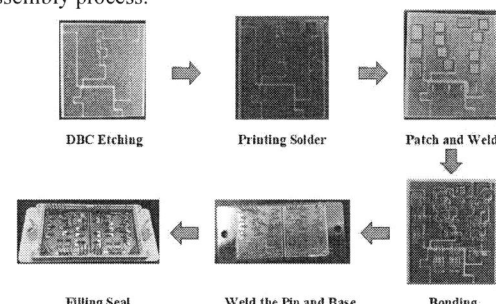

Fig 9. Module assembly process.

IV. EXPERIMATNAL RESULTS

A. Electical charactristics

Static test is mainly used to measure the electrical parameters of the power module under steady state conditions. Static test is mainly used to measure the electrical parameters of the power module under steady state conditions. The static working test equipment is a static parameter tester for power devices, and its model is HUSTEC-2000A-MTZ. Figure 10 shows the equipment for static testing, and Figure 11 shows the result of the static test. Figure 11(a) shows the V_{ce}-I_c characteristic curve and Figure 11(b) shows the I_{ce}-V_{ce} characteristic curve.

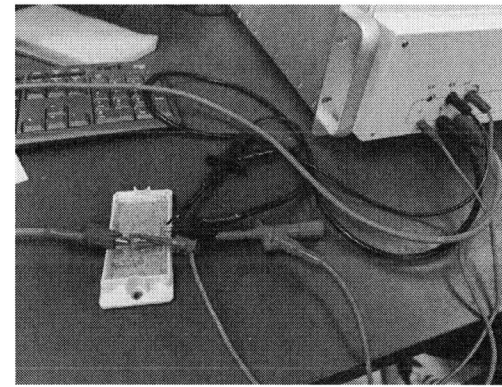

Fig 10. Static testing equipment.

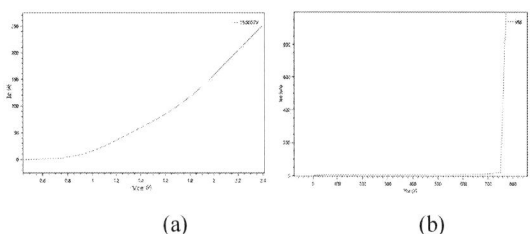

(a)   (b)

Fig 11. Static test results.

B. Thermal characrisitcis

In this paper, the overall thermal characteristics will be tested by thermal imager. Figure 12 is the thermal imaging image when the chip is working. It can be seen from the thermal imaging image that the temperature of the chip reached 62.8℃ at its highest point.

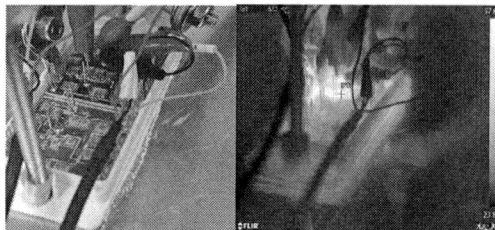

Fig 12. Thermal imaging diagram.

C. The reliablility test results

During the assembly process, the void ratio in the solder layer of the module was monitored using X-ray imaging to ensure the reliability of the soldering, as shown in Fig 13.

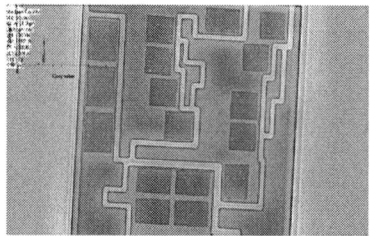

Fig 13. X-Ray image of the chip after welding.

Furthermore, the designed module will undergo power cycling tests at both second-level intervals using power cycling equipment to validate its reliability. The test conditions of the power cycle are as follows:

Test mode: Constant current mode

Test current: 160A

Initial temperature: 30℃

△T: 100℃

Over-limit values: Von offset ±5%, thermal resistance offset ±20%

Activation and shutdown time: 2 seconds for activation and 4 seconds for shutdown.

Figure 13 shows the results of the power cycling test. Figure 1(a) shows the Von curve, and Figure 13(b) shows the thermal resistance curve. The module completed 35,371 power cycles during the test.

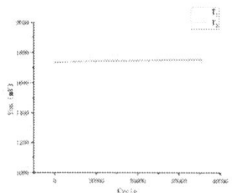 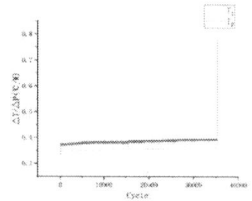

(a)　　　　　　　　　(b)

Fig 13. Power cycle test results.

V. CONCLUSION

Through the experimental results of this paper, it can be shown that in the design module, multiple variables such as inductance, junction temperature and stress are optimized at the same time, and the layout mode with appropriate values of the three values is selected as a compromise. Not only the parasitic inductance value of the three-level circuit is reduced, but also the junction temperature of the chip is reduced during operation, and the electric heating performance of the module is effectively increased.

VI. REFERENCES

[1] RENN,HUH,LYUXF,etal.Investigation on single pulse avalanche failure of SiC MOSFET and Si IGBT [J].Solid StateElectronics,2019,152:33-40

[2] Dbeiss M, Avenas Y, Zara H. Comparison of the electro-thermal constraints on SiC MOSFET and Si IGBT power modules in photovoltaic DC/AC inverters[J]. Microelectronics Reliability, 2017, 78: 65-71.

[3] LI S N , TOLBERT L M , WANG F , et al. P-cell and N-cell based IGBT module : Layout design , parasiticextraction , and experimental verification [C] //2011 Twenty-Sixth Annual IEEE Applied Power Electronics Conferenceand Exposition (APEC) . Fort Worth : IEEE , 2011 : 372-378.

[4] LI S N , TOLBERT L M , WANG F , et al. Stray inductancereduction of commutation loop in the P-cell and N-cell-based IGBT phase leg module [J] . IEEE Transactions onPower Electronics , 2014 , 29 (7) : 3616-3624.

[5] SUN X N , CHEN M , LI B D , et al. Design and evaluationof a face-down embedded SiC power module with lowparasitic inductance and low thermal resistance [J] . IEEETransactions on Power Electronics , 2023 , 38 (3) : 2799-2804.

[6] Bouguezzi S, Ayadi M, Ghariani M. Developing a Simplified Analytical Thermal Model of Multi-chip Power Module[J]. Microelectronics Reliability, 2016, 6664-77.

[7] G. Tang, T. c. Chai and X. Zhang, "Thermal Optimization and Characterization of SiC-Based High Power Electronics Packages With Advanced Thermal Design," in IEEE Transactions on Components, Packaging and Manufacturing Technology, vol. 9, no. 5, pp. 854-863, May 2019

[8] C. Ding, H. Liu, K. D. T. Ngo, R. Burgos and G. -Q. Lu, "A Double-Side Cooled SiC MOSFET Power Module With Sintered-Silver Interposers: IDesign, Simulation, Fabrication, and Performance Characterization," in IEEE Transactions on Power Electronics, vol. 36, no. 10, pp. 11672-11680, Oct. 2021.

[9] Górecki, Górecki P, K. SPICE-Aided Nonlinear Electrothermal Modeling of an IGBT Module[J]. Electronics ,2023,12(22):

[10] Jing, W. (2011). Research on power device losses in high-power three-level inverters . China University of Mining and Technology

A Novel Zero-pressure Packaging Structure of High-voltage SiC IGBT Device

Tang Xinling, Wangliang, Du yujie, Wei Xiaoguang
Beijing Huairou laboratory
Beijing, China
e-mail: duyujie@neps.hrl.ac.cn

Abstract—The press-pack device can fully leverage the advantages of silicon carbide (SiC) devices in high voltage, heat dissipation, and low inductance; however, the existing press-pack scheme cannot meet the requirements of high-voltage SiC devices due to the small area of SiC chips. Therefore, considering the pressure balancing design of SiC chips and the requirements of high field strength, low sensing, and heat dissipation, a zero-pressure packaging scheme for high-voltage SiC devices is designed in this paper. An 18 kV SiC insulate gate bipolar transistor (IGBT) press pack module is developed based on SiC IGBT chips independently. The static and switching characteristics of the module are tested, and the test results show that the module exhibits low leakage current and high blocking characteristics. At the same time, the SiC IGBT module verifies the feasibility of the zero-pressure packaging structure.

Keywords—press pack device, SiC IGBT, zero-pressure

I. INTRODUCTION

Silicon carbide (SiC) power devices above 10 kV, with advantages such as high blocking voltage, high frequency, high junction temperature, and lower switching loss, have become the core components for the development of power electronics in new power system [1-2].

As a bipolar device, SiC insulate gate bipolar transistor (IGBT) will make better leverage the high voltage characteristics of SiC materials. In recent years, many world-class universities and research institutions at home and abroad have carried out research on SiC IGBT, a frontier hot spot. In 2015, the US Army Research Laboratory (ARL) and Cree jointly developed a 27 kV/20 A SiC IGBT single-chip microcomputer module [3-5], which is the highest voltage class SiC IGBT module reported at present; In 2019, the US Army Research Laboratory cooperated with Wolfspeed (Cree) to develop the 18 kV/20 A SiC IGBT module [6] which realized the verification of the packaging structure at high temperatures above 200 ℃ and extremely low thermal resistance at 0.1 ℃/W; In 2021, Global Energy Internet Research Institute Co., LTD. (GEIRI) cooperated with North China Electric Power University (NCEPU) and Nanjing Institute of Electronic Devices to jointly develop a 15 kV/5 A SiC IGBT compression encapsulation module [7]; In 2023, GEIRI once again cooperated with NCEPU, NEDI, etc., proposed the 18 kV SiC IGBT single-chip module and ten-chip parallel package design scheme, and developed 18 kV/125 A SiC IGBT module [8]. However, due to the pressure balance design, terminal protection and packaging insulation difficulties in crimped SiC IGBT device packaging, it still cannot meet the needs of device industrialization and grid scale applications.

In this paper, considering the pressure balancing design of SiC chips and balancing the requirements of high field strength, low sensing, and heat dissipation, a zero-pressure packaging scheme for high-voltage SiC devices is designed.

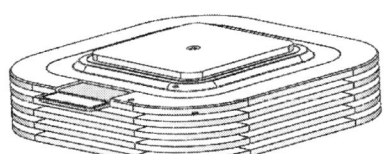

Fig. 1. Zero-pressure packaging scheme for high-voltage SiC devices

The module structure is shown in Fig. 1. Based on the chips independently developed by the research group, an 18 kV SiC IGBT module was prepared, and the static and switching characteristics of the module were obtained, which provided important data for the preparation and application of high-voltage and high-power SiC IGBT devices.

II. CHALLENGES IN PRESS PACK SIC

With the continuous development of SiC chips, its voltage has reached more than 10 kV, and the junction temperature has reached more than 200 ℃. Therefore, fully leveraging the advantages of SiC devices presents higher challenges for packaging design.

(1) High-voltage insulation design: the voltage level of the SiC chip and the terminal electric field strength of the chip far exceed the existing devices, which brings challenges to the external insulation and internal insulation design of the device; At the same time, the higher insulation design leads to the increase of the height of the device, which is contradictory to the control and heat dissipation of the parasitic inductance of the device.

(2) Pressure balance design: the active area of the SiC chip is much smaller than that of the silicon chip, which puts forward higher requirements for the design of the crimping structure. SiC materials differ from Si materials in terms of hardness and elastic modulus. The pressure range that SiC chip can withstand and the change of SiC chip characteristics under different pressures need to be further studied. At the same time, with the continuous improvement of the voltage level of flexible DC engineering, the explosion-proof characteristics of devices have gradually become one of the main factors limiting the application of devices.

(3) Multi-chip parallel electrical balance design: Similar to Si devices, SiC devices also have electrical balance problems when multi-chip parallel use, and because of faster switching speed, SiC devices are more sensitive to chip parameter dispersion and package parasitic parameters under high di/dt and dv/dt conditions. In order to realize multi-chip parallel co-flow of SiC devices, on the one hand, it is necessary to study the influence of chip parameter dispersion on device characteristics and screen chip parameters. On the other hand, reduce the parasitic parameters of packaging as much as possible, and adopt symmetrical layout and other solutions;

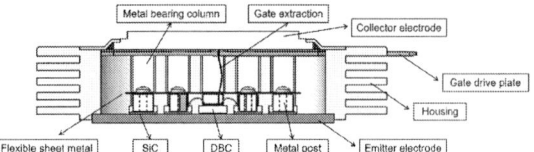

Fig. 2. Section view of SiC press-pack module

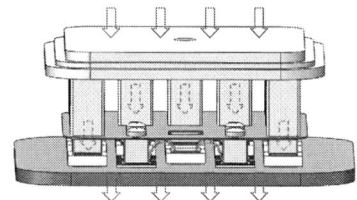

Fig. 3. Pressure propagation path of Module

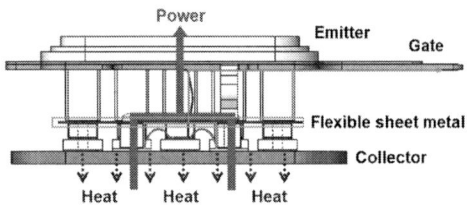

Fig. 4. Power, Drive, and Heat dissipation path of Module

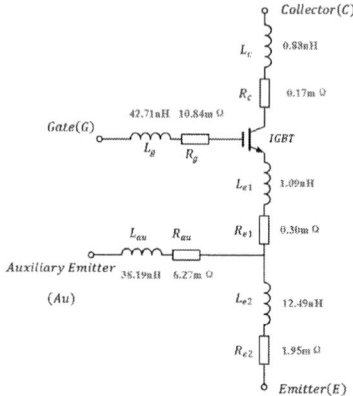

Fig. 5. Simulation results of module parasitic parameters

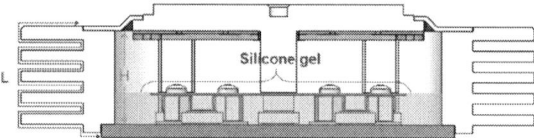

Fig. 6. Cross section of the module housing

(4) Packaging material selection and process optimization: high-reliability packaging material selection and packaging process is the key to give full play to the advantages of high pressure and high temperature SiC devices. At present, the highest voltage level of silicon IGBT devices on the market are 6.5 kV, and the voltage level of thyristors can reach 8.5 kV. The existing packaging material selection and technology cannot meet the packaging requirements of SiC IGBT above 15 kV. Therefore, it is necessary to study the selection of packaging materials and the development of supporting packaging technology.

III. A SOLUTION TO PRESS PACK SiC

A. Package design

A press pack SiC package structure is proposed, and the cross-section diagram is shown in Fig.2. The package structure has the technical characteristics of zero pressure, low parasitic parameter, high thermal conductivity, short circuit failure and explosion proof.

In order to solve the problems of small bearing area, strong pressure and easy mechanical damage in Press pack of silicon carbide chip, a zero-pressure design idea was proposed in this scheme, as shown in Fig.3. The pressure of the module is achieved by the direct bonding copper (DBC) and the pressure metal column, and the emitter of the power chip is drawn through the metal column to the flexible metal plate. Flexible metal plate, bearing metal column and module emitter general pressure contact to achieve electrical extraction.

As shown in Fig.4, a flexible metal sheet is used in the middle of the device as the connecting medium for the parallel connection of chips and current collection, and the device is divided into the upper and lower regions. In the

lower part of the device, aluminum nitride (AlN) insulation sheet and insulation potting material are combined to solve the field intensity concentration and realize the pressure resistance requirement. The upper part of the device is filled with insulating gas to achieve insulation. The chip collector is directly connected with the metal electrode through the silver sintering process, and the heat dissipation of the module is mainly borne by the collector side. According to the principle of parallel heat transfer, the thermal resistance is less than the minimum value of parallel thermal resistance, so that the overall thermal resistance of the module is greatly reduced. The laminated printed circuit board (PCB) is welded to the driver PCB and the grid copper column, the Kelvin link is realized. At the same time, the parasitic inductance of the packaged power loop of the module is 14.46 nH, the simulation results are shown in Fig.5.

Fig. 6 shows the cross section of the module housing. The material selection is a composite material composed of glass fiber, resin and filler, which is connected with the metal electrode by silicone bonding. The insulation distance H of the housing and outer surface creep age distance L shall meet the insulation requirements of the module's 18 kV voltage level. In addition, compared with the traditional ceramic housing, the composite material has better toughness, has the function of absorbing energy buffer when subjected to impact load, and the adhesive seal can also provide an easier detonation channel; From the point of view of structural protection, it has better explosion-proof performance, and a large number of fragments will not be formed after breaking, and the application safety of the module is better.

B. Simulation Result

The preparation of the module consists of 8 process steps, as shown in Fig. 7: (1) The SiC chip, DBC and collector electrode are connected by reflow welding process; The collector electrode is molybdenum copper alloy, and the

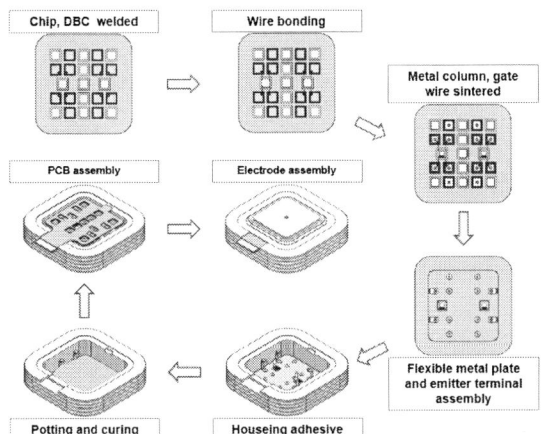

Fig. 7 Module packaging process diagram

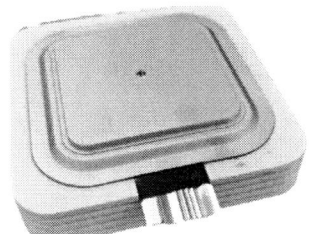

Fig. 8 Module packaging process diagram

solder is high temperature brazing metal (melting point, 317 °C); (2) The SiC chip gate is connected with DBC through aluminum wire bonding; (3) The pressure metal column is connected to the DBC, the emitter metal column is connected to the chip, and the gate electrode is connected to the DBC by the silver sintering process; (4) The flexible metal plate, the emitter terminal and the metal column are connected by bolts; (5) The module housing is connected with the collector electrode through adhesive; (6) The silicone gel is poured to flood the flexible metal plate; The use temperature of silicone gel is more than 150 °C, and the curing process requires multiple vacuum degassing to ensure that the silicone gel is fully filled; (7) PCB board and power terminal connected by brazing; (8) The module housing is connected with the emitter electrode through adhesive.

The 18 kV SiC module was prepared based on press pack structure mentioned above. The module consists of 8 SiC IGBTs and 4 SiC Fast recovery diodes (FRDs). The electrical parameters of the SiC IGBT chip are 18 kV/12.5 A, and the electrical parameters of the SiC FRD chip are 18 kV/25 A. The physical appearance of the prepared 18 kV/125 A SiC IGBT module is shown in Fig. 8.

IV.   MODULE CHARACTERISTICS

A. Static characteristics

The output and transfer characteristic curves of the prepared SiC IGBT module were tested using Agilent B1505A. Among them, the output characteristic curve test obtained the relationship between collector current I_C and collector-emitter voltage V_{CE} under gate emitter voltage V_{GE} ranging from 10 V to 20 V, as shown in Fig. 9. It can be concluded that when V_{GE} = 20 V, I_C = 100 A, the V_{CE} is

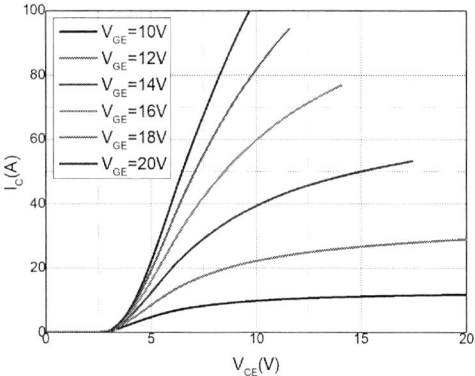

Fig. 9 Output characteristic curve of SiC IGBT

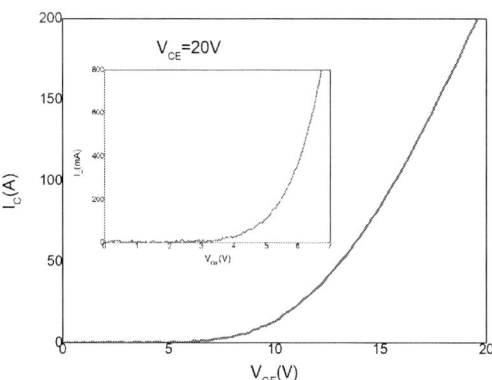

Fig. 10 Transfer characteristic curve of SiC IGBT

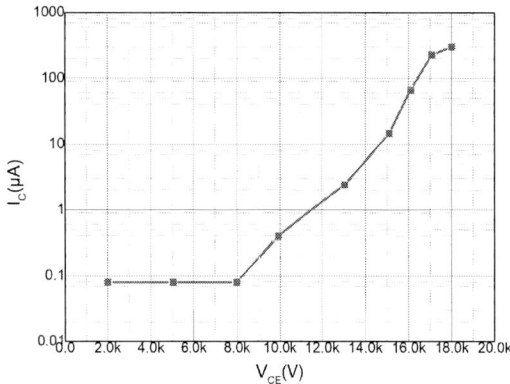

Fig. 11 Blocking characteristic curve of SiC IGBT

approximately 9.64 V. The transfer characteristic curve represents the relationship between I_C and V_{GE} under a fixed V_{CE}, as shown in Fig. 10. It can be concluded that when V_{CE} = 20 V, I_C = 80 mA, the threshold voltage $V_{GE(th)}$ of SiC IGBT is 4.69 V.

In order to conduct blocking characteristic testing of SiC IGBT modules, a high-voltage testing platform was independently built. The test results is shown in Fig. 11, which show that the leakage current of SiC IGBT module at 18 kV is about 298.9 μA, and the module exhibits low leakage current and good blocking characteristics.

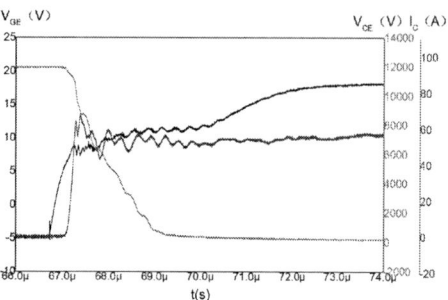

Fig. 12 The turn on waveform of SiC IGBT

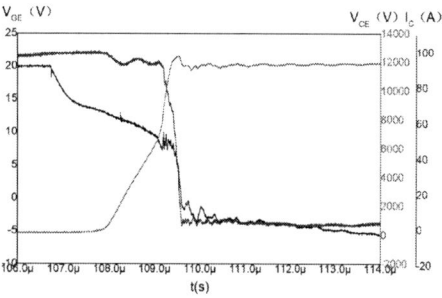

Fig. 13 The turn off waveform of SiC IGBT

B. Switching characteristics

The switching waveform of the SiC IGBT module is obtained through the high-voltage SiC dynamic test platform test. The test conditions are V_{CC} = 12 kV, V_{GEon} = -5 V, V_{GEoff} = -20 V, R_{Gon} = 20 Ω, R_{Goff} = 33 Ω, C_{GE} = 200 nF. The second turn on waveform in double pulse of SiC IGBT module is shown in Fig.12, and the test result show that the turn on current is 50 A, the turn on delay time $t_{d(on)}$ is approximately 214.5 ns, the rise time t_r is 120 ns, and the turn on energy E_{on} is 550 mJ.

The second turn off waveform in double pulse of SiC IGBT module is shown in Fig. 13, it can be seen that the turn off current is 100 A, the turn off delay time $t_{d(off)}$ is approximately 2357 ns, the fall time t_f is approximately 380 ns, and the turn off energy E_{off} is 748.8 mJ. The switch transient current and voltage overshoot of the SiC IGBT module are relatively small, and the switch loss has significant advantages.

V. CONCLUSION

This paper presents a high insulation, low thermal resistance, short-circuit failure and explosion-proof functions 18 kV press pack solution, and an 18 kV/100 A SiC IGBT module is manufactured to study the static and switching characteristics. Static characteristics test results show that when V_{GE} = 20 V, I_C=100 A, the on-state voltage drop $V_{CE(sat)}$ is about 9.64 V, and the leakage current under 18 kV is 298.9 μA. The zero-pressure packaging structure proposed in this article provides a new method for solving the pressure bonding packaging of high-voltage SiC devices.

VI. ACKNOWLEDGMENTS

This work was supported by the Beijing Nova Program, Grant Number: 20220484094.

REFERENCES

[1] T. M. Do, O. Lesaint and J. .-L. Auge, "Streamers and partial discharge mechanisms in silicone gel under impulse and AC voltages," in *IEEE Transactions on Dielectrics and Electrical Insulation*, vol. 15, no. 6, pp. 1526-1534, December 2008, doi: 10.1109/TDEI.2008.4712654.

[2] Wang, L., et al.: Modelling and optimization of SiC MOSFET switching voltage and current overshoots in a half-bridge configuration. *IET Power Electronics*. 14, 1684-1699 (2021). doi:10.1049/pel2.12146

[3] E. V. Brunt *et al.*, "22 kV, 1 cm2, 4H-SiC n-IGBTs with improved conductivity modulation," *2014 IEEE 26th International Symposium on Power Semiconductor Devices & IC's (ISPSD)*, Waikoloa, HI, USA, 2014, pp. 358-361, doi: 10.1109/ISPSD.2014.6856050.

[4] M. Hinojosa, H. O'Brien, E. Van Brunt, A. Ogunniyi and C. Scozzie, "Solid-state Marx generator with 24 KV 4H-SIC IGBTs," *2015 IEEE Pulsed Power Conference (PPC)*, Austin, TX, USA, 2015, pp. 1-5, doi: 10.1109/PPC.2015.7297038.

[5] Van Brunt, Edward, et al. "27 KV, 20 A 4H-SiC n-IGBTs." Materials Science Forum, vol. 821–823, Trans Tech Publications, Ltd., June 2015, pp. 847–850. Crossref, doi:10.4028/www.scientific.net/msf.821-823.847.

[6] M. Hinojosa, A. Ogunniyi and H. O'Brien, "Performance of 18-kV Silicon Carbide High-Voltage Boost-Chopper Modules," *2019 IEEE Pulsed Power & Plasma Science (PPPS)*, Orlando, FL, USA, 2019, pp. 1-4, doi: 10.1109/PPPS34859.2019.9009740.

[7] Yujie Du, Xinling Tang, et al.: 15 kV Press Pack SiC IGBT. *2021 IEEE Workshop on Wide Bandgap Power Devices and Applications in Asia (WiPDA Asia)*. (2021). doi:10.1109/WiPDAAsia51810.2021.

[8] Yufeng Qiu, Xinling Tang, et al.: Research on Key Technologies for Development and Series Application of 18 kV/125A SiC IGBT Device. *Proceedings of the CSEE*. 43(17), 6765-6775 (2023). doi : 10.13334/j.0258-8013. pcsee.222711.

Design and Characterization of A Novel Embedded SiC Power Module with Double-sided RDL

1st Yifei DU
College of Electrical Engineering
Zhejiang University
Hangzhou, China
22310190@zju.edu.cn

2nd Min CHEN
College of Electrical Engineering
Zhejiang University
Hangzhou, China
calim@zju.edu.cn

3rd Xinnan SUN
College of Electrical Engineering
Zhejiang University
Hangzhou, China
sxnan@zju.edu.cn

4th Jie LI
College of Electrical Engineering
Zhejiang University
Hangzhou, China
12210100@zju.edu.cn

5th Yucheng WU
College of Information Science and Electronic Engineering
Zhejiang University
Hangzhou, China
22431134@zju.edu.cn

6th Han ZHOU
College of Electrical Engineering
Zhejiang University
Hangzhou, China
3210104140@zju.edu.cn

7th Feng JIANG
College of Electrical Engineering
Zhejiang University
Hangzhou, China
jiangfeng@zju.edu.cn

Abstract—**Existing packaging technologies struggle to meet the demands of high-frequency and high-temperature applications for SiC devices, making embedded packaging a promising solution. This study designs a novel double-sided RDL embedded package structure and proposes a new "Molding-Drilling-RDL" process route, which not only reduces the process complexity and cost of embedded packaging but also improves package performance. By employing a double-sided direct plating RDL method, the solder required for connecting the back of the die to the copper layer in traditional embedded packages is eliminated. This not only reduces the junction-to-case thermal resistance but also avoids potential reliability issues associated with the solder layer after long-term operation. Thermal simulation results show that the junction-to-case thermal resistance of the proposed package structure is reduced by at least 68% compared to the traditional lead frame-based embedded package. Furthermore, by utilizing a compact layout and copper pad design, this paper reduces the total parasitic inductance of the module to 0.41 nH, an 81% reduction compared to the traditional embedded package with through-vias. Double-pulse testing results demonstrate that the proposed solution reduces switching losses by 13% compared to a commercial TO-247 package.**

Keywords—*embedded packaging, double-sided RDL, copper spacer, SiC MOSFET, parasitic inductance, thermal resistance*

I. INTRODUCTION

Compared to traditional silicon-based devices, SiC devices can operate at higher temperatures, voltages, and switching frequencies while significantly reducing energy losses, making them highly valuable in applications such as electric vehicles and renewable energy [1]. However, the full potential of SiC is limited by packaging performance. Current SiC power devices predominantly use substrate wire bonding technology, which has shortcomings in both electrical and thermal performance, severely constraining the capabilities of SiC devices [2].

The parasitic inductance introduced by bonding wires (>10 nH) exacerbates voltage spikes, switching losses, and electromagnetic interference (EMI) [3], [4]. For SiC devices with faster switching speeds, the impact of parasitic parameters is further amplified. Therefore, there is an urgent need to explore new interconnection methods to eliminate bonding wires and reduce loop inductance. In recent years, with advancements in packaging technology, lead-free

packaging technologies such as planar interconnects (Cu-Clip), Press-Pack, and embedded package have emerged [5]-[7]. Among these, in embedded package, the die is embedded in an insulating molding compound and connected to external circuits through blind vias and re-distribution layers (RDLs). This lead-free three-dimensional interconnection structure significantly shortens the electrical interconnection length within the package, achieving low parasitic inductance and a compact layout, thus demonstrating advantages in SiC power module applications. Fraunhofer IZM was the first to conduct research on embedded packaging, proposing a process route involving vacuum lamination pre-embedded dies and copper via interconnections [8]. Subsequently, D. J. Kearney and S. Bensebaa improved the process, enhancing die positioning accuracy and electrical interconnection performance [9], [10]. Additionally, F. Hou proposed an embedded packaging solution based on pre-processed BT substrates to improve mechanical strength, but the parasitic inductance and thermal resistance performance of this solution were not ideal [11]. Overall, these studies have been limited to packaging processes and have overlooked in-depth research on electrical and thermal performance. Due to the ultra-compact layout and minimal size of lead-free packaging structures, the thermal flux density within the package is significantly increased. Particularly for power dies, achieving efficient heat dissipation in such a small space has become a new challenge.

This research designs a novel double-sided RDL embedded package structure and proposes a novel "Molding-Drilling-RDL" process route, which not only reduces the complexity and cost of embedded packaging but also enhances packaging performance. The proposed solution involves directly electroplating RDLs on both sides of the die, eliminating the need for solder to connect the die's backside pad to the copper layer in traditional embedded packaging. This not only reduces the junction-to-case thermal resistance but also avoids reliability issues associated with solder layers after long-term operation. Furthermore, the process eliminates the high-temperature soldering step, reducing processing difficulty and cost. The RDL on the backside of the package serves not only as an electrical connection but also as a thermal dissipation layer. Since there are no interface materials between the die and the backside RDL, heat generated by the die can be efficiently transferred to the copper layer surface and dissipated through air or liquid cooling. In terms of electrical performance, through compact layout and copper spacer design, the overall parasitic inductance of the module is reduced to 0.41 nH, representing an 81% reduction compared to traditional embedded package with through-vias. Double-pulse test (DPT) results show that the proposed package reduces switching losses by 13% compared to commercial TO-247 package.

II. DESIGN AND FABRICATION OF THE DOUBLE-SIDED RDL EMBEDDED POWER MODULE

A. Packaging Structure and Layout

Fig. 1 illustrates the cross-sectional view of the proposed double-sided RDL embedded power module. The module consists of two MOSFET bare dies, connected in series to form a half-bridge circuit. The two MOSFET bare dies are embedded in epoxy resin molding compound, with the dies featuring dual-sided pads that establish electrical connections to external circuits through the top and bottom fan-out RDL layers. Considering the layout characteristics of the pads, the pads on the front side are connected to the top RDL through blind vias, while the pads on the back side are directly connected to the bottom RDL. Fig. 2 shows the RDL layout of the designed power module. It can be observed that several copper spacers are embedded in the power module, located in the same layer as the SiC MOSFETs, to ensure robust interconnection between the upper and lower RDL layers. Compared to existing embedded packaging solutions that use through-vias spanning the upper and lower layers [10], this approach enhances current-carrying capacity, improves electrical connection reliability, and contributes to better thermal performance. Additionally, the bottom RDL of the power module is directly exposed to the exterior, enabling facile integration with cold plates or microchannel heat sinks on the RDL surface to achieve a highly compact thermal management system integrated with the package.

B. Fabrication of the Embedded Power Module

The first stage includes Step 2 and Step 3. First, a BT substrate with pre-processed grooves is attached to a carrier board coated with thermal release adhesive. Then, the dies and copper spacers are placed in their respective positions, and all structures are covered by filling with epoxy molding compound (EMC) or laminating with sheet molding compound (SMC). The carrier board is removed by heating and releasing the adhesive, exposing the backside of the dies and copper spacers, while their front sides remain covered.

The second stage includes Step 4. Blind vias are prepared above the pads of the dies and copper spacers using laser drilling to expose their surfaces. To avoid damaging the internal structure of the dies, flat-top laser beams should be used instead of Gaussian beams.

The third stage includes Step 5 and Step 6. First, seed layers (Ti/Cu) are sputtered on both the front and back sides of the package. Then, a circuit pattern mask is created on the seed layers, and the copper layer is thickened through electroplating while simultaneously filling the blind vias. Subsequently, an inverse mask pattern is created, and the circuit pattern is formed through etching, resulting in double-sided fan-out RDL. Fig. 4 shows the top view and cross-sectional view of the blind vias above the dies. Fig. 4(a) illustrates the exposed pads of the dies after laser drilling, and Fig. 4(b) shows the connection effect between the blind vias and the pads after electroplating.

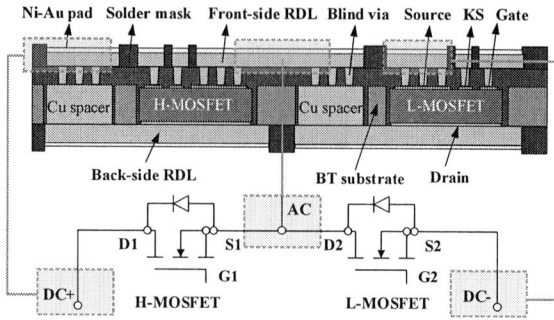

Fig. 1. Cross-section and circuit schematic of proposed module.

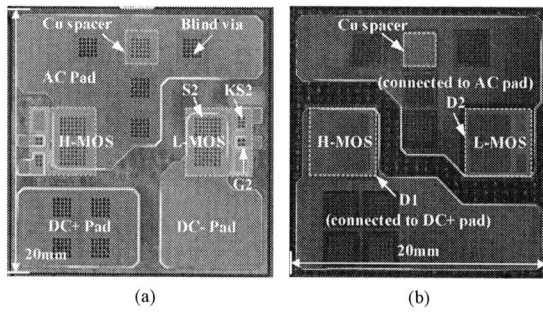

Fig. 2. Layout of proposed module. (a) Front side. (b) Back side.

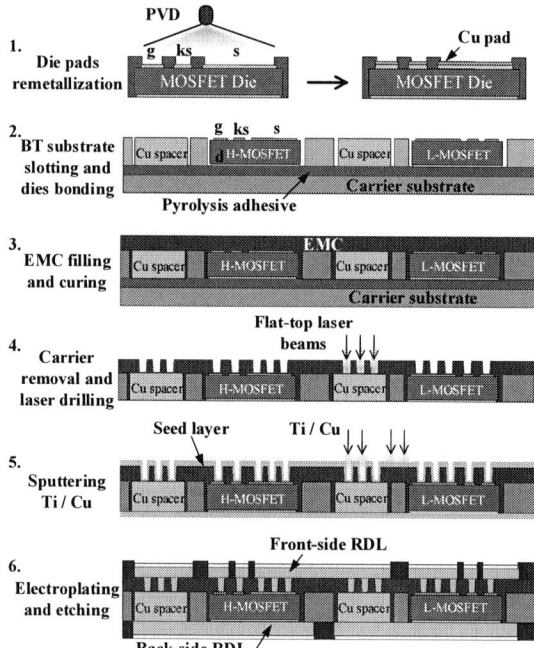

Fig. 3. Fabrication process flow of proposed module.

Finally, solder mask ink is sprayed on the front and back surfaces of the package, and the exposed copper layers are chemically plated with nickel and gold to enhance stability and solderability. Fig. 5 shows the embedded packaging after Ni-Au plating. By slicing, a half-bridge embedded SiC power module can be obtained.

979-8-3315-1110-4/25 $31.00 © 2025 IEEE

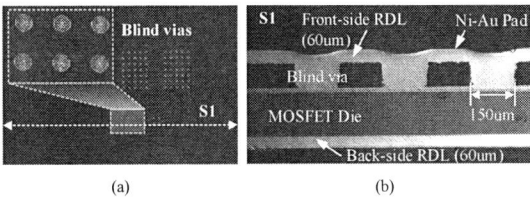

(a)   (b)

Fig. 4. Blind vias above the MOSFET die. (a) Top (pre-fill). (b) Cross-section (post-fill).

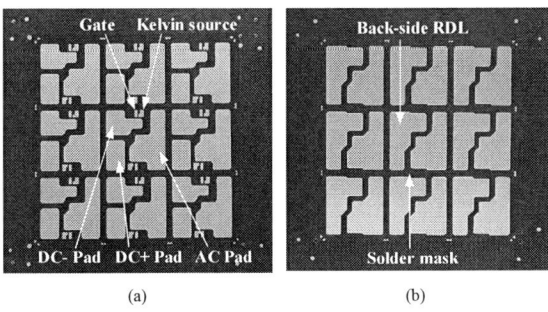

(a)   (b)

Fig. 5. Photograph of the double-sided RDL substrate embedded packaging. (a) Front side. (b) Back side.

The proposed "Molding-Drilling-RDL" process flow eliminates the need for lead frame fabrication and die attachment (soldering) required in conventional embedded packaging, significantly reducing both the process complexity and cost of embedded modules.

III. EXPERIMENTAL VERFICATION

A. Electrical Characterization

The parasitic inductance of the proposed double-sided RDL embedded power module was extracted using ANSYS Q3D Extractor. Fig. 6 shows the parasitic inductance network model of the proposed embedded power module. Table I summarizes the extracted parasitic inductance results at 1 MHz frequency. Benefiting from optimized die placement and the "copper spacer + blind via" vertical interconnection scheme, the proposed module achieves a power loop parasitic inductance below 330 pH, with drive loop parasitic inductances of 44 pH and 39 pH for the upper and lower switches, respectively. Additionally, Table II lists the parasitic inductance values of several advanced package solutions. It can be observed that the proposed package solution reduces the overall parasitic inductance by 81% compared to traditional embedded package with through-vias. Furthermore, compared to other package solutions, the proposed solution reduces the parasitic inductance by at least one order of magnitude.

To further evaluate the switching characteristics, this work compares the dynamic switching performance of the proposed embedded package with the traditional TO-247 package using double-pulse testing (DPT). The setup of the DPT platform is shown in Fig. 7. To ensure fairness in the experiment, the layouts of the DPT boards for both package types are nearly identical, and the pins of the two discrete TO-247 devices are placed as close as possible.

Fig. 8 compares the switching waveforms of both packages, demonstrating significantly enhanced switching

speed in the proposed package compared to the TO-247 counterpart. The figure inset further reveals that the proposed package exhibits lower turn-off voltage spikes and reduced oscillation even under higher current slew rate (di/dt) conditions. Fig. 9 compares the key parameters of the switching process. Compared to the TO-247 package, the proposed package has 23.98% and 7.62% lower turn-on and turn-off times, and 11% and 17.8% lower turn-on and turn-off losses, respectively.

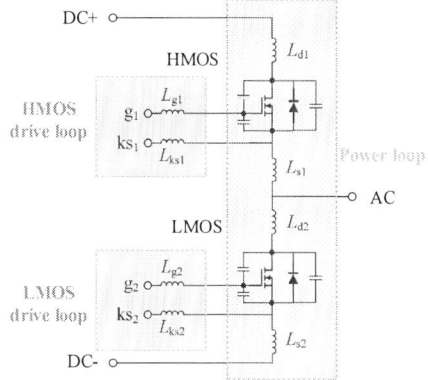

Fig. 6. Parasitic inductance network model for embedded SiC half-bridge module

TABLE I.   RESULTS OF PARASITIC INDUCTANCE EXTRACTION

Parasitic inductance	Value (pH)
HMOS drive loop ($L_{g1}+L_{ks1}$)	44
LMOS drive loop ($L_{g2}+L_{ks2}$)	39
Power loop ($L_{d1}+L_{s1}+L_{d2}+L_{s2}$)	324
Total	407

TABLE II.   PARASTIC INDUCTANCE COMPARISON

Package	Total parasitic inductance(nH)
Proposed embedded module	0.41
Through-via embedded module [10]	2.1
Press-pack package power module [4]	4.3
SKiN package power module [5]	1.4
Wire-bonded package [3]	≥10

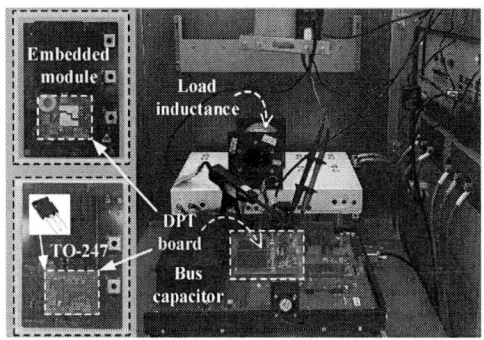

Fig. 7. DPT setup.

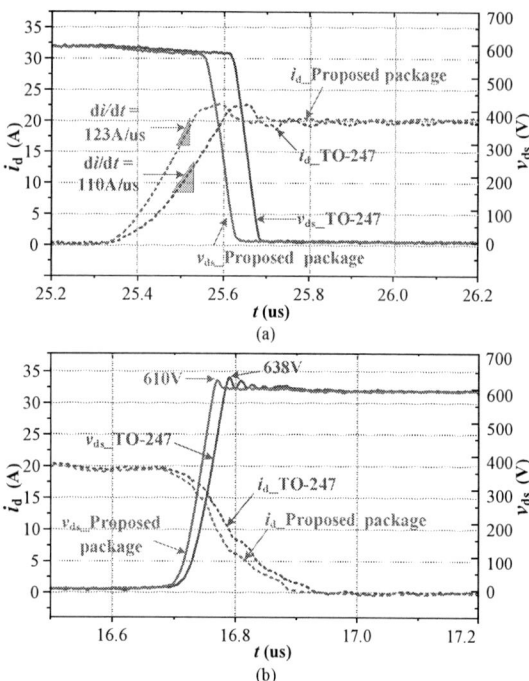

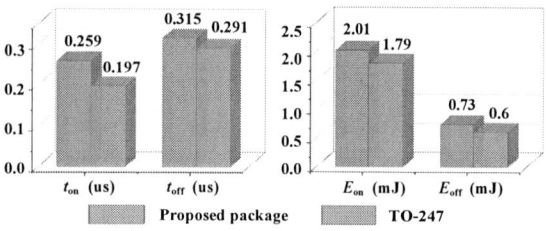

Fig. 8. Switching waveforms of proposed package and TO-247 package. (a) Turn-on. (b) Turn-off.

Fig. 9. Key parameters comparison of the DPT.

B. Thermal Characterization

The thermal resistance of the proposed module was analyzed using the ANSYS Steady-State Thermal platform. The thermal resistance network model of the half-bridge module is shown in Fig. 10, including the junction-to-case thermal resistance (R_{JC}) and the junction-to-board thermal resistance (R_{JB}), which can be described by

$$\begin{cases} R_{JC} = (T_J - T_C)/P \\ R_{JB} = (T_J - T_B)/P \end{cases} \quad (1)$$

Furthermore, a traditional embedded package model based on a lead frame and lamination process was established for comparison, as shown in Fig. 10 [7]. To ensure fairness, both packages adopted the same package size (20 mm × 20 mm) and die size (5 mm × 5 mm). The material and dimensions of other components in the package were configured with reference to actual processing conditions. Table III lists the thickness settings of each component in the two packages. Compared to the proposed package, the die in the traditional embedded package is connected to the pre-processed lead frame via solder, resulting in a longer heat dissipation path. Limited by the process, the thickness of the lead frame cannot

be designed too thin; otherwise, severe warpage will occur due to thermal stress during processing. Considering both thermal performance and thermal stress, the lead-frame thickness is typically suitable in the range of 150 μm to 250 μm. In the model established in this paper, the thickness of the lead frame was selected as 150 μm. In addition, the die in the traditional embedded package also needs to be fixed to the lead frame by soldering. The thermal conductivity of solder is usually not very high, and the introduction of a solder layer will increase two thermal interfaces, including the die-to-solder interface and the solder-to-lead frame interface. In this simulation model, SAC305 was used as the solder layer.

Table IV lists the material parameter settings for the thermal simulation. Among these, the thermal conductivities of EMC and BT laminate were referenced from the models used in the actual process. The SiC MOSFET die material was approximated as pure 4H-SiC in the thermal simulation. The thermal resistance extraction results are shown in Fig. 11. The thermal resistance of both packages was measured under a single die power dissipation of 20 W. As can be seen, benefiting from the elimination of the solder layer and the direct interconnection between the back copper layer and the die, the R_{JC} of the proposed double-sided RDL embedded package is significantly reduced, showing a decrease of at least 68% compared to the traditional embedded package. The optimization of thermal resistance in the proposed package structure can be attributed to two aspects. First, directly plating RDL on the bottom of the die can effectively eliminate the interfacial thermal resistance between the solder, die, and copper layer. Second, the electroplating method eliminates the solder layer, which has a relatively low thermal conductivity, allowing heat to be directly dissipated to the outside through the copper layer with high thermal conductivity.

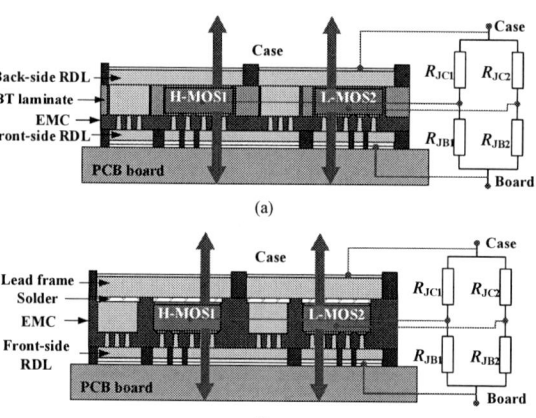

Fig. 10. Thermal resistance network models of half-bridge embedded power modules. (a) Proposed packaging. (b) Conventional embedded packaging.

TABLE III.   COMPONENT THICKNESS SETTINGS OF TWO PACKAGING STRUCTURES

Component	Proposed package	Conventional embedded package
Front-side RDL	60um	60um
SiC MOSFET	150um	150um
Back-side RDL	60um	/
Solder	/	30um
Lead frame	/	150um

TABLE IV. THERMAL SIMULATION PARAMETER CONFIGURATION

Component	Material	Thermal conductivity
EMC	CV8511CUB	0.8W/m.K
BT laminate	BT	0.8W/m.K
RDL	Cu	400W/m.K
SiC MOSFET	4H-SiC	490W/m.K
Cu spacer	Cu	400W/m.K
Solder	SAC305	67W/m.K
Blind via	Cu	400W/m.K
Lead frame	Cu	400W/m.K

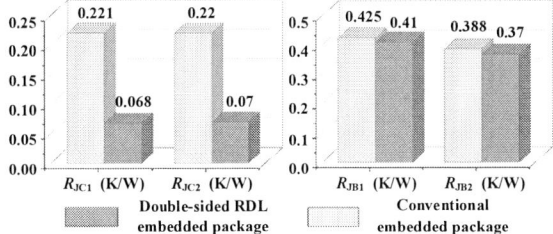

Fig. 11. Comparison of thermal resistance parameters.

IV. CONCLUSION

This paper proposes a compact embedded silicon carbide power module utilizing double-sided RDL, aiming to address the thermal performance limitations of existing embedded packaging technologies. Furthermore, the adopted "Molding-Drilling-RDL" process flow reduces both cost and fabrication complexity. The double-sided direct electroplating of RDLs eliminates the need for solder layers traditionally required for connecting the chip backside to copper layers in conventional embedded packaging. This approach not only reduces junction-to-case thermal resistance but also mitigates long-term reliability concerns associated with solder layers. Thermal simulation results demonstrate that the proposed double-sided RDL embedded package achieves at least a 69% reduction in R_{JC} compared to conventional embedded packaging. Electrically, the module exhibits an ultra-low parasitic inductance of 0.41 nH, representing an 81% reduction compared to conventional embedded packaging with through-vias, owing to its compact layout and copper block design. DPT results reveal that the proposed solution reduces turn-on and turn-off losses by 11% and 17.8%, respectively, when benchmarked against commercial TO-247

packages. In conclusion, the proposed embedded packaging structure offers superior low parasitic inductance and thermal resistance while reducing process cost and complexity, demonstrating significant potential for high-power-density and high-frequency applications.

REFERENCES

[1] Dimitrijev S, Han J, Moghadam HA, Aminbeidokhti A. Power-switching applications beyond silicon: status and future prospects of SiC and GaN devices. MRS Bull 2015;40(5):399–405.

[2] I. Josifović, J. Popović-Gerber and J. A. Ferreira, "Improving SiC JFET Switching Behavior Under Influence of Circuit Parasitics," in *IEEE Transactions on Power Electronics*, vol. 27, no. 8, pp. 3843-3854, Aug. 2012.

[3] C. Chen, F. Luo and Y. Kang, "A review of SiC power module packaging: Layout, material system and integration," in *CPSS Transactions on Power Electronics and Applications*, vol. 2, no. 3, pp. 170-186, Sept. 2017.

[4] J. Noppakunkajorn, D. Han, and B. Sarlioglu, "Analysis of high-speed PCB with SiC devices by investigating turn-off overvoltage and interconnection inductance influence," *IEEE Trans. Transport. Electrific.*, vol.1, no.2, pp.118–125, Aug. 2015.

[5] N. Zhu, H. A. Mantooth, D. Xu, M. Chen and M. D. Glover, "A Solution to Press-Pack Packaging of SiC MOSFETS," in *IEEE Transactions on Industrial Electronics*, vol. 64, no. 10, pp. 8224-8234, Oct. 2017.

[6] P. Beckedahl, S. Buetow, A. Maul, M. Roeblitz and M. Spang, "400 A, 1200 V SiC power module with 1nH commutation inductance," *CIPS 2016; 9th International Conference on Integrated Power Electronics Systems*, 2016, pp. 1-6.

[7] X. Sun et al., "Design and evaluation of a face-down embedded SiC power module with low parasitic inductance and low thermal resistance," *IEEE Trans. Power Electron.*, vol. 38, no. 3, pp. 2799–2804, Mar. 2023.

[8] A. Ostmann, L. Boettcher, D. Manessis, S. Karaszkiewicz and K. -. Lang, "Power modules with embedded components," *2013 Eurpoean Microelectronics Packaging Conference (EMPC)*, 2013, pp. 1-4.

[9] D. J. Kearney, S. Kicin, E. Bianda, and A. Krivda, "PCB embedded semiconductors for low-voltage power electronic applications," *IEEE Trans. Compon., Packag. Manuf. Technol.*, vol. 7, no. 3, pp. 387–395, Mar. 2017.

[10] S. Bensebaa, M. Berkani, S. Lefebvre, M. Petit and N. Schmitt, "Experimental And Numerical Characterization Of PCB-Embedded Power Dies Using Solderless Pressed Metal Foam" *2020 22nd European Conference on Power Electronics and Applications (EPE'20 ECCE Europe)*, 2020, pp. 1-10.

[11] F. Hou et al., "Fan-Out Panel-Level PCB-Embedded SiC Power MOSFETs Packaging," in *IEEE Journal of Emerging and Selected Topics in Power Electronics*, vol. 8, no. 1, pp. 367-380, March 2020.

Low-Profile Negative Coupled Inductor for AI Chip Vertical Power Delivery Voltage Regulator

1st Ruibo Cao
Microsystem Packaging Research Center
Institute of Microeletronics of The Chinese Academy of Sciences
Beijing, China
caoruibo@ime.ac.cn

2nd Xiangan You
Microsystem Packaging Research Center
Institute of Microeletronics of The Chinese Academy of Sciences
Beijing, China
youxiangan@ime.ac.cn

3rd Fengze Hou
Microsystem Packaging Research Center
Institute of Microeletronics of The Chinese Academy of Sciences
Beijing, China
houfengze@ime.ac.cn

4th Na Yan
Microsystem Packaging Research Center
Institute of Microeletronics of The Chinese Academy of Sciences
Beijing, China
yanna@ime.ac.cn

5th Jian Song
Microsystem Packaging Research Center
Institute of Microeletronics of The Chinese Academy of Sciences
Beijing, China
songjian2024@ime.ac.cn

6th Qidong Wang
Microsystem Packaging Research Center
Institute of Microeletronics of The Chinese Academy of Sciences
Beijing, China
wangqidong@ime.ac.cn

7th Liqiang Cao
Microsystem Packaging Research Center
Institute of Microeletronics of The Chinese Academy of Sciences
Beijing, China
caoliqiang@ime.ac.cn

Abstract—The rapid advancement of artificial intelligence (AI) technology has led to increasing electricity consumption in data centers, necessitating improved power delivery efficiency. Vertical power delivery, placing the voltage regulator modules (VRMs) on the processor's backside, is gaining attention due to its lower conduction losses compared to lateral power delivery. This study focuses on reducing inductor height while increasing current density. It explores the relationship between inductance, core footprint, and flux structures using vertical flux designs with ferrite materials to optimize the inductor magnetic field. This paper proposes a height of 2.2mm and a peak current density of 1.9A/mm². The prepared negative coupling inductor is expected to achieve a current handling capability of up to 110 A.

Keywords—Coupled inductor, low-profile, vertical power delivery

I. Introduction

With the rapid development of artificial intelligence (AI) technology, data centers are consuming more and more electricity[1-3]. It is essential to improve the efficiency of power delivery. With the continuous increase in processor input currents[4] and the reticle limits of EUV[5], these trends are placing stricter requirements on the efficiency and current density of voltage regulator modules (VRMs)[6]. Compared with lateral power delivery, vertical power delivery, which places the VRM on the backside of the processor, has attracted much attention due to its low conduction loss characteristics, as shown in Fig. 1. For this power delivery architecture, NVIDIA proposed that the new generation of VRM should achieve a current density higher than 1A/mm² while having a height of less than 5mm due to interconnect

lengths[7]. Unfortunately, no product currently meets this requirement. Some commercial products[8, 9] can achieve a current density higher than 1 A/mm², but their height is above 7.5 mm and magnetic components account for nearly 60% of the height. Therefore, reducing the height of VRM magnetic components while maintaining the current density is a challenge.

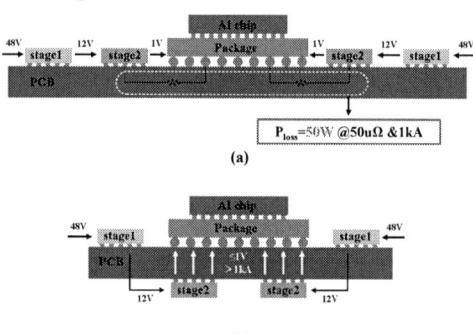

Fig.1. Processors power delivery architectures: (a) lateral power delivery (b) vertical power delivery.

For low-profile inductors, Li Qiang et al.[10] first demonstrated that inductors with lateral flux structures can save more footprint and improve the power density of VRMs. In 2013, they proposed a VRM with a power density of 1.1 kW/in³[11], and in 2022 they proposed a printed circuit board embedded inductor with a thickness of only 0.54 mm and a current density of 0.7 A/mm²[12]. Youssef at el.[13] combined the linear-extendable group operated point-of-load (LEGO-POL) architecture with negatively coupled inductors to improve circuit response and proposed a VRM with a magnetic core height of 8.5 mm and a current density of 1 A/mm². Building on this, Baek et al.[14] reduced the magnetic core height to 5.25 mm and the VRM height to 16.65 mm. They further reduced the core height to 2.5 mm and the VRM height to 8.4 mm, but the current density decreased to 0.71 A/mm²[15]. Wang et al.[16] proposed a VRM based on the multistack switched-capacitor point-of-load (MSC-PoL) architecture and coupled inductors, with a magnetic core height of less than 4 mm, a VRM height of less than 6 mm, and a current density of 0.55 A/mm². It is evident that current research results still fall short of NVIDIA's requirements for VRMs.

It is evident that reducing the inductor height while increasing current density is an attractive option and a key focus of this study. This work analyzes the relationship between inductance and core footprint with different flux structures, using vertical flux structures to construct the inductor magnetic field and gap distribution with ferrite materials with a relative permeability of 900. In addition, the proposed inductor is a common EI structure, which is convenient to manufacture and ideal for 3D integrated VRM integration. Therefore, the proposed inductor structure is suitable for the VPD architecture of high current density and low-profile 3D integration VRM. Section II presents a comparison between vertical and horizontal magnetic flux structures, revealing the quantitative relationship between inductance and physical dimensions for specific magnetic path configurations. For ferrite inductors, vertical flux structures should be prioritized when aiming to achieve high current density. Section III presents the design method of a two-phase coupled inductor with a saturation current of 110 A. In Section IV, the proposed two-phase negatively coupled inductor is tested. Finally, Section V concludes this paper.

II. MODELING AND COMPARISON FOR DIFFERENT FLUX STRUCTURES

A. Modeling For Different Flux Structures

For vertical flux inductor structures, rectangular windings are typically used, while for lateral flux inductor structures, cylindrical windings are commonly used[10, 17], this paper will only focus on these two shapes of windings, as shown in Fig. 2. The method assumes that the entire magnetic flux path is composed of an infinite number of ring cores with varying radius r. When the width of each ring core Δr is sufficiently small, the magnetic flux density within each ring core can be considered uniform. The total energy stored in the inductor is then obtained by summing the energy stored in all such ring cores.

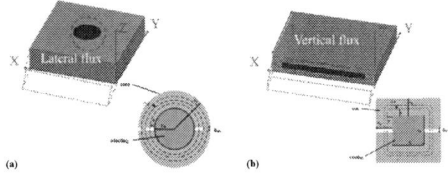

Fig. 2. Different planar inductor structures (a) lateral flux structure (b) vertical flux structure.

Considering the similarity between the two models, this section presents only the derivation process for the circular winding; the process for the rectangular winding is similar. For a single-turn inductor as shown in the Fig. 2(a), the

magnetic path with the same inner radius r can be divided into two segments: one within the core and the other within the air gap. The magnetic field intensity H along different paths can be calculated using Ampere's circuital law:

$$H_C \cdot l_C + H_g \cdot l_g = I \tag{1}$$

In the equation, H_C and Hg represent the magnetic field intensities in the core and air gap regions of the loop, respectively, while l_c and l_g are the magnetic path lengths in the core and the air gap. The inductor is excited by a dc current I. Since the size of the air gap is much smaller than the total magnetic path length, l_c and l_g can be calculated as follows:

$$l_C \approx 2\pi \cdot (r_0 + r) \tag{2}$$

$$l_g = 2 \cdot g_{air} \tag{3}$$

The inductor in this work uses ferrite material, which is approximated to have a constant relative permeability μ_r before magnetic saturation, μ_0 is the permeability of free space. By summing the magnetic flux of each toroidal core section, the total magnetic flux within the core can be calculated. Neglecting fringing effects at the air gap, the total flux and the inductance can be expressed as follows:

$$\phi = \int_0^{r_a - r_0} B_r \cdot dr \cdot h = \int_0^{r_a - r_0} \frac{\mu_0 \cdot \mu_r \cdot dr \cdot h \cdot I}{2\pi \cdot (r_0 + r) + \mu_r \cdot l_g} \tag{4}$$

$$L = \int_0^{r_a - r_0} \frac{\mu_0 \cdot \mu_r \cdot dr \cdot h}{2\pi \cdot (r_0 + r) + \mu_r \cdot l_g} \tag{5}$$

The footprint and height of the lateral flux structure can be calculated as follows:

$$S_{lateral} = \pi \cdot r_a^2 \tag{6}$$

$$Height = h \tag{7}$$

For the vertical flux structure, the formulas are as follows:

$$L = \int_0^{r_w} \frac{\mu_0 \cdot \mu(H) \cdot dr \cdot h}{2 \cdot \mu_0 \cdot (r_c + r_d + \pi \cdot r) + \mu(H) \cdot l_g} \tag{8}$$

$$S_{vertical} = (r_a + 2w_c) \cdot h \tag{9}$$

$$Height_{vertical} = r_a + 2w_c \tag{10}$$

B. Comparison of Lateral and Vertical Flux Structures

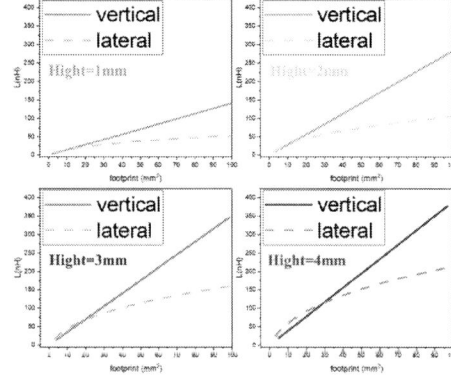

Fig. 3. The inductance of vertical magnetic flux and lateral magnetic flux at different heights varies with the footprint.

Fig. 3 illustrates the variation of inductance with footprint area for both vertical and lateral flux structures at different heights. It can be seen that the inductance in the vertical flux structure increases linearly with the occupied area when the inductance height is the same. In contrast, in the lateral flux structure, the increase in inductance gradually decreases as the footprint increases. Therefore, in the inductance structure made of high permeability material air gap, the vertical magnetic flux structure has a higher inductance density and is more suitable for integrating high current density VRMs. After balancing the trade-off between inductor height and footprint area, a rectangular coil inductor with a height of 2.5 mm and a footprint of 28.8 mm² was selected.

III. PROPOSED VERTICAL FLUX NEGATIVE COUPLED INDUCTOR

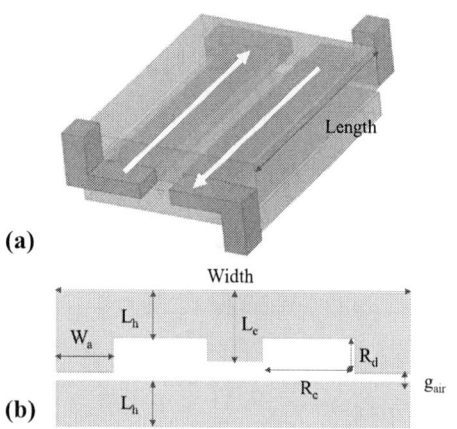

(a)

(b)

Fig. 4. Proposed negative coupled inductor: (a)3D-view (b) side-view.

Based on the analysis presented in Section II, a proposed negative coupled inductor with vertical flux is shown in Fig. 4. Compared with uncoupled inductors and forward coupled inductors, negative coupled inductors have better steady-state and transient responses[13-16, 18]. A two-phase inverse coupled inductor was designed. The design objectives of the coupling inductor are steady-state inductance L_{ss} and transient inductance L_{tr}, and its expression is determined by the equation (11)-(12), where L is self-inductance, M is mutual inductance, and D is duty cycle.

$$L_{ss} = \frac{L^2 - M^2}{L + \frac{D}{1-D} \cdot M} \quad (11)$$

$$L_{tr} = L + M \quad (12)$$

Based on the designed process, the proposed inductor with dimensions of 9.0 mm × 6.4 mm × 2.2 mm was fabricated using the ferrite material, as shown in Fig. 5 and Fig. 6. The performance of the inductor is simulated by 3D finite element analysis simulation tool. As shown in Fig. 7, when the air gap is 0.05mm, he proposed inductor has a saturation current of 110A, DCR of 0.24mΩ, self-inductance of 131nH and mutual inductance of -98nH. The inductor has a steady-state inductance of 60nH and a transient inductance of 33nH when operating on a 12V input, 1V output, 1MHz frequency buck converter.

Fig. 5. Customized magnetic components.

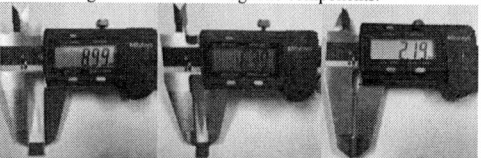

Fig. 6. Proposed inductor with a dimension of 9.0 mm × 6.4 mm × 2.2 mm

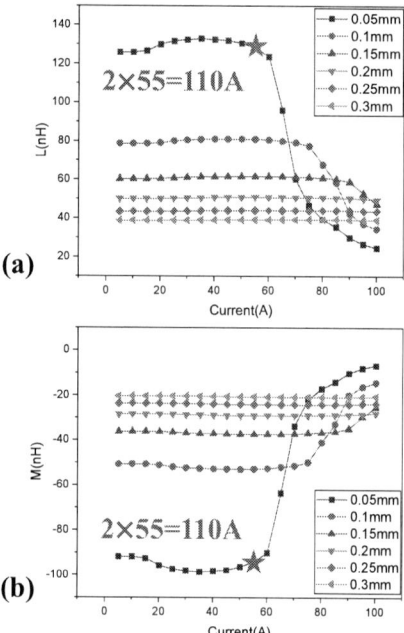

Fig. 7. Influence of excitation current on (a) self-inductance and (b) mutual inductance with different air gap.

IV. EXPERIMENTAL RESULTS AND DISCUSSION

Fig. 8. Measurement setup for the proposed negative coupled inductor using TI TPS536C7EVM-051 buck converter evaluation board.

As shown in Fig. 8, to evaluate the inductance and current handling capability of the inductor, the TPS536C7EVM-051 buck converter evaluation board from Texas Instruments was employed, where the proposed inductor structure replaced the discrete inductor. The input voltage was set to 12 V, the output voltage to 1 V, and the switching frequency to 1 MHz. To facilitate inductor current measurement, the windings were extended via additional wires to the pads. The inductance introduced by these connection wires was characterized using an LCR analyzer and determined to contribute approximately 30 nH of self-inductance per phase. Additionally, due to assembly tolerances, the measured air gap was approximately 70 µm. The inductor current waveforms i_{L1} under load currents of 10 A and 110 A are shown in Fig. 9(a) and Fig. 9(b), respectively. Based on these current waveforms, the L_{ss} and L_{tr} of the coupled inductor were calculated using the method described in[18]. Fig. 10 compares the measured inductance with 3D FEA results, showing excellent agreement. Therefore, the proposed inductor is capable of supporting a load current of 110 A, achieving a current density of 1.9 A/mm² with a height of only 2.2 mm.

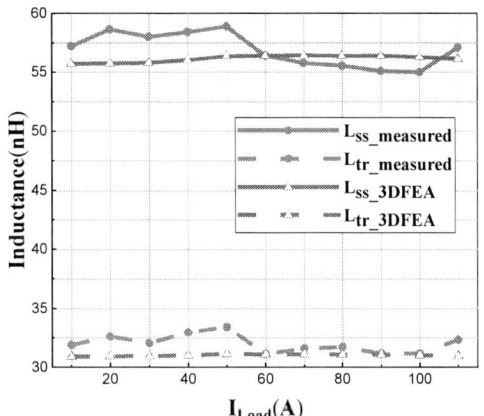

Fig. 9. Proposed inductor current waveform with: (a) I_{Load} = 10 A and (b) I_{Load} = 110 A.

Fig. 10. Comparison of steady-state and transient inductances obtained from measurements and 3D FEA simulations at different load currents.

CONCLUSION

In this work, planar inductors with air gaps were modeled, and the quantitative relationships between inductance and three-dimensional geometry for both lateral and vertical flux structures were revealed. The analysis indicates that vertical flux structures offer greater advantages for high current density ferrite inductors. Based on the proposed theory, a high current density inductor with a saturation current of 110A was designed. Experimental results show that the inductor achieves a peak current density of 1.9 A/mm² with a compact height of only 2.2 mm. The experimental results are consistent with FEM simulations, validating the accuracy of the proposed model. Therefore, the inductor is well suited for high current density vertical power delivery VRMs.

ACKNOWLEDGMENT

The authors would like to thank the support of the National Natural Science Foundation of China under Grant 62174177.

REFERENCES

[1] N. Jones, "THE INFORMATION FACTORIES," *Nature,* vol. 561, pp. 163-166, 2018.

[2] J. An, W. Ding, and C. Lin, "ChatGPT: tackle the growing carbon footprint of generative AI," Nature, vol. 615, p. 586, 2023.

[3] M. Studer, "The energy challenge of powering AI chips," (June 11, 2023). [Online]. Available: https://www.robeco.com/en-int/insights/2023/11/the-energy-challenge-of-powering-ai-chips.

[4] B. Nauta, "1.2 Racing Down the Slopes of Moore's Law," in *2024 IEEE International Solid-State Circuits Conference (ISSCC)*, 18-22 Feb. 2024 2024, vol. 67, pp. 16-23, doi: 10.1109/ISSCC49657.2024.10454417.

[5] D. Patel, "Die Size And Reticle Conundrum – Cost Model With Lithography Scanner Throughput," (June 19, 2022). [Online]. Available: https://www.semianalysis.com/p/die-size-and-reticle-conundrum-cost.

[6] K. Radhakrishnan, M. Swaminathan, and B. K. Bhattacharyya, "Power Delivery for High-Performance Microprocessors— Challenges, Solutions, and Future Trends," *IEEE Transactions on Components, Packaging and Manufacturing Technology,* vol. 11, no. 4, pp. 655-671, 2021, doi: 10.1109/TCPMT.2021.3065690.

[7] S. S. K. M. Mosa, "Challenges to Enabling Vertical Power Delivery in High-Power GPU Applications," in *IEEE Applied Power Electronics Conference and Exposition (APEC)*, Long Beach, CA, USA, 2024.

[8] "BMR510," Flex,[Online].Available: https://flexpowermodules.com/products/bmr510?model=BMR510x034%2F002.

[9] "MPC22161-120," MPS,[Online]. Available: https://www.monolithicpower.com/en/products/power-management/data_center/mpc22161-120.html.

[10] Q. Li and F. C. Lee, "High Inductance Density Low-Profile Inductor Structure for Integrated Point-of-Load Converter," in *2009 Twenty-Fourth Annual IEEE Applied Power Electronics Conference and Exposition*, 15-19 Feb. 2009 2009, pp. 1011-1017, doi: 10.1109/APEC.2009.4802786.

[11] Y. P. Su, Q. Li, and F. C. Lee, "Design and Evaluation of a High-Frequency LTCC Inductor Substrate for a Three-Dimensional Integrated DC/DC Converter," (in English), *Ieee Transactions on Power Electronics,* vol. 28, no. 9, pp. 4354-4364, Sep 2013, doi: 10.1109/Tpel.2012.2236359.

[12] F. Y. Zhu and Q. Li, "A Novel PCB-Embedded Coupled Inductor Structure for a 20-MHz Integrated Voltage Regulator," (in English), *Ieee Journal of Emerging and Selected Topics in Power Electronics,* vol. 10, no. 6, pp. 7452-7463, Dec 2022, doi: 10.1109/Jestpe.2022.3194133.

[13] Y. Elasser, J. Baek, and M. Chen, "A Merged-Two-Stage LEGO-PoL Converter with Coupled Inductors for Vertical Power Delivery," in *2020 IEEE Energy Conversion Congress and Exposition (ECCE)*, 11-15 Oct. 2020 2020, pp. 916-923, doi: 10.1109/ECCE44975.2020.9236294.

[14] J. Baek et al., "Vertical Stacked LEGO-PoL CPU Voltage Regulator," *IEEE Transactions on Power Electronics,* vol. 37, no. 6, pp. 6305-6322, 2022, doi: 10.1109/tpel.2021.3135386.

[15] Y. Elasser et al., "Mini-LEGO CPU Voltage Regulator," *IEEE Transactions on Power Electronics,* vol. 39, no. 3, pp. 3391-3410, 2024, doi: 10.1109/TPEL.2023.3337171.

[16] P. Wang, Y. Chen, G. Szczeszynski, S. Allen, D. M. Giuliano, and M. Chen, "MSC-PoL: Hybrid GaN–Si Multistacked Switched-Capacitor 48-V PwrSiP VRM for Chiplets," *IEEE Transactions on Power Electronics,* vol. 38, no. 10, pp. 12815-12833, 2023, doi: 10.1109/tpel.2023.3293022.

[17] Q. Li, Y. Su, M. Mu, F. C. Lee, and D. Gilham, "Modeling of planar inductors with non-uniform flux distribution and non-linear permeability for high-density integration," in *2012 Twenty-Seventh Annual IEEE Applied Power Electronics Conference and Exposition (APEC),* 5-9 Feb. 2012 2012, pp. 1002-1009, doi: 10.1109/APEC.2012.6165941.

[18] W. Pit-Leong, X. Peng, P. Yang, and F. C. Lee, "Performance improvements of interleaving VRMs with coupling inductors," *IEEE Transactions on Power Electronics,* vol. 16, no. 4, pp. 499-507, 2001, doi: 10.1109/63.931059.

979-8-3315-1110-4/25 $31.00 © 2025 IEEE

Shape Optimization of High-Stress-Buffering Bridge-Typed Spacers for Double-Sided Power Modules

Gaojia Zhu
School of Electrical Engineering
Tiangong University
Tianjin, China
zhugaojia@tiangong.edu.cn

Ke Xu
School of Electrical Engineering
Tiangong University
Tianjin, China
307076182@qq.com

Bingru Li
School of Electrical Engineering
Tiangong University
Tianjin, China
1426145902@qq.com

Youzheng Wang
School of Electrical Engineering
Tiangong University
Tianjin, China
youzhengwang@tiangong.edu.cn

Longnv Li
School of Electrical Engineering
Tiangong University
Tianjin, China
lilongnv@tiangong.edu.cn

Yun-Hui Mei*
School of Electrical Engineering
Tiangong University
Tianjin, China
meiyunhui@tiangong.edu.cn

Abstract—Owing to their superior characteristics of high thermal dissipation and low parasitic inductance, double-side-cooled (DSC) power modules have garnered extensive research attention. Optimizing the spacer structures has emerged as a focus to mitigate the stresses induced by thermal expansion mismatches and prolong the modules' life cycles. In this paper, a topology optimization (TO) design for the shapes of bridge-typed spacers (BTSs) in a DSC module is proposed, considering simultaneously thermal and stress impacts. The thermo-mechanical stress distributions in the module are investigated to decide the co-related and decoupled thermal and stress efforts of the BTSs. A topology optimization (TO) approach is then developed for the BTSs' shapes from both thermal and mechanical perspectives, including the TO design considerations and settings decided based on the analyses. The TO designs are conducted to identify the optimal spacer shapes, with the proper objective determined.

Keywords—Double-side-cooled (DSC) module, thermo-mechanical stress, bridge-typed spacer (BTS), topology optimization (TO), multi-physical problem

I. INTRODUCTION

Double-side-cooled (DSC) power modules are characterized by their low parasitic inductance and superior thermal dissipation efficiency [1]. These attributes render them highly suitable for applications in electric vehicles, rail traction systems, and more- or all-electric aircraft [2-3]. However, operational challenges arise due to stresses induced by thermal expansion mismatches [4].

To mitigate this issue, the insertion of metal or alloy spacers between the chips and one (or both) substrates has emerged as an effective strategy [5-7]. Among these materials, copper (Cu) is highly favored for its excellent electrical and thermal conductivity, making it the commonly used material for spacers [5]. However, copper's thermal expansion

This work was supported by the National Natural Science Foundation of China under Grant 52177189, the Tianjin Municipal Science and Technology Bureau under Grants 24JCZXJC00130 and 21JCJQJC00150, and also by the China Postdoctoral Science Foundation under Grant 2024M761259.

properties still exhibit significant differences compared to those of power chips. To enhance the buffering effect, in [6], Ding *et al.* proposed sintered nano-silver spacers, where the interconnection layer and the spacer material were identical, thereby showing superior thermo-mechanical stress characteristics. In [7], Wang *et al.* used molybdenum (Mo), which has properties similar to those of the chips, to fabricate spacers, significantly reducing the thermo-mechanical stress in the interconnection layer.

To further maximize the stress buffering effects, the spacer shapes are designed [8-9]. Jeon *et al.* [8] fabricated several spacers with distinct morphologies and evaluated their efforts on thermo-mechanical stress. In [9], copper spacers with 2×2 and 3×3 notched shapes were employed replacing traditional configurations to mitigate thermal stress caused by CTE mismatches. Finite element analysis (FEA) revealed that the 3×3 notched structure reduced the maximum stress and plastic strain in the solder layer by 18.7% and 67.8%, respectively. However, the notches decreased the contact area between the spacers and the chips, leading to an increase in the junction temperature. Ref. [10] introduced horizontally placed cylindrical and hollow cylindrical spacers. Both configurations demonstrated the ability to reduce thermo-mechanical stress. Nevertheless, compared to traditional structures, the thermal dissipation capability was found to be inferior and in need of improvement. Ref. [11] proposed an interconnected parallel plate structure, utilizing slender cylindrical copper rods instead of solid copper blocks as spacers. This structure significantly reduced the stresses. However, the larger spacing between the copper rods led to a decline in the module's thermal dissipation performance, resulting in elevated junction temperatures. Building on the work in [11], Ref. [12] replaced traditional spacers with copper columns and analyzed the impact of the columns' position, size, and height on thermal and thermo-mechanical stress performances. The results indicated that a smaller spacing between the columns, a larger diameter, and a shorter height of the copper columns led to lower thermal resistance and a significant reduction in stress. Ref. [13] introduced a pyramidal copper pad, which optimized heat diffusion to reduce chip temperature rise and thereby enhance the module's reliability. Ref. [14] compared the thermal performance of four different spacer shapes, offering a new optimization pathway for the thermal reliability design of DSC modules. To enhance both stress buffering and heat dissipation, Cao et al. [15] proposed a novel bridge-typed spacer (BTS). This structure features a vertically connected columnar spacer with extensions connecting upper and lower substrates, to not only facilitate heat dissipation but also provide additional mechanical support, thereby mitigating the thermo-mechanical stresses.

Based on the BTS proposed in [15], this study develops topology optimization (TO) designs for the shapes of BTSs in DSC modules. First, the efforts of the BTS in thermal and stress aspects are systematically analyzed, providing a foundation for latter TO designs. Subsequently, a BTS TO approach is established, with the optimization objectives of maximizing heat dissipation and minimizing thermal stress. Two distinct BTS shapes are derived according to the two objectives. Finally, the thermo-mechanical stresses of the module with the two BTSs are evaluated to decide the proper objective.

II. THERMAL AND MECHANICAL EFFORTS ANALYSES OF BTSs

A. Coupled Thermal-Stress Efforts of BTSs

The prototype module studied in this paper is a DSC one proposed in [15], as illustrated in Fig. 1(a). In the prototype module, BTSs are used to not only connect the IGBTs and Diodes but also provide additional heat transfer paths, as shown in Fig. 1(b). During simulations, the physical properties utilized are tabulated in Table I.

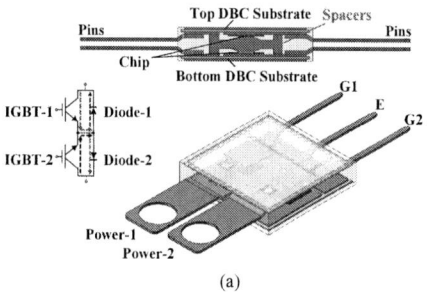

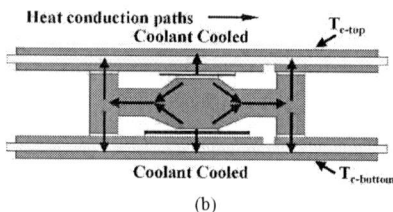

Fig. 1. Prototype module structure, where (a) basic structure, and (b) heat conduction paths.

TABLE I. MATERIAL PROPERTIES

Material	Properties					
	Density (kg/m³)	Isobaric Heat Capacity (J/kg·K)	CTE (10^{-6}/K)	Thermal conductivity (W/m·K)	Poisson ratio	Young's modules (GPa)
Cu	8930	385	16.5	388	0.34	128
Si	2330	712	4.2	148	0.28	162.7
Al₂O₃	3960	850	6.5	35	0.28	270
Sintered-Ag	8580	234	19.5	240	0.37	9
Epoxy	1160	550	11	0.84	0.3	13.5

When the module is double-sided cooled, the temperature-stress coupled fields are numerically investigated. The results are compared with those obtained using a traditional cuboid-typed spacers (CTSs) to validate the effectiveness. The temperature distribution of the module with CTSs is shown in Fig. 2(a), while that with BTSs is depicted in Fig. 2(b). Comparison results show that, the use of BTSs can reduce the module's maximum temperature by approximately 3.80 °C (5.78%), demonstrating its effectiveness in assisting heat dissipation. The thermo-mechancial stress distribution of the module with CTSs is illustrated in Fig. 2(c), and that with BTSs is shown in Fig. 2(b). The computation results indicate that the use of BTSs reduces the thermo-mechanical stress in the attachments by 5.20 MPa (33.12%), confirming its effectiveness in mitigating thermal stress.

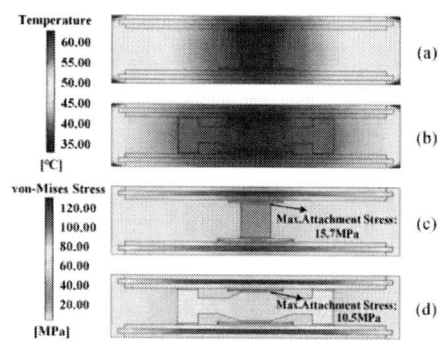

Fig. 2. Temperature and thermo-mechanical stress distributions of the module, where (a) temperature distribution with CTSs, (b) temperature distribution with BTSs, (c) stress distribution with CTSs, and (d) stress distribution with BTSs.

B. Decoupled Stress Efforts of BTSs

The aforementioned simulations show that the BTS not only enhances the thermal dissipation efficiency but also reduces the accumulation of thermal stress. To eliminate the indirectly reduced stress caused by temperature decrease, the thermal conductivity of the connecting extensions of the BTS is adjusted to that of epoxy, while the stress-strain-related parameters (Young's modulus and Poisson's ratio) are maintained at the values of the spacer material (copper). The temperature and thermal stress distributions of the module are recalculated under these conditions, as shown in Figs. 3(a) and 3(b), respectively, to decouple the thermal effect on stress. The results indicate that even when the BTS no longer contributes to enhanced thermal dissipation (with the calculated temperature being nearly identical to that of the module using a brick-typed spacer), the maximum stress in attachments is still reduced by approximately 3.10 MPa (19.75%). This finding confirms the effectiveness of the inherent structure of the BTS in mitigating thermal stress.

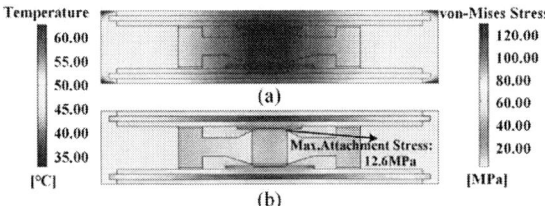

Fig. 3. Temperature and thermo-mechanical stress distributions of the module using BTS with low-thermal conduction extensions, where (a) temperature distribution, and (b) stress distribution.

C. Discussions

Based on the aforementioned analysis, it is evident that even without considering the enhanced heat dissipation efforts of the BTSs (where the thermal conductivity of the connections is set to the value of the surrounding epoxy), the bridge-like structure of the BTS and its connections to the upper and lower substrates can significantly reduce the thermo-mechanical stress in attachments by 19.75%. When the effects of BTS in enhancing thermal conductivity and reducing thermal concentration are taken into account, the thermo-mechanical stress within the attachments can be further reduced by an additional 16.67%, resulting in a total reduction of 33.12% compared to that with conventional CTSs. This finding confirms the significant efforts of the BTSs in mitigating modules' thermo-mechanical stresses. In the TO designs of the BTS profiles, the design considerations will be explored from both thermal and stress perspectives to enhance their performances.

III. SPACER SHAPE TOPOLOGICAL OPTIMIZATION

A. TO Settings and Considerations

To achieve optimal thermo-mechanical stress buffering, this paper establishes a TO model for the BTS, as shown in Fig. 4(a). In order to optimize the spacer shape while maintaining its fundamental electrical and thermal conductivity, the peripheral frame of the spacer is defined as the spacer frame and is excluded from the TO processes. Two regions on the left and right sides of the spacer frame are designated as spacer fillings and are included in the TOs. Considering the symmetry of the model, the TOs are performed based on a half-model, as depicted in Fig. 4(b). During TOs, the overall volume (weight) of the spacer is maintained.

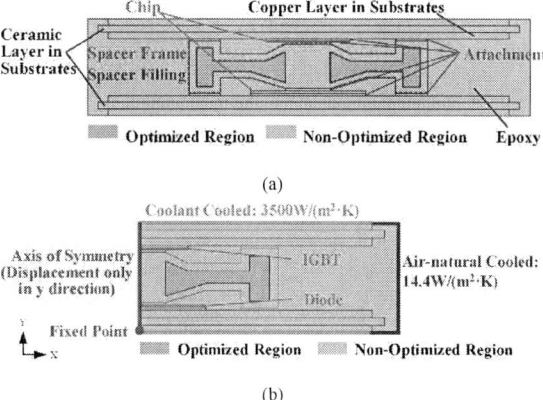

Fig. 4. TO model and settings, where (a) component illustration, and (b) TO domains and settings.

In addition, the BTS achieves thermo-mechanical stress mitigation in two ways: first, by enhancing the heat dissipation paths, which indirectly reduces the stresses, and second, by its own supporting structure, which suppresses the overall stresses. In light of this, in the TOs, two objectives are settled as:

① The maximum of heat dissipation capacity.

② The minimum of thermo-mechanical stress.

For the boundaries in TOs, the thermal ones are:

① The boundaries the module connect to the heat sinks are defined as heat convection one with the coefficient of 3500 W/(m²·K).

② A low convective heat transfer coefficient of 14.4 W/(m²·K) is applied to the outer surface of the epoxy-covered portion to simulate air natural convection.

③ The thermal sources are 38 W for the IGBT and 27 W for the Diode, with the corresponding values for the half-model being 19 W and 13.5 W, respectively.

The stress and strain related boundaries are:

① Thermal expansion of each component is considered as the source for stresses and deformations.

② Since the thermal expansion conditions on both sides of the symmetry axis are identical, the symmetry axis is set as a fixed displacement constraint (with displacement allowed only in the vertical y-axis direction).

③ To prevent translation of the model along the y-axis, the bottom point of the symmetry axis is set to a fully fixed constraint.

B. TO Results and Discussions

Based on the TO model and settings, the BTS shapes are optimized. When the objective is set to minimize the module's maximum temperature, the TO processes are illustrated in Fig. 5(a), with convergence achieved after 186 iterations. The resulting temperature distribution of the module is shown in Fig. 5(b), and the corresponding thermo-mechanical stress distribution is depicted in Fig. 5(c). Following the TO, the module temperature is reduced by only 0.24 °C (0.38%) compared to that with conventional BTSs of the same volume (weight). Furthermore, the stress is also very close to that with the conventional BTS (reduced by 0.3 MPa). Therefore, the results indicate that, under the conventional BTS configuration, the thermal performance of the module has already approached its limits. Further optimization of the BTS achieves minor improvements in heat dissipation characteristics. Similarly, its improvement in the stress-buffering effect is also negligible.

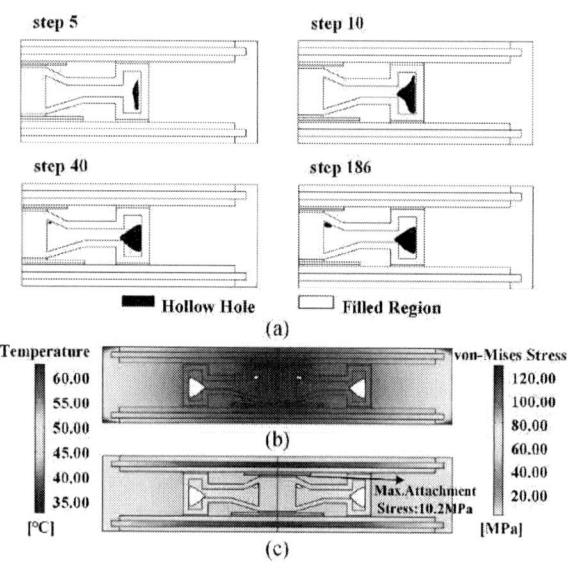

Fig. 5. TO design with minimum temperature as objective, where (a) iteration processes, (b) temperature distribution, and (c) stress distribution.

The TO processes with the objective of minimizing stress is illustrated in Fig. 6(a), where convergence is achieved after 287 iterations. Under this objective function, the temperature and stress distributions are shown in Figs. 6(b) and 6(c), respectively. In this scenario, the BTS is optimized to maximize the buffering effect. As a result, the temperature is even slightly higher than that of a module using conventional BTSs of the same volume (weight), but the difference is minimal (0.40°C, 0.64%). However, the stress at the attachments is reduced by 2.31 MPa (19.41%), indicating that the module exhibits superior thermo-mechanical stress mitigation.

The analyses indicate that when the temperature is used as the optimization objective, the improvement potentials in heat dissipation and stress-buffering are minor. Conversely, when minimizing stress is set as the objective, the temperature difference relative to the non-optimization one is negligible, while the thermal stress is significantly reduced. Therefore, in this case with BTSs, stress-related objectives can yield more effective shape designs. In addition, the TO-suggested structure , from the spacer-depth perspective, features perforated slots compared to the original BTS, as shown in Fig. 6(d). As the minimum width of the slots is larger than 0.3 mm , the TO-designed BTSs can be manufactured through precise wire cutting, thereby ensuring the manufacturability of the design.

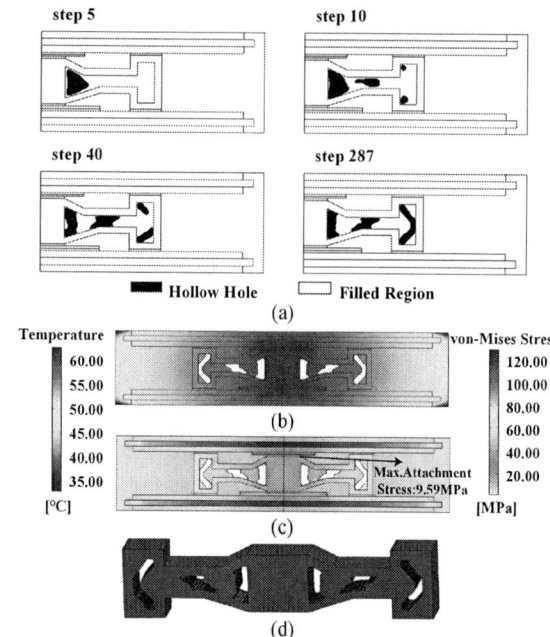

Fig. 6. TO design with minimum stress as objective, where (a) iteration processes, (b) temperature distribution, (c) stress distribution, and (d) 3D structure of the BTS TO design.

IV. CONCLUSION

This paper proposes the TO designs of BTSs for DSC power modules. The thermo-mechanical efforts of BTSs on a DSC module are analyzed. The results show that the BTSs can reduce thermo-mechanical stresses of DSC modules from both heat-dissipation and stress-buffering aspects, which provides the perspectives for TOs. The TO design settings and considerations are then established. With temperature and stress objectives, the BTS shapes are output by TO iterations. Numerical analyses of the module with the BTSs indicate that the TO design targeting temperature has minimal impacts, while that targeting stress demonstrates a more pronounced buffering effect with a minor influence on the temperature. This study offers valuable insights and references for the BTS design in DSC power modules.

In terms of future improvements, the prototype module with the TO-designed BTSs will be manufactured, and experimental validations will be conducted.

REFERENCES

[1] Y. Y. Yan, C. Chen, Z. H. Wu, J. J. Guan, J. W.i Lv, and Y. Kang, "A High Power Density Double-Side-End Double-Sided Bonding SiC Half-Bridge Power Module," IEEE Trans. Transport. Electrific., vol. 9, no. 2, pp. 3149-3163, June 2023.

[2] C. Ding, S. C. Lu, Z. C. Zhang, K. Zhang, T. Nguyen, K. D. T. Ngo, R. Burgos, and G. Q. Lu, "Double-Side Cooled SiC MOSFET Power Modules With Sintered-Silver Interposers for a 100-kW/L Traction Inverter," IEEE Trans. Power Electron., vol. 38, no. 8, pp. 9685-9694, Aug. 2023.

[3] R. Paul, R. Alizadeh, X. L. Li, H. Chen, Y. Y. Wang, and H. A. Mantooth, "A Double-Sided Cooled SiC MOSFET Power Module for EV Inverters," IEEE Trans. Power Electron., vol. 39, no. 9, pp. 11047-11059, Sept. 2024.

[4] W. Mu, A. Janabi, B. R. Hu, L. Shillaber, and T. Long, "Liquid Metal Fluidic Connection and Floating Die Structure for Ultralow

Thermomechanical Stress of SiC Power Electronics Packaging," IEEE Trans. Power Electron., vol. 39, no. 7, pp. 7808-7814, July 2024.

[5] S. Q. Liu, Y. H. Mei, J. Li, X. Li, and G. Q. Lu, "Copper-wire stress buffers for extending lifetime of double-sided bidirectional SiC modules," IEEE Trans. Power Electron., vol. 38, no. 6, pp. 7118-7127, June 2023.

[6] C. Ding, H. Z. Q. Liu, K. D. T. Ngo, R. Burgos, and G. Q. Lu, "A double-side cooled SiC MOSFET power module with sintered-Silver interposers: I-design, simulation, fabrication, and performance characterization," IEEE Trans. Power Electron., vol. 36, no. 10, pp. 11672-11680, Oct. 2021.

[7] M. Y. Wang, Y. H. Mei, W. Liu, Y. J. Xie, S. C. Fu, X. Li, and G. Q. Lu, "Reliability improvement of a double-sided IGBT module by lowering stress gradient using molybdenum buffers," IEEE J. Emerg. Sel. Topics Power Electron., vol. 7, no. 3, pp. 1637-1648, Sep. 2019.

[8] J. Jeon, J. Seong, J. Lim, M. K. Kim, T. Kim, and S. W. Yoon, "Finite element and experimental analysis of spacer designs for reducing the thermomechanical stress in double-sided cooling power modules," IEEE J. Emerg. Sel. Topics Power Electron., vol. 9, no. 4, pp. 3383–3391, Aug. 2021.

[9] X. Cao, G. Q. Lu, and K. D. T. Ngo, "Planar Power Module With Low Thermal Impedance and Low Thermomechanical Stress," in IEEE Trans. Compon., Packag., Manuf. Technol., vol. 2, no. 8, pp. 1247-1259, Aug. 2012.

[10] J. Li, A. Castellazzi, T. Dai, M. Corfield, A. K. Solomon, and C. M. Johnson, "Built-in Reliability Design of Highly Integrated Solid-State

Power Switches With Metal Bump Interconnects," IEEE Trans. Power Electron., vol. 30, no. 5, pp. 2587-2600, May 2015.

[11] S. Haque, K. Xing, C. Sushicital, D. J. Nelson, and G. Q. Lu, "Thermal management of high-power electronics modules packaged with interconnected parallel plates," in Proc. Fourteenth Annual IEEE Semiconductor Thermal Measurement and Management Symposium (Cat. No.98CH36195), San Diego, CA, USA, 1998, pp. 111-119

[12] H. P. Chen, C. Peng, G. L. Sun, Y. D.Shi, W. H. Zhu, and L. C. Wang, "Structure Optimization and Reliability Investigation on Copper Pillar Interconnections for Double-sided Cooling Power Module," in Proc. 2022 23rd International Conference on Electronic Packaging Technology (ICEPT), Dalian, China, 2022, pp. 1-5.

[13] R. Paul, A. Hassan, and H. A. Mantooth, "A Double-Sided Cooled Power Module With Embedded Decoupling Capacitors," IEEE J. Emerg. Sel. Topics Power Electron., vol. 12, no. 2, pp. 1813-1821, April 2024.

[14] Y. Y. Yan, B. H. Liu, Y. F .Zhang, J. X.Liu, C. Chen, and Y. Kang, "Thermal Performance Comparisons of Different Spacer Structures in Double-Sided Cooling Power Modules," in Proc. 2023 IEEE Energy Conversion Congress and Exposition (ECCE), Nashville, TN, USA, 2023, pp. 5895-5899.

[15] J. L. Cao, J. Li, and Y. H. Mei, "A double-sided bidirectional power module with low heat concentration and low thermomechanical stress," IEEE Trans. Power Electron., vol. 36, no. 9, pp. 9763-9766, Sept. 2021.

A Study of Multi-Chip Carrier Structure for SiC PCB-Embedded Power Module

Jia-jun Lai
Zhejiang Geener Microelectronics Co.,Ltd
Zhejiang, China
Jiajun.Lai@geener.cn

Xuelun Zhang
Zhejiang Geener Microelectronics Co.,Ltd
Zhejiang, China
Xuelun.Zhang @geener.cn

Hao Sun
Zhejiang Geener Microelectronics Co.,Ltd
Zhejiang, China
Hao.Sun23@geener.cn

Ye Zhu
Zhejiang Geener Microelectronics Co.,Ltd
Zhejiang, China
Ye.Zhu2@geener.cn

Yunbin Pan
Zhejiang Geener Microelectronics Co.,Ltd
Zhejiang, China
Yunbin.Pan @geener.cn

Mingfu Li
Viridi E-Mobility Technology (Ningbo) Co., Ltd
Zhejiang, China
minfu.li@zeekrlife.com

Ziliang Shi
Viridi E-Mobility Technology (Ningbo) Co., Ltd
Zhejiang, China
Ziliang.Shi@zeekrlife.com

Xiong Yang
Viridi E-Mobility Technology (Ningbo) Co., Ltd
Zhejiang, China
Xiong.Yang02@zeekrlife.com

Xunjin Xu
Viridi E-Mobility Technology (Ningbo) Co., Ltd
Zhejiang, China
Xunjin.Xu@zeekrlife.com

Min Chen
College of Electrical Engineering, Zhejiang University
Zhejiang, China
calim@zju.edu.cn

Abstract—**This work proposes a new chip carrier structure to enable the integration of multiple chips. Compared with the solution proposed by Schweizer, the new structure significantly reduces size of PCB embedded power module. To evaluate the proposed concept, a power module was created. Based on this module, the performance differences of by using of single-chip and multi-chip carrier were evaluated in terms of the size, thermal resistance, power loop inductance, and current balancing. Meanwhile, the effect of chip distance in multi-chip carrier on the above performance is also systematically analyzed to find out a balance between electrical preparties and thermal management. This multi-chip carrier is investigated to provide a practical reference for the development of compact and high-performance PCB-embedded power modules.**

Keywords—*SiC, PCB-Embedded Power Module, multi-chip carrier*

I. INTRODUCTION

The application of SiC (Silicon Carbide) power modules is continuously increasing in the field of the traction inverter in electrical vehicle. This rise is primarily due to the advantages of SiC devices [1], such as higher voltage withstands capability, lower power loss, and faster switching frequency than IGBT (Insulate-Gate Bipolar Transistor) devices. However, conventional housing or molding modules cannot fully maximize the performance advantages of SiC devices due to high parasitic parameters value and high thermal resistance [2]. Embedded power modules, owing to their small size [3], low parasitic parameters, and low thermal resistance [4], have emerged as a key research focus in the packaging of next-generation SiC power modules.

The main approaches for chip embedding in PCB (Printed Circuit Board) can be categorized into three types, as shown in Figure 1, including Fraunhofer IZM's chip-substrate process [5], AT&S's Embedded Component Packaging (ECP) technology [6], and Schweizer Electronic AG's cavity embedding approach [7]. The main differences between the three solutions are as follows: (a) The chip is attached to a copper substrate; then, prepreg and copper foil are stacked, followed by drilling to establish electrical connections; (b) Copper plating on the chip's backside, bonding to a thin copper sheet using nonconductive adhesive, and creating connections through laser-drilled micro-vias ; (c) The chip is mounted into a copper carrier, followed by lamination of prepreg and copper foil to complete the packaging.

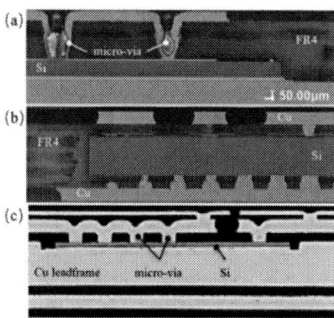

Fig. 1. Cross-sectional view of embedded solutions[5-7]: (a) IZM solution; (b) AT&S solution; (c) Schweizer Electronic AG solution

In comparison, in the first approach the structured prepreg is required, which is imposing complexity of raw materials. The nonconductive bonding materials in the second approach leads to reliability risk of the micro-vias in concerns. Under external stress, the bonding strength of the micro-vias could degrade, causing shear stress or fractures and shortening the module's lifetime. Using chip carrier instead of using substrate could avoid to use the structured prepreg and

This work was financially supported by the Key R&D plan of Zhejiang Province (No.2025C01048).

increase the reliability performance. The chip carrier is designed as a mechanical supporting for the chip during the packaging process. Additionally, the carrier is a large copper block, greatly increases the thermal interface area, thereby enabling more effective heat transfer to the next layer of the structure. However, the larger size of the chip carrier, especially in multi-chip parallel applications, the larger module size after packaging in third approach.

Consequently, this paper proposed a structure of the multi-chip carrier based on Schweizer's solution (single-chip carrier), allowing more than one chips placed in one carrier. An PCB embedded power module concept was also introduced to support this carrier design. Based on this layout, the performance differences between single-chip and multi-chip carrier were evaluated. The influence of chip distance in the multi-chip carrier on the electrical and thermal performance was analyzed. The trade-off between electrical and thermal performance was finally obtained.

II. NEW CHIP CARRIER DESIGN

This chapter introduces a structure of PCB embedded power module and describes the design of the multi-chip carrier for PCB embedded package and the simulation methods.

A. Structure of PCB embedded power module

The multi-layer PCB is the main platform for connecting circuits. Figure 2 shows its cross-sectional structure, which has six routing layers. Each layer is divided into multiple regions [2] The materials properties used for each layer are listed in Table I. The device driving circuits are located on the top layers. The power loop is mainly in L2 and L5. The chip carriers' structural layers are L3 and L4. The bottom layer L6 connects to heatsink with soldering layer. The chip carrier is embedded within the layers L3 and L4. Interlayer connections are achieved through micro-vias. The processing precision of micro-vias is less than 150 μm [8]. In this study, the micro-via diameter is set to 400 μm with a pitch of 800 μm.

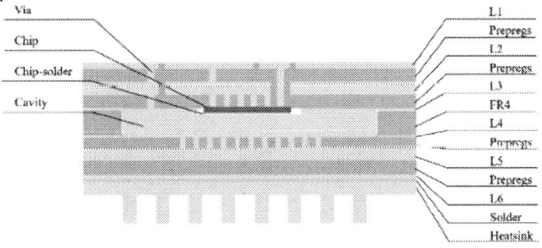

Fig. 2. Structure of PCB embedded power module [2]

TABLE I. MATERIALS PROPERTIES

	Thermal Coounductivity (W/mK)	CTE (ppm)	Young's moduls (MPa)	Poisson's ratio
Cu	385	17	127000	0.33
Chip solder	200	29	24700	0.35
FR4	0.4	12.5	20400	0.14
SiC	400	4.4	425000	0.15
Solder	52	54	13700	0.34
Prepreg	0.64	10	36000	0.30

B. Structural design of the chip carrier

The dimensions of the SiC chip set as an example is 5mm × 5mm × 0.16mm with the copper plating on surface. A typical chip carrier structure is shown in Figure 3(a) [8], where the carrier length is denoted as d1, the width as $d2$, the distance from the edge of the carrier cavity to the edge of the carrier as d3, and the distance from the chip to the edge of the carrier cavity as d4. The structural dimensions of the carrier are detailed in Table II, where the chip soldering layer thickness measures 0.03 mm, the cavity depth is 0.19 mm, and the overall carrier thickness is 1.2 mm.

The multi-chip carrier in this paper proposed as illustrated in Figure 3(b), where maximal six chips are mounted. Considering the poor flow-ability of the prepreg and its inability to fully fill the carrier cavity due to excessively large chip spacing, the chip distance is defined as 1 mm. Other parameters are specified in Table II.

TABLE II. DETAILED DIMENSIONS OF THE CARRIER

	d1	d2	d3	d4
Single-chip carrier	11mm	11mm	2mm	1mm
Multi-chip carrier	17mm	24mm	2mm	1mm

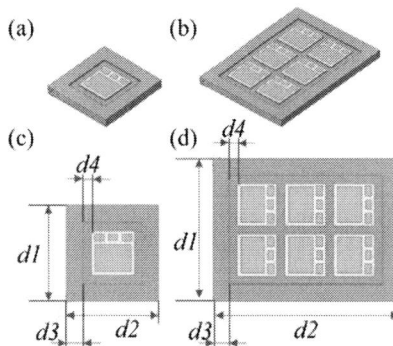

Fig. 3. (a) single-chip carrier, (b) multi-chip carrier, (c) front view of single-chip carrier, (d) front view of multi-chip carrier

C. Simulation model equivalence

In embedded power modules, interlayer interconnects are achieved through micro-vias [10]. To enhance current-carrying capacity and inter-connect reliability, the micro-vias array are designed to be as dense as possible. However, the dense micro-vias result in an exponential increase in the number of solid elements during numerical simulation, leading to excessively long simulation times. Therefore, it is necessary to apply equivalent model of the through-hole to improve computational efficiency. To validate the accuracy of the equivalent model, a simplified carrier model is established, as shown in Figure 4(a). The top of the model features the chip electrode source micro-via array, while the bottom includes the chip electrode drain through-hole array. In the geometric equivalent model, as shown in Figure 4(b), using of cubes replaces all micro-vias, with the same thickness as the replaced micro-vias.

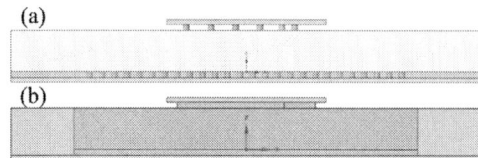

Fig. 4. (a) micro-vias model, (b) equivalent model

Numerical simulations were performed based on the two models. In the parasitic inductance simulation, the bottom-up inductances of both models were calculated. At 10 MHz, the micro-vias model was an inductance of 0.33 nH, while the equivalent model was 0.34 nH, showing minimal difference.

In the thermal simulation, thermal resistance was also used to evaluate the equivalent model. The volume of copper in the cubic model was 22.99 mm³, while it was only 3.84 mm³ in the micro-via model. The thermal conductivity of the cubic model was converted. Assuming the pre-impregnated material is isotropic and homogeneous, and the micro-vias are parallel cylindrical structures, forming the composite material as shown in Figure 5, referencing the Rayleigh model[10], the effective thermal conductivity in the Z-direction, $k_{eff,ZZ}$, satisfies the following relationship:

$$\frac{k_{eff,ZZ}}{k_m} = 1 + \left(\frac{k_1 - k_m}{k_m}\right)\phi \qquad (1)$$

where km represents the thermal conductivity coefficient of the pre-impregnated material, k_1 is the thermal conductivity coefficient of copper, and ϕ denotes the volume fraction. Based on the above formula, the calculated thermal conductivity coefficient is incorporated into the equivalent model, which results in a thermal resistance of 0.149 K/W. The thermal resistance of micro-vias model is 0.152 K/W, with an error of 1.9 % compared between the two methods. Therefore, this model is suitable for equivalent simulation of micro-vias in subsequent embedded modules during electro-thermal simulation.

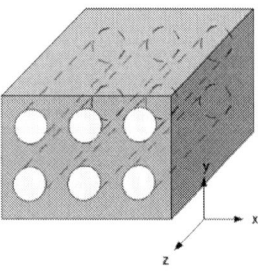

Fig. 5. The schematic of the second type composite medium considered by Rayleigh, consisting of parallel cylinders embedded in a continuous matrix [11].

III. EVALUATION OF INFLUENCE OF CARRIER STRUCTURE ON POWER MODULE PERFORMANCE

This chapter investigated the power module performance as a function of carrier structure of single-chip and multi-chip, including thermal and electrical performance.

A. Layout design of PCB power module

To investigate the performance differences between single-chip and multi-chip carrier, this paper proposed a layout scheme. Figure 6 (a) shows the circuit topology. In Figure 6 (b), the blue arrows represent the current flow in layer L5, and the red arrows represent the current flow in layer L2.

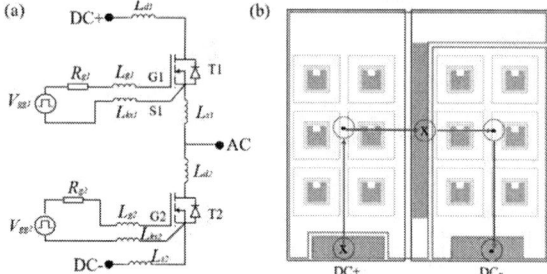

Fig. 6. (a) half-bridge topology, (b) Layout

The symbol " $\otimes$ " indicates that the current flows vertically into the layer, and "$\odot$" indicates that the current flows vertically out of the layer. They display the power loop of a half bridge. The high side bridge is on the left, and the low side bridge is on the right, each side has six chips. Electrode Darin of each chip connects to L5 through carrier and micro-vias, electrode source of each chip connects to L2 through micro-vias. The power loop is from the DC+ terminal (L1) through the micro-vias into layer L5 and from layer L5 through the micro-vias and high side chips back to layer L2, then through micro-vias from layer L2 back to layer L5, and subsequently layer L5 through the micro-vias and low side chips back to layer L2, finally to DC- terminal (L1).

As shown in Figure 7, the size of power module with six chips in parallel in using of multi-chip carrier is significant smaller by multi-chip carrier.

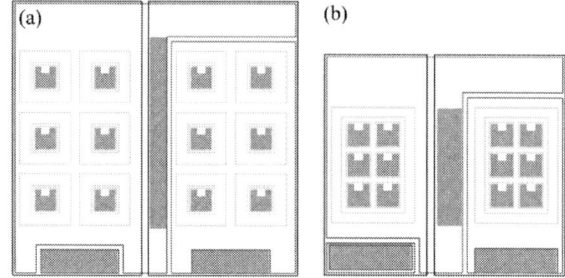

Fig. 7. Based on Layout I: (a) single-chip carrier, (b) multi-chip carrier

B. Thermal performance

This section evaluates the thermal resistance differences between single-chip and multi-chip carriers based on thermal simulations under the same conditions. According to Equation (1), the equivalent thermal conductivity of the vias is calculated to be approximately 78.7 W/m·K. The inlet liquid temperature is set to 65 °C, with a flow rate of 10 L/min. Under these conditions, the thermal resistance is calculated.

The simulation results indicate that the thermal resistance of the single-chip carrier is 0.102 K/W, with minimal thermal coupling between individual carriers. In the case of the multi-

979-8-3315-1110-4/25 $31.00 © 2025 IEEE

chip carrier, when the chip distance is 1 mm, the thermal resistance increases to 0.170 K/W, with increasing of the distance of chip, the thermal resistance is decreasing. As shown in Table III, when the distance reaches 5 mm, the thermal resistance decreases to 0.114 K/W.

TABLE III.   THERMAL RESISTANCE AND PCB AREA

Distance	Single-chip carrier	Multi-chip carrier				
		1mm	2mm	3mm	4mm	5mm
$R_{junction-liquid}$ (K/W)	0.102	0.170	0.149	0.137	0.122	0.114
PCB size (mm²)	40.76	24.83	26.86	28.98	31.2	33.45

The thermal coupling effects are compared between the single-chip and multi-chip carrier. Based on simulation it's significantly observed that the thermal coupling between each chip in multi-chip carrier situation, as show in Figure 8.

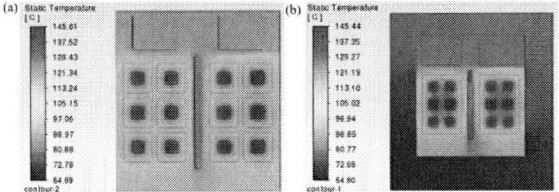

Fig. 8.   Thermal coupling effects on (a) single-chip carrier (b) multi-chip carrier with chip distance 1 mm

C. Electrical performance

This section evaluates the electrical performance of PCB embedded power module by using of different chip carrier schemes. The power loop paths of the power module is illustrated in figure 9. This loop inductance calculations at 10 MHz is presented in Table IV. In the satiation of the multi-chip carrier with 1 mm chip distance results in the loop inductance of 4.26 nH, while increasing the chip distance to 5 mm slightly reduces the loop inductance to 4.21 nH. Compared to the single-chip carrier, the loop inductance achieves an approximate reduction of 1 nH. The shorter Loop path achieves the lower inductance.

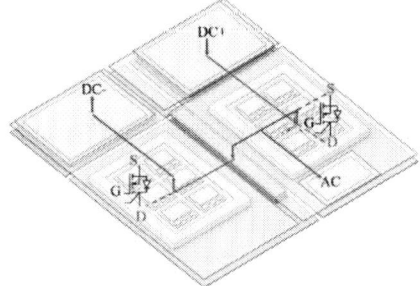

Fig. 9.   Power loop path in PCB embedded power module

TABLE IV.   POWER LOOP INDUCTANCE

Distance	Single-chip carrier	Multi-chip carrier				
		1mm	2mm	3mm	4mm	5mm
Power-loop-inductance /nH	5.26	4.26	4.23	4.22	4.21	4.21

A double-pulse simulation circuit is constructed with a DC bus voltage of 800 V, a load inductance of 40 µH, and a gate resistance of 6 Ω, to compare the peak current and steady-state current during the turn-on and turn-off transitions. The simulated waveform for the multi-chip carrier with a chip spacing of 4 mm is shown in Figure 10. The turn-off time is 167.75 ns with a turn-off loss of 34.12 mJ, while the turn-on time is 280.89 ns with a turn-on loss of 63.15 mJ. For the single-chip carrier, the turn-off time is 169.17 ns with a turn-off loss of 34.57 mJ, and the turn-on time is 281.5 ns with a turn-on loss of 64.1 mJ. These results indicate that in the satiation of multi-chip carrier with chip distance 4 mm, the switching losses are comparable to the single-chip carrier.

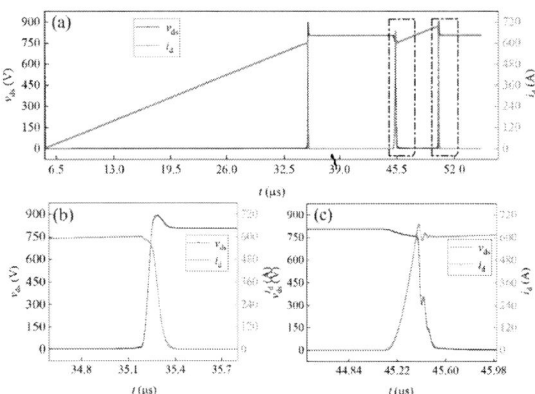

Fig. 10. Double-pulse waveforms: (a) complete waveform, (b) turn-on (c) turn-off.

Further investigation into the current balancing performance in using of single-chip and multi-chip carrier. The deviation of the Ls among the chips, denoted as ε, are the critical influencing factors. By extracting the Ls values of the chips from both schemes and utilizing the statistical formula (2), the deviation of individual chip Ls values from the overall mean can be quantified:

$$\varepsilon = \frac{\sqrt{\sum_{i=1}^{N}(X_i - \bar{x})^2}}{\bar{x}} \qquad (2)$$

Among them, Xi represents the Ls of each chip, and $\bar{x}$ denotes the mean. Specific data are shown in Table V.

In the satiation of single-chip carrier, the deviation of Ls for the high side bridge is 98.6%, while those for the low side bridge is 98.5%.

In the satiation of multi-chip carrier layout, with a chip distance of 5 mm, the high side bridge shows the deviation of Ls is 99.9%, while the low side bridge exhibits 100.3%. At this chip distance, the deviation of Ls is a little bigger than the single-chip carrier.

As the chip distance increases, the deviation of Ls increases (as shown in Figure 11), leading to larger disparities in switching response times across the chips, which may adversely affect current balancing performance.

TABLE V.   EACH CHIP'S L_S INDUCTANCE VALUE

distance	Single-chip carrier	Multi-chip carrier				
		1mm	2mm	3mm	4mm	5mm
High	98.6%	50.9%	51.7%	76.3%	83.5%	100.3%
Low	98.5%	49.5%	51.8%	77.4%	83.6%	99.9%

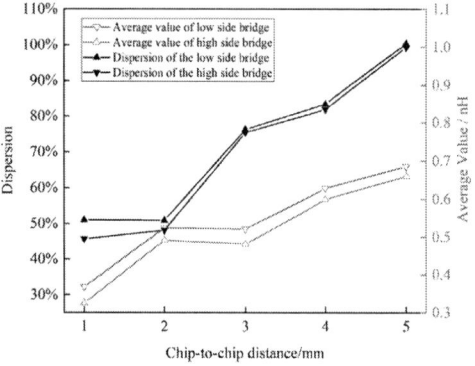

Fig. 11. Mean and dispersion of L_S

The current balancing performance in multi-chip carriers is presented in Table VI. As the chip distance increasing, both the static and dynamic current balancing of the high and low bridges gradually display a decreasing trend. Meanwhile, the thermal coupling effect between the chips is also reduced.

TABLE VI.   CURRENT BALANCING PERFORMANCE

Cavity	distance	High side bridge			Low side bridge		
State		Static	Dynamic		Static	Dynamic	
			On	Off		On	Off
Single-chip carrier		0.96%	9.98%	17.25%	1.78%	13.18%	16%
Multi-chip carrier	D=1mm	0.70%	4.05%	5.55%	0.80%	4.2%	4.2%
	D=2mm	0.75%	6.79%	7.64%	1.01%	6.5%	7.4%
	D=3mm	0.80%	7.73%	9.40%	1.21%	7.7%	8.4%
	D=4mm	0.84%	8.92%	13.05%	1.39%	10.1%	11.5%
	D=5mm	0.87%	11.76%	16.50%	1.51%	12.5%	13.4%

As illustrated in Figure 12, a favorable balance between thermal and electrical performance of the power module with the multi-chip carrier with chip distance 4 mm.

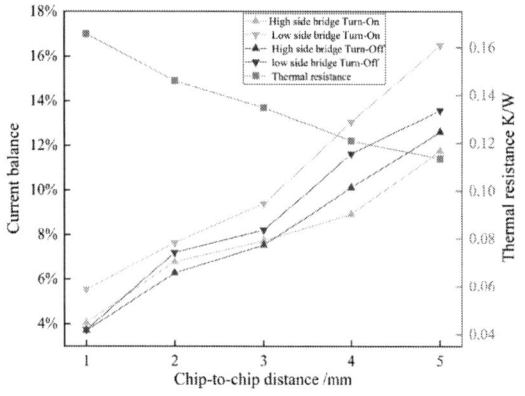

Fig. 12. Effect of chip spacing on thermal resistance and current-sharing performance.

IV. CONCLUSIONS

Through the above simulation analysis, the following conclusions can be drawn:

(1) By using of multi-chip carrier can effectively reduce the size of PCB embedded power module.

(2) By using of single-chip carrier, the size of the PCB embedded power module is larger than using of multi-chip carrier. Relatively, the power loop path is longer, leading a larger parasitic inductance. However, thermal performance is better than using multi-chip carrier.

(3) By using of multi-chip carrier, the chip distance affects the electrical and thermal performance of the power module. As the chip distance increases, the thermal coupling effect between chips is reduced, resulting in low thermal resistance. However, the electrical performance, specifically, the deviation of inductance (L_S) and current balancing becomes worse. When the chip distance is set to 4 mm, a balance between thermal and electrical performance is achieved.

(4) Simulation results indicate that by using of multi-chip carrier with chip distance 4 mm, PCB power module display a more comprehensive performance than by using of single-chip carrier.

References

[1] La Via F, Alquier D, Giannazzo F, et al. Emerging SiC applications beyond power electronic devices[J]. Micromachines, 2023, 14(6): 1200.

[2] Huesgen T. Printed circuit board embedded power semiconductors: A technology review[J]. Power Electronic Devices and Components, 2022, 3: 100017.

[3] Buttay C, Martin C, Morel F, et al. Application of the PCB-embedding technology in power electronics–State of the art and proposed development[C]//2018 Second International Symposium on 3D Power Electronics Integration and Manufacturing (3D-PEIM). IEEE, 2018: 1-10.

[4] Sun X, Chen M, Li B, et al. Design and evaluation of a face-down embedded SiC power module with low parasitic inductance and low thermal resistance[J]. IEEE Transactions on Power Electronics, 2022, 38(3): 2799-2804.

[5] Jung, E., Ostmann, A., & Landsberger, C. (2000). Verfahren zum Integrieren eines Chips innerhalb einer Leiterplatte und Integrierte Schaltung. European Patent, EP1230680B1.

[6] Stahr H. ECP Technology for Packaging[J]. AT AG, ECPE Cluster-Leiterplatten-Einbetttechnologien für die Leistungselektronik, 2021.

[7] Gottwald T, Roessle C. Minimizing Form Factor and Parasitic Inductances of Power Electronic Modules: The p^2 Pack Technology[C]//2018 7th Electronic System-Integration Technology Conference (ESTC). IEEE, 2018: 1-5.

[8] Reiner R, Weiss B, Meder D, et al. PCB-embedding for GaN-on-Si power devices and ICs[C]//CIPS 2018; 10th International Conference on Integrated Power Electronics Systems. VDE, 2018: 1-6.

[9] Gottwald T, Roessle C. P2 Pack-the paradigm shift in interconnect technology[C]//PCIM Europe 2014; International Exhibition and Conference for Power Electronics, Intelligent Motion, Renewable Energy and Energy Management. VDE, 2014: 1-9.

[10] Feix G, Hoene E, Zeiter O, et al. Embedded very fast switching module for SiC power MOSFETs[C]//Proceedings of PCIM Europe 2015; International Exhibition and Conference for Power Electronics, Intelligent Motion, Renewable Energy and Energy Management. VDE, 2015: 1-7.

[11] Pietrak K, Wiśniewski T S. A review of models for effective thermal conductivity of composite materials[J]. Journal of Power Technologies, 2015, 95(1)

979-8-3315-1110-4/25 $31.00 © 2025 IEEE

CNN-Based Rapid Co-Optimization of BV and $R_{ON,sp}$ for 4H-SiC SJ MOSFET

Tiefu Wang[1]
College of Engineering
Zhejiang University
Hangzhou, China
12460062@zju.edu.cn

Haoyuan Cheng[2]
College of Electrical Engineering
Zhejiang University
Hangzhou, China
chenghaoyuan@zju.edu.cn

Chi Zhang[3]
College of Electrical Engineering
Zhejiang University
Hangzhou, China
czhang00@zju.edu.cn

Hengyu Wang[*]
College of Electrical Engineering
Zhejiang University
Hangzhou, China
wanghengyu@zju.edu.cn

Kuang Sheng[+]
College of Electrical Engineering
Zhejiang University
Hangzhou, China
shengk@zju.edu.cn

Abstract—The structural design of power devices tends to be more complex, which leads to the problem of gradient explosion in the high-dimensional parameter design space. It is more difficult to discover the inherent physical laws among numerous design parameters. Using emerging artificial intelligence technologies to assist the structural design of power devices has become a research hotspot. Taking the 4H-SiC super junction metal oxide semiconductor field effect transistor (4H-SiC SJ MOSFET) power device as an example, this paper uses the Sentaurus TCAD software to simulate the breakdown voltage (BV), special on-resistance ($R_{ON,sp}$)of the SJ MOSFET. Based on the convolutional neural network (CNN), an multi-scale one- dimensional convolutional neural network (MS-1DCNN) model suitable for 4H-SiC SJ MOSFET is further proposed, which can assist the structural design of the SJ MOSFET power device and the multi-objective optimization of BV and $R_{ON,sp}$. The MS-1DCNN model can not only achieve a prediction accuracy of over 97% for BV, $R_{ON,sp}$, but also automatically and intelligently provide optimized device design schemes, thereby significantly reducing the device design cost and improving the design efficiency.

Keywords—Multi-scale one-dimensional convolutional neural networks (MS-1DCNN), SiC SJ MOSFET, breakdown voltage, co-optimization, pareto front

I. INTRODUCTION

Due to the excellent characteristics of the high breakdown electric field and high mobility of SiC materials, SiC power devices have a higher breakdown electric field, lower on-resistance, and better thermal conductivity, which are very efficient and durable in high-voltage, high-frequency, and high-temperature environments[1][2][3]. The Super Junction (SJ) technology can further enhance the device performance and is one of the most promising methods at present[4]. Compared with the traditional SiC MOSFET, the SiC SJ MOSFET utilizes the charge compensation effect, with the P pillars and N pillars alternately arranged in the drift region, significantly reducing $R_{ON,sp}$ while maintaining a high BV[5]. This is to improve the trade-off between BV and $R_{ON,sp}$ in unipolar power devices[6].

Currently, the design methods of power devices mainly rely on using finite element simulation software such as Sentaurus TCAD for modeling, simulation, and subsequent experimental verification[7].The simulation calculation process is limited by computing resources, has convergence problems, and requires a long computing time[8]; the experimental verification requires specialized equipment and consumes a large amount of manpower and material resources[9].With the continuous development of new device structures, the number of adjustable structural parameters has increased sharply, making the device design more complex. However, machine learning-assisted TCAD simulation can fully explore the intrinsic connections between data, achieve accurate analysis and optimization of structural parameters, thereby improving the device performance[10].

Fan et al.[11] proposed a graph attention network (RelGAT)-based TCAD surrogate modeling framework using a unified graph encoding scheme to accelerate semiconductor device simulation and design technology co-optimization (DTCO) through transfer learning. Hong et al.[12] developed an inverse design method for electron guns using a cascaded convolutional neural network combined with transfer learning, achieving fast and accurate structural prediction with a beam waist radius error below 5%.Cao et al.[13] used the CNN model to optimize the BV of the LDMOS structure with multiple floating buried layers. Han et al.[14] used the CNN model to generate the electrostatic potential required for the simulation of semiconductor devices to accelerate the simulation of semiconductor devices. Li et al.[15] used the neural network (NN) to optimize the design of the BV of the field limiting ring structure. Yee et al.[16] used the neural network (NN) to design the GaN-on-GaN diode with the protection ring termination to achieve the joint optimization of the breakdown voltage (BV) and the forward conduction voltage (VFQ). In this paper, the MS-1DCNN model and the Pareto front are used to achieve the multi-objective optimization of the BV and $R_{ON,sp}$ of the SiC SJ MOSFET device.

This paper combines machine learning with Sentaurus TCAD simulation, using multi-scale one- dimensional convolutional neural network (MS-1DCNN) and Pareto front to collaboratively optimize the $R_{ON,sp}$ and BV of SiC SJ MOSFET devices, which can achieve the highest figure of merit (FOM) while saving the simulation time. After setting a series of process limitation conditions (such as BV $\geq$ 6500V, $H_p \leq 40\mu m$, the highest FOM, and cell pitch $\geq$ $5\mu m$), the model can still intelligently optimize and automatically recommend the combination of the optimal

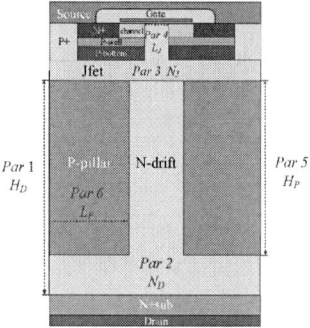

Fig.1 Cross-sectional schematic of SJ MOSFET active area cell structure and six parameters

TABLE 1 SiC SJ MOSFET SIMULATION STRUCTURE ARAMETER RANGE

Parameters	Physical Meaning	Ranges
$H_D\,(\mu m)$	Drift region height	[20,50]
$N_D\,(cm^{-3})$	N-drift region concentration	[5×10^{15}, 5×10^{16}]
$N_J\,(cm^{-3})$	JFET region concentration	[4×10^{16}, 1×10^{17}]
$L_J\,(\mu m)$	JFET region length	[2,6]
$H_P\,(\mu m)$	P-pillar region height	[$0.5H_D$, H_D]
$L_P\,(\mu m)$	P-pillar region length	[$0.4L_{All}$, $0.6L_{All}$]

input parameters, and the recommended combination breaks the one-dimensional limit of unipolar 4H-SiC devices.

II. DEVICE MODEL AND NEURAL NETWORK PREDICTION MODEL

A. SiC SJ MOSFET Structure and Parameter Selection

This paper uses the Sentaurus TCAD to simulate the active area cell structure of SiC SJ MOSFET. Six key structural parameters that can significantly affect the $R_{ON,sp}$, and BV of SJ MOSFET are selected as variables. Table 1 shows the six parameter variables and their physical meanings, as well as the range of variable values. Fig.1 shows the cross-sectional schematic diagram of the SiC SJ MOSFET cell and the annotations of the six parameters.

B. MS-1DCNN Neural Network Modeling

Through TCAD simulation,670 sets of data on BV and $R_{ON,sp}$ of SJ MOSFET devices with different structural parameters were obtained, of which 80% of the data (536 groups) were used as the training set and 20% of the data (134 groups) were used as the test set. Firstly, the data set is preprocessed, normalized, and converted into PyTorch Tensor. Then, the MS-1DCNN model is defined and initialized. Subsequently, the training of the model is started and the Loss curve is displayed in real time. Finally, the model evaluation indicators such as mean absolute error (MAE), root mean squared error (RMSE) and R-squared (R^2) are output, the predicted BV and $R_{ON,sp}$ are output, and the Pareto front is used for multi-objective optimization to find the optimal structural parameters in the parameter space.

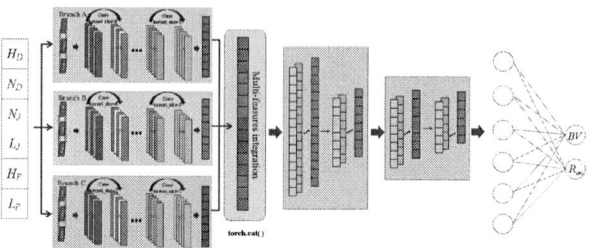

Fig.2 Schematic diagram of the MS-1DCNN model for predicting BV and $R_{ON,sp}$ for SJ MOSFET

TABLE 2 DETAILED NETWORK ARCHITECTURE OF MULTI-SCALE 1DCONVNET

Layer	Layer Name	Input Size	Output Size
1,2,3	Conv1D(kernel=3,5,7)+ReLU	(16,6,8)	(16,64,8)
4	Concatenation	(16,64×3,8)	(16,192,8)
5	Conv1D(kernel=1)+ReLU	(16,192,8)	(16,128,8)
6	MaxPool1D(kernel=2)	(16,128,8)	(16,128,4)
7	Conv1D(kernel=1)+ReLU	(16,128,4)	(16,256,4)
8	MaxPool1D(kernel=2)	(16,256,4)	(16,256,2)
9	Global Average Pooling	(16,256,2)	(16,256)
10	Linear + ReLU	(16,256)	(16,64)
11	Linear	(16,64)	(16,2)

Fig.2 is a schematic diagram of the MS-1DCNN model used for SiC SJ MOSFET devices, mainly including data preprocessing, multi-scale feature extraction, feature fusion, two convolution and pooling processes, and fully connected processing to finally output the predicted parameters. The trained model can accurately predict the BV and $R_{ON,sp}$ corresponding to different structural parameters. The detailed network architecture of the MS-1DCNN is shown in Table 2.

III. RESULT ANALYSIS

A. Training Results of the MS-1DCNN Model

Fig.3(a) (b) respectively show the comparison between the BV, $R_{ON,sp}$ (x-axis) simulated by TCAD and the BV, $R_{ON,sp}$ (y-axis) predicted by the MS-1DCNN model. The blue dots represent the 536 sets of training set data, and the pink dots represent the 134 sets of test set data. The closer the data points are to the 45-degree line, the better the model's prediction effect. And the closer R^2 is to 1, and the smaller the MAE and RMSE are, the better the model effect. The prediction accuracy can reach 97%.The MAE and RMSE formulas are as follows:

$$MAE = \frac{1}{n}\sum_{i=1}^{n}\left| y_{pre} - \hat{y}_{true} \right| \tag{1}$$

$$RMSE = \sqrt{\frac{1}{n}\sum_{i=1}^{n}\left(y_{pre} - \hat{y}_{true} \right)^2} \tag{2}$$

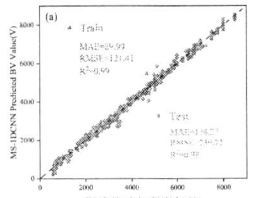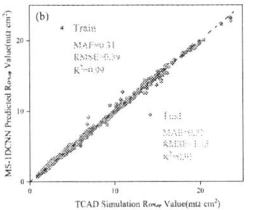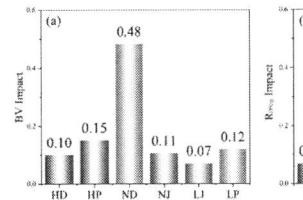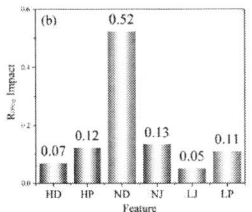

Fig.3 (a) Comparison of TCAD simulated BV and MS-1DCNN model predicted BV. (b) Comparison of TCAD simulated $R_{ON,sp}$ and MS-1DCNN model predicted $R_{ON,sp}$.

Fig.4 (a) Impact weights of six features on BV. (b) Impact weights of six features on $R_{ON,sp}$.

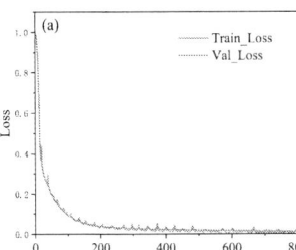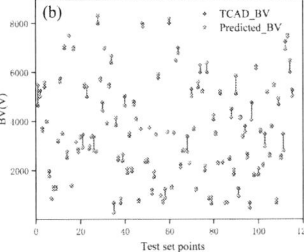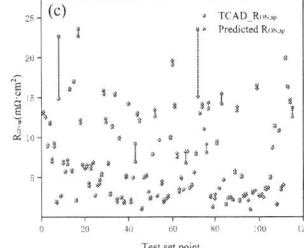

Fig.5 (a) MS-1DCNN model train and validation loss curve. (b) Regression fitting of predicted and true $R_{ON,sp}$ values on the test set. (c) Regression fitting of predicted and true BV values on the test set.

TABLE 3 COMPARISION OF OPTIMAL PARAMETERS UNDER SEVERAL CONDITIONS

Group	Structure Parameters						Output Parameters		
	H_D (μm)	N_D (cm^{-3})	N_J (cm^{-3})	L_J (μm)	H_P (μm)	L_P (μm)	BV (V)	$R_{ON,sp}$ ($m\Omega \cdot cm^2$)	FOM (MW/cm^2)
1	47.58	6.1e15	4.7e16	1.04	29.14	$0.501L_{All}$	7109	6.41	7.8
2	45.51	1.4e16	7.7e16	1.07	38.36	$0.501L_{All}$	7250	5.63	9.3
Simulation Best	50	5e16	1e17	1	50	$0.5L_{All}$	6474	6.74	7.3

Fig.4(a) (b) respectively represent the weights of six parameters on the influence of BV and $R_{ON,sp}$. It can be seen that the doping concentration in the drift region has the greatest influence on BV and $R_{ON,sp}$. The visualization of the influence weights is helpful for the further optimization design of the device structure.

Fig.5(a) shows the loss curves of the MS-1DCNN model on the training set and the validation set. The loss curve of the training set stabilizes at 0.005 after 800 epochs, and the loss curve of the validation set stabilizes at 0.02 after 800 epochs, indicating that the model training effect is good and there is no overfitting. Fig.5(b) (c) respectively show the differences between the predicted values of the models $R_{ON,sp}$, and BV on the test set and the simulated values on TCAD. The blue dots represent the TCAD simulation values, the pink dots represent the model prediction values, and the purple line represents the difference between the two.

B. Simulation Verification Comparison

The MS-1DCNN model can also achieve the function of recommending the optimal parameter combination under a series of artificially set constraints. First, let the model recommend the possible optimal parameter combinations under a series of constraints, and then input the parameter combinations into TCAD for simulation verification. Table 3

shows the verification results in TCAD after the MS-1DCNN model recommends the best parameter combinations under different constraints. All the recommended combinations exceed the theoretical limit of silicon carbide (SiC) unipolar devices after TCAD simulation. Due to the limitations of equipment and process conditions, a series of constraints are imposed on the device parameters and performance. The constraints of the first group are breakdown voltage (BV) $\geqslant$ 6500V, epitaxial layer thickness (H_p) $\leqslant$ $30\mu m$, and cell pitch $\geqslant$ $5\mu m$, which can achieve the highest figure of merit (FOM). The second group relaxes the H_p constraint to $H_p \leqslant 40\mu m$. "Simulation Best" represents the parameter combination with the highest figure of merit (FOM) value among 670 groups of data.

C. Comparison of Simulation Time

This study combines neural networks with the design of semiconductor devices to assist in the design of the two-dimensional structure of the active region of SiC SJ MOSFET, which greatly improves the simulation speed (TABLE 4). The time consumed by using the MS-1DCNN model to predict BV and R_{ON} is reduced by five orders of magnitude compared with TCAD simulation.When facing complex model designs such as terminal design and three-dimensional structure in the future, it can still indicate the simulation direction, reduce the amount of experiments required for simulation, and shorten the research and development cycle of power devices.

TABLE 4 COMPARISION OF TIME BETWEEN TCAD SIMULATION AND MS-1DCNN MODEL PREDICTION

Time	TCAD Simulation(s)	MS-1DCNN(s)
	4.8E3	0.01

IV. CONCLUSION AND INNOVATION

This study first constructs a MS-1DCNN model based on CNN, which can accurately predict the BV and $R_{ON,sp}$ of SiC SJ MOSFET, and the prediction accuracy reaches 97%. The accuracy rate is superior to other models. The model can obtain the influence weights of each input parameter on the output parameter, providing a direction for the subsequent precise simulation optimization. Meanwhile, the time consumed is reduced by five orders of magnitude.

Traditional methods often adopt single-objective optimization or simple empirical parameter tuning. In this work, multi-objective optimization of BV and $R_{ON,sp}$ is achieved. It can effectively search for different trade-off solutions in a high-dimensional optimization space. By changing different limiting conditions, the optimal solution that breaks through the theoretical limit can be obtained. This study proposes a data-driven optimization framework by combining machine learning and optimization algorithms. This framework is not only applicable to SiC SJ MOSFET devices, but can also be extended to the optimal design of other new semiconductor devices, more complex terminal structures, and more difficult to simulate three-dimensional device structures.

ACKNOWLEDGMENT

This work was supported by the National Key Research and Development Program of China under Grant 2023YFB3610001, and in part by the Young Elite Scientists Sponsorship Program by CAST under Grant 2022QNRC001.

REFERENCES

[1] T. Kimoto et al., "Physics and Innovative Technologies in SiC Power Devices," 2021 IEEE International Electron Devices Meeting (IEDM), San Francisco, CA, USA, 2021, pp. 36.1.1-36.1.4.

[2] J. A. Cooper, M. R. Melloch, R. Singh, A. Agarwal, and J. W. Palmour, "Status and prospects for SiC power MOSFETs," IEEE Trans. Electron Devices, vol. 49, no. 4, pp. 658–664, Apr. 2002.

[3] M. Bhatnagar and B. J. Baliga, "Comparison of 6H-SiC, 3C-SiC, and Si for power devices," IEEE Trans. Electron Devices, vol. 40, no. 3, pp. 645–655, Mar. 1993.

[4] H. Cheng, H. Wang, C. Wang, J. Wan, C. Zhang and K. Sheng, "4.15 kV/4.6 m$\Omega \cdot$ cm^2 4H-SiC Epi-Refilled Super-Junction Schottky Diode With Ring Assisted Super-Junction Termination Extension," in IEEE Electron Device Letters, vol. 45, no. 12, pp. 2311-2314, Dec. 2024.

[5] Y. Duan, Y. -L. Zhang, J. Q. Zhang and P. Liu, "Development of SiC Superjunction MOSFET: A Review," 2022 19th China International Forum on Solid State Lighting & 2022 8th International Forum on Wide Bandgap Semiconductors (SSLCHINA: IFWS), Suzhou, China, 2023, pp. 13-17.

[6] F Udrea, G Deboy et al., "Superjunction Power Devices History Development and Future Prospects", IEEE Transactions on Electron Devices, vol. 64, no. 3, pp. 713-727, 2017.

[7] X. Cheng, J. K. O. Sin, J. Shen, Y.-J. Huai, R.-Z. Li, Y. Wu, et al., "A general design methodology for the optimal multiple-field-limiting-ring structure using device simulator", IEEE Trans. Electron Devices, vol. 50, no. 11, pp. 2273-2279, Nov. 2003.

[8] N. El Baradai, C. Sanfilippo, R. Carta, F. Cappelluti and F. Bonani, "An improved methodology for the CAD optimization of multiple floating field-limiting ring terminations", IEEE Trans. Electron Devices, vol. 58, no. 1, pp. 266-270, Jan. 2011.

[9] T. Hirao, H. Onose, K. Yasui and M. Mori, "Edge termination with enhanced field-limiting rings insensitive to surface charge for high-voltage SiC power devices", IEEE Trans. Electron Devices, vol. 67, no. 7, pp. 2850-2853, Jul. 2020.

[10] T. S. Rawat et al., "A Reinforcement-Learning Based Approach for Designing High-Voltage SiC MOSFET Guard Rings," in IEEE Open Journal of Power Electronics, vol. 5, pp. 1853-1861, 2024.

[11] G. Fan, T. Ma, X. Sun, L. Shao and K. L. Low, "Graph Attention Network-Based Unified TCAD Modeling Enabling Fast Design Technology Co-Optimization Through Transfer Learning," in IEEE Transactions on Electron Devices, vol. 72, no. 1, pp. 474-481, Jan. 2025.

[12] W. Hong et al., "Inverse Design of Electron Gun With Transfer Learning Based on Neural Network," in IEEE Transactions on Electron Devices.

[13] Zhen Cao et al. "Machine Learning-Based Modeling and BV Optimization for LDMOS With Multifloating Buried Layers" IEEE TED, 550-556, Feb. 2025.

[14] Seung-Cheol Han et al. "Acceleration of Semiconductor Device Simulation With Approximate Solutions Predicted by Trained Neural Networks", IEEE TED, 5483-5489, Nov. 2021.

[15] Jingyu Li et al. "An Innovative and Efficient Approach for Field-Limiting Ring Design of Power Devices Based on Deep Neural Networks" IEEE EDL, 488-491, Mar. 2024.

[16] Nathan Yee et al. "Rapid Inverse Design of GaN-on-GaN Diode with Guard Ring Termination for BV and (VFQ)-1 Co-Optimization" ISPSD, 143–146, 2023.

Thermal Dissipation in GaN-on-Sapphire Power HEMTs: Synergy of Substrate Thinning and High-Conductivity Thermal Interface Materials

Chang Liu
School of Materials Science and Engineering,
Anhui University
Hefei, China
ms2234118@stu.ahu.edu.cn

Junsong Jiang*
School of Electronic and Information Engineering,
Anhui University
Hefei, China
24751@ahu.edu.cn

Kun Tan*
Anhui Province Engineering Research Center for Advanced Power Electronics and Energy Conversion (APEEC),
Anhui University
Hefei , China
k.tan@ahu.edu.cn

Jie Lu
Institutes of Physical Science and Information Technology,
Anhui University
Hefei, China
lujie@stu.ahu.edu.cn

Suxia Guo
Institutes of Physical Science and Information Technology,
Anhui University
Hefei, China
guosx2503@ ahu.edu.cn

Cungang Hu
Anhui Province Engineering Research Center for Advanced Power Electronics and Energy Conversion (APEEC),
Anhui University
Hefei, China
hcg@ ahu.edu.cn

Zhaofu Zhang
The Institute of Technological Sciences,
Wuhan University
Wuhan, China
zhaofuzhang@whu.edu.cn

Xi Tang*
Anhui Province Engineering Research Center for Advanced Power Electronics and Energy Conversion (APEEC),
Anhui University
Hefei, China
xitang@ahu.edu.cn

Wenping Cao
Anhui Province Engineering Research Center for Advanced Power Electronics and Energy Conversion (APEEC),
Anhui University
Hefei, China
wpcao@ ahu.edu.cn

Abstract—**Despite their significant potential for high-power applications, GaN-on-sapphire HEMTs face critical thermal management challenges due to the inherently low thermal conductivity of sapphire substrates. This study investigates the thermal performance of GaN-on-sapphire and GaN-on-Si HEMTs through finite element simulations of flip-chip ball grid array (FC-BGA) packaging. Simulation results reveal that substrate thinning effectively mitigates the impact of thermal conductivity on junction temperature differences between GaN-on-sapphire and GaN-on-Si devices. Reducing the substrate thickness to 0.1 mm narrows the junction temperature gap to merely 0.355°C. Furthermore, optimization of thermal interface materials (TIMs) demonstrates unique advantages for GaN-on-sapphire configurations. The electrically insulating nature of sapphire substrates permits the use of high-thermal-conductivity metal-based TIMs, whereas silicon substrates are restricted to insulating TIMs (≤ 10 W·m^{-1}·K^{-1}) to avoid electrical risks. At a 0.1 mm substrate thickness, GaN-on-sapphire HEMTs exhibit thermal performance equivalent to GaN-on-Si HEMTs when using a 15 W·m^{-1}·K^{-1} TIM. Increasing the TIM's thermal conductivity to 50 W·m^{-1}·K^{-1} further reduces the junction temperature to 91.977°C. Transient thermal cycling analysis confirms the robustness of this synergistic design approach. These results highlight promising applications for high-power-density systems, such as electric vehicle fast-charging infrastructure.**

Keywords—*GaN-on-Sapphire, thermal management, thermal interface material, finite element simulation*

I. INTRODUCTION

Gallium nitride (GaN) high electron mobility transistors (HEMTs) utilize heterojunction structures in GaN-based semiconductors to achieve high electron mobility, offering significant advantages including high-frequency operation and high-power handling capabilities. Nevertheless, conventional GaN-on-Si HEMTs face fundamental limitations in simultaneously achieving both large-scale production and high-voltage operation [1].

GaN-on-sapphire technology has emerged as a promising solution, leveraging sapphire's intrinsic advantages as substrate material [2-5]. The superior mechanical robustness and electrical insulation properties of sapphire enable substantial reduction in GaN buffer layer thickness, effectively mitigating thermal expansion coefficient mismatch and lattice dislocation issues prevalent in GaN-on-Si architectures. This configuration demonstrates remarkable performance enhancements, including: (1) breakdown voltage exceeding 1200V, (2) enhanced reliability via parasitic channel suppression [6], and (3) improved thermal dissipation characteristics. These merits collectively contribute to lower on-state resistance, higher breakdown voltages [6,7], and manufacturing cost advantages through compatibility with large-diameter wafer processes. Such technological advancements render GaN-on-sapphire particularly suitable for high-power applications such as electric vehicle traction systems, high-speed charging solutions, and 5G radio frequency infrastructure [8].

Nevertheless, GaN-on-sapphire devices face substantial implementation challenges in practical applications. While demonstrating exceptional electrical characteristics, these devices are fundamentally constrained by sapphire's limited thermal conductivity (~35 W·m^{-1}·K^{-1}), creating a thermal management bottleneck that critically impedes power density enhancement [9].

This study employs finite element modeling to analyze the thermal performance of flip-chip ball grid array (FC-BGA) packaged GaN HEMTs. Through controlled heat source application at the chip layer and comprehensive temperature field analysis, we systematically compare the thermal dissipation characteristics between GaN-on-sapphire and GaN-on-Si configurations. By leveraging sapphire's unique

979-8-3315-1110-4/25 $31.00 © 2025 IEEE

material properties through substrate thickness optimization and high-thermal-conductivity TIM implementation, the results demonstrate GaN-on-sapphire HEMTs' expanded application potential.

II. STRUCTURAL DESCRIPTION AND SIMULATION

The GaN HEMT operates as a field-effect transistor utilizing an AlGaN/GaN heterojunction structure, where current conduction and switching are controlled through polarization effects and gate modulation. In steady-state

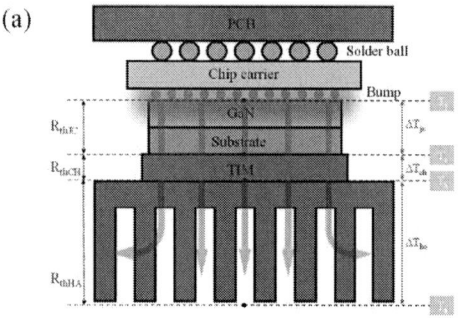

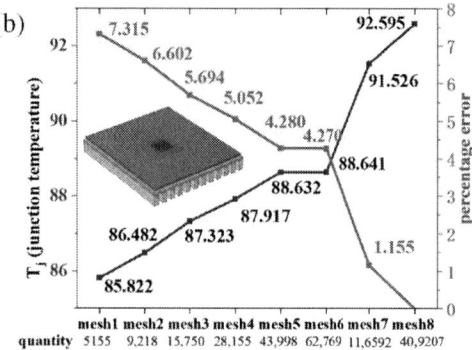

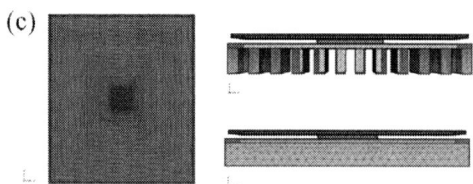

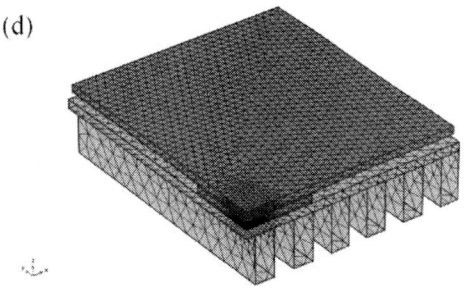

Fig. 1. (a) Geometric structure of a GaN-on-Sapphire/Si power HEMT with FC-BGA packaging and its key parameters. (b) Mesh independence verification of the junction temperature of GaN-on-Sapphire HEMTs. (c) Three-view diagrams of the mesh division of the packaging model. (d) 1/4 view after mesh division.

thermal analysis, thermal contributions from both the AlGaN barrier layer and two-dimensional electron gas (2DEG) are negligible. The dominant thermal factors are instead determined by the substrate's thermal resistance distribution and the heat dissipation pathways through both the substrate and GaN layers.

The FC-BGA packaging configuration achieves substantial improvements in I/O density and signal transmission efficiency through two key design features: (1) an optimized solder ball array layout, and (2) direct bump connections between the chip active surface and chip carrier. This architecture effectively addresses the critical industry demands for miniaturization and higher integration levels [10]. Unlike conventional BGA packaging that primarily relies on bottom-side thermal conduction to the PCB, FC-BGA implements superior top-side heat dissipation pathways. Fig. 1(a) presents a schematic representation of the GaN HEMT's FC-BGA package structure, highlighting its simplified geometry and critical temperature monitoring points.

This investigation systematically examines the thermal characteristics of GaN HEMTs, particularly analyzing how substrate structural parameters and thermal interface material (TIM) properties influence heat dissipation performance. To ensure controlled single-variable analysis, we maintain identical geometric configurations between GaN-on-Si and GaN-on-sapphire HEMTs (Fig. 1(a)), varying only substrate material parameters. Table I specifies the geometric dimensions and material properties for each layer, where the substrate thickness and TIM thermal conductivity values serve as baseline parameters for subsequent parametric studies. The packaging architecture follows JCET's FC-BGA_2021_12_22 specifications [11].

For computational efficiency, we simplify the modeling of solder balls and bumps as rectangular blocks, as their microscopic features negligibly affect thermal analysis outcomes. Accounting for the complex structure inside the PCB, the thermal conductivity of the simplified interconnect blocks and PCB are set to be anisotropic [12].

TABLE I.   MODEL DIMENSIONS AND MATERIAL PARAMETERS

	Size	Material	Thermal Conductivity ($W \cdot m^{-1} \cdot K^{-1}$)
PCB	101.5mm×114.5mm ×1.6mm	FR4	95,95,0.6
Solder ball	number:792, pitch:0.5mm, thickness:0.5mm	SAC Alloy	0.1,0.1,35
Chip carrier	31mm×31mm×1.5mm	ABF	0.48,0.48,17
Bump	number:5329, pitch:130μm, thickness:60μm	Cu	0.5,0.5,35
GaN	9.6mm×9.6mm×2.5 μm	GaN	185
Substrate	9.6mm×9.6mm minimum thickness: 100 μm	Sapphire/Si	35/135
TIM	12mm×12mm×0.1mm	Thermal grease	10
heatsink	104.5mm×117.5mm ×12.7mm	Al	160

Using COMSOL Multiphysics, three-dimensional thermal models of a GaN-on-Sapphire HEMT and a GaN-on-Si HEMT are separately constructed. A steady-state thermal simulation is performed by applying a constant heat source of

25W to the top of the GaN layer. Convective heat transfer is set on the side of the PCB in contact with air, with a convective heat transfer coefficient h = 8 W·m⁻¹·K⁻¹. The ambient temperature was set at 45°C. The emissivity of the bump layer and solder ball layer was set to 0.9, while that of the other layers was 0.7. To save computational time and ensure calculation quality, a mesh independence verification of the junction temperature was conducted. The relationship between the junction temperature and the number of mesh elements is shown in Fig. 1(b). When the mesh density reaches the mesh7 level, the junction temperature error is only 1.155%, which is considered acceptable. Therefore, this mesh configuration was adopted for subsequent analyses. The three-view and quarter-sectional diagrams of the entire meshed model are presented in Fig. 1(c) and Fig. 1(d), respectively.

III. HEAT DISSIPATION STUDY ON SYNCHRONIZED THINNED SUBSTRATES

Substrate thinning, also referred to as backgrinding, is a critical process in semiconductor manufacturing that involves the precision grinding of the wafer's backside to achieve a reduced thickness. This thinning process offers several key advantages, including improved thermal resistance, which enhances heat dissipation and contributes to superior device performance and reliability. Furthermore, thinner wafers enable the fabrication of more compact devices, supporting the ongoing trend toward miniaturization in electronics. Additionally, the reduction in wafer thickness can decrease thermal conductivity, thereby accelerating signal transmission speeds. Due to these benefits, thinned wafers have become indispensable in modern industrial applications, particularly in the development of ultra-thin electronic devices [13].

Since the thermal resistance differences in the thermal management models of GaN-on-Si HEMTs and GaN-on-Sapphire HEMTs are confined to the substrate layer and the thermal interface material layer, and given that the thermal conductivity of sapphire is 35 W·m⁻¹·K⁻¹, while that of silicon is 135 W·m⁻¹·K⁻¹. The thermal conductivity of silicon is nearly four times higher than that of sapphire, the heat dissipation of GaN-on-Sapphire HEMTs is significantly impacted by the substrate's thermal conductivity.

To further investigate the distinct impacts of packaging processes such as substrate thinning on the thermal management effectiveness, it is essential to systematically analyze the thinning effects of both types of substrates using finite element thermal analysis methods. Based on the aforementioned model structure, mesh division, and boundary conditions, three-dimensional thermal management models for both GaN-on-Sapphire HEMTs and GaN-on-Si HEMTs have been developed.

As illustrated in Fig. 2, the simulation results demonstrate that, with a constant thermal conductivity of 10 W·m⁻¹·K⁻¹ for the thermal interface material and all other parameters held constant, applying a uniform substrate thinning process significantly lowers the steady-state operating temperatures of both device types. This aligns with the theoretical predictions of Fourier's law of heat conduction. Given sapphire's lower thermal conductivity, the junction temperature of GaN-on-Sapphire HEMTs is marginally higher than that of GaN-on-Si HEMTs. Nevertheless, as substrate thickness decreases, the temperature gap between the two narrows, underscoring the pivotal role of precise substrate thinning in improving the thermal management of GaN-on-Sapphire HEMTs.

When the substrate is sufficiently thinned, the temperature difference becomes negligible, allowing the thermal impact on GaN-on-Sapphire applications to be disregarded. Specific comparisons of the packaging junction temperatures are detailed in Table II.

TABLE II. VARIATION OF T_j DURING SUBSTRATE THINNING

	0.30mm	0.25mm	0.20mm	0.15mm	0.10mm
GaN-on-Si (degC)	91.981	91.939	91.899	92.008	92.145
GaN-on-Sapphire (degC)	93.547	93.134	92.858	92.651	92.500
ΔT_j (degC)	1.566	1.195	0.959	0.643	0.355

IV. HEAT DISSIPATION STUDY ON CO-TURNED TIM THERMAL CONDUCTIVITY

Thermal interface materials (TIMs), including thermal paste, phase change materials, and thermal shims, are essential for reducing contact thermal resistance and optimizing heat transfer in electronics. The advancements in nanocomposite systems, along with increasing thermal management demands in high-power density devices, have led to significant improvements in TIM characteristics, particularly in thermal conductivity, interfacial compliance, and operational stability [14].

TABLE III. COMPARISON OF THERMAL CONDUCTIVITY AND INSULATION PARAMETERS OF THERMAL INTERFACE MATERIALS

TIM	Thermal Conductivity (W·m⁻¹·K⁻¹)	Volume Resistivity (Ω·cm)	Insulativity
Honeywell TGP 6000PT [15]	6.0	4.0×10^{15}	✔
DOWSIL TC-4083 [16]	8.2	1.6×10^{13}	✔
CoolZorb-Ultra Hybrid [17]	11.5	6.0×10^{16}	✔
Mesophase asphalt-based carbon fiber TIM [18]	29.0	/	✘
Tflex HP34 Series [19]	34.0	1×10^1	✘
Graphene fiber-based TIM [20]	82.4	/	✘

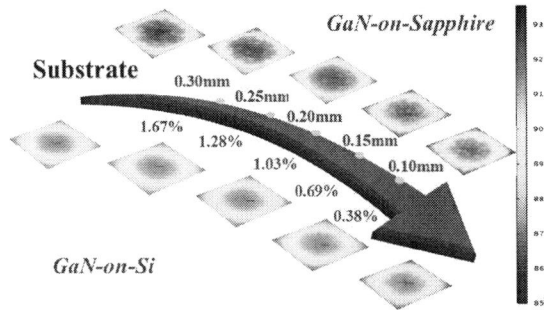

Fig. 2. Temperature distribution on the GaN layer surface of the finite element thermal model under gradient substrate thickness, and the difference in junction temperature between the two types of substrate packaging.

Table III lists key parameters of commercial and experimental TIMs. The analysis indicates that insulating TIMs exhibit a maximum thermal conductivity of 10 W·m^{-1}·K^{-1}, while most TIMs exceeding this exhibit electrical conductivity. Importantly, the intrinsic insulation of sapphire substrates allows for direct integration with ultrahigh-conductivity TIMs to enhance thermal management. In contrast, the conducting nature of silicon substrates imposes significant limitations on their combination with highly conductive TIMs due to electrical compatibility concerns.

Integrating conductive TIMs with silicon substrates introduces electrical risks. during switching or high bias, conductive paths may form between the substrate and heatsink, leading to device failure. Parasitic source-drain currents can increase leakage currents and alter the threshold voltage and switching behavior of HEMTs. Sustained voltage differences can also lead to electrode corrosion and interface degradation.

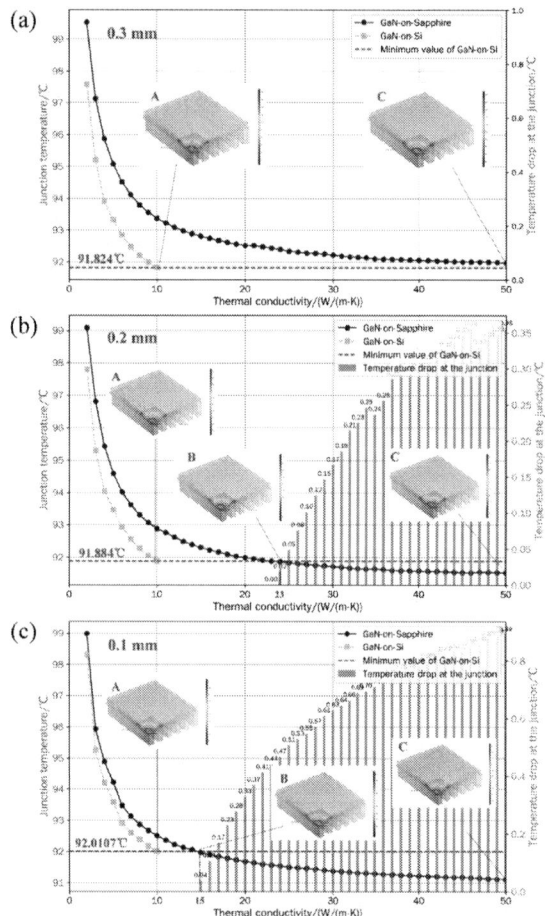

Fig. 3. Comparison of the variation trends and differences in junction temperature between GaN-on-Sapphire and GaN-on-Si as a function of TIM thermal conductivity at different substrate thicknesses. Here, (a), (b), and (c) correspond the cases of substrate thicknesses of 0.3 mm, 0.2 mm, and 0.1 mm, respectively. The red dashed line indicates the minimum junction temperature of GaN-on-Si, while the grey bar chart illustrates the difference between GaN-on-Sapphire and this value. The thermal conductivities at the intersection points in (b) and (c) are 23 W·m^{-1}·K^{-1} and 15 W·m^{-1}·K^{-1}, respectively, whereas there is no intersection point in (a).

To gain a deeper understanding of the relationship between the packaging temperature distribution and the thermal conductivity of the TIM, the following simulation studies were conducted, building on previous work. With the substrate thickness fixed at 0.3 mm, a parametric sweep of the TIM's thermal conductivity was performed in COMSOL Multiphysics, while other parameters remained consistent with those in the earlier experiments. The scanning range for GaN-on-Sapphire was set from 1 to 50 W·m^{-1}·K^{-1}, while for GaN-on-Si, it was set from 1 to 10 W·m^{-1}·K^{-1}. The experimental junction temperature data are shown in Fig. 3(a). Analysis of the figure reveals that throughout the scanning process for GaN-on-Sapphire, the junction temperature was always higher than the lowest junction temperature of GaN-on-Si, which is 91.824°C (point A in Fig. 3(a)). The thermal management performance of the GaN-on-Sapphire HEMTs packaging structure is severely limited by the low thermal conductivity of the sapphire material.

However, when the application of high-thermal-conductivity TIM is synergistically regulated with substrate thinning, such as reducing the substrate thickness to 0.2 mm as shown in Fig. 3(b), the junction temperature distribution undergoes a significant change. It can be observed that the lowest junction temperature of the GaN-on-Si HEMTs packaging is 91.884°C (point A in Fig. 3(b)). When the thermal conductivity of the TIM for the GaN-on-Sapphire HEMTs is increased to 23 W·m^{-1}·K^{-1} (point B in Fig. 3(b)), the same junction temperature can be achieved. Subsequently, further optimization of the TIM thermal conductivity will also enhance the optimization degree of the packaging junction temperature. This phenomenon, which stems from the potential electrical risks associated with applying high-thermal-conductivity TIMs to silicon materials, fully demonstrates the significant thermal management potential of sapphire substrate device packaging. Similarly, as shown in Fig. 3(c), when the substrate thickness is reduced to 0.1 mm, the GaN-on-Sapphire HEMTs can achieve the lowest junction temperature of 92.011°C for the GaN-on-Si HEMTs when the TIM thermal conductivity is 15 W·m^{-1}·K^{-1}. Table IV presents the junction temperatures at critical points A, B, and C for the three aforementioned substrate thicknesses, together with the corresponding thermal conductivities of the employed TIMs.

TABLE IV. KEY POINT TEMPERATURES AND TIM THERMAL CONDUCTIVITY IN FIG. 3

Substrate thickness	A (Si)		B (Sapphire)		C (Sapphire)	
	T_j (degC)	TIM (W·m^{-1}·K^{-1})	T_j (degC)	TIM (W·m^{-1}·K^{-1})	T_j (degC)	TIM (W·m^{-1}·K^{-1})
0.3mm	91.284	10	—	—	91.118	50
0.2mm	91.884	10	91.884	23	91.528	50
0.1mm	92.011	10	92.011	15	91.977	50

The curves in Fig. 3 indicate that, under a given substrate thickness constraint, the device temperature can be categorized into two distinct working regions: the low-thermal-conductivity region, where GaN-on-Si is dominant, and the high-thermal-conductivity region, where GaN-on-Sapphire HEMTs exhibit superior thermal management performance. The thermal conductivity at the intersection point of the curves serves as a crucial indicator. By comparing a large amount of simulation data, it can be clearly seen that this intersecting thermal conductivity shifts towards the low-thermal-conductivity region as the substrate thickness decreases. This discovery holds significant practical implications for the thermal design of packaging and the selection of TIMs.

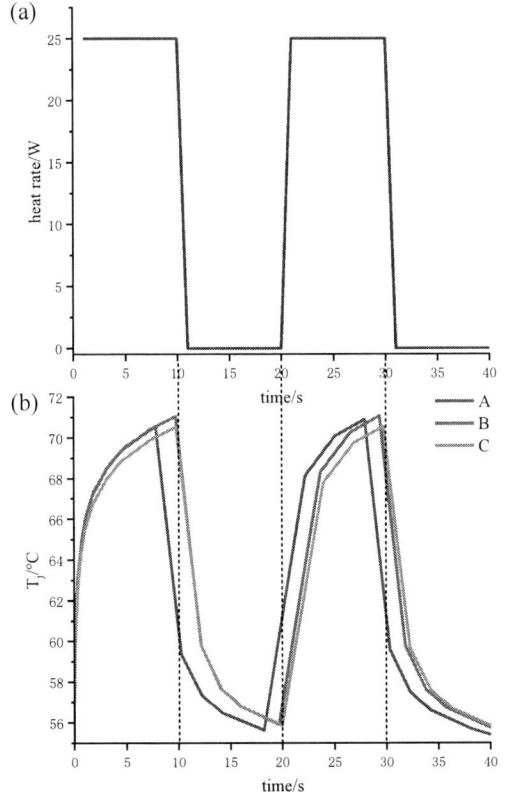

Fig. 4. (a) The cyclic power load applied to the packaging model by power cycling. (b) The junction temperature variations for three typical cases at substrate thicknesses of 0.1 mm. Points A, B, and C correspond one-to-one with those in Fig. 3(c), representing GaN-on-Si using 10 W·m⁻¹·K⁻¹ TIM, GaN-on-Sapphire at the intersection point, and GaN-on-Sapphire using 50 W·m⁻¹·K⁻¹.

To further validate the feasibility of the aforementioned synergistic design and to gain insights into the transient thermal behavior and long-term thermal stability of the packaging, transient thermal analysis was performed using COMSOL Multiphysics for the scenario with a substrate thickness of 0.1 mm. A cyclic load, as shown in Fig. 4(a), was applied to the packaging structure following substrate thinning. A power of 25 W was applied for 10 s and then shut off for 10 s, with this cycle repeated twice to simulate the repeated on-and-off conditions of power devices in practical applications. Unlike temperature cycling, power cycling better reflects the impact of each packaging layer on the junction temperature. Fig. 4(b) shows the case with a substrate thickness of 2 mm, where points A, B, and C correspond to the three points in Fig. 3(c).

As illustrated in Fig. 4, during the heating phase, the temperature rise response and response speed at points B and C are slower than those at point A. Additionally, cooling

begins earlier and the overall duration of high-temperature periods is shorter. These observations indicate the feasibility of replacing GaN-on-Si HEMTs with GaN-on-Sapphire HEMTs packaging in practical applications. Point C exhibits even better thermal performance, demonstrating that high thermal conductivity TIMs can provide solutions for achieving extreme thermal optimization effects.

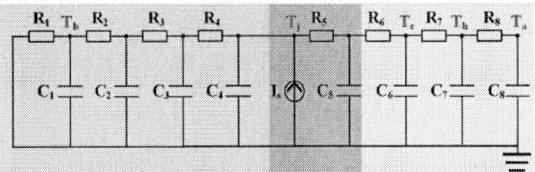

Fig. 5. The Cauer bidirectional thermal network model with critical node temperatures, the primary heat dissipation path for the packaging using FC-BGA is to the right.

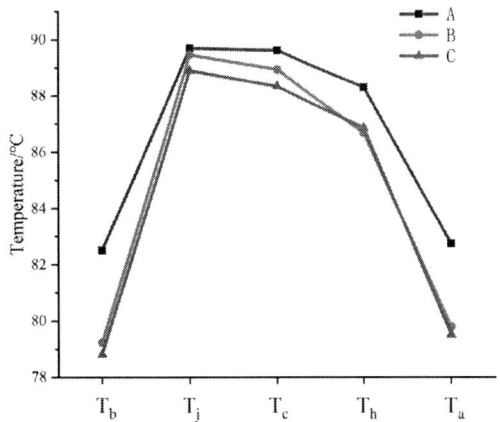

Fig. 6. Critical node temperatures for GaN-on-Sapphire with a substrate thickness of 0.1 mm.

The model can be equivalently represented by the Cauer bidirectional thermal network model depicted in Fig. 5 [21]. For the GaN-on-Sapphire HEMTs packaging with a substrate thickness of 0.1 mm, the temperatures of the critical nodes under conditions A, B, and C are illustrated in Fig. 6. Additionally, Table V lists the thermal resistance of each key layer. After calculation and analysis, it is concluded that for HEMTs packaging using FC-BGA, most of the heat is dissipated through the backside, specifically via the GaN-substrate-TIM-heat sink path, with only a small portion of the heat being dissipated from the top, which can be approximately neglected. This flip-chip configuration efficiently conducts heat away and enhances the thermal stability of the chip during high-speed operation.

TABLE V. THERMAL RESISTANCE OF THE CRITICAL LAYER FOR GAN-ON-SAPPHIRE WITH A SUBSTRATE OF 0.1 MM

	R_{bj}	R_{jc}	R_{ch}	R_{ha}
A (K/W)	0.183	0.00819	0.0694	0.00647
B (K/W)	0.183	0.0312	0.0463	0.00647
C (K/W)	0.183	0.0312	0.0139	0.00647

V. CONCLUSION

The synergistic combination of sub-millimeter substrate thinning and high-thermal-conductivity interface materials enables GaN-on-sapphire HEMTs to achieve thermal performance comparable to GaN-on-Si counterparts, with superior junction temperature characteristics at thinner substrate thicknesses. Experimental results conclusively demonstrate that packaging optimization significantly enhances thermal management efficiency. Despite technical challenges such as the high processing cost of sapphire substrates, the difficulty of ultra-thinning processes (<100 μm), and the high cost for development and application of TIMs, the superior thermal performance of GaN-on-Sapphire HEMTs has been experimentally confirmed, laying a theoretical foundation for the development of high-power-density devices based on heterogeneous integration packaging technology.

ACKNOWLEDGMENT

This work was supported by the National Natural Science Foundation of China (52377035 and 62474001), the Universities Collaborative Innovation Project of Anhui Province (GXXT-2023-001).

REFERENCES

[1] J. Wang et al., "Report of GaN HEMTs on 8-in Sapphire," in IEEE Transactions on Electron Devices, vol. 71, no. 7, pp. 4429-4432, July 2024.

[2] Z. Cheng et al., "Achieving ≥ 1200-V High-Performance GaN HEMTs on Sapphire With Carbon-Doped Buffer," in IEEE Transactions on Electron Devices, vol. 71, no. 12, pp. 7689-7695, Dec. 2024.

[3] G. Gupta et al., "1200V GaN Switches on Sapphire Substrate," 2022 IEEE 34th International Symposium on Power Semiconductor Devices and ICs (ISPSD), Vancouver, BC, Canada, 2022, pp. 349-352.

[4] W. Wang et al., "Suppression of Dynamic Resistance Degradation in 1200-V GaN-on-Sapphire E-Mode GaN HEMTs by Drain-Side Thin p-GaN Design," in IEEE Transactions on Electron Devices, vol. 72, no. 3, pp. 1537-1540, March 2025.

[5] F. Zhou et al., "Ultrathin-Body GaN-on-Sapphire HEMT With Megahertz Switching Capability Under Prompt Irradiation Dose Rate Exceeding 1010 rad(Si)/s," in IEEE Electron Device Letters, vol. 45, no. 8, pp. 1433-1436, Aug. 2024.

[6] Z. Han et al., "p-GaN Gate HEMTs on 6-Inch Sapphire by CMOS-Compatible Process: A Promising Game Changer for Power Electronics," IEEE Electron Device Lett., vol. 45, no. 7, pp. 1257–1260, Jul. 2024.

[7] J. Wang et al., "High-Performance GaN HEMTs on Single Crystalline AlN Templates with a 230 nm Ultra-Thin Buffer and Al_2O_3 /SiO_2

Passivation," in 2024 36th International Symposium on Power Semiconductor Devices and ICs (ISPSD), Bremen, Germany: IEEE, Jun. 2024, pp. 522–525.

[8] X. Li et al., "1700 V High-Performance GaN HEMTs on 6-inch Sapphire With 1.5 μm Thin Buffer," in IEEE Electron Device Letters, vol. 45, no. 1, pp. 84-87, Jan. 2024.

[9] J. Lu et al., "Scalable Compliant Graphene Fiber-Based Thermal Interface Material with Metal-Level Thermal Conductivity via Dual-Field Synergistic Alignment Engineering," in ACS Nano, vol. 18, no. 28, pp. 18560–18571, Jun. 2024.

[10] G. Pascariu, P. Cronin and D. Crowley, "Next generation electronics packaging utilizing flip chip technology," in IEEE/CPMT/SEMI 28th International Electronics Manufacturing Technology Symposium, 2003. IEMT 2003., San Jose, CA, USA, 2003, pp. 423-426.

[11] JCET Group Co., Ltd., "Flip Chip BGA," https:// www.jcetglobal.com /uploads/fcBGA_22Dec2021.pdf.

[12] H. Dang et al., "A Detailed Thermal Resistance Network Analysis of FCBGA Package," Journal of Thermal Science, vol. 33, no. 1, pp. 18–28, Oct. 2023.

[13] S. Seki, T. Funaki, J. Arima, M. Fujita, J. Hirabayashi and K. Hanabusa, "Evaluation of Thermal Resistance Reduction by Thinning Substrate of β-Ga2O3 SBD," 2023 International Conference on Electronics Packaging (ICEP), Kumamoto, Japan, 2023, pp. 193-194.

[14] N. Nagabandi et al., "Metallic nanocomposites as next-generation thermal interface materials," 2017 16th IEEE Intersociety Conference on Thermal and Thermomechanical Phenomena in Electronic Systems (ITherm), Orlando, FL, USA, 2017, pp. 400-406.

[15] Honeywell International Inc., "Honeywell PTM5000/PTM5000-SP Phase Change Thermal Interface Materials," https://www.honeywell.com.cn/content/dam/honcn/documents/advanced-materials/electrical-materials/thermal-interface-materials/EM-TIMs%20Brochure-202206%20CN.pdf.

[16] The Dow Chemical Company, "DOWSIL™ TC-4083 Dispensable Thermal Pad," https://www.ellsworth.com.cn/pdf/11-4272-01-dowsil-tc-4083-dispensable-thermal-pad.pdf.

[17] Laird Technologies, Inc., "CoolZorb-Ultra Hybrid Thermal/EMI Absorber," https://www.laird.com/sites/default/files/2021-08/MFS-DS- COOLZORB%20ULTRA%2020210430.pdf.

[18] B. Li et al., "Preparation, Microstructure and Thermal Properties of Aligned Mesophase Pitch-Based Carbon Fiber Interface Materials by an Electrostatic Flocking Method," Nanomaterials, vol. 14, no. 5, pp. 393–393, Feb. 2024.

[19] Laird Technologies, Inc., "Tflex HP34 Series," https://www.laird.com/ sites/default/files/2024-08/A1833200%20Tflex%20HP34%20Data%2 0Sheet%2008272024.pdf.

[20] Lu, Jiahao, et al. "Scalable Compliant Graphene Fiber-Based Thermal Interface Material with Metal-Level Thermal Conductivity via Dual-Field Synergistic Alignment Engineering.," ACS Nano, vol. 18, no. 28, 28 June 2024, pp. 18560–18571.

[21] Q. Li, S. Cheng, Y. Chen, J. Ye, X. Cui and P. Li, "Study on Cauer Thermal Network Model Considering Bidirectional Heat Transfer," in IEEE Access, vol. 12, pp. 90525-90534, 2024.

979-8-3315-1110-4/25 $31.00 © 2025 IEEE

Analysis of Influencing Factors on the Thermal Performance of SiC Multi-Chip Parallel Packaging Structure

Lei Wang
Beijing Institute of Smart Energy
Beijing, China
wanglei@bise.hrl.ac.cn

Rui Jin
Beijing Institute of Smart Energy
Beijing, China
jinrui@bise.hrl.ac.cn

Xuebao Li
North China Electric Power University
Beijing, China
lxb08357x@ncepu.edu.cn

Peng Zhang
Beijing Institute of Smart Energy
Beijing, China
zhangpeng@bise.hrl.ac.cn

Chao Li
Beijing Institute of Smart Energy
Beijing, China
zhangpeng@bise.hrl.ac.cn

Abstract—With the advancement of chip and packaging technology, silicon carbide device voltage and current levels, power density, operating junction temperature and other performance indicators are also increased accordingly, the device packaging, especially packaging heat dissipation poses a higher challenge. Device packaging miniaturization, integration so that the device heat flow density increased significantly, research shows that more than 50% of the power device failure is caused by the high temperature, and the temperature rises 10°C, the device failure rate will be nearly doubled. Especially for multi-chip parallel package devices, the package internal chip thermal coupling leads to high chip temperature and temperature distribution is extremely uneven, exacerbating the formation of hot spots. Silicon carbide material itself has the ability to withstand high temperatures, but due to the existing silicon carbide device packaging is still using the traditional silicon-based device packaging route, so that the existing silicon carbide power devices can not give full play to its potential. In order to enhance the high-temperature resistance of silicon carbide devices, the use of high thermal conductivity packaging materials, optimize the thermal resistance at the packaging level and improve the external heat dissipation have become two hot topics in the research of silicon carbide device packaging, and the heat dissipation problem of silicon carbide devices has become one of the key technical bottlenecks affecting the operational reliability of the devices. Based on the above problems, this paper firstly establishes a three-dimensional heat conduction model of silicon carbide devices based on Fourier's law of thermal conductivity. Secondly, based on the finite element numerical method, the thermal performance of silicon carbide multi-chip parallel device packaging is studied, and the influence of layout parameters such as the number of chips and spacing, the type of encapsulation material and parameters on the thermal performance of the device packaging is obtained, so as to clarify the optimization direction of the thermal performance of the device at the encapsulation level. Finally, the correctness of the simulation results is verified based on experimental tests. The results of the study can provide a theoretical basis for the design of silicon carbide device packaging layout, packaging material selection and performance enhancement.

Keywords—silicon carbide, multi-chip parallel, package structure，thermal performance，finite element analysis

This work was financially supported by the science and technology project of state grid (NO. 5500-202499380A-3-3-ZH).

I. INTRODUCTION

Silicon carbide due to its high breakdown voltage [1], so that the size of silicon carbide chips can be much smaller than the same voltage level of silicon-based chips, which also makes the package size of silicon carbide devices is getting smaller and more compact, which in turn makes the heat flux density of silicon carbide devices increased dramatically [2], which also brings a huge challenge to the packaging of silicon carbide devices, the high power density produces higher heat flux, which in turn leads to chip The high power density generates higher heat flux, which in turn leads to higher junction temperature, reducing device reliability and efficiency and even leading to device failure [3].

Although the silicon carbide material itself has the ability to withstand high temperatures [4-7], the packaging of existing silicon carbide devices still basically follows the traditional silicon-based device packaging technology and connecting materials, so that the operating temperature of existing silicon carbide power devices is generally still limited to below 200°C [8-11], which greatly limits the potential of the application of silicon carbide devices [1, 12].

In addition, the current-carrying capacity of silicon carbide chips decreases under high-temperature conditions [13], and in order to realize applications with larger current ratings, it is usually necessary to package multiple chips in parallel to achieve the desired current rating. However, in the limited package space, multi-chip paralleling often exists due to the small chip spacing, resulting in serious thermal coupling between the chips [13-16], making the chip junction temperature in the middle of the position to be higher than the edge of the chip temperature, resulting in uneven distribution of the chip junction temperature inside the package, and the chip junction temperature distribution is in turn affected by the uneven current-carrying capacity of the chip. Therefore, in order to maximize the performance advantages of high-temperature resistant silicon carbide devices, to ensure the reliability of silicon carbide devices in high-temperature conditions, it is necessary to enhance the thermal performance of silicon carbide devices, usually used in two ways to achieve this purpose, one is to reduce the thermal resistance of the power device junction, that is, the development of low-thermal resistance of the encapsulation structure, high-temperature-resistant high-thermal conductivity of the packaging materials and interconnection process; the second is to reduce the thermal resistance of the power device junction to the environment thermal resistance, i.e., to enhance external heat dissipation [13]. In recent years, with the development of silicon carbide technology, the application scenarios of silicon carbide devices have been broadened [6, 17], and heat dissipation has become one of the key technologies that cannot be bypassed in the packaging and application of silicon carbide devices. This study explores the packaging factors affecting the thermal performance of silicon carbide devices from the perspective of improving the thermal performance of multi-chip parallel packaging structure of silicon carbide devices.

This study establishes a numerical thermal conduction model for power devices using heat transfer theory. Through finite element discretization, the thermal distribution of the packaging structure is analyzed. The research explores the thermal performance of SiC multi-chip parallel packaging, identifying the impact of key parameters like chip layout, material type, and thickness. The findings provide a theoretical basis for optimizing SiC device packaging design, material selection, and thermal performance enhancement.

II. FINITE ELEMENT SIMULATION OF SILICON CARBIDE DEVICES

A. Physical Modeling of Silicon Carbide Multi-Chip Parallel Packaging Structures

In order to analyze the influence of chip layout and packaging material parameters on the thermal performance of silicon carbide devices, this paper carries out a study based on a commercial 62 mm silicon carbide power device, and obtains the factors affecting the thermal performance of silicon carbide device packaging by analyzing the parameters of the encapsulation process, so as to provide theoretical support for the optimal design of the encapsulation performance of silicon carbide devices and provide a basis for the selection of materials. The structural modeling of the commercial 62 mm silicon carbide power device is shown in Fig. 1, in which the thickness of the ceramic layer is 0.32 mm, the thickness of the base plate is 3 mm, the chip spacing is 1 mm, the structural parameters of the finned heat sink are shown in Table 1, and the relevant parameters of the device encapsulation materials are shown in Table 2.

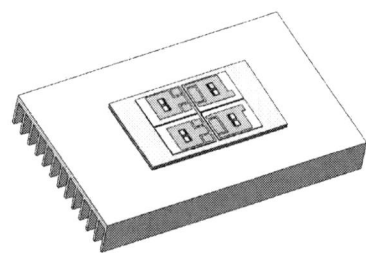

Fig. 1. Layout of a commercial 62 mm silicon carbide power device package with finned heat sinks.

TABLE I. FINNED HEAT SINK STRUCTURE PARAMETERS

Material	Dimensions (mm)	Thickness (mm)	Height (mm)	Spacing (mm)	quantity
Al	200×122 ×5	2	20	8	13

TABLE II. 62 MM SILICON CARBIDE POWER DEVICE PACKAGING MATERIALS AND THERMOPHYSICAL PARAMETERS

Material	Material type	Thermal conductivity ((W/m·K))	Specific heat capacity (J/(kg·K))	Coefficient of Thermal Expansion (10⁻⁶/K)	Density (g/cm3)
Chip	SiC	370	1200	3.4	3.2
Chip solder	PbSnAg	27	126	29	10.75
DBC copper	Cu	398	385	17.2	8.96
Ceramic	Si_3N_4	60	710	3.0	3.2
DBC solder	SnSb10	46.2	223	23.97	7.24
Base plate	Cu	398	385	17.2	8.96
TIM	Thermal grease	2.0	1.0	18	2.5
Heat sink	Al	230	880	26.4	2.7

B. Mathematical Model

Thermal transient simulation is performed for the above structure, the thermal physical parameters of thermal conductivity λ, density ρ and specific heat capacity c of the encapsulation material are constant and the encapsulation material is isotropic, and its differential equation of heat transfer can be expressed by (1) as follows:

$$\frac{\partial T}{\partial t} = \frac{\lambda}{\rho c}\left(\frac{\partial^2 T}{\partial x^2} + \frac{\partial^2 T}{\partial y^2} + \frac{\partial^2 T}{\partial z^2}\right) + \frac{q_v}{\rho c} \qquad (1)$$

where q_V is the internal heat source.

C. Initial Conditions

The SiC chip serves as the heat source, with the heat sink employing forced convection cooling. The single SiC chip exhibits a thermal power of 125 W, and with a source terminal volume of approximately 1.494 mm3, its power density reaches 83.67 W/mm3. The initial ambient temperature is assumed to be 22°C.

D. Boundary Conditions

According to the actual test conditions, the device is connected to the heat sink through the thermally conductive silicone grease, and the heat sink is analyzed with the third type of boundary conditions, i.e., given the convective heat transfer intensity and ambient temperature, the forced convection cooling analysis is carried out for the lower surface and the side of the heat sink, and it is given that the convective heat transfer intensity on the lower surface of the heat sink is 10,000 W/(m2·°C), and that on the end surface is 25 W/(m2·°C) [18, 19], and the ambient temperature is 22°C; the upper surface of the chip and the surrounding of each layer of the package material in the actual module are encapsulated by the potting silicone gel, which can be regarded as adiabatic around the upper surface of the chip and the surrounding sides of each layer of the package internal material due to the relatively low thermal conductivity of the silicone gel [16, 20].

III. EFFECT OF CHIP LAYOUT ON THERMAL PERFORMANCE OF SILICON CARBIDE PACKAGING STRUCTURES

A. Number of chips

Based on the characteristics of multi-chip parallel packaging in existing commercial silicon carbide (SiC) devices, modeling and analysis were conducted for both dual-chip and triple-chip parallel configurations on a single DBC substrate to investigate the impact of chip quantity on the thermal performance of SiC device packaging. The chip spacing was maintained at 1 mm for all configurations, with comparative results shown in Fig. 2.

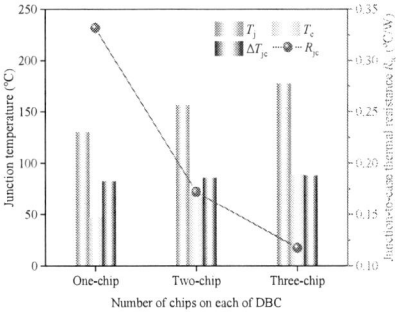

Fig. 2. The impact of the number of chips on the thermal characteristics of SiC device packaging.

The simulation results show that as the number of parallel chips on the DBC substrate increases, both the chip junction and the device case temperature gradually increase. However, the temperature difference between the device junction and the case remains relatively stable. This indicates that as the number of parallel-connected chips in the silicon carbide device increases, the thermal resistance from the junction to the case decreases, but the junction temperature of the silicon carbide chip increases.

B. Chip spacing

According to the literature, the smaller the chip distance is, the more serious the chip is affected by the neighboring chips. In order to analyze the effect of chip spacing on the thermal performance of the device, the simulation analysis of the thermal performance of the silicon carbide device is carried out under different chip spacing layouts, and the results are shown in Fig. 3.

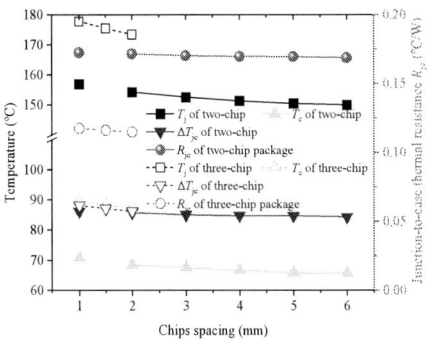

Fig. 3. The impact of chip spacing on the thermal characteristics of power device packaging.

From the results shown in Fig. 3, it can be seen that a single DBC substrate with 2 chips in parallel, when the chip spacing is 1 mm, the maximum junction temperature of the chip is 156.9°C, and the case temperature is about 70.7°C. As the chip spacing gradually increases to 5 mm, the chip junction temperature gradually decreases and stabilizes near 150°C, with a decrease of about 4.4%, and the case temperature gradually decreases and stabilizes at 65.8°C, with a decrease of about 6.9%. The temperature difference between the junction and case of the device decreases slightly with the increase of chip spacing, and remains stable at 85°C. Further analysis shows that the junction to case thermal resistance of the devices corresponding to different chip spacing layouts is basically stable at about 0.085 °C/W, indicating that the junction to case thermal resistance is independent of the chip spacing and is dominated by the interfacial properties of the package material. From the above analysis, it can be found that, for a specific device package structure, increasing the chip spacing can reduce the thermal coupling between the chips, thus reducing the chip junction temperature, but the junction to case thermal resistance is basically unchanged. 5 mm is the critical point of the spacing optimization, and there is no significant gain by continuing to increase the spacing.

IV. EFFECT OF PACKAGING MATERIALS ON THE THERMAL PERFORMANCE OF PACKAGING STRUCTURES

A. Ceramic Thickness

Studies have shown that in power devices, the base plate and DBC ceramics have the greatest influence on device heat dissipation. Therefore, in this study, the Si_3N_4 ceramic substrate is used as an example, and the thermal dissipation power of a single chip is set to 125 W, with a chip spacing of 1 mm, to study the effect of the thickness of DBC ceramics on the thermal performance of silicon carbide devices.

As can be seen from Fig. 4, the thickness of Si_3N_4 ceramic substrate has a significant effect on the thermal characteristics of SiC power devices. When the ceramic thickness increases from 0.32 mm to 2 mm, the chip junction temperature shows a monotonically increasing trend from 156.9°C to 199.6°C, and the corresponding device junction-to-case thermal resistance ($R_{th, j-c}$) increases linearly from 0.172 °C/W to 0.270 °C/W, with a relative increase of 13.1%-36.3%, which indicates that the increase in thickness significantly deteriorates the thermal conduction performance. The case temperature, on the other hand, shows a decreasing trend, and the junction-case temperature difference increases significantly with increasing thickness.

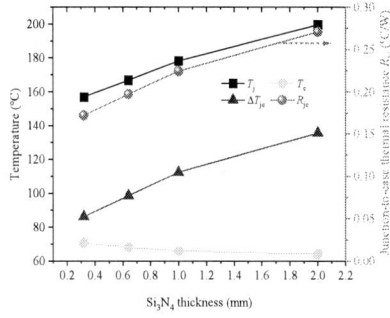

Fig. 4. The impact of Si_3N_4 ceramic thickness on the thermal characteristics of power device packaging.

In the SiC device maximum operating junction temperature of 175°C under the limiting conditions, 0.32 mm thickness of the Si₃N₄ ceramic substrate shows optimal thermal performance, the junction temperature of 156.9°C with a safety margin of 18.1°C. The thermal resistance of 0.172 °C/W in this thickness configuration is 22.2%-57.0% lower than other thickness schemes, verifying the thermal management advantages of thinned ceramic substrates in power device packaging.

B. Types of DBC ceramics

In the field of power electronic device packaging, the selection of ceramic substrate materials has a decisive impact on the thermal management performance of devices. This study focuses on the comparative analysis of the thermal performance of three mainstream ceramic substrate materials (Al_2O_3, AlN and Si_3N_4) in direct copper bonding (DBC) technology, aiming to provide a theoretical basis for the selection of packaging materials for silicon carbide (SiC) power devices.

As shown in Table 3, the three ceramic materials present significant differences in the above performance parameters: Al_2O_3 has a cost advantage but low thermal conductivity (24-28 W/(m·K)); AlN exhibits excellent thermal conductivity (140-180 W/(m·K)) but insufficient mechanical strength; and Si_3N_4 achieves the thermal conductivity (70-90 W/(m·K)) and mechanical properties (650-800 MPa) in an optimized balance.

Fig. 5 reveals the strong correlation between the thermal conductivity of the ceramic substrate and the thermal performance of SiC power devices. Quantitative analysis shows that Al_2O_3 ceramic substrate has the highest junction temperature of 171.6°C ($\Delta T = 101.6$°C), corresponding to $R_{th,j-c} = 0.204$ °C/W, which is close to the reliability threshold of 175°C for SiC devices. The AlN ceramic substrate achieves the optimal thermal performance ($T_j = 141.7$°C, $R_{th,j-c} = 0.144$ °C/W), while the Si_3N_4 ceramic substrate exhibits a compromise performance ($T_j = 155.4$°C, $R_{th,j-c} = 0.170$ °C/W). The case temperatures of the three ceramic-substrate SiC devices were maintained in the range of 69.4-70.0°C. The thermal resistance composition analysis shows that the

substrate thermal conductivity enhancement can reduce the junction temperature by as much as 17.4% ($Al_2O_3 \rightarrow$ AlN). Considering that Si_3N_4 combines excellent mechanical strength and moderate dielectric constant, its comprehensive advantages in power cycle reliability and high-voltage insulation make it the preferred solution for high-power SiC device packaging.

C. Thickness of DBC copper layer

The variation of the thermal characteristics of the device package with the thickness of the upper and lower copper layers of the DBC is shown in Fig. 6. The increase of the upper copper layer thickness causes the junction temperature to show a "U-shaped" trend, the critical thickness of 3 mm to reach a very small value ($T_{j,min}$). And the lower copper layer thickness causes the junction temperature to decrease monotonically and then increase slowly, the optimal thickness range is 2-3 mm, and the global optimal configuration is [3,3] mm (T_j=142.3°C), which is lower than that of the benchmark [0.3,0.3] mm configuration by ΔT=28.7°C. The thickness of the upper copper layer has a quadratic relationship with $R_{th,j-c}$, and the thermal resistance is lower in the range of 1.5-5 mm, and the thermal resistance of junction-to-case is more sensitive to the increase of the thickness of the lower copper layer. Thickening of the copper layer increases the lateral heat diffusion efficiency, and the direct heat dissipation contribution of the upper copper layer to the hot spot on the chip is significantly higher than that of the lower copper layer.

D. Thickness of base plate

The influence of the thickness of the base plate on the thermal performance of the power device package shows counter-intuitive heat conduction characteristics. Through the numerical simulation results in Fig. 7, it can be found that when the thickness of the base plate is gradually increased, the chip junction temperature and case temperature show monotonically decreasing characteristics, while the junction-to-case thermal resistance show an abnormal positive correlation trend.

The variation of thermal resistance due to the increase of the thickness of the substrate is significantly different from the conventional heat conduction theory, and this phenomenon breaks through the linear prediction of the traditional thermal resistance model, implying the existence of a multi-dimensional heat conduction mechanism coupling, which can be explained by the following dimensions of the heat conduction mechanism. First, the lateral thermal diffusion enhancement effect: the increase of the base plate thickness makes the thermal diffusion angle larger, and the equivalent heat dissipation area increases; second, the heat capacity regulation effect: the increase of the base plate thickness increases the volumetric heat capacity of the base plate, which slows down the transient temperature rise, and prolongs the thermal time constant; third, the interfacial effect undergoes a transformation: when the thickness of the base plate is >3 mm, the thermal resistance of the packaging interface decreases, and the thermal resistance of the base plate material body becomes the dominant one.

V. EXPERIMENTAL VERIFICATION

In order to verify the accuracy of the thermal simulation model of the multichip power module, this study establishes a

TABLE III.   THE PERFORMANCE PARAMETERS OF THE THREE CERAMIC MATERIALS

Types of Ceramics	Thermal Conductivity (W/m·K)	Specific Heat Capacity (J/(kg·K))	Density (g/cm3)
Al_2O_3	29	790	3.8
AlN	170	780	3.24
Si_3N_4	60	710	3.2

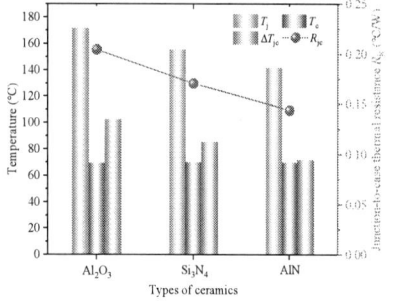

Fig. 5.   The impact of Si₃N₄ ceramic thickness on the thermal characteristics of power device packaging.

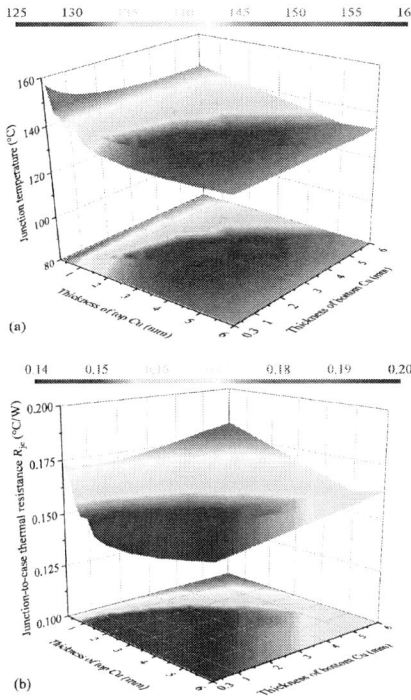

Fig. 6. (a) The variation of the junction temperature and (b) the junction-to-case thermal resistance of the SiC device with the thickness of the upper and lower copper layers of the DBC.

thermal simulation model of the SiC 62 mm multichip parallel package and designs a comparative experimental verification scheme. The experimental platform shown in Fig. 8 uses a 1200 V/21 mΩ SiC MOSFET chip to construct the test module, and applies a 200 A bus current (corresponding to 100 mA inductance current) for transient thermal impedance testing under the gate driving condition of V_G=+15/-5 V, and the on/off times are both set to 120 s. Based on the measured electrical parameters (R_{on}=21 mΩ and I_{load}=200 A) the steady-state power consumption of the single chip is calculated to be 52.5 W [21], and this value is input into the simulation model as a heat source boundary condition. The simulation calculation results and thermal resistance experimental test results are shown in Table 4, and the two results are in good agreement, which verifies the reliability of the simulation model in transient thermal analysis.

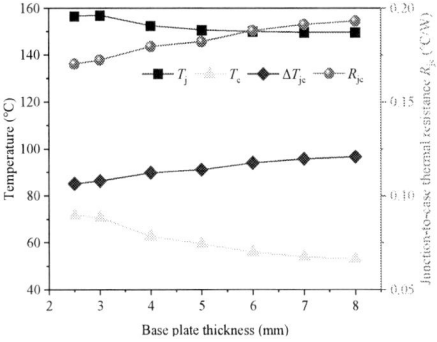

Fig. 7. The impact of metal baseplate thickness on the thermal characteristics of power device packaging.

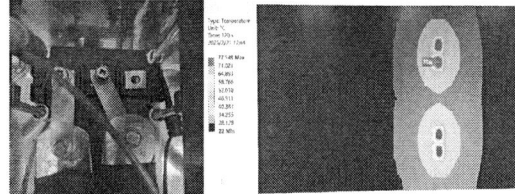

Fig. 8. Photos of experimental tests and simulation results

TABLE IV. COMPARISON OF SIMULATION AND EXPERIMENTAL JUNCTION TEMPERATURE OF MULTI-CHIP PARALLEL PACKAGE MODULE.

	Input Power (W)	Loading Current (A)	Junction Temperature (°C)
Simulation	210	-	77.1
Experiment	206.2	200	82.4
Relative error (%)	1.8	-	6.4

VI. CONCLUSION

In this paper, based on Fourier's law of thermodynamics, firstly, the physical and mathematical models of SiC multi-chip parallel devices are established. Secondly, the thermal performance of SiC multi-chip parallel package is studied and explored through finite element discretization, and the influence of key parameters such as chip layout, material type and thickness on the thermal performance of SiC multi-chip parallel package is determined, and specific parameters to optimize the thermal resistance of the package are obtained through the study of related parameters. Finally, the correctness of the simulation model and method is verified based on the thermal performance test of the silicon carbide multi-chip parallel device. The conclusions of the study can provide theoretical support for the optimized design of the thermal performance of the silicon carbide multichip parallel package. The main conclusions of this paper are:

(1) Thermal analysis reveals a trade-off in parallel chip configurations: while thermal coupling intensifies and junction temperature rises with increasing chip count, the effective thermal resistance decreases due to parallel heat dissipation paths. Optimal chip spacing demonstrates enhanced thermal management efficacy, particularly in high-density configurations.

(2) Material thickness optimization shows contrasting effects: DBC ceramic thickness increment degrades thermal performance, whereas baseplate thickening achieves a better thermal compromise-reducing junction temperature despite junction-to-case increase. The optimum base plate thickness is about 5 mm.

(3) Copper layer thickness optimization in SiC devices exhibits non-monotonic behavior: the [3,3] mm configuration yields minimum T_j, with sensitivity analysis showing upper Cu thickness dominates thermal performance. The Pareto-optimal design space is defined by the thickness of upper copper at [1.5,5] mm and the thickness of lower copper at [0.3,2] mm, achieving a lower $R_{th,j-c}$.

REFERENCES

[1] X. She, A. Q. Huang, Ó. Lucía, and B. Ozpineci, "Review of Silicon Carbide Power Devices and Their Applications," in IEEE T. Ind. Electron., vol. 64, no. 10, pp. 8193–8205, October 2017.

[2] J. Schuderer, U. Vemulapati and F. Traub, "Packaging SiC power semiconductors—challenges, technologies and strategies," 2014 IEEE

Workshop on Wide Bandgap Power Devices and Applications, Knoxville, TN, USA, pp. 18–23, October 2014.

[3] H. Lee, V. Smet and R. Tummala, "A review of SiC power module packaging technologies: challenges, advances, and emerging issues," IEEE J. Em. Sel. Top. P. vol. 8, no. 1, pp. 239–255, March 2020.

[4] D. R. M. Woo, H. H. Yuan, J. A. J. Li, H. S. Ling, L. J. Bum, Z. S. bai, Z. H. yun, S. L. Selvaraj, S. D. Velez, and R. P. Singh, "High power SiC inverter module packaging solutions for junction temperature over 220°C," 2014 IEEE 16th Electronics Packaging Technology Conference (EPTC), Singapore, pp. 31–35, December 2014.

[5] K. Puschkarsky, T. Grasser, T. Aichinger, W. Gustin, and H. Reisinger, "Review on SiC MOSFETs high-voltage device reliability focusing on threshold voltage instability," in IEEE T. Electron. Dev. vol. 66, no. 11, pp. 4604–4616, November 2019.

[6] M. Chen, H. Wang, D. Pan, X. Wang, and F. Blaabjerg, "Thermal characterization of silicon carbide MOSFET module suitable for high-temperature computationally efficient thermal-profile prediction," in IEEE J. Em. Sel. TOP. P., vol. 9, no. 4, pp. 3947–3958, August 2021.

[7] B. J. Baliga, "Silicon carbide power devices: progress and future outlook," in IEEE J. Em. Sel. TOP. P., vol. 11, no. 3, pp. 2400–2411, June 2023.

[8] C. Ding, H. Liu, K. D. T. Ngo, R. Burgos, and G. -Q. Lu, "A double-side cooled SiC MOSFET power module with sintered-silver interposers: I-design, simulation, fabrication, and performance characterization," in IEEE T. Power Electr. IEEE Transactions on Power Electronics, vol. 36, no. 10, pp. 11672–11680, October 2021.

[9] G. Tang, L. C. Wai, T. G. Lim, Z. Chen, Y. L. Ye, R. P. Singh, L. Bu, B. L. Lau, T. C. Chai, K. Yamamoto, and X. Zhang, "Development of SiC chip based power package for high power and high performance application," 2018 IEEE 20th Electronics Packaging Technology Conference (EPTC), Singapore, pp. 83–87, December 2018.

[10] G. Tang, L. C. Wai, T. G. Lim, Y. L. Ye, P. S. Ravinder, L. Bu, B. L. Lau, T. C. Chai, K. Yamamoto, and X. Zhang, "Development of high power and high junction temperature SiC based power packages," 2019 IEEE 69th Electronic Components and Technology Conference (ECTC), Las Vegas, NV, USA, pp. 1419–1425, May 2019.

[11] T. Sugioka, S. Nagao, S. Ogawa, T. Fujibayashi, Y. Sumida, Z. Hao, K. Suganuma, "High thermal stability of SiC packaging with thermosetting imide-based nanocomposite encapsulating materials combined with sintered Ag paste die-attach," 2016 IEEE 16th International Conference on Nanotechnology (IEEE-NANO), Sendai, Japan, pp. 184–187, August 2016.

[12] H. Lee, V. Smet, and R. Tummala, "A review of SiC power module packaging technologies: challenges, advances, and emerging issues," in IEEE J. Em. Sel. Top. P., vol. 8, no. 1, pp. 239–255, March 2020.

[13] F. Yang, L. Jia, L. Wang, C. Zhao, J. Wang, T. Zhang, Y. Gan, and H. Zhang, "The study on thermal coupling effect for SiC power module design guidelines," 2020 IEEE Workshop on Wide Bandgap Power Devices and Applications in Asia (WiPDA Asia), Suita, Japan, pp. 1–7, September 2020.

[14] A. Stippich, M. Neubert, A. Sewergin, and R. W. De Doncker, "Significance of thermal cross-coupling effects in power semiconductor modules," 2016 IEEE 2nd Annual Southern Power Electronics Conference (SPEC), Auckland, New Zealand, pp. 1–6, December 2016.

[15] H. Wang, Z. Zhou, Z. Xu, X. Ge, Y. Yang, Y. Zhang, B. Yao, and D. Xie, "A thermal network model for multichip power modules enabling to characterize the thermal coupling effects," in IEEE T. Power Electr., vol. 39, no. 5, pp. 6225–6245, May 2024.

[16] W. Li, W. Chen, J. Jiang, W. Wang, X. Fan, G. Zhang, and J. Fan, "Double-sided heat dissipation numerical modeling of an embedded half-bridge power module with multiple chips," 2024 25th International Conference on Thermal, Mechanical and Multi-Physics Simulation and Experiments in Microelectronics and Microsystems (EuroSimE), Catania, Italy, pp. 1–8, April 2024.

[17] A. O. Adan, D. Tanaka, L. Burgyan, and Y. Kakizaki, "The current status and trends of 1200-V commercial silicon-carbide MOSFETs: deep physical analysis of power transistors from a designer's perspective," in IEEE Power Electron., vol. 6, no. 2, pp. 36–47, June 2019.

[18] R. Paul, R. Alizadeh, X. Li, H. Chen, Y. Wang, and H. A. Mantooth, "A double-sided cooled SiC MOSFET power module for EV inverters," in IEEE T. Power Electr., vol. 39, no. 9, pp. 11047–11059, September 2024.

[19] B. P. Singh, S. S. Ghahfarokhi, K. Kostov, H. -P. Nee, and S. Norrga, "Analysis of the thermo-mechanical performance of double-sided cooled power modules," 2024 25th International Conference on Thermal, Mechanical and Multi-Physics Simulation and Experiments in Microelectronics and Microsystems (EuroSimE), Catania, Italy, pp. 1–8, April 2024.

[20] Y. Zhang, E. Deng, Z. Zhao, S. Fu, and X. Cui, "A Physical Thermal Network Model of Press Pack IGBTs Considering Spreading and Coupling Effects," in IEEE T. Comp. Pack. Man., vol. 10, no. 10, pp. 1674–1683, October 2020.

[21] Wu, Y., Wang, Z., Yan, X., Xin, G., Shi, X., and Kang, Y., "Conduction thermal runaway of SiC MOSFET under natural convection heat dissipation," 2023 IEEE 2nd International Power Electronics and Application Symposium (PEAS), Guangzhou, pp. 125-129, November 2023.

A Synchronously Implanted Termination for 1200V SiC Trench-Gate MOSFET

Zhengyun Zhu
Nanjing NARI Semiconductor Co., Ltd.,
NARI Group Corporation (State Grid
Electric Power Research Institute
Nanjing, China
zhuzhengyun@sgepri.sgcc.com.cn

Jian Luo
Nanjing NARI Semiconductor Co., Ltd.,
NARI Group Corporation (State Grid
Electric Power Research Institute
Nanjing, China
luojian2@sgepri.sgcc.com.cn

Kai Huang
College of Integrated Circuits,
Zhejiang University
Hangzhou, China
huangk@zju.edu.cn

Qian Wang
Nanjing NARI Semiconductor Co., Ltd.,
NARI Group Corporation (State Grid
Electric Power Research Institute
Nanjing, China
wangqian12@sgepri.sgcc.com.cn

Mingyang Gao
Nanjing NARI Semiconductor Co., Ltd.,
NARI Group Corporation (State Grid
Electric Power Research Institute
Nanjing, China
gaomingyang@sgepri.sgcc.com.cn

Aoxue Xu
Nanjing NARI Semiconductor Co., Ltd.,
NARI Group Corporation (State Grid
Electric Power Research Institute
Nanjing, China
xuaoxue@sgepri.sgcc.com.cn

Liang Zhou
Nanjing NARI Semiconductor Co., Ltd.
NARI Group Corporation (State Grid
Electric Power Research Institute
Nanjing, China
zhouliang1@sgepri.sgcc.com.cn

Abstract—**This work addresses the feasibility challenges of conventional float limiting ring (FLR) termination structures in Silicon Carbide (SiC) trench-gate MOSFETs. Through high-energy aluminum ion implantation, the narrow spacing between float rings is difficult to achieve. To avoid critical dimension (CD) constraints, we propose a novel termination structure fabricated synchronously with standard low-energy P-well and P+ implantation steps. The proposed methodology eliminates additional implantation processes while maintaining compatibility with conventional SiC manufacturing flows. Compared to traditional FLR approaches, the new structure demonstrates enhanced process tolerance, providing a practical solution for scalable production of next generation SiC power devices.**

Keywords—SiC; trench; MOSFET; termination

I. INTRODUCTION

Relying on the inherent advantage of wide-bandgap semiconductor material property, SiC power devices have emerged as transformative solutions in high power electronic converters[1]. Particularly, SiC metal-oxide-semiconductor field effect transistor (MOSFET) is regarded as a promising switching device owing to its superior voltage blocking capability, high current conduction density, low switching energy loss and reliable high-temperature operation[2]. Thanks to these advantages, SiC MOSFET plays a critical role across electric vehicles, high-speed railway, fast-charging infrastructure and renewable energy conversion systems [3].

In the past decade, SiC planar-gate MOSFETs have achieved commercial maturity. To enhance device power density for increasing profits, extensive research efforts have been devoted into SiC trench-gate MOSFETs due to their higher channel mobility and reduced specific on-resistance

(Rsp)[4-8]. However, the development of SiC trench-gate MOSFETs confronts inherent constraints imposed by the sub-micron CD scale, which brings challenges on both design and fabrication.

Focusing on the termination design of SiC trench-gate MOSFET, this article is arranged from an overall consideration on both device design and process flow. Firstly, an ideal FLR termination formed by high-energy aluminum ion implantation is simulated. Then, the CD issues bonded with the high-energy implantation is discussed. Next, an optimized termination with wider CD is proposed. Finally, the design tolerance of the proposed termination is simulated and discussed.

II. TERMINATION DESIGN IN SiC TRENCH-ATE MOSFET

A. Elementary termination structure in SiC power devices

Termination is necessary in power devices as it lessens the peak electric field at the edge of the main junction. In SiC power devices, junction termination extension (JTE) and field limiting ring (FLR) are two common termination structure for SiC devices. [9-10]

JTE consists of one or more P-type rings of precise dopant concentration surrounding the main junction. As shown in Fig. 1(a), before avalanche breakdown occurs at the outer edge of the ring, the low-doped ring will be completely depleted and terminate the field lines. In terms of termination design, JTE is efficient in shrinking termination length but needs additional masks for implantation.

FLR consists of a series of isolated P+ rings surrounding the main junction. Fig. 1(b) illustrates a FLR termination with equipotential lines under reverse bias. As the reverse voltage increases, the depletion region of the main junction gradually expands towards the floating rings. FLR is usually formed in the same implantation step with the main junction so that no additional processing is required. Except occupying more wafer area at higher rated voltage, the heavily doped rings of FLR are not sensitive to the activation rate of dopants.

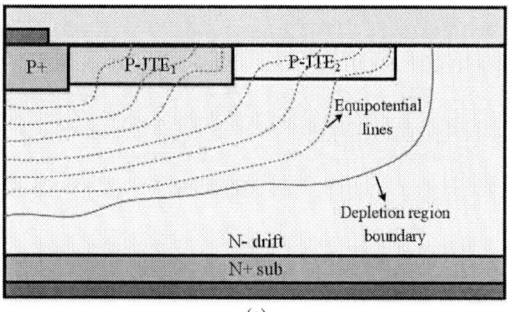

(a)

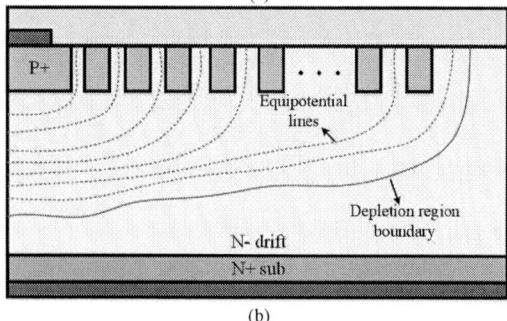

(b)

Fig. 1 Two elementary termination structure in SiC power device. (a) JTE (b) FLR

B. FLR termination for SiC trench-gate MOSFET

In SiC planar-gate MOSFETs, FLR tends to be formed simultaneously with the main junction whose junction depth is below 1μm. However, in SiC trench-gate MOSFETs, building deep P+ shielding regions is necessary to protect the gate oxide from high electric field. As a result, the main junction in SiC trench-gate MOSFET usually extends with a maximum depth of 2-3μm.

Fig. 2 shows the cell structure of a SiC trench-gate MOSFET with deep P+ shielding regions. With a two-layer epi design, the device achieves a blocking voltage of 1610V. The first epitaxy layer is a 11μm thick drift layer with a doping concentration of 9×10^{15}cm^{-3}. And the second epitaxy layer is a 2μm current spreading layer (CSL) which alleviates JFET effect between the gate trench and the deep P+ shielding region to improve current conduction.

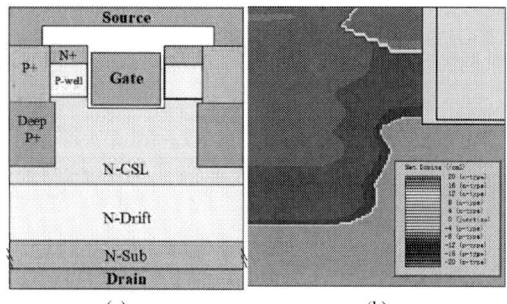

(a)   (b)

Fig. 2 (a) The cell structure of a SiC trench-gate MOSFET with deep P+ shielding regions. (b) The cell structure built with process simulator in TCAD simulation.

To design a termination for the SiC trench-gate MOSFET in Fig. 2, first of all the ideal breakdown voltage of the epitaxial layers should be examined. According to the TCAD simulation result on a P-i-N diode (in which the epi design is the same as that in MOSFET) shown in Fig. 3, the ideal breakdown voltage of the epitaxy layer is proved to be 1768V.

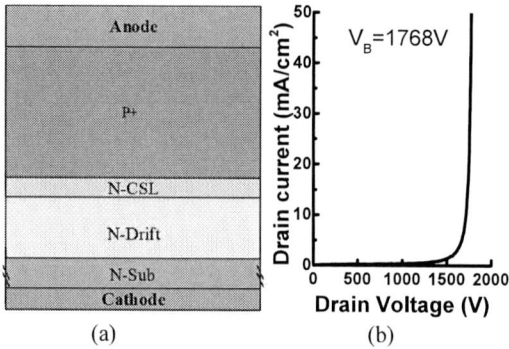

(a)   (b)

Fig. 3 (a) The P-i-N structure used for simulation, in which the epi design is the same as that in MOSFET. (b) The reverse blocking curve of the P-i-N structure.

From the view of process convenience, FLR and derivative structures are studied in this work. Within TCAD simulation, the FLR termination structure demonstrates a theoretical viability. Fig. 4 presents a fundamental 25-ring FLR termination whose simulated breakdown voltage is 1715V, reaching 97.0% of the ideal value. As formed by the same implantation process, the depth of each ring is identical to that of the main junction. Each ring maintains a fixed width (Rw) of 2.8μm. The spacing between the main junction and the first ring (Sw$_{1st}$) is 1.2μm, and the spacing between neighboring rings increases progressively outwards, reaching a maximum gap of 3.4μm at the most outer rings (Sw$_{25th}$). However, such narrow line width is hard to realize in wafer processing, especially within the deep P+ implantation which requires a thick hard mask.

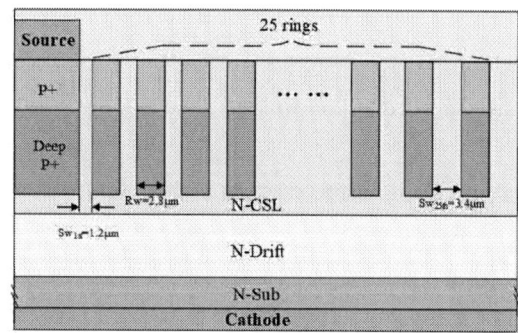

Fig. 4 A 25-ring FLR termination formed simultaneously with the main junction in SiC trench-gate MOSFET.

III. THICK HARD MASK CD ISSUES

Fig. 5 shows the ideal formation process of hard mask for ion implantation into SiC wafer. Firstly, a SiO$_2$ or multi-layer hard mask is deposited. Then, photoresist (PR) is deposited and patterned through lithography. When the thickness of PR is thin, it is easy to acquire a subvertical tilt angle (θ_{PR}) and an opening width in PR (W$_{PR}$) equivalent to that on mask (W$_{mask}$=W$_{PR}$). Subsequently, the hard mask is etched, and the

979-8-3315-1110-4/25 $31.00 © 2025 IEEE

pattern is transferred from PR to the hard mask. ($W_{OX}= W_{PR} =W_{mask}$)

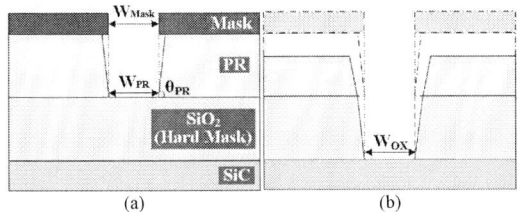

(a)   (b)

Fig. 5 The ideal formation process of hard mask. (a) lithography (b) hard mask etching

For most sub-MeV implantation process in SiC, the thin hard mask hardly brings about CD issues. However, the deep P+ implantation, which is essential for SiC trench-gate MOSFET, is born with puzzling CD issues. Attribute to its high implantation energy which may surpass 3MEV, deep P+ implantation requires a thick hard mask. When the opening width is narrow, the high aspect ratio of the etched part in hard mask will bring obstacles in practical implementation. During the etching process, the remained SiO_2 at the bottom of the opening part receives gradually decreasing plasma. As a result, the etching rate decreases continuously.

A possible issue is illustrated in Fig. 6(a). As the etching rate decreases, the hard mask has not been etched thoroughly when the opening width reaches W_{PR}. If the etch continues, as shown in Fig. 6(b), a wider opening width (W_{OX}') will emerge on the SiC/SiO_2 interface. In another possible case, when the PR is intensively consumed, the tilt angle of PR (θ_{PR}') degrades and influences the tilt angle of hard mask. As Fig. 6 (c) illustrates, this corrupts the morphology of hard mask .

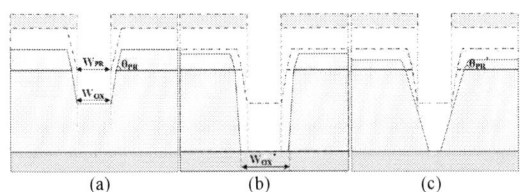

(a)   (b)   (c)

Fig. 6 Possible issues occurring during the etching process of thick hard mask. (a) Thick hard mask hasn't been thoroughly etched when the opening width reaches the designed value. (b) When the hard mask is etched completely, the opening width exceeds the designed value at the interface. (c) When the PR is intensively consumed, the tilt angle of PR and hard mask degrades.

IV. AN OPTIMISTIC TERMINATION DESIGN CONSIDERING CD ISSUES

To avoid CD issues and simplify processing for SiC trench-gate MOSFET, we propose a novel termination structure without either deep P+ implantation or extra implantation steps. The proposed termination structure consists of two parts. As illustrated in Fig. 7, the inner part is synchronously formed by P-well and P+ implantation. The P-well implantation works as a JTE and is filled with 8 P+ rings for dose compensation. The width of each filled ring is 3μm and the total length of P-well implantation is 34μm. The outer part contains 17 float rings formed by P+ implantation. The ring width ranges from 4μm to 2.5μm while the spacing between rings ranges from 1μm to 2.2μm.

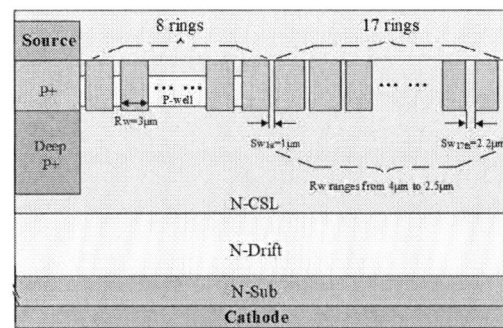

Fig. 7 A novel termination structure without either deep P+ implantation or extra implantation steps, consisting inner JTE-like structure and outer float rings.

Fig. 8 presents the TCAD simulation results of the proposed termination structure. Fig. 8 (a) shows the dopant distribution of the termination structure. And Fig. 8 (b) reveals the two peaks of electric field under 1200V reverse voltage bias. One peak emerges at the main junction, and the other emerges at the 12th float ring in the outer part of the termination. The electric field distribution on cutline A-A' and B-B' is shown in Fig. 8(c). The maximum electric field at the main junction is 2.85MV/cm, and the maximum electric field

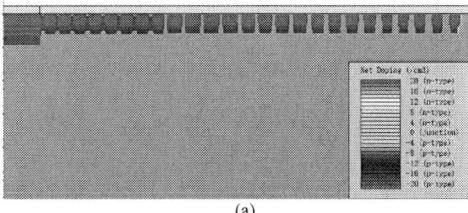

(a)

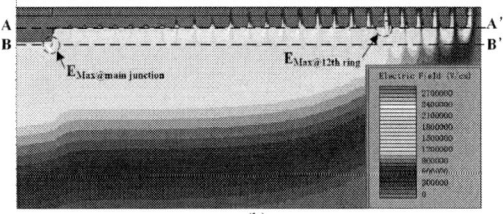

(b)

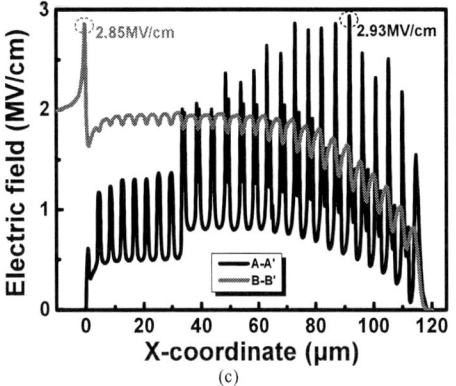

(c)

Fig. 8 (a) The TCAD simulation structure of the proposed termination. (b) The electric field distribution of the proposed termination under 1200V reverse voltage bias. (c)The electric field distribution on cutline A-A' and B-B'.

at float rings is 2.93MV/cm. When the reverse voltage bias increases to the blocking voltage of 1625V, a maximum electric field of 3.30MV/cm is located at the main junction.

To examine the tolerance on processing, extending simulations are conducted on nonideal situations. Firstly, the dose variation of dopants is studied. As Fig. 9(a) shows, when the dose of P-well and P+ implantation varies within 20%, the blocking voltage of the proposed termination structure remains over 1620V in most cases. Merely when the P+ dose decreases 20%, the blocking voltage degrades to 1581V. This demonstrates that the structure keeps a good robustness while the dopant activation rate is not steady.

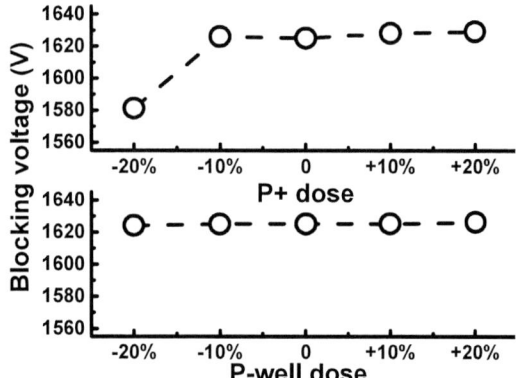

Fig. 9 The influence of P+ and P-well dose variation on blocking voltage.

Then, the influence of CD deviation on P+ float rings is studied. As Fig. 10 shows, when the ring width expands 0.1μm or shrinks within 0.2μm, the blocking voltage remains over 1620V. Nonetheless, when the ring width expands 0.2μm, the narrowed ring spacing leads to an imbalanced distribution of electric field. As a result, the blocking voltage degrades to 1606V.

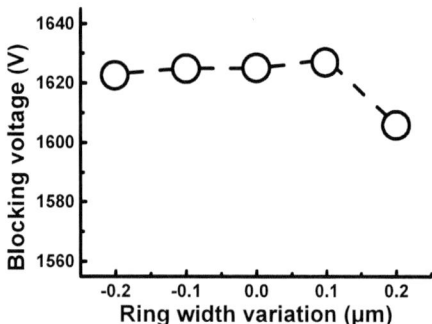

Fig. 10 The influence of P+ ring width on blocking voltage.

In practical wafer processing, an activation rate deviation within 10% and a CD deviation within 0.1μm are both obtainable. Thus, the simulation demonstrates a stable voltage blocking capability of the proposed termination structure. Nevertheless, the blocking voltage merely reaches 91.9% of the ideal value. To enhance blocking voltage, more efforts need to be devoted into optimization work, especially focusing on the modification of rings and spacings.

V. CONCLUSIONS

In this work, a novel termination structure is proposed to break the constraint arising from the CD issues in high energy deep P+ implantation. Synchronously formed with standard low-energy P-well and P+ implantation steps, the proposed termination structure achieves a 91.9% blocking voltage of the ideal value. Considering practical feasibility, the tolerance on processing is discussed. The proposed termination structure demonstrates robustness towards potential dopant and CD deviations.

ACKNOWLEDGMENT

This work was supported by the self-funded project named 'Research and Application on Key Technology of 1200V SiC Power Device' from Nanjing NARI Semiconductor Co., Ltd.. And this work was supported and funded by 'the Outstanding Postdoctoral Program of Jiangsu Province'.

REFERENCES

[1] T. Kimoto, "Material science and device physics in sic technology for high-voltage power devices", Japanese Journal of Applied Physics, Vol.54, No.040103, 2015.

[2] Baliga B J. Gallium Nitride and Silicon Carbide Power Devices. 2017.

[3] Power SiC-Manufacturing 2024, Yole Intelligence, 2024.

[4] T. Kimoto, H. Yoshioka and T. Nakamura, "Physics of SiC MOS interface and development of trench MOSFETs," The 1st IEEE Workshop on Wide Bandgap Power Devices and Applications, Columbus, 2013, pp. 135-138.

[5] Peters D, Siemieniec R, Aichinger T, et al. Performance and ruggedness of 1200V SiC-Trench-MOSFET, Proceedings of the 2017 29th International Symposium on Power Semiconductor Devices and IC's. IEEE, 2017.

[6] T. Nakamura et al., "High performance SiC trench devices with ultra-low ron," 2011 International Electron Devices Meeting, 2011, pp. 26.5.1-26.5.3.

[7] Y. Kobayashi et al., "Body PiN diode inactivation with low on-resistance achieved by a 1.2 kV-class 4H-SiC SWITCH-MOS," 2017 IEEE International Electron Devices Meeting (IEDM), San Francisco, 2017, pp. 9.1.1-9.1.4.

[8] Fukui Y, Sugawara K, Tanaka R, et al. Effects of grounding bottom oxide protection layer in trench-gate sic-mosfet by tilted al implantation. Materials Science Forum. 2020, 1004:764-769.

[9] M. Bhatnagar, H. Nakanishi, S. Bothra, P. K. McLarty and B. J. Baliga, "Edge terminations for SiC high voltage Schottky rectifiers," Proceedings of the 5th International Symposium on Power Semiconductor Devices and ICs, 1993, pp. 89-94.

[10] Temple, V.A.K. (1977) Junction termination extension (JTE): a new technique for increasing avalanche breakdown voltage and controlling surface electric fields in p-n junctions. International Electron Devices Meeting Technical Digest, pp. 423–426.

2025 IEEE Workshop on Wide Bandgap Power Devices and Applications in Asia (WiPDA Asia)

Research on a Crosstalk Parasitic Oscillation Suppression Method Based on Series Resonance Principle

1st Tiancong Shao
School of Electrical Beijing Jiaotong University
Beijing, China
shaotc@bjtu.edu.cn

2nd Pei Hu
School of Electrical Beijing Jiaotong University
Beijing, China
23121417@bjtu.edu.cn

3rd Yuhan Sun
School of Electrical Beijing Jiaotong University
Beijing, China
22121506@bjtu.edu.cn

4th Kai Wang
School of Electrical Beijing Jiaotong University
Beijing, China
23126355@bjtu.edu.cn

5th Bin Qi
School of Electrical Beijing Jiaotong University
Beijing, China
23126282@bjtu.edu.cn

6th Fangwei Zhao
Huaneng Clean Energy Research Institute
Beijing, China
fw_zhao@qny.chng.com.cn

7th Zhe Zhang
Eaton Corporation
USA,
zhezhang2@eaton.cnm

Abstract—This paper proposes a crosstalk parasitic oscillation suppression method based on the series resonance principle for SiC MOSFETs. A P-channel MOSFET and an auxiliary capacitor are connected in parallel to drive the SiC MOSFETs' gate. The gate drive's auxiliary capacitors and parasitic inductance create a low impedance loop for the parasitic oscillation, while the P-channel MOSFET provides negative feedback regulation. Theoretical analysis and experimental verification show the method can effectively suppress crosstalk parasitic oscillation during switching transients without slowing the switching speed dv/dt and keeping the gate-source voltage stable.

Keywords—Crosstalk parasitic oscillation; Series resonance; Negative feedback regulation

I. INTRODUCTION

Wide band gap semiconductor devices represented by silicon carbide (SiC) provide an opportunity for technological innovation in the field of power electronics [1,2]. However, with the increased switching speed, SiC MOSFETs' gate voltage gradually generates more prominent crosstalk parasitic oscillation due to the influence of parasitic parameters [3,4]. It affects the efficiency, power density and lifetime of power electronic converters, limiting further cost reductions, and even reducing their reliability [5].

To suppress the crosstalk parasitic oscillation, researchers have investigated both the passive and the active suppression methods [6]. Passive suppression methods change the external gate drive resistance and external capacitor between the gate source [7,8]. However, these methods slow the device's switching speed, causing additional losses [9,10]. Active Miller Clamp (AMC) technology [11] adds a clamp device to the SiC MOSFET gate drive and detects the gate voltage through the logic circuit inside the driver IC. When the SiC MOSFET gate voltage fluctuation exceeds the clamp device threshold, the clamp device is activated, and the

current through the miller capacitor flows through the clamp device, thus suppressing the crosstalk parasitic oscillation and improving the gate voltage stability. However, the research results have shown that the AMC technique can significantly restrain gate parasitic oscillations at dv/dt below 20V/ns. Still, at a higher dv/dt over 20V/ns, the AMC has a limited effect on mitigating gate parasitic oscillations than the conventional gate drive (CGD) [11]. Our previous research in reference [12] proposed an active gate drive based on the negative feedback mechanism. However, the auxiliary inside the gate driver is very large so the drive loss would be enlarged. The suppression of crosstalk parasitic oscillation in high dv/dt around or above 20V/ns is still worth researching.

This paper analyzes the mechanism of crosstalk parasitic oscillation and proposes a method to suppress crosstalk parasitic oscillation based on the series resonance principle to create a low impedance loop for oscillation.

II. THE MECHANISM OF CROSSTALK PARASITIC OSCILLATION OF SiC MOSFET

Fig.1 illustrates the crosstalk parasitic oscillation issue resulting from the coupling between the Miller capacitor and total parasitic inductance L_i. In this phase-leg configuration, R_g and $R_{g.int}$ represent the external and internal gate drive resistance; D is the parasitic diode; C_{gs}, C_{gd} and C_{ds} stand for gate-source capacitance, gate-drain capacitance and drain-source capacitance respectively. For the above symbols, the upper and lower devices are denoted as 1 and 2, respectively. In this analysis, Q_2 is the active device, and Q_1 is the passive device. Fig.1 (a) and (b) show the Miller current generation with positive and negative dv/dt, respectively.

In the phase-leg configuration, during the high positive slew rate of drain voltage in one switch, such as Q_1 in Fig.1 (a), current with oscillation is injected towards the gate, charging the device's Miller capacitance (C_{gd1}). A positive spike appears on the gate (v_{GS1}) due to the voltage rise caused by the Miller current across the overall path impedance, including the total parasitic inductance L_i. Then the positive

---

This work was supported in part by the National Natural Science Foundation of China under Grant 52377165.

spike gradually decreases, featuring a damping crosstalk parasitic oscillation. If the positive spike exceeds the threshold voltage during the rapid rise of the v_{DS1} transient, a shoot-through may occur across the phase-leg.

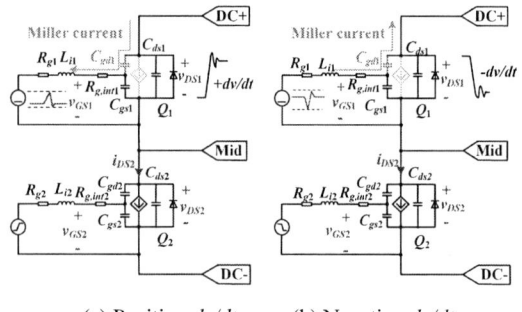

(a) Positive dv/dt (b) Negative dv/dt

Fig. 1 Mechanism of crosstalk parasitic oscillation

Similarly, during the high negative slew rate of drain voltage in one switch, such as Q_1 in Fig.1 (b), oscillation currents are injected into the gate by discharging the device's Miller capacitance (C_{gd1}). A negative spike appears on the gate (v_{GS1}) due to the voltage drop caused by the Miller current across the overall path impedance, including the total parasitic inductance L_i. Then the negative spike gradually decreases, featuring a damping crosstalk parasitic oscillation. The negative gate voltage peak has to be kept within the Amplitude Maximum Rating (AMR) to avoid any possible gate oxide damage.

III. OPERATION PRINCIPLE OF THE RT-NFAGD

To ensure the stable operation of SiC MOSFETs with limited crosstalk parasitic oscillation at high-speed switching dv/dt around or above 20V/ns, it is necessary to reduce the impact of the drive loop parasitic parameters. Hence, this paper proposes a novel gate drive method based on the series resonance principle: the resonant type negative feedback active gate drive (RT-NFAGD), as shown in Fig.2.

The RT-NFAGD only requires adding an auxiliary MOSFET Q_P and a capacitor C_a connected in parallel with Q_P in the drive loop. The crosstalk parasitic oscillation caused by total parasitic inductance L_i can be significantly suppressed. This is mainly because of the function of Q_p and Ca, which we will discuss separately.

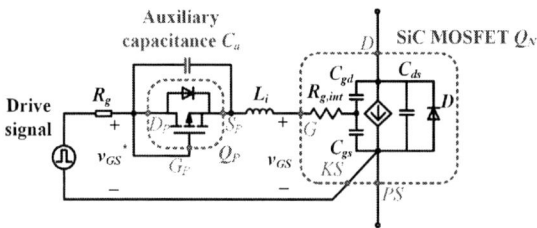

Fig. 2 Resonance Type Negative Feedback

Active Gate Drive (RT-NFAGD)

A. The function of auxiliary MOSET Q_P

The auxiliary MOSFET Q_P will show its function outside the crosstalk parasitic oscillation duration. The function of auxiliary MOSET Q_P includes two aspects.

Firstly, the primary function of the auxiliary MOSFET Q_P is to trigger the negative feedback mechanism, and it typically uses a P-channel device to share the same gate drive signal with the N-channel power MOSFET Q_N through the external drive resistance R_g. By sensing the potential difference between the D_P and S_P, Q_P can automatically manage the charging and discharging of Q_N's input capacitance C_{iss}, following with the drive signal v_{GS*} by the negative control regulation.

Secondly, another function of the auxiliary MOSFET Q_P is to reset the auxiliary capacitor C_a after the capacitor is enabled during the crosstalk parasitic oscillation duration.

B. The function of auxiliary capacitor C_a

The auxiliary capacitor C_a is mainly used to form a series resonance with the total parasitic inductance (L_i) in the drive loop during the crosstalk parasitic oscillation duration.

During the crosstalk parasitic oscillation, C_a and L_i create a low-impedance path for the oscillatory component around the resonant frequency. This low-impedance path will reduce the voltage across the drive path. Hence, it reduces the crosstalk parasitic oscillation amplitude.

Moreover, the auxiliary capacitor C_a is embedded in the forward loop of the gate drive. It is only enabled during the crosstalk parasitic oscillation duration and consumes the reactive power without significant active power loss.

IV. EXPERIMENTAL VERIFICATION

To verify the effectiveness of the proposed RT-NFAGD, this paper builds the double-pulse platform, as shown in Fig. 3. The phase-leg configuration is illustrated in Fig. 1.

Fig. 3 Experimental platform of RT-NFAGD

To minimize the parasitic influence of package parasitic parameters, this platform uses SiC MOSFET bare die, which connects with bonding wires. It should be noted that the proposed RT-NFAGD is not only suitable for embedding inside the device package. It can be composed of discrete devices. This paper will not show the discrete device version of RT-NFAGD because of the space limit.

In this experimental verification, the auxiliary MOSFET Q_P is chosen based on the principle in reference [12], and the value of auxiliary capacitor C_a is determined by engineering trial and error. Thus, the auxiliary capacitor Ca =44nF is used and AGM30P18E bare die is used as the auxiliary MOSFET Q_P. An 1200V 40mΩ SiC MOSFET bare die is used as the Q_N.

The DC bus voltage between DC+ and DC- in Fig. 1 is V_{dc}=800V, and load current I_d=30A. Fig. 4 and Fig. 5 illustrate the waveforms of the conventional gate drive (Fig.

1) and the proposed RT-NFADG (Fig. 2 connected as phase-leg configuration) with a drive resistance R_g is 2Ω and 1Ω, respectively.

When R_g=2Ω, by comparing the crosstalk parasitic oscillation voltage at the Q_2 turning-on moment given in Fig. 4 (a) and (b) at the left column, the crosstalk parasitic oscillation amplitude at v_{GS1} is reduced by 23% from 8.3 to 6.4V. When R_g=1Ω, by comparing the crosstalk parasitic oscillation voltage at the Q_2 turning-on moment given in Fig. 5 (a) and (b) at the left column, the crosstalk parasitic

oscillation amplitude at v_{GS1} is reduced by 14.14% from 9.9V to 8.5V.

Especially in a very high dv/dt around 40V/ns, see to the turning-off moment in Fig. 5 at the right column, the crosstalk parasitic oscillation amplitude at v_{GS1} still reduced by 4.26% from 4.7V to 4.5V.

In sum, the experimental results in Fig. 4 and Fig. 5 demonstrate that the proposed RT-NFAGD can significantly reduce the crosstalk parasitic oscillation without affecting the high-speed switching (dv/dt close to 40V/ns).

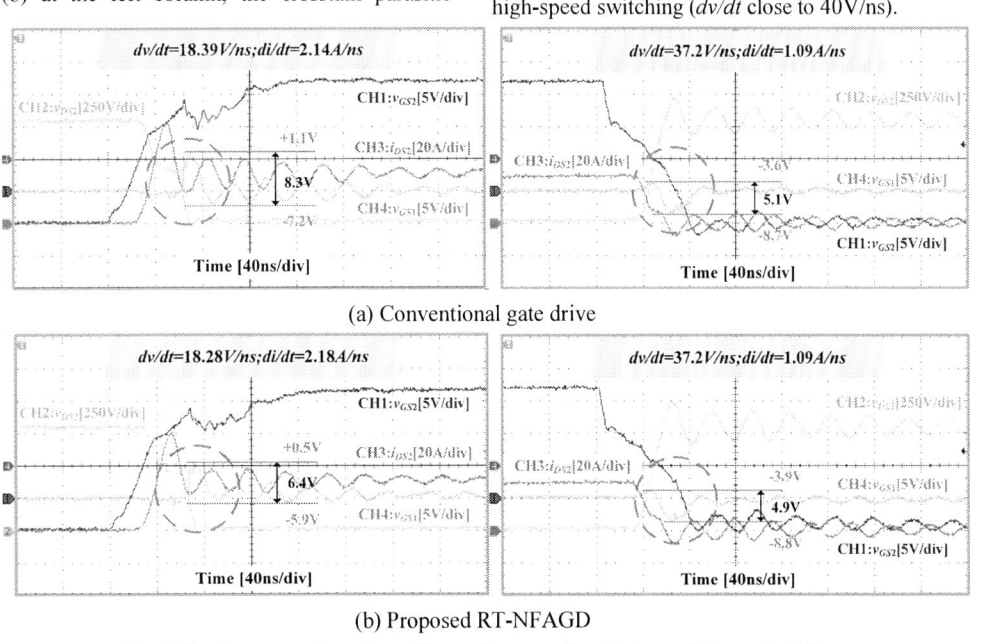

(a) Conventional gate drive

(b) Proposed RT-NFAGD

Fig. 4 Turning-on and turning-off waveforms under driving resistance R_g=2Ω

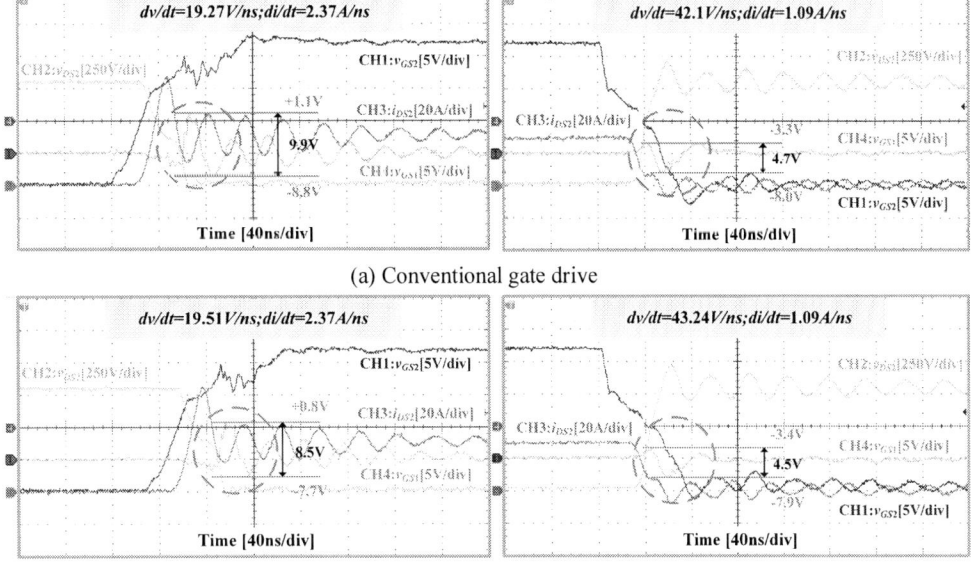

(a) Conventional gate drive

(b) Proposed RT-NFAGD

Fig. 5 Turning-on and turning-off waveforms under driving resistance R_g=1Ω

V. CONCLUSION

To address the issue of crosstalk parasitic oscillation in SiC MOSFET operating at high-speed switching, this paper proposes a suppression method based on the series resonance principle. The auxiliary capacitor is embedded in the forward loop of the gate drive. It is only enabled during the crosstalk parasitic oscillation duration and consumes only the reactive power without significant active power loss. Finally, the experimental comparison between the proposed RT-NFAGD and the conventional gate drive, the experimental results show that the proposed RT-NFAGD can significantly reduce the crosstalk parasitic oscillation without affecting the high-speed switching (dv/dt close to 40V/ns).

REFERENCES

[1] Z. Qian, J. Zhang and K. Sheng, "Status and Development of Power Semiconductor Devices and Its Applications," Proceed-ings of the CSEE, vol. 34, pp. 5149-5161, 2014.

[2] F. Wang and Z. Zhang, "Overview of Silicon Carbide Technology: Device, Converter, System, and Application," CPSS Transactions on Power Electronics and Applications, vol. 1, pp. 13-32, 2016.

[3] J. Chen, X. Du, Q. Luo, X. Zhang, P. Sun, and L. Zhou, "A Review of Switching Oscillations of Wide Bandgap Semiconductor Devices," IEEE Transactions on Power Electronics, vol. 35, no. 12, pp. 13182-13199, Dec. 2020.

[4] C. Li, Z. Lu, Y. Chen, C. Li, H. Luo, W. Li, and X. He, "High Off-State Impedance Gate Driver of SiC MOSFETs for Crosstalk Voltage Elimination Considering Common-Source Inductance," IEEE Transactions on Power Electronics, vol. 35, no. 3, pp. 2999-3011, March 2020.

[5] T. Xiong, Y. Yang, W. Xu, Y. Bi, X. Lin, T. Cao, and X. Yang, "An Active Crosstalk Suppression Driving Circuit for SiC MOSFET," Semiconductor Technology, vol. 49, no. 12, pp. 1135-1143, Dec. 2024.

[6] H. Li, Z. Qiu, H. Du, T. Shao, and Z. Wang, "Research on active gate driver for improving turn-off performance and gate voltage stability of SiC MOSFET in bridge circuit," Proc. CSEE, vol. 42, no. 21, pp. 7922–7933, Nov. 2022.

[7] B. Yang, F. Ding, C. Shen, M. Wang, S. Yang, P. Sun, and Z. Zhao, "Impact of Gate Resistance on Crosstalk in SiC MOSFET Half-Bridge Circuit," Semiconductor Technology, vol. 49, no. 12, pp. 1114-1120, Dec. 2024.

[8] H. Qin, W. Wang, F. Bu, Z. Peng, A. Liu and S. Bai, "Analysis of Crosstalk and Suppression Methods for Enhancement-Mode GaN HEMTs in A Phase-Leg Topology, " 2021 IEEE Workshop on Wide Bandgap Power Devices and Applications in Asia (WiPDA Asia),Wuhan,China,2021,pp. 147-152.

[9] D. Woldegiorgis, M. M. Hossain, Z. Saadatizadeh, Y. Wei, and H. A. Mantooth, "Hybrid Si/SiC Switches: A Review of Control Objectives, Gate Driving Approaches and Packaging Solutions," IEEE Journal of Emerging and Selected Topics in Power Electronics, vol. 11, no. 2, pp. 1737-1753, Apr. 2023.

[10] Z. Zeng and X. Li, "Comparative Study on Multiple Degrees of Freedom of Gate Drivers for Transient Behavior Regulation of SiC MOSFET," IEEE Transactions on Power Electronics, vol. 33, pp. 8754-8763, 2018.

[11] STMicroelectronics AN-5355. Mitigation Technique of the SiC MOSFET Gate Voltage Glitches with Miller Clamp, [Online]. Available: https://www.st.com (accessed on 18 March, 2020).

[12] T. Shao, T. Q. Zheng, H. Li, J. Liu, Z. Li, B. Huang, and Z. Qiu, "The Active Gate Drive Based on Negative Feedback Mechanism for Fast Switching and Crosstalk Suppression of SiC Devices," IEEE Transactions on Power Electronics, vol. 37, pp. 6739-6754, 2022.

979-8-3315-1110-4/25 $31.00 © 2025 IEEE

Effect of argon-oxygen ratios on crystallization and surface morphology of gallium oxide thin films by low-pressure chemical vapor deposition

Zhihao Yang
Department of Electronic Engineering
Xi'an University of Technology
Xi'an, China
18292469545@163.com

Jichao Hu*
Department of Electronic Engineering
Xi'an University of Technology
Xi'an, China
jchu@xaut.edu.cn

Jinhui Yu
Department of Electronic Engineering
Xi'an University of Technology
Xi'an, China
18292084680@163.com

Xiaodong Yang
Department of Electronic Engineering
Xi'an University of Technology
Xi'an, China
15239300254@163.com

Xiaomin He
Department of Electronic Engineering
Xi'an University of Technology
Xi'an, China
hexiaomin@xaut.edu.cn

Bo Peng
Key Laboratory of Wide Bandgap
Semiconductor Materials
Xidian University
Xi'an, China
boopeng@xidian.edu.cn

Abstract—β-Ga$_2$O$_3$ thin films were heteroepitaxially grown on AlN(0001) substrates via low-pressure chemical vapor deposition (LPCVD). A comprehensive investigation was conducted to characterize the crystalline structure, surface morphology, and chemical composition of both the deposited films and heterostructures. X-ray diffraction (XRD) analysis revealed that the β-Ga$_2$O$_3$ films exhibited phase purity with a highly preferred (-201) growth orientation. The surface topography evolution under various growth conditions was systematically examined using scanning electron microscopy (SEM) and atomic force microscopy (AFM). X-ray photoelectron spectroscopy (XPS) measurements demonstrated an optimal Ga/O atomic ratio of 1.5, confirming the formation of stoichiometric Ga-O bonding configurations. These findings substantiate the feasibility of fabricating high-quality β-Ga$_2$O$_3$ films with superior physical properties through LPCVD heteroepitaxy on AlN substrates, thereby establishing a crucial foundation for developing next-generation Ga$_2$O$_3$-based power electronic devices.

Keywords—LPCVD, heteroepitaxy, β-Ga$_2$O$_3$, AlN

I. INTRODUCTION

Boasting an ultra-wide bandgap (4.6-4.9eV) and exceptional Baliga's figure-of-merit (3444), β-Ga$_2$O$_3$ has emerged as a promising candidate for next-generation high-power electronic devices [1,2]. However, this material faces a critical challenge in thermal management due to its relatively low thermal conductivity (27 W/m·K) compared to conventional semiconductors like Si (150 W/m·K), 4H-SiC (490 W/m·K), and GaN (230 W/m·K) [3]. To address this limitation, aluminum nitride (AlN) presents an ideal heteroepitaxial substrate solution, offering dual advantages: First, the (-201) β-Ga$_2$O$_3$ and (0001) AlN planes exhibit remarkably low lattice mismatch (2.4%), facilitating high-quality epitaxial growth. Second, AlN's superior thermal conductivity (340 W/m·K) effectively compensates for β-Ga$_2$O$_3$'s thermal limitations [4,5]. In this study, we employed low-pressure chemical vapor deposition (LPCVD) for β-Ga$_2$O$_3$ film growth on AlN substrates, motivated by its high deposition rate and cost-effectiveness. While process parameters crucially influence film quality, the argon-to-oxygen ratio - a critical yet underexplored growth parameter -

remains insufficiently studied. We systematically investigated the effects of varying argon-oxygen ratios on the structural characteristics and surface morphology of β-Ga$_2$O$_3$ films. Through comprehensive XRD, AFM, SEM and XPS characterization, our work determines optimal process conditions that balance crystalline quality with surface uniformity.

II. EXPERIMENTAL DETAILS

Commercial AlN substrates were sequentially cleaned in acetone and anhydrous ethanol through ultrasonic agitation to remove surface contaminants. Residual liquids on substrates and crucibles were purged using high-purity nitrogen gas. Metallic gallium (99.999% purity) and oxygen served as precursors, with argon as the carrier gas. The evaporation temperature of Ga source and growth temperature of Ga$_2$O$_3$ were maintained at 950°C and 750°C, respectively. Under a constant Ar flow rate of 200 sccm, oxygen flow rates were systematically varied at 5, 10, 15, and 20 sccm. Film growth was conducted for 90 minutes under each Ar/O$_2$ ratio, followed by natural cooling to room temperature prior to sample extraction for characterization. The crystalline orientation and quality of β-Ga$_2$O$_3$ films were characterized using X-ray diffraction (XRD, SHIMADZU XRD-7000) with Cu-Kα radiation (λ = 0.154 nm). Surface morphology was examined by atomic force microscopy (AFM, Dimension Icon) and scanning electron microscopy (SEM, JEOL JSM-6700F). Surface chemical composition analysis was performed via X-ray photoelectron spectroscopy (XPS).

III. CHARACTERIZATION AND ANALYSIS

XRD tests were performed on the β-Ga$_2$O$_3$ thin films prepared at different argon-oxygen ratios, and the results are shown in Figure 1(a). The diffraction peaks with significant intensities appearing near 18.87°, 38.34°, 59.00° and 36.02° which correspond to the (-201), (-402), (-603) crystal surface of β-Ga$_2$O$_3$ and (0002) crystal plane of the AlN substrate,

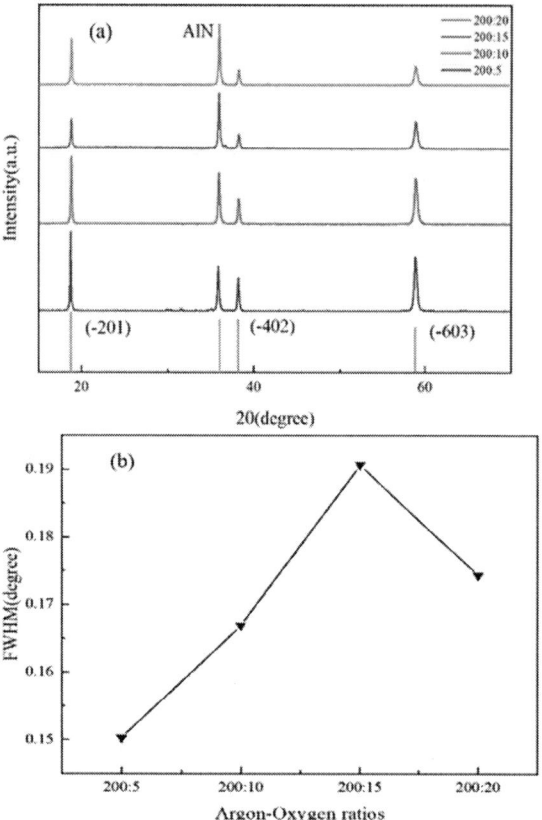

Fig. 1. (a)XRD curve of β-Ga₂O₃ thin film grown at different argon-oxygen ratio, (b)FWHM values of β-Ga₂O₃ thin film at argon-oxygen ratio.

respectively. No impurity peaks were detected. It demonstrates that the pure β-Ga$_2$O$_3$ films optimally grown along the (-201) crystal surface was obtained at different argon-oxygen ratio. The intensity of diffraction peak of β-Ga$_2$O$_3$ (-201) decreases and then increases with the increase of the oxygen flow rate increases from 5sccm to 20sccm. From Figure 1(b), it can be seen that a maximum FWHM value of 0.191 ° is obtained at an oxygen flow rate of 15sccm.

As comprehensively illustrated in Figure 2 through systematic scanning electron microscopy (SEM) investigations, the surface evolution mechanisms of β-Ga$_2$O$_3$ thin films exhibit pronounced dependence on oxygen flow rate variations. Under the initial deposition condition with an oxygen flux of 5 standard cubic centimeters per minute (sccm) (Figure 2a), discrete island-like nucleation domains are observed to initiate coalescence across the substrate surface[6]. This early-stage growth behavior results in the formation of discontinuous film structures characterized by incomplete inter-island connectivity. Progressive augmentation of oxygen supply induces measurable improvements in surface attributes. Specifically, when the oxygen flow rate is elevated to the range of 10–15 sccm (Figures 2b and 2c), a distinct transition toward enhanced morphological regularity becomes evident. The initially isolated nucleation islands undergo gradual planarization, accompanied by the development of continuous film matrices.

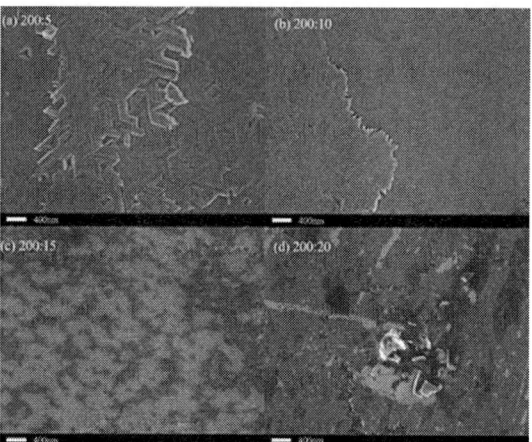

Fig. 2. SEM images of β-Ga₂O₃ thin film grown at different argon-oxygen ratio: (a)200:5, (b) 200:10, (c) 200:15, (d) 200:20.

Concurrent optimization of crystallographic alignment is achieved during this phase, as later corroborated by X-ray diffraction (XRD) crystallographic orientation analysis. Notably, surpassing the critical oxygen flux threshold of 15 sccm (Figure 2d) instigates detrimental growth phenomena. Aberrant crystalline development manifests as randomly distributed polyhedral grain clusters with pronounced geometric irregularity[7]. These structural anomalies lead to substantial degradation of surface uniformity, where aggregated particulate formations become visually predominant.

The results of AFM test the range from a 5 μm × 5 μm area of the film surface, which have been prepared under different argon-oxygen ratio from 200:5 to 200:20 is shown in Figure 3. Correspondingly, the surface RMS roughness of the four samples, the values are 21.45 nm, 1.79 nm, 4.202 nm, 5.776 nm respectively. As shown in the Figure 3, the surface roughness of the film changes with the change of oxygen flow rate, which is consistent with the SEM observation results. This indicates that increasing or decreasing the oxygen flow rate is not conducive to improving the uniformity of the film surface.

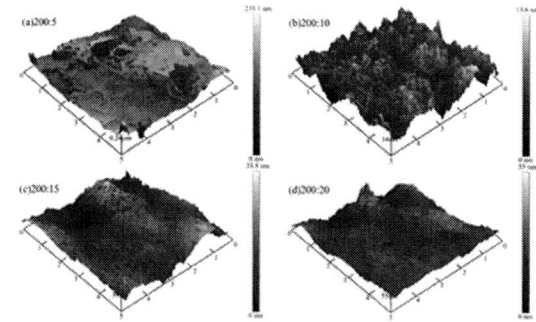

Fig. 3. AFM images of β-Ga₂O₃ thin film grown at different argon-oxygen ratio: (a)200:5, (b) 200:10, (c) 200:15, (d) 200:20.

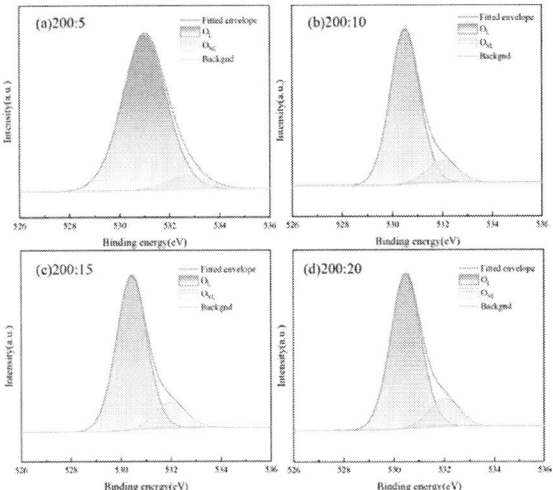

Fig. 4. Scanned fine spectra of O 1s of β-Ga₂O₃ thin films with different argon-oxygen ratio.

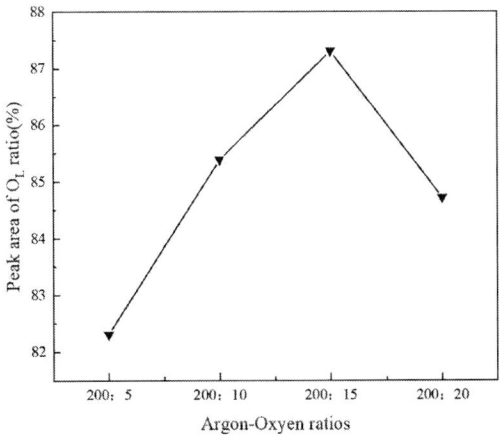

Fig. 5. The acquired XPS spectrum for the peak area ratio of O_L.

The surface chemical composition of β-Ga₂O₃ thin films was characterized by wide-scan X-ray photoelectron spectroscopy (XPS), as presented in Figure 4. The C 1s spectral line at 284.8 eV was used as the reference for all spectrum calibrations. The XPS spectra were used to label signals such as C 1s, O 1s, and Ga 3d, and the peak areas of the O 1s and Ga 3d signals were calculated. The Ga and O atomic ratios were obtained as 4.47, 1.43, 1.50, and 1.41, respectively. Further analysis of O 1s spectra using fine scanning. As shown in Figure5 (a), (b), (c) and (d), processing the asymmetric O 1s signal peak can be divided into lattice O and non lattice O. The signal peak with lower binding energy (530.1eV~531.03eV) belongs to lattice O bound to Ga atoms, while the signal peak with higher binding energy (531.52eV~532.56eV) belongs to non lattice O[8]. The presence of non-lattice O atoms indicated that the as-prepared β-Ga₂O₃ thin films were oxygen deficient, which confirmed the presence of oxygen defects and surface-adsorbed oxygen such as O–H or C–O[9]. This indicates that reducing the argon oxygen ratio appropriately can provide more oxygen elements and improve the problem of oxygen vacancies in the film. However, a too low argon oxygen ratio provides more oxygen to participate in the reaction, but the content of oxygen vacancies does not decrease, thus affecting the quality of the film.

IV. CONCLUSION

β-Ga₂O₃ thin films were deposited on AlN substrates via low-pressure chemical vapor deposition (LPCVD) under four Ar/O₂ flow ratios: 200:5, 200:10, 200:15, and 200:20. Characterization results demonstrate that the Ar/O₂ ratio significantly influences the crystal structure, surface morphology, and chemical composition of β-Ga₂O₃ films. X-ray diffraction (XRD) analysis reveals maximum diffraction peak intensity at the 200:5 ratio with a full width at half maximum (FWHM) of 0.151°. Furthermore, the crystalline quality of β-Ga₂O₃ films deteriorates with increasing oxygen proportion. Scanning electron microscopy (SEM) and atomic force microscopy (AFM) results indicate that the film surface becomes progressively smoother with higher oxygen ratios, accompanied by significantly reduced surface roughness. At 200:10 and 200:15 ratios, the root mean square (RMS) roughness values reach 1.79 nm and 4.21 nm, respectively, corresponding to relatively flat surface morphologies. However, excessive oxygen content (200:20) leads to decreased film densification. X-ray photoelectron spectroscopy (XPS) analysis confirms the coexistence of two Ga oxidation states and non-lattice oxygen in the films, evidencing oxygen vacancy formation. The proportion of Ga³⁺ increases steadily with oxygen enrichment, while non-lattice oxygen content decreases correspondingly. These observations suggest gradual quality improvement with oxygen ratio elevation. Notably, the calculated stoichiometric ratio reaches 1.50 at the 200:15 condition, approaching the optimal value. This work provides fundamental guidelines for fabricating high-quality β-Ga₂O₃ films on AlN substrates, laying a foundation for future device applications.

ACKNOWLEDGMENT

This work was supported by the National Natural Science Foundation of China (Grant No.62474139 and 61904146); Science and technology planning project of Xi'an (No.2023JH-GXRC-0122).

REFERENCES

[1] Hiroyuki Nishinaka, Tatsuji Nagaoka, Yuki Kajita, et al. Materials Science in Semiconductor Processing. 2021.Volume 128. 105732. p.2

[2] Michele Baldini, Zbigniew Galazka, Günter Wagner. Materials Science in Semiconductor Processing. 2018. Volume 78. p.2

[3] Tao Zhang, Yifan Li, Qian Feng, Yachao Zhang, Jing Ning, Chunfu Zhang, Jincheng Zhang, Yue Hao. Materials Science in Semiconductor Processing. 2021. Volume 123. 105572. p.1

[4] P. Zhang et al. IEEE Transactions on Electron Devices. 2024. doi: 10.1109

[5] Li Y, Zhang Y, Zhang J, Zhang T, Xu S, Feng L, et al. I Semiconductor Science and Technology. 2022.Volume 37. 095004.

[5] Wheeler V D, Nepal N, Boris D R, et al. Chemistry of Materials. 2020. Volume 32. 1140

[6] W. Li, K. Nomoto, Z. Hu, D. Jena, H.G. Xing, Fin-channel orientation dependence of forward conduction in kV-

class Ga_2O_3 trench Schottky barrier diodes, APEX 12 (2019) 061007.

[7]Zhang. T, Li. Y, Cheng. Q, et al., Research on the Crystal Phase and Orientation of Ga_2O_3 Hetero-Epitaxial Film, Superlattice. Microstruct. 159(2021) 107053.

[8]Hu. J, Zhang. K, Yang. X, et al., Effect of Growth Temperature on Properties of β- Ga_2O_3 Films Grown on AlN by Low-Pressure Chemical Vapor Deposition. Journal of Luminescence, 274(2024) 120709.

[9] W. Yu, Q. Gui, X. Wan, J. Robertson, et al., High-throughput interface prediction and generation scheme: The case of β- Ga_2O_3/AlN interfaces, Appl. Phys. Lett. 123 (2023) 161601.

Design of an Active Gate Driver IC for E-Mode GaN HEMTs Based on Adaptive Phased Regulation

Wei Liu
Department of Automation and
Information Engineering
Xi'an University of Technology
Xi'an,710048, China
liuwei331226@outlook.com

Yuan Yang
Department of Automation and
Information Engineering
Xi'an University of Technology
Xi'an,710048, China
yangyuan@xaut.edu.cn

Wenbin Xing
Department of Automation and
Information Engineering
Xi'an University of Technology
Xi'an,710048, China
xing_wen_bin@163.com

Abstract—This paper proposes an active gate driver IC design for enhancement-mode gallium nitride high-electron-mobility transistors (GaN HEMTs), specifically addressing the false turn-on issue induced by cross-talk between bridge arms in high-frequency switching applications. To tackle this challenge, two innovative techniques are introduced: an adaptive drain-source voltage slew rate(dV_{DS}/dt) regulation method and a negative voltage self-recovery scheme. The former dynamically adjusts the gate-source voltage(V_{GS}) to align with the dV_{DS}/dt variation, effectively reducing the dV_{DS}/dt from 244 V/ns to 124 V/ns, thereby suppressing cross-talk effects. The latter applies a negative voltage to the lower switch during the upper switch's turn-on transient to block forward cross-talk and automatically restores a zero-voltage state after the upper switch is fully activated, achieving a 104.1 ps reduction in turn-on propagation delay. The proposed design provides an effective solution to enhance the reliability and performance of GaN power devices in high-frequency applications.

Keywords—Enhancement-mode GaN HEMT; active gate driving; segment-independent control; adaptive dV_{DS}/dt

I. INTRODUCTION

In recent years, with growing global attention on third-generation wide-bandgap(WBG) semiconductor technologies, GaN HEMT driver ICs and high-power-density power modules have emerged as critical enablers for green energy production, efficient power distribution, automotive electrification, and energy consumption reduction in manufacturing[1,2].Under this trend, the importance of GaN power device driver ICs has become increasingly prominent.

Conventional gate drive(CGD) employs fixed resistors to apply voltage step functions to the gate [3], whereas active gate drive(AGD) dynamically modulates the gate resistance, gate voltage, or gate current during turn-on/turn-off transients. This driving paradigm can be categorized into open-loop and closed-loop control. As demonstrated in [4,5],open-loop methods predict the gate voltage(V_{GS}) of GaN HEMTs at the Miller plateau's initiation and termination, dividing the turn-on process into multiple phases driven by varying current intensities. However, open-loop control requires a trial-and-error process to determine optimal driving patterns, which is laborious and often lacks precision. Closed-loop approaches in [6,8] utilize real-time operational state detection for adaptive regulation. Unfortunately, existing closed-loop systems demand extremely high bandwidth for nanosecond-level switching in enhancement-mode GaN HEMT applications. Due to inherent loop delays (difficult to achieve sub-nanosecond resolution),timely

and effective control of enhancement-mode GaN HEMTs remains challenging.

This paper introduces a discrete-time feedback technique that enables segmented cycle-by-cycle adaptive regulation by modulating the V_{GS} of enhancement-mode GaN HEMTs. This method avoids the need for high-speed feedback loops, achieves timely control of the drain-source voltage slew rate (dV_{DS}/dt), effectively suppresses voltage overshoot, and employs negative-voltage turn-off to mitigate cross-talk.

II. BRIEF REVIEW OF PRIOR ART

Current research on active gate driving circuits for GaN devices has proliferated, with several prominent methodologies emerging. These include closed-loop control approaches that detect and feedback the switching states of GaN devices to the driving stage for regulation, and open-loop methods that employ pre-optimized driving strategies stored in memory for direct GaN device control.

Reference [9] presents a 600V GaN active gate driver incorporating Dynamic Feedback Delay Compensation (DFDC) technology to optimize switching transient control, reducing energy losses and electromagnetic interference (EMI). The design features a Miller Plateau (MP) detector for real-time dv/dt transient monitoring, coupled with DFDC technology for dynamic phase adjustment of control signals to compensate for feedback delays. A segmented output stage and phase modulation circuit inject high driving current during the initial switching phase to minimize delay, while dynamically adjusting driving current during the MP phase to suppress dv/dt slope and reduce voltage-current overlap losses.

Reference [10] introduces a 6.7 GHz active gate driver designed to optimize switching waveforms of GaN, addressing issues of overshoot, ringing, and EMI in high-frequency switching operations. The driver features high temporal resolution (150ps) and a wide output resistance range (0.12-64Ω) for dynamic gate resistance adjustment, enabling sub-nanosecond resolution optimization of GaN gate waveforms. While the article presents an active gate driving strategy, practical implementation of real-time gate resistance sequence optimization faces challenges, particularly under dynamic load and operating conditions, necessitating the development of more sophisticated algorithms and control logic for diverse operational scenarios.

Reference [11] proposes an intelligent gate driving `generator to address the trade-off between gate voltage overshoot and switching losses in E-mode GaN HEMTs. The design simplifies control processes by systematically generating dynamic gate driving strength patterns through external bias resistors that adjust the timing of current-controlled delay chains, overcoming the complexity of trial-and-error methods in existing active driving technologies.

Through this analysis, it becomes evident that crosstalk suppression and high dv/dt control are critical aspects in the design of active drivers for GaN half-bridge configurations during switching transitions. In response to these challenges, this paper proposes an adaptive segmented control scheme for enhanced-mode GaN HEMT active gate driver ICs, aiming to achieve timely control of dV_{DS}/dt, effective overshoot suppression, and crosstalk mitigation through negative voltage turn-off. The proposed architecture maintains the intrinsic advantages of GaN technology while addressing the critical challenges associated with high-speed switching operations in power electronic systems.

III. PROPOSED GATE DRIVER ARCHITECTURE AND SWITCHING TRANSIENT ANALYSIS

This paper proposes an adaptive multi-level voltage driving scheme. During the dV_{DS}/dt phase of turn-on for enhancement-mode GaN HEMTs, the scheme detects the rate of change of the power device's drain-source voltage and sends this information back to the control terminal. This enables adaptive adjustment of the driving voltage to regulate dV_{DS}/dt. During the turn-off phase, the scheme eliminates the need for an additional negative power rail by generating negative voltage to turn off GaN power devices, thereby preventing false turn-on caused by crosstalk.

The overall architecture shown in Fig 1 comprises three main components: a drive circuit, a dV_{DS}/dt detection circuit, and a logic control circuit. The main driver employs 5V PMOS and NMOS power transistors, while the auxiliary driver utilizes 2V PMOS and NMOS power transistors. Subsequent sections will detail the transient analysis of the proposed drive circuit operation and device design considerations.

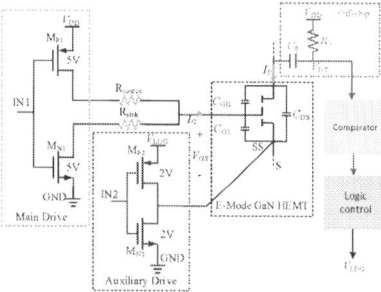

Fig 1. Schematic diagram of the adaptive multi-stage voltage drive scheme

The turn-on process of enhancement-mode GaN HEMTs is divided into four stages, which will be explained below in conjunction with the control strategy proposed in this paper. Fig 2 illustrates the waveforms of the driving voltage V_{DR}, gate-source voltage V_{GS}, drain current I_D, and drain-source voltage V_{DS} during the staged turn-on process, demonstrating the implementation of the control strategy.

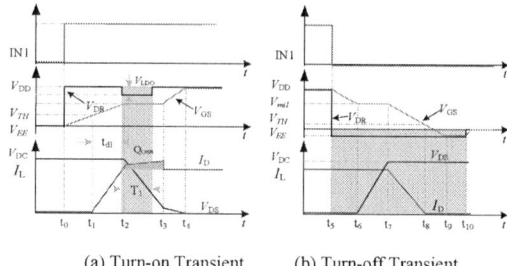

(a) Turn-on Transient   (b) Turn-off Transient

Fig 2. Transient signal waveforms during conduction

A. Turn-on Transient (Fig 3(a)(b))

- t_0–t_1: This is the turn-on delay stage. The enhancement-mode GaN HEMT has not yet activated, and no switching occurs. Since there are no di/dt or dV_{DS}/dt effects caused by excessive driving strength, the goal is to minimize losses by shortening this stage's duration. A maximum gate driving voltage of 5 V is applied to accelerate the turn-on of the GaN power device.

- t_1–t_3: This is the current rise and voltage fall stage. To avoid rapid charging of the GaN power device due to a high driving voltage V_{DR}, the proposed strategy introduces a delay period td1td1 after the gate-source voltage V_{GS} reaches the threshold voltage V_{TH}. During this period, the source voltage of the enhancement-mode GaN HEMT is adaptively adjusted based on the detected dV_{DS}/dt. This dynamic regulation lasts for T1, reducing the gate-source voltage V_{GS} to control dV_{DS}/dt.

- t_3–t_4: This is the full turn-on stage. The gate-source voltage V_{GS} rises to the supply voltage VDD (5 V) and remains constant. During this stage, the drain current I_D and drain-source voltage V_{DS} remain relatively stable with no voltage or current overshoot. To minimize turn-on losses, a high driving voltage (5 V) is maintained to fully activate the GaN power device, thereby reducing the duration of this stage.

B. Turn-off Transient (Fig 3(c)(d))

- t_5–t_9: This is the device turn-off stage. When the input signal IN1 transitions to a low level at t5, a negative voltage is generated to initiate the discharge of the input capacitance C_{ISS}. The gate-source voltage V_{GS} begins to drop from 5 V until the GaN power device is fully turned off at t9. To achieve rapid turn-off of the enhancement-mode GaN HEMT, a negative voltage is applied to enhance the sink current capability.

- t_9–t_{10}: This is the crosstalk immunity stage. The negative voltage is maintained to keep the enhancement-mode GaN HEMT in the off state, ensuring that V_{GS} remains significantly below the threshold voltage V_{TH}. This provides a sufficient margin to prevent unintended turn-on caused by crosstalk, thereby improving gate robustness.

- t_{10}–: This is the negative voltage self-recovery stage. After the upper switch Q_H completes its turn-on, the negative voltage of the lower switch Q_L is reset to 0 V, as shown in Fig 3(d). By activating transistors M_{N1} and M_{N2}, a 0 V driving voltage is generated, placing the enhancement-mode GaN HEMT in a pre-turn-on state.

This work was supported in part by the National Natural Science Foundation of China under Grants 62174134, in part by the Shaanxi Innovation Capability Support Project under Grant 2021TD-25, and in part by the Xi'an key industrial chain key core technology research projects under Grant 23LLRH0044.

This reduces turn-on propagation delay and increases the safety margin against negative crosstalk.

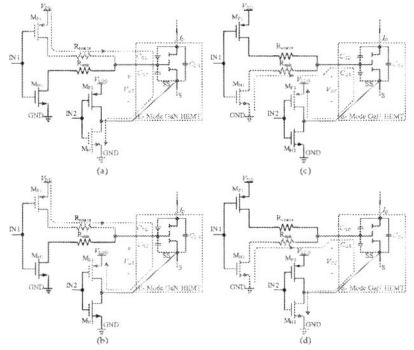

Fig 3. Operating principle of the proposed AGD circuit

This work addresses the false turn-on issue caused by crosstalk in enhancement-mode GaN HEMTs through two integrated strategies. First, an adaptive dV_{DS}/dt-based regulation dynamically adjusts the gate-source voltage V_{GS} based on the detected dV_{DS}/dt, directly controlling dV_{DS}/dt to suppress false triggering during switching transitions. Second, a negative voltage self-recovery mechanism is implemented: during the turn-on of the QH, a negative voltage is applied to the QL to block positive crosstalk from QH; once QH is fully turned on, QL's negative voltage is autonomously reset to 0 V. This approach not only effectively mitigates crosstalk-induced false turn-on but also reduces turn-on propagation delay and enhances the safety margin against negative crosstalk. By combining real-time V_{GS} adjustment with intelligent voltage recovery, the proposed solution eliminates the need for additional negative power rails while optimizing switching performance, loss reduction, and reliability in high-frequency GaN-based power systems.

IV. SIMULATION AND EXPERIMENT

The proposed circuit was simulated and experimentally tested to verify the suppression effectiveness of the driving scheme on half-bridge configurations. The GaN HEMT device selected for simulation and testing was the GS66504B from GaN Systems. For the high-side switch, a commercial driver IC (SI8271GB-ISR) was used to drive the enhancement-mode GaN HEMT, while the low-side switch employed the custom-designed driver IC proposed in this work to drive the same device.

A. Simulation Verification

Fig 4 compares the crosstalk effects on the upper switch under CGD and AGD schemes when the upper switch is turned off with voltages of 0 V and -1.2 V, respectively, while the lower switch operates normally. The results demonstrate how the proposed AGD effectively mitigates crosstalk-induced voltage disturbances on the upper switch by dynamically adjusting the driving strategy and optimizing the negative turn-off voltage.

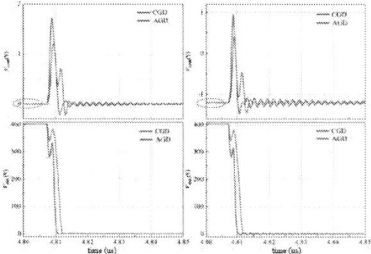

Fig 4. Conventional gate drive and active gate drive upper tube for 0V, -1.2V shutdown voltage

As shown in the Fig 4, when the upper switch is turned off with voltages of 0 V and -1.2 V, Proposed AGD scheme exhibits a significantly lower peak in the upper switch's V_{GSH} compared to CGD. Furthermore, the AGD achieves superior control over the dV_{DS}/dt relative to the CGD, as quantitatively detailed in Table 1. These results validate the effectiveness of the proposed adaptive regulation and negative voltage recovery mechanisms in suppressing crosstalk-induced disturbances and optimizing switching dynamics.

TABLE I. CONVENTIONAL GATE DRIVE AND ACTIVE GATE DRIVE UPPER TUBE FOR 0V, -1.2V SHUTDOWN VOLTAGE

	CGD		Proposed AGD	
Q_H Shutdown voltage(V)	0	-1.2	0	-1.2
dV_{DS}/dt (V/ns)	244	230	124	122
V_{GSH}(V)	1.726	0.9405	1.203	0.4035
Turn-on or off	Turn-on	Soon to be connected	Turn-on	Turn-off

Experimental results demonstrate that the proposed AGD achieves 30.3% and 57.1% improvement in crosstalk suppression under 0 V and -1.2 V turn-off voltages, respectively, compared to CGD, further validated by DPT platform using the GS66504B GaN HEMT.

B. Experimental Validation

The entire circuit layout was designed using the TSMC 0.18-μm process and fabricated. The test circuit for the custom-designed driver IC is shown in Fig 5, where V_{DC} represents the bus voltage and C_{DC} denotes the bus capacitor. For simulation and testing, the GaN HEMT device selected was the GS66504B from GaN Systems. The high-side switch employed a commercial driver IC (SI8271GB-ISR) to drive the enhancement-mode GaN HEMT, while the low-side switch utilized the custom-designed driver IC proposed in this work. A photograph of the fabricated test chip is provided in Fig 6.

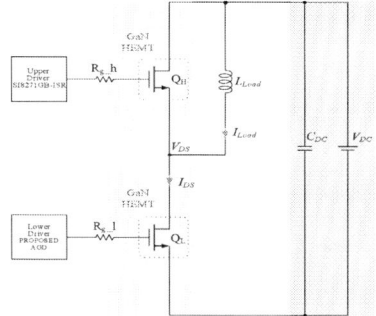

Fig 5. Switching waveform test diagram of proposed AGD

Fig 6. Chip application test diagram

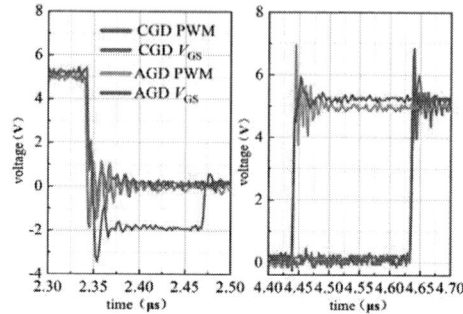

Fig 9. Comparison of active gate drive and conventional gate drive gate source voltage V_{GS} testwaveforms

a) Switching Transient Test. Fig 7 shows the switching waveform test diagram of the active gate drive. It can be seen that during the turn-off process, negative voltage can be used for turn-off, and it can return to zero voltage after a certain period of time, so the self-recovery negative voltage function works normally. Fig 8 shows the amplified waveform diagram of the active gate drive turn-on and turn-off. It can be seen that the signal propagation delays are all less than 15ns, meeting the design specifications.

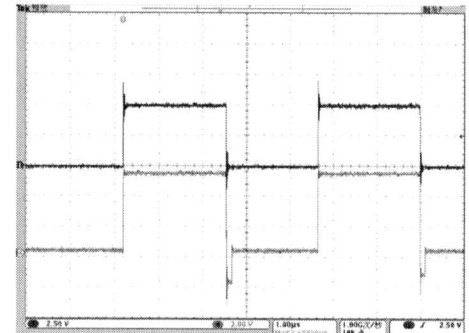

Fig 7. Switching waveform test diagram of proposed AGD

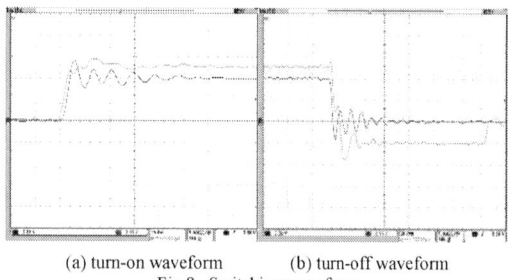

(a) turn-on waveform   (b) turn-off waveform
Fig 8. Switching waveform

b) Crosstalk Suppression Test. Fig 9 presents the comparative test waveforms of the V_{GS} for AGD and CGD. The left - hand side shows the comparison during turn - off, and the right - hand side during turn - on. It is evident that during turn - on, the V_{GS} overshoot of AGD is smaller than that of CGD. Thus, the driver IC designed in this paper can suppress crosstalk generated during switching.

From the above test results, it can be seen that the adaptive dV_{DS}/dt regulation technology and the negative voltage self - recovery method proposed in this paper can reduce crosstalk without sacrificing performance.

V. Conclusion

This paper addresses the issue of spurious turn-on caused by crosstalk in high-frequency applications of enhanced GaN HEMT, and proposes an innovative design scheme for an active gate driver chip. Through the dynamic regulation technology of the drain-source voltage change rate dV_{DS}/dt, combined with the negative voltage self-recovery method, the crosstalk effect between the upper and lower bridge arms is effectively suppressed. Based on the TSMC 0.18μm CMOS process, the chip integrates key modules such as a bandgap reference, a tunable LDO, and a logic control unit, and its performance is verified through double-pulse testing. The experimental results show that the proposed method significantly reduces the dV_{DS}/dt from 244 V/ns in the traditional driver to 124 V/ns, and at the same time shortens the transmission delay by 104.1 ps. Moreover, the overall area of the chip is only 770μm×751μm. This design provides a reliable solution for high-power density power supply systems, and has important application potential especially in high-frequency scenarios such as electric vehicles and data centers. Future work will further optimize the integrated protection function and explore the expansion of the full-bridge driver architecture to enhance the robustness and versatility of the system.

References

[1] Ramachandran R, Nymand M. Experimental demonstration of a 98.8% efficient isolated DC–DC GaN converter[J].IEEE Transactions on Industrial Electronics, 2016, 64(11):9104-9113.

[2] GaN Systems.The 2020 top Technology Trends in Power[EB/OL]. https://gansystems.com/newsroom/2020-trends-data-and-power/,2019-12-09.

[3] Dymond H C P, Wang J, Liu D, et al. A 6.7-GHz active gate driver for GaN FETs to combat overshoot, ringing, and EMI[J]. IEEE Transactions on Power Electronics, 2018,33(1): 581-594.

[4] Ke X, Sankman J, Chen Y, et al. A Tri-Slope Gate Driving GaN DC-DC Converter With Spurious Noise Compression and Ringing Suppression for Automotive Applications[J]. IEEE Journal of Solid-State Circuits, 2018, 53(1): 247-260.

[5] J. Wang et al., "Infinity Sensor: Temperature Sensing in GaN Power Devices using Peak di/dt," 2018 IEEE Energy Conversion Congress and Exposition (ECCE), Portland, OR, USA, 2018, pp. 884-890, doi: 10.1109/ECCE.2018.8558287.

[6] Zhang W J, Yu J, Cui W T, et al. A Smart Gate Driver IC for GaN Power HEMTs With Dynamic Ringing Suppression[J]. IEEE Transactions on Power Electronics, 2021, 36(12):14119-14132.

[7] Chen Y, Ma D B. A 10-MHz Closed-Loop EMI-Regulated GaN Switching Power Converter Using Emulated Miller Plateau Tracking and Adaptive Strength Gate Driving[J].IEEE Journal of Solid-State Circuits, 2021, 56(2): 531-540.

[8] Sun B, Burgos R, Zhang X, et al. Active dv/dt control of 600V GaN transistors[C]//2016 IEEE Energy Conversion Congress and Exposition (ECCE). IEEE, 2016: 1-8.

[9] J. Zhu et al., "33.2 A 600V GaN Active Gate Driver with Dynamic Feedback Delay Compensation Technique Achieving 22.5% Turn-On Energy Saving," 2021 IEEE International Solid-State Circuits Conference (ISSCC), San Francisco, CA, USA, 2021, pp. 462-464, doi: 10.1109/ISSCC42613.2021.9365974.

[10] H. C. P. Dymond et al., "A 6.7-GHz Active Gate Driver for GaN FETs to Combat Overshoot, Ringing, and EMI," in IEEE Transactions on Power Electronics, vol. 33, no. 1, pp. 581-594, Jan. 2018, doi: 10.1109/TPEL.2017.2669879.

[11] W. J. Zhang, J. Yu, Y. Leng, W. T. Cui, G. Q. Deng and W. T. Ng, "A Segmented Gate Driver for E-mode GaN HEMTs with Simple Driving Strength Pattern Control," 2020 32nd International Symposium on Power Semiconductor Devices and ICs (ISPSD), Vienna, Austria, 2020, pp. 102-105,doi:10.1109/ISPSD46842.2020.9170108.

A Vertical Ultra-Thin Voltage Regulator Module (VRM) with Dual-Phase Integrated Inductor for High-Performance Computing Applications

Nianzheng Wang
School of Electrical Engineering
Xi'an Jiaotong University
Xi'an, China
2204411648@stu.xjtu.edu.cn

Hang Kong
School of Electrical Engineering
Xi'an Jiaotong University
Xi'an, China
konghang@stu.xjtu.edu.cn

Laili Wang
School of Electrial Engineering
Xi'an Jiaotong University
Xi'an, China
llwang@mail.xjtu.edu.cn

Abstract—**As future high-performance computing systems trend toward requiring substantially higher power levels, vertical power delivery architectures have been recognized as a compelling solution for achieving next-generation power density specifications. This article presents the development of a vertical ultra-thin 6-1 V voltage regulator module (VRM) with dual-phase integrated inductor, achieving compact dimensions of 9 mm × 10 mm × 4.8 mm and demonstrating a remarkable power density of 180 W/cm³. The implemented design features per-phase operational parameters of 108 nH inductance and 40 A current capacity. Simulation analyses validate the module's enhanced electrical characteristics and thermal management capabilities, confirming its effectiveness in high-density power conversion applications. Experimental measurements of the prototype show a stable output voltage of 720 mV, though noticeable output current ripple persist.**

Keywords—vertical power delivery architectures, VRMs, dual-phase integrated inductor, high-performance computing applications

I. INTRODUCTION

As future high-performance computing systems trend toward requiring substantially higher power levels, the implementation of high-power-density and high-efficiency power delivery architecture is becoming increasingly critical. This necessitates that advanced voltage regulator modules (VRMs) transform the input voltage of the motherboard (more than 32V) to sub-1V powering CPUs or GPUs, which require extreme current delivery ranging from hundreds to thousands amperes within a chip area of mere square centimeters. Vertical VRMs have emerged as a promising approach for achieving next-generation power density specifications, which offer distinct advantages in reducing power losses and enhancing transient response, satisfying the demands of contemporary CPUs or GPUs [1], [2].

Conduction and switching losses inherent in traditional horizontal layouts are significantly mitigated by vertical power delivery. In conventional designs, elongated PCB traces between VRMs and load introduce large parasitic resistance and inductance, which exacerbate high I^2R losses during high-current operation. Vertical integration shortens current pathways, minimizing trace resistance and associated joule heating in the power delivery network [3]. This efficiency gain is critical for energy-sensitive applications, such as mobile devices and data centers.

The dynamic response of a VRM—its ability to maintain voltage stability during rapid load transients—is directly influenced by layout optimization. Vertical configurations

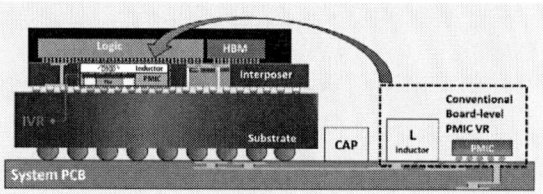

Fig. 1. Conceptual diagram of the integrated voltage regulator (IVR) proposed by TSMC.

Fig. 2. Vertical profile comparison of state-of-the-art VRMs (Infineon TDM22545D, MPS MPC22161-120, ADI LTM4664).

reduce loop inductance in the power delivery network (PDN), which improves the control loop bandwidth. Lower inductance decreases the voltage droop during sudden current demands, enabling faster correction with minimal voltage droop and overshoot, ensuring stable operation of the computing system [4].

As shown in Fig. 1, compared with the conventional board-level VR, integrated voltage regulator (IVR) needs to meet height constraint to be integrated into the interposer. Minimizing the vertical height of VRM, whose dominant factor is the thickness of inductors in interleaved multiphase buck converters, is imperative to enable seamless integration within the interposer layer. However, limiting inductor height typically constrains its inductance value, which directly governs transient and steady-state ripple performance. Consequently, optimizing both the inductor's design and its interconnection layout with substrates to achieve minimal thickness while satisfying inductance value and DC resistance (DCR) requirements is essential for compact, high-efficiency power delivery.

As illustrated in Fig. 2, the TDM22545D from Infineon exhibits a profile height of 5 mm, while the MPC22161-120 from Monolithic Power Systems (MPS) measures 7.65 mm in vertical dimension, and Analog Devices Inc.'s (ADI) LTM4664 demonstrates a height of 7.72 mm. In these state-of-the-art power modules, the thickness of magnetic components constitutes the critical design parameter that fundamentally constrains the overall vertical profile of the system [5].

Magnetic integration and the design of coupled inductors are currently the focus of research efforts aimed at reducing the thickness of magnetic components, specifically displacing discrete inductors with coupled magnetic assemblies in interleaved multiphase buck converters. Such coupled inductors demonstrate dual operational advantages: enhanced di/dt capability during transient phases for rapid load regulation, coupled with attenuated current ripple in steady-state operation [6]-[9]. For instance, W. Ping et al. proposed a multistack switched-capacitor point-of-load (MSC-PoL) VRM module. By stacking multiple H-bridge switched-capacitor units and designing coupled magnetic components, the module achieves a height of 6 mm, enabling it to be embedded in CPU or Chiplet sockets [10].

The rest of this paper is organized as follows. Section II details the circuit topology, the VRM module and the integrated inductor configuration. Section III presents comprehensive simulation results encompassing electrical, magnetic and thermal analyses. Section IV describes the vertical VRM with dual-phase integrated inductor prototype and electrical experimental results. The paper concludes with Section V.

II. TOPOLOGY AND STRUCTURE

A. Circuit Topology

This article demonstrates a vertical ultra-thin 6-1 V voltage regulator module featuring a dual-phase integrated inductor, with multi-module integration capability as illustrated in Fig. 3. The proposed converter utilizes a dual-phase buck topology comprising two half-bridge units (S11-S12 and S13-S14) operating with 180° phase-shifted switching sequences. Voltage regulation is achieved through duty cycle adjustment, while parallel module operation enables scalable power delivery under heavy loads.

B. VRM Module

As depicted in Fig. 4, the PCB layout incorporates critical components including input capacitors, DrMOS chips (MP86936), output capacitors, bonding pads, and signal traces. Input and output capacitors are placed near chip terminals to minimize parasitic effects, with inductor windings symmetrically positioned on both sides of the DrMOS chips. Considering the overall module dimensions of 9 mm × 10 mm × 4.8 mm and an output power of 80 W, the module achieves a remarkable power density of 180 W/cm³.

C. Integrated Inductor Configuration

As illustrated in Fig. 5, the dual-phase integrated planar inductor assembly maintains a total vertical profile of 4 mm. The magnetic core is fabricated by compacting magnetic powder core material, with two copper windings embedded within its central structure. The magnetic powder core material used to form the core is a high flux alloy composed of iron-nickel powder (DH060 [11]). DH060 exhibits superior DC bias capability, achieving a saturation flux density of up to 1.6 T.

The inductor's configuration demonstrates significant advantages compared to current state-of-the-art VRMs. By employing a single-turn copper winding and magnetic

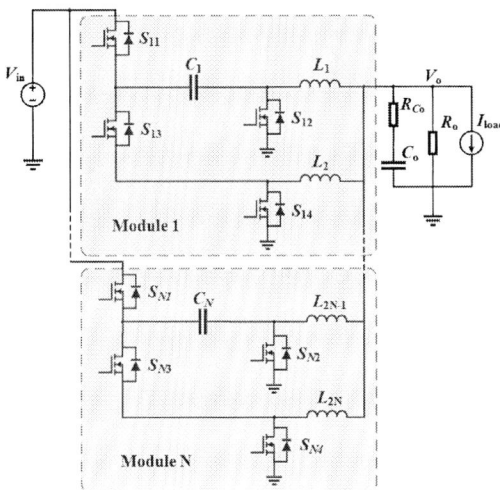

Fig. 3. Topology of multi-module dual-phase buck converter.

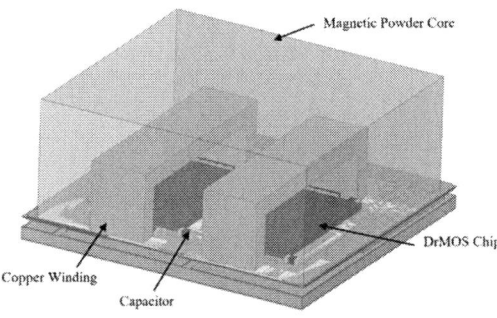

Fig. 4. 3-D stacked packaging of the vertical VRM. The PCB area is 10 mm × 9 mm, and the total VRM height is only 4.8 mm.

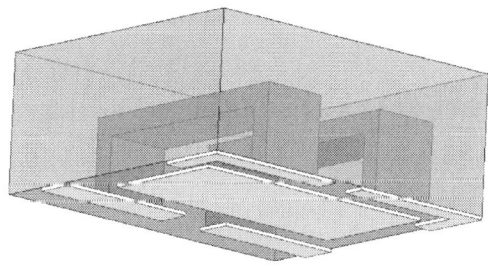

Fig. 5. Dual-phase integrated planar inductor schematic: copper windings (dark gray); magnetic powder core (light gray); solder pads (white) with 4 mm vertical profile.

powder core that can be tightly integrated through compression molding process, the design achieves an inductance value of 108 nH at merely 4 mm height. Furthermore, the innovative structure allows press-formed thermal pads on the inductor's exterior surface, effectively establishing additional vertical thermal dissipation pathways.

III. SIMULATION AND ANALYSIS

This work validates the proposed structure through LTspice, ANSYS Maxwell, and ANSYS Icepak simulations, with the analysis results presented below.

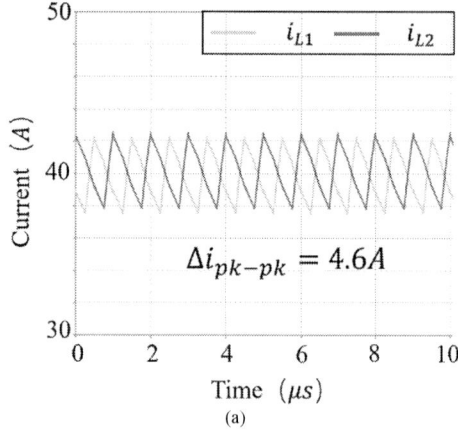

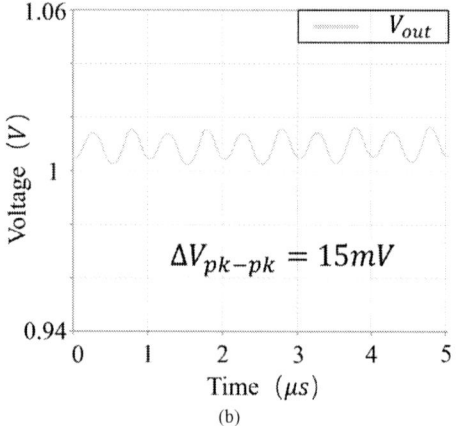

Fig. 6. (a) Steady-state phase-current waveform for the dual-phase integrated inductor. (b) Steady-state output voltage waveform. $V_{in} = 6$ V, $V_{out} = 1$ V, $I_{out} = 80$ A, $f_s = 1$ MHz, $C_{out} = 30$ μF, $L_1 = L_2 = 108$ nH and $k = 0.08$.

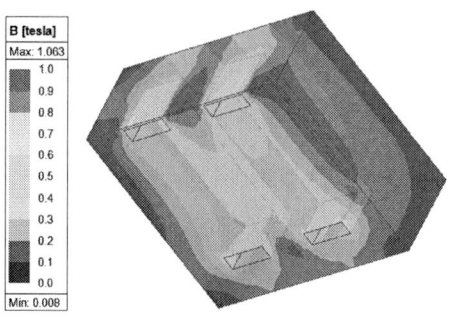

Fig. 7. ANSYS Maxwell 3D Magnetostatic simulation of the DC flux density distribution when supporting 40 A average current per phase (80 A in total).

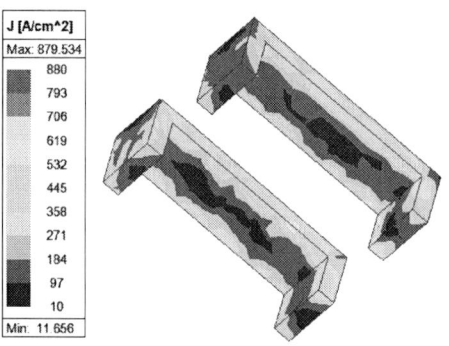

Fig. 8. ANSYS Maxwell 3D Eddy Current simulation of the current density distribution of two windings. Each winding is excited with a sinusoidal AC current of 4.6 A peak-to-peak at a frequency of 1 MHz.

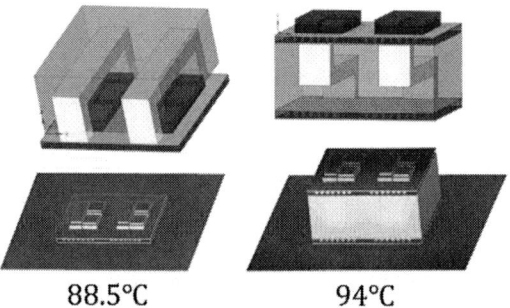

Fig. 9. ANSYS Icepak thermal simulation of two vertical magnetically integrated configurations (left: integrated inductor positioned above, right: integrated inductor positioned below) under forced air cooling. $P_{high-side} = 0.9$ W, $P_{low-side} = 1.1$ W and $P_{winding} = 0.5$ W.

The LTspice simulation validates stable converter operation under design specifications, achieving a regulated output voltage of 1 V with an input voltage of 6 V, a load current of 80 A, and a switching frequency of 1 MHz. The simulation configuration employs an output capacitance of 30 μF, a single-phase inductance of 108 nH, and an inter-phase coupling coefficient k of 0.08.

As shown in Fig. 6 (a), the steady-state phase-current waveform demonstrates that the current ripple per phase is effectively constrained to 4.6 A. Fig. 6 (b) illustrates the steady-state output voltage waveform, where the output voltage ripple is measured at 15 mV. These results indicate that the vertical VRM exhibits favorable steady-state performance, with both current and voltage ripples maintained at minimal levels.

ANSYS Maxwell simulation resultes indicate that per-phase inductance value measures 108 nH accompanied by a DCR of 0.143 mΩ. In Fig. 7, an ANSYS Maxwell 3D magnetostatic simulation was performed to display the DC flux distribution when each phase conducts 40 A average current per phase (80 A in total). The max DC flux density is 1.063 T, but it is still much lower than the saturation flux density of DH060 (1.6 T). Therefore, this inductor can

support 80 A current, which is sufficient for the vertical VRM designed in this article.

In Fig. 8, the current density distribution resulting from an eddy current simulation of the coupled inductor structure, conducted in ANSYS Maxwell 3D under excitation conditions of 4.6 A peak-to-peak sinusoidal alternating current at 1 MHz applied to individual windings. The simulation results demonstrate pronounced skin effect characteristics in the conductor windings, which provides

Fig. 10. Vertical VRM module prototype.

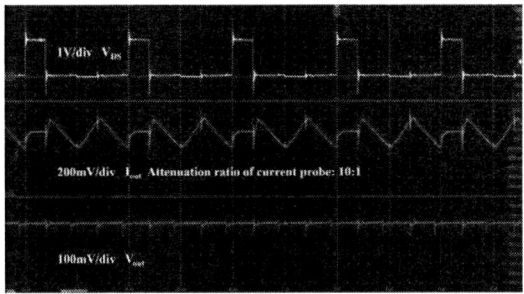

Fig. 11. Experiment results of the drain-source voltage, output voltage and output current of the vertical VRM module prototype, with an input voltage of 6 V.

critical insights for optimizing geometric parameters to mitigate AC losses through strategic dimensional adjustments.

The DC flux distribution and the AC current density distribution are relatively balanced across the two phases, indicating that the dual-phase integrated planar inductor is quite symmetric.

Fig. 9 shows the divergent thermal performance of two vertical magnetically integrated configurations under identical operating conditions. The integrated inductor positioned above (ILPA) exhibited a hotspot temperature of 88.5 °C, while the integrated inductor positioned below (ILPB) showed a higher hotspot temperature of 94 °C. The ILPA configuration adopted in this work demonstrates better thermal performance.

Simulation results indicate that the vertical VRM module architecture presents significant development potential, demonstrating effective performance across electrical, thermal and magnetic characteristics.

IV. EXPERIMENTAL RESULTS

Following the VRM module architecture detailed in Section II, a prototype featuring a custom-designed inductor mounted above the substrate is fabricated for experimental

validation, with compact dimensions of 9 mm × 10 mm × 4.8 mm, as illustrated in Fig. 10.

Voltage probes and current probes are employed to measure the drain-source voltage, output voltage, and output current, respectively, with results captured by a Tektronix oscilloscope (see Fig. 11). Experiment results show an output voltage of 720 mV with a single-phase output current approaching 20 A, aligning closely with theoretical predictions. While the output voltage remained stable, the observed current ripple highlights the need for further optimization of the inductor and prototype design.

V. CONCLUSION

In this paper, a vertical ultra-thin VRM with dual-phase integrated inductor is introduced. An analysis on dual-phase inductor's vertical integration and selection of magnetic powder core materials is conducted. Both simulation and experimental results confirm the VRM module's effective electrical, thermal and magnetic characteristics. Compared to commercial modules, the vertical VRM not only realizes high power density but also possesses a reduced overall thickness, which is advantageous for high-performance computing applications. The dual-phase inductor configuration enhances magnetic core utilization, contributing to improved power density.

REFERENCES

[1] F. C. Lee and Q. Li. High-frequency integrated point-of-load converters: Overview. IEEE Trans. Power Electron., 28(9):4127–4136, Sep. 2013.

[2] Y. Zhu, J. Zou, and R. C. Pilawa-Podgurski. A 1500-A/48-V-to-1-V switching bus converter for next-generation ultra-high-power processors. IEEE Trans. Power Electron., 39(9):11340–11355, Sep. 2024.

[3] J. Baek et al. Vertical stacked LEGO-PoL CPU voltage regulator. IEEE Trans. Power Electron., 37(6):1244–1245, Jun. 2022.

[4] C. Kong et al. A vertical power delivery architecture for high-performance computing. IEEE Trans. Power Electron., DOI 10.1109/TPEL.2025.3531414.

[5] J. Baek, Y. Elasser, and M. Chen. MIPS: multiphase integrated planar symmetric coupled inductor for ultrathin VRM. *IEEE Trans. Power Electron.*, 38(5):5609–5614, May 2023.

[6] P. L. Wong, P. Xu, B. Yang, and F. C. Lee. Performance improvements of interleaving VRMs with coupling inductors. *IEEE Trans. Power Electron.*, 16(4):499–507, Jul. 2001.

[7] Y. Elasser et al. Mini-LEGO CPU voltage regulator. *IEEE Trans. Power Electron.*, 39(3):3391–3410, Mar. 2024.

[8] M. Chen and C. R. Sullivan. Unified models for coupled inductors applied to multiphase PWM converters. *IEEE Trans. Power Electron.*, 36(12):14155–14174, Dec. 2021.

[9] D. Hou, F. C. Lee, and Q. Li. Very high frequency IVR for small portable electronics with high-current multiphase 3-d integrated magnetics. *IEEE Trans. Power Electron.*, 32(11):8705–8717, Nov. 2017.

[10] P. Wang et al. MSC-PoL: Hybrid GaN–Si multistacked switched-capacitor 48-V PwrSiP VRM for chiplets. *IEEE Trans. Power Electron.*, 38(10):12815–12833, Oct. 2023.

[11] DMEGC, DH060 Mater. Characteristics, 2022. Accessed: Feb. 10, 2025. [Online]. Available: https://dongyangdongci.oss-cn-hangzhou.aliyuncs.com/uploads/20230104/43c992b3b8a7f270677a04 7527b00645.

Design of a GaN 1-MHz/48-V-to-1-V/320-A Series-Capacitor Buck Converter

Wendi Fan, Yenan Chen

College of Electrical Engineering, Zhejiang University, Hangzhou, China
ZJU-Hangzhou Global Scientifc and Technological Innovation Center, Hangzhou, China
{wendif,yenanc}@zju.edu.cn

Abstract—This paper presents the design of a single-stage switched-capacitor converter based on the series-capacitor buck (SCB) topology, suitable for 48-V/MHz VRM. The SCB circuit features reduced voltage stress on switching devices, a device count consistent with multi-phase buck circuits, and the ability to achieve self-balancing of voltage and current. All switching devices utilize GaN transistors, leveraging their advantages such as low parasitic inductance due to lateral structure packaging and high switching speed. An 8-phase SCB prototype was constructed, capable of delivering a full-load output of 320 A. At 1 MHz, the prototype achieves a peak efficiency of 92.07% and a full-load efficiency of 82.06% at 1-V output.

Index Terms—Gallium Nitride, switched capacitor, series-capacitor buck, loss calculation

I. INTRODUCTION

The surge in data center power usage has significantly driven up operational costs. In 2024, the global electricity consumption of data centers was approximately 415 TWh, accounting for roughly 1.5% of the world's total electricity usage [1]. To reduce resistive losses on the input side, data centers are gradually adopting a 48-V bus voltage architecture to replace the traditional 12-V bus voltage architecture [2].

The 48-V Voltage Regulator Module (VRM) solutions are primarily divided into two-stage and single-stage approaches. The two-stage converter structure [3]–[5] consists of a front-stage step-down circuit with a fixed conversion ratio and a second-stage voltage regulation circuit. The second-stage voltage regulation circuit can directly leverage mature multi-phase buck circuit solutions, enabling a smooth transition from the 12-V architecture to the 48-V architecture. In addition to the two-stage solution, series-capacitor buck is a widely used single-stage topology. Series-capacitor buck (SCB) converter [6], [7] has the same number of semiconductor devices as multi-phase buck topology. Meanwhile the voltage stress on switches is reduced, enabling higher voltage conversion ratios. Additionally, higher output currents can be achieved by increasing the number of phases. Many two-stage and single-stage 48-V VRM solutions are based on this topology [8]–[10]. In low-voltage, high-current applications, vertical Trench MOSFETs with low on-resistance are commonly used. For 12-V VRM or the second-stage multi-phase buck circuit in 48V-VRM typically employs DrMOS with low parasitic inductance, but they are limited to inputs below 12 V. For 48-V VRM, discrete Trench MOSFETs are typically used,

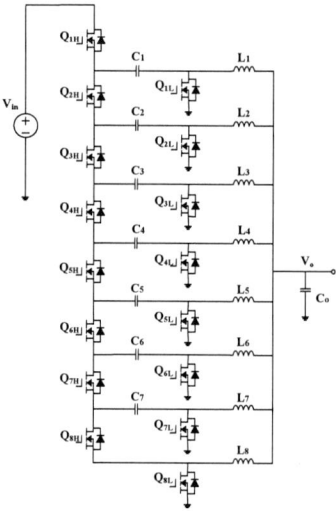

Fig. 1. The 8-phase SCB converter.

with relatively high parasitic inductance, affecting high-frequency performance. In contrast, LGA-packaged lateral GaN HEMTs offer lower parasitic inductance compared to Si devices, making GaN a promising candidate for high-frequency, high-efficiency power conversion.

This paper presents the design of a GaN-based 48-V/MHz-level 8-phase series-capacitor buck converter. By leveraging the advantages of GaN transistors, such as the low parasitic inductance and high switching speed, the converter achieves high power density and high efficiency. A 1-MHz/48-V-to-1-V/320-A prototype is built to validate the performance of the GaN transistors.

II. TOPOLOGY AND OPERATIONAL PRINCIPLE ANALYSIS

Fig. 1 illustrates an 8-phase SCB converter topology. Each series capacitor serves as the input voltage source for the next phase and provides a DC bias for its respective phase. In the 8-phase SCB, the DC bias of capacitors C_1 to C_7 decreases linearly from $\frac{7}{8}V_{in}$ to $\frac{1}{8}V_{in}$, thereby reducing the voltage stress on the switching devices. In this design, a two-phase modulation scheme is adopted. Fig.2 illustrates the operational states of the 8-phase SCB converter under two-phase modulation, while Fig. 3 shows the driving signals, inductor current waveforms, currents of

979-8-3315-1110-4/25 $31.00 © 2025 IEEE

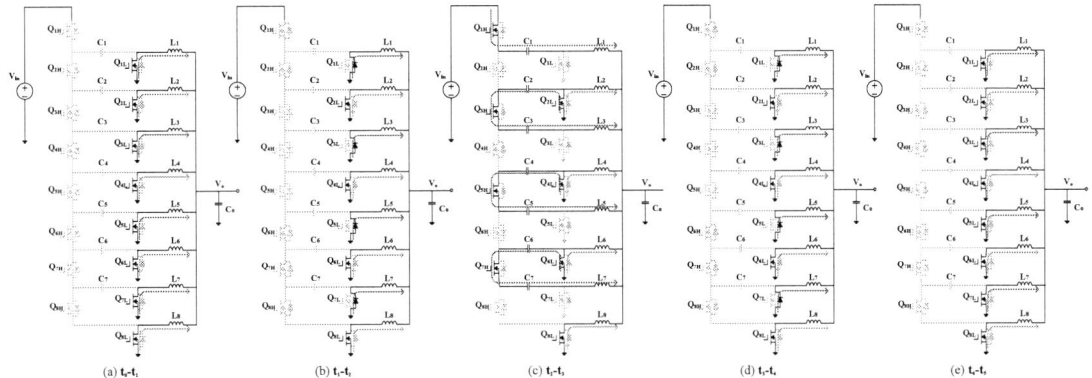

Fig. 2. The operation states of 8-phase SCB converter in two-phase modulation.

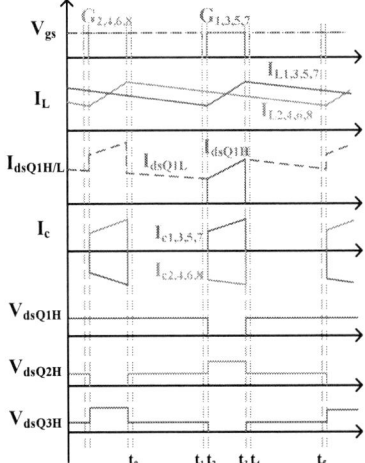

Fig. 3. The operation waves of 8-phase SCB converter.

the odd-phase switches, flying capacitor currents, and the voltages of the first three high-side switches.

At $t_0 - t_1$, the circuit operates in freewheeling mode, with low-side switches conducting. The inductor current decreases, and the low-side switch currents match the inductor current. No current flows through the capacitors, so voltages remain unchanged. The voltages of all high-side switches are $\frac{1}{8}V_{in}$.

At $t_1 - t_2$, the low-side switches of the odd phases turn off and conduct in reverse, while the low-side switches of the even phases remain in freewheeling mode. The currents of all low-side switches continue to match the inductor current, which keeps decreasing. The capacitor voltages remain unchanged, and the voltages of the high-side switches stay at $\frac{1}{8}V_{in}$.

At $t_2 - t_3$, the high-side switches of the odd phases turn on. The currents through these switches match the inductor current, which continues to rise. The currents of the low-side switches in phases 2, 4 and 6 are the sum of the currents of their respective phase and the next phase, while the current of the low-side switch in phase 8 equals the inductor current. At this time, the flying capacitors are charging and discharging respectively. The drain voltage of the high-side switch in phase 2 is the input voltage V_{in},

and its source voltage is the voltage of capacitor C_2, $\frac{6}{8}V_{in}$. Therefore, the drain-source voltage of the high-side switch in phase 2 is $\frac{2}{8}V_{in}$, and the situation is similar for the other phases.

At $t_3 - t_4$, the low-side switches of the odd phases turn off and enter reverse conduction mode, similar to the t_1-t_2 interval. The $t_4 - t_5$ interval is identical to the $t_0 - t_1$ interval. Afterward, the circuit enters the conduction state of the even phases, which is analogous to the conduction state of the odd phases.

III. LOSS CALCULATION AND ANALYSIS

A. Loss Calculation

In this section, a loss model for the 8-phase SCB converter based on GaN devices is established. The parameters are defined as follows: input voltage V_i, output voltage V_o, switching frequency f_{sw}, period T, number of phases N, inductance L, inductor current i_L, and duty cycle D. The peak inductor currents are $i_{Lmin} = i_L - \frac{\Delta i_L}{2}$ and $i_{Lmax} = i_L + \frac{\Delta i_L}{2}$.

When the high-side switch is conducting, the current flowing through it equals the inductor current. The RMS current of the high-side switch in the first phase is calculated as follows:

$$I_{Q1Hrms} = \sqrt{f_{sw} \cdot \int_0^{\frac{D}{f_{sw}}} \left(i_{Lmin} + \frac{\Delta i_L \cdot f_{sw}}{D} \cdot t \right)^2 \, dt} \quad (1)$$

Thus, the total conduction loss of the high-side switches is: $P_{conH} = N \cdot R_{DSonH} \cdot i_{Q1Hrms}^2$. Next, the switching

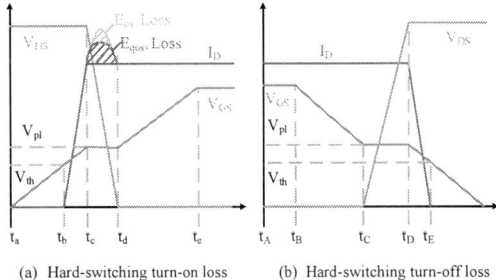

(a) Hard-switching turn-on loss   (b) Hard-switching turn-off loss

Fig. 4. The hard-switching process of the high-side switch [12].

losses of the high-side switches are calculated. The turn-on process is illustrated in Fig.4(a). According to the literature [11], during $t_b - t_c$, V_{GS} rises to the Miller plateau voltage V_{pl}, and the channel current increases, with the current rise time denoted as t_{ri}. During $t_c - t_d$, V_{GS} remains at the Miller plateau, and the driving current i_{Gon} discharges C_{GD}, causing V_{DS} to fall, with the voltage fall time denoted as t_{fu}. For the turn-off losses, as shown in Fig.4(b), similar to the turn-on process, with the rise time denoted as t_{ru} and the current fall time denoted as t_{fi}. The formulas for calculating the turn-on and turn-off losses are as follows:

$$
\begin{aligned}
P_{onH} &= N \cdot V_{DD} \cdot i_{Lmin} \cdot \frac{t_{ri} + t_{fu}}{2} \cdot f_{sw} \\
P_{offH} &= N \cdot V_{DD} \cdot i_{Lmax} \cdot \frac{t_{ru} + t_{fi}}{2} \cdot f_{sw}
\end{aligned}
\tag{2}
$$

The rise and fall times of voltage and current are detailed in the literature [11]. The switching losses caused by the overlap of voltage and current are given by $P_{sw} = P_{onH} + P_{offH}$.

During the turn-on process, the output capacitance of the GaN transistor also contributes to losses [12]. During the commutation process of Q_{1H} turning on, the output capacitance of Q_{1H} discharges, while the output capacitances of Q_{1L} and Q_{2H} charge. The output capacitance losses are:

$$
\begin{aligned}
P_{ossQ1H} &= f_{sw} \cdot \int_0^{V_{DD}} V_{ds} \cdot C_{OSS(V_{ds})} \mathrm{d}V_{ds} \\
P_{qossQ1L} &= f_{sw} \cdot \int_0^{V_{DD}} (V_{DD} - V_{ds}) \cdot C_{OSS(V_{ds})} \mathrm{d}V_{ds} \\
P_{ossQ2H} &= f_{sw} \cdot \int_{V_{DD}}^{2 \cdot V_{DD}} (2 \cdot V_{DD} - V_{ds}) \cdot C_{OSS(V_{ds})} \mathrm{d}V_{ds}
\end{aligned}
\tag{3}
$$

Finally, the total output capacitance loss is calculated as $P_{C_{oss}} = N \cdot (P_{ossQ1H} + P_{ossQ2H} + P_{qossQ1L})$.

Based on the I_{dsQ1L} waveform shown in Figure 3, it is divided into three stages to compute the total RMS current i_{Q1Lrms}. Note that this RMS current is only applicable to the low-side switch currents of the first 7 phases. For the low-side switch current of the 8th phase, it equals the inductor current throughout the conduction period, and its RMS value i_{Q8Lrms} is calculated. Thus, the conduction loss of the low-side switch is:

$$
P_{conL} = (N - 1) \cdot R_{DSonL} \cdot i_{Q1Lrms}^2 + R_{DSonL} \cdot i_{Q8Lrms}^2
\tag{4}
$$

During the dead time, the low-side switch conducts in reverse. Although GaN transistors do not have a body diode, they can conduct in reverse in a manner similar to a body diode. The reverse conduction loss is calculated as:

$$
\begin{aligned}
P_{dead1} &= V_{SD(i_{Lmin})} \cdot i_{Lmin} \cdot t_{dead} \cdot f_{sw} \\
P_{dead2} &= V_{SD(i_{Lmax})} \cdot i_{Lmax} \cdot t_{dead} \cdot f_{sw} \\
P_{dead} &= N \cdot (P_{dead1} + P_{dead2})
\end{aligned}
\tag{5}
$$

Inductor losses are divided into resistive losses and core losses, expressed as $P_{Lcon} = R_L \cdot i_{Lrms}^2 + P_{Lcore}$. When calculating capacitor losses, it is important to note that the capacitor undergoes both charging and discharging within one cycle, so the loss is doubled: $P_C = 2 \cdot R_{C_{ESR}} \cdot (i_{Q1Hrms})^2$.

Using Ansys software, the parasitic inductance and resistance on the PCB are simulated, and the losses associated with these parasitic parameters are estimated:

$$
\begin{aligned}
P_{ESL} &= \frac{1}{2} \cdot ESL \cdot i_L^2 \cdot f_{sw} \\
P_{ESR} &= ESR \cdot (i_L \cdot N)^2 \\
P_{Loop} &= P_{ESL} + P_{ESR}
\end{aligned}
\tag{6}
$$

B. Device Selection and Loss Analysis

Based on the analysis of the previous working process, the high-side switch operates with hard switching and has a short conduction time, while the low-side switch operates with soft switching and has a longer conduction time. Therefore, the high-side switch should be selected with a device that has a small parasitic capacitance, and the low-side switch should be chosen with a device that has a low on-resistance. Additionally, the maximum drain-source voltage stress on the switching transistors is 12 V. Each phase has a rated current of 40 A. The required minimum value of the flying capacitor is determined by equation (7):

$$
C = \frac{P_o}{V_{in} \cdot f_{sw} \cdot \Delta V}
\tag{7}
$$

The $\triangle V$ is taken as 10% of the capacitor's voltage. The selected capacitor must have a capacitance value that exceeds the calculated value under the DC bias, and its ESR should be as small as possible. In order to reduce winding losses and decrease the size of the inductor, a two-phase coupled inductor was selected for the experiment. Table 1 lists the main component selections for the prototype.

After device selection, the ratios of individual power loss components to total power are calculated separately at different output power levels under 1MHz, as shown in Figure 5. Under light-load condition, switching losses and output capacitance losses constitute dominant portions. Under heavy-load condition, the conduction loss of the low-side switch becomes predominant due to the current doubling effect in the SCB topology. Furthermore, GaN devices exhibit approximately 1.5V higher reverse conduction voltage drop compared to silicon-based devices, resulting in increased dead-time losses.

TABLE I
MAIN COMPONENTS IN 8-PHASE SCB PROTOTYPE

Component	Parameter
$Q_{1H} - Q_{8H}$	EPC2067, 40 V, 1.55 mΩ
$Q_{1L} - Q_{8L}$	EPC2066, 40 V, 1.1 mΩ
C_1	C3216X6S2A106K160AC1206 ×8
$C_2 - C_5$	GRM21BZ71H475KE15 ×8
$C_6 - C_7$	C2012X7S1E106K125AC ×8
$L_1 - L_8$	Two-phase coupled inductor, 85 nH, 0.15 mΩ
Gate driver	NCP51810, Half-bridge driver

IV. EXPERIMENTAL RESULTS

Figure 6 shows the prototype. In Figure 7, the voltage waveforms of the high-side and low-side switches for phases 7 and 8 at 48 V-1 V/833 kHz/160 A are shown. Analyzing phase 8, when V_{8H} is turned off, $V_{8HDS} = 6$ V. After half a cycle, when V_{7H} is turned on, $V_{8HDS} = 12$ V, which aligns with the theoretical analysis. In Figure 8, efficiency curves at different frequencies ranging from 625

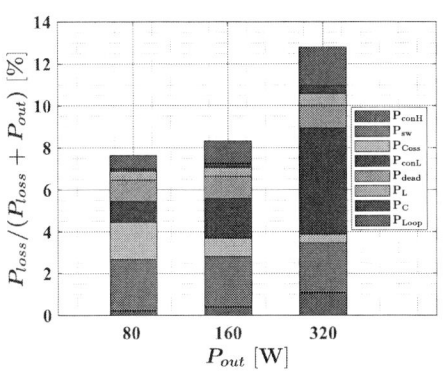

Fig. 5. Power loss distribution under different output loads at 1MHz.

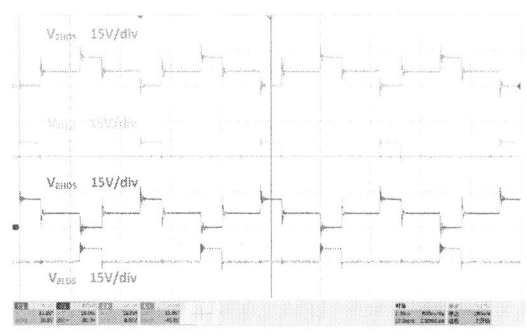

Fig. 7. The waveforms of $V_{7HDS}, V_{7LDS}, V_{8HDS}, V_{8LDS}$ at 833 kHz/160 A.

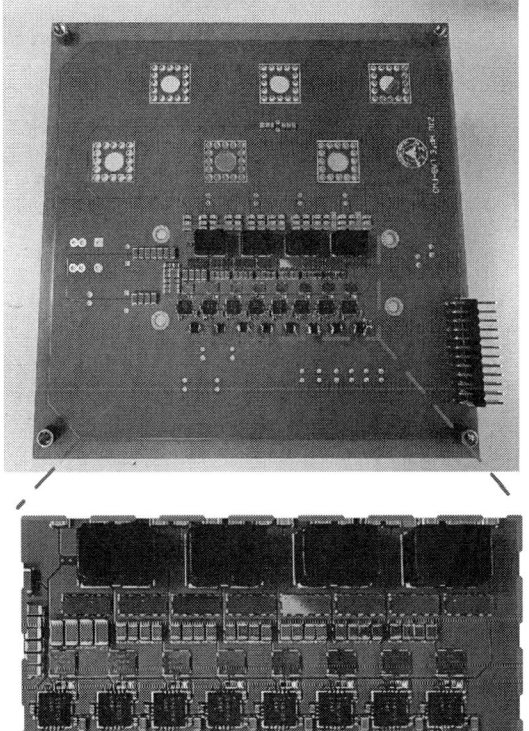

Fig. 6. Prototype of 8-phase SCB.

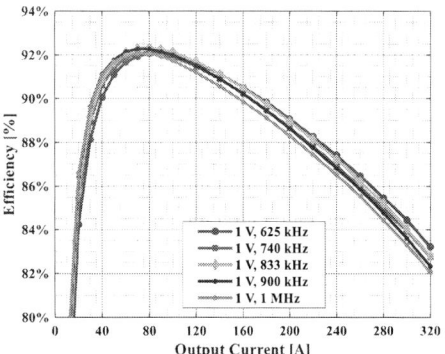

Fig. 8. The efficiency curves for 48 V to 1 V.

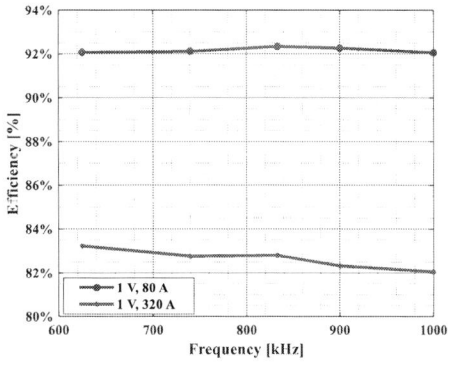

Fig. 9. The efficiency curves at 80 A/320 A.

kHz to 1 MHz are compared. With an output voltage of 1 V, the peak efficiency at 625 kHz is 92.06%, and the full-load efficiency is 83.23%. At 1 MHz, the peak efficiency is 92.07%, and the full-load efficiency is 82.06%. In Figure 9, it can be observed that as the frequency increases, the efficiency decreases slightly. At a light load of 80 A, the efficiency remains almost unchanged, while at full load, the efficiency drops by only 1.17% as the frequency increases from 625 kHz to 1 MHz. Based on the loss analysis, it can be concluded that the switching losses of GaN transistors constitute a relatively small portion of the total losses. As

a result, the efficiency does not decrease significantly as the frequency increases.

V. CONCLUSIONS

This paper presents the design of an MHz-level 8-phase SCB converter based on GaN devices, achieving high step-down ratio conversion from 48 V to 1 V. By utilizing GaN transistors to improve efficiency, and greater power density can be realized. A prototype was constructed and tested under conditions of 1 MHz/320 A output. The prototype demonstrated a peak efficiency of 92.07% and a full-load efficiency of 82.06% at 1-V output. These results validate the superior performance of GaN devices.

VI. ACKNOWLEDGMENT

This paper and its related research are jointly supported by the Natural Science Foundation of China under Grant No. 52307225, Zhejiang Provincial Natural Science Foundation of China under Grant No. LQ23E070004, and grants from the Delta Power Electronics Science and Education Development Program of Delta Group.

REFERENCES

[1] IEA (2025), Energy and AI, https://www.iea.org/reports/energy-and-ai

[2] Y. Chen, K. Shi, M. Chen and D. Xu, "Data Center Power Supply Systems: From Grid Edge to Point-of-Load," in IEEE Journal of Emerging and Selected Topics in Power Electronics, vol. 11, no. 3, pp. 2441-2456, June 2023.

[3] Y. Chen, H. Cheng, D. M. Giuliano and M. Chen, "A 93.7% Efficient 400A 48V-1V Merged-Two-Stage Hybrid Switched-Capacitor Converter with 24V Virtual Intermediate Bus and Coupled Inductors," 2021 IEEE Applied Power Electronics Conference and Exposition (APEC), Phoenix, AZ, USA, 2021, pp. 1308-1315.

[4] Y. Elasser et al., "Vertical Stacked 48V-1V LEGO-PoL CPU Voltage Regulator with 1A/mm2 Current Density," 2022 IEEE Applied Power Electronics Conference and Exposition (APEC), Houston, TX, USA, 2022, pp. 1259-1266.

[5] M. Ahmed, C. Fei, F. C. Lee and Q. Li, "High efficiency two-stage 48V VRM with PCB winding matrix transformer," 2016 IEEE Energy Conversion Congress and Exposition (ECCE), Milwaukee, WI, USA, 2016, pp. 1-8.

[6] Nishijima, K. Harada, T. Nakano, T. Nabeshima and T. Sato, "Analysis of Double Step-Down Two-Phase Buck Converter for VRM," INTELEC 05 - Twenty-Seventh International Telecommunications Conference, Berlin, Germany, 2005, pp. 497-502.

[7] Y. Jang, M. M. Jovanovic and Y. Panov, "Multiphase buck converters with extended duty cycle," in Proc. IEEE Appl. Power Electron. Conf. Expo., Dallas, TX, 2006, pp. 38-44.

[8] P. Wang, Y. Chen, G. Szczeszynski, S. Allen, D. M. Giuliano and M. Chen, "MSC-PoL: Hybrid GaN–Si Multistacked Switched-Capacitor 48-V PwrSiP VRM for Chiplets," in IEEE Transactions on Power Electronics, vol. 38, no. 10, pp. 12815-12833, Oct. 2023.

[9] Y. Zhu, T. Ge, N. M. Ellis, L. Horowitz and R. C. N. Pilawa-Podgurski, "The Switching Bus Converter: A High-Performance 48-V-to-1-V Architecture With Increased Switched-Capacitor Conversion Ratio," in IEEE Transactions on Power Electronics, vol. 39, no. 7, pp. 8384-8403, July 2024.

[10] X. Xu and Q. Li, "Analysis of Parasitic Stored Energy Loss and PCB Layout Optimization for 48V-to-1V Series-Capacitor Buck," 2024 IEEE Applied Power Electronics Conference and Exposition (APEC), Long Beach, CA, USA, 2024, pp. 898-905.

[11] Dušan Graovac, Marco Pürschel and Andreas Kiep, "MOSFET Power Losses Calculation Using the Data-Sheet Parameters," Infineon, July. 2006. [Online]. Available: https://application-notes.digchip.com/070/70-41484.pdf

[12] R. Hou, J. Lu and D. Chen, "Parasitic capacitance Eqoss loss mechanism, calculation, and measurement in hard-switching for GaN HEMTs," 2018 IEEE Applied Power Electronics Conference and Exposition (APEC), San Antonio, TX, USA, 2018, pp. 919-924.

GaN-based Low-Voltage, High-Current DC-DC Converter for Aerospace Applications

Xiao Chen
School of Electrical and Information Engineering
Tianjin University
Tianjin, China
chen_xiao@tju.edu.cn

Zhihao Zhang
School of Electrical and Information Engineering
Tianjin University
Tianjin, China
dz_hao@tju.edu.cn

Zhongjie Wang
School of Electrical and Information Engineering
Tianjin University
Tianjin, China
zhongjie@tju.edu.cn

Mingming Ji
Beijing Spacecraft
Beijing, China
877307162@qq.com

Lingxiao Xue
School of Electrical and Information Engineering
Tianjin University
Tianjin, China
xuel@tju.edu.cn

Abstract— **Power supplies are critical to the stable operation of spacecrafts such as satellites. Currently, aerospace DC-DC converters designed for 100 V bus have been seldom studied. Due to the unique requirement of device voltage derating in aerospace applications, the converter design including topology selection and efficiency optimization under high bus voltage condition presents significant challenges. This paper focuses on a 120 W converter with an input voltage range of 70 to 120 V and a 5 V output. The space power supply was optimized through detailed modeling of different topologies. Specifically, power losses caused by PCB-winding vias at high frequency were simulated. The high-frequency via losses were mitigated by employing an interleaved layout. Finally, an experimental prototype was developed, achieving an efficiency of up to 92.4%.**

Keywords—Asymmetrical half-bridge flyback Converter, Aerospace power, current RMS modeling, DC-DC converter

I. INTRODUCTION

The space power supply system, as a critical component of spacecraft such as satellites, plays a pivotal role in ensuring the stable operation of these platforms. Currently, the efficiency levels of space power supplies vary considerably. Traditional silicon-based DC-DC converters typically have efficiencies around 82%[1]. Gallium Nitride (GaN)-based technology offers improved performance. For example, a high-frequency GaN DC-DC converter utilizing a Buck-Boost and LLC hybrid topology, operating at a switching frequency of 3 MHz and delivering 200 W output power, has been reported to achieve an efficiency of 85.7%[2]. Additionally, a 100 W spacecraft DC-DC converter based on GaN technology has been developed and tested, demonstrating an efficiency of 90%, thereby validating the potential of GaN-based power supplies to reduce size, improve efficiency, and enhance power density[3].

To meet the diverse power requirements of onboard equipment, low Earth orbit (LEO) communication satellites utilize a variety of bus voltage levels. The unregulated power bus voltage typically ranges from 23 V to 42 V. For example, the Iridium NEXT satellite operates within a bus voltage range of 23 V to 37.5 V, Globalstar ranges from 23 V to 38 V, and OneWeb operates between 28 V and 37 V[4]. While the platform system voltage is commonly set at 5 V, the evolving demands of aerospace power systems have driven the need for higher bus voltage levels. The 100 V high-voltage bus architecture is particularly advantageous for meeting the high-power requirements of spacecraft and reducing cable losses [5]. However, research on aerospace DC-DC converters designed for 100 V high-voltage busbars remains limited. The selection of topologies, derating design, and efficiency optimization under high-voltage bus architectures present significant technical challenges.

II. TOPOLOGY SELECTION AND DESIGN

A. Comparison of Topologies

In the design of isolated DC-DC converters with high gain and wide input voltage range, commonly adopted topologies include active clamp flyback and active clamp forward converters. These converters achieve zero-voltage switching (ZVS) through clamp circuits, enhancing efficiency under high-voltage and high-frequency conditions. However, in aerospace applications, switching device voltage derating is typically set at 40% to ensure system reliability.

Although the ACF flyback topology enables wide-range voltage regulation, it imposes excessive voltage stress on the primary-side switch. Similarly, while the ACF forward converter supports wide input voltage operation, the voltage stress on its primary-side switch varies significantly with the transformer turns ratio (N) and input voltage, potentially resulting in even higher stress in certain designs. In contrast, the asymmetric half-bridge (AHB) flyback topology not only maintains wide voltage regulation capability but also offers superior voltage stress reduction. Specifically, the primary-side switch in the AHB flyback experiences significantly lower voltage stress compared to both ACF flyback and ACF forward converters, while the secondary-side switch also operates under reduced stress relative to the ACF flyback.

Furthermore, in low-voltage, high-current scenarios, the secondary-side synchronous rectifier turn-off current contributes significantly to power losses. The ACF forward converter inherently exhibits large secondary-side turn-off currents, whereas the AHB flyback topology, through optimized design, can reduce these turn-off currents, thereby improving conversion efficiency.

As summarized in TABLE I and illustrated in Fig. 1(a)–(c), a comprehensive comparison of voltage stresses across primary-side switches, secondary-side switches, and resonant capacitors has been conducted for these three topologies. The analysis considers voltage stress characteristics under varying transformer turns ratios and input voltages. The results demonstrate that the AHB flyback topology exhibits the lowest voltage stress levels. Therefore, the AHB flyback topology has been selected.

TABLE I. VOLTAGE STRESS COMPARISON OF DIFFERENT TOPOLOGIES

	VOLTAGE STRESS COMPARISON OF DIFFERENT TOPOLOGIES		
	ACF Forward	*ACF Flyback*	*AHB Flyback*
$V_{ds(pri)}$	$\dfrac{V_{in}^2}{V_{in}-N\cdot V_o}$	$V_{in}+NV_o$	V_{in}
$V_{ds(sec)}$	$\dfrac{V_{in}}{N}$	$\dfrac{V_{in}}{N}+V_o$	$\dfrac{V_{in}}{N}$
$V_{ds(cr)}$	$\dfrac{V_{in}^2}{V_{in}-N\cdot V_o}$	NV_o	NV_o

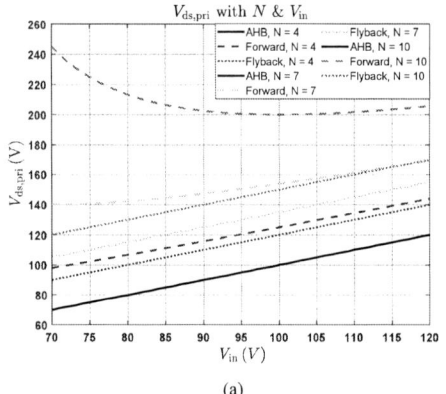

(a)

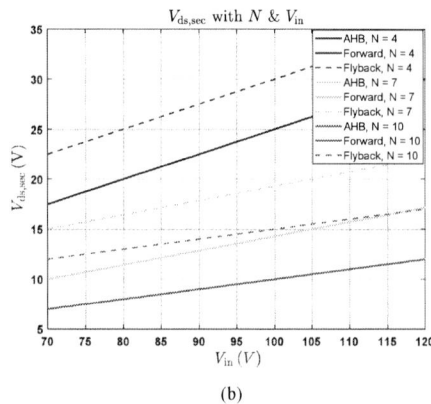

(b)

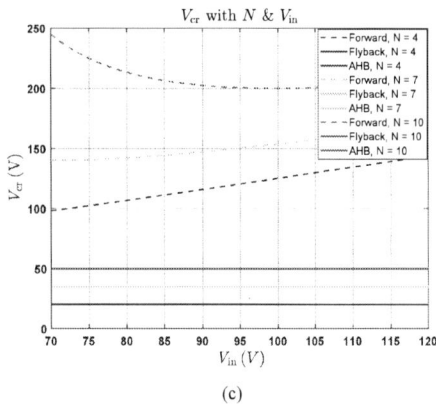

(c)

Fig. 1. Voltage stress comparison of ACF-Forward, ACF-Flyback, AHB-Flyback

B. AHB Design Optimization

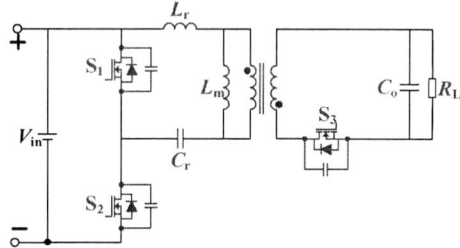

Fig. 2. AHB topology

Fig. 2 illustrates the main structure of the asymmetric half-bridge topology. S_1 and S_2 are the two primary-side main power switches, while S3 is the secondary-side synchronous rectifier switch. Through careful design, the converter operates in discontinuous conduction mode (DCM) and utilizes the negative current of the excitation inductance to achieve zero-voltage switching turn-on, according to:

$$(V_{in}-nV_o)=L_m\frac{\Delta i}{\Delta t} \tag{1}$$

The average field current of the primary side is:

$$I_{Lm\_avg}=I_{Lr\_avg}+I_{s\_avg}=\frac{I_o}{n} \tag{2}$$

The minimum value of the excitation current can be analyzed as:

$$I_{Lm\_min}=I_{Lm\_avg}-\frac{nV_o(1-D)(T_s-t_d)}{2L_{m\_max}} \tag{3}$$

The energy required by the primary side switch tube to achieve ZVS:

$$I_{Lm\_min}t_d=2C_{oss}V_{in}+C_{ps}\frac{V_{in}}{n} \tag{4}$$

The maximum limit of excitation inductance can be resolved:

$$L_{m\_max}=\frac{n^2V_o(1-D)(T_s-t_d)t_d}{(4nC_{oss}+2C_{ps})V_{in}+2I_ot_d} \tag{5}$$

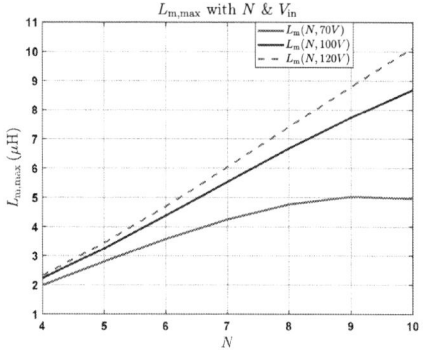

Fig. 3. Maximum magnetizing inductance requirement to achieve ZVS

It is clear from Fig. 3 that in order to achieve full range ZVS, the magnetizing inductance L_m should be designed at lowest input voltage. As long as ZVS can be achieved at the lowest input voltage of 70 V, full range ZVS is guranteed at other input voltages.

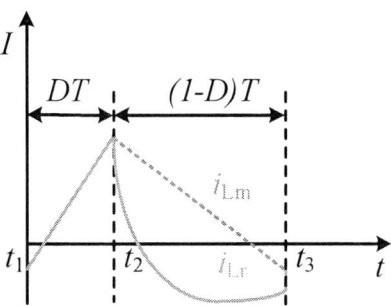

Fig. 4. Waveforms of AHB-Flyback

With ZVS achieved, most of the switching loss of the switches can be ignored, so the design for low voltage and high current needs to focus on the conduction loss of the switch, for which the analytical expression of the key currents need to be derived.

According to [6] and combined with Fig.4, the expression of current on the primary side of t_1-t_2 is shown as

$$i_{Lm}(t) = i_{Lr}(t_1) + \frac{V_{in} - V_{Cr}}{L_r + L_m} t \quad (6)$$

Combined with Fig.4, the expression of primary resonant current during t_2-t_3 is shown in formula (7)

$$i_{Lr}(t) = I_{Lr\_ini} \cos(\omega t) + \frac{(nV_o - V_{Cr\_ini})}{Z} \sin(\omega t) \quad (7)$$

where

$$I_{Lr\_ini} = I_{Lm\_max} - \frac{\dfrac{V_{in} - V_{Cr}}{L_m + L_r} L_m + V_{Cr}}{Z_{r1}} \quad (8)$$

$$V_{Cr\_ini} \approx nV_o + \frac{(I_{Lm\_max} + I_{Lm\_min})DT_s}{2C_r} \quad (9)$$

$$\omega = 1/\sqrt{L_r C_r}, Z = \sqrt{\frac{L_r}{C_r}} \quad (10)$$

$$i_{Lm}(t) = I_{Lm\_max} - \frac{nV_o t}{L_m} \quad (11)$$

The expression of the secondary side current can be expressed by combining (7) and (11)

$$i_{sec}(t) = n \times (i_{Lm}(t) - i_{Lr}(t)) \quad (12)$$

Fig. 5(a) and Fig. 5(b) illustrate the modeled primary-side and secondary-side current waveforms, respectively, under different resonant capacitor values (Cr = 1 μF, 1.5 μF, and

2 μF) and transformer turns ratios (4, 6, and 8). By calculating the RMS values of these current waveforms, a suitable design point can be selected.

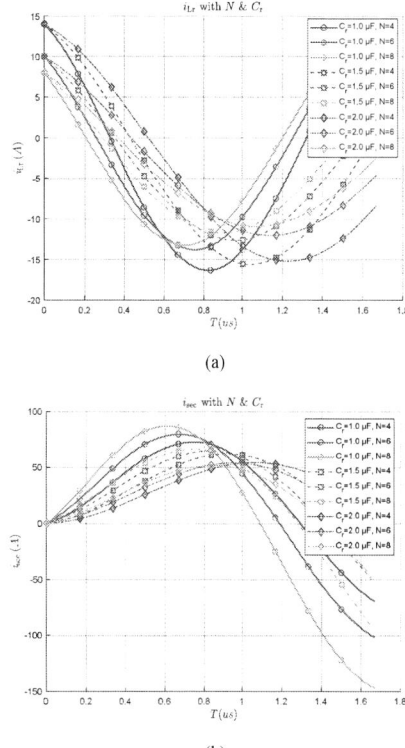

(a)

(b)

Fig. 5. Modeling of current waveforms: (a) primary current; (b) secondary current

C. Transformer Design

Based on the optimal design point, a turns ratio of 6:1 has been selected as the design baseline. While increasing the number of parallel turns on the secondary side generally helps reduce winding losses, it also raises manufacturing costs. As a result, the final winding configuration was selected as six turns in series on the primary side and two turns in parallel on the secondary side.

For the chosen 6:1 turns ratio, two candidate winding arrangements—SPPPPPPS and PPSPPSPP—have been identified. The magnetomotive force (MMF) distribution characteristics of these configurations are comparatively analyzed in Fig.6.

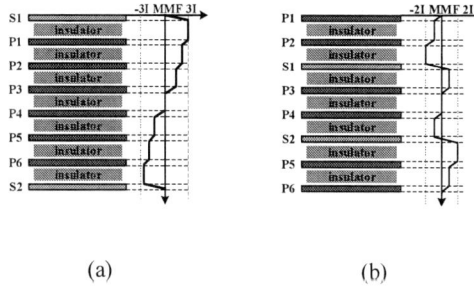

(a)   (b)

Fig. 6. MMF distribution in different winding arrangement: (a) SPPPPPPS; (b) PPSPPSPP.

We can see that the magnetomotive force value of the first winding arrangement is higher, indicating greater leakage inductance and increased winding losses. In contrast, the second arrangement exhibits a significantly lower MMF value, resulting in reduced leakage inductance and lower winding losses. However, in the second arrangement, the secondary-side winding requires via-connections in the middle layer and the outer-layer. Those through-hole vias can lead to significant power losses at high frequency. Fig. 7 presents a simulation result of the current density distribution in the secondary-side winding at 600 kHz. It is clearly seen that the current density through the vias is very high, indicating significant power losses at high frequency.

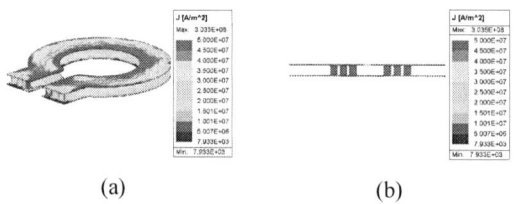

(a)   (b)

Fig. 7. Simulated secondary winding current density: (a)Secondary winding current density perspective view; (b) Secondary-side vias current density front view

In order to obtain a smaller leakage inductance and avoid the high-frequency via losses from the secondary side, a PCB-winding structure comprised of three PCBs are proposed to integrate the windings, SRs and output capacitors, as shown in Fig.8. The upper and lower winding plates are two turns of primary side winding respectively, and the middle winding plate contains two turns of primary side winding and two turns of secondary side winding. At the same time, SR and output capacitors are directly connected to the winding to avoid high frequency via losses.

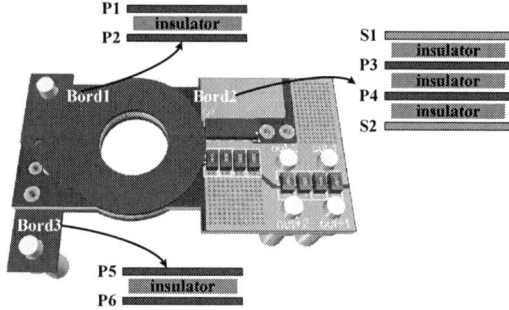

Fig. 8. 3D tomography image of the PCB winding

III. EXPERIMENTAL RESULTS

In order to verify the feasibility of the converter, we built a prototype with 600kHz, 120W, 70-120V input and 5V output, as shown in Fig.9:

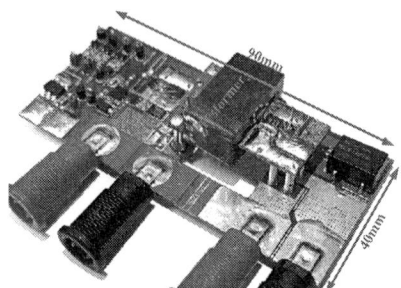

Fig. 9. 600 kHz 120W prototype.

Fig. 10 shows the experimental waveforms of V_{gs} and V_{ds} of S_2, as well as the resonant current waveform i_{Lr}, at full load of 5V/120W output and different input voltages of 70V, 100V, and 120V. The experimental waveforms prove the functionality of the prototype. Moreover, all three figures show that ZVS of S_1 and S_2 have been achieved, as indicated by the fact that V_{gs} rises after V_{ds} falls to zero.

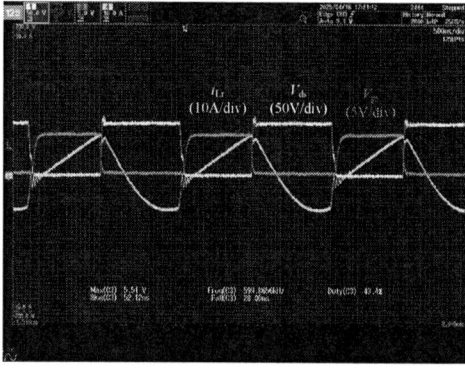

(a)

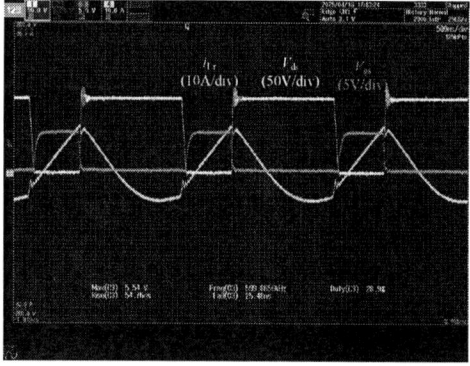

(b)

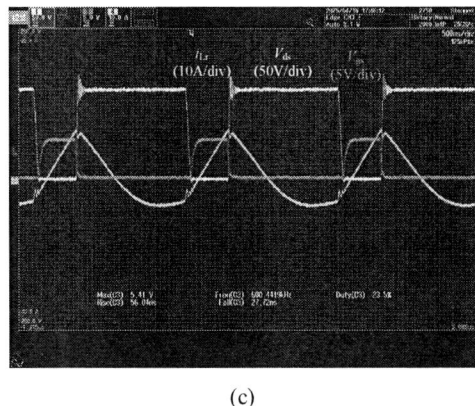

(c)

Fig. 10. Experimental waveforms at full-load and different input voltages: (a) 70V; (b) 100V; (c) 120V

Fig.11 shows the influence of different resonant capacitance C_r values on the efficiency at 100V input. When the resonant capacitance C_r is 2.2uF, the converter achieves a peak efficiency of 92.4%.

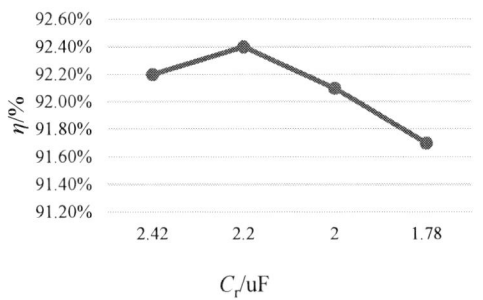

Fig. 11. Full-load efficiency against different resonant capacitance at 100V

A comparison of the proposed solution with some literatures was conducted and the results are listed in Table II. The peak efficiencies of the converters in [7] and [8] are 88% and 89.4%, respectively, whereas the peak efficiency of the proposed converter is as high as 92.4%.

TABLE II.   EFFICIENCY COMPARISON

Efficiency comparison				
	Frequency	*Power*	*Output voltage*	*Efficiency*
[7]	550kHz	120W	5V	88%
[8]	600kHz	120W	5V	89.4%
This work	600kHz	120W	5V	92.4%

Fig.12 shows the loss breakdown of the proposed converter at 100V input and 5V/120W output. The power loss is obtained based on both the tested current waveforms and analytical equations, including primary and secondary switch losses, as well as transformer loss (core and winding). The winding loss accounts for 72.1% of the total losses. In future work, we will focus winding loss analysis and reduction to further improve the converter efficiency.

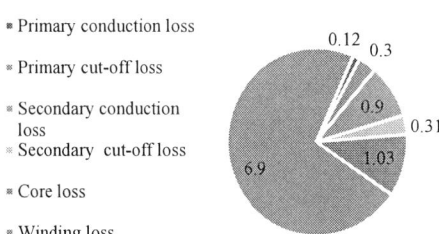

Power Loss Breakdown

- Primary conduction loss
- Primary cut-off loss
- Secondary conduction loss
- Secondary cut-off loss
- Core loss
- Winding loss

Fig. 12. Converter Loss Breakdown at 120W (5V/24A)

IV. CONCLUSION

This paper presents a converter design of for a low-voltage, high-current application delivering 120 W at 5 V, operating over a wide input bus voltage range of 70 V to 120 V. Through systematic evaluation of different topologies, the asymmetric half-bridge (AHB) flyback topology was selected for this high-voltage, wide-range input bus applications. The design process incorporates detailed modeling of the converter's current, along with optimized magnetic component design and winding layout considerations.

A 120 W prototype based on the proposed design methodology was developed and experimentally validated, achieving a peak efficiency of 92.4%. This performance significantly exceeds that of existing solutions available on the marketplace, demonstrating the effectiveness of the proposed design approach. Loss breakdown analysis revealed that winding losses constitute a major portion of total losses; therefore, future work will focus on further optimizing winding losses to enhance overall efficiency.

REFERENCES

[1]   Q. Tong, "Design of a Synchronous Rectifier Flyback DC-DC Converter Using GaN FETs for Commercial Aerospace Applications," 2020 IEEE 3rd International Conference on Electronics Technology (ICET), Chengdu, China, 2020, pp. 308-313.

[2]   L. Yun, Z. Pang, D. Luo, F. He, S. Jiang and X. Zhang, "Research on GaN Digital DC/DC Power Supply for Aerospace Applications," 2023 3rd International Conference on Electrical Engineering and Control Science (IC2ECS), Hangzhou, China, 2023, pp. 1082-1088.

[3]   C. Parmar, V. Jani and A. Kumar, "Design consideration of low to medium power dc-dc converter using enhancement-mode gallium nitride (e-GaN) HEMT device at MHz switching frequency for satellite payload power supply," 2021 National Power Electronics Conference (NPEC), Bhubaneswar, India, 2021, pp. 01-05.

[4]   M. G. Maqueda and V. Svikovic, "High Efficiency GaN Based Resonant Reset Forward Converter with Synchronous Rectification for Space Applications," 2023 13th European Space Power Conference (ESPC), Elche, Spain, 2023, pp. 1-8.

[5]   D. Koch et al., "Highly-Integrated, Low-Noise, Dual-Output GaN DC/DC for GaN Solid State Power Amplifier Supplies in Space Applications," 2023 IEEE 10th Workshop on Wide Bandgap Power Devices & Applications (WiPDA), Charlotte, NC, USA, 2023, pp. 1-6.

[6]   M. Li, Z. Ouyang and M. A. E. Andersen, "Analysis and Optimal Design of High-Frequency and High-Efficiency Asymmetrical Half-Bridge Flyback Converters," in IEEE Transactions on Industrial Electronics, vol. 67, no. 10, pp. 8312-8321, Oct. 2020.

[7]   SynQor. "MQFL-28-05S DC-DC Converter". [EB/OL]. [Accessed 10 October 2024], https://www.synqor.com/products/hi-rel/mqfl-28-05s.

[8]   Vicor Corporation. "DCM™ DC-DC Converter DCM3623xA5N06A 2y7z CE Isolated, Regulated DC Converter." [EB/OL]. [Accessed 10 March 2024], https://www.vicorpower.com/documents/datasheets/DCM3623xA5N06A2y7z_ds.pdf.

A New Method for Estimating the DC-DC Converter Performance under Low Voltage and High Current Application

Maohang Qiu[a], Hongbo Ma[a]
[a] Southwest Jiatong University, Chengdu, Sichuan

Abstract— Under low voltage and high current application with soft switching operation, the conduction loss usually dominates the devices' power loss(I^2R) at heavy load. Moreover, it's found that different converter shows significant differences on the heavy load. Traditional method adopts Total Devices Power Ratings (TDPR, TDPR=ΣIV) to explain the differences. However, the essential relationship between devices' stress and power loss is not revealed. In order to reveal the devices stress and power loss, this paper presents a new method called Circuit Level Figure of Merit(*CLFOM*) for explanation. It can reveal the devices' stress and power loss in a derivable way, and the minimum power loss at heavy load among different converters are compared in this paper.

Key Words: Circuit Level Figure of Merit(CLFOM), Soft switching.

I. Introduction

With the development of high power density ceramic capacitors, the Resonant Switched Capacitor DC-DC Converter (ReSC) has received more and more attention. However, the circuit topologies varies a lot among them, so it's important to find a method to evaluate the performance of all the ReSC. Except adopting the peak efficiency as a standard to evaluate the

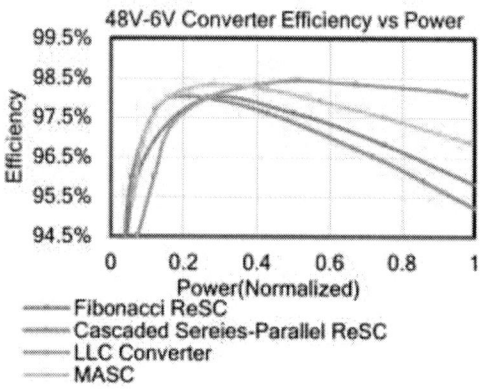

Figure 1 The 48V-6V DC-DC converter with soft switching

performance of the converter, it's more important to estimate the efficiency at heavy load since it directly decides the heatsink size or the converter's load capability. Moreover, the conduction loss (I^2R) dominates the devices power loss at heave load, especially under high current application. The Series-Parallel ReSC can reach peak efficiency as 98% at $0.21P_{full}$ but shows 2.9% drop at P_{full}[2], and the Fibonacci ReSC can reach peak efficiency as 98% at $0.3P_{full}$ but shows 2.1% drop at P_{full}[3].Compared to Fibonacci ReSC and Series Parallel ReSC, the LLC circuit can reach peak efficiency as 98.1% at $0.5P_{full}$ but only shows 0.5% drop at P_{full}[4], and the MASC also shows smaller efficiency drop on the heavy load[5]. The detailed efficiency curves are shown in Figure 1. Therefore, it's obvious that LLC converter shows the better performance on the heavy load, thus the conduction loss is less than the two other converters. To explain this, traditional method adopts circuit level evaluation method TDPR[1] to make comparison, but this method doesn't show the inner relationship between it and the conduction loss. Therefore, in this paper, a new method called Circuit Level Figure of Merit(CLFOM) is proposed to evaluate the ReSC's performance at heavy load, and this method is based on considering the semiconductor's die size.

II. The Derivation of the Circuit Level Figure of Merit: CLFOM

For the low voltage and high current application under the soft switching condition，the conduction loss dominates the majority of the full loss. For a specific device, the R_{ds} can be expressed in (1), where S_{die}, α, V_B represent the die area, the technology of the semiconductor devices and its blocking voltage respectively. Then the conduction loss can be expressed in (2), where I_k represents the RMS current through the kth devices. By substituting (2) into (1), then the (3) can be obtained. Then the question becomes how to distribute the die area to minimize the whole converter's conduction loss. Suppose the whole die area is S_{total}, then the minimum conduction loss P_{Con_Loss} can be expressed in (5) under the condition that the mth device can be expressed in (6).

$$R_{ds}S_{die} = \alpha(V_B)^2 \qquad (1)$$

$$P_{Con\_Loss} = \sum_{k-1}^{n}(I_k)^2 R_{ds(k)} \qquad (2)$$

$$P_{Con\_Loss} = \sum_{k=1}^{n} \frac{\alpha(I_k)^2(V_{B(k)})^2}{S_{die(k)}} \qquad (3)$$

$$\sum_{k=1}^{n} S_{die(k)} = S_{total} \qquad (4)$$

$$\min(P_{Con\_Loss}) = \frac{n\alpha}{S_{total}}\sum_{k=1}^{n}[I_k V_{B(k)}]^2 \qquad (5)$$

$$S_{die(m)} = \frac{[I_m V_{B(m)}]^2}{\sum_{k=1}^{n}[I_k V_{B(k)}]^2} S_{total} \qquad (6)$$

Then by adopting the (5), the minimum conduction loss at a specific power point can be compared through different circuits. For the low voltage and high current application, the heat dissipation is the key feature that influences the converter's load capacity. Therefore, the loss at full power should be optimized, which means I_1, $I_2,..., I_n$ should be the current at maximum power rating. For a fair comparison among different converters, the maximum power rating, the total die area and the technology of the semiconductor devices are assumed the same, which means the parameters P_{out}, S_{total} and α are the same. Then the new circuit-level figure of merit(CLFOM) that eliminates the S_{total} and α from (5) is defined in (7), and it is also normalized by P^2.

$$CLFOM = \frac{n\sum_{k=1}^{n}[I_k V_{B(k)}]^2}{P_{OUT}^2} \qquad (7)$$

III. The Calculation Example with CLFOM

Firstly, let's take the Switched Tank Converter(STC) shown in Figure 2 with conversion ratio as n into consideration, then the counts of the total devices should be 3n-2. The counts of the devices voltage stress as V_{out} and $2V_{out}$ should be $2n$ and (n-2) respectively, and the current stress of all the devices is I_{in}. As a result, the *CLFOM* of STC can be expressed in (8).

$$CLFOM_{STC}$$
$$= \frac{(3n-2)[2n(I_{in}V_{OUT})^2 + (n-2)(2I_{in}V_{OUT})^2]}{(I_{out(ave)}V_{OUT})^2}$$
$$= \frac{2.4674(3n-2)(6n-8)}{n^2} \qquad (8)$$

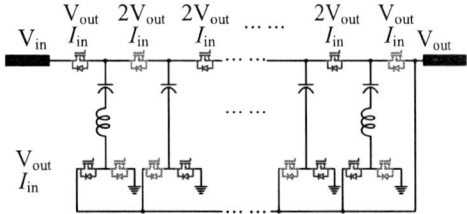

Figure 2 The STC with conversion ratio *n*

Similarly, the Series Parallel Switched Capacitor Converter(SPSC), Buck, LLC and MASC with conversion ratio as *n* can be expressed in (9), (10), (11) as well as (12) respectively.

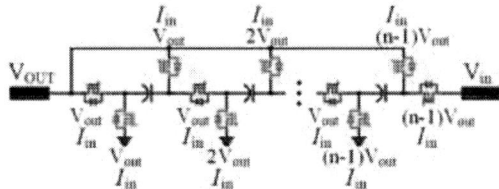

Figure 3 The Series-Parallel Switched Capacitor with conversion ratio as *n*

$$CLFOM_{SPSC} = \frac{1.6449(3n-2)(n^2-1)}{n} \qquad (9)$$

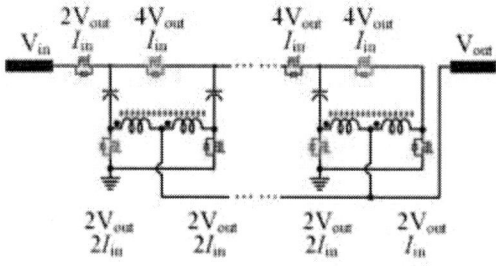

Figure 4 The Buck converter with conversion ratio as *n*

$$CLFOM_{Buck} = \frac{2V_{IN}^2 I_{OUT}^2}{V_{OUT}^2 I_{OUT(ave)}^2} = 4.9348n^2 \qquad (10)$$

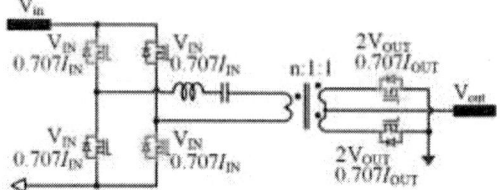

Figure 5 The LLC converter with conversion ratio as *n*

$$CLFOM_{LLC} = \frac{6(2V_{IN}^2 I_{IN}^2 + 4V_{OUT}^2 I_{OUT}^2)}{V_{OUT}^2 I_{OUT(ave)}^2} = 44.4 \qquad (11)$$

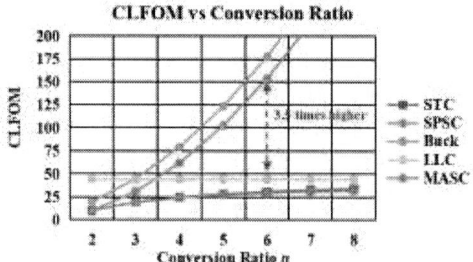

Figure 6 The MASC with conversion ratio as *n*

$$CLFOM_{MASC} = \frac{19.7393(2n-3)}{n} \qquad (12)$$

Figure 7 shows the CLFOM of the different converter versus the conversion ratio *n*. For example, when the conversion ratio as 6, the conduction loss of SPSC's at heavy load is theoretically 3.5 times than LLC's if the total die area of both are equal. Therefore, the performance of LLC is obviously better than SPSC. Besides that, the CLFOM of STC and MASC are the same, which means the devices' power loss of STC and MASC are the same.

Figure 7 The CLFOM versus conversion ratio *n*

The Efficiency Verification with MASC

In order to verify the efficiency, the 8x (48V to 6V) MASC proposed by the author is used for comparison[4]. The prototype is shown in Figure 8.

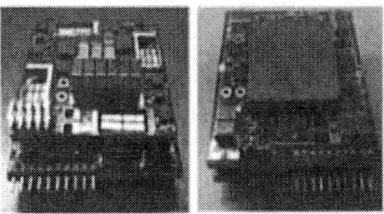

Figure 8 The 48V-6V, 500W MASC

Conclusion

This paper presents a new method called CLFOM to estimate the devices' power loss with soft switching under heavy load. It can be found that the conduction loss of the LLC and STC are around the same, while the devices' total conduction loss of SPSC is far higher than LLC and STC's under the consumption that the total die area is the same. The analysis in this paper complies with the efficiency on the heavy load of the converters shown in Figure 1.

Reference

[1] M. Shen, F. Z. Peng and L. M. Tolbert, "Multilevel DC–DC Power Conversion System With Multiple DC Sources," in IEEE Transactions on Power Electronics, vol. 23, no. 1, pp. 420-426, Jan. 2008.

[2]R. A. Abramson, Z. Ye, T. Ge and R. C. N. Pilawa-Podgurski, "A High Performance 48-to-6 V Multi-Resonant Cascaded Series-Parallel (CaSP) Switched-Capacitor Converter," *2021 IEEE Applied Power Electronics Conference and Exposition (APEC)*, Phoenix, AZ, USA, 2021, pp.

1328-1334.

[3]Z. Ye, R. A. Abramson and R. C. N. Pilawa-Podgurski, "A 48-to-6 V Multi-Resonant-Doubler Switched-Capacitor Converter for Data Center Applications," *2020 IEEE Applied Power Electronics Conference and Exposition (APEC)*, New Orleans, LA, USA, 2020, pp. 475-481.

[4]M. H. Ahmed, F. C. Lee, Q. Li and M. d. Rooij, "Design Optimization of Unregulated LLC Converter with Integrated Magnetics for Two-Stage 48V VRM," *2019 IEEE Energy Conversion Congress and Exposition (ECCE)*, Baltimore, MD, USA, 2019, pp. 521-528.

[5]M. Qiu, M. Wei, X. Liu, H. Meng and D. Cao, "A Matrix Autotransformer Switched-Capacitor Converter for Data Center Application," in *IEEE Transactions on Power Electronics*, vol. 38, no. 12, pp. 14982-14999, Dec. 2023,

2025 IEEE Workshop on Wide Bandgap Power Devices and Applications in Asia (WiPDA Asia)

A Totem Pole PFC Control Based on Dynamic Root Locus with Wide Output Voltage Range and Low THD

Hao Chu
School of Electrical and Electronic
Huazhong University of Science and
Technology
Wuhan,China
M202472392@hust.edu.cn

Xiangan Xiao
School of Electrical and Electronic
Huazhong University of Science and
Technology
Wuhan,China
U202112335@hust.edu.cn

Xuchen Sun
School of Electrical and Electronic
Huazhong University of Science and
Technology
Wuhan,China
M202372289@hust.edu.cn

Jiajia Guan
School of Electrical and Electronic
Huazhong University of Science and
Technology
Wuhan,China
jiajiaguan@hust.edu.cn

Cai Chen
School of Electrical and Electronic
Huazhong University of Science and
Technology
Wuhan,China
caichen@hust.edu.cn

Yong Kang
School of Electrical and Electronic
Huazhong University of Science and
Technology
Wuhan,China
ykang@hust.edu.cn

Abstract—The single-stage totem pole power factor correction (PFC) circuit is widely adopted as the front-end topology for on-board chargers (OBC) due to its high power factor (PF) and simplified structure. However, traditional control methods are inadequate for PFC control under a wide range of operating conditions. To address this challenge, this paper establishes a small signal model of the totem pole PFC current inner-loop and introduces a PI control method based on dynamic root locus (DRL-PI). Experimental results on a 3 kW platform show that compared with traditional fixed parameter control, the proposed control method reduces the estimated total harmonic distortion (THD) by 0.74% in constant resistance experiment and 3.36% in constant voltage experiment.

Keywords—Totem pole PFC, wide voltage range, dynamic root locus, low THD.

I. INTRODUCTION

With the development of the global economy, the energy crisis has become one of the most significant challenges facing humanity, resulting in an increasing demand for electrified transportation systems [1]. As a key component of electric vehicles and plug-in hybrid electric vehicles, the on-board charger (OBC) holds substantial application potential [2]. The typical structure of two-stage OBC is illustrated in Fig. 1. The single-stage bidirectional totem pole power factor correction (PFC) circuit has garnered extensive research and practical application as the pre-stage AC/DC converter due to its high power factor, simple structure, and low switching losses [3], [4], [5].

The voltage of a single cell in the electric vehicles typically ranges from 2.5 V to 4.2 V [6], [7], leading to a significant variation in the required charging voltage for the battery pack. To accommodate this wide voltage range and achieve higher overall efficiency, the two-stage OBC imposes requirements on the voltage regulation capabilities of its preceding totem pole power factor correction circuit [8], [9]. The traditional fixed parameters PFC control can not achieve accurate control and fast response at the same time in a wide range [10].

In order to realize the wide voltage range control of totem pole PFC with low current THD, this paper establishes large

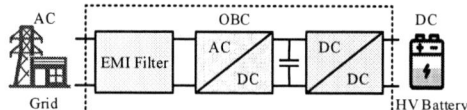

Fig. 1. Typical structure of OBC.

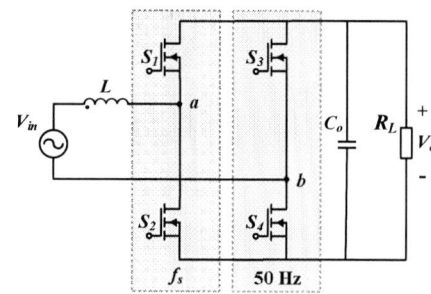

Fig. 2. Totem pole PFC topology.

and small signal models for totem pole PFC, simplifies the current transfer function, and proposes a totem pole PFC control based on dynamic root locus (DRL-PI). A 3 kW experimental platform was built to compare with the traditional fixed PI control. It is proved that the proposed method can effectively reduce total harmonic distortion (THD) in a wide voltage range.

II. LARGE AND SMALL SIGNAL MODEL ANALYSIS OF TOTEM POLE PFC

The topology of totem pole PFC is shown in Fig. 2, where V_{in} is the input voltage, V_o is the output voltage, L is the inductance, C_o is the output capacitance, and R_L is the output load. The switching tube S_1 and S_2 are high-frequency switching tubes. The switching tube S_3 and S_4 constitute a low-frequency synchronous rectification bridge, which operates at the utility frequency. Fig. 3 shows the switching waveform of the switching tubes in one cycle.

When the AC input voltage is under positive half cycle, the operating state shown in Fig. 4(a) and (b). During this stage, switching tube S_3 is off and S_4 is on. In the operating

979-8-3315-1110-4/25 $31.00 © 2025 IEEE

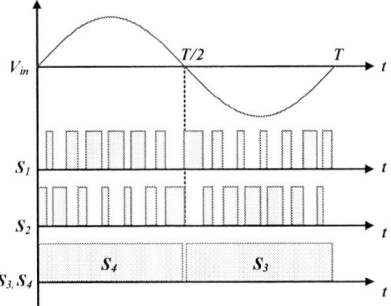

Fig. 3. The switching waveform of the totem pole PFC.

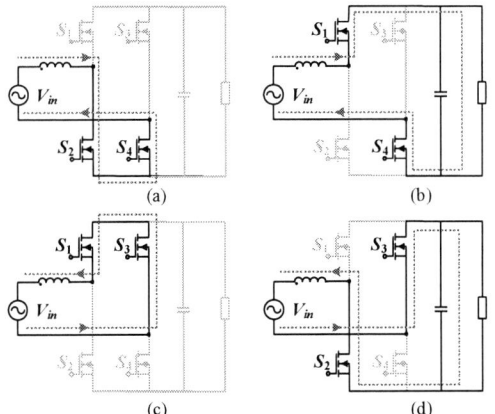

Fig. 4. Operation process stages of totem pole PFC. (a) Inductive charging under positive half cycle. (b) Inductive discharge under positive half cycle. (c) Inductive charging under negative half cycle. (d) Inductive discharge under negative half cycle.

state shown in Fig. 4(a), S_1 is off and S_2 is on, allowing the AC input to charge the inductor. At the same time, the capacitor supplies power to the output load, causing the output voltage to decrease. In the operating state shown in Fig. 4(b), S_1 is on and S_2 is off, causing the inductor to discharge. The input voltage and the inductor voltage are superimposed, simultaneously charging the capacitor and powering the load, resulting in an increase in the output voltage. When the AC input voltage is under negative half cycle, the operating states are shown in Fig. 4(c) and (d), where the switching tubes S_3 and S_4 operate in the opposite way to that described above. The analysis of these two operating states is similar to the states analysis above. During the negative half-cycle, the inductor is charged when S_1 is conducting and discharged when S_1 is not conducting.

Given the symmetrical working characteristics of totem pole PFC in both the positive and negative half cycles of the input voltage, this paper examines the positive half.

$$\dot{X} = AX + BU, X = \begin{bmatrix} i_L \\ V_o \end{bmatrix}, U = \begin{bmatrix} V_{in} \end{bmatrix} \quad (1)$$

where i_L is the inductive current. For a complete cycle, the state equation of the totem pole PFC large signal model within a switching period is:

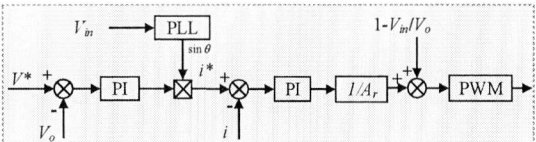

Fig. 5. Traditional control scheme.

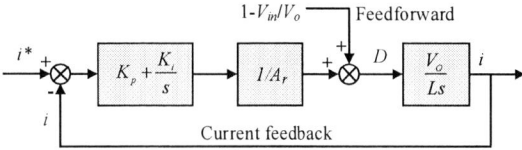

Fig. 6. Control block diagram of the current inner-loop.

$$\dot{X} = AX + BU, A = \begin{bmatrix} 0 & -\dfrac{1-D}{L} \\ \dfrac{1-D}{C} & -\dfrac{1}{R_L C} \end{bmatrix}, B = \begin{bmatrix} 1/L \\ 0 \end{bmatrix} \quad (2)$$

where D is the duty cycle.

Small signal model is a common analysis method in power electronics and other fields. By linearizing nonlinear circuits, the analysis of circuits is more intuitive. In this paper, perturbation method is used to build a small signal model of totem pole PFC. On the basis of the large signal model, add small perturbations. Disregarding second-order small quantities, the small-signal model (3) is separated:

$$\dot{\hat{x}} = A\hat{x} + B\hat{u} + \left[(A_1 - A_2)X + (B_1 - B_2)U \right]\hat{d} \quad (3)$$

where the lower case letters represent the small perturbations of the signal corresponding to the upper case. In double closed-loop control, the inner current loop ensures that the input current is close to the sinusoidal waveform, which is the main factor affecting the input current THD. The following analysis in this paper will focus on the control of the inner current loop.

Considering that the system is in a steady state, the effect of the feedforward loop is ignored. The transfer function from duty cycle to input inductor current can be derived through Laplace transform of the small-signal model equation:

$$G_{id} = \frac{V_o CRs + 2V_o}{LCRs^2 + Ls + (1-D)^2 R}. \quad (4)$$

The totem pole PFC converter designed in this paper works at a switching frequency of 75 kHz. As mentioned in [11], the analysis of the transfer function of the current loop focuses on the response characteristics of high frequency. The transfer function is simplified as follows:

$$G_{id} = \frac{V_o}{Ls}. \quad (5)$$

III. WIDE RANGE PI CONTROL BASED ON DYNAMIC ROOT LOCUS

Fig. 5 shows the traditional control scheme, where PLL denotes the phase-locked loop and A_r represents the comparison value for sinusoidal pulse width modulation. This scheme uses double closed-loop control of the voltage

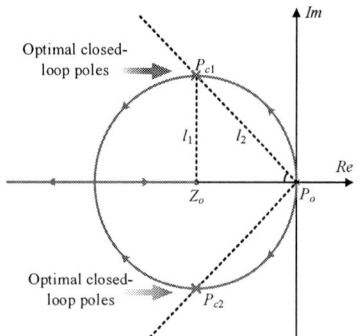

Fig. 7. The root locus curve of the inner current loop.

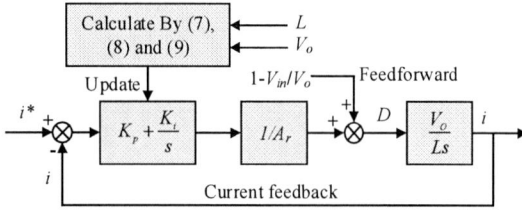

Fig. 8. The proposed DRL-PI control block diagram.

outer loop and the current inner-loop, and introduces feedforward in the current inner-loop to accelerate the response and stabilize the system.

The control block diagram of the current inner-loop can be obtained as shown in Fig. 6. The open-loop transfer function expressed is:

$$GH(s) = \left(\frac{s + K_i / K_p}{s^2}\right) \cdot \frac{K_p V_o}{A_r L} \qquad (6)$$

where K_p is the proportional coefficient, K_i is the integral coefficient, and the Evans' gain value is expressed as $K_E = K_p V_o / A_r L$. The root locus curve of the inner current loop is shown in Fig. 7, where there are two open-loop poles P_0 and one open-loop zero $Z_0(-K_i/K_p, 0)$, and the directions of the two root locus branches are marked in the figure. For a particular value of Evans' gain, P_{c1} and P_{c2} are two closed-loop poles, l_1 and l_2 are respectively the distances from the closed-loop pole to the open-loop pole and the open-loop zero on the root locus.

For closed-loop systems, $\xi = 0.707$ is often used as a design guideline. Under this damping ratio, the system responds faster and has an overshoot of less than 5% [12]. At this time, $\theta = 45°$, $l_1 = -K_i/K_p = l_2/\sqrt{2}$, and the closed-loop poles are within the optimal response interval. The PI controller parameters satisfy:

$$\frac{K_p^2}{K_i} = \frac{2C_p L}{V_o}. \qquad (7)$$

The larger the absolute value of the X-coordinate of the closed-loop pole is, the larger the attenuation index of the system, and thus the faster response can be obtained. In this paper, the attenuation index is selected as $\xi\omega_n = 10^4$, which satisfies:

$$K_i / K_p = \xi\omega_n. \qquad (8)$$

Fig. 9. Prototype of the experimental OBC.

TABLE I. PROTOTYPE PARAMETERS

Parameters	Values
Input AC voltage (V_{in})	220 V@50 Hz
Input AC current (i_L)	0~14 A
Output voltage (V_o)	350~700 V
Max Output power (P_{max})	3.0 kW
Switching frequence	75 kHz
Inductance (L)	200 μH
Output capacitance (C_o)	260 μF
Differential mode inductance	50 μH
Differential mode capacitance	4.4 μF
Controller	TMS320F28379S

The relationship between PI parameters in the digital controller and the continuous time domain transfer function derived is as follows:

$$\begin{cases} K_p' = K_p \\ K_i' = TK_i \end{cases}. \qquad (9)$$

(7), (8) and (9) illustrate the approximate relationship between the discrete PI control parameters and the circuit parameters in the current inner-loop of the discrete control system. It is obvious that the control parameters are mainly affected by the inductance and output voltage of the totem pole PFC circuit. Fig. 8 shows the PI control based on dynamic root locus (DRL-PI) proposed in this paper. By substituting the inductance value and the target output voltage into the function, the PI parameters can be adjusted continuously, so that the current inner-loop can achieve a better control effect in a larger output voltage working range.

IV. EXPERIMENTAL RESULTS AND ANALYSIS

In order to evaluate the performance of the proposed control scheme, a three-phase vehicle OBC circuit of the 900 V platform was built as a test prototype, as shown in Fig. 9. In the experiment, only the 3 kW single-phase totem pole PFC is tested and analyzed. The main parameters of the circuit are shown in Table I.

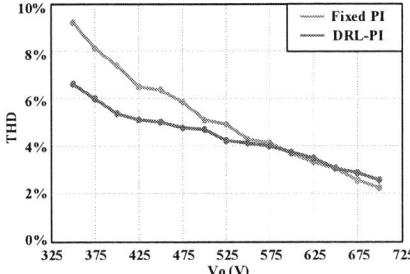

Fig. 10. 160 Ω constant resistance experiment.

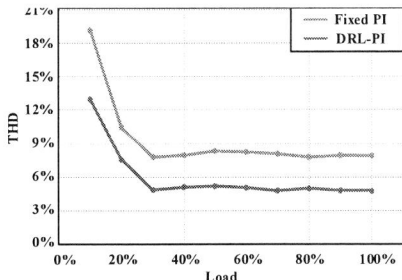

Fig. 11. 400 V constant voltage experiment.

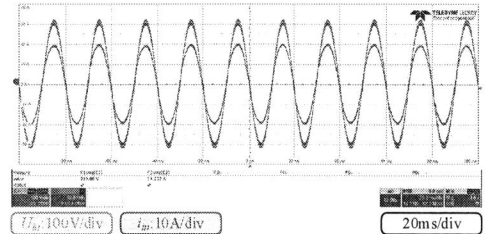

U_b: 100 V/div   i_B: 10A/div   20ms/div

Fig. 12. The measured waveform of the DRL-PI at 400 V output full load.

Urms1	0.2197	kV	Urms2	0.4014	kV
Irms1	14.169	A	Irms2	7.614	A
P1	3.1064	kW	P2	3.0552	kW
Ithd1	4.54	%		98.35	%
λ1	0.9980	C	fU1	50.005	Hz
Idc1	-0.025	A	Cfi1	1.46	

Fig. 13. Screenshot of the power analyzer.

In the comparison experiment, the PI parameter set at 700 V voltage output was used as a fixed PI parameter for control, which was compared with the DRL-PI control designed in this paper. The input current THD data was collected by the power analyzer PA5000H to evaluate the control effect. Fig. 9 shows the results of constant resistance experiment by changing the output voltage. Fig. 11 shows the results of constant voltage experiment by changing the output load resistance when the output voltage is constant 400 V. DRL-PI can effectively reduce THD, compared with the fixed PI control, the average THD reduction of 160 Ω constant resistance experiment is 0.74%, and the average THD reduction of 400 V constant voltage experiment is 3.36%. Fig. 12 shows the measured waveform of the proposed algorithm at 400 V output full load. Fig. 13 shows a screenshot of the power analyzer.

V. CONCLUSION

In this paper, a control method for totem pole PFC based on dynamic root locus is proposed, which can achieve low current THD over a wide output voltage and load range. Compared with the traditional fixed parameter control, this control method shows good adaptive characteristics under a wide output voltage range.

Experimental results have verified that using the proposed method, THD decreases by 0.74% on average in constant resistance experiment and 3.36% on average in constant voltage experiment.

REFERENCES

[1] V. I. Silaev, O. A. Gavrina and A. V. Kuzina, "The Problem of Renewable Energy Sources and Market Mechanisms in Various Energy Systems in the Era of Global Crises," 2023 International Ural Conference on Electrical Power Engineering (UralCon), Magnitogorsk, Russian Federation, 2023, pp. 83-88.

[2] H. Wouters and W. Martinez, "Bidirectional Onboard Chargers for Electric Vehicles: State-of-the-Art and Future Trends," in IEEE Transactions on Power Electronics, vol. 39, no. 1, pp. 693-716, Jan. 2024.

[3] T. Li et al., "Design and optimization of a high efficiency energy router," 4th Energy Conversion and Economics Annual Forum (ECE Forum 2024), Beijing, China, 2025, pp. 1154-1158.

[4] T. Liu, C. Chen, K. Xu, Y. Zhang and Y. Kang, "GaN-Based Megahertz Single-Phase Inverter With a Hybrid TCM Control Method for High Efficiency and High-Power Density," in IEEE Transactions on Power Electronics, vol. 36, no. 6, pp. 6797-6813, June 2021.

[5] J. Guan, Z. Wang, Z. Tang, J. Lv, C. Chen and Y. Kang, "Design of a High Power Density Bidirectional AC/DC Converter Based on GaN," 2021 IEEE Workshop on Wide Bandgap Power Devices and Applications in Asia (WiPDA Asia), Wuhan, China, 2021, pp. 393-397.

[6] P. Mohseni, O. Husev, D. Vinnikov, R. Strzelecki, E. Romero-Cadaval and I. Tokarski, "Battery Technologies in Electric Vehicles: Improvements in Electric Battery Packs," in IEEE Industrial Electronics Magazine, vol. 17, no. 4, pp. 55-65, Dec. 2023.

[7] Y. Wei and P. Sun, "Review of Techniques for Resonant Converters With Wide Voltage Gain Range Applications," in IEEE Transactions on Transportation Electrification, vol. 10, no. 3, pp. 5544-5569, Sept. 2024.

[8] H. Wang, S. Dusmez and A. Khaligh, "Maximum Efficiency Point Tracking Technique for LLC-Based PEV Chargers Through Variable DC Link Control," in IEEE Transactions on Industrial Electronics, vol. 61, no. 11, pp. 6041-6049, Nov. 2014.

[9] A. Kazemtarghi, N. Ishraq, P. Rathod, A. Mallik and N. Johnson, "DC Link Voltage Optimization for Efficiency Enhancement of Electric Vehicle Onboard Chargers," 2023 IEEE 2nd Industrial Electronics Society Annual On-Line Conference (ONCON), SC, USA, 2023, pp. 1-6.

[10] C. Song and H. Li, "Evaluation of Efficiency and Power Factor in 3-kW GaN-Based CCM/CRM Totem-Pole PFC Converters for Data Center Application," 2023 11th International Conference on Power Electronics and ECCE Asia (ICPE 2023 - ECCE Asia), Jeju Island, Korea, Republic of, 2023, pp. 2663-2668.

[11] L. Huber, M. Kumar and M. M. Jovanović, "Performance Comparison of PI and P Compensation in DSP-Based Average-Current-Controlled Three-Phase Six-Switch Boost PFC Rectifier," in IEEE Transactions on Power Electronics, vol. 30, no. 12, pp. 7123-7137, Dec. 2015.

[12] G. C. Goodwin, S. F. Graebe, and M. E. Salgado, Control System Design, 1st ed. USA: Prentice Hall PTR, 2000.

A SiC/Si Hybrid Pulsed Power Supply for High-Current Laser Diode Driving

Yihui Tang
School of Electrical Engineering And Automation
Hefei University of Technology
Hefei, China
yihui.tang@mail.hfut.edu.cn

Xingye Chen
School of Electrical Engineering And Automation
Hefei University of Technology
Hefei, China
xingye.chen@mail.hfut.edu.cn

Yan Su
School of Electrical Engineering And Automation
Hefei University of Technology
Hefei, China
suyan@mail.hfut.edu.cn

Helong Li
School of Electrical Engineering And Automation
Hefei University of Technology
Hefei, China
helong.li@mail.hfut.edu.cn

Zhiqing Yang
School of Electrical Engineering And Automation
Hefei University of Technology
Hefei, China
zhiqing.yang@mail.hfut.edu.cn

Xianbin Qi
School of Electrical Engineering And Automation
Hefei University of Technology
Hefei, China
xianbin_qi@hfut.edu.cn

Abstract—In response to the development trends toward lightweight and digitalization of semiconductor laser pump driving power supplies, this paper proposes a switched constant-current-driven pulsed discharge network based on SiC (silicon carbide) devices, capable of flexibly adjusting current amplitude, pulse width, and repetition frequency. Compared to traditional linear constant-current driving methods, the switched constant-current driving approach significantly reduces conduction losses, making it more suitable for long-pulse applications. A diode is used instead of a traditional resistor as the load, making it closer to actual operating conditions. A modeling analysis and control design for the constant-current pulsed discharge network are presented, achieving precise pulse current control at the microsecond scale. A prototype power supply was developed, demonstrating a maximum output of 100 A pulsed constant current with a rise time under 10 μs, a maximum pulse width of 300 μs, and a repetition frequency up to 1000 Hz.

Keywords—SiC, switched-mode power supply, pulsed power supply, interleaved buck, semiconductor pumping

I. INTRODUCTION

Pulse lasers play a significant role in modern technology and defense applications. Laser diode-pumped solid-state lasers (DPSSL) combine the advantages of both semiconductor lasers and solid-state lasers, offering benefits such as compact size, low weight, high efficiency, excellent beam quality, high reliability, and long lifespan. As a result, they represent an important development direction for solid-state laser technology [1-3].One of the key components of a pulse laser system is the pulsed power supply, which provides the necessary energy.

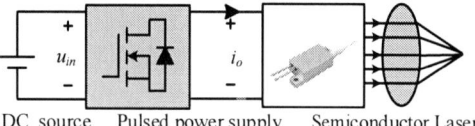

DC source   Pulsed power supply   Semiconductor Laser

Fig.1. Laser system brief architecture.

This paper focuses on the research of pulse power supplies in the directions of large-current output, high repetition frequency, and low current ripple. Semiconductor pumping is highly sensitive to minute current variations, where even subtle changes can lead to a reduction in output laser energy,

This work was supported in part by Fundamental Research Funds for the Central Universities under Grant PA2024GDSK0085, in part by Anhui Province Key Laboratory of Semiconductor Packaging and Reliability (Hefei University of Technology)

thus imposing strict requirements on current ripple. Additionally, it is required that the pulse current has no overshoot, as excessive current overshoot can directly damage the semiconductor pump. The output pulse current of high-power pulse power supplies is required to have a steep rising edge, which needs to be controlled within 20 μs. This paper analyzes the discharge network of the pulsed power supply, and a common power supply architecture is shown in Fig. 1.

Generally speaking, the driving modes of semiconductor lasers can be divided into two types: linear mode driving and switching mode driving. The driver power supply adopting the linear mode features a simple structure, low cost, and the ability to produce sufficiently fast pulse rise times. The linear control method operates the switching device in the linear mode by controlling the gate voltage, resulting in a fast response. However, under such conditions, the conduction loss of the device is relatively large, making it unsuitable for achieving long pulses. Reference [4] proposed a laser pulse power supply with multiple energy storage units, which achieves high repetition frequency and high-current output. However, it adopts a linear constant-current mode for control, which results in lower efficiency, significant heat generation, and the requirement for large heat dissipation components, making it unfavorable for device miniaturization. In contrast, switching mode drivers offer higher efficiency. A research team from the University of North Carolina developed a pulse power supply based on capacitor energy storage, adopting switching mode driving. By combining digital and analog control methods, they achieved a pulse current of 50 A. Nevertheless, the output current is entirely borne by a single device, leading to excessive current stress on the device and making it difficult to achieve high-current pulsed outputs [5]. A typical topology uses a traditional Buck circuit with a purely resistive load, which results in relatively simple control. However, due to the inability of the inductor current to change instantaneously, the pulse current's edge and ripple are significantly affected by circuit parameters [6].

To address the aforementioned issues, this paper proposes a low-voltage, wide-output-current-range semiconductor pulsed drive current source, adopting a switch-mode discharge topology. By utilizing wide-bandgap semiconductor devices (SiC) to increase the switching frequency and employing a multiphase interleaved Buck converter, the system effectively reduces output current ripple. Since the pulsed power supply requires microsecond-scale constant current control, this paper conducts modeling and control design for the pulse-forming network with a diode load. To overcome the slow response speed of traditional switch-mode power supplies, a high-energy flow rapid switching mechanism is implemented by controlling two high-power IGBTs, enabling a faster pulse rise time. The technical specifications of this power supply are as follows: input voltage of 80~200 V, adaptive output voltage based on load, pulse current of 100 A, pulse repetition frequency of 10~1000 Hz, pulse width of 100~300 μs, and pulse rise time within 20 μs. Ultimately, this paper presents a prototype of a pulsed power supply capable of delivering a maximum pulse current of 100 A, a repetition frequency of up to 1000 Hz, and a current pulse width of 300 μs.

II. SYSTEM ANALYSIS AND MODELING

This paper presents an analysis and modeling of the topology's operating principles. The proposed topology employs a switching transistor drive scheme to modulate both high- and low-frequency switching operations. In this paper, the output current ripple is suppressed by adopting wide - bandgap semiconductor devices (SiC devices) to increase the switching frequency and using the interleaved parallel connection of inductor branches. A prototype of pulse power supply is designed in this paper, which has a maximum output pulse current of 100 A, a maximum repetition frequency of 1000 Hz, and a current pulse width of 300 μs. Since semiconductor lasers are costly and their load characteristics are similar to those of diodes, multiple series - connected diode loads are used in this paper to simulate the semiconductor pump load. Specifically, SiC MOSFETs (S_1 and S_2) are utilized for high-frequency switching, while IGBTs (Q_1 and Q_2) handle low-frequency switching. The detailed analysis proceeds as follows.

A. Inductive Storage

At this stage, the switching transistors S_1, S_2, and Q_1 are turned on, while Q_2 is turned off. The inductors L_1 and L_2 are charged by the input voltage u_{in}, and their currents increase linearly. Where L represents the inductance, i_L denotes the single-channel current, and R_L is the parasitic resistance of the inductor. The equivalent circuit diagram and modeling are as follows.

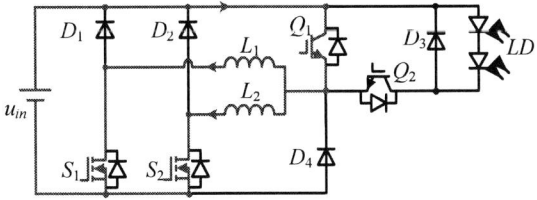

Fig.2. Equivalent circuits at inductive storage stages.

$$L \frac{di_L}{dt} = u_{in} - i_L R_L \qquad (1)$$

B. Pulse Forming

At this stage, the switching transistor Q_1 is turned off, while Q_2 is turned on. At this moment, the circuit is equivalent to a two-phase interleaved buck converter. The driving signals of the two parallel buck circuits are controlled by interleaving technology to achieve lower current ripple [7], [8].The switching transistors S_1 and S_2 operate in a high-

frequency mode with a 180° phase shift to maintain a constant inductor current. Where u_o denotes the output voltage. The equivalent circuit diagram and modeling are as follows.

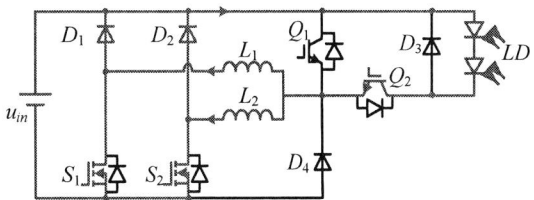

Fig.3. Equivalent circuits at pulse forming stages.

$$\begin{cases} L \dfrac{di_L}{dt} = u_{in} - u_o - i_L R_L, & \text{S ON} \\[2mm] L \dfrac{di_L}{dt} = -u_o - i_L R_L, & \text{S OFF} \end{cases} \qquad (2)$$

C. Energy Recovery

At this stage, all switching transistors are turned off. The remaining energy stored in inductors L_1 and L_2 is fed back to the power supply through D_1, D_2, and D_4. The residual load energy continues to flow through the freewheeling diode D_3. The equivalent circuit diagram and modeling are as follows.

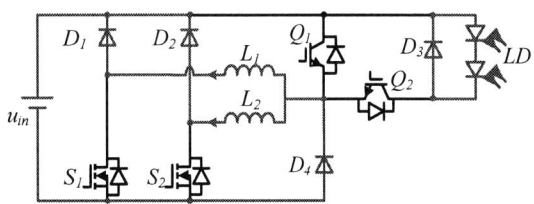

Fig.4. Equivalent circuits at energy recovery stages.

$$L \frac{di_L}{dt} = -u_{in} - i_L R_L \qquad (3)$$

III. CONTROL AND HARDWARE DESIGN

A. Pre-Charging Control

The interleaved parallel technique adopted in this paper can reduce both the required inductance value and physical size of the inductor components. Typically, the inductance values range from several tens to several hundreds of microhenries (μH), which makes the precise control of the inductor energy more challenging. For example, when the input voltage u_{in} is 150 V and the storage inductor L is 93 μH, neglecting the inductor's internal resistance, the time required for the inductor current to rise to 50 A can be calculated using the following equation:

$$\Delta t = \frac{L di_{Lref}}{u_{in}} \qquad (4)$$

According to Equation (4), the inductor charging time is 31 μs. Thus, the inductor charging rate is 1.61 A/μs, which means that if PWM mode were used for charging, the digital controller's computation frequency would need to reach 1610 kHz to achieve precise control of the inductor current within 1 A-an obviously impractical requirement. Therefore, directly charging the inductor is not advisable. Instead, this study adopts a fixed pre-charging time while adjusting the pre-charging duty cycle d_{pre} to charge the inductor, enabling more accurate current control. The inductor charging current and PWM signal waveform for a single branch are shown in Fig. 5. The internal resistance of the inductor is neglected, and the inductor current is assumed to remain nearly loss-free during the low-level period of the driving signal.

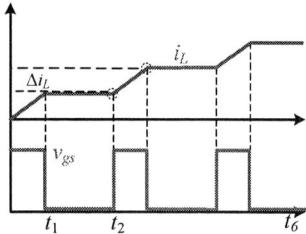

Fig.5. Inductive precharge waveform.

$$d_{pre} = \frac{Li_{Lref}}{u_{in}\Delta t} \tag{5}$$

In the equation, Δt represents the total pre-charging time, and i_{Lref} denotes the target current value to which the inductor needs to be charged. Based on these parameters, the required pre-charging duty cycle d_{pre} can be calculated.

B. Pulsed Current Regulation

When transistor Q_2 is turned on, the pulsed current rapidly reaches the flat-top phase. Since the semiconductor-pumped load has stringent requirements for current stability, a stable flat-top output current must be maintained. Therefore, closed-loop control is employed to ensure current stability. Considering that the diode load can be equivalently modeled as a voltage source in series with a small AC resistance , this paper applies the average switching modeling method to mathematically model the interleaved parallel Buck circuit [9], [10]. The equivalent circuit model is shown in Fig. 6. From the equivalent circuit model, the small-signal AC model can be derived and is expressed as [11]:

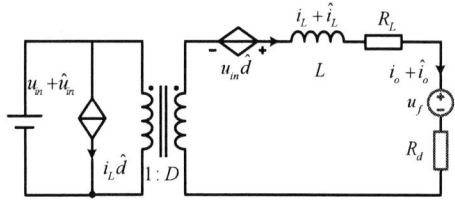

Fig.6. Interleaved buck converter average switching model.

$$\begin{cases} \dfrac{Ld\hat{i}_L}{dt} = -R_L\hat{i}_L - \hat{u}_o + D\hat{u}_{in} + u_{in}\hat{d} \\ \hat{u}_o = \hat{i}_o R_d + u_f \end{cases} \tag{6}$$

In the equation: L is the inductance value; i_L represents the single-phase inductor current; R_L denotes the inductor's parasitic resistance; i_o indicates the output current; R_d stands for the equivalent load resistance; u_f corresponds to the diode's threshold voltage; u_{in} and u_o refer to the input voltage and output voltage, respectively. Under the assumptions that the inductance and parasitic resistance are constant and identical across all phases, and that the inductor currents are balanced, the following expression can be derived:

$$L\frac{d\hat{i}_L}{dt} = -R_L\hat{i}_L - \hat{i}_L R_d + u_{in}\hat{d} \tag{7}$$

The transfer function from output current to duty cycle is:

$$G_{id} = F_{plant} = \frac{\hat{i}_o}{\hat{d}} = \frac{u_{in}}{Ls + R_L + R_d} \tag{8}$$

Upon deriving the transfer function of the topology, a proportional-integral (PI) control strategy is adopted to ensure system stability. The PI parameters obtained using the pole-zero cancellation method are as follows:

$$\begin{cases} K_p = \dfrac{\omega L}{u_{in}} \\ K_i = \dfrac{\omega\left(R_L + R_d\right)}{u_{in}} \end{cases} \tag{9}$$

$$F_{cc} = K_p + \frac{K_i}{s} \tag{10}$$

As evidenced by the preceding equations,the current loop control block diagram is shown below.

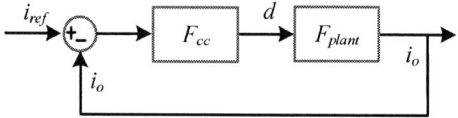

Fig.7. Current loop control block diagram.

Consequently, the closed-loop transfer function of the system is obtained as:

$$G_{cl} = \frac{G_{id}}{1+G_{id}} = \frac{\omega}{s+\omega} = \frac{2\pi f_i}{s+2\pi f_i} \tag{11}$$

In actual circuit implementations, the inductance exhibits current-dependent characteristics rather than maintaining a constant value. In order to analyze the controller design after inductor saturation, a Bode plot analysis was conducted for three sets of inductors.

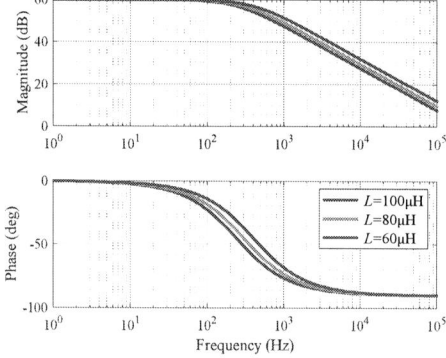

Fig.8. Topology transfer function under different inductances.

As illustrated in the graph, an increase in inductance results in a reduction of the cutoff frequency, which in turn causes the amplitude curve to decay earlier at low frequencies. A larger inductance leads to a significant phase lag at lower frequencies, and the entire phase transition region correspondingly shifts towards lower frequencies.

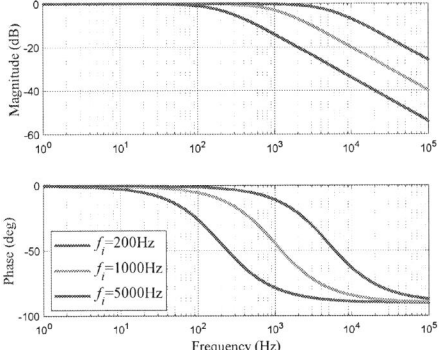

Fig.9. Closed-loop transfer function under different bandwidths.

To verify the feasibility of the topology modeling and the design performance of the controller, this paper builds a model in the PLECS simulation software for verification.

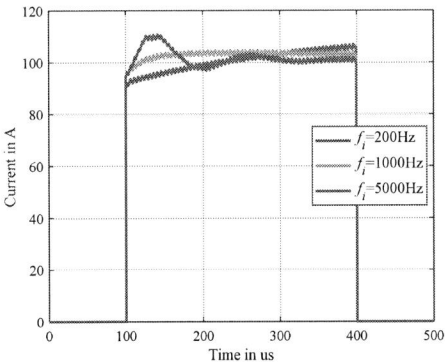

Fig.10. Output Current Simulation Waveforms Under Different Bandwidths

By comparing the Bode plots, it is evident that the bandwidth directly determines the system's ability to retain signal gain within the frequency domain as well as the rate of phase change. Ultimately, a bandwidth of 1000 Hz is chosen for the design of the controller.

Fig. 10 illustrates the impact of current loop bandwidth on system response. When the current loop bandwidth is set to 200 Hz, the system exhibits a relatively slow response, requiring a longer duration to reach the steady state. This is attributed to the limited bandwidth restricting the speed at which the controller can react to changes in the current. Increasing the bandwidth to 1000 Hz significantly improves the system's dynamic performance. As shown in Figure 10, the current achieves a more stable output from the initial stage, indicating a faster and more precise response to the control command. However, an excessively large bandwidth, such as 5000 Hz, leads to undesirable oscillations in the current waveform and substantial overshoot. This is because a very high bandwidth can make the system more susceptible to noise and disturbances, potentially causing instability.

Therefore, the simulation results presented in Figure 10 validate the design considerations for the current loop controller, demonstrating the importance of selecting an appropriate bandwidth to achieve a balance between response speed and stability.

C. Energy Storage Inductor Parameter Design

The operational modes of interleaved parallel converters can be classified into three categories based on whether the inductor current drops to zero during each switching cycle: Continuous Conduction Mode (CCM), Boundary Conduction Mode (BCM), and Discontinuous Conduction Mode (DCM). In the proposed topology, the system operates in CCM mode (interleaved parallel Buck configuration) during the flat-top phase of output current. The ripple current (Δi_L) of a single inductor is given by:

$$\Delta i_L = \frac{u_{in}d(1-d)}{Lf_s} \quad (12)$$

In the equation, d represents the duty cycle of the MOSFET drive signal during the flat-top current phase. The formula indicates that the inductor current ripple is inversely proportional to the inductance value-larger inductance results in smaller current ripple. For interleaved Buck converters, the inductor significantly affects both the output current ripple and power density, making its practical design particularly critical. According to Reference [12], the total output current ripple of the interleaved parallel configuration is given by:

$$\begin{cases} \Delta i_o = \dfrac{u_{in}}{Lf_s} \cdot (d - \dfrac{m}{n}) \cdot (1+m-nd) \\ k(d,n) = (d - \dfrac{m}{n}) \cdot (1+m-nd) \end{cases} \quad (13)$$

where: $k(d,n)$ is the current ripple coefficient, n represents the number of interleaved phases , $m = \text{floor}(dn)$. When the current ripple coefficient $k(d,n)$ reaches its maximum value, there exist n corresponding duty cycles d . Based on the above formula, the value of the energy storage inductance can be determined as:

$$L = \frac{u_{in}k_m(d,n)}{\Delta i_o f_s} \quad (14)$$

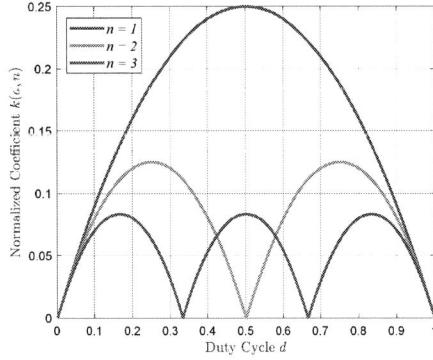

Fig.11. Normalized coefficient of total current ripple.

Taking the two-phase interleaved buck converter in this paper as an example, the circuit parameters are as follows: u_{in}=150 V, f_s=100 kHz, output current ripple Δi_o=2 A, and as shown in Fig. 11, the output current ripple coefficient $k_m(d,n)$=0.125. According to the following formula, the inductance L=100 µH.

D. Design of Bus Capacitor Parameters

Since the topology in this paper needs to generate high-frequency pulse current, frequent charging and discharging of the bus capacitor is required. To minimize the voltage drop across the capacitor after the pulse occurs, a large-capacitance capacitor must be used. The design formula for the capacitor's capacitance is as follows:

$$\frac{1}{2}C(u_1^2 - u_2^2) = i_o u_o t_p \tag{15}$$

The energy stored in the capacitor is depleted after each pulse discharge, resulting in a voltage drop. Therefore, an appropriate capacitance value must be calculated to ensure that the voltage fluctuation remains within an acceptable range after each pulse. Here, u_1 and u_2 represent the initial voltage of the capacitor and the remaining voltage after a single pulse discharge, respectively, with $u_1 = 150$ V and $u_2 = 145$ V. The pulse current is approximated as a rectangular waveform, where i_o is the flat-top pulse current, u_o is the output voltage, and t_p is the duration of a single pulse. In this design, the output topology generates a flat-top pulse current $i_o = 100$ A, with an output voltage $u_o = 18$ V and a pulse duration $t_p = 300$ μs. Based on these parameters, the calculated capacitance C is 732 μF. Considering design margin, a capacitor with a final capacitance value of 800 μF is selected.S

IV. EXPERIMENTAL VERIFICATION

To verify the feasibility of the proposed discharge forming network and control strategy, an experimental prototype has been built. The device topology employs 650 V, 120 A high-switching-frequency SiC MOSFETs for switches S_1 and S_2, and 650 V, 150 A low-switching-frequency IGBTs for switches Q_1 and Q_2. Since S_1 and S_2 operate at a switching frequency of $f_s = 100$ kHz, the diodes in series with them must also work at the same frequency. To minimize losses, Schottky diodes (which have no reverse recovery characteristics) were selected.

To enhance control flexibility, the prototype in this study adopts digital control. The controller is implemented using Texas Instruments' LAUNCHXL-F28379D DSP platform for digital control. In this design, a constant DC source is directly connected across the bus capacitor for power supply, so the bus capacitance is relatively small, primarily serving for voltage stabilization and filtering. Since semiconductor lasers are costly and their load characteristics are similar to those of diodes, multiple series - connected diode loads are used in this paper to simulate the semiconductor pump load. The remaining hardware circuit parameters are listed in Table 1.

TABLE I. PARAMETERS FOR THE CIRCUIT

Parameter	Symbol	Value
Switching frequency	f_s	100 kHz
DC voltage	u_{in}	150 V
Inductor	L	100 μH
Output current	i_o	100 A
Experimental load	*Diode*	DSEI2X101-06A
Pulse frequency	f_p	1000 Hz
Pulse width	T_p	300 μs

To verify the feasibility of the proposed topology, this paper designed an interleaved parallel Buck-converter-based pulsed laser power supply hardware test platform, as shown in Fig. 12.

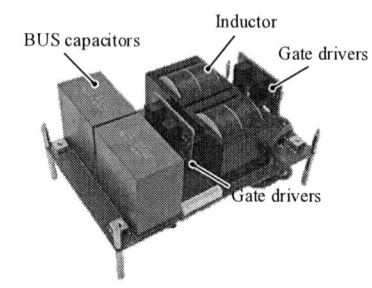

Fig.12. Pulsed power supply prototype.

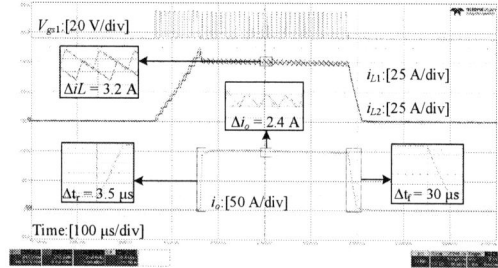

Fig.13. Discharge experiment waveform. $i_o = 100$ A,f_p=1 kHz,T_p= 300 μs.

As shown in Fig. 13, the current distribution between the two inductor branches remains well balanced. The total output pulse current ripple is significantly smaller than the inductor current ripple, indicating that the interleaved parallel structure can effectively reduce the output current ripple. In addition, the rise time of the output pulse current is less than 10 μs, confirming that the topology has a fast current response speed. The fall time of the pulse current is approximately 30 μs, indicating that the freewheeling circuit of pulse discharge has a strong ability to consume excess energy.

Fig. 14 shows the output current waveforms at different pulse frequencies when the pulse current is 50 A. By comparing the waveforms at different current levels, it can be found that: as the output current increases, the rise edge of the pulse current becomes slower. This is because when the switching device is turned on, the current flowing through the device does not reach the maximum value instantaneously, but gradually increases with the formation of the device's conductive channel. The larger the current, the longer the establishment time required. Similarly, the fall edge of the pulse current relies on the freewheeling path of the diode for energy dissipation, and a larger current also requires a longer decay time. The above experiments verify the feasibility of the pulse laser power supply topology based on the interleaved parallel architecture.

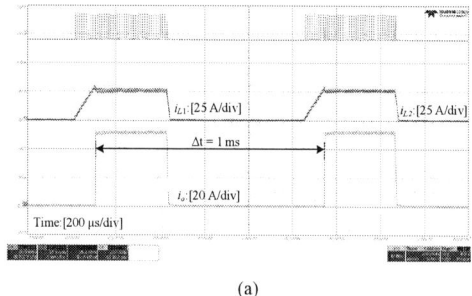

(a)

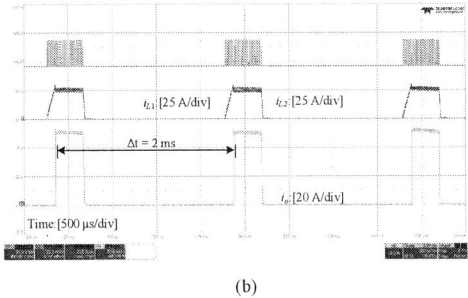

(b)

Fig.14. Discharge waveform under different operating conditions (a) $i_o = 50$ A, f_p=1 kHz, T_p = 300 μs.(b) $i_o = 50$ A, f_p=500 Hz, T_p = 300 μs.

V. CONCLUSION

This paper primarily focuses on the modeling and analysis of an interleaved parallel Buck-based pulsed power supply operating with a diode load. A control method for precharging the energy storage inductor is proposed, with a detailed analysis of the control strategy during the pulsed constant current phase. Additionally, the hardware parameters of the key components are designed. Finally, a pulsed power supply prototype was developed, achieving an output pulse current of 100 A, a pulse width of 300 μs, and a repetition frequency of 1000 Hz.

REFERENCES

[1] D. Morton et al.Pulsed power design for a small repetitively pulsed electron beam pumped KrF laser, Digest of Technical Papers[C]// PPC-

[2] Platz, René et al. Progress in high duty cycle, highly efficient fiber coupled 940-nm pump modules for high-energy class solid-state lasers[J]. LASE (2018).

[3] Li Ji, Chen Jiexiang, Zhang Yi, et al. High-power diode laser drive power supply[J]. Journal of Quantum Electronics,2003,(01):30-34

[4] Zhou Jinlang, Shen An, He Jiaojiao, et al. Research on a high-power and high-repetition frequency pulsed laser power supply[J]. Electronic Design Engineering ,2022,30(06):151-155.

[5] R. K. Kokkonda, S. Bhattacharya, V. Veliadis et al. A SiC based Two-Stage Pulsed Power Converter System for Laser Diode Driving Applications[C]//2022 IEEE Energy Conversion Congress and Exposition (ECCE), Detroit, MI, USA, 2022.

[6] S. M. Granat,.Switched mode current source for pulsed loads[P] Ed: Google Patents, 2010.

[7] I. -O. Lee, S. -Y. Cho and G. -W. Moon. Interleaved Buck Converter Having Low Switching Losses and Improved Step-Down Conversion Ratio[J]. IEEE Transactions on Power Electronics, vol. 27, no. 8, pp. 3664-3675, Aug. 2012.

[8] A. Priyadarshi, P. K. Kar and S. B. Karanki. An inductor-less bidirectional DC-DC converter topology for high voltage gain applications[C]// TENCON 2017 - 2017 IEEE Region 10 Conference, Penang, Malaysia, 2017, pp. 303-308.

[9] G. Gkizas, C. Yfoulis, C. Amanatidis, et al. Digital state-feedback control of an interleaved DC–DC boost converter with bifurcation analysis[J], Control Engineering Practice,Volume 73,2018,Pages 100-111.

[10] R. W. Erickson and D. Maksimovic. Fundamentals of power electronics[M]. Springer Science & Business Media, 2007.

[11] A. Villarruel-Parra and A. J. Forsyth, "Modeling Phase Interactions in the Dual-Interleaved Buck Converter Using Sampler Decomposition," in IEEE Transactions on Industrial Electronics, vol. 66, no. 5, pp. 3316-3322, May 2019, doi: 10.1109/TIE.2018.2854585.

[12] X. Yang, S. Zong and G. Fan. Analysis and validation of the output current ripple in interleaved buck converter[J]. IECON 2017 - 43rd Annual Conference of the IEEE Industrial Electronics Society, Beijing, China, 2017, pp. 846-851

2025 IEEE Workshop on Wide Bandgap Power Devices and Applications in Asia (WiPDA Asia)

Interleaved LLC Resonant Converter with Hybrid PSM/PFM Control for Wide Voltage Gain Applications

1st Jintong Dong
School of Electrical Engineering & Automation
Harbin Institute of Technology
Harbin, China
23s136642@stu.hit.edu.cn

2nd Yijie Wang
School of Electrical Engineering & Automation
Harbin Institute of Technology
Harbin, China
wangyijie@hit.edu.cn

3rd Xiaohui Xu
Wuhan Institute of Marine Electric Propulsion
Wuhan, China
346564867@qq.com

4th Kai Ji
Wuhan Institute of Marine Electric Propulsion
Wuhan, China
jikai712@yeah.net

Abstract—This paper proposes a dual-phase interleaved LLC resonant converter with hybrid control strategy. By integrating phase shift modulation (PSM) and pulse frequency modulation (PFM) slope modulation techniques, the scheme achieves smooth mode transition without added switches or topology modification, effectively suppressing dynamic current impact during switching. Continuous wide voltage gain regulation from 0.55 to 1.83 times is realized within narrow frequency range (around 100kHz resonant frequency). A phase-shifting current-sharing strategy is adopted for the two-phase interleaved configuration to ensure balanced current distribution. Experimental verification is conducted on a 200W prototype with 35-64V input/output voltage range, demonstrating the LLC's zero-voltage switching (ZVS) characteristics, wide gain regulation capability and current-sharing performance under the proposed hybrid control strategy.

Keywords—LLC Resonant Convertert , Hybrid Control Strategy , Current Sharing , Wide Voltage Gain

I. INTRODUCTION

In recent years, LLC resonant converters have gained significant attention in renewable energy systems due to their soft-switching characteristics enabling high efficiency and power density [1]. However, with the increasing power capacity of renewable energy installations, there emerges a growing demand for LLC topologies that simultaneously achieve wide voltage gain regulation and high-power processing capability. Conventional pulse frequency modulation (PFM)-controlled LLC converters, while capable of delivering tens of kilowatts, suffer from limited gain adjustment range and aggravated switching losses in high-frequency regions [2-4]. As systematically categorized in [2], four technical routes exist for expanding LLC gain range: resonant network optimization, primary/secondary topology reconstruction, and hybrid control strategies. Notably, in high-power multi-module series-parallel systems, topology modification approaches may exacerbate parameter mismatches and control complexity [3], making control-based gain extension methods more practical.

Existing LLC control strategies exhibit inherent limitations in single modulation dimensions: PFM shows insufficient gain adjustment margin in high-frequency zones leading to efficiency degradation [4]; phase shift modulation

(PSM) achieves step-down operation but risks losing primary-side zero-voltage switching (ZVS) at excessive phase angles [5]; pulse width modulation (PWM) requires secondary-side active devices, substantially increasing system complexity [6-7]. As a compromise, PFM-PSM hybrid control avoids topology modification costs but introduces surge currents during dynamic transitions, aggravating output current ripple and compromising efficiency/reliability [8-10].

To address these challenges, this paper proposes a dual-phase interleaved parallel LLC resonant converter for high-power wide-gain applications. By integrating PFM-PSM hybrid control with multi-phase current sharing technology (scalable to n-phase), the proposed strategy breaks single modulation limitations: PFM executes boost control when voltage gain exceeds 0.8, while fixed-frequency PSM handles buck regulation below 0.8 gain. Continuous control domain coverage from 0.55 to 1.83 times full gain range eliminates dynamic stress from discrete mode transitions. Experimental validation is performed on a 200W prototype with 35-64V input/output voltage range.

The paper proceeds as follows: Section II analyzes the PFM-PSM hybrid control mechanism and proposes a slope-compensated transition method. Section III quantifies the impact of resonant parameter deviations (ΔL_r, ΔC_r, and ΔL_m) on gain M, subsequently developing a phase-shifting current-sharing strategy. Section IV validates the gain regulation, current balancing performance, and ZVS characteristics through a 35-64V/200W experimental prototype. Section V concludes the research contributions.

II. PFM-PSM HYBRID CONTROL

A. An Analysis of LLC PFM and PSM

As illustrated in Fig. 1, the full-bridge LLC resonant converter topology comprises a DC input source, full-bridge inverter, LLC resonant tank, center-tapped rectifier, and load:

Full-bridge inverter employs four MOSFETs (Q_1-Q_4) for DC-AC conversion;

LLC resonant tank integrates series resonant inductance Lr, resonant capacitor Cr, and parallel magnetizing inductance L_m to determine system resonance characteristics;

979-8-3315-1110-4/25 $31.00 © 2025 IEEE
128

Secondary rectification stage adopts full-wave rectification with fast recovery diodes D1-D4 for AC-DC conversion;

In the interleaved parallel configuration, Q_{X1}/Q_{X2} and Q_{X3}/Q_{X4} form complementary switching pairs with 180° phase-shift driving. The shaded MOSFETs act as phase-shifted lagging legs in PSM mode, enabling step-down control through conduction phase adjustment [11].

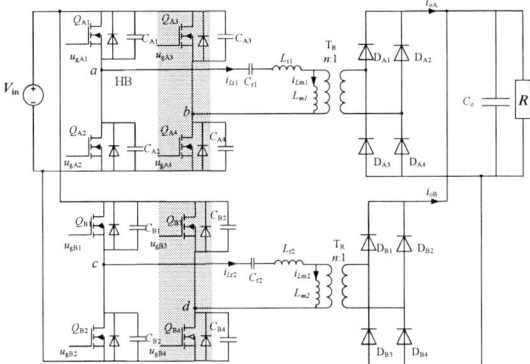

Fig. 1. Two-Phase Interleaved Hybrid-Controlled LLC Circuit Topology Diagram

Based on First Harmonic Approximation (FHA) analysis, the voltage gain expressions of LLC resonant converter under dual operating modes are derived:

PFM Mode:

$$M_{PFM} = \frac{1}{2\sqrt{\left(1 + k - \frac{k}{f_s}f_r\right)^2 + Q^2\left(\frac{f_s}{f_r} - \frac{f_r}{f_s}\right)^2}} \tag{1}$$

PSM Mode:

$$M_{PSM} = \frac{1 + \cos\theta}{2\sqrt{\left(1 + k - \frac{k}{f_{max}}f_r\right)^2 + Q^2\left(\frac{f_{max}}{f_r} - \frac{f_r}{f_{max}}\right)^2}} \tag{2}$$

Where $k = \frac{L_m}{L_r}$, denotes inductance ratio, $Q = \frac{1}{R}\sqrt{\frac{L_r}{C_r}}$ represents quality factor, $f_r = \frac{1}{2\pi\sqrt{L_r C_r}}$ is resonant frequency.

Under hybrid control strategy, assuming constant load, the converter initially operates at PFM mode with minimum frequency (f_{min}) to deliver maximum gain. As input voltage increases, the required gain decreases, leading to linear frequency increment until reaching upper limit $f_{max} = 120kHz$, then switching to PSM mode. In PSM mode, f_s is clamped at f_{max} while phase shift angle θ increases from 0 to further reduce gain. Fig. 2 illustrates gain curves under different Q values.

B. PFM-PSM HYBRID CONTROL

As shown in the single-phase control block diagram (Fig. 3), the output voltage V_o is compared with the reference V_{ref} to generate an error signal ΔV, which feeds both PFM and PSM controllers:

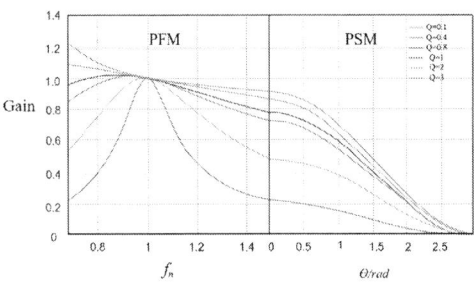

Fig. 2. Gain Curve Diagram of PFM and PSM Control

PFM Control Mode: When steady-state gain M≥0.8, the PFM controller is selected with phase shift angle $\theta_{PFM}=0$. The PFM controller integrates a voltage-controlled oscillator (VCO) and a PI module, regulating switching frequency f_s as:

$$f_s(t) = f_{s0} + K_{VCO} \cdot V_{ctrl}(t) \tag{3}$$

where the PI control law is:

$$K_{VCO} = f_{s0} + K_{VCO} \cdot V_{ctrl} \tag{4}$$

PSM Control Mode: When M<0.8, the PSM controller is engaged, locking f_s at $f_{max}=120kHz$ while adjusting θ_{PSM} via:

$$\theta_{PSM}(t) = K_{P\_PSM} \cdot \Delta V + K_{i\_PSM} \int \Delta V \, dt \tag{5}$$

To prevent frequent mode switching around the gain threshold of 0.8, a hysteresis band is implemented with state-dependent transition criteria:

PSM→PFM Transition: When operating in PSM mode, switching to PFM is triggered only if the gain exceeds M≥0.82, During this transition, θ_{PSM} is ramped down to zero with a controlled slope $k_\theta = \frac{0 - \theta_{PSM}(t_0)}{T_{ramp}}$, where T_{ramp}=5-8ms, to avoid abrupt gain changes and mitigate current surges. Simultaneously, f_s gradually decreases to the PFM minimum frequency.

PFM→PSM Transition: In PFM mode, transition occurs when M<0.78. The switching frequency f_s ramps up to f_{max} with a slope $k_f = \frac{f_{max} - f_s(t_0)}{T_{ramp}}$, while θ_{PSM} linearly increases from zero.

Fig. 4. illustrates the dynamic trajectories of θ_{PSM} and f_s during bidirectional transitions at M=1.8(PFM) and M=0.55(PSM), validating the switching mechanism.

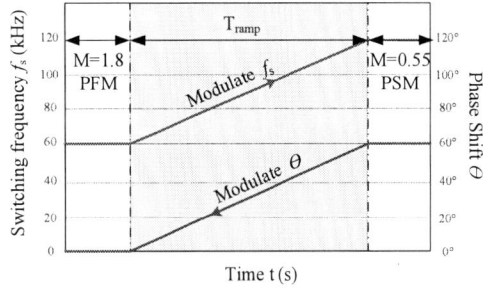

Fig. 3. Dynamic Trajectories of θ PSM and fs During Bidirectional Transitions at M=1.8 (PFM) and M=0.55 (PSM)

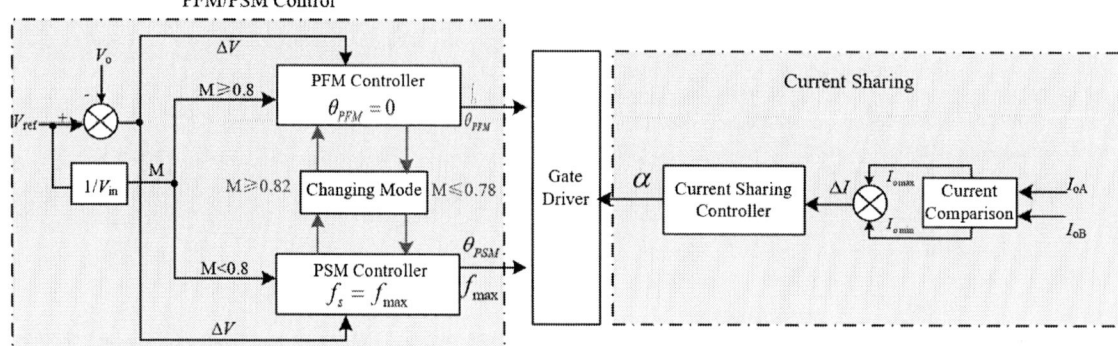

Fig. 4. LLC PFM and PSM Control Current-Sharing Control

III. CURRENT-SHARING STRATEGY

A. Phase-Shifting Current Sharing Theoretical Analysis

In practical engineering, the IPOP interleaved parallel LLC system's resonant components (L_r, C_r, L_m) in each phase have manufacturing tolerances, which can cause voltage gain mismatch between parallel modules.

Assume that in phases A and B, the resonant tank parameters L_{m1} and L_{m2}, C_{r1} and C_{r2}, L_{r1} and L_{r2} all have deviations. Then, the two phases have different gains $M_1 \neq M_2$. If the system is in PFM control mode, the voltage gain expressions for phases A and B are given by (6) and (7).

$$M_{PSM\_1} = \frac{1}{2\sqrt{\left(1+k_1-\frac{k_2}{f_s}f_{r1}\right)^2 + Q_1^2\left(\frac{f_s}{f_{r1}}-\frac{f_{r1}}{f_s}\right)^2}} = \frac{1}{D_1} \quad (6)$$

$$M_{PSM\_2} = \frac{1}{2\sqrt{\left(1+k_2-\frac{k_2}{f_s}f_{r2}\right)^2 + Q_2^2\left(\frac{f_s}{f_{r2}}-\frac{f_{r2}}{f_s}\right)^2}} = \frac{1}{D_2} \quad (7)$$

There will be a gain difference $\Delta M = M_1 - M_2$ (assuming $M_1 > M_2$), which needs to be adjusted by PSM. The calculation formula of the current-sharing angle α is given by (8) and (9).

$$M_{PSM\_1}' = \frac{1+\cos\alpha}{D_1} = \frac{1}{D_2} = M_{PSM\_2} \quad (8)$$

$$\cos\alpha = \frac{D_1}{D_2} - 1 \quad (9)$$

Similarly, when the two phases are in the PSM control stage (assuming $M_1 > M_2$), the same principle applies.

$$M_{PFM\_1}' = \frac{1+\cos\alpha}{D_1} = \frac{1+\cos\theta_2}{D_2} = M_{PFM\_2} \quad (10)$$

$$\cos\alpha = \frac{D_1(1+\cos\theta_2)}{D_2} - 1 \quad (11)$$

The total differential expansion of the gain expression quantifies parameter deviation effects:

$$\Delta M = \frac{\partial M}{\partial k}\Delta k + \frac{\partial M}{\partial Q}\Delta Q + \frac{\partial M}{\partial f_n}\Delta f_n \quad (12)$$

The partial derivatives of the intermediate variables Δk, ΔQ, and Δf_n with respect to the component parameters.

$$\Delta k = \frac{\partial k}{\partial L_r}\Delta L_r + \frac{\partial k}{\partial L_m}\Delta L_m = -k\frac{\Delta L_r}{L_r} + \frac{1}{L_r}\Delta L_m \quad (13)$$

$$\Delta Q = \frac{\partial Q}{\partial L_r}\Delta L_r + \frac{\partial Q}{\partial C_r}\Delta C_r = \frac{Q}{2L_r}\Delta L_r - \frac{Q}{2C_r}\Delta C_r \quad (14)$$

$$\Delta f_n = \frac{\partial f_n}{\partial L_r}\Delta L_r + \frac{\partial f_n}{\partial C_r}\Delta C_r = \frac{f_n}{2L_r}\Delta L_r + \frac{f_n}{2C_r}\Delta C_r \quad (15)$$

The relationship between gain error and parameter deviations is:

$$\Delta M = F_{L_r}\frac{\Delta L_r}{L_r} + F_{C_r}\frac{\Delta C_r}{C_r} + F_{L_m}\frac{\Delta L_m}{L_m} \quad (16)$$

$$F_{L_r} = -\frac{M}{D}\left(\frac{1}{k} + \left(\frac{1}{k^2}+Q^2 f_n^2\right)\left(1-\frac{1}{f_n^2}\right)\right) \quad (17)$$

$$F_{C_r} = \frac{MQ^2}{D}\left(f_n - \frac{1}{f_n}\right)^2 \quad (18)$$

$$F_{L_m} = \frac{MQ^2}{D}\left(f_n - \frac{1}{f_n}\right)^2 \quad (19)$$

Sensitivity coefficients F_{Lr}, F_{Cr}, F_{Lm} reveal:

Dominant F_{Lr}: Affects k, Q, and f_n simultaneously, with high-frequency term $Q^2 f_n^2$.

Secondary C_r: Indirectly coupled through Q and f_n, sensitivity increases with f_n deviation from resonance.

Weakest L_m: Directly impacts only k, scaled by $1/k^2$.

B. Phase-Shifting Current Sharing Control Strategy

To suppress current imbalance caused by resonant parameter mismatch, a phase-shifting impedance compensation mechanism is proposed:

Dynamic Compensation: The current difference $\Delta I = I_1 - I_2$ is detected and fed into a PI controller to generate a compensation phase shift α, which is superimposed on the PSM phase angle θ_{PSM}, resulting in the final gate signal $\theta = \theta_{PSM} + \alpha$.

Ripple Optimization: A 90° interleaving angle (180°/n for $n=2$) is implemented to minimize output ripple.

IV. EXPERIMENTAL VERIFICATION

A 35-64V input/output, 200W-rated prototype was developed to validate the proposed dual-phase interleaved LLC resonant converter with PSM-PFM hybrid control strategy. Key parameters of the single-phase LLC resonant tank are listed in Table 1, where the primary side utilizes HSP15810C SiC MOSFET (V_{DS}=100V, $R_{DS(ON)}$=4.7mΩ, C_{oss}=609pF) to minimize conduction and switching losses, while the secondary side employs C3D10060A SiC diodes. The gate driver circuit, implemented with NSI6602B-DSPNR isolated gate driver IC, configures a dead time of t_{dead}= 100ns. A STM32G474RERET6 executes the hybrid control algorithm with switching frequency range f_s=60k-120kHz.

The experimental setup (Fig. 5) demonstrates the dual-phase interleaved system, and Fig. 6 shows the hardware realization of a single-phase 100W full-bridge LLC submodule.

TABLE I.   PARAMETERS OF THE SINGLE-PHASE LLC RESONANT

Parameter	Value	Unit
f_r	100	kHz
V_{in}	35-64	V
V_{out}	35-64	V
P_{out}	100	W
L_r	8	μH
C_r	300	nF
L_m	36	μH

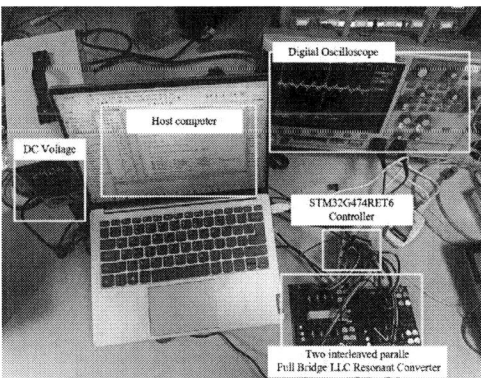

Fig. 5.   Experimental Environment

Fig. 6.   A single-phase 100W full-bridge LLC submodule

To validate the mode transition capability of the hybrid control strategy and LLC resonant tank's zero-voltage switching (ZVS) characteristics, experimental analyses were conducted:

Fig. 7 shows the resonant inductor current i_{Lr} during bidirectional PFM → PSM → PFM transitions. Smooth transition without overshoot is observed in PFM → PSM direction, while a transient current overshoot of approximately 3A (15% below device maximum rating) occurs during PSM → PFM transition, remaining within safe operational margins.

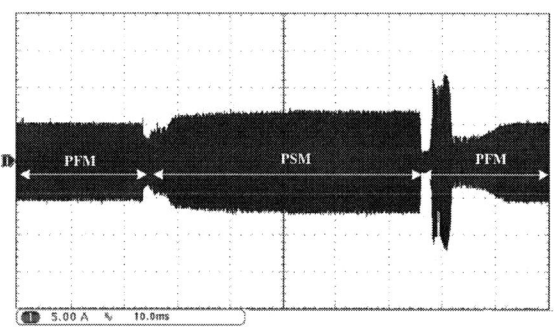

Fig. 7. i_{Lr} Waveform During PFM→PSM→PFM Transition

Fig. 8 displays expanded waveforms of i_{Lr} in PSM mode(f_s=120kHz), demonstrating phase modulation effects of shift angles on current profiles. Fig. 9 presents PFM-mode (f_s=80kHz)resonant current

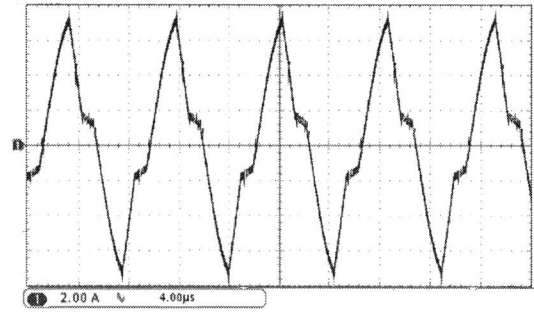

Fig. 8. i_{Lr} Waveform During PSM Mode

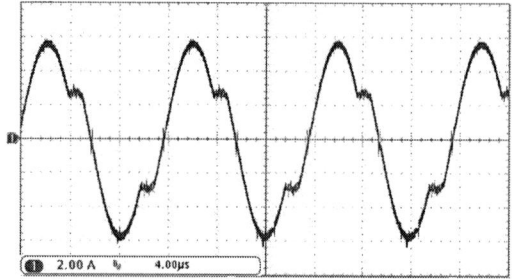

Fig. 9. i_{Lr} Waveform During PFM Mode

Fig. 10 illustrates dual-phase i_{Lr} waveforms at f_s=100kHz, $V_{in}=V_o$=50V. (near resonant frequency), showing less than 2% amplitude deviation and under 1µs phase misalignment, which confirms the effectiveness of the phase-shift compensation strategy. The phase lag of i_{Lr} relative to input voltage validates LLC's ZVS operation, enabling complete body diode reverse recovery before switch turn-on, thereby minimizing switching losses.

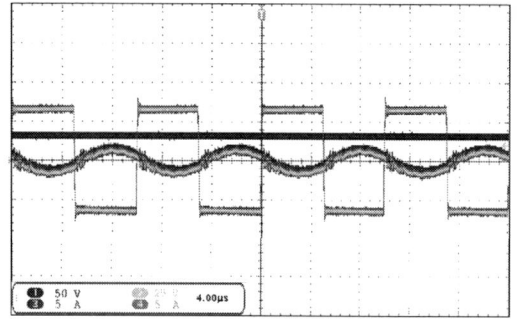

Fig. 10. Two-Phase LLC iLr and Input/Output Voltage Waveforms at f_s = 100kHz

V. CONCLUSION

This paper proposes a dual-phase interleaved LLC resonant converter with PSM-PFM hybrid control strategy for high-power wide-gain applications. By integrating dynamic slope compensation technology, seamless mode transition between PFM and PSM is achieved at the 0.8 gain threshold, eliminating surge current issues at switching boundaries. The converter realizes continuous wide voltage gain regulation from 0.55 to 1.83 times within a narrow frequency range (60kHz-120kHz). To address current imbalance caused by resonant parameter mismatches in parallel modules, an adaptive current-sharing strategy based on phase-shifting impedance compensation is developed, achieving ±2%

current balance accuracy through real-time phase angle adjustment. Experimental validation on a 35-64V input/output, 200W prototype demonstrates: 1) Mode transition current overshoot below device safety limits; 2) Less than 2% amplitude deviation and 1µs phase synchronization error between dual-phase currents at resonant frequency (100kHz). This work provides theoretical and practical foundations for high-power-density and high-reliability LLC converters.

REFERENCES

[1] Author, F.: Article title. Journal 2(5), 99–110 (2016). A Structure-Reconfigurable Isolated Bidirectional DC–DC.

[2] Q. Cao, Z. Li and H. Wang, "Wide Voltage Gain Range LLC DC/DC Topologies: State-of-the-Art," 2018 International Power Electronics Conference (IPEC-Niigata 2018 -ECCE Asia), Niigata, Japan, 2018, pp. 100-107.

[3] Z. Xu, J. Yang, Q. Wang and G. Zhang, "A Wide Input Range High-Frequency LLC Resonant Converter With Hybrid Control Strategy," 2023 3rd International Conference on Electrical Engineering and Mechatronics Technology (ICEEMT), Nanjing, China, 2023, pp. 51-58.

[4] Y. Liu et al., "A Multimode Wide Output Range High-Voltage Power Supply for Magnetrons," IEEE Trans. Ind. Electron., vol. 70, no. 11, pp.11153-11162, Nov. 2023.

[5] L. Shi, B. Liu and S. Duan, "Burst-Mode and Phase-Shift Hybrid Control Method of LLC Converters for Wide Output Range Applications," IEEE Trans. Ind. Electron., vol. 67, no. 2, pp. 1013-1023, Feb. 2020.

[6] X. Tang, Y. Xing, H. Wu, and J. Zhao, "An improved LLC resonant converter with reconfigurable hybrid voltage multiplier and PWM-plus-PFM hybrid control for wide output range applications," IEEE Trans. Power Electron., vol. 35, no. 1, pp. 185–197, Jan. 2020.

[7] Z. Li, B. Xue, and H. Wang, "An interleaved secondary-side modulatedLLC resonant converter for wide output range applications," IEEE Trans.Ind. Electron., vol. 67, no. 2, pp. 1124–1135, Feb. 2020.

[8] R. Moriyasu, M. S. Hassan, H. Funaki, M. Shoyama and Y. Noge, "Surge Current Reduction in LLC Resonant Converter with PSM under Light Load Condition," 2023 IEEE International Future Energy Electronics Conference (IFEEC), Sydney, Australia, 2023, pp. 173-178.

[9] R. Moriyasu, H. Funaki, M. Shoyama, Y. Noge and M. S. Hassan, "Surge Current Analysis and Reduction in LLC Resonant Converter With a New Hybrid Control Strategy of Pulse-Frequency Modulation and Phase-Shift Modulation," in IEEE Transactions on Power Electronics, vol. 39, no. 11, pp. 14448-14464, Nov. 2024.

[10] R. Moriyasu, Y. Noge, M. Shoyama and M. S. Hassan, "Surge Current Reduction in LLC Resonant Converter with a Hybrid Control Strategy of PFM and PSM for Expansion of Output Voltage Range," 2023 IEEE Applied Power Electronics Conference and Exposition (APEC), Orlando, FL, USA, 2023, pp. 2747-2752

[11] Z. Xiao, Z. He, R. Guan, and A. Luo, "Piecewise-approximate time domain analysis of LLC resonant converter considering parasiti capacitors and deadtime," IEEE Trans. Power Electron., vol. 38, no. 1,pp. 578–592, Jan. 2023.

2025 IEEE Workshop on Wide Bandgap Power Devices and Applications in Asia (WiPDA Asia)

A Reconfigurable Half-Bridge Buck Based on Dickson SC Converter with DCX-LLC for Wide Input Voltage Range Applications

Lichang Man, Yijie Wang, Shanshan Gao, Hongqi Ben, Dianguo Xu

School of Electrical Engineering and Automation, Harbin Institute of Technology,Harbin,China

23s106130@stu.hit.edu.cn；wangyijie@hit.edu.cn；gaoshanshan@hit.edu.cn；benhq@hit.edu.cn；xudiang@hit.edu.cn

Abstract—**With the rapid development of electric vehicles, on-board DC-DC converters need to achieve efficient power conversion at high frequency and wide input voltage range. In this paper, a two-stage DC-DC converter is proposed, the front stage adopts a Dickson switched-capacitor reconfigurable half-bridge buck converter with wide-range voltage regulation through dynamic configuration of the number of charging cells, which serves as the bus voltage and enables soft charging. The second stage is a DCX-LLC resonant converter, designed with optimised parameters to ensure Zero Voltage Switching (ZVS) over the entire load range. Finally, the correctness of the theoretical analysis is verified by simulation and experiment results with an input of 250-420 V and an output of 200 W/12 V laboratory prototype.**

Keywords—Wide input voltage range, reconfigurable half-bridge buck converter, LLC converter，Zero Voltage Switching

I. INTRODUCTION

With the growing global energy crisis and environmental challenges, electric vehicles (EVs), as representatives of clean energy transportation, have developed rapidly in recent years. However, the on-board DC-DC converters in EVs face numerous challenges, one of which is how to efficiently and stably convert the electrical energy from high-voltage power batteries to the lower voltage needed for auxiliary systems[1]. Therefore, maintaining high conversion efficiency and a wide voltage adjustment range under high-frequency conditions are the main challenges in the topology and control research of on-board DC-DC converters.

The LLC resonant converter based on GaN devices is expected to achieve high conversion efficiency at megahertz frequency. However, when the input voltage varies widely, the switching frequency or duty cycle will change widely with the input voltage[2]. To overcome this problem, literature [3]-[6] proposed two stage structures, LLC converter works as a DC transformer (DCX) with constant voltage conversion ratio operating at the optimum point for maximum efficiency. A pre-regulation module is placed before the DCX-LLC to achieve voltage adjustment for a wide input voltage range. For instance, buck-boost converters or trapezoidal switched-capacitor converters were merged with LLC resonant converter. Literature [7] adopted a cascaded Boost LLC converter, but the front-stage Boost converter faces difficulties in achieving ZVS of the switching devices, and the back-stage LLC resonant

converter's switching devices are subject to higher voltage stress. Literature [8] proposed a new topology by cascading interleaved parallel Boost converters with a dual-resonant LLC resonant converter, but when the input voltage range is wide, the frequency range of the LLC resonant converter also widens, and the dual-resonant cavity structure becomes more complex and costly. Literature [9] shows that reconfigurable switched capacitor (RSC) converters SRC can be soft-charged by adjusting the number of active charging units to control the output voltage amplitude. However, for achieving too high dropout ratios more switching tubes are required, making it difficult to achieve high efficiency and high integration.

This paper proposes a two-stage DC-DC converter, as shown in the Fig. 1. A Dickson switched capacitor reconfigurable half-bridge buck converter proposed in [9] is used for the front stage and a DCX-LLC circuit is used for the second stage. In this paper, the following aspects are discussed: the topology and working principle of Dickson switched capacitor reconfigurable half-bridge buck converter are presented in Section II. Section III optimises the parametric design of the LLC resonant converter. Finally, simulation results are given in Section IV to verify all the analyses.

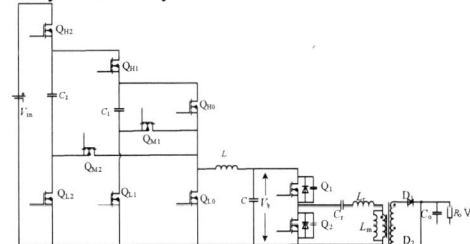

Fig. 1. The proposed two-stage DC-DC converter

II. OPERATION PRINCIPLE OF THE DICKSON SWITCHED-CAPACITOR RECONFIGURABLE HALF-BRIDGE

Fig. 4(b) shows the schematic of the two-stage DC-DC converter. The front stage consists of two charging units in cascade, and each charging unit contains a flying capacitor $C_i(i=1,2)$ and the corresponding power switches (high-side switching Q_H, mid-side switching Q_M, and low-side switching Q_L). Inductor L, capacitor C form the filter circuit. The back stage LLC consists of switching circuits composed of Q_1 and Q_2; resonant inductor L_r, resonant capacitor C_r, and excitation inductance L_m constitute a

979-8-3315-1110-4/25 $31.00 © 2025 IEEE   133

resonant circuit and are connected to the primary side of the transformer; the secondary side of the transformer is a full-wave uncontrolled rectifier circuit composed of diodes D_1 and D_2, which is connected to the output capacitor C_o and then connected to the load R_o. The reconfigurable half-bridge buck converter maintains constant pulse amplitude under input voltage fluctuations by dynamically configuring the number of effective-charging-cell, thereby significantly reducing switching losses. The converter operates in two distinct modes based on input voltage ranges: (1) When $2V_b < V_{in} < 4V_b$, the converter works in two-charging-cell mode to achieve a 3:1 conversion ratio; (2) When $4V_b < V_{in} < 6V_b$, the converter works in one-charging-cell mode to achieve a 2:1 conversion ratio. V_b is the bus voltage. This dual-mode

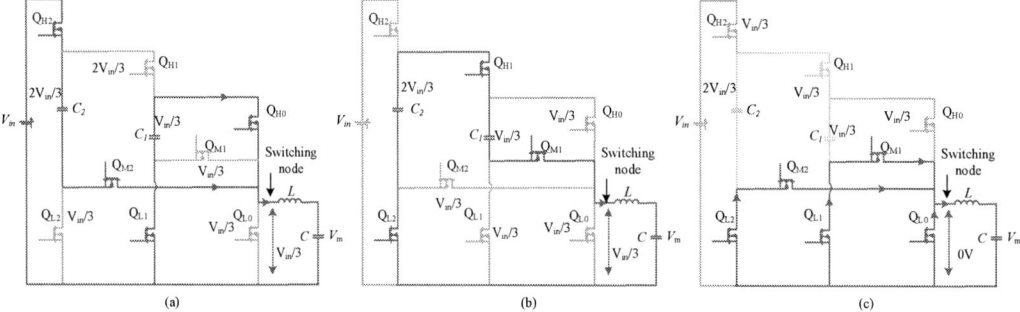

Fig. 2. The operation of two-effective-charging-cell duiring (a) first charge transfer phase,(b) second charge transfer phase,and (c) resting phase.

reconfiguration mechanism dynamically adapts to wide input voltage variations.

A. Two-charging-cell Operation

Fig.2 illustrates the process of transferring charges in a two-charging-cell-based half-bridge, in which a 3:1 conversion ratio is produced by the half-bridge. In Fig.2, the substage following the reconfigurable half-bridge is an LC filter, which forms a buck converter design with the reconfigurable half-bridge. A pulse voltage is generated at the switching node.

During the first charge transfer phase, as shown in Fig.2(a), two charge transfer paths are established simultaneously. In path 1, switches Q_{H2} and Q_{M2} are turned on, transferring charges from V_{in} to C_2. In path 2, switches Q_{L1} and Q_{H0} are turned on, transferring charges out of C_1. The two charge transfer currents converge at the switching node and flow into the output inductor. During the second charge transfer phase, as shown in Fig. 2(b), switches Q_{L2}, Q_{H1}, and Q_{M1} are turned on, transferring charges from C_2 to C_1. Under the steady-state operation, the voltages across C_2 and C_1 are $2/3V_{in}$ and $1/3V_{in}$ respectively. During both charge transfer phases, the voltage produced at the switching node is $1/3V_{in}$. There is also a resting phase operation, as shown in Fig. 2(c), where low-side power switches Q_{L2} and Q_{L1}, and middle-side power switches Q_{M2} and Q_{M1} are turned on. The switching node is essentially tied to the ground, producing 0 V at the switching node during the resting phase operation. The voltage stresses of all power switches are also marked in Fig. 2.

To identify the required operation to achieve charge balance with C_2 and C_1 and produce a standard rectangular pulse waveform at the switching node, an operation pattern for two-active-charging-cell configuration is depicted in Fig. 3. There are two occurrences of the first charge transfer phase and one occurrence of the second charge transfer phase. The duration of the first charge transfer phase and the second charge transfer phase is the same.

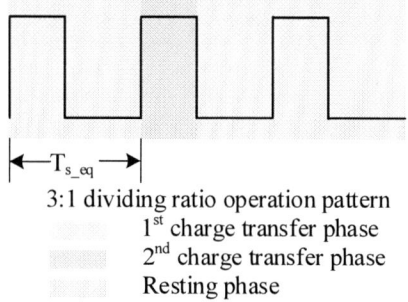

3:1 dividing ratio operation pattern
1st charge transfer phase
2nd charge transfer phase
Resting phase

Fig. 3. The operation pattern under two-active-charging-cell configurations (3:1 conversion ratio operation).

B. One-charging-cell Operation

Fig. 4 illustrates the process of transferring charge under a one-charging-cell implementation, in which a 2:1 conversion ratio is produced. One-charging-cell implementation is reconfigured from two-charging-cell implementation by deactivating the first charging cell, which is the one connected to the input voltage. The high-side power switch of the charging cell is permanently turned on, and both the middle-side and low-side power switches are permanently turned off. Under the one-active-charging-cell configuration, C_2 are disengaged from the rest of the system, so only C_1 functions as a flying capacitor. The operation with one active charging cell is essentially the same as a three-level buck converter.

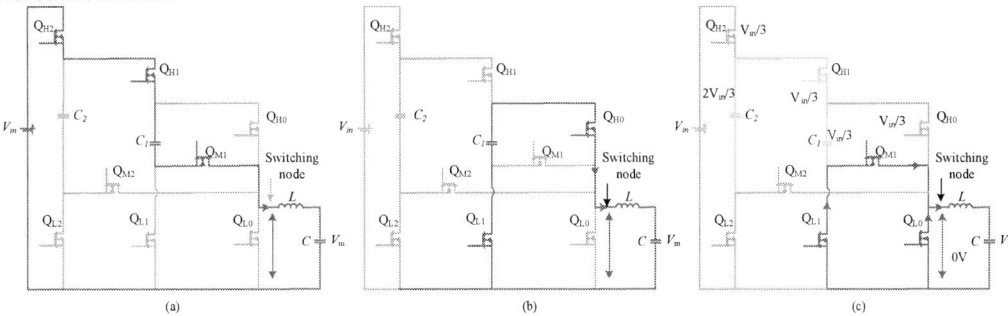

Fig. 4. The operation of One-effective-charging-cell duiring (a) first charge transfer phase,(b) second charge transfer phase,and (c) resting phase.

III. LLC RESONANT CONVERTER PARAMETER DESIGN

A. Gain analysis of LLC Resonant Converter

The LLC resonant converter, as a nonlinear system, is approximated as a linear circuit using fundamental wave analysis, which yields the voltage gain characteristics. The relationship between the voltage gain M and the switching frequency f_s, resonant frequency f_r, λ and Q is obtained by fundamental wave analysis as follows.

$$M = \frac{1}{\sqrt{\left[1+\frac{1}{\lambda}\left(1-\frac{1}{f_n^2}\right)\right]^2 + \left[Q\left(f_n - \frac{1}{f_n}\right)\right]^2}} \quad (1)$$

$$M = \frac{1}{\sqrt{\left[1+\frac{1}{\lambda}\left(1-\frac{1}{f_n^2}\right)\right]^2 + \left[\frac{1}{\lambda}\sqrt{\frac{(1+\lambda)f_n^2-1}{1-f_n^2}}\left(1-\frac{1}{f_n^2}\right)\right]^2}} \quad (2)$$

The pure resistive gain curves and DC voltage gain curves drawn according to Eqs. (1) and (2) for the same and different Q values are given in Fig.5. According to Fig.5(a), it can be seen that M is always equal to 1 when $f_n = 1$ regardless of the increase or decrease of Q. Fig.5(b) shows the voltage gain curve corresponding to different values of λ with fixed Q. It can be seen that the larger λ is, the flatter the curve is and the more stable the voltage gain is.

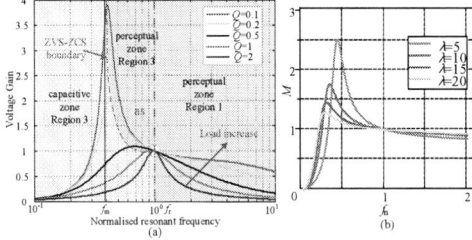

Fig. 5. Voltage Gain Curve. (a) Voltage gain versus f_n curve for different Q values. (b) Voltage gain versus λ value curve for the

same Q value.

B. Excitation inductor design

The LLC resonant converter requires that the i_{Lr} needs to discharge the voltage of the junction capacitor to 0 during the dead time and that the i_{Lr} cannot flow in the reverse direction in order to achieve the ZVS of the switching tube, therefore the resonant inductance needs to satisfy Eq.(3).

$$L_m \leq \frac{T_d}{16C_{oss}f_r} \quad (3)$$

Losses in the converter's parts are tied to the RMS values of currents: i_{rmsp}, i_{rmss}, i_{rmsSi} and i_{offSi}. Therefore, to minimize losses, the converter's parameter design is optimized. Fig. 6 gives the relationship between the currents and the excitation inductance L_m. From the figure, it can be seen that the increase of the excitation inductance reduces the currents at various key points in the circuit, especially the reduction of i_{offSi} is more obvious, which is conducive to reducing the losses and improving the efficiency. Therefore, in combination with Eq.5, it is more appropriate to choose around 50μH in the project.

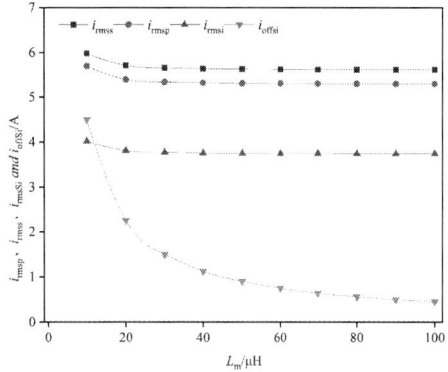

Fig. 6. Relationship between key point current and excitation

inductance L_{m}.

C. Resonant Cavity Parameter Design

According to Eq.(4), when the excitation inductance L_{m} is determined, λ determines the size of the resonant inductance L_{r}.

$$L_{\mathrm{r}} = L_{\mathrm{m}}/\lambda \qquad (4)$$

The output voltage of the converter can be expressed as:

$$V_{\mathrm{o}} = \frac{1}{n} \cdot \frac{V_{\mathrm{b}}}{2} \cdot M = \frac{V_{\mathrm{b}}}{2n} \cdot \frac{1}{\sqrt{\left[1+\frac{1}{\lambda}\left(1-\frac{1}{f_{\mathrm{n}}^2}\right)\right]^2 + \left[Q\left(f_{\mathrm{n}}-\frac{1}{f_{\mathrm{n}}}\right)\right]^2}} \qquad (5)$$

The LLC resonant converter has maximum voltage gain with $Q = 0$ at no load and minimum voltage gain with $Q = Q_{\max}$ at full load. Therefore, taking the output voltage variation within ±1% of the rated output voltage. According to the fact that the larger λ is, the flatter the curve is and the more stable the voltage gain is, the transformer leakage inductance can be used as the resonant inductor, and at the same time the power density of the system can be improved.

IV. RESULTS OF SIMULATION

The proposed topology is verified using the circuit simulation software. The main simulation parameters are shown in Table 1.

TABLE I. KEY CIRCUIT PARAMETERS

Parameters	Value
Input voltage	250V-420V
Output voltage	12V
Bus voltage	84V
Equivalent frequency at the switching node	1MHz
Flying capacitor C2&C1	100uf
LLC Converter Excitation Inductor	50uH
LLC Converter Resonant Inductor	1.1uH
LLC Converter Resonant Capacitor	20.9nF
LLC Converter switching Frequency	1MHz

The input voltage range of the converter is 250 V to 420 V, and the critical input voltage is 336 V. When the input voltage is in the range of 250 to 336 V, the front-stage converter operates in one-charging-cell mode; when the input voltage is in the range of 336 to 420 V, the converter operates in two-charging-cell mode.

From Figs. 7(a) and 8(a), it can be seen that under steady state operation, the voltages across Q_{H2} are $1/3\,V_{\mathrm{in}}$ and $1/2\,V_{\mathrm{in}}$, respectively. From Fig. 7(b) and Fig. 8(b), it can be seen that the output bus voltage of the pre-stage is 84V with low voltage ripple.

From Figs.9(a) and 10(a), which are the waveforms of the pole-to-pole voltage and current

during zero-voltage turn-on of switch Q_1, respectively, it can be seen that the primary-side MOSFET switches

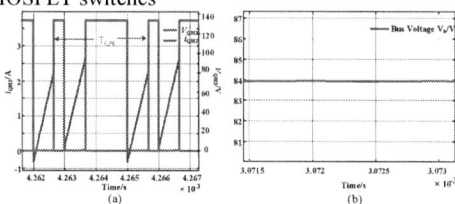

Fig. 7. The key waveforms for the pre-stage operating with one-charging-cell at 252 V input voltage. (a) Voltage and current of Q_{H2}. (b) Output bus voltage V_{b}.

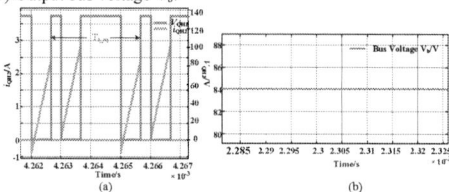

Fig. 8. The key waveforms for the pre-stage operating with two-charging-cell at 378V input voltage. (a) Voltage and Current of Q_{H2}. (b) Output bus voltage V_{b}.

achieve ZVS. Fig. 9(b) and Fig. 10(b) show the output voltage waveforms, and it can be seen that the output voltage ripple is small and the output stability is high.

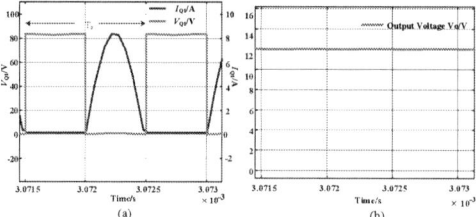

Fig. 9. The steady state waveform of LLC converter at 252 V input voltage. (a) Voltage and current of Q_1. (b) Output voltage V_{o}.

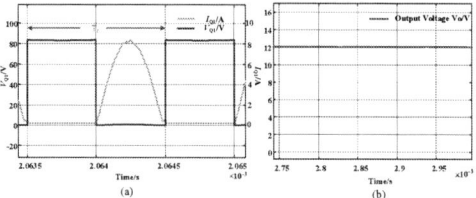

Fig. 10. The steady state waveform of LLC converter at 378V input voltage. (a) Voltage and current of Q_1. (b) Output voltage V_{o}.

V. CONCLUSION

This paper proposes a two-stage DC/DC converter with a reconfigurable half-bridge buck converter in the front stage and a DCX-LLC converter in the second stage. According to different input voltage ranges, the front stage operates in the one-charging-cell mode and the two-charging-cell mode, and the resonant parameters of the LLC resonant converter are optimally designed to achieve the soft switching of all switching tubes. Finally, the above analysis is verified by simulation.

REFERENCES

[1] G. Jagadan, U. K. Sourav, V. Chathayil, R. S. Kumar, and M. V, "Design of LLC Resonant Converter for High Efficiency EV Charging," in *IEEE Glob. Conf. Comput., Power Commun. Technol (GlobConPT).*, Sep. 2022, pp. 1–7.

[2] Y. Zuo, X. Pan, and C. Wang, "A Reconfigurable Bidirectional Isolated LLC Resonant Converter For Ultra-Wide Voltage-Gain Range Applications," *IEEE Trans Ind Electron*, vol. 69, no. 6, pp. 5713–5723, Jun. 2022.

[3] R. C. N. Pilawa-Podgurski and D. J. Perreault, "Merged Two-Stage Power Converter With Soft Charging Switched-Capacitor Stage in 180 nm CMOS," *IEEE J Solid-State Circuits.*, vol. 47, no. 7, pp. 1557–1567, Jul. 2012.

[4] S. Lim, J. Ranson, D. M. Otten, and D. J. Perreault, "Two-Stage Power Conversion Architecture Suitable for Wide Range Input Voltage," *IEEE Trans Power Electron.*, vol. 30, no. 2, pp. 805–816, Feb. 2015.

[5] Y. Shen, H. Wang, Z. Shen, Y. Yang, and F. Blaabjerg, "A 1-MHz Series Resonant DC–DC Converter With a Dual-Mode Rectifier for PV Microinverters," *IEEE Trans Power Electron.*, vol. 34, no. 7, pp. 6544–6564, Jul. 2019.

[6] J. Deng and H. Wang, "A Hybrid-Bridge and Hybrid Modulation-Based Dual-Active-Bridge Converter Adapted to Wide Voltage Range," *IEEE J Emerg Sel Top Power Electron.*, vol. 9, no. 1, pp. 910–920, Feb. 2021.

[7] F. Musavi, M. Craciun, D. S. Gautam, and W. Eberle, "Control Strategies for Wide Output Voltage Range LLC Resonant DC–DC Converters in Battery Chargers," *IEEE Trans Veh Technol.*, vol. 63, no. 3, pp. 1117–1125, Mar. 2014.

[8] F. SHI, R. LI, J. YANG, and W. YU, "High Efficiency Bidirectional DC-DC Converter with Wide Gain Range for Photovoltaic Energy Storage System Utilization," in *Proc.-IEEE Int. Power Electron. Appl. Conf. Expo (PEAC).*, Nov. 2018, pp. 1–6.

[9] P. Fang and R. Rice, "Reconfigurable Half-Bridge Based on Dickson SC Converter for Wide Input Voltage Range Applications," *IEEE J Emerg Sel Top Power Electron.*, vol. 12, no. 5, pp. 4447–4462, Oct. 2024.

Degradation Evaluation of Single-Event Effects Induced by Heavy-Ion Irradiation on E-mode p-GaN Gate HEMT Devices

Yujie Cheng
Center for More Electric Aircraft
Power System (Nanjing University of
Aeronautics and Astronautics)
Nanjing Electronic Devices Institute
Nanjing, Jiangsu Province, China
zhaolumuzhu@163.com

Dengjun Wang
Nanjing Electronic Devices Institute
Nanjing, Jiangsu Province, China

Haifeng Cheng
Nanjing Electronic Devices Institute
Nanjing, Jiangsu Province, China

Zhangzhe Yan
Nanjing Electronic Devices Institute
Nanjing, Jiangsu Province, China
15757161112@163.com

Zhuangzhuang Hu
Nanjing Electronic Devices Institute
Nanjing, Jiangsu Province, China

Jianjun Zhou
Nanjing Electronic Devices Institute
Nanjing, Jiangsu Province, China

Hongfei Wu
Center for More Electric Aircraft
Power System (Nanjing University of
Aeronautics and Astronautics)
Nanjing, Jiangsu Province, China
wuhongfei@nuaa.edu.cn

Abstract—To assess the single-event effects (SEE) resistance of enhancement-mode (e-mode) p-GaN gate HEMT devices for aerospace applications, heavy-ion irradiation tests were conducted on the devices, featuring a 25 A conduction current and a 350 V breakdown voltage. Post-irradiation degradation were evaluated through electrical analyses under varying bias and linear energy transfer (LET) conditions. The devices exhibited zero Single Event Burnout (SEB) under off-state conditions at a drain voltage of 250 V, even when exposed to heavy-ion irradiation with a LET of 82.8 MeV $\cdot$ cm^2/mg and fluence of 1×10^7 ions $\cdot$ cm^{-2}. Degradation severity showed pronounced sensitivity to bias conditions and LET, with threshold voltage (Vth) shifts recognized as the main variation. These findings affirm the SEE resistance of the e-mode p-GaN gate HEMT device.

Keywords—GaN, HEMT, p-GaN gate, single-event effect, heavy-ion irradiation

I. INTRODUCTION

In space applications, electronic devices are required not only to have high power, high efficiency, and a compact size, but also to be able to withstand high and low temperatures and resist radiation. As a wide-bandgap semiconductor device, the Gallium Nitride (GaN) High Electron Mobility Transistor (HEMT) device has the characteristic of high power density due to the extremely high electron mobility of the two-dimensional electron gas (2DEG) near the interface of the AlGaN and GaN heterostructure. At the same time, because the bond energy of the compounds in its crystal structure is relatively large, it also has excellent radiation resistance [1]. GaN HEMT devices have broad application prospects in the fields of aerospace, satellites, and so on.

There are many types of radiation that the device may be exposed to. High-energy protons, neutrons, electrons, etc. will all cause different types and degrees of damage to the device. Although GaN HEMT devices have relatively good radiation resistance [2, 3], they can still be damaged by radiation [4, 5]. The single-event effect of the device refers to the effect caused by heavy ions with high kinetic energy incident into the

semiconductor material during the irradiation process. Due to their high energy, a large number of electron-hole pairs will be generated through collisions along the incident trajectory during the incident process [6]. In GaN HEMT devices, these electron-hole pairs will change the charge distribution in the depletion region, which is likely to cause the device to burn out. Therefore, it is necessary to conduct heavy ion radiation tests to explore the working conditions of the device under radiation and to evaluate and analyze the radiation degradation of the device.

At present, a great deal of research and analysis have been carried out within the industry on various radiation damage mechanisms of GaN HEMT devices [7]. However, in the aspect of high-voltage power devices, there are relatively few existing experiments and analyses. This paper conducts heavy ion irradiation experiments under multiple conditions in this direction, providing reliable support for the research on the single-event effect of GaN HEMT devices and the reinforcement design of the devices.

II. DPEVCE CHARACTERIZATION AND MEASUREMENT

The cross-sectional structure schematic and the top-view of the fabricated GaN HEMT device are illustrated in Figure 1. On the Si substrate, a GaN buffer layer, a GaN channel layer, and an AlGaN barrier layer are sequentially grown. Subsequently, a p-GaN gate cap layer is formed via Mg doping. The device exhibits a threshold voltage (Vth, defined as the voltage when the drain current density JD reaches 10 mA/mm) of 2.2 V and an on-resistance of 30 mΩ. Under a gate voltage of 5 V, the drain saturation current of the device attains 25 A. Moreover, when the device is in the off-state, its breakdown voltage exceeds 350 V.

979-8-3315-1110-4/25 $31.00 © 2025 IEEE

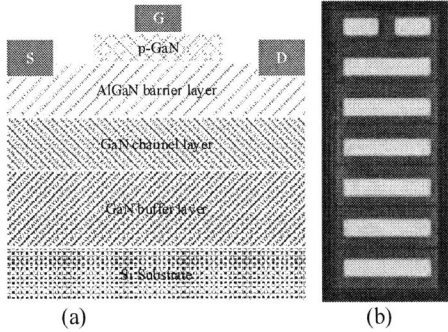

Fig. 1. (a) Cross - sectional view of the fabricated GaN HEMT device; (b) Top view of the device

The heavy-ion single-event effect simulation experiments utilized a cyclotron (HIRFL) as the radiation source. This cyclotron offers the capability to generate ions with high Linear Energy Transfer (LET) values, while its fluence rate can be precisely tuned over a continuous range. The maximum energy of the accelerated ions can surpass 1 GeV, and their penetration depth in silicon exceeds 30 μm [8].

The cyclotron operated in a 12-second cycle, delivering accelerated ions during a 1-second interval. The average fluence rate was maintained within the range of 1×10^4 to 1.5×10^4 ions $\cdot$ cm^{-2} $\cdot$ s^{-1}, resulting in an accumulated ion flux of 1.5×10^5 to 1.8×10^5 ions $\cdot$ cm^{-2} per cycle. This pulsed delivery mechanism subjects the devices to intense particle bombardment over short durations, inducing pronounced performance degradation.

For power devices, test termination criteria were defined as single-event burnout (SEB). In the absence of burnout, experiments concluded once the total ion fluence reached 1×10^7 ions $\cdot$ cm^{-2} —a benchmark commonly adopted as the standard for assessing single-event irradiation tolerance in semiconductor devices.

The device was subjected to extreme biasing conditions, with high-voltage stress applied to the drain terminal during the off-state operation (VGS = 0, -2, -4 V). Under such circumstances, upon incidence of high-energy heavy ions, the penetration depth extended into the GaN buffer layer, as illustrated in Figure 2. Owing to their substantial kinetic energy, these incident ions induced copious electron-hole pair generation via inelastic collisions along their trajectories. The generated electron-hole pairs induced significant perturbations to the charge distribution in the outer regions of the depletion layer. Specifically, electrons were swept towards the high-potential drain terminal by the electric field, while holes accumulated within the device structure [9].

In Si-based GaN HEMT devices, hole accumulation predominantly occurred at two critical regions: the channel and buffer layers in the vicinity of the gate, and the heterointerfaces between the buffer layer and the substrate. The p-GaN gate, structured as a pin junction, remained in a reverse-biased state under off-state drain biasing conditions. This reverse-biasing configuration facilitated the accumulation of holes primarily in the channel and buffer layers adjacent to the gate [10]. The Si-based GaN fabrication process relies on heteroepitaxial growth, which inherently exhibits a higher lattice mismatch compared to homoepitaxial growth techniques. This elevated lattice mismatch results in a

greater density of interfacial defects [11, 12]. Generated holes interacted with these defects, leading to their trapping and subsequent accumulation at the heterointerfaces.

These microscopic alterations in charge distribution manifested as observable changes in the macroscopic electrical characteristics, specifically variations in the drain and gate currents. Among these, the evolution of the drain current served as a direct and sensitive indicator of the device degradation induced by heavy-ion irradiation.

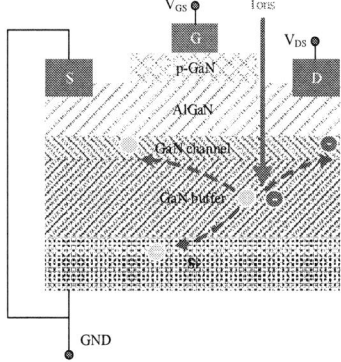

Fig. 2. Schematic Diagram of Heavy Ions Incident on GaN HEMT Device

The experiment utilized Kr and Ta ions with perpendicular incidence onto the device. The Linear Energy Transfer (LET) values of Kr and Ta ions were 36 MeV $\cdot$ cm^2/mg and 82.8 MeV $\cdot$ cm^2/mg, respectively, both exhibiting ranges exceeding 30 μm within silicon substrates [8]. This investigation was meticulously designed to systematically explore the influence of ion LET magnitudes and diverse biasing conditions on the performance degradation mechanisms of GaN HEMT devices. The study sought to elucidate the complex interplay between these critical parameters and the resultant changes in device functionality under heavy - ion irradiation.

III. RESULT AND DISCUSSION

Table 1 presents the experimental conditions for several heavy-ion irradiation tests. Experiments will be conducted under these conditions, and the degradation of the devices will be evaluated.

Table 1 Simulation Test Conditions for Single-Event Effects

Test Serial Number	Ion Species	V_{DS}	V_{GS}
1	Kr	250	0
2	Kr	300	0
3	Ta	250	0
4	Ta	300	0
5	Ta	200	0
6	Ta	200	-4
7	Ta	200	-6

Upon exposure to periodic high-energy heavy-ion bombardment, the outer region of the device's depletion layer experiences a rapid generation of numerous electron-hole pairs, which is macroscopically manifested as pulsed increases in the drain current. Prior to the arrival of heavy ions in the subsequent cycle, holes accumulated near the gate are depleted by the combined action of the AlGaN barrier layer and the p-GaN gate cap layer. This self-recovery process is reflected in the decrease of the drain current. However, complete depletion of all generated holes is not always achieved. Under the impact of high-LET ions, stringent biasing conditions, or cumulative effects, residual holes persist, preventing the drain current from fully reverting to its pre-irradiation baseline. Over successive ion bombardment cycles, this phenomenon can lead to a stepwise increase in the drain current, ultimately culminating in device burnout.

As illustrated in Figure 2(a), the drain current demonstrates three distinct response patterns under varying bias conditions. In the first scenario, exemplified by Test 1, heavy-ion bombardment induces periodic pulsed fluctuations in the drain current, which subsequently recovers to a value within the same order of magnitude as its pre-irradiation state. The second scenario, observed in Tests 2 and 3, involves incomplete recovery of the drain current following ion impact. With continuous degradation accumulation, the drain current escalates by three to four orders of magnitude. The third scenario, as evidenced in Test 4, is characterized by a rapid, pulsed increase in the drain current under ion bombardment, ultimately resulting in device failure due to burnout.

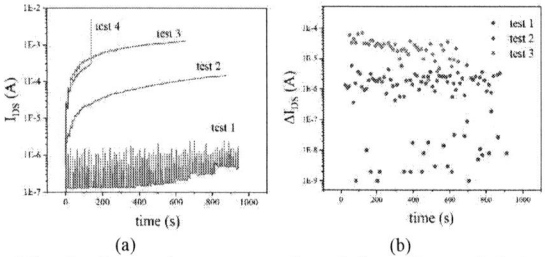

(a)   (b)

Fig. 3. Comparison test results of the effects of drain voltage and ion LET on device degradation. (a) Variation of drain current; (b) Variation of drain current in a single radiation cycle

Figure 2(b) presents the statistical analysis of the variation in drain current within each cycle of the experiment. It is evident that under the conditions of Test 1, the device exhibits the weakest degradation effect. In contrast, Test 2 employed a higher drain voltage, while Test 3 utilized ions with a higher Linear Energy Transfer (LET) value. Both modifications led to a significant exacerbation of device degradation.

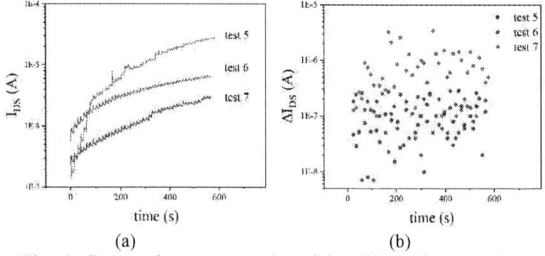

(a)   (b)

Fig. 4. Comparison test results of the effect of gate voltage on device degradation. (a) Variation of drain current; (b) Variation of drain current in a single radiation cycle

In this experiment, the strength of the barrier depletion effect was systematically varied by applying different gate voltages, and the resultant findings are illustrated in Figure 3. As the gate voltage VGS was incrementally adjusted from 0

V to -4 V and then to -6 V, a progressive enhancement in the barrier depletion effect was observed. Concurrently, there was a corresponding stepwise reduction in the magnitude of drain current variations. These results unequivocally demonstrate that the depletion mechanism at the device's gate plays a pivotal role in modulating the extent of device degradation under experimental conditions.

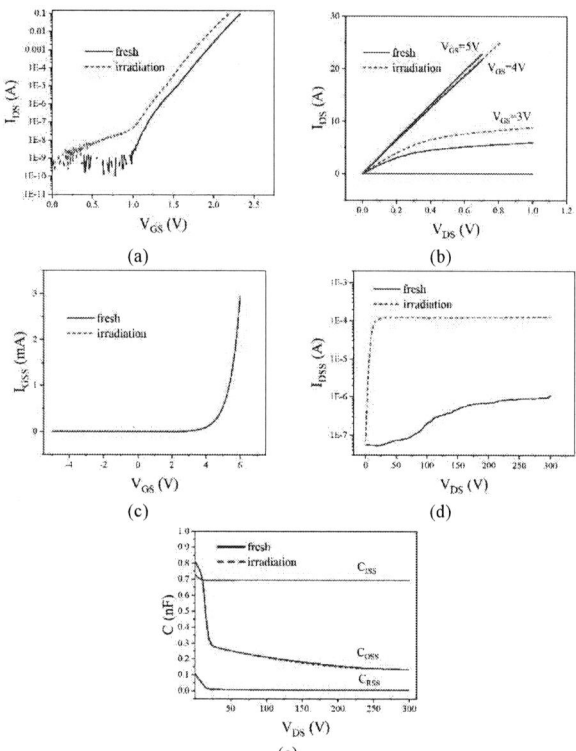

(a)   (b)

(c)   (d)

(e)

Fig. 5. Comparison of the characteristics of the device in Test 2 before and after irradiation(a) trans, (b) output (c) Igss, (d) BV (e) C-V

A comprehensive comparative analysis was performed on the characteristics of the device in Test 2 before and after irradiation, with the results depicted in Figure 4. Post-irradiation, the threshold voltage of the device exhibited a notable reduction of 0.15 V. When a gate voltage equivalent to the threshold voltage was applied, a substantial alteration in the on-resistance was observed. Conversely, minimal variation in on-resistance occurred when the device reached full conduction (VGS = 5 V). In the off-state configuration, the drain current experienced a significant increase, with the magnitude of change aligning closely with that recorded during the irradiation test. Notably, the gate current, along with input, output, and reverse transfer capacitances, remained largely unchanged. These findings strongly suggest that the extent of degradation differs among various regions of the device, with the barrier and channel layers in the vicinity of the gate undergoing more severe degradation.

IV. CONCLUSION

In a nutshell, the focal point of this paper lies in the degradation effects of E-mode p-GaN HEMTs when exposed to heavy-ion radiation, with a primary emphasis on the extent of device degradation under diverse conditions. The obtained results clearly indicate that both a higher drain voltage and ions with elevated Linear Energy Transfer (LET) values contribute to the exacerbation of device degradation. Moreover, the bias voltage applied to the gate also exerts an influence on the degradation of the device. Among the characteristics of the device, the most significant changes before and after irradiation are observed in the threshold voltage and the on-resistance when the gate voltage is at the threshold level. It is evident that the fabricated E-mode p-GaN HEMTs possess commendable radiation resistance capabilities.

REFERENCES

[1] S. Pearton, A. Haque, A. Khachatrian, A. Ildefonso, L. Chernyak, and F. Ren, "Opportunities in single event effects in radiation-exposed SiC and GaN power electronics," ECS Journal of Solid State Science and Technology, vol. 10, no. 7, p. 075004, 2021.

[2] Chen et al., "Effects of Applied Bias and High Field Stress on the Radiation Response of GaN/AlGaN HEMTs," Nuclear Science, IEEE Transactions on, 2015.

[3] S. Yang, A. Barchowsky, G. Allen, S. Vartanian, and K. Smedley, "Single Event Effects Testing for Integrated GaN Power Devices," in 2024 IEEE Radiation Effects Data Workshop (REDW) (in conjunction with 2024 NSREC), 22-27 July 2024 2024, pp. 1-6, doi: 10.1109/REDW61286.2024.10759221.

[4] Z. Zhen, C. Feng, Q. Wang, D. Niu, X. Wang, and M. Tan, "Single Event Burnout Hardening of Enhancement Mode HEMTs With Double Field Plates," IEEE Transactions on Nuclear Science, vol. 68, no. 9, pp. 2358-2366, 2021, doi: 10.1109/TNS.2021.3102980.

[5] M. Zafrani, J. Brandt, R. Strittmatter, B. Sun, S. Zhang, and A. Lidow, "Radiation Results for Modern GaN-on-Si Power Transistors," in 2022 IEEE Radiation Effects Data Workshop (REDW) (in conjunction with 2022 NSREC), 18-22 July 2022 2022, pp. 1-4, doi: 10.1109/REDW56037.2022.9921699.

[6] D. M. Fleetwood, E. X. Zhang, R. D. Schrimpf, and S. T. Pantelides, "Radiation Effects in AlGaN/GaN HEMTs," IEEE Transactions on Nuclear Science, vol. 69, no. 5, pp. 1105-1119, 2022, doi: 10.1109/TNS.2022.3147143.

[7] S. Kuboyama, A. Maru, H. Shindou, N. Ikeda, and T. Tamura, "Single-Event Damages Caused by Heavy Ions Observed in AlGaN/GaN HEMTs," IEEE Transactions on Nuclear Science, vol. 58, no. 6, pp. 2734-2738, 2011.

[8] X. Zhou, J. Yang, and H. P. Team, "Status of the high-intensity heavy-ion accelerator facility in China," AAPPS Bulletin, vol. 32, no. 1, p. 35, 2022.

[9] C. Abbate et al., "Experimental study of Single Event Effects induced by heavy ion irradiation in enhancement mode GaN power HEMT," Microelectronics Reliability, vol. 55, no. 9, pp. 1496-1500, 2015/08/01/ 2015, doi: https://doi.org/10.1016/j.microrel.2015.06.139.

[10] J. Wei et al., "Charge Storage Mechanism of Drain Induced Dynamic Threshold Voltage Shift in p-GaN Gate HEMTs," IEEE Electron Device Letters, vol. 40, no. 4, pp. 526-529, 2019, doi: 10.1109/LED.2019.2900154.

[11] Y. Puzyrev et al., "Gate Bias Dependence of Defect-Mediated Hot-Carrier Degradation in GaN HEMTs," IEEE Transactions on Electron Devices, vol. 61, no. 5, pp. 1316-1320, 2014, doi: 10.1109/TED.2014.2309278.

[12] H. Xinwen et al., "The energy dependence of proton-induced degradation in AlGaN/GaN high electron mobility transistors," IEEE Transactions on Nuclear Science, vol. 51, no. 2, pp. 293-297, 2004, doi: 10.1109/TNS.2004.825077.

979-8-3315-1110-4/25 $31.00 © 2025 IEEE

2025 IEEE Workshop on Wide Bandgap Power Devices and Applications in Asia (WiPDA Asia)

Development of a High-Power-Density SiC Module

1st Jiajun Yang
University of Chinese Academy of Sciences
Key Laboratory of High Density Electromagnetic Power and Systems (Chinese Academy of Sciences),Institute of Electrical Engineering, Chinese Academy of Sciences
Beijing, China
yangjiajun@mail.iee.ac.cn

1st Xiaoshuang Hui
University of Chinese Academy of Sciences
Key Laboratory of High Density Electromagnetic Power and Systems (Chinese Academy of Sciences),Institute of Electrical Engineering, Chinese Academy of Sciences
Beijing, China
hui00@mail.iee.ac.cn

1st Puqi Ning
University of Chinese Academy of Sciences
Key Laboratory of High Density Electromagnetic Power and Systems (Chinese Academy of Sciences),Institute of Electrical Engineering, Chinese Academy of Sciences
Beijing, China
npq@mail.iee.ac.cn

Abstract—With the rapid development of power electronics technology, power modules are increasingly applied in fields such as electric vehicles, renewable energy generation, and power systems. These applications demand higher efficiency and reliability from power modules. However, heat generated during operation causes temperature rise, which compromises performance and longevity. Additionally, parasitic inductance and thermal resistance within power modules are critical factors contributing to power losses. To address these challenges, researchers continue to explore innovative design and material technologies to minimize losses in power modules.

This paper proposes a novel double-layer Direct Bonded Copper (DBC) layout design applied to a low-power module integrating sixteen chips in parallel. By arranging chips and interconnection structures on a double-layer DBC substrate, the design effectively reduces parasitic inductance and thermal resistance. Experimental results demonstrate that compared to traditional single-layer DBC layouts, the proposed design reduces parasitic inductance by approximately 40%, suppresses voltage and current oscillations, and significantly improves module efficiency and reliability. Furthermore, the double-layer DBC layout enhances thermal management capabilities, extending the operational lifespan of the module.

Through comparative analysis of different chip applications, we find that the high-temperature suppression characteristics and stability of chips significantly influence module performance. Under elevated temperatures, The multi-chip integrated modules of some chips may explode. Therefore, in practical applications, the performance of a single chip and multiple chips needs to be comprehensively considered.

This research provides new ideas for the design and optimization of power modules, which has important theoretical significance and practical application value. By improving the DBC layout, parasitic inductance and thermal resistance can be effectively reduced, the efficiency and reliability of power modules can be enhanced, and the overall performance of power electronic systems can be promoted.

Keywords—power module, DBC, parasitic inductance

I. BACKGROUND

With the rapid development of power electronics technology, power modules are increasingly widely used in electric vehicles, new energy power generation, power systems and other fields[1]. These applications place higher demands on the performance of power modules, especially in terms of efficiency and reliability. However, the power module generates a lot of heat during operation, resulting in an increase in temperature, which in turn affects its performance and life. In addition, the parasitic inductance and thermal resistance in the power module are also important factors leading to increased loss.

To address these challenges, researchers are constantly exploring new design and material technologies to reduce the loss of power modules. For example, by optimizing the DBC layout and using high-performance materials, parasitic inductance and thermal resistance can be effectively reduced. In addition, double-sided heat dissipation structures and advanced packaging technologies are also widely studied to improve heat dissipation efficiency and reduce switching losses. These studies not only help to improve the efficiency and reliability of power modules, but also promote the overall performance of power electronics systems to meet the needs of modern energy systems for efficient and reliable power conversion.

II. INTRODUCTION

In this paper, a novel double-layer direct bonded copper (DBC) layout design is presented. This design is applied to a small power module, which consists of eight chips in parallel and a total of 16 chips. The design effectively reduces the parasitic inductance and thermal resistance by altering the DBC layout and rationally arranging the chip and interconnection structure.

The experimental results of the double-pulse test on the power module composed of the DBC substrate demonstrate that the new DBC substrate exhibits excellent performance in the power module. It significantly improves the reliability of the module and reduces the power module's losses. The results show that the parasitic inductance of the newly designed module is reduced by approximately 40% compared to the module with the traditional single-layer DBC layout. Notably, the voltage and current oscillations are smaller, which significantly enhances the efficiency and reliability of the module. Furthermore, the double-layer DBC layout enhances the thermal management capabilities of the modules, thereby helping to extend their service life.

This research provides a new approach to the design and optimization of power modules, offering significant theoretical and practical application value.

III. BRIEF REVIEW OF PRIOR ART

In power modules, Direct Bonded Copper (DBC) substrates play a crucial role due to their high current-carrying capacity and excellent thermal conductivity. They have been widely used in power modules[2-3]. Therefore, conducting in-depth research on DBC substrates is significant for enhancing the fatigue life of the entire module. Currently, to improve the reliability of DBC and reduce losses in power modules, many studies have been carried out focusing on improving DBC

979-8-3315-1110-4/25 $31.00 © 2025 IEEE 142

materials, optimizing manufacturing processes, and enhancing layouts.

For instance, Literature [4] reviews the challenges, advances, and emerging issues in silicon carbide (SiC) power module packaging technology, and discusses how to improve module performance by improving package layout and material systems. Literature [5] proposes a new wire-bond package layout, which effectively reduces parasitic inductance by adjusting the position of the chip and the interconnect structure. In literature [6], the influence of edge tail length on the microstructure and mechanical strength of the substrate was studied by means of microstructure analysis and mechanical property testing. In literature [7], a power module is made by using the method of stacked DBC substrates, so that the current of the power loop flows in two opposite planes respectively, the distance is only the thickness of the ceramic layer.

In previous work, the effects of DBC materials, layouts, and bonding wires on parasitic inductance were discussed. This paper further investigates the impact of DBC layout on the module based on stacked DBC, proposing a novel DBC layout design that reduces losses in the power module during the switching process.

IV. MODEL BUILDING AND EXPERIMENTAL TESTING

A. Comparative analysis of two different chip applications

Fig. 1. Manufacturer 1 Single Chip Module Schematic Diagram

In the double-pulse tests conducted at 25°C and 150°C, the modules composed of chips from Manufacturer 1 and Manufacturer 2 were compared. The results indicate that, in the case of single-chip modules, the loss of the module using Manufacturer 1 chips is significantly lower than that of the module using Manufacturer 2 chips. However, when multiple chips are combined into a module, the loss of the module using Manufacturer 1 chips is only approximately 20% lower than that of the module using Manufacturer 2 chips.

Fig. 2. Manufacturer 1 Multi-Chip Module Schematic Diagram

In our comparative study of the integrated modules from two manufacturers, we observed that the single-chip integrated module from Manufacturer 1 exhibited

significantly lower loss compared to that of Manufacturer 2, demonstrating superior performance in this regard. However, when eight chips were connected in parallel to form a module comprising a total of 16 chips, the difference in loss between the multi-chip integrated modules of Manufacturer 1 and Manufacturer 2 became less pronounced.

Further analysis revealed that the multi-chip parallel integrated module from Manufacturer 1 suffered from uniformity issues, which likely accounted for the insignificant difference in loss. Although the single multi-chip integrated module from Manufacturer 1 exhibited lower loss than that of Manufacturer 2 at low temperatures, the module from Manufacturer 1 was prone to explosion under high pressure at high temperatures. This issue was consistently observed across multiple experimental tests. Even when a more balanced bond line layout was adopted (compared to the standard layout), the problem of high-temperature and high-pressure explosion persisted during testing. In contrast, the multi-chip integrated module from Manufacturer 2 did not experience explosion under the same conditions. The primary reason for this difference may be attributed to the inferior high-temperature inhibition capability of Manufacturer 1's chips.

Therefore, in actual module experiments, it is crucial not only to consider the quality of single-chip integrated modules but also to address the issues related to high-temperature inhibition and stability in multi-chip configurations.

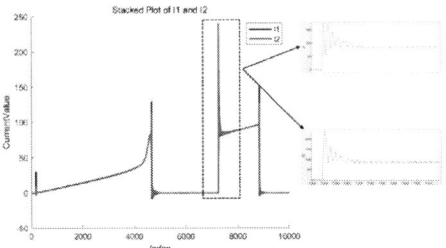

Fig. 3. Manufacturer 1 simulation current distribution diagram

Figure 3 above illustrates the simulated current distribution of two chips from Manufacturer 1 during a double-pulse test within the module. The horizontal axis (Index) indicates the sequence number of each test point, while the vertical axis (Current Value) represents the magnitude of the current measured at each corresponding point in the sequence.

By comparing and analyzing the current distribution curves shown in the figure, it is evident that there are significant differences in the current variation between the two chips during the double-pulse test. Specifically, the chip from Manufacturer 2 exhibits a relatively stable current change throughout the test, whereas the chip from Manufacturer 1 shows a sharp increase in current at the same sequence point. This inconsistency in current distribution can lead to unstable module performance in practical applications, potentially causing damage or even explosion of the module. Therefore, the chip from Manufacturer 2 was selected for use in this experiment.

979-8-3315-1110-4/25 $31.00 © 2025 IEEE

B. New DBC model

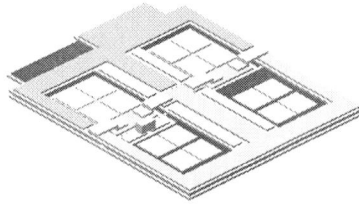

Fig. 4. DBC model diagram

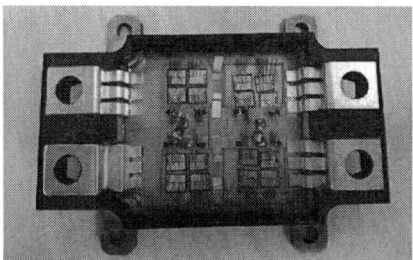

Fig. 5. Test power module physical diagram

In this experiment, we used a new type of direct bonded copper (DBC) substrate, whose structure is shown in Figure 4. The integrated test power module diagram is shown in Figure 5 for excellent heat transfer and electrical connections. The DBC layout was changed in the design to reduce parasitic inductance and improve electrical performance and reduce power module losses.

TABLE I.   STATIC TEST DATA

Static test data		
Thermal resistance of upper bridge arm	Thermal resistance of lower bridge arm	stray inductance
0.122K/W	0.125K/W	11.82nh

Before this experiment, the thermal resistance and stray inductance of the new DBC layout power module were tested. The data are shown in Table 1. The thermal resistance of the upper bridge arm is 0.122K/W, the thermal resistance of the lower bridge arm is 0.125K/W, and the stray inductance is 11.82nh.

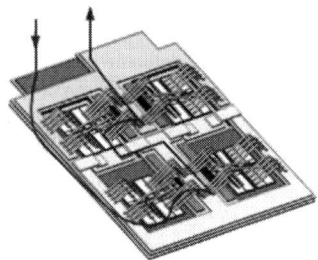

Fig. 6. DBC model current flow diagram

As depicted in Figure 6, the current path is as follows: The current originates from the positive DC terminal and flows downward through the lower layer of the DBC substrate. It then enters the drain of the SiC MOSFET and returns via the source electrode of the SiC MOSFET to the upper copper layer of the upper DBC through the bonding wire.

Subsequently, the current is conducted back to the upper copper layer of the lower DBC via an aluminum strip. It then flows through the drain, source, and bonding lines of another SiC MOSFET, returning to the upper copper layer of the upper DBC. Finally, the current completes its path by returning to the negative DC terminal.

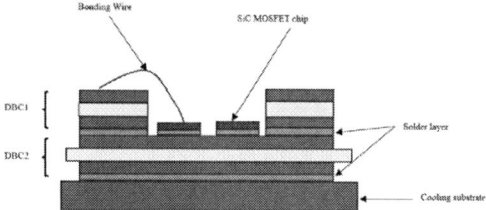

Fig. 7. Cross-Section Diagram of the Power Module

The cross-section diagram of its power module is shown in Figure 7. The chip is located on the upper copper of the lower DBC, and the thermal conductivity path of the chip has only one layer DBC.

C. Double pulse circuit test

To comprehensively evaluate the performance of the power module based on the novel DBC substrate, a double-pulse test platform was established in the laboratory. The schematic diagram of the double-pulse test is illustrated in Fig. 8. In this configuration, the switching devices S1 and S2 initially remain in the off-state, with no current flowing through the inductor L. During the first pulse, S2 is activated, allowing current to charge the inductor L from the power supply U, resulting in a rising current. Upon termination of the first pulse, S2 turns off, but the current in L freewheels through the parallel-connected diode. After a brief delay, a second pulse reactivates S2, enabling evaluation of its turn-off characteristics, including turn-off time and tail current. Following the second pulse, S2 turns off again, and the current in L discharges through the diode.

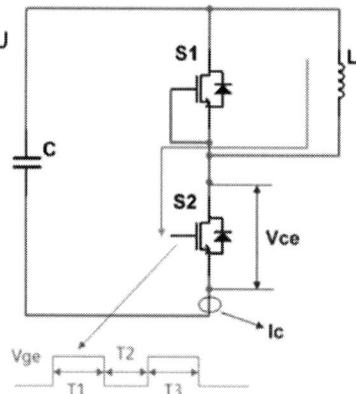

Fig. 8. The Schematic Diagram of the Double Pulse Test

To comprehensively assess the performance of the power module incorporating the new DBC substrate, we constructed a double-pulse test platform in the laboratory. The experimental test platform is illustrated in Figure 8.

Fig. 9. Double-Pulse Test Platform

In the experiment, the chip of manufacturer 2 was used to test the switching loss under the condition of low temperature 25°C and high temperature 150°C, DC bus voltage 800V and current 300A.

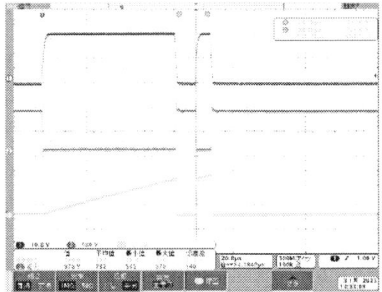

Fig. 10. Lower Tube 25°C Power Module Double-Pulse Waveform

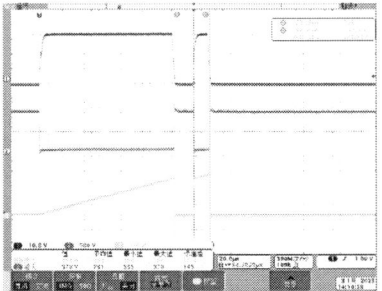

Fig. 11. Lower Tube 150°C Power Module Double-Pulse Waveform

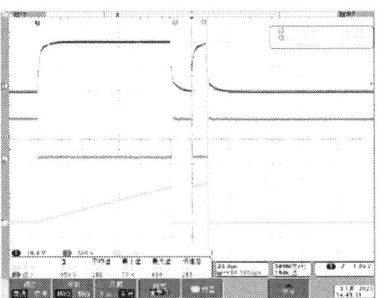

Fig. 12. Upper Tube 25°C Power Module Double-Pulse Waveform

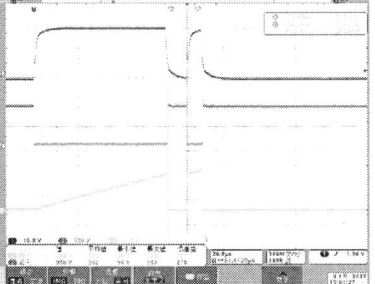

Fig. 13. Upper Tube 150°C Power Module Double-Pulse Waveform

As shown in the figure, the reverse recovery current of the diode is larger at high temperatures.

The formula of Turn-on Loss E_{on} is:

$$E_{on} = \int_0^{t_{on}} V_{DS} * I_{DS} dt \qquad (1)$$

The formula of Turn-off Loss E_{off} is:

$$E_{off} = \int_0^{t_{off}} V_{DS} * I_{DS} dt \qquad (2)$$

V_{DS} is Drain-Source Voltage, I_{DS} is Drain-Source Current, t is Integration time. The turn-on loss at 150°C is lower than the turn-on loss at 25°C, and the turn-off loss at 150°C is higher. The waveforms of the lower pipe at 25°C and 150°C are shown in Figure 9 and 10. The waveform of the upper pipe at 25°C and 150°C is shown in Figure 11 and 12.

V. CONCLUSION

This novel double-layer direct bonded copper (DBC) layout design effectively reduces parasitic inductance. Experimental tests show that compared with the traditional single-layer DBC layout, the parasitic inductance of the newly designed module is reduced by about 40%, and the voltage and current oscillation are smaller, which significantly improves the efficiency and reliability of the module, while enhancing the thermal management capability, helping to extend the service life, and providing a new idea and method for the design and optimization of the power module.

ACKNOWLEDGMENT

This work was .supported by the National Key R&D Program of China (2021YFB2500600).

REFERENCES

[1] A. Q. Huang, Power semiconductor devices for smart grid and renewable energy systems, Proceedings of IEEE, vol. 105, no. 11, pp. 2019-2047, 2019.

[2] N. Iwase, K. Anzai, K. Shinozaki, O. Hirao, T. Thanh, Thick Film and Direct Bond Copper Forming Technologies for Aluminum Nitride Substrate, IEEE Transactions on Components Hybrids & Manufacturing Technology, vol. 8, no. 2, pp. 253-258, 2003.

[3] J. Schulz-Harder, K. Excel. "Advanced DBC Substrates for High Power and High Voltage Electronics" Proc. EPE, Dresden, 2005

[4] C. Chen, F. Luo, Y. Kang, A review of SiC power module packaging: Layout, material system and integration, CPSS Transactions on Power Electronics and Applications, vol. 2, no. 3, pp. 170-186, 2017.

[5] B. Zhang, S. Wang, Parasitic inductance modeling and reduction for wire-bonded half-bridge SiC multichip power modules, IEEE Transactions on Power Electronics, vol. 36, no. 5, pp. 5892-5903, 2021.

[6] G. Dong, G. Lei, X. Chen, K. Ngo, G.-Q. Lu, Edge Tail Length Effect on Reliability of DBC Substrates under Thermal Cycling, Soldering and Surface Mount Technology, vol. 21, no. 10-15, 2009.

[7] H. Zhuzhao, C. Chen, X. Yue, et al., A high-performance embedded SiC power module based on a DBC-stacked hybrid packaging structure, IEEE Journal of Emerging and Selected Topics in Power Electronics, vol. 8, no. 1, pp. 351-366, 2020.

Minimizing Sensor Usage in TSEPs-based Junction Temperature Estimation of SiC Power Transistor through Time-Series Analysis

Valentyna Afanasenko
Institute of Robust Power
Semiconductor Systems
University of Stuttgart
Stuttgart, Germany
valentyna.afanasenko@ilh.uni-stuttgart.de

Oleksandr Solomakha
Institute of Robust Power
Semiconductor Systems
University of Stuttgart
Stuttgart, Germany
oleksandr.solomakha@ilh.uni-stuttgart.de

Kevin Muñoz Barón
Institute of Robust Power
Semiconductor Systems
University of Stuttgart
Stuttgart, Germany
kevin.munoz-baron@ilh.uni-stuttgart.de

Ingmar Kallfass
Institute of Robust Power
Semiconductor Systems
University of Stuttgart
Stuttgart, Germany
ingmar.kallfass@ilh.uni-stuttgart.de

Abstract—This paper presents a machine learning-based sensor fusion approach for non-invasive junction temperature estimation in SiC power modules, integrating multiple temperature-sensitive electrical parameters (TSEPs). By incorporating both current and previous TSEP values through Long Short-Term Memory (LSTM) networks, the junction temperature can be accurately predicted using a minimal number of sensors. Optimal TSEP combinations are explored to enhance accuracy and efficiency of the temperature estimation model and to consider the device degradation effect. Experimental results are presented on the dataset, acquired from two commercial SiC power modules under different operating conditions, with reference temperature measured by an optical sensor for verifying the proposed approach. Results demonstrate the method's effectiveness, offering a sensor-efficient solution for an accurate and degradation-aware temperature estimation in SiC MOSFET devices, achieving a mean estimation error of less than 1 K and a maximum error below 8 K.

Keywords— *Junction temperature estimation, Power Semiconductor, Temperature Sensitive Electrical Parameter (TSEP), Deep Learning (DL), SiC MOSFET*

I. INTRODUCTION

Monitoring the junction temperature (T_j) of power semiconductor devices is crucial for health management and remaining useful lifetime prediction, ensuring their reliability. Temperature Sensitive Electrical Parameters (TSEPs) enable a non-invasive, real-time temperature estimation of packaged SiC power semiconductor devices during operation [1], [2], with a time step shorter than 100 μs [2]. However, TSEPs are dependent on operational parameters such as current and voltage, which introduce uncertainty in the relationship between TSEP and T_j, making a single TSEP insufficient for accurate temperature estimation without calibration and additional measurements of operational parameters. Previously, to address this challenge, multiple TSEPs have been utilized jointly in a sensor fusion approach to estimate junction temperature in SiC MOSFETs in real time, compensating for variations in operational conditions [2], [3], [4], [5]. Machine Learning (ML) methods offer a practical solution to learn the complex and nonlinear relationship

between T_j and multiple multi-dimensional TSEPs directly from measured data, without the need of explicit modeling. Previous studies have demonstrated the effectiveness of using multiple TSEPs as inputs to Neural Networks (NNs) for T_j estimation in SiC power MOSFETs [2], [3], [5]. In [6], a similar NN-based sensor fusion approach was applied for the junction temperature estimation in SiC power MOSFET, integrating parameters such as load current, applied voltage, and device resistance, achieving a maximum estimation error of 2.5°C. In [4], a load-independent junction temperature model was developed using three TSEPs: the peak drain-source voltage $V_{DS,peak}$, the peak drain current $I_{D,peak}$, and the turn-on delay time $t_{d,on}$. The model employed multiple linear regression to capture the relationship between T_j, the TSEPs, and load conditions, achieving a Mean Absolute Percentage Error (MAPE) of 6.31%. However, achieving high precision in T_j estimation with a minimal number of sensors is essential for practical applications, as the cost of such monitoring systems must remain manageable.

Moreover, most TSEPs are sensitive to a device's health condition, with degradation from different failure mechanisms causing data drift in the TSEPs [7][8]. This drift can reduce the accuracy of data-driven models, which are typically valid only within the range of the data they were trained on. Several studies have addressed this challenge for SiC MOSFETs. In [5], a data-driven T_j estimation model was developed, with the training dataset incorporating on-state drain-source voltage $V_{DS,on}$, threshold voltage V_{th}, and turn-on transient parameter dI_D/dt, and I_D, measured across various levels of gate oxide degradation, to ensure accurate T_j estimation across different State of Health (SoH) of device. In [9], to address device degradation, the fatigue accumulation number was used as an additional input to the NN, along with Tj, wich was calculated from $R_{DS,on}$, I_{DS}, and V_{DS}, based on the T_j-$R_{DS,on}$ relationship calibrated for a healthy SiC MOSFET. However, TSEPs within the same transistor model and production batch can exhibit varying degradation evolution due to differences in production variations and load profiles. These factors may cause certain degradation mechanisms to dominate, while chip and package degradation can progress at different rates for individual Devices Under Test (DUTs). Consequently, the relationship between the offset due to aging for one TSEP-T_j

pair and the offset for another TSEP-T_j pair can be non-linear, or even unpredictable, potentially leading to a reduction in accuracy when applying a trained data-driven model to other degraded DUTs. In [10], the authors propose a real-time T_j measurement method based on td,on, using an additional V_{th} measurement circuit for aging compensation. Based on the sensitivity of V_{th} over T_j which was found to be consistent over aging, the percentage of the Vth offset is derived over aging and then incorporated into the $t_{d,on}$-T_j curve to compensate for the impact of data drift.

This paper focuses on improving the accuracy of TSEP-based temperature estimation in SiC power MOSFET while reducing the number of sensors. To achieve this, the selection of optimal TSEP combinations for data-driven modeling is first explored. Next, current and previous TSEP values are used as inputs to a memory-enabled machine learning model, namely a Long Short-Term Memory (LSTM) network, enabling additional information to be extracted from TSEP time series data. This approach enhances estimation accuracy with a reduced number of sensors. Finally, V_{th} and $t_{d,on}$, both dependent on gate oxide degradation, measured across various levels of SOHs, are jointly used in the LSTM model to account for device degradation.

II. JUNCTION TEMPERATURE ESTIMATION BASED ON TEMPERATURE-SENSITIVE ELECTRICAL PARAMETERS

A. TSEPs Datasets

Two datasets with different TSEPs measured from SiC MOSFET power modules under various operational conditions are used in this work. An optical sensor is used for ground truth T_j measurements:

Dataset 1: Referred as "Dataset 1", this dataset is collected from a commercial SiC power module using the TSEPs acquisition concept described in [2]. It includes the following six TSEPs: on-state voltage $V_{DS,on}$, quasi threshold voltage $V_{th,on}$ and $V_{th,off}$, quasi turn-on and turn-off delay $t_{d,on}$ and $t_{d,off}$, and peak gate voltage $V_{rg,peak}$.

Dataset 2: Referred as "Dataset 2", this dataset is obtained from a PWM full-bridge three-phase converter using a real-time oscilloscope. The measurements are taken from two SiC MOSFET 120 A, 1200 V modules from the same batch, a new module and a degraded device that had been in use for three years, and include eight TSEPs: threshold voltage V_{th}, turn-on/-off delay $t_{d,on}$, $t_{d,off}$, peak gate current $I_{g,peak}$, on-state voltage $V_{DS,on}$, on-state resistance $R_{DS,on}$, fall time t_f, and current slew rate di/dt. The experimental conditions cover a range of temperatures of 25°C, 40°C, 60°C, and 80°C. The DC voltage VDS is varied up to 400 V, and the AC current ID is adjusted up to 130 A.

B. Junction Temperature Estimation

Since NNs can effectively model nonlinear dependencies, the amount of information a TSEP contains about T_j is more important than its linearity or sensitivity for an accurate NN-based temperature prediction. Mutual Information (MI) is used to quantify this information, as it captures overall dependence, including both linear and nonlinear relationships, and thus providing a more comprehensive measure of the TSEP-T_j dependency compared to correlation coefficients, that might miss important nonlinear dependencies. Fig. 1 illustrates the MI and correlation coefficients between TSEPs and T_j, while Fig. 2 shows the Mean Absolute Error (MAE) of NN-based T_j predictions on the test dataset as a function of the

total MI of the TSEPs used as NN inputs for Dataset 1 and Dataset 2 when different combinations of available TSEPs were utilized. Fig. 2 demonstrates that higher MI between inputs and the target of NN leads to improved estimation accuracy.

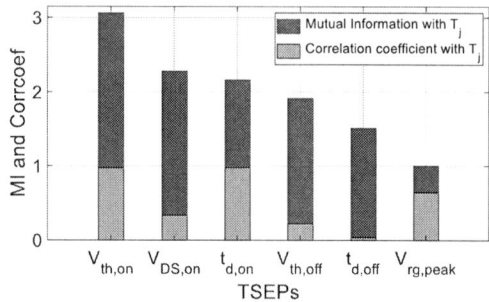

Fig. 1. Mutual information and correlation coefficient between each TSEP and junction temperature T_j (Dataset 1)

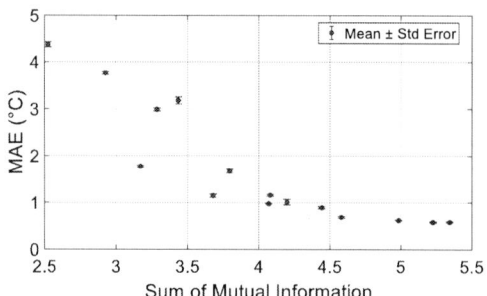

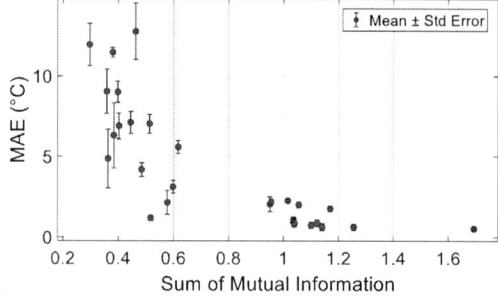

Fig. 2. MAE of T_j predictions on test dataset in dependence of sum of Mutual Information between TSEPs and T_j when using Dataset 1 (above) and Dataset 2 (below)

Another way to improve the learning of a data-driven model, and consequently its estimation accuracy, is ensuring training data consistency, meaning that there should not be different target values (T_j) for the same input values (TSEPs). As depicted in Fig. 3 (above), there are overlapping regions in $I_{g,peak}$ and $V_{DS,on}$ combination of values corresponding to different T_j values, specifically 40°C and 60°C, which can make learning even more challenging if significant noise is present. Fig. 3 (below) shows the results of T_j estimation using an NN trained on this data, demonstrating that the estimation error is higher for T_j values where the TSEPs tuples exhibit such overlap. To address this, adding another TSEP to the TSEPs combination can reduce the overlap. As shown in Fig. 4, incorporating V_{th} alongside $I_{g,peak}$ and $V_{DS,on}$ eliminates the overlap, thus enhancing estimation accuracy by improving dataset consistency on one hand, and providing more information about T_j on the other hand. Incorporating additional sensors increases the cost of temperature estimation, though.

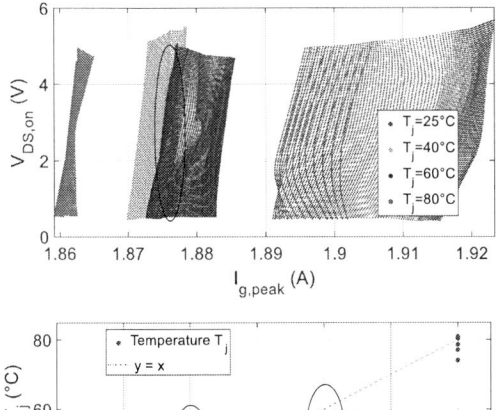

Fig. 3. Above: Combination of 2 TSEPs $V_{DS,on}$ and $I_{g,peak}$ for different T_j values (Dataset 2). Below: T_j estimation error of the NN model based on the $V_{DS,on}$ and $I_{g,peak}$ data shown above

One way to reduce the number of sensors without reducing model performance is through hybrid modeling, which integrates analytical knowledge with ML, also occurs in the literature as physics-informed ML, but this approach requires human engineering effort to develop the physics-based knowledge and to integrate it with the data-driven part.

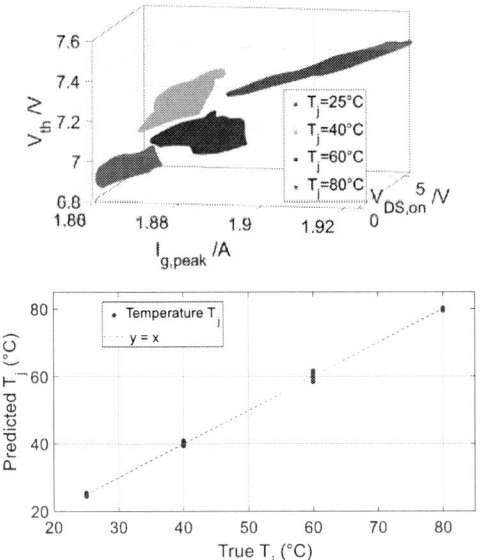

Fig. 4. Above: Combination of 3 TSEPs for different T_j values (Dataset 2). Below: T_j estimation error of the NN model based on the data shown above

Another way to enhance T_j estimation accuracy in the NN without adding additional TSEP is by treating TSEPs as time-series data. Since temperature changes gradually rather

than in discrete steps, previous TSEP values can provide additional information about the current T_j, reducing uncertainty. Instead of using only the current value of a TSEP, both current and past values are fed as inputs to the NN model. For this, we employ LSTM networks.

TABLE I. T_j ESTIMATION ERROR ON TEST DATA (20 TRAINING RUNS)

(Dataset) TSEPs	MSE NN	MAE NN	MSE LSTM	MAE LSTM
(1) $V_{th,on}$, $t_{d,on}$	3.119 ±0.240	1.285 ±0.083	0.374 ±0.072	0.378 ±0.037
(1) $V_{th,on}$, $V_{DS,on}$	1.388 ±0.042	0.826 ±0.019	0.332 ±0.044	0.376 ±0.034
(1) $V_{th,on}$, $V_{rg,peak}$	3.802 ±0.167	1.432 ±0.050	1.117 ±0.076	0.650 ±0.027
(1) $V_{th,on}$, $t_{d,on}$, $V_{DS,on}$	0.736 ±0.035	0.607 ±0.019	0.047 ±0.008	0.162 ±0.013
(1) $V_{th,on}$, $t_{d,on}$, $V_{rg,peak}$	2.550 ±0.277	1.163 ±0.084	0.233 ±0.011	0.308 ±0.009
(2) V_{th}, $V_{DS,on}$	1.732 ±0.760	0.722 ±0.177	0.854 ±0.545	0.474 ±0.154
(2) V_{th}, $I_{g,peak}$	2.069 ±0.328	0.567 ±0.088	0.477 ±0.149	0.346 ±0.057

Table I presents the Mean Squared Error (MSE) and MAE of T_j estimation using NN and LSTM for different TSEPs combination. Each model is trained 20 times with random initialization to ensure better comparability. The feedforward NN architecture consists of two hidden layers with 48 and 32 neurons respectively, followed by a single output neuron (48-32-1). The LSTM model employs a single LSTM layer with 64 hidden units, followed by a single output neuron. The dataset is split into training, validation, and test sets, and all features are scaled using min-max normalization. Models are trained with early stopping to prevent overfitting, and hyperparameter optimization is performed using random search. The input sequence length for the LSTM ranges from 8 to 15 values, depending on the specific combination of TSEPs used. Tendentially, TSEPs that are noisier or have lower mutual information with the target variable T_j required longer input sequences to achieve optimal performance. The results (Table I) clearly demonstrate that using TSEPs as time-series data in LSTM significantly improves T_j estimation accuracy compared to using single-valued TSEPs in NN models. For example, when using TSEPs $V_{th,on}$ and $t_{d,on}$, the MAE was reduced from 1.285±0.083 in NN to 0.378±0.037 in LSTM and the MSE decreased from 3.119±0.24 in NN to 0.374 ±0.072 in LSTM. The disadvantage of LSTM is longer training times and that it is a more complex model compared to NN. For the NN model, the total number of parameters is 1793, and the total memory usage is 7172 bytes (7.00 KB). In contrast, for the LSTM model, the total number of parameters is 17473, with a model size of 69892 bytes (68.25 KB). Due to its recurrent nature, the LSTM requires significantly more computation, also depending on the input length. This increase in computational cost and memory can be justified by significantly higher prediction accuracy, illustrating a trade-off between efficiency and performance.

Fig. 5 illustrates an example of the T_j estimation error distribution using NN and LSTM for combination of three TSEPs from Dataset 1 ($V_{th,on}$, $t_{d,on}$ and $V_{rg,peak}$), showing a reduction in the maximal absolute error from 8°C in NN to 5°C in LSTM. A cost assessment of the TSEPs acquisition hardware, which depends on the specific measurement equipment used, is not conducted in the present work, and we refer the reader to [2] for further technical details related to the acquisition of the Dataset 1, which is used in this work.

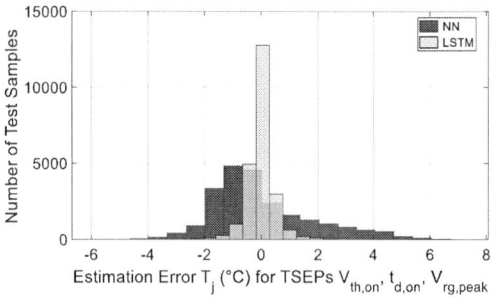

Fig. 5. Temperature estimation error with NN and LSTM when using combination of TSEPs as input data of models (Dataset 1)

C. Junction Temperature Estimation Considering Device Degradation Effect

To account for the device degradation in T_j estimation, two TSEPs, V_{th} and $t_{d,on}$, are used together, as in [10], since both are sensitive to gate oxide degradation. Thus, a relatively similar relationship between the degradation-induced offsets in the V_{th}-T_j and $t_{d,on}$-T_j relationships across different DUTs can be expected. This suggests that the variation in degradation evolution across different DUTs will have a smaller impact on T_j estimation accuracy when applying the trained model to other devices, compared to TSEPs influenced by multiple degradation mechanisms. The data-driven temperature model is developed using TSEPs measured across different health statuses of the DUT (Dataset 2), and an LSTM model is employed to improve the estimation accuracy.

TABLE II.   MSE AND MAE ON TEST DATA FOR DATASET 2 USING DATA FROM NEW AND DEGRADED DEVICES (50 TRAINING RUNS)

TSEPs	MSE NN	MAE NN	MSE LSTM	MAE LSTM
V_{th}, $t_{d,on}$	3.436 ±0.602	0.917 ±0.133	0.475 ±0.917	0.301 ±0.220

Table II presents the MSE and MAE of T_j estimation for SiC MOSFET under different health statuses and various operational conditions, using both NN and LSTM models with Dataset 2. The same model architectures as before were employed. The degradation-aware LSTM model shows significantly low estimation error, with a mean error of approximately 0.6 K.

III. CONCLUSION

This paper presents a degradation-aware and sensor-efficient approach for junction temperature estimation in SiC MOSFET devices. Two TSEPs, threshold voltage and turn-on delay, which have similar degradation dependencies, are measured across different health statuses of the device and used together in LSTM as time-series data to compensate for operational condition variations and consider degradation-induced data drift in T_j estimation. The mean evaluating error is less than 1 K, with the maximum estimation error below 8 K across 50 training runs.

ACKNOWLEDGMENT

This research is accomplished within the project "AUTOtech.agil" (FKZ 01IS22088). We gratefully acknowledge the financial support for the project by the Federal Ministry of Research, Technology and Space of Germany (BMFTR, formerly BMBF).

REFERENCES

[1] H. Yu, X. Jiang, J. Chen, Z. J. Shen and J. Wang, "Comparative Study of Temperature Sensitive Electrical Parameters for Junction Temperature Monitoring in SiC MOSFET and Si IGBT," *2020 IEEE 9th International Power Electronics and Motion Control Conference (IPEMC2020-ECCE Asia)*, Nanjing, China, 2020, pp. 905-909, doi: 10.1109/IPEMC-ECCEAsia48364.2020.9367830.

[2] K. M. Barón, D. K. Melgar, V. Afanasenko, R. Schnitzler and I. Kallfass, "Machine Learning Based Sensor Fusion for Junction Temperature Estimation," *2024 IEEE 10th International Power Electronics and Motion Control Conference (IPEMC2024-ECCE Asia)*, Chengdu, China, 2024, pp. 4076-4081, doi: 10.1109/IPEMC-ECCEAsia60879.2024.10567929.

[3] K. Sharma, S. Kamm, K. M. Barón and I. Kallfass, "Characterization of Online Junction Temperature of the SiC power MOSFET by Combination of Four TSEPs using Neural Network," *2022 24th European Conference on Power Electronics and Applications (EPE'22 ECCE Europe)*, Hanover, Germany, 2022, pp. 1-8.

[4] M. Luo *et al.*, "Load-Independent Junction Temperature Estimation via Combined TSEPs Modeling for SiC MOSFETs," in *IEEE Transactions on Power Electronics*, vol. 40, no. 1, pp. 851-861, Jan. 2025, doi: 10.1109/TPEL.2024.3473529.]

[5] Q. Xu, B. Zhang, X. Wang, J. Kang and Z. Xin, "A Novel Junction Temperature Estimation Method for SiC MOSFET Considering Device Aging Effect," *2023 IEEE 2nd International Power Electronics and Application Symposium (PEAS)*, Guangzhou, China, 2023, pp. 429-433, doi: 10.1109/PEAS58692.2023.10395720.

[6] O. Solomakha, V. Afanasenko and I. Kallfass, "Online Junction Temperature Estimation of Power Semiconductor Devices using Neural Network and Model-Based Design," *IEEE EUROCON 2023 - 20th International Conference on Smart Technologies*, Torino, Italy, 2023, pp. 187-192, doi: 10.1109/EUROCON56442.2023.10198878.

[7] P. Heimler, M. Alaluss, C. Schwabe, X. Liu, J. Lutz and T. Basler, "Online Threshold Voltage Monitoring at SiC Power Devices during Power Cycling Test and Possible Consequences," *2023 25th European Conference on Power Electronics and Applications (EPE'23 ECCE Europe)*, Aalborg, Denmark, 2023, pp. 1-10, doi: 10.23919/EPE23ECCEEurope58414.2023.10264579.

[8] Q. Zhang and P. Zhang, "A Novel Model of the Aging Effect on the ON-State Resistance of SiC Power MOSFETs for High-Accuracy Package-Related Aging Evaluation," in *IEEE Transactions on Industrial Electronics*, vol. 70, no. 9, pp. 9495-9504, Sept. 2023, doi: 10.1109/TIE.2022.3212405.

[9] Q. Liu, X. Du, Z. Wang, H. Ren, J. Zhou and R. Chen, "SiC MOSFETs Junction Temperature Estimation Method Considering Aging Influence," *2023 6th Asia Conference on Energy and Electrical Engineering (ACEEE)*, Chengdu, China, 2023, pp. 71-75, doi: 10.1109/ACEEE58657.2023.10239581.

F. Yang, S. Pu, C. Xu and B. Akin, "Turn-on Delay Based Real-Time Junction Temperature Measurement for SiC MOSFETs With Aging Compensation," in *IEEE Transactions on Power Electronics*, vol. 36, no. 2, pp. 1280-1294, Feb. 2021, doi: 10.1109/TPEL.2020.3009202

Molecular dynamics simulations of AgCu core-shell nanoparticle sintered with nanoflake

Yifeng Chen,
School of Nuclear Science, Energy and
Power Engineering
Shandong University
Jinan, China
chenyifeng@mail.sdu.edu.cn

Xin Lan*,
School of Nuclear Science, Energy and
Power Engineering
Shandong University
Jinan, China
lanxin@sdu.edu.cn

Ziyang Zhang
School of Nuclear Science, Energy and
Power Engineering
Shandong University
Jinan, China
ziyang@mail.sdu.edu.cn

Xin Li
School of Nuclear Science, Energy and
Power Engineering
Shandong University
Jinan, China
lixin2@mail.sdu.edu.cn

Abstract—Nanoscale interconnect materials enable high-power devices to efficiently dissipate heat and serve at high temperatures. AgCu nanoparticles with core-shell structure(Ag shell covering Cu core) not only leverage the cost-effectiveness of Cu materials but also exhibit enhanced resistance to oxidation. This study investigates the sintering of nanoparticle(NP) with nanoflake(NF) and analyzes the potential energy evolution throughout the sintering process. Besides, the sintering performance under different particle sizes and core-shell thickness ratios is investigated. The results reveal that the variation of potential energy occurs in four distinct stages in NP-NF sintering, it has an additional plateau phase compared to that of NP-NP sintering. A reduction in particle size shortens the plateau phase, furthermore, a larger core-shell thickness ratio reduces the centroid distance between NP and NF which facilitates the sintering process.

Keywords—core-shell, nanoparticle, nanoflake, molecular dynamics, sintering

I. INTRODUCTION

Traditional Pb-free solder is facing significant challenges in high reliability, high-heat dissipation capability and high-temperature environments due to its low melting point and poor thermal conductivity [1]. Nanoscale interconnect materials have excellent performances as low conductivity and high thermal conductivity, as well as their small size effect. It can be prepared at lower temperatures and has high reliability for high-temperature applications. Therefore, it exhibits excellent potential in the interconnection application of high power density devices [2]. However, the nanoscale interconnect interfaces significantly complicate the study of sintering mechanisms.

Molecular dynamics (MD) simulations have been extensively used to study the sintering behavior of materials at the atomic level. It can reveal the details of NP sintering, such as neck growth, displacement vectors, and so on[3]. Ye et al. discussed the effect of sintering temperature on the bonding quality between Ag NPs and SiC/Cu substrates based on MD [4], finding that the width of the sintering neck increased with higher sintering temperatures. Hu et al. [5] analyzed the effects of sintering pressure and temperature on Mean Square Displacement (MSD) and neck width during the sintering of Cu NPs. The results indicated that in pressure-assisted sintering, the dominant factor in the sintering of Cu

NPs was pressure rather than temperature. The cost of nano-Ag is relatively high, while low-cost nano-Cu is prone to oxidation during sintering. Cu@Ag NPs combine the excellent electrical and thermal conductivity of both Cu and Ag, effectively overcoming issues such as the migration of Ag and oxidation of Cu. Wang et al. [6] investigated the sintering mechanism of Cu@Ag NPs, revealing that the Cu core enhanced the mobility of the Ag shell.

Existing work mostly focuses on sintering between NPs. There is still a lack of investigations into the sintering of other shapes such as NP-NF and its sintering mechanism. The synthesized particles are not only spherical but also other shapes such as polygons, flakes, etc. Therefore, the work of this study is to explore the sintering of NP-NF particles under the Cu@Ag CS model by MD and to analyze the effects of particle sizes and core-shell(CS) thickness ratios on the sintering performance. It can provide a theoretical basis for optimizing the sintering process of Cu@Ag, and improve the material performance of nanoscale interconnect materials. The findings will promote the application of nanomaterials in electronic packaging.

II. METHODOLOGY

A. Model Construction

In this study, the LAMMPS (Large-scale Atomic/Molecular Massively Parallel Simulator, a widely used molecular dynamics simulation software) was utilized to construct two distinct sintering processes based on the CS models: nanoparticle-nanoparticle(NP-NP) and nanoparticle-nanoflake(NP-NF). These configurations were selected to mimic realistic scenarios in electronic packaging. In reality, the sintering of Cu@Ag NPs is usually carried out at around 550K to connect the die to the substrate. Therefore, the sintering temperature is selected as 550 K.

The effects of different particle sizes and shell thicknesses on the sintering performance are investigated. The particle sizes of NP were set to 6 nm, 8 nm, and 10 nm, respectively. The length and width of NF were set to be the same as the particle size of NP, and the thickness was set to be a fixed value of 4 nm. Furthermore, the core-shell thickness ratios were set to be 9:1, 4:1, and 2:1, respectively.

979-8-3315-1110-4/25 $31.00 © 2025 IEEE

TABLE I.   CONSTRUCTED NPS MODEL

	d=6nm	d=8nm	d=10nm
9:1			
4:1			
2:1			

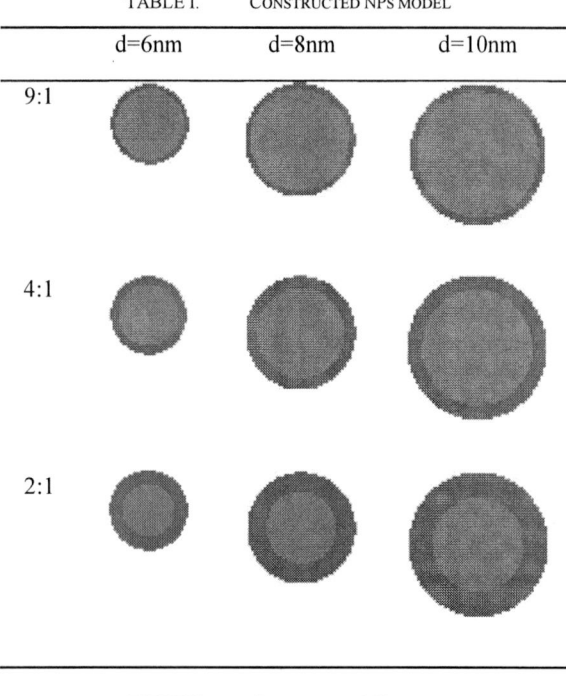

TABLE II.   CONSTRUCTED NFS MODEL

	L=6nm	L=8nm	L=10nm
9:1			
4:1			
2:1			

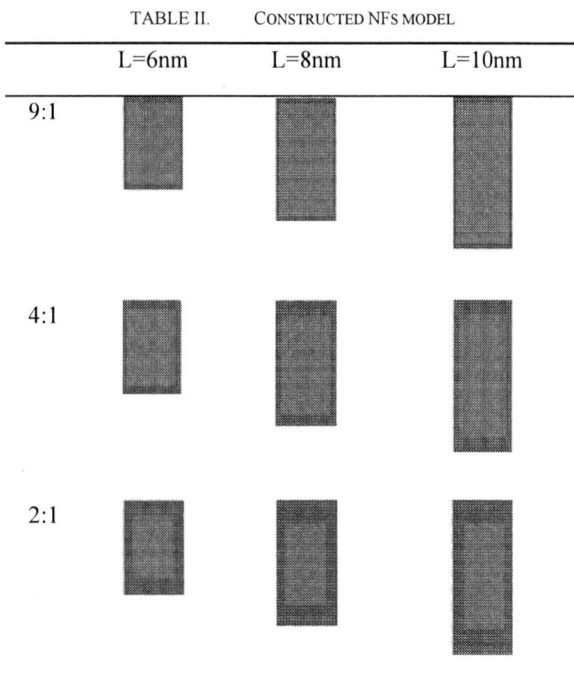

Atomic configurations of the NPs and NFs are visualized in Tables 1 and 2 (Ag shell atoms: blue; Cu core atoms: red). Before the sintering simulation began, all the nanoparticles and nanoflakes were minimized in energy to ensure the accuracy of the simulation. Simulations utilized an NVT ensemble with a Langevin thermostat (damping parameter: 100 fs) to maintain 550 K. A time step of 1 fs ensured numerical stability, and trajectories were recorded every 10 ps for post-processing.

B. Validation of the Potential Function

The accuracy of the EAM.ALLOY potential function was rigorously verified prior to sintering simulations. In this study, the EAM.ALLOY potential function was used. The validation was carried out by creating Ag NP with a 6 nm particle size and Cu NP with a 6 nm particle size. The temperature varies from 300 K to 1500 K.

The monatomic potential energy (PE) was recorded and compared with that measured by Ye et al [4]. The results indicated that the temperature dependence of the monatomic potentials of Cu and Ag were consistent with those measured by Ye et al. The maximum error of the monatomic potentials of Cu was kept within 1.81%, as shown in Fig.1. While the maximum error of the monatomic potentials of Ag was kept within 1.96%, as shown in Fig.2. The potential function is validated to be applicable in the sintering simulations.

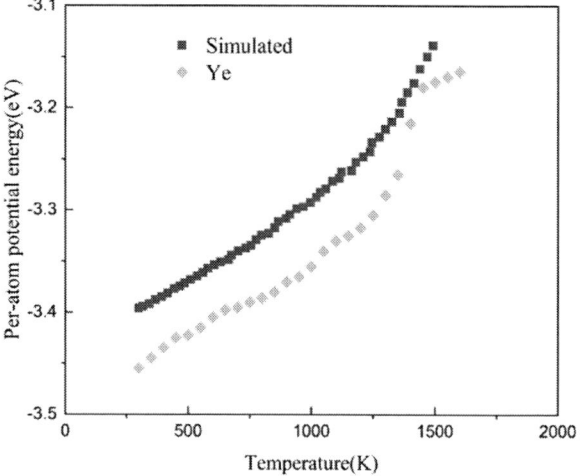

Fig.1. Monatomic PE of Cu

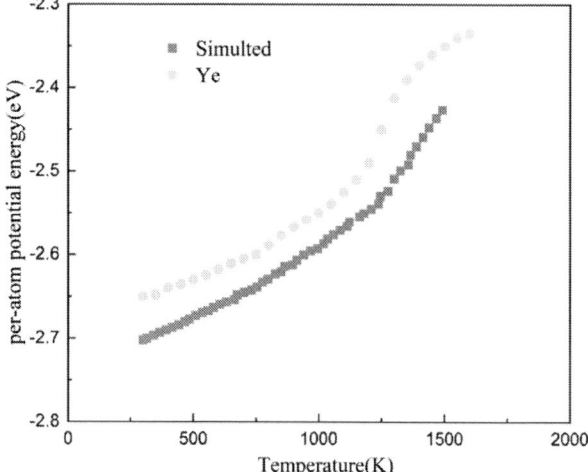

Fig.2. Monatomic PE of Ag

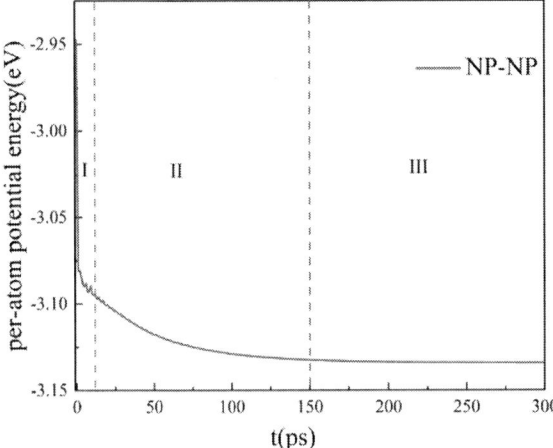

Fig.3. PE-t of NP-NP sintering

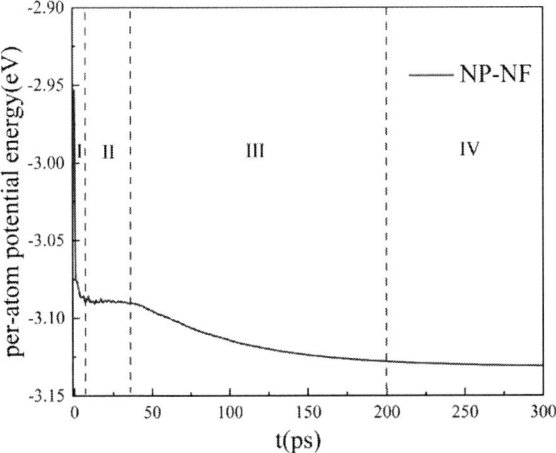

Fig.4. PE-t of NP-NF sintering

III. RESULTS AND DISCUSSION

A. PE during sintering of Cu@Ag

The sintering processes were simulated for both NP-NP and NP-NF. The trends of PE with time for the two sintering processes are different. The PE-t plot of NP-NP and NP-NF at a particle size of 10 nm, a sintering temperature of 550 K, and a core-shell thickness ratio of 4:1 are given in Fig.3 and 4, respectively.

For NP-NP, the change of PE can be divided into three stages: (I) a zone of rapid decrease of PE; (II) a slowing down of the rate of decrease of PE; and (III) an equilibrium of PE. For NP-NF, the change of PE is divided into four stages: (I) the PE decreases rapidly; (II) the PE remains unchanged for a period of time after the rapid decrease of the PE, which is called the plateau period; (III) the potential energy continues to decrease, but at a slower rate compared with that of the first stage; and (IV) the PE stays unchanged.

In stage (I), the PE decreases rapidly for both NP-NP and NP-NF sintering because the particles are just in contact with each other and the interatomic interactions at the interface are strong. The PE decreases by 0.14 eV in the first 10 ps for NP-NF sintering, while the PE decreases by 0.15 eV

in only 7 ps for NP-NF sintering. This is due to the fact that NF provides a larger contact area and the interfacial atoms are more likely to bond, resulting in a rapid reduction of the potential energy of the system. In contrast, the initial stage of NP-NP sintering is more stressful due to the smaller contact area, leading to disruption of the lattice structure in localized regions of the interface. As shown in the locally enlarged view of Fig.5, the content of the FCC phase decreases faster in the early stage of NP-NP sintering, and more FCC lattice structure is converted to an amorphous structure due to the larger stress.

In stage (II) of NP-NF sintering, a plateau period occurs. The PE is relatively stable at this stage. After an initial rapid release of PE, the system enters a locally stable state where the diffusion of atoms at the interface contact slows down, resulting in the PE remaining stable in the short term. As shown in the region contained by the blue dashed line in Fig.6, the atomic diffusion rate of NP-NF goes up and down but remains basically constant during the plateau period. In contrast, the MSD of NP-NP rises rapidly during the same time period, so there is no plateau period for NP-NP sintering.

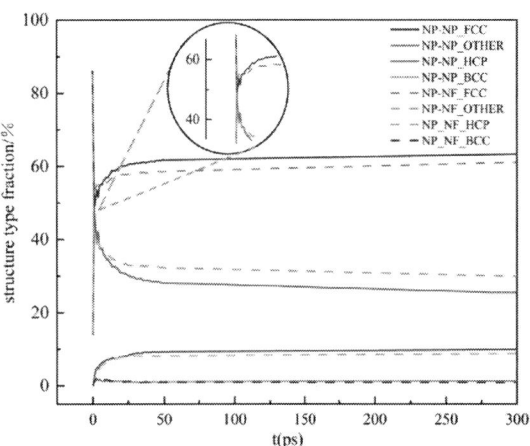

Fig.5. Changes in crystal structure content during sintering of NP-NP and NP-NF

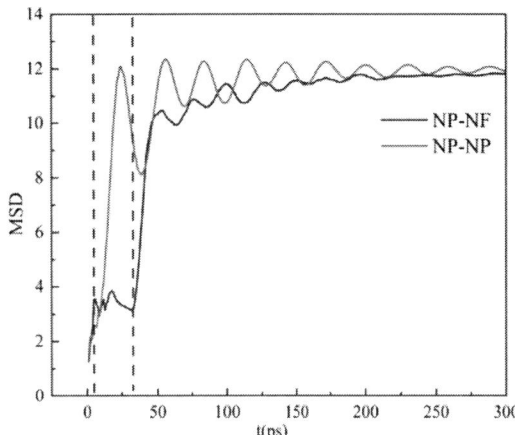

Fig.6. MSD of NP-NP or NP-NF with a particle size of 10nm and a core-shell thickness ratio of 4:1 sintered at 550K.

As the temperature continues to increase, the atoms continue to diffuse and the PE decreases further. When the temperature is not sufficient to promote diffusion of a sufficient number of atoms, the sintering process stagnates, as evidenced by the fact that both the PE and the MSD remain essentially unchanged.

B. Effect of particle size and core-shell thickness ratio on the sintering process

The sintering dynamics exhibit strong dependence on nanoparticle size and shell architecture. It was found that the plateau phase stage of NP-NF became longer and slower to enter stage (III) as the particle size increased, as shown in Fig.7. This size-dependent behavior arises from two competing factors: smaller NPs (6 nm) possess higher surface-to-volume ratios, which amplify atomic mobility at the Ag shell surface, while larger NPs require longer diffusion timescales to redistribute mass across their greater interfacial contact area with the nanofilm. Thus, the small-size NP with NF sintering has better sintering performance.

The thickness of the shell also affects the sintering performance of NP-NF. In practical applications, even small changes in shell thickness can lead to noticeable differences in the final material properties. By systematically testing core-shell ratios (9:1, 4:1, 2:1), we reveal a non-monotonic relationship between shell thickness and sintering efficiency. These ratios were carefully selected to cover a range of thicknesses commonly encountered in nanomaterial synthesis. As shown in Fig.8, the distance between the center of mass undergoes the maximum change when the core-shell thickness ratio is 9:1. This is because the Ag shell is thinner and the proportion of Cu nuclei is larger. The thinner shell layer makes the diffusion of Cu atoms and the mutual contact between particles easier. While sintering is carried out at a core-shell thickness ratio of 2:1. A thicker shell layer will result in slower changes in the center of mass during the sintering process, and the distance between the center of mass will be maximized, which is not conducive to the sintering process.

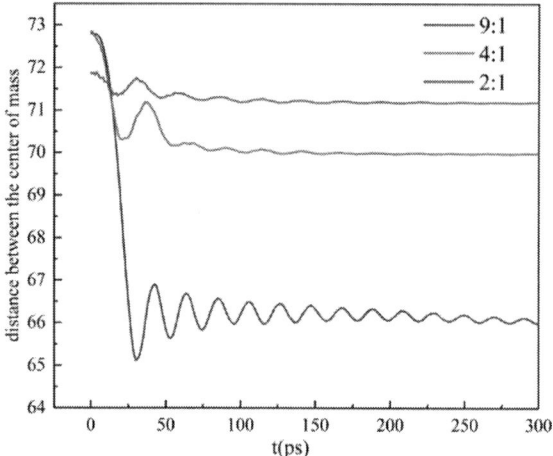

Fig.8. Center of mass distance of NP-NF at 10 nm particle size and sintering temperature of 550K

IV. CONCLUSION

This study mainly explores the sintering mechanism of NP-NF under the Cu@Ag CS model. The variation of PE with time during the sintering of NP-NP and NP-NF was analyzed, and the effects of different particle sizes and core-shell thickness ratios on the sintering performance of NP-NF were investigated. There are the following results:

1. The variation of NP-NP potential with time under the Cu@Ag CS model is divided into three phases, while the variation of NP-NF potential with time can be divided into four phases, with an additional plateau period compared to NP-NP.

2. When small-sized Cu@Ag NP-NF is sintered, the plateau period becomes shorter and the sintering performance is better.

3. When the core-shell thickness ratio takes the maximum value of 9:1, the distance between the center of NP-NF is the smallest, indicating that the larger the core-shell thickness ratio is more favorable to the sintering of Cu@Ag NP-NF.

ACKNOWLEDGMENT

This work was supported by the National Natural Science Foundation of China (No. 52376118).

REFERENCES

[1] Dele-Afolabi T T, Ansari M N M, Hanim M A, Oyekanmi A A, Ojo-Kupoluyi O J, Atiqah A, "Recent advances in Sn-based lead-free solder interconnects for microelectronics packaging: materials and technologies". Journal of Materials Research and Technology, 2023, 25: 4231-4263.

[2] Yu F, Cui J, Zhou Z, Fang K, Wayne Johnson R, Hamilton H C, "Reliability of Ag sintering for power semiconductor die attach in high-temperature applications". IEEE Transactions on Power Electronics, 2016, 32(9): 7083-7095.

[3] Hu Y, Wang Y, Yao Y, "Molecular dynamics on the sintering mechanism and mechanical feature of the silver nanoparticles at different temperatures". Materials Today Communications, 2023, 34: 105292.

[4] Ye G, Zhang J, Zhang P, Meng K, "Sintering mechanism between silver nanoparticles and SiC/Cu plates: A molecular dynamics simulation". Powder Technology, 2024, 439: 119695.

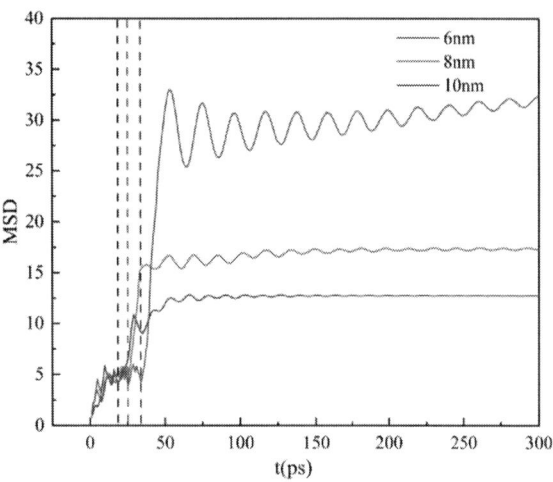

Fig.7. MSD of NP-NF sintered at different particle sizes (particle size of 6 nm and core-shell thickness ratio of 9:1)

[5] Hu D, Cui Z, Fan J, Fan X, Zhang G, "Thermal kinetic and mechanical behaviors of pressure-assisted Cu nanoparticles sintering: A molecular dynamics study". Results in Physics, 2020, 19: 103486.

[6] Wang J, Shin S, Hu A, "Geometrical effects on sintering dynamics of Cu–Ag core-shell nanoparticles". The Journal of Physical Chemistry C, 2016, 120(31): 17791-17800.

2025 IEEE Workshop on Wide Bandgap Power Devices and Applications in Asia (WiPDA Asia)

The Impact of Power Module Layout on Stray Inductance

Yuhui Kang
Key Laboratory of High Density
Electromagnetic Power and Systems
(Chinese Academy of
Sciences),Institute of Electrical
Engineering, Chinese Academy of
Sciences Beijing, China
kangyuhui@mail.iee.ac.cn

Puqi Ning
Key Laboratory of High Density
Electromagnetic Power and Systems
(Chinese Academy of
Sciences),Institute of Electrical
Engineering, Chinese Academy of
Sciences Beijing, China
npq@mail.iee.ac.cn

Xiaoshuang Hui
University of Chinese Academy of
Sciences
Key Laboratory of High Density
Electromagnetic Power and Systems
(Chinese Academy of
Sciences),Institute of Electrical
Engineering, Chinese Academy of
Sciences Beijing, China
hui00@mail.iee.ac.cn

Jiajun Yang
University of Chinese Academy of
Sciences
Key Laboratory of High Density
Electromagnetic Power and Systems
(Chinese Academy of
Sciences),Institute of Electrical
Engineering, Chinese Academy of
Sciences Beijing, China
yangjiajun@mail.iee.ac.cn

Abstract—**With the rapid development of power electronics technology, SiC power modules are increasingly being used in various electric vehicles. The performance of SiC power modules directly affects the efficiency and reliability of the entire system. Stray inductance, as an important parasitic parameter in power modules, has a significant impact on the module's switching characteristics, electromagnetic compatibility, and overall performance. The layout design of power modules is one of the key factors affecting stray inductance. SiC devices are more sensitive to module stray inductance, and to achieve parallel operation of multiple chips, it is necessary to minimize stray inductance as much as possible. Optimizing the layout can effectively reduce stray inductance, optimize gate waveforms, and improve module performance. This paper analyzes and compares the stray inductance of modules under different layout structures through simulation, and compares the impact of different layouts on stray inductance. The optimized scheme is then tested through packaging, ultimately seeking the optimal layout.**

Keywords—stray inductance, layouts, simulation

I. INTRODUCTION

In high power density and high frequency power electronic applications, the presence of stray inductance can lead to significant voltage spikes and current oscillations during the switching process of power devices, increasing switching losses and reducing the efficiency and reliability of the module[1-2]. It may even cause thermal failure of the device and electromagnetic interference issues. When SiC modules are paralleled with multiple chips, a large stray inductance can affect the gate waveform of the SiC module, and in severe cases, it can cause gate breakdown, making it impossible to parallel multiple chips. Therefore, in-depth research on the impact of power module layout on stray inductance is of great theoretical and practical significance for optimizing module design and enhancing module performance.

The factors that mainly affect stray inductance in power modules include the layout of the DBC (Direct Bonded Copper), the number of parallel chips, the wiring method, and the size of the power terminals. Among these, the number of

parallel chips, the wiring method, and the terminal size are generally less likely to be changed due to the constraints of packaging and application scenarios. The layout of the DBC, however, can be optimized by altering the layout structure and improving the routing path to reduce the area of the current loop, thereby lowering the stray inductance[3-4].

In this paper, the DBC layout structure is optimized through simulation comparison to determine the layout orientation. The results are then validated through packaging and experimental verification. Ultimately, based on the established optimization plan, a new type of low-inductance, low-loss SiC power module is designed and developed.

II. THE IMPACT OF PARASITIC INDUCTANCE ON SiC POWER MODULES

The structure of the SiC power module is shown in Fig. 1, which includes power terminals, copper-clad substrates, DBC (Direct Bonded Copper), chips, bonding wires, solder layers, copper substrates, housing, and encapsulant. Among these components, the power terminals, copper-clad substrates, DBC, chips, and bonding wires form the power loop of the module. The layout of the power terminals and DBC, as well as the chips and bonding wires within the loop, are the main factors affecting stray inductance[5-6].

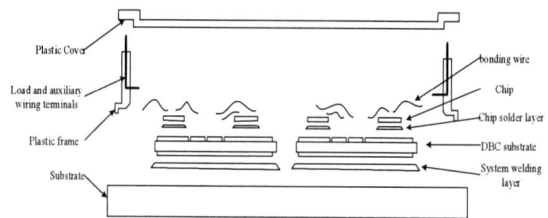

Fig.1 Power Module Structure Diagram

In SiC modules, stray inductance is primarily distributed in the power terminal connection paths, bonding wire interconnection structures, DBC (Direct Bonded Copper) connected copper layers, gate loops, and commutation loops. The stray inductance distribution diagram of a half-bridge SiC module is shown in Figure 2. As illustrated in Figure 2, LS represents the stray inductance generated in the connection paths of the positive and negative power terminals, which mainly includes the connection from the bus capacitor to the module and the length of the terminals. Optimizing the connection paths can reduce this portion of stray inductance.

Ls1-Ls6 are the stray inductances generated by the bonding wires and DBC copper layers. The interconnection method of bonding wires is one of the main contributors to stray inductance in traditional packaging. The longer the bonding wire, the smaller its diameter, and the fewer the number of parallel wires, the greater the stray inductance. Optimizing the bonding wires can reduce stray inductance. Meanwhile, the layout of the DBC and the design of the copper layer can reduce the commutation loop area through optimized routing, thereby reducing stray inductance.

LL is the stray inductance generated by the gate loop. The gate commutation inductance has a significant impact on the switching characteristics of the module. The larger the gate loop inductance, the worse the gate waveform, making it more difficult to parallel multiple chips.

The aforementioned LS and Ls1-Ls6 collectively form the stray inductance of the commutation loop. The commutation loop inductance is a key factor causing voltage overshoot. By optimizing the commutation loop path and reducing the loop area, stray inductance can be reduced, thereby minimizing voltage spikes.

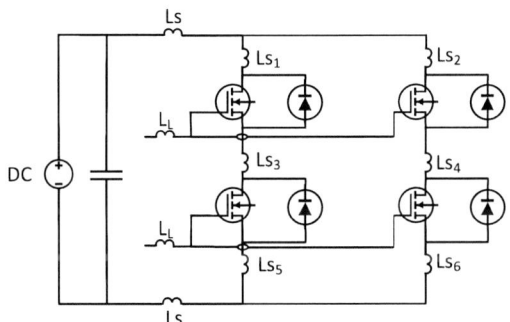

Fig.2 Stray Inductance Distribution Diagram

Stray inductance is a critical factor in SiC power modules. Due to the high switching speed and frequency characteristics of SiC devices, stray inductance can lead to more significant issues such as voltage overshoot, oscillations, increased losses, and even device failure. The stray inductance in the main circuit of a SiC power module can cause voltage overshoot and oscillations during switching, resulting in a drain-source voltage significantly exceeding the device's rated withstand capability, which may ultimately lead to device breakdown. Stray inductance prolongs the voltage rise time during turn-off, increasing switching overlap duration and consequently raising turn-off losses.Similarly, the energy stored in the inductance during turn-on current rise slows down current establishment, leading to higher conduction losses. The stray inductance in the gate path forms an LR circuit with the gate resistor, slowing down the gate voltage transition and reducing switching speed. This may even induce gate oscillation[7-8].

III. SIMULATION ANALYSIS AND VALIDATION

The stray parameters were extracted for various layout structures using the ANSYS Q3D software. The extraction

results are shown in Table 1. The various layout structures are illustrated in Figure 1.

Table 1   Table of Parasitic Inductance for Different Structures

Name	DC-side stray inductance
Conventional Design	23.7 nH
Conventional Design with Fewer Chips	25.9 nH
Conventional Planar Design	20.8 nH
Conventional Design with Doubled Area	12.75 nH
Conventional Axially Symmetric Design	10 nH
Conventional Design with Central Terminal	8.8 nH
Conventional Design with Extended Central Termina	20.32 nH

The stray inductance extraction result for the conventional design is 23.7 nH, while that for the conventional design with fewer chips is 25.9 nH. Comparing these two results reveals that, with the loop size unchanged, the fewer the parallel chips, the higher the stray inductance.

The stray inductance for the conventional planar design is 20.8 nH. Comparing the conventional design with the conventional planar design indicates that replacing bond wires with copper sheets or other structures to connect the chips to the DBC can reduce stray inductance, but only by 12%, which is a limited improvement.

When the area is doubled and an axially symmetric structure is adopted, the stray inductance is reduced to 12.75 nH, a reduction of 46%. This demonstrates that symmetric structures can effectively lower stray inductance. Based on this principle, the conventional design was modified to an axially symmetric layout, resulting in a stray inductance of 10 nH, a reduction of 57.8%.

By relocating the power terminal to the center to create a more symmetric layout, the conventional design with a centrally located power terminal was developed. This structure has a stray inductance of 8.8 nH, further reducing the stray inductance.

Considering the central power terminal, to facilitate connections, the terminal was extended, resulting in the conventional design with an extended central terminal. This structure has a stray inductance of 20.32 nH, an increase of 130% compared to the conventional terminal. This shows that an overly long power terminal can significantly increase stray inductance.

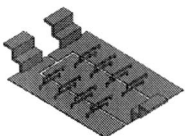   

Conventional Design   Conventional Design with Fewer Chips

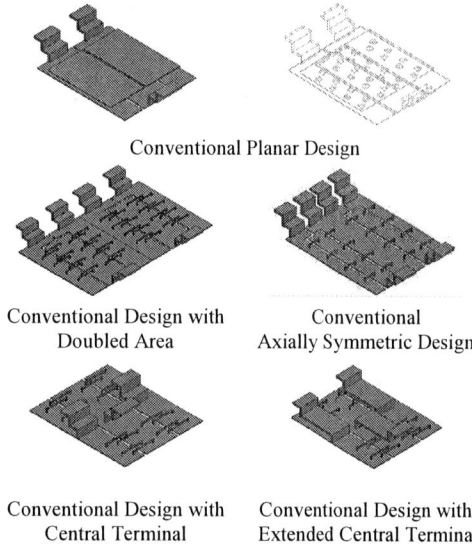

Conventional Planar Design

Conventional Design with   Conventional
Doubled Area   Axially Symmetric Design

Conventional Design with   Conventional Design with
Central Terminal   Extended Central Termina

Fig.2 Layout Diagram of Different DBC Structures

From the simulation analysis of the above seven structural layouts, it is evident that the layout structure has a significant impact on stray inductance, and an axially symmetric structure can effectively reduce stray inductance. To achieve a layout structure with even lower stray inductance, it is necessary to first validate the stray inductance of the aforementioned packaging structures. Based on the aforementioned simulation results, the optimal design solution was selected for module packaging. The physical image of the packaged power module is shown in Figure 3.

Fig.3 Power module packaging diagram

In accordance with the stray inductance measurement principles, comprehensive testing was conducted to characterize the parasitic inductance of the packaged power module.

The schematic diagram of the parasitic inductance test for the power module is shown in Fig. 4. In this figure, the DUT (Device Under Test) represents the power module being evaluated. M1 functions as an auxiliary switching device. The double pulse waveform is connected to the gate of the auxiliary switch M1 through driving, and the gate of the test module is short circuited. The probe is connected to the positive and negative terminals of the test module, and the Roche coil is connected in series to the negative terminal of the power module. The switch waveform of the test bridge arm is observed through an oscilloscope. By using the above connection method, adjust the bus voltage and finally obtain the test waveform shown in Fig.5.

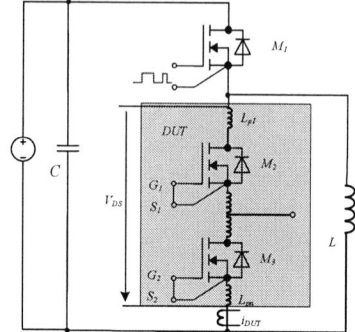

Fig.4 Schematic diagram of parasitic inductance testing

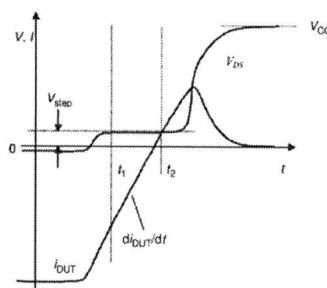

Fig.5 Parasitic inductance test waveform diagram

After obtaining the above test waveform through an oscilloscope, the final result is calculated using the oscilloscope according to the formula (1) for the value of stray inductance.

$$L_p = V_{step} / \left| \left(di_{DUT} / dt \right) \right| \qquad (1)$$

Stray inductance test platform was established, as shown in Fig. 6. The stray inductance test results of two power modules packaged with axisymmetric structure are8.6nH and 9.2nH, respectively. The test results are basically consistent with the simulation results

Fig.6 Stray Inductance Test Platform Diagram

IV. CONCLUSION

Conclusions were drawn from the simulation analysis of DBC (Direct Bonded Copper) with different layout structures as follows:

1）Increasing the number of chips can reduce stray inductance. Parallel connection of multiple chips generates mutual inductance, which in turn cancels out part of the stray inductance.

2）Compared with the bonding wire structure, the planar structure has a smaller impact on stray inductance.

3）Axially symmetric structures can effectively reduce stray inductance. Symmetric structures also reduce stray inductance by increasing mutual inductance.

4）Moving the terminal position to the center of the DBC to form a structure that is symmetrical in all directions (up, down, left, and right) can further reduce stray inductance.

5）The size of the terminals has a significant impact on stray inductance.

Based on the above conclusions, the axially symmetric structure with terminals located in the center is currently the optimal solution for minimizing stray inductance. This paper will package the module based on this solution and build a stray inductance test platform in accordance with the IEC60747-15 standard to conduct stray inductance packaging tests and double-pulse tests on the new module. This will lead to the development of a low-inductance, low-loss SiC power module.

ACKNOWLEDGMENT

This work was supported in part by National Key R&D Program of China (2021YFB2500600).

REFERENCES

[1] Tan Yifan. Research on the Reliability of Bonding Wires in Power Modules Based on Novel Stacked DBC Hybrid Packaging [D]. Huazhong University of Science and Technology, 2019. DOI:10.27157/d.cnki.ghzku.2019.001344

[2] He Mingzhuang. Design and Optimization of Stacked DBC Hybrid Packaging Structure for SiC Power Devices [D]. Guilin University of Electronic Technology, 2023.

[3] Wang Siyuan, Liang Yuqian, Sun Peng, et al. Research on Low-Inductance Bidirectional Switch SiC Power Modules Based on 3D Packaging [J]. Journal of Power Supply, 2024, 22(3): 87-92. DOI: 10.13234/j.issn.2095-2805.2024.3.87.

[4] Li Dongrun, Ning Puqi, Kang Yuhui, et al. Design of High Power Density SiC Power Modules Using Large Chips [J]. Journal of Power Supply, 2024, 22(3): 93-99. DOI: 10.13234/j.issn.2095-2805.2024.3.93.

[5] Chen, K., Zhao, Z., Li, H., & Lu, T. (2020)."Parasitic Inductance Reduction and Layout Optimization for High-Power SiC MOSFET Modules."IEEE Transactions on Power Electronics, 35(5), 4613-4626.

[6] Huang, X., Li, Q., & Lee, F. C. (2018)."Analysis and Optimization of Stray Inductance in High-Frequency SiC Power Modules Considering DBC Layout and Bonding Wire Arrangement."IEEE Journal of Emerging and Selected Topics in Power Electronics, 6(2), 870-881.

[7] Wang, J., Zhao, T., Li, J., Huang, A. Q., & Callanan, R. (2017)."Study and Handling Methods of Parasitic Inductance in High-Power SiC MOSFET Modules."IEEE Transactions on Power Electronics, 32(9), 7083-7096.

[8] Zhang, Z., Wang, F., Tolbert, L. M., & Blalock, B. J. (2014)."Impact of Parasitic Inductance on Switching Performance of SiC Devices in a Phase-Leg Configuration."IEEE Transactions on Industry Applications, 50(4), 2715-2725.

2025 IEEE Workshop on Wide Bandgap Power Devices and Applications in Asia (WiPDA Asia)

Cu/diamond Composite Electrodes for Low Thermal Resistance Packaging in High-Power IGCTs

Tang Xinling, Wei Xiaoguang, Lin Zhongkang, Yang Guang, Yu Kefan, Wang Jingfei

Beijing Huairou Laboratory
Beijing, China
e-mail: yangguang1@neps.hrl.ac.cn

Abstract—This study demonstrates the superior thermal management capabilities of Cu/diamond composite electrodes in 6-inch press-pack insulated gate commutated thyristors (IGCTs). Through systematic material characterization and device-level validation, the Cu/diamond composite exhibited an exceptional thermal conductivity of 580-680 W/(m·K) and a low coefficient of thermal expansion (<6×10⁻⁶ K⁻¹). Experimental results from a thermal resistance test platform revealed a 39.2% reduction in junction-to-case thermal resistance (1.24 K/kW improvement) compared to conventional edge-gate IGCTs, enabling a rated current capacity of 4246 A with 20% design margin. The high surface flatness achieved by a thin Mo-Cu coating further facilitated reliable interfacial contact without compromising the intrinsic thermal advantages of the Cu/diamond matrix.

Keywords — Cu/diamond composite, press-pack IGCTs, thermal management, thermal resistance test platform

I. INTRODUCTION

The escalating power density of modern power electronics necessitates advanced electrode materials combining high thermal conductivity and dimensional stability. Cu/diamond composite (Fig. 1), synergize the exceptional properties of both diamond and copper, has emerged as a promising solution [1], [2], [3], offering:

1) Ultrahigh thermal conductivity (500–800 W/(m·K)) surpassing pure copper (~400 W/(m·K)[2], [4]).

2) Tailorable thermal expansion (4–8×10⁻⁶ K⁻¹) matching semiconductor materials, e.g., Si (2.6×10⁻⁶ K⁻¹[5]).

3) Mechanical robustness for press-pack assembly.

This work presents the first implementation of Cu/diamond composites as the electrodes in the insulated gate commutated thyristors (IGCTs), with a thin Mo-Cu surface coating for better surface conditions and lower thermal contact resistance. Finally, significant enhancement of the IGCT's thermal performance comparing to the conventional ones was revealed, which primarily originated from the Cu/diamond matrix's intrinsic thermal conductivity.

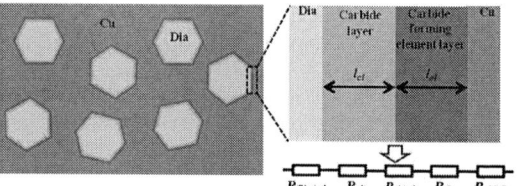

Fig. 1. Schematic diagram of the typical Cu/diamond composite and the thermal resistance network.

II. MATERIAL DEVELOPMENT AND CHARACTERIZATION

A. Composite Fabrication

The Cu/diamond electrodes were fabricated via the hot-pressing sintering processes [2]. The Cu/diamond composite was fabricated using commercial MBD8-type synthetic single-crystal diamond particles (average size ~200 μm) and spherical copper powder (purity: 99.9%, diameter: 10 μm). Prior to composite synthesis, a chromium interlayer (thickness ~200 nm) was deposited on the diamond surfaces via magnetron sputtering under the following conditions. Energy-dispersive X-ray spectroscopy (EDS) confirmed the chromium-coated diamond surfaces exhibited a C/Cr mass ratio of 95.06:4.94. The coated diamond particles were then blended with copper powder at a 7:3 volume ratio in anhydrous ethanol using a magnetic stirrer (mixing time ~3 h). The slurry was subsequently dried in a vacuum oven to remove residual solvent. The mixed powder was filled in a graphite mold. To prevent the sintered diamond/copper composite from sticking to the graphite mold, graphite paper was placed between the powder mix and the graphite mold. The diamond/copper hybrid powder was sintered using a thermal simulation machine. The vacuum was 7 Pa. An argon atmosphere of 500 Pa filled the chamber. The minimum sintering temperature was 800 °C and the maximum was 1000 °C. Then, the Cu/diamond electrode was obtained after the furnace was cooled. A schematic diagram of the Cu/diamond electrode preparation is shown in Fig. 2.

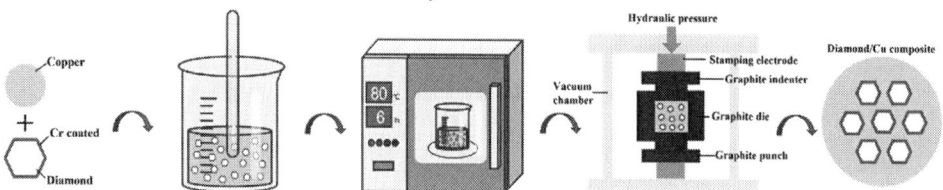

Fig. 2. Schematic diagram of the Cu/diamond electrode preparation process.[2]

This work was supported by the Beijing Nova Program, Grant Number: 20220484094.

979-8-3315-1110-4/25 $31.00 © 2025 IEEE

Then, a Mo-Cu alloy film with ~150 μm thicknesses were bonded to both the surfaces of Cu/diamond electrodes with sintering processes, which achieves better surface flatness, roughness and thermal contact resistance. The appearance of the Cu/diamond electrodes are shown in Fig. 3. Furthermore, the surface flatness of the electrode was tested by the pressure measurement film. The color of pressure measurement film appears red where pressure is applied, and the color density varies according to the amount of pressure. Thus, the flatness of the third sample shown in Fig. 3 appears to be the best.

B. Composite Properties Characterization

The coefficient of thermal expansion (CTE) of the Cu/diamond composite was measured by a thermal expansion meter (DIL402 Expedis Supreme, Netzsch, Germany), while the used rectangular small samples with sizes of 25×5×3 mm are shown in Fig. 4. The four samples were placed at different positions in the furnace during material preparation process. The CTE results of the four samples are listed in Table I, which are all below 6×10^{-6} K^{-1}, which are significantly lower than that of Cu ($\sim1.7\times10^{-5}$ K^{-1}) [2], [6].

The density of the composites (ρ) was measured by the Archimedean drainage method. The specific heat capacity (c) of the diamond/Cu composite was measured by a differential scanning calorimeter (DSC214, Netzsch, Germany) at room temperature. The thermal diffusion coefficient of the composites (α) was measured at room temperature using a laser flash meter (LFA447, Netzsch, Germany) with a disk sample size of $\Phi12.65$ mm × 3 mm. Then, the thermal conductivity of the composite (κ) is calculated by [2]

$$\kappa = \alpha\rho c \qquad (1)$$

The density and thermal conductivity of twenty disk samples with two different specifications (ten different products per batch located at different areas in the furnace) are measured in this study, which are ranging within 580–680W/(m·K), and results are listed in Table II.

Besides thermal performance, the electrical conductivity (σ), is another essential property for high current capability demand of the IGCTs. Using the eddy current conductance meter, the electrical conductivity of Cu/diamond composite was ~20% IACS, which could meet the demand of electrical conduction as well.

III. DEVICE-LEVEL THERMAL PERFORMANCE

A. Thermal Test Platfrom

To precisely measure the device-level thermal resistance, a test platform (Fig. 5) was developed in this work. The thermal resistance test platform was constructed based on the steady-state heat flux method, complying with multiple standards including: National Standard GB/T 15291 (Semiconductor Devices - Part 6: Thyristors), International Standard IEC 60747-6, Industry Standard JB/T 7626 (Test Methods for Reverse Blocking Triode Thyristors). Moreover, three additional temperature sensors (PT-100) were inserted into the upper and lower aluminum-based meter bars which contact to both the sample surfaces, improving the precision of measuring device temperatures.

This method is based on the idealized heat conduction between two parallel, isothermal surfaces separated by a test specimen of uniform thickness. The thermal gradient imposed on the specimen by the temperature difference between the two contacting surfaces causes the heat flow through the specimen. This heat flow is perpendicular to the test surfaces and is uniform across the surfaces with no lateral heat spreading. Accordingly, the thermal resistance is calculated by the temperature difference between the two isothermal surfaces divided by the heat flow through them.

As shown in Fig. 5, the platform features the following specifications: Pressure range 0-200 kN with automated pressure control/holding, testing area: 145 mm diameter, compatible with ≤6-inch press-pack devices, temperature resolution: 0.01°C, heating/cooling power: 2 kW, testing efficiency: 30 minutes per sample, measurement relative uncertainty: <2%.

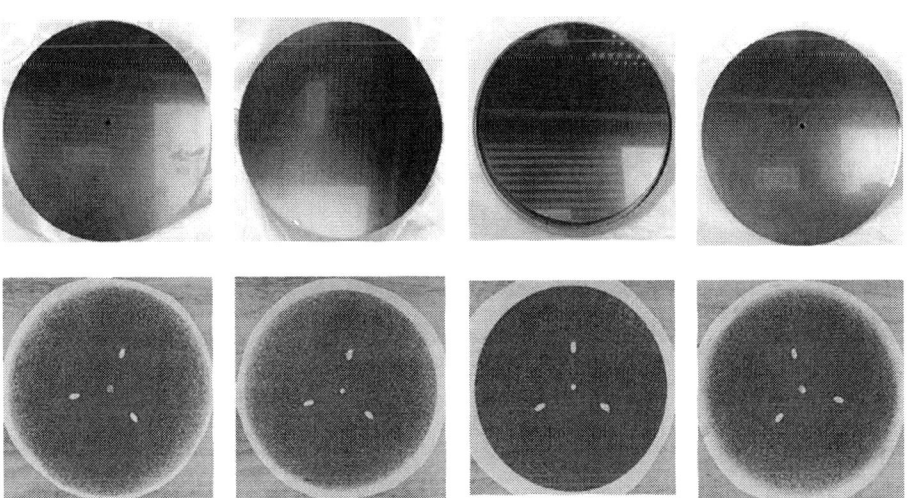

Fig. 3.   The Cu/diamond electrode coated with Mo-Cu alloy, and the flatness of surface was characterized by pressure measurement film.

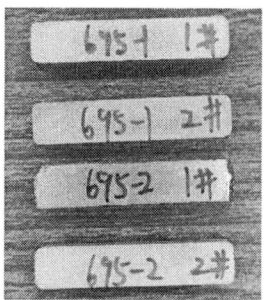

Fig. 4. Rectangular samples for CTE measurements.

TABLE I.   CTE MEASUREMRNT RESULTS

Sample	CTE (10^{-6} K^{-1})
695-1-1	5.55
695-1-2	5.62
695-2-1	5.74
695-2-2	5.77
Si [5]	2.6
Diamond [2], [7]	2.3
Cu [2], [6]	17.3
Mo [8]	4.5

B. Experimental Validation

To validate the optimization effectiveness of the Cu/diamond electrode technology, two sets of Cu/diamond anodes and two sets of cathodes were fabricated. These electrodes were assembled with GCT chips to form triple-layer packaging structures. The electrode samples and testing configurations are illustrated in Fig. 6.

Testing Protocol:

- Control Group: Three 35mm-thick conventional edge-gate devices;

- Experimental Group: Two triple-layer edge-gate devices consisting Cu/diamond electrodes.

Thermal resistance testing using the heat flux method platform yielded the results summarized in Table III. The Cu/diamond configuration demonstrated a 39.2% average reduction (1.24 K/kW) in thermal resistance compared to the conventional edge-gate structure (control group). The relatively large standard deviation observed in the Cu/diamond group originated from the limited sample size and associated data dispersion.

IV. CONLUSIONS

This study demonstrates the significant thermal management capabilities of Cu/diamond composite electrodes in 6-inch insulated gate commutated thyristors (IGCTs). Comprehensive material characterization revealed that the Cu/diamond composite achieves outstanding thermal properties, including a high conductivity ranging within 580–680 W/(m·K) and a low thermal expansion coefficient below 6×10^{-6} K^{-1}. By applying the Cu/diamond composite electrodes with Mo-Cu coatings into IGCTs, a 39.2% reduction in junction-to-case thermal resistance (1.24 K/kW improvement) compared to conventional edge-gate IGCTs was achieved, enabling a rated current capacity of 4246 A.

TABLE II.   THE DENSITY AND THERMAL CONDUCTIVITY MEASUREMENTS

Batch/Specification	20240525-B695-1		20240523-B695-2	
Sample	*Density (g/cm³)*	*Thermal conductivity (W/(m·K))*	*Density (g/cm³)*	*Thermal conductivity (W/(m·K))*
1	5.33	647.1	5.46	684.6
2	5.32	611.3	5.27	622.4
3	5.30	657.7	5.36	647.4
4	5.24	651.9	5.39	663.4
5	5.31	619.2	5.16	595.4
6	5.35	650.9	5.39	650.1
7	5.24	627.0	5.43	682.3
8	5.36	650.7	5.20	658.7
9	5.30	622.6	5.32	651.4
10	5.23	583.2	5.33	635.1
Cu [2]	8.90	398	/	/
Diamond [2], [4]	3.52	1600–2200	/	/

TABLE III.   THERMAL RESISTANCE MEASUREMENTS OF IGCT DEVICES (TR: THERMAL RESISTANCE)

Sample	Thermal resistance (K/kW)	
	Edge-gate devices (control group)	*Triple-layer structures with Cu/diamond electrodes (experimental group)*
1	3.13	1.98
2	3.18	1.87
3	3.17	/
Average	3.16	1.92
TR Reduction	1.24	

979-8-3315-1110-4/25 $31.00 © 2025 IEEE

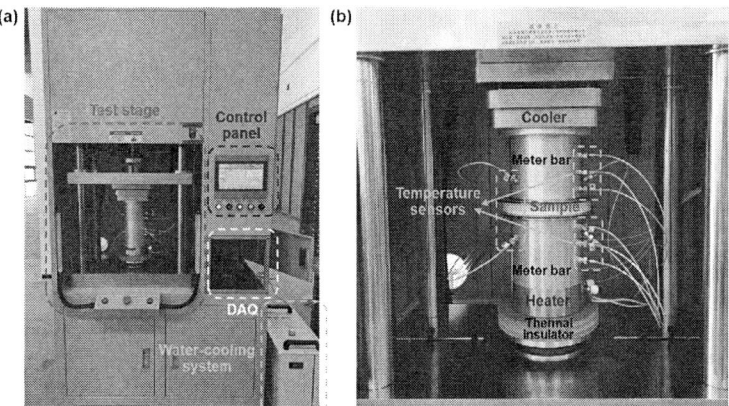

Fig. 5. The thermal test platfrom. (a) Overall platform design, and (b) the sample stage.

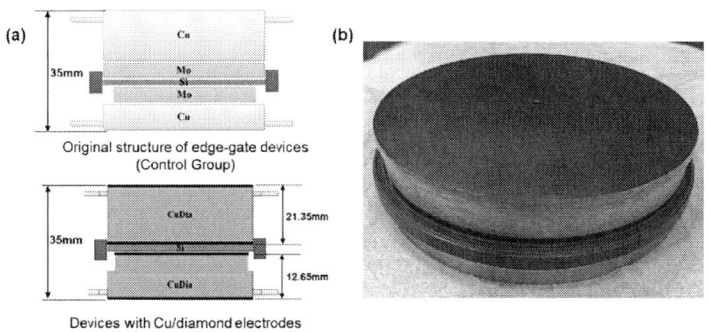

Fig. 6. Samples for device-level thermal resistance test. (a) Samples of control group and experimental group, and (b) a sample in the experimental group.

REFERENCES

[1] Y. Zhang, L. Wang, J. Hao, N. Li, X. Wang, and H. Zhang, "Effect of diamond particle size on thermal conductivity and thermal stability of Zr-diamond/Cu composite," *Diamond and Related Materials*, vol. 146, p. 111257, Jun. 2024, doi: 10.1016/j.diamond.2024.111257.

[2] K. Lu *et al.*, "Study of the hot-pressing sintering process of diamond/copper composites and their thermal conductivity," *Journal of Alloys and Compounds*, vol. 960, p. 170608, Oct. 2023, doi: 10.1016/j.jallcom.2023.170608.

[3] L. Zhou, J. Liu, R. Ding, J. Cao, K. Zhan, and B. Zhao, "A review of diamond interfacial modification and its effect on the properties of diamond/Cu matrix composites," *Surfaces and Interfaces*, vol. 40, p. 103143, Aug. 2023, doi: 10.1016/j.surfin.2023.103143.

[4] Y. Zhang *et al.*, "Interfacial Thermal Conductance between Cu and Diamond with Interconnected W−W2C Interlayer," *ACS Appl. Mater.*

Interfaces, vol. 14, no. 30, pp. 35215–35228, Aug. 2022, doi: 10.1021/acsami.2c07190.

[5] H. Watanabe, N. Yamada, and M. Okaji, "Linear Thermal Expansion Coefficient of Silicon from 293 to 1000 K," *International Journal of Thermophysics*, vol. 25, no. 1, pp. 221–236, Jan. 2004, doi: 10.1023/B:IJOT.0000022336.83719.43.

[6] T. A. Hahn, "Thermal Expansion of Copper from 20 to 800 K—Standard Reference Material 736," *Journal of Applied Physics*, vol. 41, no. 13, pp. 5096–5101, Dec. 1970, doi: 10.1063/1.1658614.

[7] Q. Y. Wang, W. P. Shen, and M. L. Ma, "Mean and Instantaneous Thermal Expansion of Uncoated and Ti Coated Diamond/Copper Composite Materials," *Advanced Materials Research*, vol. 702, pp. 202–206, 2013, doi: 10.4028/www.scientific.net/AMR.702.202.

[8] R. Chen *et al.*, "Preparation of a B4C hollow microsphere through gel-casting for an inertial confinement fusion (ICF) target," *Ceramics International*, vol. 43, no. 1, pp. 571–577, Jan. 2017, doi: 10.1016/j.ceramint.2016.09.196.

2025 IEEE Workshop on Wide Bandgap Power Devices and Applications in Asia (WiPDA Asia)

1200V SiC MOSFETs Under Dynamic Reverse Bias Testing in High-Temperature and High-Frequency Environments

1st Liudan Kong
College of Intelligent Robotics and Advanced Manufacturing
Fudan University
Shanghai, China
23210860049@m.fudan.edu.cn

2nd Jiuyang Tang
College of Intelligent Robotics and Advanced Manufacturing
Fudan University
Shanghai, China
tangjy23@m.fudan.edu.cn

3rd Jiaying Cao
College of Intelligent Robotics and Advanced Manufacturing
Fudan University
Shanghai, China
23210860002@m.fudan.edu.cn

4th Yifei Chang
College of Intelligent Robotics and Advanced Manufacturing
Fudan University
Shanghai, China
22210860070@m.fudan.edu.cn

5th Hao Guan
College of Intelligent Robotics and Advanced Manufacturing
Fudan University
Shanghai, China
23210860040@m.fudan.edu.cn

6th Yuhan Duan
College of Intelligent Robotics and Advanced Manufacturing
Fudan University
Shanghai, China
22210860002@m.fudan.edu.cn

Qingchun Zhang*
College of Intelligent Robotics and Advanced Manufacturing
Fudan University
Shanghai, China
Research Institute of Fudan University in Ningbo
Fudan University
Ningbo, China
Qingchun_Zhang@fudan.edu.cn

Pan Liu*
College of Intelligent Robotics and Advanced Manufacturing
Fudan University
Shanghai, China
Research Institute of Fudan University in Ningbo
Fudan University
Ningbo, China
panliu@fudan.edu.cn

Abstract—Silicon carbide (SiC) MOSFETs are widely used in high-voltage and high-temperature applications, however, their reliability under dynamic operating conditions remains a concern. In this work, three types of commercial 1200 V SiC MOSFETs with different structures were evaluated using a dynamic high-temperature reverse bias (DHTRB) test. The test was conducted at 25 °C and 175 °C, with switching frequencies of 50 kHz and 100 kHz for 168 hours. Real-time drain leakage current was monitored, and key static parameters were measured before and after the test. The results show that temperature had a more pronounced effect on parameter variation. The trench-gate device experienced a larger reduction in breakdown voltage. Among the two planar-gate devices, one exhibited threshold voltage shift and increased leakage current at high temperature, while the other maintained stable performance across all test conditions. This study provides useful insights for the structural design and reliability evaluation of SiC MOSFETs under dynamic stress conditions.

Keywords—SiC MOSFET, Dynamic high-temperature reverse bias, Reliability

I. INTRODUCTION

SiC MOSFETs, owing to their high breakdown voltage, excellent thermal stability, and low switching losses, have been widely adopted in high-performance and high-efficiency applications such as power electronics, photovoltaic inverters, and electric vehicle drive systems[1]. As these devices operate under complex conditions for extended periods, their reliability has garnered increasing attention. In particular, the stability of different device structures under real-world conditions, such as high temperature and high frequency,

exhibits notable variation, posing greater challenges for large-scale deployment and system integration.

Currently, standard reliability evaluation methods such as High Temperature Reverse Bias (HTRB) and High Temperature Gate Bias (HTGB) tests are widely adopted in the industry[2]. These static stress tests are effective in analyzing variations in key electrical parameters, including Gate threshold voltage (V_{th}), on-resistance ($R_{ds(on)}$), and leakage currents (I_{dss}, I_{gss}). For example, Wang et al. [3]and Yang et al. [4] reported that under prolonged high-voltage and high-temperature conditions, SiC MOSFETs exhibited slight increases in leakage current and electrical parameter drifts; Chowdhury et al. [5] investigated reliability under various bias levels and found that different JFET designs showed distinct degradation trends in breakdown voltage (BV) and gate leakage current under high-voltage stress. Wu et al. [6, 7] further analyzed the effects of temperature and bias on the static electrical parameters of the devices. These studies systematically reveal the typical degradation characteristics of SiC MOSFETs under static electrothermal stress and highlight that device structure, fabrication process, and operating conditions collectively determine long-term electrical stability.

However, in practical applications, power devices often operate under dynamic conditions characterized by frequent switching and high dV/dt, which static tests cannot fully replicate. To better simulate such real-world stress environments, Dynamic Reverse Bias (DRB) and its variants have been proposed in recent years. Lu-Wei et al. [8] developed a DRB test platform with adjustable dV/dt and found that dynamic stress could accelerate device degradation more effectively compared to static stress testing.

979-8-3315-1110-4/25 $31.00 © 2025 IEEE

Although the DRB method offers a higher evaluation value, most existing studies focus on a single device structure, lacking a comparative analysis of different structures under identical dynamic stress conditions. In practice, commercial SiC MOSFETs feature a variety of structural designs, including planar gate and trench gate. These structural differences can lead to notable variations in reliability performance under dynamic stress. However, systematic comparative studies in the literature remain limited, hindering a deeper understanding of the relationship between device structure and stress response.

In this study, a DHTRB testing method was employed to comparatively evaluate three types of commercial 1200 V SiC MOSFETs with different structures. Under uniform test conditions, the evolution of both static and dynamic parameters was monitored. The findings provide valuable insights for optimizing reliability and selecting appropriate device structures in power electronics applications.

II. EXPERIMENT AND METHOD

A. Devices Under Test(DUT)

Three types of commercial 1200 V SiC MOSFETs were selected for this study, referred to as DUT A, DUT B, and DUT C. These devices differ in gate oxide structures. As shown in Fig. 1, DUT A and DUT B adopt conventional planar gate structures, whereas DUT C adopts a trench gate structure. The specifications of the three devices are listed in TABLE I.

TABLE I.   THE SPECIFICATIONS OF THE THREE DEVICES

DUT	V_{DS}(V)	V_{th}(V)	$R_{ds(on)}$(mΩ)	I_D(A)
A	1200	1.8~4.5	30	79
B	1200	2.3~3.6	40	64
C	1200	2.8~4.8	36	84

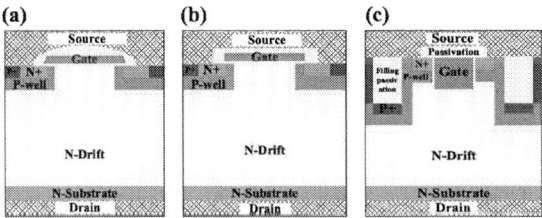

Fig. 1. Structures of the three SiC MOSFETs.

B. DHTRB Test Conditions

The DHTRB method was employed to apply stress to the devices in this study. During the test, a reverse drain voltage of 960 V was applied, and the gate voltage was maintained at -4 V (V_{GS} = -4 V) to ensure the devices remained in the off-state. The stress was applied in the form of a repetitive square wave with a duty cycle of 50%, and a voltage rise rate (dV/dt) of no less than 50 V/ns, simulating the electrical stress experienced during high-speed switching in actual circuits. The test was conducted at two temperatures, 25 °C and 175 °C, with switching frequencies of 50 kHz and 100 kHz, to explore the effect of temperature and frequency on device behavior. Additionally, the leakage current was monitored online every 30 minutes during the aging process to evaluate the dynamic degradation behavior. The test circuit is shown in Fig 2.

TABLE II.   DHTRB TEST CONDITIONS FOR 1200 V SIC MOSFETS

Test Group	Temperature (°C)	Frequency (kHz)	V_{DS} (V)	V_{GS} (V)	Duty Cycle (%)	dv/dt (V/ns)
A	25	50	960	-4	50	>50
B	25	100	960	-4	50	>50
C	175	50	960	-4	50	>50
D	175	100	960	-4	50	>50

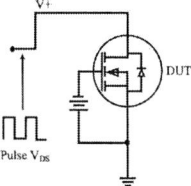

Fig. 2. Test circuit

Key static electrical parameters were measured before and after stress application to evaluate the performance changes induced by dynamic stress. All measurements were performed at room temperature using a Keysight B1505A high-voltage parameter analyzer.

C. TCAD Simulation

To support the interpretation of experimental phenomena, this study employed Synopsys Sentaurus TCAD 2019 to model and simulate a typical trench-gate MOSFET structure. The model was constructed based on representative device cross-sections and doping profiles, with reverse drain bias and negative gate voltage applied to simulate the DHTRB stress conditions. To more accurately reproduce the experimental environment, a measured pulse waveform from the actual test platform was introduced into the simulation. The simulation focused on analyzing the internal electric field distribution of the device, and mesh refinement was applied in the trench geometry and oxide regions to improve accuracy.

III. RESULTS AND DISCUSSION

This section presents the results of DHTRB reliability evaluation for the three tested devices. The discussion focuses on online degradation behavior, static parameter shifts, and their correlation with device structure.

A. Online Leakage Current and Static Parameter Degradation Trends

Under DHTRB stress with different temperature and frequency combinations, the three devices exhibited varying degrees of change in their key static electrical parameters. Fig. 3 to Fig. 5 show the online monitored I_{dss} curves of the three structures under the four stress conditions. The results indicate that DUT A exhibited a continuous increase in leakage current over time under high-temperature aging, while DUT B and DUT C showed no notable changes in online leakage current under either room temperature or high-temperature conditions. At room temperature, all three devices demonstrated consistent behavior with minimal variation across samples. However, under high-temperature stress, DUT A showed noticeable sample-to-sample variation, indicating poorer consistency compared to DUT B and DUT C.

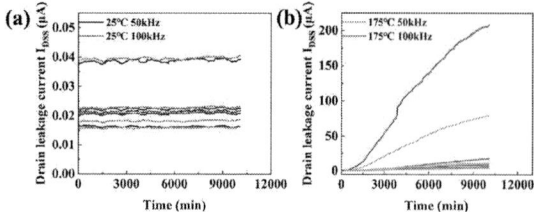

Fig. 3. Online monitored leakage current of DUT A, (a) under aging at 25 °C; (b) under aging at 175 °C.

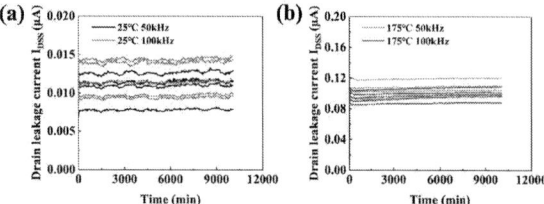

Fig. 4. Online monitored leakage current of DUT B: (a) under aging at 25 °C; (b) under aging at 175 °C.

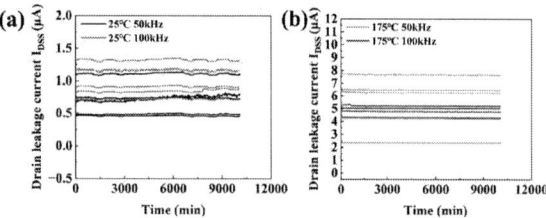

Fig. 5. Online monitored leakage current of DUT C. (a) under aging at 25 °C; (b) under aging at 175 °C.

Pre- and post-stress variations in V_{th}, I_{dss}, $R_{ds(on)}$, and BV are summarized in Fig. 6 to Fig. 8. I_{gss} is not included, as it remained within the picoampere range with negligible variation. Devices exposed to high-temperature and high-frequency stress conditions experienced more severe parameter shifts, indicating accelerated degradation and reduced stability under harsher operating environments.

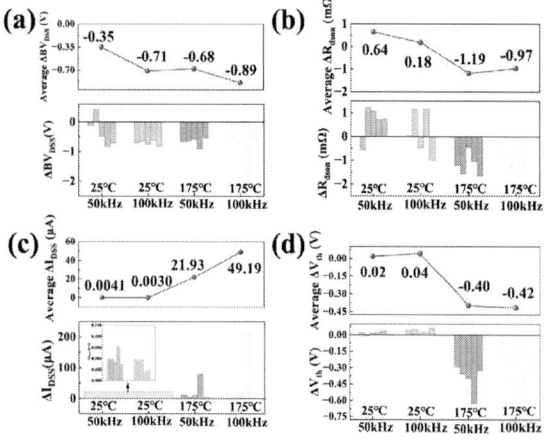

Fig. 6. DUT A Static Parameter Changes. (a) BV Variation Before and After Testing, (b) $R_{ds(on)}$ Variation Before and After Testing, (c) Online monitored Leakage Current Variation, (d) V_{th} Variation Before and After Testing

DUT A showed a clear negative shift in V_{th} and an increase in I_{dss} at 175 °C. Under high temperature and high frequency, the average I_{dss} increased by more than 40 µA. The figure shows that the increase in I_{dss} corresponds to the negative shift in V_{th}, indicating that V_{th} degradation may be the main reason for the leakage current increase. The reduction in V_{th} reduces the gate control margin at $V_{GS} = 0$ V, making the channel more susceptible to partial conduction and increasing the drain leakage current.

Notably, marked sample-to-sample variation emerged under high-temperature conditions. Although the devices had been pre-screened for consistency prior to testing, pronounced discrepancies were still observed after aging. Specifically, one sample exhibited severe leakage current degradation under high-frequency stress, while another showed similar behavior under low-frequency conditions, suggesting that elevated temperature plays a dominant role in amplifying device variability.

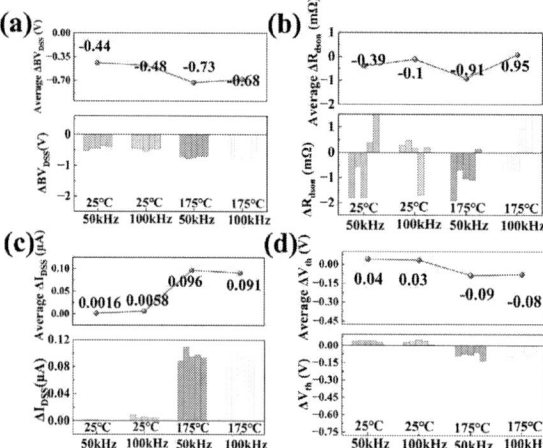

Fig. 7. DUT B Static Parameter Changes. (a) BV Variation Before and After Testing, (b) $R_{ds(on)}$ Variation Before and After Testing, (c) Online monitored Leakage Current Variation, (d) V_{th} Variation Before and After Testing

DUT B showed stable performance under all stress conditions. The values of I_{dss} and other static parameters changed very little. Especially at 175 °C, both V_{th} and BV remained almost unchanged. In addition, DUT B exhibited good sample-to-sample consistency, with minimal variation observed across different test units. This suggests that DUT B has good resistance to dynamic stress, likely due to better process control and gate oxide quality.

In contrast, DUT C showed the most severe degradation under high-temperature and high-frequency conditions. The average BV decreased by more than 150 V under the stress condition of 175 °C and 100 kHz.

By comparing the results from different test groups, it can be seen that elevated temperature (175 °C) had a greater impact on leakage-related parameters, while the effect of frequency was relatively minor. At 25 °C, even under high-frequency stress at 100 kHz, the variation in online I_{dss} remained small for all devices. In contrast, at 175 °C, leakage-related parameters increased noticeably, especially for DUT A. Additionally, DUT C exhibited a notable reduction in BV under high-temperature conditions.

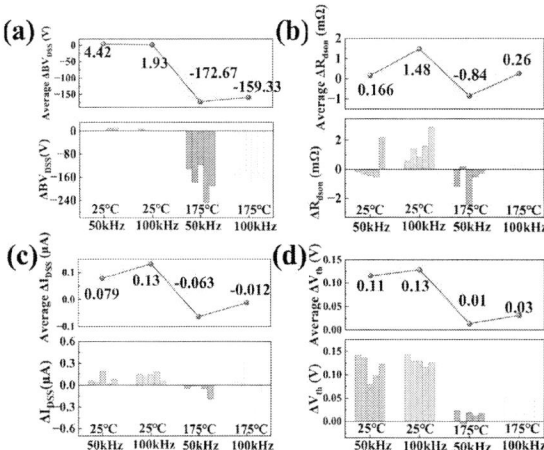

Fig. 8. DUT C Static Parameter Changes. (a) BV Variation Before and After Testing, (b) $R_{ds(on)}$ Variation Before and After Testing, (c) Online monitored Leakage Current Variation, (d) V_{th} Variation Before and After Testing

B. Structure-Dependent Reliability Differences

The three devices exhibited different degradation behaviors under the same stress conditions, highlighting the strong influence of structural design on dynamic reliability.

Among them, DUT C showed the most pronounced degradation and was therefore selected as the focus of the simulation. DUT A, which adopts a conventional planar gate structure, remained stable under mild stress, while exhibiting noticeable degradation at elevated temperatures. Parameter analysis indicated that the degradation was primarily caused by a negative shift in V_{th}, which led to an increase in drain leakage current. Since the cause of degradation in DUT A is relatively well understood, the simulation focused on DUT C to analyze its internal electric field distribution from a structural perspective. DUT B shares a similar structure with DUT A, while benefiting from gate oxide quality and better process control. As a result, it remained stable even under the harsh condition of 175 °C / 100 kHz, with only minimal parameter variation. In contrast, DUT C employs a trench-gate structure and exhibited decreased BV after aging, indicating higher sensitivity to high-temperature and high-frequency stress. Although the trench structure offers advantages in on-resistance and integration density, its geometric complexity may lead to localized field enhancement, which can affect device stability under harsh conditions.

To further investigate the internal electric field distribution in DUT C, a 2D cross-sectional model was developed and simulated using the Sentaurus TCAD platform, with emphasis on identifying the field distribution characteristics in different structural regions under reverse bias stress.

Simulation results show that under all four tested conditions— 25 °C / 50 kHz, 25 °C / 100 kHz, 175 °C / 50 kHz, and 175 °C / 100 kHz—electric field concentration consistently appeared near the trench sidewalls and at the bottom of the P-well region. These simulation setups correspond directly to the experimental test conditions in the DHTRB study, ensuring consistency between field simulation and measured stress scenarios.

At elevated temperatures (Fig. 9c and 9d), the electric field intensity at the P-well bottom increased compared to the 25 °C cases, even though the overall field distribution shape remained largely unchanged. This indicates that temperature is the primary factor enhancing local electric field magnitude. When comparing different frequencies at the same temperature (Fig. 9a vs. 9b, and Fig. 9c vs. 9d), increasing frequency from 50 kHz to 100 kHz had only a minor impact on the spatial distribution and peak field strength. In contrast, when holding frequency constant and increasing the temperature (Fig. 9a vs. 9c, and Fig. 9b vs. 9d), a noticeable rise in electric field magnitude was observed, particularly in the P-well junction area.

Overall, the simulation results suggest that temperature plays a more dominant role than switching frequency in aggravating internal electric field stress. This aligns with the experimental findings, where high-temperature conditions led to more pronounced parameter degradation and earlier breakdown in the trench-gate device.

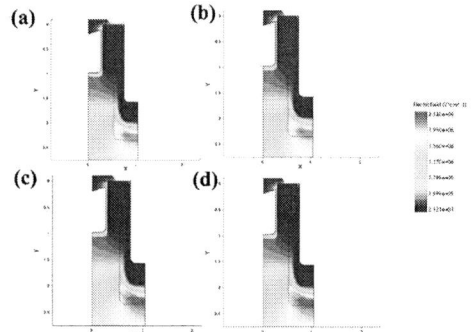

Fig. 9. Simulated electric field distribution in DUT C under different temperature and frequency conditions: (a) 25 °C / 50 kHz, (b) 25 °C / 100 kHz, (c) 175 °C / 50 kHz, (d) 175 °C / 100 kHz.

IV. CONCLUSION

This study evaluated the reliability of three 1200 V SiC MOSFETs under DHTRB stress. The results showed that temperature had a greater impact on device degradation than frequency. DUT A exhibited V_{th} shift and increased leakage current under high-temperature conditions; DUT B remained stable, while DUT C, which adopts a trench-gate structure, showed degradation in BV. To support the analysis, TCAD simulation was performed for DUT C. The results showed localized electric field enhancement under reverse bias, consistent with experimental trends, indicating that the device structure has a certain influence on stress response.

Future research will focus on the following three aspects: (1) Expansion of stress conditions: Further exploring a wider range of gate voltage, switching frequency, and bias combinations to determine critical conditions for degradation; (2) Analysis of structural differences: Introducing more types of device structures to systematically compare the influence of design factors on dynamic reliability; (3)Refinement of simulation models: Incorporating charge trapping effects and multiphysics coupling, along with terminal modeling, to establish more accurate degradation simulation models.

ACKNOWLEDGEMENT

The work is supported by Shanghai SiC Power Device Engineering and Technology Research Center (19DZ2253400), Zhejiang Provincial Science and Technology Program (2024C01247(SD2)), and industrial cooperation project (KCH2310169).

REFERENCE

[1] B. Shi *et al.*, "A review of silicon carbide MOSFETs in electrified vehicles: Application, challenges, and future development," *IET Power Electronics,* vol. 16, no. 12, pp. 2103-2120, 2023.

[2] Y. Zhong, X. He, Z. Wang, J. Luo, B. Wang, and Q. Li, "Review of HTRB and HTGB reliability of SiC MOSFETs," in *2024 7th International Conference on Energy, Electrical and Power Engineering (CEEPE)*, 2024: IEEE, pp. 1161-1168.

[3] G. Wang, L. Yuan, X. Wang, Y. Zhang, and R. Jia, "High temperature reliability and performance evaluation of 1200 V SiC MOSFETs," *Journal of Crystal Growth,* vol. 606, p. 127086, 2023.

[4] L. Yang and A. Castellazzi, "High temperature gate-bias and reverse-bias tests on SiC MOSFETs," *Microelectronics Reliability,* vol. 53, no. 9-11, pp. 1771-1773, 2013.

[5] S. Chowdhury, K. Matocha, B. Powell, G. Sheh, and S. Banerjee, "Next generation 1200V, 3.5 mΩ. cm 2 SiC planar gate MOSFET with excellent HTRB reliability," in *2018 IEEE 30th International Symposium on Power Semiconductor Devices and ICs (ISPSD)*, 2018: IEEE, pp. 427-430.

[6] W. Peifei, T. Guangfu, Y. Fei, D. Zechen, D. Yujie, and W. Junmin, "Influence of high temperature reliability test on threshold voltage and on resistance of 1200V SiC MOSFET," in *2021 IEEE 1st International Power Electronics and Application Symposium (PEAS)*, 2021: IEEE, pp. 1-6.

[7] P. Wu, G. Tang, F. Yang, Z. Du, Y. Du, and J. Wu, "Influence of high temperature reliability test of 1200V SiC MOSFET on static parameters," in *Journal of Physics: Conference Series*, 2021, vol. 2033, no. 1: IOP Publishing, p. 012096.

[8] Z. Lu-Wei, M. Hui, Z. Ze, Y. Bin, L. Haoze, and X. Zhen, "Dynamic reverse bias test circuit for SiC MOSFET with adjustable dV ds/dt," in *2024 IEEE 10th International Power Electronics and Motion Control Conference (IPEMC2024-ECCE Asia)*, 2024: IEEE, pp. 3688-3693.

Investigation of Electrical Property and Reliability in AlGaN/GaN MIS-HEMTs Fabricated by Using N₂/O₂ Composite Plasma Treatment

1st Qingyuan Zuo
School of Optoelectronic Engineering and Instrumentation Science, Dalian University of Technology
Dalian 116024, People's Republic of China
zuoqy98@mail.dlut.edu.cn

2nd Huolin Huang*, Senior Member, IEEE
School of Optoelectronic Engineering and Instrumentation Science, Dalian University of Technology
Dalian 116024, People's Republic of China
hlhuang@dlut.edu.cn

3rd Yun Lei
School of Optoelectronic Engineering and Instrumentation Science, Dalian University of Technology
Dalian 116024, People's Republic of China
22242022@mail.dlut.edu.cn

4th Jiayu Zhang
School of Optoelectronic Engineering and Instrumentation Science, Dalian University of Technology
Dalian 116024, People's Republic of China
zhangjiayu.330@mail.dlut.edu.cn

5th Jianyu Zhao
School of Optoelectronic Engineering and Instrumentation Science, Dalian University of Technology
Dalian 116024, People's Republic of China
32342022@mail.dlut.edu.cn

6th Jianxun Dai
School of Optoelectronic Engineering and Instrumentation Science, Dalian University of Technology
Dalian 116024, People's Republic of China
jianxundai@dlut.edu.cn

Abstract—Metal–insulator–semiconductor–high electron mobility transistors (MIS-HEMTs) devices using recessed gate technology typically exhibit higher threshold voltage (V_{th}) and are commonly employed in the fabrication of E-mode devices. A stable and reliable gate scheme is crucial for the development of enhancement-mode (E-mode) AlGaN/GaN MIS-HEMTs. This study reports a plasma etching approach using a nitrogen and oxygen composite plasma to repair and passivate the gate region. The fabricated devices demonstrate a high V_{th} of 3.4 V and a gate breakdown voltage of 26 V. Furthermore, the device reliability is significantly improved compared to untreated device.

Keywords— Metal‐insulator‐semiconductor-high electron mobility transistors (MIS-HEMTs), enhancement-mode (E-mode), nitrogen and oxygen composite plasma treatment, reliability

I. INTRODUCTION

Gallium Nitride (GaN), as a third-generation semiconductor, possesses a wide bandgap of 3.4 eV and exhibits superior characteristics such as high critical breakdown field, high thermal conductivity, and high electron saturation drift velocity[1]. These properties make GaN widely applicable in fields such as power electronics, microwave radio-frequency (RF) devices, and optoelectronics, with significant applications in the medium and low voltage ranges, and its presence is gradually expanding into the high-voltage market[2]. Due to the spontaneous and piezoelectric polarization effects at the AlGaN/GaN heterojunction interface, a two-dimensional electron gas (2DEG) with high electron concentration and mobility is generated at the interface[3], [4]. The resulting GaN-based High Electron Mobility Transistors (HEMTs) offer advantages such as low on-resistance, low power loss, and high switching speed. However, traditional AlGaN/GaN HEMTs are depletion-mode devices. To ensure safe operation, reduce power consumption, and simplify drive circuitry, E-mode AlGaN/GaN devices are necessary[5]. Among the various methods for achieving normally-off AlGaN/GaN devices, recessed gate etching technology offers

advantages such as high gate breakdown voltage, large gate voltage swing, and low gate leakage current[6], [7]. However, it still faces reliability issues, including an increase in interface states caused by etching, channel scattering, and unstable or non-uniform threshold voltage[8]–[10].

In this study, devices with over-etched gate regions were fabricated, where the AlGaN barrier layer in the gate region was completely etched, and part of the GaN channel layer was also over-etched. Before growing the gate dielectric, a N₂/O₂ composite plasma treatment was applied to the interface to repair the damage and passivate the surface[11]. Previous reports have utilized various plasma atmospheres for interface treatment, such as N₂[12], O₂[13], O₃[14], N₂O[11], NH₃[15], H₂/N₂[16], SF₆[17], etc. However, the combined effects of N₂/O₂ plasma on over-etched gate interfaces have rarely been studied. This experiment investigated the impact of N₂/O₂ composite plasma treatment on etching damage repair and performance enhancement in over-etched gate devices. The results show that N₂/O₂ composite plasma treatment effectively improves the interface quality between the recessed gate and the gate dielectric, reduces the interface trap density, and enhances electrical performance. As a result, the devices treated with the composite plasma exhibited higher source-drain saturation current, better threshold voltage stability, and lower on-resistance.

II. DEVICE STRUCTURE AND FABRICATION PROCESS

AlGaN/GaN epitaxial structures were fabricated on a 6-inch silicon substrate using the metal organic chemical vapor deposition (MOCVD) technique. The structure comprises a 60-nm in situ Si_3N_4 passivation layer, a 25-nm $Al_{0.25}Ga_{0.75}N$ barrier, and a 4.0-μm undoped GaN buffer layer, as shown in Fig. 1(a). Hall measurements determined a room-temperature 2DEG density of 9.2×10^{12} cm⁻² and an electron mobility of 1500 cm²/V·s. The source-to-gate spacing, gate-to-drain spacing, gate length, and gate width are 4.5, 24.0, 2.0, and 100 μm, respectively.

After the mesa isolation process，a shallow etching process for ohmic contact was carried out.，The ohmic contact pads for source and drain were composed of Ti/Al(30/500) from bottom to top. The ohmic contact resistance was 0.5 Ω·mm after rapid thermal annealing at 560 ∘C in N_2 for 300 s. The gate trenches were formed by Cl_2/BCl_3-based plasma using inductively coupled plasma reactive ion etching (ICP-RIE) system and the optimized ICP and bias power values were kept at 30 and 150 W, respectively. Subsequently, N_2/O_2 plasma treatment was performed using an Inductively Coupled Plasma Chemical Vapor Deposition (ICP-CVD) system. After optimizing the gas flow rates, they were set to 30 sccm, with the ICP power set at 300 W. No bias power was applied to minimize damage, and the treatment temperature was maintained at 100°C for a duration of 300 s. The treated sample was labeled as Sample A, while the untreated sample was labeled as Sample B. The gate dielectric was grown by Atomic Layer Deposition (ALD), with a thickness of 30 nm Al_2O_3. The Ni/Al (50 nm/150 nm) gate metal stack was deposited by magnetron sputtering, followed by a 5min annealing at 400°C in a N_2 atmosphere. The typical device top view is shown in Fig. 1(b).

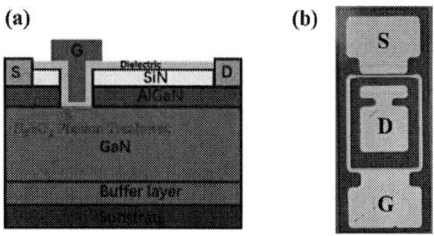

Fig . 1 (a) Schematic cross-sectional diagram of the fabricated E-mode MIS-HEMT device after recessed gate etching and interface treatment, (b) typical device photograph.

III. Prepare Your Paper Before Styling

A. Chemical composition characterization analysis

To further investigate the effect of N_2/O_2 plasma treatment on the chemical composition of the recessed gate etching interface, X-ray photoelectron spectroscopy (XPS) measurements were conducted after the shallow etching of the gate region. The XPS was performed using an Al Kα X-ray source with a kinetic energy of 1486.6 eV.The binding energy (BE) was calibrated to the adventitious hydrocarbon peak position at 284.8 eV. Fig. 2 (a) and (b) present the Ga 3d core-level spectra of the GaN surface after gate recess. The Ga 3d signals split into two peaks corresponding to Ga–N and Ga–O bonds, which are fitted using Gaussian distributions[18]. Although Sample B did not undergo N_2/O_2 plasma treatment, its surface still exhibited a relatively high oxygen content, nearly identical to that of Sample A. This may be attributed to the exposure of the etched interface to air, leading to the oxidation of the surface post-etching. Upon fitting analysis of the O1s peak, it was observed that the N_2/O_2 plasma treatment significantly reduced the presence of H-O and C-O bonds at the interface[11], as shown in Fig. 2 (c) and (d). This reduction in interface states not only enhances device reliability but also results in a more uniform and smoother interface morphology, thereby facilitating the subsequent growth of the gate dielectric.

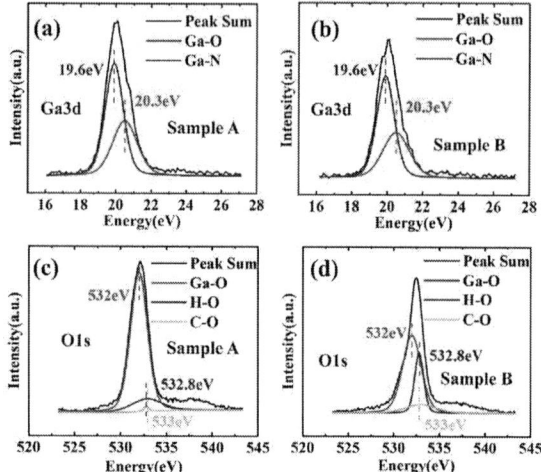

Fig. 2 XPS spectra, Peak fitting of the Ga 3d peak for (a) Sample A and (b) Sample B, and peak fitting of the O 1s peak for (c) Sample A and (d) Sample B.

B. Electrical Properties

Fig. 3 presents the transfer characteristic curves of the devices. Both Sample A and Sample B exhibit a high on-off ratio of 10^8, with the gate leakage current stable at 10^{-4} mA. The V_{th} is defined as the gate-source voltage (V_{GS}) at which the drain current (I_{DS}) reaches 1 mA/mm. Sample A shows a V_{th} of 3.4 V, which is the same as the V_{th} of Sample B, which did not undergo plasma treatment. Additionally, the plasma treatment improved the interface quality[19], reducing channel scattering, which led to an increase in the source-drain saturation current from 141 mA/mm to 315 mA/mm,as shown in Fig. 3 (a) and (b). Moreover, Sample A exhibited a transconductance (G_m) of 58 mS/mm and a subthreshold swing (SS) of 166 mV/dec, while Sample B had a gm of 34 mS/mm and an SS of 450 mV/dec, as shown in Fig. 3(c) and (d). Due to the over-etching of the gate region, the device performance is more sensitive to changes in interface quality. The N_2/O_2 plasma treatment significantly enhanced the interface quality, thereby improving the electrical performance of the devices.

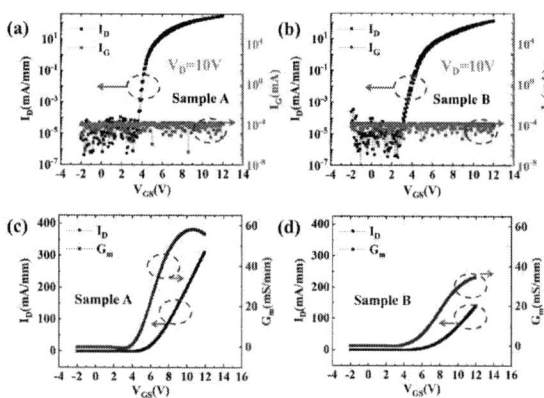

Fig. 3 Transfer characteristics and gate leakage current curves of (a) Sample A and (b) Sample B in a logarithmic scale, and transfer characteristics and transconductance curves of (c) Sample A and (d) Sample B in a linear scale.

Fig. 4 presents the output characteristics of the fabricated devices. The measurements were conducted with V_{GS} swept from 0 to 12 V in 3 V steps, while V_D was scanned from 0 to

10 V. At $V_{GS} = 12V$, Sample A exhibited a 302 % increase in maximum saturation current compared to Sample B. Additionally, the on-resistance (R_{on}) of Sample A was 19.5 $\Omega\cdot$mm, whereas that of Sample B was 40.2 $\Omega\cdot$mm. This substantial performance enhancement further underscores the critical role of N_2/O_2 plasma treatment in mitigating etching-induced damage and improving device characteristics.

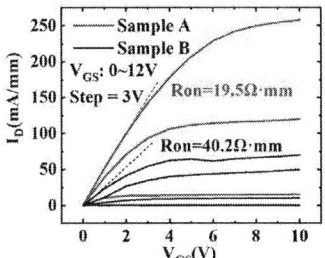

Fig. 4 Comparison of the output characteristics of Sample A and Sample B.

Fig. 5 illustrates the gate breakdown characteristics of the devices. Although the gate dielectric is 30 nm thick Al_2O_3, the gate breakdown voltage of Sample B is only 14 V due to etching-induced damage at the interface, which is significantly lower than the theoretical breakdown voltage of 30 nm Al_2O_3 (~ 30 V)[20]. In contrast, after N_2/O_2 plasma treatment, Sample A exhibited an improvement in the GaN/Al_2O_3 interface quality, resulting in a gate breakdown voltage of 26 V, which is close to the theoretical breakdown voltage of Al_2O_3.

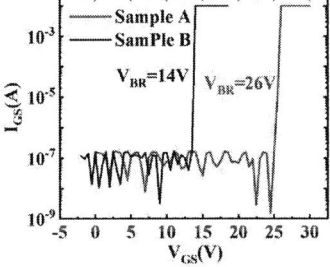

Fig. 5 I_{GS}–V_{GS} characteristics of Sample A and Sample B.

C. Reliability

Transfer characteristic measurements were conducted at varying temperatures to analyze the temperature dependence of the threshold voltage (V_{th}). As shown in Fig. 6 (a) and (b), both samples exhibited a negative shift in V_{th} with increasing temperature. For Sample A, which underwent plasma treatment, the negative shift at 523 K was 0.9 V compared to room temperature, demonstrating improved reliability relative to Sample B, which exhibited a larger negative shift of 2.1 V, as illustrated in Fig. 6 (c). Furthermore, based on the transfer characteristics at different temperatures, the temperature-dependent subthreshold swing (SS) was extracted, and the interface state density at various temperatures was calculated using Equations (1) and (2)[19], [21]:

$$SS = \left(1 + \frac{C_{it} + C_q}{C_o}\right) \cdot \frac{k_B T}{q} \cdot \ln 10 \qquad (1)$$

with

$$D_{it} = \frac{C_{it}}{q^2} \qquad (2)$$

where C_{it}, C_o, and C_q represent the interface trap capacitance, the dielectric capacitance, and the quantum capacitance, respectively. In the deep subthreshold region, C_q is very small, and thus it can be neglected in the calculation process. It is worth noting that the barrier layer capacitance does not appear in Equation (1), as the AlGaN in the gate region was fully etched during the device fabrication process, and therefore, the contribution of the barrier layer capacitance is disregarded in the calculations. k_B is the Boltzmann constant, T is the Kelvin temperature, q is the elementary charge, D_{it} is the interface trap density. As shown in Figure 6d, Sample A, which underwent plasma treatment, exhibited a one-order-of-magnitude reduction in the D_{it} compared to Sample B, resulting in significantly improved gate control. This explains why Sample B, with a higher density of trap interface states, experiences a positive shift in threshold voltage under the application of electric stress, as these traps capture electrons[22]. At elevated temperatures, these traps de-capture electrons more rapidly, leading to a larger negative shift in the threshold voltage.

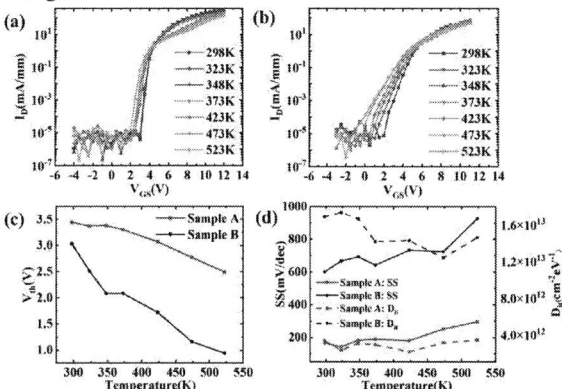

Fig. 6 Temperature-dependent transfer characteristics of the devices (a) Sample A and (b) Sample B, and (c) comparison of the Vth variation for both samples, (d) comparison of the extracted SS and the calculated D_{it} for Sample A and Sample B at different temperatures.

IV. CONCLUSION

A novel plasma treatment strategy for E-mode MIS-HEMTs has been proposed to enhance device performance. The characterization results demonstrate significant improvements in both static electrical properties and reliability under high temperatures and stress conditions. Devices treated with N_2/O_2 plasma exhibited a high threshold voltage of 3.4 V, a gate breakdown voltage of 26 V, a threshold voltage shift of only 0.9 V at 523 K, all of which represent substantial improvements over untreated devices. This plasma treatment approach provides a simple and effective method with promising applications in the development of MIS-HEMTs technology.

REFERENCES

[1] M. Ishida, T. Ueda, T. Tanaka, and D. Ueda, "GaN on Si Technologies for Power Switching Devices," IEEE Trans. Electron Devices, vol. 60, no. 10, pp. 3053–3059, Oct. 2013, doi: 10.1109/TED.2013.2268577.

[2] K. J. Chen, O. Haberlen, A. Lidow, C. L. Tsai, T. Ueda, Y. Uemoto, and Y. Wu, "GaN-on-Si Power Technology: Devices and Applications," IEEE Trans. Electron Devices, vol. 64, no. 3, pp. 779–795, Mar. 2017, doi: 10.1109/TED.2017.2657579.

[3] W. Yang, and J.-S. Yuan, "Substrate Bias Enhanced Trap Effects on Time-Dependent Dielectric Breakdown of GaN MIS-HEMTs," IEEE Trans. Electron Devices, vol. 68, no. 5, pp. 2233–2239, May 2021, doi: 10.1109/TED.2021.3067615.

[4] S. Zhang, X. Liu, K. Wei, S. Huang, X. Chen, Y. Zhang, Y. Zheng, G. Liu, T. Yuan, X. Wang, H. Yin, Y. Yao, and J. Niu, "Suppression of Gate Leakage Current in Ka -Band AlGaN/GaN HEMT With 5-nm SiN Gate Dielectric Grown by Plasma-Enhanced ALD," IEEE Trans. Electron Devices, vol. 68, no. 1, pp. 49–52, Jan. 2021, doi: 10.1109/TED.2020.3037888.

[5] M. Li, J. Wang, H. Wang, Q. Cao, J. Liu, and C. Huang, "Improved performance of fully-recessed normally-off LPCVD SiN/GaN MISFET using N_2O plasma pretreatment," SOLID STATE ELECTRON, vol. 156, pp. 58–61, Jun. 2019, doi: 10.1016/j.sse.2019.03.067.

[6] H. Wang, J. Wang, M. Li, Q. Cao, M. Yu, Y. He, and W. Wu, "823-mA/mm Drain Current Density and 945-MW/cm^2 Baliga's Figure-of-Merit Enhancement-Mode GaN MISFETs With a Novel PEALD-AlN/LPCVD-Si_3N_4 Dual-Gate Dielectric," IEEE Electron Device Lett., vol. 39, no. 12, pp. 1888–1891, Dec. 2018, doi: 10.1109/LED.2018.2879543.

[7] T. Oka, and T. Nozawa, "AlGaN/GaN Recessed MIS-Gate HFET With High-Threshold-Voltage Normally-Off Operation for Power Electronics Applications," IEEE Electron Device Lett., vol. 29, no. 7, pp. 668–670, Jul. 2008, doi: 10.1109/LED.2008.2000607.

[8] M. Hua, J. Wei, Q. Bao, Z. Zhang, Z. Zheng, and K. J. Chen, "Dependence of V_{th} Stability on Gate-Bias Under Reverse-Bias Stress in E-mode GaN MIS-FET," IEEE Electron Device Lett., vol. 39, no. 3, pp. 413–416, Mar. 2018, doi: 10.1109/LED.2018.2791664.

[9] J.-H. Lee, C. Park, K.-W. Kim, D.-S. Kim, and J.-H. Lee, "Performance of Fully Recessed AlGaN/GaN MOSFET Prepared on GaN Buffer Layer Grown With AlSiC Precoverage on Silicon Substrate," IEEE Electron Device Lett., vol. 34, no. 8, pp. 975–977, Aug. 2013, doi: 10.1109/LED.2013.2265351.

[10] Qi Zhou, Bowen Chen, Yang Jin, Sen Huang, Ke Wei, Xinyu Liu, Xu Bao, Jinyu Mou, and Bo Zhang, "High-Performance Enhancement-Mode Al2O3/AlGaN/GaN-on-Si MISFETs With 626 MW/cm^2 Figure of Merit," IEEE Trans. Electron Devices, vol. 62, no. 3, pp. 776–781, Mar. 2015, doi: 10.1109/TED.2014.2385062.

[11] J. H. Park, S.-K. Hwang, J. Kim, W. Jeon, I. Hwang, J. Oh, B. Kim, Y. Park, D.-C. Shin, J.-B. Park, and J. Kim, "N_2O plasma treatment effect on reliability of p-GaN gate AlGaN/GaN HEMTs," Applied Physics Letters, vol. 120, no. 13, p. 132103, Mar. 2022, doi: 10.1063/5.0082165.

[12] F. Qian, T. Yuan, B. Zhi-Wei, Y. Yuan-Zheng, N. Jin-Yu, Z. Jin-Cheng, H. Yue, and Y. Lin-An, "The improvement of Al_2O_3 /AlGaN/GaN MISHEMT performance by N_2 plasma pretreatment," Chinese Phys. B, vol. 18, no. 7, pp. 3014–3017, Jul. 2009, doi: 10.1088/1674-1056/18/7/066.

[13] A.-C. Liu, Y.-W. Huang, H.-C. Chen, Y.-J. Dong, P.-T. Tu, L.-H. Hsu, Y.-Y. Lai, P.-C. Yeh, I.-Y. Huang, and H.-C. Kuo, "Investigating the effect of O_2 plasma treatment on the operational characteristics of Schottky-gate AlGaN/GaN HEMT," Semicond. Sci. Technol., vol. 39, no. 8, p. 085002, Aug. 2024, doi: 10.1088/1361-6641/ad54e6.

[14] J. He, K. Wen, P. Wang, M. He, F. Du, Y. Jiang, C. Tang, N. Tao, Q. Wang, G. Li, and H. Yu, "Interface charge engineering on an in situ SiN_x /AlGaN/GaN platform for normally off GaN MIS-HEMTs with improved breakdown performance," Appl. Phys. Lett., vol. 123, no. 10, p. 103502, Sep. 2023, doi: 10.1063/5.0169944.

[15] S. Dogar, W. Khan, F. Khan, and S.-D. Kim, "Effect of NH_3 plasma treatment on the transient characteristics of ZnO nanorod-gated AlGaN/GaN high electron mobility transistor-based UV sensors," THIN SOLID FILMS, vol. 642, pp. 69–75, Nov. 2017, doi: 10.1016/j.tsf.2017.09.022.

[16] C. Huang, J. Wang, M. Wang, J. He, M. Li, B. Zhang, X. Wang, J. He, Z. Liu, and Y. He, "Investigation of the Trap States and V_{th} Instability in Normally-Off GaN MIS-FETs With LPCVD SiN_x /PEALD AlN Gate Dielectric Stack and In Situ H_2 /N_2 Plasma Pretreatment," IEEE Trans. Electron Devices, vol. 70, no. 11, pp. 5563–5569, Nov. 2023, doi: 10.1109/TED.2023.3309618.

[17] N. Sun, H. Huang, Z. Sun, R. Wang, S. Li, P. Tao, Y. Ren, S. Song, H. Wang, S. Li, W. Cheng, and H. Liang, "Improving Gate Reliability of 6-In E-Mode GaN-Based MIS-HEMTs by Employing Mixed Oxygen and Fluorine Plasma Treatment," IEEE Trans. Electron Devices, vol. 69, no. 1, pp. 82–87, Jan. 2022, doi: 10.1109/TED.2021.3131118.

[18] R. Yin, Y. Li, Y. Sun, C. P. Wen, Y. Hao, and M. Wang, "Correlation between border traps and exposed surface properties in gate recessed normally-off Al_2O_3/GaN MOSFET," Appl. Phys. Let, vol. 112, no. 23, p. 233505, Jun. 2018, doi: 10.1063/1.5037646.

[19] R. Wang, P. Saunier, X. Xing, C. Lian, X. Gao, S. Guo, G. Snider, P. Fay, D. Jena, and H. Xing, "Gate-Recessed Enhancement-Mode InAlN/AlN/GaN HEMTs With 1.9-A/mm Drain Current Density and 800-mS/mm Transconductance," IEEE Electron Device Lett., vol. 31, no. 12, pp. 1383–1385, Dec. 2010, doi: 10.1109/LED.2010.2072771.

[20] C. Lin, J. Kang, D. Han, D. Tian, W. Wang, J. Zhang, M. Liu, X. Liu, and R. Han, "Electrical properties of Al_2O_3 gate dielectrics," MICROELECTRON ENG, vol. 66, no. 1–4, pp. 830–834, Apr. 2003, doi: 10.1016/S0167-9317(02)01007-9.

[21] Z. Sun, H. Huang, R. Wang, Y. Liu, N. Sun, F. Li, P. Tao, Y. Ren, S. Song, H. Wang, S. Li, W. Cheng, J. Gao, and H. Liang, "Effects of SiON/III-nitride interface properties on device performances of GaN-based power field-effect transistors," J. Phys. D: Appl. Phys., vol. 54, no. 2, p. 025109, Jan. 2021, doi: 10.1088/1361-6463/abbf79.

[22] X. Xie, Q. Wang, M. Pan, P. Zhang, L. Wang, Y. Yang, H. Huang, X. Hu, and M. Xu, "Improved Vth Stability and Gate Reliability of GaN-Based MIS-HEMTs by Employing Alternating O_2 Plasma Treatment," NANOMATERIALS-BASEL, vol. 14, no. 6, p. 523, Mar. 2024, doi: 10.3390/nano14060523.

979-8-3315-1110-4/25 $31.00 © 2025 IEEE

Investigation on Surge Current Failure in 1200V SiC MOSFETs with Varied Gate Biases in the Third Quadrant

Chengyuan Zhou
Zhejiang University
College of Electrical Engineering
Zhejiang, China
zhou_chengyuan@zju.edu.cn

Hengyu Wang
Zhejiang University
College of Electrical Engineering
Zhejiang, China
wanghengyu@zju.edu.cn

Yang Zou
Zhejiang University
College of Electrical Engineering
Zhejiang, China
zouyang@zju.edu.cn

Kuang Sheng
Zhejiang University
College of Electrical Engineering
Zhejiang, China
shengk@zju.edu.cn

Abstract—In this work, third-quadrant I-V characteristics and surge current tests are conducted on three commercial SiC MOSFETs (Device A, B and C) under varying gate bias conditions (V_{GS}). Experimental results indicate that under a -5V gate bias, the maximum surge current the device can tolerate decreases compared to operation at 0V and +15V gate biases. This suggests that for high-current devices, both 0V and +15V gate biases enhance surge reliability. The findings highlight the importance of gate bias optimization in improving surge tolerance, emphasizing that certain gate voltages significantly contribute to better surge resilience in high-power devices. Decapsulation analysis identifies localized metal layer melting on the chip surface as the primary failure mechanism at V_{GS}=0V. The findings highlight gate-bias-dependent electro-thermal dynamics as a factor in SiC MOSFET surge reliability optimization.

Keywords—SiC MOSFET, parasitic body diode, surge current test, gate bias

I. INTRODUCTION

Silicon carbide (SiC) MOSFETs have gained prominence in electric vehicles and energy systems due to their high-voltage capability, high-frequency performance, and power density advantages [1]. While commercial planar/trench-structured devices show superior performance, critical reliability challenges persist, particularly under surge currents 3-5 times the rated I_{DS}, causing gate oxide degradation and parasitic diode-induced bipolar degradation [2]. Recent studies reveal dual failure modes in body diodes and demonstrate a 71.4% surge tolerance improvement with V_{GS} adjustment from 0V to -5V [3]. However, current research focuses primarily on low-current devices (<50A), with insufficient investigation of high-current variants and standardized failure features.

This study systematically investigates three commercial 1200V high-current SiC MOSFETs under varied V_{GS}(15/0/-5V). Through surge testing and decapsulation, we quantify maximum bearable surge current, maximum bearable surge energy, and establish gate-bias-dependent failure features. In addition, The decapsulation results elucidate distinct damage mechanisms: aluminum interconnect failure at V_{GS}=0V, providing critical references for high-current device optimization.

TABLE I.   DUTs STUDIED IN THIS WORK

DUT	Forward Rated Current@300K	Diode Rated Current@300K	Package
A	55A	55A	TO-247-3
B	60A	60A	TO-247-3
C	88A	88A	TO-247-3

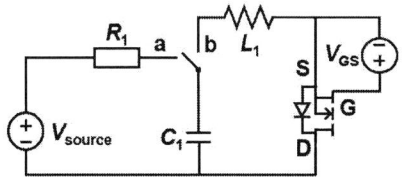

Fig. 1. Surge test circuit

II. SURGE TEST

The schematic diagram of the surge test platform utilized in this study is depicted in Fig. 1. By leveraging the resonance between the capacitor and inductor, the system generates a sinusoidal half-wave surge current, which is mathematically expressed in (1):

$$I(t) = I_{max} \sin(\frac{\pi}{t_w}t) \ (0 \le t \le t_w) \qquad (1)$$

Where t_w denotes the sinusoidal pulse width which is fixed at **10 ms** in this experiment), and I_{max} represents the peak surge current [4].

The testing protocol begins at 50A, with each subsequent surge increasing by 10A per pulse. Each test interval lasts for 2 minutes, allowing the device to cool down adequately and preventing any heat buildup.

Partial data sheets for three available commercial 1200V MOSFETs, named DUT A (Cree Inc.), DUT B (Onsemi Corp.) and DUT C (NARI Group), are shown in TABLE I. All DUTs share identical TO-247 packaging

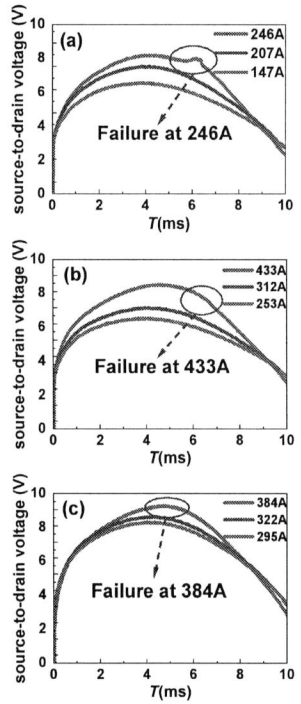

Fig. 2. Single-pulse current tests of voltage at V_{GS}=**-5V** (a) DUT A. (b) DUT B. (c) DUT C.

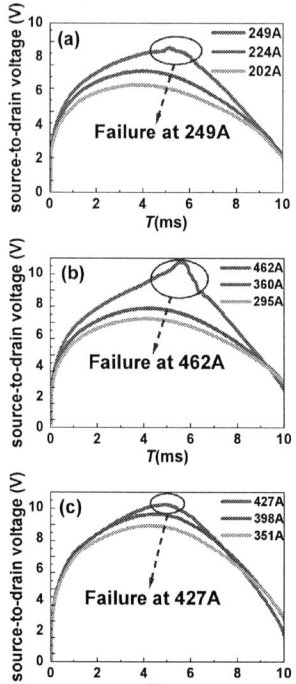

Fig. 3. Single-pulse current tests of voltage at V_{GS}=**0V** (a) DUT A. (b) DUT B. (c) DUT C.

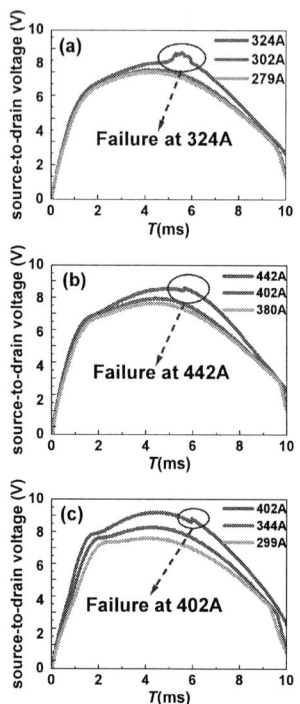

Fig. 4. Single-pulse current tests of voltage at V_{GS}=**15V** (a) DUT A. (b) DUT B. (c) DUT C.

and room-temperature specific on-resistance ($R_{on,sp}$) of 40 mΩ. While the forward rated current and the diode rated current are different.

III. RESULTS AND DISCUSSIONS

Fig. 2, Fig. 3 and Fig. 4 show the single-pulse current tests of voltage at V_{GS}=-5V/0V/15V. It can be observed that despite different gate voltages, all DUTs exhibit significant distortion in their surge voltage curves after a single surge failure event compared to those prior to failure [5].

Under the gate voltage of -5V, as illustrated in Fig. 2(a) and (b), the surge voltage profiles of DUTs A and B exhibit distortion after failure. However, as shown in Fig. 2(c), although the surge voltage profile of device C does not undergo distortion post-failure, a noticeable shift in the peak surge voltage is observed relative to the pre-failure curve.

Under the gate voltage of 0V, as shown in Fig. 3, the surge voltage curves of all three DUTs exhibit distortion following failure. As depicted in Fig. 3(a) and (b), the distortion in the surge voltage curves of DUTs A and B is particularly pronounced. However, in Fig. 3(c), a noticeable shift in the peak surge voltage is observed post-failure.

In addition, under the gate voltage of 15V, as shown in Fig. 4, the surge voltage curves of all three DUTs exhibit significant distortion following failure. This distortion occurs with a delay relative to the peak of the surge voltage.

Therefore, the experimental results indicate that the surge voltage curves of high-current SiC MOSFETs under different gate voltages exhibit distortion at the point of surge failure. Moreover, the peak surge voltage is delayed relative to the pre-failure surge voltage peak.

Fig. 5, Fig. 6 and Fig. 7 show the single-pulse current tests of surge trajectory. It can be clearly observed in Fig. 7 that the surge trajectory under 15V gate voltage shows significant differences compared to those under -5V and 0V gate voltages. Under gate voltages of -5V or 0V, the surge trajectories exhibit counterclockwise rotation prior to failure, indicating that the surge voltage peak leads the surge current peak. Following surge failure, the trajectories transform into figure-eight loop patterns with clockwise rotation near the current peak region, demonstrating a phase reversal where the voltage peak now lags behind the current peak.

The 15V gate voltage condition maintains a persistent figure-eight loop pattern throughout the testing. Before failure, the surge trajectory displays a sharp peak feature with minimal voltage differential around the current peak. Post-failure, however, the loop becomes flattened with significantly enlarged voltage variations in the current peak region, accompanied by complete disappearance of the sharp peak characteristics.

Based on the failure features of DUT A, B and C proposed above, the surge-related parameters for the three DUTs are shown in Table. 2. It can be observed that for SiC MOSFET devices with high-current capabilities, the application of negative gate voltages results in a certain degree of suppression on their surge resistance. Specifically, at a gate voltage of 15V, DUT A was able to withstand surge energy significantly higher than that at 0V and -5V gate voltages. This indicates that a positive gate

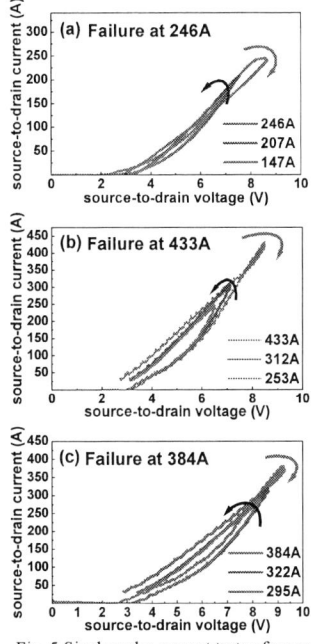
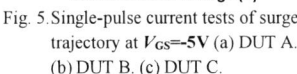
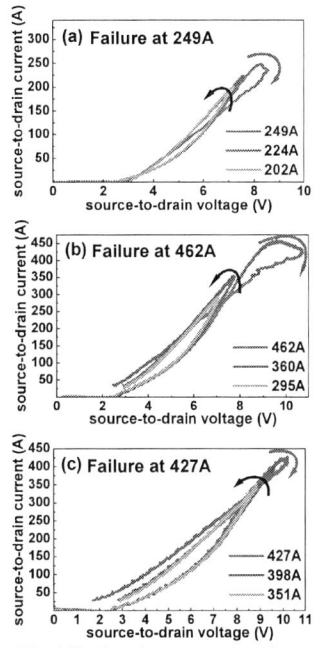
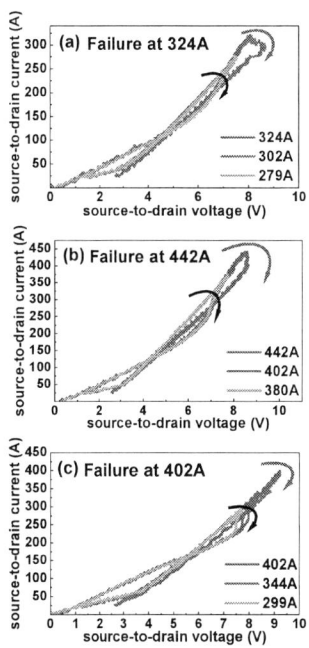

Fig. 5. Single-pulse current tests of surge trajectory at V_{GS}=-5V (a) DUT A. (b) DUT B. (c) DUT C.

Fig. 6. Single-pulse current tests of surge trajectory at V_{GS}=0V (a) DUT A. (b) DUT B. (c) DUT C.

Fig. 7. Single-pulse current tests of surge trajectory at V_{GS}=15V (a) DUT A. (b) DUT B. (c) DUT C.

voltage enhances the surge energy tolerance of DUT A. DUT C demonstrated the highest surge energy tolerance when operated at a 0V gate voltage, highlighting its inherent surge resistance under such conditions. In contrast, DUT B exhibited almost identical surge capabilities across all three gate voltages, suggesting that its surge performance is relatively unaffected by changes in the gate voltage. Among the three DUTs, DUT B exhibited the best surge robustness, maintaining stable performance across varying gate bias conditions.

The 15V gate voltage condition maintains a persistent figure-eight loop pattern throughout the testing. Before failure, the surge trajectory displays a sharp peak feature with minimal voltage differential around the current peak. Post-failure, however, the loop becomes flattened with significantly enlarged voltage variations in the current peak region, accompanied by complete disappearance of the sharp peak characteristics.

Based on the failure features of DUT A, B and C proposed above, the surge-related parameters for the three DUTs are shown in TABLE. II. It can be observed that for SiC MOSFET devices with high-current capabilities, the application of negative gate voltages results in a certain degree of suppression on their surge resistance. Specifically, at a gate voltage of 15V, DUT A was able to withstand surge energy significantly higher than that at 0V and -5V gate voltages. This indicates that a positive gate voltage enhances the surge energy tolerance of DUT A. DUT C demonstrated the highest surge energy tolerance when operated at a 0V gate voltage, highlighting its inherent surge resistance under such conditions. In contrast, DUT B exhibited almost identical surge capabilities across all three gate voltages, suggesting that its surge performance is relatively unaffected by changes in the gate voltage. Among the three DUTs, DUT B exhibited the best

TABLE II. SURGE CAPABILITY OF DUTs

DUT	Applied Gate Source Bias	Maximum Bearable Surge Current	Maximum Bearable Surge Energy
A	-5V	246A	10.9J
	0V	249A	11.0J
	15V	324A	14.7J
B	-5V	433A	20.3J
	0V	462A	24.6J
	15V	442A	23.1J
C	-5V	384A	19.7J
	0V	427A	23.4J
	15V	402A	20.5J

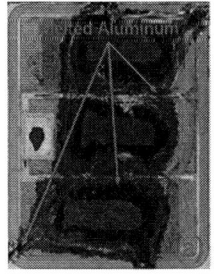
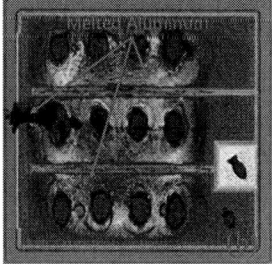

Fig. 8. Decapsulation after Single-pulse current tests at V_{GS}= 0V (a) DUT B. (b) DUT C.

surge robustness, maintaining stable performance across varying gate bias conditions.

To further investigate the surge failure mechanisms, decapsulation was performed on DUT B and DUT C after surge-induced failure at a 0V gate bias. The results are illustrated in Fig. 8. In both DUT B and C, substantial melting of aluminum metal was observed in the active regions of the devices. This melting of aluminum was identified as the primary cause of the short circuit formation between the gate and source electrodes [6], which subsequently led to the failure of the devices. The localized thermal stress and resultant metal degradation appear to play a critical role in triggering the catastrophic failure observed in these devices.

IV. CONCLUSION

This study investigates surge failure mechanisms in 1200V SiC MOSFETS under varied gate biases. The results show that higher gate voltages enhance surge tolerance in DUT A, while DUT C exhibits the best performance at 0V. Decapulation analysis reveals that aluminum interconnect melting causes short circuits between the gate and source, leading to device failure. These findings offer valuable insights for optimizing SiC MOSFETS for improved surge resistance and reliability in high-current applications.

ACKNOWLEDGMENT

This work was supported by the Science and Technology Project of State Grid (5108-202218280A-2-335-XG).

REFERENCES

[1] K. Han and B. J. Baliga, "Comprehensive physics of third quadrant characteristics for accumulation and inversion-channel 1.2kV 4H-SiC MOSFETs," IEEE Transactions on Electron Devices, vol. 66, no. 9, pp. 3916-3921, Sept. 2019.

[2] Huang W, Deng X, Li X, Wen Y, Li X and Li Z, "Investigation of surge current reliability of 1200V planar and trench SiC MOSFET," in 2020 IEEE 15th International Conference on Solid-State & Integrated Circuit Technology (ICSICT), Kunming, China, 2020, pp. 1-3.

[3] Z. Zhu, H. Xu, L. Liu, N. Ren and K. Sheng, "Investigation on surge current capability of 4H-SiC trench-gate MOSFETs in third quadrant under various VGS biases," IEEE Journal of Emerging and Selected Topics in Power Electronics, vol. 9, no. 5, pp. 6361-6369, Oct. 2021.

[4] H. Xu, N. Ren, Z. Zhu, J. Wu, L. Liu and Q. Guo, "Methodology for enhanced surge robustness of 1.2kV SiC MOSFET body diode," IEEE Journal of Emerging and Selected Topics in Power Electronics, vol. 10, no. 5, pp. 5039-5047, Oct. 2022.

[5] N. Ren, L. Liu, J. Wu and K. Sheng, "Plasma spreading layers: an effective method for improving surge and avalanche robustness of SiC devices," IEEE Transactions on Electron Devices, vol. 68, no. 11, pp. 5687-5694, Nov. 2021

[6] X. Zhan, Y. He, J. Lai, H. Yuan, Q. Song and X. Tang, "Investigation on degradation of 1200V planar and trench SiC MOSFET under surge current stress of body diode," IEEE Transactions on Electron Devices, vol. 71, no. 1, pp. 709-714, Jan. 2024.

Study on SiC Power Module Cooling Technologies Suitable for High-Junction-Temperature and High-Power-Density Applications

Baihan Liu
School of Electrical and Electronic Engineering
Huazhong University of Science and Technology
Wuhan, China
loubeckham@hust.edu.cn

Jianwei Lv
School of Electrical and Electronic Engineering
Huazhong University of Science and Technology
Wuhan, China
jianweil@hust.edu.cn

Yipeng Liu
School of Electrical and Electronic Engineering
Huazhong University of Science and Technology
Wuhan, China
d202480992@hust.edu.cn

Yifan Zhang
School of Electrical and Electronic Engineering
Huazhong University of Science and Technology
Wuhan, China
z_yifan@hust.edu.cn

Siqi Dai
School of Electrical and Electronic Engineering
Huazhong University of Science and Technology
Wuhan, China
dsq0921@hust.edu.cn

Zexiang Zheng
School of Electrical and Electronic Engineering
Huazhong University of Science and Technology
Wuhan, China
zzx_huster@hust.edu.cn

Suhang Wei
School of Electrical and Electronic Engineering
Huazhong University of Science and Technology
Wuhan, China
shwei@hust.edu.cn

Cai Chen
School of Electrical and Electronic Engineering
Huazhong University of Science and Technology
Wuhan, China
caichen@hust.edu.cn

Yong Kang
School of Electrical and Electronic Engineering
Huazhong University of Science and Technology
Wuhan, China
ykang@hust.edu.cn

Abstract—The emergence of wide-bandgap semiconductors with superior performance has elevated expectations for the high-temperature operation and power density of power electronics systems, while simultaneously presenting more stringent challenges for cooling technologies related to system reliability and lifespan. This study, based on Multiphysics coupling simulation, reveals that conventional forced water cooling and forced air cooling are unable to meet the demands of high-temperature and high-power-density applications. Key principles for selecting related cooling technologies were summarized. Considering these principles and implementation challenges, forced oil cooling and water-based two-phase cooling technologies were selected. Simulation validation confirmed their appropriate coolant properties and heat dissipation effects. Additionally, this study addressed the challenge of simulating phase change processes in two-phase cooling by introducing an extended RPI model in the simulation.

Keywords—*wide bandgap semiconductor, silicon carbide device, cooling technology, power module packaging, high temperature, high power density application*

I. INTRODUCTION

Wide-bandgap semiconductor power devices (e.g., SiC and GaN), leveraging their high critical breakdown field strength, superior thermal conductivity, and high carrier mobility, have surpassed the performance boundaries of silicon-based devices, providing technical support for power electronics (PE) systems operating under high voltage, high temperature (HT), and high-frequency conditions. Theoretical studies indicate that the stable operation of SiC power devices at HT (>200°C) can significantly reduce the size of the cooling

This work was supported by the National Natural Science Foundation of China under Grant 52077094.

system, thereby achieving an increase in power density. However, current research predominantly employs natural or forced air convection cooling methods, which introduce bulky and heavy cooling components, leading to persistently low system power density [1], [2], [3]. While double-sided cooling packaging technology has alleviated some of these challenges by reducing module thermal resistance, the overall optimization space remains limited due to the dominant effect of cooling system thermal resistance. Therefore, the development of novel cooling methods with high heat transfer coefficients (HTC) represents a critical breakthrough point. In electric vehicle (EV) traction systems and aircraft actuators, advanced cooling techniques such as jet impingement cooling [4], spray cooling [5], and phase-change cooling [6] have effectively reduced the size of the cooling components, but their adaptability to SiC devices' HT thermal management capabilities requires comprehensive validation and target-ed improvements.

This study focuses on the thermal management design of SiC power modules with a highly compact layout for high-temperature and high-power-density (HT&HPS) applications. Using a self-designed double-sided cooling power module as the reference heat source, as shown in Fig. 1, the study employs simulation analysis to investigate the feasibility of water cooling (commonly used in conventional PE) and air cooling (commonly used in high-temperature PE systems) in HT&HPS applications. Key characteristics of cooling technologies suitable for HT&HPS PE systems were summarized, and oil cooling and two-phase cooling methods, which possess significant HT application potential, were validated and evaluated. The research results provide important guidance for the thermal management design of HT&HPS PE systems.

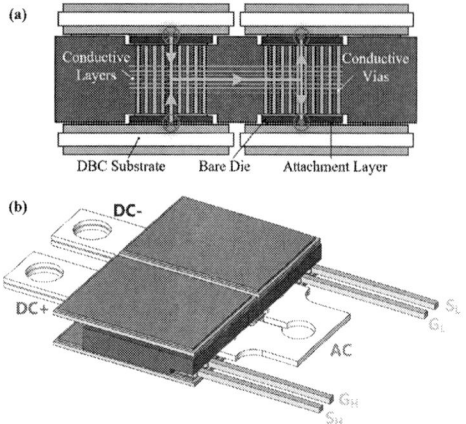

Fig. 1. The self-designed power module. (a) Internal structure; (b) 3D simulation model.

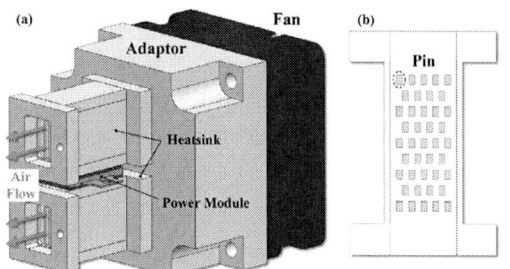

Fig. 2. (a) The customized air cooling system; (b) Pin layout of pin-fin heatsinks.

Fig. 3. Thermal simulation result of the air cooling model.

II. RESEARCH ON TRADITIONAL POWER MODULE COOLING TECHNOLOGIES

This section employs finite element simulation to investigate the application of commonly used air cooling and water cooling technologies in HT&HPS PE systems, evaluating their feasibility and potential risks, thereby providing a foundation for subsequent research into cooling technologies suitable for HT&HPS applications.

A. Evaluation of conventional cooling technologies

Fig. 1 illustrates the packaging structure and appearance of the power module. This is a two-chip parallel chip-on-chip power module, designed in a highly compact layout to meet the demands of high-power-density applications. This half-bridge power module integrates four SiC chips. Chips' heat is primarily transferred from the adjacent DBC substrate to the exterior.

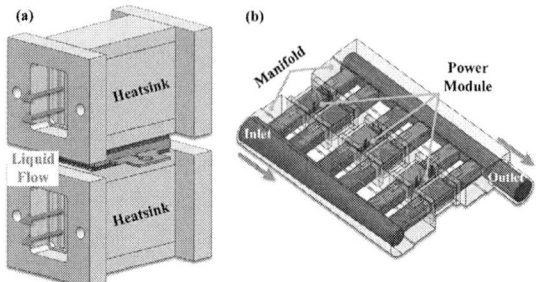

Fig. 4. (a) Liquid cooling unit model; (b) Liquid cooling system diagram of the 3-phase inverter.

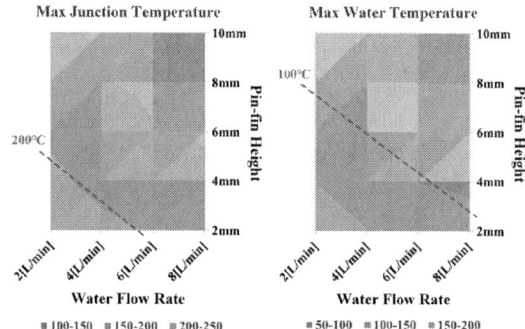

Fig. 5. Thermal simulation results of the water cooling model.

Fig. 2 illustrates the air cooling system designed for the power module. The heat sink features a square pin fin structure, matching the module's dimensions. The fan and heat sink are connected via a custom adapter to align with the module's size and facilitate air flow. The selected fan is the most suitable high-temperature type among the limited options. Simulation under the power loss of 100W/die, the ambient temperature of 20 °C, and the fan curve provided in the datasheet revealed a maximum junction temperature of 676 °C, as shown in Fig. 3. This outcome is anticipated, as air cooling is ineffective for HT applications, and most existing HT power electronics research targets low-power, space-sufficient cases. However, since this study focuses on HT&HPS applications, air cooling is no longer suitable.

The water cooling unit model of the power module, as shown in Fig. 4(a), uses the same heatsink as before. Based on the model, a parameter sweep simulation was performed to investigate the effects of pin-fin height and water flow rate (with power loss and fluid temperature maintained the same, boiling ignored). In the 3-phase inverter, the incoming water flow is distributed to the three water cooling units, as shown in Fig. 4(b). Hence, the flow rate for each heatsink was set to 1/6 of the inlet flow rate. Fig. 5 illustrates the maximum junction and water temperature obtained under different parameters. The results indicate that the power module's junction temperature cannot achieve HT operation above 200 °C to maintain water unboiling.

In summary, conventional forced air and water cooling methods fail to meet the demands of HT&HPS applications. Based on the above findings, appropriate cooling techniques for HT&HPS applications should possess the following characteristics: cooling performance lies between air and water cooling, no concerns about liquid boiling, and ease of implementation.

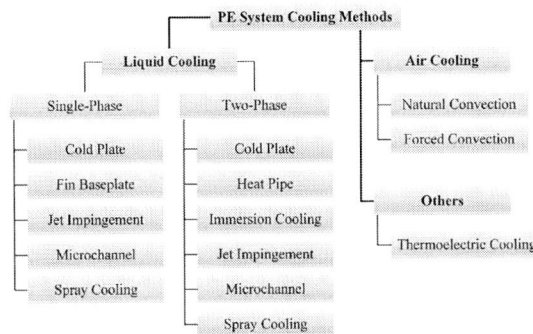

Fig. 6. Classification of power module cooling methods.

B. Selection of Electronic Cooling Technologies

To select suitable cooling technologies, this study summarizes and categorizes PE system cooling technologies based on existing research, as illustrated in Fig. 6. Liquid cooling is the most suitable option due to its superior cooling efficiency and maturity. Two-phase liquid cooling eliminates the concern of boiling under HT due to its inherent phase change process; single-phase liquid cooling requires a high-boiling-point coolant to accommodate HT applications. For the self-designed power module and heatsink, forced oil cooling and water-based two-phase cooling are believed to have high potential for HT applications. They can directly utilize existing EV components (such as motor oil cooling systems and air conditioning systems) to build a cooling loop, facilitating the achievement of high-density integrated motor drive [7], [8]. Hence, they have gradually received attention in the PE thermal management. Other liquid cooling technologies in Fig. 6 also have potential but were not studied in this work due to their incompatibility with the already designed cooling system in Fig. 4.

III. EVALUATION OF POTENTIAL COOLING TECHNOLOGIES FIT FOR HTHPS APPLICATIONS

This section evaluates the feasibility of forced oil cooling and water-based two-phase cooling applied in the HT&HPS PE system cooling.

A. Forced Oil Cooling

In EV, aerospace, and warship applications, oil cooling is used to cool motor bearings and winding due to its excellent thermal properties and stability. However, motor drives typically employ extra water/ethanol cooling, leading to bulky systems. To achieve an integrated motor drive, many studies utilize shared oil cooling systems, such as the hub motor cooling system shown in Fig. 7. This approach is easy to achieve and can directly utilize the engine oil cooling system to achieve HT&HPS PE system cooling.

Based on the liquid cooling unit model in Fig. 4, oil cooling simulations were conducted. The simulation used the commercial automatic transmission fluid (ATF), commonly used in motor cooling, with a flash point of 229 °C. The simulation results in Fig. 8 demonstrate that ATF enables the power module to operate safely at a junction temperature of 250 °C, proving the feasibility of its HT&HPS applications. It is noteworthy that the coolant choice should consider the operating temperature, and the available coolant's flash point currently can reach up to 318 °C.

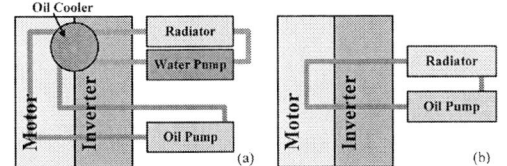

Fig. 7. EV in-wheel motor cooling system. (a) conventional; (b) all oil cooling.

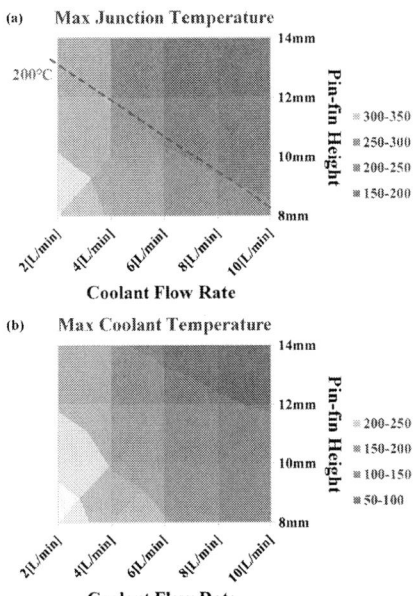

Fig. 8. Thermal simulation results of the oil cooling model.

B. Water-Based Two-Phase Cooling

The two-phase cooling system comprises an evaporator, condenser, compressor, and expansion valve, as shown in Fig. 9. The refrigerant absorbs heat in the evaporator, evaporates, is compressed into high-temperature, high-pressure gas by the compressor, condenses in the condenser, and then flows through the expansion valve to reduce pressure before returning to the evaporator to complete the cycle. Since the commonly used refrigerant R134a cannot withstand HT, this study uses deionized water as the refrigerant to evaluate the application potential of two-phase cooling technology in HT&HPS applications.

The simulation of two-phase cooling presents significantly greater challenges, introducing extra phase change heat transfer and refrigerant phase change issues based on the original forced liquid cooling problem. Existing research often avoids simulation due to the high complexity involved. This study introduced the extended RPI wall boiling model in the simulation to describe the water boiling process, comprising a mass transfer model and a heat transfer model [9]. The former model describes the bubble generation and growth by calculating the mass flux caused by evaporation and condensation. The equations for mass flux during evaporation and condensation are as follows:

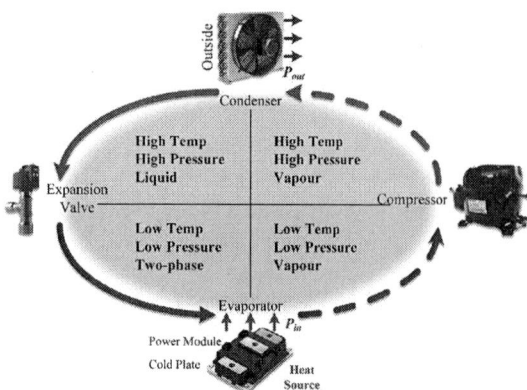

Fig. 9. Schematic diagram of the two-phase cooling loop.

$$m_{evap} = \rho_g \phi_{evap} \text{ with } \phi_{evap} = \max\left(0, \frac{\bar{\alpha}_g - \alpha_g}{t_{evap}}\right)$$

$$m_{cond} = \rho_g \phi_{cond} \text{ with } \phi_{cond} = \min\left(0, \frac{\bar{\alpha}_g - \alpha_g}{t_{cond}}\right) \tag{1}$$

They consider key parameters such as bubble detachment frequency, detachment diameter, and nucleation site density to accurately simulate the boiling process.

The heat transfer model describes the heat transfer during boiling by calculating the heat fluxes in evaporation, quenching, and convection independently, which greatly improves the accuracy, as shown in the equation below:

$$q_w = f\left(\alpha_{l,y_f^+}\right)\left(q_{c,l} + q_q + q_e\right) + \left(1 - f\left(\alpha_{l,y_f^+}\right)\right)q_{c,g} \tag{2}$$

The simulation set the initial water temperature 5 °C below the boiling point and a fluid flow rate of 3.3 L/min (equivalent to 10 L/min in an inverter cooler). Simulation results show that the water-based two-phase cooling can provide effective cooling while supporting the power module to operate at a high T_j of 228 °C. Additionally, local overheating of the power module caused by excessive vaporization of the refrigerant at HT is a necessary concern. Fig. 10 shows the gas volume fraction distribution, indicating that no complete vaporization was observed in the flow under the given conditions, ensuring system safety.

IV. CONCLUSION

Through multi-physics coupled simulation, it was found that conventional air and water cooling are unsuitable for HT&HPS applications due to poor cooling efficiency and, notwithstanding HT, respectively. Then, the following cooling technology selection criteria are summarized: cooling effect lies between air and water cooling; no concerns about liquid boiling; and ease of implementation. Considering the criteria and the designed heatsink, forced oil cooling and water-based two-phase cooling were selected. Simulation results validated their feasibility for HT&HPS applications. To address the simulation challenges in the phase change process of two-phase cooling, the extended RPI model was introduced in the simulation.

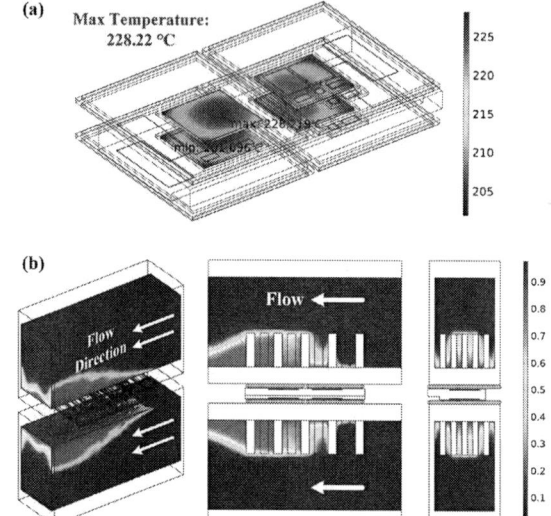

Fig. 10. Thermal simulation results of two-phase cooling. (a) SiC chip temperature; (b) Gas phase volume fraction.

REFERENCES

[1] J. M. Homberger, E. Cilio, R. M. Schupbach, A. B. Lostetter, and H. A. Mantooth, "A high-temperature multichip power module (MCPM) inverter utilizing silicon carbide (SiC) and silicon on insulator (SOI) electronics," in 2006 37th IEEE Power Electronics Specialists Conference, 2006, pp. 1–7. doi: 10.1109/pesc.2006.1711732.

[2] R. Wang et al., "A High-Temperature SiC Three-Phase AC - DC Converter Design for > 100/spl deg/C Ambient Temperature," IEEE Transactions on Power Electronics, vol. 28, no. 1, pp. 555–572, 2013, doi: 10.1109/TPEL.2012.2199131.

[3] P. Ning et al., "Development of a 10 kW high temperature, high power density three-phase AC-DC-AC SiC converter," in 2011 IEEE Energy Conversion Congress and Exposition, 2011, pp. 2413–2420. doi: 10.1109/ECCE.2011.6064089.

[4] S. Jones-Jackson, R. Rodriguez, and A. Emadi, "Jet Impingement Cooling in Power Electronics for Electrified Automotive Transportation: Current Status and Future Trends," IEEE Transactions on Power Electronics, vol. 36, no. 9, pp. 10420–10435, 2021, doi: 10.1109/TPEL.2021.3059558.

[5] H. Bostanci, D. Van Ee, B. A. Saarloos, D. P. Rini, and L. C. Chow, "Thermal Management of Power Inverter Modules at High Fluxes via Two-Phase Spray Cooling," IEEE Transactions on Components, Packaging and Manufacturing Technology, vol. 2, no. 9, pp. 1480–1485, 2012, doi: 10.1109/TCPMT.2012.2190933.

[6] I. Aranzabal, I. M. de Alegría, N. Delmonte, P. Cova, and I. Kortabarria, "Comparison of the Heat Transfer Capabilities of Conventional Single- and Two-Phase Cooling Systems for an Electric Vehicle IGBT Power Module," IEEE Transactions on Power Electronics, vol. 34, no. 5, pp. 4185–4194, 2019, doi: 10.1109/TPEL.2018.2862943.

[7] S. H. Hong, D. S. Jang, S. Park, S. Yun, and Y. Kim, "Thermal performance of direct two-phase refrigerant cooling for lithium-ion batteries in electric vehicles," Applied Thermal Engineering, vol. 173, p. 115213, 2020, doi: https://doi.org/10.1016/j.applthermaleng.2020.115213.

[8] T. Chen et al., "Double-Side Direct Oil-Cooling Automotive Power Module: from Material Compatibility to Thermal Management," in 2023 11th International Conference on Power Electronics and ECCE Asia (ICPE 2023 - ECCE Asia), 2023, pp. 33–38. doi: 10.23919/ICPE2023-ECCEAsia54778.2023.10213969.

[9] X. Sun, Z. Han, and X. Li, "Simulation study on cooling effect of two-phase liquid-immersion cabinet in data center," Applied Thermal Engineering, vol. 207, p. 118142, 2022, doi: https://doi.org/10.1016/j.applthermaleng.2022.118142.

An Enhanced Over-Current Fault Detection System for a 7.2kV Power Module Using Series-Connected SiC MOSFETs

Chunyao Hou
College of Electrical Engineering
Zhejiang University
Hangzhou, China
houcy@zju.edu.cn

An Lou
College of Electrical Engineering
Zhejiang University
Hangzhou, China
12310070@zju.edu.cn

Yue Wu
Electric Power Research Institute
China Southern Power Grid
Guangzhou, China
wuyue@csg.cn

Aozu Luan
College of Electrical Engineering
Zhejiang University
Hangzhou, China
eelaz@zju.edu.cn

Shuai Shao
College of Electrical Engineering
Zhejiang University
Hangzhou, China
shaos@zju.edu.cn

Abstract—**This paper presents a high-speed overcurrent protection system for a 7.2kV power module using series-connected SiC MOSFETs. The system integrates a tunnel magneto-resistance (TMR) current sensor for non-invasive current measurement and an optical splitter. In the event of over-current, the central controller sends the shutdown signal to each series-connected SiC MOSFET simultaneously through the 1-to-6 optical splitter. Through finite element analysis (FEA), the TMR sensor placement is optimized to mitigate skin-effect-induced measurement errors, achieving a current detection delay of 30 ns. The optical splitter enables simultaneous shutdown commands to six sub-modules with 33 ns latency and 230ps inter-module deviation. Testing this prototype on a half-bridge arm demonstrated short-circuit protection within 81.6 ns, demonstrating the system's capability to enhance reliability for the 7.2kV SiC power module.**

Keywords—overcurrent protection, tunnel magnetoresistance (TMR), optical splitter, series-connected SiC MOSFET modules

I. INTRODUCTION

In power systems, industry, and transportation, there exist numerous applications involving medium and high voltage applications. The voltage levels in these applications far exceed the breakdown voltage of individual semiconductor power devices. While silicon carbide devices with a 15 kV rating have been demonstrated in laboratory settings [1], their commercial adoption remains distant due to cost constraints and performance limitations. More importantly, the specific ON-resistance of power devices tends to increase in proportion to the 2.3 to 2.5 power of their breakdown voltage [2], leading to a significant decline in device performance as the breakdown voltage rises. Series-connecting standardized low-voltage power devices has emerged as a promising alternative.

The voltages of series-connected SiC MOSFETs can be effectively balanced using the active clamping module (ACM) proposed in [3][4]. This method ensures that the voltage imbalance remains below 5%, without compromising the switching speed of the power devices. Moreover, the power loss associated with the ACM is negligible. Building on this ACM technology, a 7.2kV SiC MOSFET module has been developed as shown in Fig. 1, utilizing 1.2kV SiC MOSFET

dies. This module is well-suited for applications in medium voltage DC/DC converters or inverters.

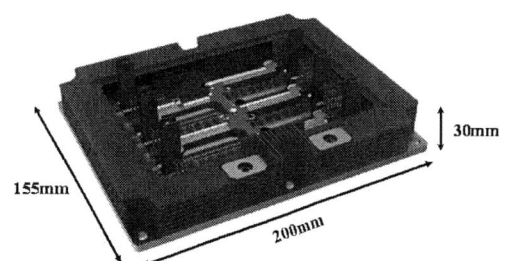

Figure 1: Photograph of 7.2 kV series-connected SiC MOSFET module

In the 7.2kV SiC MOSFET module, it is crucial to detect overcurrent and short-circuit faults within an extremely short time frame. Subsequently, a turn-off signal must be simultaneously sent to each series-connected device to prevent further damage. However, neither conventional desaturation detection nor di/dt detection schemes are capable of detecting over-current in SiC MOSFET modules. The former requires a long blanking time (approximately 500ns), which results in an extended detection time (see Fig. 2)[5]-[8]. While the latter is prone to false alarms due to switching noise coupling into the gate drive circuit during normal turn-on transients, a consequence of the high voltage ratings and ultra-fast switching speeds characteristic of SiC devices.

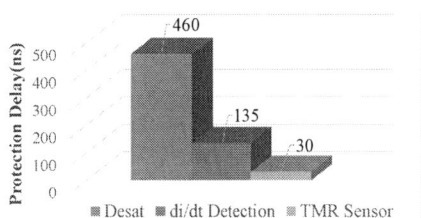

Figure 2: Comparison of fault response times for different detection methods

This paper proposes an overcurrent protection system for the 7.2kV SiC MOSFET power module, consisting of a

TMR current sensor and a splitter-based communication device (see Fig.3). The TMR sensor, integrated into the series-connected module, compares the measured current with a predefined threshold to achieve overcurrent protection. Finite Element Analysis (FEA) determines the TMR's position to mitigate skin effect interference, enabling an overcurrent protection response within 30 ns. Subsequently, the shutdown signal is directly transmitted to each MOSFET's drive board via optical fiber and a splitter within 33 ns, facilitating rapid and simultaneous shutdown. The entire system can thus execute a shutdown command in 81.6 ns. Section II will first introduce the operating principles and detail the optimization of TMR sensor locations. Section III will then describe the fault signal transmission through the optical splitter and the whole system. Finally, Section IV concludes the paper.

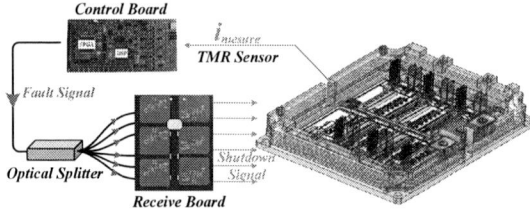

Figure 3: Frame diagram of overcurrent protection system

II. AN OVERCURRENT PROTECTION DEVICE BASED ON TMR

A TMR current sensor consists of a Wheatstone bridge with four magnetoresistances and a differential amplifier, as shown in Fig.4. The tunnel magneto-resistance exhibits a proportional relationship with the magnetic flux density (B) within a defined operational range, thereby enabling non-invasive current measurement. This sensor is integrated into the 7.2 kV power module, as shown in Fig.5, where it is mounted above a copper clip carrying the module's current.

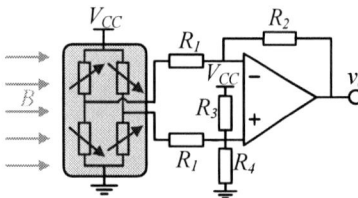

Figure 4: TMR sensor circuit diagram

While the copper clip's dimensions are specified in Fig. 5, the placement of the TMR sensor along the clip is critical. The skin effect, concentrating current flow towards the conductor's surface and edges, means that an improperly located sensor will experience a magnetic flux density (B) that does not scale linearly with the total current (I)[9]. This deviation particularly affects the I-to-B gain stage of the TMR sensor (the overall gain also includes the B-to-output stage). Such non-uniform field exposure can result in measurement delays and distortion of the sensed current waveform. To address this challenge, FEA simulation was conducted to characterize the current distribution and identify optimal sensor locations.

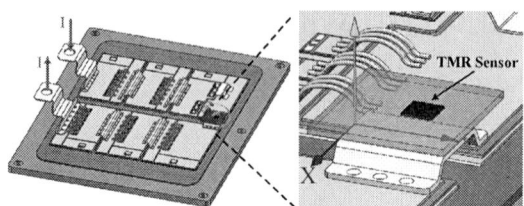

Figure 5: The model of Series-Connected Power module

A coordinate system is established as shown in Fig.5, with its origin located at the midpoint of the upper edge of the clip. Considering the TMR sensor is much smaller compared to the clip and it is only sensitive to the magnetic field in the Y direction, the TMR is approximated as a point on the YZ plane at X = 0. The primary objective is to identify a sensor location where the magnetic flux density in the Y-direction (B_y) should ideally converge to a single value for a given current across the operational frequency spectrum (1 Hz to 1 MHz). This ensures the B/I ratio effectively forms a 'straight line' across varying frequencies, thereby preventing measurement delays and waveform distortion. The skin effect inherently challenges this by causing frequency-dependent current distributions. For example, if the TMR sensor is placed at Z = 0.6 mm, simulations (Fig.6) demonstrate that B_y values diverge significantly across different frequencies.

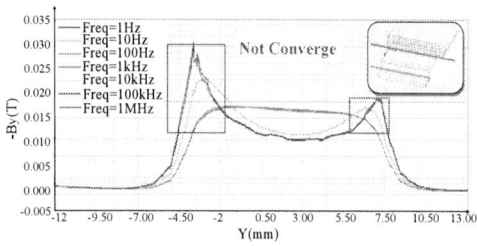

Figure 6: Simulation results for B_y vs. Y at Z = 0.6 mm across various frequencies

It is speculated that the value of Z also affects convergence. Taking the Y and Z coordinates as independent variables and B_y as the dependent variable, simulations were conducted, and the results are shown in Fig. 7. These simulations revealed a significant skin effect influence at low Z values for high-frequency currents. As the value of Z increases, the distortion caused by the skin effect gradually decreases. It is inferred that when Z is large enough, there might be a point of convergence for different-frequency currents, as the red line in Fig. 8 shows.

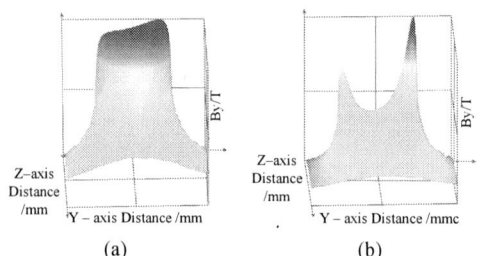

(a)   (b)

Figure 7: (a) Magnetic field distribution at f=1Hz
(b) Magnetic field distribution at f=100kHz

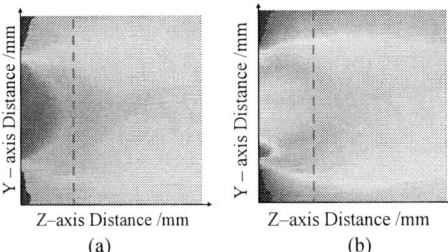

(a)   (b)

Figure 8: (a) Magnetic YZ field distribution at f=1Hz
(b) Magnetic YZ field distribution at f=100kHz

Re-simulating with a higher Z value (3 mm), yielded results (Fig.9) showing consistent gain for different frequencies at a Y-axis origin distance of 6.543 mm. The optimal TMR sensor position is marked by a star in Figure 10, with four comparison positions selected.

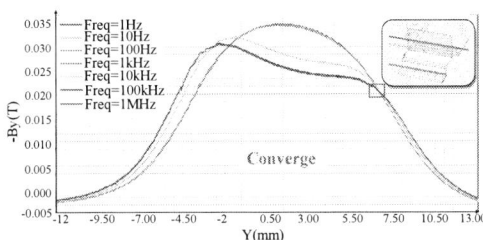

Figure 9: Simulation results for B_y vs. Y at Z = 3 mm across various frequencies

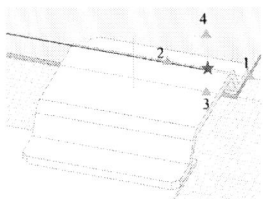

Figure 10: The selected position(star) and the comparison positions (1-4)

A prototype was developed to validate the optimal installation location of the TMR sensor, as shown in Fig. 11. The current through the 7.2 kV module is emulated using a PCB board, whose current path is identical to that of the original power module. The current direction is shown in Fig. 11. The TMR sensor is precisely positioned at the optimal location determined through prior analysis.

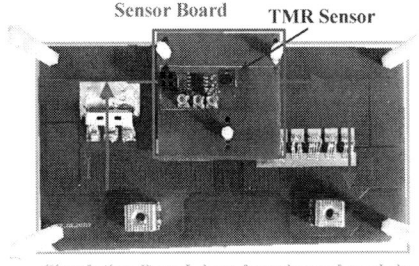

Figure 11: The prototype of TMR-based overcurrent protection device

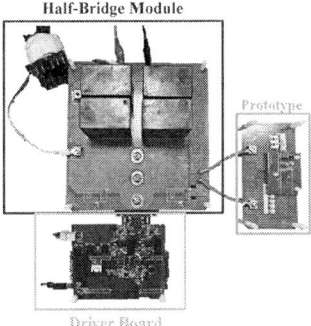

Figure 12: The Double-Pulse Test platform

A Double-Pulse Test is conducted on this prototype to assess the current measurement performance of the TMR sensor. The test platform is shown in Fig. 12. As shown in Fig. 13, a comparison was made between the TMR output at the optimal location and the measurements obtained from a commercial Rogowski coil (CWTUM/1/B). The sensor at the optimal position exhibited good linearity, low oscillation, and a 30 ns delay, outperforming other positions, as detailed in Table I and illustrated with waveforms in Fig. 14. Several sensor deficiencies were noted: low gain at Position 1, significant ripple at Position 2, poor linearity at Position 3, and a long delay at Position 4.

TABLE I   COMPARISON OF TMR OUTPUT AT DIFFERENT POSITIONS

	Gain(mV/A)	Delay(ns)	Linearity	Oscillation(mV)
Optimal location	4	30	Good	20
Location 1	1.5	N/A	Medium	340
Location 2	5.92	200	Medium	400
Location 3	4.1	N/A	Bad	100
Location 4	3.25	50	Good	10

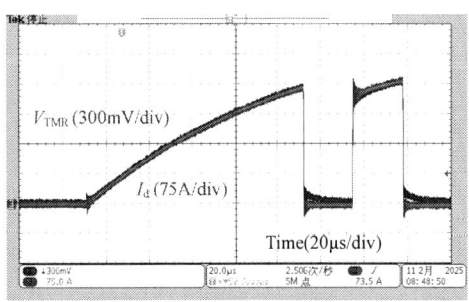

Figure 13: (a)Double-pulse test waveforms (b) Delay measurement

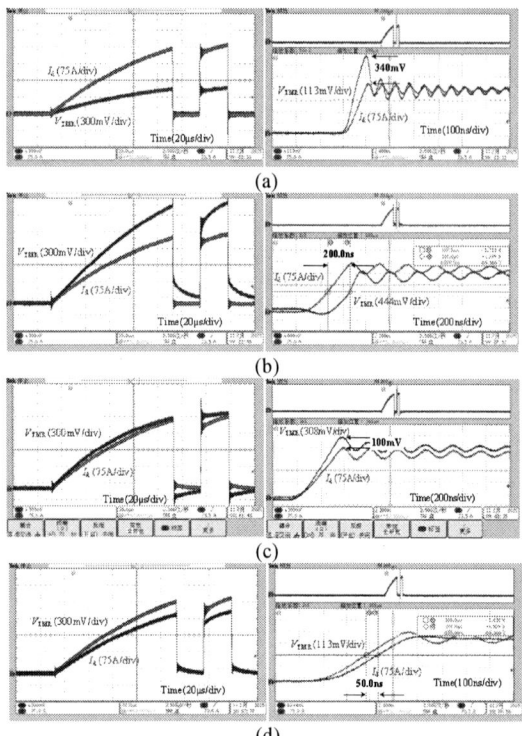

(a)

(b)

(c)

(d)

Figure 14: TMR output and current in different positions, positions 1-4 correspond to (a)-(d) respectively.

III. FAULT TRANSMISSION DEVICE BASED ON OPTICAL SPLITTER

Upon detecting an overcurrent fault via the TMR sensor, the central controller must immediately send a turn-off signal to each SiC MOSFET simultaneously. Given the insulation requirements and the necessity to suppress EMI noise, an optical cable serves as an ideal medium for signal transmission. An optical splitter distributes the incoming optical signal to multiple output ports in a predetermined ratio, as illustrated in Fig.15. This device is commonly employed in internet fiber-optic communication systems.

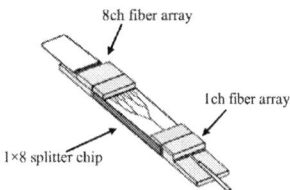

Figure 15: Schematic of 1×8 optical splitter

In this configuration, the optical splitter is interfaced with photoelectric converters at both ends. As depicted in Fig.16, the main control board can directly relay fault signals to six drive boards.

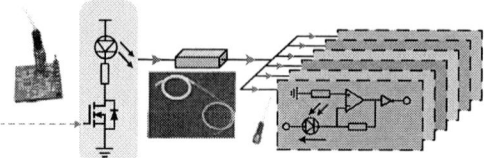

Figure 16: Structure of fault transmission device

To shorten the response time of the signal receiving circuit, a transimpedance amplifier circuit is employed, as shown in Fig. 17. Simulations were performed with V_{ref} = 2V and V_{bios} = 0V, yielding a derived R_f = 200 kΩ and R = 190 kΩ. An inverter was added in the subsequent stage to prevent incorrect protection logic.

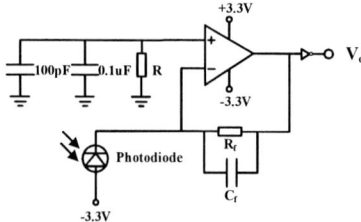

Figure 17: Structure of transimpedance amplifier circuit

Experimental results of the signal transmission are shown in Fig. 18 and 19. A DSP sent a turn-off signal through the optical splitter. The output of three receiving units was measured. The overall response time was 32.9 ns, with 230ps output asynchrony between units, demonstrating excellent consistency.

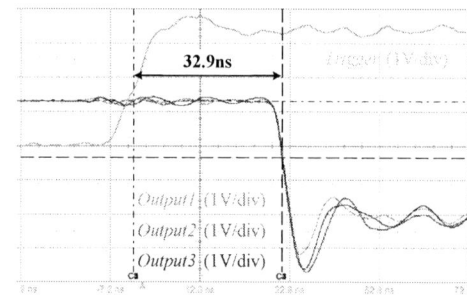

Figure 18: Fault signal response of the prototype

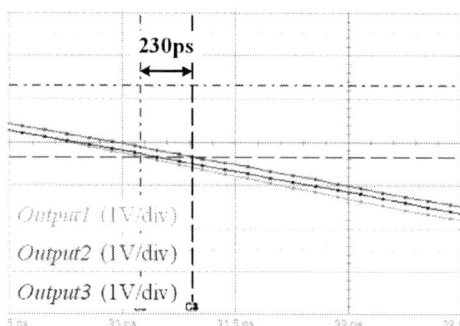

Figure 19: Output delay asynchrony between units

By integrating a TMR-based overcurrent protection device with an optical splitter-based fault transmission device, the complete enhanced overcurrent fault detection system was developed.

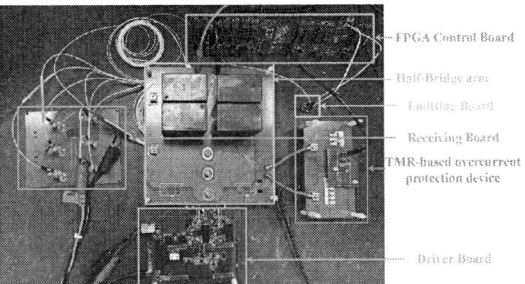

Figure 20: A short-circuit test bench equipped with the enhanced overcurrent fault detection system.

A short-circuit test bench equipped with this system was then built to test its capabilities(see Fig. 20). A short-circuit signal is issued by the DSP via the Driver board. If the short-circuit current on the half-bridge arm reaches its threshold, it triggers the TMR-based overcurrent protection device to report a fault signal to the FPGA Control Board. Subsequently, this signal is distributed to six different receiving board units through the emitting board using an optical splitter, achieving protection within 81.6ns. (see Fig. 21)

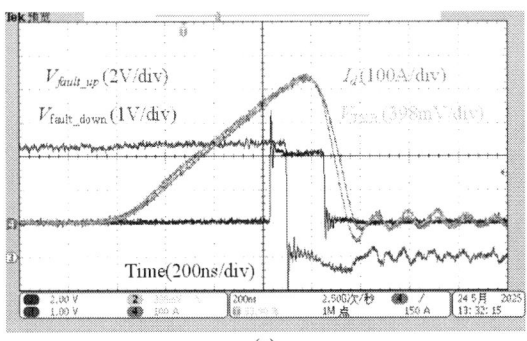

(a)

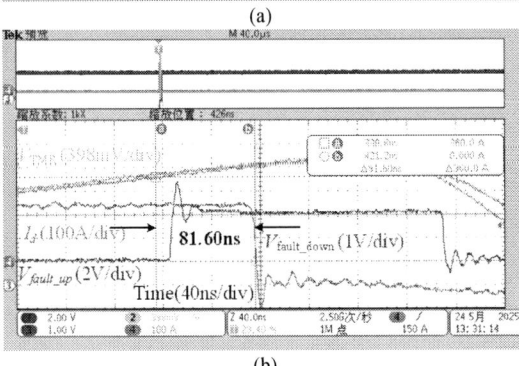

(b)

Figure 21: TMR output and current (a) Short-Circuit test (b) Delay measurement

IV. CONCLUSION

This paper demonstrates an overcurrent protection system for a 7.2kV power module using series-connected SiC MOSFETs. The TMR sensor non-invasively measures module current, with finite element simulation overcoming skin effect bandwidth limitations for a 30 ns measurement delay. The optical-splitter-based communication device issues fault responses within 33 ns, sending fault signals to six drive boards simultaneously and directly. The entire system executes a shutdown command within 81.6 ns, significantly enhancing the reliability of the 7.2kV power module using series-connected SiC MOSFETs.

ACKNOWLEDGMENT

This work was supported by the research project of China Southern Power Grid Co., Ltd. under Grant 030400KK52222006 (GDKJXM20222138).

REFERENCES

[1] V. Pala et al., "10 kV and 15 kV silicon carbide power MOSFETs for next-generation energy conversion and transmission systems," 2014 IEEE Energy Conversion Congress and Exposition (ECCE), Pittsburgh, PA, USA, 2014, pp. 449-454

[2] R. P. Zingg, "On the specific on-resistance of high-voltage and power devices," in IEEE Transactions on Electron Devices, vol. 51, no. 3, pp. 492-499

[3] J. Zhang, S. Shao, Y. Li, J. Zhang and K. Sheng, "A Voltage Balancing Method for Series-Connected Power Devices in an LLC Resonant Converter," in IEEE Transactions on Power Electronics, vol. 36, no. 4, pp. 3628-3632

[4] Z. Gao, S. Shao, W. Cui, J. Zhang, X. Chen and K. Sheng, "A Voltage Balancing Method for Series-Connected Power Devices Based on Active Clamping in Voltage Source Converters," in IEEE Transactions on Power Electronics, vol. 37, no. 9, pp. 10620-10632

[5] Y. Wu, C. Li, J. Liu and Z. Zheng, "An Ultrafast and Low-Invasive Short-Circuit Current Limiting Method by Gate Voltage Control for SiC MOSFETs With Kelvin-Source," in IEEE Transactions on Power Electronics, vol. 39, no. 12, pp. 15696-15708, Dec. 2024, doi: 10.1109/TPEL.2024.3441329.

[6] Y. Wen, Y. Yang, H. Ning, Y. Zhang and Y. Gao, "Reivew on short-circuit protection technology of SiC MOSFET", Trans. China Electrotechnical Soc., vol. 37, no. 10, pp. 2538-2548, 2022.

[7] W. Ouyang, P. Sun, M. Xie, Q. Luo and X. Du, "A Fast Short-Circuit Protection Method for SiC MOSFET Based on Indirect Power Dissipation Level," in IEEE Transactions on Power Electronics, vol. 37, no. 8, pp. 8825-8829, Aug. 2022, doi: 10.1109/TPEL.2022.3161741.

[8] Y. Feng, S. Shao, J. Du, Q. Chen, J. Zhang and X. Wu, "Short-Circuit and Over-Current Fault Detection for SiC MOSFET Modules Based on Tunnel Magnetoresistance With Predictive Capabilities," in IEEE Transactions on Power Electronics, vol. 37, no. 4, pp. 3719-3723, April 2022, doi: 10.1109/TPEL.2021.3121572.

[9] H. Chen, W. Lin, S. Shao, X. Wu and J. Zhang, "Application of Tunnel Magnetoresistance for PCB Tracks Current Sensing in High-Frequency Power Converters," in IEEE Transactions on Instrumentation and Measurement, vol. 72, pp. 1-11

[10] A. Aratake, "Field reliability of silica-based PLC splitter for FTTH," 2015 Optical Fiber Communications Conference and Exhibition (OFC), Los Angeles, CA, USA, 2015, pp. 1-3

Modeling of Static Current Sharing of Direct-Paralleled 1.7kV 800A SiC Power Modules

Zhongjie Wang
School of Electrical and Information Engineering
Tianjin University
Tianjin, China
zhongjie@tju.edu.cn

Hui Liao
School of Electrical and Information Engineering
Tianjin University
Tianjin, China
liaohui1895@tju.edu.cn

Lingxiao Xue
School of Electrical and Information Engineering
Tianjin University
Tianjin, China
xuel@tju.edu.cn

Xiangqian Zhang
Tianjin Research institute of Electric Science Co.,Ltd.
Tianjin, China
zhangxiangqian@tried.com.cn

Liwen Zou
School of Electrical and Information Engineering
Tianjin University
Tianjin, China
zouliwen@tju.edu.cn

Xiaonan Dong
School of Electrical and Information Engineering
Tianjin University
Tianjin, China
dongxiaonan@tju.edu.cn

Abstract—**High-power applications often require paralleling of power modules. Due to lower on-resistance and parasitic inductance than that of silicon IGBTs, SiC power module paralleling is found to be more challenging. This paper analyzes the impedance characteristics of three paralleled SiC modules and develops a time-domain model to quantify the parasitic's impact on current distribution. Double-pulse tests verified the model and demonstrated the parallel operation of three SiC modules with satisfactory current sharing. By optimizing power loop impedance, current-sharing deviation is reduced to less than 4% at 1100V/1200A switching conditions.**

Keywords—SiC Power Module, Current Sharing, Double Pulse Test

I. INTRODUCTION

SiC power semiconductor has emerged as a competitive technology replacing silicon IGBT counterparts, thanks to their advantageous performance such as high switching frequency and low on-resistance [1-2]. In order to expand power level, device paralleling is typically used. In this case, the steady-state and dynamic current sharing among devices constitutes one of the key research areas. Existing research on SiC paralleling primarily focuses on die-level or discrete-device-level [3]. At higher power level, multiple power modules need to be paralleled to fulfill the current handling capacity. While it is common to have silicon IGBT power modules paralleled in industry practice [4], SiC power module paralleling has been rarely reported. Due to higher switching speed, lower parasitic inductance and on-resistance, more compact system layout, it is expected that SiC paralleling at the module level will exhibit distinguishable difference in comparison with that of IGBT devices.

Unlike the current sharing issue among parallel dies within a single power module, when multiple power modules are connected in parallel, greater emphasis is placed on the impact of the parasitic impedance parameters of the overall system's power loop on current sharing among the modules. Fig. 1 illustrates a comparison of parasitic inductance parameters between IGBT-based and SiC-based systems[5-8], encompassing the inductances from the DC capacitor, power module, and the interconnecting busbar. The total parasitic inductance of the converter based on SiC modules (approximately 50 nH) is generally lower than that of the IGBT-based converter (approximately 150 nH). Moreover, as the packaging technology for SiC modules continues to advance, their parasitic inductance keeps decreasing. For example, a commercial 1700V, 800A SiC half-bridge power module depicted in Fig. 2 has a packaging parasitic inductance of nearly 5 nH. Consequently, the proportion of parasitic parameters introduced by the system's mechanical structure further increases, emerging as a significant factor influencing the current sharing of parallel modules.

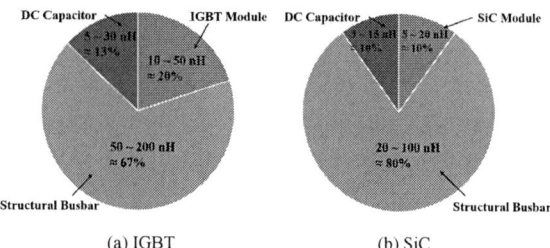

(a) IGBT   (b) SiC

Fig. 1. Parasitic impedance differences between IGBT and SiC systems

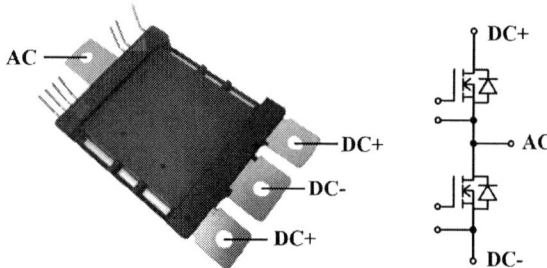

Fig. 2. 1700V, 800A SiC half-bridge power module

To investigate the current-sharing effect of directly paralleled SiC modules, this paper models the impedance parameters of the parallel circuit under steady-state conduction conditions and elucidates the impact of each impedance parameter on the steady-state current behavior of each module. The current-sharing effect among paralleled modules is experimentally verified through a double-pulse test. By reasonably controlling the impedance parameters of the power loop, the steady-state current sharing error can be maintained within 4% under the conditions of a 1100V switching voltage and a 1200A switching current.

II. MODELING OF POWER LOOP IMPEDANCE

Fig. 3 illustrates the steady-state conduction current paths of three paralleled SiC half-bridge modules in a double-pulse test circuit. The currents flowing through the lower switches of the three modules are i_1, i_2, and i_3, respectively. Considering the influence of parasitic parameters on the return path of the DC-side current, Fig. 4 presents a simplified lumped-circuit model. Here, L_1, L_2, L_3, and L_4 represent the lumped parasitic inductances of each parallel module and the current return path. R_1, R_2, R_3, and R_4 represent the lumped parasitic resistances of their corresponding paths. M_{12}, M_{13}, M_{14}, M_{23}, M_{24}, and M_{34} represent the mutual-inductance parameters between each path.

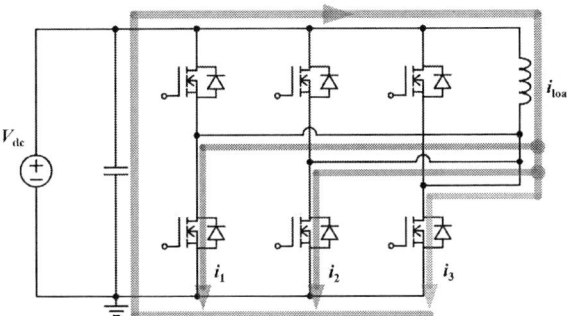

Fig. 3. Current paths of three-paralleled SiC half-bridge modules

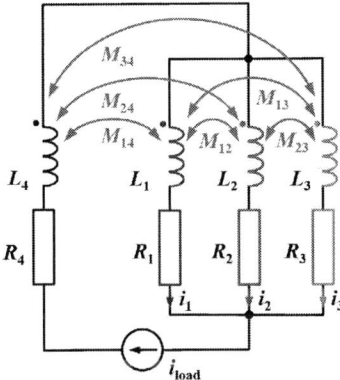

Fig. 4. Simplified lumped - circuit model

Utilizing the Laplace transform, the node-voltage equations and loop-current equations of the circuit loop depicted in Fig. 4 are presented in equations (1)-(3).

$$sL_1i_1 + sM_{12}i_2 + sM_{13}i_3 - sM_{14}(i_1 + i_2 + i_3) + R_1i_1 = \\ sL_2i_2 + sM_{12}i_1 + sM_{23}i_3 - sM_{24}(i_1 + i_2 + i_3) + R_2i_2 \quad (1)$$

$$sL_2i_2 + sM_{12}i_1 + sM_{23}i_3 - sM_{24}(i_1 + i_2 + i_3) + R_2i_2 = \\ sL_3i_3 + sM_{13}i_1 + sM_{23}i_2 - sM_{34}(i_1 + i_2 + i_3) + R_3i_3 \quad (2)$$

$$i_1 + i_2 + i_3 = k / s^2 \quad (3)$$

Here, k denotes the rate of change of the total loop current i_{load}, and s represents the Laplace operator.

$$i_1(t) = \frac{R_2R_3kt}{R_1R_2 + R_1R_3 + R_2R_3} + C_1 + A_1e^{(\lambda+\gamma)t} + B_1e^{(\lambda-\gamma)t} \quad (4)$$

$$i_2(t) = \frac{R_1R_3kt}{R_1R_2 + R_1R_3 + R_2R_3} + C_2 + A_2e^{-Dt} \quad (5)$$

$$i_3(t) = \frac{R_1R_2kt}{R_1R_2 + R_1R_3 + R_2R_3} + C_3 + A_3e^{Qt} \quad (6)$$

By applying the inverse Laplace transform, the time-domain expressions for the currents in the three paralleled modules can be derived, as shown in equations (4)-(6). Among them, A_1, A_2, A_3, B_1, C_1, C_2, C_3, D, Q, λ, and γ are constant coefficients related to the impedance parameters depicted in Fig. 4.

From the results in equations (4)-(6), the time-domain current model reveals three key observations. Firstly, a linearly increasing term related to the parasitic resistances of the three parallel branches indicates the impact of parasitic resistances on current distribution among the modules. Conversely, the exponentially varying term reveals the influence of resistances, self-inductances, and mutual-inductances among the conducting paths on the current. Once the mechanical structure is determined, the constant coefficients associated with the exponential term are uniquely determined. The sign (positive or negative) of these coefficients determines whether the exponential terms will diverge or converge over time. Lastly, the constant terms in the formula represent the initial conditions of the current expressions.

III. SIMULATION AND EXPERIMENTAL VERIFICATION

Fig. 5 illustrates the double-pulse test setup, which comprises three paralleled 1700V, 800A SiC modules, a DC capacitor busbar, an AC busbar, and a power inductor. Fig. 6 depicts the structure of the parallel drive board for the three SiC modules. Modules 1, 2, and 3 are horizontally arranged in sequence. The overall design employs a centralized drive mode.

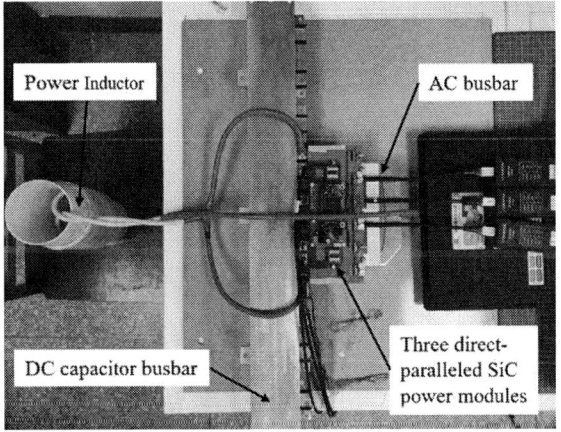

Fig. 5. Double pulse test setup with a symmetrical loop structure

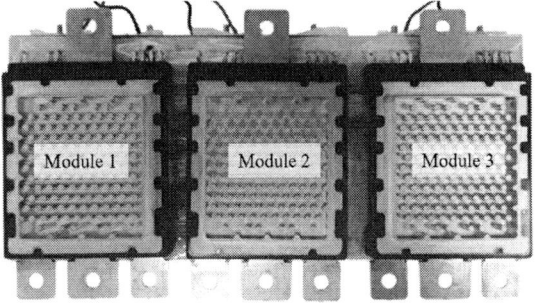

Fig. 6. Direct-parallel connection of three SiC modules

To verify the accuracy of the current time-domain model, the parasitic impedance parameters of the power loop are first extracted using the Q3D software. Fig. 7 illustrates the structural models of the DC and AC busbars. To balance the equivalent impedance among the three ports of the AC busbar, a cutout is introduced at its middle section. The Q3D simulation results indicate that each port has a parasitic resistance of approximately 0.1 mΩ and a parasitic inductance of 20 nH.

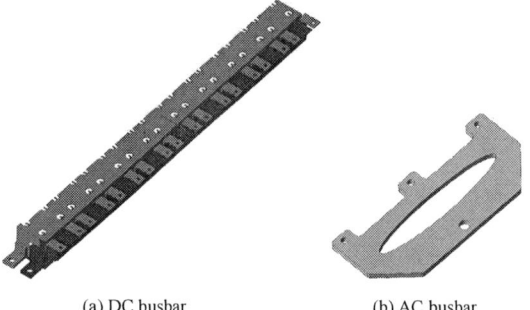

(a) DC busbar   (b) AC busbar

Fig. 7. Structural simulation model in Q3D

The parasitic parameters of the SiC half-bridge power module are measured using an impedance analyzer. The parasitic inductance is 6.4 nH, and the on-state resistance is 7 mΩ. The equivalent parasitic inductance of the DC capacitor is 20 nH, and the parasitic resistance is 1 mΩ. Based on these measurements, the specific parameter values of the simplified parallel impedance circuit depicted in Fig. 4 are obtained, thereby determining the specific numerical values of equations (4)-(6). Fig. 8 presents the theoretical calculations and Simplorer simulation results for the current distribution among the three parallel modules.

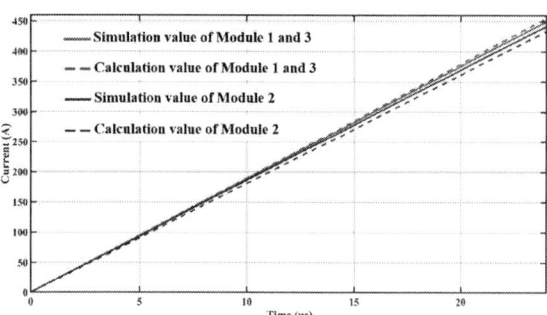

Fig. 8. Comparison between the theoretical model and simulation

As shown in Fig. 8, when the three paralleled SiC modules are arranged horizontally, Module 1 and 3 on the left and right sides have nearly identical impedance parameters due to geometric symmetry. Theoretically, the conducting currents of Modules 1 and 3 are the same. However, for Module 2 in the middle position, its equivalent parasitic resistance is larger than that of Modules 1 and 3, resulting in a smaller current compared to the currents on the two sides. Additionally, after substituting the extracted parasitic parameters into equations (4)-(6), the coefficients of the exponential terms are all negative. As time progresses, the exponential terms gradually decay, and ultimately the current distribution is primarily influenced by the linearly increasing terms dominated by the on-state resistance. There is a certain deviation between the theoretical calculation and the simulation. One of the main reasons is that there is a certain deviation between the parasitic parameters extracted by Q3D and the simplification in the equivalent circuit model.

To verify the influence of the loop parasitic impedance parameters on current sharing, Fig. 9 illustrates a double-pulse test setup with an asymmetrical loop structure. Unlike the setup in Fig. 5, the cable connecting the AC busbar and the power inductor exhibits significant asymmetry. Fig. 10 presents the double-pulse test waveforms under the condition of 1100V and 1200A. With this asymmetrical loop structure, the turn-off currents of Modules 1, 2, and 3 are 540 A, 380 A, and 340 A, respectively. The absolute current-sharing error among the modules reaches 200 A, severely affecting the current stress balance among the modules.

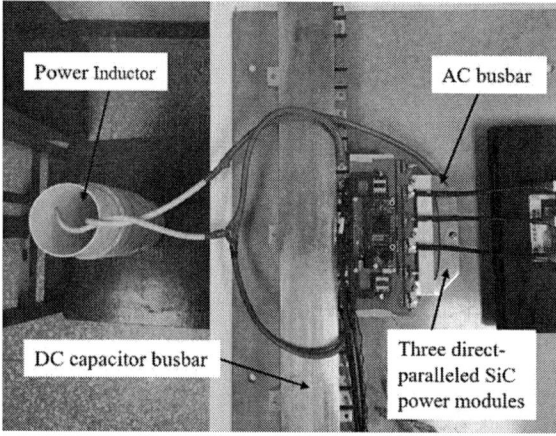

Fig. 9. Double pulse test setup with an asymmetrical loop structure

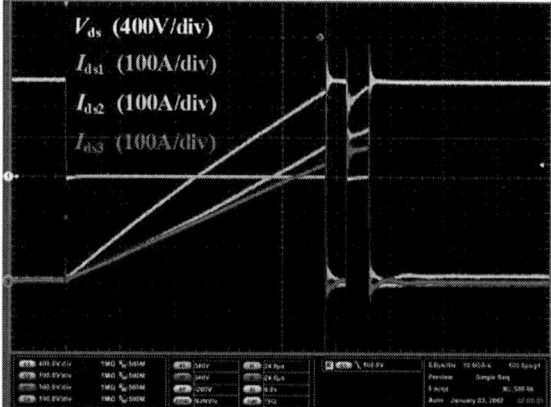

Fig. 10. DPT waveforms at 1100V, 1200A with asymmetrical structure

For comparison, Fig. 11 presents the double-pulse test waveforms under a symmetrical loop structure, as shown in Fig. 5. Under the same condition of 1100V and 1200A, Modules 1 and 3 exhibit better current-sharing performance, with their turn-off currents both approximately 428 A. The steady-state current value of Module 2 in the middle position is relatively smaller, at approximately 404 A. The experimental results are consistent with the theoretical and simulation analyses. Fig. 12 shows the corresponding turn-on and turn-off processes. It can be seen that Modules 1 and 3 exhibit relatively high dynamic current sharing consistency, while the deviation in the dynamic current of Module 2 is primarily determined by the difference in steady-state current sharing.

Equation (7) defines the calculation method for the current-sharing deviation among the three SiC modules.

979-8-3315-1110-4/25 $31.00 © 2025 IEEE

$$\Delta I_x = (I_x - I_{ave}) / I_{ave}, \quad I_{ave} = (\sum_{x=1}^{3} I_x) / 3 \qquad (7)$$

Table 1 presents a comprehensive analysis of the current-sharing error among three paralleled SiC modules under the condition of 1100 V and 1200 A.

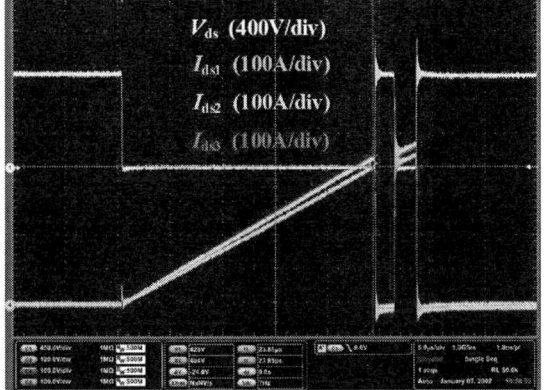

Fig. 11. DPT waveforms at 1100V, 1200A with symmetrical structure

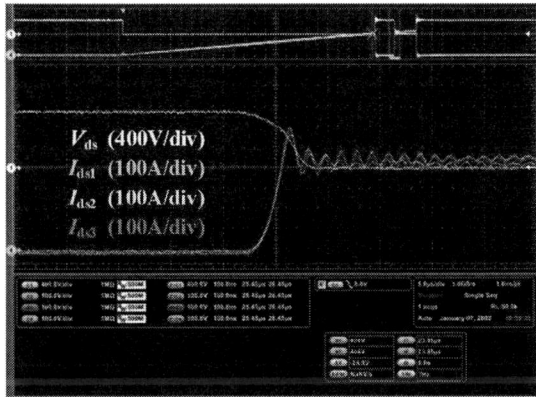

(a) Turn on process

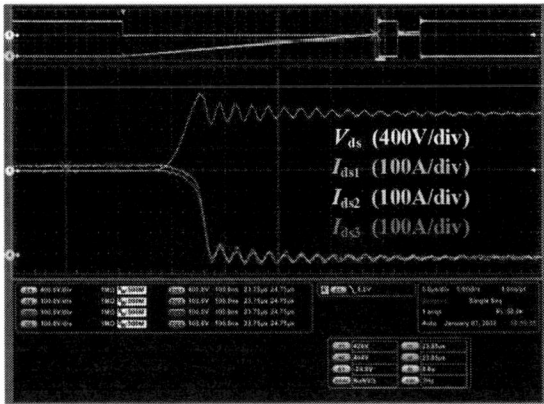

(b) Turn off process

Fig. 12. Turn-on and turn-off processes at 1100V, 1200A

TABLE I. COMPREHENSIVE ANALYSIS OF CURRENT-SHARING ERROR

	Calculation	Simulation	Experiment
Module 1	1.5 %	0.6 %	1.9 %
Module 2	3.1 %	1.16 %	3.8 %
Module 3	1.5 %	0.58 %	1.9 %

As shown by the experimental results, for three paralleled SiC modules, when the parasitic impedance of the power loop is symmetrically controlled, the steady-state current-sharing error among the paralleled modules can be maintained within 4%, thereby achieving a relatively good current balance among the three paralleled modules. It is worth noting that further analysis of the circuit's impedance model and optimization of the mechanical structure are still required for Module 2 to further reduce its current-sharing error compared to Modules 1 and 3.

IV. CONCLUSIONS

In this paper, the current-sharing problem among three directly paralleled SiC half-bridge modules is analyzed through mathematical modeling of the loop impedances. With extracted parasitic parameters, the derived model can predict the current-sharing behavior with high accuracy. Finally, the double-pulse test experiment shows that when the power loop is symmetrical, good current sharing among the three SiC modules can be achieved. Under a switching voltage of 1100 V and a switching current of 1200 A, the current-sharing deviation among the three SiC power modules is controlled within 4%, achieving a satisfactory current-sharing performance.

REFERENCES

[1] Y. Cao, M. Ngo, N. Yan, D. Dong, R. Burgos and A. Ismail, "Design and Implementation of an 18-kW 500-kHz 98.8% Efficiency High-Density Battery Charger With Partial Power Processing," in IEEE Journal of Emerging and Selected Topics in Power Electronics, vol. 10, no. 6, pp. 7963-7975, Dec. 2022, doi: 10.1109/JESTPE.2021.3108717.

[2] F. Diao et al., "A Megawatt-Scale Si/SiC Hybrid Multilevel Inverter for Electric Aircraft Propulsion Applications," in IEEE Journal of Emerging and Selected Topics in Power Electronics, vol. 11, no. 4, pp. 4095-4107, Aug. 2023, doi: 10.1109/JESTPE.2023.3266197.

[3] Y. Mao, Z. Miao, C. -M. Wang and K. D. T. Ngo, "Balancing of Peak Currents Between Paralleled SiC MOSFETs by Drive-Source Resistors and Coupled Power-Source Inductors," in IEEE Transactions on Industrial Electronics, vol. 64, no. 10, pp. 8334-8343, Oct. 2017, doi: 10.1109/TIE.2017.2716868.

[4] J. Cai, X. Du, L. Zhou, P. Sun, X. Du and J. Zhang, "Investigation on the Influence Factors of Thermal Resistance Monitoring of Parallel Connected IGBT Modules," 2020 IEEE 9th International Power Electronics and Motion Control Conference (IPEMC2020-ECCE Asia), Nanjing, China, 2020, pp. 3424-3427, doi: 10.1109/IPEMC-ECCEAsia48364.2020.9367898.

[5] K. Takao and S. Kyogoku, "Ultra low inductance power module for fast switching SiC power devices," 2015 IEEE 27th International Symposium on Power Semiconductor Devices & IC's (ISPSD), Hong Kong, China, 2015, pp. 313-316, doi: 10.1109/ISPSD.2015.7123452.

[6] Y. Yan et al., "A Novel Double-Sided Cooling Silicon Carbide Power Module With Ultralow Parasitic Inductance Based on an Interleaved Power Loop," in IEEE Transactions on Power Electronics, vol. 39, no. 10, pp. 12570-12588, Oct. 2024, doi: 10.1109/TPEL.2024.3410509.

[7] Z. -F. Li, N. Nishida, H. Aoki, H. Shibata, C. -C. Liao and M. -S. Huang, "Simulation-Assisted Design of a Power Stack for Improving Static Current Sharing Among Three IGBT Modules Connected in Parallel," in IEEE Access, vol. 10, pp. 10079-10093, 2022, doi: 10. 1109/ ACCESS.2022.3144575.

[8] G. Zou, Z. Zhao and L. Yuan, "Study on DC busbar structure considering stray inductance for the back-to-back IGBT-based converter," 2013 Twenty-Eighth Annual IEEE Applied Power Electronics Conference and Exposition (APEC), Long Beach, CA, USA, 2013, pp. 1213-1218, doi: 10.1109/APEC.2013.6520453.

A 500 W 13.56 MHz Amplifier Realized by Two Combining Class-E Circuits

Yanfei Ji
College of Electrical and Control Engineering
North China University of Technology
Beijing, China
Jiyanfei123@outlook.com

Mei Liang
College of Electrical and Control Engineering
North China University of Technology
Beijing, China
Liangmei@ncut.edu.cn

Jiwen Chen
College of Electrical and Control Engineering
North China University of Technolog
line 4: Beijing, China
Chenjiwen@ncut.edu.cn

Pengyu Jia
College of Electrical and Control Engineering
University of Technology
Beijing, China
line 4: Beijing, China
08117338@bjtu.edu.cn

Yihang Zhang
College of Electrical and Control Engineering
North China University of Technology
Beijing, China
1763794465@163.com

Abstract—**This paper presents a 500 W,13.56 MHz power amplifier, which uses a simple method to combine the output power from two Class-E circuits based on GaN HEMT devices. The output combining circuits are implemented by the resonated inductors and capacitors, the principle and design method of which are presented. The simulation verification for the 500 W, 13.56 MHz power amplifier is made based on Plecs software. Finally, an experimental prototype is built to verify the effectiveness of this power amplifier realized by two combining Class-E using direct power circuits.**

Keywords—Class-E circuit, combining method, GaN

I. INTRODUCTION

Gallium nitride (GaN) has superior physical properties compared to silicon (Si) material. GaN has a relatively wide band gap, lower intrinsic carrier concentration, higher electron saturation velocity, and higher thermal conductivity [1]. Therefore, the GaN power devices have the low on-resistance, the small parasitic capacitors, and the short switching time, which results in high efficiency and high power density, The designs of MHz power amplifiers based on GaN power devices have been introduced in many literature [2-4]. The literature [5] designed and fabricated a prototype of a Class-E power amplifier with 200 W and 13.56 MHz, and its efficiency is 81%. The output power of a single-channel Class-E power amplifier cannot reach the target, so power combining is required to obtain a sufficiently large output power. At present, transformers are widely used for power combining [6-7], but they have the disadvantages of additional core losses and an increase in the overall system size. In this work, we propose another method to solve the problem of power combining. This method eliminates the additional core losses associated with the power combining network and reduces the system size and weight.

In this paper, the conditions that the resonant components of the LC power combining network need to meet and the parameter calculation methods are introduced in detail. The simulation is set up in Chapter Three to verify the correctness of the design method. In Chapter Four, a 500

This work was sponsored by the Beijing Natural Science Foundation under Grant 3232044, and sponsored by the Youth Research Special Project of NCUT (Project No.2025NCUTYRSP003).

W, 13.56 MHz power amplifier is presented, which combines the output power from two GaN Class-E circuits.

II. DESIGN METHODOLOGY

A. Inductor-Capacitor Power Synthesis Network Research

The circuit of the inductance-capacitance power combining network is shown in Fig. 1. Among them, the parallel capacitors C_{p1}, C_{p2}, the series resonant networks C_{s1}, L_{s1}, C_{s2}, L_{s2} and the extra inductors L_{ext1}, L_{ext2} constitute the Class-E power amplifier network. L_{com1}, C_{com1}, L_{com2}, C_{com2} are the power combining networks, and R_{comp} is the rated load at the port of the two-way Class-E power amplifier power combining network.

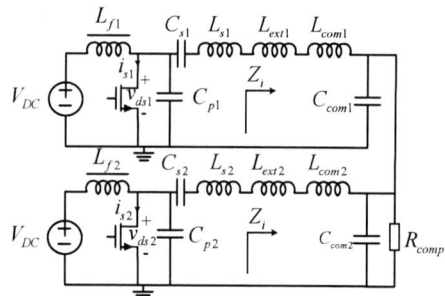

Fig. 1. Two-way LC resonant power combining circuit schematic

The basic power unit of this paper is a Class-E power amplifier, and a power combining circuit with a resonant network is used to directly combine the output power of two Class-E power amplifiers. This scheme needs to satisfy:

(1) The capacitor C_{com1}, inductor L_{com1}, and the extra inductor L_{ext1} in the Class-E power amplifier resonate at the switching frequency point, which is expressed by the following equation:

$$X_{Ccom1} = X_{Lcom1} + X_{Lext1} \tag{1}$$

In the formula, X_{Ccom1} is the impedance of capacitor C_{com1}, X_{Lcom1} is the impedance of inductor L_{com1}, and X_{Lext1} is the impedance of inductor L_{ext1}.

(2)The equivalent impedance seen from the output port of the resonant network C_{s1}、L_{s1} of the single-channel Class-E power amplifier towards the load side is:

$$Z_i = R_i + j\omega L_{ext1}$$

$$= \frac{R_{KL}}{1+\omega^2 C_{com1}^2 R_{KL}^2} + j\left(\omega L_{sum} + \frac{\omega C_{com1} R_{KL}^2}{1+\omega^2 C_{com1}^2 R_{KL}^2}\right) \quad (2)$$

In the formula, R_{KL} is the equivalent resistance seen from the output end of the single-channel power combining network towards the load side, R_i is the rated load resistance of the single-channel Class-E power amplifier, and L_{sum} is the sum of L_{ext1} and L_{com1}. The second Class-E power amplifier also satisfies the above conditions.

Based on the above circuit characteristics, for simplified analysis, when the first channel is operating, the fundamental wave of the drain voltage of the switching device S_1 is regarded as an AC voltage source v_1, as shown in Fig. 2. According to the superposition theorem, the voltage source of the second channel is considered a short circuit, and the second channel is regarded as the load of the first channel. Since the circuit parameters satisfy the resonance of capacitor C_{com1}, inductor L_{com1}, and extra inductor L_{ext1} at the switching frequency point, and the resonance of capacitor C_{s1} and inductor L_{s1} at the switching frequency, the impedance of the gray network is infinite (equivalent to an open circuit). The same logic applies when the second channel is operating.

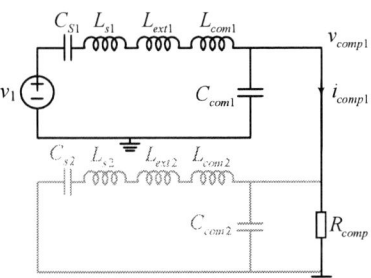

Fig. 2. Power combining circuit with v_1 acting alone

Therefore, the output current i_{comp1} of the first channel will all flow to the load. Similarly, the output current i_{comp2} of the second channel will also all flow to the load. After combining the power of the two channels, the expressions for the current and voltage of the load are as follows:

$$i_{comp} = i_{comp1} + i_{comp2} \quad (3)$$

$$v_{comp} = (i_{comp1} + i_{comp2})R_{comp} \quad (4)$$

After applying the superposition theorem, the current flowing into the load R_{comp} increases after the superposition of the two channels, while its voltage remains unchanged. Therefore, the output impedance after power combining will decrease, as shown in (5).

$$R_{comp} = \frac{1}{2}R_{KL} \quad (5)$$

According to the circuit analysis of the LC-based power combining scheme and the conditions to be satisfied, by combining (1), (2), and the calculation equations of Class-E power amplifiers, it can be derived that:

$$C_{com1} = \frac{1}{1.152\omega R_{KL}} \quad (6)$$

$$R_i = \frac{R_{KL}}{1.753} \quad (7)$$

$$L_{com1} = \frac{1}{\omega^2 C_{com1}} - L_{ext1} \quad (8)$$

Where L_{ext} is determined by the design equations of the Class-E power amplifier. The calculation for the second power combining network follows the same logic.

B. Impedance Matching Network Research

Since the load R_L is 50Ω and the rated load for the two-way power combined output is 2Ω, it is necessary to add an impedance matching network to match 50Ω to 2Ω. As shown in Fig. 3.

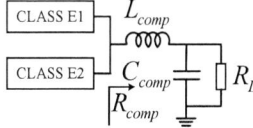

Fig. 3. Schematic Diagram of Impedance Matching

As shown in Fig. 4, the two-element L-type matching network is the simplest and most widely used impedance matching circuit. The L-type matching network has two structures. When the load is a resistive load, it can be designed through the following equations.

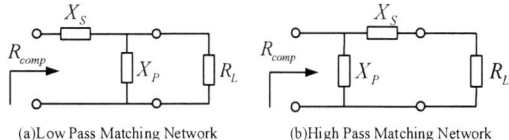

(a)Low Pass Matching Network   (b)High Pass Matching Network

Fig. 4. The diagram of high pass and low pass matching networks

When $R_L > R_{comp}$, for the L-type impedance matching using Structure (a), the quality factor Q_s of the series part and the quality factor Q_p of the parallel part are calculated using (9).

$$Q_S = Q_P = \sqrt{\frac{R_L}{R_{comp}} - 1} \quad (9)$$

When $R_L < R_{comp}$, the L-type impedance matching uses Structure (b). Calculate Q_s and Q_p using (10).

$$Q_S = Q_P = \sqrt{\frac{R_{comp}}{R_L} - 1} \quad (10)$$

After calculating Q_s and Q_p, the values of the reactive components can be determined using (11) and (12).

$$X_S = Q_S \cdot R_{comp} \quad (11)$$

$$X_P = \frac{R_L}{Q_P} \quad (12)$$

In the equations:
- R_L denotes the load resistance,
- X_P represents the parallel matching reactance,
- R_{comp} signifies the source matching resistance,
- X_S stands for the series matching reactance.

Here, X_P and X_S can each be either a capacitor or an inductor, but they must be of opposite types. Specifically, if X_P is a capacitor, X_S must be an inductor; conversely, if X_P is an inductor, X_S must be a capacitor. This complementary configuration ensures effective impedance matching.

III. SIMULATION AND ANALYSIS

A. Simulation Parameter Design

The parameter design of the single-channel Class-E power amplifier is calculated based on an input voltage of 30V and an output power of 260W. According to the design formulas (13), (14), (15) , (16) and (17) for Class-E power amplifiers, the values of each parameter in the Class-E power amplifier can be obtained[8].

$$R_i = \frac{8V_{DC}^2}{(\pi^2+4)P_o} \tag{13}$$

$$L_s = \frac{R_i(Q\text{-}1.1525)}{\omega} \tag{14}$$

$$L_{ext} = \frac{1.1525R_i}{\omega} \tag{15}$$

$$C_p = \frac{8}{\pi(\pi^2+4)\omega R_i} \tag{16}$$

$$C_s = \frac{1}{\omega R_i(Q\text{-}1.1525)} \tag{17}$$

The value of the quality factor Q generally ranges between 2 and 10. The curve of voltage gain plotted using MATLAB simulation software is shown in Fig. 5. A larger Q value implies better frequency selection characteristics of the circuit, which can more effectively filter signals of a specific frequency while strongly suppressing interference signals of other frequencies. For ease of calculation, ($Q = 9.152$) is adopted in this paper.

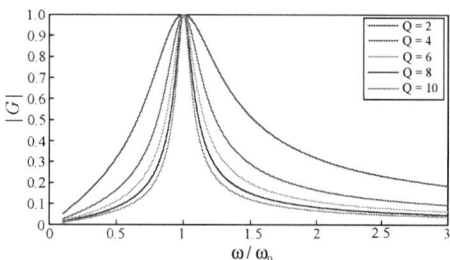

Fig. 5. Series Resonance Voltage Gain Curve

All design parameters of the circuit can be obtained according to the design formulas of the Class-E power amplifier in Section 2, and their values are shown in TABLE I.

TABLE I. Parameters of resonant components

Device designator	Value	Device designator	Value
R_i	2 Ω	C_{com1}、C_{com2}	2904 pF
C_{p1}、C_{p2}	1078 pF	L_{com1}、L_{com2}	20 nH
C_{s1}、C_{s2}	734 pF	C_{comp}	1232.59 pF
L_{s1}、L_{s2}	188 nH	L_{comp}	107.85 nH
L_{ext1}、L_{ext2}	27 nH	R_L	50 Ω

B. Simulation Analysis of Two-Channel Class-E Power Amplifier Combined Output

Use PLECS simulation software to carry out simulation analysis on the combined output of a 500W two-channel Class E power amplifier. The simulation circuit diagram is shown in Fig. 6.

In the simulation circuit diagram of the two-channel Class E power amplifier combined output in Fig. 6, the switching device uses the GS66516B model from the PLECS official website. A small resistor is connected in series with the capacitor C_{p1}、C_{p2} branch to avoid numerical calculation convergence issues caused by ideal capacitors during simulation.

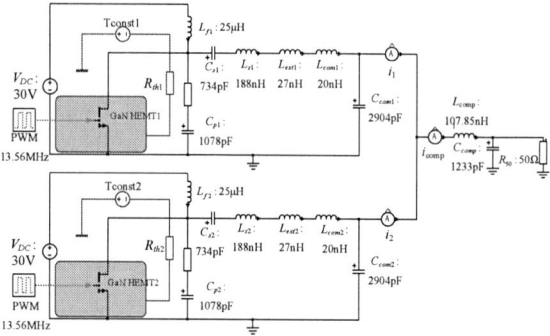

Fig. 6. Simulation circuit diagram of two-channel Class E PA combiner

The input DC voltage V_{DC} of the two-channel Class E power amplifiers is 30V. The voltage and current of key nodes in the main circuit are shown in Fig. 7, which presents the driving signal of the switching tube (GaN HEMT), the drain voltages v_{ds1}, v_{ds2}, and the waveform of the voltage v_{out} across R_{50}.

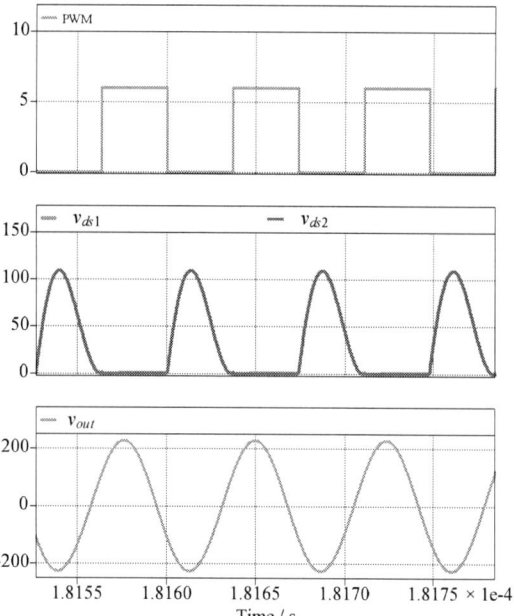

Fig. 7 Simulation results

As shown in the simulation results of Fig. 7, the voltage waveforms of the switching tubes (v_{ds1} and v_{ds2}) in the two-channel Class E power amplifiers are identical and consistent with the ideal waveforms of Class E power amplifiers, both meeting the conditions of ZVS (Zero Voltage Switching) and ZVDS (Zero Voltage Derivative Switching), with the output voltage being a standard sine wave. The amplitude of the drain voltage of the switching device is 107V, which is consistent with the theoretically calculated voltage amplitude of 3.55 V_{DC}; the peak value of the output voltage v_{out} is 228.1V, and the output power is 520W. Under the condition of a 30V input, the output power of each amplifier is 260W, which is consistent with the calculation of the single-channel Class E power amplifier in Chapter 3. The simulation results verify the correctness of the scheme.

Fig. 8 shows the simulation results of the power combiner section. Here, i_1 represents the current measured by ammeter i_1 in the main circuit simulation (Fig. 6), i_2 corresponds to the reading of ammeter i_2, and i_{comp} denotes the current from ammeter i_{comp}. The voltages across capacitors C_{com1} and C_{com2} are the output voltages v_{com1} and v_{com2} of the single-channel power combining networks, respectively. The waveforms of i_1, i_2, i_{comp}, v_{com1} and v_{com2} are presented in Figure 8. Evidently, the combined current i_{comp} equals the sum of the single-channel output currents i_1 and i_2, while the single-channel output voltages v_{com1} and v_{com2} are identical. These simulation results validate (3) and (4).

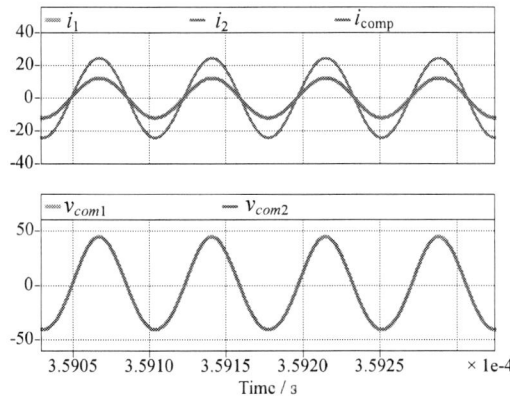

Fig. 8. Simulation results of power combining

IV. EXPERIMENTAL RESULTS

In this paper, based on GaN devices and using an inductive-capacitive resonant power combining method, a radio-frequency (RF) power supply with a switching frequency of 13.56 MHz and an output power of 500 W is designed and fabricated. Fig. 9 shows the physical diagram of the RF power supply manufactured using GaN devices.

Fig. 9. 500 W, 13.56 MHz amplifier Circuit

A voltage-controlled oscillator (LTC1799) is used to generate a 13.56 MHz driving signal. The load R_{50} is a 50 Ω high-power RF load SQ-SJ21W, which has a characteristic impedance of 50 Ω and is suitable for a wide frequency range covering 0–3 GHz. It can withstand a power of 500 W.

Fig. 10 shows the waveform of the driving signal generated by the driver chip (LMG1020), with an amplitude of 5.5V and relatively stable shape. Although minor oscillations exist during the high-level period, they have little impact on the driving circuit. A larger amplitude of the driving voltage reduces the switching loss of the switching tube, so the gate driving voltage should be as high as possible. However, since the maximum voltage withstand capability of the driver chip LMG1020 is 6V, the amplitude of the driving waveform is set to 5.5V in this paper to prevent the driver chip from being broken down.

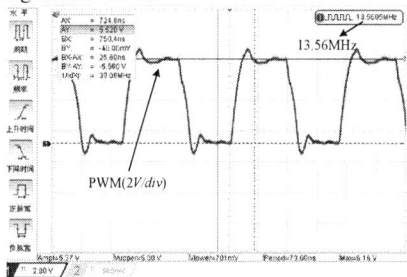

Fig. 10. Driving Waveform

Fig. 11 shows the measured drain voltage waveforms v_{ds1} and v_{ds2} of the switching tubes in two Class E power amplifier units when the input voltage is 42V. The drain voltages exhibit the characteristics of soft-switching Class E waveforms and are mutually dual. However, after the capacitor C_p completely releases its stored voltage, the reverse channel of the GaN device turns on, resulting in slight differences from the ideal waveforms. Nevertheless, the circuit still achieves a high efficiency of 86% at an output power of 502W.

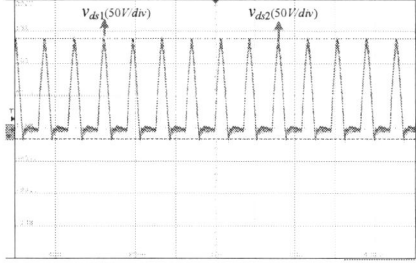

Fig. 11. Switches' drain voltages in two circuits

Fig. 12 shows that when the input voltage is 42V, the input power is 598W. The output voltage v_{out} across the load R_{50} has an amplitude of 224V, a frequency of 13.56MHz, and an output power of 502W. As shown in Fig. 13, the temperatures of the two GaN chips GS66516B are 102.4°C.

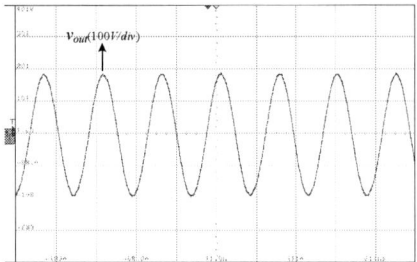

Fig. 12. Load voltage v_{out}

Fig. 14 shows the power output characteristics, where the horizontal axis represents the DC input voltage V_{DC} and the vertical axis represents the output power. Fig. 15 is the efficiency curve of the combined output of the two-channel Class E power amplifiers. As the input voltage increases, the output power continues to rise. When the input voltage is 42V, the maximum output power reaches 502W with an efficiency of 86%.

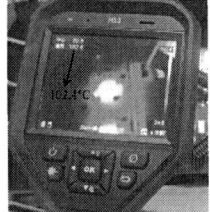

Fig. 13. Switching Tube Temperature

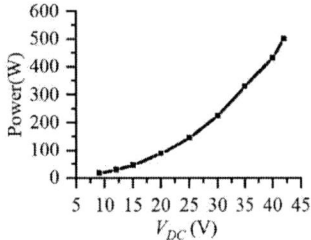

Fig. 14. Power output characteristics

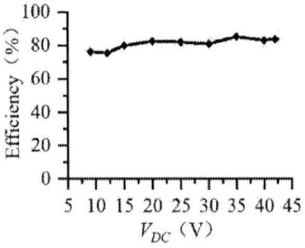

Fig. 15. Output efficiency curve

V. CONCLUSIONS

This paper presents a power combining design method for high frequency Class-E inverters based on GaN devices. An impedance matching network is used to convert the load into the rated load of the Class-E inverter. PLECS is utilized to simulate and verify the main circuit parameters. Finally, with GaN as the basis and Class-E power amplifiers as units, a 13.56 MHz Class-E inverter prototype is built. The prototype can achieve a maximum power of 500 W and an efficiency of 86% at rated output.

REFERENCES

[1] J. Millán, P. Godignon, X. Perpiñà, A. Pérez-Tomás and J. Rebollo, "A Survey of Wide Bandgap Power Semiconductor Devices," in IEEE Transactions on Power Electronics, vol. 29, no. 5, pp. 2155-2163, May 2014.

[2] T. A. Filipek, "Design and optimization of high efficiency GaN HEMT class-E power amplifiers," TENCON 2015 - 2015 IEEE Region 10 Conference, Macao, China, 2015, pp. 1-4.

[3] M. Okamoto, T. Tanaka, K. Matuzaki, T. Hashizume and H. Yamada, "13.56-MHz Class-E RF power amplifier using normally-on GaN HEMT," IECON 2014 - 40th Annual Conference of the IEEE Industrial Electronics Society, Dallas, TX, USA, 2014, pp. 982-987.

[4] M. Kim and J. Choi, "High-Frequency Self-Driven Push-Pull Class E Rectifier using Capacitive Voltage Divider," 2021 IEEE 22nd Workshop on Control and Modelling of Power Electronics (COMPEL), Cartagena, Colombia, 2021, pp. 1-6.

[5] F. H. Raab, "200-W 13.56-MHz Class-E PA with Gate-Driver ICs," 2023 IEEE/MTT-S International Microwave Symposium - IMS 2023, San Diego, CA, USA, 2023, pp. 705-708.

[6] A. Al Bastami, H. Zhang, A. Jurkov, A. Radomski and D. Perreault, "Comparison of Radio-Frequency Power Architectures for Plasma Generation," 2020 IEEE 21st Workshop on Control and Modeling for Power Electronics (COMPEL), Aalborg, Denmark, 2020, pp. 1-8.

[7] H. Zhang, A. Al Bastami, A. S. Jurkov, A. Radomski and D. J. Perreault, "Multi-Inverter Discrete Backoff: A High-Efficiency, Wide-Range RF Power Generation Architecture," 2020 IEEE 21st Workshop on Control and Modeling for Power Electronics (COMPEL), Aalborg, Denmark, 2020, pp. 1-8.

[8] E. Court and V. Balyan, "A High Efficiency Class-E Power Amplifier with Tuned Output Network," 2023 9th International Conference on Signal Processing and Communication (ICSC), NOIDA, India, 2023, pp. 17-21.

A Method to Enlarge the Soft-switching Range for the Dual Active Bridge Series Resonant Converter

Pengyu Jia
College of Electrical and Control Engineering
North China University of Technology
Beijing, China
jiapengyu@ncut.edu.cn

Yimei Xing
College of Electrical and Control Engineering
North China University of Technology
Beijing, China
2023312080119@mail.ncut.edu.cn

Kai Qiu
Beijing HyperStrong Technology Corporation
Beijing, China
qiukai@hyperstrong.com

Mingjun Liu
College of Electrical and Control Engineering
North China University of Technology
Beijing, China
2023312080102@mail.ncut.edu.cn

Abstract—The dual-active-bridge series-resonant-converter (DAB-SRC) has been widely used due to its ability to achieve natural synchronous rectification. At light load conditions, the small turn-off current leads to long charging and discharging times of the switching junction capacitor, making it difficult to achieve soft switching. In order to extend the range of soft switching for light load conditions, a very high switching frequency can be used to increase the harmonic component. This approach often results in an extremely wide switching frequency range. In order to extend the soft-switching region for very light loads and reduce the range of switching frequency without increasing the number of devices, the effects of magnetizing inductance and transformer turns ratio on the soft-switching region are considered, and a hybrid control strategy are proposed. In this paper, variable frequency phase shift (VFPS) control strategy is adopted at 20%~100% load and constant frequency phase shift (CFPS) is adopted at 5%~20% load. The feasibility and correctness of the proposed control strategy are verified by a 1.2kW experimental prototype.

Keywords—*Dual-active-bridge, series-resonant-converter, soft switching, variable frequency control, phase-shifted control*

I. INTRODUCTION

Due to the high voltage drop of the reverse conduction of the enhanced GaN (E-GaN) switch, generally around 2V-4V, designers often adopt the complementary driving pulse for the bridge leg when using the E-GaN switch as the power switches in the converter. As known, this modulation method is always employed in a dual active bridge (DAB). Since a series resonant converter (SRC) can be controlled as a DAB by replacing the secondary-side diodes with switches, which can realize a natural synchronous rectification in this way, DAB-SRC becomes more and more popular [1-4].

Variable frequency phase shift (VFPS) and constant frequency phase shift (CFPS) are two conventional control strategies for the DAB-SRC. The advantage of variable frequency (VF) control can achieve high efficiency, but the switching frequency varies over a wide range when the load changes, CFPS or duty cycle control strategies are easier to design magnetic components, but these methods may lead to a small soft-switching range especially when the normalized gain deviates from unity, resulting in low efficiency [4, 5]. Ref.

[4] proposes a VFPS control strategy based on the frequency-load linear function to achieve soft switching over the full voltage range. However, this approach results in extremely wide switching frequency range, increasing the complexity of the transformer design. According to the model of the DAB-SRC, it can be concluded from [4, 6-8] that it is hard to realize a soft-switching for the light load condition when employing the VFPS control.

In order to further extend the soft-switching region under very light load conditions and reduce the range of the switching frequency at the same time, [9] proposes a hybrid modulation strategy to improve the soft-switching region at light loads, which applies VFPS control in the range from intermediate power to rated power but applies asymmetrical modulation for the light load condition. However, the control strategy is a little bit complicated.

Without increasing the number of devices, a hybrid control strategy of VFPS combined with CFPS is proposed in this paper. Different from [9], symmetrical modulation is always ensured. The magnetizing inductance is utilized to extend the soft-switching range, especially for the light-load condition. A prototype is built to verify the correctness of the theoretical results, where the soft-switching region can cover a load variation range of 5% to 100%, and the switching frequency variation range is only 22kHz by this method.

II. THE ANALYSIS OF THE SOFT-SWITCHING REALIZATION

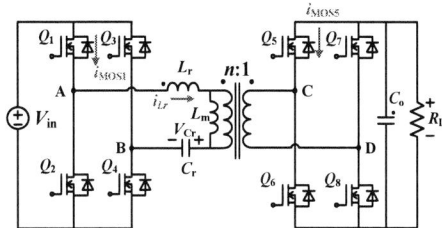

Fig 1. Topology diagram of DAB-SRC

Fig 1. shows the topology of the DAB-SRC. Ref. [4] derives a time domain analytical model and gives the soft switching boundary conditions. On this basis, a control strategy with a linear function (f_n=k*Q+b) is proposed, where f_n is defined as the normalized frequency, Q is the quality

This work was supported in part by the Beijing Natural Science Foundation under Grant 3232044, and in part by the R&D Program of Beijing Municipal Education Commission under Grant KM202110009011.

factor, besides, k and b are the slope and intercept of the linear function, respectively. Fig. 2 shows the soft-switching region and contours of different turn-off currents expressed in the Q-f_n coordinate system for the same gain, and the boundary contours of the turn-off currents form a cluster of nonlinear curves (①~⑥ in Fig. 2). As the load power decreases, the turn-off current decreases, which will have to prolong the charging and discharging time of the switch junction capacitor. In Fig. 2, straight line, which is named as track 1, shows the operation trajectory of the converter proposed, where $\varphi_{\text{deadtime}}$ is defined in (1). It is clear that the turn-off current inevitably becomes very tiny at light load with a high switching frequency.

$$\omega_r t_{\text{Dead}} = \omega_s t_{\text{Dead}} \frac{\omega_r}{\omega_s} = \frac{\varphi_{\text{deadtime}}}{f_n} \tag{1}$$

In order to solve this problem, a control strategy is proposed in this paper, which utilizes the feature of the magnetizing inductance and the soft-switching region with respect to the transformer turns ratio in the DAB-SRC. CFPS is adopted under 20% rated load. From Fig.2, it is clear that CFPS control strategy indeed increases the turn-off current, but under heavy load (Q is small) a large turn-off current will always exist, resulting in large losses and affecting the efficiency of the converter. Therefore, it only adopted below 20% rated load.

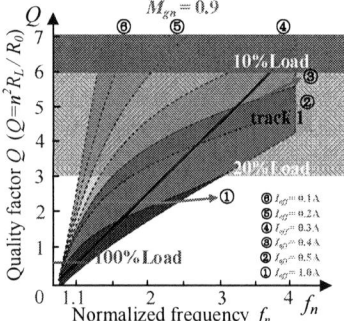

Fig. 2. Contours of different turn-off currents in the Q-f_n soft-switching region at fixed gain (M_{gn} =0.9, $\varphi_{\text{deadtime}}$ =0.12).

The model of DAB-SRC shows it is difficult to realize soft switching on the low voltage side, both at M_{gn} >1 and M_{gn} <1, where M_{gn} denotes the normalized voltage gain as defined in [4]. Therefore, by adjusting the transformer ratio, the converter can work with the buck mode in the forward operation, and with boost mode when operating reverse mode. By this means, the secondary-side is always operated as the low-voltage side, and correspondingly, it is hard to realize the soft-switching. However, it should be noted that the magnetizing exists and as a result, it provides the additional circulating current, which is helpful to realizing the soft-switching for the secondary-side-switch. In this way, the determination of the soft switching is changed, as shown in (2) and (3).

$$\begin{cases} \sin\left[\frac{\pi - 2\varphi_{deadtime}}{2f_n}\right] > \frac{1}{M_{gn}} \sin\left[\arccos\left[\frac{\left(M_{gn}\pi + 2f_n Q\right)}{2f_n Q \sec(\frac{\pi}{2f_n})}\right] - \frac{\varphi_{deadtime}}{f_n}\right] \;, M_{gn} \geq 1 \\[12pt] \sin\left[\frac{\pi - 2\varphi_{deadtime}}{2f_n}\right] > M_{gn} \sin\left[\arccos\left[\frac{\left(M_{gn}\pi + 2f_n Q\right)}{2f_n Q \sec(\frac{\pi}{2f_n})}\right] + \frac{\varphi_{deadtime}}{f_n}\right] \;, M_{gn} < 1 \end{cases}$$

$$(2)(3)$$

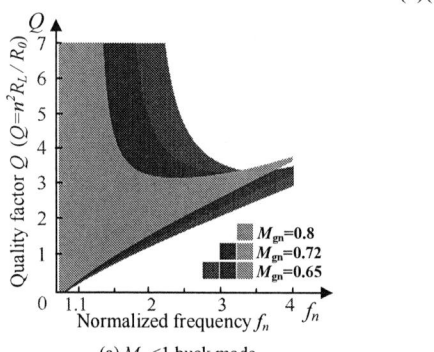

(a) M_{gn}<1 buck mode

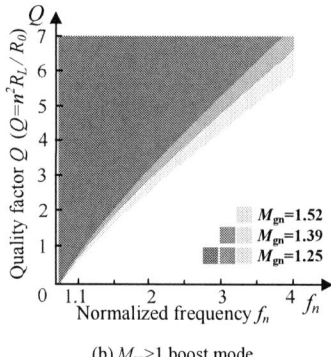

(b) M_{gn}>1 boost mode

Fig. 3. Soft switching regions of converter only considering high voltage side (t_{dead}=300ns).

Since the low voltage side is always paralleled with a magnetizing inductor, only the soft-switching range of the high-voltage side is considered as shown in Fig. 3. The range of the soft switching has been significantly increased, the further M_{gn} is away from 1, the larger the corresponding soft-switching margin. Selecting appropriate frequency points for CFPS control can achieve soft switching under extremely light load or even no-load conditions for all switching .

With fixed gain, the relationship between turn-off current and Q at different frequencies is shown in Fig. 4 and Fig. 5. The gain variation range is from 0.6 to 1.5. When Q equals 20, it corresponds to an extremely light load condition.

From Fig.5, it can be analyzed that when the gain is fixed, selecting different fixed-frequency f_n results in different turn-off currents. The larger the f_n, the smaller the corresponding turn-off current.

Therefore, taking M_{gn} =0.9 as an example, the turn-off current contour lines only considering the high voltage side are given as shown in Fig 6. It can be seen that the larger the f_n, the smaller the turn-off current. The fixed frequency f_n = 1.3 was finally selected through several iterations in the experiment. CFPS control is used at 5%~20% rated power,

979-8-3315-1110-4/25 $31.00 © 2025 IEEE

and Q - f_n linear function is used for VFPS control at 20%~100% rated power. The function trajectory is: $f_n =0.0986* Q +1.0594$. As shown in Fig. 6, track 2, which is the selected operation trajectory, is also applicable to the reverse condition.

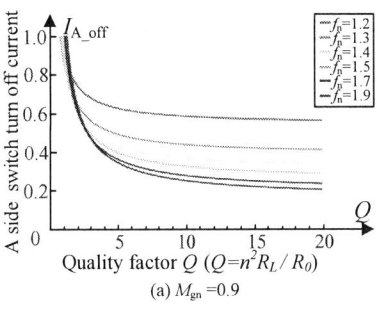

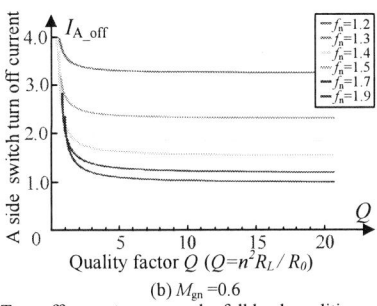

(a) M_{gn} =0.9

(b) M_{gn} =0.6

Fig. 4. Turn off current curves under full load conditions at different frequencies during forward buck mode.

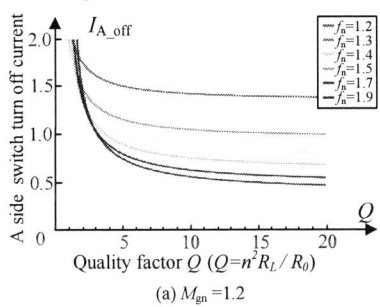

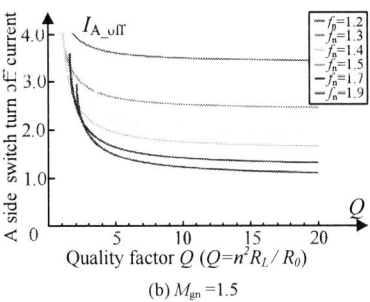

(a) M_{gn} =1.2

(b) M_{gn} =1.5

Fig. 5. Turn off current curves under full load conditions at different frequencies during boost mode.

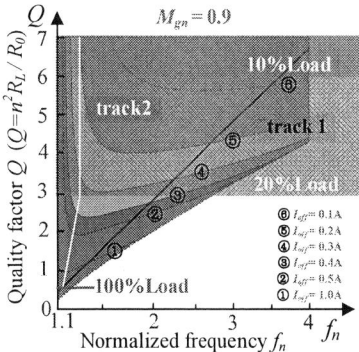

Fig. 6. Different turn-off current contours at high voltage side in Q - f_n soft-switching region under fixed gain (M_{gn} =0.9, $\varphi_{deadtime}$ =0.12).

III. EXPERIMENTAL VERIFICATION

TABLE I.   MAIN CIRCUIT PARAMETERS OF THE PROTOTYPE

Parameter	Symbol	Value
Input voltage	V_{in}	360V-440V
Output voltage	V_o	400V
Nominal load	$P_{nominal}$	1.2kW
Resonant inductance	L_r	351μH
Resonant capacitance	C_r	7.05nF
Resonant frequency	f_r	101kHz
Control frequency	f_c	50kHz
Transformer turns ratio	n :1	18:25
Excitation inductance	L_m	93.3μH
Power Switches	Q_1-Q_8	GS66508T
Deadtime	T_{dead}	300ns

A 1.2kW prototype is built and the main circuit parameters are shown in TABLE I. The experimental waveforms of the ZVS realization of the switches are observed at 5% and 100% rated load, respectively. The experiments in forward direction are shown in Fig.7 to Fig.10, and the reverse experiments are shown in Fig. 11 to Fig. 14. It can be found that the soft-switching can be always ensured from 5% to 100% rated power during the whole voltage range, with the frequency range only from 108kHz to 130kHz.

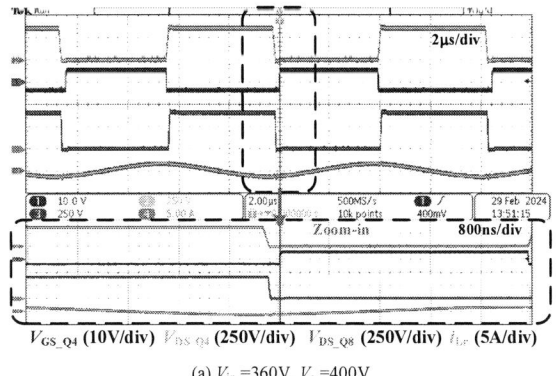

V_{GS_Q4} (**10V/div**) V_{DS_Q4} (**250V/div**) V_{DS_Q8} (**250V/div**) i_{Lr} (**5A/div**)

(a) V_{in} =360V, V_o =400V

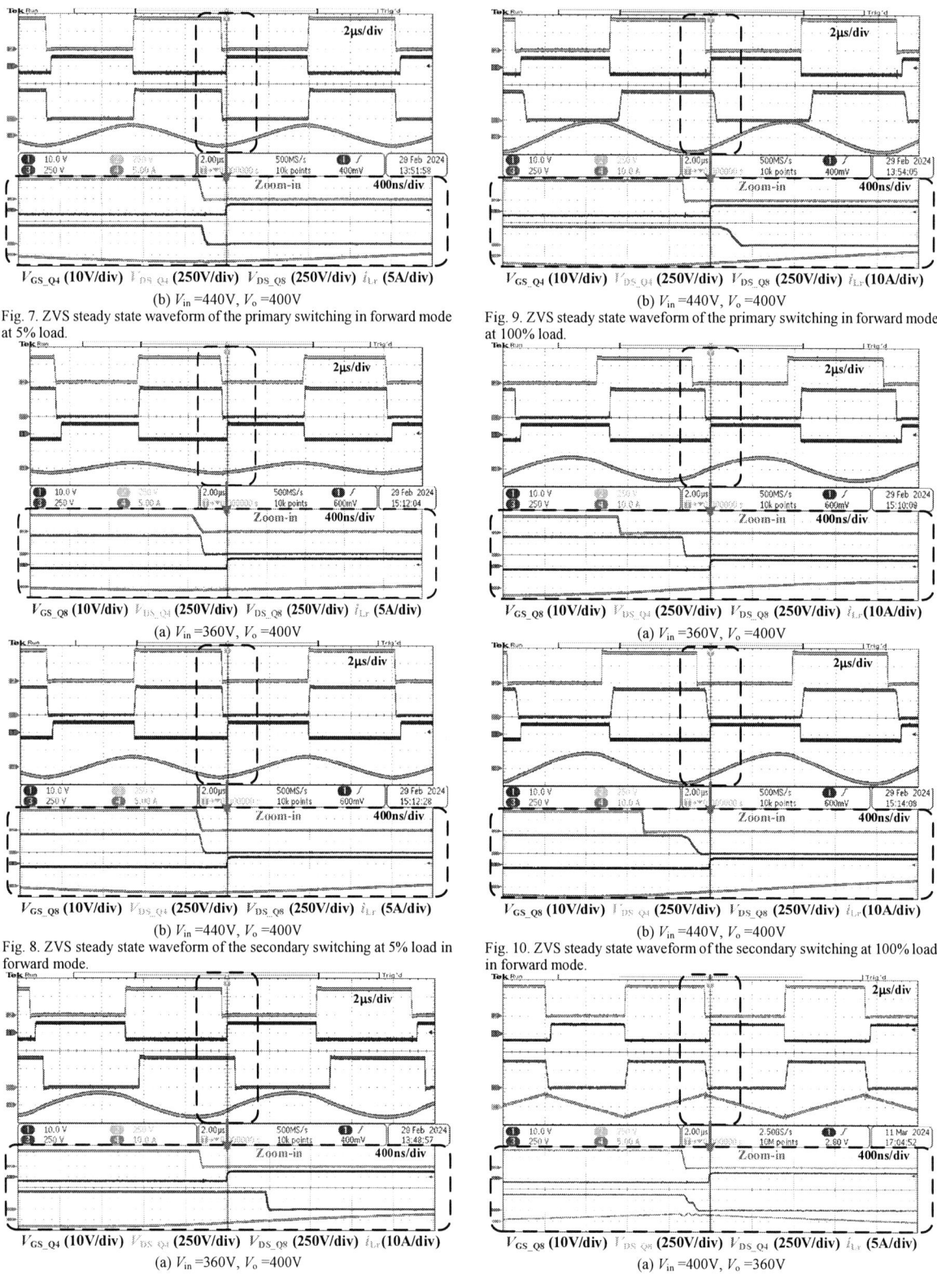

V_{GS_Q4} **(10V/div)** V_{DS_Q4} **(250V/div)** V_{DS_Q8} **(250V/div)** i_{Lr} **(5A/div)**

(b) V_{in} =440V, V_o =400V

Fig. 7. ZVS steady state waveform of the primary switching in forward mode at 5% load.

V_{GS_Q8} **(10V/div)** V_{DS_Q4} **(250V/div)** V_{DS_Q8} **(250V/div)** i_{Lr} **(5A/div)**

(a) V_{in} =360V, V_o =400V

V_{GS_Q8} **(10V/div)** V_{DS_Q4} **(250V/div)** V_{DS_Q8} **(250V/div)** i_{Lr} **(5A/div)**

(b) V_{in} =440V, V_o =400V

Fig. 8. ZVS steady state waveform of the secondary switching at 5% load in forward mode.

V_{GS_Q4} **(10V/div)** V_{DS_Q4} **(250V/div)** V_{DS_Q8} **(250V/div)** i_{Lr} **(10A/div)**

(a) V_{in} =360V, V_o =400V

V_{GS_Q4} **(10V/div)** V_{DS_Q4} **(250V/div)** V_{DS_Q8} **(250V/div)** i_{Lr} **(10A/div)**

(b) V_{in} =440V, V_o =400V

Fig. 9. ZVS steady state waveform of the primary switching in forward mode at 100% load.

V_{GS_Q8} **(10V/div)** V_{DS_Q4} **(250V/div)** V_{DS_Q8} **(250V/div)** i_{Lr} **(10A/div)**

(a) V_{in} =360V, V_o =400V

V_{GS_Q8} **(10V/div)** V_{DS_Q4} **(250V/div)** V_{DS_Q8} **(250V/div)** i_{Lr} **(10A/div)**

(b) V_{in} =440V, V_o =400V

Fig. 10. ZVS steady state waveform of the secondary switching at 100% load in forward mode.

V_{GS_Q8} **(10V/div)** V_{DS_Q8} **(250V/div)** V_{DS_Q4} **(250V/div)** i_{Lr} **(5A/div)**

(a) V_{in} =400V, V_o =360V

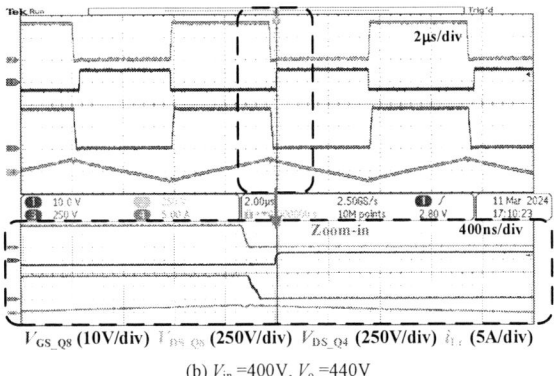

(b) V_{in} =400V, V_o =440V

Fig. 11 ZVS steady state waveform of the primary switching in reverse mode at 5% load.

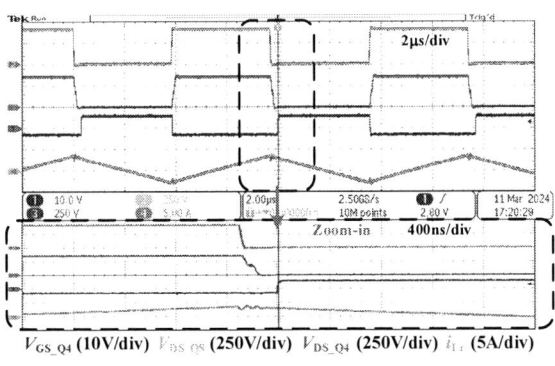

(a) V_{in} =400V, V_o =360V

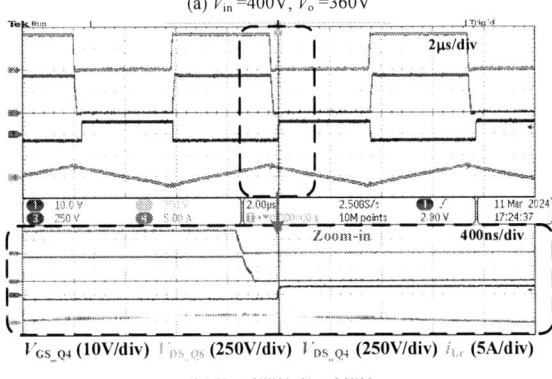

(b) V_{in} =400V, V_o =440V

Fig. 12 ZVS steady state waveform of the secondary switching in reverse mode at 5% load.

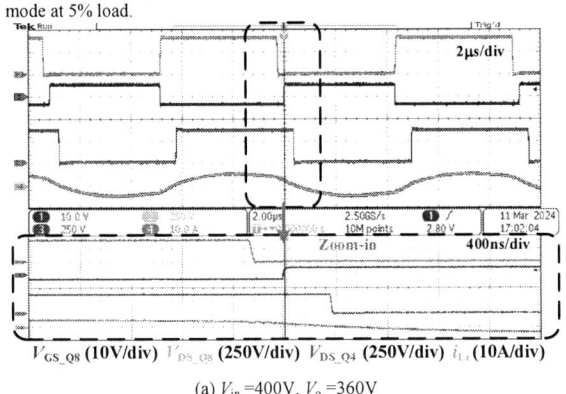

(a) V_{in} =400V, V_o =360V

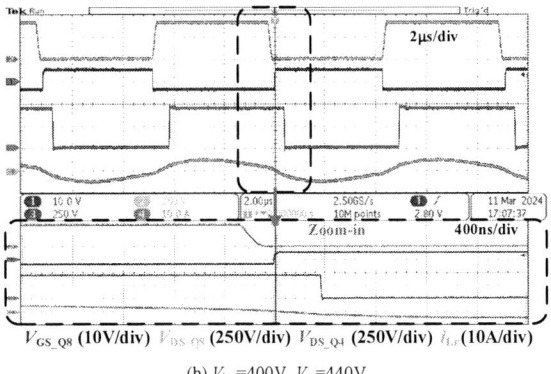

(b) V_{in} =400V, V_o =440V

Fig. 13 ZVS steady state waveform of the primary switching in reverse mode at 100% load.

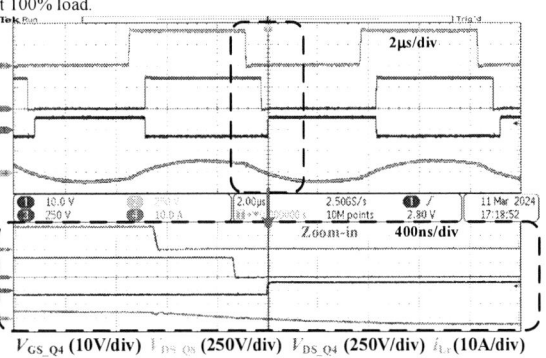

(a) V_{in} =400V, V_o =360V

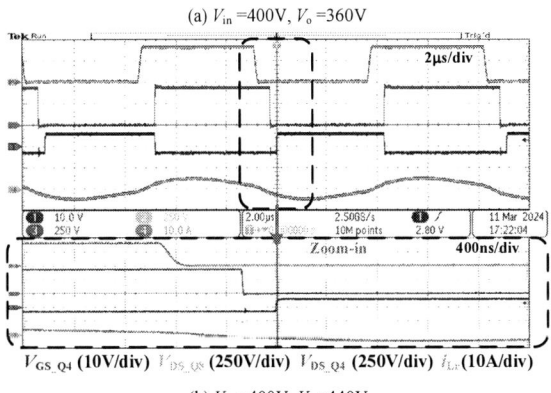

(b) V_{in} =400V, V_o =440V

Fig. 14 ZVS steady state waveform of the secondary switching in reverse mode at 100% load.

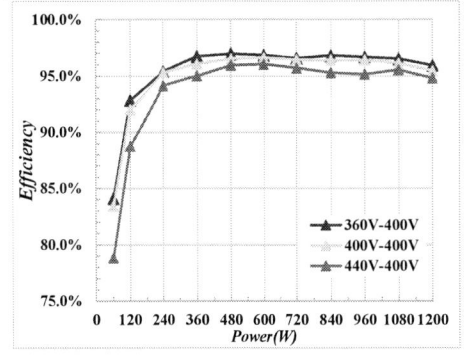

(a) forward mode

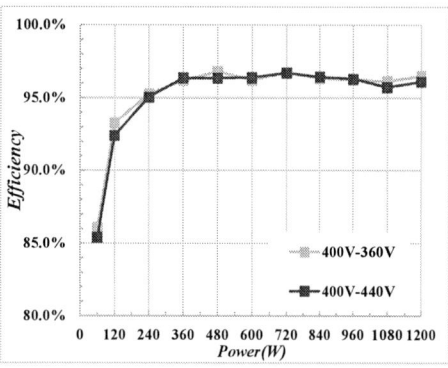

(b) reverse mode

Fig. 15 Efficiency curve of the experimental prototype.

The efficiency curves are shown in Fig 15. The highest efficiency of the converter in forward and reverse modes is 97% and 96%.

IV. CONCLUSION

In this paper, A hybrid control strategy of VFPS and CFPS is proposed for the DAB-SRC. By this means, the soft-switching region can be significantly enlarged especially in the light load condition. The effectiveness of the method is verified by the 1.2kW experimental platform.

REFERENCES

[1] K. Siebke, M. Giacomazzo and R. Mallwitz, "Design of a Dual Active Bridge Converter for On-Board Vehicle Chargers using GaN and into Transformer Integrated Series Inductance," 2020 22nd European Conference on Power Electronics and Applications (EPE'20 ECCE Europe), Lyon, France, 2020, pp. 1-8

[2] R. Yamada, A. Hino and K. Wada, "Improvement of Efficiency in Bidirectional DC-DC Converter with Dual Active Bridge Using GaN-HEMT," 2022 International Power Electronics Conference (IPEC-Himeji 2022- ECCE Asia), Himeji, Japan, 2022, pp. 1398-1403

[3] T. Chen, R. Yu and A. Q. Huang, "A Bidirectional Isolated Dual-Phase-Shift Variable-Frequency Series Resonant Dual-Active-Bridge GaN AC–DC Converter," in IEEE Transactions on Industrial Electronics, vol. 70, no. 4, pp. 3315-3325, April 2023.

[4] P. Jia, K. Qiu, M. Liang, T. Shao and Y. Zhang, "A Simple Variable Frequency and Single-Phase-Shifted Control for the Dual Active Bridge Series Resonant Converter," in IEEE Journal of Emerging and Selected Topics in Industrial Electronics, vol. 6, no. 1, pp. 308-326, Jan. 2025.

[5] L. Corradini, D. Seltzer, D. Bloomquist, R. Zane, D. Maksimovi´c, and B. Jacobson, "Minimum current operation of bidirectional dual-bridge series resonant DC/DC converters," IEEE Trans. Power Electron., vol. 27, no. 7, pp. 3266–3276, Jul. 2012.

[6] W. Han and L. Corradini, "General Closed-Form ZVS Analysis of Dual-Bridge Series Resonant DC–DC Converters," in IEEE Transactions on Power Electronics, vol. 34, no. 9, pp. 9289-9302, Sept. 2019.

[7] M. Yaqoob, K. H. Loo, and Y. M. Lai, "A four-degrees-of-freedom modulation strategy for dual-active-bridge series-resonant converter designed for total loss minimization," IEEE Trans. Power Electron., vol. 34, no. 2, pp. 1065–1081, Feb. 2019.

[8] Y. Gao and H. Ma, "Analysis and design of a control strategy for wide range ZVS of isolated bidirectional dual-bridge series resonant DC/DC converters," in Proc. 10th Int. Conf. Power Electron. ECCE Asia, 2019, pp. 3116–3121.

[9] Y. Gao et al., "Hybrid-level modulation scheme for dual-bridge series resonant converter," IEEE Trans. Ind. Electron., vol. 70, no. 11, pp. 11205–11215,Nov.2023.

Bidirectional Energy-Storage Converter Based on Partial Power Processing

Zhihao Zhang, Xiao Chen, Fan Zhai, Zhongjie Wang, Lingxiao Xue
School of Electrical and Information Engineering
Tianjin University
Tianjin China
dz_hao@tju.edu.cn

Abstract—This paper proposes a high-efficiency bidirectional converter based on partial power processing for energy storage applications. This converter combines an unregulated LLC converter with a bidirectional active-clamp flyback converter. The topology achieves regulation across wide voltage range and demonstrates active power decoupling in simulation, reducing input current ripple by 96% compared to traditional solutions. Experimental results confirm the feasibility, with soft switching achieved in the DC-DC stage. The design offers a compact and efficient solution for residential photovoltaic systems.

Keywords—Bidirectional DC-AC Converter, Active Power Decoupling, Partial Power Processing, Soft-Switching

I. INTRODUCTION

To address the energy crisis and promote carbon neutrality, residential energy storage systems have undergone rapid development. Bidirectional converters play a crucial role as the bridge connecting energy storage batteries with household appliances and even the grid[1-2].

Isolated bidirectional DC-AC converters can be mainly categorized into single-stage, quasi-single-stage, and two-stage topologies[3]. The single-stage topology achieves DC-AC conversion through a single power conversion stage, offering advantages such as simple structure and low cost. However, it suffers from high control complexity and limited voltage gain[4-6]. The quasi-single-stage topology employs a two-stage power conversion process: a front-stage DC-DC converter transforms the DC voltage into a 100Hz sinusoidal half-wave, and a second-stage line-frequency inverter that further converts the 100Hz half-wave into a 50Hz sinusoidal wave. This topology maintains high efficiency while reducing control difficulty, but at the cost of increased component count[7-9]. The two-stage topology consists of an isolated DC-DC converter cascaded with a DC-AC inverter, enabling independent voltage regulation and control. It offers advantages such as flexible control and high power quality, but typically exhibits lower efficiency and power density[10-12]. From the above analysis, it is evident that the single-stage topology has potential advantages in terms of efficiency and power density. However, the presence of double-line-frequency ripple significantly limits its performance, posing a major challenge for the practical application of single-stage topologies.

Double-line-frequency ripple is a common issue in single-phase DC-AC converters, fundamentally caused by the fluctuation of instantaneous power on the AC side at twice the line frequency (100Hz), which leads to significant double-line-frequency ripple in the DC-side current or voltage if not controlled properly [13-16]. Research indicates that double-line-frequency ripple has multiple negative impacts on system performance: firstly, ripple current causes additional copper and iron losses, reducing system efficiency [17]; secondly, ripple voltage accelerates the capacity degradation of

batteries, shortening their lifespan [18]. Notably, although two-stage topologies are generally considered to have lower power density, their ability to suppress double-line-frequency ripple is significantly better than that of single-stage topologies. Therefore, under the same ripple suppression requirements, the power density of two-stage topologies may not be inferior to that of single-stage topologies. This finding provides a new perspective for topology selection.

To address the suppression of double-line-frequency ripple, existing solutions can be mainly categorized into passive and active approaches. Passive methods absorb ripple energy by adding capacitors or inductors. While simple and reliable, these methods significantly increase the volume and cost of the converter. Active solutions typically employ active power decoupling techniques, whose core idea is to transfer double-line-frequency ripple energy to auxiliary energy storage components (such as capacitors or inductors) by introducing auxiliary circuits and advanced control strategies, thereby significantly reducing reliance on passive components[19-20]. This approach can effectively suppress double-line-frequency ripple, greatly enhancing the power density of the converter. However, the introduction of active power decoupling techniques increases system complexity and control difficulty. Therefore, designing active power decoupling solutions with simple structures, high efficiency, and low cost has become a current research hotspot.

Based on the above analysis, this paper proposes a DC-AC converter scheme based on partial power processing, aiming for high power density and high efficiency. Section II provides a detailed introduction to the proposed topology and its power distribution scheme for partial power processing. Section III focuses on analyzing the mechanism of double-line-frequency ripple suppression in the proposed topology and compares it with traditional passive power decoupling solutions. Simulation results demonstrate that the proposed scheme can reduce input current ripple by 96%. Section IV preliminarily verifies the feasibility of the scheme through experiments. Section V provides a summary of the entire paper.

II. PROPOSED PARTIAL POWER PROCESSING CONVERTER

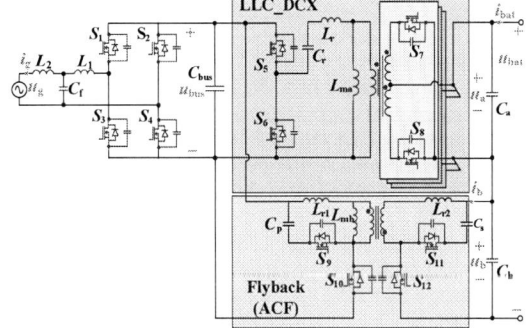

Fig. 1. Proposed Topological Structure

As shown in Fig. 1, this paper proposes a bidirectional DC-AC converter topology. The topology adopts a two-stage structure, where the DC-DC stage consists of a non-regulated LLC converter (LLC_DCX) and a bidirectional active-clamp flyback converter (Flyback_ACF), forming a partial power processing structure. The LLC converter is known for its high efficiency and high-power density. To achieve its extremely high efficiency, the LLC converter must always operate at the resonant point. The flyback converter exhibits strong voltage regulation capability, which provides the topology with a wide voltage regulation range, compensating for the LLC converter's limited voltage regulation capability. Furthermore, analysis reveals that if the output voltage (u_b) ripple of the flyback converter can fully compensate for the output voltage (u_a) ripple of the LLC converter, effective suppression of the double-line-frequency ripple on the DC side of the DC-AC converter can be achieved. This significantly reduces the required bus capacitance, thereby further enhancing the power density of the converter.

In the partial power topology, the design of power distribution is critical. Since the LLC converter and the flyback converter are connected in series on the low-voltage side, the ratio of the power processed by the LLC converter to that processed by the flyback converter is related to their series voltage division ratio, as expressed by:

$$P_{LLC} / P_{Flyback} = u_a / u_b \qquad (1)$$

where $P_{LLC} + P_{Flyback} = P_{in}$, P_{in} is the total input power; and $u_a + u_b = u_{bat}$, u_{bat} is the DC input voltage of the entire converter.

Based on the above relationship, the overall efficiency of the DC-DC stage can be expressed as:

$$\eta_{DC\_DC} = \eta_{LLC} P_{LLC} / p_{total} + \eta_{Flyback} P_{Flyback} / p_{total}$$
$$= \eta_{LLC} u_a / u_{bat} + \eta_{Flyback} u_b / u_{bat} \qquad (2)$$

where η_{LLC} is the efficiency of the LLC converter, and $\eta_{Flyback}$ is the efficiency of the flyback converter. According to (2), the overall efficiency of the DC-DC stage depends not only on the efficiencies of the LLC converter and the flyback converter but also on their voltage distribution. When u_a / u_{bat} is larger, the efficiency of the DC-DC stage approaches that of the LLC converter. conversely, it approaches that of the flyback converter. Typically, the efficiency of the LLC converter is higher than that of the flyback converter. Therefore, in designing the operating mode, u_a / u_{bat} should be made as close to 1 as possible.

The designed DC-AC converter has a DC input voltage range of 32-62V, with a rated operating voltage of 48V. To ensure the highest efficiency under rated conditions, a voltage distribution scheme for the partial power topology is designed, as shown in Fig. 2, which maximizes the input voltage of the LLC converter under rated conditions. As previously analyzed, a smaller u_b is generally preferred. However, when $u_{bat} < 48V$, u_b cannot be reduced indefinitely due to the presence of power imbalance, which leads to ripple in u_b. Therefore, the average value u_b should not be smaller than the ripple amplitude. Experimental results have verified that setting $u_{b(avg)}$ to approximately 3.5 V is feasible. When $u_{bat} > 48V$, keeping $u_{b(avr)}$ constant would cause the bus voltage to increase proportionally with u_a, which is constrained by the voltage ratings of the bus capacitor and switching devices. Thus, when $u_{bat} > 48V$, we choose to maintain a constant u_a and increase u_b.

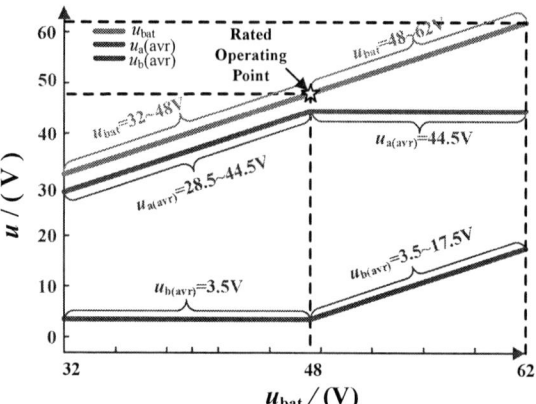

Fig. 2. DC-DC Stage Voltage (Power) Distribution Scheme

III. OPERATING PRINCIPLE OF ACTIVE POWER DECOUPLING

A. Power Decoupling Analysis

As shown in Fig. 3, a simplified schematic of the proposed scheme is presented. When the inverter load is linear, the output voltage u_g and current i_g of the inverter are both 50Hz sinusoidal waves. Since the output current i_g of the inverter is entirely derived from the DC bus current i_r, the DC bus current i_r should consist of a 100Hz sinusoidal half-wave and switching ripple. As most of the switching ripple in i_r is absorbed by the LCL filter, we focus only on the 100Hz low-frequency ripple. For the DC-DC stage, the inverter stage can be equivalently represented as a current source, leading to the simplified circuit shown in Fig. 3(b). Next, we focus only on the ripple power, i.e., the AC component in i_r, and establish a simplified sinusoidal circuit model for the circuit in Fig. 3(b), as shown in Fig. 3(c). Since the switching frequency of the converters is much higher than 100 Hz, the LLC and Flyback converters are each modeled as equivalent transformers [20]. Here, I_r is the equivalent AC source of the inverter, reflecting the magnitude of the unbalanced power. C_{bus} is the intermediate bus capacitor, R_{L1}, R_{L2} represent the losses of the LLC converter; R_{F1}, R_{F2} represent the losses of the flyback converter, C_a, C_b are the DC-side capacitors; and R_s is the internal resistance of the DC source. Since this is an AC circuit model, the DC source is equivalent to a short circuit.

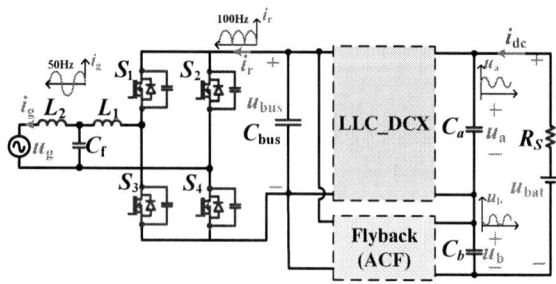

(a) Simplified Diagram of the Proposed Topology

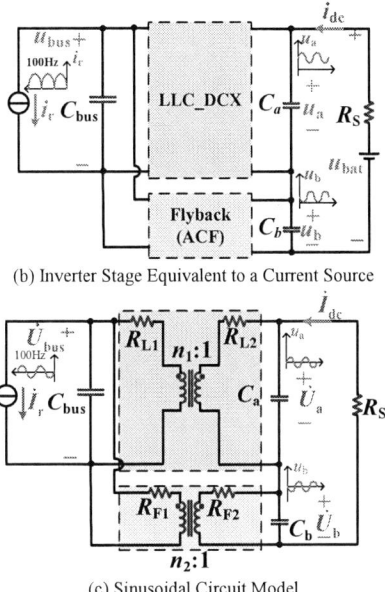

(b) Inverter Stage Equivalent to a Current Source

(c) Sinusoidal Circuit Model

Fig. 3. Power Decoupling Analysis Model Development Diagram

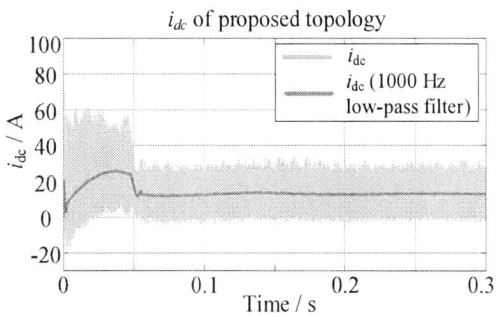

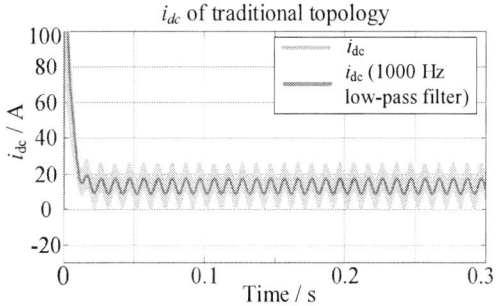

Fig. 5. Simulation Comparison of DC Input Current Waveforms

The sinusoidal circuit model shown in Fig. 3(c) is solvable. During the solving process, we find that the input current ripple satisfies the following loop current equation:

$$\dot{I}_{dc} R_s = \dot{U}_a + \dot{U}_b \qquad (3)$$

LLC converter, and U_b is the input voltage on the low-voltage side of the flyback converter. If we control the sum of U_a and U_b to be close to zero, the input current ripple on the DC side can be minimized, thereby achieving power decoupling. A more detailed theoretical analysis will be presented in the final manuscript.

B. Comparison of Decoupling Performance

To validate the power decoupling advantages, the proposed topology is compared via simulation with the conventional cascaded structure of an LLC converter and a full-bridge inverter, as shown in Fig. 4. Under DC-AC mode, the input current ripple on the DC side is compared under the same conditions: output power (500W), DC input voltage (48V), bus capacitance (100μF), and DC source internal resistance Rs (0.1Ω). As shown in Fig. 5, the input current ripple rate of the traditional topology is 65.3%, while that of the proposed topology is only 2.5%, representing a 96% reduction.

IV. EXPERIMENTAL VERIFICATION

To further verify the feasibility and power decoupling advantages of the proposed topology, a prototype was built and tested up to the rated power 500W. Fig. 6 shows the experimental prototype, along with thermal images of the LLC converter at 230 W and the flyback converter at 20 W, respectively. All switching devices in the DC-DC stage have achieved soft switching, as shown in Fig. 7. Fig. 8 presents the operating waveforms of the bidirectional DC-AC converter in two modes. Fig. 8(a) corresponds to the DC-AC mode, in which the RMS value of the input current ripple is approximately 23.8% of the average input current. Fig. 8(b) corresponds to the AC-DC mode, where the total harmonic distortion of the input current (THDi) on the AC side is approximately 7.43%. The utilization of wide bandgap semiconductors in the prototype is detailed in Table 1. The converter achieved a power density of 21W/in³.

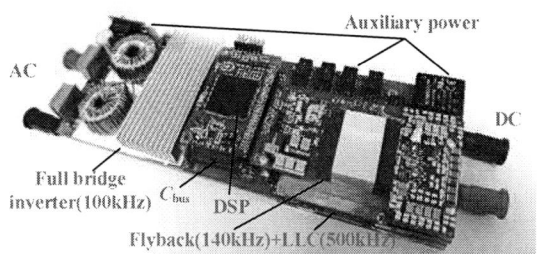

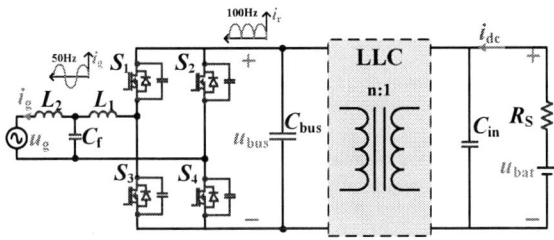

Fig. 4. Traditional DC-AC Topology

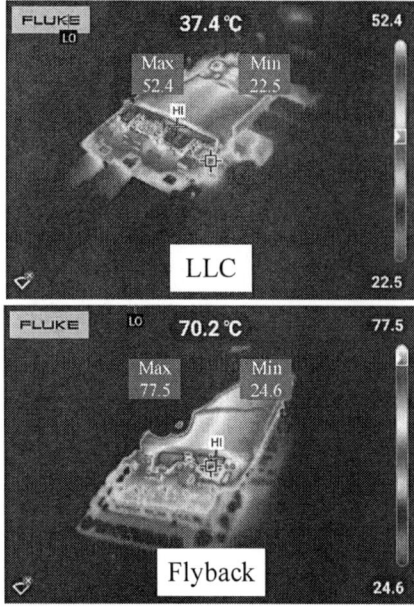

Fig. 6.   Prototype (500W) Diagram and Thermal Imaging

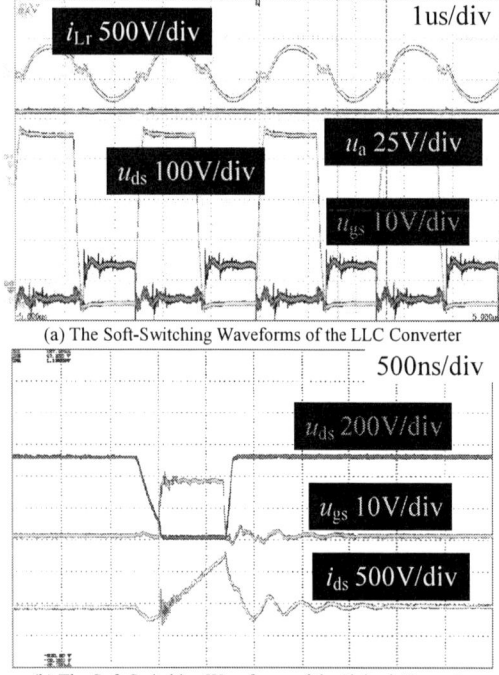

(a) The Soft-Switching Waveforms of the LLC Converter

(b) The Soft-Switching Waveforms of the Flyback Converter

Fig. 7.   Soft-Switching Implementation Waveforms

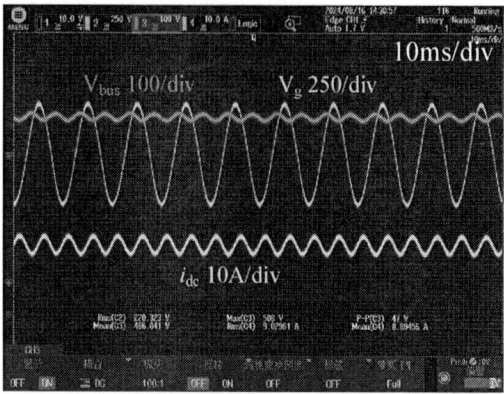

(a)DC-AC mode

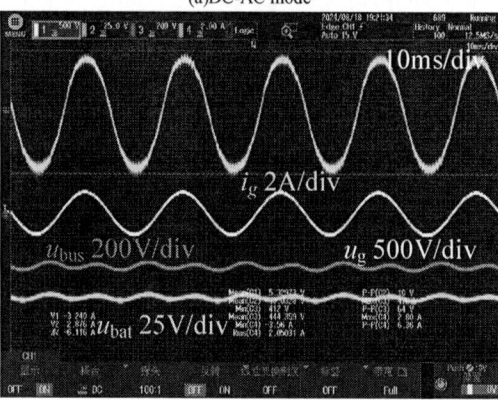

(b)AC-DC mode

Fig. 8.   The Operating Waveform Curves of the Bidirectional DC-AC Converter

TABLE I.   WIDE BANDGAP DEVICES USED IN THE PROTOTYPE

Topology Name	Switching Frequency	Type	Model Number
LLC	500 kHz	GaN	INN700D140C
	500 kHz	GaN	INN150EQ070A
Flyback	140 kHz	GaN	GS61004B
	140 kHz	SiC	P3M12080G7
Full bridge inverter	100 kHz	GaN	INN650TA070A H

V. CONCLUSION

A bidirectional DC-AC converter with partial power processing is proposed, integrating an LLC converter and a bidirectional-flyback converter for high efficiency and power density. The proposed topology offers double-line-frequency ripple suppression capability and can reduce the input current ripple by approximately 96% compared to traditional solution, according to simulation results. An experimental platform was built to preliminarily verify the feasibility of the proposed topology. Future work will focus on optimizing the control algorithm, experimentally validating power decoupling performance, and improving the overall efficiency of the converter, aiming to provide a promising solution for residential photovoltaic systems.

REFERENCES

[1] H. Wu, X. Tang, J. Zhao and Y. Xing, "An Isolated Bidirectional Microinverter Based on Voltage-in-Phase PWM-Controlled Resonant Converter," in IEEE Transactions on Power Electronics, vol. 36, no. 1, pp. 562-570, Jan. 2021.

[2] Q. Huang, A. Q. Huang, R. Yu, P. Liu and W. Yu, "High-Efficiency and High-Density Single-Phase Dual-Mode Cascaded Buck–Boost Multilevel Transformerless PV Inverter With GaN AC Switches," in IEEE Transactions on Power Electronics, vol. 34, no. 8, pp. 7474-7488, Aug. 2019.

[3] S. B. Kjaer, J. K. Pedersen and F. Blaabjerg, "A review of single-phase grid-connected inverters for photovoltaic modules," in IEEE Transactions on Industry Applications, vol. 41, no. 5, pp. 1292-1306, Sept.-Oct. 2005.

[4] S. Jain and V. Agarwal, "A Single-Stage Grid Connected Inverter Topology for Solar PV Systems With Maximum Power Point Tracking," in IEEE Transactions on Power Electronics, vol. 22, no. 5, pp. 1928-1940, Sept. 2007.

[5] Y. Zhang et al., "A Boost-Inductorless Electrolytic-Capacitorless Single-Stage Bidirectional Isolated AC–DC Converter," in IEEE Transactions on Power Electronics, vol. 38, no. 4, pp. 5469-5478, April 2023.

[6] X. Zhou, J. Xu and S. Zhong, "Single-Stage Soft-Switching Low-Distortion Bipolar PWM Modulation High-Frequency-Link DC–AC Converter With Clamping Circuits," in IEEE Transactions on Industrial Electronics, vol. 65, no. 10, pp. 7719-7729, Oct. 2018.

[7] F. Peng, G. Zhou, N. Xu and S. Gao, "Zero Leakage Current Single-Phase Quasi-Single-Stage Transformerless PV Inverter With Unipolar SPWM," in IEEE Transactions on Power Electronics, vol. 37, no. 11, pp. 13755-13766, Nov. 2022.

[8] T. Chen, R. Yu and A. Q. Huang, "Variable-Switching-Frequency Single-Stage Bidirectional GaN AC–DC Converter for the Grid-Tied Battery Energy Storage System," in IEEE Transactions on Industrial Electronics, vol. 69, no. 11, pp. 10776-10786, Nov. 2022.

[9] T. Chen, R. Yu and A. Q. Huang, "A Bidirectional Dual-Phase-Shift Variable-Frequency Series Resonant Dual-Active-Bridge GaN AC–DC Converter," in IEEE Transactions on Industrial Electronics, vol. 70, no. 4, pp. 3315-3325, April 2023.

[10] T. -F. Wu, C. -H. Chang, L. -C. Lin and C. -L. Kuo, "Power Loss Comparison of Single- and Two-Stage Grid-Connected Photovoltaic Systems," in IEEE Transactions on Energy Conversion, vol. 26, no. 2, pp. 707-715, June 2011.

[11] J. -S. Kim, J. -M. Kwon and B. -H. Kwon, "High-Efficiency Two-Stage Three-Level Grid-Connected Photovoltaic Inverter," in IEEE

Transactions on Industrial Electronics, vol. 65, no. 3, pp. 2368-2377, March 2018.

[12] B. Zhao, X. Zhang and J. Huang, "AI Algorithm-Based Two-Stage Optimal Design Methodology of High-Efficiency CLLC Resonant Converters for the Hybrid AC–DC Microgrid Applications," in IEEE Transactions on Industrial Electronics, vol. 66, no. 12, pp. 9756-9767, Dec. 2019.

[13] H. Li, K. Zhang and H. Zhao, "Active DC-link power filter for single phase PWM rectifiers," 8th International Conference on Power Electronics - ECCE Asia, Jeju, Korea (South), 2011, pp. 2920-2926.

[14] R. Wang, F. Wang, R. Lai, P. Ning, R. Burgos and D. Boroyevich, "Study of Energy Storage Capacitor Reduction for Single Phase PWM Rectifier," 2009 Twenty-Fourth Annual IEEE Applied Power Electronics Conference and Exposition, Washington, DC, USA, 2009, pp. 1177-1183.

[15] S. Li, W. Qi, S. -C. Tan and S. Y. Hui, "Enhanced Automatic-Power-Decoupling Control Method for Single-Phase AC-to-DC Converters," in IEEE Transactions on Power Electronics, vol. 33, no. 2, pp. 1816-1828, Feb. 2018.

[16] H. Hu, S. Harb, N. Kutkut, I. Batarseh and Z. J. Shen, "A Review of Power Decoupling Techniques for Microinverters With Three Different Decoupling Capacitor Locations in PV Systems," in IEEE Transactions on Power Electronics, vol. 28, no. 6, pp. 2711-2726, June 2013.

[17] P. T. Krein and R. S. Balog, "Cost-Effective Hundred-Year Life for Single-Phase Inverters and Rectifiers in Solar and LED Lighting Applications Based on Minimum Capacitance Requirements and a Ripple Power Port," 2009 Twenty-Fourth Annual IEEE Applied Power Electronics Conference and Exposition, Washington, DC, USA, 2009.

[18] T. Shimizu, T. Fujita, G. Kimura and J. Hirose, "A unity power factor PWM rectifier with DC ripple compensation," in IEEE Transactions on Industrial Electronics, vol. 44, no. 4, pp. 447-455, Aug. 1997.

[19] R. Wang, F. Wang, D. Boroyevich and P. Ning, "A high power density single phase PWM rectifier with active ripple energy storage," 2010 Twenty-Fifth Annual IEEE Applied Power Electronics Conference and Exposition (APEC), Palm Springs, CA, USA, 2010.

[20] C. Liu and J. -S. Lai, "Low Frequency Current Ripple Reduction Technique With Active Control in a Fuel Cell Power System With Inverter Load," in IEEE Transactions on Power Electronics, vol. 22, no. 4, pp. 1429-1436, July 2007.

2025 IEEE Workshop on Wide Bandgap Power Devices and Applications in Asia (WiPDA Asia)

3kV/42kW Pump-back Test for Medium Voltage DC Transformer with Series-Connected SiC MOSFETs

Yujian Zong
College of Electrical Engineering
Zhejiang University
Hangzhou, China
22310172@zju.edu.cn

Shuai Shao
College of Electrical Engineering
Zhejiang University
Hangzhou, China
shaos@zju.edu.cn

Chaojun Wang
College of Electrical Engineering
Zhejiang University
Hangzhou, China
3210104166@zju.edu.cn

Wentao Cui
College of Electrical Engineering
Zhejiang University
Hangzhou, China
pes@zju.edu.cn

Junming Zhang
College of Electrical Engineering
Zhejiang University
Hangzhou, China
zhangjm@zju.edu.cn

Abstract—The medium voltage DC transformer (DCX) plays a crucial role in DC grids. To increase the power density of medium voltage DCX, series-connected SiC MOSFETS are utilized to construct a DCX prototype. Active clamping modules (ACMs) with a corresponding voltage balancing strategy are employed to achieve even voltage sharing among series-connected SiC MOSFETS while minimizing the induced power loss. A critical aspect is the ACM capacitor energy recovery during deadtime; with an uncontrolled rectifier, this process degrades DCX load regulation. Consequently, this paper employs an active rectifier, substantially improving load regulation. To validate the voltage balancing control under higher voltage and power conditions, a pump-back test topology with 2DCXs and an additional boost converter is employed, keeping both the forward and reverse LLC converters behaving like a DCX. Experimental results demonstrate excellent voltage sharing and a peak efficiency of 98.6%.

Keywords—series-connected power devices, active clamping circuit, pump-back test, DC transformer

I. INTRODUCTION

The medium voltage direct current transformer (DCX) is used to interconnect medium voltage (e.g., 10 kV) and low voltage (e.g., 750 V) DC grids. Solutions to meet the medium voltage requirements include using SiC MOSFETs with 10 kV blocking voltage, series-connected power devices, multilevel topologies, and the input-series-output-parallel (ISOP) topology [1]. The ISOP topology is widely employed due to its reliability and maturity; however, its power density is relatively low, leading to high land usage costs. The high module count of ISOP topology causes massive reduction of overall power density [2].To increase power density, series-connected SiC MOSFETs are utilized in this work. Considering that parameter deviations can cause unequal voltage sharing among the SiC MOSFETs, a voltage balancing method is necessary [3].

Currently, various voltage balancing methods for series-connected devices have been proposed in academia. These include: on the power side, paralleling snubber circuits [4] or RCD clamp circuits with the devices [5]; and on the drive side, closed-loop regulation of drive delay or drive voltage [6].

Snubber circuits, despite their simplicity, increase equivalent device capacitance, thereby reducing switching speed and increasing switching losses. RCD clamp circuits maintain switching speed but the absorbed energy from the clamp capacitor is still dissipated. Closed-loop control strategies, such as drive delay regulation, offer precise control but are often complex to implement and necessitate high

closed-loop bandwidth. A significant drawback is their potentially slow response during transient events outside of nominal steady-state operation.

The active clamp voltage balancing method proposed in previous literature [7][8] maintains the good voltage balancing performance of RCD clamp circuits while reducing additional losses by recovering capacitor energy during dead time. Based on this voltage balancing method, this paper constructs a high power density DC transformer (DCX) prototype.

This paper proposes and experimentally validates a medium voltage DCX design utilizing series-connected SiC MOSFETs and an active voltage balancing technique. To ensure consistent commutation speeds and true bidirectional DCX operation while avoiding undesirable DAB modes, active bridge modulation is implemented on both medium and low voltage sides. The experimental validation is conducted using a pump-back test setup, which includes an additional Boost converter to maintain open-loop DCX operation and regulate circulating power. Furthermore, this work reports on the experimental verification of the DCX prototype at 3 kV/42 kW, presenting detailed analyses of its efficiency, loss distribution, and voltage balancing performance.

II. VOLTAGE BALANCING AND ENERGY RECOVERY BY THE ACTIVE CLAMPING CIRCUIT

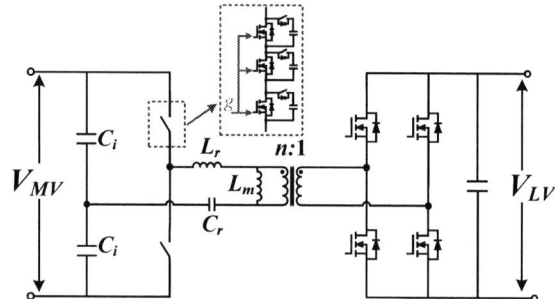

Fig. 1. LLC-DCX topology with series-connected MOSFETs.

979-8-3315-1110-4/25 $31.00 © 2025 IEEE

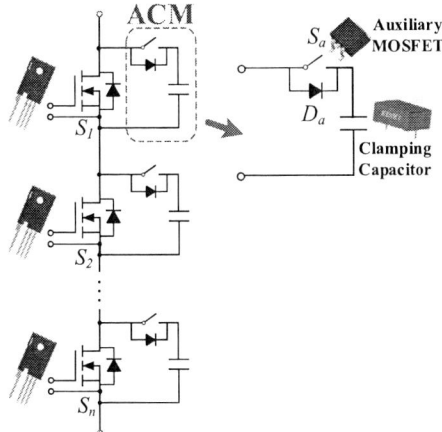

Fig. 2. The circuit structure of active clamping modules.

To achieve voltage sharing of series-connected SiC MOSFETs while minimizing the induced additional power loss, the topology and control methods proposed by [7][8] are employed in this paper. This method is suitable for DCX topologies that include a freewheeling period.

The DCX topology shown in Fig. 1 can be viewed as a resonant converter in which MV-side switches are replaced by series-connected SiC MOSFETs. Fig. 2 shows the circuit structure of the proposed active clamping modules (ACMs) which are connected in parallel across each SiC MOSFET (S_1-S_n). Each ACM consists of an auxiliary MOSFET S_a (with a body diode D_a) and a clamping capacitor C_a in the microfarad (μF) range.

Consider a scenario where S_1, one of the series-connected SiC MOSFETs, turns OFF prematurely or with a higher dv/dt compared to the others. Assuming each clamping capacitor C_a initially holds a voltage of V_{dc}/N (where N is the number of series-connected MOSFETs and V_{dc} is the DC bus voltage). The clamping process is shown in Fig. 3. At t_1, when the drain-source voltage v_{ds1} reaches V_{c1} (V_{dc}/N), D_{a1} starts to conduct, inserting C_{a1} into the circuit. Consequently, v_{ds1} along with v_{c1} increases slowly, thereby protecting S_1 from overvoltage. After t_2, v_{ds1} to v_{dsn} equal steady-state voltage V_{dc}/N, while v_{c1} is slightly higher than the other capacitor voltages due to excessive charge during t_1-t_2. C_{a1} is charged only when v_{ds1} reaches v_{c1}, hence the ACM will not affect the switching speed of the SiC MOSFET.

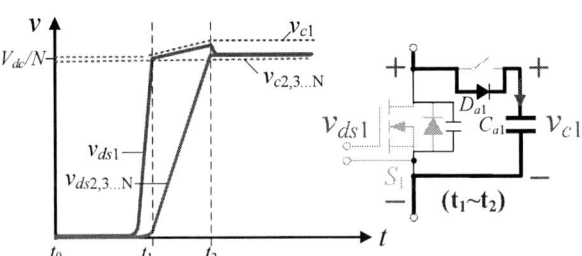

Fig. 3. Waveform of voltage clamping during turn-OFF period.

To achieve voltage balancing of SiC MOSFETs, the voltage of each C_a should be maintained at V_{dc}/N. Considering that the excessive charge across each C_a is uncertain and time-variant, a selective energy recovery strategy is employed. To recover energy, C_a is only discharged back to the DC bus

during the freewheeling period, as part of the deadtime, of S_1-S_n when the load current flows from source to drain.

The employed voltage balancing method is executed in every switching cycle. All the ACM capacitor voltages ($V_{c1} \sim V_{cN}$) are measured and sorted by an FPGA. Then, only k of the N auxiliary MOSFETs S_a with higher ACM capacitor voltages will be turned ON during the deadtime in this switching cycle.

Fig. 4 illustrates the ACM's energy recovery process. Assuming only auxiliary switch S_{a1} is activated by its gate signal g_{a1} from the FPGA, the process unfolds as follows: First, as shown in Fig. 4(a), during a designated period t_a within the deadtime t_d (prior to S_1's turn-on), S_{a1} conducts. This action connects C_{a1} to the DC bus, allowing its stored energy to be recovered. Fig. 4(b) presents the equivalent circuit for this energy recovery phase. Following this, the remaining deadtime interval (t_d - t_a) is dedicated to achieving Zero-Voltage Switching ON for S_1, as shown in Fig. 4(c).

Fig. 5 shows simulation results of ACM capacitor voltages V_c before and after enabling the energy recovery at rated voltage. Through voltage clamping, sorting, and energy recovery, voltage balancing can be achieved with an imbalance degree of less than 5%. The degree of voltage imbalance d_{im} can be defined by (1).

$$d_{im} = \frac{V_{c\max} - V_{c\min}}{\sum_{i=1}^{N} V_{ci} / N} \tag{1}$$

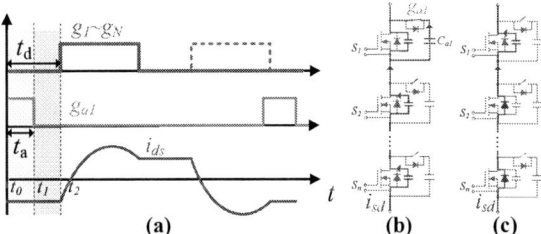

Fig. 4. (a) Waveforms of energy recovery. (b) Equivalent circuit when recovering energy from C_{a1} (c) Equivalent circuit when S_1 achieve ZVS-ON.

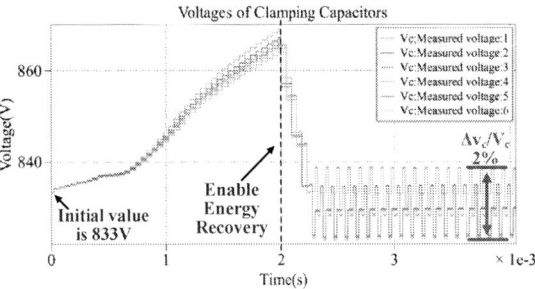

Fig. 5. Simulation of the voltage across the clamping capacitors when energy recovery is activated at 2 ms.

III. PUMP-BACK TEST OF THE MEDIUM VOLTAGE DC TRANSFORMER

In the DCX, the switching frequency equals to the resonant frequency as shown in Fig. 1. Due to the insertion of clamp capacitor C_a during the dead time, the commutation speed on the medium voltage (MV) side of the resonant tank is slower compared to the low voltage (LV) side. Consequently, as illustrated in Fig. 6 for power flow from the LV to the MV side (with the MV side operating in passive rectification mode), the change in the MV-side resonant tank voltage (v_{rMV}) lags behind the LV-side resonant tank voltage (v_{rLVT}). This portion of volt-second excitation applied to the resonant tank causes i_{rMV} to change in the negative direction as depicted in the figure. This change is particularly significant when the resonant inductance L_r in the DC transformer is relatively low.

The phase difference between the voltages on both sides of the resonant tank affects the magnitude of the transmitted power, similar to the operating mode of a Dual Active Bridge (DAB) converter. This causes the DC transformer to lose some of its load-independent constant voltage ratio transmission characteristics, leading to poorer voltage regulation. To mitigate this effect and restore DCX-like operation, active bridge modulation can be implemented on both the MV and LV sides, similar to the approach in [9]. By lagging v_{rLV}, the net volt-second excitation during commutation can be reduced. This, in turn, minimizes the phase difference between the voltages on both sides of the resonant tank, thereby achieving a voltage regulation closer to unity. The experimental test comparison results are shown in Fig. 7.

A 10kV/167kW DCX prototype is designed and constructed, as shown in Fig. 8, where each arm consists of 6 series-connected SiC MOSFETs. In order to eliminate the requirement for high-power power supplies and loads in testing, a pump-back test circuit is proposed, as illustrated in Fig. 9, where 2 DCXs are connected in parallel on the MV side.

In the conventional pump-back testing setup for DC/DC converters, one converter acts as a current source while the other serves as a voltage source [10]. However, since DC transformers typically operate at a fixed frequency and voltage transfer ratio without control, this pump-back testing method is not easily applicable. To solve this problem, an external boost converter is used to compensate for the voltage drop in the circulating loop and control the circulating power of the test. Both DCXs are devices under test (DUT), representing forward and backward modes respectively. Thus, 2 DCXs can maintain the behavior of DCX and be tested simultaneously.

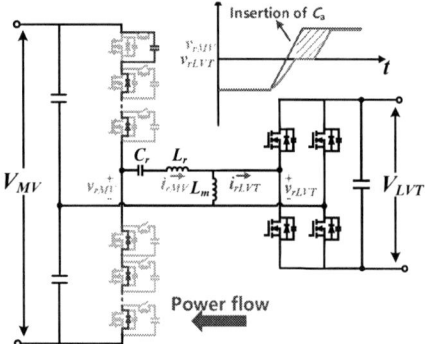

Fig. 6. LLC when MV side acts as the rectifier.

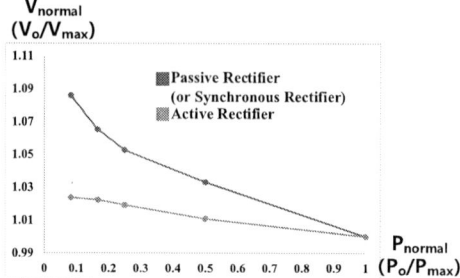

Fig. 7. The load regulation of DCXs with uncontrolled and active rectifier.

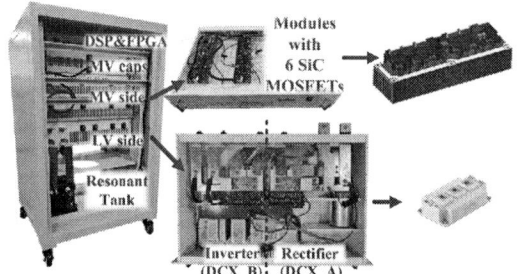

Fig. 8. The prototype of 2 medium voltage DCXs with series-connected devices modules.

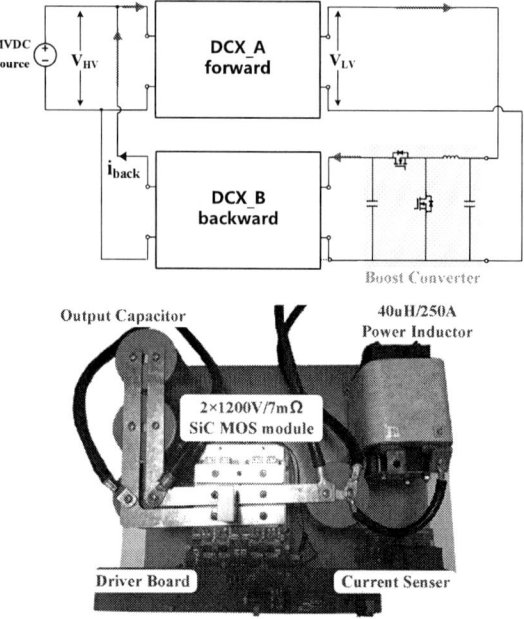

Fig. 9. The topology of the pump-back test with an additional boost converter.

When only considering the losses that primarily cause voltage drop in the converter, the efficiency η of the converter is also equal to the voltage transfer ratio from output to input side. To satisfy Kirchhoff's voltage law, the Boost converter must elevate a certain voltage, as described by the following equation (2), where d represents the duty cycle of the main power device in the Boost converter. Since the efficiency of DCX and Boost converters can reach around 98% to 99%, the Boost converter only requires a duty cycle of less than 5%, and the circulating power is regulated through closed-loop control of the duty cycle.

$$V_{MV} \times \eta_{DCXA}\eta_{DCXB}\eta_{Boost} \times \frac{1}{1-d} = V_{MV}$$

$$d = 1 - \eta_{DCXA}\eta_{DCXB}\eta_{Boost} \tag{2}$$

In the section where the efficiency of each converter decreases with increasing circulating power, the lower switch duty cycle d of the Boost converter has a positive correlation with the circulating power P_c. Therefore, closed-loop control of the circulating power can be established. Fig. 10 shows the dynamic response of the actual operating current for the pump-back test when the circulating current reference is increased from 4 A to 10 A. By designing the PI controller, the pump-back test circulating current can track the set reference value.

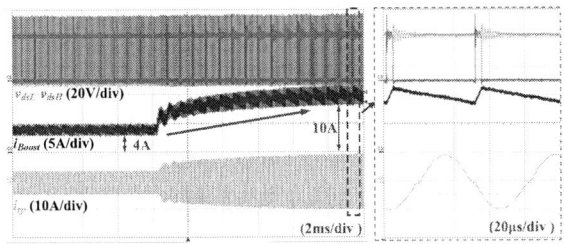

Fig. 10. Pump-back test circulating current increase from 4 A to 10 A with current loop control of Boost converter.

IV. EXPERIMENTAL RESULTS

The voltage balancing capability of the medium voltage DCX with series-connected SiC MOSFETs was tested under the conditions of a 3 kV input voltage and 42 kW circulating power. The parameters of the pump-back test are shown in Table I.

TABLE I.   THE PARAMETERS OF THE PUMP-BACK TEST

Parameter	Value
MV/LV Voltage	3 kV / 225 V
Resonant Frequency	21 kHz
Transformer Ratio	20:3
Magnetizing Inductance	4.5 mH
Resonant Inductance	65 µH
Circulating Power	Up to 42 kW

The voltage and current waveforms of the primary and secondary sides of the resonant tank are shown in Fig. 11, illustrating the converter's behavior at the resonant point. The voltage waveforms of three out of six clamping capacitors and the MV-side resonant tank current i_{rMV} are depicted in Fig. 12. Since the clamping capacitors are well balanced, the V_{ds} voltages also show good balance, as illustrated in Fig. 13. The imbalance degree of the power devices and clamping capacitors is less than 5% under different conditions.

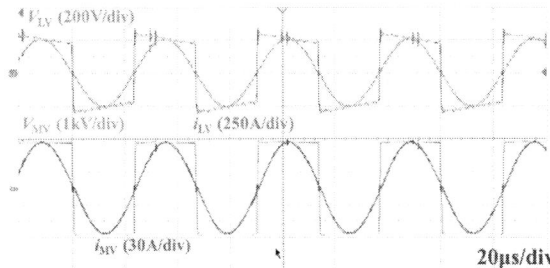

Fig. 11. i_{rMV}, V_{rMV}, i_{rLV}, and V_{rLV} at 3kV/42kW.

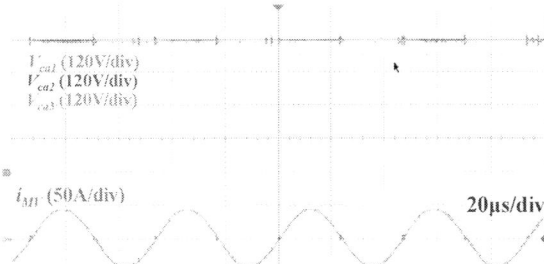

Fig. 12. Voltages of the three out of six clamping capacitors with MV side resonant tank current.

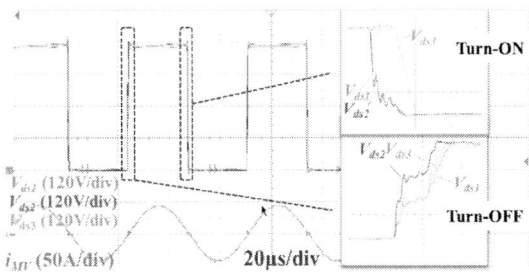

Fig. 13. Voltages of the three out of six SiC MOSFETs with MV side resonant tank current.

The excess energy stored in the clamping capacitor can be recovered back to the DC bus through reactive power, and the auxiliary switch only conducts during part of the dead time and is turned on at zero voltage. The additional losses introduced by the voltage balancing method are relatively small. Fig. 14 shows the simulation comparison of the losses of a single main power switch and the auxiliary switch under different cycling power levels with a 3 kV input. At full load, the loss of the auxiliary switch is only 1.6% of that of the main switch.

Fig. 15 illustrates the loss distribution of each component under a 3 kV input. The primary and secondary side devices can achieve ZVS-ON and ZCS-OFF, respectively, resulting in relatively low switching losses. However, under heavy load conditions, the secondary side experiences significant current stress, leading to higher conduction losses, which can be optimized by increasing the capacity of the power module.

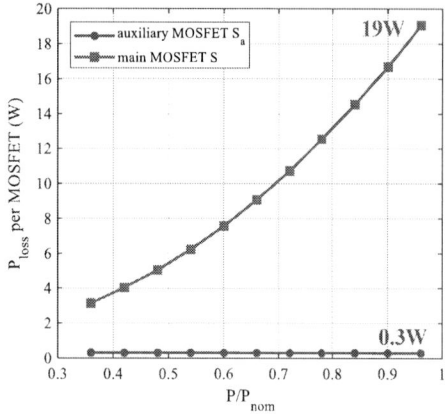

Fig. 14. Simulation loss comparison of auxiliary and main power switches.

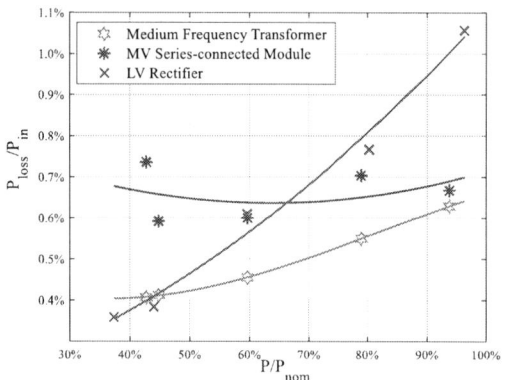

Fig. 15. Loss breakdown of the series-connected device modules, transformer, and low-side power devices.

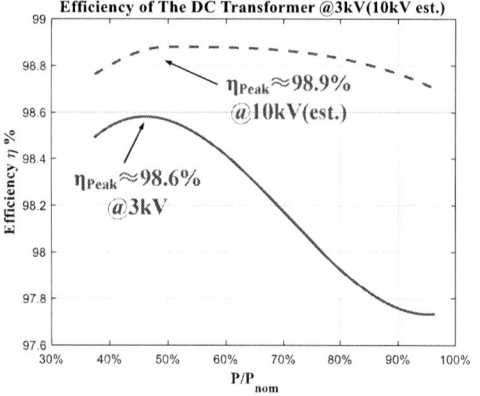

Fig. 16. The Efficiency of the DCX under different load conditions at 3kV

The efficiency of the medium voltage DCX prototype is calculated under different loads at a 3 kV input. Fig. 16 shows that the peak efficiency is approximately 98.6%. It is estimated that the peak efficiency can reach 98.9% under the designed conditions of 10 kV / 166 kW.

V. CONCLUSION

This paper detailed a high-power-density DC transformer (DCX) built with directly series-connected SiC MOSFETs and an active clamping voltage balancing scheme. The DCX load regulation can degrade due to the ACM capacitor energy recovery process during deadtime. This paper employs an active rectifier, substantially improving load regulation. To validate the voltage balancing control under higher voltage and power conditions, a pump-back test topology with 2DCXs and an additional boost converter is employed, keeping both the forward and reverse LLC converters behaving like a DCX.

A 3 kV, 42 kW pump-back test successfully validated the design. Key findings include consistent voltage balancing (imbalance < 5%) among series devices with minimal associated losses. The prototype achieved a peak efficiency of 98.6% around half load. These results suggest that scaling this DCX to 10 kV/166 kW could yield efficiencies approaching 98.9%, highlighting the effectiveness of the proposed techniques for high-power DC conversion.

ACKNOWLEDGMENT

This work was supported by the research project of China Southern Power Grid Co., Ltd. under Grant 030400KK52222006 (GDKJXM20222138).

REFERENCES

[1] S. Castagno, R. D. Curry, and E. Loree, "Analysis and comparison of a fast turn-on series IGBT stack and high-voltage-rated commercial IGBTS," IEEE Trans. Plasma Sci., vol. 34, no. 5, pp. 1692–1696, Oct. 2006.

[2] Kolar, Johann W., and Jonas Huber. "Solid-State Transformers-Fundamentals, Industrial Applications, Challenges." 14th Annual IEEE Energy Conversion Congress and Exposition (ECCE 2022). ETH Zurich, Power Electronic Systems Laboratory, 2022.

[3] A. Marzoughi, R. Burgos, and D. Boroyevich, "Active gate-driver with dv/dt controller for dynamic voltage balancing in series-connected SiC MOSFETs," IEEE Trans. Ind. Electron., vol. 66, no. 4, pp. 2488–2498, Apr. 2019.

[4] Trochimiuk, Przemysław, et al. "Medium voltage power switch based on 1.7 kV SiC MOSFETs connected in series inside power modules." 2019 21st European Conference on Power Electronics and Applications (EPE'19 ECCE Europe). IEEE, 2019.

[5] Zhang, Fan, et al. "A high input voltage auxiliary power supply utilizing automatic balanced MOSFET stack." IEEE Transactions on Industrial Electronics 68.11 (2020): 10654-10665.

[6] Lin, Xiang, et al. "Hybrid voltage balancing approach for series-connected 10 kV SiC MOSFETs for DC-AC medium-voltage power conversion applications." 2020 IEEE Energy Conversion Congress and Exposition (ECCE). IEEE, 2020.

[7] J. Zhang, S. Shao, Y. Li, J. Zhang and K. Sheng, "A Voltage Balancing Method for Series- Connected Power Devices in an LLC Resonant Converter," in IEEE Transactions on Power Electronics, vol. 36, no. 4, pp. 3628-3632, April 2021.

[8] Wu, Tong, et al. "Medium Voltage DC Transformer with Series-Connected Power Devices Based on Active Clamping Circuit." 2023 IEEE Energy Conversion Congress and Exposition (ECCE). IEEE, 2023.

[9] Guillod, Thomas, Daniel Rothmund, and Johann Walter Kolar. "Active magnetizing current splitting ZVS modulation of a 7 kV/400 V DC transformer." IEEE Transactions on Power Electronics 35.2 (2019): 1293-1305..

[10] Heinig, Stefanie, et al. "Experimental insights into the MW range dual active bridge with silicon carbide devices." 2022 International Power Electronics Conference (IPEC-Himeji 2022-ECCE Asia). IEEE, 2022.

2025 IEEE Workshop on Wide Bandgap Power Devices and Applications in Asia (WiPDA Asia)

Thermal Network of Double-sided Cooling Power Module Based on Cauer Model

1st Jie LI
College of Electrical Engineering
Zhejiang University
Hangzhou, China
12210100@zju.edu.cn

2nd Min CHEN
College of Electrical Engineering
Zhejiang University
Hangzhou, China
calim@zju.edu.cn

3rd Xinnan SUN
College of Electrical Engineering
Zhejiang University
Hangzhou, China
sxnan@zju.edu.cn

4th Jun HUANG
Shanghai Institute of Space Power Sources
Shanghai, China
huju1981@163.com

5th Yucheng WU
College of Information Science and Electronic Engineering
Zhejiang University
Hangzhou, China
22431134@zju.edu.cn

6th Yifei DU
College of Electrical Engineering
Zhejiang University
Hangzhou, China
22310190@zju.edu.cn

7th Han ZHOU
College of Electrical Engineering
Zhejiang University
Hangzhou, China
3210104140@zju.edu.cn

8th Feng JIANG
College of Electrical Engineering
Zhejiang University
Hangzhou, China
jiangfeng@zju.edu.cn

Abstract—As the volume of power modules decreases and power density increases, the problem of efficient heat dissipation of high-power modules needs to be solved urgently. However, existing power module analytical models have poor portability, a complex establishment process, and low accuracy. Therefore, establishing a universal and high-precision heat conduction model is significant in guiding module thermal design. This paper demonstrates a three-dimensional Cauer thermal circuit model with clear physical meaning for the double-sided heat dissipation power module used in electric vehicle motor controllers. First, based on the module's internal temperature and heat flux density distribution rules, the concepts of longitudinal thermal resistance and transverse diffusion thermal resistance are proposed, and the traditional thermal resistance calculation formula is modified based on the principle of thermal diffusion. On this basis, a wide range of working conditions was established to verify the accuracy and high robustness of the dynamic and static models of the thermal network model, providing a solid basis for the thermal design of power modules and multi-chip parallel thermal decoupling.

Keywords—Power module, double-sided cooling, thermal design, Cauer thermal network

I. INTRODUCTION

Module thermal capacity is not only related to the device switching performance but also seriously affects the aging/degradation speed of insulation and welding materials, affecting the module's reliability [1], [2].

Currently, theoretical research methods for module thermal design primarily involve establishing numerical thermal network models and finite element analysis (FEA) models[3], [4], [5], [6]. While numerical computation methods offer advantages such as model simplicity and fast calculation speed, their accuracy is compromised by necessary model simplifications when addressing actual complex module structures and operating conditions. Conversely, finite element simulation provides high accuracy but suffers from complex modeling processes, prolonged computation time, significant computational demands, and a lack of clear physical interpretation.

During the preliminary design phase, multiple iterations of the package structure are typically required to optimize the

This work was financially supported by the Key R&D plan of Zhejiang Province (No.2025C01048).

module's electrothermal performance. The time cost increases substantially if each iteration involves redesigning the package structure and conducting FEA simulations. Establishing a high-precision, high-robustness numerical model early in the design process, combined with later-stage FEA validation, can reduce design costs and shorten development cycles. Therefore, the primary objective of this paper is to develop a high-precision and high-robustness analytical computational model for package thermal analysis to enhance module design efficiency.

Thermal network models are categorized into continuous and partial network models, also known as Cauer models and Foster models, respectively, as illustrated in Fig. 1. The Cauer model accurately represents the equivalent thermal impedance of each material layer within the module. Each layer (e.g., die, die-attach solder layer, insulating substrate, substrate solder layer, and baseplate) corresponds to a specific set of thermal resistance (R) and capacitance (C) parameters. When the material properties and structural parameters of each layer in the thermal system are known, the Cauer model can be theoretically derived. Solving the nodal temperatures in the continuous thermal network can determine the temperature distribution across each material layer.

In contrast, the Foster model does not directly correlate its R/C parameters with physical material layers, nor do its network nodes carry explicit physical meaning. However, the R/C parameters in the Foster model can be extracted by curve-fitting the transient thermal impedance (Z_{th}) obtained from

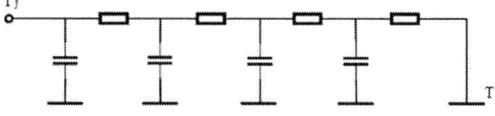

(a) Continuous network thermal model (Cauer model or T-Model)

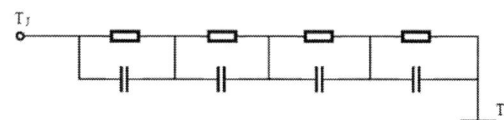

(b) Partial thermal circuit network (Foster model or π-Model)

Fig. 1. Typical thermal network models.

experimental measurements. As a result, the Foster model is widely adopted for practical power module modeling.

Since the Foster model requires physical module testing parameters for establishment, its implementation is not feasible during the packaging design phase. In contrast, the Cauer thermal circuit model, with its clear physical significance and significantly reduced computational effort compared to finite element simulation models, plays a crucial role in initial thermal design and optimization of modules. However, existing Cauer thermal circuit models employ excessive simplifications to reduce modeling complexity, resulting in generally low computational accuracy. Moreover, current Cauer models are predominantly applied to single-side cooling modules, leaving a research gap in dual-side cooling modeling [7], [8], [9].

In this paper, we first validate the equivalent heat transfer coefficient (HTC) model and liquid cooling model, as well as the substitution effect of equal-area circular and rectangular dies. Building on this, we introduce the concepts of longitudinal and transverse diffusion thermal resistance for the single heat source and double-sided heat dissipation paths. Finally, we verify the accuracy and robustness of the proposed improved Cauer thermal network model for double-sided heat dissipation power modules by comparing it with the FEA simulation model under a wide range of operating conditions.

II. VALIDATION OF SIMPLIFIED EQUIVALENT FEA MODULE

This paper uses a substrate-embedded double-sided liquid-cooled module as an example (Note: the thermal network modeling method proposed here is also applicable to other double-sided heat dissipation package designs) to study thermal circuit modeling for double-sided heat dissipation power modules. The sectional structure of the module is shown in Fig. 2.

Existing high-power modules typically employ forced liquid cooling for heat dissipation, and high simulation accuracy can be achieved using a liquid-cooled fluid FEA simulation model. However, the actual fluid modeling process is complex and computationally intensive. An equivalent HTC can be set on the module surface to reduce the computational effort while maintaining accuracy.

Therefore, this paper focuses on the substrate-embedded double-sided cooling power module and sets up a double-sided liquid-cooled fluid model to obtain the equivalent HTC. This model, along with the liquid-cooling model and the equivalent heat transfer model for the single die with double-sided heat dissipation, is shown in Fig. 3.

An equal-area circular die replaces the rectangular die to reduce computational complexity further and improve

efficiency. A rectangular die has two variables—length and width—but by replacing it with a circular die of equal area, the number of variables is reduced to just one, the radius, thus significantly simplifying the computation.

This section takes the ROHM automotive-grade chip model TK-S4661 (1200V/122A) as an example, combining its embedded package structure and actual operating conditions, to research the equivalent heat transfer coefficient model of dual-side cooling modules and the validation of equivalent substitution effects using equal-area circular chips.

The study establishes: A dual-side liquid-cooled fluid heat transfer model (rectangular chip), an equivalent heat transfer model (rectangular chip), and an equivalent heat transfer model (equal-area circular chip), to verify the consistency between the computational results of the equivalent substitution model and the actual model.

The embedded package module's structural, material, and operating condition parameters are detailed in Tables I, II, and III. The simulation parameters are configured based on the actual operating conditions and specifications of automotive-grade modules in current electrical vehicle motor controllers.

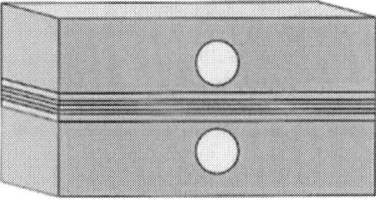

(a) Liquid-cooled fluid modeling.

(b) Equivalent heat transfer model.

Fig. 3. FEA model of double-sided liquid cooling power module.

TABLE I. MATERIAL PARAMETER OF SUBSTRATE EMBEDDED POWER MODULE

Material	Thermal conductivity (W/(m^2·°C))	Specific heat (J/(kg·k))	Density (kg/m³)
Copper	387.6	195	8933
FR₄	0.35	1300	1250
Solder	56.9	239	8690
AlN Ceramics	180	720	3300
Silicon Carbide	490	677.8	3210

TABLE II. STRUCTURAL PARAMETERS OF THE EMBEDDED POWER MODULE

Thermal Path	Material Thickness (mm)
Chip	0.2
Chip-to-interconnect solder	0.05
Epoxy Molding Compound	0.3
Interconnect Layer	0.3
Metal-Ceramic Bonding Layer	0.05
Ceramic Substrate	0.5
Sink	3
Inlet/Outlet Port Area (mm²)	30

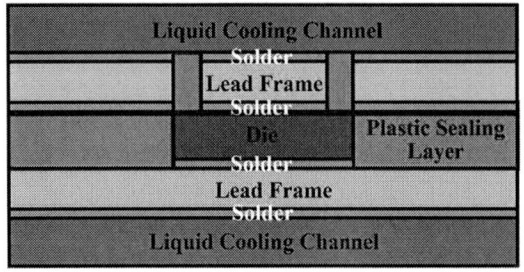

Fig. 2. Diagram of the cross-sectional of the double-sided cooling power module.

TABLE III. OPERATING CONDITION SIMULATION PARAMETERS

Condition	Parameter Settings
Ambient Temperature / °C	65
Coolant Inlet Temperature / °C	65
Die Power Dissipation / W	150
Coolant Type	50% Ethylene Glycol
Inlet Flow Velocity / (m/s)	3

The simulation results, shown in Fig. 4, reveal that when the equivalent HTC for the upper and lower surfaces of the module is set to 62,000 W/(m^2·°C), the maximum junction temperatures for all three equivalent models are 94°C. The temperature distribution within the module is virtually identical, which confirms the consistency of the equivalent heat transfer model.

III. DOUBLE-SIDED COOLING PATH THERMAL NETWORK MODEL

A. Steady-State Thermal Network Model of Symmetric Double-Sided Cooling Path

In this section, the study begins with an example of a double-sided symmetric heat dissipation path, where the heat dissipation from both the upper and lower surfaces of the die to the top and bottom shells of the module is assumed to be identical. Traditional Cauer modeling calculates thermal resistance for the heat sink path on the same plane using (1).

$$R_i = \frac{h_i}{(\lambda_i A_i)} \qquad (1)$$

According to (2), it can be seen that its application condition is assumed that the thermal resistance is uniformly distributed in the same plane, however, according to the actual

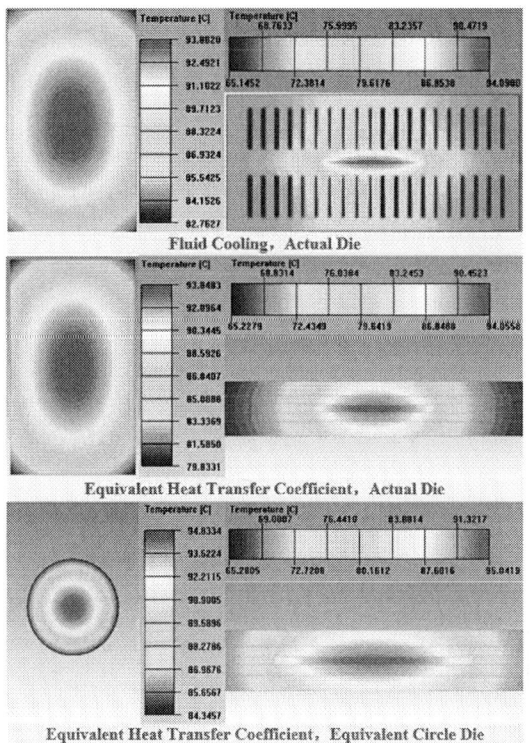

Fig. 4. Simulation results comparison between liquid cooled fluid model, equivalent heat transfer coefficient model, and equivalent die model.

test results and simulation results, the temperature gradually decreases from the center to the peripheral edge of the same plane, that is to say, the density of the heat flow along the radial direction shows a gradual decrease, and at this time, along with the use of (2) will result in a significant deviation. Therefore, this section proposes the concept of gradient thermal resistance, the thermal resistance of the heat conduction path is divided into longitudinal thermal resistance and transverse thermal resistance, as shown in Fig. 5 where R_{i_m} is the longitudinal thermal resistance of the region directly below the circle of the equivalent area of the die, and R_{i_a} is the transverse thermal diffusion thermal resistance of the heat conduction path, the longitudinal thermal resistance and the transverse thermal resistance in parallel to form the i-th layer of the die conduction path conduction thermal resistance.

At the same time, the traditional thermal resistance calculation formula shown in (1) is corrected. The thermal resistance calculation formula that varies with the thermal diffusion radius is proposed, as shown in (2) and (3), which utilizes the thermal diffusion radius to correct the traditional thermal resistance calculation formula, and with the increase of the thermal diffusion radius, its equivalent thermal resistance is also increased.

$$R_{i\_a} = \frac{h_i}{\lambda_i A_{ia}} \cdot \frac{r_0 + \sum_{j=1}^{j=i} \Delta r_j}{\sum_{j=1}^{j=i} \Delta r_j} \qquad (2)$$

$$R_{i\_m} = \frac{h_i}{\lambda_i A_{im}} \cdot \frac{r_0 + \sum_{j=1}^{j=i} \Delta r_j}{r_0} \qquad (3)$$

Based on the proposed concept of gradient thermal resistance, the traditional double-sided heat dissipation Cauer continuous thermal path model is modified as shown in Fig. 6. The longitudinal thermal resistance and transverse thermal resistance of each layer are connected in parallel to form the total thermal resistance of each layer in the heat dissipation path.

B. Transient Continuous Thermal Circuit Modeling of Double-Sided Symmetric Heat Dissipation Paths

Due to the differences in the specific heat capacity of different materials' intrinsic properties, the ability of heat

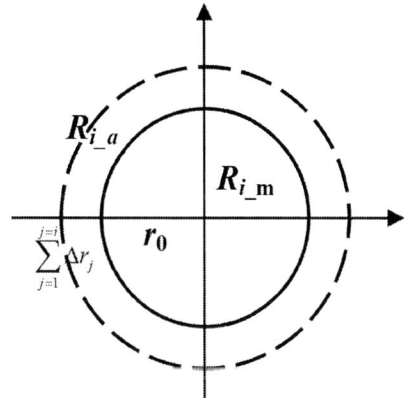

Fig. 5. Vertical and horizontal gradient thermal resistance model.

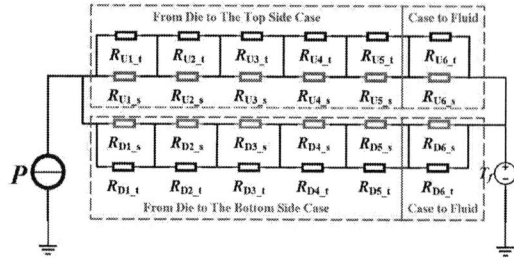

Fig. 6. Steady-state thermal network of double-sided cooling power module with the single heat source.

absorption and release between different encapsulation materials will also be different. Therefore, in the module design process, it is necessary to research the influence of the heat capacity of the packaging material on the module temperature change process. The heat capacity is treated the same way as thermal resistance, divided into longitudinal heat capacity and transverse heat diffusion heat capacity, and the heat capacity is composed of longitudinal heat capacity and transverse heat capacity in parallel. The longitudinal and transverse heat capacity calculation formulas are shown in (4) and (5).

$$C_{U/D\_i\_m} = c_{U/D\_i} \cdot \rho_{U/D\_i} \cdot A_{U/D\_i\_m} \cdot h_{U/D\_i} \cdot \frac{r_0 + \sum_{j=1}^{j=i} \Delta r_j}{r_0} \quad (4)$$

$$C_{U/D\_i\_a} = c_{U/D\_i} \cdot \rho_{U/D\_i} \cdot A_{U/D\_i} \cdot h_{U/D\_i\_a} \cdot \frac{r_0 + \sum_{j=1}^{j=i} \Delta r_j}{\sum_{j=1}^{j=i} \Delta r_j} \quad (5)$$

C_{U/D_i_m} is the longitudinal conduction heat capacity from the die to the upper/lower surface of the module, C_{U/D_i_a} is the transverse diffusion heat capacity from the die to the upper/lower surface of the module.

According to (4) and (5), combined with the equivalent heat transfer model shown in Fig. 5, the consistency of the transient change process of the junction temperature between the continuous heat path model and the FEA model is verified. The dynamic change process of the junction temperature of the die under the heat transfer conditions of pulse period 0.4 s, duty cycle 50 %, die heating power 150 W, ambient temperature 65 °C, and HTC 62000 W/(m^2·°C) is shown in Fig. 7.

According to the die junction temperature dynamic change process shown in Fig. 7, a comparison between the improved Cauer thermal network and the FEA model calculation results can be found. The improved Cauer thermal network junction temperature dynamic response curve and the FEA model are consistent with the dynamic response of the temperature, indicating that the ability to predict the change in accuracy is high, instead of the FEA model achieving the accurate prediction of steady state and transient junction temperature.

IV. CONCLUSION

This paper proposes an improved Cauer thermal network with clear physical significance, high accuracy, and high robustness. Based on the module's internal temperature and heat flow density law, the concept of gradient thermal resistance is proposed, and the parallel connection of

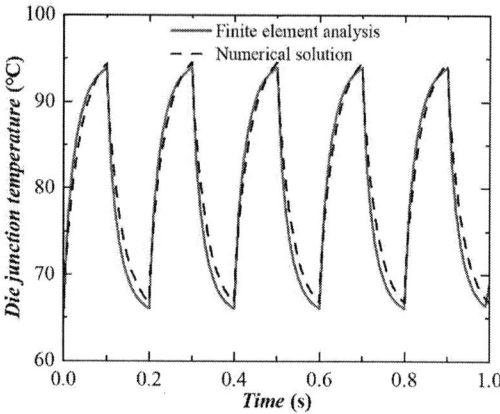

Fig. 7. Results comparison of FEA and thermal circuit model for transient temperature changes in die junction equations.

longitudinal and transverse thermal resistance is used to significantly improve the modeling accuracy of Cauer's three-dimensional thermal circuit model. Given the inadequacy of the equivalent modeling of the existing double-sided cooling thermal network, an equivalent thermal resistance calculation method for the heat dissipation path is proposed, and the model accuracy is closer to the actual working conditions in the heat dissipation path, as circles of the same area.

REFERENCES

[1] M. Liu, A. Coppola, M. Alvi and M. Anwar, "Comprehensive Review and State of Development of Double-Sided Cooled Package Technology for Automotive Power Modules," in IEEE Open Journal of Power Electronics, vol. 3, pp. 271-289, 2022.

[2] J. Knoll, C. DiMarino and C. Buttay, "A Guide for Accurate and Repeatable Measurement of the RTH, JC of SiC Packages," 2022 IEEE Energy Conversion Congress and Exposition (ECCE), Detroit, MI, USA, 2022, pp. 1-7.

[3] R. Wu, F. Iannuzzo, H. Wang and F. Blaabjerg, "An Icepak-PSpice co-simulation method to study the impact of bond wires fatigue on the current and temperature distribution of IGBT modules under short-circuit," 2014 IEEE Energy Conversion Congress and Exposition (ECCE), Pittsburgh, PA, USA, 2014, pp. 5502-5509.

[4] R. Wu, F. Iannuzzo, H. Wang and F. Blaabjerg, "Fast and accurate Icepak-PSpice co-simulation of IGBTs under short-circuit with an advanced PSpice model," 7th IET International Conference on Power Electronics, Machines and Drives (PEMD 2014), Manchester, UK, 2014, pp. 1-5.

[5] Z. Hu, M. Du and K. Wei, "An Adaptive Thermal Equivalent Circuit Model for Estimating the Junction Temperature of IGBTs," IEEE Journal of Emerging and Selected Topics in Power Electronics, vol. 7, no. 1, pp. 392-403, March. 2019.

[6] Y. Zhang, E. Deng, Z. Zhao, S. Fu and X. Cui, "A Physical Thermal Network Model of Press Pack IGBTs Considering Spreading and Coupling Effects," in IEEE Transactions on Components, Packaging and Manufacturing Technology, vol. 10, no. 10, pp. 1674-1683, Oct. 2020.

[7] A. S. Bahman, K. Ma and F. Blaabjerg, "A Lumped Thermal Model Including Thermal Coupling and Thermal Boundary Conditions for High-Power IGBT Modules," in IEEE Transactions on Power Electronics, vol. 33, no. 3, pp. 2518-2530, March 2018.

[8] Z. Wang and W. Qiao, "A Physics-Based Improved Cauer-Type Thermal Equivalent Circuit for IGBT Modules," IEEE Transactions on Power Electronics, vol. 31, no. 10, pp. 6781-6786, Oct. 2016.

[9] X. Sun et al., "Design and Evaluation of a Face-Down Embedded SiC Power Module With Low Parasitic Inductance and Low Thermal Resistance," in IEEE Transactions on Power Electronics, vol. 38, no. 3, pp. 2799-2804, March 2023.

Multi-Objective Automatic Design of Power Module Packaging Based on Artificial Neural Network and Deep Reinforcement Learning

Weina Mao
School of Electrical Engineering and
Automation
Hefei University of Technology
Hefei, China
2023170557@mail.hfut.edu.cn

Jianing Wang
School of Electrical Engineering and
Automation
Hefei University of Technology
Hefei, China
jianingwang@hfut.edu.cn

Shaolin Yu*
The Institute of Energy, Hefei
Comprehensive National Science
Center(Anhui Energy Laboratory)
Hefei, China
yusl@ie.ah.cn

Zhenchun Xia
School of Electrical Engineering and
Automation
Hefei University of Technology
Hefei, China
2023170526@mail.hfut.edu.cn

Honghong Li
School of Electrical Engineering and
Automation
Hefei University of Technology
Hefei, China
2023170551@mail.hfut.edu.cn

Xiang Pan
School of Electrical Engineering and
Automation
Hefei University of Technology
Hefei, China
2023110402@mail.hfut.edu.cn

Xiahao Wang
School of Electrical Engineering and
Automation
Hefei University of Technology
Hefei, China
Huainan, China
2023170534@mail.hfut.edu.cn

Abstract—Traditional design methods for power module packaging heavily rely on manual experiences. Although the computer-aided optimization (CAO) methods utilizing metaheuristic algorithms have accelerated the design process, they are still affected by lengthy iterative cycles when including 3D numerical simulations. Regarding these issues, this paper proposes a multi-objective automated design (MOAD) method for power modules based on machine learning, combining Artificial Neural Networks (ANN) and Deep Reinforcement Learning (DRL). The proposed ANN model can save massive time of iterative 3D numerical calculations in multi-objective optimization (MOO) design of the power module packaging, meanwhile avoiding the complex interaction of different software. The DRL algorithm can output the optimal design variables efficiently, even when design requirements change after the agent are well trained. A 1200V/300A double-sided cooled (DSC) module is used as a case study to validate the proposed method.

Keywords —power module, multi-objective automated design (MOAD), machine learning, artificial neural networks (ANNs), deep reinforcement learning (DRL)

I. INTRODUCTION

With the rapid advancement of new energy generation, electrified transportation, etc., the industrial applications have imposed increasingly higher requirements on the power modules[1][2].

The packaging design of power modules is crucial as it significantly impacts their performance. In recent years, multi-objective optimization design has gradually become a research focus in power module packaging design.

This work was supported by Anhui Province Key Research and Development Program Project, Project Number: JZ2024AKKG0057.

Traditional optimization methods rely heavily on trial-and-error procedures, where human experience and extensive testing lead to issues such as long design cycles and high costs.

Currently, computer-aided multi-objective optimization design methods are widely used in power module packaging[3][4]. However, existing multi-objective optimization methods still have certain limitations. Firstly, parasitic parameters, thermal resistance, and other objectives are often mathematically modeled using analytical formulas. However, for complex power module structures, such as dual-side water-cooled modules, the accuracy of analytical models remains limited. Secondly, although some literature has developed finite element models (FE models) for stress, they require extensive parameterized simulations, which lead to significant computational time.

To improve design efficiency, some studies have simplified the power module structure using graphical models or DNA strings to achieve multi-objective automated layout[5]-[7]. However, this approach greatly simplifies the module's physical structure into points, lines, and rectangular surfaces, resulting in significant errors in the calculations of parasitic inductance, junction temperature, and other parameters. To achieve a higher precision model, multi-field coupled FE simulations are necessary. Yet, another problem arises: the computational load is large and time-consuming[8]-[13]. More importantly, when design requirements change, such as voltage, current, or power density, the above methods require re-execution, leading to significant manual effort and time consumption.

To address the issues, this paper proposes a novel multi-objective automated design method for the packaging of power modules based on machine learning, combining Artificial Neural Networks (ANN) and Deep Reinforcement Learning (DRL). The main innovations of this paper are as follows:

1) A unified ANN model is proposed to replace numerical simulations in different physical domains, which can save massive time of iterative 3D numerical calculations in MOO design of the power module packaging and avoid the complex interaction of different software.

2) DRL embedding the above ANN models is firstly proposed for the MOO design of the packaging of power modules, which significantly improves design efficiency while increasing accuracy. With the DRL-based design, not only a MOO can be achieved, but also new optimal design variables can be obtained within very short time when a new design requirement are input.

The remaining parts of this paper are as follows: Section II introduce the design case and the basic framework of the prosed method. Section III presents the specific methodology and design process. Section IV evaluates the optimization performance of the proposed method and compares it with existing MOO methods to demonstrate its effectiveness. Section V presents the assembly of power module and experimental results. Section VI concludes this paper.

II. FRAMEWORK OF THE PROPOSED AUTOMATED DESIGN METHOD

To address the MOO challenges in power module packaging design, this paper proposes a multi-objective automated design method based on the combination of ANN and DRL. The framework of the method is shown in Fig. 1. Three basic concepts are defined as follows:

1) Design requirements: defined as the changing requirements from the client for a power module, e.g., the rating current I here in our simplified case.

2) Design variables: defined as key parameters of power module to be designed by engineers, the key layout dimensions of the substrate (l_1, l_2, l_3, l_4).

3) Design objectives: defined as the design objectives of the power module, e.g., parasitic inductance L, junction temperature T_j, power density ρ here.

Overall, the MOAD method contains two steps that are the modeling and the self-learning based on DRL.

III. A DESIGN CASE OF MOAD FOR DSC POWER MODULE

In this section, the proposed MOAD of the DSC power module is presented as a case study. The modeling and self-learning are introduced respectively.

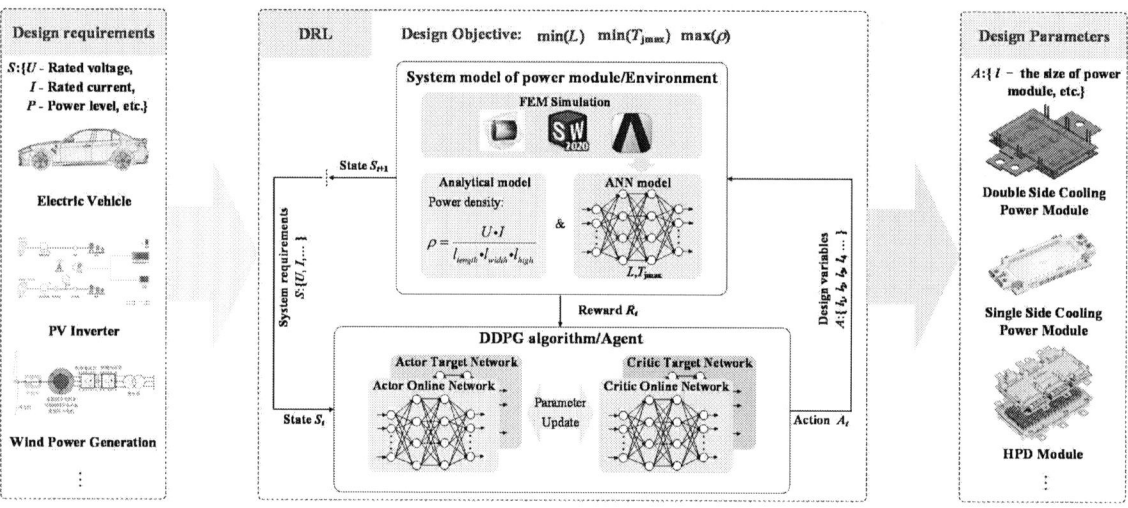

Fig. 1 The framework of proposed MOAD method for power module.

A. Modeling of design objectives

1) Parasitic Inductance and Junction temperature

The modeling process is shown in Fig. 2. Firstly, the range and discretization step of the design variables are defined based on the DSC module, which is shown in Table I. The possible design requirement is also set, which is the rating current I here from 200A to 300A as a simplified example. The total number of data points based on the discretization step size can be obtained by multiplying $N_1 \times N_2 \times \cdots \times N_i$. These data points are then used for FE simulation, from which parasitic inductance and junction temperature are extracted as training samples for the ANNs.

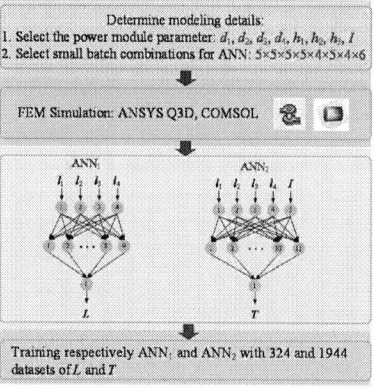

Fig. 2 The steps of the modeling of parasitic inductance and junction temperature based on ANN.

TABLE I
VARIABLE RANGE AND THE DISCRETIZATION STEPS

Parameter Variables	Constraint Range	N_i
l_1	[14, 24]	6
l_2	[14, 24]	6
l_3	[9, 13]	3
l_4	[9, 13]	3
Current	[200, 300]	6

2) Power density

The power density of the power module can be calculated by the ratio of power P to volume V. I The analytical model for power density ρ is shown in (1).

$$\rho = \frac{P}{V} = \frac{P}{l \times w \times h} \qquad (1)$$

B. The verification of the ANN models

The predictions of the ANN model are compared with the actual simulation results to verify the accuracy of the ANN model. Fig. 3 shows the actual simulation data versus the ANN predictions based on the test set, where the red represents the actual simulation values and the black represents the predicted values. It can be seen that the predicted values closely match the actual values for most design combinations.

Overall, the training results are satisfactory.

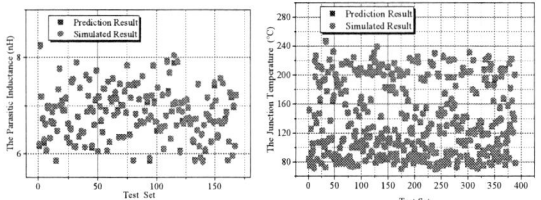

Fig. 3 Comparison between ANN predictions and simulation values.

C. The Self-learning based on DRL

In various DRL-based algorithms, DDPG can output continuous actions and performs well in high-dimensional action spaces, making it suitable for solving complex design problems with many continuous variables[14]. Therefore, it is selected as the self-learning DRL algorithm in this study.

1) DDPG Algorithm

The process of self-learning of the agent is an iterative interaction of data between the agent and the environment.

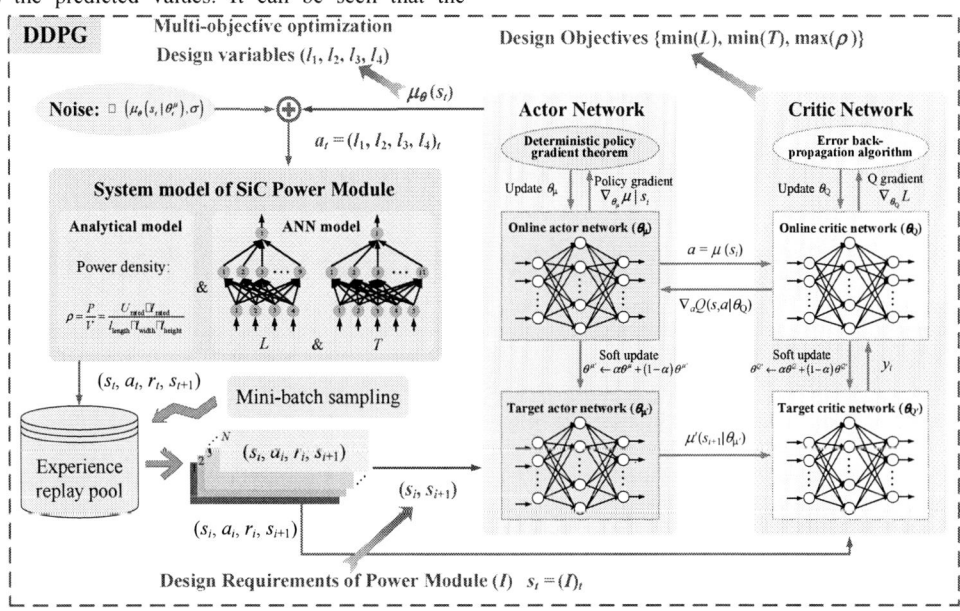

Fig. 4 Design framework of L- Tj- - ρ based on DDPG algorithm for the power module.

The design framework of DDPG algorithm is shown in Fig. 4.The DDPG agent consists of four neural networks: online actor network μ, target actor network μ', online critic network Q and target critic network Q'. Firstly, the online networks are initialized randomly. Then, in each episode, the state is randomly set. At each step t in an episode, under the state $s_t = (I)_t$, the action $a_t = (l_1, l_2, l_3, l_4)_t$ is obtained by the actor network and noise N. According to the interaction of MDP, the actor network μ of the agent gives the action a_t according to the state s_t, which is input into the power module system environment to obtain the reward r and the next state s_{t+1}. The states, actions and rewards during the interaction will be transformed into a sequence (s_t, a_t, r, s_{t+1}) and stored in the experience replay pool. The online critic network $Q^{\theta_Q}(s, a)$ is approximates to the action-value function of the action a_t. The network parameters of the online critic network Q and the online actor network μ are updated by the error back propagation algorithm and the deterministic strategy gradient theorem $\nabla_{\theta_\mu}\mu \mid s_i$, respectively[15]. The target actor network μ' and the target critic network Q' are updated by soft update method[15].

Based on this, the updating speed of the target values can be restricted effectively, and the stability of the training procedure can be improved significantly.

After successful training, the trained actor network can promptly provide the optimal design parameters (l_1, l_2, l_3, l_4) for any input system requirement (I), achieving the design goals of L-T_j-ρ.

2) The Training Result of DDPG Agent

As shown in Fig. 5, After training, the optimized actor network is obtained, and the optimal action is obtained by inputting any state in the state space to the strategy network. Since the maximum reward that can be obtained in various states is different, the reward tends to oscillate in the later stages of learning.

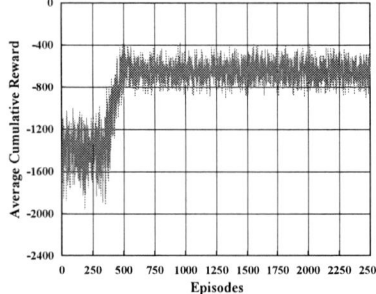

Fig. 5 Average cumulative reward during the training process of the DDPG algorithm.

IV. NUMERICAL VALIDATION OF THE PROPOSED METHOD

This section covers validation of the calculation speed and optimization result of the proposed DRL method, which is compared with that of the Brute Search (BS) and Genetic Algorithm (GA) method.

A. Comparison of Computation Speed

In this paper, two modeling methods-simulation method (SM) and ANN and three optimization algorithms-BS, GA and DRL are discussed. Thus, there are six combinations, and their overall runtimes (including data gathering and networks training) are obtained respectively.

The relationship between runtime and N_{req} is depicted in Fig. 6. For both modeling methods, ANN is significantly faster than SM because it requires only a few simulations to gather data. For the three algorithms, when N_{req} increases, the runtime of BS and GA grows proportionally, while DRL almost holds the line. This is because $T_{DRL\_res} \approx 0.1$ ms and can be almost ignored. Regarding the preparation of DRL training, if $N_{req} \leq 10$, the runtime of GA and DRL is comparative. However, as N_{req} increases, as more design requirements are given, the proposed ANN-DRL combination is much faster than GA due to the absence of running the search program again.

B. Comparison of Optimization Results

Using the optimization results obtained from ANN-GA as a benchmark, the optimization results based on ANN-DRL can be compared to prove the optimization ability of the proposed method. Table II presents the design variable values optimized by ANN-DRL and ANN-GA under different current design requirements, with the weight ratios w_1:w_2:w_3 set to 1:1:1. The current specifications are 210A, 230A, 250A, 270A, and 290A. It can be observed that the optimization results of the two algorithms are almost identical.

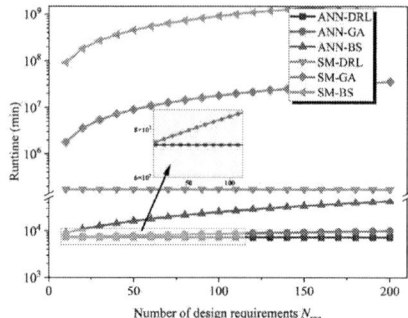

Fig. 6 Runtime of six modeling-optimization combinations.

TABLE II
COMPARISON OF OPTIMIZATION RESULTS BETWEEN GA AND DRL METHODS UNDER DIFFERENT DESIGN REQUIREMENTS

Current I (A)		Optimum size parameters (mm)			
		l_1	l_2	l_3	l_4
210A	ANN-GA	17.5	14.0	12.2	9.0
	ANN-DRL	17.3	14.0	11.7	9.0
	Error	1.1%	0	4.1%	0
230A	ANN-GA	17.9	16.0	9.1	9.0
	ANN-DRL	17.2	16.0	9.2	9.0
	Error	3.9%	0	1.1%	0
250A	ANN-GA	17.5	14.7	12.1	9.0
	ANN-DRL	17.9	15.1	11.7	9.0
	Error	2.3%	2.7%	3.3%	0
270A	ANN-GA	17.6	14.0	12.9	9.0
	ANN-DRL	17.5	14.0	12.4	9.0
	Error	0.6%	0	3.9%	0
290A	ANN-GA	9.1	14.0	9.1	9.0
	ANN-DRL	9.3	14.0	9.3	9.0
	Error	2.2%	0	2.2%	0

V. EXPERIMENTAL VALIDATION OF THE POWER MODULE

In this section, according to the optimization results in Section IV, the optimized size parameters of the power module are selected when the system requirement I is 250A. A half-bridge DSC power module is fabricated and evaluated for electrical, thermal and power density characteristics.

A. Fabrication of power module

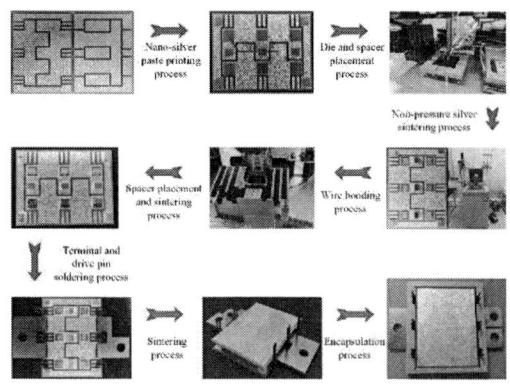

Fig. 7 Assembly process of optimized DSC power module.

TABLE III
BILL OF MATERIALS FOR THE DESIGNED POWER MODULE

Part	Material	Specification
Bare Die	EPM3-1200-0017D1	1200V,134A
Substrate	AlN DBC	0.32mm Cu;0.64mm Ceramic;0.32mm Cu
Wire Bonding	Al	5mil
Solder	Nano sliver paste	Thickness:100um
Terminal	Cu	Thickness:1mm
Spacer	Mo alloy	Molybdenum surface plating with Ag

The main fabricating progress and components are shown in Fig. 7 and Table III.

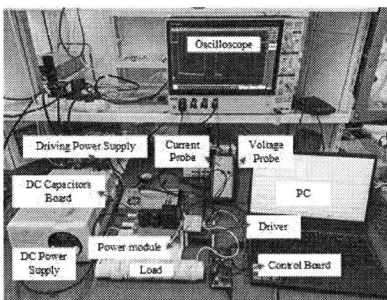

Fig. 8 Platform of DPT test.

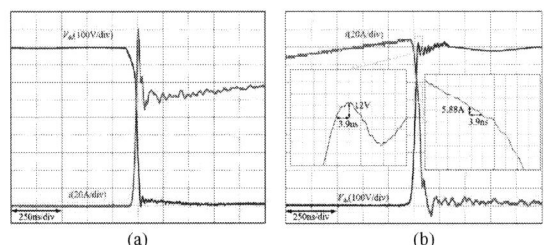

(a)   (b)

Fig. 9 Experimental waveforms of DPT. (a) Turn-On switching waveform. (b) Turn-off switching waveform.

B. Dynamic characteristics

In this subsection, a double-pulse test (DPT) is performed on the half-bridge power module at room temperature. The DPT platform is shown in Fig. 8. The low bridge is taken as the measured object. The gate and Kelvin source of the high bridge are connected. The inductance of the load is 36 uH. Drive signals are input from the control board to the drive board, which is located below the power module. The voltage probe is used to measure the voltage between the AC terminal and the DC negative terminal, while Rogowski coil is inserted at the DC negative terminal. The DC terminal voltage is provided by the DC voltage source, and the 15V drive voltage is provided by the switching power supply. The voltage and current waveform are obtained through the Tek oscilloscope.

The fabricated power module is tested at 500V/120A, the double pulse time is 5us and 1us respectively, and the turn-on and turn-off resistance is 10 Ω. Measured turn-on and turn-off waveforms are shown in Fig. 9 (a) and (b) respectively. At the moment of turn-off, the drain-source voltage rises to 500V, accompanied by a 12V voltage overshoot. At the same time, the measured current drop rate is 1.51A/ns. The parasitic inductance of the module can be calculated as 7.9nH by (2).

$$L = \frac{\Delta V}{di/dt} = \frac{12}{1.51} = 7.9 \text{nH} \qquad (2)$$

ΔV and L denote the overshoot of the drain-source voltage, and the parasitic inductance of the power module separately. As shown in Fig. 10, the test value of parasitic inductance is larger than the simulation value due to some test error. The test error is still within the acceptable range of 2nH.

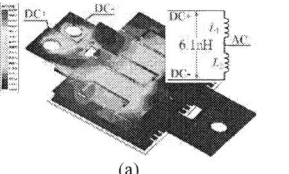 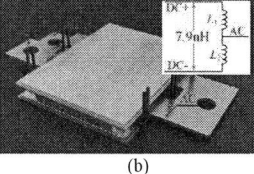

(a)   (b)

Fig. 10 Comparison of simulation and experiment for parasitic inductance. (a) Simulation result by Ansys Q3D. (b) Test result by DPT.

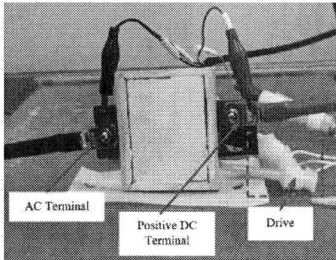

Fig. 11 Platform of thermal characteristics test.

C. Thermal characteristics

Two measurement methods, namely the K-factor measurement method and the thermal imager measurement method are considered for the thermal characteristic test. Since the K-factor measurement method cannot accurately measure the temperature distribution of chips, the thermal imager measurement method is selected for the thermal characteristics test. The high bridge is used as the test objects due to the symmetrical layout. As shown in Fig. 11, the DC positive and AC terminals are connected to the power cycle tester, while the gate terminal and Kelvin terminal of the low bridge are shorted. The high bridge was driven with 15V to maintain the normally open state, and a current of 150A was used for the test due to the power and current limitations of the experimental platform.

An infrared thermal imager (FLIRT650sc) was used to accurately measure the thermal characteristics of the lower substrate. Due to the silver-plating process of the DBC substrate, the temperature measured by the infrared imager deviates significantly compared to the real temperature. Therefore, the back of the DBC substrate is painted with a uniform and thin layer of thermally conductive silicone grease to reduce the error. The high emissivity of thermal conductive silicone grease can help the infrared thermal imager to obtain more accurate test values.

The Fig. 12 shows the simulated and experimental comparison of temperature distribution. Fig. 12 (a) shows the measured temperature distribution of the power module. It can be observed that the parallel MOSFETs of the high bridge can maintain good temperature uniformity, and the temperature difference is less than 2°C.

The actual measured temperature is higher than the

simulation due to the cooling of the power module failing to reach the value set in the simulation. In addition, the actual thickness of the silver sintered layer is different from the design caused by the placement force and the tolerance of spacer can also cause the actual measured temperature to be high than the simulation. In summary, the experimental results can verify that the designed multi-chip parallel power module has the characteristics of uniform temperature.

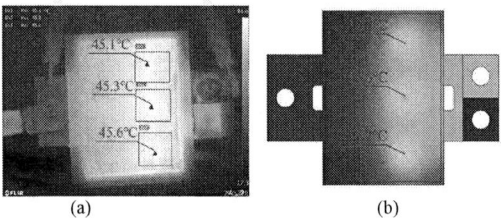

(a)   (b)

Fig. 12 Comparison of simulation and experiment for temperature distribution. (a) Test result. (b) Simulation result by COMSOL.

Fig. 13 Dimensions of the fabricated power module.

TABLE IV
COMPARISON OF DESIGNED AND FABRICATED POWER MODULE

Parameter	Designed	Fabricated	Error
l (mm)	53.8	53.87	0.07
w (mm)	38.5	38.53	0.03
h (mm)	5.94	6.17	0.23

D. Power density characteristics

The physical dimensions of the power module are shown in Fig. 13 and summarized in Table IV. The module length, width, and height are measured using vernier calipers and found to be 53.87mm, 38.53mm, and 6.17mm, respectively. The error in length, width, and height are attributed to the substrate manufacturing process and the manual printing process of the silver paste.

VI. CONCLUSION

This paper proposes a novel multi-objective automated design method for the packaging of power modules based on machine learning, combining Artificial Neural Networks (ANN) and Deep Reinforcement Learning (DRL), which can significantly improve the design efficiency of power modules. The use of ANN facilitates the rapid establishment of complex nonlinear relationships between design variables and objectives, thereby reducing reliance on lengthy iterative numerical simulations. Meanwhile, DRL enables intelligent exploration of the design space, allowing for quick identification of optimal design variables under varying input conditions. Validation through a case study involving a 1200V/300A DSC power module demonstrates the method's effectiveness.

REFERENCE

[1] X. She, A. Q. Huang, Ó. Lucía and B. Ozpineci, "Review of Silicon Carbide Power Devices and Their Applications," in IEEE Transactions on Industrial Electronics, vol. 64, no. 10, pp. 8193-8205, Oct. 2017.

[2] Puqi Ning, Xiaoshuang Hui, Yuhui Kang, Tao Fan, Kai Wang, Yunhui Mei, Guangyin Lei. Review of Hybrid Packaging Methods for Power Modules*[J]. Chinese Journal of Electrical Engineering, 2023, 9(4): 23-40.

[3] Ji, X. Song, E. Sciberras, W. Cao, Y. Hu and V. Pickert, "Multiobjective Design Optimization of IGBT Power Modules Considering Power Cycling and Thermal Cycling," in IEEE Transactions on Power Electronics, vol. 30, no. 5, pp. 2493-2504, May 2015.

[4] Z. Zeng, K. Ou, L. Wang and Y. Yu, "Reliability-Oriented Automated Design of Double-Sided Cooling Power Module: A Thermo-Mechanical-Coordinated and Multi-Objective-Oriented Optimization Methodology," in IEEE Transactions on Device and Materials Reliability, vol. 20, no. 3, pp. 584-595, Sept. 2020.

[5] P. Ning, F. Wang and K. D. T. Ngo, "Automatic layout design for power module," in IEEE Transactions on Power Electronics, vol. 28, no. 1, pp. 481-487, Jan. 2013.

[6] P. Ning, X. Wen, L. Li and H. Cao, "An improved planar module automatic layout method for large number of dies," in CES Transactions on Electrical Machines and Systems, vol. 1, no. 4, pp. 411-417, December 2017.

[7] Y. Zhou, Y. Jin, Y. Chen, H. Luo, W. Li and X. He, "Graph-Model-Based Generative Layout Optimization for Heterogeneous SiC Multichip Power Modules With Reduced and Balanced Parasitic Inductance," in IEEE Transactions on Power Electronics, vol. 37, no. 8, pp. 9298-9313, Aug. 2022.

[8] T. M. Evans et al., "PowerSynth: A Power Module Layout Generation Tool," in IEEE Transactions on Power Electronics, vol. 34, no. 6, pp. 5063-5078, June 2019.

[9] Al Razi, Q. Le, T. M. Evans, S. Mukherjee, H. A. Mantooth and Y. Peng, "PowerSynth Design Automation Flow for Hierarchical and Heterogeneous 2.5-D Multichip Power Modules," in IEEE Transactions on Power Electronics, vol. 36, no. 8, pp. 8919-8933, Aug. 2021.

[10] Al Razi, Q. Le, T. M. Evans, H. A. Mantooth and Y. Peng, "PowerSynth 2: Physical Design Automation for High-Density 3-D Multichip Power Modules," in IEEE Transactions on Power Electronics, vol. 38, no. 4, pp. 4698-4713, April 2023

[11] Y. Jia, F. Xiao, Y. Duan, Y. Luo, B. Liu and Y. Huang, "PSpice-COMSOL-Based 3-D Electrothermal–Mechanical Modeling of IGBT Power Module," in IEEE Journal of Emerging and Selected Topics in Power Electronics, vol. 8, no. 4, pp. 4173-4185, Dec. 2020.

[12] X. Li et al., "EM-Electrothermal Analysis of Semiconductor Power Modules," in IEEE Transactions on Components, Packaging and Manufacturing Technology, vol. 9, no. 8, pp. 1495-1503, Aug. 2019.

[13] Y. Yang, Y. Ge, Z. J. Wang and Y. Kang, "An Automated Electro-Thermal-Mechanical Co-Simulation Methodology Based on PSpice-MATLAB-COMSOL for SiC Power Module Design," 2021 IEEE Workshop on Wide Bandgap Power Devices and Applications in Asia (WiPDA Asia), Wuhan, China, 2021, pp. 499-503.

[14] Silver et al., "Deterministic policy gradient algorithms," in Proc. 31st Int. Conf. Mach. Learn., 2014, pp. 387–395.

[15] T. P. Lillicrap et al., "Continuous control with deep reinforcement learning," 2015, arXiv:1509.

979-8-3315-1110-4/25 $31.00 © 2025 IEEE

A High-Performance Double-Sided Cooling SiC Power Module Packaging Design for EV Inverters

Haobin Chen[a], Haidong Yan[a,b], Maosheng Zhang[b], Ji Cheng[b], Yakun Zhang[c], Chaohui Liu[c]

[a]School of Electrical Engineering, Zhejiang University, Hangzhou, China
[b]ZJU-Hangzhou Global Scientific and Technological Innovation Center, Hangzhou, China
[c]Powertrain Department, National New Energy VehicleTechnology Innovation Center (NEVC), Beijing, China
haidong_yan@zju.edu.cn

Abstract—**Wide bandgap (WBG) devices, particularly SiC devices, are increasingly being applied in electric vehicle (EV) inverters due to their excellent performance. To fully leverage the advantages of WBG devices, specialized packaging designs are required. Currently, the trend in packaging for WBG devices is aimed at reducing parasitic inductance, achieving higher heat dissipation capabilities, and enhancing reliability. In this work, a high-performance double-sided cooling (DSC) package using SiC MOSFETs suitable for EV inverters is studied. A novel 1200 V/600 A half-bridge module with 10 SiC MOSFETs in parallel is proposed. The module has a compact layout with external dimensions of only 50 mm x 50 mm x 4.3 mm. The electromagnetic behavior of the module is analyzed using the finite element method (FEM) in ANSYS Q3D, and the parasitic inductance of each component is extracted. The simulation results of the proposed design indicate that the total parasitic inductance is only 50% of that of commercial devices at the same rated power level. Thermal simulation analysis is conducted in ANSYS to verify the low thermal resistance of the designed module. A module prototype is fabricated with Cu sintering technology at its core, and the high reliability of the interconnections is verified.**

Keywords—*packaging, double-sided cooling (DSC), SiC MOSFETs, high performance, Cu sintering*

I. INTRODUCTION

In recent years, the demand for electric vehicles (EVs) has increased significantly. Power modules are the core components of EV inverters, and their packaging technology has a crucial impact on system performance and reliability. The traditional single-sided cooling (SSC) power module has become one of the most commonly used packaging structures for EV inverters. However, due to parasitic inductance and thermal dissipation issues, this design limits the performance of SiC power modules. Compared with Si devices, SiC devices have lower switching losses, higher operating frequencies, smaller sizes, and higher operating temperatures. Compared with IGBTs, MOSFETs have no tail current and can operate at higher frequencies. These advantages have driven the application of SiC MOSFETs in the EV industry [1]. Despite the superior performance of SiC devices compared to Si devices, the high switching frequency leads to performance issues related to electromagnetic parasitics in the devices and packaging. Parasitic inductance in the packaging structure and high di/dt during switching can cause voltage overshoot, resulting in severe stress, high power losses, and high EMI emissions [2]. This currently poses a barrier to high-performance switching in SiC power modules. Moreover, for the same rated power, SiC devices are much smaller than Si devices, leading to

higher heat flux density. Therefore, it is necessary to design new power module packaging to address the thermal issues associated with the use of SiC devices.

To fully leverage the high switching speed of SiC MOSFETs, parasitic inductance should be minimized as much as possible. Literature indicates that if the switching speed of SiC MOSFETs reaches 100 V/ns, the parasitic inductance should theoretically be below 4 nH [3]. To reduce the parasitic inductance in power module packaging, many novel power module structures have been designed, including SkiN technology [4], hybrid packaging [5], power overlay package [6], PCB-embedded packaging [7], and double-sided cooling (DSC) packaging [8]. Among these packaging structures, DSC power modules offer the best thermal performance, as heat can be dissipated from both sides.

To meet the requirements of 800V DC links, power modules rated at 1200V are preferred. This has become a research hotspot for double-sided cooling (DSC) power modules. In recent years, several DSC SiC power modules with parasitic inductance not exceeding 4 nH have been developed [9, 10, 11]. However, these power modules have limited power ratings and do not reach the 1200 V/600 A level. Commercial 1200 V/600 A power modules have a total parasitic inductance higher than 8 nH. Additionally, the interconnect layer is a critical part of DSC power modules that requires special attention. Sintered Cu, with its high electrical conductivity, thermal conductivity, and reliability, is highly suitable for interconnecting components within DSC power modules [12]. However, there are few reports on DSC power modules based on Cu sintering.

This paper proposes a structurally ultra-compact 1200 V/600 A double-sided cooling (DSC) SiC module. Considering both self-inductance and mutual inductance, a parasitic inductance model of the module is derived. The electromagnetic behavior of the module is analyzed using the finite element method (FEM) in ANSYS Q3D. Simulation results show that the total parasitic inductance of the proposed DSC power module is only 3.88 nH, which is just 50% of that of a commercial module at the same rated power level. Thermal simulation analysis is performed in ANSYS, calculating a thermal resistance (Rth_jf) of only 0.118 K/W for the DSC power module. Using formic acid-assisted low-temperature pressureless Cu sintering technology as the core, the designed module is fabricated and its high reliability is preliminarily verified. The shear strength of the sintered Cu joints reaches as high as 43 MPa, exceeding the requirements of the MIL-STD-883J standard.

This work was supported by the National Key Research and Development Program of China under Grant 2021YFB3602300.

II. STRUCTURAL DESIGN OF THE DSC POWER MODULE

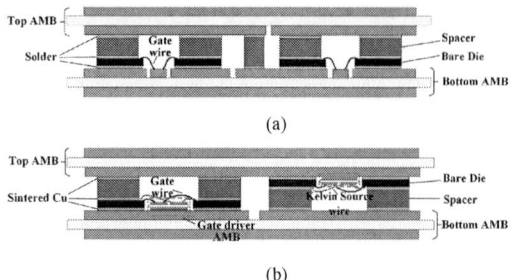

(a)

(b)

Fig. 1. Two different basic DSC structures. (a) Mainstream commercial DSC power module layout. (b) Proposed DSC power module layout.

A sandwich structure composed of a double-layer substrate, chips, spacers, and soldering layers is widely used in DSC power modules. Currently, commercially available DSC power modules mainly adopt the structure shown in Fig. 1(a). In this structure, two bare chips are placed on the same bottom substrate, and an additional intermediate spacer is required to achieve the electrical connection between the upper and lower devices of the half-bridge. The drive circuit shares the upper copper layer of the AMB with the power circuit, which reduces its current-carrying capacity and mechanical strength. Moreover, DSC power modules have very high reliability requirements for the interconnect layer, which traditional solders cannot meet. In particular, as the rated power increases, these issues become more pronounced.

To address the issues with the DSC power module structure in Fig. 1(a), this paper proposes a new structure as shown in Fig. 1(b). In this new structure, the two bare chips are placed face-up and face-down on the top and bottom AMB substrates, respectively. This eliminates the need for an additional intermediate spacer to achieve the electrical connection between the upper and lower devices of the half-bridge. As a result, not only is material usage reduced, but the layout can also be made more compact, achieving higher cost-effectiveness and power density. Moreover, parasitic parameters and thermal coupling are minimized, and heat dissipation efficiency is improved. Additionally, an integrated gate drive AMB is incorporated for the drive circuit. This enhances the utilization rate of the copper layers on the top and bottom AMBs, significantly improving current-carrying capacity, heat dissipation, and mechanical strength. Furthermore, Cu sintering technology is used for interconnection. This not only meets the high reliability requirements for interconnects in DSC power modules but also provides excellent electrical and thermal performance.

Based on the basic structure shown in Fig. 1(b), a novel 1200 V/600 A DSC power module is designed, as shown in Fig. 2. The layout of the top AMB, the layout of the bottom AMB, the complete structure, and the equivalent circuit diagram are shown in Fig. 2(a)-(d), respectively. Both the top and bottom AMBs feature large-area copper layers that handle high currents, high heat dissipation, and high mechanical strength. The high-side and low-side of the half-bridge module are each composed of 10 parallel SiC MOSFETs. The parallel SiC MOSFETs are symmetrically placed on both sides of the gate drive AMB, which is conducive to current sharing. Additionally, each SiC MOSFET employs a Kelvin source connection and a configured gate resistor, which is beneficial for the efficient and stable operation of the power module. Sintered Cu is

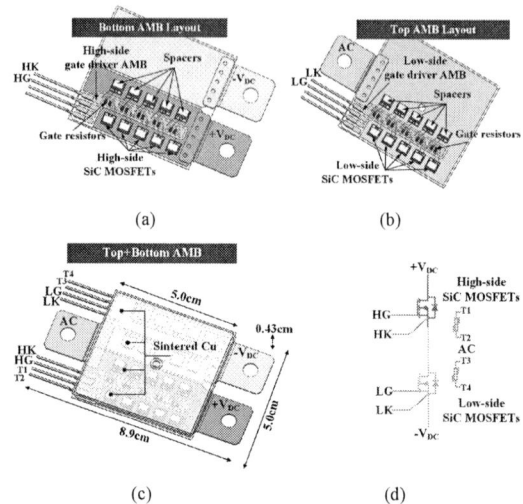

(a)   (b)

(c)   (d)

Fig. 2. The proposed 1200V 600A DSC SiC power module structure. (a) The bottom AMB layout. (b) The top AMB layout. (c) The combination of the top AMB with changed transparency and the bottom AMB. (d) Circuit diagram.

used for the reliable connection of each interconnect layer. The module has a compact layout, with external dimensions of only 50 mm × 50 mm × 4.3 mm.

III. DSC POWER MODULE SIMULATION ANALYSIS

A. Electromagnetic Simulation Analysi

Fig. 3 compares the commutation loops of two power module layouts. Compared to mainstream commercial DSC modules, the power module designed in this paper has a shorter commutation loop. Moreover, the chips of the upper and lower bridge arms are laid out more compactly, enhancing the cancellation effect of mutual inductance. As a result, the new structure is conducive to reducing the total parasitic inductance of the power module.

Fig. 4 presents the parasitic inductance model of the commutation loop. The terminal layout of the DSC power module designed in this paper is shown in Fig. 4(a). Fig. 4(b) shows the circuit model of the module considering both self-inductance and mutual inductance. This includes the self-inductances L_{a1a1}, L_{a2a2}, L_{a3a3} for the current path from $+V_{DC}$ to AC, and the self-inductances L_{b1b1}, L_{b2b2}, L_{b3b3} for the current path from AC to $-V_{DC}$. The mutual inductances between the two current paths are represented as M_{ab1}, M_{ab2} and M_{ab3}. The model is further simplified into an equivalent inductance model, as shown in Fig. 4(c). In this model, the equivalent inductances for each current path, L_{a1}, L_{a2}, L_{a3}, L_{b1},

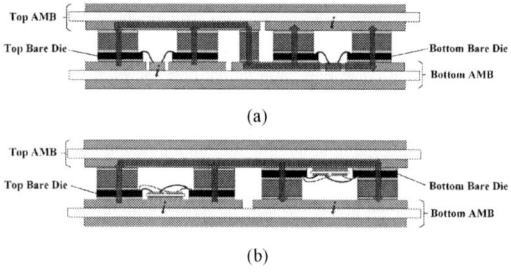

(a)

(b)

Fig. 3. Comparison of the current paths in the commutation loop. (a) Mainstream commercial DSC power module. (b) Proposed DSC power module.

979-8-3315-1110-4/25 $31.00 © 2025 IEEE

L_{b2}, L_{b3}, and the total loop inductance L_{module}, can be expressed as follows:

$$\begin{cases} L_{a1(b1)} = L_{a1a1(b1b1)} - M_{ab1} \\ L_{a2(b2)} = L_{a2a2(b2b2)} - M_{ab2} \\ L_{a3(b3)} = L_{a3a3(b3b3)} - M_{ab3} \\ L_{\text{module}} = \sum_{i=1}^{3} L_{ai} + \sum_{j=1}^{3} L_{bi} \end{cases} \quad (1)$$

The parasitic inductances of the current paths are extracted using ANSYS Q3D. The $+V_{DC}$ and $-V_{DC}$ of the commutation loop are set as Sources, and AC is set as the Sink. The self-inductances and mutual inductances of the two current paths are calculated, as shown in Fig. 5 and Table I. The self-inductances of the two current paths are 12.912 nH and 12.888 nH, respectively, while the mutual inductance is as high as 10.982 nH. According to (1), the total parasitic inductance of the power module is calculated to be 3.836 nH. Thanks to the low self-inductance and high mutual inductance, the total parasitic inductance is relatively small.

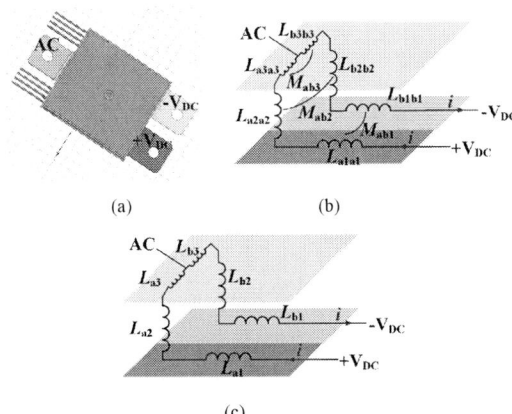

Fig. 4. Parasitic inductance model of the commutation loop. (a) Terminal layout of the main power circuit. (b) Circuit model considering the self and mutual inductance. (c) The equivalent inductance model.

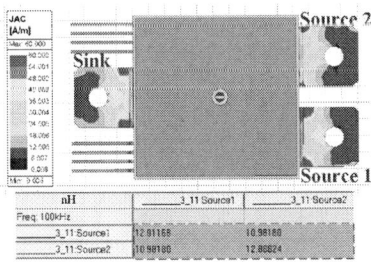

Fig. 5. The self-inductances and mutual inductances of the two current paths.

TABLE 1. PARASITIC INDUCTANCE VALUES OF THE COMMUTATION LOOP.

Parameter	Inductance (nH)
$L_a = L_{a1a1} + L_{a2a2} + L_{a3a3}$	12.912
$L_b = L_{b1b1} + L_{b2b2} + L_{b3b3}$	12.888
$M_{ab} = M_{ab1} + M_{ab2} + M_{ab3}$	10.982
$L_{\text{module}} = L_a + L_b - 2M_{ab}$	3.836

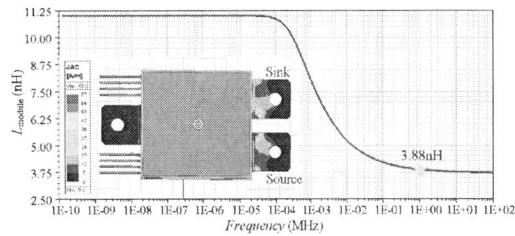

Fig. 6. Total parasitic inductance of ANSYS Q3D simulation.

The total inductance of the DSC power module is calculated at different frequencies by directly setting the $+V_{DC}$ and $-V_{DC}$ of the commutation loop as the Source and Sink, respectively. Due to the skin effect, the inductance varies with frequency, and the frequency dependence of the proposed structure is shown in Fig. 6. The simulation results indicate that the total inductance decreases rapidly below 1 MHz and then remains almost constant in the rest of the range. At 1 MHz, the total inductance Lmodule is only 3.88 nH.

B. Thermal Simulation Analysis

Fig. 7 compares the heat dissipation paths of two power modules. Mainstream commercial DSC modules can achieve double-sided cooling. However, the heat distribution is uneven, with most of the heat (65%) being transferred downward, as shown in Fig. 7(a). Moreover, the concentrated heat can also exacerbate the phenomenon of thermal coupling. By adopting the DSC module designed in this paper, the heat distribution is more uniform, with 50% of the heat dissipated upwards and 50% downwards. The uniform heat distribution reduces the occurrence of thermal coupling. Therefore, it helps to improve the overall thermal management efficiency.

Thermal simulations are conducted in ANSYS to analyze the thermal performance of the proposed power module. The boundary conditions for the thermal simulation are shown in Fig. 8. The losses for each high-side SiC MOSFET are set at 50 W, the ambient temperature is set at 60°C, and the heat transfer coefficient is set at 7000 W/(m²K). During the thermal resistance simulation, simulations are performed under three different cooling conditions: top side cooling, bottom side cooling, and both sides cooling.

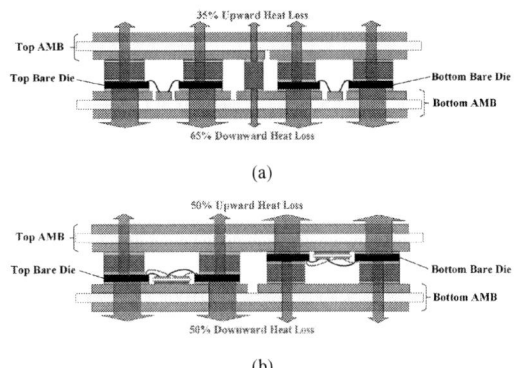

Fig. 7. Comparison of heat dissipation paths. (a) Mainstream commercial DSC power module. (b) Proposed DSC power module.

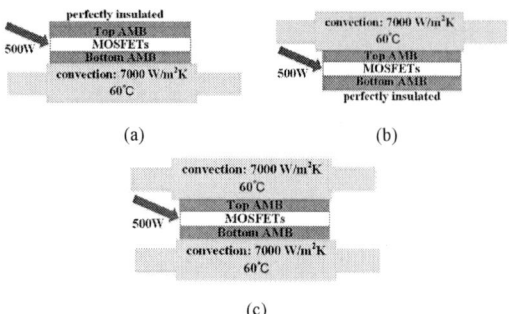

Fig. 8. Boundary conditions of thermal simulation of DSC module under different cooling conditions. (a) Top side cooling. (b) Bottom side cooling. (c) Both sides cooling.

The thermal simulation results are shown in Fig. 9. The thermal resistance of the DSC power module can be calculated using the following formula:

$$R_{\mathrm{th\_jf}} = \frac{T_j - T_f}{P} \tag{2}$$

where T_j is the average junction temperature of the high-side SiC MOSFETs, T_f is the ambient temperature, and P is the power loss of the high-side SiC MOSFETs.

Therefore, under the top side cooling condition, the calculated thermal resistance $R_{\mathrm{th\_jf\text{-}top}}$ is 0.281 K/W. Under the bottom side cooling condition, the calculated thermal resistance $R_{\mathrm{th\_jf\text{-}bot}}$ is 0.184 K/W. Under the both sides cooling condition, the thermal resistances calculated for each side are 0.295 K/W and 0.197 K/W, respectively. In this case, due to the presence of thermal coupling between the two sides, the thermal resistance of each side is slightly higher than that under single side cooling conditions. However, effective heat dissipation can occur on both sides, equivalent to the parallel connection of thermal resistances. Ultimately, under the both sides cooling condition, the thermal resistance $R_{\mathrm{th\_jf}}$ of the DSC power module is only 0.118 K/W.

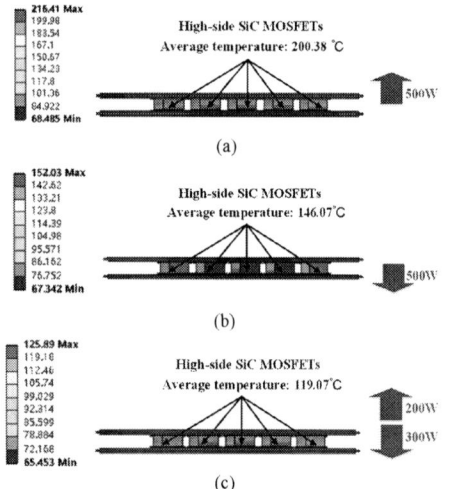

Fig. 9. Thermal simulation results of the proposed power module. (a) under top side cooling condition. (b) under bottom side cooling condition. (c) under the both sides cooling condition.

IV. FABRICATION OF THE DSC POWER MODULE

In the fabrication process of the DSC power module, obtaining a reliable interconnect layer is crucial for achieving electrical connections and thermal conduction between the chips and the substrates. Moreover, the interconnect layer is the primary weak point in the module packaging structure. The performance of the interconnect layer directly affects the electrical, thermal, and reliability characteristics of the module. Sintered Cu offers high electrical and thermal conductivity. Additionally, sintered Cu can operate effectively in high-temperature and high-current density environments without causing electromagnetic (EM) issues. It also has advantages in terms of cost and coefficient of thermal expansion matching. With excellent mechanical properties and high fatigue resistance, sintered Cu is more suitable for high-power-density DSC power module packaging.

To achieve these objectives, this paper has developed a formic acid-assisted low-temperature pressureless Cu sintering process. The self-made Cu paste used in the experiment is shown in Fig. 10(a). The distribution of the Cu paste is precisely controlled using syringes. The sintering process curve is shown in Fig. 10(b). The pre-assembled samples are first preheated for 20 minutes at 150°C in a formic acid atmosphere. Then, the temperature is increased to 240°C, and formic acid is introduced again for 30 minutes of pressureless Cu sintering. Both the preheating and sintering processes are carried out in a VADU 100.

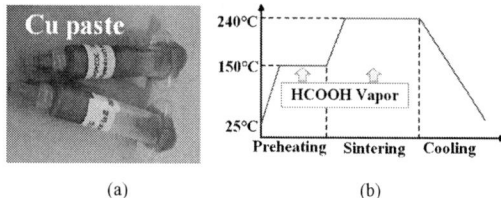

Fig. 10. Formic acid-assisted low-temperature pressureless Cu sintering process. (a) Self-made Cu paste. (b) Sintering process curve.

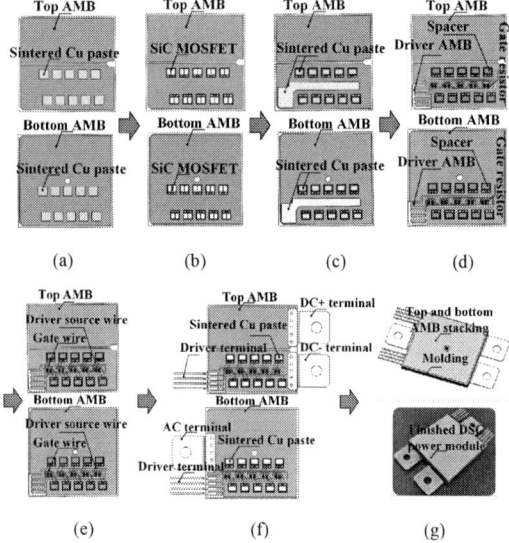

Fig. 11. Fabrication process of the DSC power module. (a) First Cu paste printing. (b) Die-attach. (c) Second Cu paste printing. (d) Spacer-attach and Driver AMB-attach. (e) Wire bonding. (f) Third Cu paste printing and add terminals. (g) Spacer-AMB-attach and molding.

Based on the aforementioned process, the DSC power module is fabricated, with the detailed process shown in Fig. 11. The formic acid-assisted low-temperature pressureless Cu sintering process is employed for die attachment (Fig. 11(a)-(b)), Mo-Cu post attachment and gate drive AMB attachment (Fig. 11(c)-(d)), as well as the attachment of the top and bottom AMBs (Fig. 11(e)-(g)). Fig. 11(g) shows a half-bridge module sample fabricated using Cu sintering.

To illustrate the effectiveness of the process, scanning electron microscopy (SEM) is performed on the cross-sectional morphology of the sintered Cu regions before and after fabrication, as shown in Fig. 12. Before sintering, the Cu nanoparticles observed by SEM are spherical in shape, as shown in Fig. 12(a). The surface of the Cu nanoparticles is covered with tiny nanoparticles. It is inferred that oxidation has occurred on the surface of the Cu nanoparticles, forming an oxide layer. Fig. 12(b) indicates that after sintering at 240°C, distinct connections appear between the Cu nanoparticles. Neck growth and pore contraction occur between the Cu nanoparticles, resulting in significant morphological changes. After formic acid-assisted low-temperature pressureless Cu sintering, the porosity of the sintered Cu nanoparticle layer is significantly reduced. This has an essential impact on its mechanical properties. The shear strength is tested to be as high as 45 MPa. The shear test results show that the Cu paste sintered in a formic acid atmosphere can achieve high-strength sintering performance. Therefore, using the formic acid-assisted low-temperature pressureless Cu sintering process, a DSC power module with highly reliable interconnects can be obtained.

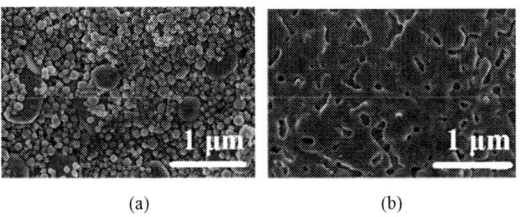

Fig. 12. Cross-sectional SEM micrographs of sintered Cu regions. (a) Initial state. (b) After sintering process.

V. CONCLUSION

This paper proposes a compact, high-performance 1200 V/600 A DSC SiC power module. Unlike mainstream commercial DSC modules, this structure does not require an additional spacer for the electrical connection between the upper and lower devices. Additionally, an auxiliary substrate is used for the gate circuit, enhancing the current-carrying and heat dissipation capabilities of the copper layer on the main substrate. This facilitates a compact layout. The external dimensions of the module are only 50 mm × 50 mm × 4.3 mm. Considering both self-inductance and mutual inductance, a parasitic inductance model of the module is derived. The electromagnetic behavior of the module is analyzed using the finite element method (FEM) in ANSYS Q3D, and the parasitic inductance of each component is extracted to calculate the total parasitic inductance. Simulation results show that the total parasitic inductance of the proposed DSC power module is only 3.88 nH. Thermal simulation analysis in Ansys verifies the low thermal resistance of the designed module. Under both sides cooling condition, the thermal resistance R_{th_jf} of the DSC power module is only 0.118 K/W. Using formic acid-assisted low-temperature pressureless Cu sintering technology as the core, the designed module is fabricated and its high reliability is preliminarily verified. The shear strength of the sintered Cu joints reaches as high as 43 MPa, exceeding the requirements of the MIL-STD-883J standard. Thus, the high reliability of the Cu sintered power module is demonstrated. Experiments are currently underway to prove the high performance of the proposed module.

REFERENCES

[1] Y. Yang, L. Dorn-Gomba, R. Rodriguez, C. Mak, and A. Emadi, "Automotive power module packaging: Current status and future trends," IEEE Access, vol. 8, pp. 160126–160144, 2020.

[2] Y. Zhang et al., "Comprehensive Analysis and Optimization of Parasitic Capacitance on Conducted EMI and Switching Losses in Hybrid-Packaged SiC Power Modules," IEEE Transactions on Power Electronics, vol. 38, no. 11, pp. 13988-14003, Nov. 2023.

[3] M. Meisser, M. Schmenger and T. Blank,"Parasitics in Power Electronic Modules: How parasitic inductance influences switching and how it can be minimized," Proceedings of PCIM Europe 2015; International Exhibition and Conference for Power Electronics, Intelligent Motion, Renewable Energy and Energy Management, Nuremberg, Germany, 2015, pp. 1-8.

[4] P. Beckedahl, I. Bogen and J. Steger, "SiC automotive power module with laser welded, ultra low inductive terminals and up to 900Arms phase current," CIPS 2022; 12th International Conference on Integrated Power Electronics Systems, Berlin, Germany, 2022, pp. 1-5.

[5] Z. Huang, C. Chen, Y. Xie, Y. Yan, Y. Kang and F. Luo, "A High-Performance Embedded SiC Power Module Based on a DBC-Stacked Hybrid Packaging Structure," IEEE Journal of Emerging and Selected Topics in Power Electronics, vol. 8, no. 1, pp. 351-366, March 2020.

[6] Y. Nishihara, K. Bando, S. Hayashibe, T. Yumoto, T. Ikeda and T. Gomyo, "Evaluation of electrical and thermal properties of POL-kW by simulation and actual measurement," 2022 IEEE CPMT Symposium Japan (ICSJ), Kyoto, Japan, 2022, pp. 150-153.

[7] J. S. Knoll, G. Son, C. DiMarino, Q. Li, H. Stahr and M. Morianz, "A PCB-Embedded 1.2 kV SiC MOSFET Half-Bridge Package for a 22 kW AC–DC Converter," IEEE Transactions on Power Electronics, vol. 37, no. 10, pp. 11927-11936, Oct. 2022.

[8] A. P. Pai, M. Ebli, T. Simmet, A. Lis and M. Beninger-Bina, "Characteristics of a SiC MOSFET-based Double Side Cooled High Performance Power Module for Automotive Traction Inverter Applications," 2022 IEEE Transportation Electrification Conference & Expo (ITEC), Anaheim, CA, USA, 2022, pp. 831-836.

[9] Y. Yan, C. Chen, Z. Wu, J. Guan, J. Lv and Y. Kang, "A High Power Density Double-Side-End Double-Sided Bonding SiC Half-Bridge Power Module," IEEE Transactions on Transportation Electrification, vol. 9, no. 2, pp. 3149-3163, June 2023.

[10] F. Yang et al., "Interleaved Planar Packaging Method of Multichip SiC Power Module for Thermal and Electrical Performance Improvement," IEEE Transactions on Power Electronics, vol. 37, no. 2, pp. 1615-1629, Feb. 2022.

[11] R. Paul, A. Hassan and H. A. Mantooth, "A Double-Sided Cooled Power Module With Embedded Decoupling Capacitors," IEEE Journal of Emerging and Selected Topics in Power Electronics, vol. 12, no. 2, pp. 1813-1821, April 2024.

[12] T. Matsuda et al., "Reduction Behavior of Surface Oxide on Submicron Copper Particles for Pressureless Sintering Under Reducing Atmosphere," Journal of Electronic Materials, vol. 51,pp. 1–7, 2022.

Optimal Design of High-temperature SiC Power Module Based on Gene Algorithm

Zhenchun Xia
School of Electrical Engineering and Automation
Hefei University of Technology
Hefei, China
2023170526@mail.hfut.edu.cn

Jianing Wang
School of Electrical Engineering and Automation
Hefei University of Technology
Hefei, China
jianingwang@hfut.edu.cn

Shaolin Yu
The Institute of Energy, Hefei Comprehensive National Science Center(Anhui Energy Laboratory)
Hefei, China
yusl@ie.ah.cn

Weina Mao
School of Electrical Engineering and Automation
Hefei University of Technology
Hefei, China
2023170557@mail.hfut.edu.cn

Xiahao Wang
School of Electrical Engineering and Automation
Hefei University of Technology
Hefei, China
2023170534@mail.hfut.edu.cn

Honghong Li
School of Electrical Engineering and Automation
Hefei University of Technology
Hefei, China
1504021691@qq.com

Runze Wang
School of Electrical and Information Engineering
Anhui University of Science and Technology
Huainan, China
2023201831@aust.edu.cn

Abstract—Silicon carbide (SiC) devices possess outstanding high-temperature performance, allowing them to operate for extended periods in high-temperature environments while maintaining high reliability. However, most existing SiC modules directly adopt packaging technologies designed for silicon chips, which limits their ability to fully utilize the superior high-temperature performance of SiC devices. To reduce thermomechanical stress in high-temperature applications and enhance the reliability of SiC modules, an optimization method combining artificial neural networks (ANN) and Genetic Algorithm (GA) is proposed. The genetic algorithm is used to automatically solve the multi-objective model of maximum principal stress and junction temperature, thereby obtaining the optimal layout structure. Finally, a multi-chip parallel half-bridge power module with a rated voltage of 1200V and a rated current exceeding 300A is fabricated. The module's high-temperature performance is evaluated through power cycling tests.

Keywords —High-temperature module, SiC MOSFET, Low stress, Chip layout.

I. INTRODUCTION

Silicon carbide (SiC) devices have broad application prospects in fields such as aerospace and oil exploration[1]-[4]. High-temperature modules based on SiC devices not only adapt to extreme environments but also offer many advantages in conventional applications. For example, the higher rated operating temperature can reduce the requirements for module heat dissipation systems, thus contributing to a reduction in the size of power systems and lowering costs. However, most current SiC modules directly adopt packaging technologies used for silicon modules, which results in a junction temperature that is lower than the 300°C rated junction temperature of SiC devices. To enhance the high-temperature performance of these modules, packaging solutions specifically tailored to the characteristics of SiC devices need to be developed.

Current research on high-temperature packaging mainly focuses on the development of high-temperature materials and advanced packaging structure designs. The primary materials for high-temperature packaging are high-temperature potting compounds and high-melting-point solders. However, commercially available potting compounds generally exhibit poor high-temperature performance, with operational temperatures not exceeding 200°C, which makes them unsuitable for high-temperature modules[5]. To address the issue of materials lacking high-temperature resistance, researchers have investigated high-temperature epoxy molding materials and silicone potting compounds[6]-[8]. In [7], detailed high-temperature tests were conducted on silicone potting compounds R-2188, SEMICOSIL 915HT, and Duraseal 1533. Reference [9] studies a high-temperature package based on low-melting-point glass. Test results showed that SiC MOSFETs packaged with glass exhibited normal static and dynamic performance, and the module demonstrated good thermal stability at 250°C.

Traditional power module design methods rely on the experience of the designer, with single optimization objectives and low power ratings in the module design. Reference [10] proposed a 900V/196A double-side cooled wire-bond-free SiC MOSFET module, which features very low parasitic inductance and good heat dissipation capabilities. Reference [11] introduced a compact interleaved packaging module with parasitic inductance as low as 0.337nH, offering a lower junction-to-case thermal resistance, and capable of operating for 30 minutes at 250°C. Reference [12] developed a 900A high-temperature SiC half-bridge power module, optimizing the layout and terminals for multi-chip parallel configurations. Reference [13] proposed cutting the double-side cooled substrate separately, replacing the metal pad with a multilayer

---

This work was supported by Anhui Province Key Research and Development Program Project, Project Number: JZ2024AKKG0057.

979-8-3315-1110-4/25 $31.00 © 2025 IEEE

ceramic substrate with embedded metal vias, effectively reducing stress concentration in the etched slots.

Currently, computer-aided multi-objective optimization (MOO) design methods are also being applied to power module design. Reference [14] constructed thermal resistance and mechanical stress models for a double-side water-cooled module, using the second-generation Non-Dominated Sorting Genetic Algorithm (NSGA-II) to obtain optimal structural parameters. However, existing multi-objective optimization methods still have certain limitations. Complex structures and objectives are difficult to model mathematically using approximate analytical formulas, resulting in limited accuracy. Additionally, while some studies have developed finite element (FE) models to simulate stress, these methods require extensive simulations. Artificial neural networks (ANN) have been widely applied in power electronics. Reference [15] used ANN to predict the relationship between solar radiation and the lifetime of power converters. To improve design efficiency, this paper proposes an optimization method for high-temperature power modules based on genetic algorithms, combining artificial neural networks (ANN) and genetic algorithms to optimize the chip layout structure, reduce stress in the interconnection layers, and extend the module's lifespan under high-temperature conditions.

II. POWER MODULE PACKAGING DESIGN

A. High-Temperature Material System

One of the challenges of high-temperature modules is the high-temperature materials. Fig. 1 shows the basic structure of a wire-bonded single-side-cooled power module, which includes power terminals, SiC chips, bond wires, substrates, and potting materials. Considering factors such as cost and thermal performance, the R-2188 potting compound was selected as the high-temperature encapsulation material. Traditional Sn-Pb-based solders, such as Sn-Pb90, Sn-Pb95, and Sn-Pb98, have melting points of 268°C, 300°C, and 316°C, respectively. Although these solders meet the high-melting-point requirements, their thermal and electrical conductivity properties are insufficient for the needs of high-temperature modules, and they are also harmful to the environment and human health. In this study, a mature silver sintering technology is used. This technology, also known as Low-Temperature Joining Technology (LTJT), enables the formation of high-melting-point (>900°C) and high-thermal-conductivity (~240 W/mK) silver chip interconnects under low temperature and low pressure conditions. The primary component of the sintered layer is silver, which has good electrical and thermal conductivity and avoids the typical fatigue effects seen in soft solder connections, greatly improving the reliability of the power module[16]-[20].

The other materials used in the packaging structure are summarized in Table I. The chip used is the CREE EPM3-1200-0017D1, with a rated voltage of 1200V, rated current of 134A, and a R_{dson} of 17mΩ at room temperature. The substrate ceramic layer is made of AlN, and the bonding process between the copper foil and ceramic layer is Active Metal Brazing (AMB). Compared to silicon nitride (Si_3N_4) and alumina (Al_2O_3), AlN has a thermal conductivity of 200-300 W/mK, which is 5 times higher than Al_2O_3 and 2 times higher than Si_3N_4.

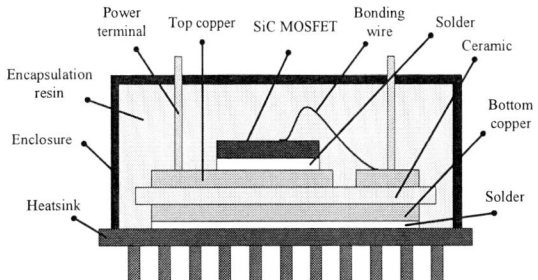

Fig. 1. Power module structure.

TBALE I. PACKAGING MATERIAL

Component	Material	Characteristic
Die	SiC MOSFET	1200V/134A
Ceramic	AlN AMB	Cu/AlN/Cu 0.3mm/0.63mm/0.3mm
Encapsulation resin	R-2188	Long term work temperature >200°C
Enclosure	30% Glass fiber reinforced PPS	Long term work temperature 200-240°C
Bonding wire	Aluminium	Gate/Kelvin source: 8mil Power source: 12mil

B. Modeling of the Multi-Objective Optimization Design Model

The silver sintering layer, as the interconnect interface between the chip and the substrate, frequently bears thermal cycling loads during the operation of the module. Due to the differences in the coefficients of thermal expansion (CTE) of different materials, the maximum stress on the solder layer increases with temperature, which accelerates the fatigue and failure of the solder layer [21]. The thermal coupling between multiple chips causes uneven temperature distribution within the module, leading to localized overheating, and consequently, excessive stress on the solder layer.

Fig. 2 shows a simplified model of the power module structure. The optimization objectives, variables, and constraints for the high-temperature module are as follows: the junction temperature and the thermal-mechanical stress on the solder layer, which mainly affects the module's lifetime, are chosen as the optimization objectives. The variables to be optimized are: d1 (longitudinal spacing between chips), d2 (lateral spacing between chips on the same arm), and d3 (spacing between the two arms). The subsequent routing space and substrate size are considered as constraints. Parametric simulations were carried out using COMSOL, with the specific structural variable ranges and step sizes provided in Table II. The chip was set to a thermal loss of 100W, and the module's single-side convective heat transfer coefficient was 2000W/m²K. Fig. 3 shows the results of the parametric simulation. Considering both the junction temperature and stress, point A represents the best chip layout in the simulation data, with a maximum temperature of 215.75°C and a maximum stress of 893.86MPa. The corresponding values for the variables d1, d2, and d3 are 15mm, 11mm, and 11mm, respectively. Due to the large step size in the simulation, point A may not represent the optimal chip layout.

The simulation of 180 data sets took 2.5 hours. To obtain a better layout by reducing the simulation step size and improving accuracy would significantly increase the simulation time. Therefore, an Artificial Neural Network

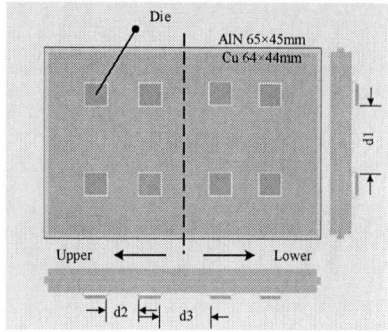

Fig. 2. Variable structure variables and their positions.

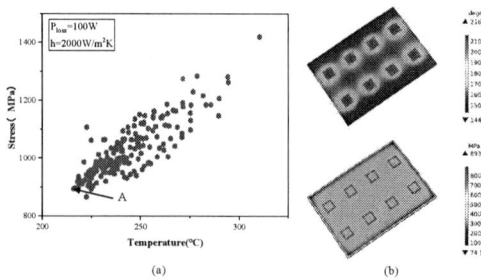

Fig. 3. (a) Parametric scanning simulation results. (b) Point A Temperature and Stress Distribution.

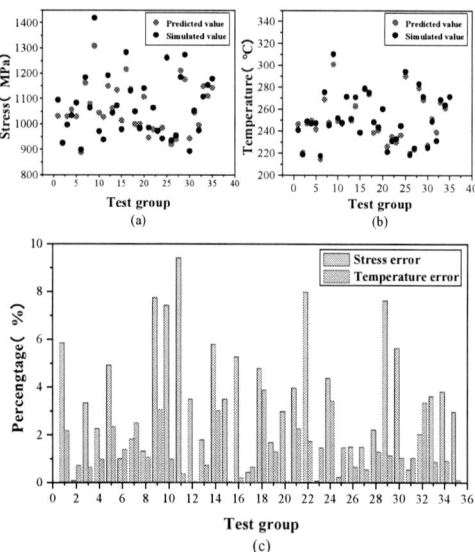

Fig. 4. Comparison of Simulation Values and Predicted Values from ANN Training Test Set. (a) Stress. (b) Junction Temperature. (c) Error Percentage.

(ANN) was used to model the relationship between the structural parameters and the objectives. The ANN training data came from the aforementioned 175 simulation data sets. The number of neurons was set to 8, the population size of the genetic algorithm was 100, with 10,000 iterations and a mutation rate of 0.2. Fig. 4 compares the prediction results of the ANN model with the COMSOL multi-physics simulation results. The fitting accuracy for the maximum junction temperature is quite good, with errors within 4%. Although the fitting accuracy for the solder layer stress is not as good as

TBALE II. VARIABLE STRUCTURE VARIABLES AND THEIR RANGE OF VARIATION

Variable	definition	Range and step
d1	Longitudinal spacing	[5mm, 5mm, 25mm]
d2	Lateral spacing	[1mm, 2mm, 11mm]
d3	Arm spacing	[5mm, 3mm, 20mm]

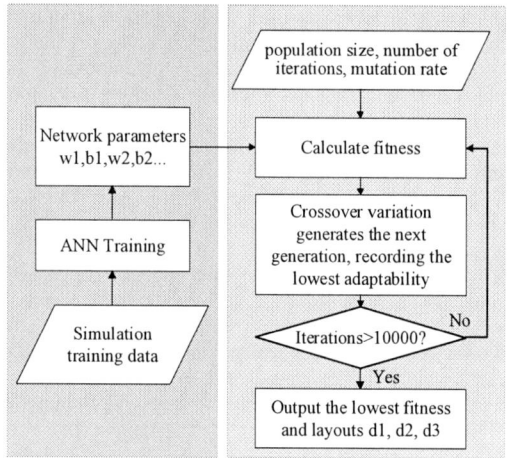

Fig. 5. Process of ANN Combined with Genetic Algorithm.

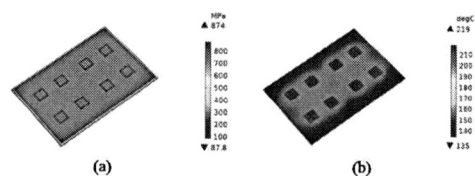

Fig. 6. Simulation at the Optimal Point. (a) Stress. (b) Junction Temperature.

that for the junction temperature, the majority of errors are around 7%. Overall, the ANN-trained model demonstrates relatively high accuracy.

C. Solution to the Multi-Objective Model

To address the issue of excessive simulation time, and considering that the relationship between the optimization objectives and variables is nonlinear and multivariable, making it difficult to perform mathematical modeling and obtain an analytical solution, a genetic algorithm (GA) is employed to solve the model. Fig. 5 illustrates the operational logic of the genetic algorithm. The GA is run for 10 cycles, and with an appropriate population size and elitism strategy, the optimization is completed in just 2 minutes and 20 seconds. The final optimal layout is obtained with d1=13.16 mm, d2=8.5 mm, and d3=11.75 mm, resulting in a maximum stress of 870.3 MPa and a temperature of 210.7°C. Fig. 6 shows that, under this layout, the simulated maximum stress and junction temperature are 874 MPa and 218.9°C, respectively. The optimized results are consistent with the multi-physics simulation outcomes.

III. MODULE ASSEMBLY

The module assembly process is divided into four parts: chip mounting, wire bonding, terminal welding, and shell installation and sealing.

TABLE III. PROCESS FLOW AND PROCESS PARAMETERS

Process	Process parameters
Silver paste printing	Scraper pressure, printing speed
Drying	Oven temperature, drying time
Chip patch	Tip temperature, pressure, table temperature
Sinter	Sintering pressure, time, temperature

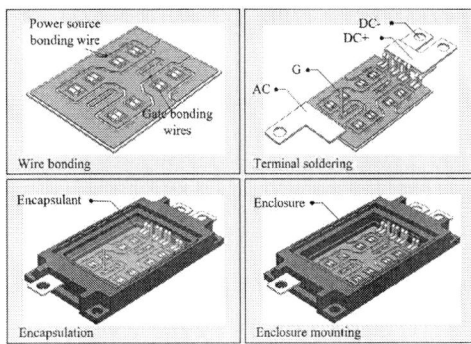

Fig. 7. Chip mounting process. (a) Silver paste printing. (b) Patch. (c) Silver paste drying. (d) Sintering.

Fig. 7 is the chip mounting process, includes silver paste printing, drying, chip placement, and sintering. During the assembly process, the process parameters of this process were repeatedly adjusted to ensure a good sintering effect. The process flow and process parameters involved in the chip mounting process are listed in Table III.

Semi-automatic printing equipment requires manual addition of silver paste, and the scraper pressure and printing speed are adjusted according to the characteristics of the silver paste. The dry method needs to set the time and temperature to make the organic solvent in the silver paste fully volatilize, reducing the porosity of the sintering layer and improving the bonding strength of the chip. When placing the chip, the operation table and the suction head also need to be set to a certain temperature, and pressure needs to be applied to the suction head through a pressure device to ensure a good placement effect. Sintering parameters directly affect the sintering quality. Silver sintering is performed under low-temperature and low-pressure conditions. The key sintering parameters include sintering pressure, sintering temperature, and sintering time.

The module's wire bonding, terminal welding, and shell installation and sealing are shown in Fig. 8. Wire bonding is a mature top-interconnection technology, which uses the high-frequency vibration of the dicing blade to weld the aluminum wire to the aluminum electrode on the top of the substrate and the copper layer on the substrate. Terminal welding uses vacuum reflow welding instead of ultrasonic welding, because the strength of the AlN ceramic is low and it is easily damaged during ultrasonic welding. The subsequent shell installation and glue filling are also routine process flows and will not be described in detail. The assembled module is shown in Fig. 9.

Researchers conducted multiple experiments to optimize the silver sintering process to improve the module's performance and finally obtained the optimal sintering process parameters. The sintering quality can be characterized by the shear strength and porosity of the sintering layer. To verify the sintering effect of the chip, a batch of samples was sintered using the same process parameters, and their shear strength and porosity were tested. Although the samples used a substrate with a different size from the half-bridge module,

Fig. 8. Module assembly process.

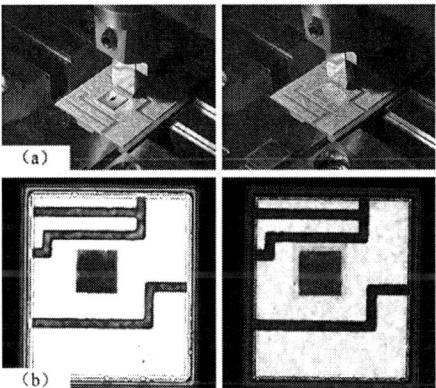

Fig. 9. Assembled Module.

Fig. 10. Sintering quality testing. (a) shear strength testing. (b) porosity testing.

the substrate material, copper layer thickness, ceramic layer thickness, and surface treatment process were completely the same.

The shear strength test is shown in Fig. 10(a), and the test results were 270.83 kg and 255.174 kg. The chip size was 5 × 5 mm, and the calculated shear strength of the two samples was 106.17 MPa and 100.03 MPa, respectively. The SEM results of the sintering layer of the samples are shown in Fig. 10(b), and the porosity was 1.89% and 1.19%, respectively. These two test results show that the process parameter adjustment of the chip mounting process is reasonable, and the sintering quality can be guaranteed.

IV. PERFORMANCE TESTING

A. Electrical Characteristics Testing

The static testing of the module is illustrated in Fig. 11.

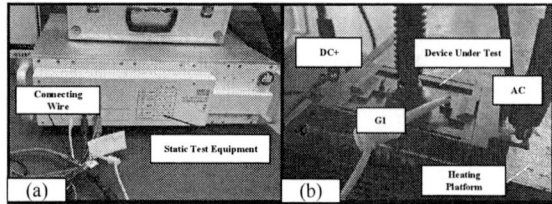

Fig. 11. Static Testing. (a) Static Testing Instrument. (b) Heating Platform and Device Under Test.

Fig. 11(a) shows the testing apparatus, while Fig. 11(b) depicts the heating platform with the module securely mounted for testing. The module was evaluated at both room temperature (25°C) and elevated temperature (200°C), with the results presented in Fig. 12. As observed, whether at ambient temperature (25°C) or high temperature (200°C), the module maintained an insulation voltage of approximately 1500V, which exceeds its rated voltage of 1200V, demonstrating the module's excellent insulating properties and its capability to operate under high-voltage conditions. Fig. 12(b) illustrates the test results for the module's on-state resistance. At room temperature, the on-state resistance is approximately 5mΩ, while at 200°C, it increases to approximately 11mΩ. This change reflects the variation in the module's current conduction capability with temperature. A lower on-state resistance indicates superior current conduction performance, facilitating smooth operation. In summary, the test results confirm that the module meets the expected insulation and on-state resistance specifications under both normal and high-temperature conditions, ensuring reliable functionality.

B. High-Temperature Reliability Testing

Power cycling tests are extensively utilized to assess the lifetime and reliability of power modules and their packaging designs. In contrast to standard modules, high-temperature modules necessitate testing at elevated junction temperatures (T_{jmax}) and broader temperature fluctuation ranges (ΔT_j). The outcomes of power cycling tests can vary significantly depending on the specific testing protocol employed. To evaluate the reliability of the module under more stringent testing conditions, the present study conducted second-level constant current power cycling tests on the power module under consideration. The initial test conditions were set as follows: $T_{jmin} = 100°C$, $T_{jmax} = 200°C$, and $\Delta T_j = 100°C$. The failure criteria for the device were defined as either an increase in the forward voltage drop (V_{on}) exceeding 5%, or a rise in the junction-to-case thermal resistance (R_{th}) greater than 20%. The results of the tests, shown in Fig. 13, indicate that after more than 17,000 cycles, the V_{on} increased by over 5%, leading to the failure of the device and the termination of the test. For comparison, researchers performed power cycling tests on SiC discrete devices from ST ($T_{jmax} = 200°C$), where the devices failed after approximately 7,000 cycles within a temperature range of 24°C to 142°C. A comparison of the test results demonstrates that the power module proposed in this study exhibits a notable level of reliability. Furthermore, the validity of the design approach and process optimization proposed in this paper has been preliminarily validated.

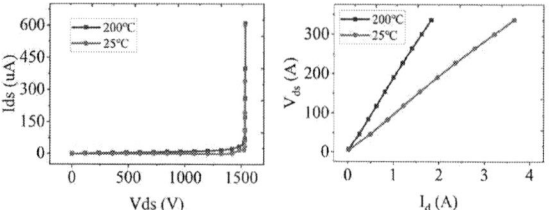

Fig. 12. Static testing result. (a) Insulation level. (b) On-state resistance.

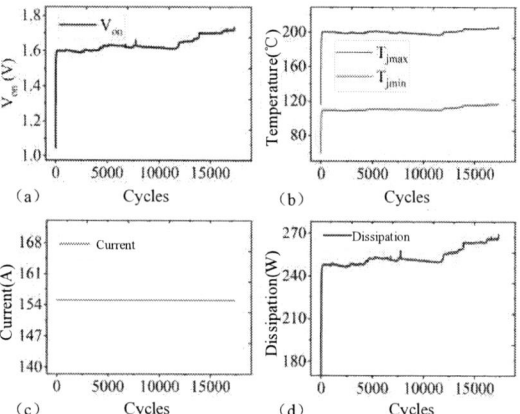

Fig. 13. Power cycle testing result. (a) Von. (b) Tjmax. (c) Current. (d) Power loss.

V. CONCLUSION

In this paper, a low-stress design method for high-temperature silicon carbide power modules based on silver sintering process is proposed. By selecting the appropriate high-temperature packaging material and optimizing the position of the chip to reduce the local high stress caused by uneven temperature distribution, the optimization method of artificial neural network combined with genetic algorithm was used to replace a large number of simulations, and the results showed that the optimal chip layout was obtained in only a short time, and the comparison error of multi-physics simulation results was very small. The reliability test results show that the module can withstand more than 17,000 cycles in high-temperature power cycles, and initially has the ability to work under high-temperature conditions, but there is still a big gap from the actual application.

REFERENCES

[1] Millan J.,Godignon P.,Perpina X.,Perez-Tomas A.. A Survey of Wide Bandgap Power Semiconductor Devices[J] IEEE Transactions on Power Electronics,2014,29(5):2155-2163.

[2] Millán J, Godignon P, Pérez-Tomás A. Wide band gap semiconductor devices for power electronics[J]. Automatika: časopis za automatiku, mjerenje, elektroniku, računarstvo i komunikacije, 2012, 53(2): 107-116.

[3] SHENG Kuang, GUO Qing, ZHANG Jun-ming, QIAN Zhao-ming. Development and Prospect of SiC Power Devices in Power Grid[J]. Proceedings of the CSEE, 2012, 32(30): 1-7,3.

[4] WANG Laili, ZHAO Cheng, ZHANG Tongyu, et al. Review of Packaging Technology for Silicon Carbide Power Modules[J]. Transactions of China Electrotechnical Society, 2023, 38(18): 4947-4962.

[5] Hou F, Wang W, Cao L, et al. Review of packaging schemes for power module[J]. IEEE Journal of Emerging and Selected Topics in Power Electronics, 2019, 8(1): 223-238.

[6] WANG Xiaolei, ZHANG Yousheng, DAI Shengwei, et al. Progress on high temperature resistant molding compounds for packaging of power chips in electrical vehicles[J]. *Insulating Materials*, 2024, 57(02): 1-9.

[7] LIN Ying, SHI Yulong, LIU Yuhao, et al. Temperature resistance of organic silicone encapsulant used in silicone carbide power devices[J]. *Insulating Materials*, 2023, 56(12): 24-33.

[8] LIU Dongming, LI Xuebao, XU Jiayu, et al. Analysis of High Temperature Wide Band Dielectric Properties of Organic Silicone Elastomer for High Voltage SiC Device Packaging[J]. *Transactions of China Electrotechnical Society*, 2021, 36(12): 2548-2559.

[9] Liu L, Nam D, Guo B, et al. Glass for encapsulating high-temperature power modules[J]. IEEE Journal of Emerging and Selected Topics in Power Electronics, 2020, 9(3): 3725-3734.

[10] Y. Chen, G. Lei, G. -Q. Lu and Y. -H. Mei. High-Temperature Characterizations of a Half-Bridge Wire-Bondless SiC MOSFET Module. IEEE Journal of the Electron Devices Society, vol. 9, pp. 966-971, 2021.

[11] Yang F, Wang L, Kong H, et al. Compact-interleaved packaging method of power module with dynamic characterization of 4H-SiC MOSFET and development of power electronic converter at extremely high junction temperature[J]. IEEE Transactions on Power Electronics, 2022, 38(1): 417-434.

[12] C. Zhang, Z. Huang, C. Chen, X. Liu, F. Luo and Y. Kang. A 900A High Power Density and Low Inductive Full SiC Power Module for High Temperature Applications based on 900V SiC MOSFETs. 2020 IEEE Applied Power Electronics Conference and Exposition (APEC), New Orleans, LA, USA, 2020, 2777-2781.

[13] B. Liu et al.A Low-Thermal-Stress Double-Sided Cooling Wire-Bondless Package Structure of SiC Power Modules for High-Temperature Applications[J].IEEE Transactions on Power Electronics, 2024 , 39(11): 14741-14757.

[14] Z. Zeng, K. Ou, L. Wang and Y. Yu, "Reliability-Oriented Automated Design of Double-Sided Cooling Power Module: A Thermo-Mechanical-Coordinated and Multi-Objective-Oriented Optimization Methodology," in IEEE Transactions on Device and Materials Reliability, vol. 20, no. 3, pp. 584-595, Sept. 2020.

[15] T. Dragičević, P. Wheeler and F. Blaabjerg, "Artificial Intelligence Aided Automated Design for Reliability of Power Electronic Systems," in IEEE Transactions on Power Electronics, vol. 34, no. 8, pp. 7161-7171, Aug. 2019,

[16] Yan H , Liang P , Mei Y ,et al.Brief review of silver sinter-bonding processing for packaging high-temperature power devices[J].Chinese Journal of Electrical Engineering, 2020, 6(3):25-34.

[17] Calabretta M, Sitta A, Oliveri S M, et al. Silver sintering for silicon carbide die attach: process optimization and structural modeling[J]. Applied Sciences, 2021, 11(15): 7012-7023.

[18] FENG Jingjing. Characterizations of A Medium-and-High Voltage IGBT Module Packaged with Nanosilver Paste[D]. Tianjin University, 2024-05-03.

[19] Chen Z, Balankura T, Fichthorn K A, et al. Revisiting the polyol synthesis of silver nanostructures: role of chloride in nanocube formation[J]. ACS nano, 2019, 13(2): 1849-1860.

[20] Wang M , Mei Y , Li X ,et al. Pressureless Silver Sintering on Nickel for Power Module Packaging[J].IEEE Transactions on Power Electronics, 2019, PP(8):1-1.

[21] YIN Zhihao, YU Dianru, ZHU Jiafeng, et al. Review of IGBT power module packaging failure mechanism and monitoring methods[J]. Advanced Technology of Electrical Engineering and Energy, 2022, 41(8): 51-70.

Monolithically Integrated Over-Temperature Protection Circuit Based on GaN HEMTs

Pingyu Cao[1,2], Kepeng Zhao[1,2], Yihao Xu[1,2], Harm Van Zalinge[2], Ping Zhang[3], Miao Cui[1]*, Fei Xue[1]

[1]Department of Electrical and Electronic Engineering, Xi'an Jiaotong-Liverpool University, Suzhou, China
[2]Department of Electrical Engineering and Electronics, University of Liverpool, Liverpool, UK
[3]Department of Communications and Networking, Xi'an Jiaotong-Liverpool University, Suzhou, China
*Miao.Cui02@xjtlu.edu.cn

Abstract

Over-temperature protection is progressively being recognised as a crucial aspect of power converter design, driven by the growing demands for higher switching frequency and power density. This paper demonstrates a monolithically integrated over-temperature protection circuit that consists of only four p-GaN HEMTs, capable of controlling the power transistors of synchronous Buck converters. The proposed work includes temperature detection and protection circuits that can avoid high-temperature damage to GaN Buck converter circuits. Due to the simple structure of this design, the protection circuit exhibits a small chip area, which is 0.16 mm². The protection circuit can accurately protect Buck converters at different temperatures by adjusting the supply voltage of the temperature detection circuit, which would be conducive to the varying demands of GaN converter circuits. The experimental results demonstrate that the proposed work achieves over-temperature protection at 200 °C.

Introduction

Low power loss, fast switching speed, and high operation frequency constitute characteristics of gallium nitride (GaN) devices, making them competitive candidates for high-efficiency, high-frequency and high-temperature power converters [1], [2]. Due to the high-temperature operating capability of GaN material, the cooling system of GaN integrated circuits (ICs) is simplified. The Buck converter represents a significant application area for GaN devices, as the low on-resistance contributes to improved efficiency. GaN-based Buck converters are widely used in various applications, such as data centres and electric vehicles [3], [4]. GaN devices under prolonged high temperature and electrical stress conditions may lead to device failure, which would result in the failure of the entire circuits [5]. Hence, over-temperature protection (OTP) is necessary to enhance the reliability of GaN ICs. A GaN integrated OTP circuit based on Fluorine implantation devices was proposed in [6], and a temperature sensor with a simple structure was introduced in [7] to achieve over-temperature protection. To prevent Buck converters from failing under high-temperature conditions, a GaN OTP circuit

for driver circuits needs to be designed. In reference [8], the over-temperature protection for GaN power converter was achieved by the temperature sensing that was based on a lateral field-effect rectifier. To achieve all GaN integration of converters, it is essential to design a simple over-temperature protection circuit to enhance the efficiency of GaN converters.

In this work, a monolithically integrated over-temperature protection circuit for the GaN Buck converter was proposed based on p-GaN HEMT technology. The proposed protection circuit consists of four devices with a small area of 0.16 mm², enhancing the power density of GaN converters. Moreover, it can accurately respond and turn off the power transistor to ensure circuit safety when the operation temperature exceeds 200 °C, thereby increasing the reliability of p-GaN circuits.

Device Fabrication and Characteristics

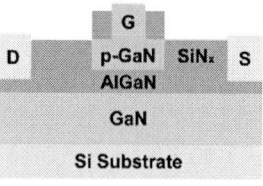

(a) E-mode device

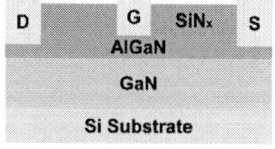

(b) D-mode device

Fig. 1. The structure of GaN devices in this work.

Fig. 1 shows a schematic cross-section of the E-mode and D-mode GaN devices in this work. The proposed GaN over-temperature protection circuit was fabricated on a commercial GaN-on-Si substrate that consists of a 70 nm p-GaN layer, a 0.8 nm AlN interlayer, a 15 nm thick AlGaN barrier layer with a 20% Al content and a 200 nm GaN channel layer. The mesa isolation and removal of the p-GaN layer were achieved by BCl₃-based inductively coupled plasma

(ICP) etching. The Ohmic contacts of the source and drain were formed using a Ti/Al/Ni/TiN metallisation, alloyed at 960 °C. Furthermore, the gate electrode of the devices consisted of Ni/TiN, and the passivation layer of the devices was a 160-nm SiN_x layer. Finally, the protection circuit was connected by the evaporated Al metal.

To evaluate the performance of p-GaN devices at room temperature (RT) and high temperatures, the electrical characteristics of the GaN devices were measured by a probe station with heating function and an Agilent B1505 semiconductor analyser. The transfer curves of E-mode and D-mode devices are demonstrated in Fig. 2(a) and Fig. 2(b). As shown in Fig. 2, the E-mode device presents a threshold voltage of 1.6 V, while the D-mode device has a threshold voltage of -2.3 V. The saturation current of the E-mode device is roughly equal to the D-mode device, which is around 215 mA/mm at RT. When the measurement increases to 200 °C, the saturation current of E-mode and D-mode devices is reduced to 73 and 70 mA/mm, respectively.

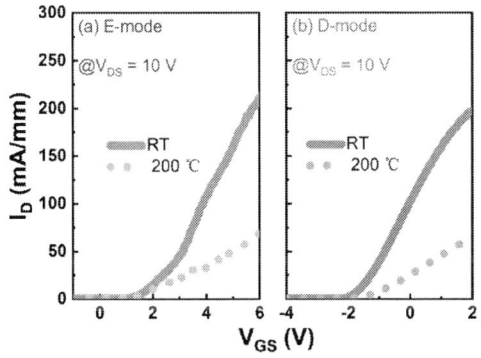

Fig. 2. The electrical characteristics of (a) E-mode device, (b) D-mode device. (dimensions: $L_{GS}/L_G/W_G/L_{GD}$=5/5/100/5 μm).

Circuit Structure

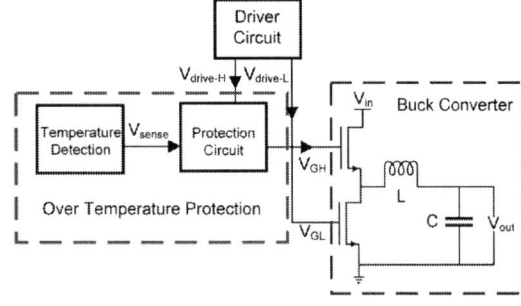

Fig. 3. Schematics of the proposed over-temperature protection for GaN power converter circuit.

The schematic of the whole circuit is shown in Fig. 3. The proposed over-temperature protection circuit can be applied to a synchronous half-bridge Buck converter to ensure that the circuit is prevented from simultaneous exposure to electrical stress and thermal conditions exceeding predefined temperature thresholds. This is achieved by cutting off the control signal of the driver circuit at high-temperature conditions.

The circuit diagram and microphotograph of the proposed work are demonstrated in Fig. 4 and Fig. 5, respectively. The area of this work is 420×380 μm² with the testing pads. The OTP circuit consists of a temperature detection circuit and a protection circuit. The detection circuit can generate a temperature-dependent bias voltage V_{sense}, which includes E-mode and D-mode devices M_1/M_2 (width=10 μm/100 μm). The protection circuit comprises D-mode and E-mode devices M_3/M_4 (width=100 μm/100 μm). At room temperature, the V_{sense} is lower than the threshold voltage of the E-mode device M_4, and the output of the protection circuit is the signal from the driver circuit V_{drive}, which would not affect the performance of the Buck converter. However, the V_{sense} increases with the elevated temperature. At high temperatures, such as 200 °C, the output of the protection circuit becomes logically low to turn off the power transistors. Moreover, the output voltage level of the temperature detection circuit can be adjusted by the supply voltage V_{DD} to accurately detect the over-temperature condition and ensure the circuit safety.

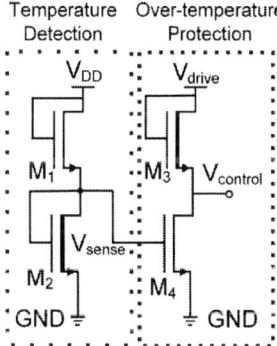

Fig. 4. The circuit schematic of the proposed over-temperature protection.

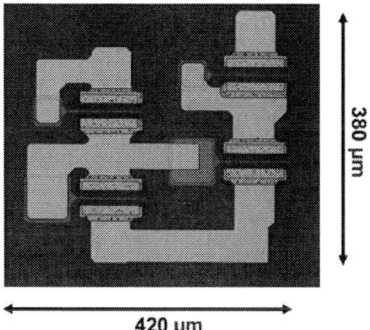

Fig. 5. The microphotograph of the proposed circuit.

Experimental Results

The current-voltage characteristics of the temperature sense circuit were measured using the Agilent B1505 semiconductor parameter analyser from RT to 200 °C, and the results are plotted in Fig. 6. As illustrated in Fig. 6, the V_{sense} increases with elevated temperatures, functioning as the control signal of the protection circuit. The V_{sense} at various temperatures are plotted in Fig. 7 with a supply voltage of 5 V. The V_{sense} is 690 mV at RT and increases to 1.64 V at 200 °C. The purpose of the temperature-dependent V_{sense} is to detect temperature changes and control the protection circuit to turn off the power transistors. The temperature threshold of the over-temperature protection can be dynamically adjusted via V_{DD} modulation, thereby enabling enhanced protection accuracy.

threshold voltage of M_4, the E-mode device M_4 will turn on, resulting in an output voltage $V_{control}$ of 0.7 V at 200 °C. Consequently, the high-side power transistor can be turned off, achieving over-temperature protection and enhancing overall system stability. The non-zero output signal at 200 °C is primarily caused by the increased on-resistance of enhancement-mode devices at high temperatures. With the progressive reduction of device dimensions, the proposed design is expected to deliver near-zero output under high-temperature conditions. The proposed work can prevent a Buck converter from simultaneous exposure to electrical stress and thermal conditions exceeding predefined temperature thresholds, thereby enhancing the reliability of the GaN converter.

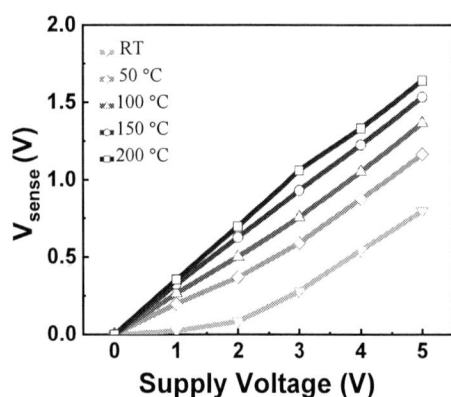

Fig. 6. The output of the temperature detection circuit under various supply voltages.

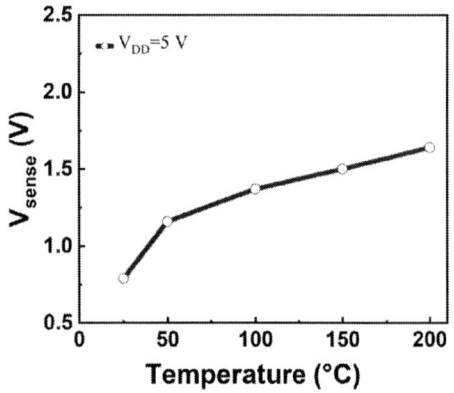

Fig. 7. The output of the temperature detection circuit at different temperatures. (V_{DD}=5 V).

Fig. 8 illustrates the output voltage of the over-temperature protection circuit at different temperatures (from RT to 200 °C). At RT, the output of the sense circuit is 0.69 V, which is lower than the threshold voltage of the E-mode device M_4 in the protection circuit. Therefore, the high-side power transistor can receive the control signal from the driver circuit. Due to the increased voltage V_{sense} and the negative shift in the

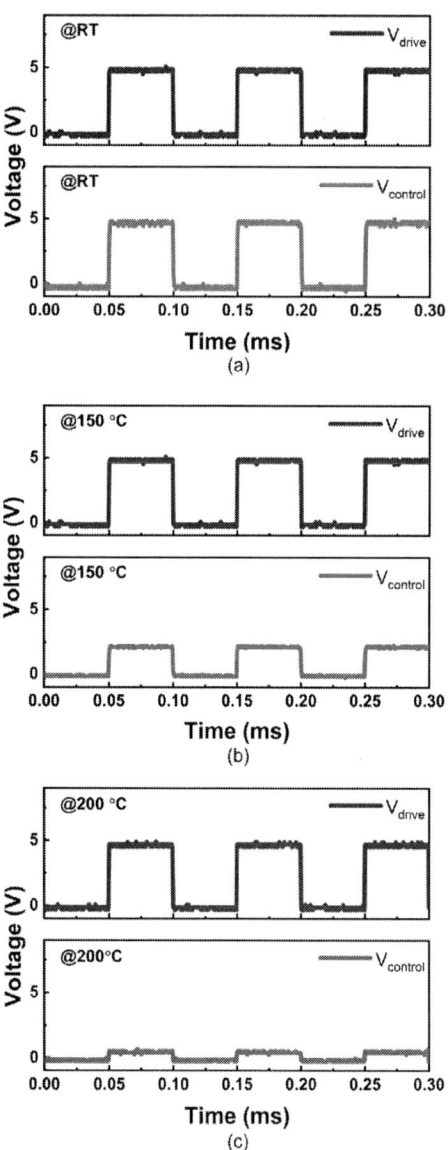

Fig. 8. The waveforms of the proposed circuit at (a) RT, (b) 150 °C and (c) 200 °C.

Conclusions

A monolithically integrated over-temperature protection circuit, consisting of a temperature detection circuit and a protection circuit based on p-GaN HEMT technology was experimentally demonstrated. The proposed protection circuit can be applied to a GaN Buck converter to turn off the power transistors at a high temperature of 200 °C. When the temperature exceeds the predefined safety threshold (200 °C), the over-temperature protection circuit is automatically activated. It achieves over-temperature protection by turning off the power transistors, thereby preventing from simultaneous exposure to electrical stress and thermal conditions, which could potentially damage the entire circuit. Furthermore, the area of this work is only 0.16 mm^2 owing to the simple structure. Therefore, the proposed circuit can enhance the overall reliability of the GaN Buck converter.

Acknowledgements

This work was supported by XJTLU Research Development Fund (RDF21-02-031, PGRS2206039).

References

[1] E. A. Jones, F. F. Wang, and D. Costinett, 'Review of Commercial GaN Power Devices and GaN-Based Converter Design Challenges', IEEE J. Emerg. Sel. Topics Power Electron., vol. 4, no. 3, pp. 707–719, Sep. 2016, doi: 10.1109/jestpe.2016.2582685.

[2] T. J. Flack, B. N. Pushpakaran, and S. B. Bayne, 'GaN Technology for Power Electronic Applications: A Review', Journal of Elec Materi, vol. 45, no. 6, pp. 2673–2682, Jun. 2016, doi: 10.1007/s11664-016-4435-3.

[3] R. Sun, Y. C. Liang, Y.-C. Yeo, C. Zhao, W. Chen, and B. Zhang, 'Development of GaN Power IC Platform and All GaN DC-DC Buck Converter IC', in 2019 31st International Symposium on Power Semiconductor Devices and ICs (ISPSD), Shanghai, China: IEEE, May 2019, pp. 271–274. doi: 10.1109/ispsd.2019.8757674.

[4] M. Cui, R. Sun, Q. Bu, W. Liu, H. Wen, A. Li, Y. Liang, C. Zhao, 'Monolithic GaN Half-Bridge Stages With Integrated Gate Drivers for High Temperature DC-DC Buck Converters', IEEE Access, vol. 7, pp. 184375–184384, 2019, doi: 10.1109/access.2019.2958059.

[5] J. He, J. Wei, S. Yang, Y. Wang, K. Zhong, and K. J. Chen, 'Frequency- and Temperature-Dependent Gate Reliability of Schottky-Type p-GaN Gate HEMTs', IEEE Trans. Electron Devices, vol. 66, no. 8, pp. 3453–3458, Aug. 2019, doi: 10.1109/ted.2019.2924675.

[6] A. M. H. Kwan, Y. Guan, X. Liu, and K. J. Chen, 'Integrated Over-Temperature Protection Circuit for GaN Smart Power ICs', Jpn. J. Appl. Phys., vol. 52, no. 8S, p. 08JN15, Aug. 2013, doi: 10.7567/jjap.52.08jn15.

[7] A. Li, F. Li, K. Chen, Y. Zhu, W. Wang, I. Z. Mitrovic, H. Wen, W. Liu, 'A Supply Voltage Insensitive Two-Transistor Temperature Sensor With PTAT/CTAT Outputs Based on Monolithic GaN Integrated Circuits', IEEE Trans. Power Electron., vol. 38, no. 9, pp. 10584–10588, Sep. 2023, doi: 10.1109/tpel.2023.3288937.

[8] L. Kang, H. Wen, Q. Bu, and W. Liu, 'Design and Evaluation of GaN-based Over-Temperature Protection Circuit', in 2019 International Conference on IC Design and Technology (ICICDT), SUZHOU, China: IEEE, Jun. 2019, pp. 1–4. doi: 10.1109/icicdt.2019.8790903.

Lifetime Modeling of IGBT Devices Based on Micro-Defect Topography Inversion

Miaomiao Shangguan
State Key Laboratory of Power Transmission Equipment Technology
Chongqing University
Chongqing, China
shangguan-mm@stu.cqu.edu.cn

Wei Lai
State Key Laboratory of Power Transmission Equipment Technology
Chongqing University
Chongqing, China
laiweicqu@126.com

Hao Wang
State Key Laboratory of Power Transmission Equipment Technology
Chongqing University
Chongqing, China
eewanghao@stu.cqu.edu.cn

Yunjie Wu
State Key Laboratory of Power Transmission Equipment Technology
Chongqing University
Chongqing, China
202311021229T@stu.cqu.edu.cn

Yu Liu
State Key Laboratory of Power Transmission Equipment Technology
Chongqing University
Chongqing, China
202411021224T@stu.cqu.edu.cn

Hang Zhao
State Key Laboratory of Power Transmission Equipment Technology
Chongqing University
Chongqing, China
202411131333@stu.cqu.edu.cn

Abstract—As a key component of traction transmission system, the reliability of power electronic devices is an important guarantee of system safety. However, the existing studies mainly focus on the life assessment of the bathtub curve failure loss period, and lack the state characterization and remaining life prediction methods of the bathtub curve random failure period. In this paper, the solder layer CT scanning results of IGBT devices are combined with the finite element simulation model to invert the fatigue degree of the alive mesh through the stress relation between the dead and alive mesh and the microscopic representation of the Coffin-Manson model. The health state of the device is evaluated according to the void distribution in the solder layer during the random failure period, and the corresponding residual life model of the IGBT device is established to realize the life cycle evaluation of the IGBT device in service. The expected results of this paper enrich the life prediction and reliability evaluation of power devices, provide data basis for the continuous use of replaced IGBT modules, and thus reduce system operation and maintenance costs, which has important strategic value and economic significance.

Keywords—Lifetime model, finit element model, reliability, insulated gate bipolar transistor (IGBT).

I. Introduction

IGBT devices are widely used in the field of power electronics, especially in high power applications such as rail transit, renewable energy, and power grid regulation. The heat dissipation and reliability of welded IGBT devices are affected by the packaging mode, and the failure modes are mostly solder layer failure and bonding line failure. However, the life evaluation of IGBT devices often neglects the fatigue aging of the solder layer, which leads to the optimistic life evaluation. When IGBT devices leave the factory, there are often small holes in the solder layer. After actual operation, the solder layer is constantly impacted by thermal stress, the growth of small holes then intensifies the concentration of thermal stress, and the holes continue to expand to form interface cracks, which seriously affects the use of IGBT modules.

In terms of IGBT failure mechanism analysis, [1] studies the degradation of the electrothermal characteristics of IGBT devices based on the study of the fracture and fall off direction of aluminum wires, and provides a test method for evaluating IGBT in operation offline or online. Reference [2] used the redesigned IGBT structure to use the on-state voltage to characterize the degradation state of the solder layer and proposed a solder layer degradation simulation method [3]. Researchers proposed a Cauer thermal network model considering the degradation of solder layer voids [4], and verified that the improved Cauer model can accurately and rapidly obtain the junction temperature of IGBT devices. The team from Chongqing University investigated the influence of low-amplitude thermal-mechanical stress and initial defects on the degradation of solder layers through power cycling experiments and microcomputer tomography [5]. Reference [6] proposed a condition monitoring method for IGBT devices.

In conclusion, the working environment where IGBT devices are located often has complex operating conditions. The growth of voids or cracks in the solder layer is inevitable, and the health status of the solder layer has a significant impact on the normal operation of IGBT devices. However, due to the randomness of the voids in the solder layer of IGBT devices and the uncontrollability of crack growth, the existing research on the solder layer of IGBT devices mostly focuses on brand-new IGBT devices, while there are relatively few studies on the expansion and growth process of the voids in the solder layer of in-service IGBT devices. IGBT devices are often tested before leaving the factory, and a large number of IGBT devices that have been in service still have the value of secondary utilization. Therefore, the remaining life assessment of IGBT devices should not be limited to brand-new devices. It is necessary to conduct health status assessment for IGBT devices that have been in service.

II. The Establishment of Finite Element Model

IGBT modules can be divided into welded IGBT and press-pack IGBT according to the difference in packaging. Since welded IGBT devices have a wider application range and are more prone to damage compared to press-pack IGBT devices, this paper focuses on the aging of the solder layer of welded IGBT devices。 The package structure of the welded IGBT module is shown in Fig. 1.

979-8-3315-1110-4/25 $31.00 © 2025 IEEE

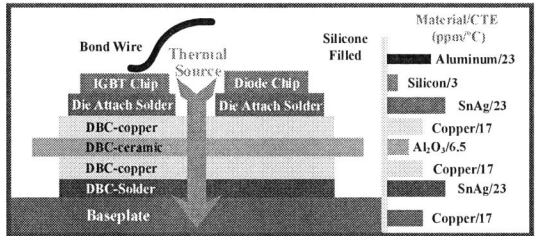

Fig. 1. Package structure of welded IGBT module.

It can be seen from the Fig. 1 that the IGBT module is stacked by multiple layers of materials, and the thermal expansion coefficient of each layer of materials is different.

In this paper, SKM50GB12T4 IGBT module produced by SEMIKRON company is selected, and the 3D model of SKM50GB12T4 IGBT module is established in COMSOL finite element simulation software, and the relevant material parameters are imported, as shown in Table 1.

TABLE I. THE ELECTROTHERMAL PARAMETERS OF EACH LAYER MATERIAL OF IGBT DEVICES

Materials	Resistivity [S·m^{-1}]	Thermal conductivity [W·(m×K)$^{-1}$]	Coefficient of thermal expansion[10^{-6}·K^{-1}]
Al	2.85×10^{-8}	238	23
Si	/	148	3
96.5Sn3.5Ag	1.5×10^{-7}	70	23
Al$_2$O$_3$	1.0×10^{-22}	55	6.5
Cu	1.67×10^{-8}	401	17

The boundary conditions of the multi-physics coupling model are set as follows: The base plate is in contact with the water-cooled plate, and the heat flux value of the substrate is set to 3000W/(m²·K) to simulate the water-cooled heat dissipation effect and simplify the calculation, while the heat flux value of the other exposed surfaces is set to 12.5W/(m²·K) under natural convection conditions. The ambient temperature was set at 298.15K, consistent with the experimental conditions. During the power cycling experiment, to prevent the component from moving in position, it is usually fastened to the heat dissipation device. Therefore, in the simulation, the installation terminal of the component is set as the displacement constraint, and the normal displacement constraint condition is applied at the perimeter of the substrate. Given that the solder layer is the weak layer of IGBT devices and is also the key research object in Chapter III, when performing mesh division, a refined mesh division was implemented for the chip and solder layer parts, while conventional mesh division was adopted for the remaining parts. The total number of meshes divided by the final finite element model is 78,415, and the established multi-field coupling simulation model is shown in Fig. 2.

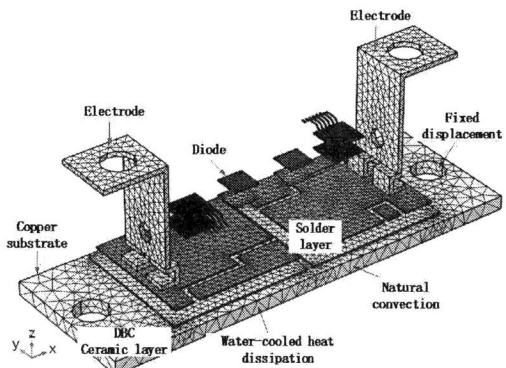

Fig. 2. Boundary constraint conditions and grid division of IGBT device.

III. THE THEORY OF INVERSION PROCESS OF FATIGUE STATE OF GRID ELEMENT

Based on the MINER linear life formula combined with the power cycling test results and the finite element simulation model, this study proposed a device health modeling method based on the surface of the solder layer micro-defect topography inversion. This method can obtain the health status of the IGBT module by introducing the stress connection between the life and death grid and the dynamic update of the grid, thus providing a more accurate theoretical basis for the reliability evaluation of the device and providing a reference for the reuse of the replaced IGBT module.

In the linear region with small thermal resistance change, the fatigue accumulation behavior usually follows a linear law, so the MINER theory can be used to analyze the linear accumulation of load periods, as shown in formula (1).

$$D = \sum_{i=1}^{n} \frac{N_i}{N_{fi}} \quad (1)$$

Common analytical calculation formulas for IGBT module life are shown in formula (2). N_f is the number of failure cycles, ΔT_j is the junction temperature difference, and T_{jm} is the average junction temperature.

$$N_f = A \square (\Delta T_j)^{\alpha_0} \quad (2)$$

Fig. 3 is a schematic diagram of the solder layer of IGBT. For an IGBT device that has been in service, at the t_0 moment of obtaining this device, the fatigue accumulation of the device is unknown. To achieve the reuse of the IGBT device that has been in service, it is necessary to determine the health status of the device to assess its remaining service life. Suppose there are multiple voids in the solder layer at time t_0. At this time, the IGBT device is subjected to a power cycling test. After ΔN power cycles, time t_1 is reached. Suppose at this time, a new void A is added in the solder layer, indicated in red. The position of the fatigue degree to be determined is x, indicated in blue. At this time, it is considered that the fatigue accumulation degree D of the new void is 1, reaching the failure state. The failure criterion is as shown in formula (3).

$$D - \sum \frac{N(i)}{N_f(i)} - 1 \quad (3)$$

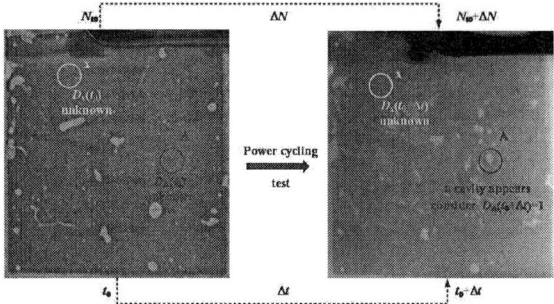

Fig. 3.   IGBT device solder layer aging diagram.

The traditional analytical Coffin-Manson life formula is shown in formula (2). Another microscopic manifestation is as shown in formula (4).

$$C = \Delta\varepsilon_{in}(N_f)^{\alpha_0} \tag{4}$$

Where $\Delta\varepsilon_{in}$ refers to the fatigue inelastic strain under a single period.

To determine the fatigue state D_x of any grid x in the non-cavity range of the solder layer at time t_0, D_x can be calculated by the correlation of stress between cavity parts due to aging of some entities after ΔN power cycles.

By combining formula (1), (2), and (4), formula (5) can be obtained as follows, where N_{t0} is the equivalent failure times under load in the early stage, ΔN is the number of power cycle tests experienced by IGBT module, N_{fx} is the failure times corresponding to load size at x, N_{fA} is the failure times corresponding to load size at A, $\Delta\varepsilon_x$ is the strain at x. $\Delta\varepsilon_A$ is the strain at A.

$$D_x(t_0 + \Delta t) = \frac{N_{t_0}}{N_{fx}} + \frac{\Delta N}{N_{fx}} = \frac{N_{t_0} + \Delta N}{\left(\dfrac{C}{\Delta\varepsilon_x}\right)^{\frac{1}{\alpha_0}}} \tag{5}$$

Similarly, the formula (6) is also consistent at A:

$$D_A(t_0 + \Delta t) = \frac{N_{t_0}}{N_{fA}} + \frac{\Delta N}{N_{fA}} = \frac{N_{t_0} + \Delta N}{\left(\dfrac{C}{\Delta\varepsilon_A}\right)^{\frac{1}{\alpha_0}}} \tag{6}$$

When there is a void in the entity at A due to aging, D_A is considered to be 1, that is, the formula (6) is equal to 1. Comparing formula (5) with formula (6), formula (7) can be obtained:

$$D_x(t_0 + \Delta t) = \frac{\left(\dfrac{C}{\Delta\varepsilon_A}\right)^{\frac{1}{\alpha_0}}}{\left(\dfrac{C}{\Delta\varepsilon_x}\right)^{\frac{1}{\alpha_0}}} = \left(\frac{\Delta\varepsilon_x}{\Delta\varepsilon_A}\right)^{\frac{1}{\alpha_0}} \tag{7}$$

Combine formula (7) with formula (1) and simplify to get formula (8) as shown below:

$$D_{xt_0} = \left(\frac{\Delta\varepsilon_x}{\Delta\varepsilon_A}\right)^{\frac{1}{\alpha_0}} - \Delta D_x = \left(\frac{\Delta\varepsilon_x}{\Delta\varepsilon_A}\right)^{\frac{1}{\alpha_0}} - \frac{\Delta N}{\left(\dfrac{C}{\Delta\varepsilon_x}\right)^{\frac{1}{\alpha_0}}} \tag{8}$$

According to formula (8), it can be found that the fatigue accumulative degree D_x at t_0 time at the solder layer x of

IGBT module is equal to the stress of cavity A generated at time t_1 is correlated, that is, we can calculate the fatigue accumulative degree D_x at x in the solder layer at time t_0 by determining the strain within the new cavity range of the solder layer $\Delta\varepsilon$.

The above method describes the calculation of the fatigue accumulative degree of the grid at t_0 time in the non-void range. It should be noted that there are still entities at t_0 time. That is, for Fig. 3, how to calculate the fatigue accumulation degree D of other voids at time t_0?

From the above analysis, it can be seen that grid aging occurs at A at time $(t_0+\Delta t)$, that is, D_A is 1. The calculation formula (9) can be obtained by combining the formulas (1), (3) and (4) as follows:

$$D_{At_0} = 1 - \Delta D_A = 1 - \frac{\Delta N}{N_{fA}} = 1 - \frac{\Delta N}{\left(\dfrac{C}{\Delta\varepsilon_A}\right)^{\frac{1}{\alpha_0}}} \tag{9}$$

Other points can also obtain D in the same way.

It can be found that to obtain the fatigue degree D at time A, B, C, etc., at t_0, the strain there $\Delta\varepsilon$ is also needed. Through $\Delta\varepsilon_A$ at place A and $\Delta\varepsilon_x$ at place x, combined with power cycle ΔN, the inversion of grid fatigue cumulant D at time x of t_0 can be realized. The strain can be obtained by the finite element model established above.

IV. DETERMINATION OF THE HEALTH STATUS OF IGBT DEVICES

This section will determine the health status of IGBT devices based on the CT scan images of the actual solder layer of IGBT devices, combined with the grid fatigue degree inversion theory proposed in Chapter III.

Select a CT image of the solder layer of an IGBT device as shown in Fig. 4 and perform binarization processing on the image. The specific steps are as follows. The image size of the solder layer obtained through CT operation may vary depending on the scanning equipment. Therefore, the scanned images need to be stretched to the actual chip size. If the dimensions of the image are not handled correctly, the actual positions of cavities or other defects will deviate due to inaccurate proportions.

The image is stretched based on the edge of the solder layer. The actual size of the IGBT device chip selected in this paper is 724mm×690mm. Therefore, the solder layer image is stretched to 724 pixels ×690 pixels through Python. After stretching the solder layer image to the same size as the actual chip solder layer, the image is binarized through the OpenCV library in Python to provide clear boundary contours and dividing lines for subsequent cavity detection and statistics. The process of image binarization is to convert color images into grayscale images, and grayscale images can perform threshold operations more easily. Meanwhile, in order to reduce the noise in the image and make the subsequent binarization effect better, Gaussian blur processing is adopted. The binarization method selected here is binarization based on the threshold, and the threshold is selected as (150,180). The binarization image is shown in Figure 5. The image after binarization processing has a more distinct hollow contour and is also more convenient for obtaining subsequent hole-related information.

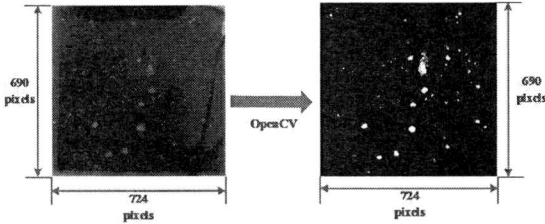

Fig. 4.   Tensile and binarized solder layer CT scan image

To construct IGBT devices, perform grid division of the model as described in the previous steps. According to the setting of the boundary conditions, the transient study of the model was conducted. A 90W heat source loss was applied to the IGBT chip, and the transient running time was controlled at (0,1,120). The results within the time range of 0-120 seconds were output, and the results were output once every 1 second. The junction temperature and shell temperature variations of IGBT devices are shown in Fig. 5. Take the difference in strain at position x during the second temperature fluctuation period, that is, 85-115s, as $\Delta \varepsilon_x$. In this paper, the vertices of the meshes in the finite element model are selected as the research objects. Through the JAVA interface in COMSOL, the position information of each mesh in the solder layer is obtained by using the getVertex function. Traverse all the grids of the solder layer, calculate the strain at all grid vertices within 85-115 seconds $\Delta \varepsilon$, obtain the maximum stress value $\Delta \varepsilon_{max}$ and the minimum stress value $\Delta \varepsilon_{min}$ within this time period by comparison, and take the difference to obtain the strain difference at each grid vertex $\Delta \varepsilon_x$. The strain information of the mesh vertices at different positions of the solder layer of the finite element model of IGBT device is obtained through the JAVA interface in COMSOL software $\Delta \varepsilon$.

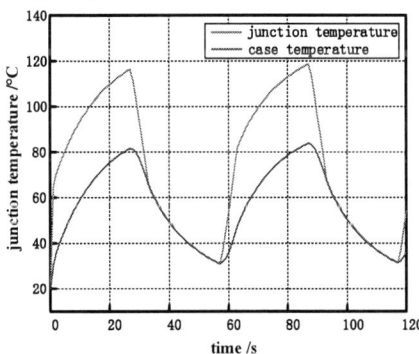

Fig. 5.   IGBT device solder layer temperature change

It is analyzed from the inversion theory that the fatigue state D of the live grid element can be obtained from the information of the raw grid element. It can be known from that the mesh of the solder layer in the finite element model of the IGBT device needs to be divided into the mesh within the range of the newly added cavity and the mesh within the range of the fatigue degree entity to be determined, that is, the dead mesh and the living mesh, which are represented by A and x respectively. As shown in Fig. 6, the CT images of the solder layer were stretched and binarized after 19,478 power cycle tests, and then the power cycle tests were continued at

ΔT_j=87°C, T_{jm}=75°C until 31286 solder layer images were recorded and processed.

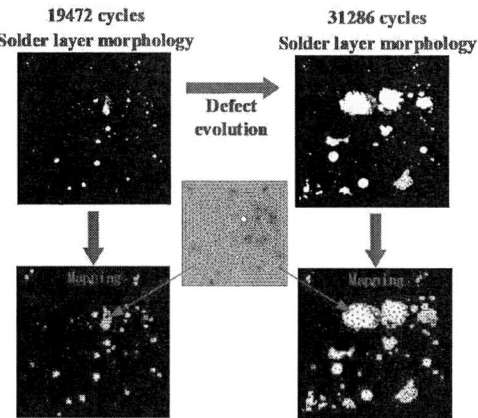

Fig. 6.   Added information acquisition of mesh vertices in the void range

As shown in Fig. 6, after ΔN power cycles, the solder layer is constantly subjected to alternating thermal stress, and the original voids in the solder layer continue to expand and new voids are formed. The OpenCV library in Python is adopted to implement contour tracing of the cavities in each binalized solder layer image, and combined with the position information of the mesh vertices of the solder layer in the IGBT finite element model, it is determined which mesh vertices are within the cavity contour. By differentiating the grid vertices within the cavity contour obtained under different power cycle times, the position information of the grid vertices within the newly added cavity range can be obtained. In Figure 7, the cavity contour is represented by green lines, and the red dots within the cavity range represent the mesh vertices in the finite element model of the corresponding solder layer.

It should be noted that according to the theory of the inversion method, in the finite element model, the solder layer is divided into multiple small grids. The fatigue state of the raw grids is inverted through the dead grid element information reference to obtain the fatigue accumulation degree D of each raw grid, and then determine the fatigue degree of the solder layer, thereby calculating the remaining life of the IGBT device. So throughout the entire calculation process, the grid position remains unchanged, that is, the grid division method remains unchanged. The mesh subdivision method is as follows: Initially, a finite element model with actual void mapping is constructed based on the CT scan pattern of the solder layer at time t_0, and then subdivision is carried out. For any subsequent operations, the mesh subdivision method remains unchanged, and the work of dividing and deleting the mesh is carried out on the basis of the original mesh.

After obtaining the specific position information of the mesh vertices within the newly added cavity range, since the stress difference of all mesh vertices within the solder layer range within 85-115s has been obtained $\Delta \varepsilon$, the position information of the vertices within the newly added cavity range in the subsequent comparison with the strain value information obtained previously $\Delta \varepsilon$ can be used to obtain the mesh strain of the solder layer in the finite element model $\Delta \varepsilon$.

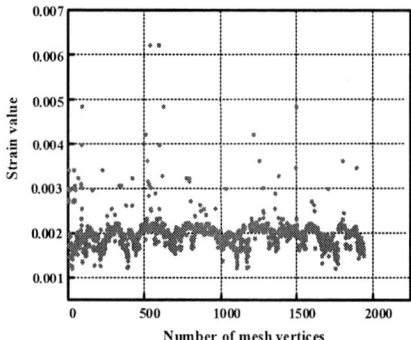

Fig. 7.   The strain values of all mesh vertices within the solid range.

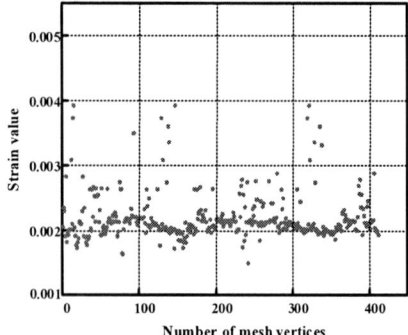

Fig. 8.   The strain values of all mesh vertices within the newly added void range.

Fig.7 shows the strain values of all mesh vertices within the solid range, and Fig. 8 shows the strain values of all mesh vertices within the newly added void range. It can be seen from the two figures that the strain values of mesh vertices are mostly concentrated in a range.

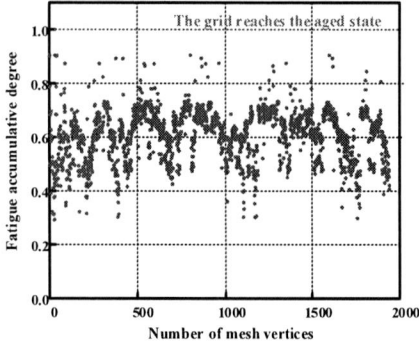

Fig. 9.   Fatigue accumulative degree D of grid vertex inversion in solid area.

The fatigue degree D of the solder layer grid in the finite element model of IGBT module at time to was inverted by formulas (8) and (9), and the results are respectively shown in Fig. 9 and 10.

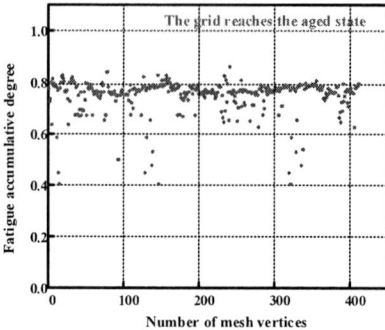

Fig. 10.   Fatigue accumulative degree D of grid vertex inversion in the add void area.

Through the inversion algorithm proposed in this paper, the fatigue state of any grid in the solder layer can be obtained at t_0 time, and the residual life of IGBT devices can be evaluated based on the fatigue state based on the finite element model.

V.   CONCLUSION

Based on the background that the existing prediction methods of the remaining life of IGBT modules are less concerned about the used devices, this paper innovatively proposes that the strain relationship between the grids in the finite element simulation model of IGBT modules can be used to evaluate the fatigue state of the devices in service and further predict the remaining life of the devices. It is of great strategic value and economic significance to provide data basis for the continuous use of replaced IGBT devices, thus reducing system operation and maintenance costs.

REFERENCES

[1]   Y. Huang, Y. Luo, F. Xiao, B. Liu and X. Tang, "Evaluation of the Degradation in Electrothermal Characteristics of IGBTs During Thermal Cycling Cocaused by Solder Cracking and Al-Wires Lifting-Off Based on Iterative Looping," in *IEEE Transactions on Power Electronics*, vol. 38, no. 2, pp. 1768-1778, Feb. 2023.

[2]   Y. Jia, Y. Huang, F. Xiao, H. Deng, Y. Duan, F. Iannuzzo, "Impact of Solder Degradation on VCE of IGBT Module: Experiments and Modeling," in *IEEE Journal of Emerging and Selected Topics in Power Electronics*, 2022, 10(4): 4536-4545.

[3]   Y. Huang, H. Deng, Y. Luo, F. Xiao, B. Liu, X. Tang, "Fatigue Mechanism of Die-Attach Joints in IGBTs Under Low-Amplitude Temperature Swings Based on 3D Electro-Thermal-Mechanical FE Simulations," in *IEEE Transactions on Industrial Electronics*, vol. 68, no. 4, pp. 3033-3043, April 2021.

[4]   M. Du, Q. Guo, H. Wang, Z. Ouyang and K. Wei, "An Improved Cauer Model of IGBT Module: Inclusive Void Fraction in Solder Layer," in *IEEE Transactions on Components, Packaging and Manufacturing Technology*, vol. 10, no. 8, pp. 1401-1410, Aug. 2020.

[5]   B. Hu, S. Konaklieva, N. Kourra, M. A. Williams, L. Ran and W. Lai, "Long-Term Reliability Evaluation of Power Modules With Low Amplitude Thermomechanical Stresses and Initial Defects," in *IEEE Journal of Emerging and Selected Topics in Power Electronics*, vol. 9, no. 1, pp. 602-615, Feb. 2021.

[6]   J. Yang, Y. Che, L. Ran, B. Hu and M. Du, "In-Situ Monitoring Solder Layer Degradation in Multichip IGBT Power Modules Using Auxiliary Emitter Voltage," in *IEEE Transactions on Power Electronics*, vol. 39, no. 10, pp. 13744-13757, Oct. 2024.

A 6.78-MHz 2.3-kW Full-Bridge GaN Inverter with Bottom-side Cooled Transistor GS-065-030-2-L for Wireless Power Transfer

Jianping Ning
School of Electrical and Electronic Engineering
Nanyang Technological University
Singapore
jianping002@e.ntu.edu.sg

Zhen Sun
School of Electrical and Electronic Engineering
Nanyang Technological University
Singapore
sunz0028@e.ntu.edu.sg

Yao Wang
School of Automation
Northwestern Polytechnical University
Xi'an, China
yao.wang@nwpu.edu.cn

Shuang Zhao
School of Electrical Engineering and Automation
Hefei University of Technology
Hefei, China
shuang.zhao@hfut.edu.cn

Abstract—**Megahertz (MHz) high-frequency wireless power transfer (WPT) system shows advantages in compactness, density, and power transfer freedom. Gallium nitride (GaN) transistors are suited for high-frequency inverter designs due to their extremely high electron mobility and fast switching speed. This paper presents the design, implementation, and experimental evaluation of a 6.78-MHz high-power GaN inverter utilizing the enhancement mode (E-mode) 650V bottom-side cooled power transistor GS-065-030-2-L for wireless power transfer. Thermal dissipation design for GaN transistors, parasitic parameters optimization, and zero-voltage-switching (ZVS) are demonstrated. A 6.78 MHz WPT prototype is set up, and the experimental results demonstrate that the GaN inverter can achieve a maximum output power of 2374 W with ZVS, and the peak DC-DC efficiency of the WPT system achieves 92.88%.**

Keywords—*Gallium nitride (GaN)-based inverter, WPT, zero-voltage-switching (ZVS), thermal management*

I. INTRODUCTION

Recent advancements in megahertz (MHz) wireless power transfer (WPT) technology have enabled its adoption in consumer electronics and medical implants, where compact size, mobility, and maintenance-free operation are critical [1]-[6]. Compared to conventional WPT systems that comply with the "Qi" or SAE standards [7]-[24], systems operating at 6.78 MHz, the lowest frequency in the industrial, scientific, and medical (ISM) band, offer significantly higher magnetic resonant coupling through high-Q resonators and support more compact coil designs [25].

Recently, advancements in wide bandgap semiconductors have enabled the inverters to work at 1 MHz, 6.78 MHz, and higher frequencies. At 6.78 MHz, a 3.3 kW full-bridge inverter achieves efficiency of 96% through optimized dead-time control and advanced thermal management [26], while a 2.28 kW full-bridge inverter attained peak efficiency of 97.22% by minimizing parasitic effects [6]. For higher frequency operation, a 13.56 MHz half-bridge resonant inverter incorporating an impedance-matching L-network delivered 471 W output power at the efficiency of 99.43%, accomplished through precise series resonance and zero-voltage-switching (ZVS) implementation [27].

This paper presents the design and implementation of a 6.78 MHz gallium nitride (GaN) high electron mobility transistor (HEMT)-based full-bridge inverter employing the enhancement mode (E-mode) 650V bottom-side cooled power transistor GS-065-030-2-L for high-power WPT systems. The presented inverter design methodology focuses on GaN device selection, gate-driver design, ZVS optimization, and printed circuit board (PCB) layout optimization. A 6.78 MHz WPT prototype is set up, and experimental results demonstrate that the WPT system can achieve a peak system efficiency of 92.88% with a 250 V input and the GaN inverter can achieve the maximum power of 2374 W at 400 V.

II. DESIGN DETAILS OF 6.78 MHZ FULL-BRIDGE INVERTER

6.78 MHz is one of the globally designated industrial, scientific, and medical (ISM) bands, making it a suitable choice for WPT applications. To simultaneously satisfy the operating requirements of multi-MHz and multi-kW, particular attention is given to parasitic parameters reduction and thermal management. The selection of power devices and the overall circuit design are optimized to mitigate the adverse impacts of parasitic parameters and to ensure effective heat dissipation under high-frequency, high-power operating conditions.

A. Selection of the Switching Device and Gate Driver

Wide bandgap (WBG) devices, such as GaN and silicon carbide (SiC), offer superior performance compared to conventional Si devices, particularly in high-frequency applications. GaN devices are favored over SiC devices at 6.78 MHz due to their superior switching characteristics, especially smaller parasitic capacitances and inductances [28]. The primary parameters affecting the switching speed of GaN transistor include the input capacitance C_{ISS} or total gate charge Q_g.

For high-power applications, the high drain-to-source blocking voltage $V_{(BL)DSS}$ and small drain-to-source on-state resistance $R_{DS(on)}$ are critical. Although a lower $R_{DS(on)}$ reduces conduction loss, it typically means a larger parasitic output capacitance C_{OSS}, which is adverse for achieving ZVS. A lower C_{OSS} enables rapid charging and discharging during short dead time, thereby facilitating ZVS and reducing switching loss. Based on these considerations, the GS-065-030-2-L GaN HEMT from *GaN Systems* is selected, balancing the $R_{DS(on)}$ and C_{OSS}. The main parameters are summarized in Table I.

To drive the GS-065-030-2-L GaN HEMT under high-current conditions, a gate drive voltage of -3.3V/+6V is recommended, as specified in the datasheet [29]. The adoption of a negative gate-source voltage $V_{GS(OFF)}$ during turn-off effectively suppresses unintended turn-on and reduces turn-off losses. However, it may increase dead-time-related losses, necessitating a careful trade-off in gate drive design. To ensure the safe and reliable operation of a full-bridge inverter under high-voltage conditions, galvanic isolation is essential for the gate drivers. Critical parameters for gate driver selection include common mode transient immunity (CMTI), propagation delay, isolation voltage, and under voltage lockout (UVLO). The Si8271 isolated gate driver is selected due to low propagation delay, high isolation capability (rated up to kilovolts), and low UVLO threshold, ensuring efficient and robust high-frequency switching. The main parameters of the Si8271 are shown in Table II.

TABLE I. MAIN PARAMETERS OF GS-065-030-2-L

Parameters	Typical Value
$V_{BL(DSS)}$	650 V
$I_{DS,max}$	30 A
$R_{DS(on)}$	50 mΩ
C_{ISS}	235 pF
C_{OSS}	60 pF
C_{RSS}	0.6 pF
Q_g	6.7 nC

TABLE II. MAIN PARAMETERS OF Si8271

Parameters	Typical Value
CMTI	200 kV/μs
Isolation Rating	2.5 kV$_{RMS}$
Output UVLO	5 V
Propagation delay	60 ns (max)

B. Design Details of GaN-Based Inverter

1) Design Details of the Gate Driver Circuit

Due to the high-frequency switching characteristic of GaN devices, minimizing parasitic inductance is essential to ensure reliable and efficient operation. Therefore, each gate driver is positioned close to its corresponding GaN transistor, reducing the parasitic inductance associated with the gate driver circuit loop. The Kelvin-source connection is employed to eliminate the adverse effects of common source inductance, as shown in Fig. 1. To enhance signal integrity, copper pours are applied between the ground terminal of the gate driver and the Kelvin source pin of the GaN transistors, as well as between the gate driver's output pin and the gate terminal of GaN transistors. These measures reduce impedance and suppress voltage oscillations during high-speed switching transients. A minimum clearance of 1.7 mm is maintained between the power and gate-drive signal traces.

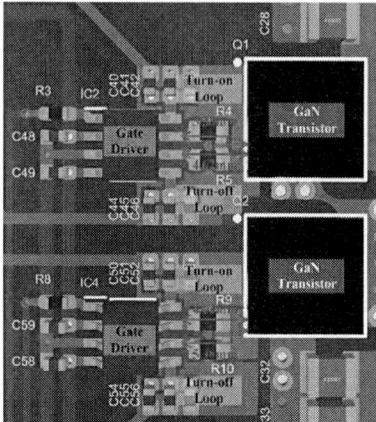
Fig. 1. Partial layout design of the gate drive circuit.

Fig. 2. Heat sink installation of the inverter.

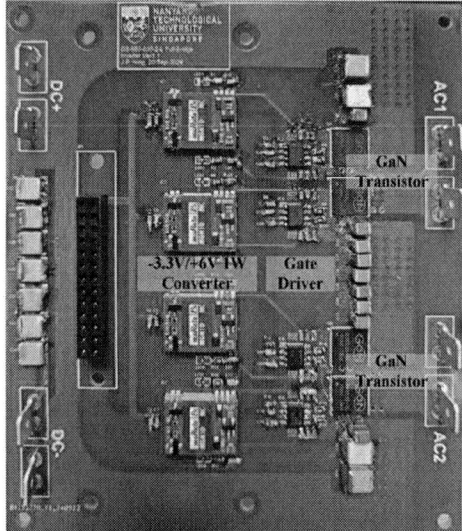
Fig. 3. The implemented GaN GS-065-030-2-L inverter.

2) Thermal Management Optimization

Effective heat dissipation is essential for the reliable operation of surface-mounted GaN transistors, as excessive junction temperatures can degrade electrical performance and compromise device longevity [30]. The GS-065-030-2-L uses bottom-side cooled passive-down-flux-on-noodle (PDFN) packaging, which necessitates a thermally optimized PCB layout. To facilitate efficient heat transfer, large copper planes are implemented beneath the GaN devices, with multiple thermal vias connecting to the bottom layer of the PCB. A heat sink is affixed to the bottom layer using thermally conductive grease to enhance thermal interface conductivity, as illustrated in Fig. 2. This assembly is further supported by active airflow via a cooling fan, ensuring adequate thermal dissipation under high-frequency, high-power operating conditions.

3) Estimation of the DC-Side Filter Capacitance

High-frequency ceramic capacitors are employed to filter switching noise and stabilize the DC bus voltage. The required total DC bus capacitance can be estimated by considering the charge variation during the capacitor charging interval, as given by:

$$C_{\text{tot}} = \frac{\Delta Q}{v_{\text{ripple}}} \tag{1}$$

$$\Delta Q = \int_{t_1}^{t_2} i(t)\mathrm{d}t \tag{2}$$

where C_{tot} is the total bus capacitance, ΔQ is the charge variation, v_{ripple} is the voltage ripple, t_1 and t_2 are the beginning and end of the capacitor charging period. Under the specified design conditions: $v_{\text{ripple}} = 4$ V (corresponding to 1% of a 400 V DC input) and a charging interval from $t_1 = 16.2$ ns to $t_2 = 57.5$ ns, a total capacitance of approximately 117.75 nF is required. To ensure stable operation at MHz-level frequencies, C0G-class ceramic capacitors are selected due to their low equivalent series resistance (ESR), minimal dielectric loss, and excellent frequency stability.

C. Hardware of the Implemented GaN Inverter

The prototype inverter is constructed based on the selected switching device GS-065-030-2-L and the proposed PCB layout, as depicted in Fig. 3. From left to right, the system integrates DC-DC converters, gate drivers, and GaN transistors. Due to the low drive voltage and small gate input capacitance of GS-065-030-2-L GaN HEMT, each gate driver-transistor pair is supplied by a 1 W isolated DC-DC converter to ensure appropriate switching voltage. To mitigate the effects of both ESR and equivalent series inductance (ESL), a capacitor network comprising 0.01 nF, 0.1 nF, and 1 nF capacitors is implemented. To balance the switching speed with the suppression of the voltage ringing, gate resistors of 4.7 Ω and 2.0 Ω are respectively inserted in the turn-on and turn-off paths. An aluminum heatsink is mounted on the underside of the PCB, covering the GaN HEMTs and gate driver section to facilitate effective thermal management.

III. Experimental Results

A. Hardware Implementation of the WPT System

To assess the inverter performance within a WPT system, an experimental setup is implemented, as illustrated in Fig. 4 (circuit diagram) and Fig. 5 (experimental setup). The system includes a DC power supply, an inverter, a resonant network, a full-wave rectifier, and a load. The ITECH IT6006C-800-25 DC power supply is employed to provide an input voltage of up to 400 V. The resonant network adopts a series–series (SS) compensation topology. The load is composed of several 4.7 Ω resistors arranged in series, enabling flexible configuration and straightforward disassembly to evaluate output power and efficiency under varying load conditions.

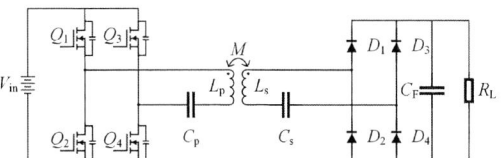

Fig.4 The circuit of the implemented WPT system

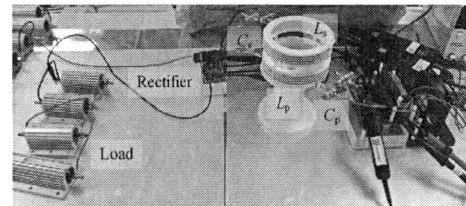

Fig. 5. Experimental setup of the WPT system.

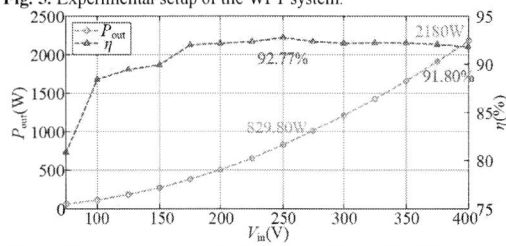

Fig. 6. The output power and efficiency at 74 Ω with different input voltages.

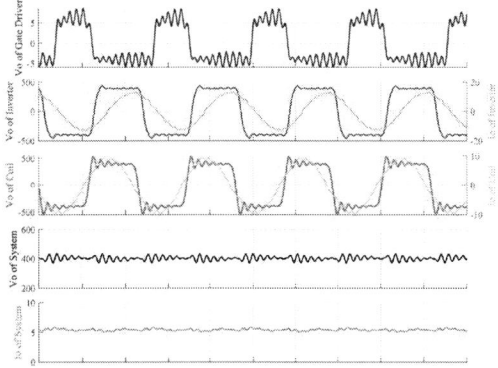

Fig. 7. The key waveforms of the system operating at $V_{\text{in}} - 400$ V and $R_{\text{L}} = 74$ Ω.

B. 6.78MHz Experimental Test

6.78MHz experimental tests are conducted to evaluate the performance of the inverter. The load resistance R_{L} is fixed at 74 Ω, while the input voltage V_{in} increases from 75 V to 400 V in steps of 25 V. As shown in Fig. 6, the efficiency increases rapidly with rising V_{in}, reaching ZVS at 175 V/381.60 W. A peak efficiency of 92.77% is achieved at 250 V/829.80 W, beyond which the efficiency gradually declined. Under the maximum input condition of $V_{\text{in}} = 400$ V, the prototype delivers an output power of 2180 W into a load impedance of 74 Ω. The corresponding input power is 2374 W, achieving a system efficiency of 91.80%. The measured phase angle of 50.13° between the output voltage and current confirms effective ZVS operation.

Fig. 7 presents the measured waveforms. Specifically, the dark blue trace is the gate driver signal, the red and green traces represent the inverter output voltage and current, the purple and light blue traces represent the resonator output voltage and current, and the black and orange traces correspond to the output load voltage and current. Since the output capacitance C_{oss} of GaN devices decreases with increasing drain-source voltage V_{DS}, the phase angle required for ZVS also varies with V_{in}. Table III summarizes the required phase angles for achieving ZVS across different input voltage conditions at 74 Ω.

Then, V_{in} is constant while the load resistance is varied from 51 Ω to 93 Ω. Fig. 8 illustrates the efficiency performance under V_{in} = 250 V. The highest efficiency, 92.88%, is achieved at R_L = 62.46 Ω, corresponding to a phase angle of 47.21° between the inverter output voltage and current.

TABLE III. ESSENTIAL ANGLE TO REALIZE ZVS UNDER DIFFERENT VOLTAGES

V_{DD}/V	θ / degree
200	53.43
250	49.69
300	49.08

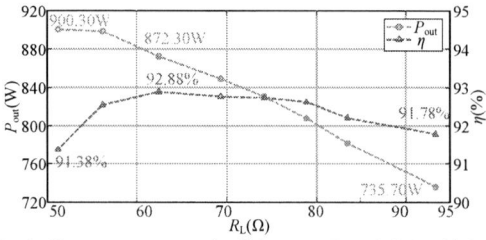

Fig. 8. The output power and efficiency at V_{in} = 250 V with load resistance from 51.21 Ω to 93.36 Ω.

IV. CONCLUSION

This paper presents the design and implementation of a 6.78 MHz GaN-based full-bridge inverter utilizing GS-065-030-2-L for WPT systems. The design focuses on improving high-frequency switching performance, including device selection, optimized gate driver circuits, effective realization of ZVS, and minimization of parasitic effects via layout optimization. Experimental validation demonstrates that the system achieves a peak efficiency of 92.88% at an input voltage of 250 V with an input power of 939.15 W. Under a 400 V input, the system delivers a maximum input power of 2374 W with an efficiency of 91.80%. These results verify the suitability of GaN technology for high-frequency, high-power WPT systems and underscore the effectiveness of the proposed design methodology.

REFERENCES

[1] S. -Y. R. Hui, Y. Yang, and C. Zhang, "Wireless power transfer: a paradigm shift for the next generation," *IEEE J. Emerg. Sel. Topics Power Electron.*, vol. 11, no. 3, pp. 2412-2427, Jun. 2023.

[2] C. Zhao and D. Costinett, "GaN-based dual-mode wireless power transfer using multifrequency programmed pulse width modulation," *IEEE Trans. Ind. Electron.*, vol. 64, no. 11, pp. 9165-9176, Nov. 2017.

[3] J. M. Arteaga, S. Aldhaher, G. Kkelis, C. Kwan, D. C. Yates, and P. D. Mitcheson, "Dynamic capabilities on multi-MHz inductive power transfer systems demonstrated with batteryless drones," *IEEE Trans. Power Electron.*, vol. 34, no. 6, pp. 5093-5104, Jun. 2019.

[4] R. S. Yang et al., "Design and performance comparison of multi-frequency inductors for megahertz wireless power transfer," in *IEEE Applied Power Electronics Conference and Exposition (APEC)*, Long Beach, CA, USA, 2024, pp. 861-868.

[5] Y. Wang, J. Yang, K. Wang, and Y. Yang, "Highly integrated hybrid inductive and capacitive power transfer system with asymmetrical

printed-circuit-board-based self-resonator," *IEEE Trans. Power Electron.*, vol. 40, no. 7, pp. 10254-10264, Jul. 2025.

[6] Y. Wang, K. Wang, Y. Yang, and S. Y. Ron Hui, "Design, implementation, and comparison of multi-MHz multi-kW H-bridge inverters based on 650V SiC and GaN devices for high-frequency WPT," in *IEEE Energy Conversion Congress and Exposition (ECCE)*, Phoenix, AZ, USA, 2024, pp. 2007-2012.

[7] X. Lu, P. Wang, D. Niyato, D. I. Kim, and Z. Han, "Wireless charging technologies: fundamentals, standards, and network applications," *IEEE Communications Surveys & Tutorials*, vol. 18, no. 2, pp. 1413-1452, 2016.

[8] Y. Yang, W. Zhong, S. Kiratipongvoot, S. -C. Tan, and S. Y. R. Hui, "Dynamic improvement of series–series compensated wireless power transfer systems using discrete sliding mode control," *IEEE Trans. Power Electron.*, vol. 33, no. 7, pp. 6351-6360, Jul. 2018.

[9] X. Liu, "Qi standard wireless power transfer technology development toward spatial freedom," *IEEE Circuits and Systems Magazine*, vol. 15, no. 2, pp. 32-39, 2015.

[10] Y. Yang, S. C. Tan, and S. Y. R. Hui, "Fast hardware approach to determining mutual coupling of series–series-compensated wireless power transfer systems with active rectifiers," *IEEE Trans. Power Electron.*, vol. 35, no. 10, pp. 11026-11038, Oct. 2020.

[11] D. van Wageningen and T. Staring, "The Qi wireless power standard," in *14th Int. Power Electron. Motion Control Conf. (EPE-PEMC)*, Ohrid, Macedonia, 2010, pp. S15-25-S15-32.

[12] Y. Yang, S. -C. Tan, and S. Y. R. Hui, "Front-end parameter monitoring method based on two-layer adaptive differential evolution for SS-compensated wireless power transfer systems," *IEEE Trans. Ind. Informat*, vol. 15, no. 11, pp. 6101-6113, Nov. 2019.

[13] E. Waffenschmidt, "Wireless power for mobile devices," in *33rd Int. Tel. Energy Conf. (INTELEC)*, Amsterdam, Netherlands, 2011, pp. 1-9.

[14] Y. Yang, "Precise modeling of nonlinear rectifier loads in wireless power transfer systems," *IEEE J. Emerg. Sel. Topics Power Electron.*, vol. 11, no. 3, pp. 3574-3585, Jun. 2023.

[15] H. Xiao, Y. Yang, K. Wang, and J. Wu, "Comparative studies of front-end model predictive control for direct inductive power transfer systems," *IEEE Trans. Power Electron.*, vol. 38, no. 10, pp. 11885-11897, Oct. 2023.

[16] Y. Yang, "A passive augmented circuit for EMI reductions of full-bridge inverters with conventional phase shift control in wireless power transfer systems," *IEEE Trans. Power Electron.*, vol. 38, no. 11, pp. 13286-13297, Nov. 2023.

[17] K. Wang, Y. Yang, H. Wang, and E. K. -W. Cheng, "Frequency regulation scheme for double-sided LCC compensated inductive power transfer systems with quasi-load-independent outputs," in *IEEE Energy Conversion Congress and Exposition (ECCE)*, Nashville, TN, USA, 2023, pp. 6383-6388.

[18] Y. Yang, S. -C. Tan, and S. Y. R. Hui, "Communication-free control scheme for Qi-compliant wireless power transfer systems," in *IEEE Energy Conversion Congress and Exposition (ECCE)*, Baltimore, MD, USA, 2019, pp. 4955-4960.

[19] K. Wang, S. Zhao, S. Shang, E. K. -W. Cheng, S. -C. Tan, and Y. Yang, "Design of wireless power transmitters for enhanced transmission distance and output power," in *IEEE Applied Power Electronics Conference and Exposition (APEC)*, Atlanta, GA, USA, 2025, pp. 3227-3231.

[20] M. Galizzi, M. Caldara, V. Re and A. Vitali, "A novel Qi-standard compliant full-bridge wireless power charger for low power devices," 2013 IEEE Wireless Power Transfer (WPT), Perugia, Italy, 2013, pp. 44-47.

[21] K. Wang, Z. Sun, X. Li, Y. Wang, and Y. Yang, "A cubic wireless charging container system with highly uniform magnetic field distribution," *IEEE Trans. Power Electron.*, vol. 40, no. 3, pp. 4613-4629, Mar. 2025.

[22] M. Treffers, "History, current status and future of the wireless power consortium and the Qi interface specification," *IEEE Circuits and Systems Magazine*, vol. 15, no. 2, pp. 28-31, 2015.

[23] K. Wang, R. Liang, X. Zhang, Y. Yang, and S. -C. Tan, "Triple-layer coil design for cubic wireless charging spaces: enhancing output power and power flow control," in *IEEE Energy Conversion Congress and Exposition (ECCE)*, Phoenix, AZ, USA, 2024, pp. 2163-2168.

[24] K. Wang, R. Liang, J. Gao, J. Wu, Y. Tang, and Y. Yang, "Optimized folded coil designs for wireless charging chambers with even

distribution of magnetic flux density," in *IEEE Applied Power Electronics Conference and Exposition (APEC)*, Long Beach, CA, USA, 2024, pp. 2859-2863.

[25] J. Nadakuduti, L. Lu and P. Guckian, "Operating frequency selection for loosely coupled wireless power transfer systems with respect to RF emissions and RF exposure requirements," in *2013 IEEE Wireless Power Transfer (WPT)*, Perugia, Italy, 2013, pp. 234-237.

[26] M. Yamaguchi, Y. Uchida, H. Watanabe, K. Kusaka and J. Itoh, "Achievement of 6.78-MHz and 3-kW single inverter in continuous operation," in *2024 IEEE 10th International Power Electronics and Motion Control Conference (IPEMC2024-ECCE Asia)*, Chengdu, China, 2024, pp. 2654-2658.

[27] Oyane, T. Senanayake, F. Hattori, J. Imaoka, M. Yamamoto and M. Masuda, "13.56MHz high power half-bridge GaN-HEMT resonant inverter achieving 99% power efficiency," in *2020 IEEE International Conference on Power Electronics, Drives and Energy Systems (PEDES)*, Jaipur, India, 2020, pp. 1-4.

[28] P. Palmer, X. Zhang, E. Shelton, T. Zhang and J. Zhang, "An experimental comparison of GaN, SiC and Si switching power devices," in *IECON 2017 - 43rd Annual Conference of the IEEE Industrial Electronics Society,* Beijing, China, 2017, pp. 780-785.

[29] GaN Systems Inc., "GS-065-030-2-L 650 V E-mode GaN transistor datasheet," GS-065-030-2-L datasheet, 2009–2022 [Revised Jul. 2022].

[30] Y. Zhang et al., "Study of heat transport behavior in GaN-based transistors by schottky characteristics method," *IEEE Trans. Electron. Devices*, vol. 64, no. 5, pp. 2166-2171, May 2017.

Two-Step Turn-off Delay Time Control for Efficiency Enhancement in Si/SiC Hybrid Switch-based Inverters

1st Zijie Zheng
College of Electrical and Information Engineering
Hunan University
Changsha, China
zijiezheng@hnu.edu.cn

2nd Jun Wang
College of Electrical and Information Engineering
Hunan University
Changsha, China
junwang@hnu.edu.cn

3rd Xuanting Song
College of Electrical and Information Engineering
Hunan University
Changsha, China
songxt@hnu.edu.cn

4th Yongzhou Zou
College of Electrical and Information Engineering
Hunan University
Changsha, China
yongzhouzou990183@hnu.edu.cn

5th Yuxing Dai
College of Electrical and Information Engineering
Wenzhou University
Wenzhou, China
daiyx@hnu.edu.cn

6th Kamal Al-Haddad
Department of Electrical Engineering
École de Technologie Supérieure
Montreal, Canada
kamal.al-haddad@etsmtl.ca

Abstract—The Si IGBT/SiC MOSFET hybrid switch offers a compelling pathway toward high-efficiency and cost-effective inverter systems, as it combines the complementary advantages of Si IGBTs and SiC MOSFETs. However, optimal control of the gate turn-off delay time between the internal Si IGBT and SiC MOSFET is required to reduce losses in the Si/SiC hybrid switch-based inverter. To address this issue, a simple two-step control method for gate turn-off delay time is proposed to improve the efficiency of the Si/SiC hybrid switch-based inverter. A relatively small gate turn-off delay time is adopted under light load conditions to leverage the fast-switching performance of the internal SiC MOSFET, while a relatively large delay time is used under heavy load to mitigate IGBT tail current losses and suppress voltage overshoot. A hardware prototype of a three-phase hybrid switch-based inverter is built. Compared to the fixed gate turn-off delay time control, experimental results show that the proposed method can achieve significant efficiency improvement without compromising switching stability.

Keywords —SiC MOSFET, IGBT, Hybrid, Drive mode, Efficiency.

I. INTRODUCTION

A Si/SiC hybrid switch consisting of a low current SiC MOSFET and a high current Si IGBT connected in parallel has attracted increasing attention as a practical approach to improving the efficiency, reliability, and cost-effectiveness of power electronic converters [1]–[3]. By combining the fast switching speed and high-temperature endurance of SiC MOSFETs with the strong current-handling capability and economic advantages of Si IGBTs, hybrid switch (HyS) effectively exploit the complementary strengths of both technologies. However, the internal Si IGBT and SiC MOSFET also exhibits inherent limitations—SiC MOSFETs are sensitive to overvoltage conditions, while IGBTs suffer from slower switching speeds and tail current losses. These constraints have led to increasing interest in coordinated switching strategies that can dynamically balance the operation of both devices for improved performance. Recent studies have proposed a variety of control methods for HyS-based converters aimed at enhancing efficiency and thermal

performance across different load conditions [4]–[7]. Some methods focus on optimization of the gate turn-off delay time in phase-leg configurations, where double-pulse testing has demonstrated notable reductions in switching energy losses [8]. Others emphasize the importance of parameter tuning to account for current overshoot, helping to suppress voltage stress and improve switching robustness [9]. Gate driver innovations have also been explored to enable multi-objective control without significantly increasing hardware complexity [10]. These efforts, along with adaptive gate timing and current distribution schemes, show potential, yet many still face challenges in practical scalability due to model dependency and computational burden. Additional developments have included Lyapunov-stable hybrid control strategies for inverter systems modeled as mixed discrete-continuous processes [11], on the device level, recent research has examined structural configurations of hybrid power switches using modern Si IGBTs and SiC MOSFETs. One such implementation demonstrated that a 1200 V hybrid module can achieve low switching losses comparable to full-SiC solutions while maintaining high current handling and cost advantages [12].

A two-step turn-off delay ($T_{\text{off-delay}}$) control strategy is proposed and experimentally validated in this work to address these limitations in hybrid Si/SiC-based three-phase inverters. The method adaptively segments the $T_{\text{off-delay}}$ according to the instantaneous load current: a shorter delay is applied under low-current conditions to capitalize on the low-loss switching of SiC MOSFETs, while a longer delay is introduced at higher currents to reduce IGBT tail losses and suppress voltage overshoot. This strategy offers a practical and hardware-friendly solution that mitigates switching losses and thermal stress without requiring additional sensors or complex control logic. The results provide a solid foundation for scalable, high-efficiency inverter design using hybrid power semiconductor technologies.

II. CONTROL STRATEGY OF HYBRID SWITCH-BASED THREE-PHASE INVERTER

A novel two-step turn-off delay ($T_{\text{off-delay}}$) control strategy is proposed in this study to effectively mitigate the aforementioned limitations in hybrid switching applications. The method divides the turn-off process into two distinct segments, where the delay duration at each step is adaptively modulated based on the instantaneous load current and inverter operating state. This dynamic segmentation enables targeted reduction of switching losses across a wide range of operating conditions.

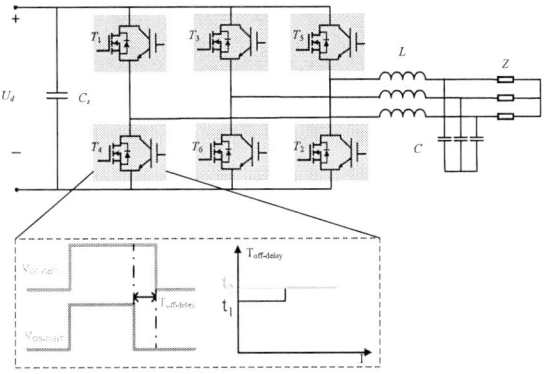

Fig. 1. Topology of three-phase full-bridge inverter based on hybrid device and switching mode of hybrid device

In light-load scenarios, the SiC MOSFET predominantly conducts the switching current. In such cases, a shorter $T_{off\text{-}delay}$ is applied to promptly suppress the current and minimize the voltage-current overlap interval, thereby reducing switching energy dissipation. Owing to the intrinsic high-speed switching capability of SiC devices, this short delay effectively leverages their performance while avoiding losses associated with abrupt transitions.

As the current level increases, the Si IGBT gradually assumes a greater portion of the conduction current. To accommodate its slower turn-off behavior, a longer $T_{off\text{-}delay}$ is introduced to ensure complete device deactivation. This delay tuning effectively suppresses voltage overshoot and mitigates reverse recovery effects by enabling a smoother dv/dt transition. Under high-current conditions, extended delays are particularly effective in reducing switching losses caused by tail current and voltage-current mismatches.

From a system perspective, the two-step $T_{off\text{-}delay}$ strategy substantially enhances inverter performance under dynamic load fluctuations and variable-frequency operation. By coordinating current distribution between Si and SiC devices through delay segmentation, the strategy provides intelligent control over energy dissipation during switching transitions. An added advantage is the mitigation of thermal stress variations, which contributes to improved long-term reliability and extended operational lifespan.

Simulation, theoretical modeling, and hardware-level experimental results collectively validate the effectiveness of the proposed approach. The strategy demonstrates robust capability in suppressing switching losses across diverse operating conditions, while simultaneously optimizing thermal behavior and overall system efficiency—highlighting its practical engineering value and scalability in high-performance inverter applications.

III. EXPERIMENT AND SIMULATION VERIFICATION

A simulation platform was developed using PLECS to evaluate the effectiveness of the proposed two-step Toff-delay strategy prior to experimental implementation. As shown in Fig. 2, the platform is based on a three-phase voltage source inverter (VSI) employing hybrid switching devices. The

inverter consists of parallel-connected Si IGBT and SiC MOSFET pairs in each leg, accurately modeled to reflect both switching and conduction characteristics. The system operates under closed-loop control with sinusoidal PWM modulation and a DC-link voltage of 800 V.

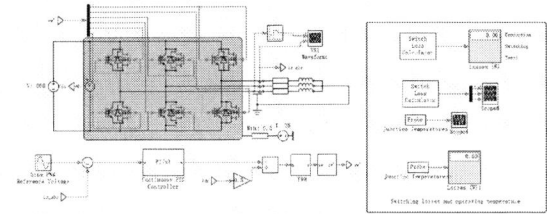

Fig. 2. Simulation model of the three-phase inverter using Si/SiC hybrid switches

Switching performance and thermal characteristics are assessed in the simulation by incorporating real-time loss models and junction temperature sensors into each device. $T_{off\text{-}delay}$ values are varied systematically across multiple load conditions. Fig. 3 presents the simulated turn-off energy loss (E_{off}) of the hybrid switching device as a function of load current, under two different $T_{off\text{-}delay}$ configurations: 0 delay, and the proposed two-step delay strategy. The results reveal that E_{off} increases nonlinearly with rising current in all cases, as higher load conditions intensify device stress during turn-off transitions. Notably, the proposed two-step $T_{off\text{-}delay}$ method (red line) consistently achieves lower turn-off losses across a wide current range compared to both fixed-delay strategies. Under low current conditions, its performance approaches that of the short-delay case, while at higher currents, it effectively limits E_{off} rise by adopting extended delay behavior similar to the long-delay strategy. This adaptive transition enables optimal energy management across varying load profiles, validating the effectiveness of the segmented delay approach in reducing switching loss while maintaining thermal balance in hybrid inverters.

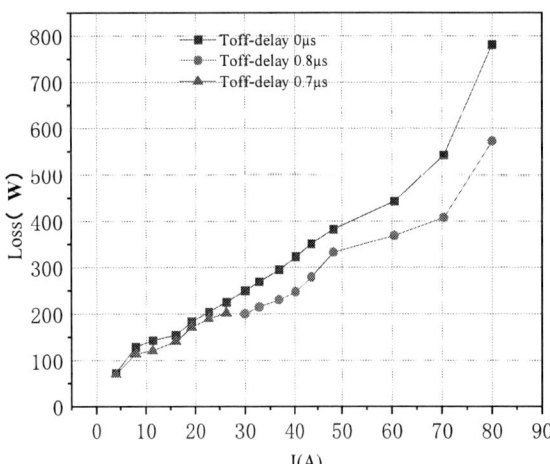

Fig. 3. Turn-off Loss Characteristics of a Three-Phase Inverter Employing Si/SiC Hybrid Switches

The experimental platform for the double-pulse test circuit based on hybrid switching devices is constructed, as shown in Fig. 4. The platform consists of a main power loop incorporating a bus capacitor, air-core inductors, and hybrid power modules composed of a Si IGBT (model IGQ120N120S7) and a SiC MOSFET (model C3M0016120K) connected in parallel. Based on this platform, the dynamic characteristics of the hybrid devices are tested，Fig. 5 shows the measured waveforms of the hybrid devices under double-pulse test conditions. In the Fig. 5, the blue waveform represents the gate-source voltage (V_{GS}), the purple waveform indicates the current through the SiC MOSFET (I_{MOS}), the green waveform shows the current through the IGBT (I_{IGBT}), and the light blue waveform represents the drain-source voltage (V_{DS}), The results clearly indicate that under low current conditions, the MOSFET predominantly conducts the load current. As the load current increases, conduction gradually shifts to the IGBT, the IGBT begins to conduct. Turn-off delay times and corresponding turn-off energy losses were tested under various bus voltage levels. The results, shown in Fig. 6, demonstrate that the optimal $T_{off-delay}$ across different test conditions at room temperature is approximately 0.6 μs, yielding the lowest observed switching loss.

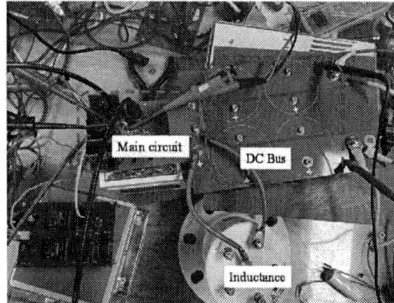

Fig. 4. Double pulse experiment platform

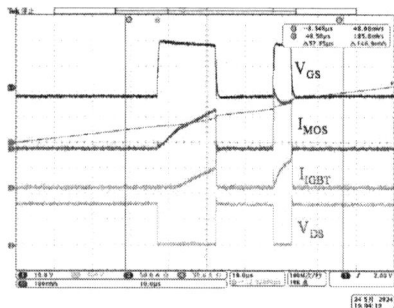

Fig. 5. Double pulse test waveform

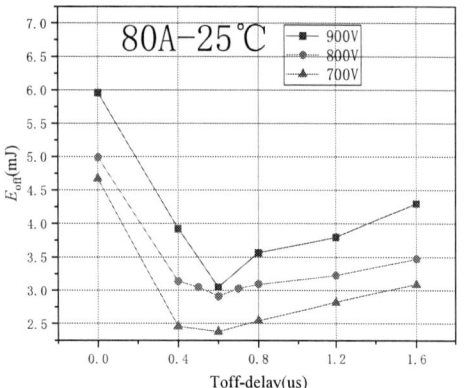

Fig. 6. Change of Eoff loss of hybrid switch with delay time

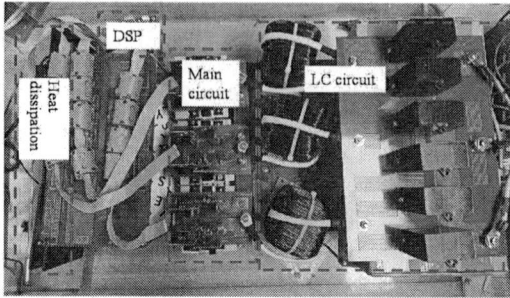

Fig. 7. Hybrid experimental platform

In addition, a three-phase inverter platform utilizing hybrid switching devices was constructed, as shown in Fig. 7. The platform comprises a bus capacitor, main power circuit, DSP controller, cooling fans, and an LC filter circuit. This setup was employed to conduct inverter performance tests and comparative analysis. The effects of different $T_{off-delay}$ values and load conditions on inverter switching losses were investigated through controlled experiments. Fig. 8 shows the output current waveform under a peak load of 80 A with a DC bus voltage of 800 V. The upper waveform (CH1) represents the line-to-line output voltage, while the lower three waveforms (CH2–CH4) correspond to the three-phase output line currents. The voltage waveform exhibits regular PWM modulation, and the output currents maintain sinusoidal waveforms with good symmetry and phase alignment, indicating stable operation of the inverter under the proposed control strategy. Some experimental parameters are shown in the table. Some experimental parameters are presented in Table 1.

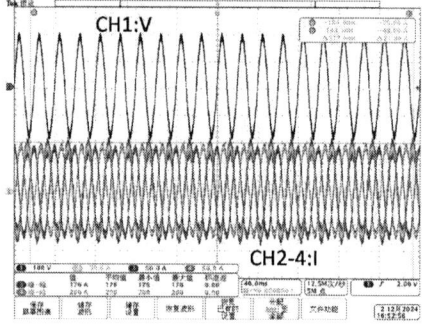

Fig. 8. Inverter experimental waveform

TABLE. I.   INVERTER PARAMETER

Parameter	Values
Supply voltage	*800V*
DC Bus capacitor	*120μF*
Output current	*0-80A*
Power factor	*1*
Switching frequency	*10kHz*
Filter inductor	*0.5mH*
Filter capacitor	*24μF*

A pure SiC MOSFET-based inverter was constructed for reference, and the efficiency comparison with the hybrid inverter is presented in Fig. 9. A noticeable difference in the optimal $T_{off-delay}$ settings is observed between the double-pulse and inverter experiments. This discrepancy arises primarily from thermal conditions: in the double-pulse test, the hybrid devices operate at room temperature, whereas in the inverter tests—particularly under high-current conditions—the junction temperature of the switching devices is significantly elevated, leading to higher switching losses. These thermal variations directly affect the loss-energy characteristics and thus shift the optimal delay configuration.

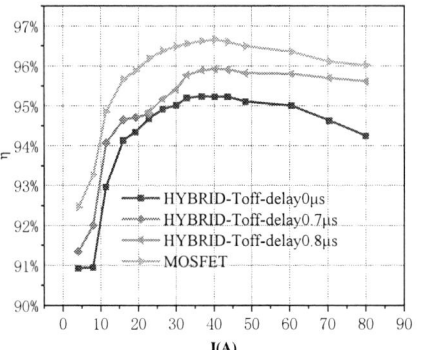

Fig. 9. Efficiency comparison diagram of hybrid inverter for optimal delay control

As illustrated in Fig. 9, the inverter employing a pure SiC MOSFET achieves a peak efficiency of 96.65% under a 0–80 A load range. In comparison, the hybrid inverter without the proposed delay strategy reaches a maximum efficiency of only 95.23%. The choice of switching strategy has a significant impact on inverter efficiency across varying load conditions. The proposed two-step $T_{off\text{-}delay}$ strategy delivers superior performance over the entire current range, particularly under high-current conditions, where it maintains an efficiency of up to 95.91%. Moreover, switching losses are reduced by as much as 209 W, indicating that the adaptive nature of the strategy effectively minimizes dynamic switching losses and ensures high efficiency under load variation. Compared to fixed-delay control, the two-step $T_{off\text{-}delay}$ approach demonstrates a clear advantage in loss reduction during high-current operation, validating its suitability for thermally constrained and efficiency-critical inverter applications.

IV. CONCLUSION

A two-step turn-off delay ($T_{off\text{-}delay}$) control strategy has been proposed for hybrid Si/SiC switches to reduce inverter switching losses under variable load conditions. In this method, the delay is dynamically shortened at low current levels and extended during high-current operation, thereby leveraging the fast-switching capability of SiC MOSFETs while mitigating IGBT tail current effects and suppressing voltage overshoot. Experimental validation conducted on a hardware three-phase inverter platform confirms that the segmented delay control achieves a measurable reduction in turn-off energy loss compared to fixed-delay configurations, without requiring additional sensing circuits or external control hardware. The strategy simplifies gate timing coordination and introduces minimal computational overhead to the control system, offering a practical and scalable solution for high-efficiency inverter applications. Future research may focus on integrating this approach with adaptive current observers or intelligent optimization algorithms to enable real-time adaptability under fast-changing operating conditions.

REFERENCES

[1] D. Woldegiorgis, M. M. Hossain, Z. Saadatizadeh, Y. Wei and H. A. Mantooth, "Hybrid Si/SiC Switches: A Review of Control Objectives, Gate Driving. Approaches and Packaging Solutions," in IEEE Journal of Emerging and Selected Topics in Power Electronics, 11(2): 1737-1753.

[2] J. He, R. Katebi and N. Weise, "A Current-Dependent Switching Strategy for Si/SiC Hybrid Switch-Based Power Converters," in IEEE Transactions on Industrial Electronics, 64(10): 8344-835.

[3] Y. Wu and D. Xu, "100kW Soft Switching Three-Phase Inverter with Hybrid Switches," 2023 IEEE 2nd International Power Electronics and Application Symposium (PEAS), Guangzhou, China, 2023, 163-169.

[4] F. Kayser, R. Baburske, P. Brandt, U. Queitsch and H. -G. Eckel, "Hybrid Switch with SiC MOSFET and fast IGBT for High Power Applications," PCIM Europe digital days 2021; International Exhibition and Conference for Power Electronics, Intelligent Motion, Renewable Energy and Energy Management, Online, 2021, 1-6.

[5] B. Xiao; Q. Guo; C. Tu; F. Xiao; P. Liu; L. Long., "A Novel Switching Strategy Based on the Driving Voltage and Switching Sequence for Si/SiC Hybrid Switch," in IEEE Transactions on Industrial Electronics, 71(11): 14265-14275.

[6] X. Chen, Y. Liu, Y. Li, Z. Yang, H. Li and L. Ding, "Evaluation of Hybrid Switch Applications in Motor Control Unit for Electric Vehicles," 2024 CPSS & IEEE International Symposium on Energy Storage and Conversion (ISESC), Xi'an, China, 2024, 287-292.

[7] A. Sheikhan and E. M. S. Narayanan, "Evaluation of a Hybrid Power Switch Based on Trench Clustered IGBT and SiC MOSFET," PCIM Europe 2024; International Exhibition and Conference for Power Electronics, Intelligent Motion, Renewable Energy and Energy Management, Nürnberg, Germany, 2024, 2601-2606.

[8] H. Qin, R. Wang, Q. Xun, W. Chen and S. Xie, "Switching Time Delay Optimization for "SiC+Si" Hybrid Device in a Phase-Leg Configuration," in IEEE Access, vol. 9, 37542-37556.

[9] H. Qin, S. Xie, Q. xun, F. Zhang, Z. Xu, L. Wang, "An optimized parameter design method of SiC/Si hybrid switch considering turn-off current spike," Energy Reports, 8(13): 789-797.

[10] Z. Li, Y. Dai, X. Jiang, F. Qi, Y. Liu, P. Ke, J. Wang, "A novel gate driver for Si/SiC hybrid switch for multi-objective optimization[J]," IET Power Electronics, 2021, 14(2): 422-431.

[11] G. E. Colón-Reyes, K. C. Stocking, D. S. Callaway and C. J. Tomlin, "Stability and Robustness of a Hybrid Control Law for the Half-bridge Inverter," 2023 European Control Conference (ECC), Bucharest, Romania, 2023.

[12] S. Lee, J. Kim, and H. Park, "Characteristics of a 1200 V Hybrid Power Switch Comprising a Si IGBT and a SiC MOSFET," Micromachines, 15(11): 1337, Nov. 2024.

2025 IEEE Workshop on Wide Bandgap Power Devices and Applications in Asia (WiPDA Asia)

6.78 MHz 2.8 kW Resonant DC-DC Power Conversion Utilizing 650 V GaN HEMT-Based H-Bridge Inverter

Zhen Sun
School of Electrical and Electronic Engineering
Nanyang Technological University
Singapore
sunz0028@e.ntu.edu.sg

Jianping Ning
School of Electrical and Electronic Engineering
Nanyang Technological University
Singapore
jianping002@e.ntu.edu.sg

Yao Wang
School of Automation
Northwestern Polytechnical University
Xi'an, China
yao.wang@nwpu.edu.cn

Yun Yang
School of Electrical and Electronic Engineering
Nanyang Technological University
Singapore
yun.yang@ntu.edu.sg

Abstract—This paper aims to investigate the challenges and feasibility of using 650V gallium nitride (GaN) high-electron-mobility transistors (HEMT) for multi-kilowatt (kW) power conversion at 6.78MHz. A transformer-based isolated resonant DC-DC converter operating at 6.78MHz is developed and driven by a GaN H-bridge inverter. The design challenges associated with high-power GaN H-bridge inverters at this frequency are thoroughly analyzed and summarized. Two inverter prototypes are implemented using GS66508T and GS-065-030-2-L HEMTs, with detailed design considerations provided. Experimental results demonstrate that the resonant DC-DC power conversion system achieves 2.8kW at 400V with a peak efficiency of 95.60% using GS66508T, and 2.06kW at 350V with a peak efficiency of 94.74% using GS-065-030-2-L.

Keywords—Gallium nitride (GaN), 6.78 MHz, DC-DC resonant power conversion, H-bridge inverter

I. INTRODUCTION

Conventional silicon (Si) semiconductors encounter significant challenges, such as narrow energy gap, slow electron motion velocity, low electric field breakdown strength, and poor thermal conductivity, limiting their performance in newly emerging industrial applications that require high power, efficiency, and density [1], such as data center power systems [2], hydrogen generation [3], electric vehicle (EV) charging, and wireless power transfer (WPT) [4].

In recent years, advancements in wide bandgap (WBG) semiconductors, such as silicon carbide (SiC) and gallium nitride (GaN) have propelled power converters toward higher performance in terms of frequencies, power, and efficiency. Compared to Si semiconductors, SiC devices offer distinct advantages, such as 10 times wider bandgaps contributing to higher blocking voltages of up to 2000 V, significantly reduced parasitic elements ($C_{iss}/C_{oss}/C_{rss}$) resulting in faster switching speeds of megahertz (MHz), and superior high-temperature tolerance [5]-[7], making them ideal for high-frequency and high-power applications. GaN is another WBG material, which features a wider bandgap and higher electron mobility than SiC, making GaN-based devices ideal for higher-frequency applications, typically ranging in MHz~tens of MHz [8]. However, due to the low thermal conductivity of GaN materials and the small package, GaN devices generally face thermal issues in high-power applications.

Recently, the GaN-based inverters have pushed the high-frequency and high-power WPT into multi-MHz and multi-

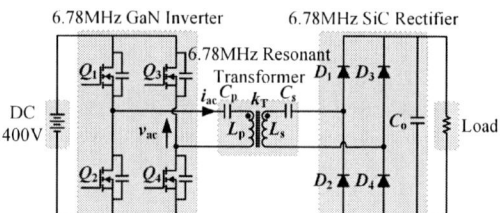

Fig. 1. 6.78 MHz 2.8 kW resonant DC-DC converter topology.

kW areas. For example, at 3 MHz, 6.6-kW WPT systems have been achieved with a GaN H-bridge inverter [9]-[10]. At 6.78 MHz, a modular inversion stage comprising two parallel-connected GaN H-bridge inverter units achieves a 4.2-kW capacitive power transfer (CPT) system [11]. Furthermore, at 13.56 MHz, [12] and [13] respectively present the 2.25-kW and 3.75-kW CPT systems excited by GaN H-bridge inverters, while [14] proposed a five-phase inverter architecture achieving a DC-AC power conversion of 6.8 kW. These results confirm the feasibility of achieving multi-kW multi-MHz power conversion with GaN H-bridge inverters; however, the design challenges of the multi-MHz and multi-kW GaN inverters, such as thermal issues and difficulty in achieving soft switching, are not fully revealed in existing works, and the corresponding inverter design details are lacking, which leaves the researchers with technical gaps.

This paper aims to investigate the challenges and feasibility of utilizing 650-V GaN-based high-electron-mobility transistors (HEMT) to achieve multi-kW power conversion at conventional cooling conditions with 6.78 MHz. As shown in Fig.1, a transformer-based isolated resonant DC-DC converter is implemented with a primary-sided GaN H-bridge inverter and a secondary-sided SiC H-bridge diode rectifier. Two GaN HEMTs, GS66508T and GS-065-030-2-L, are selected to achieve GaN inverters with a particular focus on thermal management in surface-mounted GaN devices and the realization of zero-voltage-switching (ZVS) to optimize efficiency. The power, efficiency, and switching performance of the implemented 6.78 MHz converters are evaluated under different conditions, demonstrating a maximum power conversion capability of 2.8 kW and a peak efficiency of 95.6%.

979-8-3315-1110-4/25 $31.00 © 2025 IEEE   250

II. 6.78 MHz High-Power GaN Inverter Design

A. Design Challenges

1) GaN Device Thermal Management

The power losses in GaN HEMTs, including conduction loss P_c and the switching loss P_{sw}, contribute to heat accumulation. P_c and P_{sw} inevitably increase with the power level. Low on-state resistance $R_{ds(on)}$ contributes to small conduction loss P_c while soft-switching helps to mitigate the switching loss P_{sw}.

In the market, the packages of mainstream commercial GaN devices are generally surface-mounted, such as *GaN Power Package Extended* (GaN*PX*) and *Power Dual Flat No-Lead* (PDFN), which are beneficial for high-frequency applications due to the extremely small parasitic inductance. However, these compact packages have limited heat dissipation capabilities, making thermal management of GaN devices a significant challenge in high-power applications.

2) ZVS Operation at 6.78 MHz

ZVS means that the drain-source voltage V_{ds} of the GaN HEMT drops to zero before the switch turns on, which is critical to minimize the switching loss P_{sw}, ensuring high efficiency. Furthermore, ZVS operation avoids the rapid charge or discharge of the parasitic output capacitance C_{oss}, which helps mitigate the oscillations in V_{gs} and V_{ds}, reducing electromagnetic interference and enhancing device reliability.

From the power perspective, achieving ZVS requires sufficient inductive reactive power Q_L from the load network to compensate the capacitive reactive power Q_{Coss} generated by the parasitic capacitance C_{oss} of GaN HEMTs, while from the time perspective, a positive drain-source current I_{OFF} during turn-off transient is needed to fully charge and discharge the C_{oss} of the GaN HEMTs within dead time. ZVS operation at 6.78 MHz generally requires a very high Q_L and a large I_{OFF}, which is hard to achieve.

B. Selection of GaN HEMTs

Two 650 V surface-mounted GaN HEMTs are selected for achieving 6.78 MHz high-power inverters, and the main parameters, including drain-to-source on resistance $R_{ds(on)}$, total gate charge Q_g, output capacitance C_{oss}, and internal gate resistance $R_{G(int)}$. A small $R_{ds(on)}$ contributes to low conduction loss while a lower Q_g helps reduce the driving power loss. A smaller C_{oss} helps minimize switching losses and increase the achievability of ZVS. A lower $R_{G(int)}$ accelerates the switching speed and reduces switching loss.

In practice, selecting GaN HEMTs involves trade-offs, because the low Q_g and C_{oss} that facilitate high-frequency switching always contradict the small $R_{ds(on)}$ that is beneficial to high-power conversion capability. In this paper, two GaN HEMTs GS66508T and GS-065-030-2-L with different cooling methods (i.e., top-cooled and bottom-cooled, respectively) from *GaN Systems* are selected, which balance the high-frequency switching and high-power conversion capability, as shown in Table I.

C. Design of the Gate Driver Circuit

The gate driver circuit for both GaN inverters is designed based on the Si8271 chip, and the circuit layout is shown in Fig. 2, with split turn-on and -off driving loops. The lengths of driving loops are as short as possible to minimize parasitic inductance, which are respectively 2.6 mm and 2.9 mm for GaN GS66508T and GS-065-030-2-L inverters.

TABLE I. SPECIFICATIONS OF THE SELECTED GAN HEMTS

GS66508T	GS-065-030-2-L
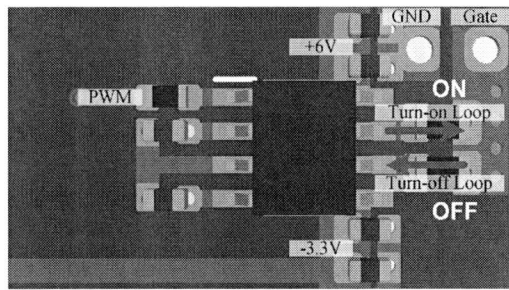	
Package Type	
GaN*PX* Top-Cooled 6.96×4.48×0.54mm³	PDFN Bottom-Cooled 8×8×0.9mm³
Static Electrical Characteristics ($V_{ds}/I_{ds}/R_{ds(on)}$)	
650V/30A/50mΩ	650V/30A/50mΩ
Dynamic Electrical Characteristics ($Q_g/C_{oss}/R_{G(int)}$)	
6.1nC/65pF/1.1Ω	6.7nC/60pF/1.3Ω

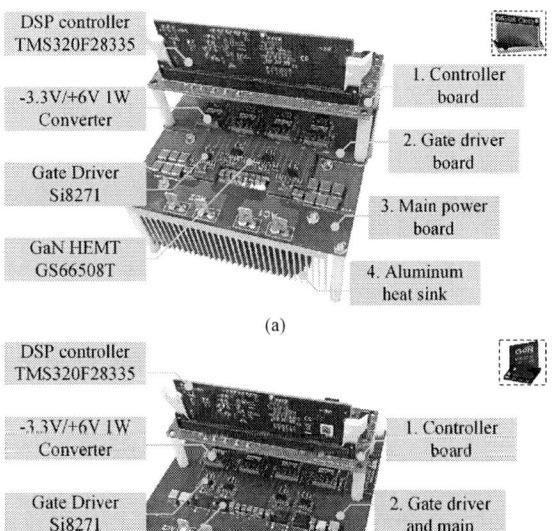

Fig. 2. Si8271-based gate driver circuit for GaN HEMT.

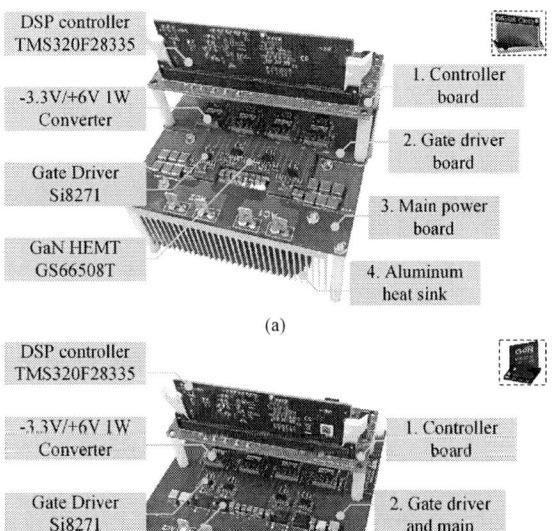

Fig. 3. Full-bridge inverters based on GaN HEMT (a) GS66508T and (b) GS-065-030-2-L.

The symmetrical and compact configuration of the driver circuit reduces impedance in the driving loop, minimizes noise and interference, and enhances circuit performance and reliability, which is important for 6.78 MHz switching operation.

D. Implementation of 650 V GaN Inverter

The GaN GS66508T inverter consists of a controller board, a gate driver board, a main power board, and an aluminum heat sink, as shown in Fig. 3(a). The split gate driver board and main power board facilitate the cooling of GaN GS66508T and simplify circuit trace routing. The four GaN switches are distributed on the back of the main power board, with their thermal pad tightly connected to the aluminum heat sink for effective heat dissipation. The GaN switches are arranged horizontally in the left and right bridges, and positioned vertically and compactly within each bridge to minimize the parasitic inductance between the upper and lower switches. Additionally, filter capacitors are placed between the drain and source of the upper and lower switches, respectively, to reduce voltage ripple and stabilize the input voltage for the inverter.

Due to the bottom-side cooling characteristic, GaN HEMT GS-065-030-2-L based inverter integrates the main power board (H-bridge inverter circuit) and the driving circuits together, as shown in Fig. 3(b). The thermal pad of GaN HEMT GS-065-030-2-L is connected to a large copper polygon of the bottom layer of the PCB, and the heat sink is tightly attached to the back side of the PCB board. The four GaN switches are arranged in parallel on the top layer side of the PCB board, and heat generated by the switches is conducted through the PCB copper to the heat sink for dissipation.

III. EXPERIMENTAL VALIDATION

A. Experimental Setup

Based on the implemented two GaN inverters, a 6.78 MHz resonant DC-DC power conversion setup is constructed. The circuit diagram is provided in Fig.1 and the hardware is demonstrated in Fig. 4. An H-bridge rectifier is implemented based on SiC diode C6D04065A. The resonant transformer is fabricated based on a fair-rite 67 magnetic core with 3 turns on the primary side and 4 turns on the secondary side. Two series capacitors C_s and C_p are used to resonate with the transformer. The main parameters of the resonant DC-DC converter are provided in Table II. The working performance of the two GaN inverters at 6.78 MHz is evaluated at different loads, input voltages, and power levels.

B. ZVS Switching Performance of GaN Inverter

Fig. 5(a), (b), and (c) demonstrate the working waveforms of the GaN GS66508T inverter at input conditions of 75 V/80 W, 200 V/640 W, and 400 V/2787 W, respectively, while Fig. 6(a), (b), and (c) present the GaN GS-065-030-2-L inverter operates at 75 V/81 W, 225 V/800 W, and 350 V/2061 W. These figures correspond to three distinct operating scenarios. Specifically, Fig. 5(a) and Fig. 6(a) represent conditions under which ZVS is not yet achieved. In Fig. 5(b) and Fig. 6(b), ZVS is critically achieved, whereas Fig. 5(c) and Fig. 6(c) demonstrate full ZVS operation. Both inverters achieve ZVS operation with inverter current i_{ac} lagging inverter voltage v_{ac} in phase. Through the comparison of the three waveform sets, the successful realization of ZVS ensures clean gate-drive signals, even under the high switching frequency of 6.78 MHz and power levels exceeding 2 kW. The implementation of ZVS thereby enables the system to operate safely and reliably under high-frequency, high-power conditions.

C. Power and Efficiency of DC-DC Converter

Fig.7 presents the waveforms of the DC-DC resonant conversion system when utilizing the GS66508T-based inverter and the GS-065-030-2-L-based inverter, respectively,

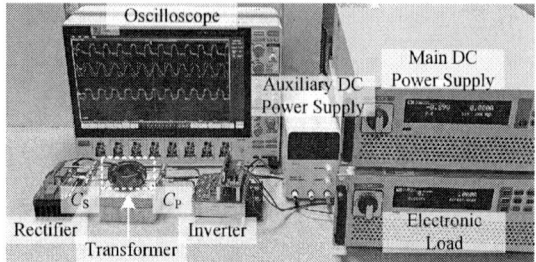

Fig. 4. Experimental setup of the resonant conversion system.

TABLE II. MAIN PARAMETERS OF THE RESONANT CONVERSION SYSTEM.

Parameter	Value	Parameter	Value
Rectifier Diode	C6D04065A	Magnetic core	Fair-Rite 67
L_p	1.74 µH	L_s	2.85 µH
C_p	466.10 pF	C_s	273.79 pF

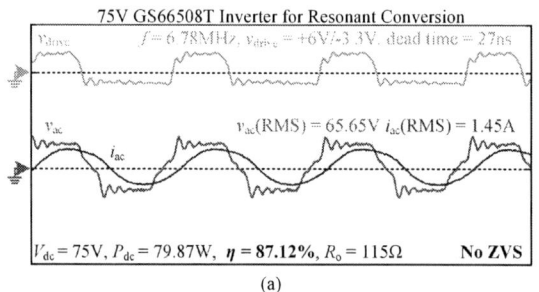

(a)

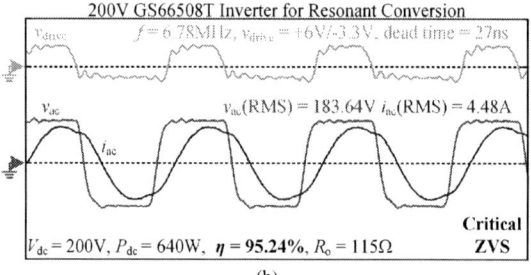

(b)

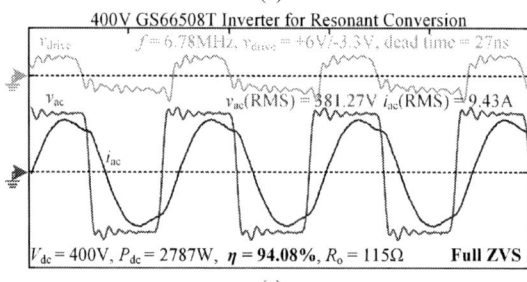

(c)

Fig. 5. Main waveforms of GS66508T inverter operating at (a) 75 V, (b) 200 V, and (c) 400 V.

demonstrating that both inverters operate effectively. In the figure, v_{gs} denotes the gate-source driving signal of the switches, while v_s and i_s represent the voltage and current delivered to the rectifier. V_o and I_o correspond to the load voltage and current, respectively. Fig. 8 shows the power levels of the 6.78 MHz DC-DC resonant conversion system under different input voltage V_{dc} and a fixed load resistance of 115 Ω. Particularly, the GaN GS66506T inverter can operate over a wide voltage range of 50 V to 400 V, achieving the maximum power output of 2.8 kW at 400 V. However, the GaN GS-065-030-2-L inverter can only operate from 50 V to

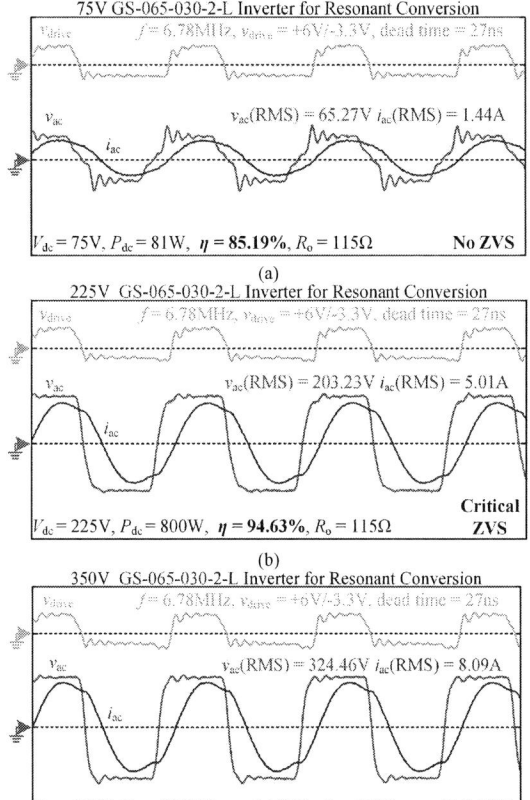

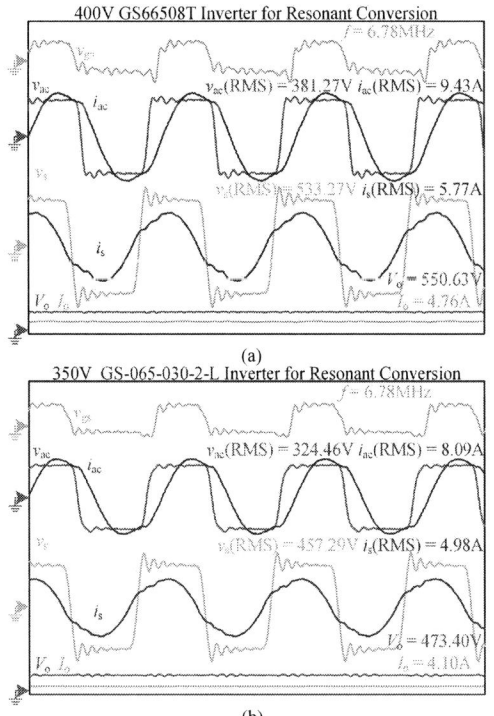

Fig. 6. Main waveforms of GS-065-030-2-L inverter operating at (a) 75 V, (b) 225 V, and (c) 350V.

Fig.7. Waveforms of the DC-DC resonant conversion systems with (a) GS66508T inverter and (b) GS-065-030-2-L inverter.

350 V, achieving a peak power of 2.06 kW at 350 V. At V_{dc} = 350 V, the GS-065-030-2-L inverter failed due to insufficient thermal dissipation. The damaged GaN HEMTs are shown in Fig. 9.

Fig. 10 depicts the DC-DC efficiency of the power conversion system, which increases with rising power at low levels and gradually stabilizes at high power levels. The efficiency improvement with power increase is mainly attributed to the realization of ZVS at higher power levels, which reduces switching losses. With the GaN GS66506T inverter, the system can achieve a maximum DC-DC efficiency of 95.60% at 480 W while with the GaN GS-065-030-2-L inverter, the maximum system efficiency only reaches 94.74% at 1007 W.

Experimental results show that the GaN GS66505T inverter supports higher voltage and power levels and can achieve slightly higher efficiency than the GaN GS-065-030-2-L inverter, which may be attributed to the difference in thermal dissipation capability.

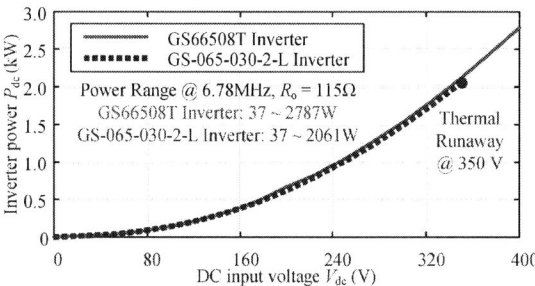

Fig.8. Inverter power levels versus V_{dc} at 6.78 MHz.

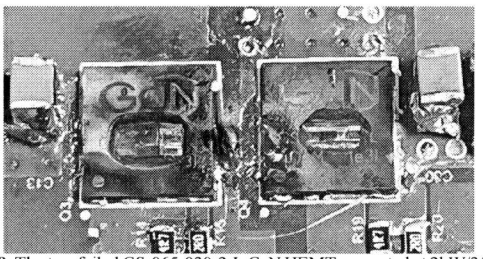

Fig. 9. The two failed GS-065-030-2-L GaN HEMTs operated at 2kW/350V.

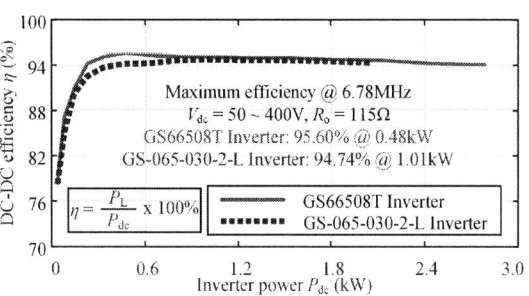

Fig. 10. The DC-DC efficiency of the two GaN inverters operating at 6.78MHz/115Ω ranges from 50 V to 400 V.

IV. CONCLUSION

This paper designs and implements a 6.78 MHz 2.8 kW DC-DC resonant power conversion system based on 650 V GaN HEMTs H-bridge inverters. The design challenges of 6.78 MHz high-power GaN inverters are summarized, and two GaN HEMTs, the GS66508T and GS-065-030-2-L, were selected for designing the 6.78 MHz high-power H-bridge inverters. The preliminary experimental results validate the 2.8 kW power conversion capability at 400 V of the GS66505T inverter and the 2.06 kW power conversion capability at 350 V of the GaN GS-065-030-2-L inverter. With different inverters, the 6.78 MHz resonant DC-DC power conversion system respectively achieves the peak efficiency of 95.60% (GS66508T based), and 94.74% (GS-065-030-2-L based).

ACKNOWLEDGMENT

The authors would like to thank the financial supports from the A*Star MTC YIRG M23M7c0115 and the Ministry of Education (MoE) Academic Research Fund (AcRF) Tier-1 Fund RG134/23.

REFERENCES

[1] A. M. S. Al-bayati, S. S. Alharbi, S. S. Alharbi, and M. Matin, "A comparative design and performance study of a non-isolated DC-DC buck converter based on Si-MOSFET/Si-Diode, SiC-JFET/SiC-schottky diode, and GaN-transistor/SiC-Schottky diode power devices," *2017 North American Power Symposium (NAPS)*, Morgantown, WV, USA, 2017, pp. 1-6.

[2] M. K. Ranjram and D. J. Perreault, "A 380-12 V, 1-kW, 1-MHz converter using a miniaturized split-phase, fractional-turn planar transformer," *IEEE Trans. Power Electron.*, vol. 37, no. 2, pp.1666-1681, Feb. 2022.

[3] D. S. Gautam and A. K. S. Bhat, "A Comparison of soft-switched DC-to-DC converters for electrolyzer application," *IEEE Trans. Power Electron.*, vol. 28, no. 1, pp. 54-63, Jan. 2013.

[4] Y. Wang, J. Yang, K. Wang, and Y. Yang, "Highly integrated hybrid inductive and capacitive power transfer system with asymmetrical printed-circuit-board-based self-resonator," *IEEE Trans. Power Electron.*, vol. 40, no. 7, pp. 10254-10264, Jul. 2025.

[5] J. Biela, M. Schweizer, S. Waffler, and J. W. Kolar, "SiC versus Si—evaluation of potentials for performance improvement of inverter and DC–DC converter systems by SiC power semiconductors," *IEEE Trans. Ind. Electron.*, vol. 58, no. 7, pp. 2872-2882, Jul. 2011.

[6] U. K. Mishra, P. Parikh, and Yi-Feng Wu, "AlGaN/GaN HEMTs-an overview of device operation and applications," *Proceedings of the IEEE*, vol. 90, no. 6, pp. 1022-1031, Jun. 2002.

[7] R. Sun, J. Lai, W. Chen, and B. Zhang, "GaN power integration for high frequency and high efficiency power applications: a review," *IEEE Access*, vol. 8, pp. 15529-15542, 2020.

[8] M. Buffolo et al., "Review and outlook on GaN and SiC power devices: industrial state-of-the-art, applications, and perspectives," *IEEE Trans. Electron. Devices*, vol. 71, no. 3, pp. 1344-1355, Mar. 2024.

[9] R. Qin, J. Li, and D. Costinett, "A 6.6-kW high-frequency wireless power transfer system for electric vehicle charging using multilayer nonuniform self-resonant coil at MHz," in *IEEE Trans. Power Electron.*, vol. 37, no. 4, pp. 4842-4856, Apr. 2022.

[10] R. Qin, J. Li, and D. Costinett, "A high frequency wireless power transfer system for electric vehicle charging using multi-layer nonuniform self-resonant coil at MHz," in *IEEE Energy Conversion Congress and Exposition (ECCE)*, Detroit, MI, USA, 2020, pp. 5487-5494.

[11] S. Maji, D. Etta, and K. K. Afridi, "A high-power large air-gap multi-MHz dc-dc capacitive wireless power transfer system for electric vehicle charging," *2023 IEEE Wireless Power Technology Conference and Expo (WPTCE)*, San Diego, CA, USA, 2023, pp. 1-6.

[12] B. Regensburger, A. Kumar, S. Sinha, J. Xu and K. K. Afridi, "High-efficiency high-power-transfer-density capacitive wireless power transfer system for electric vehicle charging utilizing semi-toroidal interleaved-foil coupled inductors," in *2019 IEEE Applied Power Electronics Conference and Exposition (APEC)*, Anaheim, CA, USA, 2019, pp. 1533-1538.

[13] B. Regensburger, S. Sinha, A. Kumar, S. Maji, and K. K. Afridi, "High-performance multi-MHz capacitive wireless power transfer system for EV charging utilizing interleaved-foil coupled inductors," *IEEE J. Emerg. Sel. Topics Power Electron.*, vol. 10, no. 1, pp. 35-51, Feb. 2022.

[14] N. K. Trung and K. Akatsu, "Design challenges for 13.56MHz 10 kW resonant inverter for wireless power transfer systems," in *2019 10th International Conference on Power Electronics and ECCE Asia (ICPE 2019 - ECCE Asia)*, Busan, Korea (South), 2019, pp. 1-7.

A Passive Cancellation Circuit Incorporating the Motor Stator Windings for Reducing the Common-Mode Noise in Motor Drive Systems

Lihong Xie
College of Automation Engineering
Nanjing University of Aeronautics and
Astronautics
Nanjing, China
xielihong@nuaa.edu.cn

Yuzhen Wu
College of Automation Engineering
Nanjing University of Aeronautics and
Astronautics
Nanjing, China
qzqgwyzq@163.com

Xinbo Ruan
College of Automation Engineering
Nanjing University of Aeronautics and
Astronautics
Nanjing, China
ruanxb@nuaa.edu.cn

Abstract—**The common-mode (CM) noise in motor drive systems is mainly caused by the CM voltage of the inverter, and it may introduce the electromagnetic interference (EMI) to the nearby equipment. With a broader applications of wide-band-gap devices such as Silicon-Carbide (SiC) and Gallium Nitride (GaN) devices, higher dv/dt are generated in the inverter, and the CM noise becomes more severe than that using Silicon devices. In this article, a passive cancellation circuit is proposed to reduce the CM noise in motor drive systems. The stator windings of a three-phase non-salient pole permanent magnet synchronous machine (PMSM) are utilized as the differential-mode (DM) inductor, and the CM voltage of the inverter is sensed for cancelling the CM noise. Compared with existing methods, magnetic components in the power stage or active devices are not required, and the proposed solution offers low volume and cost. The effect of high-frequency parasitics is analyzed, and the high-frequency range for effective CM noise cancellation is derived. Finally, a motor drive system is built and experimental results are given to verify the effectiveness of the passive cancellation circuit.**

Keywords—CM noise, motor drive systems, wide-band-gap devices, CM voltage sensing, passive cancellation circuit

I. INTRODUCTION

In motor drive systems, by applying wide-band-gap devices such as Silicon-Carbide (SiC) or Gallium-Nitride (GaN) devices, the switching frequency of pulse-width modulation (PWM) inverters can be increased to achieve high power density. Meanwhile, the reduced switching time and on-state resistance of the devices help realize high efficiency. However, a shorter switching transient indicates severe conducted electromagnetic interference (EMI), as the spectrum of the noise source in the inverter becomes higher [1]. The common-mode (CM) noise, which propagates through both the power lines and returns from the protective earth (PE), is a major EMI issue in motor drive systems. The CM noise has the potential to cause deterioration in the performance of electrical machines and its drive systems [2], leading to the degradation of system performance or even shutdown.

To ensure safe operation of motor drive systems, the attenuation of CM noise has been investigated over the past 30 years. The available methods can be classified into passive filter, active filter, modulation, topology, balancing and cancellation. The passive filter can reduce the CM noise over the entire conducted EMI frequency range, but it suffers from high volume and high cost. The active filter achieves smaller size by sensing and injecting the noise voltage or current, but the effective frequency range is limited by the bandwidth of operational amplifier [3]. Regarding the modulation schemes, the CM voltage (CMV) of the inverter, which is the source for CM noise, can be reduced by 2/3 when removing zero vectors, but the power loss and torque ripple is increased. The modulation of switching frequency is also effective for spreading the noise spectrum, and the efficiency can be improved with proper variation of switching frequency [4]. At the topology level, the CM noise can be cancelled by introducing additional phase leg, multi-level inverters, paralleled structure or dual three-phase motor, but the cost is a major concern [5]. The balancing technique eliminates the CM noise by building a balanced Wheatstone bridge [6], but additional inductor at the power line is required, degrading the power density.

In addition to the solutions above, the CM noise can be also attenuated by cancellation, where the CM voltage of the inverter is sensed for cancelling the CMV [7] [8] or CM current [9]. However, bulky CM transformers or operational amplifier with dc power supply are required. In this article, a passive circuit for cancelling the CM noise in motor drive systems is proposed. The stator windings of the motor are applied as differential-mode (DM) inductor to sense the CMV of inverters, and the magnetic components in the power stage or active devices can be omitted, so that low cost and low volume can be achieved.

This article is organized as follows. In Section II, the CM noise equivalent circuit of motor drive system is presented. In Section III, the characteristics of CM impedance for non-salient three-phase permanent magnet synchronous machine (PMSM) is analyzed, and the proposed method is introduced and explained. In Section IV, the effect of parasitics is analyzed. In Section V, experimental results are presented to validate the effectiveness of the proposed method. Finally, Section VI concludes this article.

II. CM NOISE MODEL OF MOTOR DRIVE SYSTEMS

Fig. 1 shows a typical three-phase motor drive system with line impedance stabilization network (LISN) and parasitic capacitance of the inverter. The LISN provides

---

This work was supported by the Science and Technology Planning Project of Jiangsu province under Grant BE2022048-4.

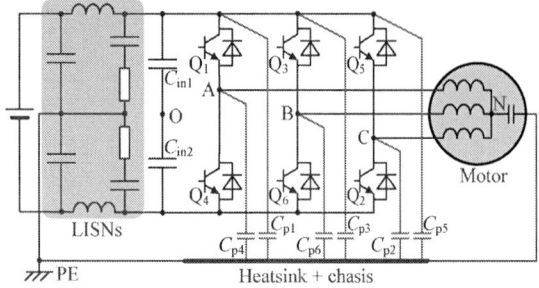

Fig. 1. The motor drive system with LISN and parasitic capacitance of the PWM inverter.

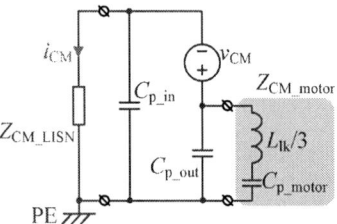

Fig. 2. The CM noise equivalent circuit for motor drive systems.

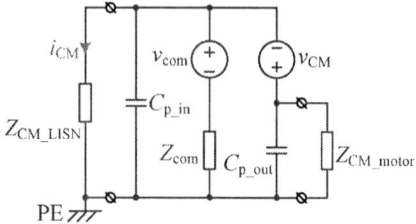

Fig. 3. The cancellation principle.

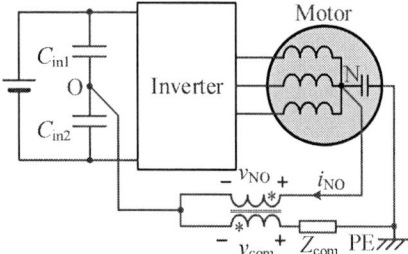

Fig. 4. The realization of the passive cancellation circuit.

III. THE PASSIVE CANCELLATION CIRCUIT

A. The cancellation principle

Fig. 3 shows the principle of CM noise cancellation. An extra branch is connected in parallel with C_{p_in}, which consists of a voltage source v_{com} in series with an impedance Z_{com}. As long as the noise current generated by v_{CM} has the same amplitude but opposite to that generated by v_{com}, the CM current i_{CM} can be cancelled by each other:

$$Z_{com} = \frac{|v_{com}|}{|v_{CM}|} \frac{Z_{CM\_motor}}{1 + sC_{p\_out}Z_{CM\_motor}} \quad (4)$$

One solution to (4) is to let v_{com} be proportional to v_{CM} by a ratio of n, and Z_{com} be proportional to the paralleled impedance of Z_{CM_motor} and C_{p_out} by the same ratio, as given in (5). As a result, v_{com} can be generated by sensing v_{CM} with a transformer, in order to avoid active devices.

$$\begin{cases} v_{com} = -nv_{CM} \\ Z_{com} = n\dfrac{Z_{CM\_motor}}{1 + sC_{p\_out}Z_{CM\_motor}} \end{cases} \quad (5)$$

B. Realization of the passive cancellation circuit

Based on (5), sensing the CMV of the PWM inverter is the key to realizing cancellation. Referring to Fig. 1, the electric potential of the neutral point N with respect to the mid-point O of input capacitors is equal to v_{CM} in (1) by assuming symmetry. For a non-salient three-phase PMSM, the inductance of stator windings are symmetrical. In specific, the self-inductance L_A, L_B, L_C of each stator winding and mutual-inductance M_{AB}, M_{BC}, M_{CA} between windings satisfy

$$\begin{cases} L_A = L_B = L_C = L_0 + L_{lk} \\ M_{AB} = M_{BC} = M_{CA} = -0.5L_0 \end{cases} \quad (6)$$

According to (6), the coupling coefficient between windings is nearly -0.5 since L_{lk} is far smaller than L_0.

specific impedance in the conducted EMI frequency range from 150 kHz to 30 MHz in standard DO 160, CISPR 32 or MIL-STD-461 [10], etc., and converts the noise current into voltage for the convenience of measurement. The CM parasitic capacitance $C_{p1} \sim C_{p7}$ are created between the power switches and the heatsink. The heatsink and motor frame are grounded to the protective earth (PE) for safety purposes. The CM current is generated by the CMV of the inverter via the CM parasitic capacitances of the inverter and motor. By assuming symmetry of three-phase paths, the equivalent CM noise circuit of the motor drive system can be derived, as shown in Fig. 2. Where, Z_{CM_LISN} and Z_{CM_motor} are the CM impedance at the LISN side and motor side, respectively.

In Fig. 2, v_{CM} is the CMV of the inverter, C_{p_in} and C_{p_out} are the total parasitic capacitance at the input and output side of inverter, respectively. v_{CM}, C_{p_in} and C_{p_out} are expressed as

$$v_{CM} = \frac{v_{AO} + v_{BO} + v_{CO}}{3} \quad (1)$$

$$\begin{cases} C_{p\_in} = C_{p1} + C_{p3} + C_{p5} \\ C_{p\_out} = C_{p2} + C_{p4} + C_{p6} \end{cases} \quad (2)$$

Where, v_{AO}, v_{BO} and v_{CO} is the voltage between the mid-point of each phase-leg and the input capacitors, respectively.

In Fig. 2, Z_{CM_motor} at low frequencies can be briefly represented by an inductance in series with C_{p_motor} between the stator winding and motor frame, and L_{lk} is the leakage inductance of each winding. Therefore, the CM impedance Z_{CM_motor} is given by

$$Z_{CM\_motor} = \frac{sL_{lk}}{3} + \frac{1}{sC_{p\_motor}} \quad (3)$$

As seen from Fig. 2, v_{CM} causes the CM noise through both the CM parasitic capacitance in the inverter and the CM impedance of motor.

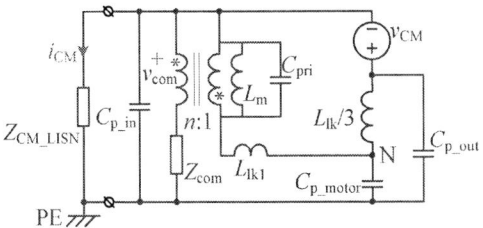

Fig. 5. The equivalent circuit with parasitics of motor and the transformer.

Therefore, the stator windings of the motor can be treated as a DM inductor for CM noise separation [11], and the CM voltage can be sensed. Referring to Fig. 3, the cancellation circuit is derived in Fig. 4. Where, the primary winding of the transformer is connected between terminals O and N, and the secondary winding is connected to the PE via Z_{com}. Since the transformer windings are not in series with the motor, its winding dimension is not determined by the current in the power stage, and a small diameter of winding can be achieved. Moreover, active devices such as bipolar junction transistors or operational amplifiers are not required, the cost is also reduced.

In Fig. 4, the primary winding of the transformer introduces magnetization current, increasing the copper loss of the stator windings. The smaller the magnetization inductance is, the high the copper loss will be. Therefore, the transformer design is a trade-off between its volume and the extra copper loss.

IV. PERFORMANCE OF CM NOISE CANCELLATION WITH PARASITICS

In this Section, the CM noise cancellation for the proposed method is analyzed, by considering the parasitics of the motor and the transformer. Based on Fig. 3 and 4, the equivalent circuit with high-frequency parasitics is given in Fig. 5(a). Where, L_m, L_{lk1} and C_{pri} are the magnetizing inductance, leakage inductance and parasitic capacitance of the primary winding.

At low frequencies, C_{pri} and C_{p_motor} can be treated as open circuit in Fig. 5. In order to generate the desired compensation voltage v_{com} in (5), the following inequality should be satisfied based on voltage division for the primary winding

$$L_m \gg L_{lk1} + \frac{L_{lk}}{3} \tag{7}$$

At high frequencies, L_m can be treated as open circuit, and v_{com} is determined by v_{CM} and the LC low-pass network composed of C_{pri}, L_{lk1}, C_{p_motor} and $L_{lk}/3$ in Fig. 5. The corner frequency f_H of the above network is given by

$$f_H = \frac{1}{2\pi\sqrt[4]{\dfrac{L_{lk1}C_{pri}L_{lk}C_{p\_motor}}{3}}} \tag{8}$$

According to (8), the effective frequency range for cancellation is far lower than f_H. In order to extend f_H, the leakage inductance and parasitic capacitance of the transformer should be kept as small as possible.

Table I. The parameters of the inverter, motor and passive cancellation circuit.

Inverter			
Input voltage	400 V	Capacitance C_{p1}~ C_{p6}	37 pF
Switching frequency	30 kHz		
Motor			
Leakage ind. L_{lk}	68 μH	Capacitance C_{p_motor}	1.6 nF
Passive cancellation circuit			
Capacitance C_{pri}	223 pF	Turns ratio n	1
Magnetizing ind. L_m	30 mH	Leakage ind. L_{lk1}	1.6 μH

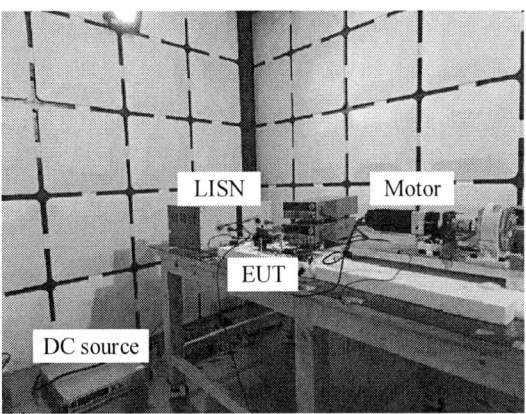

Fig. 6. Experimental setup for the conducted EMI measurement.

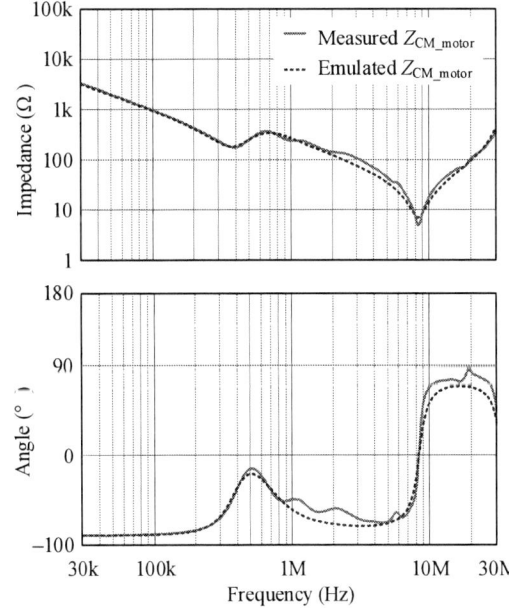

Fig. 7. The measured and emulated CM impedance of the motor.

V. EXPERIMENTAL RESULTS

To validate the proposed passive cancellation method, a motor drive system for a PMSM is built in the lab. The associated parameters of the inverter, motor and passive cancellation circuit are listed in Table I. The cancellation

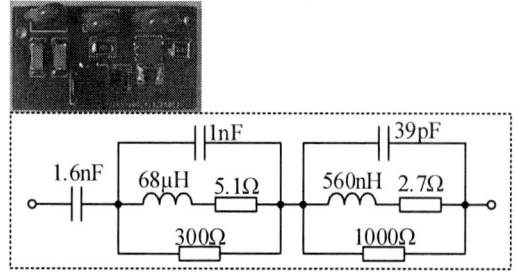

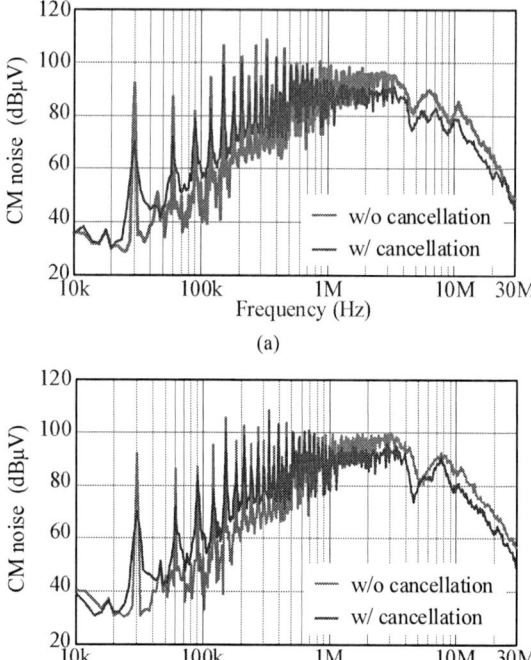

Fig. 8. The network emulating the CM impedance of the motor.

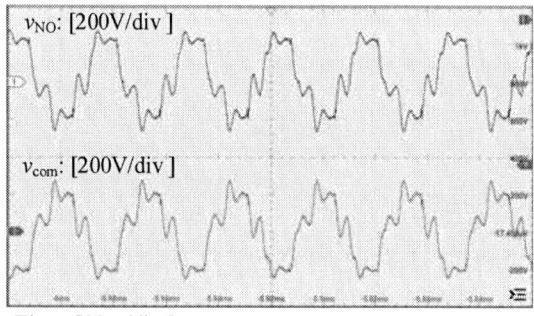

Time: [20µs/div]

Fig. 9. Time-domain waveforms of v_{NO} and v_{com}.

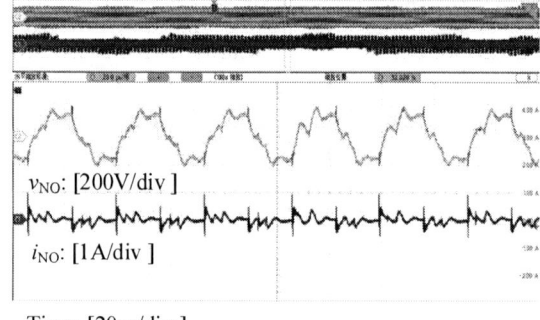

Time: [20µs/div]

Fig. 10. Time-domain waveforms of v_{NO} and i_{NO}.

transformer uses a nanocrystalline core (JK-3718) to achieve large magnetizing inductance for the transformer. Moreover, the transformer winding has 15 turns, and its winding dimension is 0.6 mm.

Fig. 6 shows the experimental setup for the conducted EMI. The LISN adopts 21115-50-TS-100-N from Solar Electronics, noise separator uses MYSE 02 from MYEMC, and EMI receiver applies ESL3 from Rohde & Schwarz. The CM impedance of the motor is measured by an impedance analyzer from 30 kHz to 30MHz, as shown by the dashed line in Fig. 7. To emulate this impedance, a network [9] in Fig. 8 is incorporated, whose impedance and phase angle curves are given by the dashed lines in Fig. 7. The emulated impedance matches well with the measured ones, which helps generate the required compensation current for CM noise reduction.

In Fig. 9, the time-domain waveforms of v_{NO} and v_{com} in several switching periods are presented. As seen, the CMV and the compensation voltage are complementary to each other, and the difference at high frequencies is caused by the parasitics of the transformer. Fig. 10 illustrates v_{NO} and the

Fig. 11. Comparison of the measured CM noise spectrum. (a) no-load condition. (b) full-load condition.

current i_{NO} through the primary winding of the transformer. The root-mean-square value of i_{NO} is 0.179 A, indicating that the power loss introduced by i_{NO} is negligible.

Fig. 11 compares the CM noise spectrum under no-load and full-load conditions, respectively. By applying the passive cancellation circuit, the CM noise is reduced by nearly 20 dB from 30 kHz to 100 kHz, alleviating the requirement for CM noise attenuation. At higher frequencies, the transformer parasitics limit the cancellation, and CM noise reduction is not obvious as expected.

VI. CONCLUSION

In this article, the CM noise cancellation in motor drive systems is presented. By connecting a transformer between the neutral of stator windings and the mid-point of input capacitors, the CMV of the inverter can be sensed for cancelling the CM noise. The leakage inductance and parasitic capacitance of the transformer should be kept small in order to have effective cancellation at higher frequencies. Experimental results indicate that the passive cancellation circuit provides 20 dB of attenuation from 30 kHz to 100 kHz, validating the proposed method.

REFERENCES

[1] B. Zhang, S. Wang, Y. Lai and Y. Yang, "Modeling and prediction of low-frequency radiated EMI for a SiC motor drive system," *IEEE Trans. Ind. Electron.*, vol. 71, no. 9, pp. 10210- 10220, Sept. 2024.

[2] Z. Zhang, Y. Hu, X. Chen, G. Jewell and H. Li, "A review on conductive common-mode EMI suppression methods in inverter fed motor drives," *IEEE Access*, vol. 9, pp. 18345– 18360, 2021.

[3] Y. Zhang and D. Jiang, "An active EMI filter in grounding circuit for dc side CM EMI suppression in motor drive system," *IEEE Trans. Power Electron.*, vol. 37, no. 3, pp. 2983-2992, Mar. 2022.

[4] D. Jiang, "Advanced Pulse-Width-Modulation Technologies in Power Electronics Converters," China Machine Press, 2018.

[5] A. L. Julian, G. Oriti, and T. A. Lipo, "Elimination of common-mode voltage in three-phase sinusoidal power converters," IEEE Trans. Power Electron., vol. 14, no. 5, pp. 982-989, Sep. 1999.

[6] L. Xing and J. Sun, "Motor drive common-mode EMI reduction by passive noise cancellation," in *Proc. 2011 14th European Conf. Power Electron. and Appl.*, pp. 1-9, 2011.

[7] S. Ogasawara, H. Ayano, and H. Akagi, "An active circuit for cancellation of common-mode voltage generated by a PWM inverter," IEEE Trans. Power Electron., vol. 13, no. 5, pp. 835–841, Sep. 1998.

[8] L. Xie, X. Yuan, "Common-mode current reduction at dc and ac sides in inverter systems by passive cancellation," *IEEE Trans. Power Electron.*, vol. 36, no. 8, pp. 9069-9079, Aug. 2021.

[9] Y. Zhang, Q. Li and D. Jiang, "A motor CM impedance based transformerless active EMI filter for dc-side common-mode EMI suppression in motor drive system," *IEEE Trans. Power Electron.*, vol. 35, no. 10, pp. 10238-10248, Oct. 2020.

[10] *Requirements For The Control Of Electromagnetic Interference Characteristics Of Subsystems And Equipment*, document MIL-STD-461G, 2015.

[11] S. Wang, F. Luo and F. C. Lee, "Characterization and design of three-phase EMI noise separators for three-phase power electronics systems," *IEEE Trans. on Power Electron.*, vol. 26, no. 9, pp. 2426-2438, Sept. 2011.

A Voltage-Sharing Cascode Switch Structure Using SiC JFET and GaN HEMT

Yin Fang, Shan Jayamaha, and Carl Ngai Man Ho

Price Faculty of Engineering, University of Manitoba, Winnipeg, Canada

fangy1@myumanitoba.ca, jayadkjs@myumanitoba.ca, Carl.Ho@umanitoba.ca

Abstract—**To maintain the advantages of high efficiency and high switching speed of the Silicon Carbide (SiC) and Gallium Nitride (GaN) semiconductor devices, this paper introduces a voltage-sharing cascode structure. The proposed structure has the capability of a higher voltage rating and simpler gate driver design compared to a series connection of devices with synchronization issues. The simulation results show a fast dv/dt at transient, enabling the implementation of high switching frequency converters. The experimental results verify the applicability of the proposed structure and validate its fast-switching speed.**

Keywords—*cascode, GaN, JFET, switching behavior, switching transient*

I. INTRODUCTION AND MOTIVATION

The Silicon Carbide (SiC) and Gallium Nitride (GaN) are two critical materials in Wide Bandgap (WBG) power semiconductor devices catalogs that illustrate a remarkable growing trend in the market by their advantages of high switching speed, lower conduction loss, and relative higher operating temperature tolerance comparing to traditional Silicon (Si) based devices [1],[2]. These characteristics of the power semiconductors allow higher power density and efficiency of their converter design [3]. Among these two semiconductor categories, GaN High-Electron-Mobility Transistor (HEMT) has the capability of withstanding extremely higher switching speed. The rising and falling times during switching in hundreds of picosecond range, hence the switching frequency of GaN HEMT is motivated to approach GHz [4], while the switching frequency of typical SiC MOSFET is below MHz [5]. Therefore, after solving the Electromagnetic Compatibility (EMC) problems driven by high switching frequency, GaN HEMT is an appropriate and available choice in the market of ultra-high switching frequency Switched-Mode Power Supplies (SMPS). Furthermore, due to the limitation of EMC and electrical field strength of the depletion layer of semiconductors [6], both enhancement and depletion mode GaN HEMTs in the market are mainly deployed for Low Voltage (LV) range, which is defined below 1000V. Various manufacturers, e.g., Infineon, TI, Renesas, Power Integration, etc., have most of their commercial products concentrated in the 400V to 700V drain-to-source voltage range. This implies that multiple devices in series connection or multilevel converter design are required for Medium Voltage (MV) or High Voltage (HV) applications. A complicated gate driver design is necessary to ensure the synchronization of the switching mechanism of each semiconductor unit that conceivably has deviations in threshold voltage and channel resistance in conduction [7]. Hence, dynamic control is recommended for current and voltage balancing by inserting an external coupled inductor in the gate driver circuit or using an RC

snubber circuit to mitigate the problem of gate signal mismatching [8],[9]. The disadvantage of this topology is that the large size of the coupled inductor significantly reduces the power density, which is the reason why it is not a popular choice in industrial applications.

Among LV and MV range devices, the SiC Junction-Field-Effect-Transistor (JFET) is popular in academic research as well as industrial applications, which takes the benefits in terms of higher voltage and current capability, low noise, high switching speed, low conduction loss, and higher junction temperature capability [13]. Furthermore, no additional antiparallel diode is required, which eliminates the reverse recovery process and further decrease the switching loss. Nevertheless, the available MV JFETs in market are normally-on devices that require negative gate driver design or implementation in a cascode configuration. Driven by the demand for high current and voltage rating of the power semiconductor, the super cascode structure is proposed in some research that uses an Si or SiC MOSFET to drive multiple High-Voltage (HV) JFETs [10]-[12]. The cascode structure, as shown in Fig. 1 (a), is a well-known topology in which the source of an LV normally-on device is connected to the gate of the HV normally-off device to make it an overall normally-off device. In such cascode configuration, the oscillations of the overall voltage during switching transient are mentioned in some research papers and solved by tuning parasitic parameters in the connection or adding an internal gate resistor [15]. Due to the principle of the cascode structure, the majority of the DC link voltage is on the High-Side (HS) transistor, and the Low-Side (LS) FET only has voltage stress below 100V. As a result of that, the breakdown voltage of the HS device is selected to be much larger than the LS driving device. Hence, any small voltage ringing in the HS transistor will impose critical oscillations across the

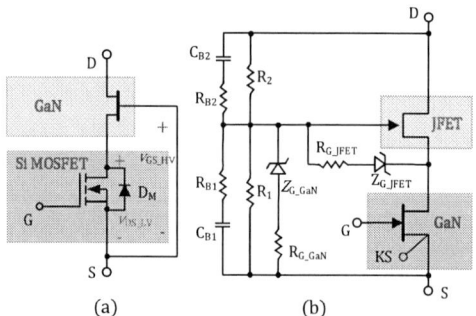

Fig. 1 Configurations of (a) conventional cascode configuration by Si MOSFET and GaN HEMT; (b) the simplified diagram of the proposed voltage-sharing cascode structure by JFET and GaN HEMT

979-8-3315-1110-4/25 $31.00 © 2025 IEEE

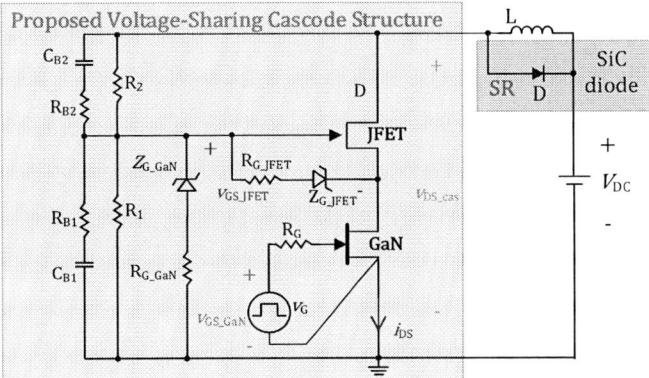

Fig. 2 Schematic of the double pulse test of proposed voltage-sharing cascode switch

LS Field-Effect-Transistor (FET), which can possibly cause the LS FET breakdown. However, MOSFETs with higher breakdown voltage normally has a higher cost and the higher channel resistance, which will increase the conduction loss. Thus, it is not efficient and cost-effective to use MOSFET with higher breakdown voltage, e.g., 900V, as the driver device in the cascode structure. Consequently, a Voltage-Sharing Cascode Structure (VSCS) is proposed in this paper to share the voltage stress of the DC link equally when the overall cascode switch is turned off for a higher voltage rating application, e.g. 1kV solar photovoltaic inverter [17]. The simulation and experimental results that are demonstrated in this paper for a 650V DC voltage show the compatibility of LV applications for grids in most countries.

II. PROPOSED STRUCTURE

The principal purpose of the cascode structure is to operate the depletion mode JFET as a normally-off device. Conventional cascode-connected GaN and Si MOSFET are shown in Fig. 1 (a) as an example, where the MOSFET is required to provide a negative v_{GS} voltage to switch off the depletion mode GaN HEMT. Consequently, the gate of the higher voltage rating GaN HEMT is connected to the source of the lower voltage rating Si MOSFET. Distinct from the example of traditional configuration, where the gate-to-source voltage (v_{GS_HV}) of the GaN is equal to the MOSFET drain-to-source voltage (v_{DS_LV}). This paper proposed a Voltage-Sharing Cascode Structure (VSCS) of enhancement GaN HEMT and depletion mode JFET that is shown in Fig. 1 (b). The most critical part of this design is sharing the voltage stress equally between the two cascode-connected devices during the off state. To reach this objective, an avalanche diode Z_{G_GaN} is connected between the gate terminals of the HS JFET and the LS GaN HEMT. The avalanche breakdown voltage of this diode is determined by half of the total breakdown voltage. Another Zener diode Z_{G_JFET} which is applied to limit the potential difference to a specific value, and an internal gate resistor R_{G_JFET} are connected between HS JFET gate and its source. The gate resistor R_G is connected between the gate driver and the gate terminal of the GaN HEMT to control its switching speed. Because of the sophisticated configuration of the VSCS, the parasitic capacitors and inductance are inevitable and can potentially cause oscillation problems in the switching

actions. The two gate resistors R_{G_GaN} and R_{G_JFET} are the damping resistor in the internal gate loop to provide damping effect on the HS JFET device [16]. The JFET and GaN HEMT have their maximum v_{DS} breakdown voltage rating of 650V and 700V respectively, hence each of the semiconductor devices are intended to carry 325V operating voltage to enough voltage breakdown margin in the design. A more concrete version of the VSCS is shown in Fig. 2 with an RC-R transient balancing circuit and the implementation into a double pulse test circuit. The parallel RC-R ladder circuit C_{B2}, R_{B2} and R_2 at HS, as well as C_{B1}, R_{B1}, and R_1 at LS is designed as a balancing circuit, which aims to equate the charge stored in the gate of both the JFET and the GaN HEMT during transient and steady state, respectively [18]. Due to the working principle of the cascode structure, the devices are switched off respectively from LS to HS in the cascode configuration. The balancing capacitors are critical for the voltage-sharing cascode configuration to provide a current path when the HS switch is still turned on, but the LS device has already been turned off by the gate driver. Different from the FREEDM Supercascode structure in [18], the series resistors R_{B1} and R_{B2} are added to improve the stability at the switching transient, and this will be explained in detail in the following section.

Additionally, the terminal Kelvin Source (KS) of the GaN HEMT directly connects to the source substrate without any accessorial wire bonding. Since the GaN is an inductance-sensitive device, the KS is fabricated to connect to the signal ground of the gate driver. Hence, a comprehensive four-terminal cascode structure is nominated in this paper that can equally share the voltage stress during the off time.

III. SIMULATION AND EXPERIMENTAL RESULTS

A. Simulation Results by LTspice

As shown in Fig. 2, the proposed VSCS is simulated in LTspice as a single switch in a double pulse test (DPT) setup. A freewheeling diode D constitutes the synchronous rectification part of the DPT and is connected in parallel with an inductor L. While the VSCS switch is turned off, the current circulates in the inductor and diode. After the switch is turned on, the current gradually transfers from the inductor to the cascode switch. Considering the current rating of GaN HEMT for safety and stability, the Schottky diode is

TABLE I SPECIFICATION OF DPT SIMULATION OF PARAMETERS

Parameter	Value	Parameter	Value	Parameter	Value	Parameter	Value
V_{DC}	650 V	R_{G_JFET}	1 Ω	C_{B2}	1 nF	R_{B1}/R_{B2}	TBD
$-v_{Gs\_GaN}$	-3 V	R_{G_GaN}	60 Ω	C_{B1}	1.5 pF	T_{on}	0.68 us
$+v_{Gs\_GaN}$	6 V	L	108 μH	R_1/R_2	5.1 MΩ	T_{off}	1.5 us

employed for the freewheeling diode to avoid a high reverse recovery current. The simulation parameters are listed in TABLE I shown below. The simulation results are demonstrated in Fig. 3 (a) when R_{B1} and R_{B2} = 0 MΩ, including the current through the cascode switch i_{DS}, and the separated drain-to-source voltage v_{DS_JFET}, v_{DS_GaN}, and the overall v_{DS_cas} of the overall structure.

Accordingly, the proposed configuration achieves the switching speed dv/dt of 62.1V/ns of 135.5V/ns at switching off and at switching on transients for the overall v_{DS_cas} in both scenarios. This fast voltage gradient verifies that the switch can be applied to high-frequency converters. The voltage sharing between two devices is more balanced when the resistance increases from 0Ω to 100Ω. Due to the circuit design principle and component selection, it will be further improved when the DC link voltage approaches 800V. The current overshoot of the overall i_{DS} was measured at the source of the LS GaN HEMT and is found to be 226% at the switching-on transient. Moreover, the current oscillation after the overshoot also indicates the risk of instability failures. It is noticeable that the HS JFET was in transconductance mode for 0.5μs before the v_{DS_JFET} decaying to 0V, which significantly increases the conduction loss. Although this phenomenon includes high overshoot amplitude and oscillations are common for the high switching speed semiconductors, but this increases the loss or even can damage the device. The reason for the high amplitude of the overshoot is addressed to the fast discharging of the balancing capacitor during the switching-on transition. To conquer this problem, the resistors R_{B1} and R_{B2} are added in series to the balancing capacitor to limit the rate of releasing the stored charge in capacitors. Consequently, slowing down the discharge rate will have an impact on the current waveform, which shows a reduction first and then an increase subsequently, not consistently rising as the conventional current behavior in DPT. The simulation results show that the series resistors effectively reduce the current overshoot amplitude from 226% to 12.6%, the oscillations are removed, and the JFET is in appropriate conduction mode after it is turned on. Another influence of adding the series the resistors is the overshoot of v_{DS_GaN} at the beginning of the switching off transient, due to the balancing capacitors charging and discharging to maintain the steady state level. In summary, there is a tradeoff between the magnitude of overshoot and capacitor discharging time. From the stability and protection point of view, the 100Ω resistors are selected to decrease the current overshoot in this research project.

B. Experimental Results Validation

The experimental setup and the photo of PCB boards is shown in Fig. 4. The DPT test consists of two individual boards: the signal connecters, power terminals, protection circuits, isolation circuits, and Digital Signal Processor (DSP) TM320F28379D to generate the pulse signals are on the

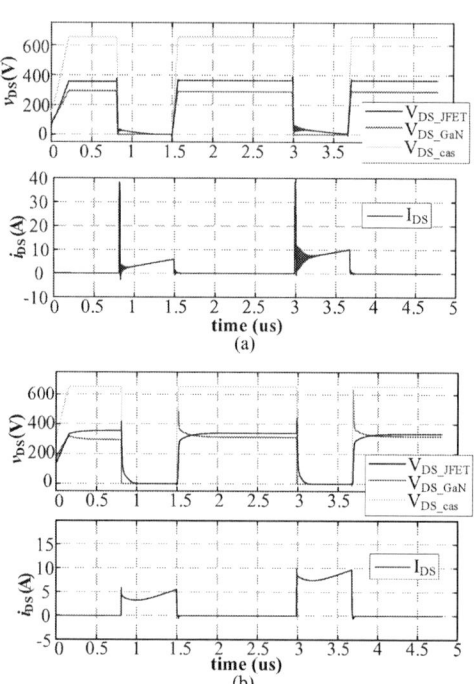

Fig. 3 Simulation results of DPT in LTspice when (a) R_{B1}, R_{B2} = 0 Ω; (b) R_{B1}, R_{B2} = 100 Ω

mother board; the VSCS circuit which is the device under test (DUT), signal power regulator, and the gate driver are on the daughter board to minimize the parasitic inductance in the gate driver loop. Following the same setup as in the simulation, the JFET UJ3N065080K3S and GaN HEMT GS-065-014-6-L are selected to configure the cascode structure with the same configuration in Fig. 2, and the series resistor values are selected as 100Ω. The experimental results of the drain-to-source voltage of each device v_{DS_JFET}, v_{DS_GaN} are measured by differential probes and the current i_{DS} is measured by a jumper wire between the source terminal and the ground and a current probe.

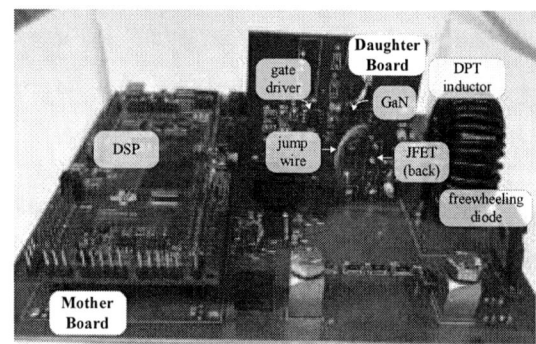

Fig. 4 Experimental setup and photos on the PCB boards

The overall switching waveform of the physical VSCS is generally smooth. There are no significant oscillations that can be observed. In the physical experiment, parasitic inductance and capacitance play a critical role in the circuit as their values will be larger in the practical prototypes rather than in the theoretical simulation, which potentially instigate oscillations and slow the switching speed. As shown in Fig. 5, the experimental results are different from the simulation. Although 100Ω resistors are applied for R_{B1} and R_{B2}, the current overshoot is 146% that is higher than the theoretical simulation outcomes. The switching speed is measured as the gradient of the voltage dv/dt, that is 14.8 V/ns at switching off transient and 27.2 V/ns at switching off transient of the overall v_{DS_cas}. The conduction loss and the switching off energy loss of the VSCS are moderate. Oppositely, the majority of the power loss is on the switching-on transition due to the current overshoot. The magnitude of the power spike is approximately 6 kW; nevertheless, the time duration is limited to 20ns. Consequently, the switching-on energy at 2nd pulse is calculated as 0.156 mJ for one single switching action. This verifies the applicability of the VSCS in higher switching frequency implementations under a 650V DC link voltage condition.

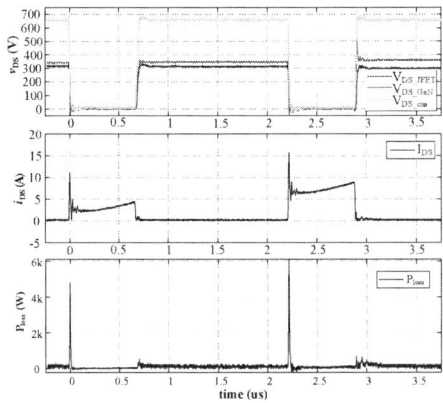

Fig. 5 Experimental results of DPT test

IV. CONCLUSION

This paper proposed a voltage-sharing cascode structure switch consisting of a high-side JFET and a low-side GaN HEMT and implemented it in a DPT in LTspice software and physical experiment. The switching speed in terms of the measured dv/dt shows that the main advantages of the SiC and GaN devices are maintained in this proposed VSCS. The simulation optimization provides a comparison of the approaches to stabilize the switching behavior, and the experimental results verify the implementation of the proposed structure in high switching speed applications.

REFERENCES

[1] R. Heckman and C. Quinn, "The Growth of Wide Bandgap Technology in Industrial Power Conversion," in IEEE Power Electronics Magazine, vol. 11, no. 4, pp. 54-60, Dec. 2024.

[2] L. Spaziani and L. Lu, "Silicon, GaN and SiC: There's room for all: An application space overview of device considerations," 2018 IEEE 30th ISPSD, Chicago, IL, USA, 2018, pp. 8-11.

[3] Y. Wang, J. Wei, Z. Zheng and K. J. Chen, "Dv/Dt-Control of 1200-V Normally-off SiC-JFET/GaN-HEMT Cascode Device," in IEEE

TPEL, vol. 36, no. 3, pp. 3312-3322, March 2021.

[4] A. Hegde, Y. Long and J. Kitchen, "A comparison of GaN-based power stages for high-switching speed medium-power converters," 2017 IEEE 5th WiPDA, Albuquerque, NM, USA, 2017, pp. 213-219.

[5] X. Xu et al., "Performance Evaluation of SiC MOSFET/BJT/Schottky Diode in A 1MHz Single Phase PFC," APEC 07 - Twenty-Second Annual IEEE APEC, Anaheim, CA, USA, 2007, pp. 1268-1272

[6] X. Wang, B. Duan, X. Yang and Y. Yang, "Novel Power MOSFET With Partial SiC/Si Heterojunction to Improve Breakdown Voltage by Breakdown Point Transfer (BPT) Terminal Technology," in IEEE J-EDS, vol. 8, pp. 559-564, 2020.

[7] S. Buonomo et al., "Series connection of power switches in high input voltage with wide range power supply for gate driving application," 2011 IEEE ECCE, Phoenix, AZ, USA, 2011, pp. 2985-2992.

[8] C. Li, S. Chen, W. Li, H. Yang and X. He, "An Active Voltage Balancing Method for Series Connection of SiC MOSFETs With Coupling Inductor," in IEEE TPEL, vol. 36, no. 9, pp. 9731-9736, Sept. 2021.

[9] W. Xu and A. Q. Huang, "15kV/50A SiC AC Switch Based On Series Connection of 1.7kV MOSFETs," 2022 IEEE ECCE, Detroit, MI, USA, 2022, pp. 1-6.

[10] A. R. Alonso, M. F. Díaz, D. G. Lamar, M. A. P. de Azpeitia, M. M. Hernando and J. Sebastián, "Switching Performance Comparison of the SiC JFET and SiC JFET/Si MOSFET Cascode Configuration," in IEEE TPEL, vol. 29, no. 5, pp. 2428-2440, May 2014.

[11] S. S. Alharbi, S. S. Alharbi and M. Matin, "Transformer-less High-Gain DC-DC Converter with SiC Cascode JFETs for High-Voltage Applications," 2018 IEEE IPMHVC, Jackson, WY, USA, 2018, pp. 37-41.

[12] P. Killeen, A. N. Ghule and D. C. Ludois, "Silicon Carbide JFET Super-Cascodes for Normally-On Current Source Inverter Switches in Medium Voltage Variable Speed Electrostatic Drives," 2019 IEEE ECCE, Baltimore, MD, USA, 2019, pp. 4004-4011.

[13] X. Song, A. Q. Huang, L. Zhang, P. Liu and X. Ni, "15kV/40A FREEDM super-cascode: A cost effective SiC high voltage and high frequency power switch," 2016 IEEE ECCE, Milwaukee, WI, USA, 2016, pp.1-8.

[14] S. Jimenez, A. Lemmon and A. Boutry, "Design and Evaluation of a 6.5 kV, 400 A Super-Cascode Power Module," in IEEE OJPEL, vol. 5, pp. 171-185, 2024.

[15] P. Xue and F. Iannuzzo, "Self-Sustained Turn-OFF Oscillation of Cascode GaN HEMTs: Occurrence Mechanism, Instability Analysis, and Oscillation Suppression," in IEEE TPEL, vol. 37, no. 5, pp. 5491-5500, May 2022.

[16] Y. Fang, C. N. Man Ho and J. Liu, "Damping Effect of Internal Gate Resistance for Cascode GaN HEMT," 2024 IEEE ECCE, Phoenix, AZ, USA, 2024, pp. 6792-6797.

[17] X. She, P. Losee, H. Hu, W. Earls and R. Datta, "Performance Evaluation of 1.5 kV Solar Inverter With 2.5 kV Silicon Carbide mosfet," in IEEE Transactions on Industry Applications, vol. 55, no. 6, pp. 7726-7735, Nov.-Dec. 2019.

[18] X. Song, A. Q. Huang, L. Zhang, P. Liu and X. Ni, "15kV/40A FREEDM super-cascode: A cost effective SiC high voltage and high frequency power switch," 2016 IEEE ECCE, Milwaukee, WI, USA, 2016, pp.1-8.

2025 IEEE Workshop on Wide Bandgap Power Devices and Applications in Asia (WiPDA Asia)

CLLLC Resonant Converter with Full-Bridge Half-Bridge Switching for Improved Voltage Gain Range and Light-Load Efficiency

Jinfeng Yu
School of Mechanical Engineering and Automation
Harbin Institute of Technology, Shenzhen
Shenzhen, China
210330330@stu.hit.edu.cn

Sinuo Liu
School of Mechanical Engineering and Automation
Harbin Institute of Technology, Shenzhen
Shenzhen, China
210330311@stu.hit.edu.cn

Yu Gu*
Education Center of Experiments and Innovations
Harbin Institute of Technology, Shenzhen
Shenzhen, China
guyu2020@hit.edu.cn

Hui Liu
Flexible Control Group of the R&D Center
CYG SUNRI
Shenzhen, China
liuhui2@cyg.com

Abstract—In this paper, a 1 kW *CLLLC* resonant converter with 400 V DC bus and 36-60 V battery voltage is designed, achieving wide voltage gain range through hybrid pulse frequency modulation (PFM) and phase-shift (PS) control, as well as full-bridge and half-bridge configurations. This paper briefly introduces the design methodology of *CLLLC* based on FHA and then analyzes the improvement in light-load efficiency provided by the HB-HB mode. Finally, the prototype with SiC MOSFETs on the high-voltage side is built and the experimental results validate the theoretical analysis.

Keywords—CLLLC resonant converter; wide voltage range; light-load efficiency

I. INTRODUCTION

With the increasing integration of renewable energy sources such as solar and wind into the power grid, the volatility of their output has become a significant challenge. Energy storage systems (ESS) are crucial in addressing this issue by stabilizing the grid and providing backup power [1]. Bidirectional isolated converters with a wide voltage range play a crucial role in energy storage applications by handling significant voltage variations caused by the charging and discharging processes. The *CLLLC* resonant converter is a promising solution due to its ability to achieve zero-voltage switching (ZVS) and zero-current switching (ZCS), which reduces switching losses and electromagnetic interference (EMI).

To expand the voltage range of the *CLLLC* resonant converter, it is typically necessary to reduce the magnetizing inductance (L_m), which can result in larger resonant currents and reduce system efficiency. A potential solution is to operate the primary-side inverter bridge in half-bridge (HB) mode instead of full-bridge (FB) [2], as shown in Fig. 1(c).

This work was support partly by Guangdong Province 2024 Educational Science Planning Project (Higher Education Special) under Grant 2024GXJK399, and partly by 2024 Harbin Institute of Technology, Shenzhen Campus Quality Engineering Project (Industry-Education Integration Practical Teaching Base) under Grant HITSZUQP24004..

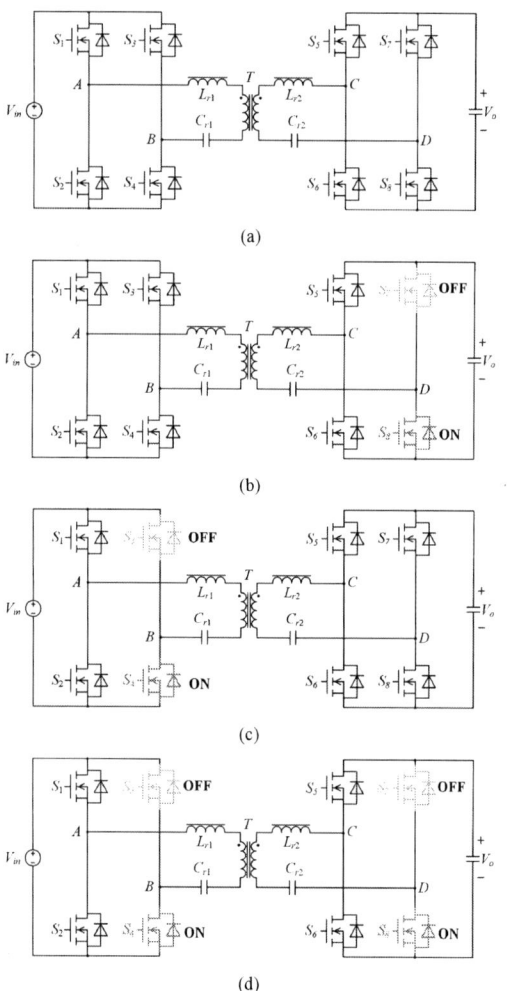

Fig. 1. Topology of *CLLLC* resonant converter: (a)FB-FB; (b)FB-HB; (c) HB-FB; (d)HB-HB.

979-8-3315-1110-4/25 $31.00 © 2025 IEEE

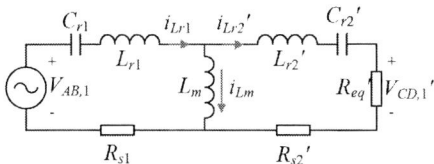

Fig. 2.   FHA model of *CLLLC* resonant converter.

This approach reduces the effective input voltage of the resonant tank $V_{AB,1}$ by half, enabling a 0.5x gain output. However, this method cannot double the gain during reverse operation.

To address this, a solution proposed in [3] involves switching the secondary-side full-bridge to half-bridge to achieve double voltage rectification. Furthermore, [4] expands the voltage range by implementing full-bridge and half-bridge configurations on both the primary and secondary sides, resulting in four operational modes—FB-FB, FB-HB, HB-FB, and HB-HB—as shown in Fig. 1.

While most converters are designed for peak efficiency near full load, they tend to suffer from relatively poor efficiency under light-load conditions, which is also a concern in energy storage applications. This paper analyzes and verifies the impact of the HB-HB mode on efficiency, demonstrating its potential for achieving higher efficiency under light-load conditions.

II. DESIGN METHODOLOGY

A. Resonant Parameters Design

Fig. 1(a) illustrates the typical topology of a *CLLLC* resonant converter operating at FB-FB mode. Neglecting higher-order harmonics, the FHA model can be obtained, as shown in Fig. 2.

The gain equation can be derived as a function of the operating frequency, resonant components, and load conditions:

$$M_{PFM} = \frac{nV_0}{V_{in}} = \frac{V_{CD,1}'}{V_{AB,1}} = \frac{\left[Z_m // \left(Z_{rs}' + R_{eq}'\right)\right]R_{eq}'}{\left(Z_{rp} + \left[Z_m // \left(Z_{rs}' + R_{eq}'\right)\right]\right)\left(Z_{rs}' + R_{eq}'\right)} \quad (1)$$

The equivalent load resistance, denoted as R_{eq}', can be expressed as:

$$R_{eq}' = n^2 \frac{V_{CD,1}}{I_{Lr2}} = n^2 \frac{\frac{2\sqrt{2}}{\pi}V_2}{\frac{\pi}{2\sqrt{2}}I_o} = n^2 \frac{8}{\pi^2} R_L \quad (2)$$

The design of the resonant parameters in a *CLLLC* converter is crucial for achieving the desired gain range, ensuring zero-voltage switching (ZVS), and maximizing efficiency. The design procedure has been thoroughly researched in [5] and can be summarized as follows:

Selection of Magnetizing Inductance L_m: The magnetizing inductance should be chosen appropriately to ensure ZVS operation:

$$L_m \leq \frac{t_{dead}}{8C_{ossmax}f} \quad (3)$$

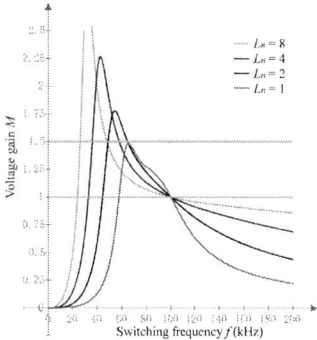

Fig. 3.   Gain curves for different L_n under full load.

In principle, L_m should be as large as possible, since the RMS value of the resonant current I_{r1} and the turn-off current of switches Q_1-Q_4 are inversely related to L_m (derivation provided later). However, increasing L_m narrows the achievable gain range, so a trade-off between efficiency and gain range must be considered when selecting L_m.

Selection of Inductance Ratio L_n: The parameter L_n is defined as:

$$L_n = \frac{L_m}{L_{r1}} \quad (4)$$

By substituting L_m into (1), a set of gain curves corresponding to different values of L_n and R_L can be plotted, as shown in Fig. 3. These curves help in selecting an appropriate L_n for desired gain range.

Once L_n and L_m are determined, the resonant capacitance C_{r1} can be calculated using the following relationships:

$$C_{r1} = \frac{1}{\left(2\pi f_r\right)^2 L_{r1}} \quad (5)$$

A symmetric resonant tank is designed to achieve a symmetric gain characteristic, so C_{r2} and L_{r2} can be calculated by:

$$L_{r1} = L_{r2}' = n^2 L_{r2}$$
$$1/C_{r1} = 1/C_{r2}' = n^2/C_{r2} \quad (6)$$

B. Expanding Voltage Gain Range via Hybrid Modulation and Bridge Configurations

In PFM mode, when $f_s > f_r$, the gain curves become flat, as shown in Fig. 3, making it difficult to achieve the desired minimum gain, especially under light load conditions. Therefore, in the region where $M_{PFM} < 1$, PS or PWM can be employed to extend the gain range. This hybrid control strategy also allows for the selection of larger L_n and L_m, thereby improving efficiency.

Fig. 4 shows the timing diagram of the *CLLLC* resonant converter operating in phase-shift control mode, where φ denotes the phase-shift angle. The introduction of φ reduces the duty cycle D of V_{AB}, thereby lowering its fundamental component $V_{AB,1}$. By Fourier analysis, the fundamental component is given by:

$$V_{AB,1} = \frac{2\sqrt{1-\cos\left(D\pi\right)}V_{in}}{\pi} \quad (7)$$

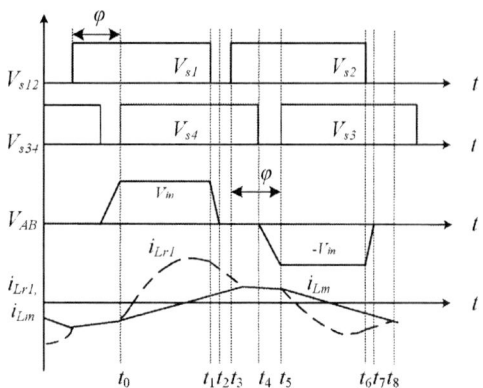

Fig. 4. Timing diagram of *CLLLC* converter under phase-shift control.

where $D = (\pi-\varphi)/\pi,\ 0 < D < 1$.

Accordingly, the voltage gain under phase-shift control is:

$$M_{\mathrm{PS}} = \sqrt{\frac{1-\cos\left(D\pi\right)}{2}}M_{\mathrm{PFM}} \qquad (8)$$

When operating at the resonant frequency, where M_{PFM} equal to 1, the gain simplifies to:

$$M_{\mathrm{PS}} = \sqrt{\frac{1-\cos\left(D\pi\right)}{2}} \qquad (9)$$

The gain range can be further extended by switching between full-bridge and half-bridge configurations, constituting 4 operation modes—FB-FB, FB-HB, HB-FB, and HB-HB—as shown in Fig. 1. Define parameters λ_{pri} and λ_{sec} as the full-bridge or half-bridge configurations on the primary and secondary sides, respectively:

$$\lambda_{\mathrm{pri}} = \begin{cases} 1, & \text{Full-Bridge} \\ 0.5, & \text{Half-Bridge} \end{cases} \qquad (10)$$

$$\lambda_{\mathrm{sec}} = \begin{cases} 1, & \text{Full-Bridge} \\ 0.5, & \text{Half-Bridge} \end{cases} \qquad (11)$$

And the gain equation of the 4 modes should be revised as:

$$M = \frac{\lambda_{\mathrm{sec}}nV_0}{\lambda_{\mathrm{pn}}V_{\mathrm{in}}} \qquad (12)$$

As shown in Fig. 5, by switching between full-bridge and half-bridge configurations on the primary and secondary

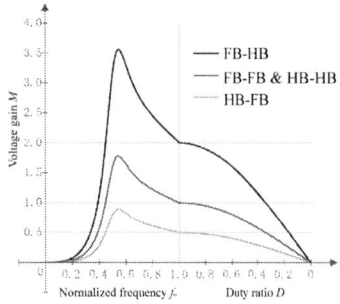

Fig. 5. Voltage gain range of the *CLLLC* converter in four operating modes with PFM+PS hybrid control.

sides, 4 operation modes (FB-FB, FB-HB, HB-FB, HB-HB) enable a wide voltage gain range without the need for any additional components.

The transition between these operating modes has been thoroughly investigated in [4]; therefore, it is not further elaborated in this paper. Instead, this work focuses on the analysis of light-load efficiency improvement enabled by FB/HB reconfiguration.

III. EFFICIENCY ANALYSIS OF *CLLLC* RESONANT CONVERTER

Similar to the approach in [6], the resonant current on the primary side, $i_{Lr1}(t)$, can be expressed as:

$$i_{Lr1}(t) = \sqrt{2}I_{Lr1}\sin(\omega_s t + \phi) \qquad (13)$$

The relationship between the currents $i_{Lr1}(t)$, $i_{Lm}(t)$ and $i_{Lr2}(t)$ is as follows:

$$i_{Lr1}(t) = i_{Lm}(t) + \frac{1}{n}i_{Lr2}(t) \qquad (14)$$

During the sufficiently short dead time, the magnetizing current i_{Lm} is approximately equal to the resonant current i_{Lr1}, leading to the relationship between the peak magnetizing current I_{Lm_peak} and I_{Lr1}:

$$I_{Lm\_peak} = \frac{V_{in}\lambda_{\mathrm{pri}}}{4L_m f_s} = i_{Lr1}(0) = \sqrt{2}I_{Lr1}\sin\phi \qquad (15)$$

The average output current I_o can be determined as:

$$I_o = \frac{V_o'}{R_L} = \frac{1}{T_s}\int_0^{T_s}\left|\lambda_{\mathrm{sec}}i_{Lr2}(t)\right|dt = \frac{2\lambda_{\mathrm{sec}}}{T_s}\int_0^{\frac{T_s}{2}}i_{Lr2}(t)dt$$
$$= \frac{2n\lambda_{\mathrm{sec}}}{T_s}\int_0^{\frac{T_s}{2}}i_{Lr1}(t) - i_{Lm}(t)dt = \frac{2\sqrt{2}n\lambda_{\mathrm{sec}}}{\pi}I_{Lr1}\cos\phi \qquad (16)$$

The RMS value of the secondary resonant current I_{Lr2} is proportional to the average output current I_o. In the half-bridge rectifier configuration, the resonant current is approximately twice that in the full-bridge rectifier configuration:

$$I_{Lr2} = \frac{\pi}{2\sqrt{2}\lambda_{\mathrm{sec}}}I_o = \frac{\pi V_o}{2\sqrt{2}R_L\lambda_{\mathrm{sec}}} \qquad (17)$$

From equations (15), (16) and (17), we can derive:

$$I_{Lr1}^2 = \frac{V_o^2\lambda_{\mathrm{sec}}^2}{8}\left[\left(\frac{\pi}{nR_L\lambda_{\mathrm{sec}}^2}\right)^2 + \left(\frac{nT_s}{2L_m}\right)^2\right]$$
$$= \frac{V_{in}^2\lambda_{\mathrm{pri}}^2}{8}\left[\left(\frac{\pi}{n^2R_L\lambda_{\mathrm{sec}}^2}\right)^2 + \left(\frac{T_s}{2L_m}\right)^2\right] \qquad (18)$$
$$= \frac{I_{Lr2}^2}{n^2} + \frac{I_{Lm\_peak}^2}{2}$$

The efficiency of the *CLLLC* resonant converter can be roughly calculated as in [7]:

$$\eta = \frac{P_o}{P_o + I_{Lr1}^2 R_{s1} + I_{Lr2}^2 R_{s2} + P_{cond} + P_{sw}} \qquad (19)$$

According to (14) and (18), we find that I_{Lr1} consists of two relatively independent components: one is related to the load and one is related to magnetization, which depend on the secondary side and the primary side of the converter, respectively. The load-related component is influenced by the load R_L (or I_{Lr2}) according to (17), while the magnetization-related component is determined by the primary-side magnetizing current I_{Lm_peak}. From (15), we know that I_{Lm_peak} is proportional to $V_{in}\lambda_{pri}$ (or V_{AB}), inversely proportional to the magnetizing inductance L_m and the operating frequency f_s, and independent of the load. Furthermore, I_{Lm_peak} is also the turn-off current of the switch, which is related to ZVS and turn-off loss.

At light load, secondary resonant current I_{Lr2}/n is much smaller compared to the magnetizing current I_{Lm_peak}, and thus, $i_{Lr1}(t)$ is primarily determined by $i_{Lm}(t)$. Under these conditions, power losses are dominated by primary side conduction and switching losses. In the HB-HB configuration, the midpoint voltage of the bridge arm V_{AB} changes from ±400V to 0 and 400V, causing I_{Lm_peak} to be half the value it is in the FB-FB configuration. A smaller I_{Lm_peak} leads to a smaller I_{Lr1}, thereby reducing conduction losses. Additionally, a smaller I_{Lm_peak} itself also reduces turn-off losses, improving performance at light load.

IV. EXPERIMENT

To validate the theoretical analysis, an experimental setup was developed using the parameters listed in Table I.

TABLE I.   PARAMETERS OF THE PROTOTYPE

Parameters	Values
Vbus	400 V
Vbat	36-60 V
Resonant inductor L_1	104 µH, PQ3220
Resonant inductor L_2	1.5 µH, PQ3535
Resonant capacitor C_1	24.2 nF, CC1206JKNPOZBN222
Resonant capacitor C_2	1700 nF, GRM31C5C2A104JA01L
Magnetizing inductance L_m	203.42 µH
Transformer	N = 25/3, EE55
Resonant frequency f_r	100 kHz
Switches $S_1 - S_4$	C3M0045065D, 650 V, 45 mΩ
Switches $S_5 - S_8$	IRFP4310ZPBF, 100 V, 4.8 mΩ

The photograph of the prototype is shown in Fig. 6. Experimental waveforms in Fig. 9 and Fig. 10 demonstrate that HB-HB configuration can reduce the magnetizing current while increasing the resonant current on the secondary side. The measured efficiency curves in Fig. 7 validate that the HB-

Fig. 6.   1kW *CLLLC* prototype.

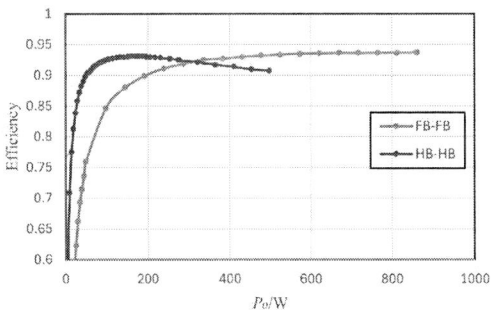

Fig. 7.   Measured efficiency curves of FB-FB mode and HB-HB mode.

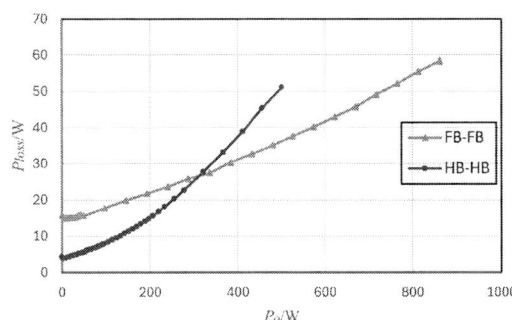

Fig. 8.   Measured efficiency curves of FB-FB mode and HB-HB mode.

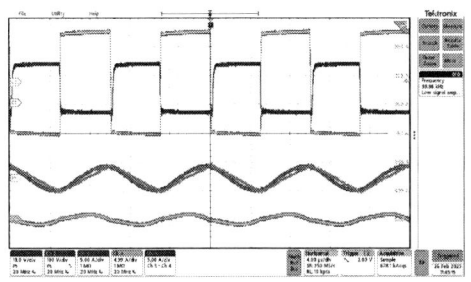

(a) HB-HB mode, $P_o = 100$W

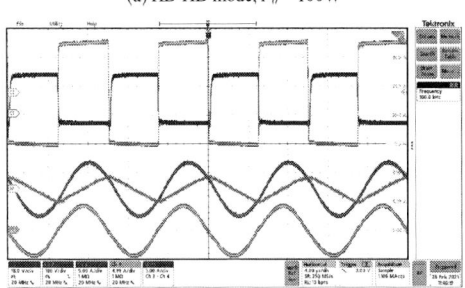

(b) HB-HB mode, $P_o = 500$W

Fig. 9.   Experimental waveforms of *CLLLC* converter – Ch1: v_{gs}; Ch2: v_{ds}; Ch3: i_{Lr1}; Ch4: $i_{Lr2}' = i_{Lr2}/n$; Math: $i_{Lm} = i_{Lr1} - i_{Lr2}'$.

HB operation mode effectively enhances light-load efficiency. Furthermore, the power loss at light load in HB-HB mode is reduced by more than 50%, as can be observed from Fig. 8.

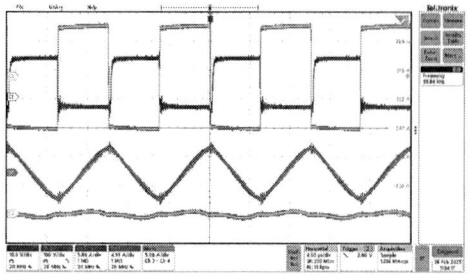

(a) FB-FB mode, P_o = 100W

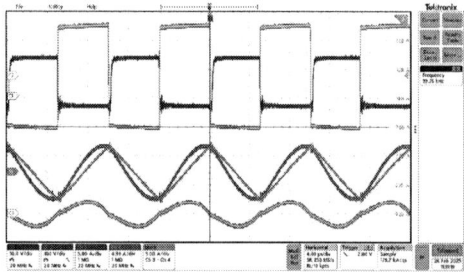

(b) FB-FB mode, P_o = 500W

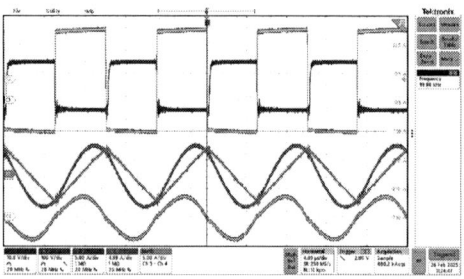

(c) FB-FB mode, P_o = 900W

Fig. 10. Experimental waveforms of *CLLLC* converter –Ch1: v_{gs}; Ch2: v_{ds}; Ch3: i_{Lr1}; Ch4: $i_{Lr2}' = i_{Lr2} / n$; Math: $i_{Lm} = i_{Lr1} - i_{Lr2}'$.

V. CONCLUSION

Apart from operating in FB-FB mode, *CLLLC* resonant converter can also operate in FB-HB, HB-FB, and HB-HB

modes, making more efficient use of the full-bridge on both primary and secondary side. FB-HB and HB-FB modes extend the voltage gain range, while HB-HB mode improves the light-load efficiency. These operation modes do not require any additional components, making them easily applicable to existing designs. However, it should be noted that the half-bridge configuration may result in higher resonant currents and DC bias voltages across the resonant capacitors, which could cause the voltages of the resonant capacitors to exceed the rated value. Therefore, the design should be checked before applying FB-HB and HB-FB modes, as they may exceed the rated voltage and current, especially under heavy load. The HB-HB mode is safer, as it operates only at light load, reducing the risk of exceeding the ratings.

REFERENCES

[1] T. S. Babu, K. R. Vasudevan, V. K. Ramachandaramurthy, S. B. Sani, S. Chemud, and R. M. Lajim, "A comprehensive review of hybrid energy storage systems: Converter topologies, control strategies and future prospects," *IEEE Access*, vol. 8, pp. 148702–148721, 2020.

[2] Z. Liang, R. Guo, G. Wang, and A. Huang, "A new wide input range high efficiency photovoltaic inverter," in *2010 IEEE Energy Conversion Congress and Exposition*, Atlanta, GA: IEEE, Sep. 2010, pp. 2937–2943.

[3] S. Zong, G. Fan, and X. Yang, "Double voltage rectification modulation for bidirectional DC/DC resonant converters for wide voltage range operation," *IEEE Trans. Power Electron.*, vol. 34, no. 7, pp. 6510–6521, Jul. 2019.

[4] Z. Xiao, F. Deng, Z. Yao, and Y. Tang, "Soft-morphing: Unlocking the potential of *CLLC* converters for ultrawide range applications," *IEEE Trans. Power Electron.*, vol. 40, no. 1, pp. 1289–1304, Jan. 2025.

[5] Z. U. Zahid, Z. M. Dalala, R. Chen, B. Chen, and J.-S. Lai, "Design of bidirectional DC–DC resonant converter for vehicle-to-grid (V2G) applications," *IEEE Trans. Transp. Electrification*, vol. 1, no. 3, pp. 232–244, Oct. 2015.

[6] J.-H. Jung, H.-S. Kim, M.-H. Ryu, and J.-W. Baek, "Design methodology of bidirectional CLLC resonant converter for high-frequency isolation of DC distribution systems," *IEEE Trans. Power Electron.*, vol. 28, no. 4, pp. 1741–1755, 2012.

[7] S. S. Queiroz and L. F. Costa, "Investigation of the influence of the dead-time on the performance of an LLC resonant converter for high-power application," in *2024 IEEE 15th International Symposium on Power Electronics for Distributed Generation Systems (PEDG)*, Luxembourg, Luxembourg: IEEE, Jun. 2024, pp. 1–6.

Analysis and Prediction Method of Gate Voltage Peak in Switching Transient of GaN E-HEMT

Yushan Liu, Yirui Hu, Xiao Li
School of Automation Science and Electrical Engineering
Beihang University
Beijing, China
yushanliu@ieee.org

Abstract—GaN HEMTs are being widely used due to their fast switching characteristics, which also cause undesirable switching oscillations and lead to voltage and current peaks related to their overshoots in fast switching operation. Gate voltage peak is a particularly severe issue in GaN devices for their narrow safe drive voltage range. This paper analyzes the synthesis of gate voltage and power oscillation as a part of interference source. Then the influence of various parameters and operating conditions are considered and a method for quantitative prediction of gate voltage peak is proposed.

Keywords—GaN E-HEMTs, voltage peak, switching oscillation, analytical model

I. INTRODUCTION

Wide band gap semiconductor devices have the advantages of high frequency, high efficiency and low threshold voltage, while fast switching also make them more susceptible to switching oscillation [1],[2], which have serious negative impacts such as overshoots [3], additional power losses [4], electromagnetic interference (EMI) noise [5] and even shoot-through faults [6] in half-bridge topologies.

For gate voltage, research efforts have been mainly devoted to crosstalk of synchronous switch [7],[8] or gate voltage of control switch concerning false turn-on and gate stability [9],[10]. However, for GaN devices with lower maximum allowable voltage, the turn-on voltage peak may lead to device failure. Studies have pointed out that the dv/dt and di/dt of the power circuit can affect the gate oscillation through the parasitic parameters [11], and qualitatively method is provided for suppression of gate oscillation [12], but there is still lack of a quantitative description of gate voltage for limiting its peak.

Since the gate oscillation mainly originates from the power oscillation [13], a second-order equivalent model is helpful in the research on power oscillation [14]. The method for obtaining the turn-off voltage overshoot by using a sloped excitation at the slew rate of dv/dt indicates that the voltage overshoot is physically a part of the sloped response oscillation [15], which is also useful to analyze the turn-on current.

Focusing on the gate voltage in the turn-on process of the control GaN HEMT in the DPT circuit shown in Fig.1, this article proposed a simplified model of gate circuit to analyze the compositions of gate voltage and their synthesis to distinguish the voltage peaks in various practical cases shown in Fig.2. Then power oscillation regarded as the source of gate oscillation is analyzed in detail to get a quantitative form considering two cases of turn-on process shown in Fig.3. As a result, the simplified quantitative form of gate voltage for peak prediction is proposed.

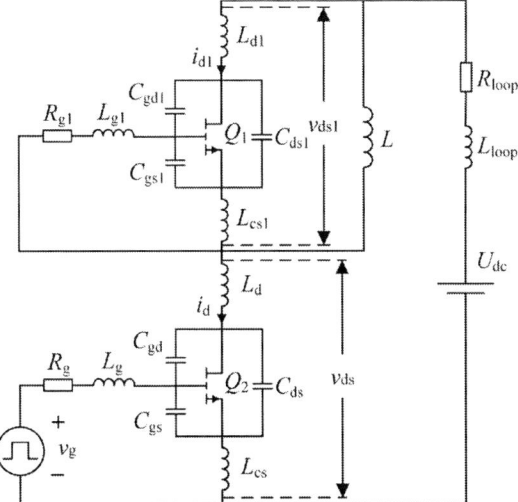

Fig. 1. DPT circuit of GaN E-HEMTs.

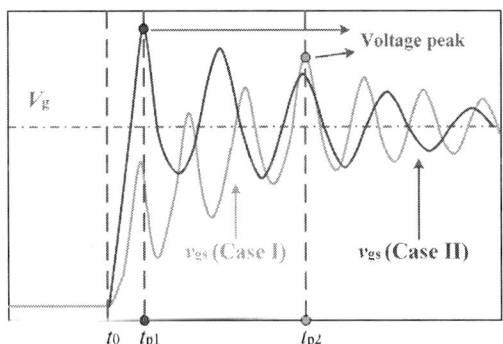

Fig. 2. Gate voltage peak in different practical cases.

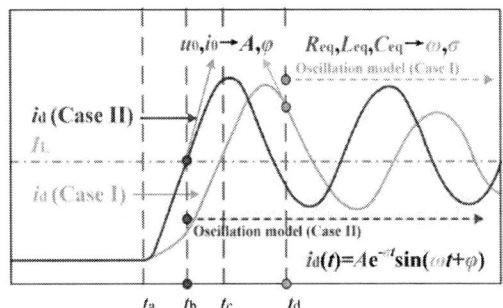

Fig. 3. Cases of drain current in turn-on process.

II. ANALYSIS OF GATE VOLTAGE COMPOSITIONS

To analysis the gate voltage of control switch Q_2, equivalent gate circuit of which is simplified from DPT circuit in Fig.1, as shown in Fig.4.

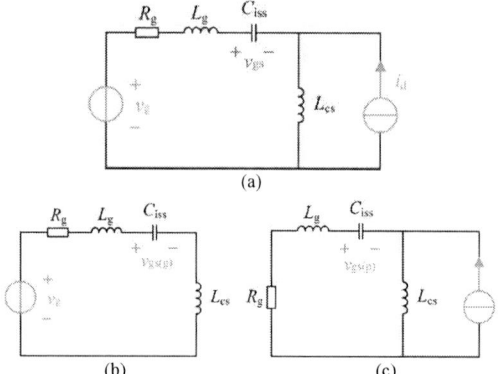

(a)

(b)   (c)

Fig. 4. Simplified equivalent circuit of gate circuit. (a)Complete circuit. (b)Decomposed circuit I. (c)Decomposed circuit II.

According to the superposition theorem, gate-source voltage v_{gs} in Fig.4(a) excited by both oscillation current i_d and drive voltage v_g can be expressed as

$$v_{gs}(t) = v_{gs(g)}(t) + v_{gs(p)}(t) \tag{1}$$

Where $v_{gs(g)}$ is the response composition of v_g as shown in Fig.4(b), and $v_{gs(p)}$ is the response composition of i_d as shown in Fig.4(c). Then the waveform of v_{gs} can be separated into two decomposed parts in two cases, as shown in Fig.5.

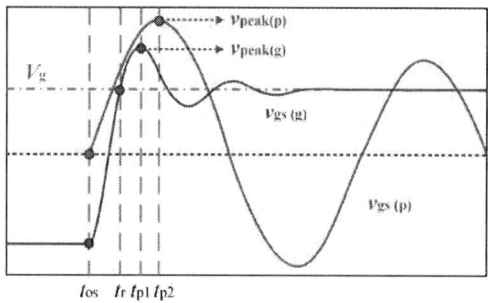

(a)

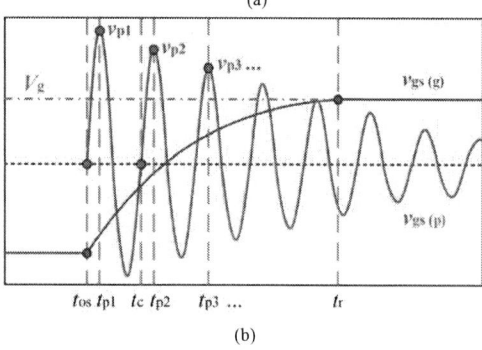

(b)

Fig. 5. Decomposed waveforms of gate voltage v_{gs}. (a)Case I. (b)Case II.

Case I that peaks appear in both decomposed waveforms is shown in Fig.5(a), which occurs when the factor $\xi < 1$, where ξ is

$$\xi = \frac{R_g}{2} \sqrt{\frac{C_{iss}}{L_g + L_{cs}}} \tag{2}$$

The synthetic result of voltage peaks $v_{peak(g)}$ and $v_{peak(p)}$ is determined by the relationship between the point in the time scale where they emerge. The worst case occurs when the two peaks appear simultaneously, the maximum voltage peak is $v_{peak(max)} = v_{peak(g)} + v_{peak(p)}$.

Case II that occurs when $\xi \geq 1$ is shown in Fig.5(b), in which there is no peak on the waveform of $v_{gs(g)}$, and the voltage peak in the synthetic result of $v_{gs(g)}$ and $v_{gs(p)}$ depends on the oscillation amplitude of $v_{gs(p)}$ and affected by the slew rate of overdamped step response $v_{gs(g)}$. The maximum voltage of v_{gs} will not exceed the value $V_g + v_{p1}$.

III. QUANTITATIVE PREDICTION OF GATE VOLTAGE PEAK

From the above discussion, the decomposed voltage part $v_{gs(p)}$, which is excited by i_d in power circuits plays a role in the synthesis of voltage peak on v_{gs} waveform. For building a quantitative relationship between $v_{gs(p)}$ and i_d, the complex frequency domain form of $v_{gs(p)}$ obtained by Laplace transform is expressed as

$$V_{gs(p)}(s) = G_p(s) \cdot I_d(s) \tag{3}$$

The frequency characteristic of G_p can be quantitatively expressed as

$$G_p(j\omega) = \frac{-j\omega L_{cs}}{1 - \omega^2 (L_g + L_{cs})C_{iss} + j\omega R_g C_{iss}} \tag{4}$$

As the excitation source in (3), i_d exhibits two cases on the characteristic of its waveform shown in Fig.3 in different operating conditions.

The voltage drop $\triangle U$ of v_{ds} before i_d reached I_L can be expressed as

$$\triangle U = \frac{g_m L_{eq}(v_g - U_{th})}{R_g C_{iss} + g_m L_{cs}} \tag{5}$$

And the judgment condition is

$$U_{dc} + V_r - \triangle U - \frac{I_L}{g_m} \geq 0 \tag{6}$$

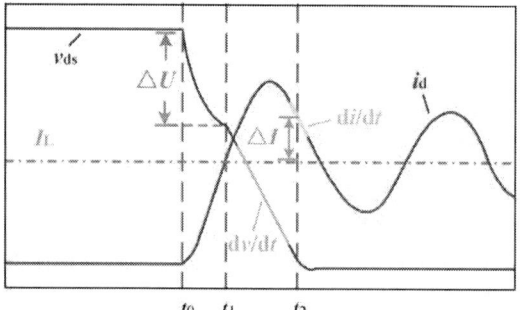

Fig. 6. Waveforms in power circuit: Case I.

If (6) is satisfied, i_d in the turn-on process will performs as case I, as shown in Fig.6. In this case, parameters of i_d can be calculated as

$$
\begin{cases}
\sigma = \xi_{on}\omega_{on} \\
\omega = \omega_{on}\sqrt{1-\xi_{on}^2} \\
A = -\sqrt{i_0^2 + \left(\dfrac{u_0 + \sigma L_{eq(on)}i_0}{\omega L_{eq(on)}}\right)^2} \\
\varphi = -arctan\left(\dfrac{\omega L_{eq(on)}i_0}{u_0 + \sigma L_{eq(on)}i_0}\right)
\end{cases}
\tag{7}
$$

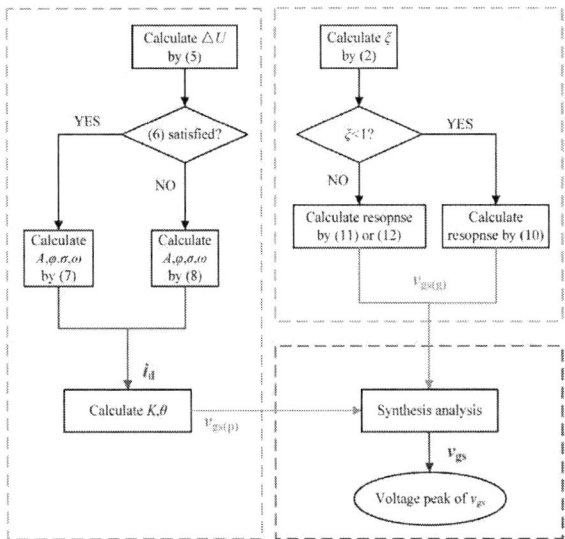

Fig. 7. Waveforms in power circuit: Case II.

If (6) is not satisfied, i_d in the turn-on process will performs as case II, as shown in Fig.7. In this case, parameters of i_d can be calculated as

$$
\begin{cases}
A = \dfrac{R_{loop}^* I_L - U_{dc} - V_r}{\omega L_{eq(on)}} \\
\varphi = 0
\end{cases}
\tag{8}
$$

Then the quantitative form of $v_{gs(p)}$ is

$$
v_{gs(p)}(t) = K \cdot A e^{-\sigma t} sin(\omega t + \varphi + \theta) \tag{9}
$$

where the ratio K and relative phase θ are derived from (5).

The quantitative form of $v_{gs(g)}$ in the cases that $\xi<1$, $\xi=1$ and $\xi>1$ can be expressed separately as

$$
v_{gs(g)}(t) = \frac{-V}{\sqrt{1-\xi^2}}e^{-\xi\omega_n t}
$$

$$
\cdot sin\left[\omega_n\sqrt{1-\xi^2}t + \beta\right] + V \tag{10}
$$

$$
v_{gs2(g)}(t) = -V e^{-\omega_n t} - \omega_n V t e^{-\omega_n t} + V \tag{11}
$$

$$
v_{gs(g)}(t) = \frac{\xi\omega_n - \omega_n\sqrt{\xi^2-1}}{2\omega_n\sqrt{\xi^2-1}}V e^{-(\xi\omega_n + \omega_n\sqrt{\xi^2-1})t}
$$

$$
- \frac{\xi\omega_n + \omega_n\sqrt{\xi^2-1}}{2\omega_n\sqrt{\xi^2-1}}V e^{-(\xi\omega_n - \omega_n\sqrt{\xi^2-1})t} + V \tag{12}
$$

From (10), voltage peak $v_{peak(g)}$ in Fig.5(a) can be calculated as

$$
v_{peak(g)} = V e^{\frac{-\xi\pi}{\sqrt{1-\xi^2}}} + V_g \tag{13}
$$

As a result, complete quantitative form of v_{gs} can be calculated by combining (1) (9) and (10), (11) or (12), then its voltage peak can be predicted considering synthesis type of compositions revealed in its time domain form. The process of proposed peak voltage prediction method is shown in Fig.8.

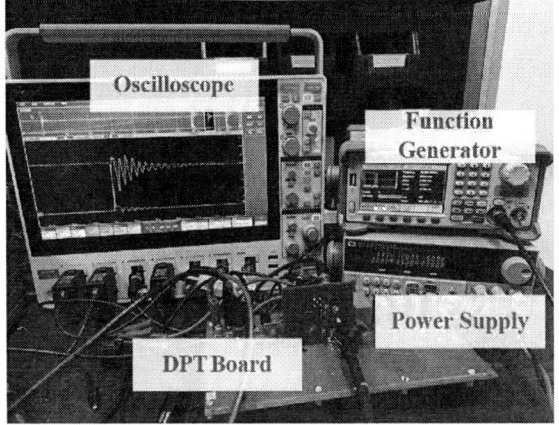

Fig. 8. Process of voltage peak prediction.

IV. VERIFICATION RESULTS

The Double Pulse Test plat is constructed as depicted in Fig.9. The DC-link voltage is generated by DC power supply IT6302. Voltage v_{gs} and v_{ds} are measured by TPP1000 probes and current i_d is acquired from a coaxial shunt. The utilized switching GaN HEMTs are EPC2065.

Waveforms of voltage v_{gs}, v_{ds} and current i_d during turn-on process in several operating conditions are measured as shown in Fig. 10.

Fig. 9. Experimental platform for DPT.

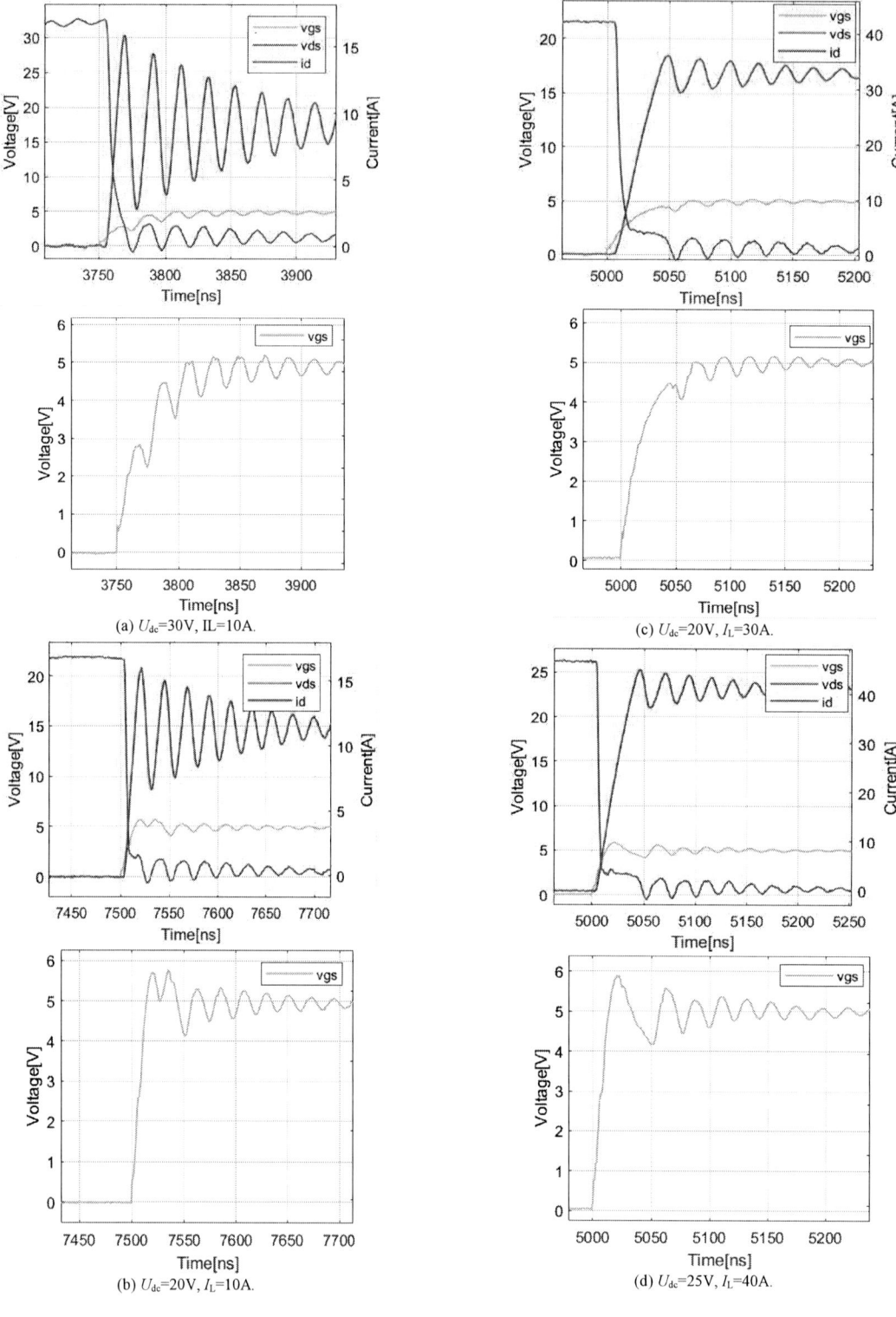

(a) U_{dc}=30V, IL=10A.

(b) U_{dc}=20V, I_L=10A.

(c) U_{dc}=20V, I_L=30A.

(d) U_{dc}=25V, I_L=40A.

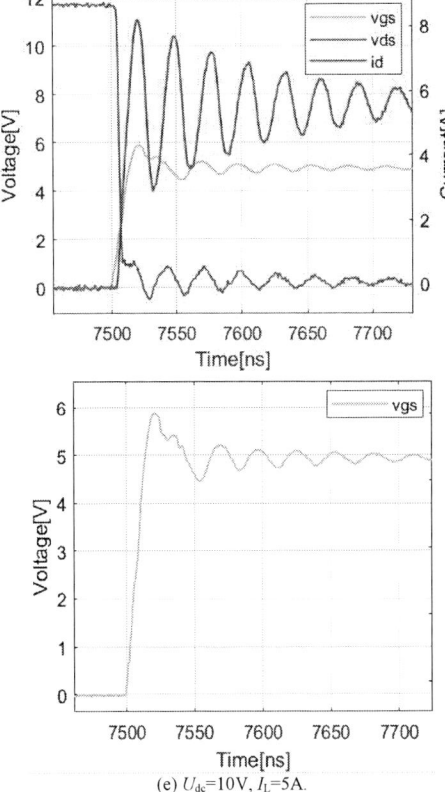

(e) U_{dc}=10V, I_L=5A.

Fig. 10. Experimental waveforms of conditions under test.

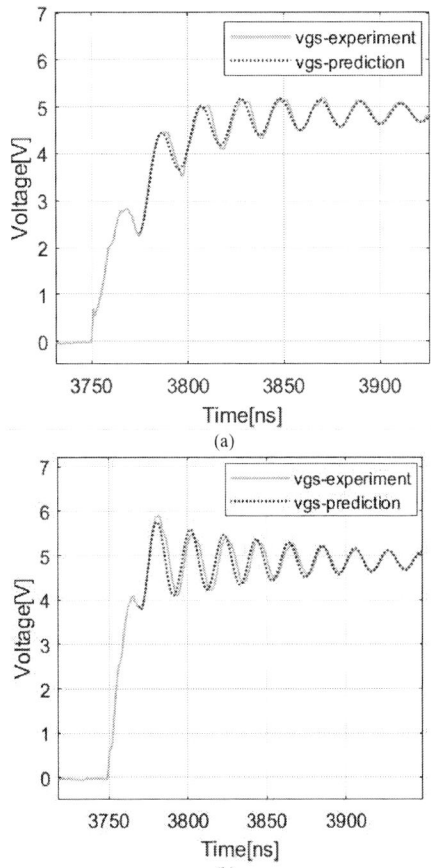

(a)

(b)

Figure 11. Experimental and prediction waveforms of i_d: (a)Case I. (b)Case II.

For i_d, subject to the judgement formula (6), Case I occurs at the high voltage, low current and slow switching conditions, such as conditions shown in Fig. 10(a)(c)(d), while Case II occurs at the relatively low voltage, high current and fast switching conditions like conditions shown in Fig. 10(b)(e). For v_{gs}, as parameters except R_g in (2) have been determined before the DPT plat was settled up, the only factor that affects the cases of v_{gs} is R_g. v_{gs} in slow switching conditions with big R_g shows the characteristic of Case II as shown in Fig. 10(a)(c). In these conditions, voltage peak is obviously lower than other conditions where v_{gs} is in Case I. The synthesis of two voltage peaks in Case I also display differently in different conditions. v_{gs} in the condition shown in Fig. 10(b) and the condition shown in Fig. 10(e) show the synthetic type that two voltage peaks of power composition and gate composition are combined, while two voltage peaks appear separately in the condition shown in Fig. 10(c), it can be explained that the power composition occurs much later than the gate composition in conditions with low di/dt or high load current I_L, where the maximum voltage peak is probably determined by the gate composition or the power composition singly.

The comparison between experimental and predicted waveforms of v_{gs} are shown in Fig.11. For v_{gs} in two typical cases, predicted waveforms obtained by the process in Fig.8 fit experimental waveforms well.

More results of experimental and predicted voltage peak of v_{gs} and comparison between them are shown in table I. Verification shows that prediction result of the proposed method is reliable.

TABLE I.   VOLTAGE PEAK PREDICTION RESULT

Operating Condition	Prediction	Experiment	Deviation
U_{dc}=30V I_L=10A	5.7549V	5.9155V	2.79%
U_{dc}=20V I_L=10A	5.7786V	5.6851V	1.64%
U_{dc}=20V I_L=30A	5.1776V	5.1486V	0.56%
U_{dc}=25V I_L=40A	5.9312V	5.8964V	0.59%
U_{dc}=10V I_L=5A	5.9153V	5.8733V	0.72%

V. CONCLUSION

For quantitatively predicting the voltage peak of gate voltage, this paper analyzed the compositions of v_{gs} and their synthetic mechanism to get a quantitative form in time domain. And during the analysis process, drain current i_d serving as the oscillation excitation source was critically analyzed due to its significant role in the synthesis of v_{gs}. Finally, a prediction method capable of providing a quantitative voltage peak value is obtained. Experimental verification has been conducted and the feasibility of the prediction method is confirmed.

ACKNOWLEDGMENT

This work was supported in part by the National Natural Science Foundation of China under Grant 52407193, and in part by Chunhui Project Foundation of the Education Department of China under Grant 202200504.

REFERENCES

[1] Tianjiao Liu, Runtao Ning, Thomas T. Y. Wong and Z. John Shen,"A Survey on Switching Oscillations in Power Converters," in *IEEE Journal of Emerging and Selected Topics in Power Electronics*, vol.8, no.1, pp.893-908, March 2020.

[2] Jian Chen, Quanming Luo, Jian Huang, Qingqing He, Pengju Sun, Xiong Du, Pengju Sun and Lin Zhou,"A Review of Switching Oscillations of Wide Bandgap Semiconductor Devices,"in *IEEE Transactions on Power Electronics*, vol.35, no.12, pp.13182–13199, December 2020.

[3] H.-T. Tang, H. S.-H. Chung, and K. J. Chen, "Adaptive level-shift gate driver with indirect gate oxide health monitoring for suppressing crosstalk of SiC MOSFETs,"in *IEEE Transactions on Power Electronics*, vol.38, no.8, pp.10196–10212, August 2023.

[4] T. Takahashi, T. Takehisa, J. Furuta, M. Shintani, and K. Kobayashi, "A three-level GaN driver for high false turn-on tolerance with minimal reverse conduction loss," in *IEEE Open J. Power Electron.*, vol. 4, pp. 357–366, 2023.

[5] J. Wang and H. Shu-Hung Chung, "Impact of parasitic elements on the spurious triggering pulse in synchronous buck converter," in *IEEE Transactions on Power Electronics*, vol.29, no.12, pp.6672–6685, December 2014.

[6] X. Wang et al., "High-frequency three-level gate driver for GaN HEMT bridge crosstalk suppression,"in *IEEE Transactions on Power Electronics*, vol. 39, no. 1, pp. 1343–1352, January 2024.

[7] Yushan Liu, Xuyang Liu, Xiao Li and Haiwen Yuan,"Analytical Model and Safe-Operation-Area Analysis of Bridge-Leg Crosstalk of GaN E-HEMT Considering Correlation Effect of Multi-Parameters,"in *IEEE Transactions on Power Electronics*, vol.39, no.7, pp.8146-8161, July 2024.

[8] X. Li, Z. Xiong and Y. Liu, "Nonsegmented Turn-On Switching Transient Analytical Model for Fast-Switching GaN HEMT With Datasheet-Driven Unified I–V Characterization," in IEEE Transactions on Industrial Electronics, 2025.

[9] Jian Chen, Quanming Luo, Jian Huang, Qingqing He, Pengju Sun and Xiong Du, "Analysis and Design of an RC Snubber Circuit to Suppress False Triggering Oscillation for GaN Devices in Half-Bridge Circuits," in *IEEE Transactions on Power Electronics*, vol.35, no.3, pp.2690–2704, March 2020.

[10] Xudong Wang, Zhengming Zhao, Yicheng Zhu, Kainan Chen and Liqiang Yuan, "A Comprehensive Study on the Gate-Loop Stability of the SiC MOSFET," in *2017 IEEE Energy Conversion Congress and Exposition (ECCE)*, 3012–3018, 2017.

[11] Junji Ke, Zhibin Zhao,Zongkui Xie ,Peng Xu and Xiang Cui,"Analytical Switching Transient Model for Silicon Carbide MOSFET under the Influence of Parasitic Parameters," in *Transactions of China Electrotechnical Society*, vol.33, no.8, pp.1762-1774,April 2018.

[12] Bochen Shi, Zhengming Zhao, Yicheng Zhu and Xudong Wang, "Time-Domain and Frequency-Domain Analysis of SiC MOSFET Switching Transients Considering Transmission of Control, Drive, and Power Pulses," in *IEEE Journal of Emerging and Selected Topics in Power Electronics*, vol.9, no.5, pp.6441–6452, October 2021.

[13] Sugihara Yusuke, Kimihiro Nanamori, Masayoshi Yamamoto and Yasuki Kanazawa,"Parasitic Inductance Design Considerations to Suppress Gate Voltage Oscillation of Fast Switching Power Semiconductor Device," in *2018 International Power Electronics Conference (IPEC-Niigata 2018 -ECCE Asia)*, 2789–2795, 2018.

[14] Tianjiao Liu, Runtao Ning, Thomas T. Y. Wong and Z. John Shen,"Modeling and Analysis of SiC MOSFET Switching Oscillations," in *IEEE Journal of Emerging and Selected Topics in Power Electronics*, vol.4, no.3, pp.747-756, September 2016.

[15] Shuang Zhao, Man Zhang, Zheng Zhang, Nan Zhu, Yue Zhao, Helong Li and Lijian Ding, "Analytic Model of the Voltage Oscillation in a High-Power Converter With Silicon Carbide Devices," in *IEEE Transactions on Industry Applications*, vol.60, no.2, pp. 3345–3358, March 2024.

979-8-3315-1110-4/25 $31.00 © 2025 IEEE

An Active Gate Charge Controlled Driver for SiC MOSFET with Zero Turn-off Loss

Yuze Zheng
School of Electrical Engineering
Xi'an Jiaotong University
Xi'an, China
zyz2001@stu.xjtu.edu.cn

Xu Cheng
Zhuhai Power Supply Bureau
Guangdong Power Grid Co., Ltd.
Zhuhai, China
664157102@qq.com

Fan Zhang
School of Electrical Engineering
Xi'an Jiaotong University
Xi'an, China
zhangfan1990@xjtu.edu.cn

Xiaolu Zhang
School of Electrical Engineering
Xi'an Jiaotong University
Xi'an, China
lu995579@stu.xjtu.edu.cn

Yukun Niu
School of Electrical Engineering
Xi'an Jiaotong University
Xi'an, China
827105370@qq.com

Xuhui Song
School of Electrical Engineering
Xi'an Jiaotong University
Xi'an, China
xuhui_song@163.com

Abstract—This paper presents an analysis of the turn-off transient in SiC MOSFETs, deriving the theoretical conditions for achieving zero turn-off loss and calculating the corresponding critical gate resistance value. Based on this analysis, an active gate driver for SiC MOSFETs utilizing gate charge control is proposed. By employing a lower turn-off voltage to enable rapid device turn-off while precisely regulating the extracted gate charge, this method achieves ultra-low turn-off losses without risking gate overvoltage breakdown. A double-pulse test circuit was implemented in LTspice to conduct comparative simulations between the proposed active gate driver and conventional gate driver. Under test conditions of 1 kV/60 A, the proposed active gate driver demonstrated over 90% reduction in turn-off losses compared to conventional gate diver, while maintaining stable ultra-low loss performance across varying load currents.

Keywords—SiC MOSFET, active gate driver, gate charge, zero turn-off loss

I. INTRODUCTION

A SiC MOSFET is a field-effect transistor fabricated from silicon carbide, which has better physical properties than Si. Compared with Si IGBTs, SiC MOSFETs exhibit virtually no tail current and can perform higher switching frequency, which can reduce the requirements for capacitors and magnetic components, and helps to reduce the size of electrical equipment [1], [2]. Switching losses have emerged as the primary limiting factor in increasing switching frequency. By adopting soft-switching circuit topologies, zero-voltage switching (ZVS) can be achieved in power devices which eliminates turn-on losses, but remains ineffective in reducing turn-off losses [3].

The inherently high switching speed of SiC MOSFETs generate high di/dt and dv/dt, rendering them highly sensitive to parasitic parameters. Therefore, quasi-zero current (QZC) turn-off can be achieved by exploiting the parasitic capacitance of load inductors [5]. Furthermore, when MOSFET generates a high dv/dt, the majority of drain current is diverted to the displacement current of C_{oss}, thereby avoiding flow through the conductive channel and preventing thermal dissipation, thereby achieving zero turn-off loss (ZTL) [4], [6], [7].

This work was supported by the China Southern Power Grid Company, Ltd., Science and Technology Project under grant 030400KC23090016 (GDKJXM20231032).

In order to achieve extremely low turn-off loss, gate driver circuit needs to provide sufficient turn-off current. Due to the high internal gate resistance of SiC MOSFETs (typically $4\Omega \sim 10\Omega$), it is difficult for CGD to further increase switching speed and reduce switching losses. By using a current-source active gate driver, driving current can be directly controlled [8]. However, the gate voltage may exceed the safe operating area during transient states, thus requiring precisely timing control to prevent gate overvoltage. A charge pump capacitor is used to provide additional turn-on voltage to quickly turn on MOSFET [9]. Noted that due to internal gate resistance and parasitic inductance, the external gate-source voltage is not equal to the actual on-chip voltage, package pin voltage exceeds safe range for a short time will not damage MOSFET.

Based on the existing research, this paper analyzes the gate charge variation during the turn-off process and proposes an active gate driver that directly controls the gate charge. The circuit structure and operating timing are simple, allowing for higher turn-off current without damaging device. Simulation results verify its effectiveness in reducing turn-off losses.

II. ANALYSIS OF ZERO TURN-OFF LOSS

During the turn-off process, as the gate-source voltage v_{gs} decreases, the device operating point progressively transitions from the triode region to the saturation region, entering the flat segment of the transfer characteristic curve. Under the clamping effect of load inductor current, the drain-source voltage v_{ds} rapidly rises in this phase while v_{gs} declines, effectively behaving as a v_{gs} controlled voltage source. When v_{ds} ascends to the DC bus voltage, further reduction of v_{gs} no longer influences v_{ds}. Instead, the clamped effect of bus voltage governs the drain current i_d, causing it to decay gradually from the load current to zero as the device enters the cut-off region.

The MOSFET characteristics in saturation and triode regions can be described by Sah's equations:

1. Triode Region, $v_{gs} - V_{th} > v_{ds} > 0$:

$$i_d = K_n \left(v_{gs} - V_{th} - \frac{v_{ds}}{2} \right) v_{ds} \qquad (1)$$

Where K_n is a constant related to the transistor structure.

2. Saturation region, $v_{gs} - V_{th} \leq v_{ds}$:

$$i_d = \frac{K_n}{2}\left(v_{gs} - V_{th}\right)^2 \left(1 + \lambda v_{ds}\right) \qquad (2)$$

Here, i_d scales quadratically with v_{gs} while exhibiting weak dependence on v_{ds} due to the channel-length modulation coefficient λ. This explains the upward tilt of saturation-region curves in transfer characteristics and accounts for the continued v_{gs} variation during the Miller plateau phase.

When drain-source voltage $v_{ds} = V_{dc}$ and $i_d = I_L$, v_{ds} is about to undergo rapid changes with the gate-source voltage, marking the onset of the Miller plateau. The corresponding gate-source voltage at this point is defined as the Miller plateau starting voltage V_{m0}. The Miller plateau concludes when the MOSFET's conductive channel is nearly fully open, i.e., when the pinch-off point of the channel is about to disappear. At this endpoint, the condition $v_{gd} = V_{th}$ (where V_{th} is threshold voltage) must hold, corresponding to the Miller plateau end voltage V_{m1}. Here, the device operates at the boundary between the triode region and saturation region. By substituting these boundary conditions into (1) and (2), V_{m0} and V_{m1} can be calculated as:

$$V_{m0} = \sqrt{\frac{2I_L}{K_n\left(1 + \lambda V_{dc}\right)}} + V_{th} \qquad (3)$$

$$V_{m1} = \sqrt{\frac{2I_L}{K_n}} + V_{th} \qquad (4)$$

When channel-length modulation effects are neglected (i.e., $\lambda = 0$), V_{m0} and V_{m1} become theoretically equivalent, implying an idealized Miller plateau region with perfectly constant gate-source voltage and an infinite Miller effect amplification factor K_m. In practice, $V_{m0} < V_{m1}$ due to non-negligible channel-length modulation. To analyze the Miller plateau interval while accounting for $\lambda > 0$, we consider the device operating in the saturation region under constant load current I_L. Under these constraints, differentiating both sides of (2) with respect to v_{gs} yields (5), which demonstrates that non-zero λ introduces a voltage-dependent K_m, deviating from the ideal infinite amplification scenario.

$$\frac{\partial v_{ds}}{\partial v_{gs}} = \frac{-2\left(1 + \lambda v_{ds}\right)}{\lambda\left(v_{gs} - V_{th}\right)} = -K_m \qquad (5)$$

Under the influence of the Miller effect, the reverse transfer capacitance C_{rss} is amplified by a factor of $(1 + K_m)$. The equivalent input capacitance $C_{iss,eq}$ and equivalent output capacitance $C_{oss,eq}$ under this condition can be expressed as:

$$C_{iss}^{eq} = C_{gs} + \left(1 + K_m\right)C_{gd} \qquad (6)$$

$$C_{oss}^{eq} = C_{ds} + \left(1 + \frac{1}{K_m}\right)C_{gd} \qquad (7)$$

By combining (4) and (5) with gate drive current equation, the analytical expression for dv/dt of an ideal MOSFET can be derived as (8), where V_{EE} is turn-off drive voltage.

$$\frac{dv_{ds}}{dt} = -K_m \frac{V_{EE} - v_{gs}}{R_g\left[C_{gs} + \left(1 + K_m\right)C_{gd}\right]} \qquad (8)$$

In practical applications, however, the parasitic capacitances of MOSFETs impose limitations on the voltage

slew rate dv/dt. For a typical half-bridge configuration and double-pulse test circuit, during the turn-off transient, displacement currents flow not only through driven device but also through parasitic capacitances of complementary device C_j. Based on the nodal current equation at the phase-leg midpoint, the load current I_L comprises the sum of displacement currents and channel current i_{ch} through the conductive channel of driven device. As dv/dt increases, the proportion of displacement currents in I_L rises significantly. When all load currents are converted into displacement currents, the theoretical maximum value of turn-off dv/dt is obtained:

$$\frac{dv_{ds}}{dt}\max = \frac{I_L}{C_j + C_{ds} + \left(1 + \frac{1}{K_m}\right)C_{gd}} \qquad (9)$$

The actual turn-off dv/dt exhibited by a MOSFET corresponds to the smaller value between (8) and (9). When the gate drive current is insufficient, the voltage slew rate is constrained by v_{gs} transition rate. Conversely, with high drive current, dv/dt becomes limited by load current, where channel current approaches zero, resulting in negligible thermal dissipation, which is zero turn-off loss (ZTL). By equating (8) and (9), the critical gate resistance $R_{g,crit}$ for achieving ZTL can be calculated as:

$$R_{g,crit} = \frac{-K_m\left(V_{EE} - v_{gs}\right)\left(C_j + C_{oss}^{eq}\right)}{I_L C_{iss}^{eq}} \qquad (10)$$

When the total gate resistance falls below this critical threshold, the turn-off loss approaches zero. However, due to the inherently large internal gate resistance $R_{g,int}$ in SiC power modules, the total gate resistance typically exceeds this critical value in most cases. Consequently, active gate drivers must be employed to increase $R_{g,crit}$ by adjusting other variables. As observed from (10), achieving ZTL is less challenging under small load current. In contrast, heavy-load scenarios, the boundary resistance can only be enhanced by either increasing the drain-source capacitance or adopting a lower negative turn-off voltage. While increasing C_{ds} reduces turn-off losses, it inherently transfers these losses to turn-on transient, resulting in no significant improvement in total switching losses. Therefore, the primary strategy for realizing ZTL should focus on optimizing the turn-off drive voltage.

III. GATE CHARGE CONTROLLED DRIVER

A. Principle of Gate Charge Control

The relationship between internal gate-source voltage $v_{gs,int}$ and gate charge Q_g during the switching process can be described in three distinct regions [10]. The variation in charge of gate-source capacitor and gate-drain capacitor during the turn-on and turn-off processes can be expressed as (11) and (12), where C_{gd0} is gate-drain capacitance when v_{ds} is 0, and C_{gd1} is capacitance when v_{ds} equals to bus voltage V_{dc}.

$$\Delta Q_{gs} = (V_{CC} - V_{EE})C_{gs} \qquad (11)$$

$$\Delta Q_{gd} = V_{CC}C_{gd0} + (V_{dc} - V_{EE})C_{gd1} \qquad (12)$$

According to (4), the Miller plateau voltage shows a strong dependence on load current. Although the position of the Miller plateau remains unchanged with increasing DC link voltage, the charge injected into gate during the plateau phase increases, which is manifested as a rightward shift of the latter part of gate charge characteristic curve. Therefore, driving the MOSFET under higher V_{dc} conditions requires a larger gate charge. Regarding the influence of load current, (11) and (12) reveal no explicit dependence on I_L, which means that the total gate charge required for MOSFET operation remains constant across varying load currents. Instead, I_L exclusively modulates the vertical positioning of Miller plateau, without changing basic gate charge requirements. Therefore, total gate charge do not change with I_L, load current only changes the position of miller plateau while bus voltage only changes the total gate charge, as shown in Fig. 1.

A simulation was conducted in LTspice using the C3M0016120K model from Wolfspeed to verify the theoretical formula of gate charge proposed above. The model was modified to enable observation of the internal gate-source voltage and channel current. The simulation conditions were configured with the DC bus voltage increasing from 100 V to 1 kV and the load current increasing from 5 A to 60 A, to observe the variation of gate charge curve.

Simulation results are shown in Fig. 2 and Fig. 3, when V_{dc} remains constant, the gate charge curves overlap in the area outside Miller plateau, a charge of 150 nC can generate a fixed Δv_{gs} of approximately 16V. In this case, directly controlling the gate charge allows for a determined internal gate-source voltage, unaffected by load variations. For voltage source converter, the DC voltage on power devices does not change while drain current varies.

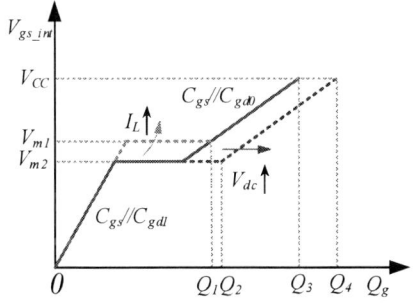

Fig. 1. Relationship between internal gate-source voltage $v_{gs,int}$ and gate charge Q_g.

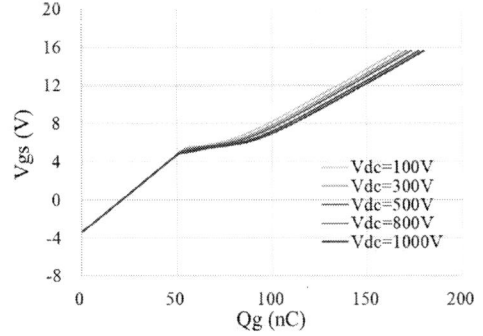

Fig. 2. Internal gate-source voltage $v_{gs,int}$ variation with gate charge Q_g under different DC bus voltage V_{dc} when load current I_L equals to 20A.

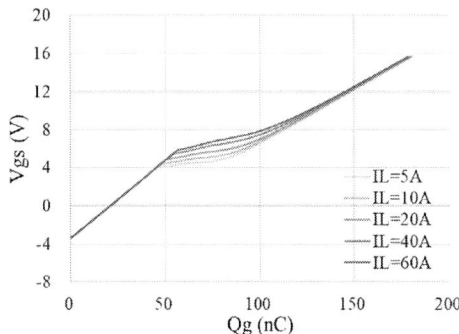

Fig. 3. Internal gate-source voltage $v_{gs,int}$ variation with gate charge Q_g under different load current I_L when DC bus voltage V_{dc} equals to 1kV.

B. Proposed Active Gate Driver

According to previous section, a gate charge control method using current mirror can be adopted to control internal gate-source voltage directly, the proposed AGD is shown in Fig. 4. A current mirror composed of transistors Q_1 and Q_2 is employed to extract charge from the MOSFET's gate. An additional negative turn-off voltage V_{neg} is applied to enhance the turn-off current. A small capacitor C_q is connected in series on the input side of the current mirror to limit the charge transfer through it. Diodes D_1 and D_2 serve as reverse-blocking diodes, while D_3 discharging the stored charge on C_q during SiC MOSFET turn-on transient. A PMOS transistor S_1 is integrated into the turn-on path of the gate driver. During turn-off transient, S_1 promptly disconnects the path to prevent a short circuit between V_{EE} and V_{neg} through Q_2.

The timing diagram of proposed AGD is shown in Fig. 5. The operation of each component of proposed AGD in different time periods is described as follows: $t_0 \sim t_5$: MOSFET turns on and conducting, the power amplifier output voltage is V_{CC}, S_1 conducts to charge C_{gs}, controller output voltage is V_{neg}, the stored charge in C_q is 0 and no current flow through Q_1 and Q_2. $t_5 \sim t_6$: AGD starts to turn off MOSFET, amplifier output voltage changed from V_{CC} to V_{EE}, S_1 turns off, and controller output voltage rises to 0 to charge C_q, charging current is amplified by current mirror, Q_2 pulling $v_{gs,ext}$ to V_{neg}, D_1 becomes reverse-biased and blocks. $t_6 \sim t_7$: C_q is fully charged and Q_2 returns to cutoff state, D_1 conducts to continue MOSFET turn-off, controller outputs V_{neg} to discharge C_q through D_3, and V_{ds} rises to V_{dc} at the end. $t_7 \sim t_8$: $V_{gs,int}$ drops to V_{EE} to form a reliable turn-off without damaging device. During the turn-off transient, the charge extracted by the current mirror from gate is $nV_{neg}C_q$, where n represents the number of parallel-connected transistors in Q_2.

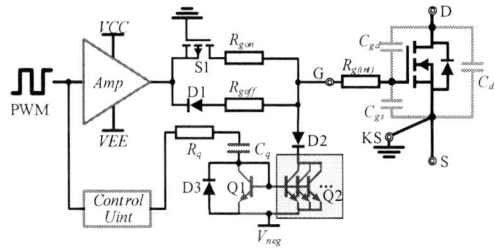

Fig. 4. Proposed gate charge controlled active gate driver.

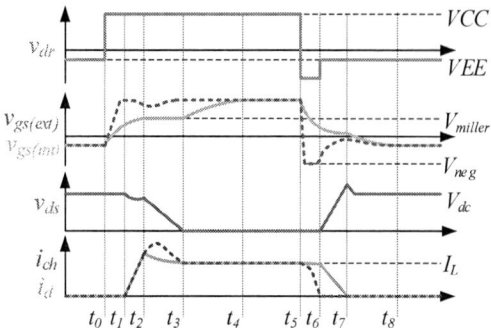

Fig. 5. Timing diagram of proposed charge controlled active gate driver.

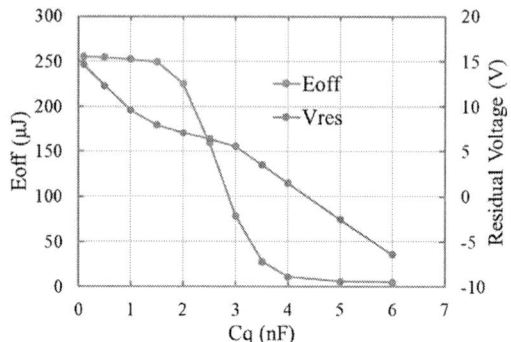

Fig. 7. Turn-off losses and residual gate-source voltage after gate charge extraction of proposed AGD with different current mirror capacitance C_q.

IV. SIMULATION VERIFICATION

A double-pulse test circuit based on C3M0016120K was constructed, with test conditions set to 1kV/60A, to validate the functionality of the proposed AGD. Two transistors are parallel-connected in Q_2, thus the charge extracted by AGD from gate should equal twice the stored charge of C_q. V_{CC}/V_{EE} are set to +15/-4 V, V_{neg} is set to -15V. Turn-off operations were performed with varying C_q capacitance values to observe the residual voltage on C_{gs} after extracting different charge quantities. Results are shown in Fig. 6, as C_q increases, the residual voltage decreases. When C_q equals to 5nF, the residual voltage measures approximately -2V, with an extracted gate charge of 150nC, consistent with the conclusions derived in previous sections.

When the extracted charge corresponds to the Miller plateau area on the gate charge characteristic curve, the change of v_{gs} with C_q slows down. The residual gate-source voltage corresponding to extracted charge in this range is almost unchanged, but turn-off loss E_{off} can be significantly reduced. However, in the range outside Miller plateau, residual v_{gs} changes significantly with C_q but E_{off} hardly changes, as shown in Fig. 7. This is because the Miller plateau area corresponds to stage of rapid change of v_{ds} in the turn-off transient. When extracted charge is larger than the gate charge required to enter Miller plateau, it indicates that there is still current flowing through current mirror during v_{ds} change. At this time, the turn-off voltage is close to V_{neg} rather than V_{EE}, and drive current provided is larger, which can significantly improve dv/dt and reduce channel current i_{ch}, thus significantly reduce the turn-off loss E_{off}.

Therefore, when selecting the value of current mirror capacitor C_q, the working voltage and maximum load current of MOSFET should be confirmed first, the maximum gate charge required for device to enter Miller plateau should be estimated, the number of parallel transistors in Q_2 and the capacitance of C_q should be reasonably configured so that the charge extracted by proposed AGD is sufficient to allow the device to pass Miller plateau, thereby minimizing turn-off loss. It should be noted that excessive C_q capacitance induce over-extraction of gate charge, driving v_{gs} below the recommended minimum turn-off voltage, and risking gate oxide reverse breakdown. The optimal value of C_q is corresponding to the gate charge that just ends Miller plateau. For example, the optimal value of C_q is 4nF in Fig. 7.

Comparative simulations were conducted between CGD with a BJT push-pull configuration and proposed AGD. Both of their turn-off resistance R_{goff} is set to 3Ω, capacitance of C_q is selected to 4nF. As shown in Fig. 8, thermal power loss are calculated as the product of channel current i_{ch} and V_{ds}, with its integral representing the energy dissipated as heat during turn-off transient. CGD produces 256.46 μJ of thermal loss, while AGD exhibits only 6.43 μJ, although faster turn-off speed induces i_d oscillations, the v_{ds} overshoot shows no significant increase.

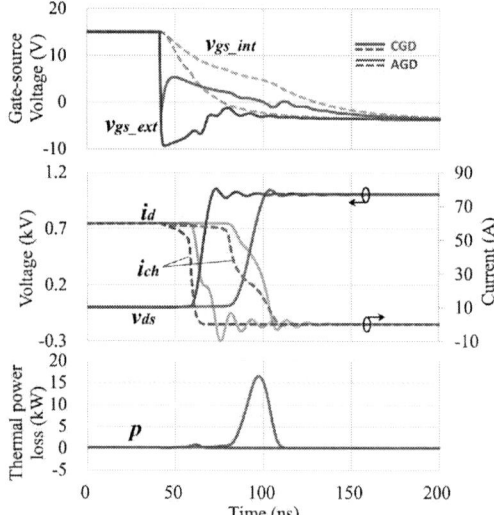

Fig. 8. 1kV/60A DPT simulation results of CGD and proposed AGD.

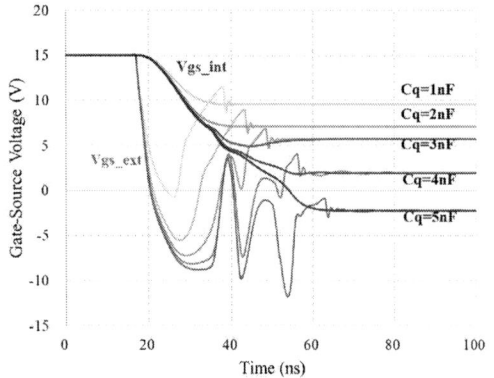

Fig. 6. External gate-source voltage $v_{gs,ext}$ and internal gate-source voltage $v_{gs,int}$ under different current mirror capacitance C_q when DC bus voltage V_{dc} is 1kV.

979-8-3315-1110-4/25 $31.00 © 2025 IEEE

Turn-off losses under varying load current were simulated, with results shown in Fig. 9. When $I_L \leq 20A$, both CGD and proposed AGD exhibit comparable turn-off losses below 10μJ. When I_L exceeds 20A, the turn-off losses of CGD increase rapidly, while the turn-off losses of AGD almost stops increasing and stabilizes at an extremely low level.

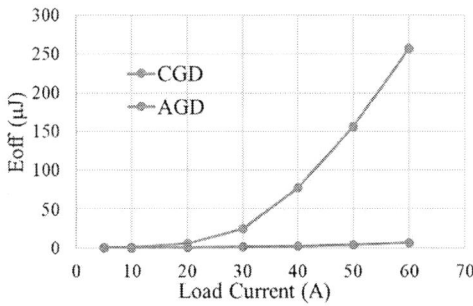

Fig. 9. Turn-off losses of proposed AGD and CGD under DC bus voltage V_{dc} equals to 1000V with varying load current I_L.

V. CONCLUSION

This paper investigates the variation trends of SiC MOSFET gate charge characteristics with bus voltage and load current, and proposes an active gate charge controlled driver, which has simple circuit topology and timing sequence. The proposed AGD achieves zero turn-off loss, eliminating over 90% of turn-off losses compared to conventional gate drivers. The proposed AGD is insensitive to load current, maintaining ZTL across different IL, thereby demonstrating exceptional adaptability to diverse application scenarios.

REFERENCES

[1] J. Biela, M. Schweizer, S. Waffler and J. W. Kolar, "SiC versus Si—Evaluation of Potentials for Performance Improvement of Inverter and DC–DC Converter Systems by SiC Power Semiconductors," IEEE Trans. Ind. Electron., vol. 58, no. 7, pp. 2872-2882, Jul. 2011.

[2] X. Li, L. Zhang, S. Guo, Y. Lei, A. Q. Huang and B. Zhang, "Understanding switching losses in SiC MOSFET: Toward lossless switching," 2015 IEEE 3rd Workshop on Wide Bandgap Power Devices and Applications (WiPDA), Blacksburg, VA, USA, 2015, pp. 257-262.

[3] S. Hazra et al., "High Switching Performance of 1700-V, 50-A SiC Power MOSFET Over Si IGBT/BiMOSFET for Advanced Power Conversion Applications," IEEE Trans. Power Electron., vol. 31, no. 7, pp. 4742-4754, Jul. 2016.

[4] S. Song, H. Peng, X. Chen, Q. Xin and Y. Kang, "Determination and Implementation of SiC MOSFETs Zero Turn-off Loss Transition Considering No Miller Plateau," IEEE Trans. Power Electron., vol. 38, no. 12, pp. 15509-15521, Dec. 2023.

[5] P. Nayak and K. Hatua, "Parasitic Inductance and Capacitance-Assisted Active Gate Driving Technique to Minimize Switching Loss of SiC MOSFET," IEEE Trans. Ind. Electron., vol. 64, no. 10, pp. 8288-8298, Oct. 2017.

[6] X. Li et al., "Achieving Zero Switching Loss in Silicon Carbide MOSFET," IEEE Trans. Power Electron., vol. 34, no. 12, pp. 12193-12199, Dec. 2019.

[7] Y. Xie, C. Chen, Y. Yan, Z. Huang and Y. Kang, "Investigation on Ultralow Turn-off Losses Phenomenon for SiC MOSFETs With Improved Switching Model," IEEE Trans. Power Electron., vol. 36, no. 8, pp. 9382-9397, Aug. 2021.

[8] H. Gui et al., "Current Source Gate Drive to Reduce Switching Loss for SiC MOSFETs," 2019 IEEE Applied Power Electronics Conference and Exposition (APEC), Anaheim, CA, USA, 2019, pp. 972-978.

[9] H. Gui, J. Sun and L. M. Tolbert, "Charge Pump Gate Drive to Reduce Turn-ON Switching Loss of SiC MOSFETs," IEEE Trans. Power Electron., vol. 35, no. 12, pp. 13136-13147, Dec. 2020.

[10] S. Yano, Y. Nakamatsu, T. Horiguchi and S. Soda, "Development and Verification of Protection Circuit for Hard Switching Fault of SiC MOSFET by Using Gate-Source Voltage and Gate Charge," 2019 IEEE Energy Conversion Congress and Exposition (ECCE), Baltimore, MD, USA, 2019, pp. 6661-6665.

A Modulation Optimization Strategy for Phase Shift Full Bridge with Low Freewheeling Loss

1st Enyou Wu
School of Electrical and
Electronic Engineering
Huazhong University of
Science and Technology
Wuhan,China
M202472394@hust.edu.cn

2nd Xuchen Sun
School of Electrical and
Electronic Engineering
Huazhong University of
Science and Technology
Wuhan,China
M202372289@hust.edu.cn

3rd Jiajia Guan
School of Electrical and
Electronic Engineering
Huazhong University of
Science and Technology
Wuhan,China
jiajiaguan@hust.edu.cn

4th Siyuan Feng
School of Electrical and
Electronic Engineering
Huazhong University of
Science and Technology
Wuhan,China
siyuanfeng@hust.edu.c

5th Tianxi Li
School of Electrical and
Electronic Engineering
Huazhong University of
Science and Technology
Wuhan,China
M202472393@hust.edu.cn

6th Cai Chen
School of Electrical and
Electronic Engineering
Huazhong University of
Science and Technology
Wuhan,China
caichen@hust.edu.cn

7th Yong Kang
School of Electrical and
Electronic Engineering
Huazhong University of
Science and Technology
Wuhan,China
ykang@hust.edu.cn

Abstract—The phase-shifted full-bridge converter is widely used in fields such as electric vehicles and renewable energy due to its advantages of high efficiency and wide input voltage range. However, conventional modulation methods may lead to large diode freewheeling loss during the freewheeling period. This paper proposes a new modulation method to reduce freewheeling loss. By increasing the duty cycle of the switching tubes in the rectifier circuit, the freewheeling time through body the diodes can be reduced or eliminated. Finally, a prototype with a switching frequency of 50kHz, 450 V-900V input, and 12-16V@2kW output was designed to verify the proposed modulation strategy, achieving an efficiency improvement of 0.2%.

Keywords—DC-DC converter, phase-shift full-bridge, modulation optimization strategy.

I. INTRODUCTION

The Phase-Shifted Full-Bridge (PSFB) converter is a highly efficient DC-DC power topology renowned for its superior performance in medium-to-high power applications PSFB converter consists of a full-bridge inverter formed by four switching devices (typically MOSFETs or IGBTs), a high-frequency transformer, and output rectification and filtering circuits.

Common secondary-side rectification circuits include the full-bridge rectifier, uncontrolled rectification [1], [2], [4] and the push-pull rectifier. This paper focuses on the optimized modulation design for the push-pull rectifier circuit. By controlling the phase shift in conduction between the Two pairs of complementary switching transistors (S_1, S_2 and S_3, S_4), the PSFB converter achieves output voltage regulation, offering a wide voltage adjustment range.

Current research efforts predominantly focus on reducing losses in primary-side switching devices [5]-[8]. To minimize freewheeling losses, improve overall converter efficiency, and reduce heat generation, this paper proposes a novel modulation method that optimizes the drive signals for the secondary-side push-pull structure.

II. ANALSIS OF VOLTAGE REGULATION

The drive signals for switches S_1-S_6 are denoted as Q_1-Q_6, respectively. In conventional modulation methods, the duty

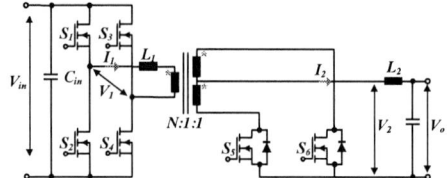

Fig.1 Phase-Shifted Full-Bridge Topological Diagram

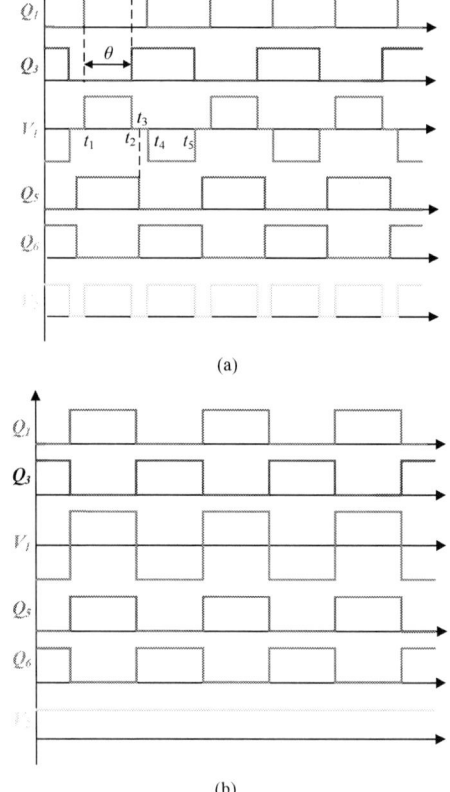

(a)

(b)

Fig.2. traditional modulation waveforms.(a) θ=150° (b) θ=180°

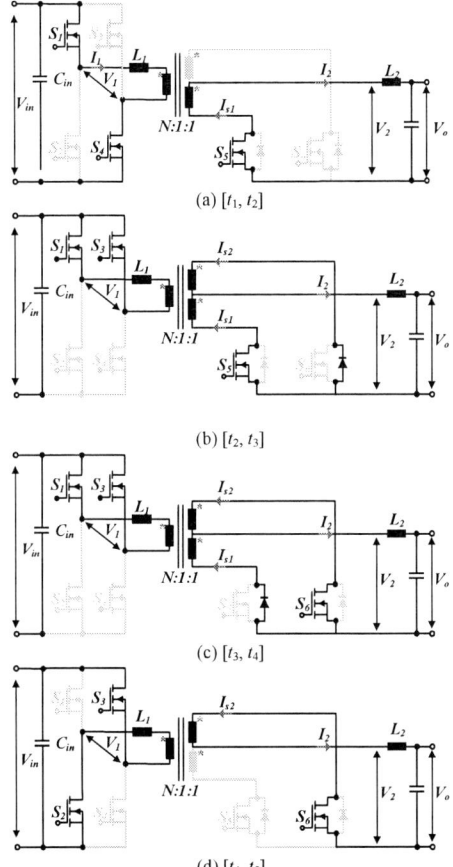

(a) [t_1, t_2]

(b) [t_2, t_3]

(c) [t_3, t_4]

(d) [t_4, t_5]

Fig.3. modal analysis of traditional modulation.

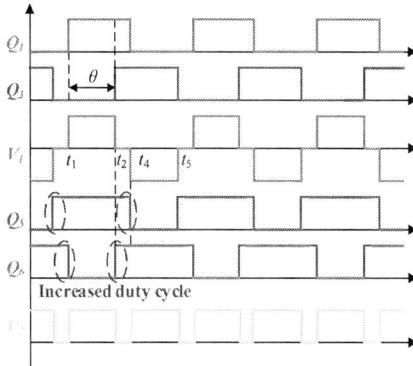

Fig.4. optimized modulation waveforms

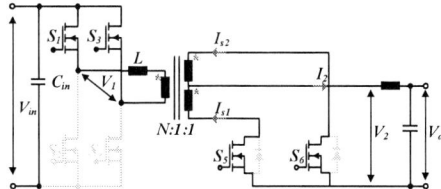

Fig.5. Modal analysis between t_2 and t_4 after optimization

the commutation processes Fig. 3(b) and Fig. 3(c) can be represented as shown in Equation 2 and 3.

$$
\begin{cases}
I_{s1} + I_{s2} = I_2 \\
-L_{s1}\dfrac{dI_{s1}}{dt} = V_o + L_2\dfrac{dI_2}{dt} + I_{s1}R_{DS(on)} \\
L_{s2}\dfrac{dI_{s2}}{dt} = V_o + L_2\dfrac{dI_2}{dt} + V_{SD}
\end{cases}
\tag{2}
$$

$$
\begin{cases}
I_{s1} + I_{s2} = I_2 \\
L_{s1}\dfrac{dI_{s1}}{dt} = V_o + L_2\dfrac{dI_2}{dt} + V_{SD} \\
-L_{s2}\dfrac{dI_{s2}}{dt} = V_o + L_2\dfrac{dI_2}{dt} + I_{s2}R_{DS(on)}
\end{cases}
\tag{3}
$$

From the modal analysis, it can be observed that during the commutation process on the transformer secondary side, the current sequentially commutates through the anti-parallel diodes of the two sets of secondary-side switches. Since the losses caused by the forward voltage drop of the diodes are typically greater than those caused by the on-resistance of MOSFETs, enabling synchronous rectification through MOSFETs can effectively reduce commutation losses.

III. MODULATION OPTIMIZATION

Chapter II introduces the voltage regulation principles of PSFB converter, analyzes the operating modes under conventional modulation methods, and identifies the root cause of high freewheeling losses in the secondary-side switches. Conventional modulation scheme also introduces voltage overshoot issues [3]. This chapter proposes an improved phase-shifted full-bridge modulation method. This

cycles of Q_1-Q_6 are all set to 50%. The phase difference between Q_1 and Q_2 is defined as the phase shift angle θ, with θ ranging from 0° to 180°. The secondary-side drive signal waveforms remain centrally symmetric with the midpoint voltage of the primary bridge arms. Fig. 2 illustrate the drive signals and voltage waveforms under θ=150° and 180°, respectively. Therefore, the relationship between the input and output voltages of the PSFB converter can be expressed as:

$$
V_o = \frac{V_{in}}{N} \cdot \frac{\theta}{180}
\tag{1}
$$

Taking θ=150° as an example, the switching modes between t_1 and t_5 are analyzed, as shown in Fig. 3. During [t_1, t_2], S_1, S_4, and S_5 are turned on, with $V_1 >0$, and the inductor current direction is as indicated in the figure, during [t_2, t_3], S_1, S_3, and S_6 are turned on, and the secondary-side winding commutates through the body diode of S_5. During [t_3, t_4], S_1, S_3, and S_5 are turned on, and the secondary-side winding commutates through the body diode of S_6. During [t_4, t_5], S_2, S_3, and S_6 are turned on, marking the end of the commutation process.

Assume that the output voltage remains constant during the commutation process, i.e., the output voltage ripple is neglected. The equations for the secondary inductor current, output voltage, and relevant parameters of the MOSFET in

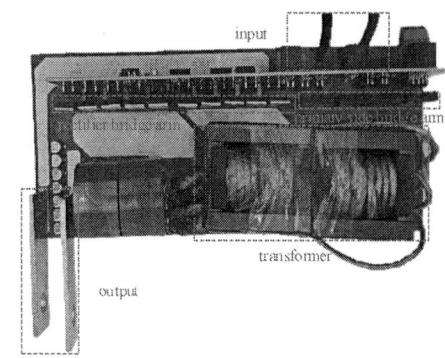

Fig.6. 2kW PSFB prototype

TABLE I.   PROTOTYPE PARAMETERS

Parameters	Values
Input DC voltage (V_{in})	450~900 V
Input DC current (I_{in})	2.2~4.4 A
Output voltage (V_o)	12~16 V
Max Output power (P_{max})	2.0 kW
Switching frequency	50 kHz
Output Inductance (L_2)	1.1 µH
Output capacitance (C_o)	400 µF
Controller	STM32G474CET6

approach effectively reduces freewheeling losses in the secondary-side switches, significantly enhances converter efficiency under high-output-current conditions, and mitigates temperature rise in the switching devices.

Since the commutation process causes freewheeling through the body diodes, the conduction time of the secondary-side switches should be extended to ensure that both switches in the secondary-side push-pull pair remain turned on during commutation. This forces the inductor current to freewheel through the switches instead of the diodes. Given that the conduction voltage drop of the switches is significantly lower than that of the diodes, the freewheeling losses can be drastically reduced under the same current. In terms of phase alignment, the secondary-side drive signals remain centrally symmetric with the midpoint voltage waveform of the primary bridge arms. The relationship between duty cycle and phase shift angle can be expressed as:

$$D_5 = D_6 = 0.5 + \lambda \frac{90° - \dfrac{\theta}{2}}{180°} \tag{4}$$

here, λ is set within the range of 0.85 to 0.95. This range is chosen to prevent both sets of synchronous rectifiers from turning on simultaneously when there is voltage across the secondary-side winding, which would cause a short circuit.

To ensure that the secondary-side drive signal waveform remains centrally symmetric with the midpoint voltage of the primary bridge leg, during the control process, when the duty cycle of the secondary drive signal varies with the primary-side phase-shift angle, its phase also shifts accordingly. The

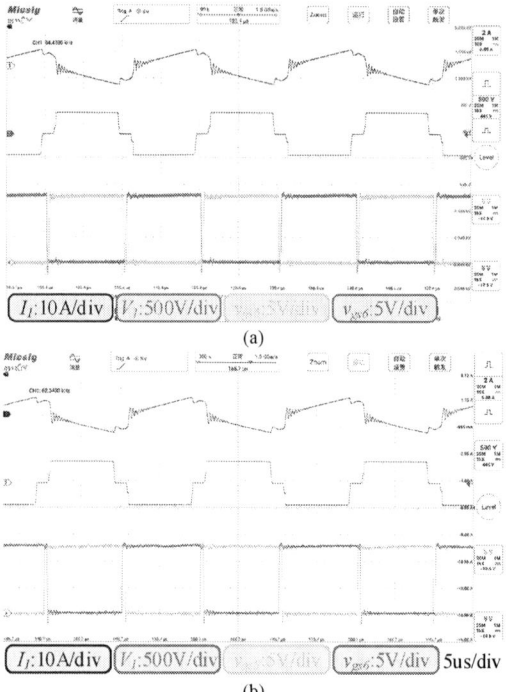

Fig.7. Experimental oscillogram at 400W (a)traditional modulation (b)optimized modulation

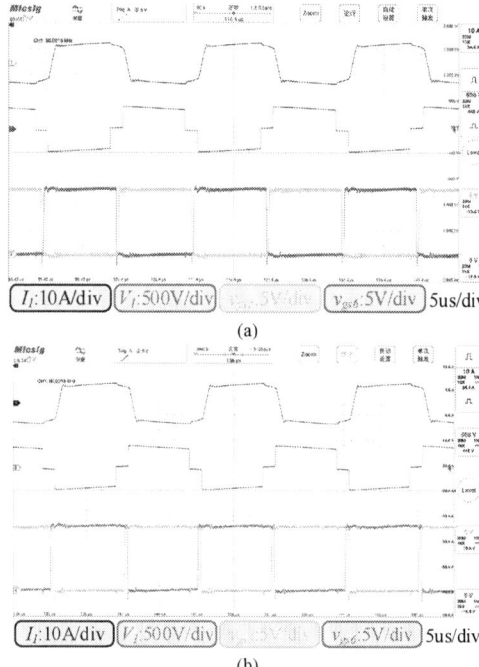

Fig.8. Experimental oscillogram at full load (a)traditional modulation (b)optimized modulation

phase of the drive signals for S_1, S_3, and S_5 can be expressed as:

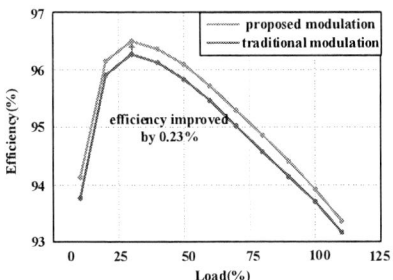

Fig.9. Efficiency curve

$$\begin{cases} \theta_1 = 90^\circ - \dfrac{\theta}{2} \\[2mm] \theta_3 = 90^\circ + \dfrac{\theta}{2} \\[2mm] \theta_5 = 0 \end{cases} \qquad (5)$$

The optimized modulation waveforms are illustrated in Fig. 4 (using θ=150° as an example).

IV. EXPERIMENTAL EVALUATION

This experiment utilizes a phase-shifted full-bridge inverter with a secondary-side push-pull rectifier, featuring a rated output current of 200 A, a rated power of 2 kW, and an input voltage range of 450–900 V. The transformer has a turns ratio of 28:1:1. Light-load and full-load tests are conducted for two modulation methods (before and after optimization). Prototype parameters are given in Table 1.

Due to the high output current, the secondary-side push-pull structure employs five parallel-connected switching devices to reduce current stress on individual switches (ignoring current imbalance issues among the switches). The analysis of converter operating modes remains consistent with the single-switch case. Fig. 7 and Fig. 8 show the experimental waveforms under the output power of 400W and full-load conditions, respectively. In the figure, I_1 is the primary winding current of the transformer, V_1 is the midpoint voltage of the primary bridge arm, and V_{gs5} and V_{gs6} are the gate-source voltages of switching tubes S_5 and S_6, respectively.

Fig. 9 shows the efficiency curves obtained from the experiments conducted with traditional modulation and the

proposed modulation respectively. The peak efficiency has increased by 0.23%.

V. CONCLUSION

This paper proposes a novel modulation method for phase-shifted full-bridge (PSFB) inverters. By increasing the duty cycle of the secondary-side switch drive signals, this method prevents the output inductor from freewheeling through the body diodes, thereby effectively reducing freewheeling losses. The proposed modulation strategy improves the overall converter efficiency and mitigates temperature rise in the secondary-side switches under high-output-current conditions. An experimental prototype implementing this method is designed, and comparative tests (half-load and full-load) are conducted under both conventional and optimized modulation schemes. Efficiency curves for both modulation methods are plotted, demonstrating a consistent efficiency improvement of over 0.2% across the entire power range, which validates the superiority of the proposed approach.

REFERENCES

[1] S. Chothe, R. T. Ugale and A. Gambhir, "Design and modeling of Phase Shifted Full Bridge DC-DC Converter with ZVS," 2021 National Power Electronics Conference (NPEC), Bhubaneswar, India, 2021, pp. 01-06.

[2] S. Mukherjee, S. S. Saha and S. Chowdhury, "Design of Duty-Ratio and Phase-Shift Control Circuits for MPPT of SPV Source using ZV-ZCS PSFB Converters," 2021 Devices for Integrated Circuit (DevIC), Kalyani, India, 2021, pp. 555-559.

[3] M. Escudero, D. Meneses, N. Rodriguez and D. P. Morales, "Modulation Scheme for the Bidirectional Operation of the Phase-Shift Full-Bridge Power Converter," in IEEE Transactions on Power Electronics, vol. 35, no. 2, pp. 1377-1391.

[4] G. Di Capua, S. A. Shirsavar, M. A. Hallworth and N. Femia, "An Enhanced Model for Small-Signal Analysis of the Phase-Shifted Full-Bridge Converter," in IEEE Transactions on Power Electronics, vol. 30, no. 3, pp. 1567-1576.

[5] H. Wang, M. Shang and A. Khaligh, "A PSFB-Based Integrated PEV Onboard Charger With Extended ZVS Range and Zero Duty Cycle Loss," in IEEE Transactions on Industry Applications, vol. 53, no. 1, pp. 585-595, Jan.-Feb. 2017, doi: 10.1109.

[6] C. -E. Kim, "Optimal Dead-Time Control Scheme for Extended ZVS Range and Burst-Mode Operation of Phase-Shift Full-Bridge (PSFB) Converter at Very Light Load," in IEEE Transactions on Power Electronics, vol. 34, no. 11, pp. 10823-10832, Nov. 2019, doi: 10.1109.

[7] X. Huang, J. Zhao and F. Lin, "The Loss Characteristics of PSFB ZVS DC-DC Converter Applied to the Auxiliary Power System," 2018 International Power Electronics Conference (IPEC-Niigata 2018 - ECCE Asia), Niigata, Japan, 2018, pp. 2051-2057, doi: 10.23919/IPEC.2018.8507987.

[8] Z. Pei et al., "Phase-Shift Full-Bridge (PSFB) Converter Integrated Double-Inductor Rectifier With Separated Resonant Circuits (SRCs) for 800-V High-Power Electric Vehicles," in IEEE Journal of Emerging and Selected Topics in Power Electronics, vol. 12, no. 1, pp. 269-282, Feb. 2024, doi: 10.1109

2025 IEEE Workshop on Wide Bandgap Power Devices and Applications in Asia (WiPDA Asia)

Optimized Trajectory Control for Soft Start-Up of Multiple-Mode Resonant Switched Capacitor Converter

1st Jingjing Qi*
Department of Electrical Engineering,
State Key Laboratory of Power System
Operation and Control
Tsinghua University
Beijing, China
jjqi@mail.tsinghua.edu.cn

2rd Kai Zhang
Department of Electrical Engineering,
State Key Laboratory of Power System
Operation and Control
Tsinghua University
Beijing, China
zhangkai22@mails.tsinghua.edu.cn

3rd Haining Zhang
Guodian Nanjing Automation Co., Ltd.

Nanjing, China
zhanghn5@chd.com.cn

Abstract—**Ensuring safe and rapid startup is paramount for improving the reliability of the resonant switched capacitor converter (ReSCC). This paper proposes a novel soft start-up strategy for a ReSCC with two-stage resonant tanks. The approach sequentially designs the state trajectories of both resonant tanks within the current-limiting band, followed by output voltage buildup through optimized switching frequency control. Rapid multi-stage capacitor voltage establishment is achieved while inrush currents and voltage spikes are effectively eliminated. The proposed strategy is elaborated in detail and its effectiveness is validated by experiments.**

Keywords—*switched capacitor, DC-DC converter, trajectory control, soft start-up*

I. INTRODUCTION

The resonant switched capacitor converter (ReSCC) has emerged as a promising solution for high-performance power conversion, offering superior energy efficiency and compactness [1]-[3]. Its resonant characteristics enable optimized energy transfer with minimized size and weight, while the start-up process presents critical challenges, including substantial inrush currents and voltage spikes that may destroy components. Moreover, the multiple resonant tanks significantly increase the complexity of the dynamic transition from initial zero-state to steady-state [4]. Consequently, investigating a reliable soft start-up methodology for ReSCC is essential to ensure system stability and operational reliability.

Hardware circuit modifications have been demonstrated to effectively facilitate capacitor voltage establishment in ReSCC. The soft-start switch approach are applied in [5], though conceptually straightforward, fails to address the critical challenge of instantaneous input voltage surges commonly encountered in practical applications. This limitation becomes particularly pronounced in high-power systems where rapid transients may generate destructive inrush currents. [6] suggests connecting a Buck converter in series at the input of the switched tank converter (STC). By gradually adjusting the duty cycle of the Buck converter, the input voltage of the STC can be progressively increased. However, the additional conversion stage not only reduces overall system efficiency due to converter losses, but also increases the footprint, impacting power density. Furthermore, these auxiliary circuits typically require additional control circuit and protection mechanisms, further

escalating system complexity and materials costs. The incorporation of auxiliary components or circuits, while enabling smooth start-up operation for ReSCC, introduces significant trade-offs in terms of system complexity and performance metrics.

The state trajectory method enables comprehensive steady-state and transient analysis of resonant converters [7],[8]. The operation and parameter design of ReSCC are inherently complex due to its multiple resonant tanks operating within each switching cycle, particularly in zero voltage switching (ZVS) mode. To address this challenge, [9] - [11] employ state trajectory analysis to mathematically model the operational behavior and parameter interrelationships in ZVS mode for different ReSCC topologies—including Fibonacci, multilevel ReSCC, and cascaded ReSCC, respectively. In addition, [12] introduces a finite-state-machine-based controller that enhances transient response by actively shaping the state trajectories. While this work also explores a soft start-up strategy, it relies on resonant capacitor voltage and inductor current sensing and does not extend to the case of multiple resonant tanks. Notably, state trajectory control methods have already enabled reliable soft start-up in isolated resonant converters, such as LLC converter [13], and CLLC converter [14], which can provide substantial insights for the soft start-up of ReSCC.

This paper extends the application of state trajectory control to optimize the soft start-up process of the multiple-mode resonant switched capacitor converter (MReSCC) [15], addressing the start-up challenges posed by its multi-resonant-tank architecture. Soft start-up stages are analyzed under a typical operating mode of MReSCC. Additionally, current bands are designed, and the state parameters, along with the duration of each circuit state, are calculated.

II. STEADY STATE ANALYSIS

The topology of the MReSCC is shown in Fig. 1, comprising three main switches S_1~S_3, two resonant tank cells Cell-1 and Cell-2, and four diodes D_a~D_d. Fig. 2 shows a typical operating mode of MReSCC. When S_1 and S_3 are turned on, Cell-1 exhibits positive series resonance and Cell-2 shows negative parallel resonance, while when S_2 is turned on, the resonance patterns in both cells are reversed. In addition, the duty cycle of each switch is 50% and the switching frequency f_s is equal to the resonant frequency f_r.

By establishing differential equations for the circuit states in Figs. 2(a) and (b) and applying normalization, the trajectory equation of the MReSCC in this operating mode can be derived as (1). Where v_{CiN} is the normalized voltage, $v_{CiN} = v_{Ci}/V_i$, i_{LiN} is the normalized current, $i_{LiN} = Z_r i_{Li}/V_i$, and

This work was supported by National Key R&D Program of China 2024YFB4206905.

979-8-3315-1110-4/25 $31.00 © 2025 IEEE

impedance $Z_r = (L_r/C_r)^{1/2}$. All variables with subscript N in the paper denote normalized variable.

Consequently, based on (1), the steady-state trajectories can be illustrated in Fig. 3. The state trajectories of v_{C1N}-i_{L1N} and v_{C2N}-i_{L2N} are circles centered at (3/7,0) and (1/7,0), respectively, indicating that the normalized capacitor voltages v_{C1N} and v_{C2N} are 3/7 and 1/7, respectively. In addition, each circuit state trajectory is semi-circular, lasting for half a switching cycle, and each switching transistor achieves zero current switching (ZCS).

$$\begin{cases} \left(v_{C1N}(t) - \dfrac{3}{7}\right)^2 + i_{L1N}^2(t) = \left(\dfrac{\pi Z_r I_o}{7 V_i}\right)^2 \\ \left(v_{C2N}(t) - \dfrac{1}{7}\right)^2 + i_{L2N}^2(t) = \left(\dfrac{2\pi Z_r I_o}{7 V_i}\right)^2 \end{cases} \quad (1)$$

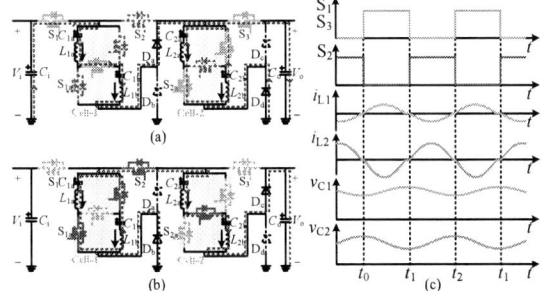

Fig. 1 Topology of MReSCC.

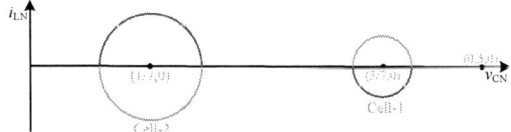

Fig. 2 A typical operating mode of MReSCC (a) and (b) circuit state (c) typical waveforms.

Fig. 3 Steady state trajectory.

III. PROPOSED START-UP STRATEGY

The proposed soft start-up strategy for MReSCC implements a three-stage voltage buildup process, sequentially establishing the capacitor voltages in Cell-1, Cell-2, and the output capacitor.

A. Soft Start-up Stage 1

During soft start-up Stage 1, the primary objective of MReSCC is to charge capacitor voltage in Cell-1 from zero to its steady-state voltage, with corresponding circuit states and state trajectories shown in Figs. 4 and 5, respectively.

When S_1 and S_3 are turned on, as shown in Fig. 4(a), the resonant tanks in Cell-1 enter positive resonance, characterized by increasing voltage v_{C1} and current i_{L1}, while

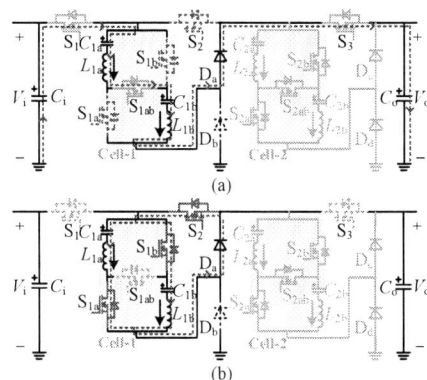

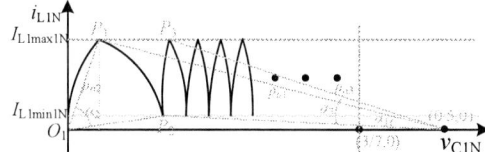

Fig. 4 Circuit states during Stage 1 (a) state1-I (b) state1-II.

Fig. 5 State trajectory of the Stage 1.

the resonant tanks in Cell-2 remain inactive due to absence of excitation. This circuit state persists until i_{L1} reaches the predefined current limit I_{L1max1}. The state trajectory equation of v_{C1N}-i_{L1N} is:

$$\left(v_{C1N}(t) - \frac{(1 - M_N)}{2}\right)^2 + i_{L1N}^2(t) = \left(v_{C11N} - \frac{(1 - M_N)}{2}\right)^2 + i_{L11N}^2 \quad (2)$$

where v_{C11N} and i_{L11N} are the initial normalized voltage and current of C_{1a}/C_{1b} and L_{1a}/L_{1b}, respectively.

When S_2 is turned on, as shown in Fig. 4(b), the resonant tanks in Cell-1 maintains positive resonance, continuing to charge C_1 with increasing v_{C1} while i_{L1} decreases. This circuit state persists until i_{L1} reaches the predefined minimum current limit I_{L1min1}. The state trajectory equation of v_{C1N}-i_{L1N} can be expressed as:

$$v_{C1N}^2(t) + i_{L1N}^2(t) = v_{C12N}^2 + i_{L12N}^2 \quad (3)$$

where v_{C12N} and i_{L12N} are the initial normalized voltage and current of C_{1a}/C_{1b} and L_{1a}/L_{1b}, respectively.

The current-limiting band can be designed as:

$$I_{L1min1N} \geq 0; \quad I_{L1max1N} = \sqrt{\frac{3}{2}} I_{L1peak1N} \quad (4)$$

where $I_{L1peak1N}$ is the normalized steady-state peak value of i_{L1}.

The output voltage V_o during Stage 1 is approximated as 0V, i.e.M_N=0. The duration $\Delta T_{\alpha1\text{-}I}$, final current $i_{P1\text{-}IEN}$, and final voltage $v_{P1\text{-}IEN}$ for State 1-I, along with the parameters $\Delta T_{\alpha1\text{-}II}$, $i_{P1\text{-}IIEN}$, $v_{P1\text{-}IIEN}$ for State 1-II, can be derived from (2) and (3), as expressed (5) and (6).

MReSCC alternately executes above two circuit states, maintaining C_{1a}/C_{1b} in continuous charging mode with the current constrained within the predefined current-limiting band. The proposed strategy allows v_{C1} to be built up as fast as possible until it reaches the steady state of $3V_i/7$ and the soft start-up Stage 1 is finished.

$$\begin{cases} \rho_{\alpha1-I} = \sqrt{i_{P1\text{-ISN}}^2 + \left(v_{P1\text{-ISN}} - \frac{1}{2}\right)^2} \\ \alpha_{\alpha1-I} = \arcsin\dfrac{I_{L1\max1N}}{\rho_{\alpha1-I}} - \arcsin\dfrac{I_{L1\min1N}}{\rho_{\alpha1-I}} \\ \Delta T_{\alpha1-I} = \dfrac{\alpha_{\alpha1-I}}{\omega_0}, i_{P1\text{-IEN}} = I_{L1\max1N}, v_{P1\text{-IEN}} = \dfrac{1}{2} - \sqrt{\rho_{\alpha1-I}^2 - i_{P1\text{-IEN}}^2} \end{cases} \quad (5)$$

$$\begin{cases} \rho_{\alpha1-II} = \sqrt{i_{P1\text{-IISN}}^2 + v_{P1\text{-IISN}}^2} \\ \alpha_{\alpha1-II} = \arcsin\dfrac{I_{L1\max1N}}{\rho_{\alpha1-II}} - \arcsin\dfrac{I_{L1\min1N}}{\rho_{\alpha1-II}} \\ \Delta T_{\alpha1-II} = \dfrac{\alpha_{\alpha1-II}}{\omega_0}, i_{P1\text{-IIEN}} = I_{L1\min1N}, v_{P1\text{-IIEN}} = \sqrt{\rho_{\alpha1-II}^2 - i_{P1\text{-IIEN}}^2} \end{cases} \quad (6)$$

B. Soft Start-up Stage 2

During soft start-up Stage 2, the primary objective of MReSCC is to charge capacitor voltage in Cell-2 from zero to its steady-state voltage. The switching sequence and corresponding inductor current variations are presented in Fig. 6. Fig. 7 shows the circuit states involved during this stage, while Fig. 8 illustrates the state trajectory characteristics of the resonant tanks. A zero-current condition ($i_{L1N}=0$) is enforced at the end of soft start-up Stage 1 to ensure seamless transition to soft start-up Stage 2 operation.

During the initial switching cycle of Stage 2, MReSCC maintains the circuit states shown in Fig. 4 until i_{L1N} decreases to zero, triggering S_2 turn-on and subsequent transition to the state 2-I illustrated in Fig. 7(a). During state 2-I($\widehat{J_{f2}J_0}, \widehat{Q_{f2}Q_0}$), energy transfers from Cell-1 to Cell-2, with i_{L1N} and i_{L2N} increasing negatively and positively respectively, maintaining $i_{L2N} = -2i_{L1N}$, until $i_{L1N}=I_{L1\min2N}$, $i_{L2N}= I_{L2\max N}$. The state trajectories equation of v_{CN}-i_{LN} can be expressed as (7). Where $V_{x1N}=4v_{C13N}+2v_{C23N}$, $V_{x2N}=2v_{C13N}+v_{C23N}$, $V_{x3N}=v_{C13N}-2v_{C23N}$.

When S_1 and S_3 are turned on, as shown in Fig. 7(b) ($\widehat{J_0J_1}, \widehat{Q_0Q_1}$), i_{L1N} and i_{L2N} decrease in the negative and positive directions, respectively, until $i_{L1N}=0$. The state trajectories equation of v_{CN}-i_{LN} can be expressed as (8).

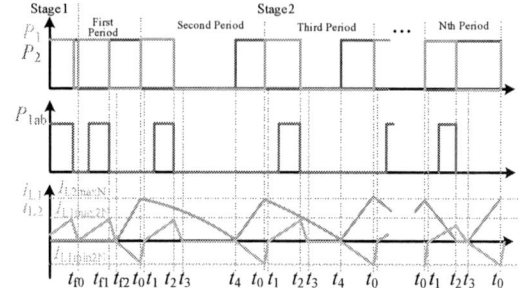

Fig. 6 Switching sequence and inductor currents during soft start-up Stage 2.

$$\begin{cases} \left(v_{C1N}(t) - \dfrac{V_{X1N}}{5}\right)^2 + i_{L1N}^2(t) = \dfrac{V_{X3N}^2}{25} + i_{L13N}^2 \\ \left(v_{C2N}(t) - \dfrac{V_{X2N}}{5}\right)^2 + i_{L2N}^2(t) = \dfrac{4V_{X3N}^2}{25} + 4i_{L13N}^2 \end{cases} \quad (7)$$

$$\begin{cases} \left(v_{C1N}(t) - 1\right)^2 + i_{L1N}^2(t) = (v_{C10N} - 1)^2 + i_{L10N}^2 \\ v_{C2N}^2(t) + i_{L2N}^2(t) = v_{C20N}^2 + i_{L20N}^2 \end{cases} \quad (8)$$

Then, as shown in Fig. 7(c) ($\widehat{J_1J_2}, \widehat{Q_1Q_2}$), i_{L2N} exhibits continuous positive decay due to low v_{C2N}, while i_{L1N} starts to positive growth until $i_{L1N}=I_{L1\max2N}$. The trajectory equation of v_{C1N}-i_{L1N} satisfies (9), while that of v_{C2N}-i_{L2N} remains (8).

$$\left(v_{C1N}(t) - \dfrac{1}{2}\right)^2 + i_{L1N}^2(t) = \left(v_{C11N} - \dfrac{1}{2}\right)^2 + i_{L11N}^2 \quad (9)$$

Notably, when v_{C2N} is high enough, i_{L1N} may not reach $I_{L1\max2N}$ before i_{L2N} decreases to zero, which cause v_{C1N} to drop slightly.

When v_{C2N} is low, all switches are turned off, as shown in Fig. 7 (d) ($\widehat{J_2J_3}, \widehat{Q_2Q_3}$), i_{L1N} and i_{L2N} decrease in the positive directions until $i_{L1N}=0$. The trajectory equation of v_{C1N}-i_{L1N} follows (3), and that of v_{C2N}-i_{L2N} is given by (8). When v_{C2N} is high enough, S_2 is turned on, as shown in Fig. 7 (e), i_{L1N} decreases in the positive direction until $i_{L1N}=0$. If $i_{L1N} = 0$ whiel $i_{L2N} > 0$, all switches remain off, as shown in Fig.7(f) ($\widehat{J_3J_4}, \widehat{Q_3Q_4}$). i_{L2N} decreases in the positive direction until $i_{L2N}=0$. The trajectory equation of v_{C2N}-i_{L2N} is still expressed as (3).

The duration $\Delta T_{\alpha2-x}$, final (or initial) currents i_{P2-xN}, and

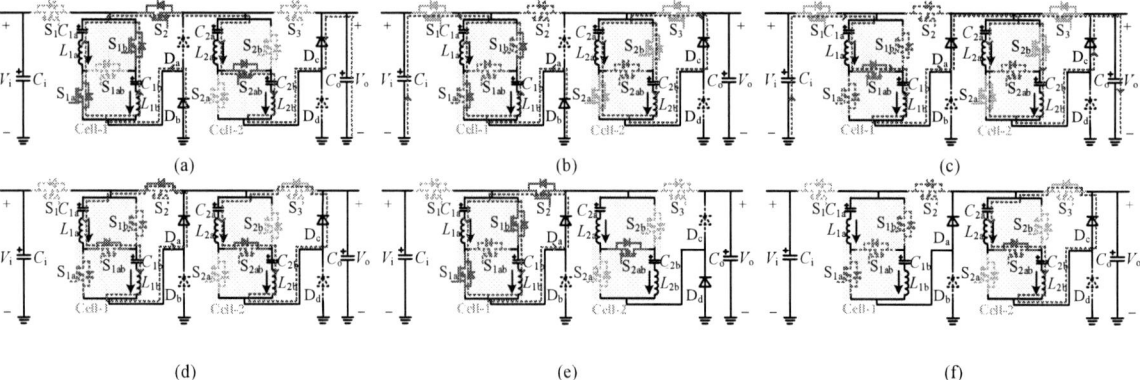

Fig. 7 Circuit states during Stage 2 (a) state2-I (b) state2-II (c) state2-III (d) state2-IV (e) state2-V (f) state2-VI.

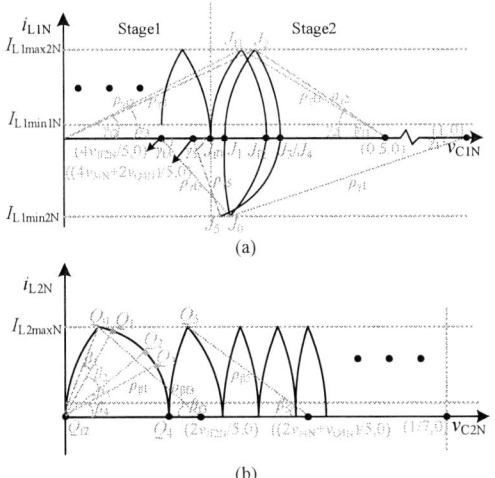

(a)

(b)

Fig. 8 State trajectory of the Stage 2 (a) v_{C1N}-i_{L1N} (b) v_{C2N}-i_{L2N}.

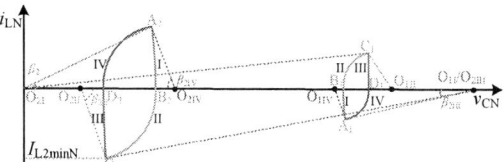

Fig. 9 State trajectory of the Stage 3-I.

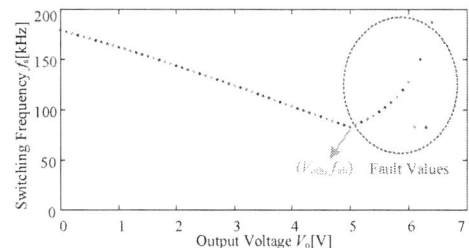

Fig. 10 Optimal trajectory control frequency f_s vs. output voltage v_o.

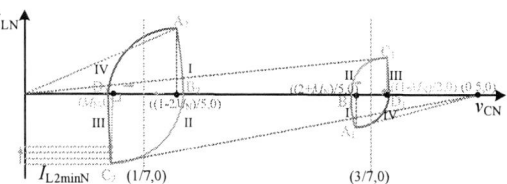

Fig.11 State trajectory of the Stage 3-II.

final (or initial) voltages $v_{\text{P2-xN}}$ for each circuit state can be derived based on the corresponding trajectory equations.

MReSCC alternately executes above circuit states, maintaining C_{2a}/C_{2b} in continuous charging mode with the current constrained within the predefined current-limiting band. The proposed strategy allows v_{C2} to be built up as fast as possible until it reaches the steady state of $V_i/7$ and the soft start-up Stage 2 is finished.

C. Soft Start-up Stage 3

The soft start-up Stage 3 focuses on establishing the output capacitor voltage to its steady state, comprising two sub-stages: stage 3-I and stage 3-II.

During Stage 3-I, optimized frequency modulation is applied to boost the output voltage v_o. The MReSCC operates through four circuit states: state 3-I, state 3-II, and state 3-IV correspond to Fig. 7(b), Fig. 2(a), and Fig.2(b), respectively, while state 3-III is similar to Fig. 7(e). The corresponding state trajectories are illustrated in Fig. 9. Combining with Fig. 9, the relationship between inductor current and capacitor voltage in Cell-2 is formulated as (10), from which the relationship between the switching frequency f_s and the output voltage v_o can be derived, i.e., $f_s = F(v_o)$.

The nonlinear relationship between switching frequency and output voltage ($f_s = F(v_o)$) can be effectively characterized through curve fitting, as demonstrated in Fig. 10. The plot reveals two distinct operational regimes: initially during Stage 3-I, the switching frequency operates above resonance ($f_r > f_r$) while exhibiting a decreasing trend with rising output voltage. This behavior continues until reaching the critical transition point (V_{oth}, f_{sth}). Beyond this threshold, the switching frequency begins increasing with output voltage as the state trajectory of state 3-II becomes tangent to the I_{L2minN} boundary, rendering the original trajectory equation (10) invalid.

Consequently, the circuit operation transitions from Stage 3-I to Stage 3-II precisely at or slightly before reaching the critical transition point (V_{oth}, f_{sth}). Fig. 11 depicts the state trajectory of v_{CN}-i_{LN} during Stage 3-II. In this stage, current limiting band is progressively reduced to further increase v_o, culminating in $|I_{\text{L2minN}}| = I_{\text{L2peak}}$ and $v_o = V_i/7$, which signifies the completion of the soft start-up process. The final state

$$
\begin{cases}
\left(\dfrac{1-2M_N}{5} - v_{\text{A2N}}\right)^2 + i_{\text{A2N}}^2 = \left(\dfrac{1-2M_N}{5} - v_{\text{D2N}}\right)^2 \\[2mm]
v_{\text{A2N}}^2 + i_{\text{A2N}}^2 = v_{\text{B2N}}^2 \\[2mm]
\left(v_{\text{C2N}} - M_N\right)^2 + I_{\text{L2minN}}^2 = \left(v_{\text{B2N}} - M_N\right)^2 \\[2mm]
\left(\dfrac{1}{2} - v_{\text{C2N}}\right)^2 + I_{\text{L2minN}}^2 = \left(\dfrac{1}{2} - v_{\text{D2N}}\right)^2 \\[2mm]
\sin\left(\beta_{2\text{I}} + \beta_{2\text{II}}\right) = \sin\left(\beta_{2\text{III}} + \beta_{2\text{IV}}\right)
\end{cases}
\tag{10}
$$

trajectory then transitions to the pattern shown in Fig. 3. Notably, the inductor current does not need to be sampled, but rather f_s is gradually reduced until $f_s = f_r$.

IV. EXPERIMENTAL VERIFICATION

Experimental validation is conducted on a 100W MReSCC prototype with 7:1 conversion ratio (48V to 6.86V). The circuit parameters are listed in Table I. The converter operates at a nominal switching frequency of 50 kHz during the steady-state condition. The soft start-up algorithm is implemented on a TMS320F28377D digital signal processor.

TABLE I. EXPERIMENTAL CIRCUIT PARAMETERS

Parameters	Specification
Input Voltage V_i/V	48
Voltage Ratio 1/M	7:1
Output Power P_o/W	100
Switching frequency f_s/kHz	50
Resonant inductor L_r/nH	660
Resonant Capacitor C_r/μF	15

When V_i= 48 V, the voltages V_{C1}, V_{C2} and V_o are approximately 20.6 V, 6.86 V and 6.86 V, respectively, as shown in Fig. 12(a). In addition, when P_o=100 W, I_{L1peak}, I_{L2peak} and I_o are about 7 A, 14 A and 14.6A, respectively, as shown in Fig. 12(b). These experimental results show excellent agreement with theoretical predictions. To ensure circuit safety, the resonant current is strictly constrained within a defined current band. Although the peak current I_{L1peak} reaches 7A, the theoretical maximum I_{L1max1} calculated from (4) is 8.61A. However, considering that switches only conduct i_{L1} during Stage 1, the devices inherently maintain sufficient current stress margin. Therefore, I_{L1max1} can be safely increased to accelerate the charging of capacitor voltage v_{C1}. In this design, by combining with I_{L2min}, the current band is set to I_{L1max1} =$|I_{L2min}|$=17A.

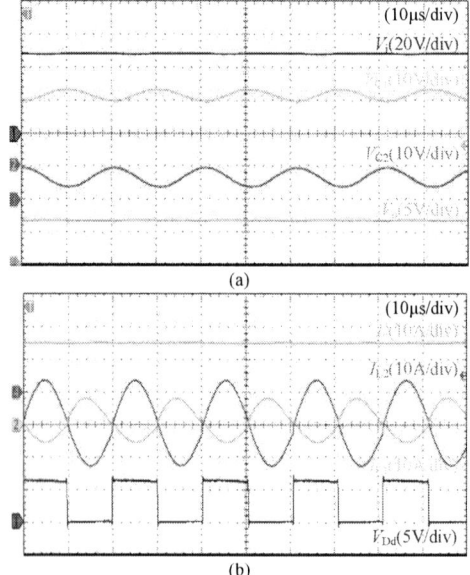

(a)

(b)

Fig. 12 Steady-state waveforms (a) Voltages, (b) Currents.

The capacitor voltages v_{C1}, v_{C2} and v_o establish sequentially and smoothly within approximately 2ms, as illustrated in Fig. 13(a). The resonant capacitor voltages and inductor currents during the soft start-up process are depicted in Figs. 13(b) and (c). During Stage 1, i_{L1} positively charges C_1 until v_{C1} reaches 20.6V. Stage 2 commences with i_{L2} charging C_2 to 6.86V. During Stage 3-I, the controller adjusts the switching frequency per (10), boosting V_o to the threshold voltage V_{oth} (5V). Finally, during Stage 3-II, the current limiting band gradually decreases to further increase V_o to 6.65V (2.9% below rated due to parasitic losses). The changes of the resonant tanks during each stage align with theoretical analysis, showing no inrush currents or voltage spikes.

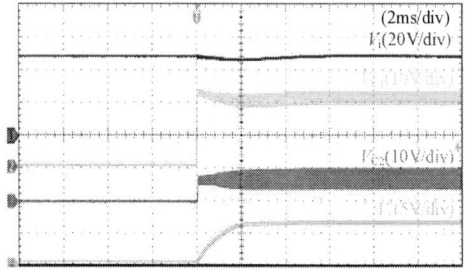

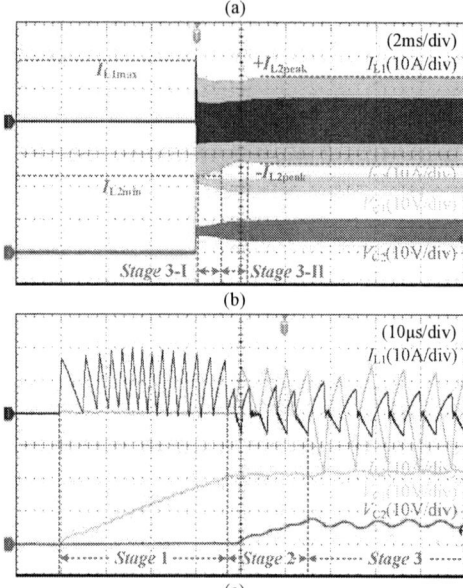

(a)

(b)

(c)

Fig. 13 Soft start-up process waveforms (a) Voltages, (b) Resonant tank voltages and currents, (c) Details.

V. CONCLUSION

In this paper, an optimized trajectory control strategy for soft start-up is proposed for MReSCC. By utilizing state trajectory analysis, precise control over each resonant tank's state trajectory during start-up is achieved, enabling smooth and rapid sequential buildup of voltages across the first-stage capacitor, second-stage capacitor, and output capacitor. In addition, the predesigned current limiting band effectively suppresses inrush currents and voltage spikes. This start-up strategy requires no additional components and relies solely on output voltage sampling, ensuring high reliability and low-cost soft start-up of the MReSCC.

REFERENCES

[1] Z. Ye, S. R. Sanders and R. C. N. Pilawa-Podgurski, "Modeling and Comparison of Passive Component Volume of Hybrid Resonant Switched-Capacitor Converters," *IEEE Trans. Power Electron.*, vol. 37, no. 9, pp. 10903-10919, Sept. 2022.

[2] R. Y. Sun, S. Webb and Y. -F. Liu, "Flying capacitor design considerations for a 48-to-12 V, 35 a split-phase dickson SC converter," *Chinese J. Electr. Eng.*, vol. 6, no. 4, pp. 28-41, Dec. 2020.

[3] P. H. McLaughlin, P. A. Kyaw, M. H. Kiani, C. R. Sullivan and J. T. Stauth, "A 48-V:16-V, 180-W Resonant Switched-Capacitor Converter With High-Q Merged Multiphase LC Resonator," *IEEE J. Emerg. Sel. Top. Power Electron.*, vol. 8, no. 3, pp. 2255-2267, Sept. 2020.

[4] S. Li, Z. Li, S. Zheng, W. Xie, J. Zheng and K. M. Smedley, "Multi-Resonance-Core-Based Dickson Resonant Switched-Capacitor Converters With Wide Regulation," *IEEE Trans. Power Electron.*, vol. 35, no. 2, pp. 1685-1698, Feb. 2020.

[5] C. Wang, Y. Lu and R. P. Martins, "A Highly Integrated 3-Phase 4:1 Resonant Switched-Capacitor Converter With Parasitic Loss Reduction and Fast Pre-Charge Startup," *IEEE Trans. Circuits Syst. II Express Briefs*, vol. 68, no. 7, pp. 2608-2612, Jul. 2021.

[6] X. Lyu, Y. Li, N. Ren, C. Nan, D. Cao and S. Jiang, "Optimization of High-Density and High-Efficiency Switched-Tank Converter for Data Center Applications," *IEEE Trans. Ind. Electron.*, vol. 67, no. 2, pp. 1626-1637, Feb. 2020.

[7] J. Zhao, L. Wu, H. Lin, X. Sun and G. Chen, "State Trajectory Control of Burst Mode for LCC Resonant Converters With Capacitive Output Filter," *IEEE Trans. Power Electron.*, vol. 37, no. 1, pp. 377-391, Jan. 2022.

[8] H. Shi, K. Sun, X. Hou, Y. Li and H. Jiang, "Equilibrium mechanism between dc voltage and ac frequency for ac-dc interlinking converters," *iEnergy*, vol. 1, no. 3, pp. 279-284, Sept. 2022.

[9] Y. Guan et al., "A High-Performance 3:1 Conversion Ratio DC–DC Converter: Analysis Method and Modular Adoption," *IEEE Trans. Power Electron.*, vol. 39, no. 4, pp. 4412-4425, Apr. 2024.

[10] X. Yang et al., "Improved Phase Shift Control for SiC-MOSFET Based Resonant Switched-Capacitor Converter With Parasitics Consideration," *IEEE Trans Ind Appl.*, vol. 56, no. 4, pp. 3995-4006, July-Aug. 2020.

[11] T. Ge, Z. Ye and R. C. N. Pilawa-Podgurski, "Geometrical State-Plane Analysis of Resonant Switched-Capacitor Converters: Demonstration on the Cascaded Multiresonant Converter," *IEEE Trans. Power Electron.*, vol. 38, no. 9, pp. 11125-11140, Sept. 2023.

[12] P.Wang, R. Ling, C. Sheng and F. Feng, "Fast Transient State Trajectory Control for Switched-Capacitor-Based Resonant Converter," *IEEE Trans. Power Electron.*, vol. 38, no. 8, pp. 9421-9435, Aug. 2023.

[13] W. Feng and F. C. Lee, "Optimal Trajectory Control of LLC Resonant Converters for Soft Start-Up," *IEEE Trans. Power Electron.*, vol. 29, no. 3, pp. 1461-1468, Mar. 2014.

[14] H. Chen, K. Zhang, L. Wang, K. Sun and Y. Li, "A Soft Startup Method With Natural Short-Circuit Tolerance Features for CLLC Converters," *IEEE Trans. Power Electron.*, vol. 40, no. 3, pp. 4008-4019, Mar. 2025.

[15] J. Qi et al., "A Multiple-Modes Resonant Switched Capacitor DC-DC Converter with Variable Voltage Ratios," *IEEE Trans. Power Electron.*, vol. 38, no. 6, pp. 7428-7443, June 2023.

Datasheet-Driven Non-segmented SPICE Model of SiC MOSFET With Improved Accuracy

Ziqi Jia
State Key Lab of Electrical Insulation
and Power Equipment
Xi'an Jiaotong University
Xi'an, China
Email: 1123248952@stu.xjtu.edu.cn

Yu Jiang
State Key Lab of Electrical Insulation
and Power Equipment
Xi'an Jiaotong University
Xi'an, China
Email: 1024854828@stu.xjtu.edu.cn

Yifan Hu
State Key Lab of Electrical Insulation
and Power Equipment
Xi'an Jiaotong University
Xi'an, China
Email: elchyf@stu.xjtu.edu.cn

Hailong He
State Key Lab of Electrical Insulation
and Power Equipment
Xi'an Jiaotong University
Xi'an, China
Email: hhlong@xjtu.edu.cn

Chunping Niu
State Key Lab of Electrical Insulation
and Power Equipment
Xi'an Jiaotong University
Xi'an, China
Email: niuyue@xjtu.edu.cn

Yi Wu
State Key Lab of Electrical Insulation
and Power Equipment
Xi'an Jiaotong University
Xi'an, China
Email: wuyic51@xjtu.edu.cn

Abstract—In recent years, SiC MOSFETs have experienced rapid advancements, driven by the growing demand for high-efficiency power devices in applications such as electric vehicles, renewable energy, and industrial automation. To achieve more accurate circuit simulations, a precise and robust SPICE model is essential. Although device manufacturers usually provide simplified SPICE models, the accuracy is insufficient to meet the simulation requirements. This paper proposes a data table-driven compact model for SiC MOSFETs, in which both static and C-V characteristics are accurately described by non-segment equations. Compared to the SPICE models provided by manufacturers, the model is characterized by the use of non-piecewise equations to describe its behavior, which enhances its convergence in simulations. Additionally, considering the different characteristics between high current and low current regions, the original EKV model is modified to improve the accuracy, and the third quadrant characteristics are modeled independently, making it more consistent with the device's actual behavior in practical circuits.

Keywords—SiC MOSFET; SPICE model; Convergence ability; modified EKV; parasitic capacitance

I. INTRODUCTION

SiC MOSFETs have become indispensable in numerous high-end applications, offering significant advantages over traditional silicon-based devices. Due to their wide bandgap, high thermal conductivity, and superior switching performance, SiC MOSFETs are increasingly used in high-efficiency applications such as electric vehicles, renewable energy systems, and industrial motor drives. However, high-precision circuit design requires device models that are accurate, efficient, and robust.

Existing SiC MOSFET models can be mainly classified into TCAD model and compact model. TCAD model can accurately predict the electrical and thermal characteristics of the device, but requires detailed physical and geometric parameters which are difficult to obtain[1-3], and the solving process is inefficient during complex circuit simulations, making it unsuitable for most circuit designers. However, compact model has gained attention in recent years due to their simpler structure and higher simulation efficiency[4-6].

McNutt et al. were the first to propose a compact model for SiC MOSFETs[7]. However, the model use a proprietary tool to model parameter extraction, making it people hard for users to build model. To enhance the accuracy, other models have been developed to improve performance. The EKV model, originally designed for Si MOSFETs, uses a single equation to describe the drain current in the weak, moderate, and strong inversion regions. While it reduces the difficulty of model extraction, its accuracy is relatively poor in applications. To adapt it for SiC MOSFETs, developers employed piecewise equations to describe the device characteristics[8]. Although this improved accuracy, the model causes convergence problems during simulation. Conventional compact models typically treat the third quadrant behavior as a mirror image of the first quadrant behavior, leading to low accuracy. Numerous attempts have been made by developers[9], but the accuracy still deviates significantly from the actual behavior.

This paper proposes a datasheet-driven SiC MOSFET SPICE model. To enhance the model's accuracy and convergence, an improved model based on the EKV model is presented, utilizing continuous equations to describe the drain current in different current regions. The model also offers a method to predict the third quadrant characteristics, but it's precise only in low voltage region before the conduction of body diode. Additionally, the model includes a diode with reverse recovery and continuous parasitic capacitance models

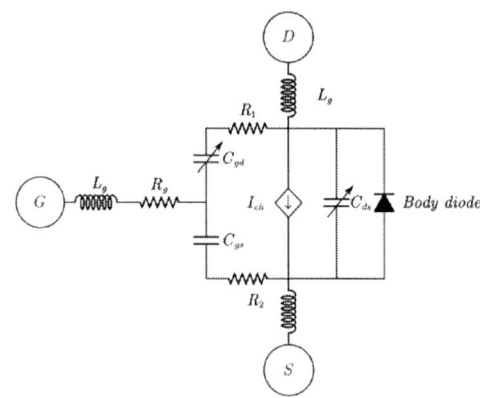

Fig. 1. SiC MOSFET SPICE Model

to accurately capture dynamic characteristics. The model's advantages are validated by comparison with the SPICE model provided by the manufacturer. Finally, dynamic characteristics are validated through double-pulse testsimulation. The novelty of this paper is described as follows:

- A continuous equation is used to describe the drain current in different channel current regions, improving model accuracy while ensuring convergence capability.

- The parameters of third quadrant part are extracted independently, which enhances the accuracy of the model during.

- The model of the parasitic capacitance are improved to ensure the convergence capability.

- The parameter extraction can be carried out simply based on the datasheet provided by manufacturers.

II. THE IMPROVED SPICE MODEL

The structure of SPICE model is shown in Fig. 1. The drain current is represented by two current sources with opposite directions, and the reverse-parallel diode represents the body diode. Parasitic capacitances are also included in the model. All unknown parameters can be extracted from datasheet.

A. Modeling the I–V Characteristics

The EKV model is a commonly used drain-to-source current model, often applied in SPICE models provided by manufacturers. This model compact model employs a superposition mechanism of dual logarithmic functions to achieve a unified analytical representation with smooth interregional transitions across operational regimes. This approach preserves derivative continuity across bias conditions, rendering it computationally advantageous for circuit simulation. The modified EKV model is presented in (1).

$$I_{ch} = 2g_m\phi_t^2 k_s \left\{ \frac{\left[\ln\left(1+e^{\frac{(V_{GS}-V_{th})}{2\kappa_s\phi_t}}\right)\right]^k}{\left[\ln\left(1+e^{\frac{(V_{GS}-V_{th})-nV_{DS}^{\dot{a}}}{2\kappa_s\phi_t}}\right)\right]^k}\right\}(1+\lambda V_{ds}) \quad (1)$$

TABLE I.   PARAMETERS USED IN EKV MODEL

Parameters	Definition
V_{DS}	Drain to source voltage
V_{GS}	Gate to source voltage
V_{th}	Threshole voltage
ϕ_t	Thermal voltage
g_m	Transconductance parameter
κ_s	Sub-threshold slope parameter
λ	Channel length modulation parameter
k	Law exponent
n and a	Triode region parameters

TABLE II.   MODEL PARAMETERS NEEDED TO BE EXTRACTED

Fitting Parameters	Initial Parameters
$K1_{xy}$	$2g_m\phi_t K_s$
$K2_{xy}$	Gate to source voltage
$K3_{xy}$	$2K_s\phi_t$
$K4_{xy}$	k
$K5_{xy}$	n
$K6_{xy}$	a
$K7_{xy}$	λ

For this function, $x = H$, L stands for the two current regions. g_m, ϕ_t and k_s represent as the transconductance, thermal voltage and the gradient in subthreshold region respectively. λ is the parameter of channel length modulation effect. To improve the accuracy of I-V characteristics, the parameters of first and third quadrants are extracted respectively. The definition of parameters are shown in Table. I.

In the model proposed by McNutt[7], the drain current is divided into low-current and high-current regions, the reason is the early turn on of the corner regions, which have lower threshold voltage. In our SPICE model, this effect is included in the form of a nonsegmented equation, which improves accuracy while ensuring convergence capability. And to improve the precision of 3rd quadrant characteristics, the parameters in this region are extracted independently. The equations are shown as follows:

$$I_{ch1} = I_{chH1} + I_{chL1} \quad (2)$$

$$I_{ch3} = I_{chH3} + I_{chL3} \quad (3)$$

where I_{chy} is the total current in the channel region, I_{chHy} and I_{chLy} represent the high-current region and the low-current. region, which can be get from (1).Y=1, 3 represent first quadrant and third quadrant respectively.

To extracted the model from datesheet provided by the manufactures, the output characteristic and transfer characteristic need to be got first. Then use curve fitting tools to extract the parameters needed in the model. It should be noted that the 3rd quadrant characteristic need to be separated from diode characteristic to ensure the accuracy of this model. So it's suggested that using the data whose V_{sd} are lower than bodydiode forward voltage. The parameters have be listed in Table. II. x and y represented different current regions and quadrant.

B. Modeling the C–V Characteristics

The parasitic capacitances of SiC MOSFETs include C_{gs}, C_{gd}, and C_{ds}. Accurate modeling of the device capacitances is crucial for accurate modeling of the dynamic characteristics. Since the variation of C_{gs} with drain voltage is minimal, it is treated as a constant in the model. The variation of C_{gd} with drain voltage is more complex. In previous physical models, modeling C_{gd} required the device's geometric parameters, which are difficult to obtain, leading to the use of various empirical models, such as tanh(x) and atan(x) piecewise functions. While these piecewise functions improved accuracy, they cause convergence problems to the model. This model proposes a non-piecewise function as follows:

$$C_{gd} = k_1\left[\tanh(k_2 V_{dg})\right] + k_3\left[\text{atan}(k_4 V_{dg} - k_4 V_{c1})\right]$$
$$+ k_5\left[\text{atan}(k_6 V_{dg} - k_6 V_{c2})\right] + k_7 \tag{4}$$

where k_n is the parameter to be extracted, and V_{c1}, V_{c2} are the points where the slope of C_{gd} changes.

The C_{ds} could be considered as p-n junction capacitance, which can be wirtten as:

$$C_{ds} = C_{j0}\left(1 + \frac{V_{ds}}{V_j}\right)^k \tag{5}$$

C_{j0} is the junction capacitance when V_{ds} is 0. V_j is the built-in potential of the diode. All the definition and the parameters needed to be extracted are listed in Table. III, IV.

C. Modeling diode Characteristics

The forward characteristics of body diode base on the shockley diode equation, which is written as:

$$I_{dio} = I_s\left(e^{\frac{q(V-IR)}{N}} - 1\right) \tag{6}$$

where I_s is the saturation current of body diode, N is the ideality factor. The reverse recovery model adopted the method presented in [10]. The parameters of bodydiode are listed as follows:

III. MODEL VALIDATION

A. Static Characteristics Validation

This study models the C3M0016120K from WolfSpeed to validate the accuracy of the model proposed. Output characteristic, transfer characteristic third quadrant characteristic and are shown in Fig. 2-4. The performance comparison between EKV and improved model is shown in Table V.

TABLE III.   PARAMETERS OF THE CAPACITANCE MODEL

Parameters	Definition
k_1 to k_6	C_{gd} parameters
C_{j0}	Junction capacitance when no bias
V_j	Built-in potential
k	Law exponent

TABLE IV.   BODYDIODE PARAMETERS NEEDED TO BE EXTRACTED

Fitting Parameters	Initial Parameters
K_1	I_s
K_2	$\frac{q}{N}$
K_3	R

TABLE V.   STATIC CHARACTERISTIC COMPARISON

MODEL	R^2	SSE	RMSE
EKV	0.9763	3284	21.63
Improved EKV	0.9992	93	0.4687
EKV in 1st quadrant	0.9998	36	0.3274

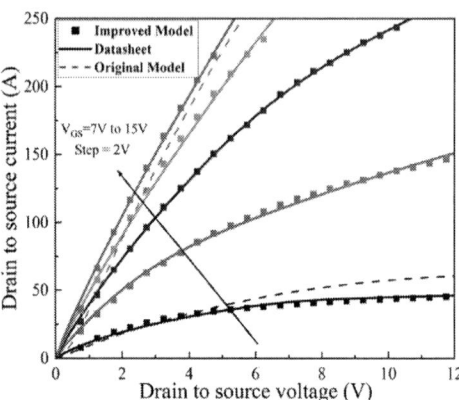

Fig. 2. Output characteristics of original model, improved model, and the data from datasheet. $T_j = 25\,°C$.

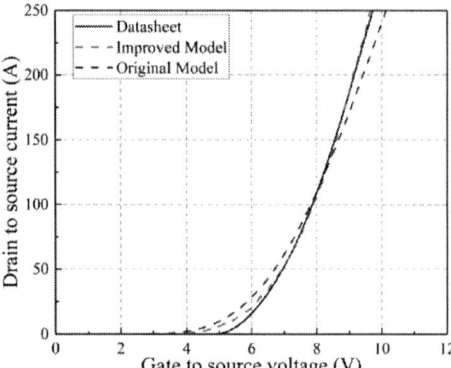

Fig. 3. Transfer characteristics. $T_j = 25\,°C$.

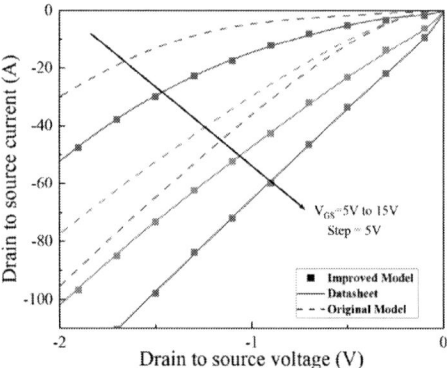

Fig. 4. 3rd quadrant characteristics.

Fig. 2, 3 show that the traditional model has poor performance on the high current region, and proposed model has high accuracy during both high-current region and low-current region. In cause of the model provided by manufacture didn't build the 3rd quadrant channel current model independently, it's almost unusable to simulate the reverse current, and Fig. 3 shows that.

B. Dynamic Characteristics

The curves of C_{gd} and C_{ds} are shown in Fig. 4. This figure shows that non-segmented model can capture the slope change point precisely.

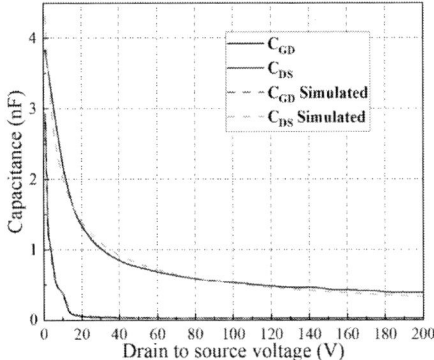

Fig. 5. Comparison of C_{GD} and C_{DS} between two models.

To further validate the reliability of the model, a double-pulse test was conducted in LTspice, with the circuit shown in Fig. 6. The waveforms are shown in Fig. 7 and Fig. 8, which provided dynamic parameters such as turn-on and turn-off loss and delay time. A comparison with the datasheet is presented in Table VI. All the characteristics in the proposed model have a better accuracy except for T_{don}, it might be caused by the inaccuracy of C_{gs}.

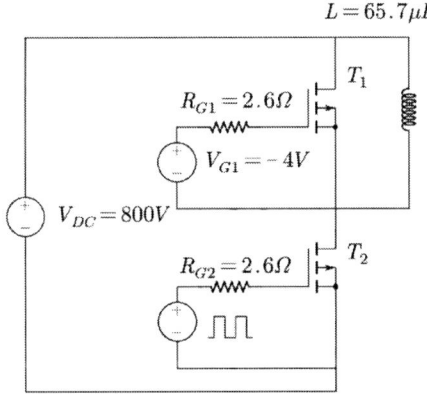

Fig. 6. Schematic of the DPT bench.

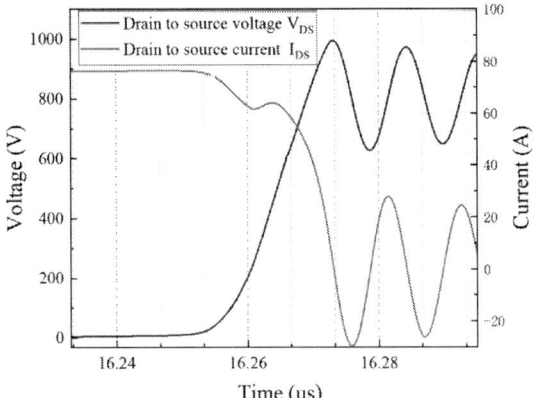

Fig. 7. Turn-OFF transient of the drain voltage and drain current.

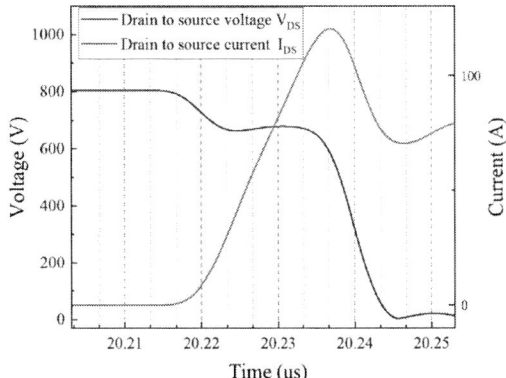

Fig. 8. Turn-ON transient of the drain voltage and drain current.

TABLE VI. COMPARISON OF DYNAMIC CHARACTERISITIC

	Datasheet	Manufacture Model	Improved Model
E_{on} (mJ)	2.3	1.56	1.06
E_{off} (mJ)	0.6	0.38	0.48
T_r (ns)	33	9.8	28.4
T_f (ns)	13	22.3	21.2
T_{don} (ns)	34	38.2	25.7
T_{doff} (ns)	65	39.7	54.1

IV. CONCLUSION

This work provides detailed insights into the modelling approach of the SiC MOSFETs in SPICE. An optimized approach was introduced to fit the model which divided the high-current and low-current regions into two parts and used a non-segmented equation to described the characteristics. To improve the accuracy in 3rd quadrant, the channel current was modeled independently. Also, the parasitic capacitance are modeled with a non-segmented equation to ensure the convergence capability. The double pules test showed that this model can predict the dynamic characteristics precisely. All the static characteristics and dynamic characteristics are compared between improved model and traditional EKV model, and the result prove that the proposed model demonstrates higher precision compared to traditional model. This study provides engineers with a reliable method for circuit design.

REFERENCES

[1] K. Han and B. J. Baliga, "The 1.2-kV 4H-SiC OCTFET: A new cell topology with improved high-frequency figures-of-merit," IEEE Electron Device Lett., vol. 40, no. 2, pp. 299–302, Feb, 2019.

[2] B. Duan, X. Yang, J. Lv, and Y. Yang, "Novel SiC/Si heterojunction power MOSFET with breakdown point transfer terminal technology by TCAD simulation study," IEEE Trans. Electron Devices, vol. 65, no. 8, pp. 3388–3393, Aug, 2018.

[3] K. Han and B. J. Baliga, "The 1.2-kV 4H-SiC OCTFET: A newcell topology with improved high-frequency figures-of-merit," IEEEElectron Device Lett., vol. 40, no. 2, pp. 299–302, Feb, 2019.

[4] Z. Ye and X. Wang, "Behavioral model of SiC MOSFET on hard-switching condition," in Proc. 21st Int. Conf. Electr. Mach. Syst. (ICEMS), Oct. 2018, pp. 811–816,

[5] A. Stefanskyi, L. Starzak, and A. Napieralski, "Universal behavioural model for SiC power MOSFETs under forward bias," in Proc. 25th Int. Conf. Mixed Design Integr. Circuits Syst. (MIXDES), Jun. 2018, pp. 343–348,

[6] Z. Zhou, Q. Ge, L. Zhao, and B. Yang, "A modified general model for sic power MOSFET in rail transportation application," in Proc. 20th Int. Conf. Electr. Mach. Syst. (ICEMS), Aug. 2017, pp. 1–5,

[7] H. A. Mantooth, K. Peng, E. Santi and J. L. Hudgins, "Modeling of Wide Bandgap Power Semiconductor Devices—Part I," in *IEEE Transactions on Electron Devices*, vol. 62, no. 2, pp. 423-433, Feb, 2015.

[8] H. Li, X. Zhao, K. Sun, Z. Zhao, G. Cao, and T. Q. Zheng, "A non-segmented PSpice model of SiC MOSFET with temperaturedependent parameters," IEEE Trans. Power Electron., vol. 34, no. 5,pp. 4603–4612, May 2019.

[9] Y. Mukunoki et al., "An improved compact model for a silicon-carbide MOSFET and its application to accurate circuit simulation," IEEE Trans.Power Electron., vol. 33, no. 11, pp. 9834–9842, Nov. 2018.

[10] P. O. Lauritzen and C. L. Ma, "A simple diode model with reverse recovery," in IEEE Transactions on Power Electronics, vol. 6, no. 2, pp. 188-191, April 1991,

Investigation and Improvement on Surge Robustness of Double-Trench SiC MOSFETs in Synchronous Rectification Mode

Hexin Zhu
*School of Integrated Circuit Science
and Engineering*
*University of Electronic Science and
Technology of China*
Chengdu, China
202322310503@ std.uestc.edu.cn

Xiaochuan Deng
*School of Integrated Circuit Science
and Engineering*
*University of Electronic Science and
Technology of China*
Chengdu, China
xcdeng@uestc.edu.cn

Haohao Dai
*School of Integrated Circuit Science
and Engineering*
*University of Electronic Science and
Technology of China*
Chengdu, China
202221020414@std.uestc.edu.cn

Qian Huang
*School of Integrated Circuit Science
and Engineering*
*University of Electronic Science and
Technology of China*
Chengdu, China
202322280211@std.uestc.edu.cn

Xuan Li
*Shenzhen Institute for Advanced Study
University of Electronic Science and
Technology of China*
Shenzhen, China
xuanli@uestc.edu.cn

Xu Li
*School of Integrated Circuit Science
and Engineering*
Southwest Jiaotong University
Chengdu, China
xuli@swjtu.edu.cn

Abstract—This paper investigates the surge capability of double-trench SiC MOSFETs under both synchronous and asynchronous rectification mode. The results indicate that in synchronous rectification mode, the terminal region fails before the cell region, which weakens the surge capability of the device. An improved surge spice model and thermal simulation model confirms that heat is concentrated in the terminal region because the conduction of the channel can suppress the activation of the body diode under positive gate bias. Subsequently, this paper utilizes TCAD simulation to optimize the device design by increasing the source trench depth to balance the surge temperatures in the terminal region and the cell region, thus achieving a compromise between surge capability and forward conduction performance.

Keywords— **SiC, MOSFET, surge, TCAD simulation, thermal model**

I. INTRODUCTION

Silicon Carbide (SiC) metal–oxide–semiconductor field effect transistors (MOSFETs) are advanced semiconductor devices that offer substantial advantages over traditional silicon MOSFETs. Due to their wider bandgap and higher breakdown voltage, SiC MOSFETs play a key role in the future of power electronics, particularly in electric vehicles, photovoltaic systems, and smart grids [1-3].

In practical applications, the body diode of a SiC MOSFET can serve as a low-cost and energy-efficient alternative to the JBS freewheeling diode. When the SiC MOSFET is under a positive gate voltage, reverse current can flow through the channel, resulting in a low conduction voltage drop and significantly reducing energy losses, which is the principle of synchronous rectification [4]. However, during the freewheeling period of the device, short-duration surge currents caused by faults in the peripheral circuitry present one of the dynamic reliability issues that need to be addressed in the practical use of SiC MOSFETs. Currently, there is amount of research on the surge characteristics of SiC devices, including studies on the degradation and failure mechanisms of SiC Schottky diodes [5] and planar SiC MOSFETs [6-8]. However, research on the surge characteristics of double-trench SiC MOSFETs, particularly under synchronous rectification, remains insufficiently explored. In addition,

there has been no research on the surge characteristics of the terminal region of SiC MOSFETs to date.

This paper presents surge testing on a double-trench SiC MOSFET under both synchronous (i.e. positive gate voltage) and asynchronous rectification mode (i.e. negative gate voltage). It is found that the surge peak current in the last test before failure under positive gate voltage was only 59% of that observed under negative gate voltage. Through device decapsulation and thermal modeling simulations, it is found that under surge conditions, the double-trench SiC MOSFET with positive gate voltage experiences thermal concentration in the terminal region, leading to device failure. This results in significantly lower surge capability under positive gate voltage compared to negative gate voltage. Furthermore, surge current analysis reveals that the opening of the channel under positive gate voltage increases the source-drain voltage V_{SD} required to turn on the body diode, preventing the body diode from discharging the surge current, thereby causing an increase in current density in the terminal region and leading to thermal failure. Finally, based on the suppression effect of reduced channel opening on the body diode, increasing the source trench depth to balance the surge temperatures in the terminal region and the cell region significantly enhances the surge capability, with only a slight increase in forward conduction resistance.

II. EXPERIMENT AND SIMULATION

The typical surge experimental waveforms of double-trench SiC MOSFET (SCT3160KL) from Rohm at gate-source bias (V_{GS}) setting to +15/-5V are shown in Fig.1. The surge current gradually increase during the test until the device fails. The maximum forward surge current (IFSM) is defined as the peak surge current at the last test before failure. It can be seen that the IFSM at V_{GS} = +15 V is 36A, which is only 59% of 61A observed under V_{GS} = -5 V. When V_{GS} = +15 V, source-drain voltage (V_{SD}) and the surge current simultaneously start to increase from zero, which is disparate in which observed for the V_{GS} = -5 V because the surge current flows through channel.

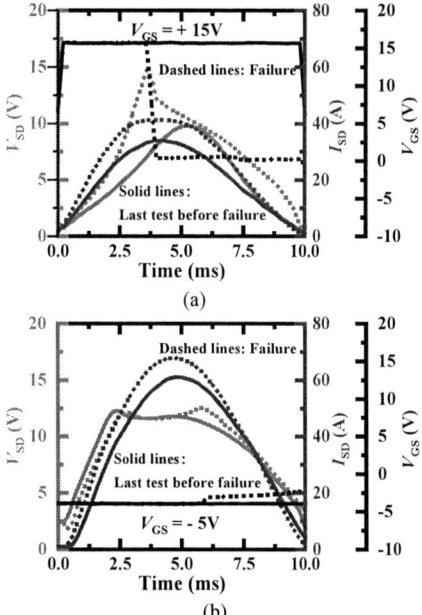

(a)

(b)

Fig.1 Surge waveforms of double-trench MOS at (a) V_{GS} = +15 V and (b) -5 V.

The failed device is decapsulated to investigate the failure mechanism of positive gate bias. As shown in Fig. 4 (a), it is evident that the hot spot is located at the edge and the surface metal does not be melted, which shows the failure mechanism of double-trench MOSFET at V_{GS} = +15 V is different from planar-gate MOSFET or double-trench MOSFET at V_{GS} = -5 V that have been reported [6]. The cross-section along the cutline verifies that the terminal is burnt through before source metal melts.

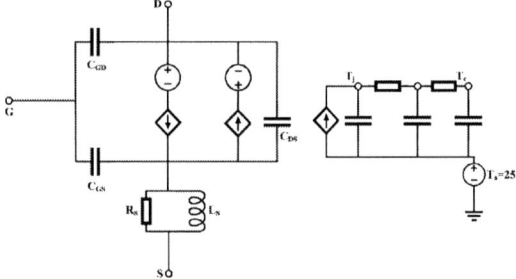

Fig. 2. The improved spice model of double-trench SiC MOSFETs

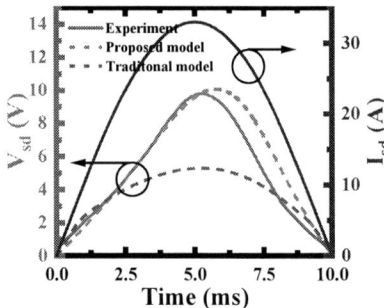

Fig. 3. Measured source-drain voltage (V_{SD}) of the device and the predicted V_{SD} obtained from the traditional simulation model and the improved model at V_{GS}=15V.

The spice model of SiC MOSFET has been continuously optimized over the years, with the aim of accurately representing characteristics under corresponding operating conditions. However, existing models cannot accurately simulate device behavior under surge conditions. Based on the measured data, the SiC MOSFET under surge conditions is fitted using an improved SPICE model, as shown in figure 2. Figure 3 shows the measured and the predicted V_{SD} obtained from the traditional simulation model and the improved model at V_{GS}=15V. The improved model provides an accurate simulation of the device's surge voltage, which allowed the distribution of the surge current between the cell region and the terminal region to be determined and incorporated into the electrothermal-force multi-physical field solution of SiC MOSFET model carried out in COMSOL as shown in Figure 4 (b). It can be observed that the terminal region experiences the highest temperature, forming a hotspot, which confirms that the device will undergo thermal damage in the terminal region, ultimately leading to a source-drain short circuit.

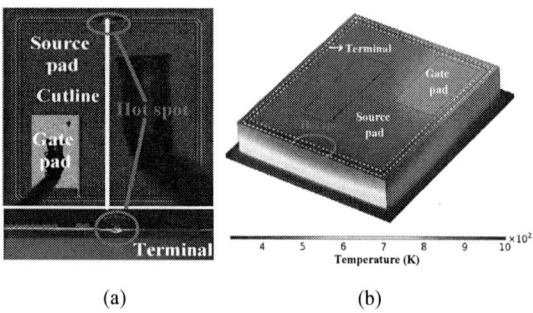

(a)   (b)

Fig.4. (a) Decapsulation image of failed double-trench SiC MOSFET at V_{GS} = +15 V. (b) COMSOL temperature distribution diagram in surge simulation of double-trench SiC MOSFET at V_{GS} = +15 V.

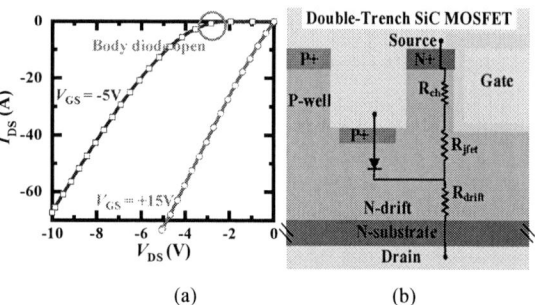

(a)   (b)

Fig.5. (a) Third-quadrant I-V characteristic. (b) Schematic of the unit-cell of double-trench SiC MOSFET.

To investigate why the surge failure mode of the double-trench MOSFET in synchronous rectification mode differs from that in asynchronous rectification mode, the third quadrant I-V test is shown in Figure 5 (a). It can be seen that when V_{GS} = -5V, the V_{SD} required to turn on the body diode, which is defined as V_{BD-ON} is 3.6V. However, when V_{GS} = +15V, the I-V curve exhibits a clear ohmic behavior, indicating that the channel opening has a suppressive effect on the body diode's turn-on. This is attributed to the electric potential distribution. Figure 5 (b) shows the schematic of the unit-cell of double-trench SiC MOSFET. The voltage fell in body diode PN junction is defined as V_{BD}. The R_{drift}, R_{ch} and

R_{jfet} are the resistance of drift, channel and JFET region respectively. V_{BD} can be given as:

$$V_{BD} = V_{SD}\left(\frac{R_{ch} + R_{jfet}}{R_{ch} + R_{jfet} + R_{drift}}\right) \qquad (1)$$

The body diode can turn on when V_{BD} is higher than V_{pn}, which is about to the bandgap of SiC. The $V_{BD\text{-}ON}$ can be written as

$$V_{BD\text{-}ON} = V_{pn}\left(1 + \frac{R_{drift}}{R_{ch} + R_{jfet}}\right) \qquad (2)$$

It can be observed that when $V_{GS} = -5V$, R_{ch} is infinite, causing $V_{BD\text{-}ON}$ to be equal to V_{pn}. However, when $V_{GS} = +15V$, the sharp decrease in R_{ch} leads to an increase in $V_{BD\text{-}ON}$ [9]. This indicates that in synchronous rectification mode, the opening of the channel suppresses the turn-on of the body diode, while the diode at the terminal is not suppressed, resulting in the surge current concentrating at the terminal.

To improve the surge capability of double-trench devices in synchronous rectification mode, the resistance of the channel or JFET region should be increased. To minimize the compromise on forward conduction capability, this paper simulates the increase in resistance of the JFET region by adjusting the trench depth. Fig.6 (a) represents the traditional double-trench device with a source trench depth of 0.8 μm, while in Fig.6 (b), the source trench depth is increased to 1.8 μm. It can be observed that, under the same surge current, the device with the deeper source trench exhibits a higher surge current, effectively reducing the concentration of surge current in the terminal region.

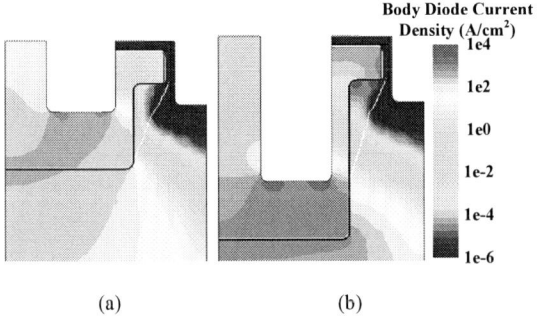

(a)   (b)

Fig.6. (a) Conventional double trench MOSFET and (b) deep source trench MOSFET simulation of body diode current density at the same surge current.

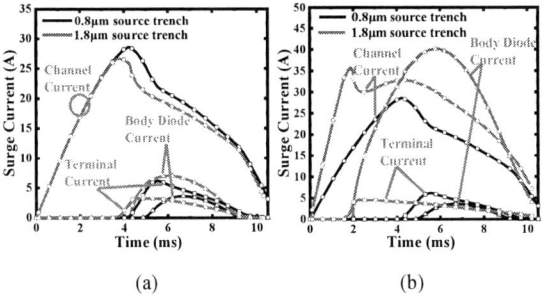

(a)   (b)

Fig.7. Surge current simulation of different source trench depth (black: 0.8 μm, red:1.8 μm) at (a) 36A surge current and (b) IFSM surge current.

Figure 7 shows the simulation surge current distribution for devices with 0.8 μm (traditional double-trench device) and 1.8 μm source trench depths. As shown in Fig.7. (a), when a 36A surge current is applied to both devices, the body diode surge current of the deep source trench device significantly increases and, at its peak, exceeds the current in the terminal region. This is attributed to the reduction in surge current in both the terminal and channel regions. Additionally, the turn-on time of the body diode occurs earlier, indicating a decrease in $V_{BD\text{-}ON}$. As the source trench depth increases, the range of the JFET region expands, and the distributed V_{SD} voltage becomes higher, promoting the turn-on of the body diode.

We define the peak surge current IFSM as the surge current value that causes the temperature in the terminal or cell region to reach the melting point of aluminum metal. According to the simulation results, the IFSM of the traditional double-trench device is 36A, while when the source trench depth is increased to 1.8 μm, the IFSM rises to 77A. Fig.7 (b) shows the surge current distribution for these two devices at their respective IFSM values. Under approximately the same surge current peak in the terminal region, the peak surge current of the body diode in the deep-source-trench device increases from 3.6A to 40.1A. The body diode becomes the primary current path for the surge current. Additionally, the time interval between the terminal turn-on and the body diode turn-on is significantly shortened. This demonstrates that the higher JFET region resistance brought about by the deeper source trench effectively reduces the suppression effect of the channel opening on the body diode.

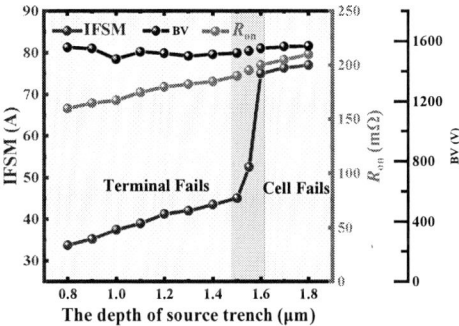

Fig.8. IFSM and forward conduction resistance at various depth of source trench.

As shown in Fig.8, with the increase in source trench depth, both IFSM and the forward conduction resistance (R_{on}) increase, while the breakthrough voltage (BV) keeps constant. Notably, there is a significant rise in IFSM between the 1.5 μm and 1.6 μm. Before this range, the device experiences failure in the terminal region, while beyond this point, failure occurs in the cell region. At a source trench depth of 1.6 μm, a good compromise is achieved between IFSM and R_{on}. The IFSM increased by 127%, while the R_{on} only increased by 25%.

III. CONCLUSION

In this study, the surge characteristics of double-trench SiC MOSFETs under both synchronous and asynchronous rectification modes were investigated. The results demonstrate that the surge capability is significantly affected by the gate voltage, with devices operating under positive gate voltage exhibiting a notably lower surge peak current compared to those under negative gate voltage. This is attributed to the thermal concentration in the terminal region, which leads to device failure. Surge current analysis further reveals that the channel opening under positive gate voltage increases the V_{SD} required to turn on the body diode, preventing it from discharging surge current and resulting in thermal failure. By increasing the source trench depth to raise the resistance of the JFET region, the suppression effect on the body diode is weakened. Simulation results validate that as the source trench depth increases from 0.8 μm (in the traditional double-trench SiC MOSFETs) to 1.8 μm, the body diode in the deep-source-trench device conducts more surge current under the same surge conditions, raising the IFSM from 36A to 77A. To balance the reduction in conduction capability caused by the increase in the JFET region, the source trench depth was adjusted. Before a depth of 1.5 μm, the terminal region fails first, while beyond 1.6 μm, the cell region fails first. At a source trench depth of 1.6 μm, IFSM increased by 127%, while the conduction resistance R_{on} increased by only 25%.

ACKNOWLEDGMENT

This work was supported in part by the National Natural Science Foundation of China under Grant 62334004 and the Natural Science Foundation of Jiangsu Province under Grant BK20232027.

REFERENCES

[1] X. She, A. Q. Huang, Ó. Lucía and B. Ozpineci, "Review of Silicon Carbide Power Devices and Their Applications," *IEEE Trans. Ind. Electron.*, vol. 64, no. 10, pp. 8193–8205, Oct. 2017, doi:10.1109/TIE.2017.2652401.

[2] J. G. Kassakian and T. M. Jahns, "Evolving and Emerging Applications of Power Electronics in Systems," *IEEE J. Emerg. Sel. Top. Power Electron.*, vol. 1, no. 2, pp. 47-58, June 2013, doi: 10.1109/JESTPE.2013.2271111.

[3] X. Deng et al., "Investigation of Failure Mechanisms of 1200 V Rated Trench SiC MOSFETs Under Repetitive Avalanche Stress," *IEEE Trans. Power Electron*, vol. 37, no. 9, pp. 10562-10571, Sept. 2022, doi: 10.1109/TPEL.2022.3163930.

[4] S. Yin, Y. Liu, Y. Liu, K. J. Tseng, J. Pou and R. Simanjorang, "Comparison of SiC Voltage Source Inverters Using Synchronous Rectification and Freewheeling Diode," *IEEE Trans. Ind. Electron.*, vol. 65, no. 2, pp. 1051-1061, Feb. 2018, doi: 10.1109/TIE.2017.2733483.

[5] X. Jiang et al., "Investigation on Degradation of SiC MOSFET Under Surge Current Stress of Body Diode," *IEEE J. Emerg. Sel. Top. Power Electron.*, vol. 8, no. 1, pp. 77-89, March 2020, doi: 10.1109/JESTPE.2019.2952214.

[6] X. Zhan et al., "Investigation on Degradation of 1200-V Planar and Trench SiC MOSFET Under Surge Current Stress of Body Diode," *IEEE Trans. Electron Devices*, vol. 71, no. 1, pp. 709-714, Jan. 2024, doi: 10.1109/TED.2023.3335892.

[7] X. Jiang et al., "Investigation on Degradation of SiC MOSFET Under Surge Current Stress of Body Diode," *IEEE J. Emerg. Sel. Top. Power Electron.*, vol. 8, no. 1, pp. 77-89, March 2020, doi: 10.1109/JESTPE.2019.2952214.

[8] A. Mihaila et al., "Surge current capability of 6.5kV-rated SiC MOSFETs," in 2020 32nd International Symposium on Power Semiconductor Devices and ICs (ISPSD), Vienna, Austria, 2020, pp. 222-225, doi: 10.1109/ISPSD46842.2020.9170186.

[9] R. Zhang, X. Lin, J. Liu, S. Mocevic, D. Dong and Y. Zhang, "Third Quadrant Conduction Loss of 1.2–10 kV SiC MOSFETs: Impact of Gate Bias Control," *IEEE Trans. Power Electron*, vol. 36, no. 2, pp. 2033-2043, Feb. 2021, doi: 10.1109/TPEL.2020.3006075.

2025 IEEE Workshop on Wide Bandgap Power Devices and Applications in Asia (WiPDA Asia)

A High-Output-Current Radiation Hardened Half Bridge DC-DC Converter for Space Applications

Dawei Li
*Power Electronics and
Renewable Energy Research
Center
Xi'an Jiaotong University*
Xi'an, P.R. China
20010312@stu.xjtu.edu.cn

Xiang Zhou
*Power Electronics and
Renewable Energy Research
Center
Xi'an Jiaotong University*
Xi'an, P.R. China
zhouxiang06@xjtu.edu.cn

Yeerzhati Nuerdebieke
*Power Electronics and
Renewable Energy Research
Center
Xi'an Jiaotong University*
Xi'an, P.R. China
1522908286@stu.xjtu.edu.cn

Abstract—A radiation-hardened DC-DC converter is proposed in this paper to meet the special requirements of space applications. Topology selection and parameter design are carried out. To ensure high efficiency and wide-range voltage regulation, two methods are introduced to suppress the impact of transformer leakage inductance. The proposed radiation-hardened prototype achieves DC-DC conversion from 95V~105V input to 3.3V output, supporting up to 80A full-load current. The prototype achieves a power density of 42.4W/in^3 and a peak efficiency of 94.74%.

Keywords—Radiation hardened, half bridge DC-DC, high efficiency, high power density

I. INTRODUCTION

With the rapid advancement of space technology, the demand for reliable and high-performance power conversion systems in space applications is increasing. In the space environment, electronic devices are exposed to intense radiation for prolonged periods, which can potentially lead to system failures.

Power converters play a critical role in space-based systems, as they are responsible for converting the high-voltage power generated by solar panels or other energy sources into appropriate low-voltage levels required by various electronic devices. For example, in lunar rovers, a stable low-voltage power supply is essential for the operation of microcontrollers, sensors, and communication modules.

However, high-energy protons, electrons, and heavy ions in space can induce Single-Event Effects (SEE), Total Ionizing Dose (TID) effects and Displacement Damage Dose (DDD) effects [1]. These phenomena may cause unintended conduction of the intrinsic P-N junctions within power MOSFETs. Additionally, they can lead to structural changes in the MOSFET lattice, resulting in electrical performance degradation [2]. Furthermore, these energetic particles can create electrical energy spikes at the input and output stages of the power management and protection circuits, adversely affecting the performance of the converter [3].

To address these challenges, this study presents a radiation-hardened converter with an input voltage range of 95V to 105V and a 3.3V/80A output. By incorporating advanced radiation-hardened components and optimizing transformer design, the proposed converter aims to deliver a stable, efficient, and radiation-hardened power solution. This paper is organized as follows: Section II covers the topology selection, parameter calculation and component selection.

Section III analyzes the impact of transformer leakage inductance on converter performance and discusses optimization methods. Section IV presents the experimental results of the prototype.

II. THE CONVERTER TOPOLOGY AND PARAMETERS DESIGN

A. Circuit Topology

The objective of this study is to design a high-efficiency and high-power-density radiation hardened converter with a high output current of 3.3V/80A. Traditional isolated DC-DC space power converters feeding from a 100 V bus use Flyback and Forward topologies [4]. However, the performance and power density of single-ended converters are limited, making them unsuitable for the power level of this study. Double ended converters allow transformer operation over two quadrants in the magnetic hysteresis B-H curve, thus making more efficient use of the transformer and achieving higher efficiency and power density.

Additionally, in noise sensitive space applications, fixed frequency pulse-width-modulation (PWM) is preferred over frequency modulated control schemes. Power converters are the main source of electromagnetic interference (EMI). In space applications, it is important to know at which frequencies power converters are running. This enables the synchronization of converter switching clocks when necessary to reduce EMI and, in addition, allows for better optimization of passive and active filters in the system [5]. Managing EMI is a key requirement for operating highly sensitive instrumentation and RF power amplifiers in space. Therefore, frequency-modulated resonant topologies that achieve ZVS DC-DC conversion, such as the LLC, are not suitable for this type of design.

Therefore, double-ended topologies with fixed-frequency pulse-width-modulation, such as half-bridge and full-bridge converters, are better suited to space applications. However, for 100V to 3.3V conversion, a full-bridge converter requires a transformer with an excessively high turns ratio, making it

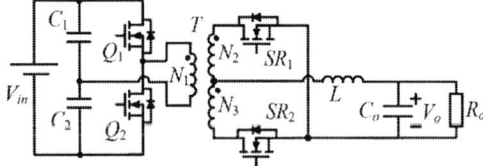

Figure 1: Main circuit of the radiation-hardened converter

TABLE I. COMPONENT CHARACTERISTICS

Component	Electrical	Radiation-hardened
Primary side switch	V_{DS}=200V, I_D=45A	100KRad for TID, 75 MeV*cm²/mg for SEE
Secondary side switch	V_{DS}=30V, I_D=75A	100KRad for TID, 75 MeV*cm²/mg for SEE

difficult to manufacture. Moreover, with a 100V bus voltage, a half-bridge converter can still achieve a low RMS current on the primary side, resulting in a simpler and more efficient circuit.

In summary, this study ultimately selects the half-bridge topology with synchronous rectification as the design solution for the radiation-hardened converter, with the main circuit illustrated in Fig. 1.

B. Parameter Design

As shown in Fig. 1, the turns ratio of the transformer is $N_1 : N_2 : N_3$ and $N_2 = N_3$. The simplified input-output voltage relationship is:

$$V_{in} * \frac{1}{n} * D = V_{out} \tag{1}$$

Where $n = N_1 : N_2$ is the turns ratio and D is the duty cycle. To enhance the utilization of the transformer while ensuring the reliability of the converter, the duty cycle is typically designed not to exceed 0.46. For the application with an input voltage of 95-105V and a 3.3V output, considering the voltage drop across the devices and the voltage drop on the line caused by the large current on the secondary side, the turns ratio is selected as 12:1:1.

The maximum RMS current of the primary side switch is:

$$I_{p,RMS,\max} = \frac{\sqrt{2}P_o}{\eta V_{in.\min}} \tag{2}$$

Where P_o is the rated output power, $V_{in,\min}$ is the minimum input voltage, $\eta = 0.86$ is the lowest efficiency of the power converter. By calculation, the maximum RMS current of the primary side switch is 4.6 A. The maximum RMS current of the primary side winding is 1.414 times the maximum RMS current of the primary side switch, which is 6.5 A.

The maximum RMS current of the secondary side switch is:

$$I_{s,RMS,\max} = \frac{\sqrt{2}I_o}{2} \tag{3}$$

Where I_o is the rated output current, the calculated maximum RMS current of the secondary side switch is 56.6A. The maximum RMS current of the secondary winding is the same as that of the secondary side switch, which is also 56.6A.

The required output filter inductor is:

$$L \geq \frac{(0.5 - D) * V_o}{f * \Delta I_o} \tag{4}$$

Where $f = 100$kHz is the switching frequency, $\Delta I_o = 10$A is the inductor current ripple. The calculated minimum required inductance is $0.34\mu H$. In practice, an $0.37\mu H$ inductor is selected.

The required output filter capacitor is:

$$C_o \geq \frac{(0.5 - D) * V_o}{16 * L * f^2 * \Delta V_o} \tag{5}$$

Where $\Delta V_o = 25$mV is the designed output voltage ripple. The calculated value shows that the output filter capacitance should be at least $230\mu F$.

C. Component Selection

The electrical and radiation-hardened characteristics of the selected MOSFETs are shown in Table I. The output filter inductor is selected with an inductance of $0.37\mu H$ and a rated current of 100A. Nine $100\mu F$ MLCC capacitors are connected in parallel as the output filter capacitor.

III. IMPACT OF LEAKAGE INDUCTANCE AND OPTIMIZATION

In isolated power converters, the transformer leakage inductance has been known to cause a voltage spike associated with the switching, which can reduce system efficiency [6]. In a half-bridge converter with synchronous rectification, increased leakage inductance not only exacerbates the drain-source voltage oscillation of the secondary-side synchronous rectifier but also generates EMI interference and additional losses [7], thereby reducing the reliability and efficiency of the converter.

To address these issues, this study adopts two improvement methods. The first is the use of interleaved winding technology for the transformer windings, as shown in Fig. 2(a), to reduce leakage inductance. At the same time, interleaved windings are also beneficial to reduce the eddy current loss [8].

The second is the addition of an RCD snubber circuit to the secondary-side synchronous rectifier, as shown in Fig.

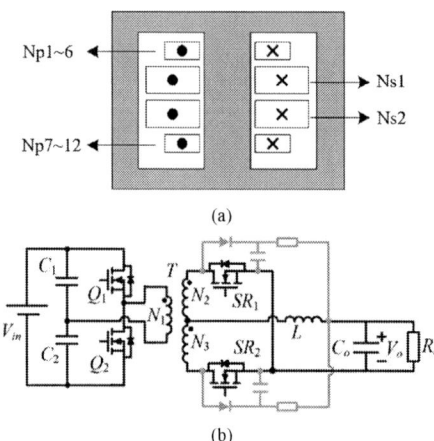

(a)

(b)

Figure 2: (a) Interleaved winding structure;
(b) Half bridge topology with an RCD snubber

2(b). The purpose of the RCD snubber circuit is to suppress the undesired oscillations at the drain - source terminal of the MOSFET. Such oscillations are caused by device imperfections, namely, the resonance between the leakage inductance of the transformer and the parasitic capacitance between the drain and source of the MOSFET [9].

During the oscillation, when the drain-source voltage of the MOSFET exceeds the voltage across the snubber capacitor, the diode conducts, and the capacitor is charged. When the drain-source voltage of the MOSFET drops below the capacitor voltage, the diode turns off, and the capacitor discharges through the snubber resistor. In this design, the snubber resistor is not connected to the drain of the MOSFET, which allows the minimum voltage after capacitor discharge to be approximately the output voltage rather than zero. As a result, the current through the diode will be smaller during the next energy absorption cycle.

The oscillation occurs during the commutation process of the two branches in the secondary full - wave rectification. At this time, the energy in the transformer leakage inductance can be expressed as:

$$E = \frac{1}{2} L_{r\_s} \left(\frac{I_o}{2} \right)^2$$

$$L_{r\_s} = \frac{1}{2} \frac{L_r}{(n/2)^2}$$

(6)

Where L_{r_s} is the leakage inductance of one branch in the secondary-side full-wave rectifier circuit, referred from the primary side L_r, and I_o is the rated output current.

Assuming that all the energy in the transformer leakage inductance can be fully absorbed by the snubber capacitor, the capacitance value of the snubber capacitor is:

$$C_{RCD} = \frac{2E}{(V_p^2 - V_o^2)}$$

$$V_p = 1.15 * \frac{V_{in}}{n}$$

(7)

Where V_o is the output voltage, and V_p is the voltage after the snubber capacitor is charged, which is set as 1.15 times the reverse voltage across the synchronous rectifier in the off state [10].

The discharge time of the snubber capacitor should be no less than four time constants [11], and the selection of the resistor should satisfy:

$$R_{RCD} \leq \frac{1}{4 C_{RCD} f}$$

(8)

The snubber diode should be a fast-recovery diode, and its rated current should be no less than 1/10 of the rated current of the MOSFET [12].

In this study, the transformer leakage inductance is designed to be no more than 1.6 µH. Based on (6)-(8), the calculated snubber capacitance and resistance values are determined to be 439 nF and below 5.7 Ω. Accordingly, a 470 nF snubber capacitor and a 5 Ω resistor are selected. The corresponding circuit configuration, as illustrated in Fig. 3, has been constructed in LTspice for simulation verification.

Without the snubber circuit, the drain-source voltage oscillation waveform of the secondary-side MOSFET is

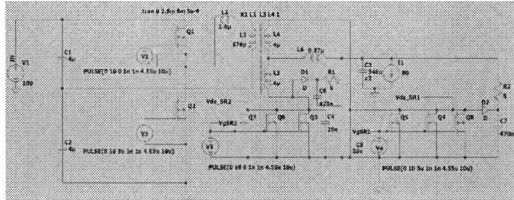

Figure 3: Circuit for simulation verification.

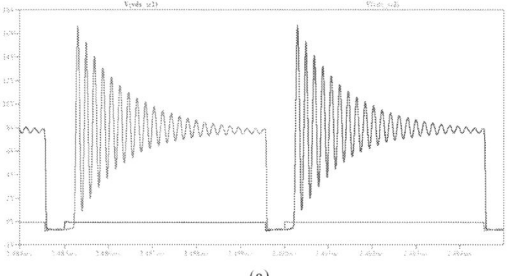

(a)

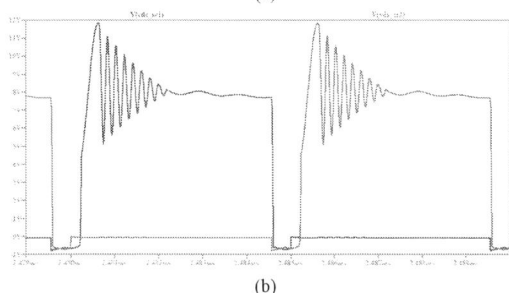

(b)

Figure 4: Comparison of drain-source voltage oscillations, (a) with snubber circuit, (b) without snubber circuit.

Figure 5: Radiation-hardened converter prototype

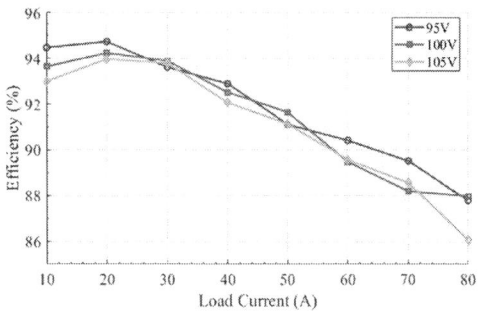

Figure 6: Efficiency curve of the prototype

shown in Fig. 4(a), where the peak drain voltage reaches 16.6 V. After implementing the snubber circuit, the oscillation waveform presented in Fig. 4(b) demonstrates a reduced peak drain voltage of 11.8 V.

IV. RESULTS

The developed radiation-hardened converter prototype is shown in Fig. 5, detailed specifications are shown in Table II.

TABLE II. PROTOTYPE SPECIFICATIONS

Parameter	Value
Input Voltage	95V-105V
Output Voltage	3.3V
Rated Current	80A
Voltage Ripple	Max = 20.4 mV
Peak Efficiency	94.74%
Line Regulation	0.03%
Load Regulation	1.3%

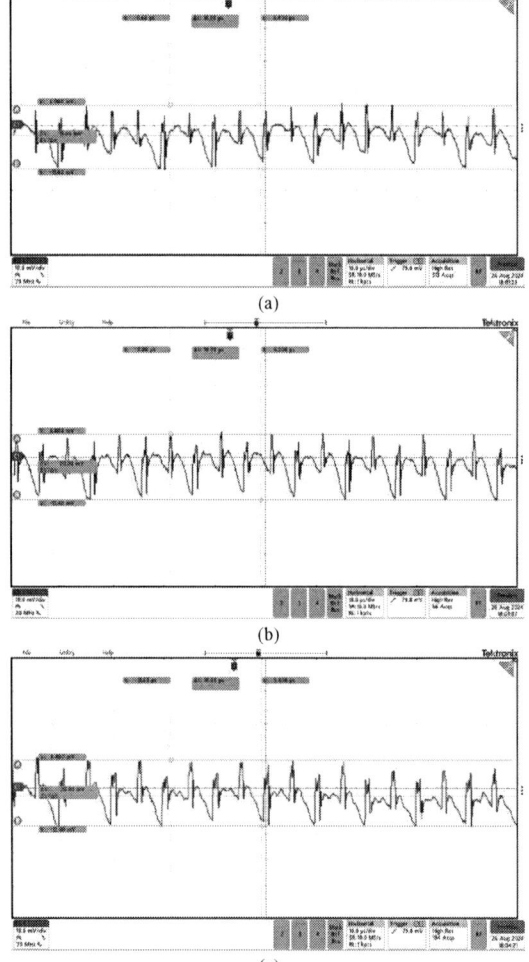

(a)

(b)

(c)

Figure 7: Measured output voltage ripple at different inputs, (a) at 95V input voltage, (b) at 100V input voltage, (c) at 105V input voltage

The prototype measures 148mm × 46mm × 15mm and delivers a 3.3V, 80A output, achieving a power density of 42.4W/inch³.

The efficiency curve of the prototype is shown in Fig. 6. The prototype achieves a peak efficiency of 94.74% at an input of 95V and a load of 20A.

In order to provide stable and accurate power supply, the output voltage ripple of the converter in this study was designed to be below 25 mV. As shown in Fig. 7, the measured results are 19.6mV, 20.2 mV and 20.4mV at 95 V, 100 V and 105 V inputs at 3.3V/80A output.

V. CONCLUSION

With the swift progress in space technology, the need for radiation-hardened converters is on the rise. To cope with the harsh conditions of space applications, this study designed a radiation-hardened converter based on a half-bridge topology with synchronous rectification. The consideration of topology selection and the design process of the main parameters are given. The design incorporates a secondary-side RCD snubber circuit and a transformer interleaved winding to improve performance and reliability. The developed prototype delivers a 3.3V, 80A output, achieving a power density of 42.4W/in³ and a peak efficiency of 94.74%.

REFERENCES

[1] S. Michelis, N. H. Van Der Blij, G. Ripamonti and P. D. Antoszczuk, "bPOL48V, a rad-hard 48V DC/DC Converter for Space and HEP Applications," 2023 13th European Space Power Conference (ESPC), Elche, Spain, 2023, pp. 1-4, doi: 10.1109/ESPC59009.2023.10298148.

[2] L. D. Edmonds, C. E. Barnes, and L. Z. Scheick, "An Introduction to Space Radiation Effects on Microelectronics," NASA Jet Propulsion Laboratory, JPL Publication 00-06, Pasadena, CA, 2000, pp. 11–59.

[3] V. Turriate, "Design and Implementation of a Radiation Hardened GaN Based Isolated DC-DC Converter for Space Applications," M.S. thesis, Virginia Tech, Blacksburg, VA, USA, 2018.

[4] V. Turriate, B. Witcher, D. Boroyevich, R. Burgos, "Design Considerations for a Gallium Nitride Based Phase Shifted Full Bridge DC-DC Converter for Space Applications," in Proc. IEEE 6th Workshop on Wide Bandgap Power Devices and Applications (WiPDA), 2018.

[5] V. Turriate, B. Witcher, D. Boroyevich and R. Burgos, "Practical Implementation and Efficiency Evaluation of a Phase Shifted Full Bridge DC-DC Converter Using Radiation Hardened GaN FETs for Space Applications," 2019 IEEE Applied Power Electronics Conference and Exposition (APEC), pp.1587-1594, doi:10.1109/APEC. 2019. 8722059.

[6] T. -J. Liang, J. -H. Lee, S. -M. Chen, J. -F. Chen and L. -S. Yang, "Novel Isolated High-Step-Up DC–DC Converter With Voltage Lift," in IEEE Transactions on Industrial Electronics, vol. 60, no. 4, pp. 1483-1491, April 2013, doi: 10.1109/TIE.2011.2177789.

[7] J. M. Zhou, X. K. Zhao, and S. Y. Tao, Kai guan dian yuan ci xing yuan jian li lun ji she ji [Theory and design of magnetic components in switching power supplies]. Beijing, China: Beijing University of Aeronautics and Astronautics Press, 2014.

[8] M. H. Ahmed, C. Fei, F. C. Lee and Q. Li, "48-V Voltage Regulator Module With PCB Winding Matrix Transformer for Future Data Centers," in IEEE Transactions on Industrial Electronics, vol. 64, no. 12, pp. 9302-9310, Dec. 2017, doi: 10.1109/TIE.2017.2711519.

[9] D. A. Rahardianto, W. Djuriatno, R. N. Hasanah and Taufik, "Design and Implementation of a RCD Snubber Circuit on Full-Bridge DC-DC Converter," 2023 International Conference on Technology and Policy in Energy and Electric Power (ICT-PEP), Jakarta, Indonesia, 2023, pp. 244-249, doi: 10.1109/ICT-PEP60152.2023.10351181.

[10] Zhang, X. B., & Li, J. Y. (2010). Simulation and Experiment of the RCD Snubber Circuit. In Proceedings of the 2010 Annual Academic Conference of Henan Metallurgical Society (pp. 65 - 73). Henan Metallurgical Society.

979-8-3315-1110-4/25 $31.00 © 2025 IEEE

[11] Z. Jiemin. Switching Power Supply Theory and Design. Beijing: Beihang University Press, Mar. 2012, ISBN 978 - 7 - 5124 - 0715 - 2.

[12] Wang, Z. A., & Liu, J. J. Power Electronics Technology. Beijing: China Machine Press, ISBN: 978 - 7 - 111 - 26806 – 2.

The Thermal Simulation Study of DC-DC Converters Based on Phase Change Cooling

Boyang Liu
School of Electrical Engineering
Dalian University of Technology
Dalian,China
1205890424@mail.dlut.edu.cn

Haoyu Zhang
School of Electrical Engineering
Dalian University of Technology
Dalian,China
zhhy@mail.dlut.edu.cn

Yu Han
State Grid Liaoning Extra-High
Voltage Company
State Grid Corporation of China
Shenyang,China
han0yu0@126.com

Jialiang Chi
School of Electrical Engineering
Dalian University of Technology
Dalian,China
chijialiang2000@mail.dlut.edu.cn

Yao Zhao
School of Electrical Engineering
Dalian University of Technology
Dalian,China
yaozhao@mail.dlut.edu.cn

Zhiqiang Wang*
School of Electrical Engineering
Dalian University of Technology
Dalian,China
wangzq@dlut.edu.cn

Abstract—DC-DC converters typically rely on liquid cooling for heat dissipation. However, these methods suffer from slow heat dissipation and large temperature fluctuations. Under short-term high-power conditions, these methods struggle to effectively limit the temperature rise of power devices, compromising the reliable operation of DC-DC converters. Phase-change heat dissipation is more promising but lacks theoretical analysis at the converter level. To this end, the flow-solid-thermal coupling model of the phase change heat sink is established by studying phase change materials (PCMs) such as paraffin wax and low-melting-point alloy (LMPA), and using numerical simulations. The heat dissipation performance of different PCMs is analyzed and compared with that of traditional cooling methods. The results show that under power losses of 45.2 W, the temperature rise of the SiC MOSFET using an LMPA is reduced by more than 50% compared to paraffin-based cooling. With power losses of 73.2 W, the phase change heat sink reduces the SiC MOSFET temperature by 23.6 °C compared to a conventional heat sink.

Keywords—*DC-DC converter, thermal management, PCMs, heat sink, computational fluid dynamics*

I. INTRODUCTION

DC-DC converters enable efficient energy conversion and are widely used in fields such as new energy generation and electromagnetic energy weaponry [1, 2]. However, in applications like battery charging, capacitor charging, and electromagnetic emission, they often operate under extreme conditions with short-term high power. The transient power impact in such conditions causes a rapid rise in the junction temperature of power devices. Conventional cooling methods, such as forced air and liquid cooling, are limited by thermal response speed, making it difficult to effectively control temperature rise within intermittent operating conditions. Phase change materials (PCMs) present a potential solution due to their high latent heat and temperature regulation capability. During the phase transition, PCMs absorb a significant amount of heat while maintaining a stable temperature, thereby mitigating fluctuations.

Many researchers have investigated PCMs for thermal management of electronic components. For instance, Borkar et al. analyzed the cooling performance of PCM-based hybrid heat sinks under different operating conditions through three-dimensional numerical modeling, revealing the effects of microchannel aspect ratio, fin spacing, and Reynolds number on PCM performance [3]. Singh et al. employed a composite structure of n-eicosane and Cu-based thermal conductivity enhancers to evaluate the cooling potential of PCMs under continuous and intermittent operation [4]. However, most existing studies focus on electronic devices, with limited analysis at the converter level.

In this paper, the circuit model of a DC-DC converter is first constructed, and power losses of the SiC MOSFET are calculated based on a theoretical analysis. On this basis, a three-dimensional coupled flow-solid-heat model of a finned phase change heat sink is developed using ANSYS Fluent software to simulate the phase change process of the heat sink. Then the thermal characteristics of the heat sink are analyzed under different PCMs, such as paraffin wax and low-melting-point alloy (LMPA). Finally, the heat dissipation effect of the LMPA heat sink is evaluated compared with that of the conventional heat sink.

II. THREE DIMENSIONAL STRUCTURE OF PHASE CHANGE HEAT SINK

A typical full-bridge LLC-type DC-DC converter is studied as an example, with its circuit topology shown in Fig. 1. The converter parameters are listed in TABLE I. The inverter circuit consists of four SiC MOSFETs. The LLC converter enables soft switching of SiC MOSFETs, with losses including conduction losses, turn-off losses and body diode conduction losses [5].

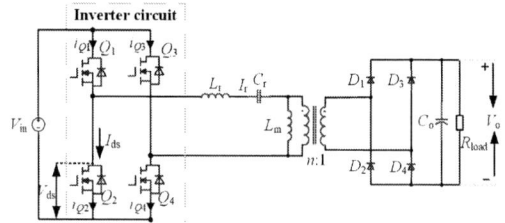

Fig. 1. Topology diagram of LLC-type DC-DC converter.

Based on the DC-DC converter topology, a three-dimensional physical model is constructed, as shown in Fig. 2. This model integrates four SiC MOSFETs as heat sources and a finned phase change heat sink as the thermal management unit. The thermal management unit consists of a phase change unit and a heat sink unit. The central phase change unit absorbs the heat generated by the SiC MOSFETs using the latent heat of the PCM, maintaining the temperature near the phase change point. The heat sink unit then transfers the stored heat to the surroundings through aluminum fins, ensuring the phase change unit operates safely and reliably. To further enhance heat transfer efficiency, a 1 mm thick copper plate is placed between the heat source and the phase change unit [6].

TABLE I.   LLC CONVERTER PARAMETERS

Symbol	Parameter	Value
V_{in}	Input Voltage	200~400 V
P_{nom}	Nominal Power	4000 W
Q_1~Q_4	Primary Switch	IMW65R048M1H
L_r	Resonant Inductance	58 µH
C_r	Resonant Capacitance	207 nF
L_m	Magnetizing Inductance	197 µH
f_{sw}	Switching Frequency	46 kHz
R_{Load}	Load	20 Ω

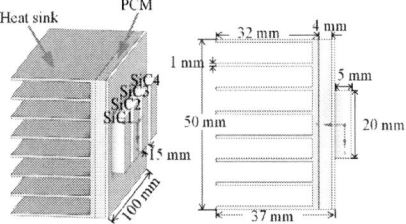

Fig. 2. Physical modeling of phase change heat sink.

The selected PCMs include paraffin RT46 and the LMPA Bi-Pb-In-Sn-Cd, both of which are commonly used for heat dissipation in electronic devices [7, 8]. The following assumptions are made for the simulation model:

1) The fluid flow is considered incompressible, laminar and Newtonian.

2) The Boussinesq approximation is applied to simulate the natural convection effects during the phase change.

3) The PCM is assumed to be ideal, isotropic, and uniformly distributed.

4) Volume changes during the phase transition are neglected, and the specific heat capacity and thermal conductivity of the PCM are considered constant.

5) Radiative heat transfer is assumed to be negligible.

TABLE II.   PHYSICAL PARAMETERS OF THE MATERIAL

Parameter	Paraffin RT46	Bi-Pb-In-Sn-Cd	Al	SiC
Density (kg/m³)	770	9495	2702	3210
Specific heat (J/(kg·K))	2200	241	903	690
Thermal conductivity (W/(m·K))	0.2	14.7	237	245
Latent heat (J/kg)	170000	32300	-	-
Phase transition temperature (℃)	41-48	41-48	-	-

The initial conditions considered in this study are as follows: the external ambient temperature $T_a = 25$ °C and heat

exchange occurs through natural convection at the side and bottom boundaries of the heat sink, with a convection heat transfer coefficient of 10 W/(m²·K). The relevant physical parameters are presented in TABLE II. Since only the phase transition of the PCM is considered, the latent heat and phase transition temperatures of other materials are not listed.

III. SIMULATION ANALYSIS

A. Selection of PCM

The power losses of SiC MOSFETs are calculated based on the Reference [9]. The total power losses of the SiC MOSFETs are approximately 45.2 W at input voltage $V_{in} = 350$ V, load $R_{load} = 20$ Ω, and switching frequency $f_{sw} = 46$ kHz. Therefore, the power losses of each SiC MOSFET are set to 11.3 W. Assuming a running time of 360 s for the LLC converter, Fig. 3 and Fig. 4 show the calculated results for the PCM temperature, SiC MOSFET temperature, heat sink temperature and liquid phase rate, as well as the temperature distribution of different PCMs over time.

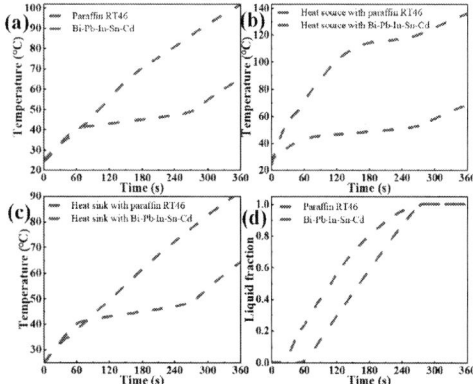

Fig. 3. Temperature comparison of (a) PCM, (b) SiC MOSFET, (c) heat sink, and (d) liquid phase fraction.

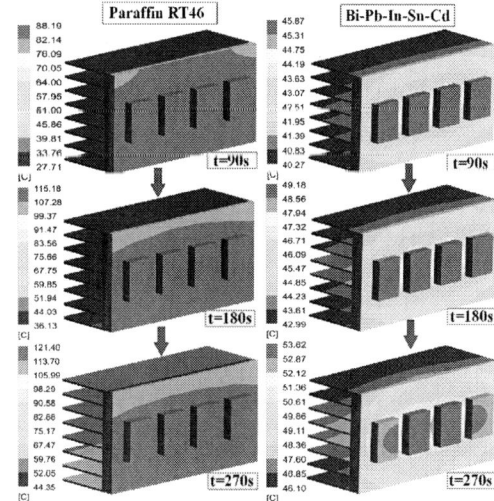

Fig. 4. Temperature distribution of paraffin RT46 and Bi-Pb-In-Sn-Cd at different time periods.

From Fig. 3(a) and (b), when paraffin RT46 is used as the PCM, its temperature reaches 101.8 °C, and the temperature of the SiC MOSFET reaches 136.1 °C. In contrast, when Bi-

Pb-In-Sn-Cd is used as the PCM, its temperature reaches 65.2 °C, and the temperature of the SiC MOSFET reaches 68.9 °C. Compared to paraffin RT46, the temperature rise of the SiC MOSFET is reduced by more than 50% when Bi-Pb-In-Sn-Cd is used, indicating that paraffin RT46 is significantly less effective in inhibiting the temperature rise of the heat source than Bi-Pb-In-Sn-Cd.

Bi-Pb-In-Sn-Cd can quickly store heat as latent heat due to its high thermal conductivity. In contrast, the heat sink using paraffin wax, as shown in Fig. 4, does not exhibit a significant reduction in the temperature rise rate during the phase transition due to the slow heat transfer rate of paraffin wax. Additionally, the delay in the temperature rise rate reduction of the SiC MOSFET in Fig. 3 is due to the fact that the paraffin wax region near the heat source melts first, while the region farther from the heat source experiences a large temperature difference between the melted and unmelted areas. This results in limited heat transfer, which prolongs the phase transition process. In contrast, LMPA, with their high thermal conductivity, rapidly releases the absorbed heat to the surrounding environment, making them better suited for the design requirements of phase change heat sinks.

B. LMPA Performance Simulation

To verify the phase transition characteristics of the Bi-Pb-In-Sn-Cd under short- duration high-power operation, the power losses of the heat source are further increased in this study. The specific conditions are as follows: T_a = 25 °C, V_{in} = 400 V, R_{load} = 13 Ω, and f_{sw} = 60 kHz. The LLC converter operates for 200 s before entering the cooling state. The power losses of each SiC MOSFET under these conditions are 18.3 W, and the convective heat transfer coefficient between the finned heat sink and the atmosphere is set to 10 W/(m²·K). The results of the SiC MOSFET junction temperature, the Bi-Pb-In-Sn-Cd temperature, and the liquid-phase rate calculations are shown in Fig. 5.

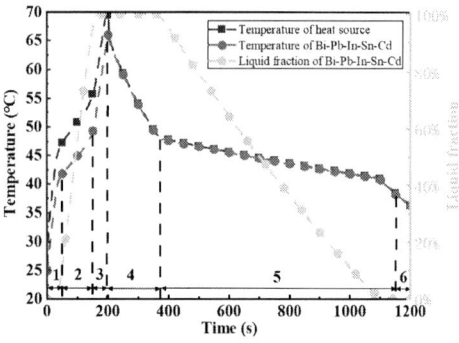

Fig. 5. The distribution chart of temperature and liquid phase fraction.

The temperature variation curve in Fig. 5 is divided into six stages, covering the two main processes of heating and solidification, with the alloy regulating the temperature by absorbing and releasing heat in an orderly manner. During the heating process, the alloy temperature rises rapidly from 25 °C to the phase transition temperature of 41 °C in stage 1, with heat absorbed by the solid sensible heat. In stage 2, the temperature stabilizes near the phase transition point, with heat mainly absorbed by the latent heat of the phase transition, lasting for about 100 s. In stage 3, the alloy completely melts, and the temperature continues to rise as heat is absorbed by the liquid sensible heat. Heating stops when the temperature

of the heat source reaches a maximum of 71.8 °C, and the solidification process begins. In stage 4, heat is rapidly released as liquid sensible heat, causing the temperature to drop quickly to the phase transition point. In stage 5, heat is released as latent heat of phase change, and the temperature changes slowly. In stage 6, the release of latent heat of phase change is completed, and the alloy solidifies completely. Solid sensible heat continues to be released, and the temperature drop rate slows down gradually. Simulation results show that the Bi-Pb-In-Sn-Cd exhibits excellent phase transition characteristics at 18.3 W, effectively coping with thermal load changes.

Finally, we compare the heat dissipation performance of the Bi-Pb-In-Sn-Cd phase change heat sink with that of a heat sink without added PCM under the same conditions. The LLC operates for 200 s, after which it enters the cooling state, with each SiC MOSFET consuming 18.3 W. The temperature changes of the heat source and heat sink during the heating and solidification phases are observed to evaluate the transient heat dissipation performance. The results are shown in Fig. 6.

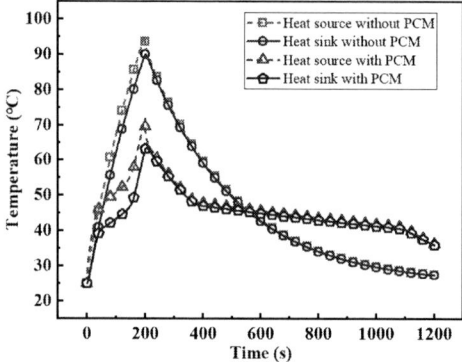

Fig. 6. Temperature comparison of the heat source and heat sink with and without PCM.

As shown in Fig. 6, under the same power losses conditions, the temperature of the heat sink with alloy-based phase change energy storage remains lower than that of the heat sink without added the PCM. At the beginning of the heating process, since the PCM has not melted, heat is transferred through thermal conduction, and the thermal resistance between the heat sink and the environment dominates the heat transfer process. Therefore, the temperature rise curves of the two heat sinks are nearly identical. As the heating process continues, the PCM melts and absorbs latent heat. At this point, the temperature rise of the alloy heat sink begins to level off. After 200 s of heating, the temperature of the heat source with the PCM reaches 71.8 °C, which is 23.6 °C lower than that of the traditional heat sink. After heating stops, the phase change heat sink cools rapidly to the phase change temperature during solidification, while the conventional heat sink takes longer to complete the heat dissipation. However, there is no advantage in dissipating heat to ambient temperature with a phase change heat sink.

IV. CONCLUSIONS

In this paper, phase change materials are used to enhance the thermal performance of DC-DC converters under short duration high power operating conditions. The thermal performance of LMPA is compared with that of paraffin wax through thermal simulation. Compared with paraffin wax, which has poorer thermal conductivity, the LMPA phase-change thermal solution can reduce the temperature rise of SiC MOSFETs by more than 50% when the power consumption is 45.2 W. When the power consumption is 73.2 W, the LMPA heat sink, with its high thermal conductivity and excellent phase change properties, significantly outperforms traditional heat sinks, making it suitable for short-term high-power operation at the converter level.

REFERENCES

[1] W. Liu, C. Zhu, Z. Lin and S. Yan, "Study on Heat Dissipation System of Liquid-cooled Lithium Battery Pack Based on Phase Change Materials," 2022 4th International Conference on Intelligent Control, Measurement and Signal Processing (ICMSP), Hangzhou: China, 2022, pp. 376-379.

[2] A. Jurkowski and R. Paluch, "Application of Phase Change Material for Thermal Management of Space Electronics," 2024 30th International Workshop on Thermal Investigations of ICs and Systems (THERMINIC), Toulouse: France, 2024, pp. 1-6.

[3] P. Borkar and V. S. Duryodhan, "Study of Phase Change Material-Based Hybrid Heat Sink for Electronics Cooling Application," IEEE Transactions on Components, Packaging and Manufacturing Technology, vol. 14, no. 10, pp. 1771-1782, October 2024.

[4] A. Singh, S. Rangarajan, L. Choobineh and B. Sammakia, "Thermal Management of Electronics During Continuous and Intermittent Operation Mode Employing Phase Change Material-Based Heat Sinks—Numerical Study," IEEE Transactions on Components, Packaging and Manufacturing Technology, vol. 11, no. 11, pp. 1783-1791, November 2021.

[5] E. S. Glitz and M. Ordonez, "MOSFET Power Loss Estimation in LLC Resonant Converters: Time Interval Analysis," IEEE Transactions on Power Electronics, vol. 34, no. 12, pp. 11964-11980, December 2019.

[6] J. Wei, H. Jiang, N. Xiao, Z. Wu, L. Wang and L. Ran, "Multiple Phase Change Materials Integrated Into Power Module for Normal and High Current Reliability Enhancement," IEEE Transactions on Device and Materials Reliability, vol. 23, no. 1, pp. 127-133, March 2023.

[7] T. Wen, J. Hu, X. Li and S. Kang, "A Wide Range of Multi-Stage Stiffness Regulation via Harnessing Phase Variability of Low-Melting-Point Alloy," IEEE Robotics and Automation Letters, vol. 9, no. 7, pp. 6664-6671, July 2024.

[8] M. Lu, X. Wang and L. Liu, "Form-Stable Composite Phase Change Material Based Cooling of Small Brushless DC Motor Windings," IEEE Transactions on Components, Packaging and Manufacturing Technology, vol. 13, no. 9, pp. 1410-1420, September 2023.

[9] H. Farooq, H. A. Khalid, M. Uzair Khalid, M. M. Farooqi and Q. Malik, "Loss Analysis of Full Bridge LLC Resonant Converter with Wide Input Range Using Si and SiC Switches," 2021 16th International Conference on Emerging Technologies (ICET), Islamabad: Pakistan, 2021, pp. 1-6 .

A 2:1 Capacitive Isolated Resonant Switched-Capacitor DC-DC Converter

1st Xinyu Zhang
School of Automation
Beijing Institute of Technology
Beijing, China
3432626243@qq.com

2nd Yu Fu
School of Automation
Beijing Institute of Technology
Beijing, China
fuyu@bit.edu.cn

3rd Yucheng Zhao
School of Automation
Beijing Institute of Technology
Beijing, China
3220231025@bit.edu.cn

4th Shouxiang Li(Corresponding author)
School of Automation
Beijing Institute of Technology
Beijing, China
lishouxiang@bit.edu.cn

Abstract—**Traditional transformers face issues such as high losses and limited power density at high switching frequencies. By utilizing capacitors for energy processing, the efficiency and power density are enhanced due to the superior energy density that capacitors provide. To reduce system weight and losses, this paper proposes a 2:1 capacitive isolated resonant switched-capacitor converter that employs capacitive isolation to block common-mode (CM) voltage and supports zero-current switching (ZCS) and zero-voltage switching (ZVS). These features significantly reduce switching losses and enhance efficiency and power density. A hardware prototype was developed, and experimental results validate the converter's soft-switching characteristics, demonstrating its potential for high-efficiency applications.**

Keywords—*Capacitive isolation, resonant switched-capacitor, DC-DC converter, ZCS, ZVS.*

I. Introduction

Isolation is required in numerous power electronics applications, such as data center power supplies, telecommunication power supplies, electric vehicle on board chargers, or dc microgrids. In these applications, both power and information need to be transmitted across the isolation barrier. There are mainly two technologies for power transmission. The first and most common technology is inductive power transfer (IPT), which utilizes magnetic transformers. Power is coupled from the primary side to the isolated side through the magnetic field within the transformer. Even when the transformer only provides isolation without additional voltage gain, it remains the main cause of the converter's volume, weight, losses, and low power density [1]. Although wide-bandgap semiconductors, such as gallium nitride (GaN) high-electron-mobility transistors (HEMT) or silicon carbide (SiC), can operate converters at MHz frequencies, increasing the frequency to reduce the size of the magnetic transformer is limited by the low permeability and core losses of magnetic core materials at MHz frequencies [2]. Therefore, the balance between power density and efficiency remains a key factor in transformer design.

The second technology for power transmission is capacitive power transfer (CPT). CPT uses capacitors as coupling elements between the input and output ports, which also bear CM voltage. Compared with IPT, energy is stored in the electric field within the capacitors, theoretically ensuring higher power efficiency and a more compact design [3].

Capacitive isolation technology based on CPT was initially mainly used for signal isolation. For example, a capacitive current isolation unit was adopted in [4] to develop a novel capacitive isolation gate driver for controlling high-power MOSFETs and IGBTs. Although CPT requires a higher operating frequency than IPT, recent advancements in power semiconductor switching speeds have made CPT a viable solution for current isolation in isolated DC-DC converters. For instance, a buck DC-DC converter with touch current limitation was investigated in [5]. By adding a capacitive coupling unit to the conventional buck converter, the touch current was reduced, electrical shock protection safety was enhanced, and advantages in efficiency and volume were demonstrated. A transformerless capacitive isolated DC-DC converter based on a full-bridge topology with phase-shift control was proposed in [6], which can achieve both 1:1 and buck conversion modes. Symmetrical current drive was employed to mitigate the DC offset across the isolation capacitor and minimize voltage stress. Additionally, a GaN-based composite hybrid switched-capacitor DC-DC converter was presented in [7], which comprised a fixed-ratio capacitive isolated resonant Dickson converter and a flying capacitor multilevel (FCML) converter. The former achieves a high step-down ratio and handles most of the system power, while the latter is used for wide-range input voltage regulation. Through optimized design, the weight of passive components is minimized.

To enhance power density, minimize volume, and improve efficiency, capacitors can serve as key components for energy processing. Additionally, to achieve high power transfer efficiency, capacitive resonant coupling is utilized, which means adding inductors to the circuit to form a characteristic frequency with the coupling capacitors at the operating frequency. To this end, this paper proposes a capacitive isolated resonant switched-capacitor DC-DC converter that provides a 2:1 step-down ratio. It offers effective isolation through capacitors to block the CM voltage and can be configured with either ZCS or ZVS.

The remainder of this paper is structured as follows. Section II provides a detailed introduction to the proposed

979-8-3315-1110-4/25 $31.00 © 2025 IEEE

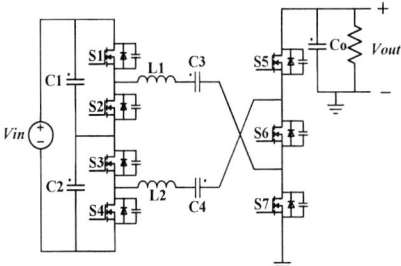

Fig. 1. Proposed capacitive isolated resonant switched-capacitor converter.

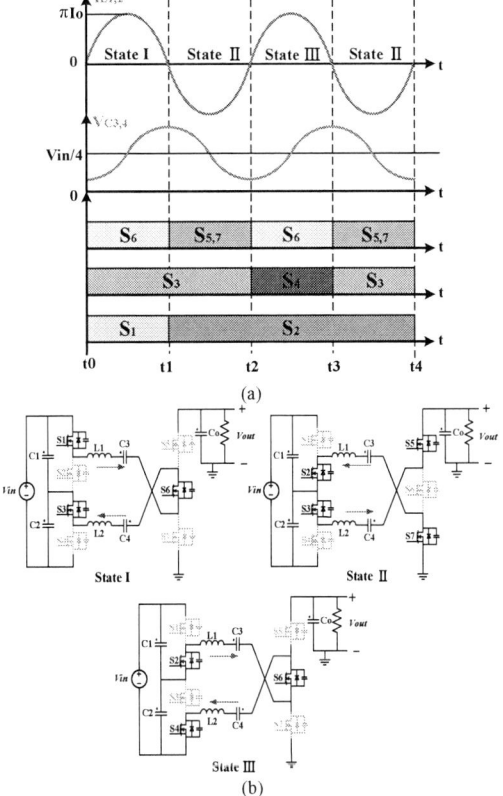

Fig. 2. Operation principles for ZCS: (a) key waveforms, (b) working states.

capacitive isolated resonant switched-capacitor DC-DC converter, elaborating on its novel topology and the operational principles for achieving ZCS and ZVS, as well as discussing component stress analysis. Section III presents the design and construction of the hardware prototype and analyzes the experimental results, which validate the soft-switching characteristics of the converter. Finally, Section IV concludes the paper with a summary of the key findings and potential applications of the proposed converter.

II. CAPACITOR-ISOLATED RESONANT SWITCHED CAPACITOR CONVERTER

A. Proposed Topology

The proposed topology of resonant switched capacitor converter is illustrated in Fig. 1. On the input side, two series-connected capacitors C_1 and C_2 are paralleled across the input voltage source to form a half-bridge structure. Two flying capacitors (isolation capacitors) are connected in series with two small inductors L_1 and L_2, respectively, and are connected to the junction points of switches $S_{1,2}$ and $S_{3,4}$, thereby forming a resonant network. To ensure circuit symmetry, the two input capacitors C_1 and C_2 are designed with identical capacitance values, and the two isolation capacitors C_3 and C_4 are also equal. Additionally, the inductance values of the two resonant inductors L_1 and L_2 are equal. The output side consists of three switches and an output capacitor, with the two flying capacitors intersecting at the wiring crosspoints of the output switches. The converter features a circulating current symmetric structure, which can eliminate common-mode interference between the input and output terminals and achieve isolation between the input and output grounds.

B. Operation Principle for ZCS

The current waveform through the resonant inductor, the voltage waveform across the isolation capacitors, and the switch modulation strategy are illustrated in Fig. 2. At this point, the converter has three working modes.

State I (t0-t1): Switches $S_{1,3,6}$ are turned on, while switches $S_{2,4,5,7}$ are turned off. During this period, the input capacitor C_1 discharges into the resonant capacitor, causing the voltage across C_3 and C_4 to increase gradually. At time t0, C_3 and C_4 transition from delivering charge to the load to absorbing charge from the input capacitor, and the voltage across the resonant capacitor reaches its minimum value, with the current through the capacitor being exactly zero.

State II (t1-t2, t3-t4): Switches $S_{2,3,5,7}$ are turned on, while switches $S_{1,4,6}$ are turned off. During this period, the resonant capacitor discharges into the load, causing the voltage across C_3 and C_4 to decrease gradually. At time t1, C_3 and C_4 complete their charging process, and the voltage across them reaches its maximum value, with the current through the capacitor being exactly zero.

State III (t2-t3): Switches $S_{2,4,6}$ are turned on, while switches $S_{1,3,5,7}$ are turned off. During this period, the input capacitor C_2 discharges into the resonant capacitor, causing the voltage across C_3 and C_4 to increase gradually. At time t2, the resonant capacitors C_3 and C_4 complete their discharging process and begin to absorb charge from the input side, with the voltage across the resonant capacitors reaching its minimum value and the current through the capacitors still being exactly zero.

Neglecting the effects of the input and output capacitors and assuming $L_1 = L_2 = L$ and $C_1 = C_2 = C$, the resonant frequency f_r, the resonant period T_r and angular resonant frequency ω_r of the inductor current can be derived:

$$f_r = \frac{1}{2\pi\sqrt{LC}} \tag{1}$$

$$T_r = 2\pi LC \tag{2}$$

$$\omega_r = \frac{1}{\sqrt{LC}} \tag{3}$$

To achieve ZCS, the switching frequency f_s should be set to half of the resonant frequency f_r.

$$f_s = \frac{1}{2}f_r = \frac{1}{4\pi\sqrt{LC}} \tag{4}$$

This ensures that the current drops to zero when the switching device is turned off, thereby realizing zero current turn-off.

Let the maximum value of the inductor current be I_{Lmax}. Its current waveform can be represented as:

$$i_{L1}(t) = i_{L2}(t) = I_{Lmax} sin(\omega_r t) \qquad (5)$$

When the inductor current is negative, the resonant capacitor discharges to the output terminal, releasing a total charge of:

$$\Delta Q = \int_{T_r}^{T_r/2} I_{Lmax} sin(\omega_r t) dt \qquad (6)$$

According to the conservation of charge, all the charge output by the circuit comes from the charge released by the resonant capacitor, thus we have the following equation:

$$\Delta Q = I_o \cdot T_r \qquad (7)$$

From this, the peak value of the inductor current can be derived as:

$$I_{Lmax} = \pi \cdot I_o \qquad (8)$$

For every two cycles of the inductor current, a half-sine wave flows through components S_1 and S_4, resulting in their RMS values being $\sqrt{1/8}$ of the peak value. Meanwhile, for S_2 and S_3, the RMS value of the carrier current is $\sqrt{3/8}$ of the peak value. This provides guidance for the subsequent selection of components.

Moreover, since the two equal-value input capacitors are connected in series across the power supply terminals, each capacitor obtains half of the supply voltage and charges/discharges the output through the isolation capacitors, it is straightforward to derive the input-output relationship as follows:

$$V_{in} = 2V_o \qquad (9)$$

C. Operation Principle for ZVS

By introducing a dead-time between the modulation signals and adjusting the turn-on and turn-off times of the switches, the resonant current is utilized to charge and discharge the parasitic capacitance within the power switches. This process ensures that the voltage across the power switches drops to zero before they are turned on, thereby achieving ZVS. The modulation signals for achieving ZVS, the current waveform through the resonant inductor, and the voltage waveform across the switch drain-source terminals are shown in Fig. 3(a). The 16 operating states of ZVS are shown in Fig. 3(b).

State 1: Switches $S_{1,3,6}$ are turned on. The input terminal charges the resonant capacitor. This phase ends while the inductor current is still positive. In the previous state 16, switch S_6 discharges through the parasitic output capacitance, dropping the drain-source voltage to zero, thus allowing S_6 to be conducted with ZVS in this state.

State 2: Switch S_1 is turned off, Switches $S_{3,6}$ continue

to conduct. The resonant current flows through the output capacitance of S_2 to continue the current path, removing charge from the transistor S_2 and charging the output capacitance of transistor S_1. This phase must end while the current is still positive; otherwise, S_2 will be recharged, and ZVS will be lost. After S_2 is fully discharged, its body diode conducts the current.

State 3: Switches $S_{2,3,6}$ are turned on. The positive resonant current continues to flow, and this phase should last until the inductor current becomes zero. S_2 conducts with ZVS.

State 4: Switch S_6 is turned off, Switches $S_{2,3}$ continue to conduct. During this phase, the current is negative, and the resonant current forms a circuit through the output capacitance of S_6, discharging the resonant capacitor towards the output terminal. The negative current removes charge from transistors S_5 and S_7 and charges the output capacitance of transistor S_6. After S_5 and S_7 are fully discharged, their body diodes conduct the current.

State 5: Switches $S_{2,3,5,7}$ are turned on. The resonant capacitor continues to discharge to the output terminal, with S_5 and S_7 conducting with ZVS.

State 6: Switch S_3 is turned off, Switches $S_{2,5,7}$ continue to conduct. The resonant current flows through the output capacitance of S_3 to continue the current path, removing charge from transistor S_4 and charging the output capacitance of transistor S_3. After S_4 is fully discharged, its body diode conducts the current. This phase must end while the current is still negative.

State 7: Switches $S_{2,4,5,7}$ are turned on. The negative resonant current continues to flow, and this phase should last until the inductor current becomes zero. S_4 conducts with ZVS.

State 8: Switches $S_{5,7}$ are turned off, Switches $S_{2,4}$ continue to conduct. The current is positive, and the input terminal forms a loop through the output capacitance of S_6 to charge the resonant capacitor. The positive current removes charge from transistor S_6 and charges the output capacitance of transistors S_5 and S_7. After S_6 is fully discharged, its body diode conducts the current.

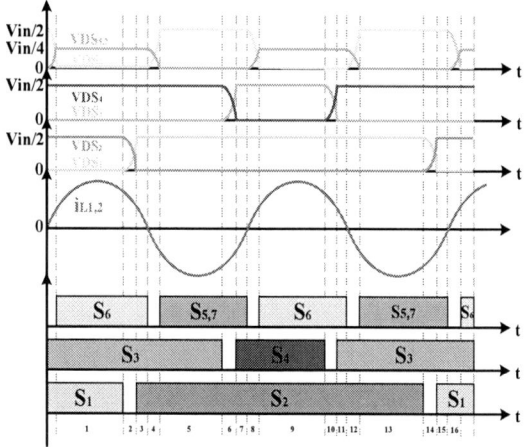

(a)

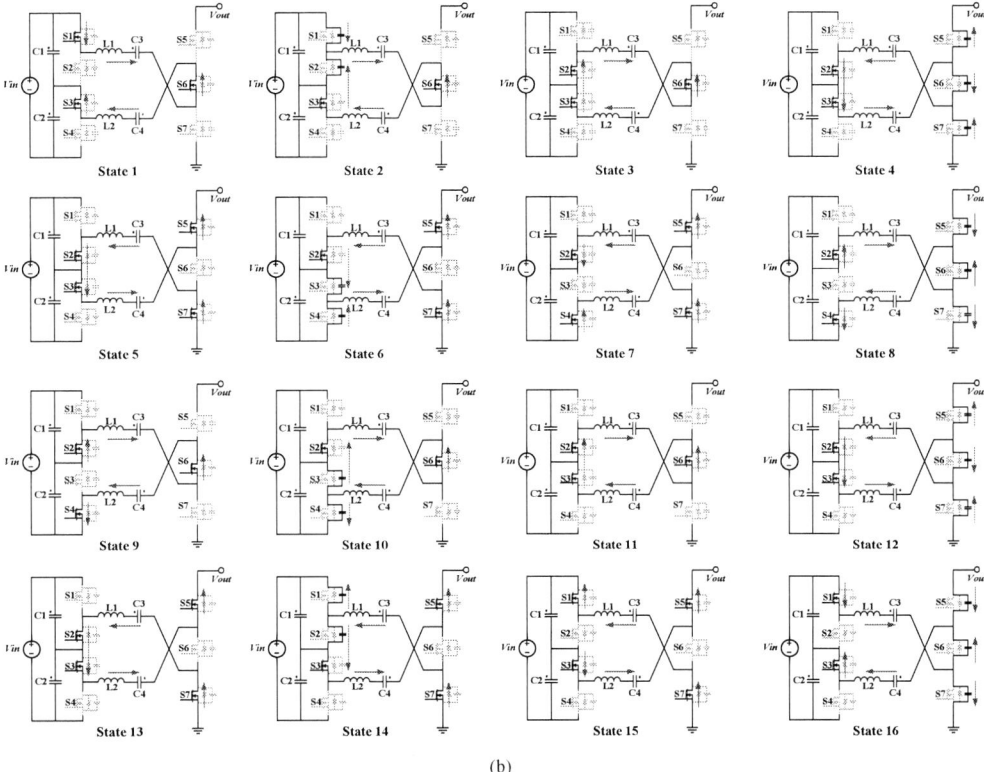

(b)

Fig. 3. Operation principles for ZVS: (a) key waveforms, (b) working states.

State 9: Switches $S_{2,4,6}$ are turned on. The input terminal continues to charge the resonant capacitor, with S_6 conducting with ZVS.

State 10: Switch S_4 is turned off, Switches $S_{2,6}$ continue to conduct. The resonant current flows through the output capacitance of S_3 to continue the current path, removing charge from transistor S_3 and charging the output capacitance of transistor S_4. After S_3 is fully discharged, its body diode conducts the current. This phase must end while the current is still positive.

State 11: Switches $S_{2,3,6}$ are turned on. The positive resonant current continues to flow, and this phase should last until the inductor current becomes zero. S_3 conducts with ZVS.

State 12: Switch S_6 is turned off, Switches $S_{2,3}$ continue to conduct. The current is negative, and the resonant current discharges through the output capacitance of S_6. The negative current removes charge from transistors S_5 and S_7 and charges the output capacitance of transistor S_6. After S_5 and S_7 are fully discharged, their body diodes conduct the current.

State 13: Switches $S_{2,3,5,7}$ are turned on.. The resonant capacitor continues to discharge to the output terminal, with S_5 and S_7 conducting with ZVS.

State 14: Switch S_2 is turned off, Switches $S_{3,5,7}$ continue to conduct. The resonant current flows through the output capacitance of S_2 to continue the current path, removing charge from transistor S_1 and charging the output capacitance of transistor S_2. After S_1 is fully discharged, its

body diode conducts the current. This phase must end while the current is still negative.

State 15: Switches $S_{1,3,5,7}$ are turned on. The negative resonant current continues to flow, and this phase should last until the inductor current becomes zero. S_1 conducts with ZVS.

State 16: Switches $S_{5,7}$ are turned off, Switches $S_{1,3}$ continue to conduct. The current is positive, and the input terminal forms a loop through the output capacitance of S_6 to charge the resonant capacitor. The positive current removes charge from transistor S_6 and charges the output capacitance of transistors S_5 and S_7. After S_6 is fully discharged, its body diode conducts the current.

It is evident that all switches can achieve ZVS.

TABLE I. COMPONENT STRESSES

V_{S1-4}	$V_{S5,7}$	V_{S6}
$\frac{1}{2}V_{in} = V_O$	$\frac{1}{4}V_{in} = \frac{1}{2}V_O$	$\frac{1}{2}V_{in} = V_O$
$I_{S1,4\_rms}$	$I_{S2,3\_rms}$	$I_{S5,6,7\_rms}$
$\frac{\pi}{\sqrt{8}}I_O$	$\frac{\sqrt{3}\pi}{\sqrt{8}}I_O$	$\frac{\pi}{2}I_O$
I_{S1-7_max}	$I_{L1,L2\_max}$	$I_{L1,L2\_rms}$
πI_O	πI_O	$\frac{\pi}{\sqrt{2}}I_O$

TABLE II. BILL-OF-MATERIAL OF THE PROTOTYPE

Component	Model number	Value
MOSFET $S_{1-4,6}$	IQE013N04LM6CG	40 V, 1.1 mΩ @ 10 V, 25 ℃
MOSFET $S_{5,7}$	IQE006NE2LM5CG	25 V, 0.5 mΩ @ 10 V, 25 ℃
Input/Output capacitors $C_{1,2,out}$	C3216X7R1H106K160AC	X7R, 50 V, 10 uF, 2.9 uF @ 30 V
Isolated capacitors $C_{3,4}$	GRM31C5C1H224GE02	C0G, 50 V, 0.22 uF
Resonant inductor $L_{1,2}$	Wurth 74431435022	220 nH, 37.6 A, 0.28 mΩ @ 100 kHz
Gate drivers	ADI LTC4440	2.4 A Peak Current
	TI UCC27212	3.7 A Peak Current

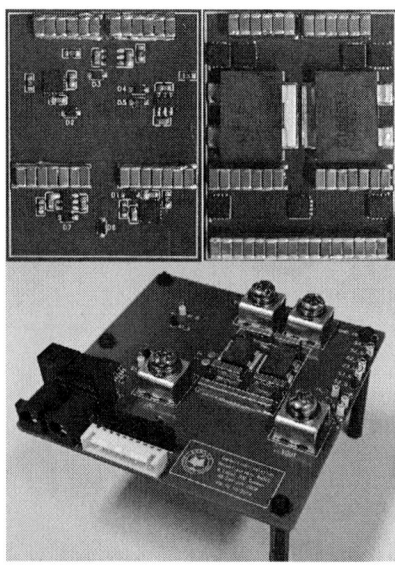

Fig. 4. Photograph of the hardware prototype.

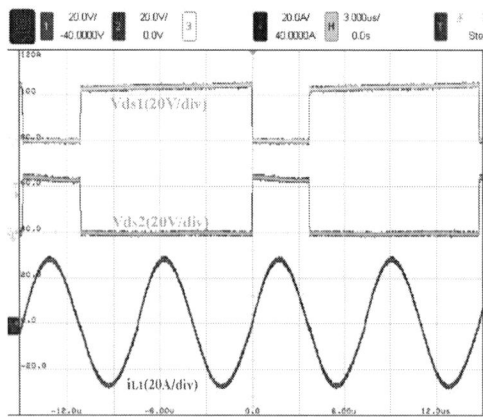

Fig. 5. Experimental waveforms of soft switching: ZCS.

D. Component Stress Analysis

The calculation of voltage and current stress on electronic components is crucial for hardware selection and loss analysis. Table I summarizes the average value of voltage stress and the RMS and maximum values of current stress borne by the switching transistors; as well as the RMS and maximum values of current stress borne by the inductors.

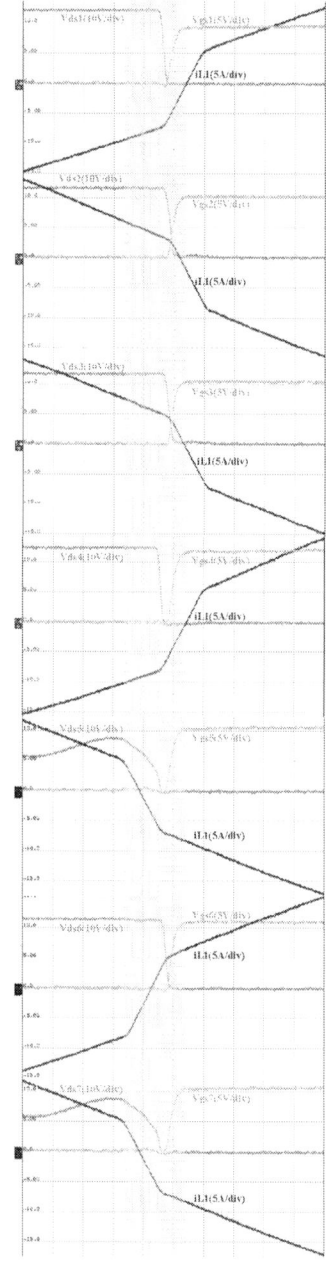

Fig. 6. Experimental waveforms of soft switching: ZVS.

III. HARDWARE PROTOTYPE DESIGN AND EXPERIMENTAL RESULTS ANALYSIS

A. Component Selection and Hardware Implementation

A 48 V-24 V, 10 A capacitive isolated resonant switched-capacitor converter was designed and fabricated. The circuit schematic is the same as Fig. 1. Based on the parameters in Table, the bill-of-material (BOM) of the prototype was listed in Table II.

Regarding the selection of power switches, since S_{1-4} and S_6 withstand the same maximum voltage stress, which is the output voltage value of 24 V, it is considered to choose the same device. For the purpose of margin protection, a device with a rated drain-source voltage of 40 V is selected. S_5 and S_7 bear the maximum voltage stress of 12 V, so a device with a rated drain-source voltage of 25 V is chosen.

In terms of capacitor selection, this design chooses ceramic capacitors that can provide higher energy density than film capacitors. Since Class I C0G multilayer ceramic capacitors (MLCC) have low capacitance, high voltage capability, and can maintain a tight tolerance on the frequency [8], they become the primary choice for the isolation/resonant capacitors. The input and output capacitors need to be designed large enough to maintain voltage stability; therefore, Class II multilayer ceramic capacitors with higher capacitance and higher volumetric energy density are selected [9].

B. Experimental Results and Analysis

The ZCS characteristics of the proposed converter are shown in Fig. 5, with the switching conditions of S_1 and S_2 exemplifying how ZCS is achieved. When the switching frequency is precisely matched with the resonant frequency, the turn-on and turn-off of each switch occur at the zero-crossing points of the inductor current.

By applying phase shift to the drive signals, the ZVS characteristics of the converter are demonstrated in Fig. 6. It should be noted that although the switching waveforms of power switches S_1 through S_7 are combined into a single diagram, their spatial arrangement does not reflect the actual phase relationships. Specifically, during the transition of power switches $S_{1,4,5,7}$, the resonant current $i_{L1}(t)$ is negative, allowing the MOSFET parasitic output capacitance to be fully discharged and the body diode to conduct, reducing the drain-source voltage to zero. Afterward, the drive voltage is fully turned on, with minimal overlap with the drain-source voltage, thereby achieving good ZVS. Similarly, during the transition of power switches $S_{2,3,6}$, the positive resonant current is utilized to reduce the drain-source voltage of the switches to zero before they are turned on, thereby achieving ZVS.

IV. CONCLUSION

This paper proposes a 2:1 capacitive isolated resonant switched-capacitor DC-DC converter that utilizes capacitance for isolation and supports both ZCS and ZVS. The operational principles of both ZCS and ZVS converters are thoroughly analyzed in this paper. The ZCS operation ensures that the current through the switching devices drops to zero at the switching instants, while the ZVS operation uses the resonant current to discharge the parasitic capacitance of the power switches, ensuring that the voltage across the switches drops to zero before they are turned on. This significantly reduces switching losses and enhances the overall efficiency and power density of the converter. A hardware prototype designed for a 48 V-24 V, 10 A application successfully validated the soft-switching characteristics of the converter through experimental results. Future work may focus on further optimizing the converter design to improve efficiency and power density, exploring different input-output voltage ratios and power levels to extend its application range in various power electronic systems. In summary, the converter represents significant progress in the field of isolated power converters, providing a viable solution for next-generation power electronic applications.

REFERENCES

[1] M. Kasper, C. -W. Chen, D. Bortis, J. W. Kolar and G. Deboy, "Hardware verification of a hyper-efficient (98%) and super-compact (2.2kW/dm3) isolated AC/DC telecom power supply module based on multi-cell converter approach," 2015 IEEE Applied Power Electronics Conference and Exposition (APEC), Charlotte, NC, USA, 2015, pp. 65-71.

[2] C. R. Sullivan, B. A. Reese, A. L. F. Stein and P. A. Kyaw, "On size and magnetics: Why small efficient power inductors are rare," 2016 International Symposium on 3D Power Electronics Integration and Manufacturing (3D-PEIM), Raleigh, NC, USA, 2016, pp. 1-23.

[3] L. Hüssen, M. -D. Wei and R. Negra, "Design and Synthesis of a Dualband Capacitive Resonant Coupler for Harmonic Backscattering Through an Isolation Barrier," 2024 54th European Microwave Conference (EuMC), Paris, France, 2024, pp. 517-520.

[4] D. A. Shevtsov, I. M. Shishov, I. V. Lukoshin, M. A. Podguzova and Y. I. Kovan, "Development of a New Gate Driver with Capacitive Isolation," 2023 IEEE 24th International Conference of Young Professionals in Electron Devices and Materials (EDM), Novosibirsk, Russian Federation, 2023, pp. 980-983.

[5] A. Toebe, R. C. Beltrame, M. Mezaroba, A. L. Batschauer and C. Rech, "Capacitive Coupled Step-Down DC–DC Converter With Touch Current Limitation," in IEEE Transactions on Power Electronics, vol. 39, no. 8, pp. 9724-9735, Aug. 2024.

[6] P. Granello, L. Schirone and P. Bauer, "A Capacitive Galvanically Isolated Full Bridge Converter," in IEEE Transactions on Transportation Electrification, early access, Jan. 09, 2025, doi: 10.1109/TTE.2025.3527724.

[7] S. Coday et al., "Design and Implementation of a GaN-Based Composite Hybrid Switched-Capacitor DC-DC Converter for Space Applications," in IEEE Open Journal of Power Electronics, vol. 6, pp. 150-161, 2025.

[8] A. Templeton, N. Reed, H. Hayes, J. Davis and J. Bultitude, "Class I Multi-Layer Ceramic Capacitors (MLCCs) Performance as Wide Band Gap (WBG) Snubbers in Hard Switching Applications," 2023 Fourth International Symposium on 3D Power Electronics Integration and Manufacturing (3D-PEIM), Miami, FL, USA, 2023, pp. 1-5.

[9] D. Zakzewski, Y. Shen, A. Hasnain, R. Resalayyan and A. Khaligh, "Class II Ceramic Capacitor Voltage Characteristic Modeling and Compensation for AC-Connected Applications," in IEEE Journal of Emerging and Selected Topics in Industrial Electronics, vol. 5, no. 4, pp. 1582-1592, Oct. 2024.

An Auxiliary Power Supply Utilizing Planar Winding Transformer with High Insulation and Low Coupling Capacitance Considerations

1st Yong Chen
Zhuhai Power Supply Bureau
Guangdong Power Grid Co., Ltd.
Zhuhai, China
35665035@qq.com

2nd Xuhui Song
School of Electrical Engineering
Xi'an Jiaotong University
Xi'an, China
xuhui_song@163.com

3rd Fan Zhang
School of Electrical Engineering
Xi'an Jiaotong University
Xi'an, China
zhangfan1990@xjtu.edu.cn

4th Yuze Zheng
School of Electrical Engineering
Xi'an Jiaotong University
Xi'an, China
zyz2001@stu.xjtu.edu.cn

5th Xiaolu Zhang
School of Electrical Engineering
Xi'an Jiaotong University
Xi'an, China
lu995579@stu.xjtu.edu.cn

6th Yukun Niu
School of Electrical Engineering
Xi'an Jiaotong University
Xi'an, China
827105370@qq.com

7th Zheyuan Yu
School of Electrical Engineering
Xi'an Jiaotong University
Xi'an, China
zheyuan.yu@smartonep.com

8th Min Wu
School of Electrical Engineering
Xi'an Jiaotong University
Xi'an, China
wumin@xjtu.edu.cn

9th Kaixiang Gong
School of Electrical Engineering
Xi'an Jiaotong University
Xi'an, China
2193512024@stu.xjtu.edu.cn

Abstract—This paper proposes an auxiliary power supply (APS) for medium-voltage SiC MOSFETs with high insulation and low coupling capacitance. Firstly, the mathematical model of the LLC resonant topology is analyzed, and the overall framework of the APS is introduced. In terms of the isolated barrier design, the advantages of printed circuit board (PCB) planar windings are elaborated, and an equivalent capacitance model based on an air-gapped core structure is established, with results indicating a model accuracy exceeding 93%. Finally, a prototype is developed with a coupling capacitance of 1.23 pF. The APS delivers stable gate driver voltages and achieves a peak efficiency of 82.547%. High-voltage testing confirms a PDIV of 24.5 kV DC and a breakdown voltage of 31 kV DC, demonstrating the design's excellent insulation performance and suitability for medium-voltage applications.

Keywords—Auxiliary power supply (APS), SiC MOSFETs, PCB planar winding, coupling capacitance

I. INTRODUCTION

Silicon carbide (SiC) Metal-Oxide-Semiconductor Field-Effect Transistors (MOSFETs) used in medium-voltage (MV) applications exhibit significant advantages, including high-voltage withstand capability, superior temperature tolerance, high frequency operation and low power losses [1]. However, the auxiliary power supply (APS), essential for ensuring the reliable operation of SiC gate drivers, faces critical challenges. In MV applications, the APS must withstand voltages exceeding 15 kV, while the high dv/dt (up to 100 kV/μs) of 10 kV SiC devices induces severe common-mode interference in the driver circuit due to parasitic capacitive couplings. As a result, the APS must have the following features: 1) high insulation ability (>15 kV); 2) low coupling capacitance (<5 pF); 3) a small footprint. Additionally, IEC 61800-5-1 standards should be met. For example, under a working voltage of 7.2~7.6 kV, the minimum air clearance is 25 mm and the minimum creepage distance is 38 mm [2]. Since the isolation barrier design is critical for APS performance,

current research efforts are predominantly focused on this, primarily through three approaches:

1) *Magnetic coupling based on transformer core.* A ferrite-core-based transformer, designed in [3] for driving 10 kV SiC MOSFETs gate drivers, achieves a 20 kV DC isolation voltage and suppress coupling capacitance to 1.2 pF through the shielding layer. A toroidal-core-based transformer [4] achieves a PDIV exceeding 15 kV RMS with a coupling capacitance of 1.03 pF. However, this APS features a complex encapsulation structure and is constrained to a power of 2.5W. Overall, the magnetic coupling method typically enables transformers to achieve high coupling coefficients and large inductance, but it often requires bulkier magnetic cores to enhance the APS insulation performance.

2) *Wireless power transfer (WPT).* This method inherently demonstrates superior dielectric performance owing to the coreless design and intrinsic isolation provided by dielectric barriers between transmitter and receiver coils [5]. The impact of transmitter and receiver coils misalignment on coupling capacitance in a WPT system is analyzed in [6] where the transformer's insulation is enhanced with polyurethane potting compound. The designed APS achieves a measured coupling capacitance of 2 pF and withstands 25 kV isolation voltage. However, the efficiency remains limited to 10% to 43%. A PCB solid dielectric-based very-high-frequency APS is proposed [7] to demonstrate a coupling capacitance of 3.2 pF and breakdown voltage exceeding 23 kV. Although the 20 MHz switching frequency enables a peak efficiency of 80%, the output power is limited to 1.8 W, restricting its applicability in high power driving applications.

3) *Optical fiber transfer.* This method provides the APS with optimal common-mode immunity, enhancing both power and signal isolation capability, which theoretically achieves zero coupling capacitance under ideal conditions. An APS based on power over fiber (PoF) is designed in [8], but it demonstrates limited performance-maximum power transfer of 0.5 W with only 24% efficiency.

This work was supported by the China Southern Power Grid Company, Ltd., Science and Technology Project under grant 030400KC23090016 (GDKJXM20231032).

979-8-3315-1110-4/25 $31.00 © 2025 IEEE

This paper presents a design of an APS based on PCB planar windings. The isolated barrier adopts a magnetic core structure with an air gap, which offers the advantages of reduced coupling capacitance and enhanced insulation capability. Meanwhile, the designed APS features high efficiency and stable output voltages, ensuring the safe and reliable operation of SiC MOSFETs gate drivers.

The rest of the paper is organized as follows. Section II analyzes and models the topology. The design of the isolated barrier is introduced in Section III, where an equivalent capacitance model is proposed. Section IV discusses the experimental results. Finally, the conclusions are drawn.

II. CIRCUIT DESIGN

In isolated DC/DC topologies, the LLC resonant converter offers distinct advantages such as high efficiency and reduced passive component count. With precise parameter design, the transformer leakage inductance referred to the primary side can serve as the resonant inductor, effectively reducing the system volume occupied by extra components.

When the switching frequency aligns with the resonant network, the LLC achieves zero-voltage switching (ZVS) and minimizes conduction and turn-off losses. This makes LLC particularly suitable for fixed-output APS applications operating in open-loop fixed-frequency mode. Furthermore, the widespread availability of commercial LLC gate drivers with mature technology ensures stable control performance and a compact footprint.

The topology of the LLC resonant half-bridge converter is shown in Fig.1, where V_{in} represents the input voltage, $Q_{1,2}$ are the power devices. L_r (resonant inductor) and C_r (resonant capacitor) form the resonant tank. I_m denotes the magnetizing inductance of the transformer. $VD_1 \sim VD_4$ constitute the full-bridge rectifier. C_o is the output filter capacitor, R_L represents the load, and the transformer features a turns ratio of 1:n between the primary and secondary windings.

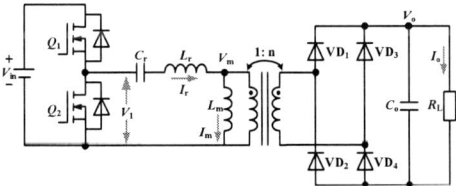

Fig. 1. LLC resonant half-bridge converter.

To establish the mathematical model of the LLC resonant converter, the equivalent circuit referred to the primary side is depicted in Fig.2.

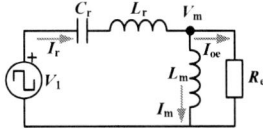

Fig. 2. Equivalent circuit of LLC resonant half-bridge converter.

V_1 represents the RMS value of the input high-frequency square wave, and its expression is as follows:

$$V_1 = \frac{\sqrt{2}}{\pi} V_{in} \tag{1}$$

R_e represents the equivalent resistance of the load R_L referred to the primary side, expressed as:

$$R_e = \frac{8}{(n\pi)^2} R_L \tag{2}$$

When the system operates at an angular frequency ω, the following expression can be derived:

$$\frac{V_m}{V_1} = \left| \frac{R_e \| (j\omega L_m)}{R_e \| (j\omega L_m) + j\omega L_r + (j\omega C_r)^{-1}} \right| \tag{3}$$

It is noted that if ω is equal to $\left(L_r C_r \right)^{-1/2}$, the ratio of V_m to V_1 becomes independent of the equivalent load resistance, which is 1 in constant. In this way, the voltage gain can be expressed as follows:

$$G = \frac{V_o}{V_{in}} = \frac{n}{2} \tag{4}$$

In this paper, the APS operates at a rated power of 20W, utilizing a self-oscillating gate driver IC (IR2085S) with a 50% duty cycle to control the SiC MOSFETs (IRFH7440TRPBF). The 24 V DC input voltage is converted into a high-frequency AC voltage (254.5 kHz) through the half-bridge and resonant tank. This AC signal is then transformed via an isolated transformer and rectified by a diode bridge (MSB56) to generate a 24 V DC output. To achieve voltage regulation, the rectified output is further processed by two buck regulator ICs (TPS5450), delivering stable gate driver voltages of +20 V and -5 V.

III. DESIGN CONSIDERATIONS FOR ISOLATED BARRIER

A. PCB Planar Winding

Transformers based on magnetic coupling inherently achieve high coupling coefficients, which facilitate compact winding designs. Compared to conventional wire-wound structures, PCB planar winding transformers exhibit significantly lower coupling capacitance due to:

1) *Minimized interwinding capacitance.* When primary and secondary windings are parallel-aligned on opposing sides of the core, their overlapping area is very small, resulting negligible interwinding capacitance.

2) *Reduced winding-to-core capacitance.* The low profile of planar windings has a smaller winding-core overlapping area than that of wire-wound designs.

Moreover, the PCB planar winding structure offers manufacturing advantages, including enhanced mechanical stability and process repeatability.

Based on these considerations, the proposed planar winding design, which is shown in Fig.3, adopts 2oz copper thickness with a trace width of 1mm, featuring 4 turns on the primary side and 8 turns on the secondary side. The primary and secondary PCBs utilize 6-layer and 10-layer FR4 dielectric stacks, respectively. This multilayer configuration ensures that surface signal layers remain free of traces and vias, thereby increasing the effective creepage distance.

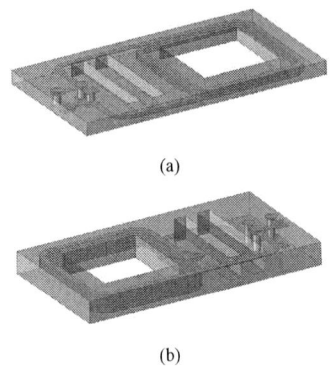

(a)

(b)

Fig. 3. Design of the planar widings. (a) primary winding. (b) secondary winding.

B. Structure of the Core

In terms of transformer cores, their volumes are often oversized to meet dielectric insulation requirements for APS. This paper proposes a split-core transformer with air gaps, where the ferrite core is divided into two segments. This design enhances the effective creepage distance and reduces coupling capacitance compared to conventional cores without air gaps.

As illustrated in Fig.4, the proposed air-gapped transformer structure is compared with a traditional non-gapped design. Simulation results in Table I quantify the improvements:

TABLE I.   COMPARISON OF TWO CORE STRUCTURES

Structure	E_{max} (kV/mm)	$C_{coupling}$ (pF)
non air gap	8.049	1.34
2mm air gap	5.577	1.15

Fig.4 illustrates an air-gapped transformer and a traditional non-gapped design of identical dimensions. The simulation results in Table I demonstrate that the transformer with a 2mm air gap exhibits lower maximum electric field strength (5.577 kV/mm) and reduced coupling capacitance (1.15 pF) compared to the non-gapped design.

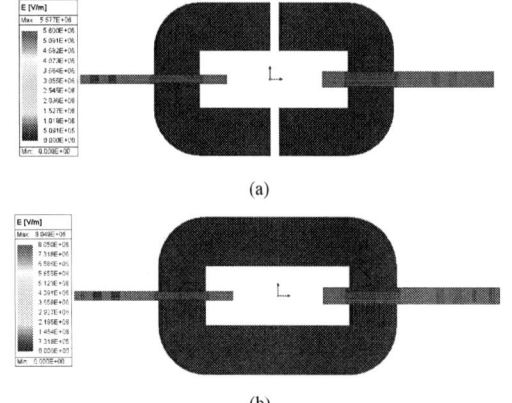

(a)

(b)

Fig. 4.   The structure of the APS transformer. (a) 2mm air gap. (b) non air gap.

Due to the air gap dividing the core into two isolated sections, the APS transformer comprises four distinct nodes.

The equivalent capacitance model schematic is depicted in Fig.5, where P represents the primary winding, T represents the primary core, R represents the secondary core, and S represents the secondary winding.

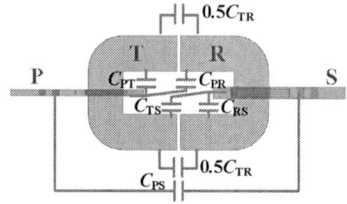

Fig. 5.   The equivalent capacitance model.

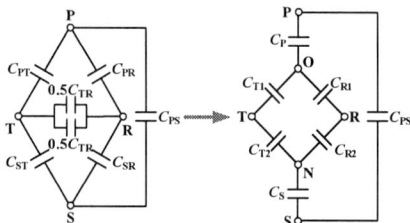

Fig. 6.   Δ-Y transformation of the equivalent capacitance model.

The equivalent capacitance model depicted in Fig.5 can undergo the Δ-Y transformation illustrated in Fig.6. The transformed capacitances are expressed as follows:

$$
\begin{cases}
C_P = C_{PT} + C_{PR} + C_{PR} \times C_{PT}/0.5C_{TR} \\
C_{T1} = C_{PT} + 0.5C_{TR} + C_{PT} \times 0.5C_{TR}/C_{PR} \\
C_{R1} = C_{PR} + 0.5C_{TR} + C_{PR} \times 0.5C_{TR}/C_{PT} \\
C_S = C_{TS} + C_{RS} + C_{RS} \times C_{TS}/0.5C_{TR} \\
C_{T2} = C_{TS} + 0.5C_{TR} + C_{TS} \times 0.5C_{TR}/C_{RS} \\
C_{R2} = C_{RS} + 0.5C_{TR} + C_{RS} \times 0.5C_{TR}/C_{TS}
\end{cases}
\quad (5)
$$

Under the condition of structural symmetry between the primary and secondary windings and the magnetic core configuration, the total coupling capacitance can be formulated as:

$$
C_{coupling} = \frac{2C_{PT}C_{PR} + C_{PT}C_{TR} + C_{PR}C_{TR}}{C_{PT} + C_{PR} + 2C_{TR}} + C_{PS} \quad (6)
$$

As illustrated in Fig.7, the analytical results achieve over 93% agreement with the finite element simulation results, validating the accuracy of the proposed coupling capacitance model.

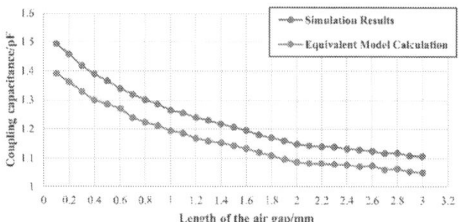

Fig. 7.   Simulated and calculated coupling capacitance.

IV. EXPERIMENTAL RESULTS

To enhance the safety and stability of the APS, an enclosure is designed in SolidWorks with comprehensive of PCB mounting, magnetic core fixation and mechanical structure. Meanwhile, five 2mm creepage grooves are incorporated to improve the insulation performance of the enclosure. The top view and assembly drawing of the 3D model of the enclosure are shown in Fig. 8.

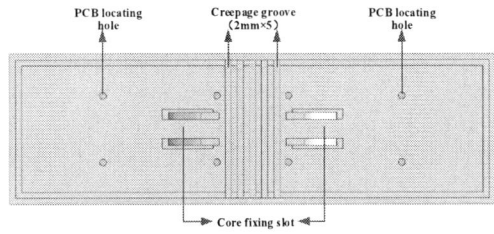

(a)

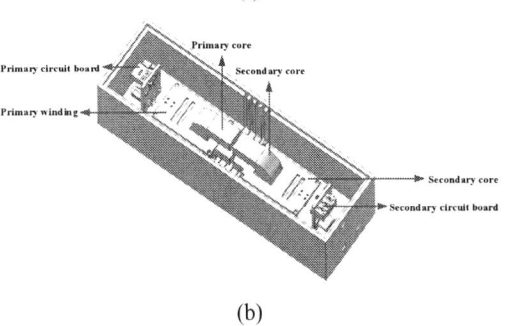

(b)

Fig. 8. 3D model of the enclosure. (a) top view. (b) assembly drawing.

The hardware prototype of the designed APS with its enclosure is shown in Fig. 9. The overall dimensions are 158.5 cm × 48 cm × 62 cm. The core material is DMR96A, and the enclosure is made of Nylon PA12, which offers high mechanical strength, compressive resistance, and excellent electrical insulation properties. In addition, it is lightweight and easy to reproduce.

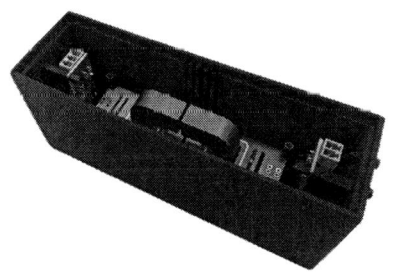

Fig. 9. The hardware prototype of the APS.

A. Coupling Capacitance Measurement

Taking insulation capability and common mode transient immunity (CMTI) into consideration, the air gap of the magnetic core is set to 1mm. The coupling capacitance of the APS is measured by a Keysight E4990A impedance analyzer. The sweep frequency is set from 6 MHz to 10 MHz and the result is 1.23 pF, which is shown in Fig. 10.

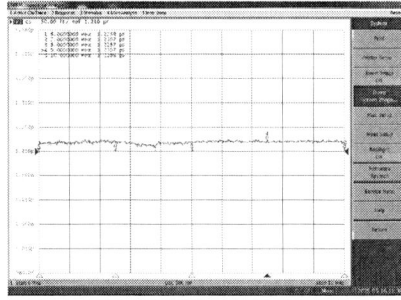

Fig. 10. Measured coupling capacitance of the APS.

B. Electrical Performance

The electrical performance of the APS is verified under 1 kV DC voltage. Fig. 11 shows the double-pulse test of the SiC device, where the gate-source voltage V_{gs} maintains stable outputs of +20 V and -5 V, indicating that the APS has excellent output voltage regulation capability.

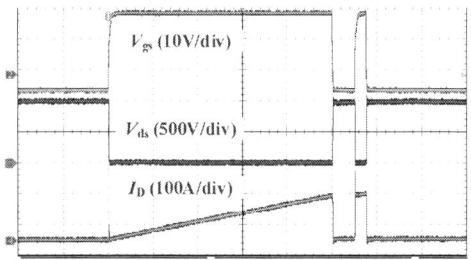

Fig. 11. Double-pulse test results.

In addition, an efficiency test is conducted. As shown in Fig. 12, the APS achieved its highest efficiency of 82.547% at an output power of 21 W.

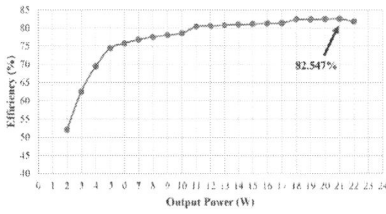

Fig. 12. Efficiency test of the APS.

C. Hi-Pot Test

The insulation capability of the APS is verified through a DC voltage testing platform. The experiment setup is shown in Fig. 13, where (a) shows the AC380 high-voltage power supply and (b) shows the high-voltage probe.

 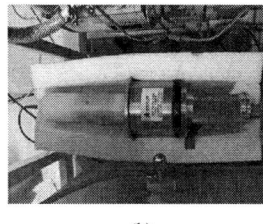

(a)   (b)

Fig. 13. Hi-pot test setup. (a) power supply. (b) high-voltage probe.

In the hi-pot test, the DC voltage is increased in steps of 0.5 kV, with each voltage level maintained for 30 seconds. Discharge behavior is monitored by using an oscilloscope, and the number of discharges exceeding 20 pC is recorded.

The partial discharge inception voltage (PDIV) is defined as the lowest voltage level at which more than 10 discharges exceeding 20 pC occur in each of three consecutive voltage steps. The PDIV test result of the APS is shown in Fig. 14. When the voltage reached 24.5 kV DC, 12 discharges are observed, and in all subsequent voltage levels the number of discharges exceeded 10. Therefore, the PDIV is determined to be 24.5 kV DC.

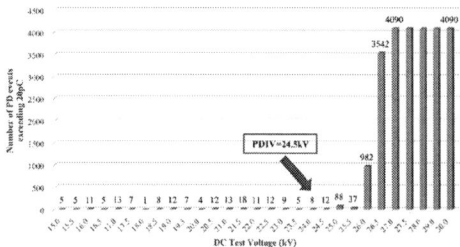

When the voltage increases to 31 kV DC, a breakdown occurs, indicating that the breakdown voltage is 31 kV DC. The hi-pot test results demonstrate that the designed APS possesses excellent insulation performance.

V. CONCLUSION

This paper proposes an auxiliary power supply (APS) based on an LLC topology and a PCB planar-winding transformer. Through optimization of the winding structure and the design of an air gap in the magnetic core, the APS achieves an ultra-low coupling capacitance of 1.23 pF, a PDIV of 24.5 kV DC, and a breakdown voltage of 31 kV DC. In addition, experimental results demonstrate that the APS provides stable output voltage regulation, maintaining consistent +20 V and −5 V outputs during the double-pulse test, and achieves a peak efficiency of 82.547%, verifying the excellent electrical performance.

REFERENCES

[1] V. Soler et al., "Reliability and Robustness Tests for Next-Generation High-Voltage SiC MOSFETs," *IEEE J. Emerg. Sel. Topics Power Electron.*, vol. 9, no. 4, pp. 4320–4329, Aug. 2021.

[2] A. Anurag, S. Acharya, N. Kolli, and S. Bhattacharya, "Gate drivers for medium-voltage SiC devices," medium-voltage SiC devices," *IEEE J. Emerg. Sel. Topics Ind. Electron.*, vol. 2, no. 1, pp. 1–12, Jan. 2021.

[3] A. Anurag, S. Acharya, Y. Prabowo, G. Gohil, and S. Bhattacharya, "Design considerations and development of an innovative gate driver for medium-voltage power devices with high dv/dt," *IEEE Trans. Power Electron.*, vol. 34, no. 6, pp. 5256–5267, Jun. 2019.

[4] H. Li, Z. Gao, and F. Wang, "Medium-voltage isolated auxiliary power supply design for high insulation capability, ultra-low coupling capacitance, and small size," *IEEE Trans. Power Electron.*, vol. 38, no. 6, pp. 7226–7240, Jun. 2023.

[5] Y. Wei, S. Zhao, R. Jimenez, A. Solangi, X. Du and H. A. Mantooth, "Design Optimization and Experimental Validation of Gate Driver for 10 kV SiC MOSFET," in *Proc. IEEE Inte. Power Electron. and Appl. Conf. and Expo. (PEAC)*, 2022, pp. 43–48,

[6] V. T. Nguyen, G. Veera Bharath, and G. Gohil, "Design of isolated gate driver power supply in medium voltage converters using high frequency and compact wireless power transfer," in *Proc. IEEE Energy Convers. Congr. Expo.*, 2019, pp. 135–142.

[7] Z. Guo, H. Li, and P. Cheetham, "A very-high-frequency isolated gate driver power supply using solid dielectrics for medium voltage SiC MOSFETs," in *Proc. IEEE Appl. Power Electron. Conf. Expo. (APEC)*, Houston, TX, USA, Mar. 2022, pp. 1394–1399.

[8] X. Zhang et al., "A gate drive with power over fiber-based isolated power supply and comprehensive protection functions for 15-kV SiC MOSFET," *IEEE J. Emerg. Sel. Topics Power Electron.*, vol. 4, no. 3, pp. 946–955, Sep. 2016.

Gate-Source Voltage Oscillation of Multi-chip SiC Power Modules Considering Differential Gate Resistance

Longnv Li
School of Electrical Engineering
Tiangong University
Tianjin, China
lilongnv@tiangong.edu.cn

Chuyuan Liu
School of Electrical Engineering
Tiangong University
Tianjin, China
liuchuyuan2023@163.com

Lu Wang
School of Electrical Engineering
Tiangong University
Tianjin, China
wanglu1073@163.com

Youzheng Wang
School of Electrical Engineering
Tiangong University
Tianjin, China
youzhengwang@tiangong.edu.cn

Gaojia Zhu
School of Electrical Engineering
Tiangong University
Tianjin, China
zhugaojia@tiangong.edu.cn

Yun-Hui Mei*
School of Electrical Engineering
Tiangong University
Tianjin, China
meiyunhui@tiangong.edu.cn

Abstract — This study investigates the application and performance characteristics of Silicon carbide (SiC) power modules, which are widely utilized in industrial settings due to their high power density and reliability. However, gate oscillations frequently manifest during the turn-off phase, particularly due to circuit asymmetry and parasitic parameters associated with the chips. To address this issue, this research establishes a dynamic model of the multi-chip turn-off process, aimed at identifying the primary sources of oscillations. The study systematically analyzes the effects of individual gate resistance and differential gate resistance, and their influence on oscillations under conditions of parasitic inductance imbalance. Experimental results demonstrate the effectiveness of the proposed approach, with further validation conducted through double-pulse testing.

Keywords: Multichip parallel, gate oscillation, gate resistance, differential gate resistance, SiC MOSFET

I. INTRODUCTION

Due to the increasing demands on power density, multichip paralleled Silicon carbide (SiC) MOSFET power modules have received extensive research attentions [1-3]. However, variations in parasitic parameters among chips often lead to current imbalance, especially during turn-off process, causing significant gate oscillations that impair switching performance and raise failure risks [4-5]. Therefore, the research on gate oscillation mechanisms and effective mitigation strategies have become a focus [6].

In [7], tests and analyses were conducted on two commercial automotive—grade power modules—one incorporating two chips and the other a single chip. The findings revealed that multi-chip parallel modules are more susceptible to gate oscillations compared to their single-chip counterparts. In [8], two multi-chip parallel configurations—parallel chips and parallel half-bridges—were proposed, and the impact of stray inductance on the current commutation process were thoroughly investigated. The study highlighted that transient current imbalances could compromise the reliability of power modules. Mismatched drain-source currents may arise from suboptimal printed circuit board (PCB) layouts or semiconductor manufacturing tolerances, with gate loop parasitic inductance or asynchronous PWM signals causing gate signal delays being the primary culprits

of transient current imbalance [9]. Despite advancements in the study of gate oscillations in multi-chip parallel configurations, there remains a notable gap in comprehensive analysis regarding the influence of gate internal resistance on gate oscillations.

To address this issue, this study investigates the impact of gate related resistance on oscillation characteristics through theoretical analysis and experimental verification. Firstly, a dynamic model of the multichip parallel turn-off process, incorporating gate resistance, is developed to identify key causes of oscillations. Then, the effects of individual gate resistance and differential gate resistance, are analyzed under parasitic inductance imbalance. Finally, a double-pulse test platform is established to validate the analysis.

II. GATE OSCILLATION MODEL DURING TURN-OFF PROCESS

Fig.1 illustrates the equivalent simplified model and the flow of induced current during the turn-off process of the parallel MOSFET. The equivalent circuit includes two MOSFETs (Q_1, Q_2), a drive circuit, a DC bus power supply V_{dc}, a load inductance L_{load}, a bus capacitance C_{dc} and line parasitic inductances. The drive circuit consists of a gate driver unit, a common gate resistance (R_G), and individual gate resistances (R_{g1} and R_{g2}). The line parasitic inductances contain source inductances (L_{s1} and L_{s2}), gate inductances (L_{g1} and L_{g2}), and Kelvin source impedances (L_{ks1} and L_{ks2})

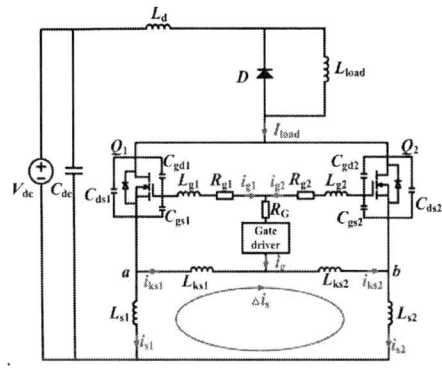

Fig.1. Equivalent circuit and induct current direction of multi-chip turn-off model.

This work was supported by the National Natural Science Foundation of China (Nos. 52177189 and 52107007), the Tianjin Municipal Science and Technology Bureau (No. 21JCJQJC00150 and 24JCZXJC00130). This work is corresponded by Yun-Hui Mei.

The primary cause of multi-chip oscillation is the coupling of the power supply loop into the driver loop through L_{ks} via the current path. Under normal conditions, when Q_1 and Q_2 are in the on state, i_{s1} and i_{s2} represent the currents flowing through the respective MOSFETs. When Q_1 and Q_2 are turned off, due to the imbalance of parasitic parameters in each parallel branch, different currents flow through $Q1$ and $Q2$, resulting in $i_{s1} \neq i_{s2}$. Assuming $i_{s1} < i_{s2}$, the imbalance between the two currents can be expressed,

$$i_{s2} - i_{s1} = \Delta i_s \qquad (1)$$

Due to the coupling of the power loop and the drive loop by the Kelvin source parasitic inductance L_{ks}, oscillations can occur in the gate loop under certain conditions [6].

$$\begin{cases} i_{g1} = i_{ks1} - i_{s1} \\ V_{gs1} = V_D - V_{g1} - V_{ks1} \\ V_{g1} = i_{g1}R_{g1} + L_{g1}\dfrac{di_{g1}}{dt} \\ V_{ks1} = L_{ks1}\dfrac{di_{ks1}}{dt} \\ i_{g1} = C_{gs1}\dfrac{V_{gs1}}{dt} \end{cases} \qquad (2)$$

Where V_D is the gate drive voltage, the condition oscillation generation in the gate circuit can be derived as follows,

$$R_{g1}{}^2 C_{gs1} - 4(L_{g1} + L_{ks1}) < 0 \qquad (3)$$

The change in Δi_s can lead to an increase in oscillation amplitude. When the source parasitic inductance and Kelvin source parasitic inductance remain unchanged, the primary factor in the circuit that can cause changes in Δi_s is the variation in gate resistance.

Due to the parasitic resistance in the gate circuit and variations in chip manufacturing processes, the gate resistances in each branch may be unbalanced. A larger gate resistance will cause the capacitance C_{gs} to charge and discharge more slowly than in branches with smaller gate resistances, leading to different MOSFET turn-off speeds across different branches. This exacerbates the imbalance in the current is and results in increased oscillation in the gate circuit.

However, an increase in the absolute value of the drive resistance in each branch will suppress the oscillation amplitude. This is because the increase in resistance in each branch reduces the difference in resistance values between the branches, balancing the currents flowing through each MOSFET. To better understand the impact of resistance difference on oscillation amplitude, we define α as the dispersion of resistances.

$$\alpha = \frac{\left|R_{g1} - \overline{R_g}\right| + \left|R_{g2} - \overline{R_g}\right| + \cdots + \left|R_{gn} - \overline{R_g}\right|}{\overline{R_g}} \qquad (4)$$

Where R_{g1} and R_{g2} represent the gate resistances in each branch, and R_g denotes the average value of the gate resistances. Under oscillating conditions, as dispersion increases, the impedance of each gate circuit becomes

unbalanced. This imbalance severely affects the discharge speed of the C_{gs} capacitance in each chip. Consequently, when multiple chips are turned off, discrepancies arise, and the amplitude of the gate oscillation becomes larger. Furthermore, a larger gate resistance further suppresses the oscillation of the gate voltage.

III. SIMULATION RESULTS AND EFFECT ANALYSIS OF TURN-OFF PROCESS

To further investigate the gate oscillation process, a simulation model based on the circuit depicted in Fig. 1 is established in LTspice. And the simulation parameters are listed in Table I.

TABLE I
MAIN PARAMETERS OF SIMULATION MODEL

Parameter		Value
Voltage	V_{dc}	500V
Drain inductance	L_d	10nH
Source inductance	L_{s1}, L_{s2}	10nH,15nH
Gate inductance	L_g	5nH
Kelvin source inductance	L_{ks1}, L_{ks1}	10nH,25nH
Gate resistance	R_G	10Ω

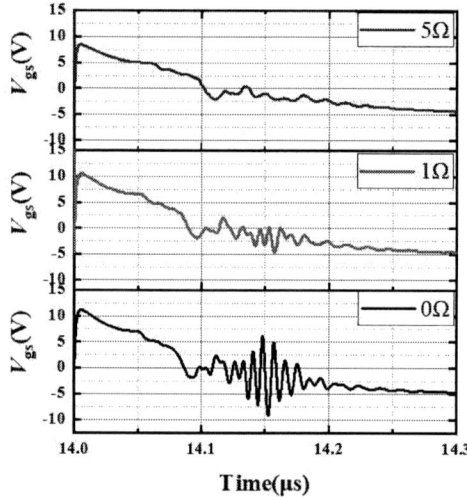

Fig.2. The influence of different individual gate resistances on gate voltage oscillation.

Fig.2 illustrates the impact of different individual gate resistances on gate voltage oscillation. It can be seen that as the individual gate resistance increases, the oscillation of gate voltage oscillation gets suppressed.

The impact of varying differential gate resistances on gate voltage oscillation is illustrated in Fig.3. As the impedance mismatch between the two gate circuits increases, the gate voltage oscillation becomes more pronounced. Even minor differences in resistance values can significantly exacerbate gate oscillations. This observation is consistent with the theoretical conclusions. The equation as a graphic and insert it into the text after your paper is styled.

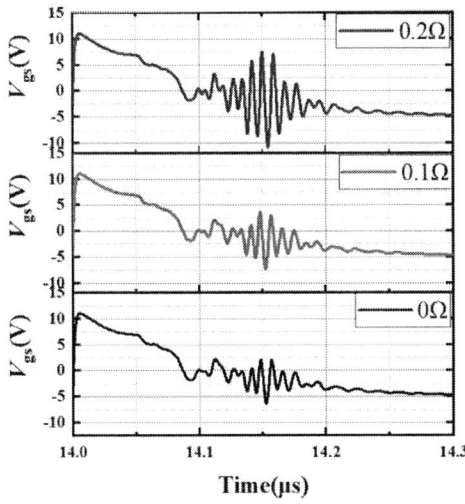

Fig.3. The influence of different differential gate resistances on gate voltage oscillation.

Fig.4 shows the influence of different differential gate resistances on gate current oscillation. As depicted in the figure, the increasing resistance difference between the two gate circuits results in a pronounced disparity in driving current. This disparity causes a substantial increase in oscillations within the gate loop.

Fig.5 illustrates the influence of divergence at different average gate resistance values. It can be observed that as the divergence increases, the gate oscillation amplitude also increases. A higher average resistance value can better suppress the effects of divergence, but it evidently also increases the turn-off loss of the MOSFET.

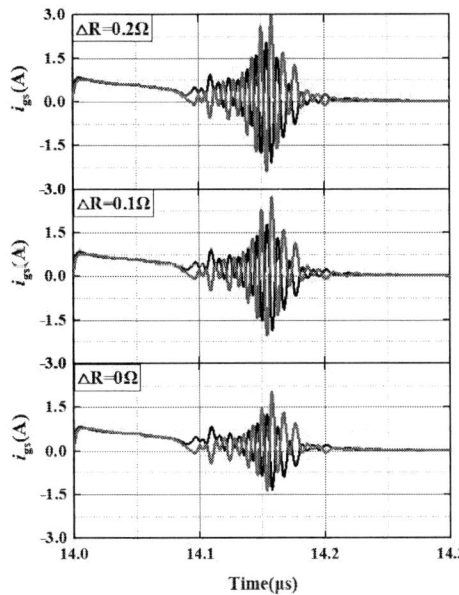

Fig.4. The influence of different differential gate resistances on gate current oscillation.

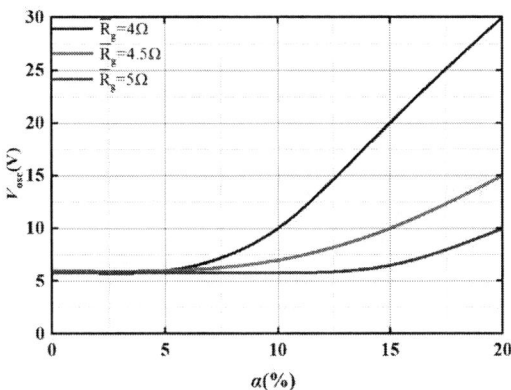

Fig.5. The influence of gate resistance dispersion α on gate voltage oscillation.

IV. EXPERIMENTAL VERIFICATION

To validate the proposed gate oscillation suppression analysis, a six-chip paralleled power module is established, as shown in Fig. 6. And a double-pulse test setup is constructed, as illustrated in Fig. 7. The setup included a bus capacitance of 1000 μ F, a main circuit bus voltage of 500V, and a 125μH hollow inductor as the load. The gate drive voltage is set to 15V/-3V. Three prototype modules, each with six parallel chips, are prepared: Module A (The chips were not screened), Module B (The chips have been screened), and Module C (Module B with a 10Ω individual gate resistance).

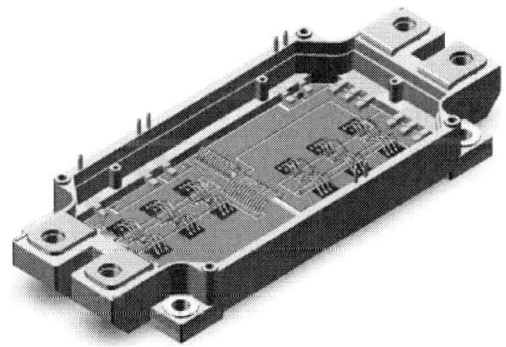

Fig.6. Prototype module.

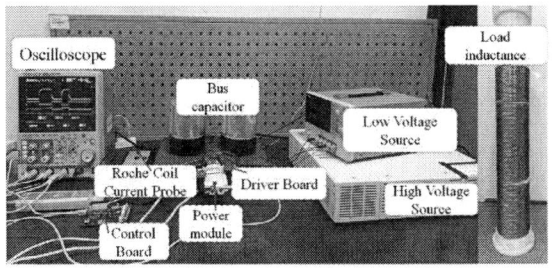

Fig.7. Double-pulse test platform.

The waveform during the turn-off process of Module A is shown in Fig. 8. As depicted, significant gate voltage oscillation occurs when the load current is 30 A. To investigate measures for suppressing gate voltage oscillations. While increasing the bus voltage, the pulse width is reduced to maintain the magnitude of the load current unchanged. Even when the bus voltage reaches 500V, no severe oscillation occurs at the gate of Module A. This indicates that the primary influence on gate loop oscillations is determined by the magnitude of the load current. When the load current exceeds a certain value, the current sharing imbalance among multiple chips will intensify. Under the condition that requirement (3) is simultaneously met, severe oscillations will occur in the gate circuit. Experimental analyses were conducted from the perspectives of reducing differential gate resistance and increasing individual gate resistance, aiming to validate the theoretical and simulation results on oscillation suppression.

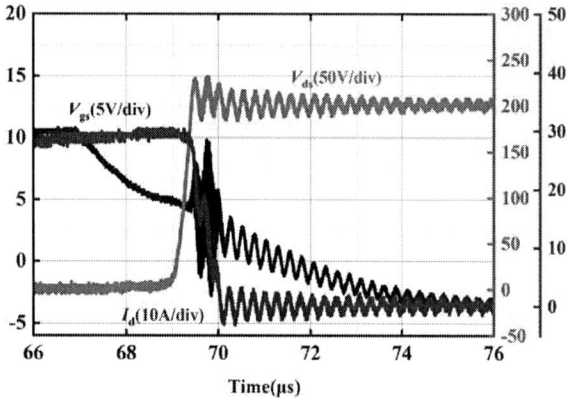

Fig.8. The test waveforms of turn-off process of Module A.

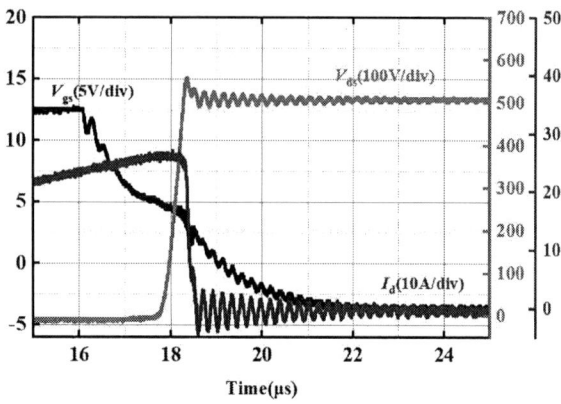

Fig.9. Test waveform of the high-voltage shutdown process for Module A

A. Reduce differential gate resistance

To validate the influence of differential gate resistance on oscillation characteristics, experimental tests are conducted on Module B, which underwent chip screening, as illustrated in Fig. 10. It can be observed that, compared to Module A, the gate voltage oscillations in Module B have been mitigated, with minor oscillations occurring when the load current reaches 40A. This improvement is attributed to the reduced differential gate resistance among the chips after screening. However, it can be noted that the gate oscillation

has not been completely suppressed at this stage. This is because, even though the difference between the gate resistances is nearly eliminated, due to the mismatch of parasitic parameters on each branch, the oscillation conditions in (3) are still satisfied. When the load current increases, severe oscillations still occur in the gate circuit.

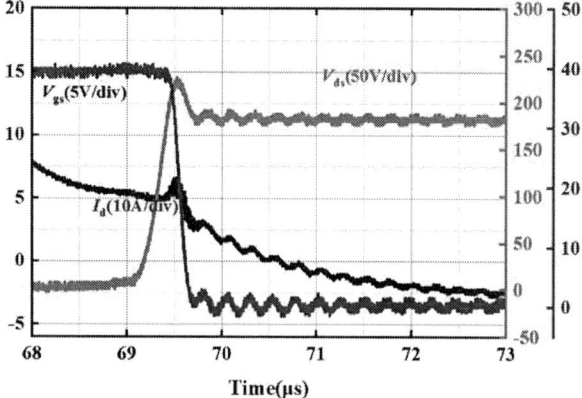

Fig.10. The test waveforms of turn-off process of Module B.

B. Increase individual gate resistance

The test waveforms of turn-off process of Module C is depicted in Fig. 11. The results indicate that the gate voltage oscillation disappears after increasing the individual gate resistance. This is primarily due to the increase in R_g, which prevents the conditions for oscillation from being met, which is shown in (3).

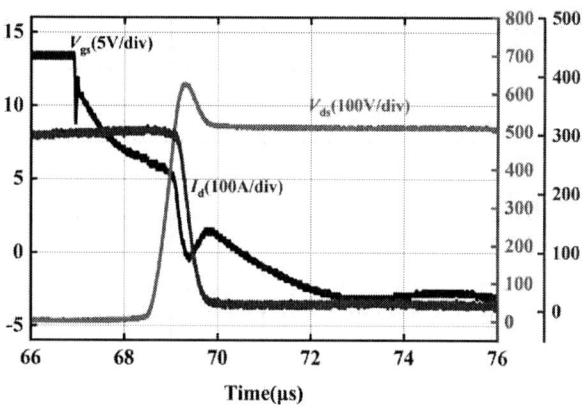

Fig.11. The test waveforms of turn-off process of Module C.

V. CONCLUSION

This paper investigates the impact of gate-related resistances on gate oscillation characteristics in multi-chip parallel power modules through theoretical, simulation, and experimental studies. By systematically optimizing chip parameters, the differential gate resistance can be effectively modulated to suppress gate oscillations. Additionally, the incorporation of appropriate individual gate resistances is shown to effectively mitigate gate oscillations. Experimental results validate the theoretical analysis and provide practical guidelines for improving the turn-off performance of multi-chip power modules. These findings offer valuable insights into the design and optimization of high-power electronic systems.

REFERENCES

[1] L. Anoldo, G. Malta, B. Mazza, G.G. Piccione, M. Calabretta, S. Russo, A. Russo, A. Sitta, A. Messina, A. Lionetto and S. Patanè,"Mult-chip SiC-based compact module for automotive applications: A high speed thermal study," *Microelectronics Reliability*, vol. 138, 2022.

[2] H. Li, S. Zhao, X. Wang, L. Ding and H. A. Mantooth, "Parallel Connection of Silicon Carbide MOSFETs-Challenges, Mechanism, and Solutions," *IEEE Transactions on Power Electronics*, vol. 38, no. 8, pp. 9731-9749, 2023.

[3] J. Lv, C. Chen, B. Liu, Y. Yan, Z. Zheng and Y. Kang, "Dynamic Current Sharing Mechanism Analysis of Paralleled SiC MOSFETs Considering Parasitic Mutual Inductances Based on an Improved Model," *IEEE Transactions on PowerElectronics*, vol. 39, no. 6, pp. 7536-7559, 2024.

[4] J Wang, S Yu, X Zhang, Z Wei, N Jiang and W Chen,"Accurate Modeling of the Effective Parasitic Parameters for the Laminated Busbar Connected With Paralleled SiC MOSFETs," *IEEE Transactions on Circuits and Systems*, vol. 68, no. 5, pp. 2107-2120, 2021.

[5] H Mao, L Ran, H Chen, X Zhou and H Jiang. "Avalanche capability degradation of the parallel-connected SiC MOSFETs", *Microelectronics Reliability*, vol. 142, 2023.

[6] X Wu, Y Wu, M Rong, C Tao, Y Xiao and J Wang. "Mitigating Gate Oscillations of Parallel SiC MOSFETs for Enhanced Performance in DC Solid-State Circuit Breaker," *IEEE Journal of Emerging and Selected Topics in Power Electronics*, vol. 12, no. 2, pp. 1822-1833, 2024.

[7] K Ma, Y Sun, X Liu, Y Song, X Li, H Shi, Z Feng, X Zhang, Y Zhou and S Liu. "The Mechanism of Short-Circuit Oscillations in Automotive-Grade Multi-Chip Parallel Power Modules and an Effective Mitigation Approach," *Sensors,* 2024.

[8] G Chang, C Peng, Y Liu, E Deng, X Li, Q Xiao and Y Huang. "Optimization and Validation of Current Sharing in IGBT Modules With Multichips in Parallel," *IEEE Transactions on Power Electronics*, vol. 39, no. 12, pp. 15672-15681, 2024.

[9] J Wang, C Wang, S Zhao, H Li, L Ding, Xun Shen and H. Alan Mantooth. "Comprehensive Analysis of Paralleled SiC MOSFETs Current Imbalance Under Asynchronous Gate Signals," *IEEE Journal of Emerging and Selected Topics in Power Electronics*, vol. 11, no. 5, pp. 4850-4866, 2023.

2025 IEEE Workshop on Wide Bandgap Power Devices and Applications in Asia (WiPDA Asia)

Development and Electrical Characteristics Study of a 10kV/125A SiC MOSFET Module

Yaodong Zhang
School of Electrical Engineering
Xi'an Jiaotong University
Xi'an, China
13613134194@163.com

Yong Chen
Zhuhai Power Supply Bureau of
Guangdong Power Grid Co.,Ltd.
DC Power Distribution and
Consumption Technology Research
Center of Guangdong Power Grid
Co.,Ltd.
Zhuhai, China
35665035@qq.com

Xu Cheng
Zhuhai Power Supply Bureau
Guangdong Power Grid Co.,Ltd.
DC Power Distribution and
Consumption Technology Research
Center of Guangdong Power Grid
Co.,Ltd.
Zhuhai, China
664157102@qq.com

Xiaotian Zhang
School of Electrical Engineering
Xi'an Jiaotong University
Xi'an, China
xiaotian@xjtu.edu.cn

Abstract—To address the needs of new power systems with large-scale integration of renewable energy sources, this paper independently developed a high-voltage silicon carbide (SiC) metal-oxide-semiconductor field-effect transistor (MOSFET) power module rated for 10 kV/125 A and investigated its electrical characteristics. First, the overall packaging structure and internal circuitry of the module were designed in detail. Subsequently, a 3D steady-state thermal finite element simulation model of the SiC MOSFET module was established in COMSOL Multiphysics to analyze its thermal performance under a constant baseplate temperature of 60°C. Additionally, ANSYS Q3D Extractor was employed to extract parasitic inductance parameters from the designed module's power loop. Simulation results revealed that the highest junction temperature occurred in the SiC MOSFET chip, reaching 123.35°C,which confirms that the existing packaging materials and structure meet the maximum allowable junction temperature requirements. The extracted parasitic inductance of the power loop was measured as = 26.1nH. The low parasitic inductance of the developed module facilitates the superior high-frequency switching performance inherent to SiC devices. This study provides critical technical support for the development and engineering application of high-voltage SiC MOSFET power modules.

Keywords—SiC MOSFET module, package,thermal simulati on,parasitic inductance

I. INTRODUCTION

SiC is a compound semiconductor material composed of silicon (Si) and carbon (C). Its insulation breakdown field strength is 10 times that of silicon, and its bandgap is nearly three times that of silicon. Its intrinsic carrier concentration is much lower than that of silicon, and its thermal conductivity is also three times that of silicon, making it more suitable for high-temperature and high-voltage applications. Due to its excellent physical and electrical properties, SiC devices have shown significant advantages in high-power, high-frequency, and high-temperature application scenarios, gradually becoming an important development direction in the field of power electronics [1]. MOSFETs have the advantages of high operating frequency, fast switching speed, low switching loss, and a simple driving system. By using 4H-SiC material to manufacture MOSFETs and packaging multiple chips into power modules, the voltage level and

capacity of SiC MOSFET modules can be further enhanced, overcoming the disadvantage of insufficient capacity of single SiC chips.

However, current commercial SiC power modules still adopt the packaging technology traditionally used for silicon IGBT modules. This leads to the generation of significant high-frequency parasitic parameters in the modules, which can cause issues such as voltage overshoot and increased device stress, turn-on current overshoot and higher losses, switching current oscillations, and degraded circuit performance. Additionally, in high-current applications, multiple chips must be connected in parallel to enhance current capacity. The presence of parasitic parameters can result in uneven dynamic and static current distribution in parallel paths, thereby reducing the operational reliability of the modules. Therefore, achieving low and symmetrical parasitic inductance during the packaging process is crucial for leveraging the high-speed switching characteristics of SiC devices [2]. How to minimize the parasitic parameters introduced by device packaging has become a key technical challenge in the application of SiC devices in high switching frequency scenarios. Furthermore, SiC chips are more compact in size compared to Si chips. While this enables higher power density, it also leads to issues such as performance degradation due to uneven heat dissipation and more pronounced thermal stress caused by mismatched coefficients of thermal expansion between different materials. Consequently, thermal management in packaging is also a critical issue that cannot be overlooked during the design and development of modules.

To this end, based on preliminary research, this paper develops a 10 kV/125 A SiC MOSFET module tailored for high-voltage and high-capacity power conversion applications. The paper proposes a packaging design scheme for this SiC MOSFET module and conducts a series of simulations on its electrical characteristics to obtain necessary parameters and evaluate its performance. The paper first introduces the module's packaging structure and design scheme, followed by thermal simulations and parasitic parameter extraction simulations of the power loop.

This work was supported by the China Southern Power Grid Company. Ltd., Science and Technology Project: Research on Insulation Design andP arasitic Parameter Optimization Technology for High Voltage SiC Devices and Development of Half Bridge Device Packaging under Grant030400KC 23090015 and Grant GDKIXM20231031.

979-8-3315-1110-4/25 $31.00 © 2025 IEEE

II. DESIGN OF MODULE

A. Overall Packaging Structure Design

The power module employs a self-designed packaging structure, which comprises a housing, a power loop, a drive loop, and a substrate. The power loop is composed of six mutually isolated and entirely symmetrical DBC sub-modules, with every two sub-modules forming a half-bridge structure. Each half-bridge structure includes a DBC substrate, power terminals, and SiC MOSFET chips, totaling 24 chips in the entire module. The DBC substrate features a double-layer structure, with the SiC MOSFET chips connected to the etched circuit on the upper surface of the top DBC substrate layer via soldering. The power terminals consist of DC+ power terminals, DC- power terminals, and AC power terminals. The drive loop includes 12 double-layer DBC substrates divided into three groups and drive terminals, with each group positioned above and below the power module respectively.

Fig. 1. Appearance of SiC MOSFET module.

Fig. 2 shows the internal structure of the SiC MOSFET module, arranged from bottom to top as follows: heat sink baseplate, direct bonding copper (DBC) ceramic substrate, SiC MOSFET chip, bonding wires, decoupling capacitor, power terminals, drive terminals, silicone gel, and housing. Among these, the DBC substrate features a dual-layer structure comprising an alumina ceramic substrate and copper layers. Its layered configuration progresses from top to bottom as: upper copper layer, upper ceramic layer, intermediate copper layer, lower ceramic layer, and bottom copper layer. The heat sink baseplate is fabricated from aluminum silicon carbide (AlSiC) material, offering enhanced temperature cycling capability.

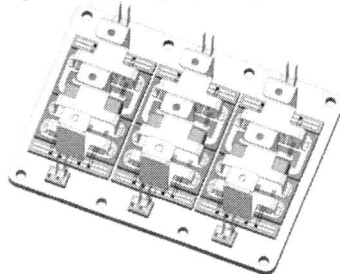

Fig. 2. Internal packaging structure of SiC MOSFET module.

B. Circuit Design

Fig. 3 illustrates the internal circuit design of the module. The module incorporates three identical half-bridge configurations. Each half-bridge configuration contains two symmetrically arranged power chip groups, forming the upper and lower arms respectively. Each group consists of four parallel-connected chips, resulting in a total of 24 chips within the entire module.

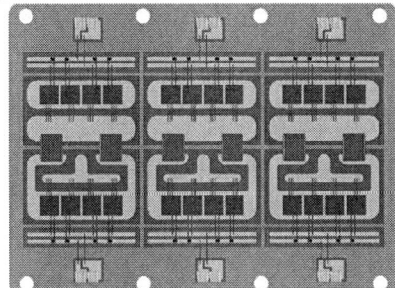

Fig. 3. Circuit design of SiC MOSFET module.

On the DBC substrate of a single half-bridge, the upper copper layer is etched into four axially symmetric sections: the DC positive layer, DC negative layer, AC layer, and an interconnection layer between the source terminals of the upper-arm chips and the AC layer. The AC layer and interconnection layer are bridged by an overhead copper structure (AC power terminal). A decoupling capacitor is positioned between the DC positive and negative layers within each half-bridge. All internal chip connections (gate, source, and copper layer interconnects) employ aluminum wire bonding.

All power chips utilize Kelvin source connections. Within the same bridge arm, the gate and Kelvin source terminals of the chips are routed via aluminum bonding wires to the driver PCB. The upper and lower bridge arms are collectively driven through dedicated driver terminals. The Kelvin source connection decouples the power loop from the driver loop, ensuring that only minimal current flows through the driver circuit, thereby mitigating parasitic parameter effects. This configuration reduces the inductive coupling of the source leads, suppressing voltage spikes and oscillations during switching transitions, which enhances switching performance and device reliability. Furthermore, by minimizing parasitic resistance, the Kelvin connection reduces joule heating during operation, thereby improving thermal management and extending module lifespan.

Fig. 4 and Fig. 5 further illustrates the terminal structure of the module. A single half-bridge structure within the module is equipped with three power terminals to interface with external circuitry, resulting in a total of nine power terminals across the entire module. These terminals are vertically mounted on the DBC substrates. The DC+ power terminal serves to interconnect the drain terminals of the upper-arm chips with the external busbar positive voltage, while the DC- power terminal links the source terminals of the lower-arm chips to the external busbar negative voltage. The AC power terminal provides connectivity to the load.

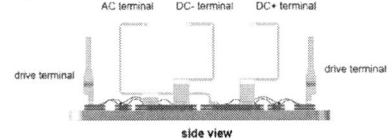

Fig. 4. Terminal schematic side view diagram.

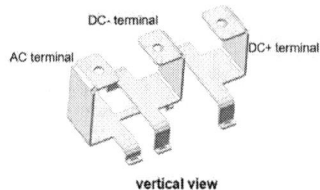

Fig. 5. Terminal schematic vertical view diagram.

The driver section incorporates six gate driver terminals and six Kelvin source driver terminals, all positioned on the driver-side DBC substrate. Each driver terminal is integrated with a gate resistor to optimize switching dynamics.

III. MODULE THERMAL SIMULATION AND PARASITIC PARAMETER EXTRACTION

A. Module Thermal Simulation Analysis

The thermal characteristics of the module are critical factors for the long-term reliability of the device, which are closely related to parameters such as chip layout, packaging material types and dimensions. The device junction temperature increases with the thickness of the DBC substrate, AlN ceramic layer, and AlSiC heat sink baseplate. Therefore, for practical packaging selection, thinner ceramic layers and heat sink baseplates facilitate module heat dissipation while meeting requirements for electrical insulation and mechanical strength.

To evaluate the overall thermal reliability and heat dissipation capability of the module under high-speed switching conditions, a three-dimensional steady-state thermal finite element model of the SiC MOSFET module was constructed in COMSOL Multiphysics for thermal simulation analysis. In the designed SiC MOSFET module, the internal structure comprises three identical half-bridge configurations. When a bridge-arm SiC MOSFET chip is in the conducting state, the complementary bridge-arm module's antiparallel diode conducts and operates in freewheeling mode. Since the module operates under high-frequency switching in power electronic converters, both the SiC MOSFET chips and their integrated antiparallel diodes alternate rapidly between conduction states. Thus, all chips were defined as heat sources in the thermal simulation.

For the thermal boundary condition setup in the thermal simulation analysis, assuming the thermophysical properties of materials remain constant regardless of temperature and maintaining a fixed temperature at the bottom of the module heat sink baseplate, this study employs Dirichlet boundary conditions (first-type boundary conditions) with the baseplate bottom temperature fixed at 60°C. In other directions, since the thermal interface materials are stationary air or silicone gel—whose thermal conductivity is orders of magnitude lower than other internal materials—these boundaries are defined as thermally insulated in the simulation. Under these conditions, the thermal simulation results of the module are shown in Fig. 6.

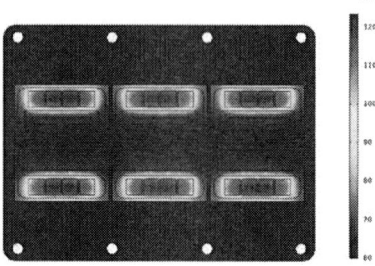

Fig. 6. Thermal simulation temperature distribution of SiC MOSFET module.

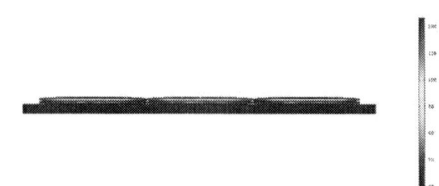

Fig. 7. Chip cross-sectional temperature distribution of SiC MOSFET module.

The simulated temperature distribution of the module indicates that the current packaging materials and structure satisfy the maximum allowable junction temperature requirements of the chips. The highest junction temperature, localized at the SiC MOSFET chip, reaches 123.35°C, confirming that the SiC MOSFET chip is the critical reliability concern under high-temperature operation.

B. Power Loop Parasitic Inductance Extraction

The main sources of parasitic inductance in power module packaging originate from the intrinsic parasitic inductance of conductors (e.g., DBC substrates, bonding wires, power terminals) and mutual inductance between these conductors. Excessive parasitic inductance can lead to critical issues such as voltage overshoot (increased device stress), turn-on current overshoot (higher switching losses), and current oscillation, making the extraction and minimization of parasitic inductance a critical focus in module packaging design.

This study employs ANSYS Q3D Extractor, a parameter extraction software based on Finite Element Method (FEM) and Method of Moments (MoM), to extract the parasitic inductance of the designed module's power loop. The extraction frequency is set to 50 MHz. Since the module contains three identical half-bridge structures, only one half-bridge is selected for parasitic inductance extraction. The simulation results show that the extracted power loop parasitic inductance is $L_{DS} = 26.1$ nH. The developed module in this paper exhibits low power loop parasitic inductance, which facilitates the full utilization of SiC devices' superior performance at high switching frequencies. Additionally, it enables more uniform dynamic and static current distribution in parallel paths, thereby enhancing the overall performance of the power module.

IV. CONCLUSION

This paper proposes a novel packaging structure for a 10 kV/125 A silicon carbide (SiC) metal-oxide-semiconductor field-effect transistor (MOSFET) module, focusing on its thermal stability and parasitic parameters. Specifically, a half-bridge SiC power module was modeled using

979-8-3315-1110-4/25 $31.00 © 2025 IEEE

multiphysics simulation software. Through simulations, the thermal stability of the module under specified temperature conditions was analyzed, and the parasitic parameters of the power loop were calculated. The results demonstrated that the module exhibits excellent thermal stability and low parasitic parameters. A physical prototype of the module will be fabricated in subsequent work, and experimental tests will be conducted to further validate its performance, aiming to provide further insights for future research and optimization.

REFERENCES

[1] L. Zhang, X. Yuan, X. Wu, C. Shi, J. Zhang and Y. Zhang, "Performance Evaluation of High-Power SiC MOSFET Modules in Comparison to Si IGBT Modules," in IEEE Transactions on Power Electronics, vol. 34, no. 2, pp. 1181-1196, Feb. 2019.

[2] H. Lee, V. Smet and R. Tummala, "A Review of SiC Power Module Packaging Technologies: Challenges, Advances, and Emerging Issues," in IEEE Journal of Emerging and Selected Topics in Power Electronics, vol. 8, no. 1, pp. 239-255, March 2020.

[3] H. A. Mantooth and S. S. Ang, "Packaging Architectures for Silicon Carbide Power Electronic Modules," 2018 International Power Electronics Conference (IPEC-Niigata 2018 -ECCE Asia), Niigata, Japan, 2018, pp. 153-156.

[4] C. Zhang, Z. Huang, C. Chen, X. Liu, F. Luo and Y. Kang, "A 900A High Power Density and Low Inductive Full SiC Power Module for High Temperature Applications based on 900V SiC MOSFETs," 2020 IEEE Applied Power Electronics Conference and Exposition (APEC), New Orleans, LA, USA, 2020, pp. 2777-2781.

[5] Y. Ge, Z. Wang, Y. Yang, C. Qian, G. Xin and X. Shi, "Layout-Dominated Dynamic Current Balancing Analysis of Multichip SiC Power Modules Based on Coupled Parasitic Network Model," in IEEE Transactions on Power Electronics, vol. 38, no. 2, pp. 2240-2251, Feb. 2023.

[6] Wei Xiaoguang, Wu Zhiwei, Tang Xinling, et al. Development and research on electrical characteristics of domestic 6.5 kV/400 A SiC MOSFET modules [J]. Proceedings of the CSEE, 2024, 44(10): 4012-4026.

[7] X. Li, Y. Chen, H. Chen, R. Paul, X. Song and H. A. Mantooth, "A 10 kV SiC MOSFET Power Module With Optimized System Interface and Electric Field Distribution," in IEEE Transactions on Power Electronics, vol. 39, no. 8, pp. 9540-9553, Aug. 2024.

2025 IEEE Workshop on Wide Bandgap Power Devices and Applications in Asia (WiPDA Asia)

Study of Thermal Stress on Devices in Si/SiC Hybrid Half-Bridge Inverter

Yulin Wang
College of Electrical and Information Engineering
Hunan University
Changsha, China
wangyl728@hnu.edu.cn

Ruixiao Dong
College of Electrical and Information Engineering
Hunan University
Changsha, China
ruixiaodong@hnu.edu.cn

Jun Wang
College of Electrical and Information Engineering
Hunan University
Changsha, China
junwang@hnu.edu.cn

Yuxing Dai
Wenzhou University

Wenzhou, China
daiyx@hnu.edu.cn

Liuchen Chang
University of New Brunswick

Brunswick, Canada
lchang@unb.ca

Abstract—Compared to full SiC-based inverters, the proposed Si/SiC hybrid half-bridge (HHB) inverter topology with partial power processing offers an effective solution. This approach significantly reduces costs while leveraging the lower operating frequency of the Si-phase to minimize the high switching loss commonly associated with Si devices. Excellent output power quality and fast dynamic response are achieved by the high frequency operation of SiC. However, the imbalance in power distribution leads to an increase in junction temperature of the Si phase devices, which produces an imbalance in thermal stress compared to the SiC devices. It negatively affects the transfer efficiency and stable operation of the inverter. To evaluate the thermal performance of power devices, this study has established a power loss and junction temperature coupling model. A 1.5-kW HHB inverter prototype has been developed. Through combining thermal simulation and offline experimental results, a comprehensive device thermal stress analysis has been conducted, thereby verifying the accuracy of the developed computational model.

Keywords—*Si/SiC HHB topology, thermal stress, temperature calculation model.*

I. INTRODUCTION

As a key carrier of power conversion, inverters are now widely used in the fields of rail transportation, automotive motor drive, and renewable energy generation. Compared with the traditional Si IGBT-based inverters, the third-generation wide-bandgap (WBG) power devices represented by SiC have shown significant improvements in dynamic response and power density [1]-[4]. However, the higher cost and limited current capacity hinders the large-scale use of SiC devices. Therefore, the concept of circuit-level mixed-parallel Si/SiC was proposed, and the improved transmission efficiency and dynamic response were achieved at a lower cost using the Si/SiC HHB topology [5], [6]. For a two-level converter, a parallel-configured inverter combining a high-current Si-based inverter and a low-current SiC-based inverter was implemented to achieve high transfer efficiency and fast dynamic response at a relatively lower cost [7]-[10]. On this basis, the concept of WBG fractional power processing (WFPP) was proposed, using low-frequency high-capacity Si devices and high-frequency small-capacity WBG devices as the base and power splitting processing, respectively [11]. However, the uneven power distribution in the Si/SiC hybrid interleaved structure leads to an imbalance in the thermal stress of the device, which makes the device junction temperature increase. According to the data, the failures caused by the power devices account

for 19% of the power converter failures, and 55% of them are due to the large junction temperature of the device [12], [13]. Reference [14] proposes an offline optimisation strategy based on an average power loss model. But the established power loss model is not applicable to Si/SiC HHB inverters. The power loss and device junction temperature are still variable under steady state operation. How to use the loss junction temperature model to analyse the thermal stress of the device without affecting the output power quality, and to optimize the thermal stress by controlling the parameter adjustment has become the focus of current research.

In this paper, the thermal stresses of devices in Si/SiC HHB inverters are quantitatively analysed by means of the established power loss junction temperature coupling model. The rest of the organisational parts are as follows, Section II derives the power loss and junction temperature coupling model, Section III fits the computational model using thermal simulation, Section IV quantitatively analyses the thermal stresses based on the offline experimental results, and Section V concludes the paper.

II. POWER LOSS AND JUNCTION TEMPERATURE CALCULATION MODEL

The topology of Si/SiC HHB inverter is presented in Fig.1, which consists of Si and SiC phase half bridge. The Si phase incorporates larger capacity Si IGBTs (S_{L1}, S_{L2}) and a main inductance L_1, the SiC phase contains smaller capacity SiC MOSFETs (S_{H1}, S_{H2}) and a auxiliary inductance L_2. Capacitors C_1, C_2 are large enough to ensure the voltage on them equal with $V_{in}/2$. The total power loss of the HHB inverter comprises two components: the Si-phase half-bridge and the SiC-phase half-bridge.

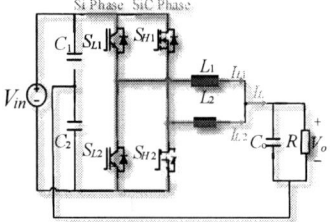

Fig. 1. Topology of Si/SiC HHB inverter.

The switching frequency f_{sH} of SiC phase is always set as a fixed and elevated value to guarantee the quality of output power. The switching frequency f_{sL} of Si phase is

979-8-3315-1110-4/25 $31.00 © 2025 IEEE

designed to be lower to reduce switching loss. To accommodate varying current loads during a line cycle, two power sharing ratios are utilized: K_1 for peak current conditions and K_2 for near-zero current conditions. The operational scheme is illustrated in Fig. 2, with the mathematical relationship expressed as follows:

$$K = \begin{cases} K_1 & (|I_{Lr}| \geq 0.5I_m) \\ K_2 & (|I_{Lr}| < 0.5I_m) \end{cases} \tag{1}$$

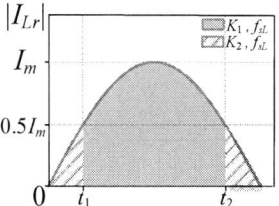

Fig. 2. The diagram of three-dimension-variant operation mode.

A. Power Loss Model of Si phase Components

The power loss of the device consists of two parts: switching loss and conduction loss. For Si-based devices, within one line cycle, the switching loss is directly determined by the sum of the product of the switching energy loss and the switching frequency. The average switching loss for the upper-arm switch during turn-on, turn-off, and reverse recovery are expressed as follows:

$$\begin{cases} P_{onSi} = \dfrac{1}{N_{0l}} \sum_{i=1}^{N_1} (a_{1Ti} I_{L1\min}(i) + b_{1Ti}) f_{sL} \\ P_{offSi} = \dfrac{1}{N_{0l}} \sum_{i=1}^{N_2} (a_{2Ti} I_{L1\max}(i) + b_{2Ti}) f_{sL} \\ P_{recSiD} = \dfrac{1}{N_{0l}} \sum_{i=1}^{N_3} (a_{3Ti} |I_{L1\max}(i)| + b_{3Ti}) f_{sL} \\ N_{0l} = \dfrac{T_0}{T_{sL}} \end{cases} \tag{2}$$

Where N_1, N_2, N_3 represent the forward-turn-on time, forward-turn-off time and reverse-recovery time of a single IGBT and its anti-parallel diode, respectively. N_{0l} denotes the total low- frequency (LF) switching times within the time period T_0. The fitting coefficients a_{1T}, b_{1T}, a_{2T}, b_{2T}, a_{3T}, b_{3T} are obtained from the switching-energy-loss-current curve, which is measured through the double pulse test under one DC bus voltage and junction temperatures. Subsequently, the switching energy loss at specific junction temperature is calculated using the linear interpolation method.

In addition to switching loss, the conduction loss of the device cannot be overlooked. For conducting Si IGBTs and diodes, the conduction loss over one line cycle are related to the on-state voltage, the root-mean-square (RMS) value of the conduction current, and the conduction duty cycle during the period T_0. These loss can be expressed as follows:

$$\begin{cases} P_{conSi} = \dfrac{1}{T_0} \int_0^{T_0} V_{ce}(I_{rms1}, T) I_{rms1}(t) D_{sL1}(t) dt \\ P_{conSiD} = \dfrac{1}{T_0} \int_0^{T_0} V_d(I_{rms1}, T) I_{rms1}(t) D_{sL2}(t) dt \end{cases} \tag{3}$$

The loss in the Si-phase inductor consist of copper loss and core loss. Based on empirical formulas, the average copper loss of the Si-phase inductor over one line cycle T_0 can be expressed as:

$$P_{L1cu} = \dfrac{1}{T_0} \int_0^{T_0} I_{rms1}(t)^2 R_{L1cu} dt \tag{4}$$

Where R_{L1cu} is the inner resistance of Si phase inductance.

In addition, the magnetic core loss of the Si-phase inductor, which encompasses hysteresis loss, eddy current loss and residual loss can be calculated using the Modified Steinmetz Equation (MSE) :

$$P_{L1core} = \dfrac{1}{T_0} \int_0^{T_0} V_{L1} K_c \left(\dfrac{2}{\Delta B^2 \pi^2} \int_0^{T_{sL}} \left(\dfrac{dB}{dt} \right)^2 dt \right)^{\alpha-1} \left(\dfrac{\Delta B}{2} \right)^{\beta} f_{sL} dt \tag{5}$$

In this context, V_{L1} denotes the winding volume, T_{sL} represents the switching period, and f_{sL} represents the switching period. K_c, α and β are the fitting coefficients obtained from the core loss curve in datasheet. ΔB is the peak-to-peak amplitude of magnetic flux generated by the current ripple.

B. Power Loss Model of SiC phase Components

Compared to Si-phase, the switching frequency of SiC-phase is higher, and the current waveform is more irregular. The switching loss of upper SiC device can be calculated as equation (6).

$$\begin{cases} P_{onSiC} = \dfrac{1}{N_{0h}} \sum_{i=1}^{N_4} (a_{4Ti} I_{L2H}(i) + b_{4Ti}) f_{sH} \\ P_{offSiC} = \dfrac{1}{N_{0h}} \sum_{i=1}^{N_5} (a_{5Ti} I_{L2H}(i)^2 + b_{5Ti} I_{L2H}(i) + c_{5Ti}) f_{sH} \\ P_{recSiCD} = \dfrac{1}{N_{0h}} \sum_{i=1}^{N_6} (a_{6Ti} I_{L2H}(i) + b_{6Ti}) f_{sH} \\ N_{0h} = \dfrac{T_0}{T_{sH}} \end{cases} \tag{6}$$

Similarly, N_{0h} represents the total number of high-frequency switching events during period T_0, while N_4, N_5 and N_6 denote the counts of the three corresponding hard-switching types. The coefficients a_{4T}, b_{4T}, a_{5T}, b_{5T}, c_{5T}, a_{6T} and b_{6T} are current-dependent switching energy fitting parameters that vary with junction temperature. It should be noted that the conduction loss of SiC MOSFETs differ from those of Si IGBTs, as there is no knee voltage effect in the on-state. Instead, the on-resistance $R_{ds(on)}$ at a specific junction temperature can be determined using one-dimensional linear interpolation. Additionally, the current in SiC devices involves smaller high-frequency current ripple and larger low-frequency reverse current ripple. For simplicity in calculation, only the dominant ripple component is considered to represent the RMS current value of SiC switching. Consequently, the conduction loss can be expressed as follows:

$$P_{conSiC} = \dfrac{1}{T_0} \int_0^{T_0} I_{rms2}(t)^2 \left(R_{dson}(T) D_{sH1}(t) + R_{sdon}(T) D_{sH2}(t) \right) dt \tag{7}$$

The calculation methods for the inductor loss of both the Si-phase and SiC-phase are identical. Given that only low-frequency current ripple is considered, we can directly substitute the current values and SiC-phase inductor parameters into Equations (4) and (5) to compute the corresponding inductor loss. And the inductor loss (P_L) consisting of two parts: copper loss (P_{Lcu}) and core loss (P_{Lcore}), i.e.:

$$P_L = P_{Lcu} + P_{Lcore} \tag{8}$$

Generally speaking, the total average power loss of an Si/SiC HHB is the sum of all individual loss components. We can reasonably assume that the upper and lower switching devices within the same half-bridge exhibit nearly identical power loss and thermal stresses. Therefore, the total power loss of the Si/SiC HHB can be expressed as:

$$P_{lossav} = 2(P_{Si,IGBT} + P_{SiC,MOS}) + P_{L1} + P_{L2} \tag{9}$$

C. Dynamic Junction Temperature Calculation Model

The junction temperature of power devices is influenced by both their thermal environment and power loss. The thermal impedance networks illustrating the heat transfer paths from the environment to the IGBT and MOSFET nodes are depicted in Fig. 3 (a) and (b), respectively. In these diagrams, $Z_{thj\text{-}c(t)}$ is the junction-to-case foster thermal impedance extracted from datasheet. T_{cI} and T_{cM} is the casing temperatures of the IGBT and MOSFET devices.

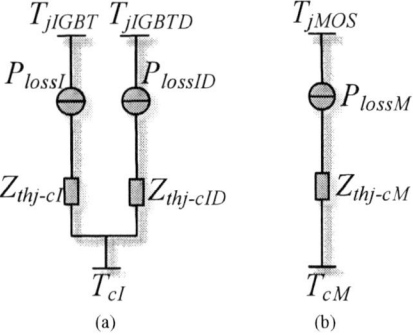

(a)   (b)

Fig. 3. The thermal impedance network of Si and SiC devices from junction to case.(a) Si IGBT.(b) SiC MOSFET.

Fig. 4 presents the detailed thermal impedance network, and the specific thermal resistance and time constant value of each layer are shown in Table I. The case temperature of device Tc varies in a steady manner without fluctuation owing to the larger heat-dissipated time constant of heatsink. The transient junction temperature $T_{jIGBT(t)}$, $T_{jIGBTD(t)}$ and $T_{jMOS(t)}$ can be expressed as following equations.

$$\begin{cases} T_{jIGBT}(t) = P_{lossI}(t)Z_{thj-cI}(t) + T_{cI} \\ T_{jIGBTD}(t) = P_{lossID}(t)Z_{thj-cID}(t) + T_{cI} \\ T_{jMOS}(t) = P_{lossM}(t)Z_{thj-cM}(t) + T_{cM} \end{cases} \tag{10}$$

$$Z_{thj-c}(t) = \sum_{i=1}^{n} R_{thi}(1 - e^{-t/\tau_i}) \tag{11}$$

τ_i is the time constant of each foster thermal impedance layer.

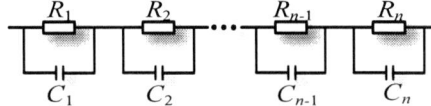

Fig. 4. N-order foster thermal network from junction to case of devices.

TABLE I. FOSTER THERMAL IMPEDANCE PARAMETERS

IGBT	R_{thi}(K/W)	0.00094	0.0568	0.07	0.17	0.008	0.0016
	τ_i(s)	2.5e-5	3.4e-4	3.1e-3	0.016	0.224	2.91
Diode	R_{thi}(K/W)	0.17605	0.298	0.293	0.0141	0.0022	
	τ_i(s)	3.1e-4	2.8e-3	0.0152	0.209	2.46	
MOSFET	R_{thi}(K/W)	0.0635	0.348	0.496	0.393		
	τ_i(s)	1.46e-5	4.6e-4	2.43e-3	1.43e-2		

The dynamic junction temperature and power loss of two devices in one line period is shown in Fig. 5. To compute the transient junction temperature, the discrete thermal-model method is employed. For an n-order foster thermal impedance, the transient temperature difference between the junction and case can be derived using the iterative algorithm given in equation (12).

$$\begin{cases} \Delta T_{j-c}^{(1)} = P_{lossd}^{(0)} \sum_{i=1}^{n} R_{thi}(1 - e^{\frac{-\Delta t}{\tau_i}}) \\ \Delta T_{j-c}^{(q)} = \sum_{i=1}^{n} \Delta T_{j-c,i}^{(q-1)} e^{\frac{\Delta t}{\tau_i}} \\ \quad + P_{lossd}^{(q-1)} \sum_{i=1}^{n} R_{thi}(1 - e^{\frac{\Delta t}{\tau_i}})(q > 1, q \in N^+) \\ T_j^{(q)} = T_c + \Delta T_{j-c}^{(q)} \end{cases} \tag{12}$$

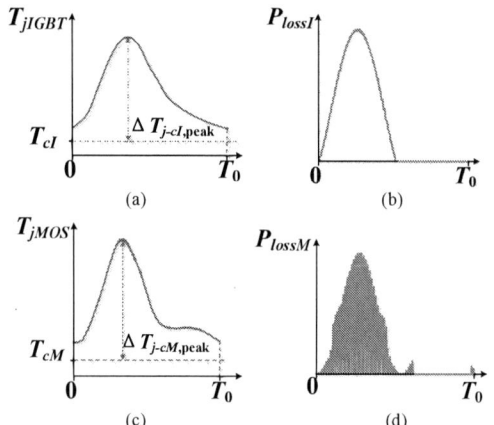

(a)   (b)

(c)   (d)

Fig. 5. The transient junction temperature and power loss of devices during one line period T_0.(a) Junction temperature of Si IGBT.(b) Power loss of Si IGBT.(c) Junction temperature of SiC MOSFET.(d) Junction temperature of SiC MOSFET.

The detailed step-shaped power loss and junction temperature of the devices are shown in Fig. 6. Based on the power loss model described earlier, the time step Δt should be set to one low-frequency period (T_{sL}). For the n-order foster thermal impedance of the device, the power loss at the first time step is calculated relative to the case temperature. Subsequently, the corresponding junction-to-case temperature difference for each order is determined. Furthermore, the power loss and temperature difference of each order at the junction temperature obtained in the first time step are used to compute the junction-to-case temperature difference at the second time step. Finally, the junction-to-case temperature difference at the current time step can be derived from the temperature differences of each order and the power loss from the previous time step.

979-8-3315-1110-4/25 $31.00 © 2025 IEEE

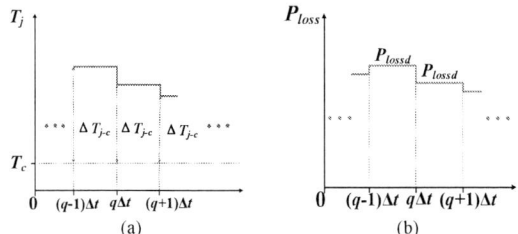

(a)   (b)

Fig. 6. The step-shaped power loss and junction temperature of devices in detail.(a) Junction temperature.(b) Power loss.

In the calculation procedure, the peak junction temperature T_{jpeak} within one line period T_0 can be derived as equation (13) and (14) :

$$T_{jpeak} = T_c + \Delta T_{j-cpeak} \tag{13}$$

$$\begin{cases} \Delta T_{jlMpeak} = | T_{jlpeak} - T_{jMpeak} | \\ T_{jlM\max} = \max(T_{jlpeak}, T_{jMpeak}) \\ T_{j,m\arg in} = T_{j,safe} - T_{jlM\max} \end{cases} \tag{14}$$

In equation (14), $\Delta T_{jlMpeak}$ represents the absolute value of the peak junction temperature difference. The maximum peak junction temperature, T_{jlMmax} is defined as the higher of the two peak junction temperatures of the Si and SiC devices. The junction temperature margin for safe operation is calculated as the difference between T_{jlMmax} and the safe-operation junction temperature $T_{j,safe}$. A smaller $\Delta T_{jlMpeak}$ indicates a more balanced thermal stress distribution across the devices.

III. SIMULATION AND MODEL VALIDATION

To verify the accuracy of the peak junction temperature calculation results based on above model. The transient junction temperature of Si and SiC devices can be both simulated in PLECS when case temperature is set as corresponding value. The thermal simulated model of devices are downloaded from Infineon official website while the simulated principle is identical as Infineon IGBT module simulation software IPOSIM. The comparison between the calculated and simulated value with variable control parameters are shown in Fig. 7 (a) , (b) and (c) . The experimental results reveal that as the power allocation ratio of the silicon phase increases, the junction temperature of the Si IGBT rises proportionally, while conversely, the junction temperature of the SiC MOSFET decreases. The results indicate that the junction temperature calculation model can offers an effective evidence for the thermal stress performance investigation.

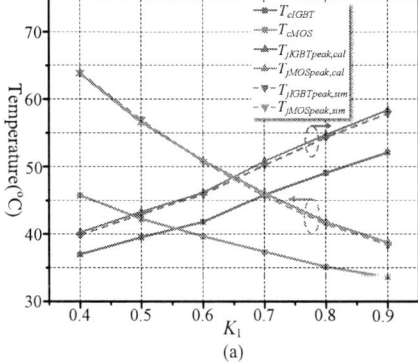

(a)

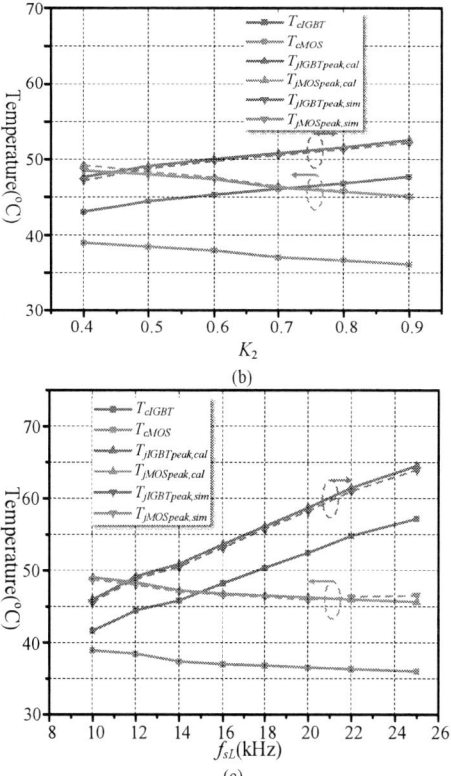

(b)

(c)

Fig. 7. Maximum junction temperature simulated and calculated value varying with single control parameter at 400V DC bus voltage and 1.5kW power. (a) K_2=0.5, f_{sL}=12 kHz. (b) K_1=0.65, f_{sL}=12 kHz. (c) K_1=0.65, K_2=0.5.

IV. EXPERIMENTAL RESULT AND DEVICES' THERMAL STRESS ANALYSIS

A. Experimental platform

An experimental platform based on a Si/SiC hybrid half-bridge inverter was built, as shown in Fig. 8. A detailed description of the inverter including electrical and component parameters is given in Table II . An FPGA board (xlinx arix-7 series AX7A200) was used to control the output power and optimise the thermal loss characteristics. The case temperature of the device is measured by two PT100 temperature sensors. In addition, the transient junction temperatures of both silicon and SiC devices can be simulated in PLECS when the case temperature is set to the corresponding value. The power analyser Zimmer LMG540 is used to measure the efficiency and total power loss of the inverter.

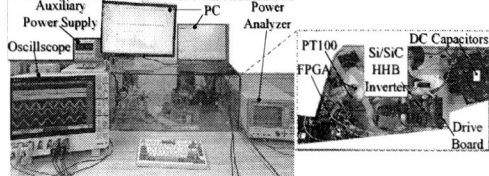

Fig. 8. Experimental platform

B. Thermal Stress Analysis

From the loss and junction temperature coupling model, both the total average power consumption and the maximum junction temperature of the device are affected by the power sharing ratio and the switching frequency under fixed operating conditions. In order to decouple these parameters, a unique control variable study is used. The 400V DC bus voltage, 17A output current amplitude, 100°C conduction limit maximum operating junction temperature are defined as the steady state operating condition, and the other two parameters are fixed at 1.5kW output power to explore the variation of the peak junction temperature and the total average power loss with respect to one control parameter. As shown in Fig. 9 (a), (b) and (c), the maximum junction temperature of two devices is positively correlated with power loss and it has opposite trends as the power distribution ratio and the Si phase switching frequency increase in the steady operating state.

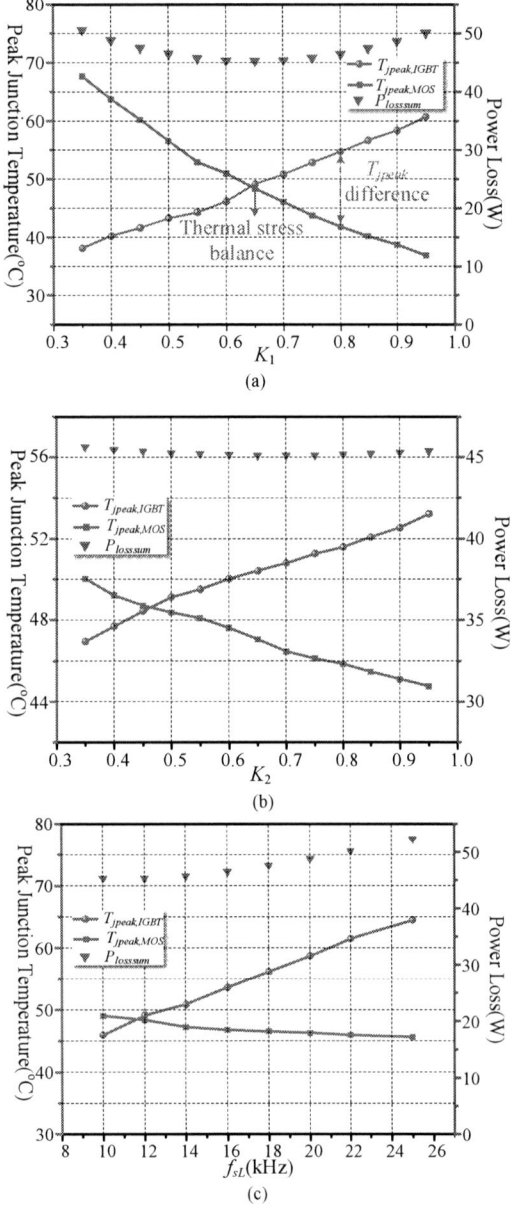

Fig. 9. Maximum junction temperature and total power loss varying with single control parameter at 400V DC bus voltage and 1.5kW output power. (a) K_2=0.5, f_{sL}=12kHz. (b) K_1=0.65, f_{sL}=12kHz. (c) K_1=0.65, K_2=0.5.

TABLE II. PARAMETERS OF COMPONENTS IN SI/SIC HHB INVERTER

Parameters		Value
Input Voltage		400V
Full Power		1.5kW
Modulation Ratio		0.55~0.85
Output Voltage THD		<2% @full load
SiC phase switching frequency		100kHz
Output filter capacitor		4uF
Si device	Model	IKW40N120CS6
SiC device	Model	IMW120R090M1H
	Secure	15A
L_1	Core	Kool Mµ60
	Turns	86
	Inductance	1mH
L_2	Core	Kool Mµ60
	Turns	72
	Inductance	400uH

For clarity in our analysis, we calculate the power loss of key components as shown in Fig. 10. Fig. 10 (a) and (b) demonstrate that the power loss trends of both Si and SiC devices align with their respective peak junction temperature variations. An increased power sharing ratio indicates greater current allocation to the Si-phase components. Accordingly, this ratio exhibits a positive correlation with inductance loss in the Si phase, while showing a negative correlation with those in the SiC phase. Notably, whether considering average power loss or peak junction temperature, the power sharing ratio parameter K_1 exerts greater influence than K_2 due to its association with higher current loads.

Fig. 10 (c) reveals that increased switching frequency leads to higher switching loss in Si IGBTs. Concurrently, the Si-phase current ripple diminishes with rising switching frequency, resulting in lower conduction loss. Additionally, the inductance loss of the Si phase decreases as its switching frequency increases. However, the switching loss of Si devices exhibits an accelerating growth rate with frequency elevation. In comparison, both the magnitude and proportion of loss variation in the SiC phase remain significantly smaller than those in the Si phase. Consequently, the overall average power loss follows a trend similar to the Si-phase loss characteristics.

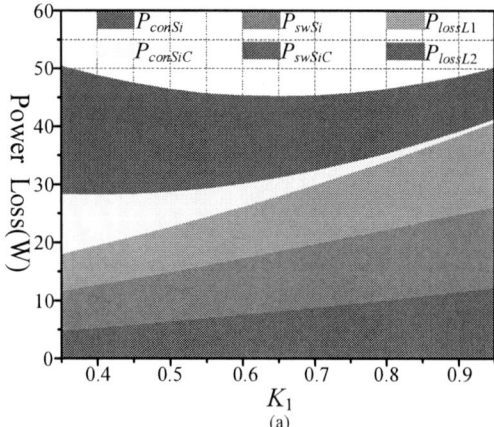

(a)

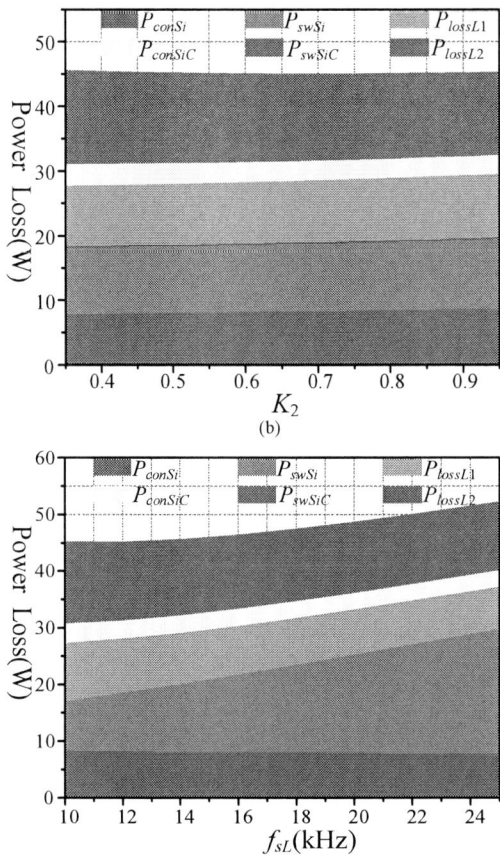

Fig. 10. The power loss of main components at a DC bus voltage of 400V, an output current amplitude of 17A and an output power of 1.5kW. (a) K_2=0.5, f_{sL}=12kHz. (b) K_1=0.65, f_{sL}=12kHz. (c) K_1=0.65, K_2=0.5.

V. CONCLUSION

This paper establishes a device loss model and junction temperature calculation model for the Si/SiC HHB inverter and proposes a thermal stress analysis method based on the coupled model of device loss and junction temperatures. Through thermal simulations, the model's calculated values are fitted with simulation results. Meanwhile, a 3.3kW experimental prototype was constructed, and quantitative analysis of thermal stress and power loss was conducted using offline measurement methods. The study obtained the variation patterns of peak junction temperatures for both Si IGBTs and SiC MOSFETs under different power sharing ratios and LF switching conditions, along with the composition of total power loss. The results verified the feasibility of the thermal stress analysis and calculation model for the devices.

REFERENCES

[1] D. Vyawahare and M. Chandorkar, "Distributed generation system with hybrid inverter interfaces for unbalanced loads," 2015 IEEE 6th International Symposium on Power Electronics for Distributed Generation Systems (PEDG), Aachen, Germany, 2015, pp. 1-7.

[2] J. Lu, M. Savaghebi, S. Golestan, J. C. Vasquez, J. M. Guerrero and A. Marzabal, "Multimode Operation for On-Line Uninterruptible Power Supply System," IEEE Journal of Emerging and Selected Topics in Power Electronics, vol. 7, no. 2, pp. 1181-1196, June 2019.

[3] J. Liu, M. Yang and T. Wang, "Impedance-Based Stability Analysis of Grid-Tied Photovoltaic System With Superconducting Magnetic

Energy Storage System," IEEE Transactions on Applied Superconductivity, vol. 31, no. 8, pp. 1-4, Nov. 2021, Art no. 5402504

[4] C. Lorenzini, L. F. A. Pereira, A. S. Bazanella and G. R. Gonçalves da Silva, "Single-Phase Uninterruptible Power Supply Control: A Model-Free Proportional-Multiresonant Method," IEEE Transactions on Industrial Electronics, vol. 69, no. 3, pp. 2967-2975, March 2022.

[5] J. Wei et al., "Review on the Reliability Mechanisms of SiC Power MOSFETs: A Comparison Between Planar-Gate and Trench-Gate Structures," IEEE Transactions on Power Electronics, vol. 38, no. 7, pp. 8990-9005, July 2023.

[6] D. Woldegiorgis, M. M. Hossain, Z. Saadatizadeh, Y. Wei and H. A. Mantooth, "Hybrid Si/SiC Switches: A Review of Control Objectives, Gate Driving Approaches and Packaging Solutions," IEEE Journal of Emerging and Selected Topics in Power Electronics, vol. 11, no. 2, pp. 1737-1753, April 2023.

[7] P. D. Judge and S. Finney, "2-Level Si IGBT Converter with Parallel Part-Rated SiC Converter Providing Partial Power Transfer and Active Filtering," 2019 20th Workshop on Control and Modeling for Power Electronics (COMPEL), Toronto, ON, Canada, 2019, pp. 1-7.

[8] T. -F. Wu, Y. -H. Huang and Y. -T. Liu, "3Φ4W Grid-Connected Hybrid-Frequency Parallel Inverter System With Ripple Compensation to Achieve Fast Response and Low Current Distortion," IEEE Transactions on Industrial Electronics, vol. 68, no. 11, pp. 10890-10901, Nov. 2021.

[9] T. -F. Wu, Y. -H. Huang, S. Temir and C. -C. Chan, "3Φ4W Hybrid Frequency Parallel Uninterruptable Power Supply for Reducing Voltage Distortion and Improving Dynamic Response," IEEE Journal of Emerging and Selected Topics in Power Electronics, vol. 10, no. 1, pp. 906-918, Feb. 2022.

[10] N. Li, M. B. Macavilca, C. Wu, S. Finney and P. D. Judge, "Converter Topology for Megawatt Scale Applications With Reduced Filtering Requirements, Formed of IGBT Bridge Operating in the 1000 Hz Region With Parallel Part-Rated High-Frequency SiC MOSFET Bridge," IEEE Transactions on Power Electronics, vol. 39, no. 1, pp. 799-813, Jan. 2024.

[11] C. Zhang, K. Qu, B. Hu, J. Wang, X. Yin and Z. J. Shen, "A High-Frequency Dynamically Coordinated Hybrid Si/SiC Interleaved CCM Totem-Pole Bridgeless PFC Converter," IEEE Journal of Emerging and Selected Topics in Power Electronics, vol. 10, no. 2, pp. 2088-2100, April 2022.

[12] Z. Peng, J. Wang, Z. Liu, Y. Dai, G. Zeng and Z. J. Shen, "Fault-Tolerant Inverter Operation Based on Si/SiC Hybrid Switches," IEEE Journal of Emerging and Selected Topics in Power Electronics, vol. 8, no. 1, pp. 545-556, March 2020.

[13] R. Han et al., "Thermal Stress Balancing Oriented Model Predictive Control of Modular Multilevel Switching Power Amplifier," IEEE Transactions on Industrial Electronics, vol. 67, no. 11, pp. 9028-9038, Nov. 2020.

[14] C. Zhang, X. Yuan, J. Wang, B. Hu, X. Yin and Z. J. Shen, "Optimization of Power Sharing and Switching Frequency in Si/WBG Hybrid Half-Bridge Converters Using Power Loss Models," IEEE Journal of Emerging and Selected Topics in Power Electronics, vol. 11, no. 3, pp. 2837-2849, June 2023.

A 8:1 Multi-Resonant Switched-Capacitor Converter

1st Yuxin Yan
School of Automation
Beijing Institute of Technology
Beijing, China
2209164970@qq.com

2nd Yu Fu
School of Automation
Beijing Institute of Technology
Beijing, China
fuyu@bit.edu.cn

3rd Yucheng Zhao
School of Automation
Beijing Institute of Technology
Beijing, China
3220231025@bit.edu.cn

4th Shouxiang Li(Corresponding author)
School of Automation
Beijing Institute of Technology
Beijing, China
lishouxiang@bit.edu.cn

Abstract—To satisfy the demands of the two-stage power distribution architectures in 48V data centers, this paper presents a multi-resonant switched-capacitor converter. This converter efficiently suppresses voltage fluctuations of the 48V data center bus, accomplishes soft capacitor charging, and guarantees high efficiency under low-voltage, high-current scenarios. The resonant frequency of the topology, the equilibrium voltage distribution across capacitors, and the phase-plane behavior are analyzed. The circuit operation, analysis, and experimental results from a 48 V-to-6 V laboratory prototype is presented, achieving a peak efficiency of 98.86% and a full-load efficiency 96.22%. The experimental results demonstrate the effectiveness of the proposed technology.

Keywords—Data center, switched-capacitor converter, state plane analysis

I. Introduction

In recent years, with the rapid advancement of information and communication technology (ICT) and the steadily escalating demand for data processing, large - scale data centers have gained increasing popularity. Correspondingly, the energy consumption of data centers is increasing annually [1]-[4], prompting extensive efforts to achieve high-power usage efficiency (PUE) [5]. In order to reduce system distribution losses, the 48V bus power architecture has been increasingly and extensively adopted.

Extensive research has been conducted on the voltage regulation module (VRM) to address the 48V bus. From a cascading perspective, the VRM solutions can be classified into two distinct categories: two-stage architectures and single-stage topologies. Single-stage VRMs evolve into various structures, including designs based on the Sigma architecture [6], the transformer based topologies [7] and series capacitor buck (SCB) converters [8]. However, single-stage solutions are limited by high control complexity and the immaturity of related research.

Compared with the single-stage architecture, the two-stage VRM demonstrates higher flexibility in terms of performance optimization and size control. The common two-stage configuration is based on the intermediate bus architecture. In the first stage, an intermediate bus converter (IBC) is employed, with its principal emphasis on maximizing power density and efficiency. The second stage utilizes a multiphase buck converter, which offers superior transient response characteristics.

Over the recent period, the demand for lower intermediate bus voltages has been continuously increasing, giving rise to the emergence of the 8:1 IBC scheme. Due to its higher power density, the resonant switched-capacitors (ReSc) architecture still receives widespread attention and has been optimized through means such as increasing the number of stages [9]-[10] and using embedded transformers [11].

In this paper, an IBC capable of 8:1 conversion ratios is proposed, and the architecture of the converter is shown in Fig.1. This converter effectively mitigates voltage fluctuations on the 48V data center bus, ensuring stable power delivery that meets the stringent requirements of modern data center operations. In this solution, the topology allows for soft switching, enabling highly efficient operation under low-voltage and high-current scenarios. The hardware experiments verified that it reaches a peak efficiency of 98.86%.

II. Multi-Resonant Switched-Capacitor Converter

A. Operation Principle

As illustrated in Figure 1, The multi-resonant switched-capacitor converter proposed in this article comprises an A-side and a B-side. The inductors and capacitors on both sides possess identical values, which means $C_{2A} = C_{2B} = C_2$, $C_{3A} = C_{3B} = C_3$, $L_A = L_B = L$. The waveforms of the drive signal and inductor current is depicted in Fig.1(b). Phase A commences with the discharge of capacitor C_1, whereas Phase B initiates with the charging of capacitor C_1. Each phase undergoes three states, resulting in the formation of three resonant cavities with varying resonant frequencies and amplitudes. The operation of Phase A is outlined as follows:

State A1: Switches S_2, S_3, S_{2A} and S_{5A} are turned on. Capacitor C_1 discharges, and capacitors C_1, C_{2A}, C_{3A} as well as the output inductor L_A form a resonant tank circuit. The resonant frequency of state A1 is given as follows:

$$
\begin{cases}
f_{A1} = \frac{1}{2\pi\sqrt{LC_{eq1}}} \\
\frac{1}{C_{eq1}} = \frac{1}{C_1} + \frac{1}{C_2} + \frac{1}{C_3}.
\end{cases}
\tag{1}
$$

979-8-3315-1110-4/25 $31.00 © 2025 IEEE

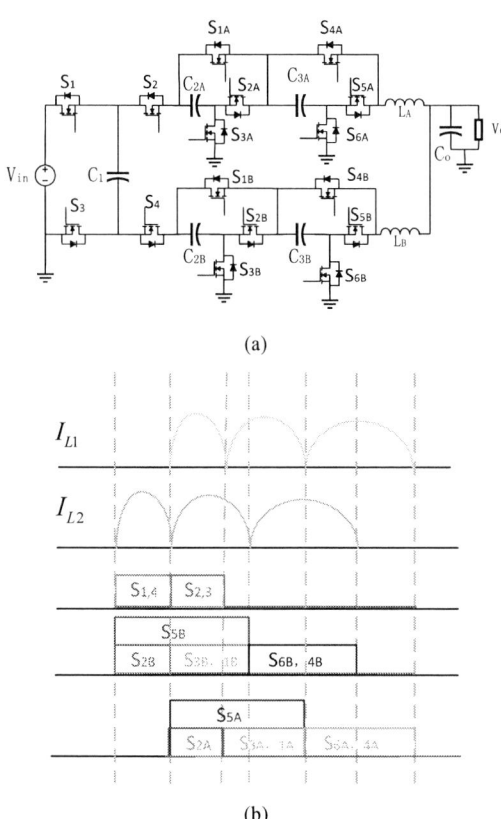

(a)

(b)

Fig. 1. (a) The proposed multi-resonant switched-capacitor converter; (b) key waveforms of 8:1 mode

State A2: Switches S_{1A}, S_{3A} and S_{5A} are operational. Capacitor C_{1A} is disconnected and capacitor C_{2A} discharges. Capacitors C_{2A}, C_{3A}, along with the output inductor L_A constitute a resonant tank circuit. In this state, the resonant frequency can be calculated as follows:

$$\begin{cases} f_{A2} = \dfrac{1}{2\pi\sqrt{LC_{eq2}}} \\ \dfrac{1}{C_{eq2}} = \dfrac{1}{C_2} + \dfrac{1}{C_3}. \end{cases} \qquad (2)$$

State A3: Switches S_{4A} and S_{6A} are activated. Capacitors C_1 and C_{2A} are excluded, leaving C_{3A} and the output inductor L_A collectively form a resonant tank circuit. In this state, the resonant frequency is:

$$\begin{cases} f_{A3} = \dfrac{1}{2\pi\sqrt{LC_{eq3}}} \\ \dfrac{1}{C_{eq3}} = \dfrac{1}{C_3}. \end{cases} \qquad (3)$$

Phase B initiates with the charging of capacitor C_1 and the conducting of S_1, S_4, S_{2B} and S_{5B}. The resonant frequency of State B is identical to that of State A, which means $f_{A1} = f_{B1}$, $f_{A2} = f_{B2}$, $f_{A3} = f_{B3}$.

According to the previous design, the duration of each state is half of the resonant period. Therefore, the period T and frequency f can be calculated:

$$T = \frac{1}{f_1} = \frac{1}{2f_{A1}} + \frac{1}{2f_{A2}} + \frac{1}{2f_{A3}} \qquad (4)$$

$$f = \frac{2}{\frac{1}{f_{A1}} + \frac{1}{f_{A2}} + \frac{1}{f_{A3}}} \qquad (5)$$

The duty cycle can be expressed as:

$$\begin{cases} D_1 = \dfrac{\frac{1}{f_{A1}}}{\frac{1}{f_{A1}} + \frac{1}{f_{A2}} + \frac{1}{f_{A3}}} \\[4mm] D_2 = \dfrac{\frac{1}{f_{A2}}}{\frac{1}{f_{A1}} + \frac{1}{f_{A2}} + \frac{1}{f_{A3}}} \\[4mm] D_3 = \dfrac{\frac{1}{f_{A3}}}{\frac{1}{f_{A1}} + \frac{1}{f_{A2}} + \frac{1}{f_{A3}}}, \end{cases} \qquad (6)$$

Where D_1 denotes the time-ratio of State 1 within a cycle, D_2 denotes the time-ratio of State 2 within a cycle, and D_3 denotes the time-ratio of State 3 within a cycle.

B. Equilibrium State Calculation

The capacitor C_1 is analyzed. In Fig.2(a), the KVL equations for the equivalent circuits are as follows:

$$x_{c1\_1} = x_{c21\_1} + x_{c31\_1} + V_{out}/V_{in} \qquad (7)$$

$$x_{c1\_2} = \frac{(V_{in} - V_{out})}{V_{in}} - x_{c22\_2} - x_{c32\_2}, \qquad (8)$$

where x_{c1_1}, x_{c1_2}, x_{c21_1}, x_{c22_2}, x_{c31_1} and x_{c32_2} are the normalized equilibrium states of v_{c1} of state 1, v_{c1} at state 2, v_{c21} at state 1, v_{c22} at state 2, v_{c31} at state 1, and v_{c32} at state 2. Due to symmetry property, there are $x_{c21\_1} = x_{c22\_2} = x_{2\_1}$ and $x_{c31\_1} = x_{c32\_2} = x_{3\_1}$. Then the middle point of x_{c1_1} and x_{c1_2} should satisfy $x_{c1m} = \frac{1}{2}$.

For fig. 2 (b), the power supply is only connected to C_1, so $I_{in} = I_{C1}$. And due to the conservation of energy, I_{out} can be calculated:

$$I_{out}V_{out} = V_{in}I_{in}. \qquad (9)$$

Therefore, the capacitor voltage ripple can be derived based on the current I_{out} as follows:

$$\Delta v_{c1} = \frac{I_{out}}{8C_1 f_s}. \qquad (10)$$

Due to the series connection of capacitors, the voltage ripples of capacitors C_2 and C_3 can be calculated as follows:

$$\begin{cases} \Delta v_{c2} = \dfrac{C_1}{C_2}\Delta v_{c1} \\[2mm] \Delta v_{c3\_1} = \dfrac{C_1}{C_3}\Delta v_{c1} \\[2mm] \Delta v_{c3\_2} = \dfrac{C_2}{C_3}\Delta v_{c1} \\[2mm] \Delta v_{c3\_3} = \Delta v_{c3\_1} + \Delta v_{c3\_2} \end{cases} \qquad (11)$$

For the capacitor C_2, the KVL equations of the circuits shown in Fig. 3(a) are :

$$x_{c1\_1} - x_{c2\_1} - x_{c3\_1} = V_{out}/V_{in} \qquad (12)$$

$$x_{c2\_2} - x_{c3\_2} = V_{out}/V_{in}, \qquad (13)$$

which indicating that capacitor C_2 approximately satisfy

$$x_{c2m} = \frac{x_{c1m}}{2} + \frac{(\Delta v_{c3\_1} + \Delta v_{c3\_2})}{4V_{in}}.$$

For the capacitor C_3, the equilibrium states in Fig. 4(a) should satisfy KVL equations:

$$x_{c1\_1} = x_{c2\_2} + x_{c3\_1} + V_{out}/V_{in} \tag{14}$$

$$x_{c2\_2} - x_{c3\_2} = V_{out}/V_{in} \tag{15}$$

$$x_{c3\_3} = V_{out}/V_{in}. \tag{16}$$

The equilibrium states of v_{c3} then should satisfy the following formula:

$$\begin{cases} x_{c3\_1} = x_{c1m} - x_{c2m} - V_{out}/V_{in} \\ x_{c3\_2} = x_{c2m} - V_{out}/V_{in} \\ x_{c3\_3} = V_{out}/V_{in} \end{cases}. \tag{17}$$

C. Phase planes

For the first-stage capacitor C_1, the equivalent circuits and state planes are illustrated in Fig. 2(a). The capacitor C_1 resonates with the inductors for a portion of the cycle, and i_{L1} represents the current through C_1. Due to symmetry of the resonator, the state trajectory is also symmetrical during state 1 and state 2. In order to obtain a circular trajectory, the inductor current and capacitor voltage are normalized. The symbols Z_1 and k_1 are defined as follows:

$$Z_1 = \sqrt{\frac{L}{C_1}}, \ k_1 = \sqrt{1 + \frac{C_1}{C_2} + \frac{C_1}{C_3}}. \tag{18}$$

The trajectories of state 1 and state 2 are combined into one coordinate and the trajectory of the switch cycle is shown in Fig. 2(b). A straight line is added to indicate the decrease of i_L during the T_{ZVS} period. The symbols I_{off1N} and I_{off2N} represent the normalized value of the solution for the turn-off current. And the capacitor voltage ripple is normalized:

$$\Delta v_{c1N} = \frac{\Delta v_{c1}}{V_{in}}. \tag{19}$$

When the switching frequency f_{sw} and the ZVS time T_{ZVS} are given, the arc angle φ_1 in Figure 2(a) is calculated through geometric conditions as follows:

$$\varphi_1 = 2\pi f_{r123}(\frac{8}{f_{sw}} - T_{ZVS}). \tag{20}$$

The arc angle φ_1' in Figure 2(b) can be expressed through geometric conditions as follows:

$$cot\varphi_1' = \frac{M_y}{\Delta v_{c1N}/2}, \tag{21}$$

where M_y is the middle point of I_{off1N} and I_{off2N}.

According to (19), (20), and (21), the average current of I_{off1} and I_{off2} can be calculated as follows:

$$\frac{I_{off1}+I_{off2}}{2} = \frac{M_y}{Z_1}V_{in}k_1 = \frac{cot\varphi_1'\Delta v_{c1N}}{2Z_1}V_{in}k_1$$

$$= \frac{I_{out}cot\varphi_1'}{16c_1Z_1f_{sw}}k_1 = \frac{\pi f_{r123}}{8f_{sw}}I_{out}cot\varphi_1' \tag{22}$$

The solution for the turn-off current I_{off1} and I_{off2} can be obtained as follows:

$$I_{off1,2} = \frac{\pi f_{r123}}{8f_{sw}}I_{out}cot\frac{\varphi_1}{2} \pm \frac{1}{2}T_{ZVS}. \tag{23}$$

Through geometric analysis of the state trajectory in Fig. 2(b), the centers and radius be calculated as follows:

$$x_{c1\_2,c1\_1} = 0.5 \pm \frac{I_{zvs}Z_1}{2V_{in}k_1}cot\frac{\varphi_1}{2} \tag{24}$$

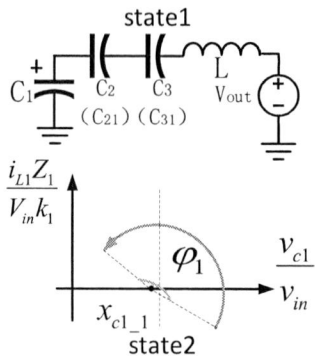

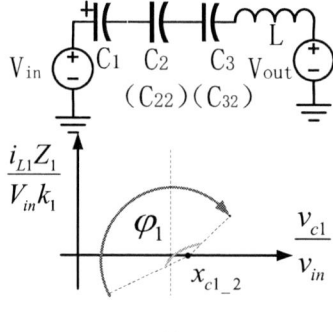

(a)

(b)

Fig. 2. (a) Equivalent circuits and corresponding state trajectories of v_{c1} and i_{L1} each operation phase; (b) Arrange the state trajectory of state 1 and state 2.

$$\rho_1 = \frac{1}{2sin\frac{\varphi_1}{2}}\sqrt{\Delta v_{c1N}^2 + (\frac{I_{zvs}Z_1}{I_{zvs}Z_1})^2} \tag{25}$$

The equivalent circuits and state planes of the capacitor C_2 are illustrated in Fig. 3(a). In state 1, C_2 is in series with C_1 and C_3, and the y-axis normalization factor is $\frac{Z_2}{V_{in}k_2}$. In state 1, C_2 is in series with C_3, and the y-axis is normalized by a factor of $\frac{Z_2}{V_{in}k_3}$, where Z_2, k_2, and k_3 are defined as follows:

$$Z_2 = \sqrt{\frac{L}{C_2}}, \ k_2 = \sqrt{1 + \frac{C_2}{C_1} + \frac{C_2}{C_3}}, k_3 = \sqrt{1 + \frac{C_2}{C_3}}. \tag{26}$$

The complete state trajectory combined state 1 and state 2 is shown in Fig. 3(b). the y-axis using the normalization factor $\frac{Z_2}{V_{in}k_3}$, so the trajectory of state 1 becomes elliptical arc.

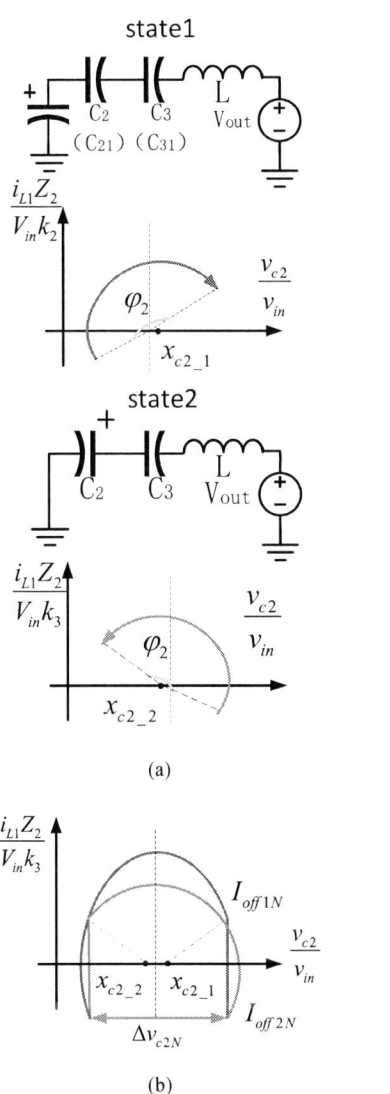

(a)

(b)

Fig. 3. (a) Equivalent circuits and corresponding state trajectories of v_{c2} and i_{L1} each operation phase; (b) Arrange the state trajectory of state 1 and state 2.

The average current of I_{off1} and I_{off2} in Fig 3(b) can be calculated as follows:

$$\frac{I_{off1}+I_{off2}}{2} = \frac{\pi f_{r23}}{8f_{sw}} I_{out} \cot \frac{\varphi_2}{2}.$$ (27)

And the angles φ_1 and φ_2 satisfy:

$$\cot \frac{\varphi_2}{2} = \frac{f_{r123}}{f_{r23}} \cot \frac{\varphi_1}{2}.$$ (28)

The arc angle φ_2 in Figure 3(a) can be calculated as follows:

$$\varphi_2 = 2\pi f_{r23}(\frac{D_2}{f_{sw}} - T_{ZVS}).$$ (29)

According to (21), (28), and (29), the duty ratio D_1 and D_2 can be expressed:

$$\cot\left[\pi f_{r23}\left(\frac{D_2}{f_{sw}} - T_{ZVS}\right)\right] =$$

Fig. 4. Photograph of the hardware prototype

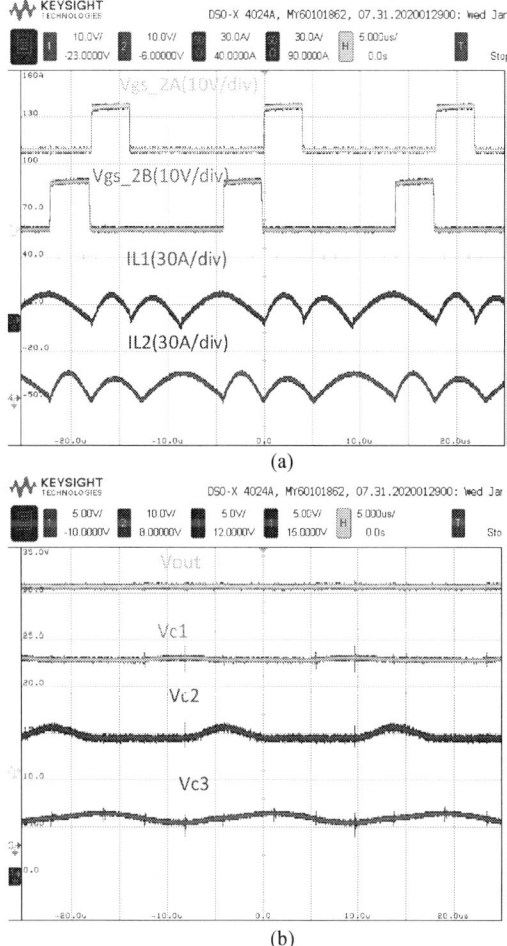

(a)

(b)

Fig. 5. Test at $Vin=48V$ and $Iout=20A$.

$$\frac{f_{r123}}{f_{r23}} \cot\left[\pi f_{r123}\left(\frac{D_1}{f_{sw}} - T_{ZVS}\right)\right].$$ (30)

For the capacitor C_3, involving two charging resonance processes and one discharging resonance process, so the phase plane contains three circles or ellipses. Similarly, the data of capacitor C_3 can be calculated.

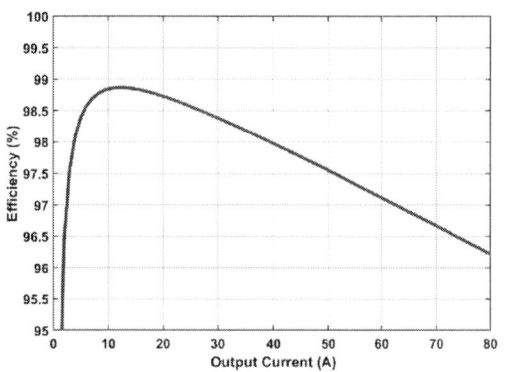

Fig. 6. The efficiency of the converter when it operates at a voltage ranging from 48V to 6V is measured.

III. HARDWARE IMPLEMENTATION AND EXPERIMENTAL RESULTS AND ANALYSIS

A hardware prototype was constructed to enable experimental validation. This chapter presents the technical specifications of the hardware prototype and the steady-state experimental results. The experimental setup conforms to the 48V data center power requirements, and the prototype's performance has been evaluated through testing.

The characteristics of the proposed inductor are demonstrated in fig. 5. Specifically, fig. 5 illustrate the operating condition of the converter when the voltage is converted from 48V to 6V and the output current is 20A. The inductor currents i_{L1} and i_{L2} shown in fig. 4(a) align closely with the theoretical predictions given in fig. 1. The output voltage v_{out}, capacitor voltage v_{c1}, v_{c2}, and v_{c3} are shown in fig. 4(b). The smooth waveforms confirm the effective realization of soft charging.

The efficiency curves are precisely measured and vividly depicted in Fig. 6. In the case of a nominal 48 V input, the peak efficiency attains an impressive 98.86%, and the full-load efficiency reaches a substantial 96.22%.

IV. CONCLUSION

This paper presents a multi-resonant switched-capacitor converter, specifically designed for 48V data center power distribution systems. The proposed converter effectively mitigates intermediate bus voltage ripple while enabling soft-switching operation through an optimized modulation strategy, thereby ensuring high conversion efficiency under low-voltage, high-current operating conditions. Key parameters of the topology, including the resonant frequency of the resonant tank, the equilibrium voltage distribution across capacitors, and the phase-plane trajectories, have been systematically analyzed to elucidate its dynamic characteristics and steady-state behavior. A hardware prototype of the MRSCC was experimentally validated, achieving a peak efficiency of 98.86% under nominal load. With its exceptional power density and energy efficiency, the MRSCC demonstrates significant competitiveness as a next-generation power delivery solution for data centers.

REFERENCES

[1] J. Liang, L. Wang, M. Fu, J. Liang and H. Wang, "Overview of Voltage Regulator Modules in 48 V Bus-Based Data Center Power Systems," CPSS Trans. Power Electron. Appl., vol. 7, no. 3, pp. 283-299, Sep. 2022.

[2] I. Khan, S. Rahman, M. A. Ayaz, M. Amir and H. Shehada, "Review of Isolated DC-DC Converters for Application in Data Center Power Delivery," IEEE Trans. Ind. Appl., vol. 60, no. 4, pp. 5436-5446, Jul./Aug. 2024.

[3] Lyu X, Li Y, Ni Z, et al. Composite modular power delivery architecture for next-gen 48V data center applications[C]//2018 1st Workshop on Wide Bandgap Power Devices and Applications in Asia (WiPDA Asia). IEEE, 2018: 343-350.

[4] Y. Zhu, Y. Yang and F. Blaabjerg, "A Comprehensive Review on Partial Power Processing-based Voltage Regulator Modules for 48-V Bus-based Data Centers," in Proc. 25th Eur. Conf. Power Electron. Appl. (EPE'23 ECCE Europe), Aalborg, Denmark, 2023, pp. 1-9.

[5] Y. Chen, K. Shi, M. Chen and D. Xu, "Data Center Power Supply Systems: From Grid Edge to Point-of-Load," IEEE J. Emerg. Sel. Topics Power Electron., vol. 11, no. 3, pp. 2441-2456, Jun. 2023.

[6] M. H. Ahmed, C. Fei, F. C. Lee and Q. Li, "Single-Stage High-Efficiency 48/1 V Sigma Converter With Integrated Magnetics," IEEE Trans. Ind. Electron., vol. 67, no. 1, pp. 192-202, Jan. 2020.

[7] S. Saggini, O. Zambetti, R. Rizzolatti, M. Picca and P. Mattavelli, "An isolated quasi-resonant multiphase single-stage topology for 48-v vrm applications", IEEE Trans. Power Electron., vol. 33, no. 7, pp. 6224-6237, July 2018.

[8] Y. Chen et al., "Virtual Intermediate Bus CPU Voltage Regulator," IEEE Trans. Power Electron., vol. 37, no. 6, pp. 6883-6898, Jun. 2022.

[9] Y. Li, X. Lyu, D. Cao, S. Jiang and C. Nan, "A 98.55% Efficiency Switched-Tank Converter for Data Center Application," IEEE Trans. Ind. Appl., vol. 54, no. 6, pp. 6205-6222, Nov./Dec. 2018.

[10] X. Li, Y. Guan, W. Yang, Y. Wang and D. Xu, "A High Performance 48-to-8 V Cascaded Switched-Capacitor Converter for Data Center Applications," in Proc. IEEE 2nd Int. Power Electron. Appl. Symp. (PEAS), Guangzhou, China, 2023, pp. 35-40.

[11] H. Wu, Y. Zhang and Z. Li, "Hybrid Resonant Converter-Based 8:1 Bus Converter With 3.5 kW/in3 and 98.6%-Efficient for 48 V Data-Center Power Systems," IEEE Trans. Power Electron., vol. 39, no. 1, pp. 36-41, Jan. 2024.

Sustainability of Power Devices: A Perspective on Design for Recycling

Jinpeng Cheng
State Key Laboratory of Power Transmission Equipment Technology
Chongqing University
Chongqing, China
Jinpeng.C@cqu.edu.cn

Shuyu Liu
State Key Laboratory of Power Transmission Equipment Technology
Chongqing University
Chongqing, China
202411131314@stu.cqu.edu.cn

Hao Feng
State Key Laboratory of Power Transmission Equipment Technology
Chongqing University
Chongqing, China
hfeng6@cqu.edu.cn

Li Ran
School of Engineering
University of Warwick
Coventry, the UK
l.ran@warwick.ac.uk

Abstract—**This paper discusses the challenges of recycling and disposal posed by the growing use of power devices to achieve net-zero carbon goals. It introduces the R-strategies and circular economy concepts, emphasizing their importance in sustainable power devices design. Existing recycling methods for electronic components are reviewed, but due to the unique characteristics of power devices, these methods are not adequately suited. Therefore, a "design for recycling" strategy is proposed, along with a recyclable design example and its potential recycling process. Finally, an evaluation framework incorporating environmental impact indicators is developed to support sustainable design decisions.**

Keywords—power devices packaging; sustainability; design for recycling; circular economy

I. INTRODUCTION

To reduce global carbon dioxide (CO_2) emissions to net zero by 2050, a fundamental transformation in the production, transportation, and consumption of energy is required [1]. Although countries have reached broad political consensus on achieving net zero emissions, significant challenges remain in realizing this goal. The key to achieving net zero lies in the unprecedented scale and speed of deploying clean energy technologies by 2030 [2]. However, the pathway to net zero is extremely narrow: staying on this trajectory requires the immediate, systematic deployment and large-scale application of all available clean and efficient energy technologies to maximize energy efficiency and significantly reduce carbon emissions.

Green energy solutions such as photovoltaic (PV) solar power, wind energy, and electric vehicles (EVs) are considered key drivers in mitigating climate change and reducing greenhouse gas emissions. As shown in Fig. 1, according to the requirements set by the International Energy Agency (IEA), the annual rate of energy intensity improvements averaging 4% to 2030, accompanied by a rapid expansion of solar and wind energy deployment during this decade. It is projected that by 2030, annual additions of 630 GW of solar PV and 390 GW of wind will be achieved, four-times the record levels set in 2020. Additionally, the market share of EVs in global vehicle sales is expected to surge from the current level of approximately 5% to over 60% by 2030 [2]. Achieving this transition necessitates the widespread adoption of high-efficiency power electronic devices, which are critical enabling

This work was supported by the National Natural Science Foundation of China under Grant 52477177.

technologies for efficient energy conversion in green energy systems, thereby supporting the transition to low-carbon energy and emission reduction goals [3].

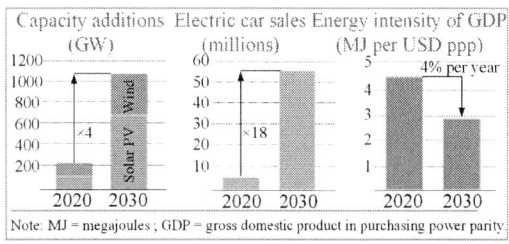

Fig. 1. Key clean technologies ramp up by 2030 in the net zero pathway.

As power electronic devices become more widely used, end-of-life disposal challenges will soon arise. Without recycling or reuse as part of a circular strategy, these devices could shift from being "green solutions" to "environmental problems". Currently, around 50 million tons of electronic waste (E-waste) are generated globally each year, with less than 20% formally recycled [4]. A large amount of discarded E-waste is ultimately shipped to developing countries for landfill or improper disposal [5]. Despite high recycling costs, sustainable management of electronic components is increasingly necessary. Many countries and organizations are promoting component reuse, waste reduction, and proper disposal of non-recyclable waste [6]-[8].

Eighty percent of the sustainability impact of power electronics is determined at the design stage [9]. For power devices, it is crucial to adopt circular economy principles early. As shown in Fig. 2, different R-strategies promote the circular use of components and materials, keeping them flowing within a closed-loop system instead of being discarded after one use [10].

To achieve an effective closed-loop, the design phase should adopt a "Design for R" approach, integrating reliability to extend the lifespan of power devices, e.g., "Design for reuse" involves reusing the entire device or its components, requiring reliability testing to assess remaining lifespan and enabling downgraded application scenarios. "Design for refurbish/remanufacture" aims to give components a second life, requiring easy disassembly and modular design to ensure that old components do not affect the performance of new ones. "Design for recycling" includes disassembling systems, separating materials, and recycling processes. Since power devices are rich in precious metals, their recyclability should be considered at the design stage, including material selection and easily detachable connections. "Design for reliability" extends the usage cycle, enhances value, reduces waste, and supports reuse and refurbishment by ensuring reliable components are easier to reuse or refurbish.

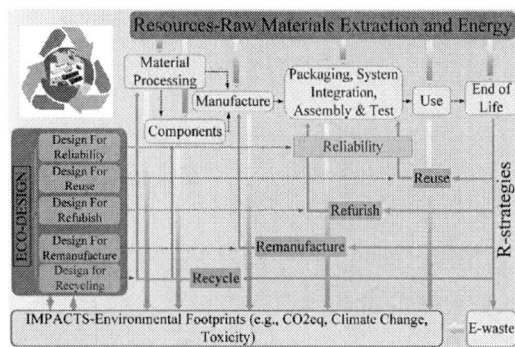

Fig. 2. Circular economy based on R-strategies life cycle for power devices (adapted from [10])

Inner-loop cycles (e.g., reuse and refurbish) are prioritized to reduce resource consumption, and minimize environmental impact. However, power devices require high reliability, ensuring performance and safety after recycling. Once recycling design matures with efficient disassembly, material separation, and reliability assessments, direct recycling becomes the preferred option.

This paper approaches the topic from the "design for recycling" perspective, summarizing and drawing from the progress and methods of recycling existing electronic components or products. It explores the ideas and understanding of recycling design, proposes potential cases and recycling processes, and briefly discusses the opportunities and challenges faced by "design for recycling".

II. APPROACHES TO RECYCLING ELECTRONIC COMPONENTS

As shown in Fig. 3, electronic components can be recycled through various methods.

Chemical recycling involves depolymerizing waste printed circuit boards (WPCB) into smaller useful molecules. Centrifugal separation separates WPCB from solder, while vacuum pyrolysis decomposes them into gases, liquids, and solid residues under low temperature and pressure, which can be reused as fuel or chemical raw materials [11].

Physical recycling uses mechanical, thermal, or other physical methods for material separation. Copper can be recovered from solid residues through crushing, sieving, and gravity separation, ensuring a clean, waste-free production process [12]. Magnetic separation extracts ferromagnetic metals, and electrostatic separation sorts metals with different conductivities [13].

Bioleaching offers a simple, energy-efficient, and eco-friendly approach by using sulfur-oxidizing bacteria to create an acidic environment that dissolves metals [14].

Recycling methods for power devices can draw on existing recycling practices. However, due to the unique structures and materials of power devices, additional design considerations are needed.

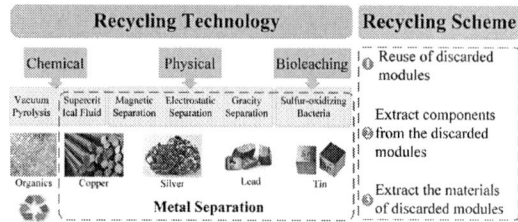

Fig. 3. Solutions for recycling electronic components

III. POWER DEVICES DESIGN FOR RECYCLING

Although the recycling methods in Section II are effective for many electronic components, they are less suitable for power modules, which consist of tightly integrated materials like chips, direct bonded copper (DBC) substrates, baseplates, solder layers, bonding wires, silicone gel, plastic housings, and terminals.

A. Recycling Design Strategies

Recycling DBC substrates involves complex, energy-intensive processes like crushing and re-sintering. Thermoset plastic housings are non-degradable and have low reuse value, while solder layers often contain hazardous substances like lead, increasing environmental processing costs. Bonding techniques can also leave craters on expensive chip surfaces, making it recycling impractical. Current technologies are unable to provide efficient, low-cost, and eco-friendly recycling for these components. To improve recyclability, it is crucial to optimize module packaging by avoiding hard-to-recycle materials and structures, using innovative designs and more sustainable materials, as shown in Fig. 4.

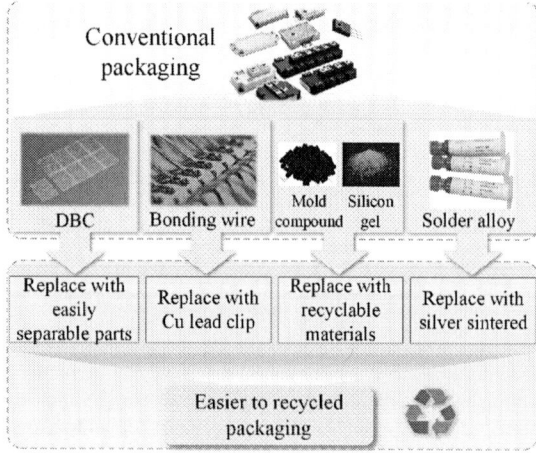

Fig. 4. Structural and material optimization strategy for power modules.

B. Recycling Design Cases

As shown in Fig. 5, the packaging with a three-section heat pipe offers strong recycling potential. The heat pipe replaces the DBC substrate, providing excellent heat dissipation and insulation, as verified in Fig. 6 [15].

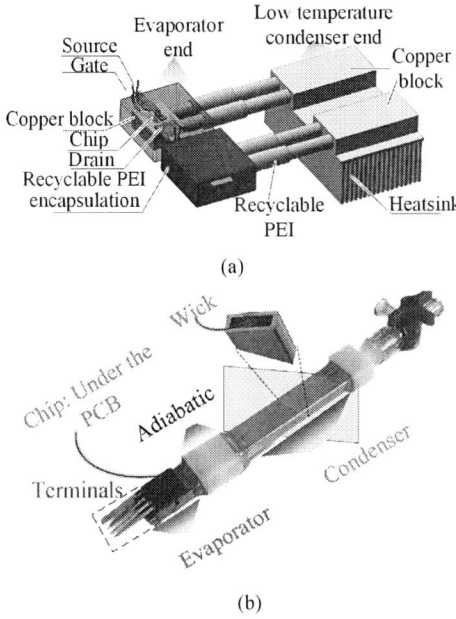

(a)

(b)

Fig. 5. Packaging structure based on a three-section heat pipe.(a) Cylindrical structure; (b) Optimized flat structure.

Polyetherimide (PEI) is a high-performance thermoplastic with excellent thermal stability, mechanical strength, electrical insulation properties, and chemical resistance, making it a promising candidate for encapsulation. Unlike conventional thermosetting plastics, PEI can be reshaped upon heating without altering its chemical structure, enhancing its recyclability. Thermoplastics offer significant advantages in recycling, including lower costs and reduced environmental impact, contributing to sustainable development and the circular economy. Notably, PEI is soluble in n-methylpyrrolidone (NMP) at 60 °C, while NMP itself is water-soluble, whereas PEI is not. This property enables efficient separation and recovery of PEI using water, facilitating its reuse in a sustainable manner [16].

The PEI insulation section is recyclable, and the copper pipes are easy to disassemble. Silver sintering is used for soldering, allowing chemical decomposition, while copper clips replace bonding wires. The PEI casing, with low

thermal conductivity, still ensures effective heat dissipation. Preliminary suggests the packaging can be easily disassembled with a simple chemical process, as shown in Fig. 7.

C. Design and Evaluation Framework

Designing recyclable power modules requires balancing performance factors (e.g., reliability, cost, power density, and efficiency) with environmental impacts (e.g., carbon footprint, resource consumption, and manufacturability). The framework in Fig. 8 integrates sustainability indicators with traditional design goals, ensuring the evaluation process incorporates both performance and environmental metrics for multi-objective optimization.

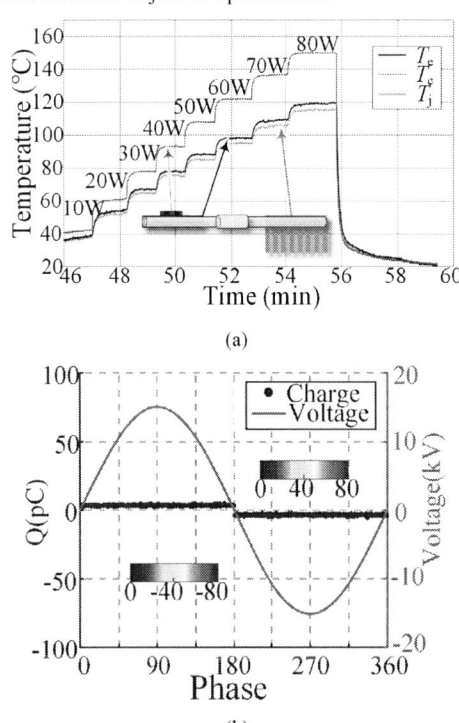

(a)

(b)

Fig. 6. Thermal and electrical insulation testing results. (a) Thermal tests; (b) Electrical insulation tests.

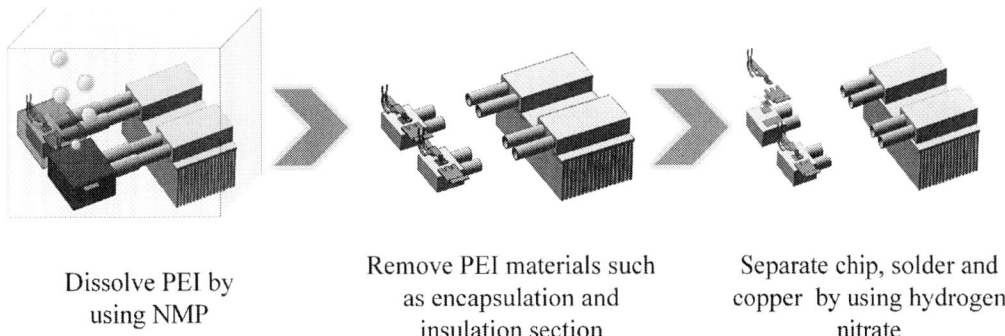

Dissolve PEI by using NMP

Remove PEI materials such as encapsulation and insulation section

Separate chip, solder and copper by using hydrogen nitrate

Fig. 7. Preliminary proposed recycling process.

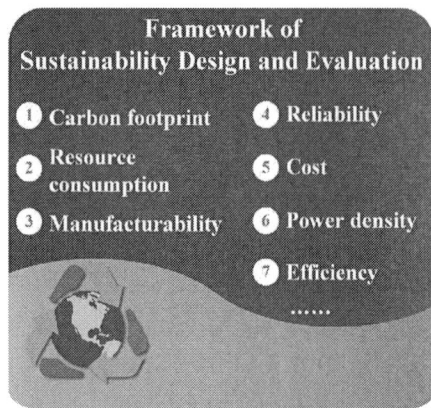

Fig. 8. Sustainability design and evaluation framework.

IV. SUMMARY AND OUTLOOK

To achieve the net-zero carbon target, the use of power devices is inevitably increasing. To avoid environmental impacts caused by improper disposal at the end of life, it is essential to incorporate "R-design strategies" and circular economy concepts at the design stage. As recycling technology matures, adopting a "design for recycling" approach will have greater potential.

Sustainable design of power devices should consider factors such as material selection, structural design, performance evaluation, and energy efficiency optimization from the early stages. Although recyclable designs may initially increase carbon footprints, this impact will gradually decrease as technology advances.

Lifecycle assessment (LCA) is a vital tool for evaluating carbon footprints and informing design decisions. However, to effectively apply it to the sustainability assessment of power devices, there is a need to enhance databases and enable data sharing across the supply chain. This will allow for a more accurate analysis of materials, manufacturing processes, and end-of-life treatments, driving the sustainable development of power devices.

Additive manufacturing reduces material waste, optimizes production processes, and lowers carbon footprints, while artificial intelligence enhances resource efficiency by optimizing designs and predicting device lifespan and energy performance, supporting circular economy strategies.

For sustainable power device designs that have not yet been adopted, the government should drive industry transformation through regulatory measures and incentives. Additionally, collaboration between universities, research institutes, and enterprises should be strengthened to foster the development and application of sustainable designs.

REFERENCES

[1] J. W. Kolar et al., "Net zero CO2 by 2050 is NOT enough (!)," in Proc. 25thEur. Conf. Power Electron. Appl. (EPE-ECCE Europe), Aalborg, Denmark, Sep. 2023. [Online]. Available: https://www.pes-publications.ee.ethz.ch/uploads/tx_ethpublications/13_EPE_Keynote_Kolar_as_presented__corr_after_Presentation_180923.pdf

[2] Net Zero by 2050 — Analysis (IEA, 2021); https://www.iea.org/reports/net-zero-by-2050.

[3] Zhang Y, Dong D, Li Q, et al. Wide-bandgap semiconductors and power electronics as pathways to carbon neutrality[J]. Nature Reviews Electrical Engineering, 2025: 1-18.

[4] C.P. Baldé. (2022). Global Transboundary E-waste Flows Monitor 2022. [Online]. Available: https://ewastemonitor.info/wp-content/uploads/2022/06/Global-TBM_webversion_june_2_pages.pdf

[5] Cornelis P. Baldé. (2024). THE GLOBAL E-WASTE MONITOR 2024. [Online]. Available: https://ewastemonitor.info/wp-content/uploads/2024/12/GEM_2024_EN_11_NOV-web.pdf

[6] https://samsolu.com/zh-hans/solutions/electronic-waste-management-solutions/

[7] Eur. Commission European Commission. (2022). Ecodesign for SustainableProducts Regulation. [Online]. Available: https://commission.europa.eu/energy-climate-change-environment/standards-tools-and-labels/products-labelling-rules-and-requirements/sustainable-products/ecodesignsustainable-products-regulation_en

[8] https://www.nea.gov.sg/our-services/waste-management/3r-programmes-and-resources/e-waste-management

[9] A. Sangwongwanich, D. -I. Stroe, C. Mi and F. Blaabjerg, "Sustainability of Power Electronics and Batteries: A Circular Economy Approach," in IEEE Power Electronics Magazine, vol. 11, no. 1, pp. 39-46, March 2024.

[10] ECS Sustainability and Environmental Footprint, Eur. Assoc. Smart Syst. Integr. (EPoSS), Brussels, Belgium, Jul. 2023.

[11] Zhou, Yihui and Ke-qiang Qiu. "A new technology for recycling materials from waste printed circuit boards." Journal of hazardous materials 175 1-3 (2010): 823-8 .

[12] Long, Laishou et al. "Using vacuum pyrolysis and mechanical processing for recycling waste printed circuit boards." Journal of hazardous materials 177 1-3 (2010): 626-32 .

[13] Veit, Hugo Marcelo, et al. "Utilization of magnetic and electrostatic separation in the recycling of printed circuit boards scrap." Waste management 25.1 (2005): 67-74.

[14] Karwowska, Ewa, et al. "Bioleaching of metals from printed circuit boards supported with surfactant-producing bacteria." Journal of Hazardous Materials 264 (2014): 203-210.

[15] J. Wei et al., "Substrate-less Power Semiconductor Packaging for the Potential of Recyclability," in IEEE Journal of Emerging and Selected Topics in Power Electronics.

[16] Alqaheem, Yousef, et al. "Preparation of polyetherimide membrane from non-toxic solvents for the separation of hydrogen from methane." Chemistry Central Journal 12 (2018): 1-8.

Cryogenic Output Capacitance Loss of GaN HD-GIT

Yudong Wang
School of Electrical Engineering
Xi'an Jiaotong University
Xi'an, China
2216110978@stu.xjtu.edu.cn

Zilong Chen
School of Electrical Engineering
Xi'an Jiaotong University
Xi'an, China
czl0928@stu.xjtu.edu.cn

Yukun Zhang
School of Electrical Engineering
Xi'an Jiaotong University
Xi'an, China
yukunzhang@stu.xjtu.edu.cn

Chong Dou
Sungrow Power Supply Co.,Ltd.
Hefei, China
douchong@sungrowpower.com

Qian Cui
Sungrow Power Supply Co.,Ltd.
Hefei, China
cuiqian@sungrowpower.com

Yuqi Wei
School of Electrical Engineering
Xi'an Jiaotong University
Xi'an, China
yuqiwei@xjtu.edu.cn

Abstract—The output capacitance (C_{oss}) loss, which is the loss generated when the output capacitor of a device is charged and discharged, has become a problem for gallium nitride (GaN) high electron mobility transistor (HEMT) in high-frequency applications. Few research has been conducted on this type of loss, especially at low temperatures. Among them, hybrid-drain gate injection transistor (HD-GIT) has lower on-resistance due to its unique conduction mechanism and is widely used in the field of power electronics. Therefore, in this paper, a cryogenic test platform for output capacitance loss of HD–GIT was built, and the non-linear resonance method was adopted to measure the output capacitance loss of GaN devices at eight temperatures, six frequencies, and six voltage amplitudes respectively. The experimental results show that: 1) The output capacitance loss of GaN devices is positively correlated with voltage and frequency; 2) As the temperature decreases, the output capacitance loss of GaN increases. The loss at 93 K is 16.8% higher than that at 293 K. Finally, the reason for the increase in loss at low temperatures is explained through semiconductor physics theory, mainly due to the decrease in electron emission rate and the increase in capture time at low temperatures. Additionally, an HD-GIT model is established on technology computer aided design (TCAD) to simulate and verify the correctness of the theory.

Keywords—GaN HEMT, non-linear resonance, output capacitance, simulation, cryogenic temperature

I. INTRODUCTION

Cryogenic power electronics technology, with its remarkable advantages in improving efficiency, power density, and reliability, is becoming a key driving force behind the innovation of power electronic systems. In the field of quantum computing, it can achieve low-noise signal amplification and precise energy control in extremely low-temperature environments, providing a guarantee for the stable operation of superconducting quantum bits [1]. In all-electric aircraft, cryogenic motor drives and distributed power management technologies can significantly reduce system weight and enhance energy utilization efficiency, helping the aviation industry transform toward low-carbon and electrification [2]. In deep-space exploration scenarios, this technology can adapt to the extreme low temperatures and high radiation environments in the universe, ensuring the long-term reliable operation of power systems and providing a solid technical foundation for humanity's exploration of the mysteries of the universe.

Power semiconductor devices are the core components of power electronic systems. Gallium nitride (GaN) high electron mobility transistors (HEMT) as a new-generation semiconductor, has a lower on-resistance [3], significantly improved over current capacity [4], faster switching speed [5], and stable breakdown voltage at low temperatures. Compared with other power semiconductor devices, GaN devices have excellent performance at low temperatures, making them ideal devices for cryogenic application fields such as the aerospace industry and deep-space exploration.

Among them, hybrid-drain gate injection transistor (HD-GIT) adopts a p-type gate, which gives it good gate stability. When a positive voltage is applied to the gate, conductivity modulation occurs, reducing the on-resistance during operation. The working principle of HD-GIT is shown in Fig. 1 [6].

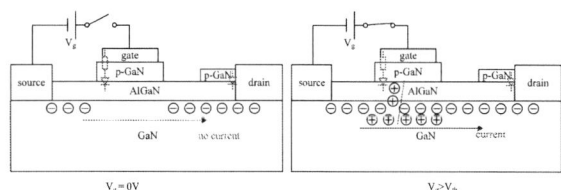

Fig. 1. The working principle of HD-GIT.

However, due to the on-resistance loss of, and the switching losses are reduced, the proportion of output capacitance loss increases significantly [7]. Therefore, in order to achieve more efficient power electronic converters, it is very important to study the output capacitance loss.

This C_{oss} loss issue first gained attention in silicon (Si) superjunction devices and recently in GaN HEMTs. After research and analysis by scholars, it has been found that the output capacitance loss of GaN devices is mainly related to factors such as switching speed and voltage amplitude [8]. Based on various experimental results, Mohammad Samizadeh Nikoo et al. proposed a GaN device models at room temperature, which can well explain the output capacitance loss of GaN at room temperature [9].

However, the above-mentioned measurements of output capacitance loss are only carried out at room temperature. Currently, the research on the cryogenic characteristics of the output capacitance loss of GaN devices is still incomplete, and there is few research on hybrid-drain gate injection transistor (HD–GIT) type GaN devices. Therefore, based on the existing research, this paper mainly focuses on the following aspects for the cryogenic measurement of the output capacitance loss of GaN: 1) Propose a set of low-temperature measurement schemes to test the output capacitance loss of GaN at low temperatures; 2) explore the cryogenic characteristics of the output capacitance loss of GaN and analyze the mechanism of its changes.

II. CRYOGENIC TEST PLATFORM

The HD-GIT device selected in this article is Infineon-IGOT65R025D2, and the key parameters of the device are shown in Table I.

TABLE I.   KEY PARAMETERS OF DUT

Parameter	Value
Manufacturer	Infineon
Model	IGOT65R025D2
Voltage/Current Rating	650 V/120 A
Static $R_{DS(ON)}$	30 mΩ
C_{oss}	130 pF (V_{DS}=400 V, f=1 MHz)
Package	PG-DSO-20
Type	HD-GIT

The measurement circuit consists of a power supply part, a device under test (DUT), and a gate drive circuit. During the off-state of the DUT, the output capacitance loss can be indirectly obtained by measuring parameters such as the voltage, current, or temperature of the DUT.

A. Test Circuit and Test Method

Currently, common methods for measuring output capacitance loss include the Sawyer-Tower circuit, the calorimetric test, the non-linear resonance-based test and the unclamped inductive switching method. The advantages and disadvantages of each test method are shown in Table II.

TABLE II.   ADVANTAGES AND DISADVANTAGES OF MEASUREMENT METHODS FOR OUTPUT CAPACITANCE LOSS

	Advantage	Disadvantage
Sawyer-tower circuit	Simple circuit	Requires a power amplifier and limits parasitic parameters
Calorimetric test	Measure the losses under soft-switching, not limited by frequency	Requires high-precision temperature measuring instruments and a power amplifier
Non-linear resonance-based test	LC resonance, no need to input high voltage, relatively simple circuit	The losses are sensitive to waveform distortion and data processing
Unclamped inductive switching method	LC resonance, relatively simple circuit	Requires high-precision resistance measurement for each device

By comparing the four measurement methods, it was finally decided to use the non-linear resonance-based method to measure the output capacitance loss. The circuit topology is shown in Fig. 2.

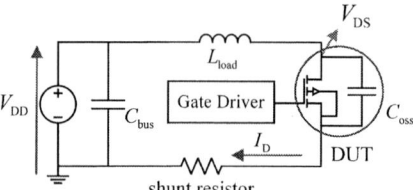

Fig. 2. Circuit topology for output capacitance loss.

where V_{DD} is the DC voltage provided by the voltage source, L_{load} is an air-core inductor (to eliminate the influence of the iron core on the inductance value), C_{oss} is the output capacitance of the DUT, C_{bus} is the bus-voltage-stabilizing capacitor, and the shunt resistor is a precision resistor with a small resistance value. The current value in the circuit can be calculated by measuring its voltage drop.

A single pulse is input to the gate of the DUT. By controlling the on-time of the GaN device, the current of the inductor is controlled. When the gate voltage falls, the switch is turned off, forming an LC resonant circuit, which causes the voltage across the DUT to rise rapidly. Solving the equations of the LC circuit shows that:

$$V_{ds} = \sqrt{\frac{C(V_{ds})}{L}} t_0 V_{DD} \sin \frac{t}{\sqrt{LC(V_{ds})}} \tag{1}$$

where $C(V_{ds})$ is the output capacitance of the device, L is the air-core inductor, and t_0 is the single-pulse width.

In the experiment, the single-pulse width was controlled to be 10 μs. By changing the bus voltage V_{DD} and the inductance L_{load}, the voltage amplitude across the DUT circuit and the charge-discharge frequency can be changed. The inductance value was selected to be small, with values of 0.5 μH-10 μH. The experimental temperature is 93 K-293 K.

An oscilloscope was used to measure the voltage across the DUT, and the shunt resistor was used to measure the current flowing through the DUT. Fig. 3 shows the main operating waveforms of the circuit during operation.

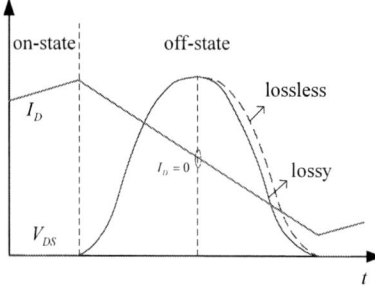

Fig. 3. Waveform of output capacitance loss.

When the device is turned off, the voltage across the DUT changes. The C-V curve of the DUT can be obtained through equation (2). Using this method to calculate the capacitance can avoid the influence of the inter-turn capacitance of the inductor on the result.

$$C(V_{DS}) = \frac{I_D}{\dfrac{dV_{DS}}{dt}} \qquad (2)$$

Then, the output capacitance loss (E_{diss}) and the energy stored in the capacitor (E_{oss}) can be obtained by equation (3).

$$E = \int_T CV_{DS}d(V_{DS}) \qquad (3)$$

B. Cryogenic Experimental Platform Optimization

The output capacitance loss of GaN devices is mainly related to factors such as switching speed and voltage amplitude. To accurately measure the relationship between the output capacitance loss and temperature, it is necessary to control the stability of the inductance value, the output of the voltage source, and the drive circuit.

The magnetic permeability of ferromagnetic materials will reduce at low temperatures. Some scholars also studied and pointed out that the output voltage of isolated power supplies will decrease significantly at low temperatures [10]. Therefore, during the experiment, the inductor, Microcontroller Unit (MCU), voltage source, and drive power supply should be placed outside the thermostatic chamber. However, in order to avoid mis-conduction, the gate driver should be placed as close to the GaN device as possible.

For the devices on the test circuit board, in addition to considering their reliability at low temperatures, it is also necessary to ensure the accuracy of the auxiliary components at low temperatures. In reference [11], summarizes the cryogenic characteristics of power electronic devices with different packaging materials. For key devices, film resistors with good cryogenic performance, as well as negative-positive zero (NP0) capacitors and film capacitors are used.

The signal is transmitted to the isolated power supply through a coaxial cable with an impedance of 50 Ω. In order to reduce the voltage overshoot caused by wave reflection and refraction during the transmission process, the coaxial cable is matched at both ends. A 50 Ω resistor is connected in series at the output end of the MCU, and an RC impedance matching circuit (a 50 Ω resistor and a 30 pF capacitor in series) is connected in parallel at the input end of the isolation chip, as shown in Fig. 4.

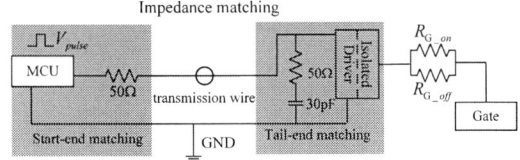

Fig. 4. Impedance matching of PWM.

At low temperatures, the test accuracy of the probes will also be affected. In order to accurately measure the voltage and current in the main circuit, a shunt resistor is used to measure the current in the main circuit, and a passive probe is used to measure the voltage in the circuit. The data measured by the probes used in the circuit are shown in Table III .

TABLE III. SELECTION OF PASSIVE DEVICES

Test object	Probe model	Characteristics
V_{ds}	LECROY PPE6KV-A	High voltage resistance: 1000 V_{rms}
		High bandwidth: 500 MHz
		High input impedance: 50 MΩ
I_d	T&M Research-SSDN-10	High bandwidth: 2000 MHz
		Resistance: 0.1 Ω
V_{gs}	LECROY PP026	High bandwidth: 500 MHz
		High input impedance: 10 MΩ

The built cryogenic test platform is shown in Fig. 5, and the details of the circuit board are shown in Fig. 6.

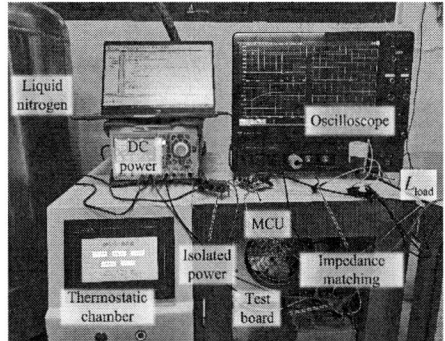

Fig. 5. Dynamic test platform for output capacitance loss.

Fig. 6. Dynamic test board for output capacitance loss.

III. EXPERIMENTAL RESULTS

Fig. 7 shows the experimental waveforms measured using a 5 μH inductor. When the gate of the DUT is at a high voltage, the device turns on, and the current in the main circuit starts to rise. After the switch is turned off, the energy in the inductor is released, causing the device voltage to rise above 600 V.

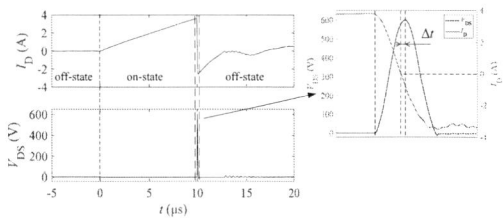

Fig. 7. Experimental waveform of output capacitance loss.

Theoretically, when the device voltage V_{ds} reaches its peak value, the current in the circuit I_d should drop to zero. Therefore, the delay between the voltage probe and the current probe can be calibrated. The delay in this experiment is shown as Δt in Fig. 7, which is approximately 8.4 ns.

From equation (2), the C-V curve of the DUT at a temperature of 293 K, a voltage amplitude of 600 V and a frequency (switching speed) of 11.2 MHz is shown in Fig. 8. The experimental data is basically consistent with the data in the data sheet.

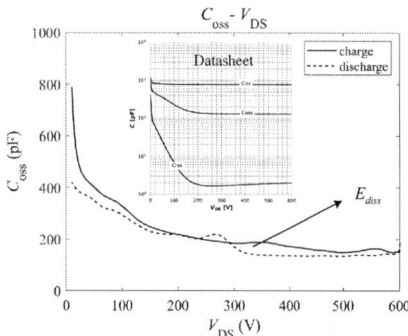

Fig. 8. The capacitance-voltage curve of DUT.

By integrating the curve using equation (3), the output capacitance loss of the DUT during a single charge-discharge cycle under these conditions can be obtained, which is 3.29 μJ.

The same method was used to process other data. To reduce experimental errors, single-pulse experiments were conducted five times under the same conditions. After each experiment, the power supply was disconnected, and enough time was allowed for the DUT to release stress. The average value of the five experimental results was taken to obtain the relationship between the single-cycle output capacitance loss and temperature, voltage, and frequency.

Fig. 9 shows the output capacitance loss at a voltage of 600 V. As the frequency increases, the output capacitance loss increases slightly, with an increase of approximately 11.7%. In the circuit, the resonant frequency is positively correlated with the maximum value of the circuit current. The resonant frequency of the same inductor also decreases as the temperature decreases, indicating that the output capacitance of the DUT increases as the temperature decreases.

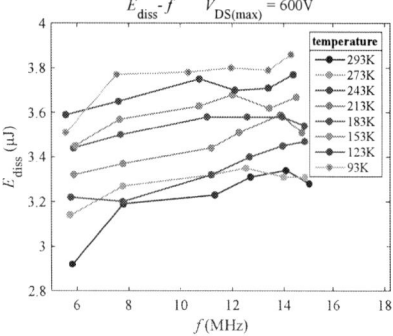

Fig. 9. The relationship between output capacitance loss and frequency and temperature.

Fig. 10 shows the relationship between the output capacitance loss and voltage at a resonant frequency of 11.2 MHz. As the voltage increases, the output capacitance loss increases rapidly. The output capacitance loss obtained at room temperature is of the same order of magnitude as that reported in [8]. Fig. 10 also shows the relationship between the output capacitance loss and temperature. As the temperature decreases, the output capacitance loss increases. At the voltage amplitude of 600 V and the frequency of 7.8 MHz, the output capacitance loss at 93 K is 16.8% higher than that at 293 K.

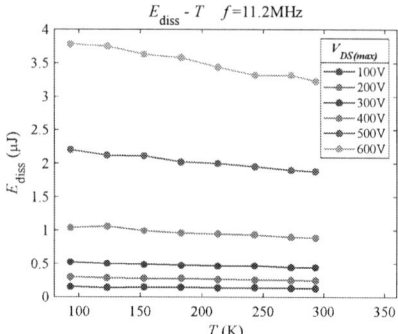

Fig. 10. The relationship between the output capacitance loss and voltage and temperature.

The relationship between the equivalent output capacitance and temperature and voltage is shown in Fig. 11. As the temperature gradually decreases, the equivalent output capacitance shows a tendency to rise. The energy of the output capacitance loss accounts for approximately 9.2% of the energy stored in the capacitor on average.

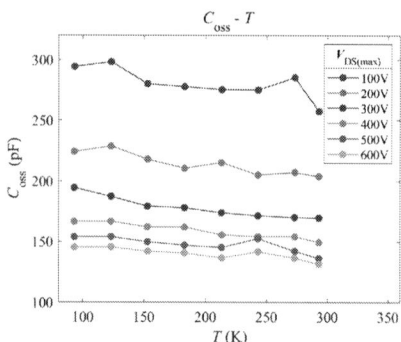

Fig. 11. The relationship between equivalent output capacitance and temperature and voltage.

IV. MECHANISM ANALYSIS

The physical origin of C_{oss} loss in GaN HEMTs remains controversial, but there is a consensus that carrier trapping/detrapping and the resulting C_{oss} hysteresis are the root causes of output capacitance loss. However, the characteristics of the main traps, including their energy levels, detrapping time constants, and energy levels, still lack a complete explanation.

In [12], the main cause of the output capacitance loss is the trapped charges in the device. James G. Rathmell et al. used statistical physics knowledge to obtain the relationship between the electron capture effect and temperature. In [13] shows that the capture time of electrons is less than 1 microsecond, but the emission time is usually in the range of milliseconds to seconds. The rate of electron capture and the rate of emission after electron capture:

$$R_{en} = c_n n p_D \tag{4}$$

$$R_{en} = c_n N_C \exp(\frac{E_D - E_C}{kT})(N_D - p_D) \tag{5}$$

The emission time constant is:

$$\tau_e = (c_n N_C \exp(\frac{E_D - E_C}{kT}))^{-1} \tag{6}$$

where c_n is the capture coefficient, N_C is the electron concentration, N_D is the doping concentration, $E_D - E_C$ is the energy difference between the donor level and the conduction band, and p_D is the concentration of empty traps. We combine the equation (4) and (5)

$$\frac{dp_D}{dt} = c_n N_C \exp(\frac{E_D - E_C}{kT})(N_D - p_D) - c_n n p_D \tag{7}$$

Free charges are proportional to displacement current, and the relationship with time is a linear function. Rewrite (8) as

$$\frac{dp}{dt} = a(1-p) - b(1-kt)p \tag{8}$$

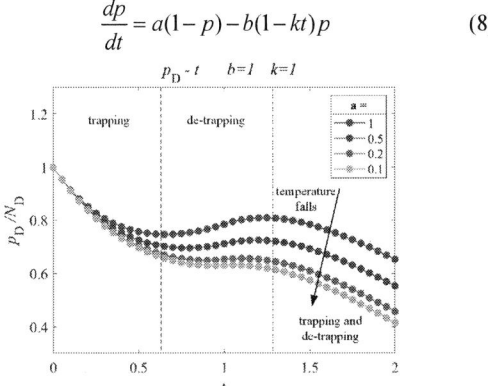

Fig. 12. The variation of the concentration of empty traps with time at different temperatures.

By means of numerical solution, the variation of the concentration of empty traps with time at different temperatures (different values of a) is obtained as shown in Fig. 12.

We can divide Fig. 12 into three parts. At the beginning, electrons are captured by traps, which is mainly controlled by R_{en} and this stage occurs when GaN is just turned off. After that, it enters the second stage, the electron emission stage. The concentration of free electrons in GaN decreases, and the capture rate reduces. In this stage, electron emission is dominant, The number of captured electrons is mainly controlled by R_{en}. However, due to the long electron emission time, the change in the number of captured charges in this stage is not significant; In the third stage, the reverse displacement current increases, the electron capture phenomenon intensifies, which is similar to the first stage. The schematic diagram of electron trapping/de-trapping in a non-linear resonant circuit is shown in Fig. 13.

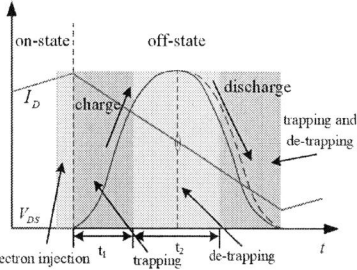

Fig. 13. The schematic diagram of electron trapping/de-trapping in a non-linear resonant circuit.

The number of captured electrons in the first stage

$$p_D = N_D \exp(-c_n n_0 t) \tag{9}$$

The number of captured electrons in the second stage

$$N_D - p_D = N_D \left[1 - \exp(-c_n n_0 t_1)\right]$$
$$\left[\exp\left(-c_n N_C \exp\left(\frac{E_D - E_C}{kT}\right)(t - t_1)\right)\right] \tag{10}$$

Energy loss

$$E_{diss} = \int q \cdot V_{DS} \cdot d\left(N_D - p_D\right)$$
$$= c_n n_0 N_D q V_m \left[\begin{array}{l} \int_0^{t_1} kt \cdot \exp(-c_n n_0 t) dt + \\ \int_{t_2}^{t_3} (2 - kt) \cdot \frac{p_D\big|_{t=t_2}}{N_D} \cdot \exp(-c_n n_0 t) dt \end{array}\right]$$
$$+ V_m N_D q \cdot \left(N_D - p_D\right)\big|_{t_1}^{t_2}$$
$$= E_1(V_m, n_0, f) - V_m N_D q\left[1 - \exp(-c_n n_0 t_1)\right]$$
$$\left(1 - \exp\left(-c_n N_C \exp\left(\frac{E_D - E_C}{kT}\right)(t_2 - t_1)\right)\right) \tag{11}$$

In equation (11), the first term represents the energy loss caused by electron capture. The value of n_0 depends on the initial current value, while V_m is the oscillation amplitude of the voltage, the oscillation frequency f affects t_1 and voltage change rate k, but has a minor impact on the loss. Therefore, the output capacitance loss mainly depends on the voltage amplitude and the magnitude of the displacement current, which is basically consistent with the conclusion drawn in Reference [14]. The second term represents the energy "generated" by electron de-trapping. It is smaller in magnitude compared to the first term and is mainly related to the oscillation time (frequency) and temperature, showing a positive correlation with both. Consequently, as the temperature decreases, the total output capacitance loss increases.

It should be noted that in Fig. 9, the positive correlation between the output capacitance loss and frequency we obtained is mainly because in the experiment, the two control factors of frequency and displacement current were not independent. When the output capacitance of the GaN HEMT remains unchanged, an increase in displacement current will also lead to an increase in frequency. In the subsequent experimental plan, a tunable capacitor will be connected in parallel with the output capacitance to achieve the independence of frequency and displacement current.

979-8-3315-1110-4/25 $31.00 © 2025 IEEE

Based on the GaN structure and doping concentration proposed [7], an HD–GIT type GaN device was built in the technology computer aided design (TCAD) simulation software. The device structure diagram and the electron capture schematic diagram are shown in Fig. 14.

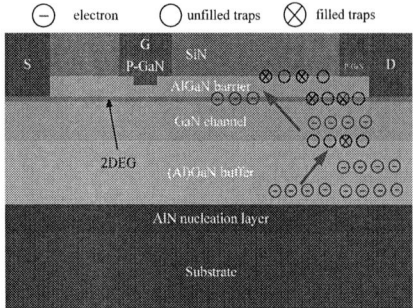

Fig. 14 Device structure diagram and electron capture schematic diagram of HD-GIT.

Among them, P-GaN is p-type doping with a doping concentration of 1×10^{19} cm^{-3}; AlGaN barrier is n-type doping with a doping concentration of 5×10^{17} cm^{-3}; GaN buffer is p-type doping with a doping concentration of 1×10^{16} cm^{-3}.

The electron capture phenomenon and the change in electron concentration of HD-GIT were simulated. During the off-state, electron capture mainly occurs in the GaN channel, (Al)GaN buffer, and the AlGaN/passivation interface. Specifically, the electron trapping that mainly affects the output capacitance loss occurs at the GaN channel.

Fig. 15 shows the charge distribution in the GaN channel at 100 K and 300 K obtained by the Silvaco TCAD simulation software. As the temperature decreases, the electron capture phenomenon is enhanced, and the charge density increases, which is the fundamental reason for the increase in capacitance at low temperatures.

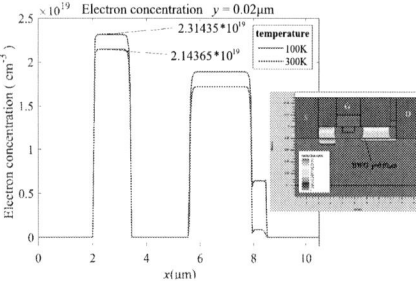

Fig. 15 The electron concentration of HD-GIT in the off-state at 100 K and 300 K by TCAD.

When the temperature drops from 293 K to 93 K, the carrier concentration increases by approximately 7.9%, which is consistent with the measured increase in capacitance in the experiment.

Therefore, the reason for the increase in output capacitance loss at low temperatures is the increase in carrier concentration and the increase in capture time caused by electron capture.

V. CONCLUSION

In this paper, a cryogenic test platform for output capacitance loss was built. A commercial HD-GIT device

was tested, and the relationship between the output capacitance loss and frequency, voltage and temperature at low temperatures were obtained. The influence of low temperatures on the output capacitance loss was explained from the perspective of electron capture, and the conjecture was verified by using TCAD simulation software.

ACKNOWLEDGMENT

This work is supported the Aeronautical Science Fund (ASF) (Grant No. ASFC-2022Z072070001) and Sungrow Power Supply.

REFERENCES

[1] Bardin J. Beyond-Classical Computing Using Superconducting Quantum Processors; proceedings of the 2022 IEEE International Solid-State Circuits Conference (ISSCC), F 20-26 Feb. 2022, 2022 [C].

[2] Schefer H, Canders W R, Hoffmann J, et al. Cryogenically-Cooled Power Electronics for Long-Distance Aircraft [J]. IEEE Access, 2022, 10: 133279-308.

[3] Ren R, Gui H, Zhang Z, et al. Characterization and Failure Analysis of 650-V Enhancement-Mode GaN HEMT for Cryogenically Cooled Power Electronics [J]. IEEE Journal of Emerging and Selected Topics in Power Electronics, 2020, 8(1): 66-76.

[4] Wei Y, Hossain M M, Mantooth H A. Cryogenic Overcurrent Characteristic of GaN HEMT and Converter Evaluation [J]. IEEE Transactions on Industry Applications, 2024, 60(4): 6479-87.

[5] Buffolo M, Favero D, Marcuzzi A, et al. Review and Outlook on GaN and SiC Power Devices: Industrial State-of-the-Art, Applications, and Perspectives [J]. IEEE Transactions on Electron Devices, 2024, 71(3): 1344-55.

[6] J. Shuo, C. Yong, W. Deliang, Z. Baoshun, K. M. Lau, and K. J. Chen, "Enhancement-mode AlGaN/GaN HEMTs on silicon substrate," IEEE Transactions on Electron Devices, vol. 53, no. 6, pp. 1474-1477, 2006, doi: 10.1109/ted.2006.873881.

[7] J. P. Kozak et al., "Stability, Reliability, and Robustness of GaN Power Devices: A Review," IEEE Transactions on Power Electronics, vol. 38, no. 7, pp. 8442-8471, 2023, doi: 10.1109/tpel.2023.3266365.

[8] Q. Song, R. Zhang, Q. Li, and Y. Zhang, "Output Capacitance Loss of GaN HEMTs in Steady-State Switching," IEEE Transactions on Power Electronics, vol. 39, no. 5, pp. 5547-5557, 2024, doi: 10.1109/tpel.2023.3279308.

[9] M. S. Nikoo, A. Jafari, N. Perera, and E. Matioli, "New Insights on Output Capacitance Losses in Wide-Band-Gap Transistors," IEEE Transactions on Power Electronics, vol. 35, no. 7, pp. 6663-6667, 2020, doi: 10.1109/TPEL.2019.2958000.

[10] M. u. Hassan, Y. Wu, F. Luo, and V. Solovyov, "Development of Gate Drive Configuration for GaN Based Cryogenic Power Electronics Converters," IEEE Transactions on Industry Applications, vol. 59, no. 6, pp. 7039-7051, 2023, doi: 10.1109/TIA.2023.3304618.

[11] H. Gui et al., "Review of Power Electronics Components at Cryogenic Temperatures," IEEE Transactions on Power Electronics, vol. 35, no. 5, pp. 5144-5156, 2020, doi: 10.1109/TPEL.2019.2944781.

[12] M. Guacci et al., "On the Origin of the C_{oss} -Losses in Soft-Switching GaN-on-Si Power HEMTs," IEEE Journal of Emerging and Selected Topics in Power Electronics, vol. 7, no. 2, pp. 679-694, 2019, doi: 10.1109/jestpe.2018.2885442.

[13] J. L. Gomes, L. C. Nunes, C. F. Goncalves, and J. C. Pedro, "An Accurate Characterization of Capture Time Constants in GaN HEMTs," IEEE Transactions on Microwave Theory and Techniques, vol. 67, no. 7, pp. 2465-2474, 2019, doi: 10.1109/tmtt.2019.2921338.

[14] Q. Song, R. Zhang, Q. Li, and Y. Zhang, "Impact of Conduction Current on Output Capacitance Loss in GaN HEMTs," in 2023 IEEE Applied Power Electronics Conference and Exposition (APEC), 19-23 March 2023, pp. 2533-2537, doi: 10.1109/APEC43580.2023.10131551.

Improved Control Strategy Based on BM Theory for Grid Forming Type VSG System

Shuanglong Li
School of Automation, Beijing Information Science & Technology University
Beijing, China
Email:2024020402@bistu.edu.cn

Yajing Zhang*
School of Automation, Beijing Information Science & Technology University
Beijing, China
Email:zhangyajing@bistu.edu.cn

Bin Liu
State Grid Economic and Technological Research Institute Co
Beijing, China
Email:
liubin@chinasperi.sgcc.com.cn

Baoying Huang
State Grid Economic and Technological Research Institute Co
Beijing, China
Email:
huangbaoying@chinasperi.sgcc.com.cn

Siyu Pan
State Grid Economic and Technological Research Institute Co
Beijing, China
Email:
pansiyu@chinasperi.sgcc.com.cn

Hao Ma
School of Automation, Beijing Information Science & Technology University
Beijing, China
Email:2022020463@bistu.edu.cn

Abstract—In view of the problem that the grid-connected system of three-phase two-level inverter using silicon carbide is susceptible to power disturbance and the system lacks frequency inertia support ability, which seriously affects the stable operation of the grid-forming system, In this paper, based on the traditional virtual synchronous generator (VSG) system, combined with the Brayton-Moser (BM) control model, a new controller composition is designed to improve the VSG system. Combined with the advantages of the BM controller, the improved VSG system has a certain power anti-interference ability and a more sufficient grid frequency inertia margin. Finally, using Matlab to verify the effectiveness and superiority of the improved VSG system.

Keywords—Three-phase inverter, Wide bandgap devices, VSG, BM theory, Power disturbance

I. INTRODUCTION

To solve the shortage of fossil energy and respond to the "double carbon" goal proposed by the state, by the end of 2024, China's installed capacity of new energy power generation, including wind, photovoltaic and biomass power generation, will reach 1.45 billion kilowatts. The scale of renewable energy power generation in China will continue to expand, and the power system will show a trend of "double high" power system [1]. Renewable energy power generation systems need to use power electronics to match the grid, but traditional silicon-based devices play a limited role in high temperature and high frequency. In this context, wide bandgap devices have become a research hotspot due to their better withstand voltage performance, higher operating temperature, and suitable for high-frequency switching applications, silicon carbide, gallium nitride and so on are commonly used [2]. With the gradual replacement of traditional synchronous machines by new energy, in the new power system, due to the limited inertial support and damping provided by the synchronous machine, and is more susceptible to power disturbance and system fault [3], and the problem of system stability gradually emerges.

Project supported by the Headquarters Technology Projects of State Grid Corporation of China (5200-202456095A-1-1-ZN).

Therefore some scholars have proposed new control theory [4], which simulates the rotor mathematical expressions of the synchronous generator to make the new energy system have inertia support and voltage support capabilities called virtual synchronous generator (VSG) control, and improve the new power system performance. The VSG system without voltage and current control loop has a lower risk of instability and a simpler structure in the strong power network, but it is difficult to limit the fault current during the fault and even cause device damage [5]. To this end, scholars have added voltage and current control to limit the short-circuit current and improve the stability of the system during faults [6]. However, the dual closed-loop regulation ability of voltage and current is limited, and the role cannot be fully exerted when the system is running stably. Therefore, some scholars have proposed virtual impedance control, and the current stability of the system combined with virtual impedance control and double closed-loop control is greatly improved [7], but the system stability still needs to be improved when dealing with power disturbances. For the sake of solving the problem that the system is susceptible to power disturbance, the combination of control methods such as neural network control [8-10], adaptive control [11-13] with traditional virtual synchronizers has become an important research direction for scholars. Such as [8-10] the neural network algorithm is used to optimize the VSG parameters or establish a nonlinear relationship between different parameters to achieve parameter cooperative control, so as to optimize the system performance. In [11-13], a mathematical model was established to analyze the relationship between moment of inertia and damping coefficient and how it affects the system work, and a reasonable parameter threshold was obtained, and finally the parameter adaptive adjustment equation was established to realize real-time parameter adjustment.

The above method only focuses on the parameter adjustment of VSG power control body, and does not consider the influence of other links in the control system, the design of its parameter is complex, and the implementation process is challenging. So this paper improves the traditional VSG control structure based on BM theory to achieve functional optimization. The Brayton-Moser control is based on the Brayton-Moser energy function, which was proposed by Brayton and Moser to analyze the stability of RLC circuits [14]. The BM controller is measured as a directly measurable physical quantity with good dynamic performance，and then gradually developed to analyze the stability of nonlinear systems to realize the control of the system[15-17].

In this paper, an improved control strategy based on BM theory for grid forming type VSG system is proposed, and the improved system has good stability and inertial support capacity. Firstly, the BM model of the three-phase inverter was established, and then the BM power forming controller was designed. Finally, the performance of the improved VSG system is verified by comparing with the traditional VSG system in Matlab, and it also verifies the effectiveness of the BM control strategy.

II. INTRODUCTION TO IMPROVED VSG SYSTEM

A. Introduction to the System Topology

In this paper, a three-phase two-level inverter using silicon carbide as a switching device is studied, and its main circuit is composed of a DC power supply and an energy storage device, an inverter, a filter, a transmission line, a load, and an AC power grid, as shown in Fig. 1. In addition to the main circuit, the system also needs a controller to control the inverter to ensure its stable operation. The control module of the system first needs to collect the voltage and current signals and calculate the power and transform the coordinates, and then generate the modulation signal through the power control loop, virtual impedance and BM controller, and finally use SPWM modulation to drive the inverter.

The schematic for the improved VSG system is shown in Fig. 1, and U_{dc} is the DC power supply voltage, the support capacitors C, e_a, e_b, e_c are the inverter phase voltage, $S1$~$S6$ are the six switch tubes, i_{abc} is the output current of the inverter, L_f and R_{Lf} are the filter inductance and equivalent resistance, C_f and R_{cf} are the filter capacitance and equivalent resistance respectively, u_{gabc} is the grid side phase voltage, i_{gabc} is the grid side current, Z is the grid side equivalent impedance, and U_g is the AC power grid.

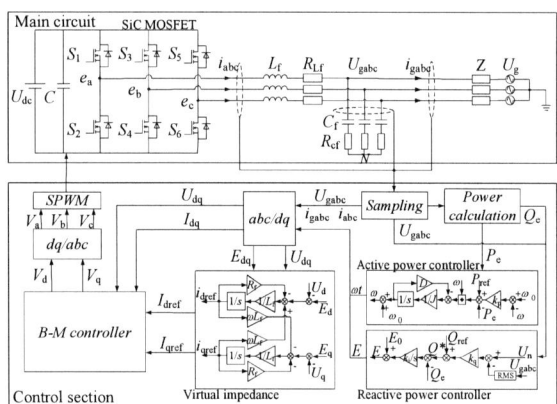

Fig. 1. Improved VSG system diagram

B. Introduction to the Active Power Control Part

The traditional synchronous generator adjusts the active output of the generator through the adjustment of mechanical torque, and realizes the response to the frequency deviation of the grid through the frequency modulator. So as to make the new energy power system have the inertia of the traditional synchronous generator, the rotor mechanical equation for simulating the synchronous generator is shown in (1):

$$\begin{cases} J\dfrac{d\omega}{dt} = T_m - T_e - T_d \\[2mm] T_m = \dfrac{P_m}{\omega_0}, T_e = \dfrac{P_e}{\omega_0}, T_d = D(\omega - \omega_0) \\[2mm] \dfrac{d\delta}{dt} = \omega \end{cases} \quad (1)$$

In Eq.(1): T_m is the mechanical torque of the synchronous generator; T_e is the electromagnetic torque of the synchronous generator; T_d is the damping torque of the synchronous generator. P_m is the mechanical power and P_e is electromagnetic power output by the synchronous generator; J is the moment of inertia of the synchronous machine; D is the damping factor; ω is the output angular frequency of VSG, ω_0 is the rotor angular frequency of the synchronous machine. The value of J has an important influence on the magnitude of inertia of the system in the dynamic process of power and frequency response. The value of D also plays an important role in suppressing the power oscillation of the grid, and these two variables are of great significance for the improvement of the operating performance of the system. Therefore, it is necessary to reasonably design the parameter values according to the actual situation.

In addition, droop control is used to adjust the active power to achieve frequency regulation. Therefore, on the basis of grid-connected power tracking, VSG can make active power adjustment for the frequency deviation of the access point to ensure its ability to cope with frequency anomalies. The active-frequency droop control expression is as follows (2):

$$P_m = P_{ref} + k_p(\omega - \omega_0) \quad (2)$$

In Eq. (2): P_{ref} is the reference active power; K_P is the active droop coefficient. Substituting (2) into (1) is sorted out to obtain the active-frequency control structure in Fig. 1.

C. The Principle of the Reactive Power Control

To simulate the reactive-voltage regulation of the traditional synchronous generator and avoid the oscillation in the process of voltage regulation of the system, the virtual electromotive force E generated by VSG was divided into three parts. They are the electric potential of reactive power regulation, the electric potential of the voltage control part and the electromotive force E_0 of the generator in the no-load state. The reactive-voltage droop control expression is as follows (3) and (4):

$$Q^* = Q_{ref} + k_q(U_n - U_{gabc}) \quad (3)$$

$$E = \left(k_{qp} + \frac{k_i}{s}\right)(Q^* - Q_e) + E_0 \quad (4)$$

In Eq. (3) (4): Q_{ref} is the set value of the reactive power, Q^* is the intermediate variable in reactive regulation, k_q is the droop coefficient of the reactive power regulation part, U_n is the terminal voltage, and U_{gabc} is the phase voltage on the grid side; k_{qp} is the reactive power proportional coefficient, which is set to 0 in this paper; k_i is the reactive power integral coefficient; Combined with (3) and (4), the VSG reactive voltage regulation control structure is obtained, as shown in Fig. 1.

From the above power control content, the VSG output virtual electromotive force E expression can be obtained:

$$E = \begin{bmatrix} e_a \\ e_b \\ e_c \end{bmatrix} = \begin{bmatrix} E\sin(\omega t) \\ E\sin\left(\omega t - \frac{2}{3}\pi\right) \\ E\sin\left(\omega t + \frac{2}{3}\pi\right) \end{bmatrix} \quad (5)$$

D. The Principle of Virtual Impedance Control

The virtual impedance module simulates the stator resistance and synchronous reactance of the synchronous generator, limiting the short-circuit current and ensuring the safety of the system during a fault. In the event of a fault, the virtual impedance module reduces the virtual potential of the VSG output to limit the current, and does not change the original control structure of the VSG, retaining the advantages of VSG control. Based on the main circuit topology in Fig. 1, a virtual impedance mathematical expression can be obtained as follows:

$$\begin{cases} i_d = \dfrac{1}{L_f s}\left(e_d - u_d - R_{Lf}i_d + \omega L_f i_q\right) \\ i_q = \dfrac{1}{L_f s}\left(e_q - u_q - R_{Lf}i_q - \omega L_f i_d\right) \end{cases} \quad (6)$$

In Eq. (6): i_{dq}, e_{dq} and u_{dq} correspond to the components of the inverter output current i_{abc}, the inverter three-phase voltage e_{abc} and the grid-side phase voltage u_{gabc} in the DQ coordinate system, respectively. ω is the power system frequency, and L_f is the filter inductance and R_{Lf} is the equivalent resistor resistance.

III. CONTROLLER DESIGN BASED ON BM MODEL

According to Fig. 1, the expression of the circuit of the three-phase two-level inverter is established, as shown in (7):

$$\begin{cases} L_f \dfrac{di_a}{dt} = u_a - R_{Lf}i_a - e_a \\ L_f \dfrac{di_b}{dt} = u_b - R_{Lf}i_b - e_b \\ L_f \dfrac{di_c}{dt} = u_c - R_{Lf}i_c - e_c \end{cases} \quad (7)$$

Assuming that the voltage amplitude of each phase of the power grid is equal, the load impedance is matched, and the phase difference of each phase is 120° in an ideal three-phase equilibrium state, and the voltage and current of each phase meet the following equations:

$$\begin{cases} u_a + u_b + u_c = 0 \\ i_a + i_b + i_c = 0 \end{cases} \quad (8)$$

The switching function is shown in (9), which represents the inverter phase voltage as (10):

$$S_k = \begin{cases} 1 \\ 0 \end{cases} \quad (9)$$

When the function value is 1, only the k-phase upper bridge arm is turned on，otherwise only the lower arm of the k phase is turned on, Combine (8) to obtain the following formula:

$$\begin{cases} e_a = U_{dc}\left(S_a - \dfrac{1}{3}\sum_{k=a,b,c} S_k\right) \\ e_b = U_{dc}\left(S_b - \dfrac{1}{3}\sum_{k=a,b,c} S_k\right) \\ e_c = U_{dc}\left(S_c - \dfrac{1}{3}\sum_{k=a,b,c} S_k\right) \end{cases} \quad (10)$$

In order to facilitate the control of voltage and current, the three-phase voltage and current are converted from the three-phase rotating coordinate system to the two-phase stationary dq coordinate system through coordinate transformation, in which the values of the d axis and the q axis change with the sinusoidal law, and the equal amplitude transformation matrix is shown in (11):

$$C_{abc/dq} = \frac{2}{3}\begin{bmatrix} \sin\omega t & \sin(\omega t - \dfrac{2\pi}{3}) & \sin(\omega t + \dfrac{2\pi}{3}) \\ \cos\omega t & \cos(\omega t - \dfrac{2\pi}{3}) & \cos(\omega t + \dfrac{2\pi}{3}) \end{bmatrix} \quad (11)$$

After the abc/dq coordinate transformation, the inverter model can be expressed as (12):

$$\begin{cases} L_f \dfrac{di_d}{dt} = u_d - R_{Lf}i_d + \omega L_f i_q - e_d \\ L_f \dfrac{di_q}{dt} = u_q - R_{Lf}i_q - \omega L_f i_d - e_q \end{cases} \quad (12)$$

$$\begin{cases} e_d = S_d * U_{dc} \\ e_q = S_q * U_{dc} \end{cases} \quad (13)$$

In Eq. (12) and (13): ω is the grid angular frequency, i_d, i_q, u_d and u_q are the components of the current output on the inverter side and the three-phase phase voltage on the grid side in the dq axis, and e_d, e_q, S_d and S_q are the components of the phase voltage of the three-phase inverter and the switching function in the dq axis.

Combined with the Brayton-Moser model from (12) and (13), it can be obtained:

$$Q\dot{X} = \frac{\partial P(x)}{\partial x} - \hat{B}u \quad (14)$$

$$\frac{\partial P(x)}{\partial x} = \begin{bmatrix} R_{Lf}i_d - \omega L_f i_q - u_d \\ R_{Lf}i_q + \omega L_f i_d - u_q \end{bmatrix} \quad (15)$$

$$Q = diag\left\{-L_f, -L_f\right\} \qquad (16)$$

$$X = \left[x_1, x_2\right]^T = \left[i_d, i_q\right]^T \qquad (17)$$

$$B = diag\left\{S_d, S_q\right\}, \ u = \left[-U_{dc}, -U_{dc}\right]^T \qquad (18)$$

To control the output current of the three-phase inverter, so that it can track the desired current $i_{abc}{}^*$, it is necessary to track the components of the output current and the expected current in the dq axis. $X^* = \left[x_1^*, x_2^*\right]^T = \left[i_d^*, i_q^*\right]^T$ is the expected current, $X_e = X^* - X = \left[x_1^* - x_1, x_2^* - x_2\right]^T$ is the error. The expected current function is as follows (19):

$$Q\dot{X}^* = \frac{\partial P^*(x)}{\partial x^*} - \hat{B}^* u \qquad (19)$$

So as to design the controller, need to inject virtual damping into the original system to reshape the system energy, and the injected damping can be connected in series with the inductor, and the mathematical model after injecting virtual damping into the system error is as follows:

$$Q\dot{X}_e = \frac{\partial P_d(x)}{\partial x_d} - \hat{B}_e u \qquad (20)$$

Among them, set the injected virtual damping factor is K.

$$R_d X = (R_{Lf} + K)X, \quad K = diag\left\{k_1, k_2\right\} \qquad (21)$$

$$\frac{\partial P_d(x)}{\partial x_d} = \begin{bmatrix} (R_{Lf} + k_1)i_d - u_d - \omega L i_q \\ (R_{Lf} + k_2)i_q - u_q + \omega L i_d \end{bmatrix} \qquad (22)$$

In Eq. (21), K is the positive definite matrix, in order to avoid the strong coupling phenomenon in the dynamic process and improve the dynamic and static performance of the three-phase DC inverter, the BM model expression of the controlled quantity can be obtained by subtracting (20) from (19):

$$\begin{cases} S_d U_{dc} = L_f\left(i_{ed} - i_d^*\right) - R_{Lf}i_d + u_d + \omega L_f i_q - k_1(i_d - i_d^*) \\ S_q U_{dc} = L_f\left(i_{eq} - i_q^*\right) - R_{Lf}i_q + u_q - \omega L_f i_d - k_2(i_q - i_q^*) \end{cases} \qquad (23)$$

Since the component of the current value in the dq coordinate system is constant at steady state, the current differential term is 0, and the BM model with S_d and S_q as the control quantity and i_d and i_q as the controlled quantity can be obtained as follows:

$$\begin{cases} S_d = \frac{1}{U_{dc}}\left[u_d - R_{Lf}i_d + \omega L_f i_q - k_1(i_d - i_d^*)\right] \\ S_q = \frac{1}{U_{dc}}\left[u_q - R_{Lf}i_q - \omega L_f i_d - k_2(i_q - i_q^*)\right] \end{cases} \qquad (24)$$

IV. CONTROLLER STABILITY ANALYSIS

According to the Lyapunov stability criterion, the energy function of the designed controller is as follows:

$$E = \frac{1}{2} L_f i_d^{2} + \frac{1}{2} L_f i_q^{2} \qquad (25)$$

And only when i_d, i_q is 0 at the same time E is 0, so $E>0$ The energy function is positively definite. Further the derivative of E is obtained, and its derivative is as follows:

$$\dot{E} = L_f i_d \dot{i}_d + L_f i_q \dot{i}_q \qquad (26)$$

Combining (10) yields the following equation:

$$\dot{E} = -R_d\left(i_d^2 + i_q^2\right) + \left(u_d - e_d\right)i_d + \left(u_q - e_q\right)i_q \qquad (27)$$

From the structure of Fig. 1, we can see that $e_{dq} > u_{dq}$ and $R_d>0$, $\dot{E}$ is less than 0, and it is negatively . So the system is stable by the Lyapunov stability criterion.

V. SIMULATION

A. Simulate System Parameter Settings

The performance of improved VSG system requires further verification, so the system model as shown in Fig. 1 was set up in Matlab compared with the traditional VSG system, and the parameter value of the model are displayed in the following table ('*' represents the normalized value of the parameters):

TABLE I. SYSTEM PARAMETERS

Parameter	Value	Parameter	Value
DC power U_{dc}/V	1000	Damping coefficient D^*/pu	103.18
Support Capacitance C/pF	5.6	Active power droop coefficient k_p/pu	30.25
Filter Capacitance C_f/uF	35	Reactive power droop coefficient k_q/pu	10.82
Filter inductance L_f/mH	3	Reactive power integral coefficient k_i^*/pu	115.47
Equivalent resistance R_{Lf}/Ω	0.5	Active power set value P_{ref}/MW	170
Terminal Voltage U_n/V	380	Reactive power set value Q_{ref}/KW	0
Power system frequence f/Hz	50	Damping coefficient k_1/pu	50
Moment of Inertia J^*/pu	3.36	Damping coefficient k_2/pu	50

B. Analysis of Simulation Result

Fig. 2 is the grid-side phase voltage and current during the operation of the improved VSG system, the improved VSG system indirectly controls the current value of the system on the grid side by controlling the given value of the power. Under the given system operating conditions in this paper, the grid voltage amplitude is about 311V, and the expected current amplitude value can be calculated to be about 365A, and according to Fig. 2 the system realizes the tracking control of the current on the network side, and is able to run smoothly.

979-8-3315-1110-4/25 $31.00 © 2025 IEEE

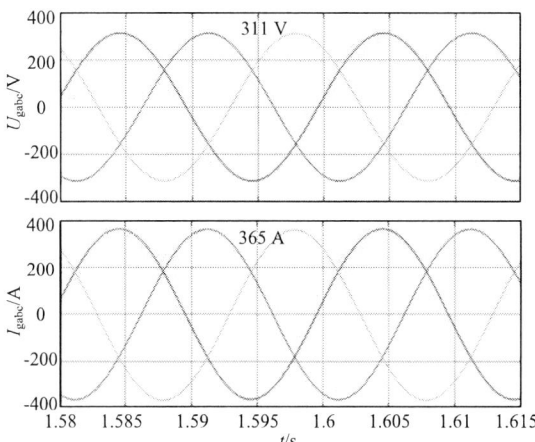

Fig. 2. System grid-side phase voltage and current

C. Simulation System Performance Verification

1) Set the grid frequency to decrease by 0.1Hz in 2-3.5 seconds

Fig. 3 is the active power response and frequency response curve of the two systems, and Fig. 3 shows that the system increases the active power to simulate the rotor to release kinetic energy to provide inertial support for the system while the power grid frequency decreases at 2 seconds. In addition, the improved VSG system has a stronger hindrance effect on its frequency change, and its frequency is reduced to 49.9Hz for the first time with a lag of 0.1 seconds compared with the traditional VSG system, providing more adequate frequency inertia support, and the overshoot of the frequency change is much smaller than that of the traditional VSG system.

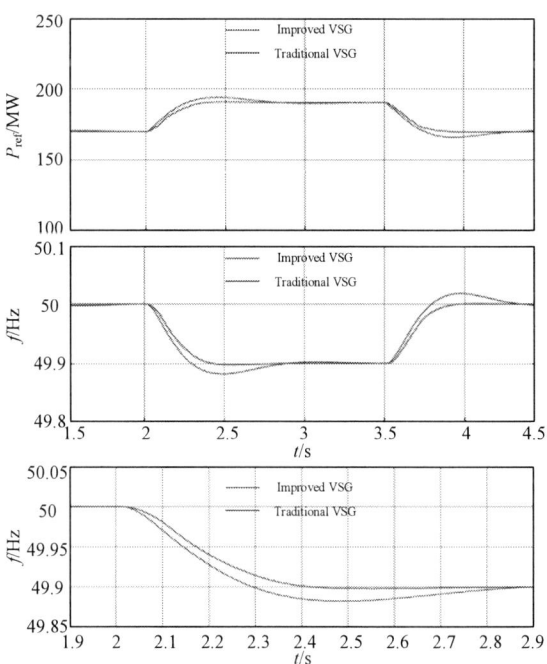

Fig. 3. System power response and frequency response

2) Set the grid voltage to drop by 20% in 1.5-1.7 seconds.

Fig. 4 is a waveform of the improved VSG system, where U_g is the grid voltage and U_{gabc} is the output voltage at the point of connection. According to the Fig. 4, when the grid voltage is reduced to 248V, the voltage at the grid-connected point is only reduced to 290V, and the system has a certain voltage support capacity to cope with voltage fluctuations to ensure voltage stability.

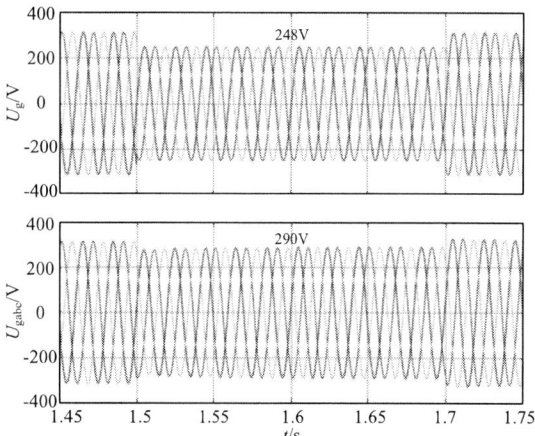

Fig. 4. Improved VSG system voltage waveform

VI. CONCLUSION

In this paper, based on BM theory an improved control strategy for grid forming type VSG system is proposed，and so as to verify the effectiveness of the improved VSG system, Compared with the traditional VSG system, the simulation results are obtained in Matlab, and the following conclusions are drawn from the analysis:

(1) The improved VSG system designed in this paper has a better effect on suppressing the rapid mutation of frequency when dealing with the power system frequency fluctuation, which provides more sufficient inertia support for the grid-forming system, and has better frequency stability and anti-interference ability.

(2) The improved VSG system not only improves the frequency stability, but also has better dynamic performance and good voltage support ability to cope with voltage mutations and disturbances, which provides a guarantee for the stable operation of the grid-forming system.

REFERENCES

[1] H. Sun, B. Wang, et al, " Research on Inertia System of Frequency Response for Power System With High Penetration Electronics," Proceedings of the CSEE, Vol.40, No.16, pp. 5179-5192, Aug. 20, 2020.

[2] W. Cai, D. Sun, M. Zhou, Q. Guo and H. Gao, "The development status of the third generation wide bandgap power semiconductor and its application," Science & Technology Review, pp. 42-55, 2021.

[3] X. Zhang, X. Zhan, et al, " Review on Grid-following/Grid-forming Hybrid Mode Control for Grid-connected Converter in High-penetration Renewable Energy Generation," Automation of Electric Power Systems, pp. 1-15, July 2024.

[4] Q. Zhong, Weiss G, "Synchronverters；Inverters That Mimic Synchronous Generators," IEEE Transactions on Industrial Electronics, Vol. 58, pp. 1259-1267, April 2011.

[5] Du W , Chen Z , Schneider K P , et al, "A Comparative Study of Two Widely Used Grid-Forming Droop Controls on Microgrid Small Signal Stability," IEEE Journal of Emerging and Selected Topics in Power Electronics, Vol.8, pp. 963-975, June 2020.

[6] Narsaiah K , Srinivas T, "A Unified Control Strategy for Three-Phase Inverter in Distributed Generation," International Journal of Innovative Science and Modern Engineering (IJISME), Vol. 29, No. 3, pp.1-5, March 2014.

[7] H. Liu, S. Gao, et al, "Small Signal Stability Analysis of Grid-Connected Photovoltaic Virtual Synchronous Generators," Journal of Solar Energy, Vol. 42, No. 2, pp. 417-424, Feb 2021.

[8] J. Guo, Y. Fan, "Adaptive control strategy for VSG parameters based on improved particle swarm optimization," Journal of Electrical Engineering & Control, Vol. 26, No. 6, pp. 72-82, June 2022.

[9] Z. Gao, J. Zhao, X. Yang, F. Yao and J. Fang, "Adaptive control strategy of VSG moment of inertia and damping coefficient based on RBF," Electric Power Construction, Vol. 43 No. 9, pp. 132-139, Sep 2022.

[10] X. Xiao, H. Deng, M. Zou, et al, "Research on Virtual Inertia Flexible Control of Photovoltaic Storage VSG Based on AHFS and RBF," Electric Power, pp. 1-12, Feb 2025.

[11] C. Wen, D. Chen, C. Hu, Z. Piao, J. Zhou, "Self-adaptive Control of Rotational Inertia and Damping Coefficient of VSG for Converters in Microgrid," Automation of Electric Power System, Vol. 42, No. 17, pp. 120-126, Sept 2018.

[12] Y. Yang, F. Mei, C. Zhang, et al, " Coordinated adaptive control strategy of rotational inertia and damping coefficient for virtual synchronous generator," Electric Power Automation Equipment, Vol. 39, No. 3, pp. 125-131, Mar 2019.

[13] S. Wang, J. Di, P. Hao, L. Kong, "Speed Feedback Based Adaptive Control Strategy for VSG Inertia and Damping," Power Grid Analysis & Study, Vol. 53 No. 3, pp. 53-61,2025.

[14] Brayton R K, Moser J K, "A theory of nonlinear networks," Quarterly of Applied Mathematics, Vol. 22, No. 2, pp. 1-33, July 1964.

[15] Liu. X, Y. Zhou, W. Zhang, S. Ma, "Stability criteria for constant power loads with multistage LC filters," IEEE Transactions on Vehicular Technology, Vol. 60, pp. 2042-2049, 2011.

[16] Marx D, Magne P, Nahid-Mobarakeh B, Pierfederici S, Davat B, "Large signal stability analysis tools in DC power systems with constant power loads and variable power loads—A review," IEEE Transactions on Power Electronics, Vol. 27, pp.1773-1787, 2011.

[17] Kumari SHIPRA, Rakesh MAURYA, Shambhu N. SHARMA, "Brayton-Moser Passivity Based Controller for Electric Vehicle Battery Charger," CPSS Transactions on Power Electronics and Applications, Vol. 6, No. 1, pp. 40-51, March 2021.

Baseplate Temperature Gradient-Based Health Status Monitoring for Power Module Bonding Wires

Yongxin Chen
School of Electrical Engineering and
Automation and Anhui Province
Engineering Research Center for
Advanced Power Electronics and
Energy Conversion (APEEC), Anhui
University
Hefei, China
z23301208@stu.ahu.edu.cn

Lei Xu
School of Electrical Engineering and
Automation and Anhui Province
Engineering Research Center for
Advanced Power Electronics and
Energy Conversion (APEEC), Anhui
University
Hefei, China
z24301270@stu.ahu.edu.cn

Kun Tan
School of Electrical Engineering and
Automation and Anhui Province
Engineering Research Center for
Advanced Power Electronics and
Energy Conversion (APEEC), Anhui
University
Hefei, China
k.tan@ahu.edu.cn

Xi Tang
Institutes of Physical Science and
Information Technology and Anhui
Province Engineering Research Center
for Advanced Power Electronics and
Energy Conversion (APEEC), Anhui
University
Hefei, China
xitang@ahu.edu.cn

Cungang Hu
School of Electrical Engineering and
Automation and Anhui Province
Engineering Research Center for
Advanced Power Electronics and
Energy Conversion (APEEC), Anhui
University
Hefei, China
hcg@ahu.edu.cn

Wenping Cao
School of Electrical Engineering and
Automation and Anhui Province
Engineering Research Center for
Advanced Power Electronics and
Energy Conversion (APEEC), Anhui
University
Hefei, China
wpcao@ahu.edu.cn

Abstract—**The widespread adoption of power modules in power electronics has elevated reliability to a critical research priority. Bonding wires, serving as pivotal components within these modules, profoundly influence performance and longevity when fractured or lifted off. Complete fracture of all bonding wires on a chip occurs after long-term operation, which may be accelerated under extreme operational conditions, inducing internal current redistribution and consequential shifts in temperature distribution. To mitigate this challenge, this study introduces a method for monitoring the health status of power module bonding wires through variations in baseplate temperature gradients. Unlike conventional approaches, the proposed method monitors the bonding-wire degradation fault via externally measuring temperatures on the baseplate; this non-invasive technique provides both effectiveness of monitoring and ease of implementation. By conducting finite-element analysis, the correlation between baseplate thermal distribution patterns and health status of bonding wires is investigated. Experimental results confirm that the proposed method effectively identifies and locates complete lift-off of all bonding wires on a chip, enabling real-time status assessment of power modules for high-reliability application scenarios and offering a practical way for power converter system maintenance.**

Keywords— temperature gradient, power module, bonding wires, health status

I. INTRODUCTION

In the domains of new-energy power generation, high-speed railways, and aerospace, power modules are pivotal components for electrical energy conversion and control, with their reliability being paramount to the stable operation of power systems. Bonding wires, which serve as critical interconnects between internal chips and external circuitry, directly govern module performance and longevity. Recent advancements in power electronics technology have driven continuous increases in operating frequency, power density, and integration levels of power modules, imposing stringent demands on bonding wire reliability. During operation, Joule

heating from active chips elevates internal module temperatures. Under prolonged operational conditions, lift-off of bonding wires from an individual chip within a power module exerts a significant influence on current density and temperature distribution across the power module. In severe scenarios, such failure mechanisms may ultimately lead to module performance degradation or catastrophic failure, potentially compromising the operational integrity of power systems. Notably, empirical studies indicate that bonding wire failures account for approximately 70% of power module malfunctions [1]. Consequently, real-time and precise monitoring of bonding wire integrity is imperative to safeguard power system reliability.

Currently, the methods used to monitor the health status of bonding wires in power modules can be primarily categorized into two approaches: thermal parameter-based methods and electrical parameter-based methods. Thermal Parameter-Based Methods evaluate the fatigue aging of bonding wires by analyzing chip junction temperature, temperature distribution, and thermal resistance. Reference [2] monitors bonding wire aging by leveraging the discrepancy between junction temperature and temperature-sensitive electrical parameters. Reference [3] explored the impact of progressive bonding wire degradation on transient thermal impedance, developed an electrothermal coupling model, and conducted numerical simulations. The study revealed that bonding wire aging not only increases the forward conduction voltage drop but also elevates the comprehensive surface temperature of the chip, with measured transient thermal resistance showing a corresponding rise. However, challenges persist in accurately assessing bonding wire conditions due to difficulties in directly measuring chip junction temperatures under operational conditions and the complexity of acquiring transient thermal impedance data [4],[5]. Electrical parameter-based methods assess module health by monitoring electrical parameters linked to bonding wire integrity. When bonding wires break, physical characteristics such as short-circuit current, turn-on/turn-off delay times, threshold voltage and on-state voltage drop will change. Reference [6] demonstrated the correlation between transconductance variations and bonding wire degradation levels, proposing a monitoring strategy for IGBT modules based on this relationship. Reference [7] investigated the influence of bonding wire faults on gate input capacitance, translating wire health status into variations in gate charge duration time, thereby introducing a fatigue monitoring method for multi-chip IGBT modules. Reference [8] analyzed magnetic field changes near bonding wires during detachment by using finite element simulations to identify regions with maximal magnetic field variation rates. By deploying sensors in these areas, an online monitoring method based on magnetic field measurements was proposed. While these studies provide valuable insights, practical applications face limitations including measurement environment constraints, equipment costs, and complexity in parameter acquisition [9]-[15].

Existing monitoring approaches exhibit inherent limitations in achieving accurate real-time assessment of bonding wire status. This paper proposes a non-invasive baseplate temperature gradient-based method for health monitoring. Based on the actual structure of the power module, a three-dimensional simulation model is constructed. Based on the established model, the effects of complete lift-off of bonding wires on the thermal distribution and characteristics of power module baseplate under extreme or prolonged operational conditions are systematically investigated and analyzed. The study demonstrates that monitoring the baseplate temperature gradient provides a viable approach for detecting the detachment of bonding wires. This methodology effectively reflects the structural integrity of bonding wires while offering practical advantages in implementation, as it requires neither specialized module encapsulation design nor destructive interventions.

II. Temperaure Gradiant Mechanism Of Power Modules

In the design and evaluation process of power modules, temperature distribution characteristics serve as a critical indicator. Temperature gradient, as a manifestation of thermal behavior in modules, offers more accessible measurement parameters compared to internal peak temperatures, particularly the temperature gradient at the baseplate's bottom plane. The fundamental principles of analyzing module thermal behavior through temperature gradients are outlined below:

Heat transfer occurs from high-temperature to low-temperature regions. Within a continuous temperature field, the direction of the most rapid temperature changes at any specific location and its rate of variation defines the temperature gradient at that point. The temperature gradient ∇T between two points within the same plane can be approximately represented as:

$$\nabla T_{ij} = \frac{T_i - T_j}{L_i - L_j} \qquad (1)$$

In equation (1-1), T_i and T_j represent the temperatures of two points, while L_i and L_j denote the spatial positions of two points.

For non-steady-state heat conduction in isotropic materials:

$$\frac{\partial^2 T}{\partial r^2} = \frac{c \cdot \rho}{k} \cdot \frac{\partial T}{\partial t} \qquad (2)$$

where T is temperature, t is time, ρ is density, k is thermal conductivity and c is the Specific Heat Capacity.

For non-steady-state electrical conduction within modules:

$$\frac{\partial^2 U}{\partial r^2} = RC \frac{\partial U}{\partial t} \qquad (3)$$

where U is voltage, C is capacitance per unit volume, R is resistance per unit length, and t is time.

By comparing Equations (2) and (3), the energy transfer equations between thermal and electrical fields exhibit structural similarity. This analogy allows thermal conduction equations to be solved using methods analogous to electric field analysis:

$$\frac{\partial^2 \phi}{\partial r^2} = a^2 \frac{\partial \phi}{\partial t} \qquad (4)$$

where a is the transmission coefficient. In the electric field, a is $\sqrt{RC}$, and in the temperature field, a is $\sqrt{cp/k}$.

The parameters in the electric field, such as voltage U, resistance R, capacitance C, charge Q, current I, and current density J, correspond to temperature T, thermal resistance R_{th}, thermal capacitance C_{th}, thermal energy Q_{th}, heat flow P, and heat flux density q in the temperature field, respectively. Similarly, the electric field strength E has a corresponding temperature parameter. Combining the following equation:

$$\begin{aligned} q &= -\lambda \nabla T \\ J &= -\lambda \nabla U \end{aligned} \qquad (5)$$

This correspondence implies that ∇T analogous to electric field strength E. Field strength E can serve as a critical parameter for characterizing thermal performance in devices. Just as E is pivotal in insulation design, temperature gradients ∇T provide insights into heat dissipation efficiency and material behavior under operational stresses.

For power modules, the thermal resistance R_{th} between any two micro-units A and B is defined as:

$$R_{th} = \frac{T_A - T_B}{P} = \frac{d\delta}{kA} \qquad (6)$$

where T_A and T_B are the temperatures of micro-units A and B, P is the power loss, and $d\delta$ represents the micro-unit size. It can also be written as :

$$\nabla T = \alpha P = P / \lambda A \qquad (7)$$

The chip region of the module serves as the primary heat source, with most of the generated heat being conducted through the baseplate to the heat sink. Since the baseplate plays a critical role in heat dissipation, the health status of internal bonding wires can be characterized by analyzing variations in the baseplate's temperature gradient.

For power modules operating under prolonged power cycling conditions, bonding wires are prone to aging or fracture due to thermal fatigue. Under prolonged or extreme operating conditions (e.g., high current loads or elevated temperatures), all bonding wires on a single chip may lift off. In such cases, the internal current redistributes, altering the temperature distribution within the module. These changes ultimately manifest as shifts in the baseplate temperature profile. Leveraging this phenomenon, real-time monitoring of temperature gradient variations between designated points on the baseplate enables effective early warning and diagnosis of bonding wire degradation, thereby enhancing fault detection capabilities in power modules.

III. FEA SIMULATION OF MODULE TEMPERATURE DISTRUBUTION

To validate the effectiveness of the proposed method, a finite element simulation study is conducted to analyze the correlation between baseplate thermal distribution patterns and the health status of bonding wires. A three-dimensional model of the power module is established, which includes bonding wires, chips, solders, copper layers, ceramic layers, and baseplate, as shown in Fig. 1(a). Mesh generation is carried out in the finite element simulation software using a multi-level meshing approach. For components with smaller dimensions (such as bonding wires, chips, and solder layers), mesh refinement is performed during the meshing process. In contrast, for larger-sized components like copper layers and ceramic layers, standard mesh division is adopted. The mesh generation is shown in Fig. 1(b).

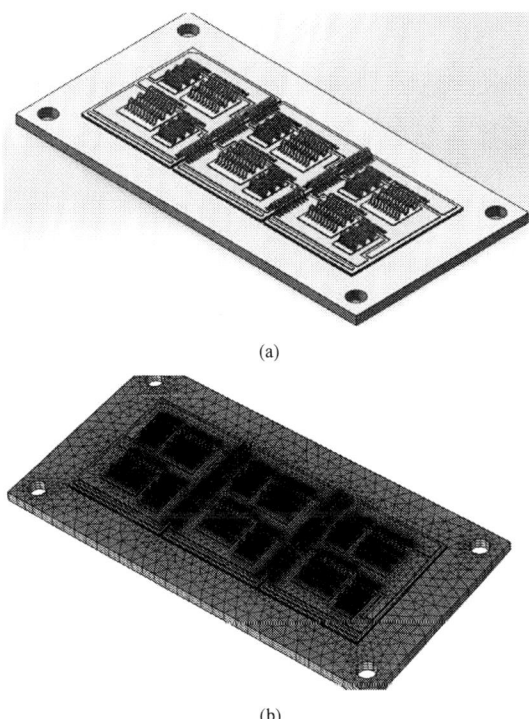

(a)

(b)

Fig. 1. Three-dimensional model of the power module: (a) Multilayer structure; (b) Mesh generation of the power module.

The chips in the power module are labeled as Chip 1 to Chip 3, with temperature monitoring points designated as P_1, P_2, and P_3, as illustrated in Fig. 2. During the simulation, variations of baseplate temperature distribution are modeled under both healthy conditions and the failure of the chip (All bonding wires in a chip are lifted off). Through finite element simulations of the power module's thermal behavior under different fault states, the following findings are obtained:

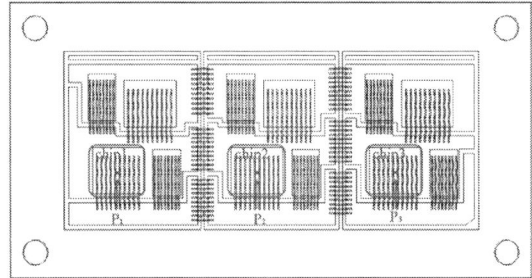

Fig. 2. chip numbering and monitoring point locations.

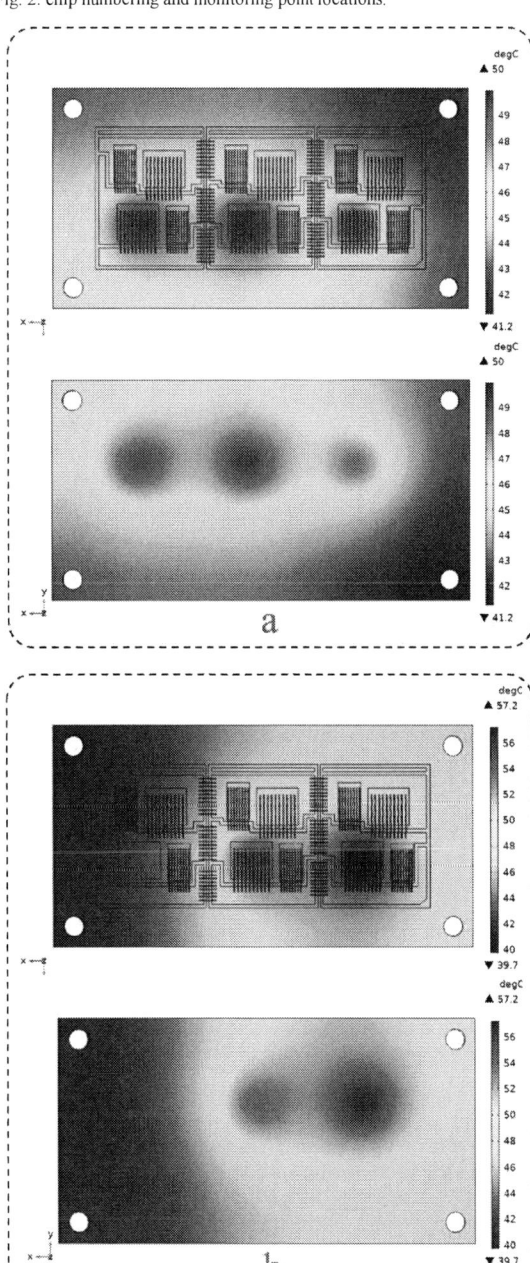

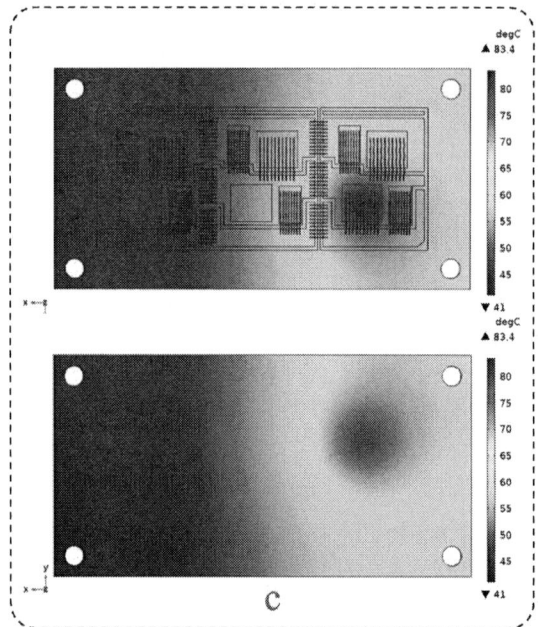

Fig. 3. Temperature distribution on upper and lower baseplate surfaces of power module under different bonding wire conditions: (a) healthy status; (b) all bonding wires on Chip 1 were lifted off; (c) all bonding wires on Chip 1 and Chip 2 were lifted off

Fig. 3 illustrates the thermal simulation outcomes of the power module. In Fig. 3(a), the temperature distribution under healthy conditions on the baseplate exhibits uniformity, providing a reference standard for the module's normal operational state.

As illustrated in Fig. 3(b), the baseplate temperature distribution undergoes significant modification when complete bonding wire detachment occurs on Chip 1. Comparative analysis of Fig. 4 and Fig. 5 reveals that the temperature at monitoring point P_1 beneath Chip 1 decreases by approximately 6°C, while temperatures at points P_2 and P_3 under Chip 2 and Chip 3 exhibit increases of 5°C and 8°C, respectively. The thermal gradient ∇T_{12} between monitoring points P_1 and P_2 demonstrates a substantial alteration, shifting from its initial value of -0.11°C/cm to -3.67°C/cm. These observations suggest that variations in thermal gradients can effectively indicate the status of bonding wires in Chip 1.

When all bonding wires on Chip 2 are lifted off, the temperature at monitoring point P_1 increased by approximately 4°C, while P_2 exhibits a temperature decrease of approximately 3°C. Notably, monitoring point P3 records a temperature rise of approximately 5°C. The temperature gradient ∇T_{12} shifts from its initial value of -0.11°C/cm to 2.33°C/cm. The status of the chip bonding wires can be effectively monitored through systematic observation of temperature gradient variations.

When all bonding wires on Chip 3 were lifted off, the baseplate temperature variation pattern demonstrates similarity to the complete bonding wire detachment scenario observed in Chip 1, and detailed descriptions are omitted here to avoid redundancy. The temperature gradient ∇T_{23} exhibits a significant increase from 0.44°C/cm to 3.99°C/cm, which can be effectively employed to monitor the bonding wire integrity of this chip.

When all bonding wires on Chips 1 and 2 become detached, the temperature at monitoring point P_1 decreases by approximately 5°C, while P_2 and P_3 exhibit temperature increases of 4°C and 32°C, respectively. The temperature gradient ∇T_{23} transforms from 0.44°C/cm to -8.55°C/cm, with its magnitude demonstrating a 19-fold amplification compared to the original value. Although the alterations in ∇T_{12} and ∇T_{13} are less substantial than those in ∇T_{23}, they still present marked deviations from normal operating conditions. The monitoring of multiple chip failure conditions can likewise be achieved through systematic observation of temperature gradient variations.

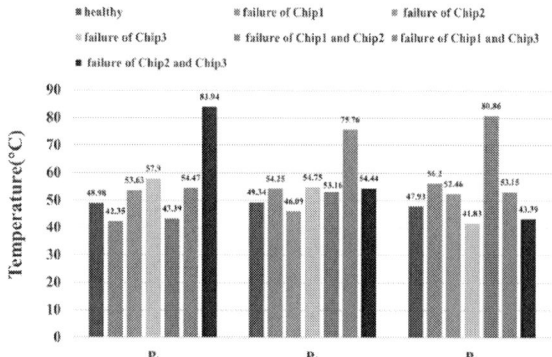

Fig. 4. Correlation between the temperature of monitoring points and the status of bonding wires.

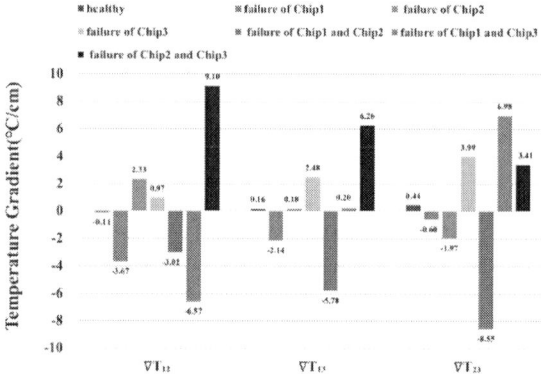

Fig. 5. Correlation between the temperature gradient of monitoring points and the status of bonding wires.

IV. EXPERIMENTAL VALIDATION PROPOSED METHOD

The thermal testing experimental platform for power modules is established, comprising essential components including an IGBT power module, constant current source, constant voltage supply, thermocouples, infrared thermal imager, and water-cooling system. Thermal characterization of the power module is conducted under various experimental conditions. For instance, by systematically varying the current levels, the temperature distribution characteristics within the power module chip and the temperature gradient across the baseplate can be comprehensively investigated.

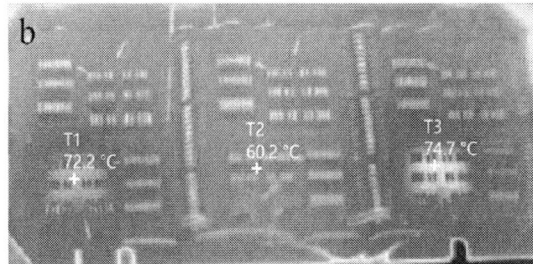

Fig. 6. Thermal distribution images of the power module when I_D=60A captured by infrared thermal imaging: (a) healthy condition; (b) all bonding wires fracture in Chip 2 are lifted off; (c) all bonding wires fracture in Chips 2 and 3 are lifted off.

As can be seen from Fig. 6, when I_D =60 A, the junction-temperature differences among the three chips in the power module under healthy conditions are very small. When all the bonding wires of Chip 2 are lifted off, the junction temperature of Chip 2 decreases by about 5°C while the junction temperatures of Chip 1 and Chip 3 increase by about 8°C and 9°C respectively. When all the bonding wires of Chip 2 and Chip 3 are lifted off, the junction temperature of Chip 1 increases by about 47°C, and the junction temperatures of Chip 2 and Chip 3 increase by about 13°C .

Meanwhile, the changes in the baseplate temperature gradient under the three conditions are shown in Fig. 7. Under healthy conditions, the baseplate temperature gradient is very small. When the bonding wires of Chip 2 are lifted off, the temperature gradient ∇T_{23} shifts from -0.12°C/cm to -0.988°C/cm, exhibiting an eightfold increase in magnitude compared to normal conditions, which are obviously different from the normal situation. The reason why ∇T_{13} changes little is that when the bonding wires of Chip 2 are lifted off, the junction temperatures of both Chip 1 and Chip 3 increase, resulting in little change in the baseplate temperature gradient ∇T_{13}. When all bonding wires on Chip 2 and Chip 3 are completely lifted off, the temperature gradients ∇T_{12}, ∇T_{13}, and ∇T_{23} all exhibit a sharp increase. Notably, ∇T_{12} shows a dramatic shift from -0.22°C/cm to 3.27°C/cm, representing a 14-fold amplification in magnitude compared to its initial value.

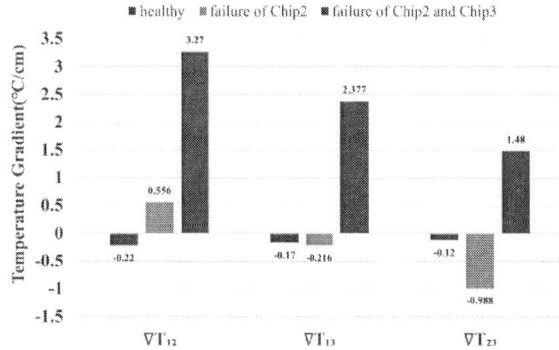

Fig. 7. Comparative analysis of baseplate temperature gradients under varying conditions at I_D=60A.

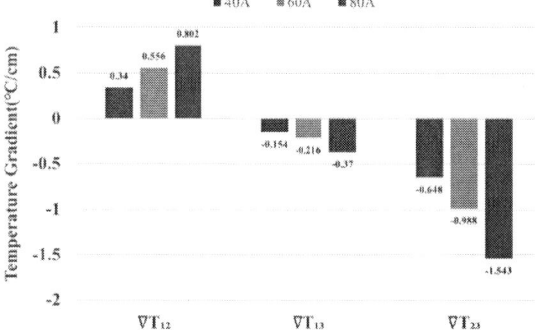

Fig. 8. Comparative analysis of baseplate temperature gradients under varying current conditions when all bonding wires are lifted off in Chip 2.

As shown in Fig. 8, the amplitude of the temperature gradient ∇T progressively increases with the elevation of current I_D. This suggests that when all bonding wires detach from a chip in the power module, the variation in the baseplate temperature gradient ∇T exhibits a dependence on the magnitude of the applied current, where enhanced current application correlates with amplified gradient magnitude.

V. CONCLUSION

This study proposes a non-invasive health state monitoring method for power module bonding wires using baseplate temperature gradients. The results demonstrate that under healthy conditions, the baseplate exhibits uniform temperature distribution with minimal temperature gradients. When all bonding wires on a specific chip are disconnected, significant changes occur in both temperature distribution and temperature gradients. For instance, all bonding wires on Chip 2 lift-off result in a junction temperature decrease of approximately 5°C for Chip 2, while junction temperatures of Chips 1 and Chip 3 increased by about 8°C and 9°C, respectively. The temperature gradients ∇T_{12} and ∇T_{23} exhibit significant variations, with the ∇T_{23} value demonstrating a particularly notable deviation from typical conditions, increasing from -0.12 °C/cm to -0.988 °C/cm.

The proposed method offers advantages of non-invasive implementation and real-time monitoring capabilities. The proposed method demonstrates the capability to effectively identify and locate defective chips exhibiting complete lift-off of bonding wires, thereby enabling real-time condition monitoring of power modules in high-reliability applications.

It provides a practical solution for maintenance strategies in power converter systems, offering critical operational insights for reliability-critical implementations.

ACKNOWLEDGMENT

This work was supported in part by Anhui Provincial Natural Science Foundation under Grant 2308085QE180, and in part by the National Natural Science Foundation of China under Grant 52377035.

REFERENCES

[1] C. Tu, H. Xu, B. Xiao, L. Long, and M. Chai, "Study on the failure of IGBT bonding wire based on temperature gradient," in IECON 2020 The 46th Annual Conference of the IEEE Industrial Electronics Society, Singapore, Singapore: IEEE, Oct. 2020, pp. 3011 – 3016. doi: 10.1109/IECON43393.2020.9255166.

[2] K. Wei, W. Wang, Z. Hu, and M. Du, "Condition Monitoring of IGBT Modules Based on Changes of Thermal Characteristics," IEEE Access, vol. 7, pp. 47525–47534, 2019, doi: 10.1109/ACCESS.2019.2909928.

[3] J. Gao et al., "Analysis of the IGBT Module with Bonding Wire Failure Based on Transient Thermal Impedance," J. Phys. Conf. Ser., vol. 2276, no. 1, p. 012033, May 2022, doi: 10.1088/1742-6596/2276/1/012033.

[4] B. Gao et al., "A Temperature Gradient-Based Potential Defects Identification Method for IGBT Module," IEEE Trans. Power Electron., vol. 32, no. 3, pp. 2227–2242, Mar. 2017, doi: 10.1109/TPEL.2016.2565701.

[5] Q. Li et al., "Review of the Failure Mechanism and Methodologies of IGBT Bonding Wire," IEEE Trans. Compon. Packag. Manuf. Technol., vol. 13, no. 7, pp. 1045–1057, Jul. 2023, doi: 10.1109/TCPMT.2023.3297224.

[6] K. Wang, L. Zhou, P. Sun, and X. Du, "Monitoring Bonding Wire Defects of IGBT Module Using Module Transconductance," IEEE J. Emerg. Sel. Top. Power Electron., vol. 9, no. 2, pp. 2201–2211, Apr. 2021, doi: 10.1109/JESTPE.2020.2973348.

[7] K. Wang, L. Zhou, P. Sun, and X. Du, "Monitoring Bonding Wires Fatigue of Multichip IGBT Module Using Time Duration of the Gate Charge," IEEE Trans. Power Electron., vol. 36, no. 1, pp. 888–897, Jan. 2021, doi: 10.1109/TPEL.2020.3005183.

[8] W. Guo, G. Xiao, and L. Wang, "Online Condition Monitoring of Bonding Wires Lift-Off in Power Modules Based on Magnetic Field Measurement," IEEE Trans. Power Electron., vol. 40, no. 3, pp. 4425–4436, Mar. 2025, doi: 10.1109/TPEL.2024.3503722.

[9] M. Jiang et al., "Finite Element Modeling of IGBT Modules to Explore the Correlation between Electric Parameters and Damage in Bonding Wires," in 2019 IEEE Energy Conversion Congress and Exposition (ECCE), Baltimore, MD, USA: IEEE, Sep. 2019, pp. 839–844. doi: 10.1109/ECCE.2019.8912236.

[10] X. Huang, M. Du, H. Fu, and S. Gao, "An Aging Monitoring Method of Bonding Wires Based on Voltage Ringing Frequency Characteristics in IGBT Modules," IEEE Trans. Device Mater. Reliab., vol. 24, no. 2, pp. 344–353, Jun. 2024, doi: 10.1109/TDMR.2024.3394517.

[11] H. Liu, F. Wang, X. Zhang, W. Xia, and L. Ren, "An Online Monitoring Method for Bonding Wire Fatigue in IGBT Module," IEEE J. Electron Devices Soc., vol. 12, pp. 440–449, 2024, doi: 10.1109/JEDS.2024.3399554.

[12] P. Sun, C. Gong, X. Du, Y. Peng, B. Wang, and L. Zhou, "Condition Monitoring IGBT Module Bonding Wires Fatigue Using Short-Circuit Current Identification," IEEE Trans. Power Electron., vol. 32, no. 5, pp. 3777–3786, May 2017, doi: 10.1109/TPEL.2016.2585669.

[13] J. Chen, E. Deng, P. Liu, S. Yang, and Y. Huang, "The Influence and Application of Bonding Wires Failure on Electrothermal Characteristics of IGBT Module," IEEE Trans. Compon. Packag. Manuf. Technol., vol. 11, no. 9, pp. 1426–1434, Sep. 2021, doi: 10.1109/TCPMT.2021.3102242.

[14] U.-M. Choi, F. Blaabjerg, S. Jorgensen, S. Munk-Nielsen, and B. Rannestad, "Reliability Improvement of Power Converters by Means of Condition Monitoring of IGBT Modules," IEEE Trans. Power Electron., vol. 32, no. 10, Art. no. 10, Oct. 2017, doi: 10.1109/TPEL.2016.2633578.

[15] S. Liu, C. Tu, L. Long, H. Xu, B. Xiao, and Z. Zhu, "A Method Monitoring Healthy State of Bonding Wires in IGBT Based on dV_{CE}/di_C," in IECON 2022 – 48th Annual Conference of the IEEE Industrial Electronics Society, Brussels, Belgium: IEEE, Oct. 2022, pp. 1–6. doi: 10.1109/IECON49645.2022.9968484.

Study on Influencing Factors and Mechanisms of Single-Event Gate Rupture in SiC STP-MOSFETs

* Ying Yang
Department of Electronic Engineering
Xi'an University of Technology
Xi'an, Shaanxi, China
Corresponding author: yangy@xaut.edu.cn

Zixuan Liu
Department of Electronic Engineering
Xi'an University of Technology
Xi'an, Shaanxi, China
2565932926@qq.com

Chen Wang
Department of Electronic Engineering
Xi'an University of Technology
Xi'an, Shaanxi, China
18392610396@163.com

Xulong Wang
Department of Electronic Engineering
Xi'an University of Technology
Xi'an, Shaanxi, China
2230321215@stu.xaut.edu.cn

Abstract—This paper investigates the impact factors, mechanism, and mitigation strategies of single-event gate rupture (SEGR) in Shallow Trench-Planar Silicon Carbide Metal-Oxide-Semiconductor Field-Effect Transistors (SiC STP-MOSFETs), which are effective in reducing specific on-resistance. Compared with conventional planar gate MOSFETs (Con-MOSFETs), STP-MOSFETs exhibit a 17.2% reduction in specific on-resistance. However, it has been observed that STP-MOSFETs are more susceptible to SEGR. Specifically, at a position 0.4 μm away from the cell center and under 90° vertical incidence, the hole accumulation effect at the bottom of the gate oxide layer is most pronounced. When the N-drift region is completely penetrated by heavy ions and the linear energy transfer (LET) value is 0.1 pC/μm, the SEGR threshold voltage of STP-MOSFETs is only 48 V, which is 10.4% lower than that of conventional structures. After reinforcement with high-k dielectric HfO_2, the SEGR threshold voltage is increased by 165%.

Keywords—STP-MOSFET, SEGR, HfO_2 hardening

I. INTRODUCTION

Silicon Carbide (SiC) MOSFETs are widely used in spacecraft DC-DC voltage converters due to their excellent electrical properties. However, they are prone to single-event gate rupture (SEGR) in space radiation environments, leading to device failure.

Conventional planar gate MOSFETs (Con-MOSFETs) experience SEGR at the bottom of the gate oxide layer, with a threshold voltage of 66 V[1]. The sensitive region for particle incidence is located at the gate oxide layer[2]. This phenomenon occurs due to the accumulation of holes at the SiC/SiO_2 interface, generating a transient electric field at the bottom of the gate oxide layer[3]. For trench gate MOSFETs, SEGR occurs at the trench corner[4], with a threshold voltage of 40 V[5]. Studies have investigated the effects of drain bias [6]and surface LET values [4,7]on SEGR and found that the sensitive positions for particle incidence are located in the trench gate region[4]. Our research group has proposed a novel Shallow Trench-Planar MOSFET (STP-MOSFET) for the first time. The introduction of shallow trenches reduces the impact of the JFET region in conventional planar MOSFETs (Con-MOS), which contributes to the high specific on-resistance. The accumulation layer formed under the gate oxide layer in STP-MOSFETs lowers the device's specific on-

resistance, combining the advantages of simple planar gate process and low specific on-resistance of trench gate. However, during the study of STP-MOSFETs, it was found that there is a significant hole charge accumulation under the shallow trench gate oxide layer, which may affect SEGR. This paper focuses on the STP-MOSFET proposed by our research group for the first time, simulates and analyzes the advantages of STP-MOSFET and conventional planar MOSFET in terms of specific on-resistance and SEGR threshold voltage, and investigates the impact factors, mechanism, and mitigation measures of SEGR in STP-MOSFETs. Follow-up work will involve tape-out experiments.

II. DEVICE STRUCTURE AND PARAMETER SETTINGS

Fig. 1 illustrates the device structure and parameters of the SiC STP-MOSFET. In the experiment, the initial incidence parameters are set as follows: heavy ions are vertically incident; the charge track radius is 0.05 μm; the incidence depth is 13 μm; the LET value is set to 0.1 pC/μm; and the incidence position is 0.4 μm away from the cell center.

The physical models used in the experiment include the Shockley-Read-Hall (SRH) recombination model, Auger recombination model, bandgap model, impact ionization model, incomplete ionization model of impurities, carrier mobility model, Okuto-Crowell model, carrier scattering model, and high-field velocity saturation mobility model.

Fig. 2 shows the output characteristic curves of Con-MOS and STP-MOS when the gate voltage (Vg) is 15 V and 20 V, respectively. It is observed that the output characteristics of STP-MOS are superior to those of Con-MOS. When STP-MOS is conducting, an accumulation layer is formed on both sides of the trench, reducing the resistance of the JFET region and thus decreasing the device's specific on-resistance, resulting in better conduction capability. The specific on-resistance of STP-MOS is calculated to be 4.85 mΩ·cm² at Vg = 20 V and V_{DS} = 2 V, while that of Con-MOS is 5.86 mΩ·cm². Therefore, the specific on-resistance of STP-MOS is reduced by 17.2% compared with Con-MOS.

III. STUDY ON THE MECHANISM OF SINGLE-EVENT GATE RUPTURE

When the gate-source voltage is 0 V, the device is in the cut-off state. Fig. 3 and Fig. 4 presents the variations of the maximum electric field strength in the gate oxide layer of

This work was supported by the Shaanxi Provincial Key Research and Development Program (Grant No.2025CY-YBXM-146) and the National Natural Science Foundation of China (Grant No.62174134)

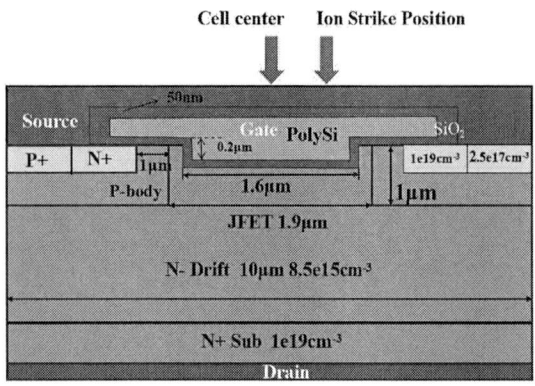

Fig. 1. Schematic of the STP-MOSFET Device Structure.

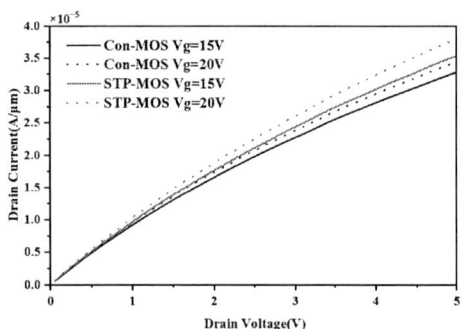

Fig. 2. Comparison of Output Characteristic Curves between STP-MOS and Con-MOS.

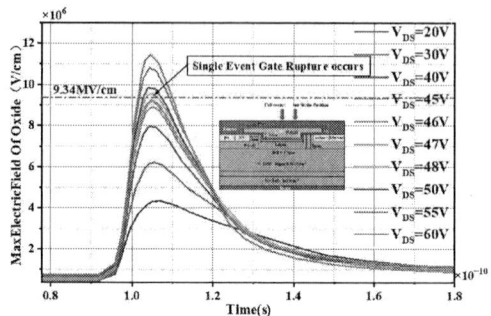

Fig. 3. Variation of Drain-Source Voltage and Maximum Electric Field in the Gate Oxide Layer of STP-MOSFET with Time.

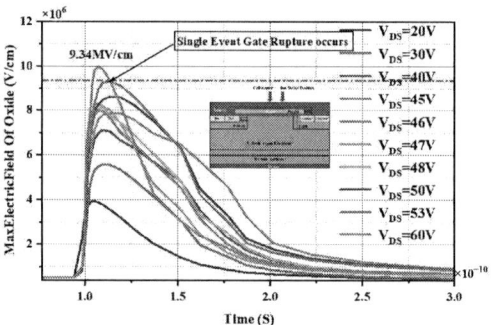

Fig. 4. Variation of Drain-Source Voltage and Maximum Electric Field in the Gate Oxide Layer of Con-MOSFET with Time.

STP-MOS and Con-MOS with time when the drain-source voltage (V_{DS}) ranges from 20 V to 60 V, respectively. As shown in Fig. 3, when $V_{DS} \geq 48$ V, the maximum electric field strength in the gate oxide layer reaches 9.40 MV/cm, which exceeds its breakdown field strength (9.34 MV/cm). This indicates that the SEGR threshold voltage of the STP-MOSFET device is 48 V. As depicted in Fig. 4, when $V_{DS} \geq 53$ V, the maximum electric field strength in the gate oxide layer is 9.35 MV/cm, surpassing its breakdown field strength (9.34 MV/cm), suggesting that the SEGR threshold voltage of the Con-MOSFET device is 53 V. The SEGR threshold voltage of STP-MOS is 10.4% lower than that of Con-MOS, making it more susceptible to SEGR under the same conditions, although the difference is not significant.

To analyze the mechanism of premature breakdown in STP-MOSFETs, Fig. 5 illustrates the distribution of hole current density in the gate oxide layer and internal electric field strength of the STP-MOSFET device at different times ($t = 9.5 \times 10^{-11}$ s, $t = 1 \times 10^{-10}$ s, $t = 1.05 \times 10^{-10}$ s, and $t = 1.48 \times 10^{-10}$ s) when $V_{DS} = 48$ V. It can be seen that as time progresses, holes gradually accumulate at the bottom of the gate oxide layer, indicating that heavy ion incidence leads to hole accumulation beneath the oxide layer. In Fig. 5(c), the hole concentration beneath the oxide layer reaches 1.25×10^8 A/cm², which is 10^6 times higher than that in the bulk. As time goes on, the maximum electric field strength in the gate oxide layer expands from the bottom of the gate oxide layer towards the trench corner. In Fig. 5(g), the electric field beneath the oxide layer reaches 9.27 MV/cm, which is 40 times higher

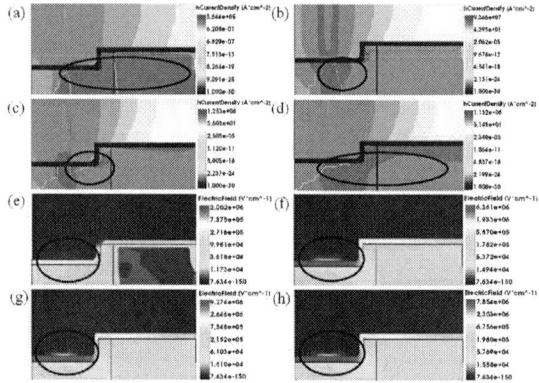

Fig. 5. The time-dependent distribution of hole current density(a) $t=9.5 \times 10^{-11}$s; (b) $t=1 \times 10^{-10}$s; (c) $t=1.05 \times 10^{-10}$s; (d) $t=1.48 \times 10^{-10}$s and electric field after the heavy ions incident (e) $t=9.5 \times 10^{-11}$s; (f) $t=1 \times 10^{-10}$s; (g) $t=1.05 \times 10^{-10}$s; (h) $t=1.48 \times 10^{-10}$s.

than that in the bulk. This is apparently a transient field caused by hole accumulation, causing the electric field strength at the bottom of the gate oxide layer to exceed 9.27 MV/cm, resulting in gate rupture. When the transient electric field generated by hole accumulation is superimposed on the original electric field strength and exceeds the critical breakdown field strength of the gate oxide layer, the gate oxide layer will be punctured, causing it to lose its insulating function.

The comprehensive analysis of the output characteristic curves of STP-MOS and Con-MOS, as well as the variations

979-8-3315-1110-4/25 $31.00 © 2025 IEEE

of drain-source voltage and maximum electric field in the gate oxide layer with time, indicates that the accumulation layer formed beneath the gate oxide layer of STP-MOS reduces the device's specific on-resistance. However, the significant hole charge accumulation under the shallow trench gate oxide layer affects the SEGR of the device. Compared with conventional structures, the threshold voltage for SEGR occurrence has slightly increased.

IV. DISCUSSION OF INFLUENCING FACTORS, MECHANISM ANALYSIS, AND HARDENING STUDY

The breakdown field strength of the gate oxide layer is primarily determined by the Linear Energy Transfer (LET) value of the incident ions. The selected LET values in the experiment are 0.02 pC/μm, 0.04 pC/μm, 0.06 pC/μm, 0.08 pC/μm, 0.10 pC/μm, 0.12 pC/μm, and 0.14 pC/μm, with corresponding gate oxide breakdown field strengths of 10.29 MV/cm, 10.03 MV/cm, 9.79 MV/cm, 9.57 MV/cm, 9.34 MV/cm, 9.14 MV/cm, and 8.94 MV/cm, respectively. Fig. 6 shows the variation of the maximum electric field strength in the gate oxide layer with time for LET values ranging from 0.02 pC/μm to 0.14 pC/μm. When the LET value is 0.08 pC/μm, the maximum electric field peak in the gate oxide layer is 8.80 MV/cm, which is lower than its critical breakdown field strength of 9.57 MV/cm. For LET values of 0.10 pC/μm, 0.12 pC/μm, and 0.14 pC/μm, the maximum electric field peaks in the gate oxide layer are 9.34 MV/cm, 9.14 MV/cm, and 8.94 MV/cm, respectively, all of which exceed the critical breakdown field strengths corresponding to the LET values of the incident ions. This indicates that as the LET value increases, the number of electron-hole pairs generated along the ion bombardment path also increases. The accumulation of more holes at the corner results in a larger transient electric field, making the SEGR effect more likely to be triggered. Therefore, it is demonstrated that the critical LET value for gate rupture in STP-MOSFETs at V_{DS} = 48 V is 0.1 pC/μm.

This thesis also investigates the impact of different incidence depths on the SEGR effect in SiC STP-MOSFET devices. Fig. 7(a) shows the variation of the maximum electric field strength in the gate oxide layer with time when the incidence depth ranges from 1 μm to 13 μm. When the incidence depth is 1 μm and 3 μm, there is no significant change in the electric field of the gate oxide layer. Fig. 7(b) displays the heavy ion charge distribution inside the device when the incidence depth is 11 μm and 13 μm, respectively. When the incidence depth is 11 μm, the particle completely passes through the N-drift region, and the maximum electric field in the gate oxide layer is 9.35 MV/cm, causing the device to experience SEGR. When the incidence depth is 13 μm, the particle reaches the N⁺ substrate. Since the doping concentration of the N⁺ substrate is much higher than that of the N⁻ drift region, the holes generated by heavy ion incidence in the N⁺ substrate will be quickly recombined, resulting in no significant change in the hole concentration near the gate oxide layer. Therefore, it is concluded that SEGR is mainly influenced by particles within the N⁻ drift region.

Fig. 8 shows the incidence of particles from 30° to 150°. Fig. 9 presents the variation of the maximum electric field strength in the gate oxide layer with time when the incidence angle ranges from 30° to 150°. When particles are incident at 90°, the path traversed by heavy ions in the device is the longest, and the number of electron-hole pairs generated along

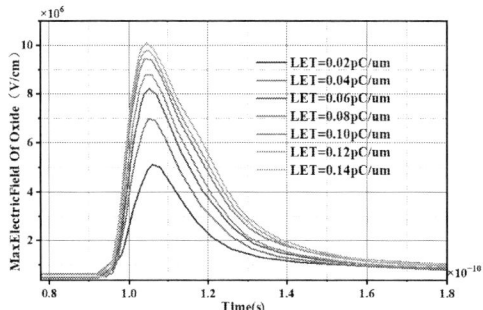

Fig. 6. The variation trend of the incident particle energy and the maximum electric field of the gate oxide layer with time.

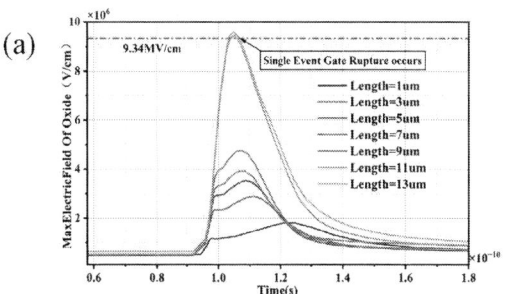

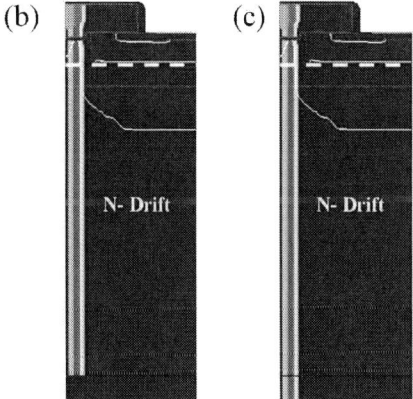

Fig. 7. The effect of particle incident depth on SEGR(a) The variation trend of particle incident depth and the maximum electric field of gate oxide layer with time; (b) The distribution of heavy ion charge density in the device with Length = 11μm; (c) The distribution of heavy ion charge density in the device with Length = 13μm.

the heavy ion bombardment path is the highest. The accumulation of more holes at the corner results in a larger transient electric field, with the maximum electric field strength in the gate oxide layer reaching 9.40 MV/cm, causingthe device to experience SEGR. At this time, the maximum electric field strength in the gate oxide layer is the highest. To clearly observe the electric field distribution in the gate oxide layer, Fig. 10 shows the electric field strength distribution at incidence angles of 30° and 90°, respectively. At an incidence angle of 30°, the maximum electric field strength in the gate oxide layer is 1.76 MV/cm. At an

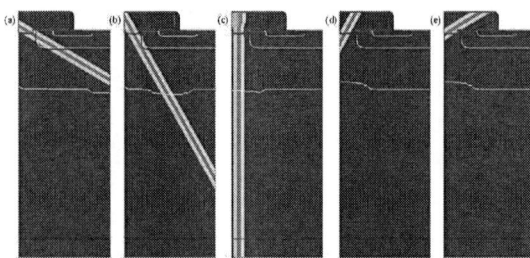

Fig. 8. Incidence results at different angles(a) Angle=30°; (b) Angle=60°; (c) Angle=90°; (d) Angle=120°; (e) Angle=150°.

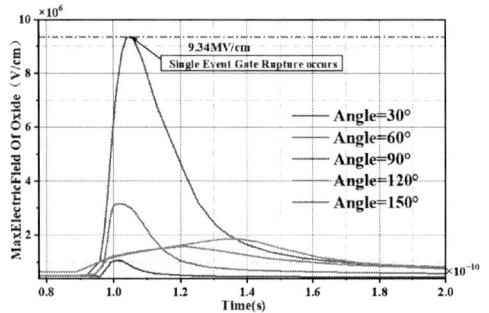

Fig. 9. Variation trend of incident angle and maximum electric field of gate oxide layer with time.

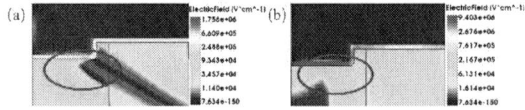

Fig. 10. The results of electric field at different angles(a)Angle=30°; (b)Angle=90°.

incidence angle of 90°, due to the generation of more electron-hole pairs, more holes accumulate at the bottom of the gate oxide layer, causing the maximum electric field strength at the bottom of the gate oxide layer to increase to 9.40 MV/cm.

Due to the uncertainty of the incidence position of heavy ions in the radiation environment, it is also crucial to investigate the impact of changing the ion incidence position on SiC STP-MOSFETs. Fig. 11 shows the incidence of particles at positions 0.2 μm, 0.4 μm, 0.8 μm, and 1.1 μm away from the cell center, respectively. Fig. 12 presents the SEGR threshold voltage of SiC STP-MOSFETs when single particles are incident from the positions shown in Fig. 11. It is found that when single particles are incident at positions 0.2 μm, 0.4 μm, 0.8 μm, and 1.1 μm away from the cell center, the SEGR threshold voltages are 93 V, 48 V, 51 V, and 60 V, respectively. Therefore, it is indicated that the most sensitive region for SEGR is 0.4 μm away from the cell center.

This paper achieves SEGR hardening of the device by employing a high-K dielectric material for the gate medium. Fig. 13 shows the variation of the maximum electric field with time under two different gate media. With the gate medium thickness set at 50 μm and the bias voltage at 48 V, the maximum electric field of the SiO_2 gate medium is 9.468 MV/cm, whereas that of the HfO_2 gate medium is only 4.664 MV/cm, thereby significantly reducing the risk of gate

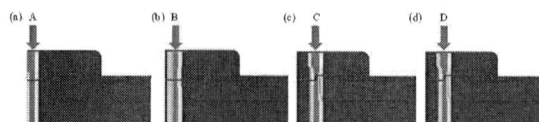

Fig. 11. The results of the incident particles at different positions(a) Position=0.2um; (b) Position=0.4um; (c) Position=0.8um; (d)Position=1.1um.

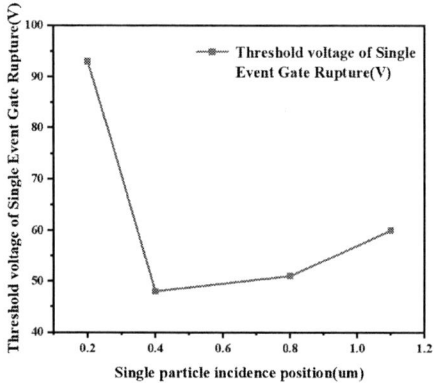

Fig. 12. The threshold voltage of gate penetration of particles incident at different positions.

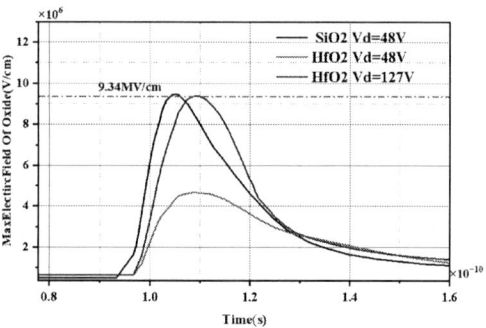

Fig. 13. Variation of Maximum Electric Field with Time for SiO_2 and HfO_2 Gate Media.

medium layer breakdown. Under a bias voltage of 127 V, the maximum electric field of the HfO_2 gate medium is 9.388 MV/cm, which exceeds the critical breakdown electric field of 4.664 MV/cm, resulting in SEGR. At this point, 127 V is the SEGR threshold voltage for HfO_2, representing a 165% increase compared to the threshold voltage with SiO_2 as the gate medium.

CONCLUSION

This paper takes STP-MOS as the research object, focusing on the impact factors, breakdown mechanism, and hardening measures of SEGR. STP-MOS, a shallow trench planar gate structure MOS proposed by our research group, is designed to effectively reduce the specific on-resistance. The study finds that the specific on-resistance of STP-MOS is reduced by 17.2% compared with Con-MOS. However, it is also found that there is a significant hole charge accumulation under the shallow trench gate oxide layer, which may affect

SRGR. Therefore, this paper investigates the impact factors, breakdown mechanism, and hardening measures of SEGR in STP-MOS and compares the SEGR threshold voltages of STP-MOS and Con-MOS. The research shows that at a position 0.4 μm away from the cell center and under 90° vertical incidence, the hole accumulation effect at the bottom of the gate oxide layer is most pronounced. When the N-drift region is completely penetrated by heavy ions and the LET value is 0.1 pC/μm, the SEGR threshold voltage is only 48 V. The larger the LET value, the more likely the device is to experience SEGR. The incidence depth for SEGR is 11 μm. The most sensitive region for SEGR is 0.4 μm away from the cell center, and when the incidence angle is 90°, the path traversed by the heavy ion in the device is the longest, making the device more susceptible to SEGR. The electric field strength distribution at different moments shows that compared with trench gate and planar gate MOSFETs, the gate rupture location in STP-MOSFETs is mainly at the bottom of the gate oxide layer and expands towards the trench corner over time. When the transient field strength is added to the original electric field strength and exceeds the critical breakdown field strength of the gate oxide layer, the gate oxide layer will be punctured, causing it to lose its insulating function and resulting in the SEGR effect. After reinforcement with high-K dielectric HfO_2, the SEGR threshold voltage is increased from 48 V to 127 V, representing a 165% improvement.

ACKNOWLEDGMENT

This work was supported by the Shaanxi Provincial Key Research and Development Program (Grant No.2025CY-YBXM-146) and the National Natural Science Foundation of China (Grant No. 62174134)

REFERENCES

[1] Zhang A, Chen W, Huang J, et al. Single-event effect hardening of the Schottky contact super barrier rectifier (SSBR) with high-k gate dielectric[J]. Journal of Computational Electronics, 2023, 22(5): 1463-1471.

[2] Yu Q, Chen W, Huang J, et al. A novel 4H–SiC power MOSFET with source-side poly-Si/SiC heterojunctions for single-event effects hardening[J]. Micro and Nanostructures, 2025, 198: 208064.

[3] Wang H, Nie Z, Huang X, et al. The impact of negative gate voltage on neutron-induced single event effects for SiC MOSFETs[J]. Microelectronics Reliability, 2024, 163: 115547.

[4] Zhou X, Tang Y, Jia Y, et al. Single-event effects in SiC double-trench MOSFETs[J]. IEEE Transactions on Nuclear Science, 2019, 66(11): 2312-2318.

[5] Wang Y, Liu T, Qian L, et al. Analysis and hardening of SEGR in Trench VDMOS with termination structure[J]. Micromachines, 2023, 14(3): 688.

[6] Martinella C, Race S, Für N, et al. Heavy-ion effects in SiC power MOSFETs with trench-gate design[J]. IEEE Transactions on Nuclear Science, 2024.

[7] Sun S, Chen F, Sun Y, et al. Single event effects hardening in SiC double-trench MOSFETs[J]. Microelectronics Reliability, 2025, 164: 115569.

SiC Power Module with Staggered Terminals Layout Design to Reduce Parasitic Inductance

Jiahang Wang
Guangzhou Institute of Technology,
Xidian University,
Guangzhou, 510555, China
23111213605@stu.xidian.edu.cn

Xi Jiang
Guangzhou Institute of Technology,
Xidian University,
Guangzhou, 510555, China
xjiang@xidian.edu.cn

Runze Ouyang
Guangzhou Institute of Technology,
Xidian University,
Guangzhou, 510555, China
22111213611@xidian.edu.cn

Song Yuan
Guangzhou Institute of Technology,
Xidian University,
Guangzhou, 510555, China
syuan@xidian.edu.cn

Yuanzhi Zhao
Guangzhou Institute of Technology,
Xidian University,
Guangzhou, 510555, China
23111213674@stu.xidian.edu.cn

Qingrong Hu
Guangzhou Institute of Technology,
Xidian University,
Guangzhou, 510555, China
23111213613@stu.xidian.edu.cn

Ying Wang
School of Electronic Engineering,
Xi'an University of Posts and
Telecommunication,
Xi'an 710121, China

Xiaowu Gong
State Key Laboratory of Wide-Bandgap
Semiconductor Devices and Integrated
Technology, School of
Microelectronics, Xidian University,
Xi'an 710071, China
xwgong@xidian.edu.cn

Abstract—This paper introduces a novel staggered terminal layout for SiC MOSFET multi-chip power modules. The proposed layout reduces parasitic inductance by canceling inductance within the power loop and between power terminals, while also enhancing current balance through an axisymmetric layout. The parasitic inductance of the proposed SiC power module is approximately 3.85 nH. Compared to commercial counterpart with the same current rating, the parasitic inductance is reduced by 4.26 nH. Furthermore, the drain voltage overshoot in the proposed power module is reduced by approximately 23 V, and the switching loss is lowered by nearly 1020 μJ under test conditions of 600 V/100 A.

Keywords—silicon carbide (SiC) ; symmetric layout. ; parasitic inductance

I. INTRODUCTION

Silicon Carbide (SiC) power devices offer several advantages over their silicon (Si) counterparts, including low specific on-state resistance, fast switching speeds, and high operating temperature capabilities [1]. These attributes have driven the rapid adoption of SiC devices in high-power applications, such as electric vehicle and industrial motor drives.

Despite their benefits, the high-frequency switching speed of SiC modules introduces significant design challenges, most notably the issue of parasitic inductance. Large inductance in the power loop leads to voltage overshoot and EMI, and exacerbates reliability concerns. Furthermore, the asymmetrical layout and internal structure of SiC modules often lead to non-uniform current distribution, resulting in localized heating and thermal stress, which degrade the performance and reliability of the power module over time.

Minimization of parasitic inductance is thus essential to improving switching performance and enhancing the reliability of SiC-based modules [1], [2]. Achieving this goal requires a comprehensive approach including power module structural optimization, advanced packaging techniques, and thermal management strategies. In particular, factors such as chip placement, interconnection design, and heat dissipation must be simultaneously considered during the module design process.

To reduce the adverse effects caused by parasitic inductance, recent researches focus on the following aspects. Techniques such as compact layout strategies, optimization of power terminal placement, and the adoption of multilayer or stacked PCB architectures have shown good results in reducing power loop inductance and improving current sharing characteristics among parallel semiconductor chips [3], [4].

In addition to power module's geometric optimization, mutual-inductance cancellation techniques have been introduced. By controlling current path geometry and adjusting magnetic field distribution, the internal mutual inductance can be significantly cancel out. One reported approach employs multidimensional coupling of self-inductance and mutual inductance has been proposed to enhance multichip power module performance [5]. Furthermore, interleaved planar packaging has been shown to mitigate the trade-off between thermal and electrical performance [6]. Compared with single-sided power modules, double-sided cooling power modules replace bonding wires with molybdenum-based connections. Combined with double-sided cooling, this approach reduces both thermal resistance and parasitic inductance. Among them, the three-dimensional layout with vertical interconnections minimizes parasitic inductance without sacrificing thermal performance[7].

This paper introduces a staggered terminal layout design for power modules, which effectively reduces the parasitic inductance of the power terminals. The proposed approach enhances the switching performance of the power module and

979-8-3315-1110-4/25 $31.00 © 2025 IEEE

ensures more uniform current distribution within the module's chips, thereby improving both the electrical and thermal performance of the module.

II. LOW-INDACTANCE LAYOUT OF POWER MODULES

A. Design of Alternately Axisymmetric Layout for DC+ and DC - Terminals

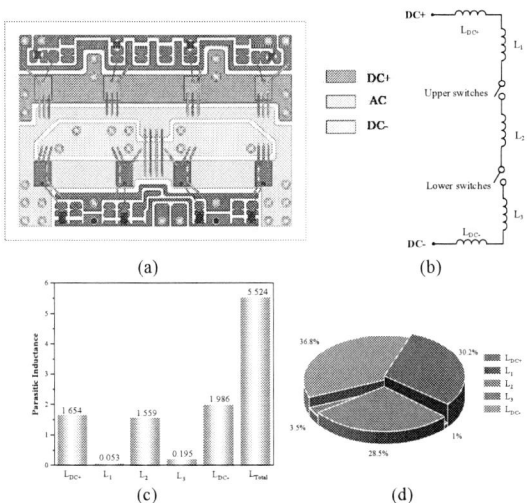

Fig.1. (a) Top view of a commercial power module. (b) Equivalent total power loop inductance. (c) Comparison of power loop inductance. (d) Proportion of power loop inductance

As shown in Fig. 1, the ANSYS Q3D s software was used to extract the parasitic inductance of the DC+ terminal of the power loop in the commercial power module CAB008M12GM3[8], the parasitic inductance from the DC+ terminal to the upper switch chip, the parasitic inductance from the upper switch chip to the lower switch chip, and the parasitic inductance of the DC- terminal. The parasitic inductances of the DC+ and DC- power terminals account for 30.1% and 36.8% of the total parasitic inductance, respectively, which constitute a significant portion of the total parasitic inductance in the power circuit. Thus, it is critical to minimize the parasitic inductance of the power terminals in order to improve the module's performance.

As can be seen from Fig. 2, the entire DBC is divided into four parts by the X-axis and the Y-axis, and each part has a power loop. These four power loops are not only axially symmetric but also centrally symmetric. To reduce mutual inductance, the DC+ terminal is placed between the DC- terminals to achieve the cancellation of the mutual inductance between the terminals. The finite element analysis software ANSYS Q3D is used to extract the self-inductance and mutual inductance matrix between the power terminals. The results is shown in Table I. It can be seen from the table that the total parasitic inductance of the power terminals at symmetric positions is similar. For example, the inductances of the DC+1 and DC+2 terminals are approximately 7.6 nH each, while the DC+3 and DC+4 terminals are around 6.5 nH each. Similarly, the inductances of the DC-1 and DC-3terminals are about 4.6 nH each, and the DC-2 and DC-4 terminals are approximately 6 nH each

However, there are still some power terminals that have larger inductance (DC+1 terminal, DC+2 terminal) due to the non-uniform distribution of mutual inductance across these terminals. To further minimize the parasitic inductance of the module, we propose an improved layout design for the power module.

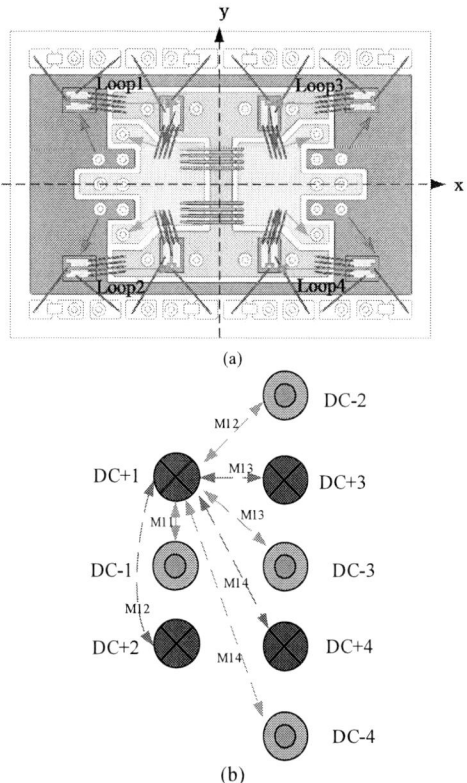

Fig. 2. (a) Distribution of four power loops. (b) DC terminal layout.

TABLE I. PARTIAL INDUCTANCE EXTRACTED FOR BASELINE TERMINALS

	DC+1	DC+2	DC+3	DC+4	DC-1	DC-2	DC-3	DC-4	Total
DC+1	12.60	4.03	5.88	3.74	-5.94	-4.88	-4.88	-2.94	7.61
DC+2	4.03	12.58	3.74	5.89	-5.87	-2.94	-4.92	-4.87	7.64
DC+3	5.88	3.74	12.55	4.00	-4.91	-5.89	-5.84	-3.05	6.48
DC+4	3.74	5.89	4.00	12.55	-4.86	-3.03	-5.89	-5.90	6.50
DC-1	-5.94	-5.87	-4.91	-4.86	12.53	3.74	5.85	3.71	4.25
DC-2	-4.88	-2.94	-5.89	-3.03	3.74	12.66	3.98	2.44	6.08
DC-3	-4.88	-4.92	-5.84	-5.89	5.85	3.98	12.52	4.03	4.85
DC-4	-2.94	-4.87	-3.05	-5.90	3.71	2.44	4.03	12.66	6.08

B. Improved power terminal layout of power module

In the power terminals of the power module proposed above, there is a relatively large enhancement of mutual inductance between DC+1 and DC+3, DC - 1 and DC - 3, as well as DC+2 and DC+4. Based on this, DC+1 and DC+3, DC - 1 and DC - 3, and DC+2 and DC+4 are staggered with each other to further enhance the cancellation of mutual inductance and weaken its enhancement. As shown in Fig 3(b), a DC+ terminal and a DC - terminal are added while maintaining the original symmetrical structure. The left, right, top, and bottom positions of each DC+ terminal are occupied by DC - terminals, and vice versa for DC- terminals. This arrangement enhances the cancellation of mutual inductance. Meanwhile, the two newly added terminals make the structure more symmetrical and the current distribution more uniform. The specific layout of the power module is shown in Fig 3(a).

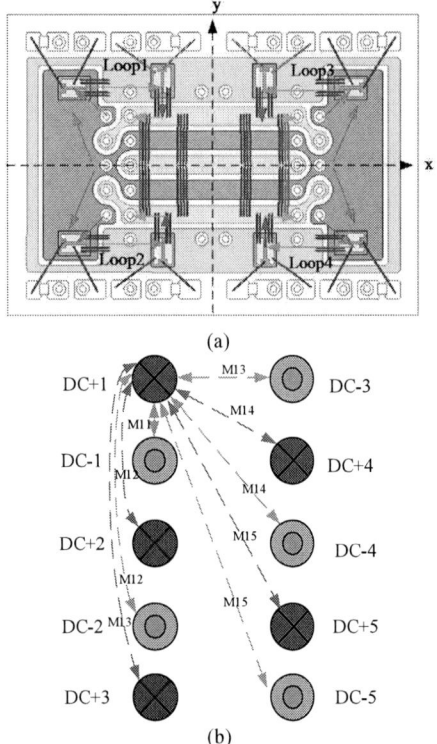

(a)

(b)

Fig. 3. (a) Distribution of four improved power loops. (b) DC terminal layout.

TABLE II. PARTIAL INDUCTANCE EXTRACTED FOR IMPROVED

	DC+1	DC+2	DC+3	DC+4	DC+5	DC-1	DC-2	DC-3	DC-4	DC-5	Total
DC+1	12.59	4.02	2.42	4.91	2.91	-5.95	-3.02	-5.89	-3.73	-2.38	5.88
DC+2	4.02	12.50	3.96	4.90	4.85	-5.87	-5.80	-3.75	-5.84	-3.70	5.27
DC+3	2.42	3.96	12.60	2.91	4.90	-3.01	-5.88	-2.38	-3.71	-5.90	5.91
DC+4	4.91	4.90	2.91	12.52	3.97	-5.86	-3.70	-5.92	-5.85	-3.01	4.87
DC+5	2.91	4.85	4.90	3.97	12.52	-3.69	-5.86	-3.02	-5.85	-5.89	4.84
DC-1	-5.95	-5.87	-3.01	-5.86	-3.96	12.51	3.95	4.94	4.86	2.90	4.78
DC-2	-3.02	-5.80	-5.88	-3.70	-5.86	3.95	12.53	2.92	4.87	4.90	4.91
DC-3	-5.89	-3.75	-2.38	-5.92	-3.02	4.94	2.92	12.59	4.01	2.43	5.93
DC-4	-3.73	-5.84	-3.71	-5.85	-5.85	4.86	4.87	4.01	12.51	3.99	5.26
DC-5	-2.38	-3.70	-5.90	-3.01	-5.89	2.90	4.90	2.43	3.99	12.59	5.93

Compared with the structure proposed above, the self-inductance and mutual- inductance of the new staggered DC+ and DC - terminals are also extracted, as presented in Table II. Compared with Table II, the self-inductance of the five DC+ terminals and five DC- terminals changes little, which is an inherent property of the material. However, changing the layout by adding two terminals alters the mutual inductance between DC+ and DC - terminals, enhancing the cancellation of mutual inductance and reducing its enhancement to lower the parasitic inductance. From the total inductance, it can be seen that the parasitic inductance of the terminals in the new structure is significantly lower than that of the power module proposed above. Moreover, the parallel connection of the two newly added terminals with the other terminals further reduces the parasitic inductance. It can also be observed from the table that the parasitic inductances of power terminals in symmetric positions are similar. Compared with the power module proposed above, the differences in parasitic inductances among terminals at different positions are smaller, and the current distribution is more uniform.

The current distribution within the proposed power module was analyzed using ANSYS Q3D Extractor. The corresponding simulation results are presented in Fig. 4(a). The current densities of the four power loops are similar, which is consistent with the inductance model established above, indicating that the current in each power loop of the proposed power module is evenly distributed.

Moreover, the thermal performance of the module was evaluated using COMSOL Multiphysics. The temperature distribution results are shown in Fig. 4(b). The heat distribution among the chips is uniform, without any temperature concentration on a single chip.

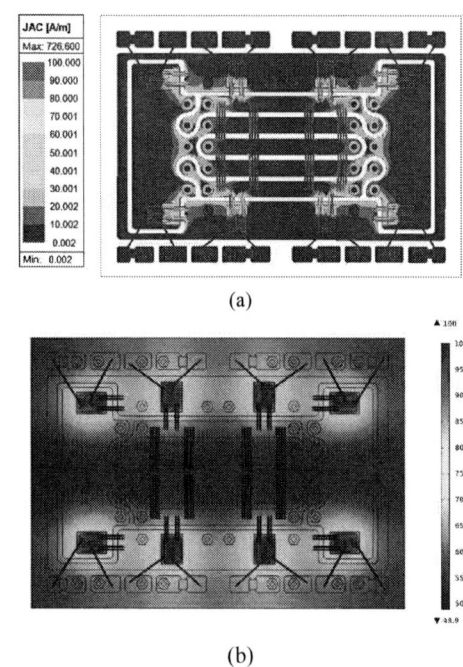

(a)

(b)

Fig . 4. (a) Circuit distribution of the power module (b) Thermal distribution Thermostable distribution of a power module.

III. POWER MODULES FABRICATION

The manufacturing process of the 1200V/100A half - bridge silicon carbide power module is shown in Fig 5, and the specific parameters are listed in Table III. In the manufacturing process, the required DBC (Direct - Bonded Copper) is fabricated first. Then, solder is applied to the corresponding positions on the DBC, and the chips are mounted on the corresponding positions of the DBC for soldering. After that, solder is applied again to the corresponding positions on the DBC, and the pin terminals are soldered to the corresponding positions on the DBC. The solder used in these two processes has different melting point, ensuring that they do not interfere with each other. Finally, the housing is assembled, and silicone gel is poured for encapsulation to form the power module.

TABLE III. COMPONENTS IN THE POWER MODULE

Part	Materials	Specifications
SiC MOSFET	IV1Q2050BD	1200V,58A
DBC	Cu/AlN/Cu	Thickness:0.3mm/0.63mm/0.3mm
Driver Bonding Wire	Aluminum	Diamete:5mil
Power Bonding Wire	Aluminum	Diameter:15mil
Terminals	Ni-plated Cu	Diameter:1mil

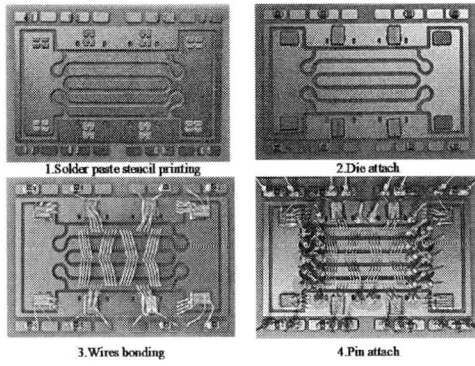

Fig. 5. Fabrication and assembly of the proposed module

IV. EXPERIMENTAL RESULTS

A. Static Characteristics

The breakdown voltage and output characteristics of the fabricated power module were measured using a Keysight B1505A power device analyzer, as depicted in Fig. 5. At room temperature conditions, when the breakdown voltage of the upper chips and lower chips is 1200 V, the maximum leakage current does not exceed 20 µA. Under the conditions of the gate-source voltage of 20 V and drain current of 150 A, the on-state resistance of the power module is 12.7 mΩ.

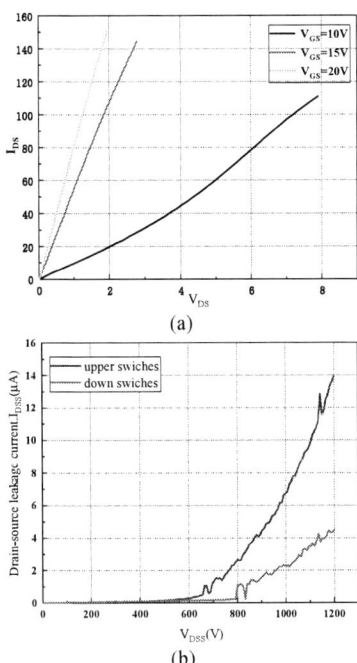

Fig. 6. Static characteristics of the proposed SiC power module. (a) Output characteristics. (b) Drain–source leakage current.

B. Switching Characteristics

During the double-pulse test, the upper switching transistors are turned off and function as freewheeling diodes. The upper switching transistors are biased with a constant off-voltage of -5 V, while the lower switching transistors operate under conditions of 600 V and 100 A. This setup is used to analyze the switching characteristics and extract the parasitic inductance of the module. The test platform and double-pulse

circuit are illustrated in Fig. 7. The specific test parameters are listed in Table IV.

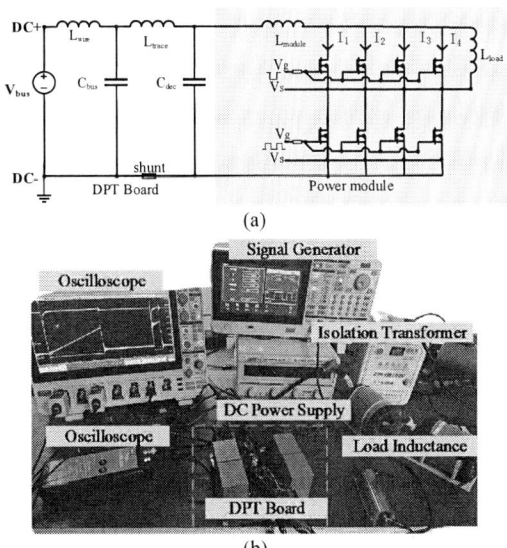

Fig. 7. Experimental setup of DPT platform. (a) Schematic of the DPT circuit. (b) Overall test platform.

TABLE IV. SIMULATIONS PARAMTERS OF DPT

Parameters	Parameter description	Value
V_{bus}	DC-link voltage	600V
I_{load}	Load current	100A
V_g	Gate driver voltage	20/-5V
C_{bus}	DC decouple capacitor	400µF
C_{des}	DC decouple capacitor	80nF

To benchmark the performance of the proposed module, a comparison is made with the commercial power module CAB008M12GM3 which has same voltage and current ratings. The same DPT circuit layout and test platform are employed for this comparison. To extract the parasitic inductance of the power module, the following formula is utilized for calculation. a hyphen for a minus sign. Punctuate equations with commas or periods when they are part of a sentence, as in:

$$f = \frac{1}{2\pi\sqrt{L_d\,C_{oss}}} \tag{1}$$

The switching waveforms are shown in Fig 8. During the turn-off transient, the voltage overshoot of the commercial module reaches 677 V, with an oscillation period of 13.86 ns. In contrast, the proposed power module shows a turn-off voltage overshoot of 654 V and an oscillation period of 10.82 ns. This represents a decrease of 23 V in the overshoot voltage. According to the datasheet, the C_{oss} of the commercial module is 600 pF, whereas the C_{oss} of the proposed module is 770 pF. Using these values, the parasitic inductance of the commercial module is estimated to be 8.11 nH, while the parasitic inductance of the proposed module is calculated to be 3.85 nH. Therefore, the parasitic inductance of the proposed module is reduced by 4.26 nH compared to the commercial module. Compared with the simulation results, the actual parasitic inductance is slightly higher because the parasitic inductance on the PCB cannot be completely eliminated.

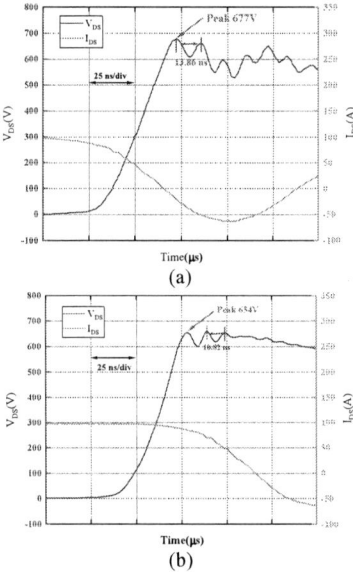

(a)

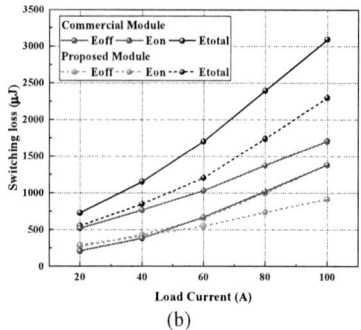

(b)

Fig. 9. (a) Switching losses under different load current (b) Turn-on transient currents of an individual chip measured at a load current of 100A

V. CONCLUSIONS

This paper presents a novel multi-loop layout for power modules that reduces parasitic inductance through the staggered arrangement of DC+ and DC- terminals. The symmetric design and staggered arrangement multi-terminal structure result in a more uniform current flow and improved thermal distribution across the chips. The parasitic inductance of the commercial module is calculated to be approximately 8.11 nH, while that of the proposed module is approximately 3.85 nH. The parasitic inductance of the proposed module is significantly reduced by 4.26 nH. Furthermore, the proposed power module demonstrates superior switching performance, with a reduction of 86 μJ in turn-off loss, 934 μJ in turn-on loss, and 1020 μJ in total switching loss compared to the commercial module.

REFERENCES

[1] Z. Zeng, W Shao, H Chen, et al. Changes and challenges of photovoltaic inverter with silicon carbide device[J]. Renewable & Sustainable Energy Reviews, 2017, 78: 624-639.DOI:10.1016/j. rser. 2017. 04. 096.

[2] H. Lee, V. Smet and R. Tummala, "A Review of SiC Power Module Packaging Technologies: Challenges, Advances, and Emerging Issues," in IEEE Journal of Emerging and Selected Topics in Power Electronics, vol. 8, no. 1, pp. 239-255, March 2020, doi: 10.1109/JESTPE.2019.2951801.

[3] Y. Wang et al., "Symmetric and Staggered Terminal Layouts for Enhanced Current Balance and Reduced Parasitic Inductance in SiC Power Modules," in IEEE Transactions on Power Electronics, vol. 40, no. 3, pp. 4100-4111, March 2025, doi: 10.1109/TPEL.2024.3499334.

[4] Z. Huang, C. Chen, Y. Xie, Y. Yan, Y. Kang and F. Luo, "A High-Performance Embedded SiC Power Module Based on a DBC-Stacked Hybrid Packaging Structure," in IEEE Journal of Emerging and Selected Topics in Power Electronics, vol. 8, no. 1, pp. 351-366, March 2020, doi: 10.1109/JESTPE.2019.2943635.

[5] J. Wang, S. Yu and W. Zhou, "A Comprehensive Design Method for Multichip Double-Sided Cooling Power Module With Multidimensional Self-and Mutual Inductances," in IEEE Transactions on Power Electronics, vol. 39, no. 8, pp. 9526-9539, Aug. 2024, doi: 10.1109/TPEL.2024.3378683.

[6] F. Yang et al., "Interleaved Planar Packaging Method of Multichip SiC Power Module for Thermal and Electrical Performance Improvement," in IEEE Transactions on Power Electronics, vol. 37, no. 2, pp. 1615-1629, Feb. 2022, doi: 10.1109/TPEL.2021.3106316.

[7] F. Yang, Z. Liang, Z. J. Wang and F. Wang, "Design of a low parasitic inductance SiC power module with double-sided cooling," 2017 IEEE Applied Power Electronics Conference and Exposition (APEC), Tampa, FL, USA, 2017, pp. 3057-3062, doi: 10.1109/APEC.2017.7931132.

[8] (Apr. 2023). Silicon Carbide Half-Bridge Module CAB008A12GM3.[Online].Available:https://assets.wolfspeed.com/upl oads/2023/05/Wolfspeed CAB008M12GM3 data sheet.pdf.

Fig. 8. Switching waveforms at 600V/100A (a) Commercial module DPT test plot. (b) Proposed module DPT test plot.

To experimentally verify the dynamic current sharing among the chips in the proposed power module, a PEM Rogowski coil was employed to measure the drain current of each chip by threading through the bonding wires connected to the lower-side switches. The measured current waveforms corresponding to the four power loops are presented in Fig. 9(a). As shown, all four lower-side MOSFETs are switched on nearly simultaneously, and the current is distributed uniformly across the parallel devices, which meets the expectations.

The switching losses of the power module were evaluated under various load current conditions. The results are presented in Fig. 9(b). Under the test conditions of 600 V/100 A, the turn -off loss of the commercial module is about 86 μJ higher than that of the proposed module, the turn - on loss is about 934 μJ higher, and the total switching loss is about 1020 μJ higher. As the load current increases, the difference in switching losses between the commercial module and the proposed module gradually increases, and the switching loss of the designed module is significantly lower than that of the commercial module.

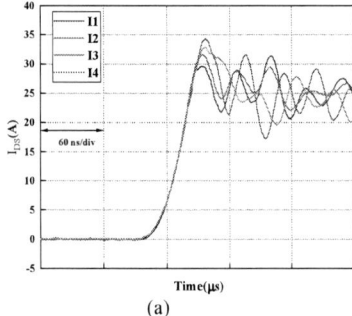

(a)

Novel Tri-gate Multichannel Device for Improved V_{th} Controllability

Quanbo He
Engineering Department
University of Cambridge
Cambridge, UK
qh243@cam.ac.uk

Hengyu Wang
College of Electrical Engineering
Zhejiang University
Hangzhou, China
wanghengyu@zju.edu.cn

Florin Udrea
Engineering Department
University of Cambridge
Cambridge, UK
fu10000@cam.ac.uk

Abstract—Gallium nitride (GaN) high electron mobility transistors (HEMTs) are considered promising candidates for next-generation power electronic applications due to their high electron mobility and wide bandgap. To address the performance limitations of conventional HEMT designs, multi-channel HEMTs have been developed. To enhance control over multiple channels, tri-gate architectures have been introduced。 Despite their advantages, achieving enhancement-mode (E-mode) operation in tri-gate multichannel HEMTs remains difficult due to the fabrication challenges associated with extremely narrow fins and the negative effects on electron concentration. This study presents a comparative analysis between a conventional tri-gate MC-HEMT and a novel design, referred to as the MOP-trigate structure. The proposed design incorporates a thin p-type GaN layer surrounding the fins, separated from the gate metal by an oxide layer, to enhance depletion of the 2DEG without reducing fin width. As an extension of the analysis, the multichannel region—serving as the primary structural component—is also examined in detail. This work innovatively designs and uses a TCAD model to analyze and develop novel multichannel devices. It provides a theoretical foundation for future fabrication and experimental efforts and contributes to the advancement of multichannel power device design.

Keywords—trr-gate, multichannel, TCAD, normally-off

I. INTRODUCTION

Gallium nitride (GaN) high electron mobility transistors (HEMTs) are widely anticipated to form the backbone of the next generation of power devices. Thanks to their exceptional electron mobility, high field strength, and wide bandgap, these devices are highly suitable for a variety of applications ranging from power converters to traction inverters [1], [2]. In response to the ever-growing need for more efficient and high-performance GaN solutions, the limitations of traditional HEMT structures have become increasingly apparent. As a result, multi-channel HEMTs (MC-HEMTs) have emerged as a revolutionary alternative, offering lower on-state resistance and significantly higher current capability [3], [4]. An innovative concept in this development is the introduction of the superjunction (SJ) concept into GaN multi-channel architectures In these undoped multi-channel configurations, the spontaneous formation of a two-dimensional electron gas (2DEG) along with a two-dimensional hole gas (2DHG) creates a natural charge balance, which is critical for the establishment of a superjunction effect [4], [5] This inherent balance not only facilitates improved device performance but also provides new possibilities for engineering advanced power systems.

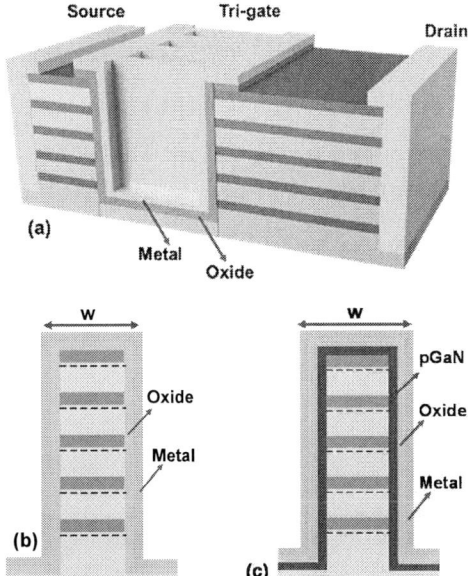

Fig.1 (a) Schematic of conventional tri-gate multi-channel MOSHEMTs. Cross section of a tri-gate fin in the multi-channel (b) conventional tri-gate and (c) MOP-trigate.

A significant challenge in advancing GaN MC-HEMT technology lies in the effective management and control of electrons within the transistor. To address this, the tri-gate design has been developed as a novel method to modulate multiple channels simultaneously, leading to enhanced electrostatic control and overall device performance. This unique technology is particularly effective in addressing the degradation in drain current ($I_{D,max}$). Research has shown that an enhancement-mode (E-mode) fin field-effect transistor (FinFET) featuring four channels can achieve a maximum transconductance that is 3.2 times greater than that of a conventional single-channel tri-gate device [6], [7]. Despite the remarkable performance, the commercialization of tri-gate GaN MC-HEMTs encounters significant challenges. One major challenge is achieving E-mode operation necessitates extremely narrow fins, typically as narrow as 40nm or even below [7], [8], [9]. However, such narrow fin widths are difficult to achieve in manufacturing and can have negative impact on the local concentration of 2DEG [10]. Moreover, the current literature lacks comprehensive studies examining how various gate parameters influence the performance of tri-gate devices.

In this paper, we will introduce a comparison between a conventional tri-gate multichannel device and a novel design multichannel device based on tri-gate. Fig. 1(a) shows the schematic of conventional tri-gate multichannel MOSHEMTs and Fig. 1(b) is the novel structure named MOP-trigate that will be discussed in this work. A thin p-type GaN layer enwraps the fins, featuring an oxide layer interposed between the gate metal and the p-GaN, effectively mitigating leakage current. Both structures play important roles in enhancing the depletion of the 2DEG in multichannel structures. This work will attempt to accelerate the depletion of the 2DEG in the channels by adding a p-type GaN layer in the gate region without changing the fin width. This comparative analysis serves as a foundation for the selection and fabrication of devices tailored for specific applications. Meanwhile, the multichannel structure serves as a fundamental component of such devices and will be discussed in this paper.

II. SIMULATION MODEL AND SETUP

Physics-based 3-D TCAD simulations have been carried in Sentaurus. TCAD Sentaurus is a set of software tools developed by Synopsys, an electronic design automation (EDA) software specialist company. Sentaurus is primarily used for modelling and simulation of semiconductor devices [11]. To provide a physical insight into the physics of these devices a 3D TCAD model has been built.

Some key analytical and physical models have been imputed into the TCAD to increase its accuracy and versatility. For example, the density of 2DEG takes into account the piezoelectric and spontaneous polarization physic model. Mobility models are specified to describe the 2DEG carrier mobility, including high-field saturation effects. Effective intrinsic density and Fermi level models are included. A Shockley-Read-Hall (SRH) model is used to characterize recombination through deep defect levels in the gap [11].

The multichannel simulation model is built on a basic multichannel block model. It can simulate devices that use a multichannel design. The UID multichannel contains n-type donors, which contribute to the 2DEG charge together with the polarization charges. In multichannel AlGaN/GaN

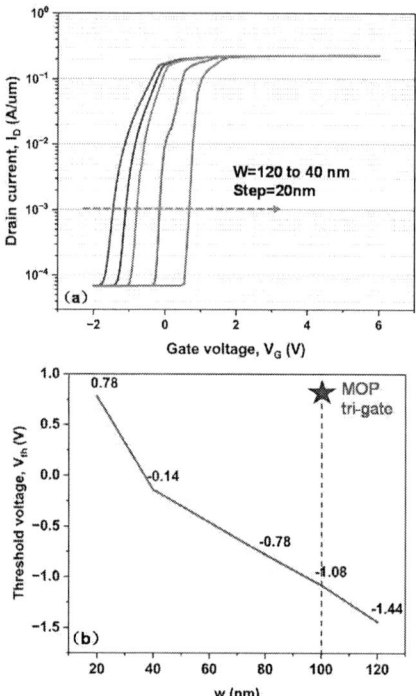

Fig. 3 (a) Transfer characteristics and (b) threshold voltage of conventional multichannel tri-gate devices with different width of w. In (b), the p-GaN in MOP tri-gate is defined as 1×10^{19} cm^{-3}.

structures, the charge composition is paramount to their operation [3]. In natural or polarization super-junctions, the charge in the two-dimensional electron gas (2DEG) compensates that in the two-dimensional hole gas (2DHG)) [12], [13]. The incorporation of donor-like traps in unintentionally doped (UID) multichannel structures, changes completely the "ideal" charge balance. It is important to highlight that, up to this point, while the stable existence of a 2DHG remains uncertain, some studies suggest that the 2DHG at the interface cannot be ignored. In 2017, Sun et al. [14] demonstrated the presence of a 2DHG in the epitaxial stack using ultra-low-frequency CV measurements, providing a promising approach for experimental verification. However, achieving a balance between 2DEG and 2DHG in undoped devices poses challenges due to the presence of unintentional doping. For the sake of simplification, the simulations presented in this paper assume an ideal, undoped condition.

Figure 2 shows the example and the experimental data from [15]. Simulation results are compared with experimental data and the close match supports the simulation models. In a previous study [16], the detailed analysis of this approach was discussed, showing that the model works well for many types of multichannel devices.

III. RESULTS AND DISCUSSION

E-mode operation is the most important consideration of designing the power tri-gate HEMT. The threshold voltage for a GaN HEMT can vary depending on the specific device and design. In most commercial GaN HEMTs devices,

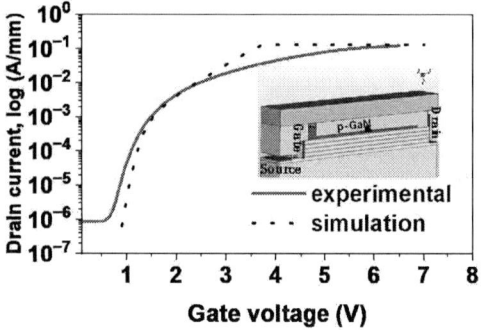

Fig. 2 3-D TCAD simulation model and experimental and simulated transfer I-V characteristics of the E-mode AlGaN/GaN multichannel HEMT.

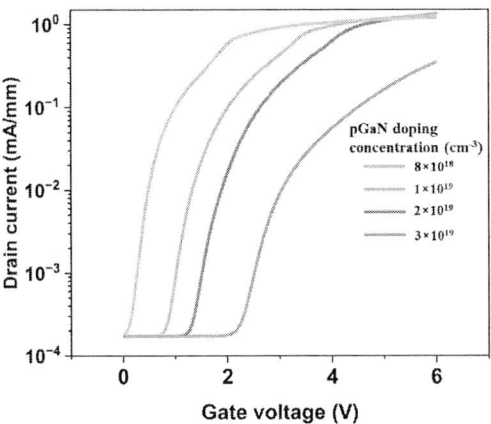

Fig. 5 Transfer curves of MOP-trigate in different p-GaN doping concentration.

threshold voltages typically range from approximately 1 to 3 volts and devices need to operate within a gate-to-source voltage range that extends up to 6V while maintaining its functionality.

Fig. 3(a) and (b) illustrate the V_{th} for different w values in conventional tri-gate, clearly showing that a narrower w results in a more positive threshold voltage. The channel control ability of the gate is directly related to the width of the fins. The fin of the tri-gate needs to be very narrow to archive a normally off operation (positive threshold voltage). This comes at the expense of a more challenging process and slight deterioration of the on-state performance. As shown in previous work on tri-gate architecture [7], [17], [18], the channel control ability is achieved by reducing the width of fin. In Fig. 3 (b), MOP-trigate shows the capability that keep the V_{th} positive when fin width in 100 nm.

To discuss the threshold voltage of novel structure MOP-trigate, defining the gate is crucial. According to the study on gate leakage in MOS-p-type semiconductor structures [19],

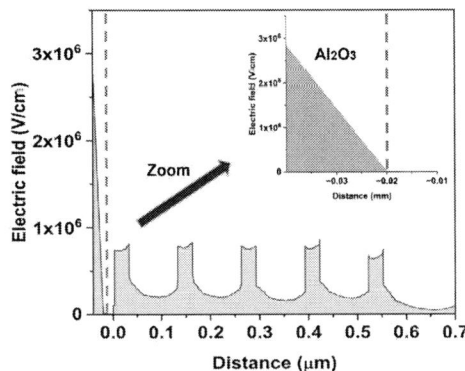

Fig. 4 Electric field of MOP-trigate and Al2O3 layer when V_D=10V, V_g=6V.

the gate tunnelling can be modelled as Fowler-Nordheim (FN) tunnelling. However, if we only discuss transfer curves, the voltage applied to the drain does not need to be large. To determine the operational state of MOP-trigate in this condition, a drain bias of 10 volts is applied. Generally, devices require working within a gate-to-source voltage range up to 6V while maintaining functionality. In this scenario, the electric field in the oxide layer is examined, as shown in Fig. 4. The electric field in the oxide is observed to be lower than 3 MV/cm, indicating that the FN tunnelling model does not apply at this operating voltage or any lower voltages [19].

Fig. 5 shows the transfer curves of novel structure MOP-trigate. In this structure, V_{th} shifts to right as the p-GaN doping concentration increases. Notably, when the p-GaN doping concentration is 2×10^{19} cm^{-3}, the device does not fully turn on at a V_g of 6V. Therefore, with the current parameters, maintaining the p-GaN doping concentration between 5×10^{18} cm^{-3} and 2×10^9 cm^{-3} keeps the threshold voltage at a suitable level.

IV. MULTICHANNEL MODEL DISCUSSION

Multichannel structure as the main block of the tri-gate devices, which worth to detailed investigation to better understand the underlying physical mechanisms. In addition to the tri-gate structure, the multichannel configuration plays a significant role in device operation. In particular, its impact on breakdown behavior is a critical aspect that deserves close attention. To achieve charge balance, a p-GaN cap is placed on top of the channels. This structural feature is considered when analyzing the device's operating mechanism. To demonstrate the reliability of the proposed model, the capacitance of the 3D multichannel structure with a p-GaN cap is compared with experimental results, as shown in Fig. 6. The close agreement between simulation and measurement confirms the credibility of the model.

Figure 7 illustrates the electric field distribution at a drain voltage (V_D) of 3000V for different length of channel. The electric field along the y-cutline (vertical direction) further elucidates this effect. This is primarily attributed to the top p-GaN cap, which aids in extending the electric field by forming a RESURF. Not only extending the electric field, the p-GaN cap also helps maintain charge balance in the multichannel region. In the 5-channel multichannel HEMT, multichannel region, consisting of UID multichannel structures serves as the main functional block. Ideally, the AlGaN and GaN layers are undoped, 2DEG and 2DHG with equal carrier concentrations

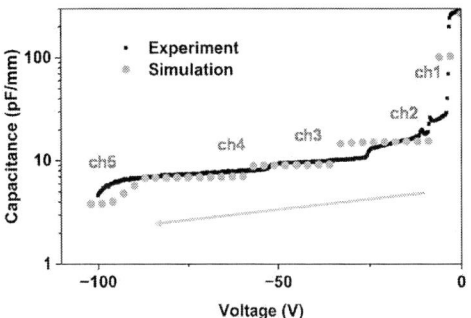

Fig.6 Experimental and simulated C-V curves for a multi-channel main building block with p-GaN cap layer, showing the successive depletion of multiple channels.

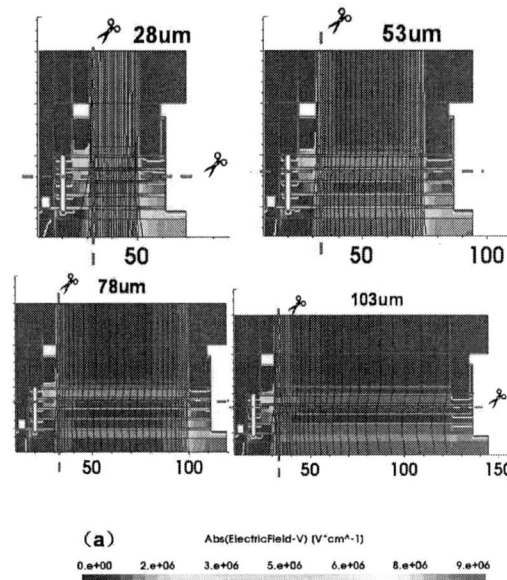

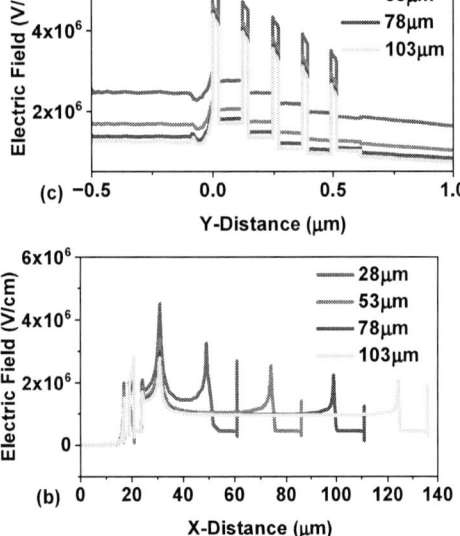

Fig. 7 (a) Simulated electric field contours and equipotential lines for different length of channel at OFF-state, V_D=3000V. (b) Electric field along x-cutline (along horizontal direction) for different length of channel. (c) Electric field along y-cutline (along vertical direction) for different length of channel, which shows the E-field across 5 channels.

(N_s and P_s, respectively) can be obtained naturally without doping due to the presence of matched polarization charges [20]. The charge balance between 2DEG and 2DHG, as well as the intrinsically matched polarization charge leads to a neutrally depleted region, resulting in a similar behavior to that of conventional doped superjunction structures.

V. CONCLUSION

This work investigates novel tri-gate multichannel structures incorporating a thin p-GaN layer surrounding the fins to enhance channel depletion. This design enables the use of relatively wide fin widths, such as 100 nm, while still maintaining effective control over the channel. Compared to conventional tri-gate devices, the proposed structure is capable of achieving a threshold voltage in the range of 1–3 V. This value is also influenced by the doping concentration of the p-GaN layer. Notably, when the p-GaN doping concentration reaches 2×10^{19} cm^{-3}, the device fails to fully turn on at a gate voltage of 6 V. Therefore, under the current design parameters, maintaining the p-GaN doping concentration within the range of 5×10^{18} cm^{-3} and 2×10^{9} cm^{-3} is essential to ensure a suitable threshold voltage and reliable device operation.

The multichannel architecture serves as a critical region in these devices, making it essential to analyze its operating behavior, particularly the distribution of the electric field. To achieve charge balance and enhance electric field management, a p-GaN cap layer is introduced on top of the multichannel region. Both simulation and experimental results indicate that the breakdown voltage is closely related to the channel length. The addition of the p-GaN cap contributes to extending the electric field through the formation of a RESURF effect, thereby improving the device's breakdown performance.

REFERENCES

[1] U. K. Mishra, P. Parikh, and Yi-Feng Wu, "AlGaN/GaN HEMTs-an overview of device operation and applications," *Proc. IEEE*, vol. 90, no. 6, pp. 1022–1031, Jun. 2002, doi: 10.1109/JPROC.2002.1021567.

[2] J. P. Kozak *et al.*, "Stability, Reliability, and Robustness of GaN Power Devices: A Review," *IEEE Trans. Power Electron.*, vol. 38, no. 7, pp. 8442–8471, Jul. 2023, doi: 10.1109/TPEL.2023.3266365.

[3] Y. Zhang, F. Udrea, and H. Wang, "Multidimensional device architectures for efficient power electronics," *Nat. Electron.*, vol. 5, no. 11, pp. 723–734, Nov. 2022, doi: 10.1038/s41928-022-00860-5.

[4] H. Ishida *et al.*, "GaN-based natural super junction diodes with multi-channel structures," in *2008 IEEE International Electron Devices Meeting*, San Francisco, CA, USA: IEEE, Dec. 2008, pp. 1–4. doi: 10.1109/IEDM.2008.4796636.

[5] O. Ambacher *et al.*, "Two-dimensional electron gases induced by spontaneous and piezoelectric polarization charges in N- and Ga-face AlGaN/GaN heterostructures," *J. Appl. Phys.*, vol. 85, no. 6, pp. 3222–3233, Mar. 1999, doi: 10.1063/1.369664.

[6] Y. Ma *et al.*, "Tri-gate GaN junction HEMT," *Appl. Phys. Lett.*, vol. 117, no. 14, p. 143506, Oct. 2020, doi: 10.1063/5.0025351.

[7] J. Ma, C. Erine, P. Xiang, K. Cheng, and E. Matioli, "Multi-channel tri-gate normally-on/off AlGaN/GaN MOSHEMTs on Si substrate with high breakdown voltage and low ON-resistance," *Appl. Phys. Lett.*, vol. 113, no. 24, p. 242102, Dec. 2018, doi: 10.1063/1.5064407.

[8] L. Nela *et al.*, "Multi-channel nanowire devices for efficient power conversion," *Nat. Electron.*, vol. 4, no. 4, pp. 284–290, Mar. 2021, doi: 10.1038/s41928-021-00550-8.

[9] Y. Zhang *et al.*, "GaN FinFETs and trigate devices for power and RF applications: review and perspective," *Semicond. Sci. Technol.*, vol. 36, no. 5, p. 054001, May 2021, doi: 10.1088/1361-6641/abde17.

[10] Y. Ma, M. Xiao, Z. Du, H. Wang, and Y. Zhang, "Tri-Gate GaN Junction HEMTs: Physics and Performance Space," *IEEE Trans. Electron Devices*, vol. 68, no. 10, pp. 4854–4861, Oct. 2021, doi: 10.1109/TED.2021.3103157.

[11] "SentaurusTM Device User Guide, Synopsys, Mountain View, CA, USA, 2015.".

[12] H. Ishida *et al.*, "Unlimited High Breakdown Voltage by Natural Super Junction of Polarized Semiconductor," *IEEE Electron Device*

Lett., vol. 29, no. 10, pp. 1087–1089, Oct. 2008, doi: 10.1109/LED.2008.2002753.

[13] L. Nela, C. Erine, A. M. Zadeh, and E. Matioli, "Intrinsic Polarization Super Junctions: Design of Single and Multichannel GaN Structures," *IEEE Trans. Electron Devices*, vol. 69, no. 4, pp. 1798–1804, Apr. 2022, doi: 10.1109/TED.2022.3151558.

[14] J. Sun *et al.*, "Substantiation of buried two dimensional hole gas (2DHG) existence in GaN-on-Si epitaxial heterostructure," *Appl. Phys. Lett.*, vol. 110, no. 16, p. 163506, Apr. 2017, doi: 10.1063/1.4980140.

[15] M. Xiao *et al.*, "Multi-Channel Monolithic-Cascode HEMT (MC ^{2}-HEMT). A New GaN Power Switch up to 10 kV," in *2021 IEEE International Electron Devices Meeting (IEDM)*, San Francisco, CA, USA: IEEE, Dec. 2021, p. 5.5.1-5.5.4. doi: 10.1109/IEDM19574.2021.9720714.

[16] Q. He, H. Wang, M. Xiao, Y. Zhang, K. Sheng, and F. Udrea, "Numerical Simulation and Analytical Modeling of Multichannel AlGaN/GaN Devices," *IEEE Trans. Electron Devices*, vol. 71, no. 3, pp. 1710–1717, Mar. 2024, doi: 10.1109/TED.2024.3359165.

[17] L. Nela *et al.*, "High-Performance Enhancement-Mode AlGaN/GaN Multi-Channel Power Transistors," in *2021 33rd International Symposium on Power Semiconductor Devices and ICs (ISPSD)*, Nagoya, Japan: IEEE, May 2021, pp. 143–146. doi: 10.23919/ISPSD50666.2021.9452238.

[18] J. Ma and E. Matioli, "High Performance Tri-Gate GaN Power MOSHEMTs on Silicon Substrate," *IEEE Electron Device Lett.*, vol. 38, no. 3, pp. 367–370, Mar. 2017, doi: 10.1109/LED.2017.2661755.

[19] J. C. Ranuárez, M. J. Deen, and C.-H. Chen, "A review of gate tunneling current in MOS devices," *Microelectron. Reliab.*, vol. 46, no. 12, pp. 1939–1956, Dec. 2006, doi: 10.1016/j.microrel.2005.12.006.

[20] C. Wood and D. Jena, *Polarization Effects in Semiconductors: From Ab Initio Theory to Device Applications*. Springer Science & Business Media, 2007.

Linear-ESO Based Control for Wide Input Range Partial Power Regulated DC-DC Converter

Dept. of Electrical Engineering, Harbin Institute of Technology, Harbin, China

Xikun Sang, Shanshan Gao, Yijie Wang, Dianguo Xu

Abstract— Traditional resonant converters suffer from resonance point shifts under frequency adjustments, leading to efficiency sacrifice and limited voltage regulation ranges. In recent years, partial power voltage regulation has been widely attracted due to advantages of low device stress, high efficiency and easy control in wide range. However, traditional fixed-parameter PI control is less effective and low robustness in a wide operating range due to the sensitivity to the system operating points. Therefore, an active disturbance rejection control (ADRC) strategy with disturbance differential observation based on high order linear ESO is proposed for partial power regulation systems with wide input range. The detailed analysis of the mathematical model, frequency-domain characteristics, the influence of control parameters for the proposed controller are presented in this paper. Finally, the effectiveness of the proposed control strategy was verified through experimental testing compared with PI control.

Keywords— *LADRC, ESO, partial power, frequency-domain analysis.*

I. INTRODUCTION

Resonant converters have been widely applied due to their advantageous soft switching characteristics and efficient transmission capabilities. However, the resonant operating point will deviate from the ideal state by frequency modulation, leading to additional reactive power and affecting the soft switching range of power devices potentially [1]. Although the DC transformer (DCX) state is conducive to enhancing the efficiency performance, it fails to accommodate the demands of a wide operating range. To address the issue of limited operating range of traditional resonant converters, numerous control schemes and topological improvements have been proposed to broaden the operating voltage range, which can roughly be divided into two categories: variable topology structure and hybrid mode control [2]. For variable topology control methods, during the transition process of different topology structures, significant transient inrush currents may occur, reducing system reliability, especially under high load power conditions. Moreover, additional auxiliary switches and other devices are required for transitions, resulting in additional losses and costs. The variable mode hybrid control method is also an effective way to achieve a wide voltage gain range by combining various control methods such as frequency regulation and phase shifting control, but the complexity of control strategies increases at the same time. Further research and optimization are needed to balance system efficiency, device cost, and soft switching characteristics [3].

In order to adapt to a wide input voltage range while ensuring high efficiency, the concept of partial power processing has attracted increasing attention in recent years. By incorporating an additional DC-DC converter into the system for dynamic voltage regulation, it only needs to transmit a portion of the total power, thereby reducing power loss of this part, known as partial power voltage regulator (PPVR). Meanwhile, the resonant converter in the system can be designed at the DCX state, transferring most of the total power, achieving efficient performance [4]-[7]. The PPVR typically can be selected as non-isolated DC-DC converters with simple structures and easy adjustment. The typical control strategies for DC-DC converters mainly include proportional-integral (PI) control, sliding mode control, active disturbance rejection control (ADRC), fuzzy control and etc. Among these, PI control is the most commonly used, improving both the response speed and steady-state accuracy by introducing proportional and integral terms simultaneously. A dual-loop PI controller based on voltage and current is applied to control a buck-boost converter in [8], but the significant voltage overshoot was observed in the experiment even over a limited voltage range. Additionally, PI control is usually designed for one nominal operation point, but the numerical small-signal model of the system will change with the operation point variations [9]. In order to achieve better dynamic responses under the operational point uncertainty and disturbances, lag-based sliding mode control was utilized as an inner loop control for a secondary boost converter in [10]. A second-order sliding mode control strategy for synchronous buck DC-DC converters was also employed in [11], achieving rapid step load, startup processes and robustness against parameter uncertainties. However, a major limitation of sliding mode control lies in its chattering phenomenon and non-constant switching frequency.

A fuzzy neural network model predictive control method is proposed in [12], which effectively overcomes instability issues and achieves rapid transient response performance. However, this controller is relatively complex and requires hardware with strong computational capabilities. In recent years, disturbance rejection control has attracted increasing attention [13]-[15], and the core idea is the directive compensation for the impact of unknown uncertain disturbances effectively. Among that, the Extended State Observer (ESO) based control requires the least amount of information, only the order of the system. On this basis, linear ESO further simplifies parameter design, making it easier to implement [16]. Therefore, automatic disturbance rejection control (ADRC) based on linear ESO has become increasingly popular and widely applied in power electronics and industrial fields.

For the partial power regulation system with a wide input range proposed in [7], an enhanced ADRC control strategy based on high-order linear ESO is presented in this paper. The limitations of traditional PI control under a wide operating range were analyzed in detail. Furthermore, the mathematical model, frequency domain characteristics and the influence of control parameters of the proposed controller were studied in this paper. Finally, the dynamic response performance of the system under this paper and PI control is compared and analyzed through experimental testing, which verifies the good dynamic regulation capability of the proposed control method under wide operating range.

979-8-3315-1110-4/25 $31.00 © 2025 IEEE

II. PI CONTROLLER DESIGN AND LIMITATION ANALYSIS

Compared with single-stage power converters, partial power regulation system contains multiple power transmission stages, including DCX power stage and the PPVR regulation stage. The DCXs operate in the open-loop state with constant switching frequency and duty cycle, achieving a constant voltage conversion ratio. Therefore, the two-port network of DCX can be considered as a voltage source in series with output impedance Z_o, which facilitates the derivation of the equivalent circuit model for the quasi-single-stage partial power regulation systems with pre-stage regulator and post-stage regulator, as illustrated in Fig. 1.

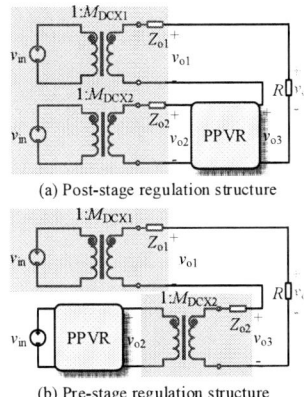

(a) Post-stage regulation structure

(b) Pre-stage regulation structure

Fig. 1 Equivalent circuit model of quasi single-stage partial power regulation system.

In order to obtain the transfer function of the proposed system, it is necessary to perform small-signal modeling for both DCX and PPVR. The system output voltage is regulated by the closed-loop PWM control of PPVR, and the small-signal model of that can be derived by state space averaging (SSA) method. In contrast, the DCXs operate at the resonant mode, so the extended describing function (EDF) modeling approach is employed. The detailed modeling process has been discussed in [7]. Based on the equivalent small-signal circuit models, the transfer function from the control duty cycle d_1 to the system output voltage v_o is expressed as (1).

$$\frac{\hat{v}_o(s)}{\hat{d}_1(s)} = \frac{I_L(D_1^2 Z_{o2} + sL) + D_1 D_1' I_L Z_{o2} - (V_{o2} - V_{o3})D_1'}{\left(\dfrac{D_1^2 Z_{o2} + sL}{R}\right)(sC_o R + sC_o Z_{o1} + 1)} \quad (1)$$

The compensation controller design will be carried out on the original system, and a closed-loop system control diagram is constructed based on the negative feedback control principle, as shown in Fig. 2. It mainly includes the system power stage $G(s)$, sampling network $H(s)$, and controller $G_c(s)$. The crossover frequency of compensated open-loop system is designed at 5kHz. A PI controller is used for compensation first, and the frequency characteristic curves of the open-loop system before and after compensation are obtained as Fig. 3.

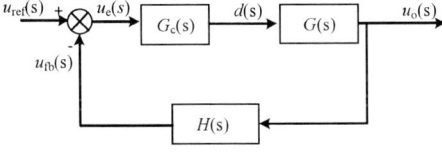

Fig. 2 Simplified closed-loop control diagram of system

From Fig. 3, it can be seen that the phase margin of the compensated open-loop system is about 60°, which improves the dynamic characteristics compared with original system. However, the numerical transfer function of the system will change as the operating point varies, so the frequency-domain characteristics of the compensated open-loop system will change accordingly. As the Bode plots of the open-loop system before and after compensation at 42V input shown in Fig. 4, it can be seen that the phase margin and control performance of compensated system reduces significantly at the same compensation parameters.

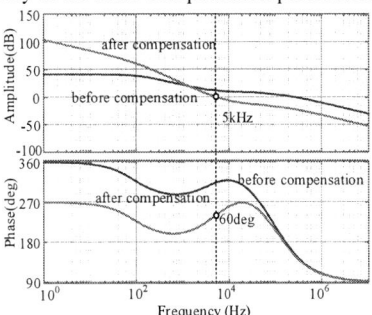

Fig. 3 Open-loop Bode diagram of the system before and after compensation (V_{in}=24V).

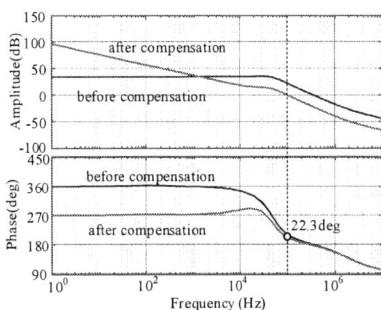

Fig. 4 Open-loop Bode diagram of the system before and after compensation (V_{in}=42V).

III. MATHEMATICAL MODEL OF HIGH ORDER LESO

The linear active disturbance rejection control (LADRC) consists of two components primarily, linear extended state observer (LESO) and linear state error feedback (LSEF), which are responsible for the total disturbance estimation and state variable errors compensation respectively. A reduced-order simplified model of the system is considered to provide a practical guideline of controller order selection, so a third-order LADRC controller is adopted here to achieve the system dynamic control.

Firstly, the transfer function of the controlled system is written in the form of differential equations as follows:

$$y^{(n)} = f_1\left(t, y, \dot{y}, \ddot{y}, ..., y^{(n-1)}, \omega\right) + b_0 u \quad (2)$$

where, n represents the controller order, t represents the time-varying state of the system, ω is the external disturbance. u and y are the input and output of the control system, b_0 is the control gain, and f_1 is the total disturbance of the system.

The state variables are introduced to represent the output quantities as well as the total disturbance of the system, as defined in (2).

$$\begin{cases} x_1 = y \\ x_2 = \dot{y} \\ \dots \\ x_n = \overset{n-1}{y} \\ x_{n+1} = f_1 \end{cases} \quad (2)$$

The equation (2) is converted into the form of state space equation as shown in (3), and $h = \dot{f}_1$.

$$\begin{cases} \dot{x} = Ax + Bu + Eh \\ y = Cx \end{cases} \quad (3)$$

where, $A = \begin{bmatrix} 0 & 1 & 0 & \dots & 0 \\ 0 & 0 & 1 & \dots & 0 \\ \vdots & \vdots & \vdots & \ddots & \vdots \\ 0 & 0 & 0 & 0 & 1 \\ 0 & 0 & 0 & 0 & 0 \end{bmatrix}_{(n+1)\times(n+1)}$, $B = \begin{bmatrix} 0 \\ 0 \\ \vdots \\ b_0 \\ 0 \end{bmatrix}$, $C = \begin{bmatrix} 1 \\ 0 \\ \vdots \\ 0 \\ 0 \end{bmatrix}^T$, $E = \begin{bmatrix} 0 \\ 0 \\ \vdots \\ 0 \\ 1 \end{bmatrix}$.

The LESO can be established according to (3), as shown in (4).

$$\begin{cases} \dot{z} = (A\text{-}LC)z + Bu + Ly \\ \hat{y} = Cz \end{cases} \quad (4)$$

where, $z = [z_1, z_2, \dots, z_{n+1}]^T$ is the state vector of LESO for the estimation of the state variables $x_1 \sim x_{n+1}$. $\hat{y}$ is the output of LESO, which is used for estimating the actual system output y. $L = [\beta_1, \beta_2, \dots \beta_{n+1}]^T$ is the gain matrix of LESO.

Further, a proportional-derivative controller control law is used to regulate the output voltage, as shown in (5).

$$\begin{cases} u = \dfrac{u_0 - z_{n+1}}{b_0} \\ u_0 = k_p(r - z_1) - k_{d1}z_2 - k_{d2}z_3 - \dots - k_{d(n-1)}z_n \end{cases} \quad (5)$$

where, r is the reference signal, k_p, k_{d1}, $k_{d2}\dots k_{d(n-1)}$ are the LESF controller parameters.

It can be seen that the control parameters for a third-order LADRC controller include b_0, $\beta_1\sim\beta_4$, as well as k_p, k_{d1} and k_{d2}. By adjusting the observer bandwidth and control bandwidth, the parameter tuning process can be simplified effectively. Assuming that the closed-loop system has three poles at $-\omega_c$, consequently, the relationship between the control parameters and control bandwidth ω_c can be derived as $k_p = \omega_c^3, k_{d1} = 3\omega_c^2, k_{d2} = 3\omega_c$.

The observation error will approach zero over time for stability, by solving for the eigenvalues of A-LC and placing the characteristic poles at locations corresponding to the observer bandwidth ω_0, the relationship between the LESO gain and observer bandwidth ω_0 can be determined as $\beta_1 = 4\omega_0, \beta_2 = 6\omega_0^2, \beta_3 = 4\omega_0^3, \beta_4 = \omega_0^4$.

According to (4), the expressions for observation z_1 regarding the control input and output quantities can be obtained as

$$\begin{cases} \dfrac{z_1(s)}{y(s)} = \dfrac{\beta_1 s^3 + \beta_2 s^2 + \beta_3 s^3 + \beta_4}{s^4 + \beta_1 s^3 + \beta_2 s^2 + \beta_3 s + \beta_4} \\ \dfrac{z_1(s)}{u(s)} = \dfrac{b_0 s}{s^4 + \beta_1 s^3 + \beta_2 s^2 + \beta_3 s + \beta_4} \end{cases} \quad (6)$$

According to (6), the amplitude-frequency characteristic curves of z_1 regarding to y and u can be obtained for different ω_0, as shown in Fig. 5. It can be seen that increasing ω_0 is

beneficial for z_1 to make real-time and accurate prediction of y and reduce the low-frequency disturbance of u. However, it will also amplify the interference of high-frequency noise at the output side.

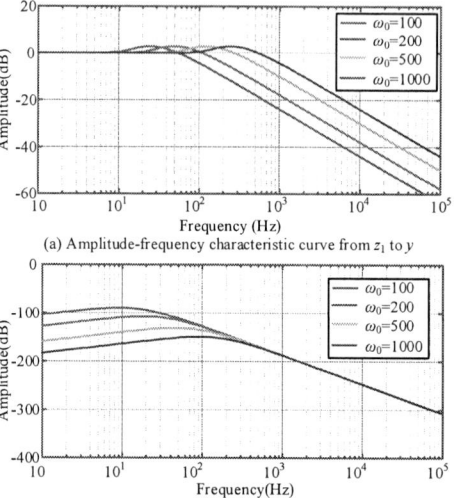

(a) Amplitude-frequency characteristic curve from z_1 to y

(b) Amplitude-frequency characteristic curve from z_1 to u

Fig. 5 Amplitude-frequency curve of observed quantity z_1 related to output quantity y and control quantity u.

III. ANALYSIS AND DESIGN OF ENHANCED LADRC

In order to better balance the coupling effects of observer bandwidth on low-frequency disturbances and high-frequency noise, an enhanced high-order LADRC controller based on disturbance differential observation is proposed for the wide input partial power regulated system. Additional state variable is further introduced as $x_5 = \dot{f}_1$ to describe the disturbance change rate. Further, a high-order disturbance observation mechanism is shown as (7) to improve the coupling contradiction between low-frequency disturbance rejection and high-frequency noise suppression.

$$\begin{cases} e = y - z_1 \\ \dot{z}_1 = \beta_1 e + z_2 \\ \dot{z}_2 = \beta_2 e + z_3 \\ \dot{z}_3 = \beta_3 e + z_4 + b_0 u \\ \dot{z}_4 = \beta_4 e + z_5 \\ \dot{z}_5 = \beta_5 e \\ u = \dfrac{u_0 - z_4}{b_0} \\ u_0 = k_p(r - z_1) - k_{d1}z_2 - k_{d2}z_3 \end{cases} \quad (7)$$

where, z_5 is the estimation of the total disturbance change rate and $\beta_1 \sim \beta_5$ is the bandwidth gain of LESO.

Further, according to the bandwidth configuration method, the improved LESO observer gain can be derived as: $\beta_1 = 5\omega_0, \beta_2 = 10\omega_0^2, \beta_3 = 10\omega_0^3, \beta_4 = 5\omega_0^4, \beta_5 = \omega_0^5$. According to (7), the expressions for the improved LESO observations can be obtained as (8).

The observation tracking errors are defined as $e_1 = z_1 - y, e_2 = z_2 - \dot{y}, e_3 = z_3 - \ddot{y}, e_4 = z_4 - f_1, e_5 = z_5 - f_1'$. The output y and control quantity u are both taken as step signals with amplitude K, i.e., $y(s)=u(s)=K/s$. According to (9), it can be indicated that $\lim_{s \to 0} se_i = 0, i = 1, 2, 3, 4, 5$, which demonstrated that the improved LESO has good convergence and unbiased estimation of state variables and generalized disturbance.

$$
\begin{cases}
z_1 = \dfrac{\beta_1 s^4 + \beta_2 s^3 + \beta_3 s^2 + \beta_4 s + \beta_5}{s^5 + \beta_1 s^4 + \beta_2 s^3 + \beta_3 s^2 + \beta_4 s + \beta_5} y + \dfrac{b_0 s^2}{s^5 + \beta_1 s^4 + \beta_2 s^3 + \beta_3 s^2 + \beta_4 s + \beta_5} u \\[2mm]
z_2 = \dfrac{\beta_2 s^4 + \beta_3 s^3 + \beta_4 s^2 + \beta_5 s}{s^5 + \beta_1 s^4 + \beta_2 s^3 + \beta_3 s^2 + \beta_4 s + \beta_5} y + \dfrac{\left(s^3 + s^2 \beta_1\right) b_0}{s^5 + \beta_1 s^4 + \beta_2 s^3 + \beta_3 s^2 + \beta_4 s + \beta_5} u \\[2mm]
z_3 = \dfrac{\beta_3 s^4 + \beta_4 s^3 + \beta_5 s^2}{s^5 + \beta_1 s^4 + \beta_2 s^3 + \beta_3 s^2 + \beta_4 s + \beta_5} y + \dfrac{\left(s^4 + s^3 \beta_1 + s^2 \beta_2\right) b_0}{s^5 + \beta_1 s^4 + \beta_2 s^3 + \beta_3 s^2 + \beta_4 s + \beta_5} u \\[2mm]
z_4 = \dfrac{\beta_4 s^4 + \beta_5 s^3}{s^5 + \beta_1 s^4 + \beta_2 s^3 + \beta_3 s^2 + \beta_4 s + \beta_5} y - \dfrac{\left(s \beta_4 + \beta_5\right) b_0}{s^5 + \beta_1 s^4 + \beta_2 s^3 + \beta_3 s^2 + \beta_4 s + \beta_5} u \\[2mm]
z_5 = \dfrac{\beta_5 s^4}{s^5 + \beta_1 s^4 + \beta_2 s^3 + \beta_3 s^2 + \beta_4 s + \beta_5} y - \dfrac{s \beta_5 b_0}{s^5 + \beta_1 s^4 + \beta_2 s^3 + \beta_3 s^2 + \beta_4 s + \beta_5} u
\end{cases} \quad (8)
$$

A comparative analysis is conducted on the disturbance rejection characteristics by adjusting the bandwidths of the original observer and improved observer to make the tracking characteristics of the output variable y consistent, and the high-frequency noise suppression effects are ensured to be the same. The transfer function of z_1 regarding to u can be obtained for the improved LESO controller, and the amplitude-frequency characteristics of that are compared with the original LADRC controller, as shown in Fig. 6. Besides, according to (2) and (8), the relationship between observation error e_4 and total disturbance f_1 can be obtained as (9), and the corresponding amplitude-frequency characteristic curves are illustrated as Fig. 7. It can be seen that the improved LADRC achieves gain attenuation in the low-frequency band, which is more conducive to reducing the influence of low-frequency interference and accurate prediction of total disturbance.

$$
e_4 = -\frac{s^5 + \beta_1 s^4 + \beta_2 s^3 + \beta_3 s^2}{s^5 + \beta_1 s^4 + \beta_2 s^3 + \beta_3 s^2 + \beta_4 s + \beta_5} f_1 \quad (9)
$$

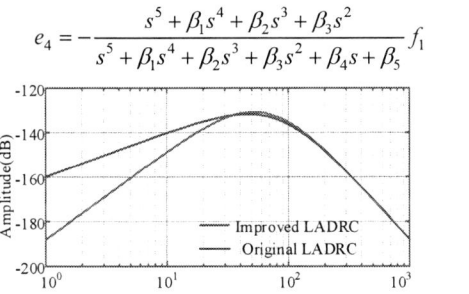

Fig. 6 Amplitude-frequency characteristic comparison of z_1 to u.

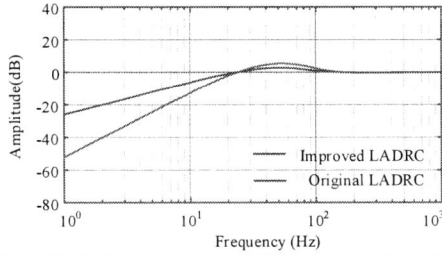

Fig. 7 Amplitude-frequency characteristic comparison of e_4 to f_1.

According to (8), the structural diagram of improved LADRC controller is illustrated as Fig. 8.

Further, the frequency domain performance and the control parameters impact of the closed-loop control system are studied. The equivalent transfer function of the closed-loop system can be obtained as (10), and the equivalent closed-loop control diagram is constructed as Fig. 9, where, $H_r(s)$ is the input filtering stage, $C(s)$ is the feedback compensation stage, and $G(s)$ is the power stage of the system.

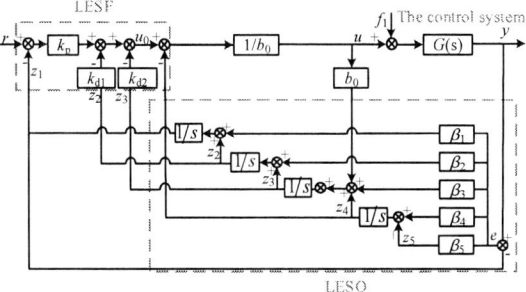

Fig. 8 Block diagram of improved LADRC control structure

$$
u(s) = H_r(s) r(s) - C(s) y(s) \quad (10)
$$

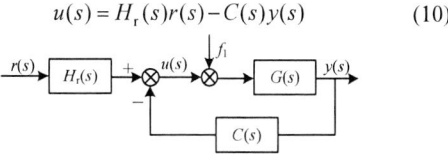

Fig. 9 Equivalent closed-loop system control diagram.

According to (7) and (8), the transfer functions of $H_r(s)$ and $C(s)$ can be obtained as (11) and (12), where, $n_0 = k_{p1}\beta_5$, $n_1 = k_{p1}\beta_4 + k_{d1}\beta_5$, $n_2 = k_{p1}\beta_3 + k_{d1}\beta_4 + k_{d2}\beta_5$, $n_3 = \beta_5 + k_{p1}\beta_2 + k_{d1}\beta_3 + k_{d2}\beta_4$, $n_4 = \beta_4 + k_{p1}\beta_1 + k_{d1}\beta_2 + k_{d2}\beta_3$, $m_4 = \beta_1 + k_{d2}$, $m_3 = \beta_2 + k_{d1} + k_{d2}\beta_1$, $m_2 = \beta_3 + k_{p1} + k_{d1}\beta_1 + k_{d2}\beta_2$.

$$
H_r(s) = \frac{k_{p1}\left(s + \omega_0\right)^5}{b_0\left(s^5 + m_4 s^4 + m_3 s^3 + m_2 s^2\right)} \quad (11)
$$

$$
C(s) = \frac{n_4 s^4 + n_3 s^3 + n_2 s^2 + n_1 s + n_0}{b_0\left(s^5 + m_4 s^4 + m_3 s^3 + m_2 s^2\right)} \quad (12)
$$

According to Fig. 9, the closed-loop transfer function from the total disturbance f_1 to the system output y is derived as (13), and the closed-loop transfer function from the system output y to the reference input r is derived as (14).

$$
G_{yf_1}(s) = \frac{b_0\left(s^5 + m_4 s^4 + m_3 s^3 + m_2 s^2\right)G(s)}{b_0\left(s^5 + m_4 s^4 + m_3 s^3 + m_2 s^2\right) + G(s)\left(n_4 s^4 + n_3 s^3 + n_2 s^2 + n_1 s + n_0\right)} \quad (13)
$$

$$
G_{yr}(s) = \frac{k_{p1}\left(s + \omega_0\right)^5 G(s)}{b_0\left(s^5 + m_4 s^4 + m_3 s^3 + m_2 s^2\right) + G(s)\left(n_4 s^4 + n_3 s^3 + n_2 s^2 + n_1 s + n_0\right)} \quad (14)
$$

According to (13) and (14), the closed-loop amplitude frequency characteristic curves of the system anti-interference and tracking performance under different parameters are shown in Figs. 10 and 11. As shown in Fig. 10(a), increasing ω_c reduces the amplitude in the low-frequency band, which is beneficial for enhancing the system disturbance rejection capability. Meanwhile, Fig. 11(a) indicates that a larger ω_c helps improve the response speed and real-time tracking ability. However, excessive ω_c may amplify high-frequency noise at the output, potentially leading to system oscillations or even instability. Additionally, according to Figs. 10 and 11, decreasing b_0 also enhances disturbance rejection but increases the impact of high-frequency noise at the output. When b_0 is too small, it may cause oscillations and disrupt the stable operation of the system. In summary, a balance should be comprehensively considered among the disturbance rejection capability, response speed, and noise suppression capability during the practical parameter design.

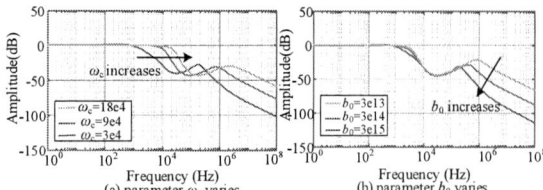

Fig. 10 The parameters impact on the system tracking characteristics.

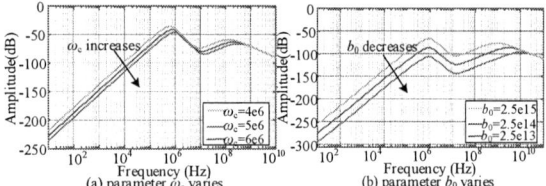

Fig. 11. The parameters impact on the system anti-disturbance characteristics.

IV. EXPERIMENTAL RESULTS AND ANALYSIS

In order to verify the correctness of the theoretical analysis in the above sections, an experimental platform is constructed for validation, and the overall structural block diagram of the closed-loop control system is presented in Fig. 12, which mainly consists of the partial power regulated converter, sampling circuit, digital controller, and driving circuit. The digital controller employs the TMS320F28377 DSP digital processor. By sampling the system output voltage, the conduction duty cycle of PPVR is regulated by LADRC controller to achieve the closed-loop control. Meanwhile, both DCXs operate in the open-loop state with constant switching frequency.

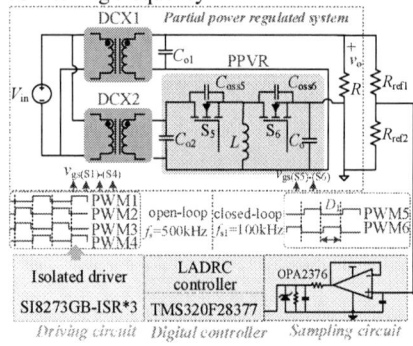

Fig. 12 Overall system structure block diagram.

As shown in Figs. 13 and 14, the experimental dynamic waveforms under LADRC and PI control are presented when the input voltage has a step variation. In the test results of Fig. 14, the waveforms from top to bottom represent the system input voltage v_{in} and total output voltage v_o. It can be observed that when v_{in} steps up, the output voltage v_o fluctuates within ±20V under PI control, leading to reduced stability and increased steady-state error. In contrast, as shown in Fig. 13, the waveforms from top to bottom represent the input voltage v_{in}, output voltage v_o, and gate-source voltage $v_{gs(S5)}$ of the main power switch in PPVR. It can be seen that the closed-loop control of the output voltage can be achieved by the duty cycle adjustment. The dynamic response performance is characterized by the overshoot level σ and settling time t_s during transient process, it can be seen that when v_{in} has an upward step change, the overshoot of v_o is approximately 9.5% with settling time 8ms. Conversely, as v_{in} steps down, the overshoot during the transition process

is 6.5% and the settling time is about 7ms. Compared with PI control, the output voltage fluctuation is reduced significantly, which improves the operation stability greatly.

Furthermore, the dynamic response test waveforms of the system when the load current abrupt changes between 0.25 A and 1 A are presented as Fig. 15. In Fig. 15(a), the waveforms from top to bottom represent the load current i_o, output voltage v_o, and gate-source drive voltage $v_{gs(S5)}$ of PPVR. It can be observed that during the sudden changes in load current, the proposed LADRC control strategy exhibits lower transient overshoot and shorter settling time. Specifically, during load current surges, the overshoot of the proposed control strategy is approximately 2%, representing a 63.6% reduction compared to PI control, with a recovery time of approximately 2.5ms, which is 54.4% shorter than PI control. While during the load current drops, the upward overshoot is 2.5%, and the recovery time is 1.75ms, both reduced by at least half compared to PI control. The experimental results confirm that the proposed control strategy exhibits strong robustness and good dynamic characteristics under different working conditions.

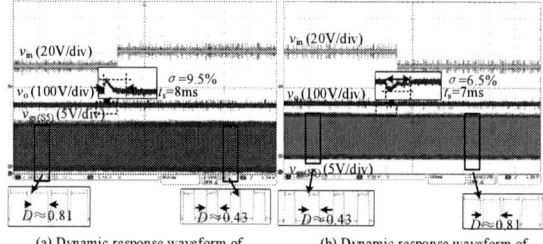

(a) Dynamic response waveform of sudden increase in input voltage   (b) Dynamic response waveform of sudden decrease in input voltage

Fig. 13 Closed-loop control test waveforms at full load in this paper.

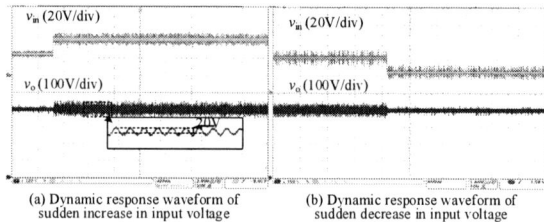

(a) Dynamic response waveform of sudden increase in input voltage   (b) Dynamic response waveform of sudden decrease in input voltage

Fig. 14 PI closed-loop control test waveforms at full load.

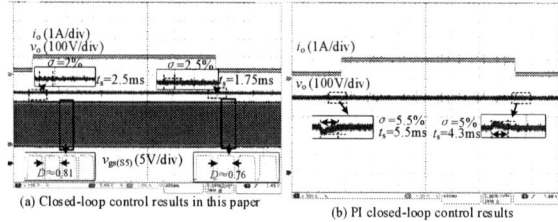

(a) Closed-loop control results in this paper   (b) PI closed-loop control results

Fig. 15 Dynamic test waveforms comparison of closed-loop control under load current abrupt.

V. CONCLUSION

In order to address the issues of low robustness and parameter sensitivity of traditional PI control under wide operating range, an enhanced high-order linear active disturbance rejection control strategy for wide input partial power regulation system is proposed in this paper. The corresponding mathematical model is established, and frequency-domain analysis and parameter characteristic analysis are conducted in detail. Finally, the effectiveness and application advantages of the proposed control method are verified through experimental results.

979-8-3315-1110-4/25 $31.00 © 2025 IEEE

REFERENCES

[1] X. Wu, H. Chen and Z. Qian, "1-MHz LLC Resonant DC Transformer (DCX) With Regulating Capability," *IEEE Trans. Ind. Electron.*, vol. 63, no. 5, pp. 2904-2912, May 2016.

[2] Y. Wei, Q. Luo and H. A. Mantooth, "LLC and CLLC resonant converters based DC transformers (DCXs): Characteristics, issues, and solutions," *CPSS Trans. Power Electron. Appl.*, vol. 6, no. 4, pp. 332-348, Dec. 2021.

[3] M. Dai, L. Cong, C. Wang, D. Sun and W. Gu, "An Adaptive Pulse-Width and Frequency Modulation Control Strategy for Light Load Efficiency Improvement in Half-Bridge LLC Converter with Wide Input Voltage Range," *2023 IEEE Energy Conversion Congress and Exposition (ECCE)*, 2023, pp. 3089-3094.

[4] R. Gu, J. Duan, D. Zhang and H. Liu, "Regulated Series Hybrid Converter With DC Transformer (DCX) for Step-Up Power Conversion," *IEEE Trans. Ind. Electron.*, vol. 69, no. 9, pp. 8961-8971, Sept. 2022.

[5] Z. Wu, Z. Wang, T. Liu, W. Xu, C. Chen and Y. Kang, "High Efficiency and High Power Density Partial Power Regulation Topology With Wide Input Range," *IEEE Trans. Power Electron.*, vol. 38, no. 2, pp. 2074-2091, Feb. 2023.

[6] T. Liu, X. Wu and S. Yang, "1 MHz 48–12 V Regulated DCX With Single Transformer," *IEEE J. Emerg. Sel. Top. Power Electron.*, vol. 9, no. 1, pp. 38-47, Feb. 2021.

[7] X. Sang, S. Gao, Y. Wang and D. Xu, "A High Step-up Partial Power Regulated DC-DC Converter with Zero Input Current Ripple and Wide Input Range," *IEEE Trans. Ind. Electron.*, vol. 71, no. 11, pp. 14097-14110, Nov. 2024.

[8] Li J, Liu J. "A Novel Buck-Boost Converter with Low Electric Stress on Components," *IEEE Trans. Ind. Electron.*, vol. 66, no. 4, pp. 2703-2713, Apr. 2019.

[9] Y. -C. Jeung and D. -C. Lee, "Voltage and Current Regulations of Bidirectional Isolated Dual-Active-Bridge DC-DC Converters Based on a Double-Integral Sliding Mode Control," *IEEE Trans. Power Electron.*, vol. 34, no. 7, pp. 6937-6946, July 2019.

[10] O. Lopez-Santos, L. Martinez-Salamero, G. Garcia, H. Valderrama-Blavi, and T. Sierra-Polanco, "Robust sliding-mode control design for a voltage regulated quadratic boost converter," *IEEE Trans. Power Electron.*, vol. 30, no. 4, pp. 2313-2327, Apr. 2015.

[11] R. Ling, D. Maksimovic and R. Leyva, "Second-Order Sliding-Mode Controlled Synchronous Buck DC-DC Converter," *IEEE Trans. Power Electron.*, vol. 31, no. 3, pp. 2539-2549, Mar. 2016.

[12] M. Gheisarnejad, A. Mohammadzadeh and M. -H. Khooban, "Model Predictive Control Based Type-3 Fuzzy Estimator for Voltage Stabilization of DC Power Converters," *IEEE Trans. Ind. Electron.*, vol. 69, no. 12, pp. 13849-13858, Dec. 2022.

[13] S. Zhuo, G. Arnaud, L. Guo, L. Xu, D. Paire, and F. Gao, "Active disturbance rejection voltage control of a floating interleaved DC-DC boost converter with switch fault consideration," *IEEE Trans. Power Electron.*, vol. 66, no. 12, pp. 12396–12406, Dec. 2019.

[14] S. Zhuo, A. Gaillard, L. Xu, D. Paire and F. Gao, "Extended State Observer-Based Control of DC–DC Converters for Fuel Cell Application," *IEEE Trans. Power Electron.*, vol. 35, no. 9, pp. 9923-9932, Sept. 2020.

[15] Y. Zuo, X. Zhu, L. Quan, C. Zhang, Y. Du, and Z. Xiang, "Active disturbance rejection controller for speed control of electrical drives using phase-locking loop observer," *IEEE Trans. Ind. Electron.*, vol. 66, no. 3, pp. 1748–1759, Mar. 2019.

[16] H. Yin, L. Cao, X. Zeng, K. H. Loo and C. Sun, "Fast Dynamic Control of Dual-Active-Bridge DC–DC Converter Based on an Adaptive Linear Extended State Observer," *IEEE Trans. Power Electron.*, vol. 39, no. 12, pp. 16019-16030, Dec. 2024.

A Multi-Objective Optimization Method Based on NSGA-II Algorithm for Electro-Thermal-Stress Collaborative Design of Intelligent Power Modules

Tao Xu
State Key Laboratory of Intelligent
Power Distribution Equipment and
System
Hebei University of Technology
Tianjin, China
2040687037@qq.com

Lei Ming
State Key Laboratory of Intelligent
Power Distribution Equipment and
System
Hebei University of Technology
Tianjin, China
minglei@hebut.edu.cn

Ningbo Li
Beijing Keytone Electronic Relay Corp
Beijing, China
ktisaac_lee@126.com

Zihang Gu
State Key Laboratory of Intelligent
Power Distribution Equipment and
System
Hebei University of Technology
Tianjin, China
lwj_qhb@sina.com

Zhiwei Jiao
State Key Laboratory of Intelligent
Power Distribution Equipment and
System
Hebei University of Technology
Tianjin, China
2572475217@qq.com

Zhen Xin
State Key Laboratory of Intelligent
Power Distribution Equipment and
System
Hebei University of Technology
Tianjin, China
xzh@hebut.edu.cn

Abstract—**Intelligent power module (IPM) has always been important for achieving efficient energy conversion due to its high integration level and high reliability. The IPM based on the insulated metal baseplate (IMB) can reduce parasitic inductance and improve its thermal cycle life. However, the low thermal conductivity of the insulating resin in the IMB makes the thermal performance of the module extremely sensitive to temperature changes. Strongly coupled with electrical-thermal-stress multi-physical characteristics, the impact of temperature sensitivity on performance is further exacerbated, limiting the heat dissipation efficiency and reliability of the module, especially in high power density applications. For that, this paper proposes a multi-objective optimization method based on the non-dominated sorting genetic algorithm II (NSGA-II), where the packaging structure and chip layout are ingeniously integrated into the Pareto frontier construction process. Simulations and experiments show that trade-offs are necessary since the three objectives cannot be optimized at once, with maximum reductions of 22.59% in loop inductance, 21.56% in chip junction temperature, and 33.43% in solder layer stress.**

Keywords—*power modules, electric-thermal-stress multi-physics, insulated metal baseplate, multi-objective optimization*

I. INTRODUCTION

The intelligent power module (IPM) primarily integrates power devices, driver circuits, protection circuits, and other components. It is widely used in fields such as home appliances, motor drives, smart grids, aerospace, and missile systems, demonstrating significant potential in high-reliability sectors like military applications [1, 2]. In recent years, IPM based on an insulated metal baseplate (IMB) has attracted interest in the industry due to its unique coefficient of thermal expansion(CTE)-matched packaging design. By adopting IMB technology, the effective chip mounting area is increased by 23%. Additionally, through the optimization of CTE in

materials, the thermal cycling lifetime of the module is enhanced [3].

However, the thermal conductivity of the insulating resin in IMB is relatively low, which makes the thermal performance of the module extremely sensitive to temperature changes. In addition, there are complex electro-thermal-stress multi-physical field coupling effects inside the power module, resulting in mutual constraints among parasitic parameters, thermal resistance, and module stress. To solve this problem, a multi-objective optimization method considering the effects of power cycling and thermal cycling is proposed in [4]. This method seeks the Pareto optimal solution between thermal resistance and mechanical stress through the non-dominated sorting genetic algorithm II (NSGA-II), thereby improving the long-term reliability and durability of the module. However, this method does not consider the influence of parasitic inductance on the module, resulting in limitations in the optimization theory. In [5], a multi-objective optimization design method for power modules based on inductance and thermal performance is proposed, and the key factors for achieving optimal performance are explored. However, this method does not consider the influence of the thickness of material layers on thermal and electrical properties, restricting its applicability. In addition, a mathematical model of thermal resistance and mechanical stress of power modules is proposed in [6] and verified using the multi-physical field finite element analysis tool COMSOL. However, this method does not consider the thermal coupling effect between chips, resulting in an incomplete optimization effect.

To address this issue, this paper proposes a comprehensive multi-objective optimization method that considers both packaging structure and chip layout from the perspective of electro-thermal-stress effects. This paper firstly establishes a set of quantitative indicators for the electro-thermal-stress multi-physics field, and then proposes a multi-objective optimization model with loop inductance, chip junction temperature, and solder layer stress as the objectives. The model is solved using NSGA-II to obtain the optimal solution

---

This study was supported in part by the National Nature Science Foundation of China under Grant 52207196 and in part by the Overseas Returnees Funding Program of Hebei Province under Grant C20230316 and funded by Hebei Natural Science Foundation under Grant E2024202265.

set. Experimental results validate the effectiveness of the proposed model.

II. STRUCTURE OF IMB-BASED IPM

A. Analysis of the IMB Structure

Compared to the traditional DBC structure, the IMB allows the top and bottom copper layers to be directly bonded to the resin insulation layer, thereby eliminating the substrate and solder layers and avoiding the failure points shown in the diagram, as illustrated in Fig. 1. The copper-clad circuit on the IMB is designed as an integrated structure, eliminating the need for traditional aluminum bonding wires for electrical connections, which effectively avoids the issue of stray inductance introduced by aluminum bonding wires. Furthermore, since the CTE of the various materials has been harmonized and aligned with that of copper, the expansion of each part remains consistent even under temperature fluctuations, thereby preventing the generation of thermal stress.

With the module volume remaining unchanged, the IMB offers greater design flexibility in the longitudinal structure compared to the traditional structure that uses a ceramic isolation substrate. For example, the copper-clad circuit above the insulating layer can be further thickened to reduce stray inductance, and the copper baseplate can also be thickened to optimize heat dissipation and mechanical stress distribution. However, the thickness of the resin insulating layer in the IMB is limited by the manufacturing process and is typically fixed at 0.12 mm, meaning it cannot be adjusted. Due to the significant changes in the IMB packaging structure, traditional optimization methods are no longer applicable. Therefore, there is an urgent need for an electro-thermal-stress multi-objective optimization method that comprehensively considers the new structure to fully leverage the technical advantages of the IMB and improve the module's overall performance.

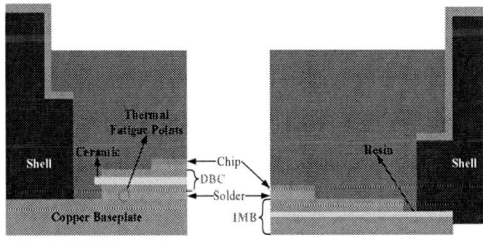

Fig. 1. Comparison between ceramic substrate and IMB structure.

B. Analysis of the IPM Layout

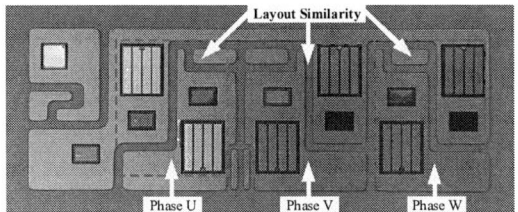

Fig. 2. Three sets of full bridge layouts with similar configurations.

This paper takes a full-bridge group of an IPM as the research object, as shown in Fig. 2. The optimization objectives are the comprehensive performance metrics across electro-thermal-stress multi-physics fields. The electrical indicator primarily focuses on the loop inductance of the

power module, the thermal indicator on the maximum junction temperature of the chips, and the stress indicator on the maximum stress in the solder layer.

III. PROPOSED ELECTRO-THERMAL-STRESS MULTI-PHYSICS QUANTIFICATION METRICS AND MULTI-OBJECTIVE OPTIMIZATION METHODOLOGY

A. Electrical Indicators

To extract the parasitic inductance of the bond wires in the power module, the inductance calculation formula for a long straight conductor with a circular cross-section can be used:

$$L = \frac{\mu_0 l}{2\pi}[\ln\frac{4l}{d} - 1] \tag{1}$$

μ_0 is the permeability of free space, l is the length of the bond wire, and d is the diameter of the bond wire.

However, since the parasitic inductance of the collector copper layer not only varies with the position of the chips but also changes with the thickness of the upper copper layer [7, 8], taking a full-bridge layout of an IPM as an example, as shown in Fig. 3, a coordinate system is established. In this system, the center coordinates of the IGBT chips are (x_{ci}, y_{ci}), and the center coordinates of the diode chips are (x_{di}, y_{di}), where $i = 1, 2$. The length of the bonding wires on the chips changes with the vertical position of the chips. The three-dimensional geometric model of the collector copper layer of the chips is imported into ANSYS Q3D software and solved using the finite element method. By employing parametric sweep analysis, the parasitic parameters of the IGBT chips at different position coordinates within the region can be extracted. The obtained results are then imported into MATLAB for data fitting and prediction. Finally, the fitting accuracy for the collector inductance of the chips is 0.9978 and 0.9975, respectively.

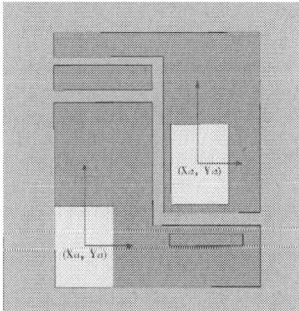

Fig. 3. Collector copper layer parasitic parameterization scan diagram.

B. Thermal Indicators

Based on the heat transfer path shown in Fig. 1, assuming the thermal diffusion angle is constant in each layer, the thermal resistance calculation expression for each material layer of the module is [8, 9]:

$$R_{thi} = \int_0^{t_i} \frac{1}{k_i A_i(z)}dz \tag{2}$$

In the formula, t_i is the thickness of each material layer in the module, k_i is the thermal conductivity coefficient of the material, and A_i is the effective thermal conductivity area. Since the isothermal lines of the square chip are distributed in a circular pattern, the effective heat conduction area A_i can be approximately expressed as

$$A_i = \pi(r + z\tan\alpha)^2 \quad (3)$$

Where r is the radius of the effective heat conduction area, and α is the thermal diffusion angle.

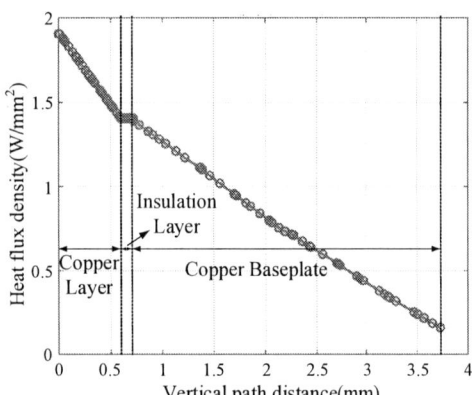

Fig. 4. Heat flux density variation curve.

The heat flux density variation curve is employed in this study to determine the thermal diffusion angle and derive the target parameter through quantitative analysis. Since the chip and solder layer have the same area shape, the effective heat transfer area is fixed, and thermal diffusion can be ignored. Therefore, starting from the copper layer on the IMB, the thermal resistance is calculated by considering the thermal diffusion angle. To achieve this, thermal simulations are performed in COMSOL to obtain a series of heat flux density data along the vertical downward path at the center of the chip, from which the variation curve is plotted, as shown in Fig. 4. The horizontal axis represents the distance from a point along the path to the top of the upper copper layer.

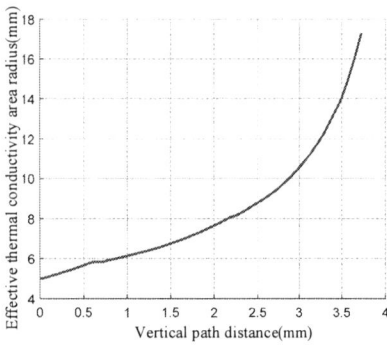

Fig. 5. Curve of effective thermal conductivity area radius variation.

The curve of the effective thermal conductivity area radius change is shown in Fig. 5, and the slope of the curve reflects the size of the thermal diffusion angle. By combining (2), the thermal resistance values of each layer can be calculated, as shown in Table I.

TABLE I. MATERIAL PARAMETERS AND THERMAL RESISTANCE OF EACH LAYER

Layer	Material	k(W/(m·K))	R_{th}(K/W)
Chip	Silicon	148	0.0199
Solder	SAC305	63.2	0.0450
Upper copper	Cu	395	0.0176
Insulation resin	Resin	10	0.1143
Copper of baseplate	Cu	395	0.0372
Convective heat transfer	-	-	0.2041

However, for a multi-chip module, the presence of multiple heat sources will result in thermal cross-coupling effects. The expression for the thermal impedance matrix model of the power module with multiple chips is:

$$\begin{bmatrix} T_{j1} \\ T_{j2} \\ \vdots \\ T_{jn} \end{bmatrix} = \begin{bmatrix} R_{11} & R_{12} & \cdots & R_{1n} \\ R_{21} & R_{22} & \cdots & R_{2n} \\ \vdots & \vdots & \ddots & \vdots \\ R_{n1} & R_{n2} & \cdots & R_{nn} \end{bmatrix} \begin{bmatrix} P_1 \\ P_2 \\ \vdots \\ P_n \end{bmatrix} + \boldsymbol{T}_a \quad (4)$$

In the thermal resistance matrix, the diagonal elements represent self-heating thermal resistance corresponding to temperature rise caused by self-generated chip losses, while the off-diagonal terms quantify cross-coupling thermal resistance characterizing inter-chip thermal interactions.

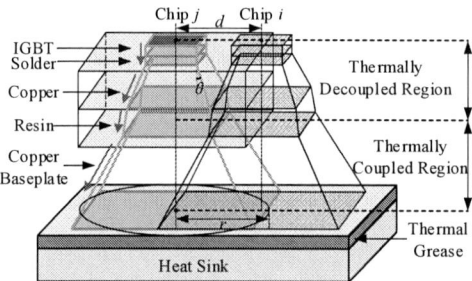

Fig. 6. Diagram of thermal diffusion and thermal coupling effects.

The intersecting cross-section divides the heat flow diffusion region into upper and lower parts, as shown in Fig. 6. Since heat flows from the high-temperature region to the low-temperature region, the coupling thermal resistance of chip j to chip i only includes the thermal resistance of the lower part of the cross-section. Specifically, it is divided into two components: the coupling conductive thermal resistance and the coupling convective thermal resistance. The coupled heat transfer thermal resistance can be directly solved by the cross-section parameter based on (2). For the convective heat transfer coupling effect between chips, the convective heat transfer thermal resistance between chip j and chip i can be expressed as:

$$R_{conv\_ij} = \frac{A_{ij}}{A_j} R_{conv\_j} \quad (5)$$

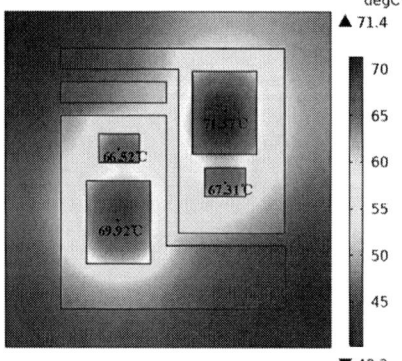

Fig. 7. Multi-chip thermal coupling simulation results.

Where R_{conv_j} is the convective heat transfer thermal resistance included in the self-heating resistance of chip j. A multiphysics simulation was performed in COMSOL under steady-state thermal analysis conditions, with a power configuration of 20 W per diode and 70 W per IGBT. Fig. 7 shows the resulting thermal field distribution and power dissipation characteristics.

Based on the chip layout shown in Fig. 7, build its thermal impedance matrix model:

$$R_{th} = \begin{bmatrix} 0.4381 & 0.0801 & 0.2241 & 0.1621 \\ 0.0801 & 0.4381 & 0.1657 & 0.2296 \\ 0.1303 & 0.0896 & 1.0848 & 0.1319 \\ 0.0891 & 0.1332 & 0.1319 & 1.0848 \end{bmatrix} \quad (6)$$

The thermal impedance matrix exhibits reciprocal symmetry in cross-coupling components: homogeneous chip pairs demonstrate identical mutual thermal impedance magnitudes, whereas heterogeneous combinations show notably distinct values.

TABLE II.   COMPARISON OF COMPUTATION AND SIMULATION

	IGBT1	IGBT2	Diode1	Diode2
simulation(°C)	69.92	71.37	66.52	67.31
calculation(°C)	69.00	69.18	64.73	64.90

As shown in Table II, the maximum error is only 3.74%, indicating that the constructed thermal impedance matrix model is accurate.

C. Stress Indicators

A parametric scan of the thickness of each material layer was performed while considering the viscoplastic behavior of the solder layer to obtain the maximum stress of the solder layer at various thicknesses [10]. Some simulation results are shown in Fig. 8, where it can be observed that significant alternating thermal stresses occur at the edges and corners of the solder layer. Finally, the data were fitted to determine the relationship between maximum solder layer stress and layer thickness.

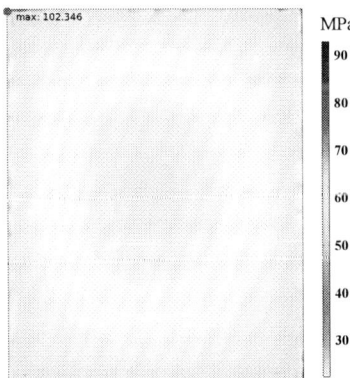

Fig. 8. Solder layer stress distribution.

D. Electro-Thermal-Stress Multi-Objective Optimization Model and Solution

Using the thickness of each material layer and the chip's horizontal and vertical coordinates as optimization variables, with loop inductance, maximum junction temperature, and maximum stress as optimization objectives, a multi-objective optimization mathematical model for the power module packaging design is established, which can be expressed as:

$$\begin{cases} \min & L = L_c + L_s \\ \min & T = R_{th} \times P + T_a \\ \min & F \\ & h_{min} \le h_i \le h_{max} \\ & X_{ci\,min} \le X_{ci} \le X_{ci\,max}, Y_{ci\,min} \le Y_{ci} \le Y_{ci\,max} \end{cases} \quad (7)$$

The NSGA-II algorithm is used to solve the problem, and the flow of the algorithm is shown in Fig. 9. As Fig. 10 shows the calculation results, the optimized solution in Region 1 minimizes the junction temperature but has higher solder layer stress and parasitic inductance, making it suitable for scenarios with high junction temperature requirements. In Region 3, the optimized solution exhibits lower parasitic inductance and solder layer stress, but a higher chip junction temperature, making it suitable for scenarios with stringent requirements on loop inductance and solder layer stress. The points in Region 2 represent a compromise between the three performance metrics, making it suitable for scenarios where all three are required.

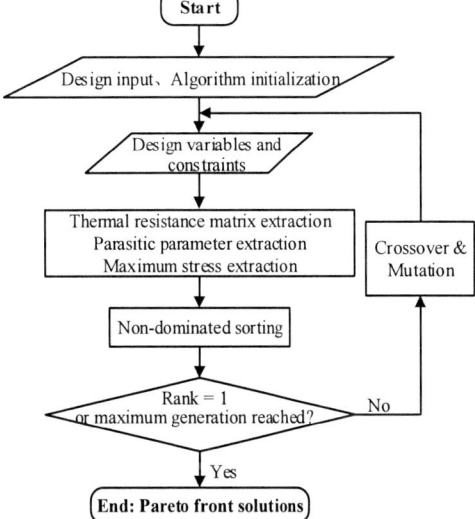

Fig. 9. NSGA-II algorithm flowchart.

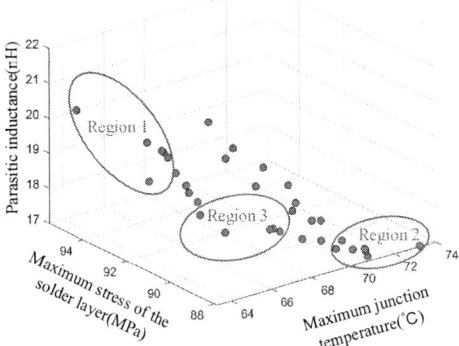

Fig. 10. Pareto-solutions of multiple-objective optimization.

Table III quantifies the optimization ratios of selected Pareto-optimal solutions, where h_1, h_2, and h_3 are the thicknesses of the solder layer, the upper copper layer, and the copper backing plate, respectively. Fig. 11(a) and (b) demonstrate the corresponding geometric configurations for distinct objective priorities. Normalized against baseline design parameters, peak reduction percentages reach 22.59% for parasitic loop inductance, 21.56% for junction temperature, and 33.43% for maximum die stress, revealing multi-physics performance enhancements through the proposed optimization framework.

TABLE III.   PARTIAL OPTIMUM SOLUTION

	h_1	h_2	h_3	L(nH)	ratios	T(°C)	ratios	F(MPa)	ratios
Initial value	0.1	0.1	1.0	21.96	-	81.08	-	132.10	-
1(L_{min})	0.1	0.6	3.7	17.00	22.59%	73.20	7.92%	87.94	33.43%
2(T_{min})	0.1	0.6	4.0	20.71	5.69%	63.60	21.56%	91.11	31.03%
3(F_{min})	0.1	0.6	3.7	17.34	21.04%	69.13	14.74%	87.94	33.43%
4	0.1	0.4	3.9	19.12	12.93%	67.54	16.70%	90.06	31.82%
5	0.1	0.5	4.0	17.62	19.76%	68.91	15.00%	89.14	32.52%
6	0.1	0.6	4.0	18.27	16.80%	65.92	18.70%	91.11	31.03%

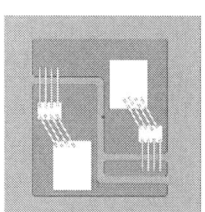
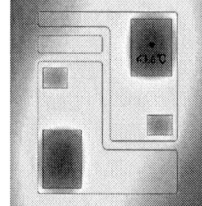

(a) L_{min} (b) T_{min}
Fig. 11. Chip layout for different optimization goals.

IV.   EXPERIMENTAL RESULTS

A. Parasitic Inductance Extraction

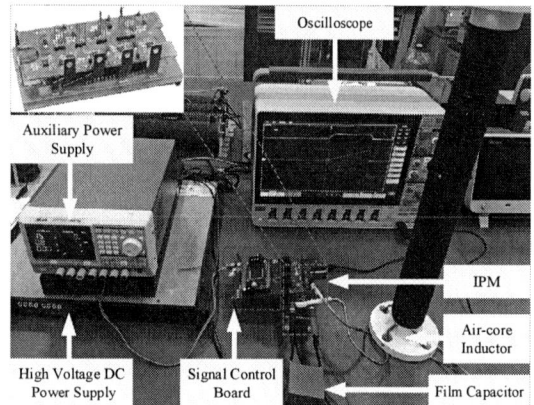

Fig. 12. Double pulse test platform.

The dual-pulse experimental test platform is shown in Fig. 12. The three-phase bridge arm power devices were tested under the conditions of 400V/120A using the dual-pulse method, and the experimental results are presented in Fig. 13. During IGBT turn-on/off transients, the changing current di/dt generates a voltage on the loop parasitic inductance, and this induced voltage will be superimposed on the bus voltage, causing a voltage spike between the IGBT collector and emitter, so the loop parasitic inductance can be calculated according to (8). The final measured parasitic inductance values show minimal deviation from the simulation results, indicating that the mathematical model for parasitic inductance constructed in this study has high accuracy.

$$L = \frac{\Delta V_{CE}}{di_c / dt} \qquad (8)$$

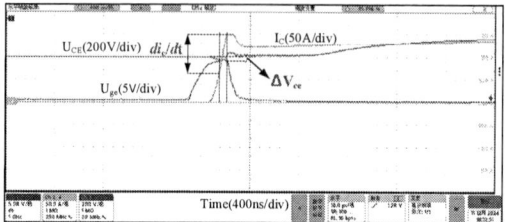

Fig. 13. IGBT turn-on waveform.

B. Thermal Resistance Testing

Thermal resistance measurement was conducted following the JESD51-14 Transient Dual Interface (TDI) method using a Simcenter Micred power tester, as illustrated in Fig. 14. The TDI methodology involves performing two junction-to-case thermal resistance measurements on the same power semiconductor device mounted on a temperature-controlled heatsink. The first measurement was taken without thermal grease application, while the second followed standard assembly specifications with a thin, uniform layer of thermal interface material.

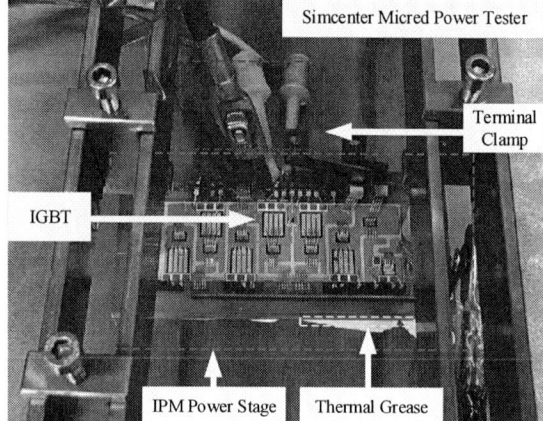

Fig. 14. Junction shell thermal resistance experimental measurement.

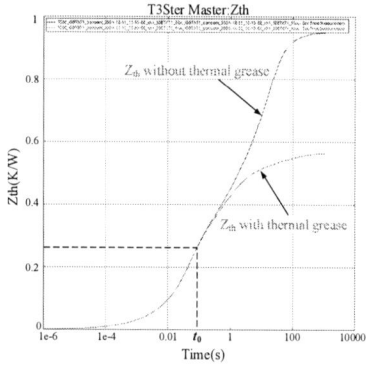

Fig. 15. Dual-interface method thermal resistance curve.

Since the difference in thermal paths only occurs at the contact surface between the package and the heatsink, the thermal impedance at the separation point t_0 is the thermal resistance of the device, as shown in Fig. 15. By analyzing the structural function, the internal structural information of the device can be obtained, as shown in Fig. 16. The measured junction-to-case thermal resistance is 0.25 K/W, while the calculated junction-to-case thermal resistance is 0.234 K/W, with an error of less than 7%.

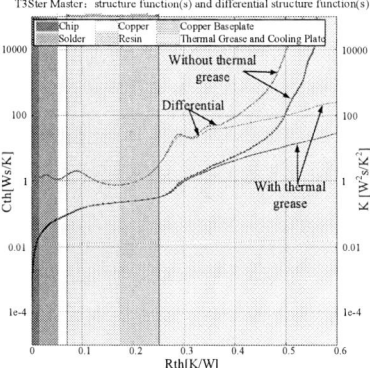

Fig. 16. Dual-interface method structure function.

V. CONCLUSION

This paper firstly analyzes the structure of IPM based on IMB packaging and the reasons for its limited application in high-power density scenarios. Then, through electro-thermal-stress multiphysics coupling analysis, quantitative characterization methods for parasitic inductance, junction temperature, and stress are investigated. Based on that, a multi-objective optimization method for IPM based on NSGA-II is proposed, targeting electro-thermal-stress performance. The results show the maximum reductions of 22.59% in loop inductance, 21.56% in chip junction temperature, and 33.43% in maximum solder layer stress, with inherent multi-objective compromises required among these parameters. Finally, experimental data validate the effectiveness of the proposed method. More detailed implementation will be presented in the full paper.

REFERENCES

[1] J. Bao, J. Hu, Y. Zhou and Y. Xu, "A Simulation Study on the Thermal Effectiveness of Graphene-Based Films In Intelligent Power Modules," 2023 China Semiconductor Technology International Conference (CSTIC), Shanghai, China, 2023, pp. 1-3.

[2] W. Zhen, F. Su, L. Xiaomin, X. Kai, H. Weiguan and S. Wenyu, "Research on Selection and Application Strategies of Intelligent Power Module based on Failure Mechanisms," 2022 IEEE International Conference on Sensing, Diagnostics, Prognostics, and Control (SDPC), Chongqing, China, 2022, pp. 389-392.

[3] K. Ohara, H. Masumoto, T. Takahashi, -. Manabu Matsumoto and Y. Otsubo, "A New IGBT Module with Insulated Metal Baseplate(IMB) and 7th Generation Chips," Proceedings of PCIM Europe 2015; International Exhibition and Conference for Power Electronics, Intelligent Motion, Renewable Energy and Energy Management, Nuremberg, Germany, 2015, pp. 1-4.

[4] B. Ji, X. Song, E. Sciberras, W. Cao, Y. Hu and V. Pickert, "Multiobjective Design Optimization of IGBT Power Modules Considering Power Cycling and Thermal Cycling," in IEEE Transactions on Power Electronics, vol. 30, no. 5, pp. 2493-2504, May 2015.

[5] A. Nakamura, M. Yoshida and T. Miyashita, "Multi-Objective Design Optimization of Power Module Performances," 2022 International Conference on Electronics Packaging (ICEP), Sapporo, Japan, 2022, pp. 157-158.

[6] Z. Zeng, K. Ou, L. Wang and Y. Yu, "Reliability-Oriented Automated Design of Double-Sided Cooling Power Module: A Thermo-Mechanical-Coordinated and Multi-Objective-Oriented Optimization Methodology," in IEEE Transactions on Device and Materials Reliability, vol. 20, no. 3, pp. 584-595, Sept. 2020.

[7] L. Ming et al., "A SiC-Si Hybrid Module for Direct Matrix Converter With Mitigated Current Spikes," in IEEE Journal of Emerging and Selected Topics in Power Electronics, vol. 10, no. 4, pp. 3805-3817, Aug. 2022.

[8] Y. Chen et al., "Layout-Dominated Electro-Thermal Optimization for Multichip Power Modules with Response Surface and Fourier Series Model," 2023 IEEE Applied Power Electronics Conference and Exposition (APEC), Orlando, FL, USA, 2023, pp. 1-6.

[9] J. Zhou, B. Li, Y. He, W. Yuan, J. Liu and H. Ni, "Electro-Thermal-Mechanical Multiphysics Coupling Failure Analysis Based on Improved IGBT Dynamic Model," in IEEE Access, vol. 7, pp. 174155-174166, 2019.

[10] E. R, G. Kavithaa, V. Samavatian, K. Alhaifi, A. Kokabi and H. Moayedi, "Reliability Enhancement of a Power Semiconductor With Optimized Solder Layer Thickness," in IEEE Transactions on Power Electronics, vol. 35, no. 6, pp. 6397-6404, June 2020.

2025 IEEE Workshop on Wide Bandgap Power Devices and Applications in Asia (WiPDA Asia)

GaN-based Partial Power DC-DC Converter with Four-quadrant Operation Capability

Chao Liu[1], Zhe Zhang[2], Shunqing Wu[3], Zeqi Yang[3] and Chuang Liu[4]

[1]Department of Wind and Energy Systems, Technical University of Denmark, Kgs. Lyngby, Denmark

[2]State Key Laboratory of Intelligent Power Distribution Equipment and System, Hebei University of Technology, Tianjin, China

[3]School of Electrical Engineering, Hebei University of Technology, Tianjin, China

[4]School of Electrical Engineering, Northeast Electric Power University, Jilin, China

E-Mail: chali@dtu.dk, zhezhangdk@gmail.com, shunqingwu1@163.com, 15031348977@163.com, victorliuchuang@163.com

Abstract—**Partial power processing is a promising technique for high-efficiency, high-density, and low-cost power conversion. In this paper, a bidirectional/bipolar partial power converter is proposed. A dedicated modulation scheme is developed and analyzed to enable smooth four-quadrant operation. A compact prototype utilizing Gallium Nitride (GaN) devices has been built and tested under 200V/1000W conditions. Experimental results demonstrate that the proposed converter achieves a peak efficiency of nearly 99%.**

Keywords—*partial power converter, four-quadrant operation, GaN devices*

I. INTRODUCTION

Partial Power Processing (PPP) technology has attracted extensive attention due to its inherent feature of dealing only with partial power since it was first proposed in [1,2]. This feature offers significant advantages in improving overall system efficiency, enabling compact converter designs, and reducing system cost. Partial Power Converters (PPCs), which are based on PPP principles, have been widely adopted in various applications such as photovoltaic (PV) systems [3-6], electric vehicle (EV) charging [7], and DC microgrids [8].

In the traditional full power converter (FPC), as illustrated in Fig. 1(a), the input and output terminals are connected in parallel with the source and load, respectively. As a result, the entire power must flow through the converter from the input to the output terminal. However, in Fig. 1(b) and Fig. 1(c), one power port of the converter forms a series loop with the source and load, while the other port is connected in parallel with either the source [Fig. 1(b)] or the load [Fig. 1(c)]. Due to the existence of a series path between source and load, only a portion of the total power flows through the converter, enabling efficient energy transfer through partial power handling capability. PPCs regulate power and voltage by adjusting the voltage difference between the source and the load, rather than directly controlling the load voltage.

Based on the voltage relationship between the source and the load, PPCs can be classified into three categories: step-up (SU-PPC) [9,10], step-down (SD-PPC) [11,12], and step-up/down types (SUD-PPC) [13-15]. In [13], the device stress factors of the three PPC configurations are compared. The results indicate that although the SUD-PPC employs more components, it offers a wider voltage regulation range and lower device stress factors. In [14], a seamless modulation strategy is proposed to address the challenge of smooth mode transition between step-up and step-down operations.

This work is supported by National Key Research and Development Program of China under Grant 2024YFB2504900.

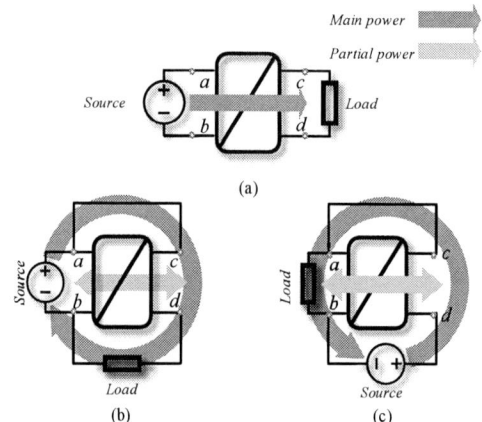

Fig. 1. Power flows in FPC and PPCs.

Meanwhile, with the rapid advancement of renewable energy technologies, energy storage systems (ESSs) featuring bidirectional power regulation capabilities have become essential for mitigating the mismatch between intermittent renewable energy sources, such as wind and solar, and dynamically varying load demands. However, most existing research has primarily focused on unidirectional power transfer, leaving four-quadrant operation less explored.

This paper proposes a four-quadrant PPC with a smooth mode switching modulation strategy and develops a compact prototype using GaN devices. The main contributions are summarized as follows:

1. A novel PPC topology is proposed that enables four-quadrant operation.

2. The use of GaN devices further increases the efficiency and power density.

3. A seamless and smooth transition modulation scheme is specifically designed, which ensures uninterrupted mode transition between SU and SD operational states.

II. TOPOLOGY AND MODULATION STRATEGY

In order to handle the four-quadrant operation, four pairs of anti-series switches are employed in the low voltage side to achieve four-quadrant operation, where S_{1-4} and S_{9-12} are the upper and lower switches in the anti-series switches, respectively.

979-8-3315-1110-4/25 $31.00 © 2025 IEEE   388

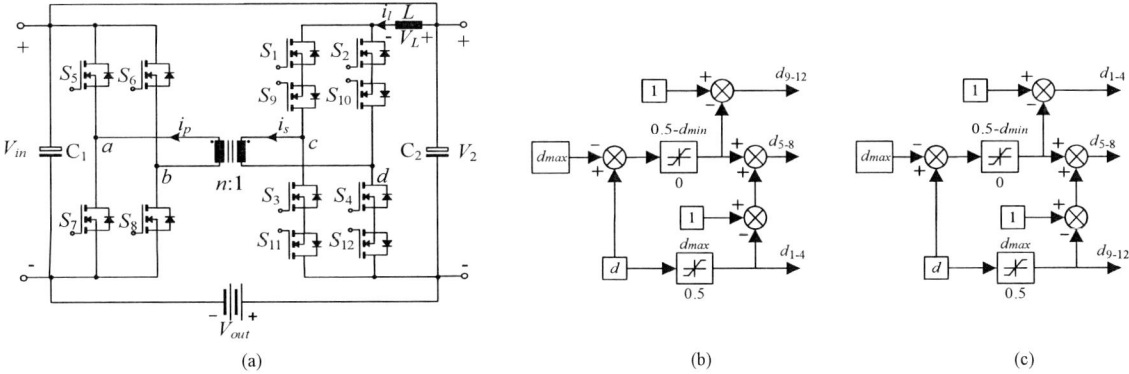

(a)　　　　　　　　　　　　　(b)　　　　　　　　　　　　　(c)

Fig. 2. Topology and modulation strategy of the four-quadrant PPC: (a) Topology of the proposed PPC; (b) Modulation strategy for charging mode; (c) Modulation strategy for discharging mode.

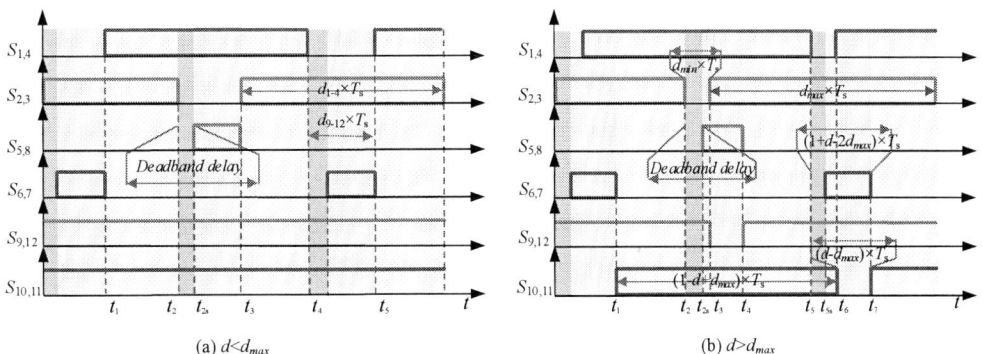

(a) $d<d_{max}$　　　　　　　　　　　　　(b) $d>d_{max}$

Fig. 3. Driver signals in charging mode.

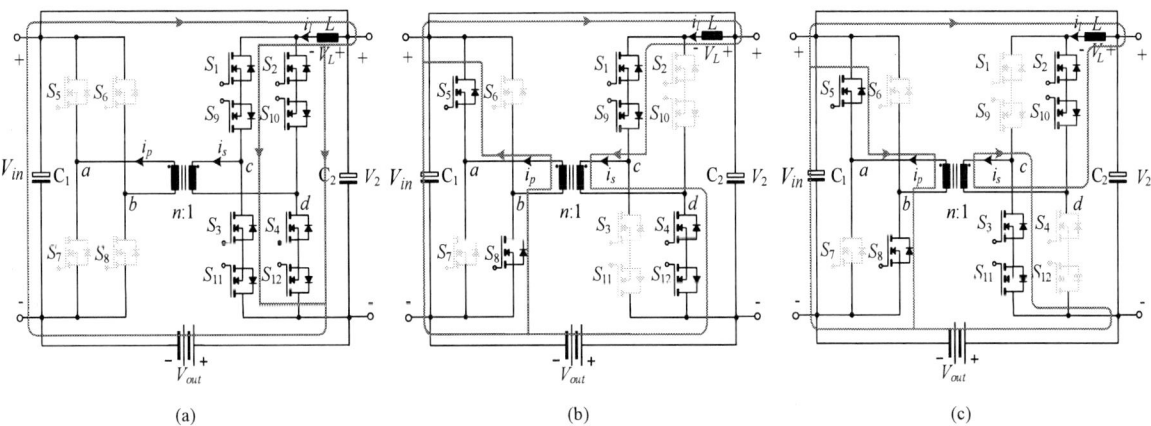

(a)　　　　　　　　　　　　　(b)　　　　　　　　　　　　　(c)

Fig. 4. Equivalent circuits of the proposed topology in charge mode: (a) Equivalent circuit during the t_1-t_2 for both $d<d_{max}$ and $d>d_{max}$; (b) Equivalent circuit during the t_2-t_3 for $d<d_{max}$; (c) Equivalent circuit during the t_3-t_4 for $d>d_{max}$.

The operational mechanisms of the charge and discharge modes are introduced in detail below. The duty cycles of S_{1-4}, S_{5-8}, and S_{9-12} are represented by d_{1-4}, d_{5-8}, and d_{9-12}, respectively.

$$d_{1-4} = \begin{cases} d & , d < d_{max} \\ d_{max} & , d \geq d_{max} \end{cases}$$
$$d_{5-8} = \begin{cases} 1-d & , d < d_{max} \\ 1+d-2d_{max} & , d \geq d_{max} \end{cases}$$
$$d_{9-12} = \begin{cases} 1 & , d < d_{max} \\ 1+d-2d_{max} & , d \geq d_{max} \end{cases} \tag{1}$$

The relationship between the duty cycle for each switch and the unit duty ratio in charging and discharging modes is given by (1) and (2), respectively.

$$d_{1-4} = \begin{cases} 1 & , d < d_{max} \\ 1+d-2d_{max} & , d \geq d_{max} \end{cases}$$
$$d_{5-8} = \begin{cases} 1-d & , d < d_{max} \\ 1+d-2d_{max} & , d \geq d_{max} \end{cases} \tag{2}$$
$$d_{9-12} = \begin{cases} d & , d < d_{max} \\ d_{max} & , d \geq d_{max} \end{cases}$$

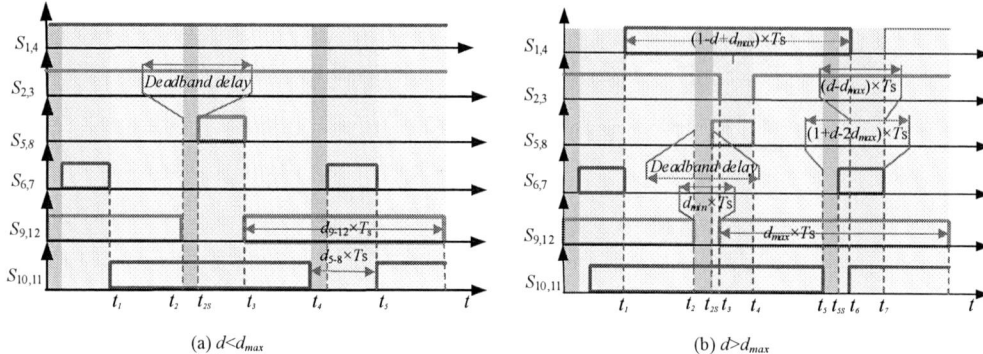

(a) $d<d_{max}$ (b) $d>d_{max}$

Fig. 5. Driver signals in discharging mode.

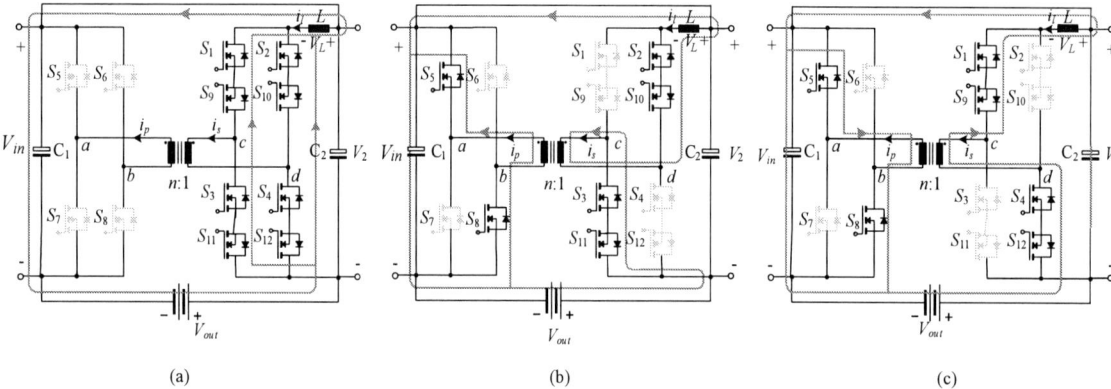

(a)   (b)   (c)

Fig. 6. Equivalent circuits of the proposed topology in discharging mode: (a) Equivalent circuit during the t_1-t_2 for both $d<d_{max}$ and $d>d_{max}$; (b) Equivalent circuit during the t_2-t_3 for $d<d_{max}$; (c) Equivalent circuit during the t_3-t_4 for $d>d_{max}$.

In both charge and discharge operation modes, the modulation strategy can be divided into two modes based on the value of d, with $d=d_{max}$ as the boundary point. The operational mechanisms of the charging and discharging modes are analyzed in detail below.

A. Charging mode

In the charging mode, the input bus serves as the power source, while the battery functions as the load. According to (1), when $d \leq d_{max}$, d_{5-8} decreases from 1, as it is defined by 1-d. In contrast, d_{1-4} increases proportionally with d. d_{1-4} and d_{5-8} are complementary. Switches S_{1-4} remain continuously on-state. One switching cycle in this case consists of four symmetric states, as illustrated in Fig. 3(a).

During the $t_1 \sim t_2$ interval, S_{1-4} and S_{9-12} remain on-state while S_{5-8} are switched off, as shown in Fig. 4(a). Under this condition, no current passes through the high frequency transformer; thus, it does not participate in power transfer and the power supply directly transmits power to the load through the power supply loop. The inductor is charged in this period; the inductor voltage is derived by (3).

$$V_L = V_{in} - V_{out} \tag{3}$$

At t_2, $S_{2,3}$ are turned off, and the inductor current discharges through $S_{1,4}$ and $S_{9,12}$. On the HV side, i_p flows through the anti-parallel diodes of $S_{5,8}$ until the switches turn on at t_{2s}, as shown in Fig. 4(b). This deadtime delay enables the ZVS switching for $S_{5,8}$. The inductor voltage in this state is

given by

$$V_L = V_{in} - V_{out} - \frac{V_{in}}{n} \tag{4}$$

At t_3, $S_{2,3}$ are turned on again, the converter returns to its initial operating state (the interval of $t_1 \sim t_2$) where the inductor voltage in this state is equal to (3). At t_4, $S_{1,4}$ are switched off, similarly, $S_{6,7}$ are switched on after a short deadtime to achieve ZVS switching. The inductor voltage in this period is equal to (4). According to (3) and (4), by applying the voltage-second balance principle, the relationship between V_{in} and V_{out} is derived in (5).

$$V_{out} = \frac{V_{in}(n + 2 \cdot d - 2)}{n} \tag{5}$$

When $d>d_{max}$, d_{5-8} increases from $1-d_{max}$ as d increases since it is defined by $1+d-2d_{max}$. Meanwhile, d_{1-4} is clamped at the maximum value d_{max}, and d_{9-12} decreases from 1 as d increases, given by $1-d+d_{max}$, as shown in Fig. 3(b). Compared to the case where $d<d_{max}$, the condition $d>d_{max}$ introduces two additional switching states. Fig. 4(c) shows the equivalent circuit during the period from t_3 to t_4 for this case. In this interval, $S_{1,4}$ are switched off, while $S_{5,8}$ and $S_{9,12}$ are turned on. Consequently, the power flows from the HV side to the LV side, and the inductor current passes through $S_{2,3}$ and $S_{10,11}$. The inductor voltage in this state is given by

$$V_L = V_{in} - V_{out} + \frac{V_{in}}{n} \tag{6}$$

979-8-3315-1110-4/25 \$31.00 © 2025 IEEE   390

TABLE I

SYSTEM SPECIFICATIONS

Symbol	Name	Value
S_{5-8}	HV side switches	GS66508T
S_{1-4} and S_{9-10}	LV side switches	EPC2034C
d_{max}	Boundary point	0.95
L	Inductance	15 μH
n	Turn ratio	16:4
V_{in}	Bus voltage	200 V
R	Load	60 Ω

According to (3), (4), and (6), by reapplying the voltage-second balance principle, the relationship between V_{in} and V_{out} for $d>d_{max}$ is the same as that for $d<d_{max}$. Therefore, both cases share the same voltage gain equation. When $d<1$, V_{out} is lower than V_{in}; whereas when $d>1$, V_{out} is exceeds V_{in}. Although the step-up and step-down operating mode transition occurs at $d=1$, to avoid the incomplete charging and discharging of the capacitor causing the voltage at both ends of the transformer to be unsatisfactory, the modulation strategy is designed to switch at $d=d_{max}$. The proposed modulation strategy avoids excessively short conduction times of the high-voltage side switches (S_{5-8}), ensuring they can fully turn on, and it enables seamless switching between step-up and step-down conversion in charging mode.

B. Discharging mode

In the discharging mode, power flows from the battery to the bus, which is the reverse of the charging mode. Consequently, the functions of switches S_{1-4} and S_{9-12} are interchanged. As a result, the corresponding duty cycles d_{1-4} and d_{9-12} in Fig. 2(c) are exchanged compared to those in Fig. 2(b) for the charging mode.

When $d \le d_{max}$, d_{5-8} continue to decrease from 1 as d increases, same to the charging mode. Meanwhile, d_{9-12} increases proportionally with d, while d_{5-8} remains clamped at 1. Fig. 5(a) shows the control waveform in the discharging mode for $d \le d_{max}$. Similar to the charging mode, the complete energy exchange cycle can be divided into four symmetrical working intervals.

During the interval from t_1 to t_2, switches S_{1-4} and S_{9-12} are all in the conducting state, while switches S_{5-8} remain turned off, as shown in Fig. 6(a). The switch configuration in this period is identical to that in the t_1-t_2 interval of charging mode, as shown in Fig. 4(a). However, since power is transferred from the battery to the bus in this case, the current on the low-voltage side flows from the bottom to the top. The inductor voltage during this phase is given by (3).

At t_2, switches $S_{9,12}$ are turned off, following a dead-time delay of $\Delta t = t_{2s} - t_2$, switches $S_{5,8}$ are activated. Similar to the charging mode, the dead time delay ensures ZVS for $S_{5,8}$. The equivalent circuit in this interval is shown in Fig. 6(b). The current path during this period is the same as that in Fig. 4(c), except that the direction is reversed. Therefore, the inductor voltage expression is given by (6).

According to (3) and (6), the relationship between V_{in} and V_{out} for $d<d_{max}$ in discharging mode can be expressed as:

$$V_{in} = \frac{n \cdot V_{out}}{n - 2 \cdot d + 2} \qquad (7)$$

Similar to charging mode, d_{5-8} increases from $1-d_{max}$ as d increases when $d>d_{max}$.

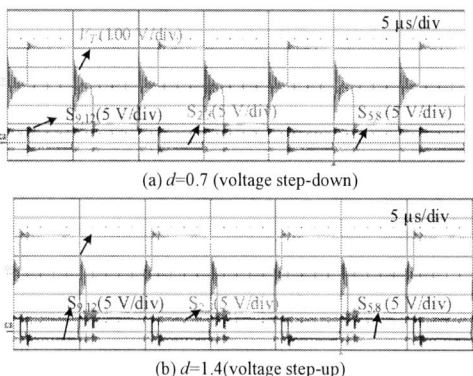

(a) d=0.7 (voltage step-down)

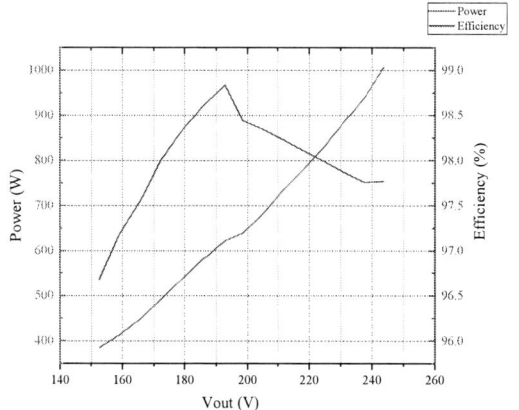

(b) d=1.4 (voltage step-up)

Fig. 7. Driver signals and transformer in charging mode.

Fig. 8. Output power and efficiency versus output voltage.

Meanwhile, d_{9-12} is clamped at the maximum value d_{max}, while d_{1-4} decreases from 1 as d increases, given by $1-d+d_{max}$, as shown in Fig. 5(b). As a result, the condition $d>d_{max}$ in discharging mode also introduces two additional switching states compared to the case where $d<d_{max}$.

Fig. 6(c) shows one of the two additional operating states, corresponding to the interval from t_3-t_4 in Fig. 5(b). It can be observed that the switch configuration in this period is identical to that in the t_2-t_3 interval of charging mode, as shown in Fig. 4(b). This implies that the current path in this phase is the same as that in Fig. 4(b), but with the opposite direction. Therefore, the inductor voltage in this state is given by (4). Both cases in discharging mode share the same voltage gain equation, similar to the charging mode.

III. EXPERIMENTAL RESULTS

A prototype is built to verify the proposed topology and modulation strategy, where the input voltage V_{in} is set to 200 V and the load is set to 40 Ω. The 650V GaN devices from Infineon [16] are used for HV side switches, while the 200V GaN devices from EPC [17] are selected for the LV side switches. The specifications of the prototype are listed in Table. I.

Fig. 7(a) and Fig. 7(b) show the gate driver signals and transformer voltage waveforms in the charging mode for d=0.7 and d=1.4, respectively. In Fig. 7(a), $S_{2,3}$ operate complementarily with $S_{5,8}$, and $S_{9,12}$ remain continuously in the on-state, which is consistent with the operation shown in

Fig. 3(a). When $S_{2,3}$ is turned off, the voltage across the high-voltage side of the transformer equals the input voltage of 200 V. In the Fig. 7(b), the duty cycle of $S_{2,3}$ is 0.95, which equals d_{max}, which is consistent with the operation shown in Fig. 3(b).

Fig. 8 illustrates the relationship between output voltage (V_{out}) and both output power and efficiency in the charging mode. The red curve represents the output power, which increases steadily from approximately 380 W at 153 V to about 1000 W at 244 V. The blue curve shows efficiency, which initially increases with V_{out}, peaking at around 98.9% when $V_{out} \approx 195$V, corresponding to $d=d_{max}$. Beyond this point, the efficiency gradually decreases, reaching about 97.8% at 244 V. This is because the power transferred by the converter is minimized when $d=d_{max}$.

IV. Conclusions

This paper presents the design of a four-quadrant PPC topology along with a corresponding modulation strategy that enables seamless switching between step-up (SU) and step-down (SD) modes in both charging and discharging operations. Based on the theoretical analysis, a GaN-based prototype was developed and tested under low-voltage conditions to validate the proposed topology and control method. Experimental results demonstrate that the prototype achieves a peak efficiency close to 99%.

References

[1] A. Jon, A. Iosu, and R. A. Arruti, "Review of architectures based on partial power processing for dc-dc applications," IEEE Access, vol. 8, DOI 10.1109/ACCESS.2020.2999062, pp. 103 405-103 418, 2020.

[2] Z. J. R. Rakoski, da Silva Martins M´ario L´urcio, P. J. Renes, and H. H. Le˜aes, "Evaluation of power processing in series-connected partialpower converters," IEEE Journal of Emerging and Selected Topics in Power Electronics, vol. 7, DOI 10.1109/JESTPE.2018.2869370, no. 1, pp. 343–352, 2019.

[3] B. Min, J. Lee, J. Kim, T. Kim, D. Yoo and E. Song, "A New Topology with High Efficiency Throughout All Load Range for Photovoltaic PCS," in IEEE Transactions on Industrial Electronics, vol. 56, no. 11, pp. 4427-4435, Nov. 2009.

[4] H. Zhou, J. Zhao and Y. Han, "PV Balancers: Concept, Architectures, and Realization," in IEEE Transactions on Power Electronics, vol. 30, no. 7, pp. 3479-3487, July 2015.

[5] B. R.M., "An advanced photovoltaic array regulator module," in IECEC 96. Proceedings of the 31st Intersociety Energy Conversion Engineering Conference, vol. 1, DOI 10.1109/IECEC.1996.552937, pp. 519–524 vol.1,

[6] J. R. R. Zientarski, M. L. d. S. Martins, J. R. Pinheiro and H. L. Hey, "Series-Connected Partial-Power Converters Applied to PV Systems: A Design Approach Based on Step-Up/Down Voltage Regulation Range," in IEEE Transactions on Power Electronics, vol. 33, no. 9, pp. 7622-7633, Sept. 2018.

[7] M. Iyer, S. Gulur, G. Gohil and S. Bhattacharya, "An Approach Towards Extreme Fast Charging Station Power Delivery for Electric Vehicles with Partial Power Processing," in IEEE Transactions on Industrial Electronics, vol. 67, no. 10, pp. 8076-8087, Oct. 2020.

[8] H. Zhang et al., "A Partially Rated Interlinking Converter With Distributed Energy Storage for Active Power Sharing in DC Microgrids," in IEEE Transactions on Power Electronics, vol. 40, no. 7, pp. 9370-9387, July 2025, doi: 10.1109/TPEL.2025.3542580.

[9] M. Shousha, T. McRae, A. ProdiA, V. Marten, and J. Milios, "Design and implementation of high power density assisting step-up converter with integrated battery balancing feature," IEEE Journal of Emerging and Selected Topics in Power Electronics, vol. 5, DOI 10.1109/JESTPE.2017.2665340, no. 3, pp. 1068–1077, 2017.

[10] R. Xu, S. Gao, Y. Wang, and D. Xu, "A high step up sepicbased partial-power converter with wide input range," in 2021 IEEE Industry Applications Society Annual Meeting (IAS), DOI 10.1109/ IAS48185.2021.9677080, pp. 1–5, 2021.

[11] M. M. C., Z. Zhe, and M. A. A. E., "Analysis and comparison of dc/dc topologies in partial power processing configuration for energy storage systems," in 2018 International Power Electronics Conference(IPEC-Niigata 2018 -ECCE Asia), DOI 10.23919/IPEC.2018.8507937,pp. 1351–1357, 2018.

[12] M. M. C., Z. Zhe, J. K. L¨uthje, and A. M. A. E., "Fractional charging converter with high efficiency and low cost for electrochemical energy storage devices," IEEE Transactions on Industry Applications, vol. 55, DOI 10.1109/TIA.2019.2921295, no. 6, pp. 7461–7470, 2019.

[13] C. Liu, Z. Zhang and M. A. E. Andersen, "Analysis and Evaluation of 99% Efficient Step-up/down Converter based on Partial Power Processing," in IEEE Transactions on Industrial Electronics, 2022, doi: 10.1109/TIE.2022.3198241.

[14] C. Liu, Z. Zhang, Z. Ouyang, J. Huang, M. A. E. Andersen and T. G. Zsurzsan, "A Seamless Modulation Strategy for Step-up/down Partial Power Processing Converter (SUD-P3C)," 2022 24th European Conference on Power Electronics and Applications (EPE'22 ECCE Europe), Hanover, Germany, 2022, pp. 1-8.

[15] N. G. F. d. Santos, J. R. R. Zientarski and M. L. d. S. Martins, "A Two-Switch Forward Partial Power Converter for Step-Up/Down String PV Systems," in IEEE Transactions on Power Electronics, vol. 37, no. 6, pp. 6247-6252, June 2022.

[16] GaN Systems, GS66508T-DS-Rev-200402.pdf https://gansystems.com/wp-content/uploads/2020/04/GS66508T-DS-Rev-200402.pdf

[17] Efficient Power Conversion Corporation (EPC), epc2034C_datasheet.pdf, https://epc-co.com/epc/Portals/0/epc/documents/datasheets/epc2034C _datasheet.pdf

UIS Ruggedness of Si/SiC Hybrid Switches

Chuanqi Zhang
College of Electrical and Information
Engineering
Hunan University
Changsha, China
E-mail: zcq0512@hnu.edu.cn

Xuanting Song
College of Electrical and Information
Engineering
Hunan University
Changsha, China
E-mail: songxt@hnu.edu.cn

Shiwei Liang
College of Electrical and Information
Engineering
Hunan University
Changsha, China
E-mail: swliang@hnu.edu.cn

Yuxing Dai
College of elecerical and electronic
engineering
Wenzhou University
WenZhou, China
E-mail:daiyx@hnu.edu.cn

Jun Wang
College of Electrical and Information
Engineering
Hunan University
Changsha, China
E-mail: junwang@hnu.edu.cn

Kamal Al-Haddad
École de Technologie Supérieure
Montreal, QC

Canada, kamal
E-mail:al-haddad@etsmtl.ca

Abstract—**Although the hybrid switch consisting of high power Si IGBT and low power SiC MOSFET achieves reduction of losses and cost, there is a severe concern of its UIS capability under single-pulse mode. The limiting factors and in-depth failure mechanism of UIS capability have not been comprehensively studied yet. In this article, we investigated UIS of the Si/SiC hybrid switch and its internal devices with various sizing ratios and gate control patterns. It is observed that the avalanche current of the hybrid device is predominantly sustained by the SiC MOSFET and increases with the increasing SiC MOSFET sizing during the avalanche process. Furthermore, the influence of gate control patterns is also analyzed. Optimizing the gate turn-off delay period in SiC MOSFET enhances avalanche ruggedness characteristics within hybrid switch. Finally, the UIS failure mechanism of the hybrid switch is discussed.**

Keywords—SiC MOSFET, Si IGBT, Hybrid switch, UIS,Avalanche.

I. INTRODUCTION

With the rapid development of new energy power generation, rail transit, and industrial frequency conversion technologies, power electronic devices must comply with rigorous operational specifications, particularly in terms of power loss and reliability. The operational environment for power devices has grown progressively more complex.Traditional silicon (Si)-based devices, such as IGBTs, struggle to meet these demands due to inherent limitations like high-frequency power losses and suboptimal high-temperature stability. Consequently, there exists an urgent imperative to develop power electronic devices characterized by low power loss, high reliability, and robust high-current capacity [1-3]. Silicon carbide (SiC) devices present a promising solution to these challenges; however, the fabrication processes for SiC materials and devices remain technically daunting and cost-prohibitive. To tackle this challenge, hybrid Si/SiC devices have been proposed and investigated in recent research[4-8]. These hybrid devices are typically configured with a high - current Si IGBT connected in parallel with a low - current SiC MOSFET. This configuration enables them to attain performance comparable to that of a SiC MOSFET with the same rated voltage and current. Subsequently, a wealth of research efforts have been dedicated to the optimization of

switching modes, gate control strategies, current - sharing mechanisms, and the refinement of simulation models.

Device reliability is a paramount concern in the field of power electronics. The unclamped inductive switching (UIS) characteristics of power devices are a pivotal factor for assessing the reliability of them. A mounts of researches have illustrated the principal factors when devices encounter UIS condition[9-12]. Recent investigations in [13] on UIS ruggedness of hybrid devices have primarily illustrated the impact of temperature and gate delay pattern. However, critical analysis of avalanche energy distribution within the hybrid switch during UIS events remains unexplored.

However, there is a notable scarcity of research addressing several critical aspects: the current distribution dynamics between the Si IGBT and SiC MOSFET throughout the entire test process, the influencing factors governing this distribution, the potential for avalanche current level augmentation, and the identification of the device that fails first during the avalanche process. Gaining a comprehensive understanding of these underlying mechanisms holds the potential to significantly enhance the robustness of hybrid devices. In this paper, an experimental investigation is conducted to meticulously analyze the current variation characteristics of Si/SiC hybrid devices operating under the single-pulse avalanche mode.

II. EXPERIMENTS SETUP

Three SiC MOSFETs with varying current ratings, two Si IGBTs, and one SiC Schottky barrier diode (SBD) were selected, all rated at 1.2 kV. Subsequently, the Si IGBTs, SiC SBD, and SiC MOSFETs were connected in parallel to fabricate a hybrid device.

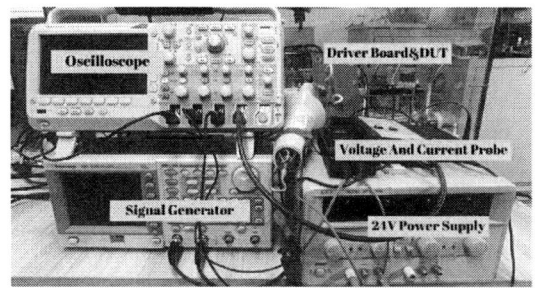

Fig.1.Experimental test platform

Table.I

Species	Hybrid Devices1 (51%SiC)	Hybrid Devices2 (44%SiC)	Hybrid Devices3 (29% SiC)	Hybrid Devices3 SiC MOSFET (29% SiC)	Hybrid Devices2 SiC MOSFET (44%SiC)	Hybrid Devices1 SiC MOSFET (51%SiC)	Hybrid Devices3 SiC MOSFET +1 SBD	IGBT	IGBT+ SBD
Failure Current (A)	106	57	30	106	140	157	111	5	6
Avalanche Energy (mJ)	561.8	162.45	45	561.8	980	1232.45	616.05	1.25	1.8

The experimental platform of the single-pulse avalanche test are shown in Figure 1 respectively. Experimental tests were conducted under the experimental conditions where the DC side voltage Vdd was 100 V, the capacitance C of the capacitor bank was composed of four capacitors in parallel, and the main inductance L was designed to be 0.1mH. Among them, the rated current of IGBT is 216A, that of MOSFET used in hybrid device 1 is 115A, that of MOSFET used in hybrid device 2 is 100A, and that of MOSFET used in hybrid device 3 is 63A.

In this experimental investigation, the hybrid Si/SiC devices were concurrently turned off. Throughout the testing procedure, the charging time (ton) was incrementally increased until device failure occurred, at which juncture the test was terminated. This testing protocol was systematically repeated for all three hybrid devices.The measured avalanche current (I_{AS}) and maximum avalanche energy (E_{AS}) values are tabulated in Table I.

III. EXPERIMENTAL RESULTS AND DISCUSSION

Figure 2 presents a three-stage analysis of the entire avalanche test. The three stages correspond to the three parts separated by four dotted lines in the figure, the first stage is called the charging stage for charging the inductor, the second stage is called the commutation stage for the commutation stage between the two devices, and the third stage is the device gate voltage shutdown device into the avalanche breakdown stage.

A. Static characteristic curve analysis

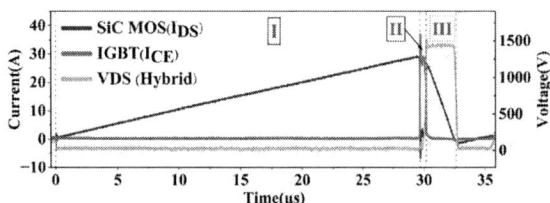

Fig.2.Hybrid device 3 single avalanche test waveform

The static characteristic curve depicted in Figure 3 reveals that, at the same device on - voltage, the hybrid device exhibits a higher conduction current than the IGBT, particularly evident at low current levels. Analysis of the hybrid device's static characteristic curve indicates that, within the low current range, the MOSFET predominantly bears the conduction current. This phenomenon results in an upward shift of the hybrid device's static curve. These findings align well with the test results, confirming that the MOSFET is the primary carrier of the conduction current under low-current conditions.

At low current levels, due to the different conduction characteristics of the devices, the current primarily flows through the SiC MOSFET. The on-state resistance ($R_{DS}(on)$) of the SiC MOSFET is lower at low currents, so the current preferentially chooses the path with lower resistance. Additionally, the Si IGBT is unable to reach its threshold at low current levels, causing it to remain in the off state.

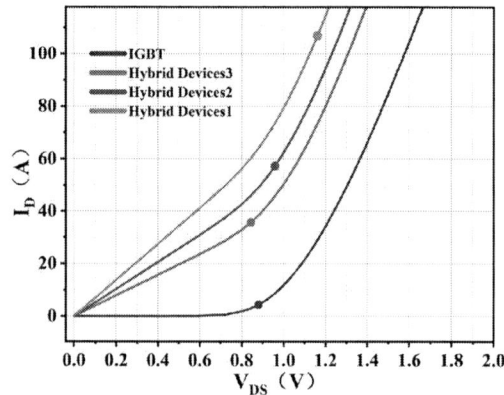

Fig.3.Static characteristic curves of three hybrid devices the point in the figure is the device avalanche failure point

A comparative analysis of four configurations—IGBT paralleled with an SBD (referred to as "Hybrid Devices1"),and three hybrid structures ("Hybrid Devices 2" "Hybrid Devices 3")—was conducted in conjunction with static characteristic curves. The results revealed that hybrid devices incorporating a SiC MOSFET with a higher rated current exhibited a greater proportion of avalanche current borne by the MOSFET during the avalanche process. This enables the IGBT in hybrid devices to operate at higher current levels and under more severe conditions. Through observations of current sharing during UIS turn-off, it is hypothesized that a specific mechanism governs current distribution in the avalanche breakdown stage.

B. Leakage current and commutation process analysis

To analyze the device behavior during the commutation stage, local amplification of Phase II facilitates easier observation of the device's variation pattern. Figure 2 presents a partial magnification of the commutation phase in the monopulse test. Given that the turn-off speed of the MOSFET is typically higher than that of the IGBT bipolar device at the same voltage level, the MOSFET tends to turn off first, with current commutation between the two devices occurring after the MOSFET is switched off.

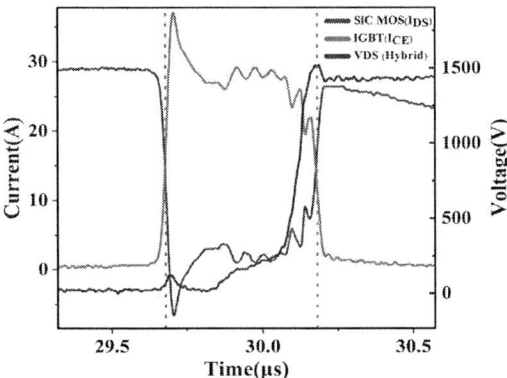

Fig. 4.Hybrid device phase II converter part

The commutation phase was measured to last 0.5 μs. During this interval, the IGBT solely carried the current until its turn-off triggered an avalanche.The induced voltage forced the device into the avalanche breakdown mode. In the avalanche breakdown phase,both parallel branches conducted current, with the MOSFET predominantly bearing the avalanche current and the IGBT carrying only a minor portion.Device failure occurred when the avalanche current borne by the IGBT during this phase matched the avalanche current measured in the single - tube IGBT avalanche test. Notably, the SBD and MOSFET remained undamaged . These findings indicate that the IGBT within the hybrid device acts as a limiting factor for the overall avalanche capability.

After conducting parallel - connection tests on IGBTs with MOSFETs and SBDs of varying rated currents, the data presented in Table 1 were obtained. An analysis of the table reveals that the critical failure current of a single - tube IGBT is 5 A, whereas the waveform current just prior to the failure of the IGBT+SBD configuration reaches 6 A, demonstrating an increase in the critical failure current. This finding strongly suggests that the shunted SBD carries a portion of the current during the avalanche breakdown stage, thereby contributing to an enhancement in the unclamped inductive switching (UIS) robustness of the IGBT. Moreover, the test results indicate that increasing the number of parallel - connected SBDs leads to improved performance, further validating the role of SBDs in bearing avalanche current.

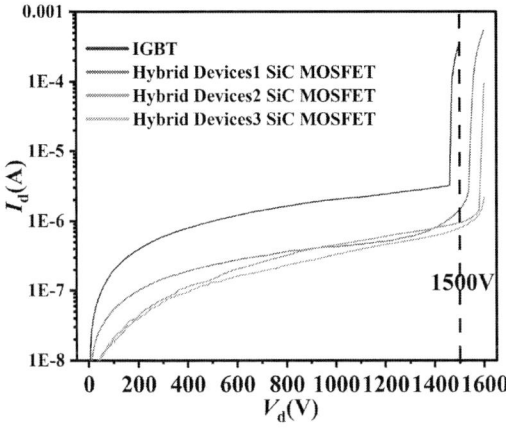

Fig. 5.Device drain leakage current characteristic curve

Figure 5 shows the trace diagram of the leakage current characteristic curve of the five devices.The leakage current of the devices was tested to infer the differences in avalanche tolerance of the parallel devices. The SBD also exhibits a relatively high leakage current, which slightly enhances the UIS avalanche tolerance of the IGBT when connected in parallel. Comparing the three hybrid devices, it can be seen from the figure that when the hybrid device reaches a 1.5 kV avalanche voltage level, the leakage current of the SiC MOSFET used in the hybrid devices increases as the chip area increases. In the UIS test, devices with higher leakage current share more of the current, and their avalanche tolerance increases with the larger SiC chip area used. This suggests that SiC MOSFET with larger chips carry more energy from the inductance, improving the overall UIS capability.

C. Influence of Turn-off Delay Time

To find out if a delayed shutdown would have an effect on the current distribution of the device during an avalanche. Therefore, relevant experiments were carried out.

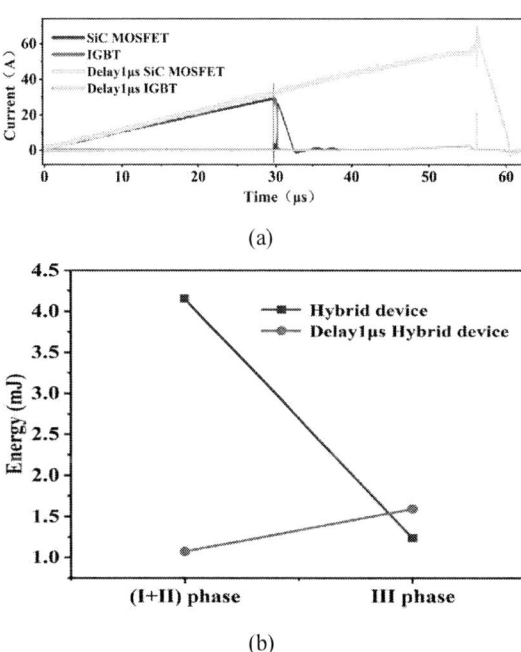

(a)

(b)

Fig. 6.(a) Hybrid device 3 with a delay of 1μs with no delay current, and (b) hybrid device 3 with a delay of 1 μs with no delay power loss at different stages

Figure 6(a) presents a comparison of the current magnitudes in Hybrid Device 3 with a 1-μs delay and without delay, while Figure 6(b) illustrates the power loss values of Hybrid Device 3 at various stages under these two conditions. In the figures, the first two stages are combined because, with a delay, Phase II does not occur. During the testing of Hybrid Device 3, the MOSFET device was turned off 1 μs later than the IGBT device. A comparison between the 1-μs delay and non-delay scenarios revealed that the avalanche current level of the hybrid device increased to 60 A. Through the observation of current distribution during the avalanche process, a preliminary analysis indicates that when avalanche breakdown occurs after the device is fully turned off, the SiC MOSFET with a 1-μs delay carries a larger share of the current during the avalanche. These findings suggest that the turn-off delay has a significant impact on the unclamped inductive switching (UIS) robustness of hybrid devices.

Analysis of power consumption with and without delay reveals that, when the SiC MOSFET is turned off without delay, the IGBT device experiences significant power consumption during the commutation stage. This leads to heat accumulation, raising the initial temperature of the device at the onset of the avalanche process in the delayed - testing scenario. Consequently, during the avalanche breakdown phase, the device under delayed - testing conditions endures higher power consumption compared to the non - delayed case. To accumulate the same amount of heat required for avalanche - induced damage, the delayed - tested device conducts a larger current than its non - delayed counterpart. As a result, the avalanche current level increases.

D. Influence of Si/SiC Sizing Ratio

In order to investigate the influence of the SiC MOSFET's chip size on the UIS capability of the HyS, 1200-V SiC MOSFETs with current rating of 115, 100, and 63A from the same manufacturer are chosen to be parallel connected to the 1200 V/216 A IGBT to comprise the HyS. Hybrid Devices1、Hybrid Devices2、Hybrid Devices3、by comparing the three situations and combining them with the static characteristic curve, it is found that the larger the rated current of the SiC MOSFET in use, the greater the current the MOSFET bears during an avalanche. It enables IGBTs in hybrid devices to operate at higher current levels and under more severe working conditions.

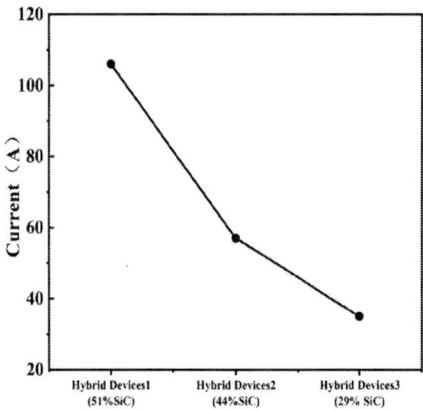

Fig. 7.Avalanche currents of three hybrid devices

IV. Conclusions

In summary, this experimental investigation systematically analyzed the avalanche robustness of Si/SiC hybrid devices under unclamped inductive switching (UIS) conditions. During the avalanche process, the study involved conducting tests by systematically varying parameters such as gate-signal turn-off delay time, drain leakage current, and static characteristics,alongside evaluating MOSFETs with different rated currents to assess their impact on device performance.The results demonstrated that the avalanche energy of the IGBT during the avalanche phase was negligible, with the avalanche current measuring

approximately 2% of the rated current.When conducting avalanche tests on hybrid devices, failure occurred when the IGBT avalanche current reached the threshold observed in single-device IGBT avalanche tests.Notably, replacing the MOSFET with a higher-rated current device significantly increased the hybrid device's avalanche energy.This phenomenon is primarily attributed to the MOSFET carrying a larger proportion of the current during avalanche, thereby enabling the IGBT to operate under more demanding current conditions.

Leakage current analysis of the three SiC MOSFETs employed in the hybrid device revealed that an increase in chip area correlated with an elevation in leakage current at the avalanche voltage inflection point.This rise in leakage current resulted in a larger proportion of avalanche current being carried during the avalanche process.Furthermore, experiments involving variations in gate-signal turn-off delay time demonstrated that delaying the turn-off of the SiC MOSFET enhanced the avalanche current level of the hybrid device.The insights derived from these findings hold significant implications for improving the single-shot robustness of Si/SiC hybrid devices.

Acknowledgment

This work was supported by Yuelushan University Science & Technology Park under Grant H202291400539.

References

[1] N. Ren, H. Hu, X. Lyu, et al., "Investigation on Single Pulse Avalanche Failure of SiC MOSFET and Si IGBT." SolidState Electronics, vol. 152, pp. 33-40, January 2019.

[2] J. G. Kassakian and T. M. Jahns, "Evolving and Emerging Applications of Power Electronics in Systems," in IEEE Journal of Emerging and Selected Topics in Power Electronics, vol. 1, no. 2, pp. 47-58, June 2013.

[3] M. S. T. McDonald, A. Murray, and T. Avram, "Power MOSFET avalanche design guidelines," International Rectifier, El Segundo, CA, USA, Tech. Rep. AN-1005, 2011.

[4] K. F. Hoffmann and J. P. Karst, "High frequency power switch - improved performance by MOSFETs and IGBTs connected in parallel," 2005 European Conference on Power Electronics and Applications, Dresden, Germany, 2005, pp. 11 pp.-P.11.

[5] R. A. Minamisawa, U. Vemulapati, A. Mihaila, C. Papadopoulos and M. Rahimo, "Current Sharing Behavior in Si IGBT and SiC MOSFET Cross-Switch Hybrid," IEEE Electron Device Letters, vol. 37, no. 9, pp. 1178-1180, Sept. 2016.

[6] Y. Zhou, M. Xu, Y. Zhang and H. Liu, "Research and Simulation of Hybrid ANPC Converter," 2023 IEEE 6th International Electrical and Energy Conference (CIEEC), Hefei, China, 2023, pp. 2399-2404

[7] J. He, R. Katebi and N. Weise, "A Current-Dependent Switching Strategy for Si/SiC Hybrid Switch-Based Power Converters," IEEE Transactions on Industrial Electronics,vol. 64, no. 10, pp. 8344-8352, Oct. 2017.

[8] Li, Zongjian;Wang, Jun;Ji, Bing;Shen, Z. John.Power Loss Model and Device Sizing Optimization of Si/SiC Hybrid Switches[J].IEEE TRANSACTIONS ON POWER ELECTRONICS,2020,Vol.35(8): 8512-8523

[9] K. Fischer and K. Shenai, "Electrothermal effects during unclamped inductive switching (UIS) of power MOSFET's," IEEE Trans. Electron Devices,vol.44,no. 5, pp. 874–878, May 1997. DOI: 10.1109/16.568052.

[10] R. R. Stoltenburg, "Boundary of power-MOSFET, unclamped inductiveswitching (UIS), avalanche-current capability," in Proc. 4th Annu. IEEE APEC, Mar. 1989, pp. 359–364. DOI: 10.1109/APEC.1989.36987.

[11] J. Wei, Z. Wei, H. Fu, et al., "Review on the Reliability Mechanisms of SiC Power MOSFETs: A Comparison Between Planar-Gate and Trench-Gate Structures," IEEE Transactions on Power Electronics, vol. 38, no. 7, pp. 8990-9005, July 2023.

[12] J. Wang, Z. Li, X. Jiang, C. Zeng and Z. J. Shen, "Gate Control Optimization of Si/SiC Hybrid Switch for Junction Temperature Balance and Power Loss Reduction," IEEE Transactions on Power Electronics, vol. 34, no. 2, pp. 1744-1754, Feb. 2019.

[13] Hangzhi Liu,Dehang He,Qicheng Guo,et al. "UIS Characterization of Hybrid Si/SiC Device under Single Pulse Avalanche Mode"IEEE 10th International Power Electronics and Motion Control Conference (IPEMC2024-ECCE Asia). 2024.

Research on Thermal Conductivity and Mechanical Properties Control of Epoxy Resin in Extreme Environments

Zhen Li
The School of Electrical Engineering
Shandong University
Jinan,China
0009-0008-7303-7302

Liang Zou
The School of Electrical Engineering
Shandong University
Jinan,China
zouliang@sdu.edu.cn

Qingsong Liu
Electric Power Research Institute
CSG
lGuangzhou,China
kyyzzbs@csg.cn

Zhiyun Han
The School of Electrical Engineering
Shandong University
Jinan,China
hanzhiyun@sdu.edu.cn

Jinyang Bai
State Grid of CHINA Technology
College
SGCC
Jinan,China
1019854711@qq.com

Abstract—In high-altitude regions with plentiful renewable energy resources, extreme environmental conditions—including substantial temperature variations, elevated humidity levels, intense radiation, and low air pressure—impose heightened demands on the insulation systems of electric power equipment. Consequently, there is an escalating imperative to engineer insulating materials that exhibit both elevated thermal conductivity and substantial dielectric strength. The fundamental physicochemical characteristics of epoxy resin-cured structures may undergo alterations in high humidity environments. Contemporary modification methodologies primarily focus on enhancing the performance of epoxy resin through the incorporation of high thermal conductivity nanomaterials. A comprehensive series of molecular simulations and analytical studies were conducted to elucidate the modification mechanism of three nanoparticle fillers, namely carbon nanotubes, graphene, and SiO_2, within the epoxy resin matrix, as well as their humidity-responsive properties. A molecular model of the nanofiller-enhanced epoxy resin was constructed, and different concentrations of water molecules were introduced for the humidity environment simulation. The changing rules of the thermal diffusion properties, mechanical properties, glass transition temperature, and dielectric properties of the composites under the humid environment were systematically explored. The study found that graphene exhibited exceptional performance in optimizing mechanical and thermal properties, with carbon nanotubes ranking second. Additionally, the dielectric constant of SiO_2 was observed to surpass that of other systems in terms of humidity adaptation.

Keywords—epoxy resin, high humidity, molecular dynamics, nano-dopant, heat diffusion coefficient, mechanical property

I. INTRODUCTION

In recent years, the rapid development of the electric power industry has caused the voltage level and network scale of the power grid system to increase continuously. The geographical distribution of energy-rich areas and load centers being mismatched further drives the transmission voltage level up[1]. High-voltage and insulation technology are the cornerstones of constructing large-scale power systems, and the level of research, development, and manufacturing of insulation materials directly constrains the development of electrical equipment. In renewable energy-rich, high-altitude

areas especially, the power equipment insulation system is subject to extreme environmental conditions such as large temperature differences, high humidity, strong radiation, and low air pressure. The development of insulating materials with both high thermal conductivity and high dielectric strength is increasingly critical in these areas[2]. Epoxy resins are widely used in gas-insulated switchgear (GIS), transformers, switchgear cabinets, cable terminals, and other electrical insulating components due to their excellent physical, chemical, and insulating properties[3].

However, the underlying physicochemical properties of epoxy-cured structures may be subject to alteration in high-humidity environments. The interface exhibits a high propensity for extrusion or stretching due to the significant disparities in thermomechanical properties and mechanical stress release between the conductor and the epoxy system. This phenomenon significantly impacts the stable operation of electrical equipment. This stress concentration frequently results in hydrolysis damage to the insulating material, which can ultimately lead to insulation failure, equipment fire, or even explosion, among other significant safety concerns. Consequently, there is an urgent need for in-depth study of the mechanical properties of epoxy curing systems under moisture attack and other physical properties of the law of change, to further reveal its moisture-induced degradation mechanism, in order to ensure that the long-term stable operation of electrical equipment in extreme environments is guaranteed. The prevailing modification methods principally augment the performance of epoxy resin through the incorporation of high thermal conductivity nanomaterials. However, the use of nanomaterials has the potential to influence various physical properties, including thermal stability, mechanical properties, dielectric properties, and insulation strength, while concomitantly enhancing thermal conductivity.

In order to optimize the selection and modification strategy of nanomaterials, molecular dynamics (MD) simulation can effectively replace traditional experimental methods and significantly shorten the research and development cycle. Clancy constructed an atomic model of epoxy resin through molecular simulation and confirmed that the reduction of cross-linking density and the increase of

979-8-3315-1110-4/25 $31.00 © 2025 IEEE

temperature and humidity will lead to the deterioration of elasticity properties[4]. Giannopoulos simulated the mechanical behaviors of graphene nanocomposites and found that the increase of graphene volume fraction can enhance the interfacial properties of epoxy resin. Giannopoulos modeled the mechanical behavior of graphene nanocomposites and found that an increase in the volume fraction of graphene could enhance the interfacial bond strength and stress transfer efficiency[5]. The research methodologies and nano-modification outcomes underpin the project, thereby providing a theoretical foundation for the development of epoxy materials that are conducive to utilization in extreme environments.

II. MODELING OF EPOXY RESIN NANOFILLER DOPING UNDER DIFFERENT MOISTURE INTRUSION

In the domain of electrical insulation application environments, glycidyl ether epoxy resins have a higher prevalence. Among these, bisphenol A type epoxy resins are particularly prominent due to their superior comprehensive performance. These resins are formed through the process of crosslinking bisphenol A diglycidyl ether (DGEBA) with a curing agent, 3,3'-diaminodiphenyl sulfone(33DDS). The structural formula for this process is depicted in Fig.1.

Fig. 1. The general structure formula of bisphenol A epoxy resin

This paper proposes a novel approach to the reinforcement of materials, employing nano-modification technology and three distinct nano-fillers: carbon nanotubes, graphene, and SiO_2. These nano-fillers possess exceptional physical properties, which have been extensively documented in the literature[6-8]. In order to investigate the effects of the introduction of three kinds of nano-fillers into an epoxy resin matrix on the improvement of the materials' mechanical and thermal properties after moisture exposure, the following steps were taken:

- The selection of Materials Studio (MS) for modeling is predicated on the consideration of interatomic interactions. The COMPASS II force field, deemed suitable for epoxy resin materials, is selected for optimization to ensure the rationality of the molecular space structure.

- The amorphous cell module in MS software was utilized to construct a three-dimensional model of the epoxy resin. This model was then used to complete the cross-linking process of the epoxy resin. Subsequently, water molecules were added to the model to establish an epoxy resin model under the invasion of different concentrations of water.

- Three types of nanomaterials were incorporated into the epoxy resin, resulting in the establishment of models of epoxy resin/carbon nanotube(EP/SWCNT), graphene/epoxy resin(EP/GR), and SiO_2/epoxy resin under the intrusion of varying concentrations of water, as illustrated in Fig.2.

- The three nanoparticle-doped epoxy resin models with varying concentrations of water intrusion were

subjected to molecular dynamics relaxation of 100 to 200ps under the constant temperature (NVT) system at 300K. This ensured that the system was fully relaxed under the given conditions, eliminated any imbalance factors, and gradually stabilized the system.

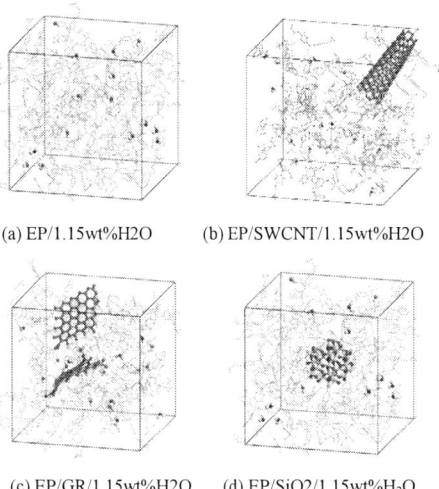

(a) EP/1.15wt%H2O   (b) EP/SWCNT/1.15wt%H2O

(c) EP/GR/1.15wt%H2O   (d) EP/SiO2/1.15wt%H_2O

Fig. 2. EP model with different water concentration invasion

III. STUDY ON THE MODIFICATION TO ENHANCE THE HEAT DIFFUSION COEFFICIENT

The thermal diffusion coefficient is a quantitative metric that quantifies the rate at which a temperature perturbation on one side is transferred to the other. That is to say, it is the rate at which the temperature difference between the two sides of a material is reduced when the material is externally heated or cooled. It has been demonstrated that an increase in the thermal diffusion coefficient leads to a decrease in the material's temperature gradient and an enhancement in its heat transfer capacity. In the context of insulating materials, the prevention of internal overheating is contingent upon the possession of a high thermal diffusion coefficient. The dimensions of this coefficient are associated with the thermal conductivity, specific heat capacity, and density, which are calculated as follows:

$$\alpha = \kappa/\rho C \qquad (1)$$

Where α is the thermal diffusion coefficient, κ is the thermal conductivity, ρ is the density of the system, and C is the specific heat capacity. To obtain the thermal diffusion coefficient of a material, its thermal conductivity and specific heat capacity need to be calculated.

A. Calculation of thermal conductivity

The thermal conductivity of the epoxy resin model with different concentrations of water intrusion under the doping of three nanoparticles can be calculated by using the non-equilibrium molecular dynamics (NEMD) method. This method is based on Fourier's law, which generates a temperature gradient by localized heat bath method. However, during the molecular dynamics simulation of water molecules contained in the system, the problem of water molecules diffusing out of the simulation space occurs. This problem can be mitigated by establishing wall conditions at both ends of the heat source and heat sinks or by employing a regional constant potential field to restrict the diffusion of the atoms.

This approach ensures that the water molecules are repelled back to the simulation region, preventing them from escaping.

The present study investigates the trend of the overall thermal conductivity with moisture absorption of the epoxy resin model doped with three types of nanoparticles and compares it with the EP/neat model. The results are plotted as shown in Fig.3.

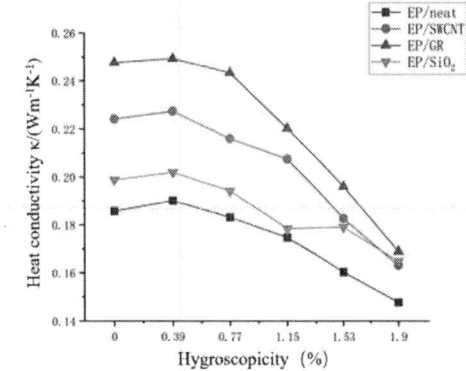

Fig. 3. The curve of thermal conductivity of epoxy resin doped with different nanomaterials with moisture absorption rate

The findings revealed substantial discrepancies in the thermal conductivity enhancement effect of diverse reinforcement materials on epoxy resin, as well as the subsequent attenuation trend following water absorption. In anhydrous conditions, the graphene-reinforced epoxy resin demonstrated the maximum overall thermal conductivity of approximately 0.2476 W·m⁻¹·K⁻¹, which was greater than that of the carbon nanotube-reinforced system (0.2241 W·m⁻¹·K⁻¹) and the silica-reinforced system (0.1988 W·m⁻¹·K⁻¹). Significant enhancements in thermal conductivity were observed for the EP/SWCNT, EP/GR, and EP/SiO2 models, with respective increases of 20.7%, 33.3%, and 7.1%, as compared with the EP/neat model. This substantial effect on the thermal conductivity of epoxy resin in a dry environment underscores the importance of these models in elucidating the underlying mechanisms and potential applications of these materials. The following text is intended to provide a comprehensive overview of the subject matter. This phenomenon can be attributed to the two-dimensional lamellar structure of GR, which facilitates a more extensive thermal conductivity path. In contrast, SWCNT, due to its one-dimensional nanotube configuration, exhibits a slightly compromised distribution uniformity compared to graphene, despite its superior thermal conductivity. Conversely, SiO₂ exhibits the least pronounced initial thermal conductivity enhancement effect, attributable to its low inherent thermal conductivity and granular composition.

B. Calculation of specific heat capacity

In order to streamline the calculation process, the dry state (0wt%) and saturated water absorption concentration (1.90wt%) models of epoxy resin under three nanoparticle dopings were utilized to calculate the specific heat capacity. Each group of epoxy model was operated at 300K for 100ps under the NVT system, and each model was subjected to five sets of repetitions. The results of the calculated specific heat capacity are shown in Fig. 4.

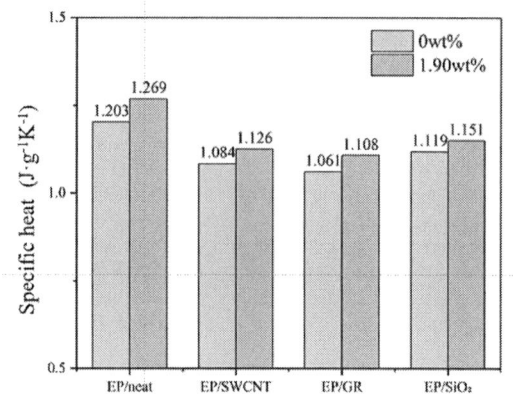

Fig. 4. The specific heat capacity of 4 groups of models before and after saturated water absorption

The specific heat capacity of all three nanoparticles decreased after doping. This phenomenon can be attributed to the lower specific heat capacities of carbon nanotubes, graphene, and SiO₂ relative to the epoxy resin matrix. Consequently, the overall specific heat capacity of the doped system is essentially equivalent to the weighted average of the matrix and the doped particles. This reduction in specific heat capacity is attributable to the lower specific heat capacities of the particles. It has been demonstrated that the doping of nanoparticles can lead to a reduction in the growth of specific heat capacity. This phenomenon can be attributed to the enhanced thermal conductivity of nanoparticles, which facilitates a more rapid and uniform distribution of heat at the microscopic level. Consequently, the contribution of heat absorption by complex chain segment dynamics is diminished, resulting in a macroscopic manifestation of reduced specific heat capacity.

C. Calculation of heat diffusion coefficient

The heat diffusion coefficient is calculated according to equation (1) and the results are shown in the TABLE I.

TABLE I.   THE DENSITY BEFORE WATER ABSORPTION AND THE THERMAL DIFFUSION COEFFICIENT BEFORE AND AFTER SATURATED WATER ABSORPTION OF THE FOUR GROUPS OF MODELS

Model	Densities (g/cm³)	0wt% Heat diffusion coefficient (m²/s)	1.90wt% Heat diffusion coefficient (m²/s)
EP/neat	1.18	0.1308	0.1021
EP/SWCNT	1.21	0.1540	0.1197
EP/GR	1.23	0.1897	0.1239
EP/SiO₂	1.26	0.1410	0.1136

The data presented in the table demonstrate a substantial alteration in the thermal diffusion coefficient of the epoxy resin following the incorporation of diverse fillers. The EP/neat system exhibited the lowest thermal diffusion coefficient prior to water absorption (0.1308m²/s), the EP/GR system demonstrated the highest (0.1897m²/s), the EP/SWCNT system exhibited the subsequent highest (0.1540m²/s), and the EP/SiO₂ system exhibited a slightly elevated thermal diffusion coefficient compared to the undoped system (0.1410m²/s). The graphene and SWCNT systems formed an efficient thermal conductivity network in the matrix due to their excellent thermal conductivity, which significantly enhanced the thermal diffusion performance.

After water absorption, the thermal diffusion coefficients of all the systems decreased, and the thermal diffusion coefficient of the undoped system decreased to 0.1021m²/s, while that of the graphene and carbon nanotube systems decreased to 0.1239m²/s and 0.1197m²/s, respectively, which showed a better resistance to humidity attenuation. The thermal diffusion coefficient of the SiO₂ system, despite of the weak thermal conductivity, decreased the smallest after water absorption, and the decrease was only 19.5%, showing better humidity adaptability.

When considered collectively, the graphene and carbon nanotube systems exhibit considerable advantages in terms of thermal performance enhancement. Conversely, the SiO₂ system is well-suited for scenarios that demand high levels of moisture absorption in the environment.

IV. STUDY ON THE MODIFICATION OF MECHANICAL PROPERTIES FOR ENHANCEMENT

The mechanical properties of the epoxy resin model doped with three kinds of nanoparticles were also calculated using the static normal strain method. The NPT simulation optimization combined with the mechanical property calculation was carried out for the physical properties of the invasion of different concentrations of moisture. Five sets of repetitive data calculations were taken for each kind of moisture concentration. The final results of the epoxy resin model doped with three kinds of nanoparticles and the EP/neat model are shown in Fig. 5.

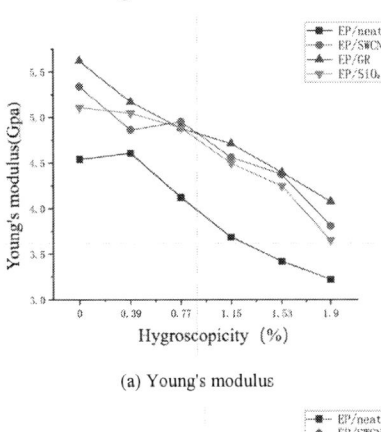

(a) Young's modulus

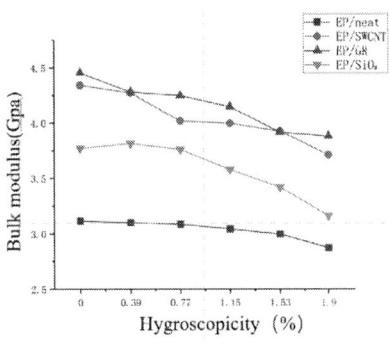

(b) Bulk modulus

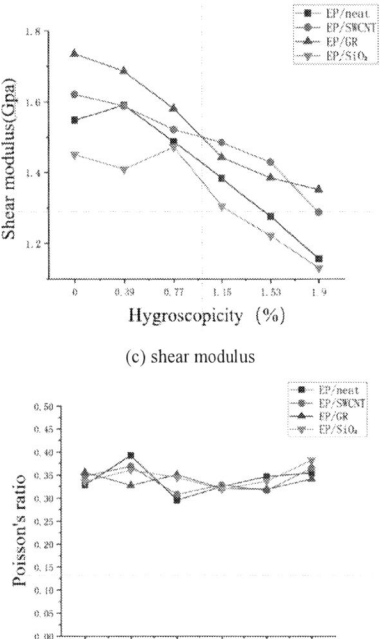

(c) shear modulus

(d) Poisson's ratio

Fig. 5. The mechanical properties of the four groups of models with humidity change curve

The clathrate modulus and bulk modulus exhibited a gradual decrease in response to the increase in moisture absorption rate. The EP/GR and EP/SWCNT systems demonstrated superior retention properties, while the EP/SiO₂ and undoped systems exhibited a faster rate of decrease, as illustrated in Figures (a) and (b). Among these, the enhancement of graphene and carbon nanotubes is associated with their exceptional interfacial binding capability and packing microstructure. The two-dimensional lamellar structure of graphene establishes a continuous rigid network, thereby effectively constraining the slippage of matrix chain segments under the influence of hydroplasticization. This, in turn, results in a retardation of modulus decay. Conversely, silica particles exhibit reduced interfacial bonding during moisture absorption and insufficient filler dispersion homogeneity, resulting in a more significant modulus decrease.

The shear modulus undergoes a more substantial decrease following moisture absorption, particularly in EP/SiO₂ and undoped systems, as illustrated in Figure (c). The modulus of the graphene system can be maintained at 1.341GPa under conditions of high humidity, indicating that it exhibits good stability in resisting shear deformation. This performance advantage is attributable to the efficient interfacial interaction between graphene and the substrate, which can uniformly disperse the moisture-induced internal stress. Conversely, the silica system exhibits a pronounced decline in modulus under conditions of high humidity. This phenomenon may be attributed to the lack of rigidity of the particles and the moisture absorption at the interfacial region, which results in the softening of the interfacial layer. This, in turn, significantly reduces the toughness properties of the material.

The Poisson's ratio, denoted by υ, exhibited minimal variation during the moisture absorption process, with values ranging from 0.32 to 0.39. No statistically significant differences were observed among the systems. This finding suggests that moisture absorption exerts a negligible influence on the transverse deformation capacity of epoxy resin. Furthermore, the filler does not modify the relationship between transverse and axial deformation of epoxy resin composites.

In summary, the absorption of moisture significantly reduced the modulus properties of the epoxy resin system. However, the presence of fillers effectively enhanced the material's moisture resistance. Graphene exhibited the optimal moisture resistance, and its two-dimensional structure exhibited a high degree of bonding to the matrix, thereby effectively constraining the deleterious effects of moisture on chain segment slip and interfacial bonding. Carbon nanotubes exhibit enhanced properties, yet their one-dimensional structure results in a network that is comparatively less homogeneous than that of graphene. Silicon dioxide demonstrated a weaker modulation of its mechanical properties in high humidity environments, attributable to its lack of hydrophilicity and rigidity. This finding indicates that carbon-based fillers, such as graphene, hold considerable promise for applications in environments necessitating high mechanical properties in the presence of high humidity.

V. CONCLUSION

This paper presents a comprehensive molecular simulation and analytical study that investigates the modification mechanism of three types of nanoparticle fillers—namely, carbon nanotubes, graphene, and SiO_2—in an epoxy resin matrix, as well as their humidity-responsive properties. A molecular model of the nanofiller-enhanced epoxy resin was constructed, and different concentrations of water molecules were introduced for the humidity environment simulation. The variation rules of the thermal diffusion properties, mechanical properties, glass transition temperature, and dielectric properties of the composites under the humid environment were systematically explored. The following conclusions were obtained:

- The doping of graphene and carbon nanotubes led to a substantial enhancement in the thermal conductivity and thermal diffusion coefficient of epoxy resin. In anhydrous conditions, the graphene system exhibits a substantial enhancement in thermal conductivity (33.3% higher than that of the undoped system), rendering it well-suited for applications in dry environments where high thermal conductivity is imperative. Following the absorption of moisture, the graphene and carbon nanotube systems demonstrated superior resistance to moisture decay. Conversely, the SiO_2 system exhibited relatively more stable thermal diffusion performance under high humidity conditions, exhibiting enhanced humidity adaptability.

- The three fillers enhanced the mechanical properties of the epoxy resin matrix. Graphene demonstrated optimal performance under both dry conditions and following moisture absorption, exhibiting remarkable modulus retention capacity. This superior performance is attributed to the inherent two-dimensional lamellar structure and robust interfacial bonding characteristics of graphene. In comparison, carbon nanotubes

enhanced the strength of the matrix through their one-dimensional network configuration. Additionally, carbon nanotubes exhibited enhanced resistance to changes in internal stresses induced by humidity, contributing to their overall performance in complex environments. In contrast, SiO_2 demonstrated enhanced modulus in dry conditions; however, its superior hydrophilicity and inadequate interfacial bonding rendered it more adept at moisture absorption. Consequently, this led to an enhancement in the mechanical performance of the epoxy resin. In contrast, while SiO_2 has been shown to enhance the modulus under dry conditions, its modulus decreases more after moisture absorption due to its higher hydrophilicity and insufficient interfacial bonding. This renders it less adaptable to moisture..

In this paper, the different modification mechanisms and advantages of carbon nanotubes, graphene and SiO_2 on the humidity adaptation of epoxy resins are revealed by molecular simulation, and it is found that graphene is outstanding in the optimization of mechanical and thermal properties, followed by carbon nanotubes, and the dielectric constant of SiO_2 is better than that of the other systems in terms of humidity adaptation. The results provide a theoretical basis for the design of functionalized epoxy composites for different environmental requirements and have important application reference value.

This paper utilizes molecular simulation to elucidate the various modification mechanisms and advantages of carbon nanotubes, graphene, and SiO_2 on the humidity adaptation of epoxy resins. The findings indicate that graphene exhibits exceptional efficacy in optimizing mechanical and thermal properties, with carbon nanotubes demonstrating a comparable performance. Additionally, the dielectric constant of SiO_2 is observed to exceed those of other systems in terms of humidity adaptation. The findings establish a theoretical framework for the design of functionalized epoxy composites tailored to diverse environmental requirements and offer significant practical application value.

ACKNOWLEDGMENT

This paper is funded by National Natural Science Foundation of China Enterprise Innovation and Development Joint Fund Project（U24B2092）.

REFERENCES

[1] Mazur, C., Hoegerle, Y., Brucoli, M., van Dam, K., Guo, M., Markides, C. N., & Shah, N. (2019). A holistic resilience framework development for rural power systems in emerging economies. Applied Energy, 235, 219-232.

[2] Shengtao Li, Shihu Yu, Feng Yang, Progress in and prospects for electrical insulating materials, High Volt 1 (3) (Oct. 2016) 122e129.

[3] Clancy T, Frankland S, Hinkley J, et al. Molecular modeling for calculation of mechanical properties of epoxies with moisture ingress[J]. Polymer, 2009, 50(12): 2736-2742.

[4] Giannopoulos G I, Kallivokas I G. Mechanical properties of graphene based nanocomposites incorporating a hybrid inter-phase[J]. Finite Elements in Analysis & Design, 2014, 90: 31-40.

[5] Kim H, Park S. Smart cure cycle with cooling and reheating for co-cure bonded steel/carbon epoxy composite hybrid structures for reducing thermal residual stress[J]. Composites Part A: Applied Science and Manufacturing, 2006, 37(10): 1708-1721.

[6] Y. Chen, H. Zhang, Y. Yang, et al. High‑Performance Epoxy Nanocomposites Reinforced with Three‑Dimensional Carbon Nanotube Sponge for Electromagnetic Interference Shielding[J]. Advanced Functional Materials, 2016, 26(3): 447-455.

[7] X. Shen, Z. Wang, Y. Wu, et al. Multilayer Graphene Enables Higher Efficiency in Improving Thermal Conductivities of Graphene/Epoxy Composites[J]. Nano letters, 2016, 16(6): 3585-3593.

[8] Allahverdi, A., Ehsani, M., Janpour, H., & Ahmadi, S. The effect of nanosilica on mechanical, thermal and morphological properties of epoxy coating[J]. Progress in Organic Coatings, 2012, 75(4): 543-548.

A 3L-ANPC SiC MOSFET Power Module with Low Thermal Coupling and Parasitic Inductance

Zedong Xue
Faculty of Integrated Circuit
Xidian University
Xi'an,China
23111213538@stu.xidian.edu.cn

Zhiyuan Qi
Faculty of Integrated Circuit
Xidian University
Xi'an,China
qizhiyuan@xidian.edu.cn

Zihao Chen
Faculty of Integrated Circuit
Xidian University
Xi'an,China
24111110081@stu.xidian.edu.cn

Hao Yuan
Faculty of Integrated Circuit
Xidian University
Xi'an,China
haoyuan@xidian.edu.cn

Qingwen Song
Faculty of Integrated Circuit
Xidian University
Xi'an,China
qwsong@xidian.edu.cn

Yuming Zhang
Faculty of Integrated Circuit
Xidian University
Xi'an,China
xidian_ic@163.com

Abstract—Three-level active neutral-point-clamped (ANPC) converters are particularly suitable for high-frequency, high-power-density applications owing to their superior voltage withstand capability and reduced harmonic distortion. However, its widespread adoption remains constrained by parasitic inductance from intricate circuit layouts and thermal management challenges stemming from uneven power loss distribution. To address the above challenges, this paper proposes a 3D-uniform layout double-sided cooled (3DUL) ANPC module. The proposed module uniformly arranges dies in three-dimensional space based on their loss distribution, utilizing the three-dimensional commutation loops to achieve mutual inductance cancellation, while simultaneously minimizing thermal coupling effects between the dies. The module additionally optimizes the thermal resistance and thermal coupling effects of the highest-loss die in ANPC converter, aiming to maximize the module's reliability.

Keywords—*ANPC module, SiC MOSFET, Double-sided Cooling, The natural frequency doubling modulation.*

I. INTRODUCTION

More-electric and all-electric aircraft, characterized by reduced carbon emissions, enhanced operational reliability, and optimized weight , represent the foremost developmental paradigms in next-generation aviation technologies[1]. The integrated starter-generator (ISG) serves as a critical enabler for next-generation aircraft electrification, combining the functions of a starter and generator into one unit, enhancing efficiency, reducing weight, and enabling more sustainable operations [2]-[4]. The growing electrical load in aircraft has increased overall energy consumption and exacerbated harmonic pollution. ISG must deliver adequate power while ensuring high-quality electrical output to meet these demands. In this context, the two-level converter utilizing Si-based power devices is no longer sufficient to meet the increasing performance demands.

The combination of silicon carbide (SiC) power devices, which offer lower switching losses and higher operating temperatures, with three-level active neutral-point-clamped (ANPC) converters, characterized by high voltage withstand capability and low harmonic output, is particularly well-suited for the electrification of future aircraft[5]. Currently, the application of ANPC converters is confronted with two critical challenges. Firstly, under high frequency switching conditions, the uneven distribution of power losses among dies becomes significantly pronounced. This disparity results in elevated thermal stress on specific die, thereby adversely affecting the overall system reliability and operational longevity. Secondly, the existing packaging architectures

exhibit high parasitic inductance, which contribute to increased switching losses and voltage spikes across the dies.

In recent years, numerous researchers have investigated the application of ANPC converters by combining various modulation strategies and device types. The ultimate goal of these studies is to reduce the maximum junction temperature of the dies within the converter and ensure the reliability of the converter system. As presented in references [6], [7] and [8], the conduction modulation effect of IGBTs under high current and the low switching losses of SiC MOSFETs are utilized by operating the two types of devices at different switching frequencies, thereby balancing conduction losses and switching losses. However, under high-frequency switching conditions, the switching losses become significantly greater than the conduction losses, rendering this balancing approach ineffective. Reference [9] implements an ANPC converter utilizing two three-phase full-bridge modules and six discrete IGBT devices. This study merely verified the improvement of uneven loss distribution by the natural frequency doubling method. The designed ANPC converter was unable to achieve high power density. Due to the low parasitic inductance and superior thermal performance, the double-sided cooling package has been recognized as potential solution for high power density SiC power module. However, most of the reported double-sided cooling power modules are half bridge configuration[10]. The research about the double-sided cooling package for ANPC module is very rare. In [11] , a ANPC single-phase module are designed in the form of a double-sided cooling package. The parasitic inductance and thermal resistance have been improved, however, it does not take into account the loss discrepancies among dies at different switching positions during actual operation, nor the thermal coupling effects caused by these loss differences.

Therefore, existing ANPC modules are not suitable for high-frequency, high-power-density applications. This paper proposes a ANPC module with 3D-uniform layout double-side cooled package (3DUL), which achieves high power density while realizing low parasitic inductance, thermal resistance and thermal coupling. In Section II , an corresponding loss model will be established, and the loss distribution among different dies based on the natural frequency doubling modulation will be calculated. In Section III，the design approach and details of the module will be provided, along with result validation using corresponding simulation software.

979-8-3315-1110-4/25 $31.00 © 2025 IEEE

II. LOSS MODEL DEVELOPMENT AND CALCULATION

A. The natural frequency doubling modulation

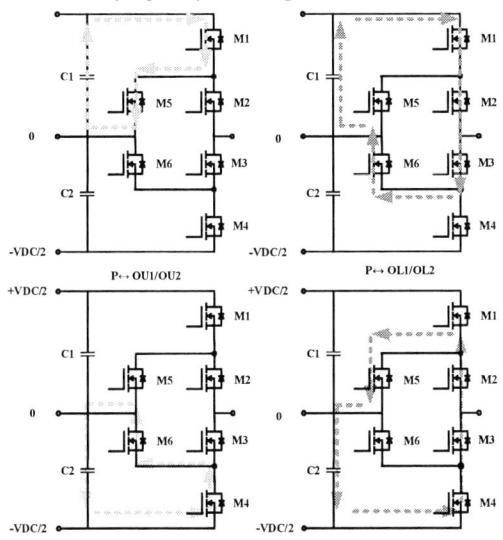

Fig 1   The basic commutation circuit loop.

TABLE I.   ANPC SINGLE-PHASE STATUS

Status	Device						Output
	M1	M2	M3	M4	M5	M6	
P	1	1	0	0	0	1	$+V_{DC}/2$
OU1	0	1	0	0	1	0	0
OU2	0	1	0	1	1	0	0
OL1	0	0	1	0	0	1	0
OL2	1	0	1	0	0	1	0
N	0	0	1	1	1	0	$-V_{DC}/2$

There are three commonly used modulation methods for ANPC converters[12]. The allowable states in ANPC single-phase arc shown in Table I, and the basic commutation circuit loop is shown in Fig. 1. Direct transitions between the P-state and N-state are prohibited, necessitating an intermediate transition through one of the four predefined zero states to ensure operational stability. M1 and M4 are referred to as outer switches, M2 and M3 as inner switches, and M5 and M6 as clamping switches. In PWM1, the outer switches and clamping switches are operated at the carrier frequency while the inner switches are operated at the fundamental frequency, utilizing the small commutation loop (LS) as the sole commutation path. In PWM2, the inner switches are operated at the carrier frequency, whereas the outer switches and clamping switches are operated at the fundamental frequency, employing the large commutation loop (LL) as the only commutation path. Using a single commutation loop will cause the switching losses to be concentrated on the outer or inner switches, and it cannot be applied under high-frequency switching.

The natural frequency doubling modulation employs two bipolar carriers of identical amplitude but with a 180° phase difference to achieve a 1:1 ratio between the large and small

commutation loops[13]. Simultaneously, this approach doubles the equivalent switching frequency of the output voltage compared to the actual switching frequency. The detailed modulation strategy is illustrated in Fig.2.

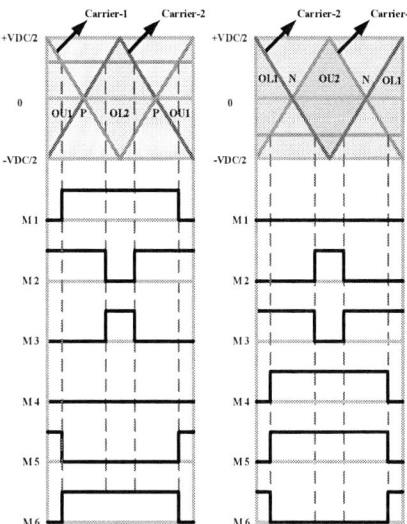

Fig 2   The natural frequency doubling modulation.

B. Loss model

The key parameters of SiC MOSFETs can be quickly obtain from the datasheets provided by the manufacturers. Using curve fitting tools, the functional relationships between these parameters can be derived. It is important to note that while the MOSFET allows bidirectional current flow, there is a difference in the current conduction capabilities in the forward and reverse directions. The establishment of conduction losses is as follows.

$$R_{DS}(I_D)=R_{DS,25}(I_D)+\frac{T_j-25}{150}\left[R_{DS,175}(I_D)-R_{DS,25}(I_D)\right] \quad (1)$$

$$R_{SD}(I_D)=R_{SD,25}(I_D)+\frac{T_j-25}{150}\left[R_{SD,175}(I_D)-R_{SD,25}(I_D)\right] \quad (2)$$

$$P_{conduction,\,forward}=\frac{1}{T}\sum[I_{max}\cdot M\cdot \sin(\omega t-\varphi)]^2\cdot R_{DS}(I_D)\cdot T_s \quad (3)$$

$$P_{conduction,\,reverse}=\frac{1}{T}\sum[I_{max}\cdot M\cdot \sin(\omega t-\varphi)]^2\cdot R_{SD}(I_D)\cdot T_s \quad (4)$$

$$P_{conduction}=P_{conduction,\,forward}+P_{conduction,\,reverse} \quad (5)$$

Where the $R_{DS,25}(I_D)$ and $R_{DS,175}(I_D)$ represent the expressions for the forward conduction resistance at 25°C and 175°C, respectively, under the specified gate voltage and gate resistance. M refers to the modulation index (MI), φ is the power factor Angle, and T_S is the sampling time. Similarly, the expression for the reverse conduction resistance can be derived. After performing linear interpolation, expressions for the forward and reverse conduction resistance at different temperatures and current levels can be obtained.

The main factors affecting the switching loss of MOSFET are the current and voltage, drive resistance and temperature, which are established as follows.

$$E_{on}(I_D) = k_{T_j,on} \cdot k_{V,\,on} \cdot k_{R_g,on} \cdot a_1 \cdot I_D^{\,b_1} \qquad (6)$$

$$E_{off}(I_D) = k_{T_j,off} \cdot k_{V,\,off} \cdot k_{R_g,off} \cdot a_2 \cdot I_D^{\,b_2} \qquad (7)$$

$$E_{rr} = \frac{I_{E_{rr}}}{I_{test}} Q_{rr} \cdot V_{DS} \qquad (8)$$

$$P_{switch} = \frac{1}{T} \sum_{i=1}^{N_{on}} E_{on}(I_{D,i}) + \frac{1}{T} \sum_{i=1}^{N_{off}} E_{off}(I_{D,i}) \qquad (9)$$

$$P_{body,diode} = \frac{1}{T} \sum_{i=1}^{N_{E_{rr}}} E_{rr,i} \qquad (10)$$

Where the coefficients $k_{T_j,on}$, $k_{V,\,on}$, and $k_{R_g,on}$ are temperature correction, voltage correction, and gate resistance correction factors for the turn-on energy function, respectively, which are extracted from the datasheet. Similarly, the expression for the turn-off energy function can also be derived. E_{rr} represents the reverse recovery energy of the body diode under different current and voltage conditions. N_{on}, N_{off}, and $N_{E_{rr}}$ denote the effective number of turn-on events, turn-off events, and reverse recovery events, respectively, during one fundamental period.

C. Loss calculation

In MATLAB/Simulink, a related model is established as shown in Fig. 3. Based on the principle of piecewise linear approximation, calculate the losses under different operating conditions, with a sampling time of 10 ns. It is important to note that during loss calculation, due to the bidirectional conduction characteristic of the MOSFET, the body diode conduction may overlap with the switching events. In such cases, the voltage between the drain and source will be clamped by the body diode, and the resulting switching losses can be ignored. A brief explanation is given for M5, as shown in Fig. 4.

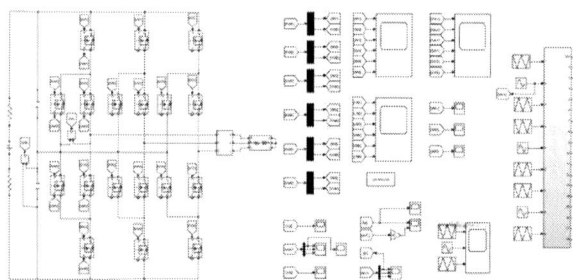

Fig 3   ANPC-simulink model.

The operational states of ISG are highly complex. During the startup phase, it functions as an inverter, while transitioning to a rectifier during steady-state operation. Additional modes such as climbing, coasting, and hybrid propulsion further diversify the operational profiles[14]. Consequently, ANPC converter exhibits multiple combinations of power factors (PF) and MI. Currently, the voltage levels of MEA are mostly 270V and 540V. For compatibility purposes, a bus voltage of 1080V and 1620V is selected for simulation. Due to the circuit symmetry, power loss calculations were selectively performed for modules M1, M2, and M5. Typical ranges of modulation index and power factor were investigated, with the resultant power loss characteristics illustrated in Fig. 5. Other simulation parameters are summarized in Table II.

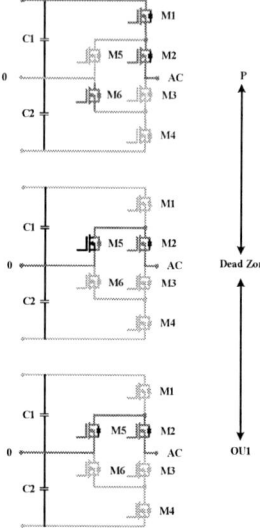

Fig 4   M5-body diode voltage clamp condition.

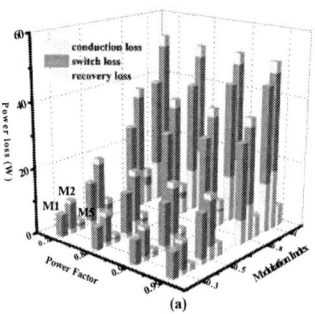

(a)

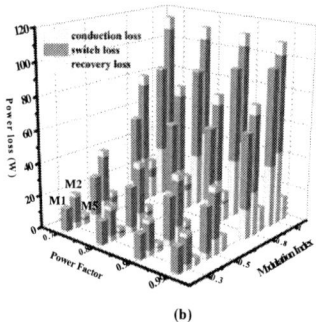

(b)

Fig 5   Loss simulation results (a) VDC=1080V (b) VDC=1620V.

TABLE II.   SIMULINK SIMULATION PARAMETERS

Parameter	Value
Load impedance	10Ω
Modulation wave frequency	400Hz
Carrier frequency	50kHz
Temperature	150°C
Sampling time	10ns

III. THE DESIGN AND SIMULATION VERIFICATION OF THE ANPC MODULE

Based on the loss characteristics illustrated in Fig. 5, three critical conclusions can be drawn. First, switching losses dominate the total power losses under high-frequency and high-voltage operating conditions. Second, in the natural double-frequency modulation strategy, the inner switches exhibit the highest power losses, followed by the outer switches, while the clamping switches demonstrate the minimal losses. Third, the power losses of the inner switches are approximately equal to the sum of the losses generated by the outer switches and clamping switches.

Therefore, in the design of power module layout, the thermal resistance of dies at the inner switch positions and their thermal coupling with dies in other switch positions should be prioritized. Additionally, the parasitic inductance in all commutation loops must be minimized to reduce switching losses under high-frequency operating conditions.

A. Overall design of the module

Based on the principles of electromagnetism, the mutual inductance cancellation principle and the reduction of loop area are recognized effective methods to reduce the parasitic inductance of the commutation loop. Double-side cooling packaging can convert the traditional two-dimensional commutation loop into a three-dimensional commutation loop, thereby effectively optimizing parasitic inductance in combination with the aforementioned two theories. As shown in Fig. 1, the track at the clamping switch position, where the die is located, serves as a common path for the two commutation loops. By positioning it between the inner and outer switches, the area of the two commutation loops can be significantly reduced. Additionally, by placing part of the path on another DBC substrate, the mutual inductance cancellation can further reduce the parasitic inductance of the commutation loop. Furthermore, this layout can also simultaneously reduce the parasitic inductance caused by the power terminals.

Building upon the double-sided cooling, each Direct Bond Copper (DBC) substrate is allocated with half of the switching devices to ensure balanced thermal distribution. Furthermore, the dies at the inner switch positions and their adjacent counterparts are positioned on independent DBC substrates. This configuration maximizes the length of the thermal coupling path between these critical heat sources, thereby significantly mitigating thermal coupling effects[15].

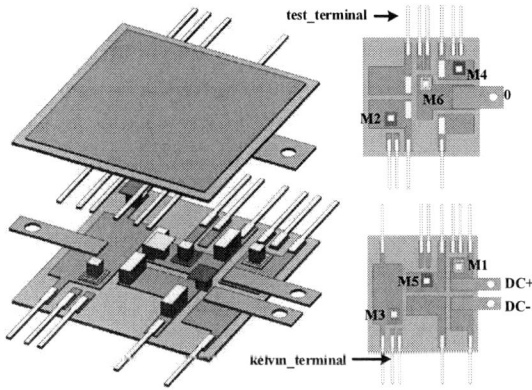

Fig 6   The explosion view of the proposed module.

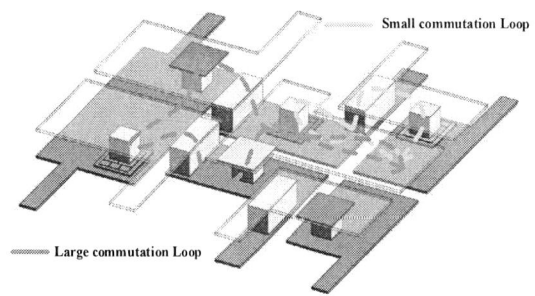

Fig 7   The commutation Loops of the ANPC module.

The overall design of the power module is illustrated in Fig. 6. Fig. 7 details the current paths of the commutation loops, where components such as power terminals and gate driver terminals are omitted for clarity.

B. The detailed optimization of the module

The height of the molybdenum block must be sufficient to accommodate bonding wire connections for both the gate and Kelvin source terminals. Additionally, the clearance between tracks of different potentials must comply with electrical safety requirements. Consequently, the design flexibility for optimizing block height and inter-track clearance is inherently limited. However, the spacing between dies significantly impacts multiple critical parameters, including thermal resistance, thermal coupling, parasitic inductance, and power density. Therefore, meticulous optimization of die-to-die spacing is essential.

In contrast to conventional packaging, when analyzing the thermal resistance of dies in double-sided cooling modules, the non-uniform temperature distribution between the top and bottom substrates prevents the thermal resistances on both sides of the dies from being simply modeled as a parallel connection[16]. Under the assumption of negligible thermal coupling, the steady-state thermal resistance model of the dual-sided cooling package is illustrated in Fig. 8. Where $R_{c\text{-}e}$ corresponds to the case-to-environment thermal resistance, $R_{j\text{-}t}$ represents the junction-to-top DBC case thermal resistance, and $R_{j\text{-}b}$ denotes the junction-to-bottom DBC case thermal resistance. The case temperature is defined as the average value to reflect practical thermal gradients.

To reduce computational complexity, the entire power module is assumed to be a linear system. By applying power to a single die, the thermal coupling between this die and others can be derived from (11). Where $k_{i,d}$ represents the thermal coupling coefficient (TCC) caused by the i-th die to the d-th die. Subsequently, Equation (12) quantifies the heat dissipation through the top and bottom DBC substrates, respectively. Where A is effective heat exchange area, h is heat transfer coefficient, and T_{sur} is the surface temperature of DBC substrate. Finally, $R_{j\text{-}t}$ and $R_{j\text{-}b}$ are determined via (13).

$$K_{I,\,d} = \frac{T_d - T_a}{P_i} \tag{11}$$

$$Q = h \cdot A \cdot (T_a - T_{sur}) \tag{12}$$

$$R_{j\text{-}x} = \frac{T_j - T_c}{Q} - \frac{T_x - T_c}{Q} \tag{13}$$

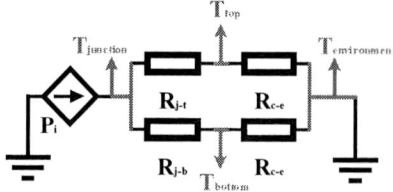

Fig 8   Steady-state thermal resistance model.

Due to the presence of power terminals, the spacing between dies at outer switch position and their distances to inner switch position dies are large, resulting in negligible thermal coupling effects. Consequently, adjusting their spacing offers limited benefits. However, dies at the inner switch position exhibit the highest power losses. To mitigate thermal coupling, clamping switch position dies adjacent to the inner switches should be placed on a another DBC substrate rather than sharing the same substrate with the inner switches, thereby avoiding exacerbating the temperature rise of the inner switch position dies. As the clamp switch position dies exhibit the lowest power losses, co-locating them with adjacent outer switch position dies on the same substrate is permissible. Additionally, reducing the spacing between clamp switch dies effectively minimizes the commutation loop area, thereby lowering parasitic inductance.

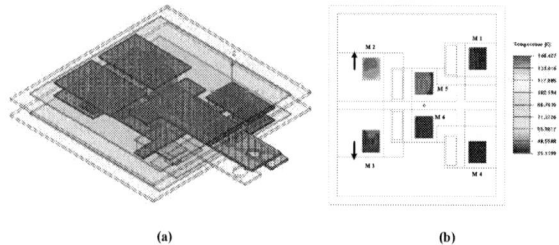

Fig 9   Examples of simulation software(a) analysis of the large commutation loop (b) analysis of thermal coupling coefficient (P₂=120w).

Therefore, optimizing the spacing between inner switch position dies is critical for enhancing module performance. Taking the configuration with the minimum commutation loop area (i.e., the closest of M2/M3 and M5/M6) as the baseline, the spacing of inner switch position dies was adjusted. Simulations were conducted using ANSYS Q3D and ANSYS ICEPAK, with test-related components omitted from the module, as illustrated in Fig. 9. The thermal simulation parameters are shown in Table III.

Fig.10(a) demonstrates the variation in parasitic inductance of the large commutation loop with different spacings, excluding the parasitic inductance contribution from power terminals. The results reveal a sharp surge in large commutation loop parasitic inductance at 3.5 mm spacing. Fig. 10(b) presents the thermal coupling trends between inner switch position dies and remaining dies across varying spacings. TCC exhibit a near-linear reduction as spacing increases. Fig. 11 shows the evolution of thermal resistances R_{j-t} and R_{j-b} under different offset distance and heat transfer coefficients. The thermal resistance decreased as the heat transfer coefficient increased, which could be attributed to alterations in the internal heat dissipation pathways within the module. However, the thermal resistance still exhibited an increasing trend with the offset distance. To

balance power density, parasitic inductance, and thermal performance, the optimal offset distance for inner switch position dies is determined to be 3.5 mm.

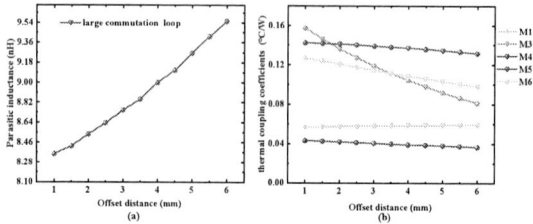

Fig 10   (a)Parasitic inductance of the loop under different offset distance(b) Thermal coupling of M2 under different offset distance, h=2000w/(m²K).

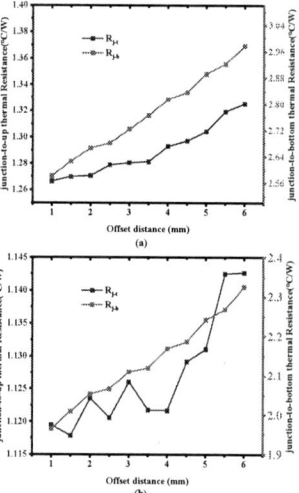

Fig 11   The changing trend of thermal resistance (a)h=2000w/(m²K) (b) h=3000w/(m²K).

M6	0.10	0.11	0.17	0.22	0.16	0
M5	0.22	0.14	0.11	0.10	0	0.16
M4	0.09	0.04	0.06	0	0.09	0.19
M3	0.04	0.11	0	0.06	0.11	0.17
M2	0.06	0	0.11	0.04	0.14	0.11
M1	0	0.06	0.04	0.09	0.20	0.09
	M1	M2	M3	M4	M5	M6
			(a)			

M6	0.05	0.07	0.10	0.15	0.11	0
M5	0.15	0.09	0.07	0.05	0	0.11
M4	0.05	0.02	0.03	0	0.05	0.13
M3	0.02	0.07	0	0.03	0.07	0.11
M2	0.03	0	0.07	0.02	0.09	0.06
M1	0	0.03	0.02	0.05	0.13	0.05
	M1	M2	M3	M4	M5	M6
			(b)			

Fig 12   The thermal coupling coefficient(°C/W) between different dies (a)h=2000W/(m²K)(b) h=3000W/(m²K).

Ultimately, the clearance between tracks is selected to be 2mm, the height of the block is chosen to be 2.5mm, and the spacing of the dies at the inner switch position is 12mm. The parasitic inductance parameters of the module are shown in Table IV. The dimensions of the positive and negative power terminals are 20mm×5mm, and the midpoint power terminal is 20mm×9mm. The thermal coupling coefficients between chips under different heat transfer coefficients are shown in Fig.12. With the increase of the heat transfer coefficient, the

TCC of the inner switch position dies and the outer switch position dies with greater loss is almost zero.

TABLE III. THERMAL SIMULATION PARAMETERS

	Materials	Thermal conductivity (W/(m²·K))	Size
Track	Copper	385	/
Solder	Nano-silver solder	200	4.5mm×5.4mm ×0.1mm
Block	Molybdenum	138	2.5mm×2.5mm ×2.5mm
Substrate	ALN	180	44mm×48mm× 0.6mm
Die	SiC	370	4.5mm×5.4mm ×0.2mm

TABLE IV. THE PARASITIC INDUCTANCE PARAMETERS OF THE MODULE

Performance	Value
LL	8.48nH
LS	4.04nH
LL(with terminals)	11.896nH
LS(with terminals)	7.184nH

IV. CONCLUSION

To address the challenges of excessive parasitic inductance and thermal management of uneven loss distribution in ANPC converters under high power density and high-frequency operation, this paper establishes a loss model to calculate the losses of a natural frequency-doubling-based all-SiC ANPC converter at an equivalent 100 kHz switching frequency. A ANPC module with 3D-uniform layout double-side cooled package was then designed. Simulations verify that the parasitic inductance in commutation loops remains below 12 nH (including power terminals), with the highest-loss die achieving ultralow thermal resistance and minimal thermal coupling coefficients through targeted thermal optimization.

REFERENCES

[1] K. Emadi and M. Ehsani, "Aircraft power systems: technology, state of the art, and future trends," *IEEE Aerospace and Electronic Systems Magazine*, vol. 15, no. 1, pp. 28–32, Jan. 2000.

[2] G. L. Calzo, P. Zanchetta, C. Gerada, A. Gaeta, and F. Crescimbini, "Converter topologies comparison for more electric aircrafts high speed Starter/Generator application," in *2015 IEEE Energy Conversion Congress and Exposition (ECCE)*, Montreal, QC, Canada: IEEE, Sep. 2015, pp. 3659–3666.

[3] J. K. Nøland, M. Leandro, J. A. Suul, M. Molinas, and R. Nilssen, "Electrical Machines and Power Electronics For Starter-Generators in More Electric Aircrafts: A Technology Review," in *IECON 2019 - 45th Annual Conference of the IEEE Industrial Electronics Society*, Oct. 2019, pp. 6994–7001.

[4] B. Sarlioglu and C. T. Morris, "More Electric Aircraft: Review, Challenges, and Opportunities for Commercial Transport Aircraft," *IEEE Transactions on Transportation Electrification*, vol. 1, no. 1, pp. 54–64, Jun. 2015.

[5] D. Han, J. Noppakunkajorn, and B. Sarlioglu, "Comprehensive Efficiency, Weight, and Volume Comparison of SiC- and Si-Based Bidirectional DC–DC Converters for Hybrid Electric Vehicles," *IEEE Transactions on Vehicular Technology*, vol. 63, no. 7, pp. 3001–3010, Sep. 2014.

[6] D. Zhang, J. He, and D. Pan, "A Megawatt-Scale Medium-Voltage High-Efficiency High Power Density 'SiC+Si' Hybrid Three-Level ANPC Inverter for Aircraft Hybrid-Electric Propulsion Systems," *IEEE Trans. on Ind. Applicat.*, vol. 55, no. 6, pp. 5971–5980, Nov. 2019.

[7] C. Wang, J. Pang, K. Wang, X. Zhang, Z. Zheng, and Y. Li, "Design and Evaluation of a SiC and Si IGBT Hybrid Three-Level ANPC Power Module," in *2023 IEEE PELS Students and Young Professionals Symposium (SYPS)*, Aug. 2023, pp. 1–7.

[8] B. Zhang, Z. Wang, and M. Gu, "An Improved Si/SiC Hybrid Three-Level ANPC Inverter with Optimized Thermal Distribution-Based Modulation Scheme," in *2024 27th International Conference on Electrical Machines and Systems (ICEMS)*, Nov. 2024, pp. 1012–1017.

[9] C. Attaianese, M. Di Monaco, and G. Tomasso, "Three-Phase Three-Level active NPC converters for high power systems," in *SPEEDAM 2010*, Jun. 2010, pp. 204–209.

[10] M. Liu, A. Coppola, M. Alvi, and M. Anwar, "Comprehensive Review and State of Development of Double-Sided Cooled Package Technology for Automotive Power Modules," *IEEE Open Journal of Power Electronics*, vol. 3, pp. 271–289, 2022.

[11] T. Wang, Y. Gan, H. Jin, L. Wang, Y. Wu, and Y. Wang, "A Novel Double-sided Cooling 3L-ANPC SiC MOSFET Power Module with Interleaved Layout," in *2024 IEEE Applied Power Electronics Conference and Exposition (APEC)*, Feb. 2024, pp. 192–196.

[12] H. Wang, X. Ma and H. Sun, "Active neutral-point-clamped (ANPC) three-level converter for high-power applications with optimized PWM strategy," PCIM Asia 2020; International Exhibition and Conference for Power Electronics, Intelligent Motion, Renewable Energy and Energy Management, Shanghai, China, 2020, pp. 1-8.

[13] D. Floricau, E. Floricau, and M. Dumitrescu, "Natural doubling of the apparent switching frequency using three-level ANPC converter," in *2008 International School on Nonsinusoidal Currents and Compensation*, Jun. 2008, pp. 1–6.

[14] J. J. Ferreira Evangelista Filho, J. P. Souza Pascon, J. F. Guerreiro, G. T. de Carvalho Ferreira, and J. A. Pomilio, "Power Electronics for All-Electric Aircraft: A Review," in *2023 IEEE 8th Southern Power Electronics Conference and 17th Brazilian Power Electronics Conference (SPEC/COBEP)*, Nov. 2023, pp. 1–6.

[15] F. Yang *et al.*, "Interleaved Planar Packaging Method of Multichip SiC Power Module for Thermal and Electrical Performance Improvement," *IEEE Trans. Power Electron.*, vol. 37, no. 2, pp. 1615–1629, Feb. 2022.

[16] L. Han, L. Liang, Z. Zhang, and Y. Kang, "Understanding Inherent Implication of Thermal Resistance in Double-Side Cooling Module," *IEEE Transactions on Power Electronics*, vol. 38, no. 2, pp. 2435–2445, Feb. 2023.

Research on Transient Electric Field Calculation in Welded Devices Considering Dielectric Relaxation of Insulating Materials

Zihan Sang
North China Electric Power University
Beijing, China
zihansang1@163.com

Zhaocheng Liu
North China Electric Power University
Beijing, China
fendou_1997@163.com

Hao Li
Beijing Institute of Smart Energy
Beijing, China
holly6865@163.com

Xuebao Li
North China Electric Power University
Beijing, China
lxb08357x@ncepu.edu.cn

Ying Cao
State Grid Zhejiang Electric Power Co.,
Ltd. Research Institute
Hangzhou, China
18868795925@163.com

Peng Shu
State Grid Zhejiang Electric Power Co.,
Ltd. Research Institute
Hangzhou, China
240979253@qq.com

Abstract—High-voltage, high-power welding-type IGBT devices are essential in high-voltage DC equipment. Accurate transient electric field distribution within these devices is crucial for effective insulation design. During operation, these devices face positive polarity repetitive square waves, leading to dielectric relaxation in insulating materials and resulting in electric field distortion. However, current methods do not consider this effect. This paper presents a transient electric field calculation model for welding-type IGBT devices that includes dielectric relaxation. Using the complex frequency domain finite element method and complex sparse matrix direct solving algorithm, the internal transient electric field distribution is analyzed, and the impact of dielectric relaxation is studied. By adjusting encapsulation insulating material parameters, the electric field intensity at weak insulation points is significantly reduced, demonstrating effective insulation design through well-matched insulating materials.

Keywords—welding IGBT, transient electric field, dielectric relaxation, complex frequency domain finite element method, direct algorithm

I. INTRODUCTION

With the continuous development of direct current (DC) transmission technology, high-voltage, high-power insulated gate bipolar transistors (IGBTs) have become core components in DC power grids and are widely utilized in various large-capacity power conversion and control equipment. To enhance the electrical stability of high-voltage, high-power IGBT devices, commonly used welded high-voltage, high-power IGBT devices ensure low contact resistance and high mechanical strength through their stable welding process. Their reliable electrical connections enable these devices to withstand the high voltage and high current generated during operation, making them one of the ideal packaging forms for high-voltage, high-power IGBT devices.

As the voltage rating of devices increases, the insulation design of device packaging has become a critical issue in the development of high-voltage IGBT devices. Inadequate structural design can lead to a decline in the insulation capability of the device, resulting in partial discharge or even insulation breakdown. Therefore, to ensure the long-term reliable operation of the devices, it is essential to obtain accurate electric field distribution within the device to achieve effective design. Existing research primarily focuses on the study of static electric field distribution characteristics in the packaging of welded devices. However, in the face of complex power system environments, sudden transient voltage waveforms can cause rapid changes in electric field distribution, potentially leading to electrical breakdown phenomena. Additionally, since the devices must withstand not only traditional alternating sinusoidal/cosine voltages but also non-sinusoidal periodic voltages, this results in transient voltage waveforms containing a rich spectrum of frequency components. The dielectric constant of the packaging material is affected by frequency variations, which is known as the dielectric relaxation effect. Current research on welded IGBT devices has not taken into account the impact of the dielectric relaxation process on the transient electric field distribution.

Moreover, traditional time-domain finite element methods for calculating transient electric fields that consider relaxation processes require convolution operations, resulting in a significant computational burden, as well as issues with iterative errors and convergence. Literature has proposed a frequency-domain finite element method based on Fourier transforms, which can address the shortcomings of time-domain methods. However, this approach cannot compute transient transition processes or zero-input responses caused by initial conditions. Research indicates that the complex frequency-domain finite element method based on discrete Laplace transforms can accurately calculate the transient electric field transition processes at a lower computational cost. However, existing solution methods employing the complex sparse matrix iterative algorithm have slow computational speeds and low efficiency.

Therefore, this paper first establishes a transient electric field calculation model for welded IGBT devices that incorporates the dielectric relaxation of materials. It employs the complex frequency-domain finite element method and a direct solution algorithm for the complex sparse matrix, achieving nearly a six-fold increase in computational speed compared to commonly used iterative algorithms. Next, it systematically analyzes the transient electric field distribution characteristics of welded IGBT devices under single shutdown conditions, identifying weak points within the devices. Then, it compares the electric field strength before and after considering the dielectric relaxation model of organic siligel, examining the impact of the dielectric relaxation process on transient electric field distribution. Finally, by adjusting the parameters of the insulating materials, the study aims to reduce the electric field strength at weak insulation points, providing new insights for the simulation and design of packaging insulation structures for welded IGBT devices.

II. MODELING AND CALCULATION OF INSULATION ELECTRIC FIELD FOR WELDED IGBT DEVICES

A. Device Electric Field Calculation Model Considering Material Dielectric Relaxation

The internal structure of the 3.3 kV welding-type IGBT device is illustrated in Fig. 1[8]. The device comprises a multi-layer structure that includes a semiconductor chip, a DBC (Direct Bond Copper) structure, a heat dissipation substrate, bond wires, power output terminals, a welding layer, and a packaging shell. Fig. 1(b) presents a cross-sectional schematic of the device, where the chip is connected to the power output terminals via bond wires. The terminals of the chip are coated with a passivation layer, which serves to protect the chip and enhance its voltage withstand capability[9]. Inside the packaging shell, siligel is injected for insulation and protection purposes.

(a) actual object image

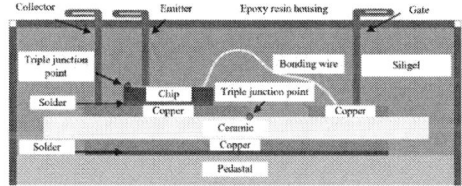

(b) cross-sectional diagram

Fig. 1. 3.3 kV Welding-Type IGBT Device[8]

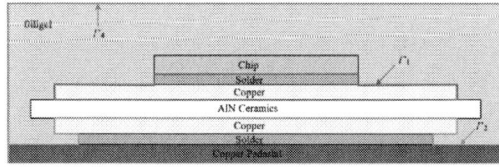

(a) IGBT device

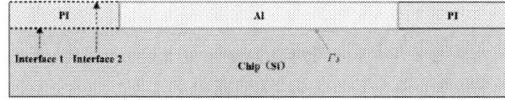

(b) IGBT chip

Fig. 2. Two-Dimensional Computational Model and Boundary Conditions of Welding-Type IGBT Device

Based on the detailed cross-sectional schematic of the device in Fig. 1(b), a simplified simulation model of the device has been constructed, as shown in Figure 2. The thickness of the aluminum layer of the chip is set to 0.005 mm, the thickness of the silicon layer of the chip is set to 0.57 mm,

the thickness of the upper and lower copper layers of the DBC is set to 0.381 mm, and the thickness of the DBC ceramic substrate is set to 0.462 mm.

The entire computational region contains four types of insulating materials. In the electric field calculation of the device packaging insulation structure, the chip is typically simplified to silicon[10]. This paper follows this simplification, with the passivation layer at the chip terminals being made of polyimide. The basic parameters of the materials involved in the model are shown in Table I.

TABLE I. MATERIAL PARAMETERS IN THE COMPUTATIONAL MODEL OF WELDING-TYPE DEVICES

Materials	ε_r	ρ/(S/m)
Siligel	2.78	1.04×10^{-12}
PI	2.96	4.00×10^{-14}
AlN	8.80	3.00×10^{-12}
Si	11.7	2.50×10^{-4}

According to literature [11], the control equation expressed in terms of scalar potential in the time domain is:

$$\nabla \cdot \left[\gamma \nabla \varphi + \frac{\partial}{\partial t}(\varepsilon \nabla \varphi) \right] = 0 \tag{1}$$

where γ and ε are the conductivity and permittivity of the material.

When dielectric relaxation is considered, the permittivity ε of the material becomes a time-dependent dielectric function $\varepsilon(t)$ and the optical-frequency permittivity ε_∞. Equation (1) is then transformed into:

$$\nabla \cdot (\gamma \nabla \varphi) + \frac{\partial}{\partial t}\left\{ \nabla \cdot \left[\varepsilon_\infty \nabla \varphi + \varepsilon(t) * (\nabla \varphi) \right] \right\} = 0 \tag{2}$$

Since Equation (2) involves convolution calculations, which are computationally intensive, and based on the assumption that the material is linear, homogeneous, and isotropic, we perform a Laplace transform on Equation (2). Thus, the control equation for transient electric field calculation considering dielectric relaxation in the complex frequency domain can be expressed as:

$$\nabla \cdot \left\{ \begin{matrix} \gamma \nabla \varphi_L(s) + \left[\varepsilon_L(s) s \nabla \varphi_L(s) + \varepsilon_\infty s \nabla \varphi_L(s) \right] \\ -\left[\varepsilon(0) \nabla \varphi(0) + \varepsilon_\infty \nabla \varphi(0) \right] \end{matrix} \right\} = 0 \tag{3}$$

Where s is the independent variable in the complex frequency domain, the subscript L denotes the complex frequency domain, and $\varepsilon(0)$ represents the initial permittivity.

Moreover, the boundary conditions and initial conditions for transient electric field calculations in the complex frequency domain can be expressed as:

$$\begin{cases} \varphi_L \big|_{\Gamma_1} = u_L(s) \\ \dfrac{\partial \varphi_L}{\partial n} \bigg|_{\Gamma_2} = \psi_L(s) \\ \varphi \big|_{t=0_-} = \varphi(0) \end{cases} \tag{4}$$

Where Γ_1 denotes the Dirichlet boundary with known potential values, Γ_2 represents the Neumann boundary with known normal derivatives of potential, and $\varphi(0)$ specifies the initial potential distribution.

During the turn-off process and in the off-state, the boundary conditions for the mathematical model of the welded IGBT device are illustrated in Fig. 2. Since the upper copper layer is subjected to a high voltage while the potentials of the lower copper layer and the chip aluminum-plated layer are set to 0, boundaries Γ_1, Γ_2, and Γ_3 are Dirichlet boundaries. Here, the high-voltage boundary is marked with a red line, and the low-voltage boundary is marked with a blue line. The outer frame of the computational model represents the device's enclosure. As only the internal electric field of the device is considered, the normal component of the electric field intensity on boundary Γ_4 is assumed to be 0, making it a Neumann boundary, which is indicated by a green line.

To accurately calculate the electric field intensity at interfaces, the transient boundary electric field constraint equation method from reference [5] is employed. This transforms the complex-frequency-domain finite element equations into Equations (5) and (6), which are used to solve for the transient potential and the normal component of electric field intensity on the boundary, respectively. Finally, conjugate symmetry operations and inverse Laplace transform are performed to obtain the time-domain nodal potential vector and the normal component of electric field intensity.

$$\boldsymbol{\Phi}_{Lk} = \boldsymbol{S}_{k11}^{-1}[\boldsymbol{T}_{11}\boldsymbol{\Phi}(0) + \boldsymbol{T}_{21}^{\mathrm{T}}\boldsymbol{U}(0) + \boldsymbol{F}_{k2} - \boldsymbol{S}_{k21}^{\mathrm{T}}\boldsymbol{U}_{Lk}] \quad (5)$$

$$\boldsymbol{E}_{\mathrm{n}Lk} = -\boldsymbol{H}_k^{-1}[\boldsymbol{S}_{k21}\boldsymbol{\Phi}_{Lk} + \boldsymbol{S}_{k22}\boldsymbol{U}_{Lk} - \\ \boldsymbol{T}_{21}\boldsymbol{\Phi}(0) - \boldsymbol{T}_{22}\boldsymbol{U}(0) - \boldsymbol{G}\boldsymbol{E}_{\mathrm{n}}(0)] \quad (6)$$

Where, $\boldsymbol{\Phi}_{Lk}$ is the column vector of node potentials to be solved in the complex frequency domain; $\boldsymbol{E}_{\mathrm{n}Lk}$ is the column vector of the normal electric field at the boundary nodes in the complex frequency domain; $\boldsymbol{U}_{Lk}$ is the column vector of known node potentials on the Dirichlet boundary; the expressions for the elements in the coefficient matrices $\boldsymbol{S}$, $\boldsymbol{T}$, $\boldsymbol{F}$, $\boldsymbol{K}$, $\boldsymbol{H}$, and $\boldsymbol{G}$ can be found in reference [5].

B. Complex Frequency Domain Finite Element Method Based on Direct Algorithm of Complex Sparse Matrix

In order to calculate the transient electric field distribution of welded devices taking into account the dielectric relaxation of insulating materials, a coupled dielectric relaxation model of the materials was employed. The complex frequency-domain finite element method and a direct solver algorithm for complex sparse matrices were used to achieve the electric field calculation of the welded IGBT devices. The calculation flowchart is shown in Fig. 3.

The core of the direct solver algorithm for complex sparse matrices is to solve the linear equation system of the complex sparse matrix using the sparse Gaussian elimination method. For the sake of clarity, equation (5) is simplified to:

$$AX = B \quad (7)$$

where A is the complex sparse matrix containing finite element information; B is the column vector containing the initial information of the boundary node potentials; and X is the column vector of node potentials in the complex frequency domain that needs to be solved.

To directly solve equation (7), it is necessary to scale and reorder matrix A to improve computational efficiency and numerical stability. Then, the processed matrix A undergoes

triangular decomposition. Finally, forward substitution and backward substitution are performed using the triangular matrices $\boldsymbol{L}$ and $\boldsymbol{U}$ obtained from the decomposition to complete the solution. The transformation and solving formulas are as follows:

$$X = A^{-1}B = (D_r^{-1}P_r^{-1}LUP_c^{-1}D_c^{-1})^{-1}B \quad (8)$$

where $\boldsymbol{D}_r$ and $\boldsymbol{D}c$ are diagonal matrices used to scale matrix A; $\boldsymbol{P}_r$ and $\boldsymbol{P}_c$ are permutation matrices used to reorder the rows and columns of matrix A.

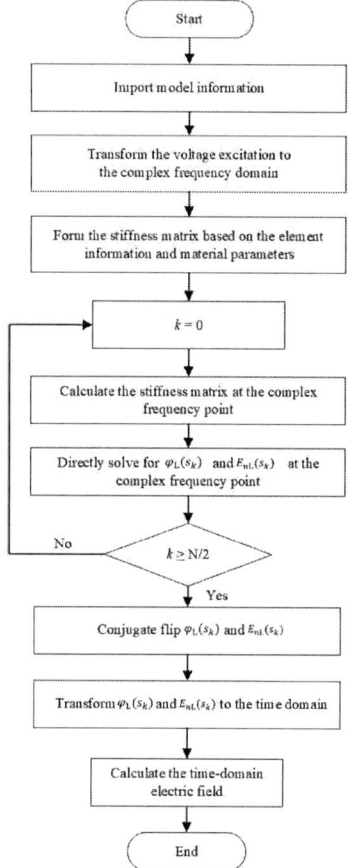

Fig. 3. Flowchart for direct solving in the complex frequency domain FEM

III. ELECTRIC FIELD ANALYSIS OF WELDING DEVICES CONSIDERING MATERIAL DIELECTRIC RELAXATION

A. Simulation Settings

To simulate the circuit breaker condition of the device[12], the excitation voltage is applied as a non-ideal step voltage with a certain rise time. The voltage waveform is shown in Fig. 4. In Fig. 4, $U_m = 3300$ V is the amplitude of the non-ideal step voltage, and $t_r = 1$s is the rise time of the non-ideal step voltage, which represents the duration of the turn-off process. A voltage of 3.3 kV is applied to the copper layer on the DBC, while a voltage of 0 V is applied to the copper layer beneath the DBC and the aluminum layer of the chip.

The welding device model is subdivided, with the maximum element size set to 0.001m. The mesh subdivision results are shown in Fig. 5.

979-8-3315-1110-4/25 $31.00 © 2025 IEEE

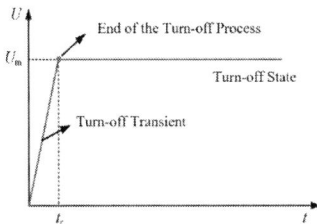

Fig. 4. Non-ideal Step Voltage in Circuit Breaker Operation

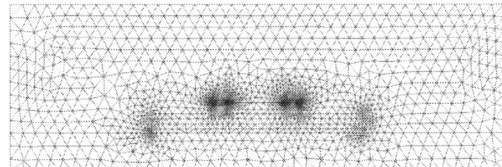

Fig. 5. Mesh Generation Diagram of Soldered Device

B. Transient Electric Field Calculation and Analysis of Welding Devices

Complete the drawing of the welding device model on the simulation platform, add medium and boundary conditions, then perform grid subdivision and set voltage excitation parameters. After outputting the simulation related files, import the files into the complex frequency domain finite element direct solver, apply the complex frequency domain finite element method and the complex sparse matrix direct solving algorithm to calculate the node potential and electric field, and finally obtain the potential and electric field distribution of the entire field through interpolation and post-processing plotting.

In order to demonstrate the transient electric field distribution of the device packaging insulation structure under circuit breaker conditions, Fig. 6 presents the electric field distribution contour maps of the device packaging insulation structure at different moments.

As shown in Fig. 6(a), when t=1s, the device has just turned off. Under the influence of the applied voltage, the electric field is locally enhanced at the junction of the aluminum layer, PI passivation layer, and chip silicon layer. This is due to the discontinuity in dielectric constant and conductivity at the junction. As shown in Fig. 6(b), when t=100s, charge begins to accumulate at the interface between the chip and the PI passivation layer (hereinafter referred to as Interface 1, marked with a black dashed line in Fig. 2(b)) and the interface between the PI passivation layer and the organic siligel (hereinafter referred to as Interface 2, also marked with a black dashed line in Fig.2(b)). This causes the electric field strength inside the PI passivation layer, as well as in the chip and organic siligel, to gradually increase. As shown in Fig. 6(c), when t=300s, charge further accumulates at the interfaces, leading to a continued increase in electric field strength within the PI passivation layer, chip, and organic siligel. In summary, under single turn-off conditions, the locations with higher electric field strength in the device appear at the PI passivation layer, Interface 1, and Interface 2, and attention should be paid to the variations in electric field strength at these locations.

To study the variation of the maximum electric field strength of the welded device over time under single turn-off conditions, and to compare the locations of the maximum

electric field strength at different moments, Fig. 7 shows the positions where the maximum electric field strength occurs and how it changes over time. In Fig.7(a), the positions of the maximum electric field strength are marked with red dots, indicating that the maximum electric field strength is found at the junction of the aluminum layer, PI passivation layer, and chip silicon layer. Fig. 7(b) presents the temporal variation of the maximum electric field strength of the device. It can be observed that the maximum electric field strength of the device gradually increases over time. When t < 1 s, due to the mismatch of dielectric parameters at the junction of the chip silicon layer, aluminum layer, and PI passivation layer, the electric field is locally enhanced at this point. When t > 1 s, as charge accumulation stabilizes gradually, the increasing trend of electric field strength slows down, and after a slow rising charge relaxation process, it tends to stabilize.

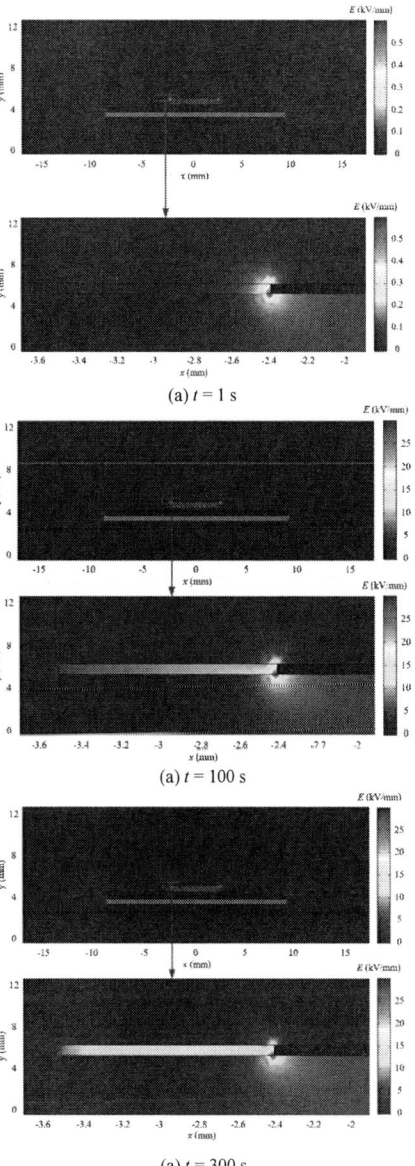

(a) t = 1 s

(a) t = 100 s

(a) t = 300 s

Fig. 6. Time-varying Electric Field Distribution Contour

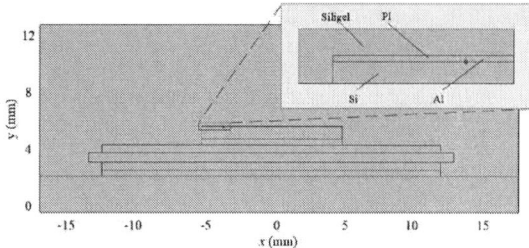

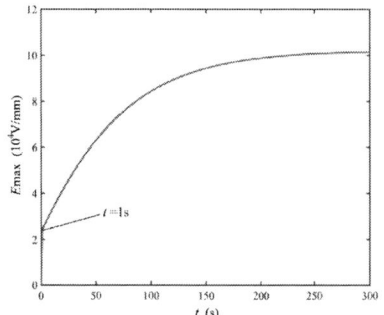

(a) The location of the maximum electric field strength in the device

(b) The temporal variation of the maximum electric field strength

Fig. 7. The distribution of the maximum electric field strength

To examine the influence of dielectric relaxation on the electric field distribution in the welded device, the Havriliak-Negami model[13] is incorporated into transient electric field calculations for the organic siligel. The model's dielectric parameters[14] are set as shown in Table II. Figure 8 compares

the time-varying electric field distribution in the welded device with and without the siligel's relaxation model.

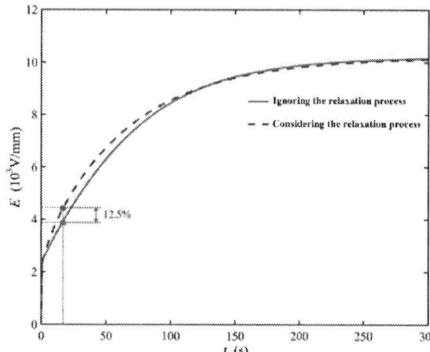

Fig. 8. Impact of Dielectric Relaxation on Electric Field Strength

Figure 8 reveals that when accounting for dielectric relaxation in the organic siligel, the electric field distribution initially displays higher values, subsequently lower values, before converging to patterns similar to non-relaxation cases. This phenomenon stems from the gel's frequency-dependent dielectric properties affecting medium polarization and consequently modifying transient field distribution. For steady-state fields, dielectric relaxation shows negligible influence. The maximum field strength difference reaches 12.5 percent at 16 seconds between relaxation and non-relaxation scenarios. Consequently, dielectric relaxation effects must be incorporated in transient field calculations for welded devices.

TABLE II. FREQUENCY-DOMAIN DIELECTRIC CHARACTERISTIC PARAMETERS OF SILIGEL

$T(°C)$	$\sigma_{dc}(S/m)$	$\Delta\varepsilon_1$	$\tau_1(s)$	α_1	β_1	α_2	$\tau_2(s)$	$\Delta\varepsilon_2$	β_2	ε_∞
20	1.72×10^{-12}	1.63	24.96	0.78	1	0.14	1.26×10^{-8}	0.11	1	2.72

C. Influence of Insulation Material Parameters on Transient Electric Field Characteristics

Due to the correlation between transient electric field strength and the differences in dielectric constant and conductivity of various insulation materials, it is observed that, in general, the variation in dielectric constants among different insulation materials is much smaller than the variation in conductivity. Therefore, the influence of conductivity on the control of transient electric fields can be studied from the perspective of matching material parameters. In this paper, siligel is used as an example, with its conductivity set to 10^{-11}, 10^{-12}, and 10^{-13} S/m. The parameters of other insulation materials and the calculation parameters remain consistent with those mentioned earlier. The transient electric field distribution of the device is calculated accordingly. When t = 300 s, the variation of the electric field strength at Interface 1 with respect to the conductivity of the filled siligel is shown in Figure 9.

As observed from Fig. 9, when the conductivity of the siligel increases from 10^{-12} S/m to 10^{-11} S/m, the electric field strength at Interface 1 significantly increases. Conversely, when the conductivity of the siligel decreases from 10^{-12} S/m to 10^{-13} S/m, the electric field strength gradually increases along the x-axis towards the right side of Interface 1.

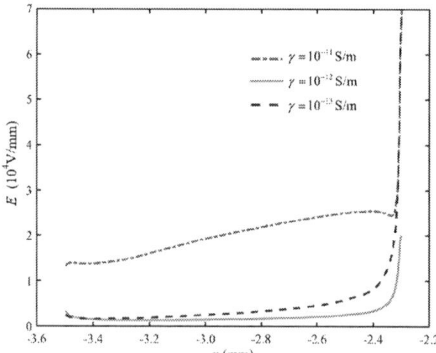

Fig. 9. The variation of electric field strength with the conductivity of insulating materials

According to the Maxwell polarization theory under the parallel plate model of two-layer dielectric materials, when the ratio of the dielectric constants and conductivities of the two media is consistent, there is no charge accumulation at the interface of the media. As shown in Table III, when the conductivity of the siligel is 10^{-12} S/m, the parameter matching between the PI passivation layer and the siligel is relatively good, resulting in less accumulated charge at Interface 1 and consequently a lower electric field strength. Therefore, in practical engineering applications, for welded devices, it is advisable to select insulation materials with better parameter matching to reduce the electric field strength at the surface of the PI passivation layer.

TABLE III. COMPARISON OF MAXIMUM ELECTRIC FIELD STRENGTH UNDER INSULATING MEDIA WITH DIFFERENT CONDUCTIVITIES

The conductivity of organic siligel / (S/m)	Maximum electric field strength
10^{-11}	2.99
10^{-12}	1
10^{-13}	2.85
10^{-11}	2.99

IV. CONCLUSION

This paper establishes a transient electric field calculation model for welded IGBT devices, taking into account the dielectric relaxation of materials, and systematically analyzes the transient electric field distribution patterns of welded IGBT devices under circuit breaker conditions. The main conclusions of this study are as follows:

1) A mathematical model for calculating the transient electric field in welded IGBT devices under single shutdown conditions has been developed. Utilizing a complex frequency domain finite element method and a direct solver based on complex sparse matrices, the distribution of the transient electric field within the device has been analyzed. The proposed calculation method improves the speed by nearly six times compared to traditional methods.

2) Under circuit breaker conditions, due to the mismatch in dielectric constant and conductivity parameters of different insulating materials, the maximum electric field strength of the 3.3 kV welded IGBT device occurs at the junction of the chip aluminum layer, the PI passivation layer, and the chip silicon layer. The maximum electric field strength of the device gradually increases over time and stabilizes after the charge relaxation process.

3) In the transient electric field calculations of welded IGBT devices, the dielectric relaxation model of siligel is considered. The frequency-dependent effect of the siligel's dielectric function can influence the variations in the transient electric field distribution. However, the dielectric relaxation process does not significantly affect the steady-state electric field distribution. When comparing the results before and after relaxation, a trend is observed where values initially overshoot, then undershoot, and eventually converge, with a maximum difference of 12.5% in electric field calculations. Therefore, it is essential to account for the impact of dielectric relaxation on the transient electric field distribution when computing the transient electric field of welded devices.

4) The effects of different conductivity insulating materials on the electric field strength at the material interface have been calculated. By selecting insulating materials with better parameter matching, the electric field strength at the interface can be reduced by nearly two times, thereby achieving improved insulation performance.

ACKNOWLEDGMENT

Acknowledgement for the funding support from Project 5500-202399657A-3-2-ZN.

REFERENCES

[1] T. Eckhardt, "Thermal management and reliability considerations in high-power IGBT modules with advanced packaging technology," IEEE Trans. Power Electron., vol. 28, no. 5, pp. 2530-2537, 2013.

[2] Y. Meng, K. Yang, Q. Zhang, X. Li, J. Hao, and Y. Kang, "Study on the characteristics of space charge in welded IGBT," in Proc. 7th Int. Conf. Smart Grid Smart Cities (ICSGSC), 2023.

[3] Z. Liu, X. Cui, X. Li, C. Ma, Z. Zhao, "A review of electric field calculation research on the insulation structure of high voltage high power IGBT devices," J. Electr. Eng., vol. 44, no. 1, pp. 214-231, 2024.

[4] T. Wen, X. Cui, X. Li, S. Liu, and Z. Zhao, "Time domain finite element method for transient electric field calculation considering the effect of dielectric relaxation," Trans. China Electrotech. Soc., vol. 37, no. 7, pp. 1735-1745, 2022.

[5] T. Wen, X. Cui, X. Li, S. Guan, S. Liu, Z. Zhao, "Complex frequency domain finite element method for transient electric field calculation in composite insulation structures," J. Electr. Eng., vol. 42, no. 7, pp. 2776-2788, 2022.

[6] S. Liu, Design and Implementation of Transient Electric Field Calculation Software Based on Complex Frequency Domain Finite Element Method Considering Relaxation Processes, M.S. thesis, North China Electr. Power Univ., Beijing, China, 2023.

[7] G. Feng, Research and Implementation of Large Sparse Matrix Direct Solving Algorithms, M.S. thesis, Northeastern Univ., Shenyang, China, 2010.

[8] J. Xu, Research and Application of the Properties of Organic Silicone Gel for High Voltage High Power IGBT Device Packaging, M.S. thesis, North China Electr. Power Univ., Beijing, China, 2021.

[9] J. Zhang and Q. Ye, "Preparation of two-component addition-type silicone rubber electronic encapsulant," Silicone Mater., vol. 23, no. 1, pp. 31-35, 2009.

[10] P. Fu, Z. Zhao, X. Cui, P. Zhang, R. Han, "Electrical field analysis of press-pack IGBTs," in Proc. 6th Asia-Pacific Conf. Antennas Propag. (APCAP), 2017.

[11] Z. Wang, Concise Numerical Calculation of Electromagnetic Fields. Beijing: China Machine Press, 2011.

[12] D. Yan, P. Liu, L. Feng, F. Zhang, S. Xiu, S. Jia, Y. Mo, "Simulation research on breaking process of hybrid DC circuit breaker based on two-level voltage source synthetic test method," High Voltage Apparatus, vol. 60, no. 5, pp. 76-83+91, 2024.

[13] L. Zhou, X. Li, D. Wang, J. Jiang, "Condition assessment of oil-paper insulation based on Havriliak-Negami dielectric relaxation model," High Voltage Engineering, vol. 42, no. 1, pp. 153-162, 2016.

[14] Y. Mao, X. Li, J. Xu, Z. Zhao, X. Cui, "Wide-band dielectric characteristics and temperature effects of silicone gel for IGBT device packaging," Electric Power, vol. 53, no. 12, pp. 45-54, 2020.

A Two-Stage Turn-On Gate Driver for SiC MOSFET with Short-Circuit Current Suppression

1st Yong Chen
Zhuhai Power Supply Bureau, Guangdong Power Grid Co., Ltd.,
Zhuhai, China
35665035@qq.com

2nd Xiaolu Zhang
School of Electrical Engineering
Xi'an Jiaotong University
Xi'an, China
lu995579@stu.xjtu.edu.cn

3rd Xu Cheng
Zhuhai Power Supply Bureau Guangdong Power Grid Co., Ltd.
Zhuhai, China
664157102@qq.com

4th Yuze Zheng
School of Electrical Engineering
Xi'an Jiaotong University
Xi'an, China
zyz2001@stu.xjtu.edu.cn

5th Xuhui Song
School of Electrical Engineering
Xi'an Jiaotong University
Xi'an, China
xuhui_song@163.com

6th Yukun Niu
School of Electrical Engineering
Xi'an Jiaotong University
Xi'an, China
827105370@qq.com

7th Kaixiang Gong
School of Electrical Engineering
Xi'an Jiaotong University
Xi'an, China
2193512024@stu.xjtu.edu.cn

8th Fan Zhang
School of Electrical Engineering
Xi'an Jiaotong University
Xi'an, China
zhangfan1990@xjtu.edu.cn

Abstract—The short-circuit tolerance of silicon carbide MOSFETs (SiC MOSFETs) is relatively limited, leading to increased demands on the response speed and protective capability of short-circuit detection and protection circuits. This paper proposes a two-stage gate drive strategy for suppressing short-circuit currents through the reduction of gate voltage, which effectively limits the development of short-circuit currents before protective measures are initiated. Simulation validation was performed using the 1700V / 310A SiC power module CAS310M17BM3. The results indicate that the proposed drive scheme can reduce short-circuit losses by 73.5\%, shorten the protection response time by 60\%, and also decrease turn-on losses under normal operating conditions.

Keywords—*SiC power module, short-circuit detection, gate driver, parasitic inductance*

I. INTRODUCTION

Short-circuit faults are one of the most common and serious problems in power electronics systems. If not detected and addressed promptly, they can lead to device damage and even safety incidents. The third-generation wide-bandgap semiconductor material silicon carbide (SiC) holds great promise in the field of power electronics because of its low conduction losses and high current density. However, when a short-circuit fault occurs, the current in SiC MOSFETs can rapidly rise to extremely high levels, with the short-circuit tolerance typically lasting only a few microseconds. Consequently, short-circuit protection measures must complete signal acquisition, fault determination, and protective actions within 2-3 µs. This presents stringent demands on the response speed and reliability of short-circuit protection systems.

Currently, the design strategies for short-circuit protection circuits in SiC MOSFETs typically follow three main processes: fault detection, signal processing, and device shutdown. Among these, short-circuit fault detection is the core component of the entire protection mechanism, as its speed and reliability directly influence the effectiveness of the short-circuit protection action. Existing fault detection methods primarily rely on monitoringtage or current values

across the device to determine whether a short-circuit fault has occurred.

Common short-circuit detection methods are as follows. Shunt resistor detection involves placing a low-value resistor in series within the power circuit to indirectly measure current by monitoring the voltage drop across the resistor, which is then compared against a threshold to trigger protection. This method is highly flexible and cost-effective; however, it incurs significant power loss, potentially affecting system efficiency [1]. Desaturation detection is widely applied, operating on the principle of monitoring the drain-source voltage of the device when it is conducting. If V_{DS} exceeds a preset threshold, a short-circuit condition is indicated. This detection circuit has a simple structure and quick response time, but due to the wide linear region of SiC MOSFETs, it is essential to appropriately set the blanking time to avoid false triggering [2]. Parasitic inductance voltage detection is executed by measuring the voltage across the parasitic inductance located between the Kelvin source and the power source of the SiC MOSFET. During a short circuit, the high di/dt in the circuit generates a considerable induced voltage across the parasitic inductance. When this induced voltage surpasses a predetermined threshold, a short-circuit condition is considered to have occurred. However, this detection method has poor noise immunity and may result in false triggering during normal device operations [3].

Generally, short-circuit protection strategies only implement appropriate protective and current-limiting measures after detecting a short-circuit fault signal. This paper proposes a two-stage gate drive strategy based on short-circuit current suppression, which can mitigate the development of short-circuit currents even before the fault signal is generated, without affecting normal device operation. Simulation results demonstrate that this strategy effectively suppresses short-circuit currents under fault conditions, significantly reducing short-circuit losses, with a response time maintained within 400ns. Additionally, this strategy also reduces turn-on losses under normal operating conditions.

This work was supported by the China Southern Power Grid Company, Ltd., Science and Technology Project under grant 030400KC23090016 (GDKJXM20231032).

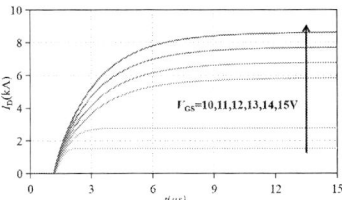

Fig. 1. Short-circuit drain current waveforms under different V_{GS}

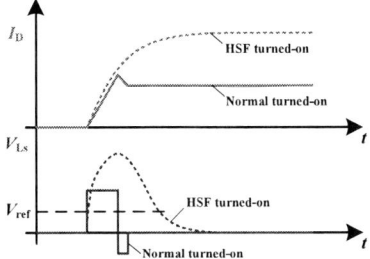

Fig. 2. Parasitic inductance voltage waveforms under normal turn-on and hard-switching fault(HSF)

II. PROPOSED TWO-STAGE GATE DRIVE STRATEGY

A. Operation Principles

When a short circuit occurs, the short-circuit current flowing through the SiC MOSFET gradually increases until it reaches saturation. The saturation value of the short-circuit current is influenced not only by the fault circuit voltage and loop parameters but also by the gate voltage. When a gate-source voltage below the recommended level is applied to the SiC MOSFET, the MOS channel does not fully turn on, resulting in increased channel resistance and limiting the device's current-carrying capability. During a short circuit, the differences in current-carrying capacity under various gate-source voltages become even more pronounced.

As shown in Fig. 1. the magnitude of the short-circuit saturation current decreases with decreasing V_{GS}, with the short-circuit saturation current at V_{GS}=10V being only about 22% of that at V_{GS}=15V.

Based on this phenomenon, this paper proposes a two-stage gate drive strategy. Upon receiving the turn-on signal, a gate-source voltage (V_{GS}) below the recommended value is first applied to the SiC MOSFET. If the device successfully turns on without any short-circuit conditions, the drive voltage is quickly switched to the recommended V_{GS} value, thereby avoiding additional conduction losses. In the event of a short circuit, the drive signal is immediately turned off, and shutdown measures are initiated. The application of a lower drive voltage across the gate-source terminals prior to protective actions effectively suppresses the development of short-circuit currents.

In the proposed scheme, the determination and switching between the two drive stages are achieved by monitoring the voltage across the parasitic inductance between the Kelvin source and the power source. As illustrated in Fig. 2., during normal operation, the induced voltage across the inductance rises above a threshold and then quickly drops. In the event of a short-circuit fault, the short-circuit current is significant, causing the di/dt to remain at a high level; thus, it takes

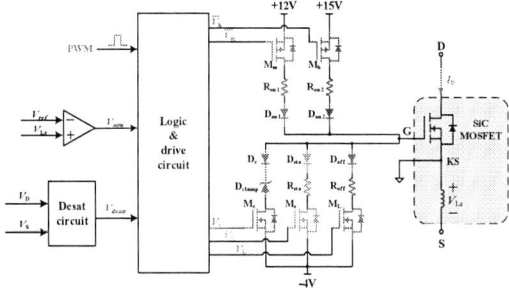

Fig. 3. Proposed two-stage gate driver circuit diagram

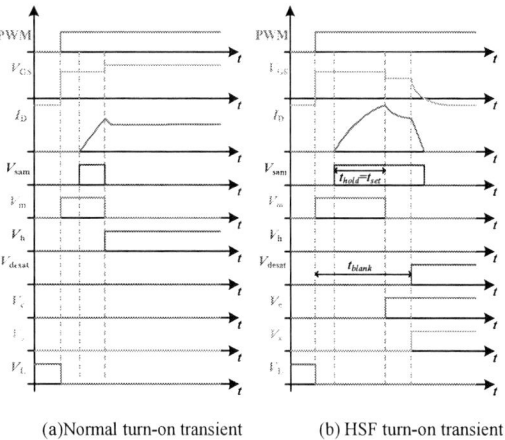

(a)Normal turn-on transient (b) HSF turn-on transient

Fig. 4. Operation waveforms of proposed two-stage gate driver

longer for the induced voltage across the inductance to fall below the threshold. Consequently, if the induced voltage is detected below the threshold within a specified time frame, it can be concluded that the device is operating normally, and the system immediately switches to the normal drive voltage. If the induced voltage remains above the threshold for longer than the set limit, a short-circuit fault is deemed to have occurred, prompting the swift implementation of protective measures.

B. Design of the Two Stage Gate Driver

The proposed two-stage gate drive scheme for SiC MOSFETs is illustrated in Fig. 3. This driving circuit consists of a common-source inductor voltage detection section, a desaturation detection circuit, a logic control circuit, and a drive and protection branch. The positive drive voltages are set at the recommended values of +15V and a slightly lower value of +12V. The negative drive voltage is set at the recommended value of -4V. The protection circuit includes a clamping branch and a high-resistance soft shutdown branch.

The operating timing diagram of the proposed two-stage gate drive circuit is shown in Fig.4. Upon receiving the turn-on PWM signal, the circuit first activates the low-drive voltage branch at +12V and monitors the circuit status. During normal turn-on conditions, the sampled voltage signal V_{sam} initially rises to a high level and then quickly returns to a low level. At this point, the circuit is considered to be in a normal state. By capturing the falling edge of the V_{sam} signal, a turn-on signal for the second stage of the drive is generated, allowing the gate voltage to be swiftly switched

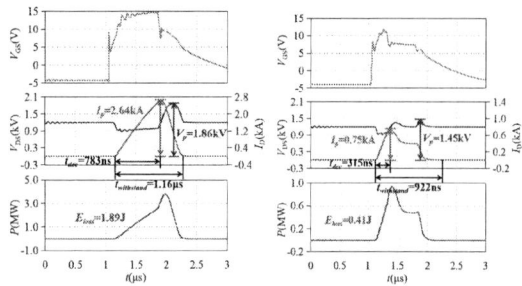
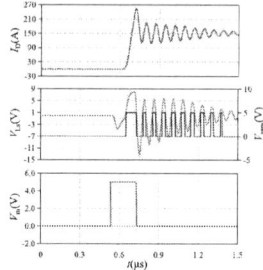

(a) Traditional gate driver   (b) Proposed two-stage gate driver

Fig. 5. Simulation results of the circuits during HSF

Fig. 6. Oscillation phenomenon during turn-on transient

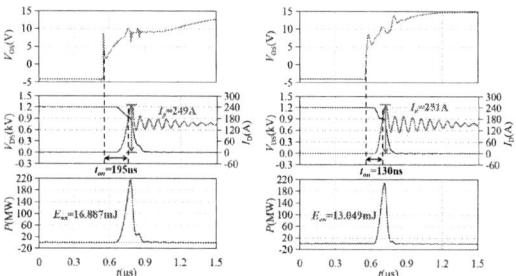

(a) Traditional gate driver   (b) Proposed two-stage gate driver

Fig. 7. Simulation Results during Normal Turn-On Transient

to +15V, fully turning on the device. Additionally, a smaller gate resistance should be selected for the +12V drive branch to enhance the gate current and accelerate the turn-on speed.

When a HSF occurs, if the comparison signal V_{sam} does not return to a low-level state within a predetermined time interval, the circuit is deemed to be in a fault condition. At this point, the logic control unit promptly disables the gate drive signal and activates a clamping branch, which limits the gate voltage to a lower level, effectively constraining the further increase of the short-circuit current. Simultaneously, after the blanking time has elapsed, the desaturation detection circuit issues a fault signal, triggering the soft shutdown path to bring the short-circuit current down to zero. This shutdown strategy helps to reduce the di/dt during the current decline, thereby lowering the voltage stress across the device during the turn-off period[4]. The gate voltage clamping is achieved using a secondary switching device alongside a Zener diode, with the selected clamping voltage being slightly lower than the drive voltage used during the first stage.

III. SIMULATION RESULTS

To validate the feasibility of the proposed two-stage gate drive scheme and its effectiveness in suppressing short-circuit currents, a simulation circuit was constructed using LTspice software. The 1700V/ 310A SiC power module CAS310M17BM3 was selected for the simulation, with the test conditions set to V_{DC}=1200V and I_L=150A.

In the simulation software, both a traditional drive circuit and the proposed two-stage gate drive circuit for the SiC power module were developed. The traditional drive circuit and the proposed two-stage drive circuit utilize the same desaturation detection circuit and soft shutdown circuit, with identical circuit parameters set for both configurations. The blanking times for both circuits are set to approximately 750 ns.

A. Simulation Results for Hard Switching Fault (HSF)

Simulation validation of the effectiveness of short-circuit protection was performed under short-circuit conditions, with the results shown in Fig. 5. The traditional drive circuit employs a desaturation detection and shutdown method, achieving a short-circuit response time of 783 ns, while the drain current peaked at 2.64 kA. After detecting a HSF, a soft shutdown strategy is employed. However, due to the high peak current, the di/dt during shutdown is also significantly high, resulting in a V_{DS} peak of up to 1.86 kV, which exceeds the rated value of 1700V for the SiC power

module. The total short-circuit loss during this process was measured at 1.89J.

The proposed two-stage gate drive circuit integrates both gate clamping and soft shutdown protection methods. When the output of the voltage comparator remains high for a duration exceeding a preset time, a HSF is recognized and protective measures are promptly implemented. The short-circuit response time is measured at 315ns, which accounts for the preset time and the delays introduced by the hardware circuitry. Following fault detection, the gate clamping branch is first activated, using a clamping voltage of 7.5V in the simulation.

While the gate voltage is clamped, the drain current begins to decrease from its peak of 0.75kA. After a brief interval, a desaturation detection signal is received, prompting the activation of the soft turn-off branch, gradually reducing the drain current to zero. The reduction in peak current and the implementation of the two-stage shutdown approach result in a significant decrease in the di/dt value within the circuit. The peak V_{DS} is measured at 1.45kV, representing a reduction 22% compared to traditional drive circuits. Throughout this process, the short-circuit loss is recorded at 0.41J, which is just 21.7% of the short-circuit losses experienced with conventional drive circuits.

B. Simulation Results during Normal Turn-On Transient

Due to the lower drive voltage implemented during the first turn-on phase of the proposed circuit, smaller drive resistor values should be employed to avoid additional turn-on losses. This adjustment ensures that the drive current is slightly higher than that of traditional drive circuits, facilitating faster switching speeds. However, the influence of parasitic parameters in the drive circuit can lead to oscillations during the turn-on process. The oscillation of the drain current I_D causes fluctuations in the voltage across the parasitic inductance, resulting in frequent voltage jumps at

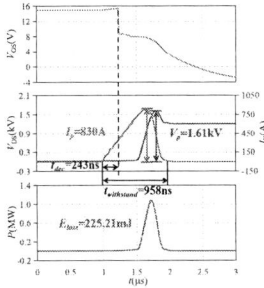

Fig. 8. Simulation results of the circuits during FUL

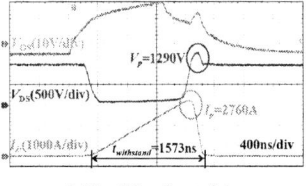

(a) Traditional gate driver

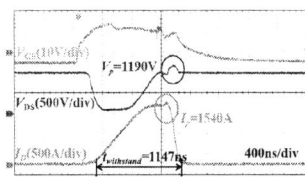

(b) Proposed two-stage gate driver

Fig. 9. Experimental results during HSF

the comparator's output, as it shown in Fig. 6. Nevertheless, as demonstrated by the simulation results under short-circuit conditions in Fig. 5, both voltage and current values during the short-circuit process do not exhibit oscillations. Therefore, even if frequent voltage jumps are detected at the comparator output, this can be interpreted as a normal turn-on process without indicating a short-circuit fault. By using a latch to capture the first falling edge of the comparator output voltage, the switching signal for the two turn-on phases can be obtained.

The simulation results comparing two circuit configurations during normal turn-on transients are shown in Fig. 7. The results indicate that, without increasing the peak current during the turn-on period, the proposed two-stage gate drive circuit exhibits a faster gate voltage rise rate. This leads to a reduction in turn-on time from 195ns to 130ns, resulting in a 22.7% decrease in turn-on losses.

C. Simulation Results for Fault Under Load(FUL)

When FUL occurs, the response time of the desaturation detection circuit is faster than that under HSF, approximately around 250ns. Since both circuits employ desaturation detection, there is no need for additional detection and protection logic. Upon detecting a desaturation signal, a method of first clamping followed by a soft turn-off is utilized for circuit protection. The simulation results during FUL are shown in Fig. 8. The fault response time is 243ns, with a maximum drain current of 830A and a voltage spike of 1.61kV. In this process, the short-circuit loss amounts to 225.21mJ.

IV. EXPERIMENTAL RESULTS

To further verify the feasibility of the proposed driving circuit and its suppression effect on short-circuit current, a prototype of the driving circuit was developed based on the proposed scheme. Short-circuit experiments were conducted using the Cree's 1700V/300A SiC power module CAS300M17BM2. The recommended driver voltage for this module is +20/-5V, and the experimental condition was set to V_{DC}=1000V.

The test results for the traditional driving circuit and the proposed driving circuit under HSF conditions are shown in Fig. 9. With the traditional driving circuit, the short-circuit current continued to increase before the protection action was taken, reaching a peak of 2760A, with a short-circuit duration of 1573ns. During the soft turn-off phase after desaturation, the voltage spike peaked at 1290V. In contrast, with the proposed two-stage turn-on driving circuit, the short-circuit current was suppressed before desaturation, as the gate-source voltage V_{GS} was lower than in the traditional circuit. The peak current was reduced to 1540A, a decrease of 44.2% compared to the traditional driving circuit. The voltage spike at turn-off was 1190V, and the short-circuit duration was shortened to 1147ns, with a reduction of 27%.

V. CONCLUSION

This paper presents a two-stage turn-on driving strategy for short-circuit current suppression. By reducing the driving voltage in the first turn-on stage, the development of short-circuit current can be suppressed before the short-circuit signal is detected. In the proposed scheme, parasitic inductance voltage is monitored to obtain the switching signals for the two-stage turn-on and the short-circuit fault signal. Simulation results indicate that the proposed two-stage driving strategy effectively reduces short-circuit losses while accelerating the normal turn-on speed, and it also lowers the voltage stress experienced by the device during faults. After developing a prototype of the driving circuit, short-circuit tests at 1000V were performed using the SiC power module CAS300M17BM2. Experimental results demonstrate that, compared to the traditional driving circuit, the proposed circuit reduces the maximum short-circuit current by 44.2% and short-circuit duration by 27%, validating the superiority of the two-stage turn-on driving circuit in suppressing short-circuit current.

REFERENCES

[1] Z. Guo and H. Li, "Dv/dt Sensing-Based Short-Circuit Protection for Medium-Voltage SiC mosfets," in IEEE Transactions on Power Electronics, vol. 38, no. 9, pp. 10554-10558, Sept. 2023.

[2] Q. Li, Y. Yang, Y. Wen, X. Tian, Y. Li and W. Xiang, "A Fast Overcurrent Protection IC for SiC MOSFET Based on Current Detection," in IEEE Transactions on Power Electronics, vol. 39, no. 5, pp. 4986-4990, May 2024.

[3] S. Lee, K. Kim, M. Shim and I. Nam, "A Digital Signal Processing Based Detection Circuit for Short-Circuit Protection of SiC MOSFET," in IEEE Transactions on Power Electronics, vol. 36, no. 12, pp. 13379-13382, Dec. 2021.

[4] T. Bertelshofer, A. März and M. . -M. Bakran, "A temperature compensated overcurrent and short-circuit detection method for SiC MOSFET modules," 2017 19th European Conference on Power Electronics and Applications (EPE'17 ECCE Europe), Warsaw, Poland, 2017, pp. P.1-P.10.

2025 IEEE Workshop on Wide Bandgap Power Devices and Applications in Asia (WiPDA Asia)

A zero-sequence injection method to reduce electromagnetic interference

1st Hui Liu
School of Electrical and Electronic Engineering
Huazhong University of Science and Technology
Wuhan, China
liuhuihui@hust.edu.cn

2nd Dong Jiang
School of Electrical and Electronic Engineering
Huazhong University of Science and Technology
Wuhan, China
jiangd@hust.edu.cn

3rd Junzhao Zhang
School of Electrical and Electronic Engineering
Huazhong University of Science and Technology
Wuhan, China
zjunzhao@hust.edu.cn

Abstract—Frequent switching will cause serious electromagnetic interference (EMI), which not only affects the performance of the inverter itself but also emits interference in the form of conduction and radiation, impacting nearby systems. This paper proposes a zero-sequence injection pulse width modulation (ZIPMW) to suppress conducted EMI. Experimental results show that compared to conventional space vector PWM, ZIPWM can reduce EMI by 5-10dB in the 300kHz-3MHz range.

Keywords—EMI, zero sequence injection, pulse width modulation

I. INTRODUCTION

Pulse width modulation (PWM) technology is widely used in railway transportation, new energy vehicles, aerospace, and ship propulsion systems. However, under the influence of PWM pulse signals, the high-frequency switching effect of power semiconductor devices resembles a step function, causing rapid changes in the voltage and current in power electronic systems. Among these, the high-frequency components of the response can cause severe electromagnetic interference (EMI) to other sensitive devices through conduction and radiation. In response to the electromagnetic compatibility (EMC) issues of power electronic devices, many governments and related organizations have established a series of strict EMC standards.

To meet EMC standards, many scholars and experts have conducted extensive research on EMI suppression. Passive EMI suppression methods can primarily be classified into grounding, shielding, and filtering. By changing the grounding method of power devices, the coupling capacitance between the devices and the ground is reduced, thereby decreasing common-mode interference [1]. A symmetrical LCL filter was designed by studying the combined magnetic integration of harmonics and EMI filters, effectively suppressing the system's EMI [2]. However, the introduction of passive components increases the system's weight and volume, which is not conducive to improving the system's power density. A new active common-mode EMI filter was proposed, based on current sampling and compensation circuits, utilizing transistor amplifiers for current compensation to achieve cancellation effects with noise signals [3]. However, this introduces additional control circuitry, increasing system complexity. Active EMI suppression technologies based on advanced PWM can suppress interference at the source. In the spectrum of constant switching frequency, there are spikes at multiples of the switching frequency. Spread spectrum modulation weakens these spikes by changing the switching frequency, thereby reducing system EMI. A randomly varying switching frequency is proposed in [4] to reduce EMI. Because the ideal random signal is difficult to obtain. [5] derived chaotic PWM

by combining chaotic sequences with variable switching frequencies and verified it through experiments. Using Fourier series, [6] derived the frequency spectrum of the output voltage of a PWM-driven inverter with periodic switching frequency variation, which more clearly reveals the impact of periodic switching frequency PWM on EMI and provides the theoretical foundation. [7] proposed a uniformly distributed spread-spectrum modulation strategy, which reduces EMI while also lowering switching losses. However, these methods generally increase the total harmonic distortion (THD) of the current, deteriorating power quality. This paper suppresses higher-frequency EMI by injecting a small zero-sequence component into the modulation wave, displacing the pulses and altering the duty cycle, without increasing the THD of the current. This modulation strategy is suitable for scenarios where high-frequency EMI needs to be suppressed while maintaining requirements on THD of the current.

Chapter II of this paper introduces the principles of zero-sequence injection PWM (ZIPWM). Chapter III presents the simulation results in Matlab/Simulink. Chapter IV provides experimental validation on the test platform. Chapter V is the conclusion of the paper.

II. THE PRINCIPLE OF ZIPWM

As shown in Fig.1, the topology used in this paper is a three-phase two-level inverter.

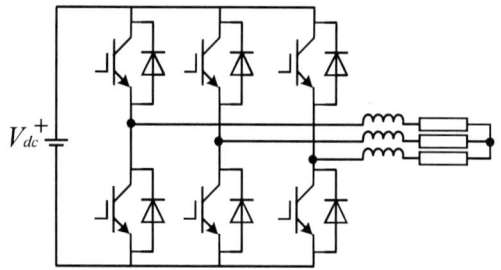

Fig.1 Three-phase two-level inverter.

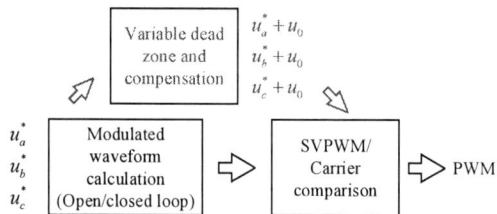

Fig.2 The method of zero-sequence component injection.

Traditional space vector PWM (SVPWM) generates three-phase modulation waves using open-loop or closed-loop control methods, and modulation is achieved through space

979-8-3315-1110-4/25 $31.00 © 2025 IEEE

vector decomposition or equivalent carrier comparison methods, which drive the switching devices to generate the required voltage. As shown in Fig.2, the method proposed in this paper injects a small zero-sequence component u_0 into the three-phase modulation wave.

Fig.3 shows the impact of zero-sequence injection on PWM, where T_s is the switching period. The green line represents the triangular carrier, the blue line represents the modulation wave of SVPWM, and the purple line represents the PWM of SVPWM. The red line represents the modulation wave of ZIPWM, which can be seen as the blue modulation wave with a small zero-sequence component superimposed. The black line represents the PWM of ZIPWM. It can be observed that the black PWM differs slightly from the purple PWM, which is precisely due to the injection of the zero-sequence component.

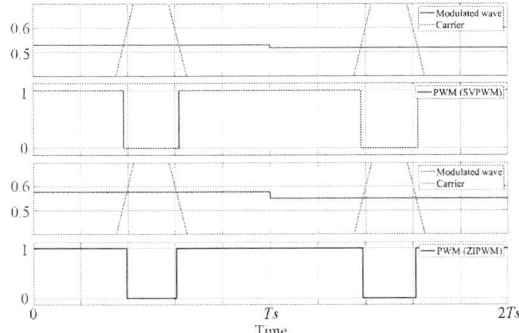

Fig.3 The impact of zero-sequence injection on PWM.

The modulation wave values range from -1 to 1. As shown in Fig4, taking the example of injecting a zero-sequence component with an amplitude of 0.02, the actual injection is a random disturbance between -0.02 and 0.02. This slightly alters the modulation wave values, thereby changing the duty cycle, with the change in duty cycle being ±0.02.

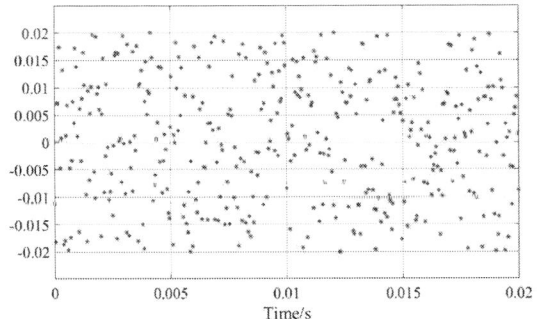

Fig.4 The value of the injected zero-order component.

A significant change in the duty cycle would theoretically degrade power quality. Therefore, the amplitude of the zero-sequence component injected in this paper is between 0.01 and 0.05, accounting for only 1-5% of the modulation wave. Additionally, the expected value of the mathematical distribution is 0, so its impact on the current THD and low-frequency components is minimal. In a closed-loop control system, where voltage and current sampling involve delays and quantization errors, the generated modulation wave is already not perfectly accurate. Thus, the impact of ZIPWM on power quality can essentially be ignored, and subsequent simulation and experimental results can further verify this.

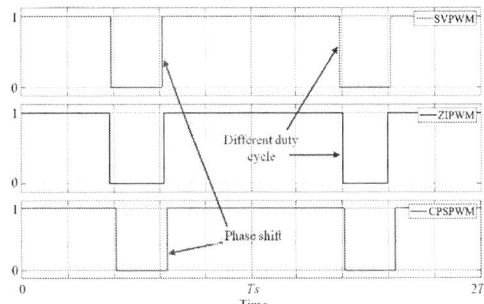

Fig.5 The difference between ZIPWM and CPSPWM.

As shown in Fig.5, the blue line represents the PWM of SVPWM, the black line represents the PWM of ZIPWM, and the red line represents the PWM of carrier phase-shift PWM (CPSPWM). It can be seen that the red line is merely a phase-shift of the blue line. In contrast, the black line shows a change in the duty cycle compared to the blue line, indicating that ZIPWM and CPSPWM are different methods.

Since the amplitude of the injected zero-sequence component accounts for 1-5% of the modulation wave's amplitude, and a PWM signal is generated for each switching period, it will affect the EMI at 20-100 times the switching frequency.

III. SIMULATION VERIFICATION

A simulation model was built in Matlab/Simulink, with parameters shown in Table I.

TABLE I SIMULATION AND EXPERIMENT PARAMETERS

Symbol	Parameter	Value
V_{dc}^*	DC voltage	200V
f_s	Switching frequency	20kHz
f	Modulated wave frequency	50Hz
m	Modulation index	0.9
L	Filter Inductor	1mH
R	Lord	30Ω
t_d	Dead zone	2us

Fig.6 shows the frequency spectrum of the phase current. It can be seen that compared to SVPWM, ZIPWM can reduce EMI by 5-10dB in the range of 500kHz to 30MHz.

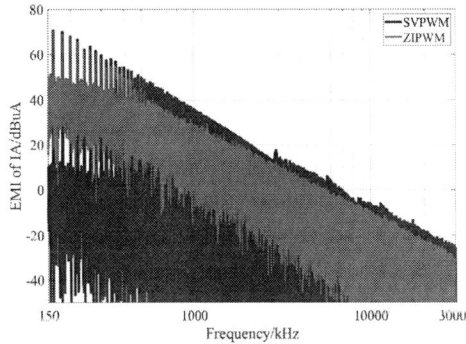

Fig.6 Frequency spectrum of the phase current (simulation).

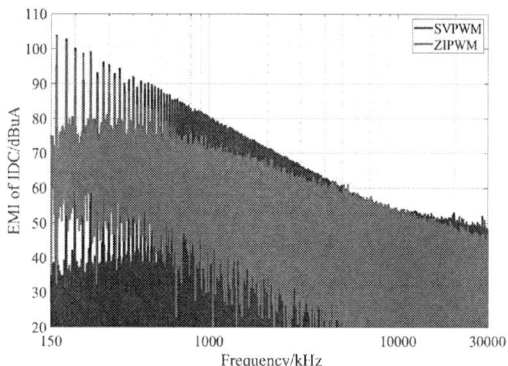

Fig.7 Frequency spectrum of the DC current (simulation).

Fig.7 shows the frequency spectrum of the DC current. It can be observed that compared to SVPWM, ZIPWM can reduce EMI by 5-10dB in the range of 400kHz to 3MHz.

Fig.8 shows the waveform and THD of the phase current. The THD of SVPWM is 11.28%, while the THD of ZIPWM is 11.33%. It can be seen that ZIPWM has little to no effect on the THD.

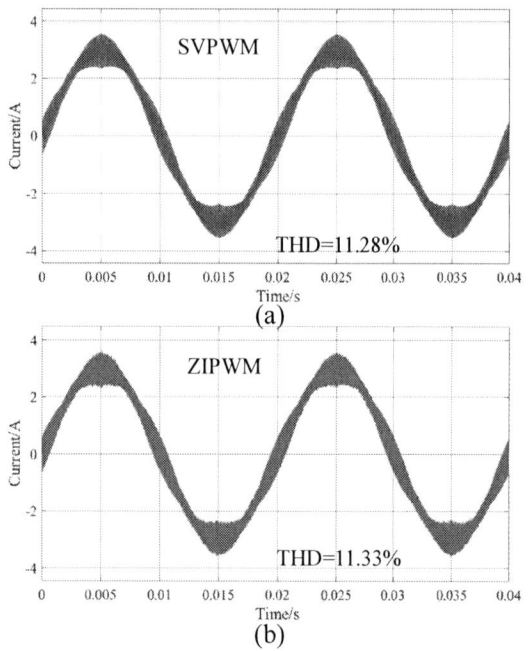

Fig.8 Waveform of the phase current (simulation)

(a)SVPWM (b)ZIPWM.

IV. EXPERIMENTAL VERIFICATION

The experimental platform is shown in Fig.9, with the experimental parameters being the same as those in the simulation, as listed in Table I. The current waveforms are recorded using an oscilloscope, and EMI is tested using an EMI receiver.

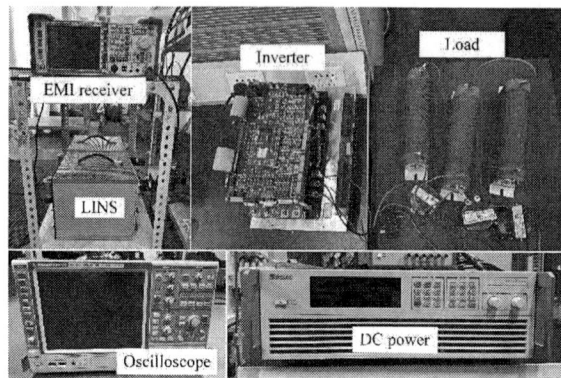

Fig.9 Experimental platform.

The frequency spectrum of the phase current is shown in Fig.10. It can be seen that compared to SVPWM, ZIPWM can reduce EMI by 5-10dB in the range of 300kHz to 3MHz. Fig.11 shows the frequency spectrum of the DC current. It can be observed that compared to SVPWM, ZIPWM can reduce EMI by 5-10dB in the range of 300kHz to 3MHz.

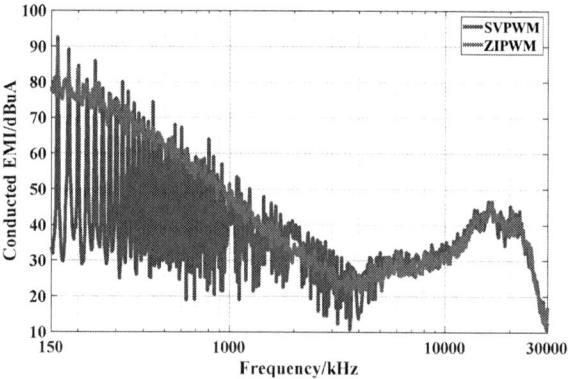

Fig.10 Frequency spectrum of the phase current.

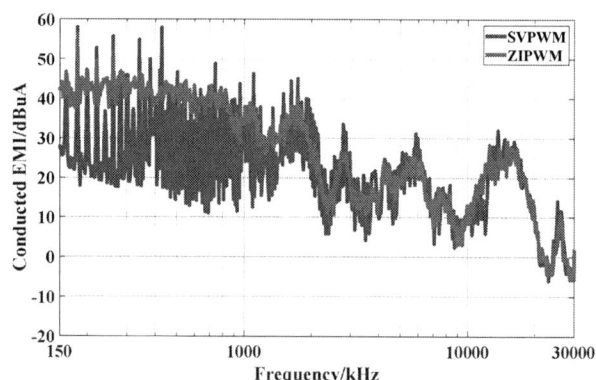

Fig.11 Frequency spectrum of the DC current.

Fig.12 shows the waveform and THD of the phase current. The THD of SVPWM is 13.46%, while the THD of ZIPWM is 13.49%. It can be seen that ZIPWM has little to no effect on the THD.

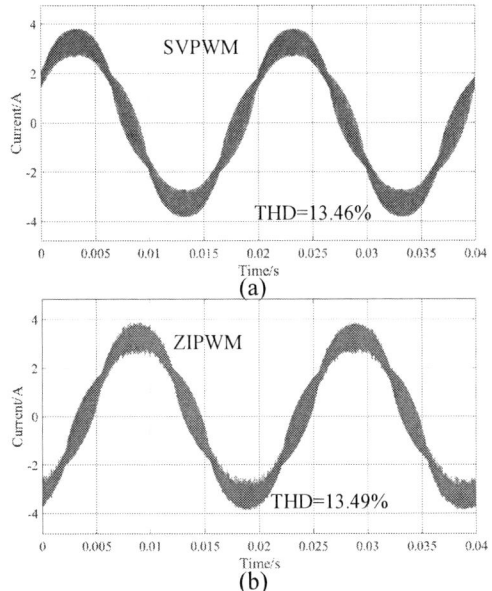

Fig.12 Waveform of the phase current (a)SVPWM (b)ZIPWM.

V. CONCLUSION

This paper proposes a modulation strategy based on zero-sequence component injection. By injecting a very small disturbance into the modulation wave, the duty cycle can be slightly altered. Simulation and experimental results show that ZIPWM has little impact on the system's power quality and can reduce EMI by 5-10dB in the range of 300kHz to 3MHz.

REFERENCES

[1] K. Li, P. Evans and M. Johnson, "Novel GaN Power Transistor Substrate Connection to Minimize Common Mode Noise," 2018 1st Workshop on Wide Bandgap Power Devices and Applications in Asia (WiPDA Asia), Xi'an, China, 2018, pp. 213-219.

[2] S. Jiang, Y. Liu, Z. Mei, J. Peng and C. -M. Lai, "A Magnetic Integrated LCL–EMI Filter for a Single-Phase SiC-MOSFET Grid-Connected Inverter," in IEEE Journal of Emerging and Selected Topics in Power Electronics, vol. 8, no. 1, pp. 601-617, March 2020.

[3] Y. -C. Son and S. -K. Sul, "A new active common-mode EMI filter for PWM inverter," in IEEE Transactions on Power Electronics, vol. 18, no. 6, pp. 1309-1314, Nov. 2003.

[4] S. Legowski and A. M. Trzynadlowski, "Advanced random pulse width modulation technique for voltage-controlled inverter drive systems," [Proceedings] APEC '91: Sixth Annual Applied Power Electronics Conference and Exhibition, Dallas, TX, USA, 1991, pp. 100-106.

[5] H. G. Li, S. D. Gong, J. W. Liu and D. L. Su, "CMOS-Based Chaotic PWM Generator for EMI Reduction," in IEEE Transactions on Electromagnetic Compatibility, vol. 59, no. 4, pp. 1224-1231, Aug. 2017, doi: 10.1109/TEMC.2016.2645784.

[6] J. Huang and R. Xiong, "Study on Modulating Carrier Frequency Twice in SPWM Single-Phase Inverter," in IEEE Transactions on Power Electronics, vol. 29, no. 7, pp. 3384-3392, July 2014.

[7] J. Chen, D. Jiang, Z. Shen, W. Sun and Z. Fang, "Uniform Distribution Pulsewidth Modulation Strategy for Three-Phase Converters to Reduce Conducted EMI and Switching Loss," in IEEE Transactions on Industrial Electronics, vol. 67, no. 8, pp. 6215-6226, Aug. 2020.

Development of A Novel Analytical Trapped Charge Model for Total Ionizing Dose Effects of SiC MOSFETs

Qingmao Hu
College of Semiconductors (College of Integrated Circuits)
Hunan University
Changsha, China
huqm@hnu.edu.cn

Xin Yang*
College of Electrical and Information Engineering
Hunan University
Changsha, China
xyang@hnu.edu.cn

Qingzhong Gui
College of Semiconductors (College of Integrated Circuits)
Hunan University
Changsha, China
qgui@hnu.edu.cn

Abstract—A novel analytical model for the total ionizing dose effect of SiC MOSFETs is proposed. The accuracy of the model has been verified by TCAD tool. The effects of the electric field across the gate oxide and temperature during the annealing process are discussed. The simulated results of these effects from the model are consistent with previous experimental results.

Keywords—total ionizing dose (TID), SiC MOSFET, trapped charge model, annealing

I. INTRODUCTION

Silicon carbide (SiC) MOSFETs are increasingly gaining attention in the aerospace field due to their wide bandgap characteristics. Compared to silicon based MOSFETs, SiC MOSFETs exhibit stronger tolerance to total ionizing dose (TID) irradiation [1]. Experimental research on the TID effect of SiC MOSFETs takes substantial cost and time, making it necessary to develop an accurate and convenient simulation method on the TID effect of SiC MOSFETs. Numerical simulation is too complex and time-consuming and difficult to apply to practical prediction. A model that can be applied in predicting the degradation of SiC MOSFETs in TID effect is needed. Campbell et al. [2] developed analytical models that account for the principal effects of total dose in MOS devices, which is Si-based. Lu et al. [3] developed a dynamic model based on Rowsey's model to describe both the mobile particles and defects in Si MOSFETs during the TID effects. Liang et al. [4] modeled for TID effect of 4H-SiC power MOSFETs under both gate bias and drain bias and proposed the feasibility of applying the models into switching-mode conditions. However, the influence of the electric field across the gate oxide on capture cross-section is not considered. Besides, there is a lack of time-dependent models for the simulation of both the radiation and annealing processes for SiC MOSFETs.

This work develops a novel analytical model for SiC MOSFETs under the TID effect based on the model used in the numerical simulation methods. The proposed model is a time-dependent model and can intuitively reveal the physical dynamics of the increase and decrease of trapped charges during the radiation and annealing process, respectively. Some important external conditions (e.g., electric field across

the gate oxide and temperature) can be completely integrated into this model. The proposed model has a maximum relative error of only 8.387% compared to the results obtained from the numerical simulation in TCAD, which takes about 1.2% of the simulation time of the latter.

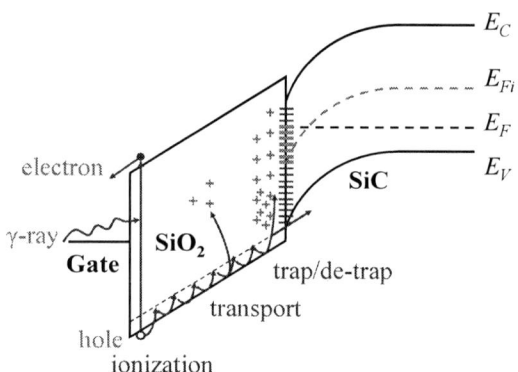

Fig. 1. The physical process of TID effect with a positive gate bias.

II. MECHANISMS OF TID EFFECT

The physical process of TID effects with a positive gate bias is described by Fig. 1. During the formation of the charged defects in the oxide, three phases are mainly included: oxide irradiation ionization, carriers transport, and carriers trapping or de-trapping. When radiation particles inject into the oxide, transferring enough energy to SiO_2, the covalent bonds of SiO_2 are broken, ionizing electron-hole pairs (EHPs), which are generated uniformly throughout the oxide. A portion of the generated EHPs by irradiation ionization will disappear within picoseconds through the initial recombination [9]. After that, driven by the electric field, electrons drift out of the gate, while holes transport slowly towards the SiO_2/SiC interface. Most of the electrons generated by irradiation ionization in the oxide will escape from being trapped due to its high mobility, while holes are captured by the traps near the interface due to relatively low mobility. Some of the captured holes can be released or neutralized through thermal emission or recombination with electrons tunnelling from the semiconductor [2]. Ultimately, the trapping and de-trapping will reach a dynamic equilibrium. The trapped holes will make a negative shift of the threshold voltage and cause the degradation of device.

This work was supported in part by the National Natural Science Foundation of China under Grant U24A20155, and in part by the Science and Technology Innovation Program of Hunan Province, China under grant 2023PT1009 and 2024RC1036.

At the interface, the acceptor-like traps capture electrons due to the rise of the Fermi level under the positive gate bias. Studies have shown that the impact of trapped holes in the oxide is much greater than that of trapped electrons at the interface [10], [12], thus the trapped electrons at the interface will be ignored in subsequent discussions.

III. MODELING APPROACH

A. Numerical Simulation Method

In the numerical simulation methods, the first phase, irradiation ionization is modeled by (1) and (2), which describe the generation rate of survived electron-hole pairs caused by irradiation. The remaining holes after initial recombination account for the proportion $Y(E)$, which is related to the electric field [11]. In (1), g_0 is the generation rate of electron-hole pairs and equals 7.88×10^{12} rad^{-1} ·cm^{-3} for SiO$_2$ [10] and dD/dt is the dose rate. In (2), for γ-ray radiation, the empirical parameters E_0, E_1, and m are 0.1 V/cm, 0.55 MV/cm, and 0.7, respectively.

$$G_\gamma = g_0 \frac{dD}{dt} Y(E) \tag{1}$$

$$Y(E) = \left(\frac{|E| + E_0}{|E| + E_1} \right)^m . \tag{2}$$

After initial recombination, the second phase, the transport of carriers in the oxide, is calculated by Poisson's equation, drift-diffusion equation, and carrier continuity equation as follows:

$$\nabla \cdot E = \frac{q}{\varepsilon_{ox}} \left(p - n + p_t - n_t \right) \tag{3}$$

$$\begin{cases} J_p = qp\mu_p E - qD_p \dfrac{dp}{dx} \\ J_n = qn\mu_n E + qD_n \dfrac{dn}{dx} \end{cases} \tag{4}$$

$$\begin{cases} \dfrac{\partial p}{\partial t} = -\dfrac{\partial J_p}{q\partial x} + g_p - r_p \\ \dfrac{\partial n}{\partial t} = \dfrac{\partial J_n}{q\partial x} + g_n - r_n \end{cases} \tag{5}$$

where E is the electric field across the gate oxide; p, n, p_t, and n_t are the density of free holes, free electrons, trapped holes, and trapped electrons, respectively; J_p and J_n are the hole and electron currents; g_p (g_n) and r_p (r_n) are the generation rate and recombination rate of holes (electrons), respectively. The finite-element simulation tools solve the above equations for every vertex of the grid. G_γ calculated by (1) contributes the generation rate of carriers in every vertex of the oxide. For the generation rate of carriers in the region of traps, the de-trapping rate of carriers should be additionally considered.

The calculation of the last phase is mainly about the concentration of the trapped charge, which means the dynamic (trapping and de-trapping) of charged traps should be mathematically described. It is obtained by calculating the trap occupation rate [11]. For holes with a considered trap level, the trap occupation rate is calculated as:

$$\frac{\partial f_p}{\partial t} = \left(1 - f_p\right) c_p - f_p e_p \tag{6}$$

where f_p is the hole trap occupation rate, c_p is the hole capture rate, and e_p is the hole emission rate. c_p and e_p are written as:

$$\begin{aligned} c_p &= c_{p,V} + c_{p,C} = c_{p,V} + e_{n,C} \\ &= \sigma_p \frac{J_p}{q} + \sigma_n v_{Tn} N_C \exp\left(\left(E_{\text{trap}} - E_C\right)/kT\right) \end{aligned} \tag{7}$$

$$\begin{aligned} e_p &= e_{p,C} + e_{p,V} = c_{n,C} + e_{p,V} \\ &= \sigma_n \frac{J_n}{q} + \sigma_p v_{Tp} N_V \exp\left(\left(E_V - E_{\text{trap}}\right)/kT\right) \end{aligned} \tag{8}$$

where the subscripts C and V represent the exchange of electrons or holes between the conduction band and valence band, respectively (as shown in Fig. 2). σ_p and σ_n are capture cross sections of the hole and electron. E_{trap} is the energy of the trap level.

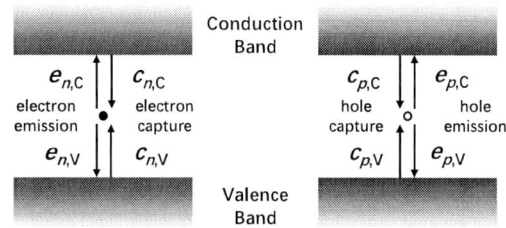

Fig. 2. Capture and emission of carriers between the conduction or valence band and trap energy [11].

Finally, the concentration of trapped holes is computed by multiplying the trap occupation by the trap concentration:

$$p_t = N_p \bullet f_p \tag{9}$$

where N_p is the concentration of hole traps. A similar calculation can be made for electrons. Mathematical iterative methods are used to solve the above equations, thus making a long simulation time and troubles to apply to circuit level simulations.

B. Proposed Analytical Model

Based on the above models, now we derive a new analytical model for trapped charges. According to (7), (8) and (9), multiplying both sides of (6) by N_p, it leads to

$$\begin{aligned} \frac{\partial p_t}{\partial t} &= \left(N_p - p_t\right) c_p - p_t e_p \\ &= \left(N_p - p_t\right) c_{p,V} - p_t e_{p,V} - p_t c_{n,C} + \left(N_p - p_t\right) e_{n,C}. \end{aligned} \tag{10}$$

The right side of (10) describes four physical pictures of trapped charges: capture, emission, recombination, and ionization (as shown in Fig. 3). For hole traps, these physical dynamics correspond to $c_{p,V}$, $e_{p,V}$, $e_{p,C}$, and $c_{p,C}$ in Fig. 2, respectively ($c_{n,C} = e_{p,C}$, $e_{n,C} = c_{p,C}$, [11]). For hole traps, $\sigma_p \gg \sigma_n$, and the trap energies are mainly distributed near the top of valence band [10], so from (7) and (8) we can conclude that the recombination and ionization of hole traps can be ignored. Thus (10) can be simplified as

$$\frac{\partial p_t}{\partial t} = -\left(c_{p,V} + e_{p,V}\right) p_t + c_{p,V} N_p \tag{11}$$

where

$$c_{p,\mathrm{V}} = \sigma_p \frac{J_p}{q} \tag{12}$$

$$e_{p,\mathrm{V}} = \sigma_p v_{Tp} N_\mathrm{V} \exp\left(\left(E_\mathrm{V} - E_{\mathrm{trap}}\right)/kT\right). \tag{13}$$

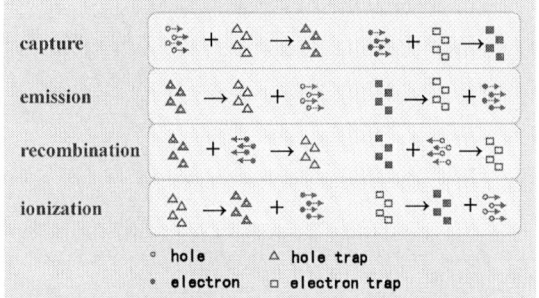

Fig. 3. Four dynamic physical processes of traps: capture, emission, recombination, and ionization.

For a stationary state, the leakage current in the oxide is time-independent, thus the trap capture rate and emission rate are time-independent. Integrating (11) gives:

$$p_t = \begin{cases} \dfrac{b}{a} - \left(\dfrac{b}{a} - p_t(0)\right)e^{-at} & p_t(0) < \dfrac{b}{a} \\[2mm] \dfrac{b}{a} + \left(p_t(0) - \dfrac{b}{a}\right)e^{-at} & p_t(0) > \dfrac{b}{a} \end{cases} \tag{14}$$

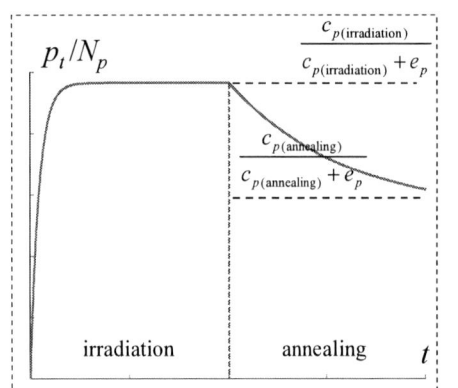

Fig. 4. Diagram of dynamic of trapped hole concentration described by (14).

where $a = c_{p,\mathrm{V}} + e_{p,\mathrm{V}}$, $b = c_{p,\mathrm{V}} N_p$. For a period of radiation-annealing, the dynamic of trapped hole concentration described by (14) is shown by Fig. 4. The increase/decrease rate and ultimate value of p_t depends on the capture and emission rate $c_{p,\mathrm{V}}$ and $e_{p,\mathrm{V}}$. For $e_{p,\mathrm{V}}$, it is determined by the trap energy E_{trap} according to (13):

$$e_{p,\mathrm{V}} = k_e \sigma_p \exp\left(\left(E_\mathrm{V} - E_{\mathrm{trap}}\right)/kT\right) \tag{15}$$

where $k_e = v_{Tp} N_V$, which is a coefficient determined by temperature and can be obtained from parameter fitting when the temperature is fixed. $c_{p,\mathrm{V}}$ switches between the processes of radiation and annealing due to different leakage currents. During the radiation process, in [2] it is assumed that the holes generated by irradiation appear immediately in the trapping region under the action of the gate bias. However, there must be some losses of holes during the transport, especially for

power MOSFETs with thicker oxides. Considering the approximate expression of flux of holes in [2], the hole current can be written:

$$J_p = k_c q G_\gamma t_{ox} \tag{16}$$

where t_{ox} is the thickness of oxide and k_c is a coefficient representing the proportion of hole loss in transport, which can obtained from parameter fitting. Substituting (12) with (1) and (16), we derive the expression of $c_{p,\mathrm{V}}$ during the radiation process:

$$c_{p,\mathrm{V(radiation)}} = k_c g_0 \sigma_p \frac{dD}{dt} t_{ox} Y(E). \tag{17}$$

During the annealing process, $c_{p,\mathrm{V}}$ is radiation-independent. Therefore, we consider it as a constant value which can be obtained from parameter fitting.

In this work, the threshold voltage of MOSFETs is used as the criterion of TID degradation. The variation of the threshold voltage is linear to the trapped charge concentration when the trapped charges are distributed near the interface, as shown in (18):

$$\Delta V_{th} \approx -\frac{qp_t}{C_{ox}} = -\frac{qp_t t_{ox}}{\varepsilon_{SiO_2}}. \tag{18}$$

IV. MODEL VERIFICATION AND EXTENSION

A. Model Verification

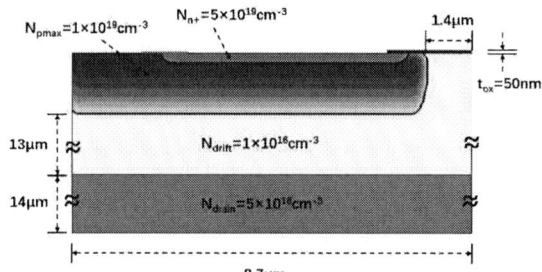

Fig. 5. Parameters of SiC MOSFET used in TCAD.

TABLE I.   FITTED PARAMETERS

Fitted parameters	Values	Units
k_c	0.166	1
k_e	3.699×10^{24}	$s^{-1} cm^{-2}$
$c_{p,\mathrm{V(anneal)}}$	2.268×10^{-8}	s^{-1}

In this work, Sentaurus TCAD tools are used to simulate the TID effect of SiC MOSFETs and verify the accuracy of the proposed model. A two-dimensional finite element model of planar-gate SiC MOSFET is built. The elementary cell parameters [12] are shown in Fig. 5. The capture cross sections of the hole and electron are 1×10^{-13} cm^2 and 1×10^{-16} cm^2, respectively [10]. A single trap level is set as 1.06 eV from the valence band. The irradiation started in 0 s and ended in 4000 s with different dose rates. The V_{gs} keeps +15V during the radiation. The fitted parameters are shown in Table I. Fig. 6 shows the results of simulated shifts of threshold voltage with the dose rate of 375 rad/s. Fig. 7 shows the results of simulated trapped charge concentration dynamics at a certain point in the oxide interface during the irradiation and annealing process. The maximum relative error of trapped charge concentration is 8.387%, which indicates the relatively high accuracy of the proposed model. Additionally, It is observed from Fig. 7(a) that with the dose rate of 100 and 200 rad/s (both of them are lower than the dose rate specified in the TID test standards), the concentration of trapped holes is linear to the dose. However, As the dose rate increases, the concentration of trapped charges tends to saturate within a short time. This saturation phenomenon under high dose conditions has also been reported in previous work [13]. Besides, the recovery of traps in the annealing process are not linear to annealing time but became slow with time, which is also reported in [14]-[16].

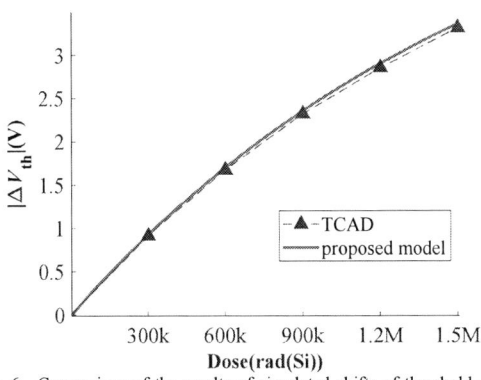

Fig. 6. Comparison of the results of simulated shifts of threshold voltage under TID effect.

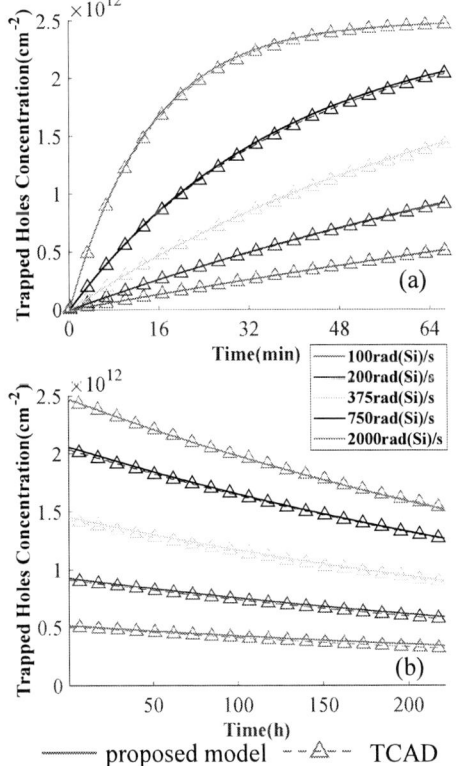

Fig. 7. Comparison of trapped holes concentration dynamics during (a)irradiation and (b)annealing process with different dose rates from TCAD simulation and proposed model.

B. Model Extension

For further application of the model, some important external factors and device parameters can be taken into consideration. We provide two aspects to develop the model according to the experiment results of previous works:

- The electric field across the gate oxide: the gate bias influences the recombination and capture of carriers by electric field during the irradiation process [17], [18].

- Temperature during the annealing process: high temperature helps to accelerate the process of annealing [14].

The electric field of gate oxide influences the TID effect mainly by the hole yield and the capture cross section. As shown in (2), the hole yield $Y(E)$ increases with increasing electric field. However, the capture cross section is also electric-field-dependent and decreases with increasing electric field [5]. A universal empirical model of the capture cross section is expressed by:

$$\sigma_p = \sigma_{p0}\left(1 + a_1 E^{p_1} + a_2 E^{p_2}\right)^{p_0} \quad (19)$$

where a_1, a_2, p_1, p_2 and p_0 are material-dependent empirical parameters [11]. Therefore, the concentration of trapped holes doesn't monotonically vary with the electric field of the gate oxide. As a result of that, as the gate voltage increases, the degradation of V_{th} first increases and then decreases for the same dose. Fig. 8 shows the simulated results of the concentration of trapped holes under different electric field values, which are consistent with the previous reports shown in Fig. 9 [18]. The empirical parameters of a_1, a_2, p_1, p_2, and p_0 are 1.3×10^{-4}, 0, 0.62, 1, -1, respectively.

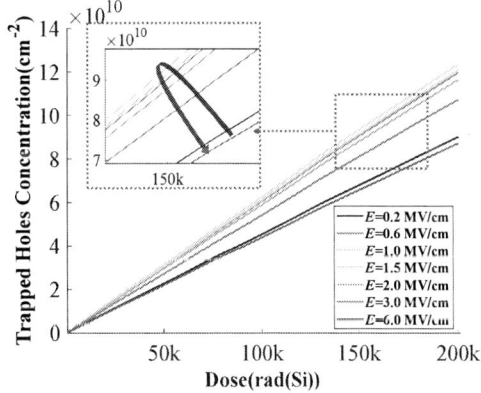

Fig. 8. Radiation induced trapped hole concentration under different electric field of gate oxide calculated by proposed model

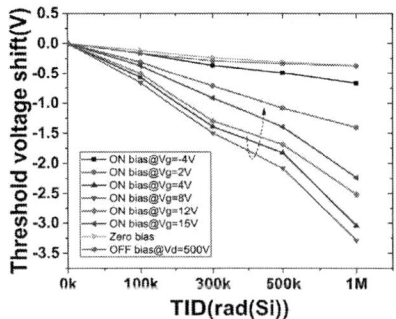

Fig. 9. The degradation of the threshold voltage dosen't monotonically vary with the positive gate bias [18].

The temperature during the annealing process mainly affects the emission rate to accelerate the process of recovery of the traps. The emission rate is determined by (13). The hole thermal velocity v_{Tp} and the effective density-of-states of the valence band N_V are temperature-dependent as shown in (20) and (21):

$$v_{Tp} = \sqrt{\frac{3kT}{m_p}} \qquad (20)$$

$$N_V = 2\left(\frac{2\pi m_p kT}{h^2}\right)^{3/2} . \qquad (21)$$

Substituting (13) with (20) and (21), we can derive that

$$e_{p,v} \propto T^2 \cdot \exp\left(-\frac{E_{trap} - E_V}{kT}\right). \qquad (22)$$

From (22) we can conclude that $e_{p,v}$ increase with T sharply, which means that high temperature may help to accelerate the process of annealing. This phenomenon is also reported in [14]. Fig. 10 shows the simulated results of the influence of temperature on the annealing process.

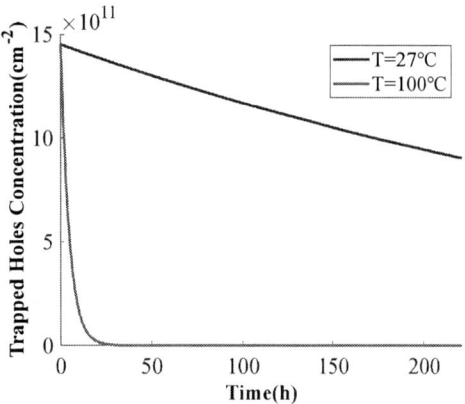

Fig. 10. High temperature helps to accelerate the process of annealing.

C. Discussion

The proposed model performs well in comparison with numerical simulation method. However, we use a single level trap to represent the distributed traps for simplicity. Besides, the trapped electrons in the interface are not considered. Actually, the energy level distribution of SiC/SiO$_2$ interface traps has been investigated [22], [23]. The proposed model can be extended by substituting N_p with the energy distribution of the hole and electron traps in the oxide and interface and integrating the energy distribution of p_t and n_t within the bandgap. In this way, the comprehensive influence of trap charges from all energy levels and locations can be obtained.

V. Conclusion

This work developed a novel analytical model for trapped charges near the interface of SiC MOSFETs under the TID effect. Compared to the results of numerical simulation in TCAD, the proposed model keeps a high precision and spends less simulation time. The model is further extended by discussion the relationship of the electric field of gate oxide and temperature during the annealing process. Further development on the new model is expected to consider muti-level of traps to simulate different kind of devices.

Acknowledgment

This work was supported in part by the National Natural Science Foundation of China under Grant U24A20155, and in part by the Science and Technology Innovation Program of Hunan Province, China under grant 2023PT1009 and 2024RC1036.

References

[1] J. Furuta, M. Mizushima and K. Kobayashi, "Measurement of Total Ionizing Dose Effects on SiC Trench MOSFETs by Gamma-ray and Alpha-particle Irradiation," 2022 22nd European Conference on Radiation and Its Effects on Components and Systems (RADECS), Venice, Italy, 2022, pp. 1-5.

[2] Campbell, Phillip Montgomery, and Carolyn W. Bogdan. "Analytical models for total dose ionization effects in MOS devices." No. SAND2008-5112. Sandia National Laboratories (SNL), Albuquerque, NM, and Livermore, CA (United States), 2008.

[3] Lu, Guangbao, et al. "Dynamic modeling of total ionizing dose-induced threshold voltage shifts in MOS devices." Chinese Physics B 32.1 (2023): 018506.

[4] S. Liang, et al., "Modeling Irradiation-Induced Degradation for 4H-SiC Power MOSFETs," IEEE Transactions on Electron Devices, vol. 70, no. 3, pp. 1176-1180, March 2023.

[5] Paillet, P., et al. "Simulation of multi-level radiation-induced charge trapping and thermally activated phenomena in SiO$_2$." RADECS 97. Fourth European Conference on Radiation and its Effects on Components and Systems (Cat. No. 97TH8294). IEEE, 1997.

[6] Fernández-Martínez, P., et al. "Simulation of total ionising dose in MOS capacitors." Proceedings of the 8th Spanish Conference on Electron Devices, CDE'2011. IEEE, 2011.

[7] I. S. Esqueda, H. J. Barnaby, and M. P. King, "Compact modeling of total ionizing dose and aging effects in MOS technologies," IEEE Trans. Nucl. Sci., vol. 62, no. 4, pp. 1501–1515, Aug. 2015, doi: 10.1109/TNS.2015.2414426

[8] Lelis, A. J., et al. "The nature of the trapped hole annealing process." IEEE Transactions on Nuclear Science 36.6 (1989): 1808-1815.

[9] Schwank, James R., et al. "Radiation effects in MOS oxides." IEEE Transactions on Nuclear Science 55.4 (2008): 1833-1853.

[10] T. Liu, et al., "Comparative Investigation on Ionizing Irradiation-Induced Threshold Voltage Degradation for 1200-V DT SiC MOSFET by Experiment and Simulation," IEEE Transactions on Nuclear Science, vol. 71, no. 11, pp. 2386-2392, Nov. 2024.

[11] Sentaurus Device User Guide, 6th ed., Synop., Mountain View, CA, USA, 2010, pp. 423-440.

[12] Romano, Gianpaolo, et al. "A comprehensive study of short-circuit ruggedness of silicon carbide power MOSFETs." IEEE Journal of Emerging and Selected Topics in Power Electronics 4.3 (2016): 978-987.

[13] Bonaldo, Stefano, et al. "Radiation-induced effects in SiC vertical power MOSFETs irradiated at ultra-high doses." IEEE Transactions on Nuclear Science (2024).

[14] Yu, Qingkui, et al. "Application of total ionizing dose radiation test standards to SiC MOSFETs." IEEE transactions on nuclear science 69.5 (2021): 1127-1133.

[15] Zhang, Yuan-Lan, et al. "The Degradation and Recovery of 1200-V SiC MOSFET with Different Total Ionizing Doses." 2023 20th China International Forum on Solid State Lighting & 2023 9th International Forum on Wide Bandgap Semiconductors (SSLCHINA: IFWS). IEEE, 2023.

[16] Li, Xinyu, et al. "Degradation of VDMOS under simultaneous and sequential stress of gamma ray irradiation and annealing process." IEEE Transactions on Electron Devices 70.6 (2023): 2947-2955.

[17] Z. Chen, et al., "Degradation Behavior of SiC Trench MOSFETs by Total-Ionizing-dose Irradiation Under Gate Voltage Stress," 2024 36th International Symposium on Power Semiconductor Devices and ICs (ISPSD), Bremen, Germany, 2024, pp. 228-231.

[18] C. Peng, Z. Lei, Z. Zhang, Y. He, T. Ma and Y. Chen, "Bias and Temperature Dependence of Radiation-Induced Degradation for SiC MOSFETs," IEEE Transactions on Nuclear Science, vol. 71, no. 5, pp. 1186-1193, May 2024.

[19] Yang, Sheng, et al. "Impact of switching frequencies on the TID response of SiC power MOSFETs." *Journal of Semiconductors* 42.8 (2021): 082802.

[20] Gao, Kexin, et al. "Degradation behavior and mechanism of SiC power MOSFETs by total ionizing dose irradiation under different gate voltages." *2021 IEEE Workshop on Wide Bandgap Power Devices and Applications in Asia (WiPDA Asia)*. IEEE, 2021.

[21] Feng, Haonan, et al. "Special Degradation Effects of 60 Co γ-Rays Irradiation on Electrical Parameters of SiC MOSFETs." *IEEE Transactions on Nuclear Science* 70.9 (2023): 2165-2174.

[22] Afanasev, Valeri V., et al. "Intrinsic sic/sio2 interface states." *physica status solidi (a)* 162.1 (1997): 321-337.

[23] I. Pesic, D. Navarro, M. Miyake, and M. Miura-Mattausch, "Degradation of 4H-SiC IGBT threshold characteristics due to SiC/SiO2 interface defects," *J. Solid-State Electron.*, vol. 101, pp. 126–130, Nov. 2014, doi: 10.1016/j.sse.2014.06.023.

[24] Luo, Runding, et al. "Total Ionizing Effects on Static Characteristics of 1200V SiC MOSFET Power Devices with Planar and Trench Structures." *2023 20th China International Forum on Solid State Lighting & 2023 9th International Forum on Wide Bandgap Semiconductors (SSLCHINA: IFWS)*. IEEE, 2023.

2025 IEEE Workshop on Wide Bandgap Power Devices and Applications in Asia (WiPDA Asia)

Method for Estimating Power Loss of IGBT Module in Wind Power Converter Based on Measured Temperature Information

Ye Tian
State Grid Fujian Electric Power
Research Institute
State Grid Fujian Electric Power Co.,
Ltd.
Fuzhou, China
tianbobing@outlook.com

Dawei Chen
State Grid Fujian Electric Power
Research Institute
State Grid Fujian Electric Power Co.,
Ltd.
Fuzhou, China
dwchen@zju.edu.cn

Zhijie Zeng
State Grid Fujian Electric Power
Research Institute
State Grid Fujian Electric Power Co.,
Ltd.
Fuzhou, China
zeng_zhijie@fj.sgcc.com.cn

Lixuan Zhu
State Grid Fujian Electric Power
Research Institute
State Grid Fujian Electric Power Co.,
Ltd.
Fuzhou, China
zlx@hhu.edu.cn

Guojun Bao
State Grid Fujian Electric Power
Research Institute
State Grid Fujian Electric Power Co.,
Ltd.
Fuzhou, China
baoguojun0216@163.com

Zhixiang Zou
School of Electronic Engineering
Southeast University
Nanjing, China
zzou@seu.edu.cn

Abstract—Online monitoring of the temperature of IGBTs represents a crucial approach for boosting the reliability of wind power converters. The junction temperature is calculated based on the device power loss in the current thermal model techniques. However, it is challenging to obtain this power loss value with high precision in real time. Consequently, this paper introduces a new type of power loss observer, aiming to accurately estimate the real-time power loss. Different from traditional methods, the proposed approach merely requires the measurement of the heatsink temperature. This paper conducts an in-depth analysis of the mathematical principle, algorithm, and implementation steps of the proposed method. Moreover, both simulation and experiment are employed to verify the effectiveness of the loss observer.

Keywords—IGBT module, thermal monitoring, junction temperature, power loss estimation

I. INTRODUCTION

A. Background

High-power converters, functioning as crucial power conversion devices, have extensive applications in wind power generation. High-power IGBT modules, which enable rapid frequency switching for power conversion processes, are the pivotal components of these converters.

Nonetheless, recent research indicates that IGBT modules are the most fragile components within high-power converters. Over half of the IGBT module malfunctions are associated with junction temperature. For instance, overheating can lead to damage, and thermal stress cycles can cause material degradation. Moreover, owing to the unpredictability of the load profile, it is arduous to estimate the temperature of the IGBT module in advance. Thus, the real-time acquisition of accurate junction temperature is of great importance for IGBT condition monitoring, converter thermal management, and life prediction. It also serves as an indispensable approach to enhancing the reliability of wind power converters. Thermal network models are capable of obtaining temperatures at specific positions within the IGBT module. They can strike a balance between model accuracy and complexity, making them appropriate for long-term

dynamic thermal analysis and online temperature monitoring of IGBT modules.

All the thermal models described above require power loss information as an input to compute the junction temperature. However, for the following reasons, the calculated loss has substantial errors in the initial phase, and these errors will further escalate as the converter's service duration lengthens.

1）A minor difference in the parasitic parameters of the commutation loop between the offline test platform and the converter can lead to a deviation in switching loss of up to 30%.[1]

2）The non-linear characteristics of the analytical model, along with the measurement inaccuracies introduced by the probe during the calibration procedure, will result in approximately a 15% error in switching loss.[2]

3）The junction temperature serves as the input for the conventional loss calculation approach. However, acquiring it precisely proves to be challenging.[3]

4）The deterioration of the chip's gate oxide layer has the potential to cause the switching loss to vary by as much as 50%.[4]

5）Owing to the aging of the bonding wires, both the conduction loss and the switching loss can experience a 10% change.

B. The novelty and contributions of this work

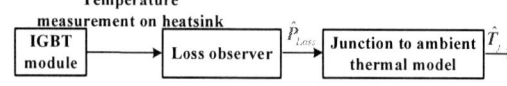

Fig. 1. Principle of the proposed technique

In response to the aforementioned problems, as depicted in Fig. 1, this paper puts forward a novel loss estimation technique for IGBT modules. This technique diverges from the traditional loss calculation approach. The proposed approach merely requires the temperatures of the heatsink and

This work is in part by the Science and Technology Program of State Grid Fujian Electric Power Co. (52130424000R).

979-8-3315-1110-4/25 $31.00 © 2025 IEEE

other relatively accessible-to-measure temperatures as inputs and is capable of accurately estimating the power loss.

The structure of this paper is as follows: Section II initially showcases the overall framework of the proposed method and analyzes the principles of the key components within the technique, with a particular emphasis on deriving the algorithm of the loss observer. Section III presents the simulation and experimental validation of the proposed methods. Section IV concludes this paper.

II. PRINCIPLE OF THE PROPOSED TECHNIQUE

A. Overview of the proposed technique

According to the analysis in Section Ⅰ, separating the loss calculation from the variables that are hard to accurately acquire, like the aging level, junction temperature, and parasitic parameters of the commutation loop, represents the crucial factor for enhancing the accuracy of loss estimation. Guided by this concept, this paper introduces a loss estimator and a junction temperature estimation approach relying on multi-point temperature measurements. The proposed approach is mainly composed of three components: the thermal resistance model of the sensor network, the loss observer, and the thermal resistance model from the junction to the ambient. Sensors S1…S5 are positioned on the surface of the heat sink, and the temperature measured is denoted as $T_s=[T_{s1}\ T_{s2}\ T_{s3}\ T_{s4}]^{\mathrm{T}}$.

The half-bridge 2MBI450VN-170 IGBT module is selected as the research case, and Fig. 2 shows its structure and circuit topology. The FEM model of the module and the heatsink is shown in Fig. 3.

As depicted in Fig. 4, within the insulated gate bipolar transistor (IGBT) module, the central nodes of each chip are designated as j_1, j_2, …, and j_{12}. Taking into account the requirements for electrical insulation, protection of the IGBT module's packaging structure from damage, and the complexity associated with the installation process, sensors S_1, S_2, S_3, S_4, and S_5 are positioned on the upper surface of the heat sink.

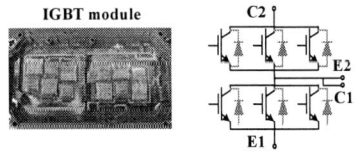

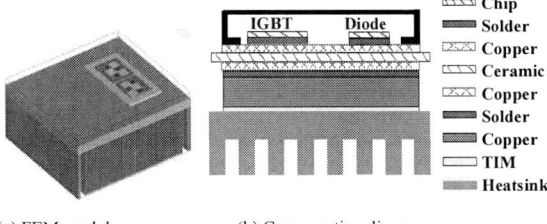

(a) Inner structure   (b)Circuit topology

Fig. 2. Prototype and circuit topology of the IGBT module

(a) FEM module   (b) Cross-section diagram

Fig. 3. FEM model and cross-section diagram of the IGBT module and the heatsink

Fig. 4. Structure of the power module

B. Sensor network thermal model

The first element of the proposed method is the thermal resistance model of the sensor network. This model is capable of predicting the temperature of the sensor nodes, and its model parameters serve as the foundation for the gain design of the loss observer.

A thermal model of the sensor network is built using a Foster-type model that considers thermal coupling effects, which serves as the foundation for the loss observer. As illustrated in Fig. 5, this model is capable of predicting the temperature at sensor nodes by leveraging loss data. The power losses of the lower-arm IGBT, lower-arm diode, upper-arm IGBT, and upper-arm diode are respectively labeled as P_1, P_2, P_3, and P_4, respectively.

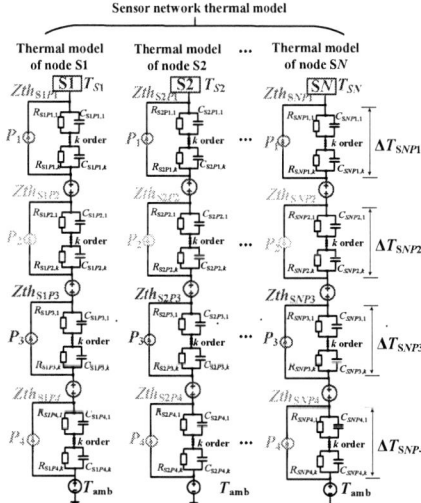

Fig. 5. Sensor network thermal model

According to [5], when the period of environmental temperature fluctuation is much larger than the maximum time constant of the thermal system dynamics, the ambient temperature (T_{amb}) can be directly added to the Foster model as a voltage source.

In application scenarios where the ambient temperature fluctuates rapidly on a second-level time scale, the accuracy of the Foster model in Fig. 5 will decrease because it does not account for the low-pass filtering of the reference temperature. In the future, in-depth research will be further carried out on this issue.

The sensor network thermal model in Fig. 5 can be represented by the thermal impedance matrix in (1).

$$\begin{bmatrix} T_{s1} \\ T_{s2} \\ \dots \\ T_{sN} \end{bmatrix} = \begin{bmatrix} Z_{thS1P1}(t) & \dots & Z_{thS1P4}(t) \\ Z_{thS2P1}(t) & \dots & Z_{thS2P4}(t) \\ \dots & \dots & \dots \\ Z_{thSNP1}(t) & \dots & Z_{thSNP4}(t) \end{bmatrix} \begin{bmatrix} P_1 \\ P_2 \\ \dots \\ P_4 \end{bmatrix} + T_{amb} \quad (1)$$

When a step loss P_j is added, Z_{thSiPj} satisfies(2). Therefore, based on the step temperature rise T_{si} of node Si, the thermal resistance R_{SiPj} and thermal capacitance C_{SiPj} ($r=1\dots k$) shown in Fig. 5 can be identified by (2) and (3).

$$Z_{thSiPj}(t) = \frac{T_{Si}(t) - T_{amb}(t)}{P_j} = \sum_{r=1}^{k} R_{SiPj,r}(1 - e^{-\frac{t}{\tau_{SiPj,r}}}) \quad (2)$$

$$\tau_{SiPj,r} = R_{SiPj,r} C_{SiPj,r} \quad (3)$$

The state space of the thermal model of the sensor network is shown in (2).

$$\begin{cases} \dfrac{dx}{dt} = Ax + DP_{Loss} \\ T_s = y = Cx \end{cases} \quad (4)$$

In (2), x is the temperature of the thermal capacitances in the thermal model. $P_{Loss\_ref} = [P_{ref\_1} \; P_{ref\_2} \; P_{ref\_3} \; P_{ref\_4}]^T$ is the vector consisting of the power loss of each device in the power module. $Ts = [T_{s1} \; T_{s2} \; T_{s3} \; T_{s4}]^T$ is the sensors' temperatures.

A, D and C are the system matrix, input matrix and output matrix of the state space, which can be calculated as shown in (5).

$$A = \begin{bmatrix} \dfrac{-1}{\tau_{S1P1,1}} & \dots & 0 \\ \dots & \dots & \dots \\ 0 & \dots & \dfrac{-1}{\tau_{SNP4,k}} \end{bmatrix} \quad D = \begin{bmatrix} \dfrac{1}{C_{S1P1,1}} & \dots & 0 \\ \dots & \dots & \dots \\ \dfrac{1}{C_{S1P1,k}} & \dots & 0 \\ \dots & \dots & \dots \\ 0 & \dots & \dfrac{1}{C_{SNP4,k}} \end{bmatrix}$$

$$C = \begin{bmatrix} 1\dots1 & \dots & 0 \\ 0 & \dots & 0 \\ 0 & \dots & 0 \\ 0 & \dots & 1\dots1 \end{bmatrix}_{N \times 4kN} \quad (5)$$

C. Junction-to-ambient thermal model

For instance, as depicted in Fig. 6, a Foster-type thermal resistance model from the junction to the ambient is constructed to estimate the temperatures of j6 and j8.

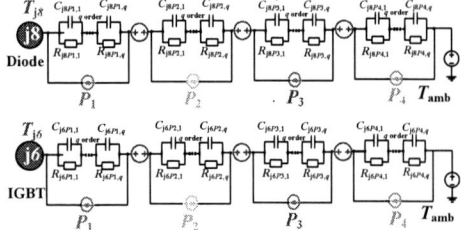

Fig. 6. Junction to ambient thermal model

Based on the method in [5], the state space of the junction-ambient thermal network model can be obtained as shown in (6).

$$\begin{cases} \theta = A_{j\text{-}a}\theta + B_{j\text{-}a}P_{Loss} \\ T_j = \begin{bmatrix} T_{j6} & T_{j8} \end{bmatrix}^T = C_{j\text{-}a}\theta \end{cases} \quad (6)$$

D. Loss observer

Take S as the left singular matrix of CD. Take U_{CD1} and U_{D1} as the first rank(D) = m^* columns of U_{CD} and U_D respectively, where U_{CD} and U_D are the left singular matrices of CD and D respectively.

Take V_{N2} as the first $n - m^*$ columns of the right singular matrix of $U_{CD1}^T C$. And let $T = [U_{D1} \; V_{N2})$.

Find the value of L_{22} such that the real parts of all the eigenvalues of $A_{22} + L_{22}C_2$ are negative.

$$T^{-1}AT = \begin{bmatrix} A_{11} & A_{12} \\ A_{21} & A_{22} \end{bmatrix}, T^{-1}D = \begin{bmatrix} D_1 \\ D_2 \end{bmatrix}, \; S^{-1}CT = \begin{bmatrix} C_{11} & 0 \\ 0 & C_{22} \end{bmatrix}$$

Solve for P_{22}, where Q_{22} is a symmetric positive-definite matrix with the same dimension as A_{22}.
$$\begin{cases} P_{22} > 0 \\ P_{22}(A_{22} + L_{22}C_{22}) + (A_{22} + L_{22}C_{22})^T P_{22} = -Q_{22} < 0 \end{cases}$$

$$\bar{P} = \begin{bmatrix} I_{m^*} & 0 \\ 0 & P_{22} \end{bmatrix} \qquad \bar{L} = \begin{bmatrix} -kC_{11}^{-1} & 0 \\ 0 & L_{22} \end{bmatrix}$$
$$\bar{Q}_{11} = A_{11} + A_{11}^T + (A_{12} + A_{21}^T P_{22})Q_{22}^{-1}(A_{12}^T + P_{22}A_{21})$$
$$k > \lambda_{max}(\bar{Q}_{11})/2$$

$$\bar{G} = \begin{bmatrix} D_1^T C_{11}^{-1} 0 \end{bmatrix}, \; L = T\bar{L}S^{-1}, \; G = \bar{G}S^{-1}$$

Fig. 7. Calculation method of L and G

The third element is the loss observer adopting the error between the sensors' temperatures predicted by the model and the measurements to derive the estimated value $\hat{d}$ of the actual power loss P_{Loss} as shown in (4).

$$\begin{cases} \dot{\hat{x}}(t) = (A + LC)\hat{x}(t) + D\hat{d}(t) - Ly(t) \\ \hat{d}(t) = rG(y - C\hat{x}(t)) \end{cases} \quad (1)$$

In (4), $\hat{x}(t)$ and $\hat{d}(t)$ is respectively the estimate of state variable x(t) and power loss. By choosing a sufficiently large r, an accurate estimation of the power loss can be made based on the observer in (4). The detailed calculation method of L and G are presented in Fig. 7.

E. Implementation steps of the proposed technique

The junction temperature monitoring method proposed in this paper, which is based on the loss observer, has its implementation steps shown in Fig. 8. Temperature sensors are placed on the heatsink surface. $Z_{\text{th}SiPj}(t)$ is utilized to identify the parameters of the sensor network thermal model. According to (2), the model state space is calculated. Next, the loss observer in (4) is designed. By merely using the temperature measurements (y) from the sensors, the estimated value ($\hat{d}$) of the power loss is obtained. Finally, this value of $\hat{d}$ is input into the junction-to-ambient thermal model in (3) to get the estimated junction temperature $\hat{T}_j$.

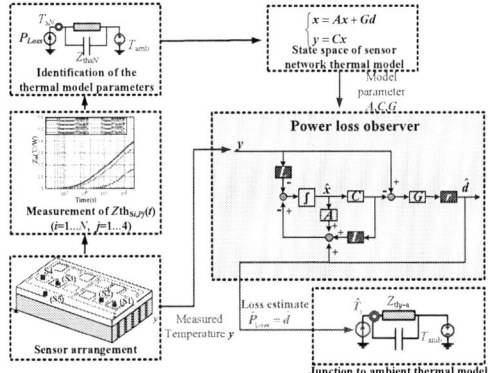

Fig. 8. Implementation steps of the proposed technique

III. VALIDATION OF THE PROPOSED TECHNIQUE

A. Identification of the parameters in thermal models

The 2MBI450VN-170 IGBT module was chosen as the device under test (DUT). It was fixed onto an aluminum forced-air cooling heatsink with a 1-mm-thick thermal pad. As shown in Fig. 9, temperature sensor 1, sensor 2, ..., and sensor 5 were installed on the surface of the heatsink. The accuracy of the estimated power loss is influenced by several key factors. Besides the rational design of the loss observer's structure and feedback gains, the modeling precision of the thermal resistance model also plays a crucial role. The parameters of the thermal resistance model were identified using the Nonlinear Least Squares algorithm. This algorithm was applied based on the cooling curves of the IGBT module.

Fig. 10 and Fig. 11 provides an example by comparing the measured and fitted transient thermal impedances of S3. It's evident that the fitting R^2 value is no less than 0.9951, and the root-mean-square error (RMSE) is no more than 0.0068. These results suggest that the model parameters were accurately identified.

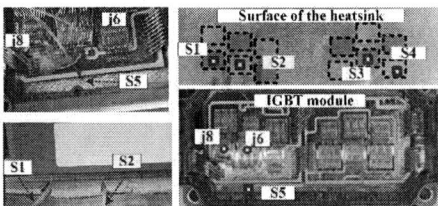

Fig. 9. Locations of the sensors

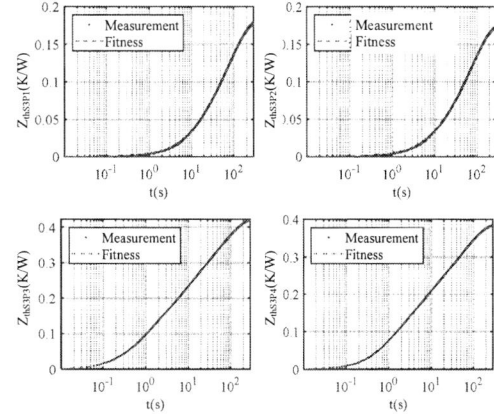

Fig. 10. Measured and fitted transient thermal impedances of the sensor network's thermal model

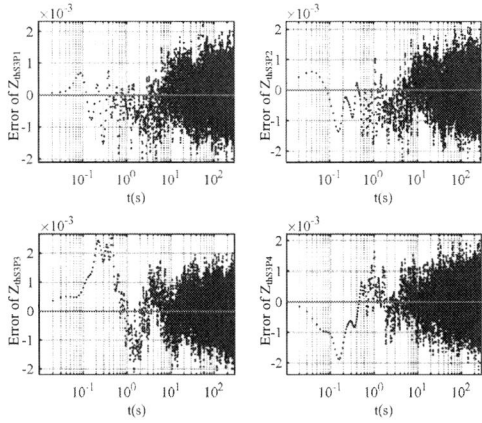

Fig. 11. Error of the measured and fitted transient thermal impedances of the sensor network's thermal model

B. Simulation validation

As shown in Fig. 12, the accuracy of the proposed method for loss observer and junction temperature estimation is verified through Simulink. A reference model is employed to mimic the thermal behavior of the actual IGBT module, which consists of two components: the loss calculation component and the reference thermal model of the DUT.

The loss calculation component serves to compute the loss reference value, denoted as P_{Ref}. The reference thermal model, relying on P_{Ref}, is utilized to calculate the temperature y of the sensor nodes and the junction temperature reference value T_{Ref}.

The loss observer uses y as its input and is capable of deriving the power loss estimation without the need for any information about P_{Ref}. The estimated power loss is then fed into the junction-to-ambient thermal model to estimate the junction temperature.

The simulation waveforms are presented in Fig.13. It is evident that the proposed technique can provide precise estimations of the power loss for each device within the IGBT module, solely based on the measured temperatures.

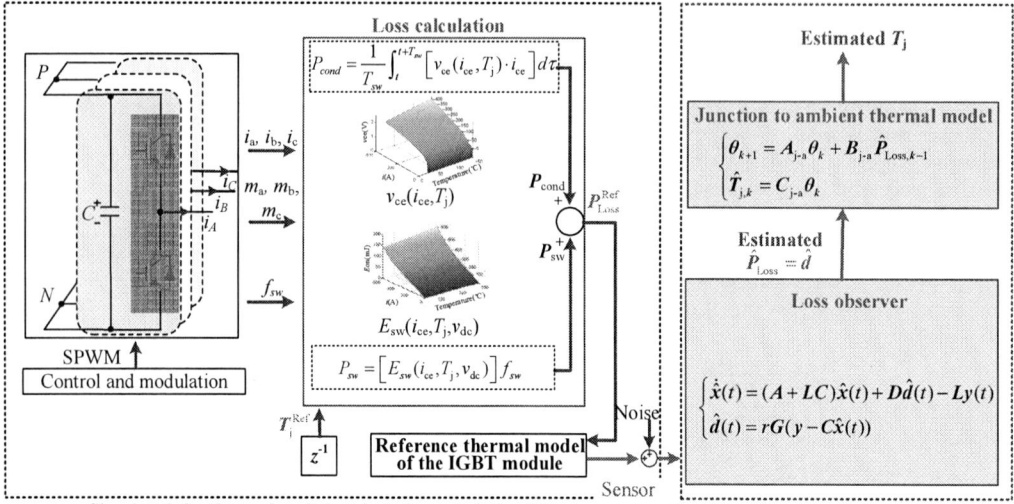

Fig. 12. Schematic diagram of simulation

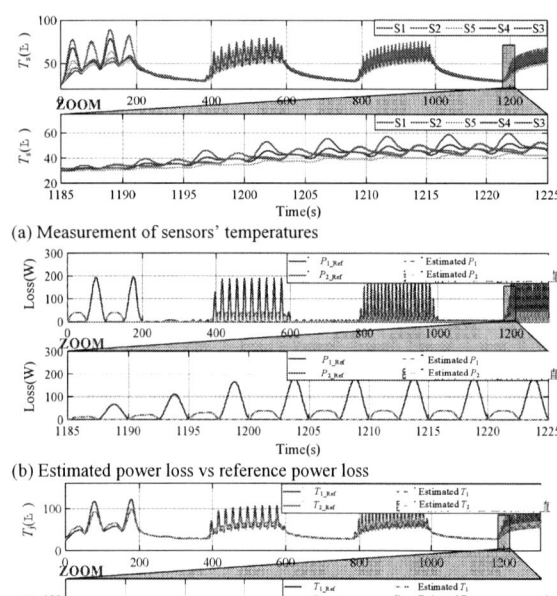

(a) Measurement of sensors' temperatures

(b) Estimated power loss vs reference power loss

(c) Estimated junction temperature vs reference junction temperature

Fig. 13. Simulation results for the proposed technique

C. Experiment evaluation

An alternating current of 56A-0A-127A-0A is fed from a DC source into the upper-bridge-arm Diode. The air velocity at the heatsink inlet is set at 5 m/s. The comparison results between the estimated loss and measured values are presented in Fig. 14.

It is observable that the estimates of the power loss and junction temperatures can effectively track the actual values. The error for the power loss estimate is less than 3.

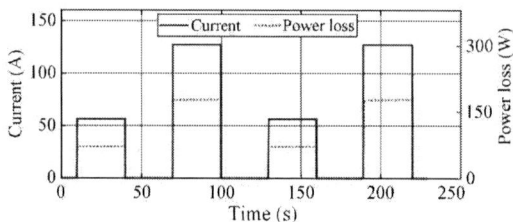

(a) Load current and power loss

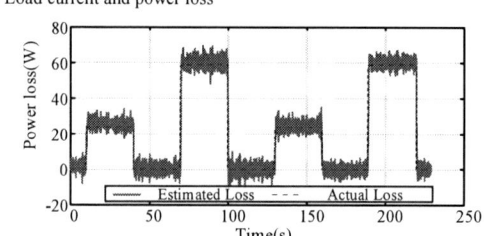

(b) Estimated and measured power loss

Fig. 14. Experiment results

The superiority of the proposed method over the traditional method is shown in TABLE 1.

TABLE I.   COMPARISON BETWEEN THE PROPOSED AND TRADITIONAL METHOD

Method	Depending on the operating conditions	Depending on the junction temperature	Affected by the degradation
Traditional method	Yes	Yes	Yes
Proposed method	No	No	No

IV. CONCLUSION

To enhance the accuracy of power loss estimation, this paper puts forward a brand-new, universal loss observer for IGBT modules in wind power converters. This observer is developed based on the measured temperature. The key work and contributions are summarized as follows:

First, leveraging the sensor network thermal model, the structure and gain of the loss observer are designed. Then, the algorithm and implementation steps of the observer are deduced.

The proposed loss observer has a distinct architecture and operating principle compared to traditional methods. It can accurately estimate the power loss by only measured temperatures.

Simulation and experimental results have verified that the error between the power loss estimated by the proposed technique and the actual values is less than 3W.

Although this paper takes the 2MBI450VN-170 IGBT module as an example to analyze and verify the proposed technology, the underlying principles and implementation procedures can also be applied to other kinds of IGBT modules.

ACKNOWLEDGMENT

The authors would thank the sponsor of this work: Research on the Dynamic Interaction Mechanism and Characteristics of "Wind Turbine-Wind Farm-Power Grid" for Offshore Wind Power Based on Grid-Forming Strategy Grant 52130424000R, funded by the Science and Technology Program of State Grid Fujian Electric Power Co.

REFERENCES

[1] C. H. van der Broeck, L. A. Ruppert, and A. Hinz et al., "Spatial electro-thermal modeling and simulation of power electronic modules," IEEE Transactions on Industry Applications, vol. 54, no. 1, pp. 404–415, January 2018.

[2] K. Ammous, H. Morel and A. Ammous et al., "Analysis of power switching losses accounting probe modeling," IEEE Transactions on Instrumentation and Measurement, vol. 59, no. 12, pp. 3218-3226, December 2010.

[3] X. Yang, S. Xu, and K. Heng et al., "Distributed thermal modeling for power devices and modules with equivalent heat flow path extraction," IEEE Journal of Emerging and Selected Topics in Power Electronics, vol. 11, no. 6, pp. 5863-5876, December 2023.

[4] X. Liu, L. Li, and D. Das et al., "An electro-thermal parametric degradation model of insulated gate bipolar transistor modules," Microelectronics Reliability, vol. 104, pp. 113559, January 2020.

[5] M. Musallam and C. M. Johnson, "Real-time compact thermal models for health management of power electronics," IEEE Transactions on Power Electronics, vol. 25, pp. 1416-1425,June 2010.

[6] A. S. Bahman, K. Ma and F. Blaabjerg et al., "A lumped thermal model including thermal coupling and thermal boundary conditions for high-power IGBT modules," IEEE Transactions on Power Electronics, vol. 33, no. 3, pp. 2518-2530, March 2018.

A method for enhancing the heat dissipation performance of power devices through graphene Coating

Xin Li
School of Electrical and Information Engineering
Anhui University of Science and Technology
Huainan, China
2023201813@aust.edu.cn

Jianing Wang
School of Electrical Engineering and Automation
Hefei University of Technology
Hefei, China
jianingwang@hfut.edu.cn

Shaolin Yu
The Institute of Energy, Hefei Comprehensive National Science Center(Anhui Energy Laboratory)
Hefei, China
yusl@ie.ah.cn

Runze Wang
School of Electrical and Information Engineering
Anhui University of Science and Technology
Huainan, China
2023201831@aust.edu.cn

Xiahao Wang
School of Electrical Engineering and Automation
Hefei University of Technology
Hefei, China
2023170534@mail.hfut.edu.cn

Baolong Yan
School of Electrical and Information Engineering
Anhui University of Science and Technology
Huainan, China
2023201832@aust.edu.cn

Zhenchun Xia
School of Electrical Engineering and Automation
Hefei University of Technology
Hefei, China
2023170526@mail.hfut.edu.cn

Abstract—With the trend of power devices towards higher power density and miniaturization, traditional cooling technologies face dual challenges in efficiency and cost. To address this issue, this paper proposes a graphene coating-based method to enhance radiative heat transfer in power devices. The effectiveness of graphene coating compared to conventional surface treatments was validated through finite element simulation. Comparative experiments were conducted using fabricated copper substrates with graphene coating and uncoated counterparts, where temperature distribution was monitored using thermal imaging and thermocouple measurements under 10A to 35A current loads. Experimental results demonstrate that at 35A operation, the graphene coating reduced the maximum substrate surface temperature by approximately 36°C (160°C without coating vs. 124°C with coating), while thermocouple measurements showed nearly 20°C reduction at substrate edges. Further analysis reveals that the cooling enhancement effect becomes particularly pronounced above 25A, showing significant temperature differentials. This study provides new insights for passive cooling design in high-power devices.

Keywords—Graphene，IGBT，Radiative heat transfer

I. INTRODUCTION

With the miniaturization and increasing power density of electronic devices, thermal management has emerged as a critical challenge affecting operational stability and reliability. As the core functional module of electronic systems, power devices generate substantial Joule heating under operational loads. If not effectively dissipated, this heat accumulation leads to persistent junction temperature rise, resulting in decreased carrier mobility, accelerated material thermal

fatigue, and eventual structural failure [1]. While active cooling solutions like forced air cooling and liquid circulation can partially mitigate these issues, they introduce significant trade-offs in system complexity and energy efficiency [3]. Therefore, developing cost-effective thermal management technologies holds urgent engineering application value.

In recent years, radiative heat transfer has gained interdisciplinary research interest as a passive thermal management solution requiring no external energy input, offering advantages of zero power consumption, compact structure, and high reliability [2]. However, conventional metallic substrates exhibit limited infrared emissivity (typically $\varepsilon<0.3$) in near-ambient temperature ranges, severely constraining their radiative cooling efficiency. The advent of graphene—a novel two-dimensional material—provides theoretical foundations for addressing this limitation: With an ultra-high in-plane thermal conductivity of 5,300 W/(m·K) and tunable surface plasmon resonance characteristics [4], it enables innovative approaches for constructing high-efficiency selective radiators. Research demonstrates that monolayer graphene films prepared via chemical vapor deposition (CVD) enhance substrate emissivity to over 0.85 in the 8-13 μm atmospheric window, significantly improving radiative heat dissipation.

This study investigates the enhancement mechanisms of graphene-functionalized coatings on radiative cooling performance in power devices. Through combined finite element numerical simulations and thermal imaging experiments, we quantitatively analyze how interfacial characteristics affect radiation heat transfer efficiency. By constructing a single-switch IGBT model, experimental results demonstrate that graphene-functionalized coatings

---

This work was supported by Anhui Province Key Research and Development Program Project, Project Number:JZ2024AKKG0057.

effectively enhance surface emissivity. Under 35A operation, coated samples exhibited a 36°C temperature reduction compared to uncoated counterparts after 120s of operation. The study concludes by evaluating the application feasibility of functionalized graphene coatings for power devices, aiming to develop commercial passive cooling solutions compliant with JEDEC reliability standards.

II. PRINCIPLE OF THERMAL RADIATION IN POWER DEVICES

In power devices, radiative heat transfer is the process of energy propagation through electromagnetic waves (independent of a medium), fundamentally governed by Planck's law and Stefan-Boltzmann law:

$$I(\lambda, T) = \frac{2\pi h c^2}{\lambda^5} \frac{1}{e^{\frac{hc}{\lambda k_B T}} - 1} \qquad (1)$$

Key parameters include:

For silicon carbide (SiC) power devices at junction temperatures $(T_j \approx 450\text{K})$, the dominant radiation band spans 3~12 μm. For gallium nitride (GaN) devices in high-temperature regimes $(T > 500\ \text{K})$, the radiation peak shifts toward shorter wavelengths (2~8μm). After applying engineering modifications to the Stefan-Boltzmann law, the actual radiative power at the device surface is expressed as:

$$Q_{rad} = \varepsilon \sigma A (T_s^4 - T_{amb}^4) \qquad (2)$$

Among them, the key parameter reflectivity is affected by three factors: surface morphology (it increases by 20% when the roughness is greater than 1μm), material (the surface of GaAs devices is approximately 0.3, and anodized aluminum oxide is approximately 0.8), and oxide layer (a SiO_2 thickness of 30 nm can cause a fluctuation of ± 0.15).

In practical thermal management of power devices, heat radiation typically operates synergistically with thermal conduction and thermal convection. Heat generated internally transfers via conduction to packaging materials, then dissipates through convection (air/liquid cooling), with a portion radiated to the external environment [4]. For high-power, compact devices, radiative heat transfer has become critical for thermal management. Traditional metal-based heat sinks exhibit low mid-infrared emissivity ($\varepsilon < 0.2$), while conventional dielectric coatings face challenges in spectral selectivity and thermal stability. Graphene offers a breakthrough: its Dirac cone electronic structure enables strong light-matter interactions in the 3~15 μm band, while its atomic-scale thickness minimizes interfacial thermal resistance.

This study employs simulation-driven design to apply differentiated surface treatments to thermal substrates, validating the superior radiative heat transfer enhancement of graphene-coated surfaces in power devices. Prototype single-chip DBC substrates were fabricated, with substrates grouped into bare (uncoated) and graphene-coated samples through reflow soldering. Synchronized comparative experiments were conducted to experimentally confirm the radiative heat transfer enhancement provided by graphene coatings. The

feasibility of applying graphene coatings in practical power device thermal management is thoroughly explored.

III. SIMULATION

To evaluate the enhancement effect of graphene coatings on radiative heat transfer in power devices, this study established a 3D thermal simulation model based on a finite element analysis platform (e.g., ANSYS Icepak). The model framework includes an insulated gate bipolar transistor (IGBT) heat source coupled with a thermal substrate, where substrate materials are pure copper (Cu), aluminum alloy (Al), aluminum silicon carbide (AlSiC), and copper-molybdenum-copper (CMC) composite. Four surface treatment process variables were designed for each substrate material: non-radiative surface, mirror-electroplated polished surface, anodized surface, and graphene-functionalized coating surface, with surface variables configured as shown in Table 1. The controlled variable approach was systematically applied to quantify the impact of different surface treatments on device thermal dissipation performance.

TABLE I.   EXPERIMENTAL PARAMETER DIAGRAM

Cu	No radiation	Plating and polishing	Surface oxidation	Graphene coating
Al	No radiatio	Plating and polishing	Surface oxidation	Graphene coating
AlSiC	No radiatio	Native surface	Graphene coating	———
CMC	No radiatio	Native surface	Graphene coating	———

The simulation boundary conditions were configured as transient thermal conduction (60W thermal dissipation operating for 120 seconds) integrated with radiative heat transfer, replicating practical thermal management scenarios in real-world applications.

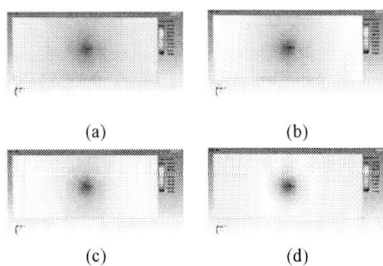

Fig. 1.   (a) Cu non-radiative heat transfer, (b) Al non-radiative heat transfer, (c) AlSiC non-radiative heat transfer, (d) CMC non-radiative heat transfer

As shown in Figure 2(a), simulation data indicate that the polished/oxidized/graphene-coated copper substrate reduced junction temperatures by 0.474°C, 8.674°C, and 13.131°C, respectively, compared to the non-radiative heat transfer condition in Figure 1. For the aluminum substrate, the corresponding temperature reductions were 1.433°C, 23.748°C, and 27.317°C. Meanwhile, composite substrates (AlSiC and CMC) achieved radiative cooling of 23.809°C and 8.697°C, respectively; applying graphene coating further lowered junction temperatures to 30.49°C (AlSiC) and 15.435°C (CMC), demonstrating its efficacy in enhancing radiative heat transfer for power devices.

Further analysis reveals that thermal performance differences stem not only from surface emissivity variations

($\varepsilon = 0.03{\sim}0.91$) but also from substrate thermal conductivity. The aluminum substrate ($k = 237$ W/(m·K)) yielded a 34.6°C higher junction temperature than copper ($k = 398$ W/(m·K)) under identical power density, enabling an additional 14.181°C radiative cooling gain with graphene coating. Variable-power testing (Figure 2(b)) demonstrates a positive correlation between temperature differential (with/without graphene coating) and Heat consumption, confirming graphene's performance advantages in high-temperature applications.

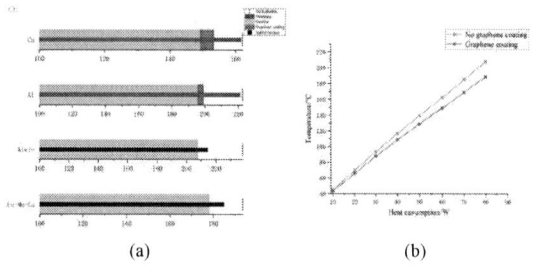

(a)   (b)

Fig. 2. Simulation results

IV. EXPERIMENT

A. Sample preparation

A DBC substrate was designed according to experimental requirements. Through three precision screen-printing trials with thickness gradients (0.10 ± 0.02 mm, 0.12 ± 0.02 mm, 0.15 ± 0.02 mm) and SEM microstructural characterization, the 0.1 mm thickness group demonstrated optimal solder ductility (coverage rate: 98.7%) and minimal porosity ($\leq$ 0.3%). After optimizing the printing process, stepwise reflow soldering (peak temperature: 185°C ± 5°C, duration: 60 s) was performed to achieve metallurgical bonding between the chip and substrate (shear strength $\geq$ 50 N).

Following chip soldering, aluminum wire bonding was conducted using a bonder to guide wires to target pads, with high-frequency ultrasonic vibration and pressure ensuring mechanical connections. Bonding parameters were calibrated to determine optimal performance. Finally, secondary reflow soldering with Sn96.5Ag3.0Cu0.5 solder established reliable electrical terminal-to-substrate connections (X-ray inspection showed void content $\leq$ 5%). The workflow is illustrated in Figure 3.

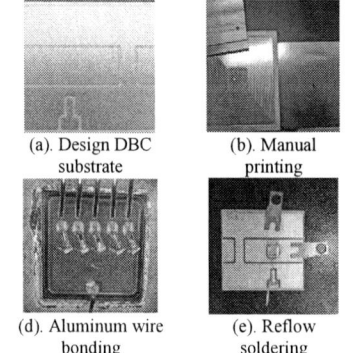

(a). Design DBC substrate   (b). Manual printing   (c). Patch

(d). Aluminum wire bonding   (e). Reflow soldering   (f). Weld the bottom plate

Fig. 3. Flow chart of sample preparation process

B. Static testing

Test project	Test conditions	Range	Measured values	Result
Igesf	Vge=25V, Gears=50nA	−200nA~200nA	0.2858	Qualified
Igesr	Vge=−25V, Gears=50nA	−100nA~100nA	−0.4404	Qualified
Vcesat	Vge=15V Ic=60A	1.6V~2.8V	1.7377	Qualified
Vgth	Vge=Vce Ic=0.25mA	4.5V~6.5V	5.0803	Qualified

(a)

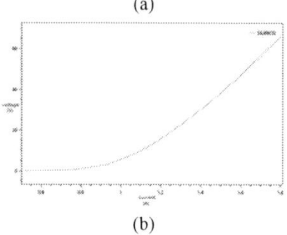

(b)

Fig. 4. (a) Test report of main parameters, (b) Vce-Ic characteristic

Static testing is performed on the fabricated samples to assess the electrical performance and packaging quality of the power devices by measuring static parameters such as voltage, current, on-resistance, and leakage current. This ensures the operational stability, reliability, and long-term performance of the module. The static test results, as shown in Figure 4, meet the specifications outlined in the technical manual.

C. Thermal testing

A thermal testing platform is established by suspending the sample on the power cycling apparatus. Insulating paper is placed at the contact points between the sample and the metal frame of the power cycling apparatus to ensure electrical insulation. A black coating is applied around the IGBT chip to facilitate accurate temperature distribution detection by the thermography camera, while thermocouples are affixed to the substrate to enable real-time temperature monitoring, The structure is shown in Figure 5.

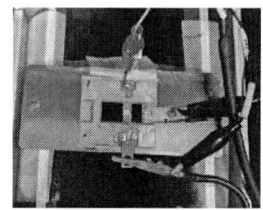

(a)Thermal test platform   (b)Coating of the base plate

Fig. 5. Thermal testing

V. CONCLUSION

(a)   (b)

(c)   (d)

Fig. 6.   (a) Experimental group 1 with graphene coating, (b) Experimental group 2 with graphene coating, (c) Experimental group 1 without graphene coating, (d) Experimental group 2 without graphene coating

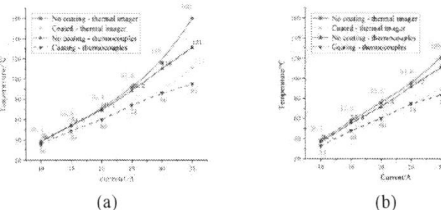

(a)   (b)

Fig. 7.   (a) Experimental temperature rise curves of experimental group 1 under different current intensities, and (b) experimental temperature rise curves of experimental group 2 under different current intensities

To minimize experimental errors, two experimental groups were established and repeated experiments were conducted. The experimental variable was the presence or absence of a graphene coating on the heat dissipation baseplate. The temperature distribution on the substrate surface was monitored using a thermography camera during a 120-second power-on period under varying current intensities, with the highest recorded temperature shown in Figure 6. Specifically, Figures 6(a) and (b) display the highest temperatures detected by the thermography camera on the substrate surface without a graphene coating under 35A current, with both experimental groups showing temperatures of 160° C. Figures 6(c) and (d) show the highest temperatures detected under the same current condition, but with a graphene coating, resulting in recorded temperatures of 124° C and 130° C, respectively. The results demonstrate that the graphene coating significantly enhances the radiation heat transfer capability of the baseplate, leading to a temperature reduction of approximately 30° C compared to the uncoated baseplate, indicating a substantial improvement in heat dissipation performance.

Based on the experimental data,As shown in Figure 7. temperature curves were plotted, where the red and yellow curves represent the temperature measurements obtained by the thermography camera and thermocouples for uncoated substrates under various current intensities, while the blue and green curves represent the corresponding measurements for substrates with graphene coating. From the curve analysis, it is evident that at low current intensities (below 25A), the difference between the graphene-coated and uncoated substrates is not significant. However, within the current range of 25A to 35A, the graphene-coated substrates exhibit lower temperatures, and a clear trend of enhanced heat dissipation with increasing power is observed. This suggests that graphene coatings hold considerable potential for application in high-power devices.

This study validates through modeling and experimental studies that graphene coating significantly enhances radiative heat transfer in power devices. Compared to traditional substrate surface treatments (e.g., polishing/oxidation), graphene-coated devices achieve notable cooling effects via improved radiative heat dissipation. Moreover, the radiative enhancement capability of graphene coatings increases with rising surface temperatures in high-power devices. These findings highlight the potential for applying graphene coatings to passively cooled power devices and those operating in harsh environments without access to external air/water-cooling systems.

REFERENCES

[1] Kuang Sheng,Wei Yu Tang,Zan Wu. Key Technologies of Silicon Carbide Power Module Packaging and Thermal Management [J]. Electric   Traction   Locomotive,2023,(05):1-9.DOI:10.13890/j.issn.1000-128X.2023.05.001.

[2] Moore A L, Shi L. Emerging challenges and materials for thermal management of electronics[J]. Materials Today, 2014, 17(4): 163-174.

[3] SHENG Kuang, GUO Qing, ZHANG Jun-ming, QIAN Zhao-ming. Development and Prospect of SiC Power Devices in Power Grid[J]. Proceedings of the CSEE, 2012, 32(30): 1-7,3.

[4] WANG Laili, ZHAO Cheng, ZHANG Tongyu, et al. Review of Packaging Technology for Silicon Carbide Power Modules[J]. Transactions of China Electrotechnical Society, 2023, 38(18): 4947-4962.

[5] ZHOU Ting. Near-field thermal radiation based on graphene multi-surface coupling[D].Nanchang University,2018.

2025 IEEE Workshop on Wide Bandgap Power Devices and Applications in Asia (WiPDA Asia)

Research on DC Characteristics of β-$(Al_{0.22}Ga_{0.78})_2O_3$/β-$Ga_2O_3$ MODFET

Haitao Zhang
Department of Electronic Engineering
Xi'an University of Technology
Xi'an, China
zhanght0226@163.com

Xiaomin He
Department of Electronic Engineering
Xi'an University of Technology
Xi'an, China
hexiaomin@xaut.edu.cn

Abstract—This research paper elucidates the significant impact of four key parameters—barrier layer thickness ($H_{Barrier}$), gate length (L_g), δ-doping layer thickness, and δ-doping layer concentration—on the transport and output characteristics of β-$(Al_{0.22}Ga_{0.78})_2O_3$/β-$Ga_2O_3$ HEMTs. Increasing the barrier layer thickness ($H_{Barrier}$) gradually decreases the device's threshold voltage, enhances the two-dimensional electron gas (2DEG) concentration in the channel, and improves the output characteristics. Decreasing the gate length (L_g) leads to a gradual decrease in the threshold voltage, an increase in the 2DEG concentration, and an enhancement of the device's output characteristics. The role of the δ-doping layer is to pull the conduction band closer to the Fermi level, thereby deepening the potential well. Increasing the thickness or concentration of the δ-doping layer further increases the 2DEG concentration, while simultaneously lowering the threshold voltage and enhancing the device's output characteristics. Notably, the barrier layer thickness has the most significant impact on the threshold voltage; when the thickness is reduced to 10 nm, the device transitions from depletion mode to enhancement mode. Gate length and δ-doping thickness have a significant impact on transconductance, on-resistance, as well as the maximum current density and carrier concentration. Furthermore, the δ-doping concentration has an important influence on all performance parameters of the device. By rationally adjusting these parameters, precise tuning and optimization of device performance can be achieved. However, due to the coupling between these parameters, caution must be exercised in their selection and adjustment in practical applications to ensure that the device operates at its optimal performance level.

Keywords—β-Ga_2O_3; Threshold Voltage; On-resistance; Transconductance; MODFET

I. INTRODUCTION

In recent years, β-Ga_2O_3 has attracted more attention because of its structural stability and ease of preparation as a single-crystal material[1], [2]. At room temperature, β-Ga_2O_3 boasts a bandgap width of 4.8 eV [3] and a breakdown electric field strength (E_{br}) of 8 MV cm^{-1} [4], which is more than double the theoretical limits of SiC and GaN. With a substantial Baliga figure of merit(2000-3400)and excellent electrical transport properties ($\mu > 200$ cm²/V·s) [5], as well as an electron saturation velocity of 2×10^7 cm/s [6], β-Ga_2O_3 finds widespread applications in the field of high-frequency devices. The modulation-doped field-effect transistor (MODFET) based on the β-$(Al_xGa_{1-x})_2O_3$/Ga_2O_3(AGO/GO) heterojunction is currently one of the β-Ga_2O_3 high-frequency devices that researchers primarily focus on. Since the first successful fabrication of a β-AGO/GO heterojunction MODFET by Ahmadi et al. [7], researchers have further enhanced the performance of β-AGO/GO MODFETs through various methods, including intentional doping with Ge [8], Si δ-doping [9], [10], [11], altering the

buffer layer thickness [12] and concentration [13], adopting a double heterojunction structure [14], changing the spacer layer thickness [15], [16], utilizing an Si_3N_4 passivation layer [17], and employing a P-type GaN gate [18]. Studies have found that the increase in surface charge density of β-AGO/GO is primarily determined by increasing the Al composition or doping concentration. However, an excessively high Al composition (>0.3) can lead to phase separation in β-$(Al_xGa_{1-x})_2O_3$ [19].In this paper, with the Al composition of 0.22, a systematic investigation is conducted into the impact of structural parameters on the direct current (DC) characteristics of AGO/GO MODFET. The findings provide valuable guidance for the design of subsequent devices.

II. METHOD

We utilized Sentaurus to simulate the electrical characteristics of β-$(Al_{0.22}Ga_{0.78})_2O_3$/β-$Ga_2O_3$ MODFET. The device structure, as illustrated in Fig. 1, is composed of the following layers from bottom to top: a semi-insulating β-Ga_2O_3 substrate, a UID (Unintentionally Doped) β-Ga_2O_3 buffer layer, a β-$(Al_{0.22}Ga_{0.78})_2O_3$ spacer layer, a δ-doped layer, and a β-$(Al_{0.22}Ga_{0.78})_2O_3$ barrier layer. The structural parameters are detailed in Table 1.

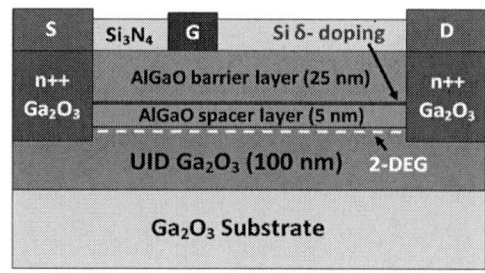

Fig.1.Structure of β-$(Al_{0.22}Ga_{0.78})_2O_3$/β-$Ga_2O_3$

III. RESULTS

This paper primarily investigates the impact of key parameters, including the barrier layer thickness ($H_{Barrier}$), gate length (L_g), δ-doped layer thickness (H_δ), and δ-doped layer concentration (N_δ),on the performance of the β-$(Al_{0.22}Ga_{0.78})_2O_3$/β-$Ga_2O_3$ MODFET device.

A. Barrier Layer Thickness

The influence of the β-$(Al_{0.22}Ga_{0.78})_2O_3$ barrier layer thickness on the electrical characteristics of the β-$(Al_{0.22}Ga_{0.78})_2O_3$/β-$Ga_2O_3$ MODFET device is illustrated in Fig. 2. When the barrier layer thickness increases in 5 nm increments from 10 nm to 30 nm, the transfer characteristic I_{DS}-V_{GS} curves of the β-$(Al_{0.22}Ga_{0.78})_2O_3$/β-$Ga_2O_3$ MODFET device are shown in Fig. 2(a). With the V_{DS}= 10 V and the

979-8-3315-1110-4/25 $31.00 © 2025 IEEE

gate scan voltage ranging from -4 V to 3 V, the threshold voltages (V_{TH}) for barrier layer thicknesses from 10 nm to 30 nm are 0.034 V, -0.37 V, -0.79 V, -1.29 V, and -1.88 V, respectively, exhibiting a negative shift with increasing barrier layer thickness (Fig. 2(b)). Analysis of the conduction band (E_C) (Fig. 2(c)) reveals that an increase in the barrier layer thickness leads to a downward shift of the conduction band at the interface, bringing it closer to the Fermi level. This deepens the electron potential well on the β-Ga$_2$O$_3$ side, increases the two-dimensional electron gas (2DEG) concentration, and thus reduces the threshold voltage. It is also observed that when $H_{Barrier}$ = 10 nm and V_{TH} = 0.034 V, the device is in the off-state at V_g = 0 V, indicating that the Schottky built-in potential under an extremely thin barrier layer can fully deplete the 2DEG, transforming the device from a normally-on type to a normally-off type. Furthermore, the peak transconductance ($g_{m,peak}$) shifts towards negative gate voltage as the barrier layer thickens, and its magnitude decreases from 334.48 mS/mm ($H_{Barrier}$ = 10 nm) to 247.60 mS/mm ($H_{Barrier}$ = 30 nm) (Fig. 2(d)). This is primarily attributed to the weakened gate control capability resulting from the increased spacing between the gate and the channel.

Table 1. Simulation Parameters

$H_{barrier}$	Thickness of β-(Al$_{0.22}$Ga$_{0.78}$)$_2$O$_3$ Barrier	25nm
H_{spacer}	Thickness of β-(Al$_{0.22}$Ga$_{0.78}$)$_2$O$_3$Spacer	5nm
N_{spacer}	Doping Concentration of β-(Al$_{0.22}$Ga$_{0.78}$)$_2$O$_3$ Spacer	1x10^{15}cm^{-3} (n-type)
N_δ	δ-Doping Concentration	5.5x10^{18}cm^{-3}
H_δ	Thickness of δ-Doped	1nm
H_{buffer}	Thickness of β-Ga$_2$O$_3$ Buffer	100nm
N_{buffer}	Doping Concentration of UID-Ga$_2$O$_3$ Buffer	1x10^{15}cm^{-3} (n-type)
N_S (N_D)	Doping Concentration of Source (Drain) Region	1x10^{20}cm^{-3}
L_g	Gate Length	1μm
L_{gs}	Gate-Source Spacing	0.5μm
L_{gd}	Gate-Drain Spacing	3μm

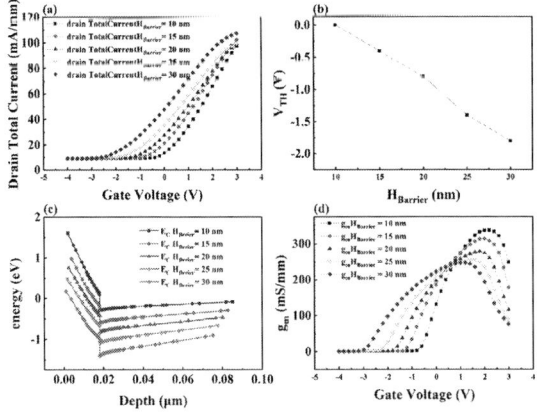

Fig.2.Different $H_{Barrier}$:(a)Transfer characteristics; (b)V_{TH} vs $H_{Barrier}$; (c) Conduction band diagrams; (d) g_m vs V_{GS}

The output characteristics for different β-(Al$_{0.22}$Ga$_{0.78}$)$_2$O$_3$ barrier layer thicknesses under the conditions of gate voltage V_{GS} = 0 V and drain voltage swept from 0 to 10 V are illustrated in Fig. 3(a). Based on the analysis of transfer characteristics, it is observed that when the barrier layer thickness is 10 nm, the device remains non-conductive at a zero gate bias voltage. Hence, this study focuses solely on the effects of barrier layer thicknesses of 15 nm, 20 nm, 25 nm, and 30 nm on the output characteristics. As the β-(Al$_{0.22}$Ga$_{0.78}$)$_2$O$_3$ barrier layer thickness increases from 15 nm to 30 nm, the saturation drain current $I_{DS,max}$ significantly rises from 21.11 mA/mm to 47.70 mA/mm (Fig. 3(b)), while the on-resistance (R_{ON}) decreases from 7.48 Ω·mm to 6.22 Ω·mm (Fig. 3(c)). The 2DEG concentration distribution (Fig. 3(d)) reveals that an increase in the barrier layer thickness leads to a notable rise in the 2DEG concentration at the heterojunction interface, indicating that the driving capability of the device is enhanced with an increase in the barrier layer thickness.

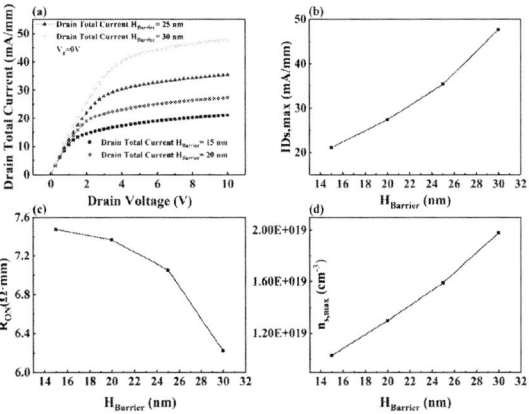

Fig.3. Different $H_{Barrier}$: (a) Output characteristics; (b)Saturation I_{DS} vs $H_{Barrier}$; (c) R_{ON} vs $H_{Barrier}$; (d) 2DEG vs $H_{Barrier}$

B. Gate Length

The gate length L_g is one of the crucial factors influencing device characteristics, and it has a significant impact on the electrical properties of the β-(Al$_{0.22}$Ga$_{0.78}$)$_2$O$_3$/β-Ga$_2$O$_3$ MODFET device. The effect of gate length L_g on the device's electrical properties, with L_g varying from 0.2 μm to 1.0 μm in increments of 0.2 μm, is illustrated in Fig. 4. With the drain voltage = 10 V, the voltage sweep from -4 V to 2 V was applied to the gate to obtain the device's transfer characteristic curve, as shown in Fig. 4(a). By extracting data from the curve, the threshold voltages of the device under different gate lengths L_g were determined to be -2.26 V, -1.83 V, -1.60 V, -1.43 V, and -1.29 V, respectively. As depicted in Fig. 4(c), when the gate length ranges from 0.4 μm to 1 μm, the device's threshold voltage varies between -1.83 V and -1.29 V, showing a relatively small change. However, when the gate length is 0.2 μm, the threshold voltage further decreases to -2.26 V. As the gate length L_g decreases, the threshold voltage of the device shifts negatively, indicating the onset of the short-channel effect in the device. Fig. 4(d) presents the conduction band E_C diagrams of the device under different gate lengths L_g.

Analysis reveals that with the reduction of gate length L_g, the conduction band of the device at the β-(Al$_{0.22}$Ga$_{0.78}$)$_2$O$_3$/β-Ga$_2$O$_3$ interface moves downward and approaches the Fermi level, leading to an increase in the 2DEG concentration in the channel. Fig. 4(b) demonstrates the influence of different gate lengths L_g on the transconductance characteristics of the β-(Al$_{0.22}$Ga$_{0.78}$)$_2$O$_3$/β-Ga$_2$O$_3$ MODFET device. The negative shift of the threshold voltage with decreasing gate length L_g also causes the position of the peak transconductance to shift towards a more negative V_{GS}. Additionally, as the gate length L_g decreases, the drain current increases because a shorter gate length helps reduce the gate resistance, thereby enhancing the current-driving capability. This enhancement in drain current leads to an increase in the magnitude of $g_{m,peak}$ as L_g decreases, rising from 254.98 mS/mm at a gate length of 1 μm to 352.53 mS/mm at 0.2 μm.

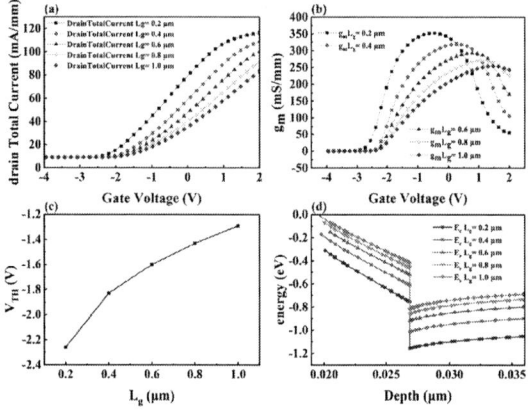

Fig.4. Different L_g :(a) Transfer characteristics; (b) g_m vs V_{GS} ; (c) V_{TH} vs L_g ; (d) Conduction band diagrams.

With gate bias = 0 V, a scanning voltage from 0 to 10 V was applied to the drain, and the output characteristic I_{DS}-V_{DS} curves of the β-(Al$_{0.22}$Ga$_{0.78}$)$_2$O$_3$/β-Ga$_2$O$_3$ MODFET device with gate lengths L_g ranging from 0.2 μm to 1 μm are shown in Fig. 5(a). Fig. 5(b) illustrates the on-state resistance of the device under different gate lengths. As the gate length decreases from 1 μm to 0.2 μm, the on-state resistance decreases from 7.01 Ω·mm to 5.59 Ω·mm. Fig. 5(c) displays the saturation drain current of the device under various gate lengths. As the gate length reduces from 1 μm to 0.2 μm, the saturation drain current increases from 35.42 mA/mm to 79.38 mA/mm. As the gate length decreases, the on-state resistance R_{ON} gradually reduces, while the maximum saturation drain current $I_{DS,max}$ gradually increases. This is because as the gate length shortens, the channel resistance under the gate progressively decreases, and the carrier transport path becomes shorter, leading to an increase in the maximum saturation drain current. Fig. 5(d) shows 2DEG distribution under different gate lengths L_g. As the gate length decreases, the 2DEG concentration rises.

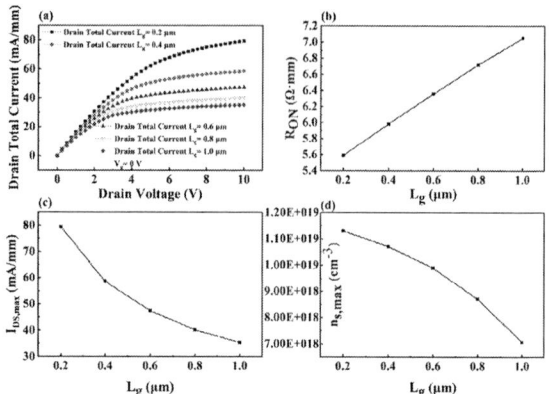

Fig.5.Different L_g: (a) Output characteristics; (b) R_{ON} vs L_g ; (c) Saturation I_{DS} vs L_g; (d) 2DEG concentration vs L_g

C. δ-doped Layer Thickness

This section primarily investigates the impact of the δ-doped layer thickness H_δ on the electrical characteristics of the β-(Al$_{0.22}$Ga$_{0.78}$)$_2$O$_3$/β-Ga$_2$O$_3$ MODFET device. The electrical property results of the device, with the δ-doped layer thickness H_δ varying from 1 nm to 7 nm in increments of 2 nm, are depicted in Fig. 6. With the drain voltage fixed at 10 V, the voltage sweep from -5 V to 3 V was applied to the gate to obtain the device's transfer characteristic curve, as shown in Fig. 6(a). When the δ-doped layer thickness H_δ values are 1 nm, 3 nm, 5 nm, and 7 nm, the corresponding threshold voltages of the device are -1.29 V, -1.88 V, -2.26 V, and -2.66 V, respectively. To better visualize this trend, Fig. 6(b) was plotted, showing that as the δ-doped layer thickness H_δ increases, the threshold voltage of the device decreases approximately linearly. The mechanism behind this change can be reasonably explained by the device's energy band structure, as illustrated in Fig. 6(c). Fig. 6(c) presents the conduction band E_C diagrams of the device under different δ-doped layer thicknesses H_δ. Analysis of this figure indicates that the δ-doped region causes a significant downward shift in the energy band on the β-(Al$_{0.22}$Ga$_{0.78}$)$_2$O$_3$ side. As the δ-doped layer thickness H_δ increases, the effect of the δ-doped region on the energy band becomes more pronounced, with the conduction band of the device moving downward and approaching the Fermi level. Consequently, the electron potential well on the Ga$_2$O$_3$ side at the β-(Al$_{0.22}$Ga$_{0.78}$)$_2$O$_3$/β-Ga$_2$O$_3$ interface deepens. This deepening of the potential well promotes electron accumulation in the channel, thereby increasing the concentration of the 2DEG and leading to a decrease in the threshold voltage V_{TH}. Fig. 6(d) displays the transconductance characteristics under different δ-doped layer thickness. As the δ-doped layer thickness increases, due to the negative shift of the threshold voltage, the position of the peak transconductance also shifts towards a more negative V_{GS}. Additionally, an increase in the δ-doped layer thickness provides more additional electrons at the heterojunction interface, meaning that the gate voltage can more effectively modulate the 2DEG, resulting in a larger peak transconductance. The peak transconductance increases from 254.61 mS/mm at a δ-doped layer thickness of 1 nm to 376.38 mS/mm at 7 nm.

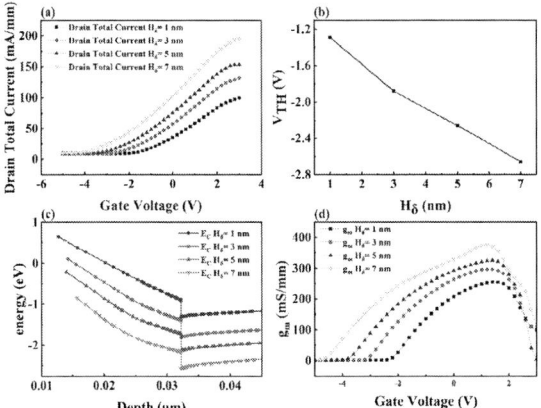

Fig.6.Different H_δ: (a) Transfer characteristics ; (b) V_{TH} vs H_δ ; (c) Conduction band diagrams; (d) g_m vs H_δ.

Under a gate bias of 0 V, a scanning voltage ranging from 0 to 10 V was applied to the drain. The output characteristic I_{DS}-V_{DS} curves of the device, with δ-doped layer thickness H_δ values of 1 nm, 3 nm, 5 nm, and 7 nm, are shown in Fig. 7(a). As the thickness of the δ-doped layer increases, the maximum saturation drain current $I_{DS,max}$ rises from 35.42 mA/mm to 102.47 mA/mm (as depicted in Fig. 7(c)), while the on-state resistance R_{ON} decreases with increasing δ-doped layer thickness, dropping from 7.05 Ω·mm to 3.28 Ω·mm (as shown in Fig 7(b)). Fig. 7(d) illustrates the 2DEG concentration under different δ-doped layer thicknesses. As the thickness of the δ-doped layer increases, the 2DEG concentration rises from 5.97×10^{18} cm^{-3} to 1.79×10^{19} cm^{-3}. This indicates that as the δ-doped layer thickness H_δ increases, the driving capability of the device is enhanced accordingly.

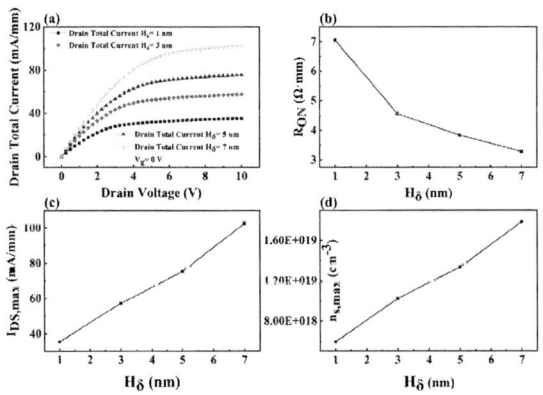

Fig.7.Different H_δ:(a) Output characteristics ; (b) R_{ON} vs H_δ ; (c)Saturation I_{DS} vs H_δ; (d) 2DEG concentration vs H_δ

D. *δ-doped Layer Concentration*

The impact of the doping concentration N_δ in the δ-doped layer on the electrical characteristics of the device is illustrated in Fig. 8. Under a fixed drain voltage of 10 V, a voltage sweep from -10 V to 2 V was applied to the gate to obtain the transfer characteristic curves of the device at five different doping concentrations in the δ-doped layer: 1×10^{18} cm^{-3}, 5×10^{18} cm^{-3}, 1×10^{19} cm^{-3}, 5×10^{19} cm^{-3}, and 1×10^{20} cm^{-3}, as shown in Fig. 8(a). By extracting data from the curves, the threshold voltages of the device under different doping concentrations in the δ-doped layer were determined to be -

1.11 V, -1.28 V, -1.50 V, -2.53 V, and -5.25 V, respectively. As depicted in Fig. 8(b), with the increase in the doping concentration of the δ-doped layer, the threshold voltage of the device gradually decreases. To explain this phenomenon, Fig. 8(c) simulates the energy band diagrams of the device under different doping concentrations in the δ-doped layer. Analysis of the energy band diagrams indicates that as the doping concentration in the δ-doped layer increases, the effect of the δ-doped layer in pulling down the conduction band E_C at the β-$(Al_{0.22}Ga_{0.78})_2O_3$/β-$Ga_2O_3$ interface becomes more pronounced. The conduction band level moves closer to the Fermi level, deepening the electron potential well on the β-Ga_2O_3 side at the β-$(Al_{0.22}Ga_{0.78})_2O_3$/β-$Ga_2O_3$ heterojunction interface and increasing the concentration of the 2DEG in the channel, thereby reducing the threshold voltage V_{TH}. Fig. 8(d) displays the transconductance characteristics under different the δ-doped layer doping concentration. As the doping concentration in the δ-doped layer increases, the position of the peak transconductance also shifts towards a more negative V_{GS}, while the magnitude of $g_{m,peak}$ changes only slightly.

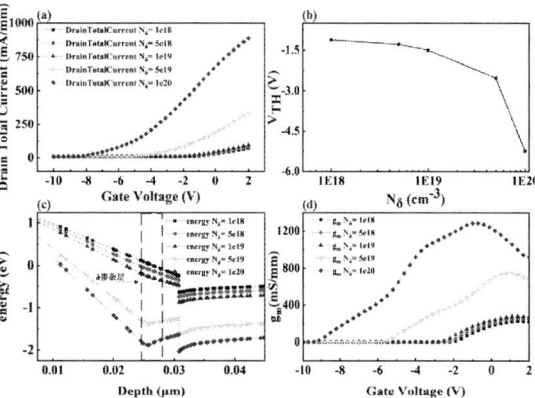

Fig.8. Different N_δ: (a)Transfer characteristics; (b)V_{TH} vs N_δ; (c) Conduction band diagrams; (d)g_m vs V_{GS}

Under a gate bias of 0 V, a scanning voltage ranging from 0 to 10 V was applied to the drain. The output characteristic I_{DS}-V_{DS} curves of the β-$(Al_{0.22}Ga_{0.78})_2O_3$/β-$Ga_2O_3$ MODFET device, with δ-doped layer doping concentrations N_δ of 1×10^{18} cm^{-3}, 5×10^{18} cm^{-3}, 1×10^{19} cm^{-3}, 5×10^{19} cm^{-3}, and 1×10^{20} cm^{-3}, are shown in Fig. 9(a). By extracting data from the curves, the on-state resistance and maximum saturation drain current of the device under different δ-doped layer doping concentrations were obtained. An increase in the δ-doped layer doping concentration led to a higher saturation drain current $I_{DS,max}$, which rose from 27.99 mA/mm to 674.18 mA/mm (as depicted in Fig. 9(c)), while the on-state resistance R_{ON} decreased from 8.16 Ω·mm to 0.67 Ω·mm (as shown in Fig. 9(b)). Fig. 9(d) illustrates the 2DEG concentration under different δ-doped layer doping concentrations N_δ. As the doping concentration in the δ-doped layer increases, the 2DEG concentration rises significantly. The aforementioned research indicates that as the doping concentration in the δ-doped layer increases, the driving capability of the device is enhanced accordingly.

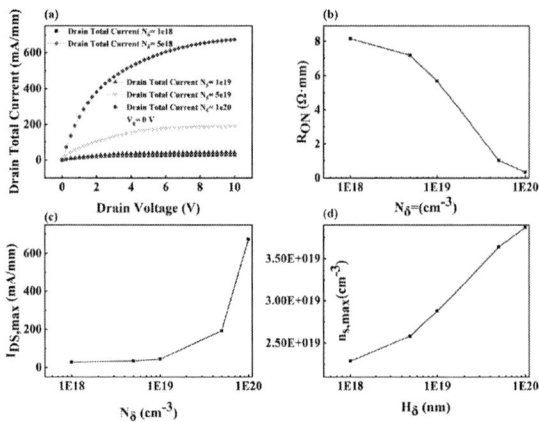

Fig. 9. Different N_δ :(a) Output characteristics vs N_δ ; (b) R_{ON} vs N_δ ; (c) Saturation I_{DS} vs N_δ ; (d) 2DEG concentration vs N_δ

This study delves into the impact of various parameters, including the barrier layer thickness ($H_{Barrier}$), gate length (L_g), and thickness and concentration of the δ-doped layer(H_δ and N_δ), on the direct current (DC) performance of the device.The findings are summarized as follows: When the barr ier layer thickness increases from 10 nm to 30 nm, the threshold voltage V_{TH} shifts negatively from 0.034 V to -1.88 V, the on-resistance (R_{ON}) decreases from 7.48 Ω·mm to 6.22 Ω·mm, and the concentration of the two-dimensional electron gas (2DEG) increases significantly. As the gate length is shortened to 0.2 μm, V_{TH} shifts negatively to -2.26 V, and the peak transconductance $g_{m,peak}$ increases to 352.53 mS/mm; however, the short-channel effect intensifies. When the δ-doping layer thickness increases from 1 nm to 7 nm, RON decreases from 7.05 Ω·mm to 3.28 Ω·mm, and the 2DEG concentration rises to 1.79×10^{19} cm^{-3}. With the δ-doping concentration increasing from 1×10^{18} cm^{-3} to 1×10^{20} cm^{-3}, R_{ON} decreases to 0.67 Ω·mm, and the maximum drain current ($I_{DS,max}$) enhances to 674.18 mA/mm.The research has unveiled the mechanism through which structural parameters influence device performance by regulating band bending and the concentration of 2DEG, providing a theoretical basis for the optimized design of β-Ga$_2$O$_3$-based MODFET.

Acknowledgment

This work was supported financially by the National Natural Science Foundation of China (Grant No. 62474139, 62104190, 61904146). This work was supported financially by Science and technology planning project of Xi'an (No.2023JH-GXRC-0122), This work was supported financially by Natural Science Basic Research Program of Shaanxi (Program No.2024JC-YBQN-0655). This work was supported financially by the Shaanxi Funds for Distinguished Young Scientists (2021JC-25).

References

[1] X. Ma and S. Luan, "Current blocking layer enables enhanced NiO/β-Ga$_2$O$_3$ heterojunction vertical MOSFET with a higher power figure of merit", *Eng. Res. Express*, vol. 7, no. 1, p. 015310, Jan. 2025.

[2] D. Herath Mudiyanselage *et al.*, "β-Ga$_2$O$_3$-Based Heterostructures and Heterojunctions for Power Electronics: A Review of the Recent Advances", *Electronics*, vol. 13, no. 7, Art. no. 7, Jan. 2024.

[3] C. Janowitz *et al.*, "Experimental electronic structure of In$_2$O$_3$ and Ga$_2$O$_3$", *New J. Phys.*, vol. 13, no. 8, p. 085014, Aug. 2011.

[4] M. Higashiwaki, K. Sasaki, A. Kuramata, T. Masui, and S. Yamakoshi, "Gallium oxide (Ga$_2$O$_3$) metal-semiconductor field-effect transistors on single-crystal β- Ga$_2$O$_3$ (010) substrates", *Applied Physics Letters*, vol. 100, no. 1, p. 013504, Jan. 2012.

[5] N. Ma *et al.*, "Intrinsic electron mobility limits in β - Ga$_2$O$_3$", *Applied Physics Letters*, vol. 109, no. 21, p. 212101, Nov. 2016.

[6] A. Kumar and U. Singisetti, "First principles study of thermoelectric properties of β -gallium oxide", *Applied Physics Letters*, vol. 117, no. 26, p. 262104, Dec. 2020.

[7] E. Ahmadi *et al.*, "Demonstration of β-(Al$_x$ Ga$_{1-x}$)$_2$ O$_3$ /β-Ga$_2$O$_3$ modulation doped field-effect transistors with Ge as dopant grown via plasma-assisted molecular beam epitaxy", *Appl. Phys. Express*, vol. 10, no. 7, p. 071101, Jul. 2017.

[8] E. Ahmadi *et al.*, "Ge doping of β-Ga$_2$O$_3$ films grown by plasma-assisted molecular beam epitaxy", *Appl. Phys. Express*, vol. 10, no. 4, p. 041102, Apr. 2017.

[9] D. Wang, D. H. Mudiyanselage, and H. Fu, "Design Space of Delta-Doped β-(Al$_x$ Ga$_{1-x}$)$_2$ O$_3$ /Ga$_2$O$_3$ High-Electron Mobility Transistors", *IEEE Trans. Electron Devices*, vol. 69, no. 1, pp. 69–74, Jan. 2022.

[10] A. Patnaik, M. Gupta, and P. Sharma, "Forward Bias Gate Leakage Mechanism in δ-doped β-(Al$_x$ Ga$_{1-x}$) $_2$ O$_3$ /Ga$_2$ O$_3$ HFET", in *2023 IEEE 33rd International Conference on Microelectronics (MIEL)*, Nis, Serbia: IEEE, Oct. 2023, pp. 1–4.

[11] A. Patnaik, N. K. Jaiswal, and P. Sharma, "Physics-Based Compact I–V Model for δ-Doped β-(Al$_x$ Ga$_{1-x}$)$_2$O$_3$/Ga$_2$O$_3$ HFET Involving Parallel Conduction", *IEEE Trans. Electron Devices*, vol. 70, no. 10, pp. 5242–5248, Oct. 2023.

[12] C. Joishi *et al.*, 'Effect of buffer iron doping on delta-doped β-Ga$_2$O$_3$ metal semiconductor field effect transistors", *Applied Physics Letters*, vol. 113, no. 12, p. 123501, Sep. 2018.

[13] Y. Zhang *et al.*, "MOCVD grown epitaxial β- Ga$_2$O$_3$ thin film with an electron mobility of 176 cm2/V s at room temperature", *APL Materials*, vol. 7, no. 2, p. 022506, Feb. 2019.

[14] Y. Zhang, C. Joishi, Z. Xia, M. Brenner, S. Lodha, and S. Rajan, "Demonstration of β-(Al$_x$Ga$_{1-x}$)$_2$O$_3$/Ga$_2$O$_3$ double heterostructure field effect transistors", *Applied Physics Letters*, vol. 112, no. 23, p. 233503, Jun. 2018.

[15] N. K. Kalarickal *et al.*, "High electron density β -(Al$_{0.17}$Ga$_{0.83}$)$_2$O$_3$/ Ga$_2$O$_3$ modulation doping using an ultra-thin (1 nm) spacer layer", *Journal of Applied Physics*, vol. 127, no. 21, p. 215706, Jun. 2020.

[16] A. Patnaik, Sachchidanand, N. K. Jaiswal, and P. Sharma, "Design considerations to enhance 2D-Electron gas density in δ-doped β-(Al$_x$Ga$_{1-x}$)$_2$O$_3$/ Ga$_2$O$_3$ HFET", *J Mater Sci: Mater Electron*, vol. 34, no. 27, p. 1853, Sep. 2023.

[17] C. N. Saha, A. Vaidya, and U. Singisetti, "Temperature dependent pulsed IV and RF characterization of β -(Al$_x$Ga$_{1-x}$)$_2$O$_3$/Ga$_2$O$_3$ hetero-structure FET with *ex situ* passivation", *Applied Physics Letters*, vol. 120, no. 17, p. 172102, Apr. 2022.

[18] A. D. Meshram, A. Sengupta, T. K. Bhattacharyya, and G. Dutta, "ON- and OFF-State Performance of Normally-OFF β-(Al$_x$Ga$_{1-x}$)$_2$O$_3$/Ga$_2$O$_3$ MODFETs with p-GaN Gate", in *2023 IEEE 20th India Council International Conference (INDICON)*, Dec. 2023, pp. 673–678.

[19] S. W. Kaun, F. Wu, and J. S. Speck, "β-(Al$_x$Ga$_{1-x}$)$_2$O$_3$/Ga$_2$O$_3$ (010) heterostructures grown on β-Ga$_2$O$_3$ (010) substrates by plasma-assisted molecular beam epitaxy", *Journal of Vacuum Science & Technology A*, vol. 33, no. 4, p. 041508, Jun. 2015.

A Structure-Reconfigurable Electronic Transformer For Renewable Energy DC Distribution Syste

Yu Feng
College of Electrical and
Information Engineering
North China University for Nationalities
Yinchuan, China
20237367@stu.nmu.edu.cn

Xianbin Qi*
School of Electrical Engineering
And Automation
Hefei University of Technology
Hefei, China
xianbin_qi@hfut.edu.cn

Zhiqing Yang
School of Electrical Engineering
And Automation
Hefei University of Technology
Hefei, China
zhiqing yang@hfut.edu.cn

Jinxiao Wei
School of Electrical Engineering
And Automation
Hefei University of Technology
Hefei, China
jxwei@hfyt.edu.cn

Peng Qin
School of Electrical Engineering
And Automation
Hefei University of Technology
Hefei, China
pengqin@hfut.edu.cn

Helong Li
School of Electrical Engineering
And Automation
Hefei University of Technology
line Hefei, China
helong.li@hfut.edu.cn

Liu Fang
College of Electrical and
Information Engineering
North China University for Nationalities
Yinchuan, China
fangliu@nmu.edu.cn

Lijian Ding
School of Electrical Engineering
And Automation
Hefei University of Technology
Hefei, China
ljding@hfut.edu.cn

Abstract — This paper focuses on DC Power Electronic Transformer (DC-PET) in renewable energy DC distribution system. According to the characteristics of variable load industrial park, a set of DC-PET which can switch topological structure according to different load conditions is designed to achieve high efficiency in the whole operation process. Through theoretical analysis and simulation verification, it is proved that the designed variable structure power electronic transformer can effectively realize the voltage conversion function, significantly improve the energy conversion efficiency, and provide technical support for the stable and efficient operation of renewable energy DC distribution system.

Keywords—Variable structure power electronic transformer, Extended phase shift modulation, Efficiency optimization

I. INTRODUCTION

With the rapid development of renewable energy, solar energy, wind energy and other large-scale access to the grid, DC distribution system because of its advantages in reducing power conversion links, reducing losses and facilitating distributed power access, has received widespread attention. DC Power Electronic Transformer (DC-PET) is the core equipment to realize voltage conversion and power control in DC distribution system, and its performance directly affects the efficiency and stability of the whole system.

Power electronic transformer uses power electronic device and high frequency transformer to realize voltage conversion. Compared with traditional transformer, power electronic transformer has the advantages of small size, light weight and flexible control. The modular DC-PET topology is shown in Fig.1. The HV side is connected to the medium-voltage DC bus in cascade mode, and the LV side is connected to the low-voltage DC bus to supply power to the park. Among them, the DC-DC part is dominated by topologies such as DAB and LLC, which have the problems of low efficiency and narrow transmission power adjustment range under light load conditions. Considering that the load of industrial parks presents obvious stepped fluctuations, these problems will significantly increase the power loss and affect the stability of the power system.

Based on this problem, many literatures have proposed improved methods, such as multiple phase-shift modulation for DAB converter[1], adaptive inductance, hybrid modulation strategy for LLC converter, etc. These schemes improve the light-load efficiency, but introduce more complex

design process, and are difficult to meet the power supply demand under variable load environment.

To solve these problems, this paper designs a DC power electronic transformer with variable structure, switches different working modes according to different load conditions, and optimizes the modulation strategy for the heavy-duty structure. Through analysis and simulation experiments, it is verified that the proposed topology structure and modulation strategy can realize the high-efficiency operation of DC-PET in the full load range.

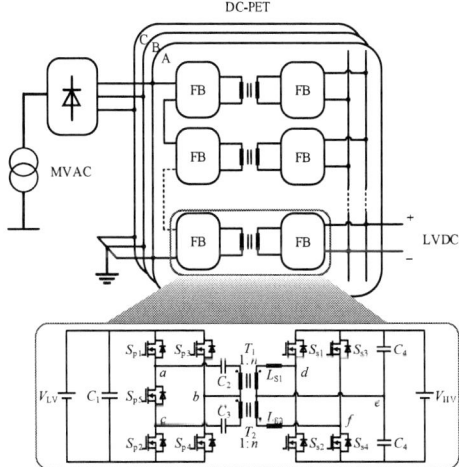

Fig. 1. Modular DC-PET architecture and topology unit.

II. STRUCTURE AND OPERATION PRINCIPLE

A. Topology of the Structure-Reconfigurable DC-EPT

The proposed reconfigurable DC-PET topology is shown in Fig.1. The V_{LV} side is the low voltage side connected to the LVDC, and the V_{HV} side is the high voltage side connected to the medium voltage bus. Inductors L_{S1} and L_{S2} are phase-shifting inductors composed of transformer leakage induction and external series inductors, and T_1 and T_2 are high-frequency transformers with a turn ratio of 1: n. The low-voltage side of the original side of the converter is composed of two half-bridge bridge arms, a half-bridge is composed of three multiplexed switches S_{P1}, S_{P2}, S_{P3}, and the midpoint of each two adjacent switches is connected to the original side of the two transformers, the other half bridge is composed of switches S_{P3}, S_{P4}, and the midpoint is connected to the midpoint of the two transformer cascades. The midpoint of the two-half bridge on the high-voltage side is connected to the secondary side of the transformer, and the cascading midpoint of the transformer secondary side is connected to the midpoint of the capacitor C_4 and C_5.

By configuring the switch to be ON or OFF, the converter can be configured to work in four configurations, as shown in Fig.2. When the voltage on both sides of V_{LV} and V_{HV} converters is fixed, these working structures can obtain different transmission power ranges, which can be divided into low transmission power structure (LTP), medium

transmission power structure (MTP) and high transmission power structure (HTP). In the process of bidirectional power transmission, the flexible switching between various operating structures can be realized by properly configuring the on or off state of the switch, so as to effectively extend the transmission power range of the converter.

The voltage waveforms of transformers in the four operating structures are all square waves. Therefore, a phase-shift modulation strategy can be applied to achieve uniform bidirectional power transmission with different operating structures.

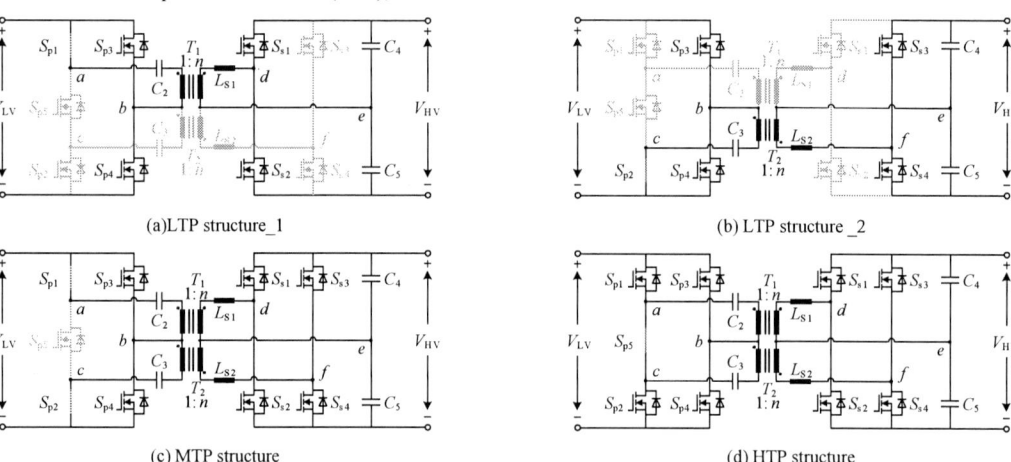

(a)LTP structure_1   (b) LTP structure _2

(c) MTP structure   (d) HTP structure

Fig. 2. Four transmission power structures of reconfigurable power electronic transformer

B. Extended phase shift modulation analysis

If the four structural modulation modes all adopt single phase shift (SPS) modulation mode, the HTP structure has the problem of low operating efficiency [2]. In order to improve the operating efficiency of HTP structure, an extended phase shift (EPS) modulation method is proposed. Compared with SPS modulation, EPS modulation increases the shift of the half-bridge on the primary side and expands the freedom of modulation. The operating state of EPS modulation is analyzed below.

In boost mode, the switching sequence of HTP structure using EPS modulation is shown in Fig.3, and there are 8 switching states in one switching cycle T_S.

The shift ratio between the two half Bridges on the primary side is $D1$, the shift ratio between the full bridge on the primary side and the full bridge on the secondary side is $D2$, and the switch S_{P5} is always on. The other switches have a constant duty cycle of 0.5. Fig.4 shows the modes of the converter under EPS modulation. Only the first four states are plotted because the operating states of each half switching cycle are symmetric.

Based on the switching timing in Fig.3 and the current path in Fig.4, the inductor current value at each moment can be listed, and i_{LS1_boost} is set as the initial current on the inductor L_{S1}. Considering the symmetry of inductor current in the whole period, the current expression of inductor L_{S1} in half a period is presented. Similarly, set i_{LS2_boost} as the initial current on inductor L_{S2}, and the current expression of inductor L_{S2} in half a cycle can be listed by Formula (2).

$$i_{LS1}(t) = \begin{cases} i_{LS1\_boost} + \dfrac{V_{HV}}{2L_{S1}}t \\ \qquad\qquad (t_0 < t < t_1) \\ i_{LS1\_boost} + \dfrac{V_{HV}}{2L_{S1}}D_1T + \dfrac{2nV_{LV}+V_{HV}}{2L_{S1}}(t-t_1) \\ \qquad\qquad (t_1 < t < t_3) \\ i_{LS1\_boost} - \dfrac{nV_{LV}}{L}D_1T + \dfrac{2nV_{LV}+V_{HV}}{2L_{S1}}D_2T + \dfrac{2nV_{LV}-V_{HV}}{2L_{S1}}(t-t_2) \\ \qquad\qquad (t_3 < t < t_4) \end{cases} \quad (1)$$

$$i_{LS2}(t) = \begin{cases} i_{LS2\_boost} - \dfrac{V_{HV}}{2L_{S2}}t \\ \qquad\qquad (t_0 < t < t_1) \\ i_{LS2\_boost} - \dfrac{V_{HV}}{2L_{S2}}D_1T + \dfrac{2nV_{LV}+V_{HV}}{2L_{S2}}(t-t_1) \\ \qquad\qquad (t_1 < t < t_3) \\ i_{LS2\_boost} + \dfrac{nV_{LV}}{L}D_1T - \dfrac{2nV_{LV}+V_{HV}}{2L_{S2}}D_2T - \dfrac{2nV_{LV}-V_{HV}}{2L_{S2}}(t-t_2) \\ \qquad\qquad (t_3 < t < t_4) \end{cases} \quad (2)$$

$$\begin{cases} i_{LS1\_boost} = \dfrac{nV_{LV}D_1 - V_{HV}D_2}{2L_{S1}}T - \dfrac{2nV_{LV}-V_{HV}}{8L_{S1}}T \\ i_{LS2\_boost} = -\dfrac{nV_{LV}D_1 - V_{HV}D_2}{2L_{S1}}T + \dfrac{2nV_{LV}-V_{HV}}{8L_{S1}}T \end{cases} \quad (3)$$

979-8-3315-1110-4/25 $31.00 © 2025 IEEE   446

Combined with the above formula, the initial values of inductors L_{S1} and L_{S2} in boost mode i_{LS1_boost} and i_{Ls2_boost} can be calculated Formula (3).

By integrating the voltage and current of V_{ab} and V_{bc} ports in a switching cycle, the power P_{boost_ab} and P_{boost_bc} transmitted by transformers T_1 and T_2 to the high voltage side can be obtained, The transmission power P_{boost} of the converter during the whole switching cycle by Formula (5).

$$
\begin{cases}
P_{boost\_ab} = \dfrac{1}{T/2}\displaystyle\int_{t_0}^{t_1} nV_{ab}i_{LS1}(t)dt \\[4pt]
\qquad = \dfrac{TV_{HV}V_{LV}n((D_1+0.5)D_2 - 0.5D_1^2 - 0.25D_1 - D_2^2)}{0.5L_{S1}} \\[10pt]
P_{boost\_bc} = \dfrac{1}{T/2}\displaystyle\int_{t_0}^{t_1} nV_{bc}i_{LS2}(t)dt \\[4pt]
\qquad = \dfrac{TV_{HV}V_{LV}n((D_1+0.5)D_2 - 0.5D_1^2 - 0.25D_1 - D_2^2)}{0.5L_{S2}}
\end{cases} \tag{4}
$$

$$
\begin{aligned}
P_{boost} &= P_{boost\_ab} + P_{boost\_bc} \\[4pt]
&= \frac{TV_{HV}V_{LV}n((D_1+0.5)D_2 - 0.5D_1^2 - 0.25D_1 - D_2^2)(L_{S1}+L_{S2})}{L_{S1}L_{S2}}
\end{aligned} \tag{5}
$$

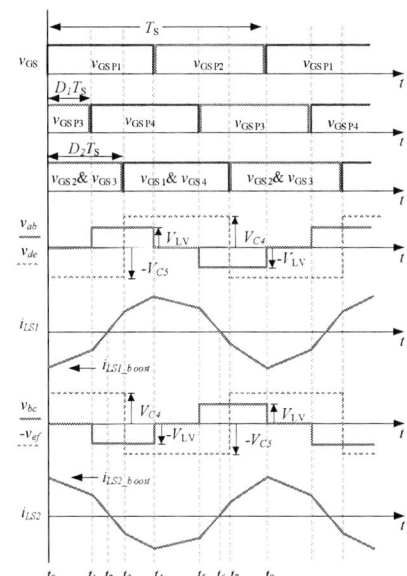

Fig. 3. EPS modulation switch sequence diagram in boost mode

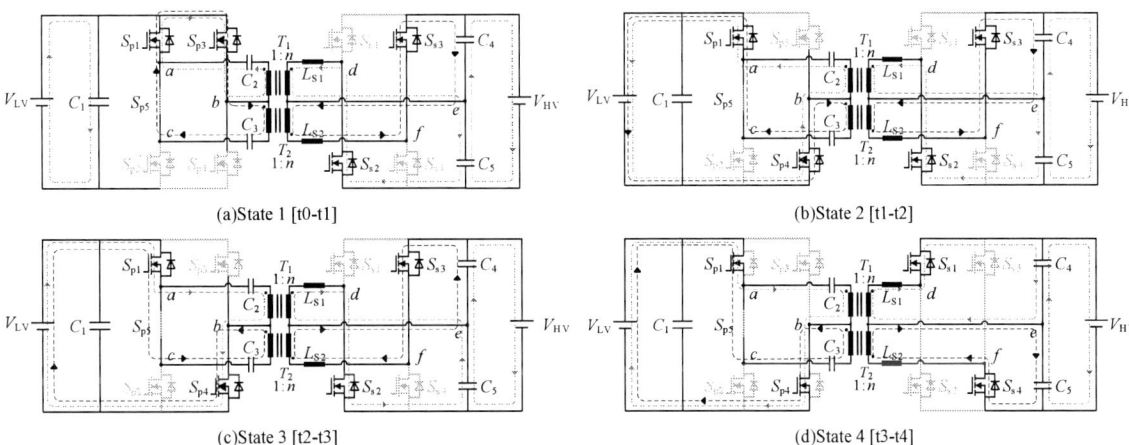

(a)State 1 [t0-t1]   (b)State 2 [t1-t2]

(c)State 3 [t2-t3]   (d)State 4 [t3-t4]

Fig. 4. The HTP structure uses EPS modulated current flow in boost mode

In boost mode, the transmission power of LTP, MTP and LTP structures under SPS modulation can be obtained in the same way, and their comparison is analyzed below.

Let the original edge shift ratio D_1 take a value between 0 and 0.5 to find the boundary of the extended shift ratio D_2 within that interval. The transmission power P_{boost} is normalized to P'. Formula (6) shows the value range of the normalized power P'.

Based on the same method, the normalized power curve of LTP, MTP and HTP (SPS modulation and EPS modulation respectively) relative shift ratio D can be obtained, as shown in Fig.5.

$$
\begin{cases}
P'_{max} = -D^2 + 0.5D + 0.0625 \\
P'_{min} = 0
\end{cases} \tag{6}
$$

It can be seen that the HTP structure has the widest power adjustment range when EPS modulation is adopted in

boost mode. Based on the same principle, the normalized power curve in step-down mode can be obtained, and the HTP structure with EPS modulation is also regulated by the widest power range.

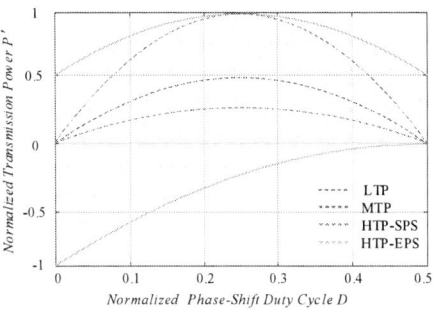

Fig. 5. Power gain curves of the proposedconverter in boost mode.

III. PARAMETER DESIGN

A. Parameter calculation

Based on the above working principle, in order to facilitate parameter calculation, the values of LS1 and LS2 are set to be the same. In order for the converter to work properly under different operating structures, the value of the inductor should be less than the maximum value in its optional range, which can be expressed as:

$$L_{S1} < \frac{TV_{HV}V_{LV}n((D_1+0.5)D_2 - 0.5D_1^2 - 0.25D_1 - D_2^2)}{P_{boost}} \tag{7}$$

In order to facilitate theoretical analysis, the leakage inductance of the transformer is ignored, and the circuit parameters are designed for the conditions mentioned above, as shown in Table I.

TABLE I.   PARAMETERS FOR THE CIRCUIT

Parameter	Value
Low-voltage side V_{LV}	400 V
High-voltage side V_{HV}	600 V
Rated Power P_{rated}	3 kW-8kW
Switching frequency f_S	100k Hz
Transformer T_1 and T_2	1:2
Capacitors C_1	470 μF
Capacitors C_2 and C_3	30 μF
Capacitors C_4 and C_5	470 μF
Inductors L_{S1} and L_{S2}	53 μH

B. ZVS Analysis

The ZVS range is analyzed by taking the operation in boost mode as an example. The ZVS range in buck mode can be derived by the same method.

As shown in Fig.3 and Fig.4, after the shutdown of Sp3 at time t1, the condition for Sp4 to achieve soft switching is that the current iLS1 and iLS2 continue to flow through the body diode during the dead time. Similarly, this is the condition for Sp2 to soft-switch after Sp1 is turned off at t4. It can be expressed as:

$$n(i_{LS1\_boost} + i_{LS2\_boost}) < 0 \tag{8}$$

Considering the convenience of parameter design, take LS1=LS2 and bring it into the formula, the ZVS condition of Sp1-Sp4 on the high voltage side under the boost mode can be deduced as follows:

$$\frac{Tn((4D_1-2)V_{LV}n + (4D_1+1-4D2)V_{HV})}{4L_{S1}} < 0 \tag{9}$$

Similarly, after SS2 and SS3 are turned off at t3 time, the condition for SS1 and SS4 to achieve soft switching is that the current iLS1 and iLS2 continue to flow through the body diode during the dead time. It can be expressed as:

$$\frac{0.5T((D_1-0.5)V_{LV}n + (0.25-D_2)V_{HV})}{L_{S1}} > 0 \tag{10}$$

According to formulas (8), (9) and (10), the ZVS range of the boost mode can be derived. $K=V_{HV}/nV_{LV}$, K=1 and K=2

respectively can be used to obtain the ZVS range of the boost mode as shown in the figure below:

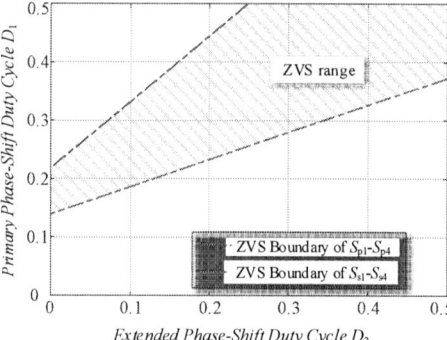

(a) K=1

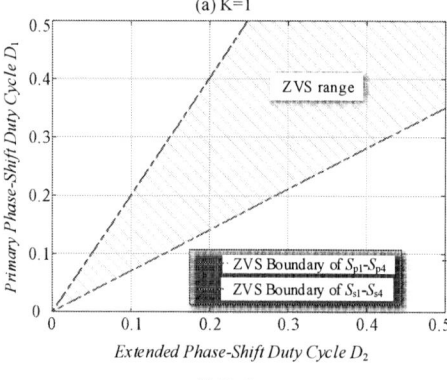

(b) K=2

Fig. 6.   ZVS range in boost mode.

IV. VERIFICATION RESULTS

The circuit model is set up in PLECS according to the parameters shown in Table I. As shown in Fig.8, the transmission power is 3.5kW under MTP structure (a), and the transmission power is 7kW by SPS modulation (b) and EPS modulation (c) under HTP structure. The transmission power characteristics of LTP structure are similar to those of MTP structure. It can be seen that the current stress is significantly reduced by EPS modulation. Based on the thermal model C3M0120090D_utorial series of Wolfspeed Company, efficiency analysis was conducted to compare the operation efficiency of SPS and EPS modulation mode in HTP mode respectively (ignoring the magnetic loss of inductors and transformers), as shown in Fig.7. It can be seen that the EPS modulation strategy significantly improves the operation efficiency of DC-PET in HTP mode.

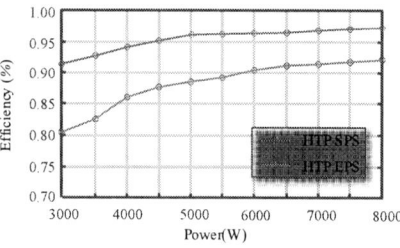

Fig. 7.   SPS and EPS efficiency curves of converter in HTP mode

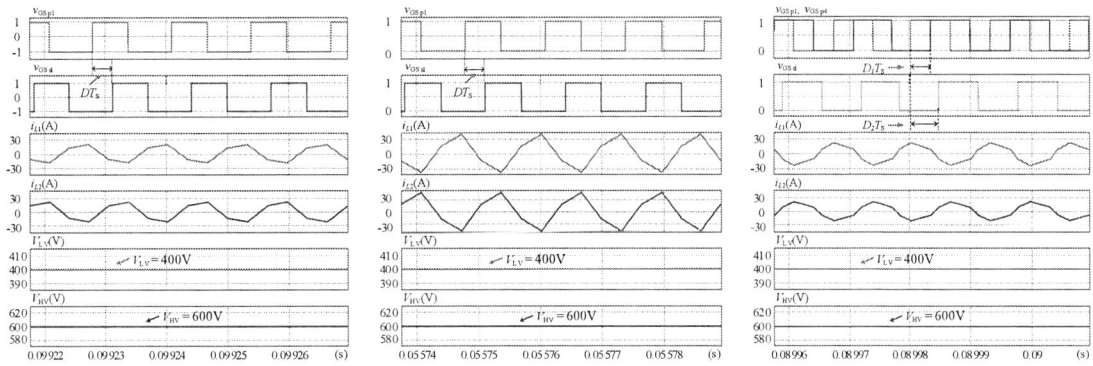

(a) Transmit 3.5kW power in MTP structure (b) Transmit 7kW power in HTP structure (SPS) (c) Transmit 7kW power in HTP structure (EPS)

Fig. 8. Simulation results of the proposed converter in boost mode.

V. CONCLUSION AND NEXT STEPS

In this paper, a reconfigurable variable structure DC-PET topology and its EPS modulation strategy are proposed to reduce the efficiency of power electronic transformer (DC-PET) in renewable energy DC distribution system under light load and high power conditions. Through theoretical analysis and simulation experiments, the high efficiency operation of the full load range is realized, the loss of high frequency switching is reduced, and the effectiveness of the proposed topology and modulation strategy is verified.

ACKNOWLEDGMENT

This work was supported in part by the Anhui Provincial Natural Science Foundation under Grant2408085QE151, in part by the Open Fund of the State Key Laboratory of High-Efficiency and High-Quality Conversion for Electric Power under Grant 2024KF010, and Graduate Innovation Project YCX24325 of North China University for Nationalities.

REFERENCES

[1] B. Zhao, Q. Yu and W. Sun, "Extended-Phase-Shift Control of Isolated Bidirectional DC–DC Converter for Power Distribution in Microgrid," in IEEE Transactions on Power Electronics, vol. 27, no. 11, pp. 4667-4680, Nov. 2012.

[2] X. Qi et al., "A Structure-Reconfigurable Isolated Bidirectional DC–DC Converter for Wide Voltage Gain Range Applications," in IEEE Transactions on Power Electronics, vol. 40, no. 4, pp. 5319-5335, April 2025.

[3] J. Wei, H. Lin, Y. Wang, H. Feng, F. Liang and L. Ran, "Harmonic Routing Between DC-link Capacitor Bank and DC/DC Stage for Reliability Enhancement in the Application of SST," 2023 IEEE 2nd International Power Electronics and Application Symposium (PEAS), Guangzhou, China, 2023, pp. 628-632.

[4] A. Dannier and R. Rizzo, "An overview of Power Electronic Transformer: Control strategies and topologies," International Symposium on Power Electronics Power Electronics, Electrical Drives, Automation and Motion, Sorrento, Italy, 2012, pp. 1552-1557.

[5] H. Fang, W. Wang, F. Jiang and Y. Du, "Back-flow Power Optimization Strategy of DAB Converter with Extended Phase Shift Control," 2024 27th International Conference on Electrical Machines and Systems (ICEMS), Fukuoka, Japan, 2024, pp. 1417-1423.

Pyrolysis Process Analysis of Polyimide at High Temperature Based on Molecular Dynamics Simulation

Yuteng Jiang
School of Electrical Engineering
Shandong University
Jinan, China
202434760@mail.sdu.edu.cn

Minglei Xie[*]
Electric Power Research Institute
CSG
Guangzhou, China
79103643@qq.com

Wenzhi He
Dongguan Power Supply Bureau
Guangdong Power Grid Co., Ltd
Dongguan, China
460478417@qq.com

Zhi Wang
Dongguan Power Supply Bureau
Guangdong Power Grid Co., Ltd
Dongguan, China
553554828@qq.com

Bingxin Chen
Dongguan Power Supply Bureau
Guangdong Power Grid Co., Ltd
Dongguan, China
392737588@qq.com

Zhiyun Han
School of Electrical Engineering
Shandong University
Jinan, China
hanzhiyun@sdu.edu.cn

Sixiao Xin
School of Electrical Engineering
Shandong University
Jinan, China
202420765@mail.sdu.edu.cn

Liang Zou
School of Electrical Engineering
Shandong University
Jinan, China
zouliang@sdu.edu.cn

Abstract—Polyimide(PI), as a high-performance material, has garnered significant attention due to its exceptional high-temperature resistance and insulation properties. and has been widely applied in fields such as electrical insulation, microelectronics, water conservancy, and power. However, PI undergoes pyrolysis at high temperatures, generating gases such as CO_2 and CN-, which may lead to insulation failure and operational safety hazards in electrical equipment. The pyrolysis behavior of PI under high-temperature conditions and its impact on the safe operation of electrical equipment remains a challenge that requires in-depth research. This study aims to reveal the pyrolysis mechanism of PI under high-temperature conditions and its influence on insulation failure through molecular dynamics simulations, thereby providing a deeper understanding of the pyrolysis mechanism of PI under high temperatures. This knowledge will guide the enhancement of the reliability and safety of power equipment.

Keywords—polyimide; molecular dynamics; high temperature; molecular model

I. INTRODUCTION

Polyimide(PI) possesses a distinctive imide ring structure and demonstrates superior thermal resistance, electrical properties, mechanical strength, and chemical stability. These exceptional characteristics have led to its widespread applications across multiple industrial sectors, including electrical insulation, microelectronics, hydroelectric power systems, and mechanical manufacturing[Error! Reference source not found.].

However, during power system operation, a prevalent issue is the thermal degradation of materials caused by equipment heating. In the lifecycle of PI materials, pyrolysis involves the transformation of macromolecular chain chemical structures[Error! Reference source not found.]. Under high-temperature conditions, PI is prone to molecular chain scission, which directly compromises its insulation properties. This degradation phenomenon poses significant challenges to the stable operation of electrical equipment and power grids, potentially leading to serious operational failures.

Current experimental research predominantly remains at the macroscopic level, investigating PI's thermal degradation through observational methods and speculative approaches,

with limited exploration of molecular-level decomposition mechanisms. The pyrolysis of PI under high temperatures constitutes an exceptionally complex chemical process involving numerous intermediates and reaction pathways. Due to experimental limitations, direct observation of the microscopic kinetic processes during PI's thermal decomposition proves particularly challenging, significantly hindering comprehensive understanding of its degradation mechanisms.

Through analysis of molecular dynamics simulation products, this study concludes that the cleavage of C-O-C and C-N bonds is responsible for the reduction in polymer chain degree of polymerization. The research also identifies CO_2 and CN- as primary pyrolysis products and establishes the correlation between CO_2 generation dynamics and temperature. Furthermore, the study elucidates the formation mechanisms of CO_2 and CN-, visually demonstrates specific bond-breaking processes during simulation, and investigates how CO_2 causes destructive effects on electrical equipment.

The paper systematically examines PI's pyrolysis degree, transient processes, and product variations under different temperatures. By integrating bond dissociation analysis with pyrolysis behavior, the study reveals the fundamental degradation mechanisms, thereby providing theoretical guidance for the design and maintenance of power equipment.

II. MODEL CONSTRUCTION AND PYROLYSIS SIMULATION

A. The principles followed in model construction

First, the constructed PI polymer model should exhibit bond lengths and angles consistent with actual materials after molecular relaxation and annealing processes, while achieving the lowest total energy state. Then, the model must incorporate all characteristic structures present in the cured material while satisfying statistical requirements for these structural features. Finally, the model scale must align with available computing resources. For reference, our laboratory servers can complete 90ps molecular dynamics simulations for models containing up to 1,000 atoms within 48 hours.

B. Model Construction Methodology

The molecular structure of the PI monomer is shown in Fig. 1. Considering the structural rationality of the PI, as well as computational cost and accuracy, we selected a PI with a polymerization degree of 10 for molecular dynamics simulations. The chosen polymer fragment was energy-minimized to obtain a reasonable three-dimensional structure. Periodic boundary conditions were applied throughout the simulation. Structural and energy optimization of the resulting periodic box was performed through relaxation and annealing processes. The relaxation process eliminates unreasonable distortions in the constructed box, followed by an annealing treatment of the relaxed model. Structural optimization was considered complete when the total energy deviation of the final few annealing cycles was less than 3%.

Fig. 1. PI monomer molecular structure diagram

C. Molecular Dynamics Simulation of Pyrolysis Process

The simulations were performed using LAMMPS software with force field parameters for C, H, O, and N systems adopted from references**Error! Reference source not found.**. Periodic boundary conditions were applied throughout the simulation. The system was maintained under isothermal-isochoric (NVT) ensemble conditions, with pyrolysis simulations conducted at 2400 K, 2600 K, 2800 K, 3000 K, and 3200 K for 90 ps. Considering the computational intensity of molecular dynamics simulations for large atomic systems, we applied the time-temperature superposition principle**Error! Reference source not found.**. This principle states that equivalent mechanical relaxation phenomena can be observed either at higher temperatures over shorter durations or at lower temperatures over extended periods. Consequently, elevating the simulation temperature accelerates chemical reaction rates and reduces computational time without compromising simulation accuracy.

III. ANALYSIS OF HIGH-TEMPERATURE PYROLYSIS PRODUCTS OF PI

During the initial stage of high-temperature pyrolysis, the atoms within the PI molecule exhibit intensified vibrations due to thermal excitation, causing torsional deformation of the entire molecular chain in three-dimensional space. However, the majority of atoms remain constrained by chemical bonds and cannot dissociate from the PI molecule. Only a minimal quantity of hydrogen and oxygen atoms detach from the molecular chain. The PI molecule maintains its continuous chain structure without fracture, as it has not yet absorbed sufficient thermal energy. The entire molecule remains at a relatively low energy state. A-D below for more information on proofreading, spelling, and grammar.

A. Analysis of Primary Chain-Scission Products at 2800K

With the increase of simulation steps, the initial backbone cleavage event occurs in the polymer chains. To ensure statistical significance and avoid experimental randomness, we conducted multiple trials on all 10 PI molecules until the primary chain-scission products reached a consistent state. The testing was terminated at this equilibrium point. The characteristic products generated from the first backbone cleavage are summarized in Table 1:

Table 1 Primary chain-scission products

Index	Products
1	$C_{44}H_{21}N_5O_{10}+C_{44}H_{21}N_5O_{10}$
2	$C_{45}H_{21}N_5O_{11}+C_{43}H_{21}N_5O_9$
3	$C_{50}H_{25}N_4O_{10}+C_{43}H_{21}N_5O_9$
4	$C_{53}H_{23}N_7O_{13}+C_{35}H_{19}N_3O_7$
5	$C_{54}H_{23}N_7O_{14}+C_{34}H_{19}N_3O_6$
6	$C_{56}H_{29}N_7O_{11}+C_{32}H_{13}N_3O_9$
7	$C_{66}H_{31}N_9O_{15}+C_{22}H_{11}NO_5$
8	$C_{66}H_{31}N_7O_{15}+C_{22}H_{11}N_3O_5$
9	$C_{66}H_{31}N_8O_{16}+C_{22}H_{11}N_2O_4$
10	$C_{75}H_{34}N_8O_{18}+C_{13}H_8N_2O_2$
11	$C_{75}H_{36}N_8O_{19}+C_{13}H_8N_2O$

For PI systems comprising just four elemental components, this relationship specifically focuses analytical attention on the bond energies formed between these constituent elements. According to literature data, the bond energies are quantified as follows: C-C (332 kJ/mol), C=C (611 kJ/mol), C-N (305 kJ/mol), C-O (326 kJ/mol), and C=O (728 kJ/mol). This reveals that the relatively weaker bonds are C-C, C-N, and C-O. Given the exceptional stability of benzene rings, our investigation should primarily focus on three structural features in PI molecules: (1) carbon atoms directly connected to benzene rings, (2) the three C-N bonds at junction regions, and (3) ether linkages bridging two benzene rings.

The widespread distribution of these vulnerable sites throughout the PI structure suggests complex decomposition behavior. The potential decomposition pathways are numerous, further complicated by the polymeric nature of the material, which significantly increases analytical challenges. This theoretical prediction aligns well with current research findings: experimental characterization struggles to identify specific decomposition products, consistent with our observation of over a dozen primary decomposition scenarios.

Through OVITO visualization, we can observe that the fracture of the main chain is primarily attributed to two factors. The first is the cleavage of the ether bond (C-O-C) linking the two benzene rings, and the second is the sequential breakage of the two C-N bonds in the imide ring, leading to the decomposition of the two cyclic structures into separate fragments. The scenario of the ether bond breaking is illustrated in Fig. 2, while the breakage of the C-N bonds is depicted in Fig. 3.

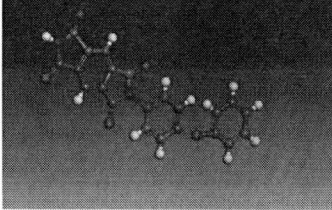

Fig. 2. Fracture occurs at the ether bond of the PI molecule

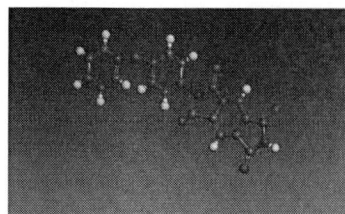

Fig. 3. Fracture occurs at the ether bond of the PI molecule

Based on the different types of bond cleavage, we can classify the 13 initial product formation scenarios according to their respective chemical bond breakage patterns, with the results summarized in Table 2.

Table 2 Correlation between products and their bond cleavage types

Index	Products	Types of chemical bond cleavage
1	$C_{44}H_{21}N_5O_{10}+C_{44}H_{21}N_5O_{10}$	C-N
2	$C_{45}H_{21}N_5O_{11}+C_{43}H_{21}N_5O_9$	C-O-C
3	$C_{50}H_{25}N_4O_{10}+C_{43}H_{21}N_5O_9$	C-O-C
4	$C_{53}H_{23}N_7O_{13}+C_{35}H_{19}N_3O_7$	C-O-C
5	$C_{54}H_{23}N_7O_{14}+C_{34}H_{19}N_3O_6$	C-N
6	$C_{56}H_{29}N_7O_{11}+C_{32}H_{13}N_3O_9$	C-N
7	$C_{66}H_{31}N_9O_{15}+C_{22}H_{11}NO_5$	C-N
8	$C_{66}H_{31}N_7O_{15}+C_{22}H_{11}N_3O_5$	C-N
9	$C_{66}H_{31}N_8O_{16}+C_{22}H_{11}N_2O_4$	C-N
10	$C_{75}H_{34}N_8O_{18}+C_{13}H_8N_2O_2$	C-O-C
11	$C_{75}H_{36}N_8O_{19}+C_{13}H_8N_2O$	C-O-C

We can conclude that during the initial reaction stage, the types of chemical bond cleavage are relatively consistent: one is the breakage of the C-N bond in the imide ring, and the other is the cleavage of the ether bond (C-O-C) linking the two benzene rings. Among these, C-O-C cleavage occurs slightly less frequently, accounting for about 45% of the total cases, while C-N bond breakage is slightly more prevalent, making up approximately 55%.

However, although the types of bond cleavage in PI are well-defined, their specific locations are not. This leads to highly diverse reaction products—up to over a dozen possible outcomes—and this is only during the initial stage, where the main chain undergoes a single cleavage. As the reaction progresses, the products become even more complex, and the variety of bond cleavage types increases. This makes it impractical to study intermediate products using traditional chemical methods, i.e., experimental approaches alone.

Additionally, through repeated experiments, we observed that C-N bond cleavage occurs more frequently than C-O-C bond breakage. Combined with the fact that C-N bond cleavage corresponds to a higher number of product variations, we can conclude that C-N bond cleavage is the primary driver of PI degradation in the early stages, leading to a reduction in its degree of polymerization..

B. Pyrolysis Product Analysis of PI at 2800K within 90ps

We first conducted experiments at 2800 K to investigate the pyrolysis products of PI molecules and the dynamic evolution of product formation under this temperature condition. In the following context, we use Cx+ to denote products with more than x carbon atoms, and Cx to represent products containing x carbon atoms. For example, C9

indicates a product containing 9 carbon atoms. The temporal evolution of C10+ molecular species is shown in the figure. 4.

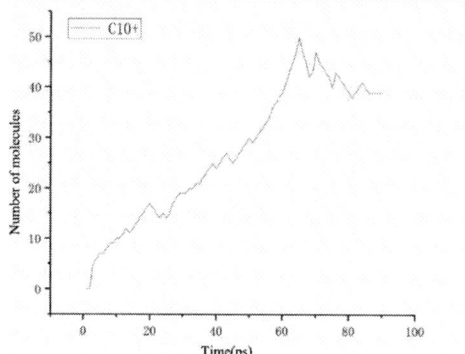

Fig. 4. Variation in the number of C10+ molecules over time.

From the above images, we can draw the following conclusions:

One is that C10+ molecules begin to form around 2 ps. Since the initial pyrolysis products of PI at 2800 K are almost exclusively fragments with main chains containing more than 10 carbon atoms, we can infer that the scission of the PI main chain and the reduction in polymerization degree occur around 2 ps at 2800 K. This suggests that, because the C-N bond in the imide ring must break twice before main chain cleavage occurs, the first C-N bond rupture in the imide ring likely takes place earlier than 2 ps.

The other is the number of C10+ molecules exhibits an inflection point—initially increasing overall, then decreasing, and finally stabilizing. In the initial phase, large molecular products (particularly those with main chains of 20+ or even 30+ carbon atoms) dominate the system, while mid-sized fragments (C10-C20) are relatively scarce. Although C10+ molecules undergo decomposition, their generation rate from the breakdown of larger precursors (C20+) exceeds their consumption rate, resulting in a net increase in C10+ population. During the later phase, the reservoir of large precursors (C20+) becomes depleted. With no new source of mid-sized fragments, the system enters a consumption-dominant regime where existing C10+ molecules progressively decompose into smaller fragments (typically C1-C9). This leads to a gradual decline in C10+ concentration until it stabilizes at equilibrium. We plotted all the data in a single graph, as shown in the figure. 5.

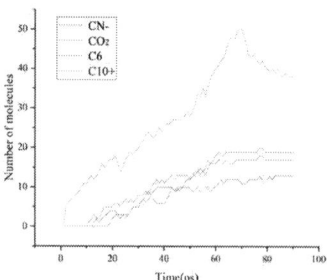

Fig. 5. The variation in the quantities of the four molecules over time

The final equilibrium quantity of C10+ molecules is significantly greater than the other three species. Moreover, the formation pattern of C10+ molecules differs from the others, exhibiting a non-monotonic trend rather than a simple monotonic increase.

Through the transient process of the products CO_2 and CN-, we can observe that the generation of CO_2 and CN- is very limited in the early stages. This is because the degree of main chain cleavage is still insufficient, and the products formed are primarily large molecules with main chains containing more than 10 carbon atoms. As a result, the growth of these three types of substances is very slow initially. However, after 10 ps, CO_2 and CN- begin to enter an orderly growth phase. Around 12 ps, CO_2 starts to precipitate. Both substances exhibit relatively rapid growth in the early to middle stages, then stabilize in the middle to later stages. Ultimately, the quantity of CO_2 exceeds that of CN-.

From the transient process of the product C6, we can deduce that the cleavage of the main chain is not a gradual progression from the head to the tail. Instead, a large molecule first breaks into relatively large fragments. Since the main chain contains multiple weak points prone to cleavage, these larger fragments remain unstable. They then gradually undergo further fragmentation into smaller subunits, eventually breaking down into small molecular units.

In addition to these primary products, some other molecules were also generated during the reaction, including products such as water and hydrogen gas. Fig. 6 illustrates the proportion of each product in the overall reaction products.

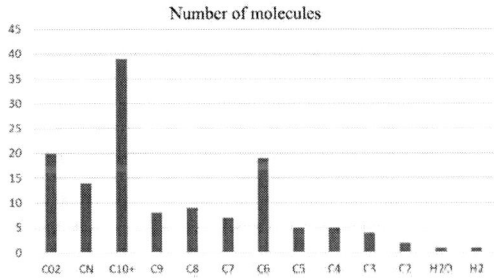

Fig. 6. The quantity of each product from the pyrolysis of PI

It can be observed that CO_2 and CN- are the primary decomposition products of PI, with small amounts of H_2O and H_2 also being generated. This aligns with the types of final products measured by H. Hatori and other researchers[Error! Reference source not found.]. Although there are some quantitative discrepancies, the differences remain within an experimentally acceptable range. These minor variations can likely be attributed to the overly idealized and confined conditions of the simulation environment.

C. Variations in PI Pyrolysis Products at Different High Temperatures

We conducted simulations with 100,000 steps at four different temperatures (2400 K, 2600 K, 3000 K, and 3200 K) and compared the resulting products. Figure 7 illustrates the production of CO_2 and the corresponding molecular quantities at each temperature.

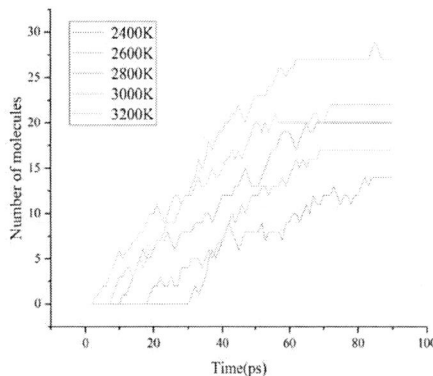

Fig. 7. CO_2 Production at Different Temperatures

From the above figure, we can draw the following conclusions:

First, regarding the onset time of CO_2 generation, the longest duration (nearly 30 ps) was observed at 2400 K, while the shortest (approximately 5 ps) occurred at 3200 K. This is because higher temperatures intensify molecular motion, allowing PI molecules to absorb more thermal energy from the heat source. Consequently, chemical bonds become more prone to destabilization and cleavage, leading to faster CO_2 release.

Second, contrary to expectations, the final yield of CO_2 did not exhibit a strictly positive correlation with temperature. Surprisingly, the CO_2 production at 3000 K exceeded that at 3200 K. This phenomenon arises because reaction selectivity depends not only on temperature but also on molecular structure. At lower temperatures, only one C-N bond in the PI's imide ring tends to break, limiting CN- formation and diverting more precursors toward CO_2 production. In contrast, at higher temperatures, both C-N bonds in the imide ring are readily cleaved, which suppresses CO_2 formation and favors CN- generation. Thus, CO_2 yield does not monotonically increase with temperature but instead peaks at an optimal value—a maximum point dictated by reaction selectivity.

Third, across all temperatures, CO_2 generation followed a consistent trend: slow initial growth, rapid acceleration in the early-middle stage, and eventual stabilization. Higher temperatures accelerated energy accumulation, significantly reducing the time required to reach quasi-equilibrium.

IV. CHEMICAL REACTION MECHANISM OF PI PYROLYSIS AT HIGH TEMPERATURES

A. Mechanism of CN- Formation in Pyrolysis Products

From the above conclusion, it is not difficult to see that the formation of CN- does not accumulate from the very beginning. Therefore, the chemical reaction process of CN- must undergo a relatively complex chemical process, involving multiple instances of bond formation and breaking as well as changes in rings.

Through visual analysis of the reaction process, we have determined the formation mechanism of CN-, which primarily consists of four steps:

Step 1: Before the main chain's degree of polymerization begins to decrease, the C-N bond in the imide ring breaks first. At this stage, the PI molecule has not yet absorbed sufficient thermal energy to fully cleave both C-N bonds. Under initial high-temperature conditions, when one C-N bond breaks, the oxygen atom connected to the carbonyl group and the original nitrogen atom each retain a free electron.

Step 2: These free electrons further enhance the electronegativity of the nitrogen atom, causing electron withdrawal from the second C-N bond due to interaction with the benzene ring. This ultimately promotes the cleavage of the second C-N bond.

Step 3: As the reaction progresses, the nitrogen atom, now carrying two free electrons, exhibits strong electronegativity and electron-attracting ability. This disrupts the π-conjugated system of the benzene ring, leading to rearrangement. The benzene ring undergoes decarbonization, forming a five-membered ring. Meanwhile, the detached carbon atom, bearing a free electron, combines with one of the nitrogen atom's free electrons to form a C=N bond. This carbon-nitrogen bond retains two unpaired electrons with strong electronegativity, providing the necessary electrons for subsequent reactions while destabilizing the ether linkage.

Step 4: Under high-temperature conditions, the five-membered ring—unlike the benzene ring—lacks a stable π-conjugated system. As a result, the carbon atom attached to it becomes unstable. Influenced by the strong electronegativity of the nitrogen atom and its lone pair, the carbon atom (bearing a lone pair), along with the rightmost carbon and nitrogen atoms, detaches from the unsaturated five-membered ring, forming an independent structure—CN-.

The four-step reaction process is illustrated in Fig. 8

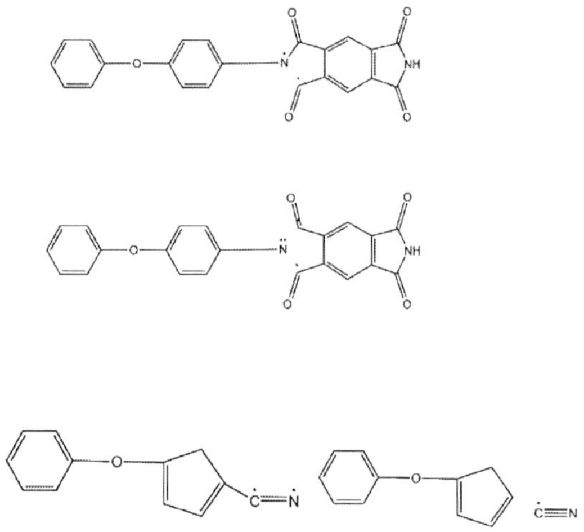

Fig. 8. Formation Process of CN-

B. Mechanism of CO_2 Formation in Pyrolysis Products

Through visual analysis of the reaction process, we have elucidated the formation mechanism of carbon dioxide (CO_2), which primarily consists of four distinct steps:

Step 1: Similar to the aforementioned process, the initial step involves the cleavage of the C-N bond in the imide ring. At this stage, the molecular chain remains intact without a reduction in polymerization degree. The oxygen atom connected to the carbonyl group and the original nitrogen atom each retain a free electron.

Step 2: Diverging from the previous pathway, instead of a second C-N bond cleavage in the imide ring, the C=O bond

adjacent to the nitrogen undergoes rupture. The liberated oxygen atom possesses one free electron, while the carbon atom from the broken bond also retains a free electron, enabling their recombination to form a C-O bond.

Step 3: The transformed C-O bond subsequently undergoes further cleavage. The carbon atom gains an additional free electron and forms a C=N bond with the nitrogen (bearing one free electron). Concurrently, this process yields a benzene ring derivative containing a carboxyl group.

Step 4: The emergence of a carboxyl group on the benzene ring introduces a reactive site. Although phenolic hydroxyl groups typically exhibit limited reactivity, elevated temperatures destabilize the structure, prompting decarboxylation. The entire carboxyl group detaches from the benzene ring while retaining a lone pair of electrons. Simultaneously, the oxygen atom also carries a lone pair, facilitating the formation of a C=O bond, ultimately generating CO_2.

The four-step reaction process is illustrated in Fig. 9

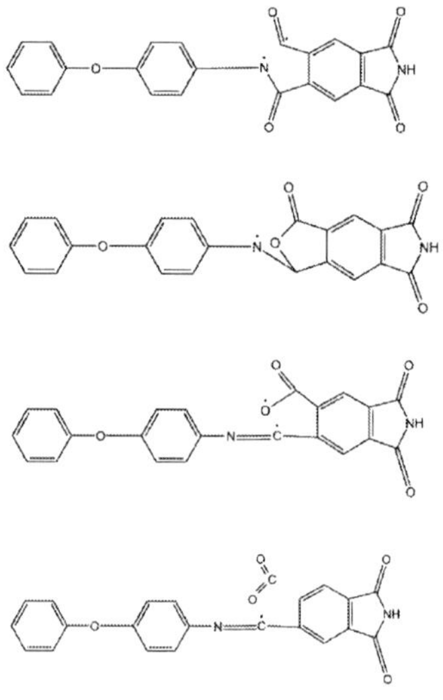

Fig. 9. Formation Process of CO_2

V. CONCLUSION

This paper employs reactive molecular dynamics simulations on PI molecular models to reveal its high-temperature pyrolysis mechanism at the atomic level and further explores the microscopic process of its insulation failure.

The simulation results indicate that the primary bond cleavage in PI molecules under high temperatures occurs at the C-N bonds on the imide ring. This finding highlights the instability of the C-N bonds on the imide ring at elevated temperatures, serving as a critical starting point for PI molecular decomposition. Further analysis reveals that the breakage of the C-N bonds connecting the two benzene rings is the main reason for the initiation of PI backbone fracture.

Our findings indicate that CO_2 and cyanide CN- are the primary products of high-temperature pyrolysis in PI molecules. At lower temperatures, only one of the C-N bonds in the imide ring of PI can break, limiting its ability to produce CN-, with more of the reaction favoring CO_2 generation. In contrast, at higher temperatures, both C-N bonds in the imide ring are easily broken, which hinders CO_2 formation and shifts the chemical reaction toward CN-production. This demonstrates that higher temperatures do not necessarily lead to a greater final yield of CO_2.

Moreover, the safe operation of electrical equipment is primarily affected by two factors related to PI pyrolysis: the CO_2 produced can alter the distribution of electric field strength, leading to partial breakdown, and the generation of flammable and explosive hydrogen gas.

In Conclusion, in practical operations, it is essential to properly manage these two gases resulting from high-temperature PI pyrolysis to ensure the safety of electrical equipment and power systems.

REFERENCES

[1] Xu Shize, Zhang Yuwen, Zhang Tao, et al. Thermodynamic properties of siloxane-functionalized polyimide via molecular dynamics simulation [J]. Plastics Science and Technology, 2022, 50(10): 74-79.

[2] Pu Chuanzhi. Molecular dynamics study on pyrolysis and thermal imidization processes of polyimide [D]. Beijing University of Chemical Technology, 2023.

[3] Van Duin AC, Zeiri Y, Dubnikova F, Kosloff R, Goddard WA 3rd. Atomistic-scale simulations of the initial chemical events in the thermal initiation of triacetonetriperoxide. [J]. Journal of the American Chemical Society,2005,127(31):11053-62.

[4] Zhang L, Zybin SV, van Duin AC, Dasgupta S, Goddard WA 3rd, Kober EM. Carbon cluster formation during thermal decomposition of octahydro-1,3,5,7-tetranitro-1,3,5,7-tetrazocine and 1,3,5-triamino-2,4,6-trinitrobenzene high explosives from ReaxFF reactive molecular dynamics simulations [J]. The journal of physical chemistry. A,2009,113(40):10619-40..

[5] Liao Ruijin, Sun Huigang, Gong Jing, et al. Kinetic model of aging and life prediction for transformer oil-paper insulation [J]. High Voltage Engineering, 2011, 37(7): 1576-1583.

[6] Hatori H, Yamada Y, Shiraishi M, et al. The mechanism of polyimide pyrolysis in the early stage[J]. Carbon, 1996, 34(2): 201-208.

[7] Goodenough J B, Hamnett A, Kennedy B J, et al. XPS investigation of platinumized carbon electrodes for the direct methanol air fuel cell[J]. Electrochimica Acta, 1987, 32(8):1233-1238.

[8] Pramoda K P, Chung T S, Liu S L, et al. Characterization and thermal degradation of polyimide and polyamide liquid crystalline polymers[J]. Polymer Degradation and Stability, 2000, 67(2): 365-374.

Analysis of SiC Output Capacitance Effects on Soft-Switching Characteristics in Three-Level Resonant Converters

Zhe Shao
Department of Electrical Engineering
Harbin Institute of Technology
Harbin, China
shaozhe@stu.hit.edu.cn

Zhiyuan Wang
Department of Electrical Engineering
Harbin Institute of Technology
Harbin, China
wangzhiyuan@stu.hit.edu.cn

Binbin Li
Department of Electrical Engineering
Harbin Institute of Technology
Harbin, China
libinbin@hit.edu.cn

Abstract—To address the high power density and high efficiency requirements of isolated DC/DC converters in novel power load systems, this paper investigates the soft-switching characteristics of a silicon carbide SiC-based three-level resonant converter. Through modal analysis, the influence mechanism of power device output capacitance parameters on soft-switching behavior is revealed. By comparing the impact of the presence and absence of secondary-side switch output capacitance on the soft-switching of primary-side switches, it is found that the magnetizing current required for soft-switching increases when considering the output capacitance, leading to higher conduction losses. The correctness of the magnetizing current calculation is validated through simulations. Experimental verification is conducted on a 2300V DC input / 900V DC output / 25kW prototype, demonstrating the effectiveness of the proposed design.

Keywords—high power density, three-level NPC, silicon carbide MOSFET, soft-switching

I. INTRODUCTION

The rapid expansion of emerging loads such as electric vehicle charging stations, data centers, and electrolytic hydrogen production systems has led to a sharp increase in power demand and growing concerns over high energy consumption [1]. These systems predominantly adopt medium-voltage AC to low-voltage DC rectification architectures, with isolated DC/DC converters serving as the core energy conversion units [2]-[4].

To enhance the power density and efficiency of DC/DC converters, resonant converter topologies and wide-bandgap semiconductor technologies have emerged as key research frontiers. Silicon carbide (SiC) power devices, representative of third-generation WBG semiconductors, exhibit significant advantages in high-power power electronics applications. Their low on-resistance and rapid switching characteristics provide critical support for high-frequency power electronic system designs [5].

However, during resonant converter operation, the interaction between the junction capacitance of SiC devices and resonant inductors induces high-frequency oscillations, posing challenges to achieving zero-voltage switching conditions. Existing studies reveal limitations: Literature [6] proposes optimizing resonant inductance parameters to adjust equivalent parasitic capacitance charging behavior, but this

This work was supported in part by the National Key Research and Development Program of China under Grant 2023YFB4204700 and in part by the Delta Power Electronics Science and Education Development Program of Delta Group.

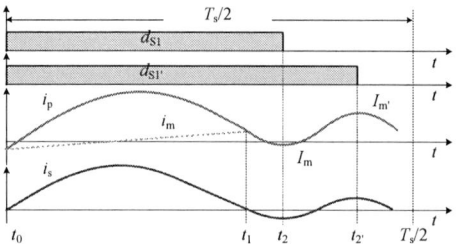

Fig. 1. Schematic diagram of the impact of parasitic capacitance on soft-switching

approach causes secondary-side junction capacitance voltage fluctuations to exceed the output voltage, thereby increasing device voltage stress. While Literature [7],[8] analyzes the transient behavior of junction capacitance during switching transitions and proposes a two-level LLC design methodology, they lack systematic analysis of three-level resonant topologies with higher voltage adaptability. Consequently, this paper focuses on three-level resonant converter topologies, employing operational mode analysis to investigate the impact mechanism of power device junction capacitance parameters on soft-switching characteristics.

II. IMPACT OF SWITCHING DEVICE OUTPUT CAPACITANCE ON SOFT-SWITCHING

During the operation of DC/DC converters, when the resonant frequency exceeds the switching frequency, the resonant current naturally crosses zero after the resonant phase. Subsequently, the switching devices undergo turn-off after a time interval Δt. The magnitude of the turn-off current at this critical moment determines the charge-discharge current characteristics of the junction capacitance in the primary-side switching devices. If the turn-off current amplitude is insufficient, the junction capacitances of both primary and secondary-side switching devices cannot complete energy exchange within the dead time, leading to Zero- Voltage Switching (ZVS) failure. Conversely, if the turn-off current amplitude is excessive, the Quasi-Zero Current Switching (QZCS) condition is violated, significantly increasing switching losses.

Under ideal conditions, the turn-off current is assumed to equal the peak magnetizing current, which is primarily governed by the converter's input-output voltage, switching frequency, and magnetizing inductance (L_m). However, in practical operation, after the resonant current crosses zero, the

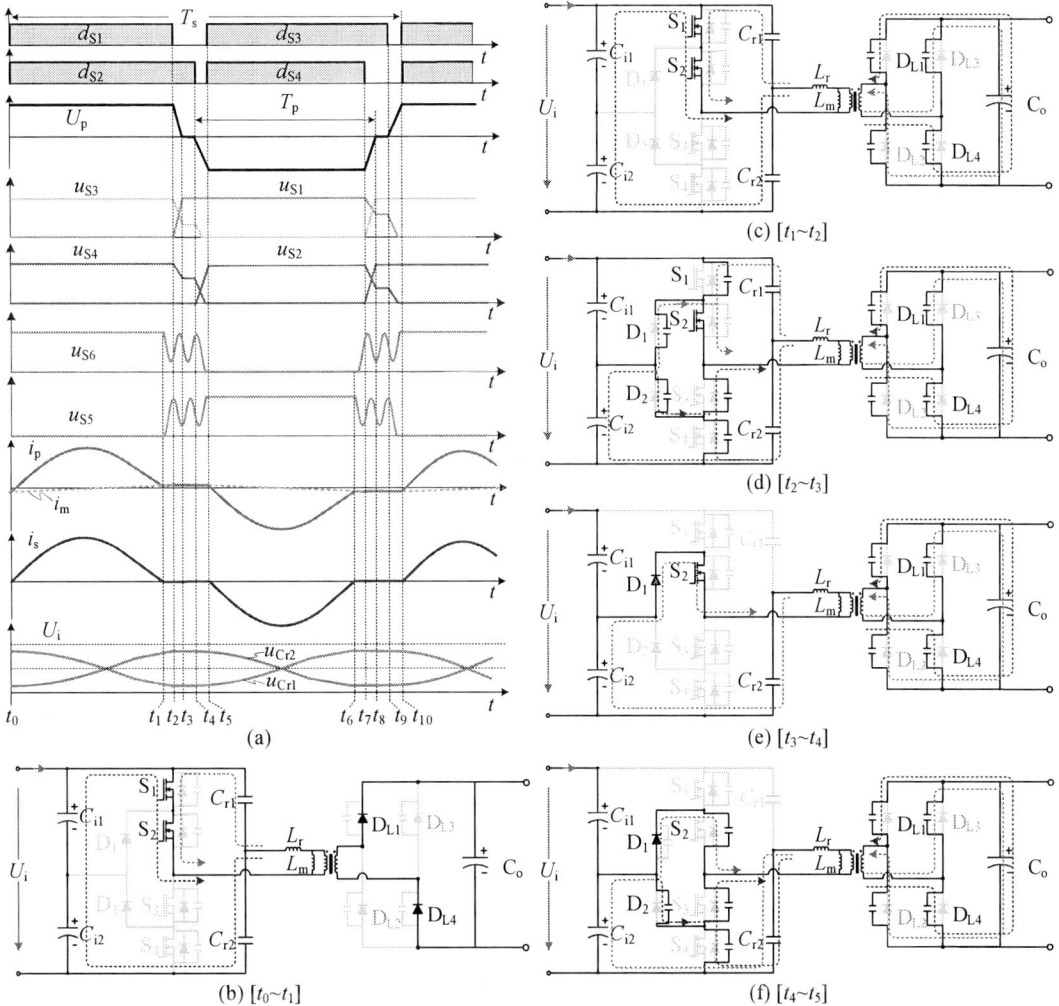

Fig. 2. Key operational waveforms and equivalent circuits of each operating mode for the converter

junction capacitances of the secondary-side switching devices initiate charge-discharge processes. These secondary-side junction capacitances, combined with parasitic capacitances in the circuit, interact with the resonant inductance (L_r) to form an oscillatory resonance in Fig. 1.

Notably, with the trend toward high-frequency operation in power electronics, the nonlinear junction capacitance characteristics of Silicon Carbide (SiC) devices exacerbate the impact on soft-switching performance. Therefore, a modal analysis of the impact mechanism of power device parasitic capacitance parameters on soft-switching characteristics clarifies the design criteria for magnetizing inductance parameters. This approach not only improves converter efficiency but also provides critical theoretical guidance for advancing high-frequency power electronics.

III. MODE ANALYSIS

Based on the operational waveforms shown in Fig.2 and the equivalent circuits of each operating mode, a modal analysis is performed for the resonant converter with a three-level half-bridge primary and a two-level full-bridge secondary. The converter consists of switching devices S_1~S_4,

diodes D_{L1}~D_{L4}, diodes D_1 and D_2, resonant capacitors C_{r1} and C_{r2}, resonant inductor L_r, input filtering capacitor C_{i1} and C_{i2}, output filtering capacitor C_o, and transformer magnetizing inductor L_m. The transformer turns ratio is defined as n:1.

The following assumptions are adopted in this analysis:

1. The magnetizing current I_m remains constant during the dead time.

2. The output capacitance C_{oss} of the SiC devices is assumed to be constant.

3. The resonant capacitor voltage U_{Cr} remains constant during the dead time.

4. Other parasitic parameters, such as the transformer's parasitic capacitance, are neglected.

Under these assumptions, five s-domain equivalent circuits of the three-level resonant converter are established and illustrated in Fig. 3 to 7.

Stage t_0 to t_1: This operational stage corresponds to the resonant phase, where the resonant inductor (L_r) and capacitor (C_r) transfer power to the load through LC oscillation. From the equivalent circuit, the KVL and KCL equations are established:

$$
\begin{cases}
i_p s L_r + i_p \dfrac{1}{s2C_r} + \dfrac{nU_o}{s} = \dfrac{U_i}{s} + I_p\left(t_0\right) L_r \\[2mm]
i_p = i_m + \dfrac{i_s}{n}
\end{cases}
\tag{1}
$$

The primary-side resonant current $i_p(t)$ can be derived as follows:

$$
\begin{aligned}
i_p\left(t\right) &= \frac{\left(U_i - nU_o\right)}{\omega_r L_r}\sin\left(\omega_r\left(t-t_0\right)\right) \\
&\quad + I_p\left(t_0\right)\cos\left(\omega_r\left(t-t_0\right)\right)
\end{aligned}
\tag{2}
$$

where the resonant frequency ω_r is expressed as:

$$
\omega_r = \frac{1}{\sqrt{2L_r C_r}}
\tag{3}
$$

Stage t_1 to t_2: At time t_1, the primary-side resonant current i_p equals the magnetizing current i_m, and the secondary-side resonant current i_s should decay to 0 A. During this interval, no current path exists in the secondary-side rectifier bridge. As a result, the output capacitances of the secondary-side diodes begin to resonate with the resonant inductor L_r.

The equivalent capacitance value of the output capacitances of the four secondary-side diodes, reflected to the primary side, is given by:

$$
C_s = C_{oss\_s} , \; C_{eq} = n^2 C_{oss\_s}
\tag{4}
$$

During this stage, the current flowing through the switching devices not only includes the constant magnetizing current but also incorporates the resonant current generated by the interaction between the output capacitances of the secondary-side diodes and the resonant inductor. Following the same method as the previous stage, determine the current magnitude:

$$
i_p\left(t\right) = I_p\left(t_1\right) + \frac{U_i - U_{Cr} - nU_o}{\omega_{r1} L_r}\sin\left(\omega_{r1}\left(t-t_1\right)\right)
\tag{5}
$$

where the frequency of parasitic capacitance ω_{r2} is expressed as:

$$
\omega_{r1} = \sqrt{\frac{n^2}{L_r C_s}}
\tag{6}
$$

Secondary-side diodes U_D is charged and it can be calculated as:

$$
U_D\left(t\right) = \left(U_i - U_{Cr} - nU_o\right)\left[1 - \cos\left(\omega_{r1}\left(t-t_1\right)\right)\right]
\tag{7}
$$

Stage t_2 to t_3: At time t_2, switching device S_1 is turned off. As shown in Fig. 2(d), five output capacitances on the primary side and four diode output capacitances on the secondary side participate in the resonance at this moment. The equivalent parasitic capacitance on the primary side is:

$$
C_{p1} = \frac{2C_{oss\_p}C_{oss\_p}}{2C_{oss\_p} + C_{oss\_p}} + 2C_{oss\_p} = \frac{8C_{oss\_p}}{3}
\tag{8}
$$

where the frequency of parasitic capacitance ω_{r3} is expressed as:

$$
\begin{cases}
\omega_{r2} = \dfrac{1}{\sqrt{L_r C_{eq1}}} \\[3mm]
C_{eq1} = \dfrac{C_{p1}C_s}{n^2 C_{p1} + C_s}
\end{cases}
\tag{9}
$$

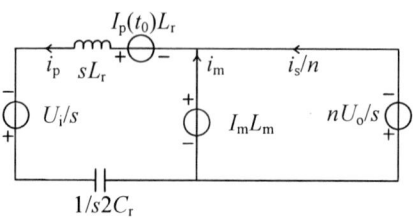

Fig. 3. t_0 to t_1 stage and resonant phase s domain equivalent circuit

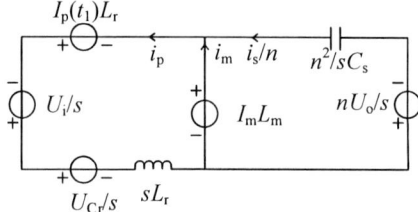

Fig. 4. t_1 to t_2 stage and magnetizing phase s domain equivalent circuit

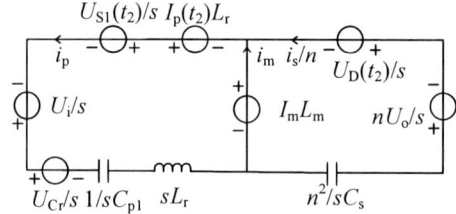

Fig. 5. t_2 to t_3 stage and soft-switching phase of the outer switches s domain equivalent circuit

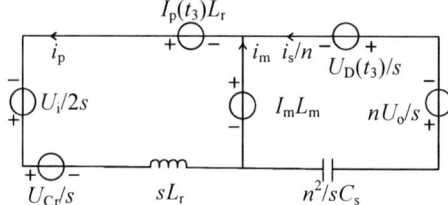

Fig. 6. t_3 to t_4 stage s domain equivalent circuit

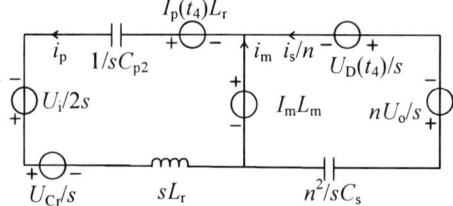

Fig. 7. t_4 to t_5 stage and soft-switching phase of the inner switches s domain equivalent circuit

Starting from time t_3, the output capacitance of switching device S_1 begins to charge gradually, reaching $U_i/2$ at time t_4.

$$
\begin{aligned}
i_p\left(t\right) &= \frac{\left(nU_{S5}\left(t_2\right) + U_i - U_{Cr} - nU_o\right)}{\omega_{r3} L_r}\sin\left(\omega_{r2}\left(t-t_2\right)\right) \\
&\quad + \left(I_p\left(t_2\right) - \frac{I_m n^2 C_{p1}}{n^2 C_{p1} + C_s}\right)\cos\left(\omega_{r2}\left(t-t_2\right)\right) + \frac{I_m n^2 C_{p1}}{n^2 C_{p1} + C_s}
\end{aligned}
\tag{10}
$$

Primary-side switching devices U_{S1} Secondary-side diodes U_D can be calculated as:

$$\begin{cases} U_{S1}(t) = \dfrac{AC_s}{n^2 C_{p1}}\left[1-\cos\left(\omega_{r2}\left(t-t_2\right)\right)\right] \\ \qquad +\dfrac{BC_s}{n^2 C_{p1}}\sin\left(\omega_{r2}\left(t-t_2\right)\right)+C\left(t-t_2\right) \\ U_D(t) = A\left[1-\cos\left(\omega_{r2}\left(t-t_2\right)\right)\right] \\ \qquad +B\sin\left(\omega_{r2}\left(t-t_2\right)\right)-C\left(t-t_2\right) \end{cases}$$

$$\begin{cases} A = \dfrac{\left(nU_D\left(t_2\right)+U_i-U_{Cr}-nU_o\right)n^2 C_{p1}}{n^2 C_{p1}+C_s} \\ B = \dfrac{n^2}{\omega_{r3}C_s}\left(I_p\left(t_2\right)-\dfrac{I_m n^2 C_{p1}}{n^2 C_{p1}+C_s}\right) \\ C = \dfrac{n^2 I_m}{n^2 C_{p1}+C_s} \end{cases}$$

$$(11)$$

Stage t_3 to t_4: During this phase, switching device S_1 is clamped at $U_i/2$, and the magnetizing current flows through D_1 and S_2. As shown in Fig. 2(e), the output capacitances of the primary-side switching devices do not participate in the resonance.

$$i_p(t) = \frac{\left(nU_D\left(t_3\right)+\dfrac{U_i}{2}-U_{Cr}-nU_o\right)}{\omega_{r2}L_r}\sin\left(\omega_{r2}\left(t-t_3\right)\right) \\ +\left(I_p\left(t_3\right)-I_m\right)\cos\left(\omega_{r2}\left(t-t_3\right)\right)+I_m \quad (12)$$

The output capacitances of the primary-side devices do not participate in the resonance, the voltage across the output capacitances of the primary-side devices remains constant, while the secondary side continues to resonate.

$$\begin{cases} U_{S1}(t) = \dfrac{n^2\left(-U_{S1}\left(t_3\right)+\dfrac{U_i}{2}-U_{Cr}-nU_o\right)}{\omega_{r1}^2 L_r C_s}\left[1-\cos\left(\omega_{r1}\left(t-t_3\right)\right)\right] \\ \qquad +\dfrac{n^2}{\omega_{r1}C_s}\left(I_p\left(t_3\right)-I_m\right)\sin\left(\omega_{r1}\left(t-t_3\right)\right)+U_{S1}\left(t_3\right) \\ U_D(t) = \dfrac{\left(U_s\left(t_3\right)-\dfrac{U_i}{2}+U_{Cr}+nU_o\right)}{2\omega_{r1}^2 L_r C_s}\left[1-\cos\left(\omega_{r1}\left(t-t_3\right)\right)\right] \\ \qquad +\dfrac{n\left(-I_p\left(t_3\right)+I_m\right)}{2\omega_{r1}C_s}\sin\left(\omega_{r1}\left(t-t_3\right)\right)+U_D\left(t_3\right) \end{cases}$$

$$(13)$$

Stage t_4 to t_5: At time t_4, switching device S_2 is turned off. As shown in Fig. 2(f), four output capacitances on the primary side and four diode output capacitances on the secondary side participate in the resonance at this moment. The equivalent parasitic capacitance on the primary side is:

$$C_{p2} = \frac{2C_{oss\_p}C_{oss\_p}}{2C_{oss\_p}+C_{oss\_p}}+C_{oss\_p} = \frac{5C_{oss\_p}}{3} \quad (14)$$

where the frequency of parasitic capacitance ω_{r4} is expressed as:

$$\begin{cases} \omega_{r3} = \dfrac{1}{\sqrt{L_r C_{eq2}}} \\ C_{eq2} = \dfrac{C_{p2}C_s}{n^2 C_{p2}+C_s} \end{cases}$$

$$(15)$$

Starting from time t_4, the output capacitance of switching device S_2 begins to charge gradually, reaching $U_i/2$ at time t_5.

$$i_p(t) = \frac{\left(nU_D\left(t_4\right)+\dfrac{U_i}{2}-U_{Cr}-nU_o\right)}{\omega_{r3}L_r}\sin\left(\omega_{r3}\left(t-t_4\right)\right) \\ +\left(I_p\left(t_4\right)-\frac{I_m n^2 C_{p2}}{n^2 C_{p2}+C_s}\right)\cos\left(\omega_{r3}\left(t-t_4\right)\right) \\ +\frac{I_m n^2 C_{p2}}{n^2 C_{p2}+C_s} \quad (16)$$

$$\begin{cases} U_{S2}(t) = \dfrac{AC_s}{n^2 C_{p2}}\left[1-\cos\left(\omega_{r3}\left(t-t_4\right)\right)\right] \\ \qquad +\dfrac{BC_s}{n^2 C_{p2}}\sin\left(\omega_{r3}\left(t-t_4\right)\right)+C\left(t-t_4\right) \\ U_D(t) = A\left[1-\cos\left(\omega_{r3}\left(t-t_4\right)\right)\right] \\ \qquad +B\sin\left(\omega_{r3}\left(t-t_4\right)\right)-C\left(t-t_4\right) \end{cases}$$

$$\begin{cases} A = \dfrac{n^2\left(nU_D\left(t_4\right)+\dfrac{U_i}{2}-U_{Cr}-nU_o\right)C_{p2}}{n^2 C_{p2}+C_s} \\ B = \dfrac{n^2}{\omega_{r3}C_s}\left(I_p\left(t_4\right)-\dfrac{I_m n^2 C_{p2}}{n^2 C_{p2}+C_s}\right) \\ C = \dfrac{n^2 I_m}{n^2 C_{p2}+C_s} \end{cases}$$

$$(17)$$

At this point, the soft-switching process of the primary-side switches is completed.

IV. COMPARATIVE ANALYSIS OF SOFT-SWITCHING CONSTRAINTS

When the primary-side soft-switching is unaffected by the secondary-side output capacitance, the constraint for achieving ZVS of the devices is expressed as:

$$I_m T_d \ge 3U_i C_{oss} \quad (18)$$

The minimum magnetizing current required to achieve ZVS on the primary side can be determined from TABLE I.

TABLE I.   SIMULATION PARAMETER

Parameter	Parameter value
Input voltage	2300V
Output voltage	900V
C_{oss}	200pF
T_d	400ns
T_m	200ns
f_s	50kHz
P	25kW

When the primary-side soft-switching is affected by the secondary-side output capacitance, to ensure ZVS of the power switches, from (11), the voltage across the outer switch must be charged to $U_i/2$ during the interval from the turn-off of the outer switch to the turn-off of the inner switch. Similarly, from (17) the voltage across the inner switch must also reach $U_i/2$. By solving the five mode equations simultaneously, the charging times for both the outer and inner switches can be derived.

$$\begin{cases} U_{S1}\left(T_{\text{outer\_ZVS}}\right) = \dfrac{U_{\text{in}}}{2} \\ U_{S2}\left(T_{\text{inner\_ZVS}}\right) = \dfrac{U_{\text{in}}}{2} \end{cases} \quad (19)$$

Therefore, the deadtime T_d must be set to at least exceed:

$$\begin{cases} T_m \ge T_{\text{outer\_ZVS}} \\ T_d \ge T_{\text{inner\_ZVS}} + T_m \end{cases} \quad (20)$$

where T_m denotes the time difference by which the outer switch turns off earlier than the inner switch. During this interval (T_m), the outer switches must achieve ZVS.

When the secondary-side output capacitance is both considered and neglected under the same deadtime, the magnetizing current magnitudes required to achieve ZVS for the primary-side switches can be obtained from (19), as shown in TABLE II.

As shown in Fig .8, the simulation results indicate that the magnetizing current required to achieve ZVS for the primary-side switches is at least 3.9 A when the secondary-side output capacitance is neglected, and increases to 5.9 A when it is considered, which agrees well with the theoretical calculations.

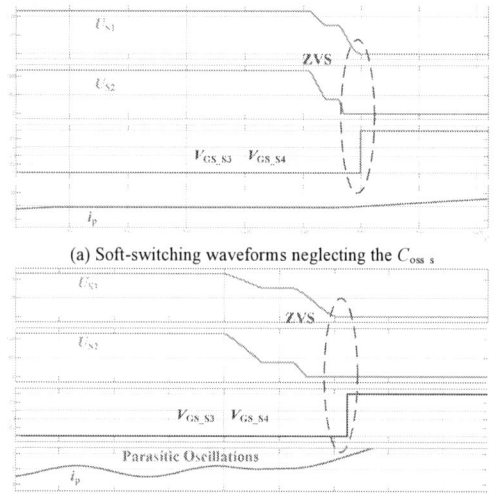

(a) Soft-switching waveforms neglecting the $C_{\text{oss\_s}}$

(b) Soft-switching waveforms considering the $C_{\text{oss\_s}}$
Fig. 8. Designed three-level resonant converter

Due to the presence of secondary-side output capacitance, the magnetizing current required to achieve soft-switching in the primary-side switches increases. This results in higher reactive current circulation in the circuit, thereby increasing conduction losses.

TABLE II.   THE COMPARATIVE RESULTS OF MAGNETIZING CURRENT MAGNITUDES

Secondary-side output capacitance	The magnetizing current magnitudes
Neglected	3.45A
Considered	6.05A

In practice, when considering parameters such as transformer parasitic capacitance, PCB parasitic capacitance,

and other stray capacitances, achieving soft-switching becomes more challenging, and the magnetizing current will further increase.

V. EXPERIMENTAL VERIFICATION

To validate the aforementioned soft-switching analysis and conclusions, a 25kW 50kHz three-level resonant converter based on SiC devices was designed. The converter prototype is shown in Fig 9, and its specifications are summarized in TABLE 1. The primary-side and secondary-side devices are rated at 1.7kV SiC MOSFET and 1.2kV SiC MOSFET.

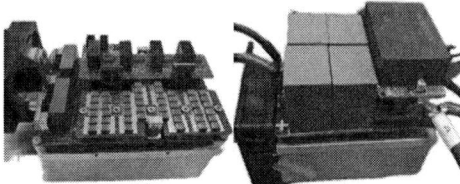

(a) Primary-Side Module   (b) secondary -Side Module
Fig. 9. Designed three-level resonant converter

TABLE III.   DESIGNED THREE-LEVEL RESONANT CONVERTER PARAMETER

Parameter	Parameter value
Input voltage	2300V
Output voltage	900V
L_m	200µH
f_s	50kHz
P	25kW
I_m	18A

As shown in Fig. 10, parasitic oscillations are present in the current during the magnetizing phase.

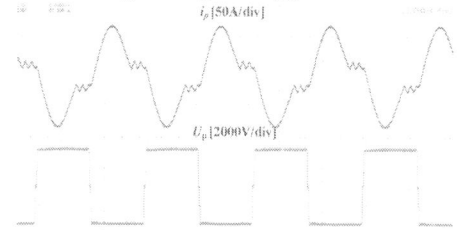

Fig. 10. Steady-state waveforms at 25 kW

Since this paper focuses solely on the impact of switch output capacitance on soft-switching, while the experimental circuit includes additional factors such as transformer parasitic capacitance, the magnetizing current required in experiments exceeds that predicted by theoretical design and simulations. The influence of transformer parasitic capacitance on soft-switching will be addressed in future studies.

VI. CONCLUSION

In this paper, the soft-switching characteristics of a silicon carbide (SiC)-based three-level resonant converter are investigated. Through modal analysis, the influence mechanism of power device output capacitance parameters on soft-switching behavior is revealed. By comparing the impact of the presence and absence of secondary-side switch output capacitance on the soft-switching of primary-side switches, it is found that a higher magnetizing current is required for soft-switching when output capacitance is considered, leading to increased losses. The correctness of the magnetizing current calculation is validated through simulations. Experimental verification is conducted on a 2300 V DC input / 900 V DC output / 25 kW prototype. Since the experimental circuit includes not only the power device output capacitance but also parasitic capacitances of the transformer and other factors, the magnetizing current required in experiments exceeds that predicted by theoretical design and simulations. The influence of transformer parasitic capacitance on soft-switching will be addressed in future studies.

REFERENCES

[1] Krein, Philip T. "Data center challenges and their power electronics." *CPSS Transactions on Power Electronics and Applications* 2, no. 1 (2017): 39-46

[2] H. Hu, Y. Shao, L. Tang, J. Ma, Z. He and S. Gao, "Overview of Harmonic and Resonance in Railway Electrification Systems," IEEE Trans. Ind. Appl., vol. 54, no. 5, pp. 5227-5245, Sept.-Oct. 2018.

[3] E. Chemali, M. Preindl, P. Malysz and A. Emadi, "Electrochemical and Electrostatic Energy Storage and Management Systems for Electric Drive Vehicles: State-of-the-Art Review and Future Trends," IEEE J. Emerg. Sel. Topics Power Electron., vol. 4, no. 3, pp. 1117-1134, Sept. 2016.

[4] M. Safayatullah, M. T. Elrais, S. Ghosh, R. Rezaii and I. Batarseh, "A Comprehensive Review of Power Converter Topologies and Control Methods for Electric Vehicle Fast Charging Applications," IEEE Access, vol. 10, pp. 40753-40793, 2022.

[5] S. Hain, M. Meiler, and M. Denk, "Evaluation of 800v traction inverter with sic-mosfet versus si-igbt power semiconductor technology," in PCIM Europe 2019; International Exhibition and Conference for Power Electronics, Intelligent Motion, Renewable Energy and Energy Management, pp. 1–6, 2019.

[6] P. Jiang, H. Feng and L. Ran, "ZVS Analysis and a Design Method for Unidirectional Medium-Voltage LLC-DCX With High Step-Up Ratio," in IEEE Transactions on Power Electronics, vol. 39, no. 3, pp. 2948-2953, March 2024.

[7] Y. Cao, M. Ngo, D. Dong and R. Burgos, "The ZVS Transition Analysis and Optimization for CLLC-Type Resonant DC Transformer," 2021 IEEE Energy Conversion Congress and Exposition (ECCE), Vancouver, BC, Canada, 2021.

[8] T. Guillod, D. Rothmund and J. W. Kolar, "Active Magnetizing Current Splitting ZVS Modulation of a 7 kV/400 V DC Transformer," in IEEE Transactions on Power Electronics, vol. 35, no. 2, pp. 1293-1305, Feb. 2020.

On the First Demonstration and Analysis of the HTGB Induced Electrical Degradation of High Voltage SiC IGBT Devices

1st Tuanzhuang Wu
National ASIC System
Engineering Research Center,
School of Integrated Circuits,
Southeast University
Nanjing 210096, China
230228400@seu.edu.cn

2nd Jiaxing Wei*
National ASIC System
Engineering Research Center,
School of Integrated Circuits,
Southeast University
Nanjing 210096, China
jiaxingwei@seu.edu.cn

3rd Junhou Cao
National ASIC System
Engineering Research Center,
School of Integrated Circuits,
Southeast University
Nanjing 210096, China
junhcao@seu.edu.cn

4th Hao Fu
National ASIC System
Engineering Research Center,
School of Integrated Circuits,
Southeast University
Nanjing 210096, China
fuhaoseu@163.com

5th Zhaoxiang Wei
National ASIC System
Engineering Research Center,
School of Integrated Circuits,
Southeast University
Nanjing 210096, China
weizhaoxiang27@163.com

6th Desheng Ding
National ASIC System
Engineering Research Center,
School of Integrated Circuits,
Southeast University
Nanjing 210096, China
dds@seu.edu.cn

7th Siyang Liu*
National ASIC System
Engineering Research Center,
School of Integrated Circuits,
Southeast University
Nanjing 210096, China
liusy2017@seu.edu.cn

8th Weifeng Sun
National ASIC System
Engineering Research Center,
School of Integrated Circuits,
Southeast University
Nanjing 210096, China
swffrog@seu.edu.cn

9th Xiaolei Yang
State Key Laboratory of Wide-Bandgap
Semiconductor Power Eletronic Devices
Naniing Electronic Devices Institute
Nanjing 210001, China
15005154193@163.com

10th Song Bai
State Key Laboratory of Wide-Bandgap
Semiconductor Power Eletronic Devices
Naniing Electronic Devices Institute
Nanjing 210001, China
13809020747@163.com

Abstract—**Silicon Carbide (SiC) possesses many superior electrical properties and is one of the most popular semiconductor materials in the field of power electronics. SiC-based power devices feature high voltage withstand capability, low ON-resistance, and excellent thermal conductivity. Due to their bipolar conduction design, Insulated Gate Bipolar Transistors (IGBTs) achieve some of the highest power levels among discrete power devices. Therefore, SiC-based IGBT devices have significant research prospects and application potential. However, limited by the reliability of SiC devices, especially the gate oxide interface, there are currently no commercial SiC IGBT products available worldwide. In this study, the degradation of threshold voltage and conduction characteristics of 20 kV/15 A SiC IGBT devices under High Temperature Gate Bias (HTGB) stress is demonstrated and investigated for the first time. Furthermore, TCAD simulation is employed to establish the relationship between defect changes at the gate oxide interface and the degradation of threshold voltage (V_{TH}) and specific differential ON-resistance ($R_{diff,sp}$) in SiC IGBT devices. The results indicate that the degradation of V_{TH} and $R_{diff,sp}$ is primarily affected by both gate voltage and** temperature. The findings provide valuable insights for the design and application of SiC IGBT devices.

Index Terms—**SiC IGBT, Electrical degradation, HTGB Sress, Interface trap, Reliability**

I. INTRODUCTION

Due to the excellent growth quality of single crystals, 4H-SiC is the mainstream substrate and epitaxial material for SiC-based power devices. Compared with silicon, 4H-SiC has a wider band gap, higher critical breakdown electric field, higher electron saturation velocity, and better thermal conductivity, among other advantages [1]–[3]. Therefore, 4H-SiC-based power devices outperform Si-based devices in terms of turn-off leakage current, blocking voltage, power density, heat dissipation capability, and radiation resistance. These devices have great potential in high-temperature and high-radiation environments such as aerospace, rail transit, new energy vehicles, and energy exploration and drilling. The IGBT

979-8-3315-1110-4/25 $31.00 © 2025 IEEE

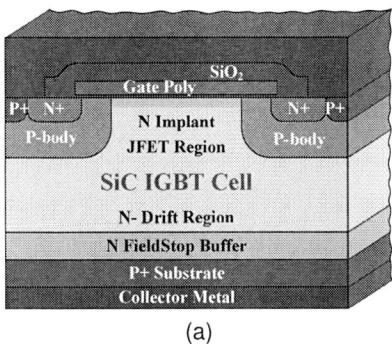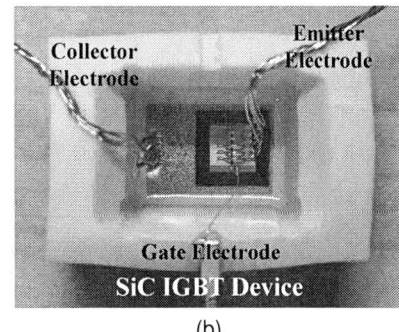

(a)   (b)

Fig. 1. Structure of SiC IGBT device. (a) Schematic diagram of SiC IGBT device structure. (b) 3D printing package of SiC IGBT device.

structure combines the high switching speed of power MOSFETs with the high current gain of bipolar devices. It features high input impedance, a simple control circuit, low driving power, and low on-state voltage, making it widely used in power conversion, smart grids, and other fields [4]–[6]. In ultra-high voltage applications such as flexible DC transmission, it is often necessary to control voltages exceeding 10 kV, which currently relies mainly on module switches [7]. Compared with modules, SiC IGBT devices are smaller in size and lower in cost, making them promising for high-voltage direct current (HVDC) transmission and related fields. However, due to complex fabrication processes, high epitaxial costs, and unresolved reliability issues, there are currently no mature commercial SiC-based IGBT chips. In this paper, the drift of V_{TH} and $R_{diff,sp}$ in SiC IGBTs under HTGB experimental conditions is investigated for the first time, and the corresponding results are presented and analyzed [8]. The degradation of V_{TH} and $R_{diff,sp}$ in SiC IGBTs is mainly caused by the combined effects of gate voltage stress and operating temperature.

In Section II, the structure and physical sample of the SiC IGBT are presented, along with the design of a specific HTGB experimental scheme. Section III provides the measured data and analysis, followed by the conclusion in Section IV.

II. DEVICE STRUCTURE & HTGB TEST METHOD

A. Device Structure

The structural schematic diagram of the SiC IGBT device is shown in Fig. 1(a), while the prepared sample with a 3D-printed package is shown in Fig. 1(b). The electrodes of the SiC IGBT are connected via leads, as illustrated in Fig. 1(b). The epitaxial N-drift layer, N+ buffer layer, and P+ collector layer are all grown on an N-type 4H-SiC substrate. On the surface, SiO_2 is used as a hard mask for each implantation. The P-well region, P+ region, and N+ region are formed by multiple different type ion implantation. Meanwhile, N-type ion implantation is carried out separately in the JFET region to reduce its resistance. The trap density at the gate oxide interface is effectively reduced by annealing in a NO atmosphere, and the final gate oxide layer

thickness is controlled to about 50 nm. After completing the front structure of the device, the N-type substrate is removed by grinding with part of the P+ layer retained as the collector of SiC IGBT. Finally, the Ohmic metal is evaporated and deposited on the backside SiC material.

B. HTGB Test Flow

The schematic circuit diagram and stress platform for the HTGB experiment on SiC IGBT devices are shown in Fig. 2. In this experiment, each SiC IGBT is subjected to a constant gate bias voltage and maintained at a high temperature for a specified duration. The detailed HTGB experimental scheme is summarized in Table I. Four fresh SiC IGBT devices were tested, with their corresponding numbers and stress conditions listed in Table I. Additionally, Table I specifies the electrical characteristics monitored during the stress process and the time points at which measurements were taken.

TABLE I
HTGB TEST DEVICES SPLIT

No.	Stress Condition	Duration	Monitors
#1	$V_G = 25$ V, $T = 150$ °C	0h, 3h, 6h, 12h, 24h, 48h, 72h	Transfer, Output
#2	$V_G = -10$ V, $T = 175$ °C		
#3	$V_G = -10$ V, $T = 150$ °C		
#4	$V_G = 25$ V, $T = 175$ °C		

III. RESULTS & DISCUSSION

This section presents the measured results of the HTGB experiment and discusses the mechanism for the electrical degradation of SiC IGBTs based on these results. Furthermore, Sentaurus TCAD simulations are used to investigate the influence of trap charges at the gate oxide/SiC interface on V_{TH} and $R_{diff,sp}$, thereby validating the discussion and analysis.

A. Transfer & Output Characteristics

The transfer characteristics of the SiC IGBT devices monitored during HTGB stress are shown in Fig. 3. Figures 3(a)–(d) present the corresponding results for devices #1–#4, respectively. The transfer characteristic

2025 IEEE Workshop on Wide Bandgap Power Devices and Applications in Asia (WiPDA Asia)

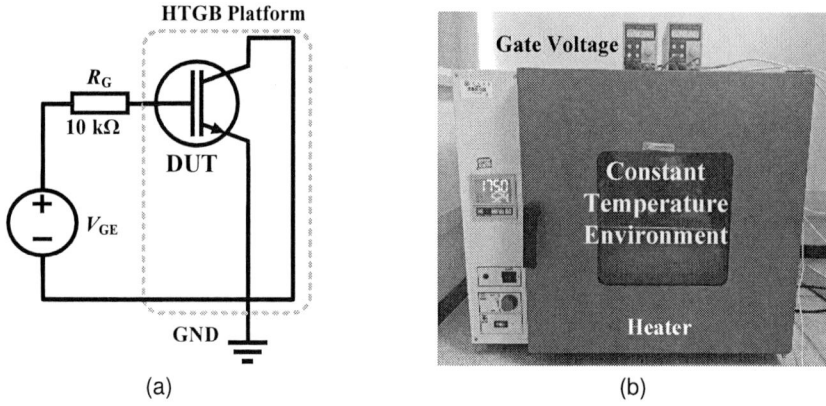

Fig. 2. (a) Schematic diagram of HTGB test circuit & (b) stress platform.

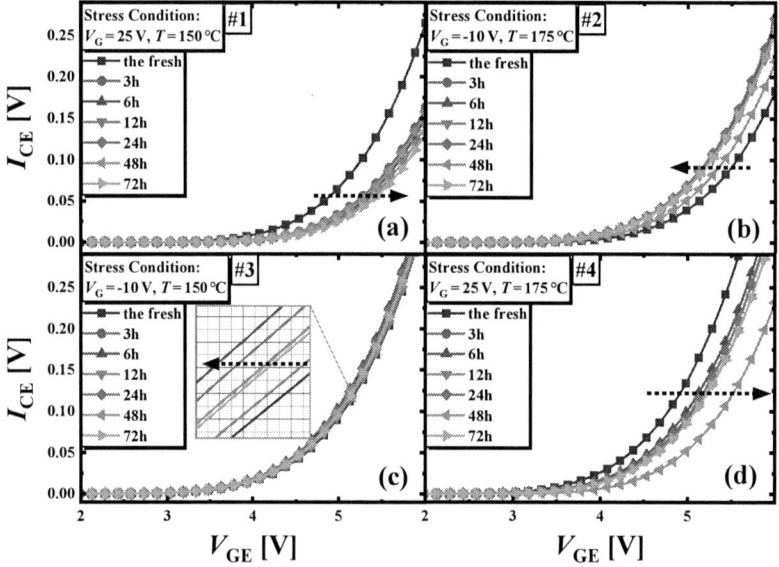

Fig. 3. (a) (d) are the variation of the transfer characteristics of each SiC IGBT device after HTGB stress, respectively.

curves of samples #1 and #4 shift to the right under the applied HTGB stress, while those of samples #2 and #3 shift to the left. Moreover, the changes in the transfer characteristics of samples #1 and #4 are more pronounced than those of samples #2 and #3. Additionally, the V_{TH} degradation for all samples is more severe during the initial stress period and gradually weakens with increasing stress duration, eventually stabilizing. It is evident that the V_{TH} of SiC IGBT devices increases under HTGB stress with a positive gate voltage, while it decreases with a negative gate voltage. A larger absolute value of the gate voltage results in more significant V_{TH} degradation. The V_{TH} of samples #1 and #4 increases, whereas that of samples #2 and #3 decreases under stress. As the stress continues, the V_{TH} shift of all samples gradually stabilizes. Furthermore, the rate of V_{TH} change is higher for samples #2 and #4, indicating that temperature also affects V_{TH} degradation, as shown in Fig. 4(a). The V_{TH} offset ratio for each sample is illustrated in Fig. 4(b),

where samples #1 and #4 exhibit larger offset ratios than samples #2 and #3, demonstrating that a higher absolute gate voltage has a more pronounced impact on the V_{TH} degradation of SiC IGBT devices.

In summary, the V_{TH} of SiC IGBT devices increases under HTGB stress with a positive gate voltage, while it decreases with a negative gate voltage. The degradation of V_{TH} becomes more pronounced as the absolute value of the gate voltage increases. Additionally, the rate of V_{TH} change is higher at elevated temperatures.

Similarly, the output characteristics of the SiC IGBT devices monitored during HTGB stress are shown in Fig. 5. Figures 5(a)–(d) illustrate the variations in output characteristics for samples #1–#4, respectively. Figure 6(a) presents the changes in specific differential ON-resistance ($R_{diff,sp}$) for each sample after HTGB stress. All samples exhibit a similar trend: $R_{diff,sp}$ decreases during the initial period of HTGB stress and then partially recovers to a stable value. The $R_{diff,sp}$ offset ratios for

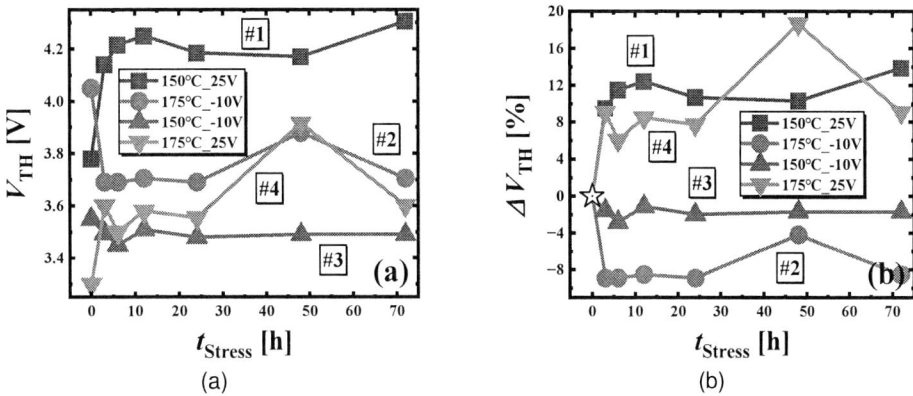

Fig. 4. (a) SiC IGBT threshold voltage drift after HTGB stress (b) V_{TH} offset ratio compared to fresh device.

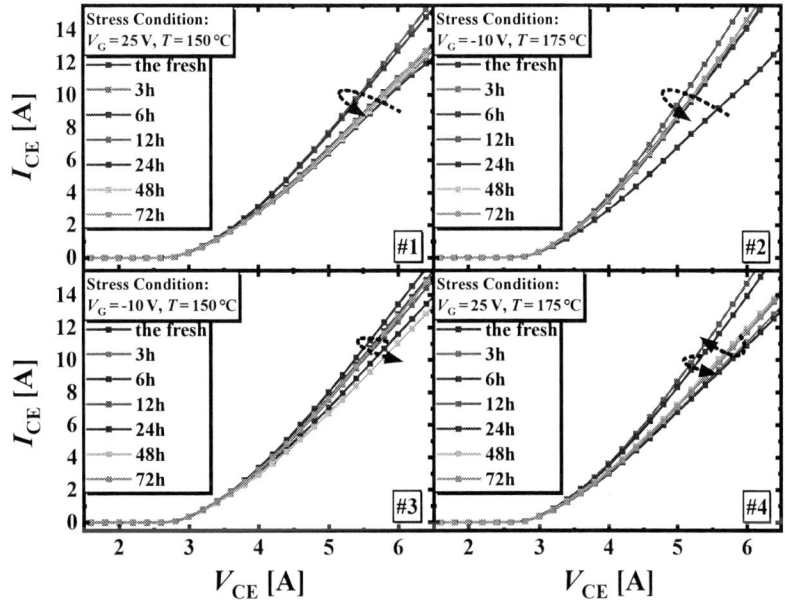

Fig. 5. (a) (d) are the variation of the output characteristics of each SiC IGBT device after HTGB stress, respectively.

each sample are shown in Fig. 6(b), where samples #1 and #4 display larger offset ratios compared to samples #2 and #3. Notably, an initial increase in $R_{\mathrm{diff,sp}}$ is observed for sample #4 at the beginning of the HTGB process, which may be attributed to pre-existing interface traps. As shown in Fig. 6(a), the degradation of $R_{\mathrm{diff,sp}}$ in samples #2 and #4 is more pronounced than in samples #1 and #3. Additionally, the recovery of $R_{\mathrm{diff,sp}}$ degradation is more evident in samples #1 and #4. In conclusion, the $R_{\mathrm{diff,sp}}$ of SiC IGBT devices generally decreases under HTGB stress, with more severe degradation occurring at higher stress temperatures. Furthermore, the recovery of $R_{\mathrm{diff,sp}}$ degradation is more apparent under positive gate voltage.

Taking into account the V_{TH} degradation of SiC IGBT devices, the channel resistance increases under HTGB stress with a positive gate voltage, indicating that negative charges are generated at the $\mathrm{SiO_2/SiC}$ interface [9], [10].

However, the $R_{\mathrm{diff,sp}}$ of SiC IGBT devices decreases under the same HTGB stress. This suggests that positive charges emerge at the interface of the JFET region, enhancing the electron density [11]. Under negative gate voltage, the V_{TH} of SiC IGBT devices decreases due to trapped holes at the channel interface, thereby causing $R_{\mathrm{diff,sp}}$ to decrease as well. In a high-temperature environment, carriers and mobile charges inside the semiconductor devices gain energy and become more active. The gate voltage attracts different types of charges depending on its polarity. Electrons near the $\mathrm{SiO_2/SiC}$ interface enter the gate oxide layer under positive gate voltage and high-temperature stress, while the effect is opposite under negative bias [12]. When the stress time becomes sufficiently long, further increases in duration do not result in significant degradation of V_{TH} due to the reduction and eventual saturation of available traps [13]. V_{TH} is more sensitive to higher absolute gate bias and

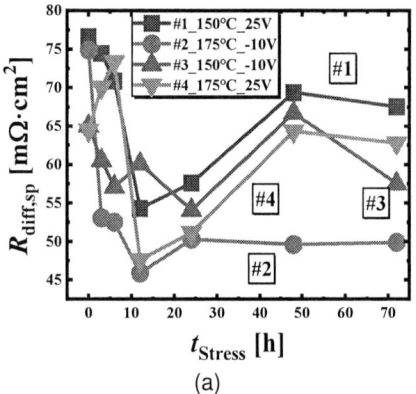
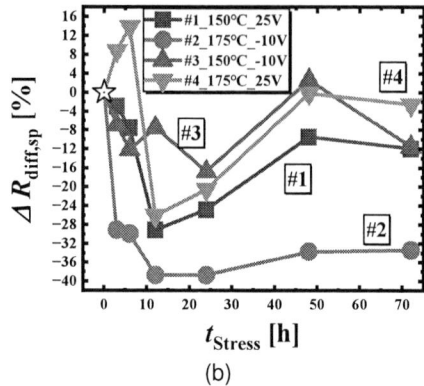

(a)   (b)

Fig. 6. (a) SiC IGBT differential ON-resistance drift after HTGB stress (b) $R_{\text{diff,sp}}$ offset ratio compared to fresh device.

higher temperatures, while $R_{\text{diff,sp}}$ is more susceptible to high temperature. Additionally, tunneling effects during this process can also contribute to the degradation [14], [15].

B. Simulation & Validation

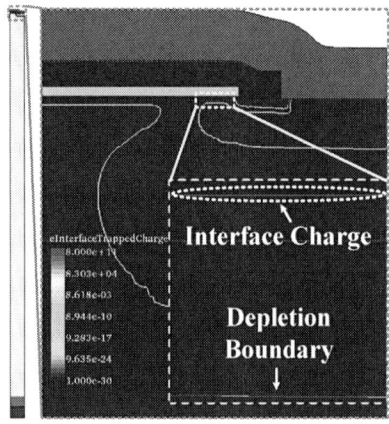

Fig. 7. Simulation structure diagram of SiC IGBT device. The small diagram shows the addition of an SiO2/SiC interface charge to its channel region.

To further validate the analysis of charge injection at the SiO_2/SiC interface after HTGB stress, a simulation structure consistent with the fabricated device was established using the Sentaurus TCAD suite, as shown in Fig. 7. Various fixed charges were introduced near the SiC IGBT channel interface to emulate the effects of HTGB stress induced damage. As illustrated in Fig. 8, the simulated transfer characteristics of the SiC IGBT device under different channel interface charge conditions are presented. The simulation results clearly support the experimental observations and analysis discussed in the previous section.

IV. CONCLUSION

In this study, the HTGB experiment on a 20 kV/15 A SiC IGBT device was carried out for the first time, and the trends of the static electrical characteristic curves

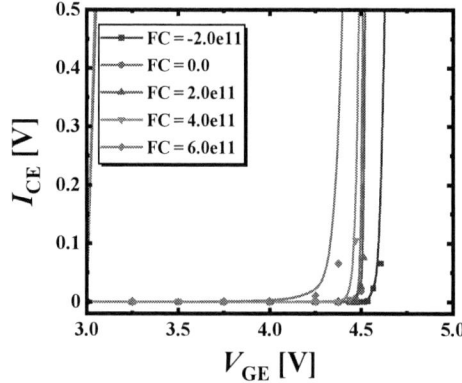

Fig. 8. The transfer characteristics of the SiC IGBT simulated with different channel SiO_2/SiC interface charges move.

were investigated. The variations of V_{TH} and $R_{\text{diff,sp}}$ under stress were obtained. Additionally, the observed electrical degradation of SiC IGBT devices was analyzed and verified by TCAD simulation. This work will facilitate the more reliable design and application of SiC IGBT devices. In subsequent studies, the mechanisms and other reliability aspects of SiC IGBT devices will be further investigated.

ACKNOWLEDGMENT

This work was supported in part by the National Natural Science Foundation of China under grant 62434002 and grant 62174029; in part by the Fundamental Research Funds for the Central Universities under grant 2242024RCB0028; in part by the Science and Technology Major Project of Jiangsu Province under grant BG 2024001; in part by the Fund for Transformation of Scientific and Technological Achievements of Jiangsu Province under grant BA2023001; in part by the Natural Science Foundation of Jiangsu Province under grant BK20232027; in part by the Distinguished Young Scientists Foundation of Jiangsu Province under grant BK20230025; and in part by the Fund for Transformation of Scientific and Technological Achievements of Wuxi City under grant C20231021.

REFERENCES

[1] J. Wang et al., "Characterization, Modeling, and Application of 10-kV SiC MOSFET," in IEEE Transactions on Electron Devices, vol. 55, no. 8, pp. 1798-1806, Aug. 2008, DOI: 10.1109/TED.2008.926650.

[2] S. Narasimhan, A. Kanale, S. Bhattacharya and J. Baliga, "Performance Evaluation of 3.3 kV SiC MOSFET and Schottky Diode for Medium Voltage Current Source Inverter Application," 2021 IEEE 8th Workshop on Wide Bandgap Power Devices and Applications (WiPDA), Redondo Beach, CA, USA, 2021, pp. 366-371, DOI: 10.1109/WiPDA49284.2021.9645089.

[3] E. Bashar et al., "Comparison of Short Circuit Failure Modes in SiC Planar MOSFETs, SiC Trench MOSFETs and SiC Cascode JFETs," 2021 IEEE 8th Workshop on Wide Bandgap Power Devices and Applications (WiPDA), Redondo Beach, CA, USA, 2021, pp. 384-388, doi: 10.1109/WiPDA49284.2021.9645092.

[4] J. Tasloglou, N. Frantzis, G. Bertolis, J. Kostandaras, J. M. Koutsoubis and C. X. Manasis, "Switching Speed Evaluation of Commercially Available IGBTs for Sub-microsecond Pulsed Power Applications," 2021 10th Mediterranean Conference on Embedded Computing (MECO), Budva, Montenegro, 2021, pp. 1-5, doi: 10.1109/MECO52532.2021.9460225.

[5] H. Xiao, W. Liu, Y. Yao, Z. Zhang, W. Hu and H. Luo, "600A/1200V S3+ IGBT Module with Fine Geometry Trench IGBT Technology for EV/HEV Application," PCIM Asia 2020; International Exhibition and Conference for Power Electronics, Intelligent Motion, Renewable Energy and Energy Management, Shanghai, China, 2020, pp. 1-4.

[6] Y. Yorozu, M. Hirano, K. Oka, and Y. Tagawa, "Electron spectroscopy studies on magneto-optical media and plastic substrate interface," IEEE Transl. J. Magn. Japan, vol. 2, pp. 740–741, August 1987 [Digests 9th Annual Conf. Magnetics Japan, p. 301, 1982].

[7] K. Rouzbehi, A. Miranian, J. I. Candela, A. Luna and P. Rodriguez, "Proposals for flexible operation of multi-terminal DC grids: Introducing flexible DC transmission system (FDCTS)," 2014 International Conference on Renewable Energy Research and Application (ICRERA), Milwaukee, WI, USA, 2014, pp. 180-184, doi: 10.1109/ICRERA.2014.7016553.

[8] M. Ren et al., "Gate Oxide Failure Mechanisms of SiC MOSFET Related to Electro-Thermomechanical Stress Under HTRB and HTGB Test," 2024 36th International Symposium on Power Semiconductor Devices and ICs (ISPSD), Bremen, Germany, 2024, pp. 208-211, doi: 10.1109/ISPSD59661.2024.10579620.

[9] G. Rescher, G. Pobegen, T. Aichinger and T. Grasser, "Preconditioned BTI on 4H-SiC: Proposal for a Nearly Delay Time-Independent Measurement Technique," in IEEE Transactions on Electron Devices, vol. 65, no. 4, pp. 1419-1426, April 2018, doi: 10.1109/TED.2018.2803283.

[10] C. Unger and M. Pfost, "Determination of the Transient Threshold Voltage Hysteresis in SiC MOSFETs after Positive and Negative Gate Bias," 2019 31st International Symposium on Power Semiconductor Devices and ICs (ISPSD), Shanghai, China, 2019, pp. 195-198, doi: 10.1109/ISPSD.2019.8757661.

[11] A. J. Lelis, R. Green, D. B. Habersat and M. El, "Basic Mechanisms of Threshold-Voltage Instability and Implications for Reliability Testing of SiC MOSFETs," in IEEE Transactions on Electron Devices, vol. 62, no. 2, pp. 316-323, Feb. 2015, doi: 10.1109/TED.2014.2356172.

[12] C. X. Zhang et al., "Origins of Low-Frequency Noise and Interface Traps in 4H-SiC MOSFETs," in IEEE Electron Device Letters, vol. 34, no. 1, pp. 117-119, Jan. 2013, doi: 10.1109/LED.2012.2228161.

[13] C. J. Cochrane, P. M. Lenahan, A. J. Lelis; "An electrically detected magnetic resonance study of performance limiting defects in SiC metal oxide semiconductor field effect transistors". J. Appl. Phys. 1 January 2011; 109 (1): 014506. https://doi.org/10.1063/1.3530600

[14] M. Sagawa, H. Miki, Y. Mori, H. Shimizu and A. Shima, "Evaluation of gate oxide reliability in 3.3 kV 4H-SiC DMOSFET with J-Ramp TDDB methods," 2018 IEEE 30th International Symposium on Power Semiconductor Devices and ICs (ISPSD), Chicago, IL, USA, 2018, pp. 363-366, doi: 10.1109/ISPSD.2018.8393678.

[15] S. Maaß, H. Reisinger, T. Aichinger and G. Rescher, "Influence of high-voltage gate-oxide pulses on the BTI behavior of SiC MOSFETs," 2020 IEEE International Reliability Physics Symposium (IRPS), Dallas, TX, USA, 2020, pp. 1-6, doi: 10.1109/IRPS45951.2020.9129232.

2025 IEEE Workshop on Wide Bandgap Power Devices and Applications in Asia (WiPDA Asia)

A New Si/SiC Hybrid Interleaved Three-level ANPC Inverter with Cost and Performance Tradeoff

Ruixiao Dong
College of Electrical and Information
Engineering
Hunan University
Changsha, China
ruixiaodong@hnu.edu.cn

Jun Wang
College of Electrical and Information
Engineering
Hunan University
Changsha, China
junwang@hnu.edu.cn

Yulin Wang
College of Electrical and Information
Engineering
Hunan University
Changsha, China
wangyl728@hnu.edu.cn

Yuxing Dai
Wenzhou University
Wenzhou, China
daiyx@hnu.edu.cn

Liuchen Chang
University of New Brunswick
New Brunswick, Canada
lchang@unb.ca

Chao Zhang
Electrical Engineering College
Guizhou University
Guiyang, China
zhangc@gzu.edu.cn

Abstract—The three-level active-neutral-point-clamped (ANPC) inverter is popular in high voltage and high power application. The different switching frequency of devices provides a situation to use full-power Si IGBTs and SiC MOSFETs for cost reduction and efficiency improving. However, the conventional hybrid ANPC topology still faces contradiction between considerable cost and performance further improvement. In this paper, a new Si/SiC hybrid interleaved three-level ANPC inverter based on wide fracrional power processing (WFPP) and Hybrid² concept is proposed towards cost and performance tradeoff. The topology and operation principle is introduced firstly, then the comparison of comprehensive performance including cost, efficiency and thermal stress on devices among proposed topology and conventional two Si/SiC hybrid topology is conducted to highlight the advantage of proposed topology.

Keywords—three-level active-neutral-point-clamped inverter, WFPP, Hybrid², comprehensive performance comparison, cost and performance tradeoff

I. INTRODUCTION

Three-level (3L) active-neutral-point-clamped (ANPC) converter is proposed to shorten the communication path and overcome the unequal power loss distribution of active devices of traditional neutral-point-clamped (NPC) converter, which is attractive at high voltage and high power application.

In reference [1] , an ANPC inverter composed with four line-frequency-low-cost Si IGBTs and two high-frequency-high-cost SiC MOSFETs is built based on specific modulation scheme. In comparison to full-SiC design, the quite cost reduction and high efficiency is achieved. However, the use of different kind devices causes the uneven power loss distribution and makes the thermal stress more concentrate on high frequency devices, which negatively affects the power capacity and reliability of converter. In [2], a megawatt-scale 3L-ANPC inverter containing two Si IGBTs and four SiC MOSFETs is developed for aircraft hybrid-electric propulsion systems, the high efficiency and power density is obtained. The quantitative evaluation about efficiency and thermal stress on devices of these two-type Si/SiC hybrid 3L-ANPC inverters is conducted in [3]. Results show that, the ANPC inverter using four SiC MOSFETs displays more even thermal stress distribution than two SiC MOSFETs while

the higher transfer efficiency at some equal power stages is derived from 2SiC/4Si hybrid use method. Moreover, four full-power SiC devices brings more cost than two used. Generally speaking, the hybrid use of Si and SiC devices is only dependent on the different operation frequency in above three-level converter. Since the 3~8 times cost of SiC material more than Si devices with similar current capacity, even though only two full-power SiC devices are used, the cost is still considerable especailly in high current application.

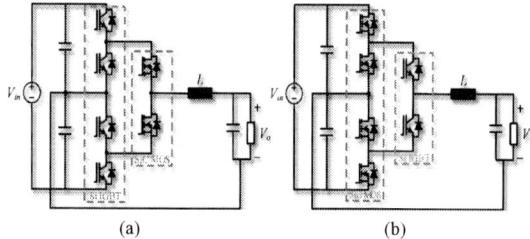

Fig. 1.Conventional Si/SiC hybrid three-level inverter topology. (a) 2SiC&4Si. (b) 4SiC&2Si.

To face this challenge, the Hybrid² concept is proposed [4], there is, replace two full-power SiC MOSFETs with Si/SiC hybrid parallel switches in three-level ANPC converter shown in Fig. 1(a) to maintain efficiency with optimal switching strategy while further reducing cost. The hybrid parallel switch(HyPS) is composed of one larger-capacity Si IGBT and one smaller-capacity SiC MOSFET. Compared with one similar large-capacity SiC MOSFET, the lower cost and nearly high efficiency is achieved through switching sequence optimization [5]. However, the switching frequency of HyPs is limited by the overcurrent problem of smaller-capacity MOSFET at each switching period, which is still a potential hidden trouble for operation reliability of power converter.

To fully highlight the excellent high-frequency performance of wide-bandgap (WBG) devices (SiC MOSFET / GaN HENT) with lower cost, the concept of WBG fractional power processing (WFPP) for power converter is proposed. A Si/SiC hybrid half-bridge(HHB) topology is built towards high power quality, fast dynamic response, high efficiency and cost tradeoff [6], which has been promoted to two-level DC-DC, PFC and inverter converters [7-9]. Si/WBG hybrid half bridge is composed of

979-8-3315-1110-4/25 $31.00 © 2025 IEEE   468

higher-power-lower-frequency Si and lower-power-higher-frequency SiC half bridge, the power between two phases could be distributed freely so that the processed current of each phase can be controlled to be lower than secure operation current of devices. Moreover, the adjustable power sharing and switching frequency brings predictable potential for more even thermal stress distribution and wider output power range of converter.

In this work, a new Si/SiC hybrid interleaved three-level ANPC inverter based on WFPP and Hybrid^2 concept is proposed towards cost and performance tradeoff. The comprehensive comparison in terms of cost and efficiency is conducted among proposed converter and other two converters in [3][4] through simulation.

The reminder of this paper is organized as following. Section II introduces the topology and operation principle of Si/SiC hybrid interleaved three-level ANPC inverter. In section III, the current ripple compensation analysis and dynamic coordinated operation is presented. Section IV shows the cirtical inductance design and control strategy. Section V shows the simulation waveform and comprehensive comparison result in terms of cost, efficiency and themal stress on devices among proposed and traditional 3L-ANPC converters. Section VI concludes this paper.

II. TOPOLOGY AND OPERATION PRINCIPLE

In this section, the topology of Si/SiC hybrid interleaved three-level ANPC inverter is introduced, while how it operates with different switching frequency is presented.

A. Topology

The topology of proposed three-level ANPC inverter based on Si/SiC HHB is presented in Fig. 2. Two series DC bus capacitors are large enough to guarantee the voltage of them equal with half of V_{in}. The full-power-fundamental-frequency operation module consists of four large-capacity Si IGBTs ($S_1 \sim S_4$). The hybrid half bridge module contains two phases: one is composed of two low-frequency-high-power(LFHP) Si IGBTs (S_{L1}, S_{L2}) and main inductance L_1, the other is composed of two high-frequency-low-power(HFLP) SiC MOSFETs (S_{H1}, S_{H2}) and small inductance L_2. In other word,the high-frequency-full-power all-SiC half bridge in fig1(a) is reconfigured as Si/SiC hybrid-frequency interleaved half bridge. When two phases coordinate working, the main power is processed by Si phase with lower switching frequency while auxiliary SiC phase operates with higher switching frequency for remaining power to obtain excellent power quality and higher effective frequency.

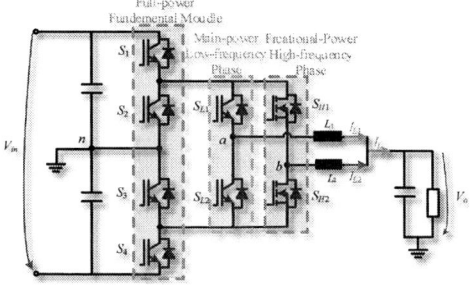

Fig. 2. Topology of proposed three-level ANPC inverter based on Si/SiC hybrid half bridge.

B. Operation Principle

The relationship between two phases' output voltages and switching states is shown in TABLE I. For sake of convenient analysis, switching function S_F, S_L and S_H(1 or 0) is introduced.The switches $S_1 \sim S_4$ operate with fundamental frequency to generate three output voltage levels. During positive half-fundamental period, switches S_1, S_3 turn on while S_2, S_4 turn off, S_F=1; During negative half-fundamental period, switches S_1, S_3 turn off while S_2, S_4 turn on, S_F=0. S_L=1 or S_H=1 means S_{L1} or S_{H1} turns on while S_{L2} or S_{H2} turns off, vice versa. When Si and SiC phase coordinate working with each other, the current communication paths of two phases are independent, so we can separately analyze each phase. For Si phase, during positive half-fundamental period, when S_L=1, the current flow though switches S_1, S_{L1} and main inductance L_1 positively, the output voltage of Si phase is $V_{in}/2$ so that the current increases; when S_L=0, the current flow though switches S_3, S_{L2} and main inductance L_1 positively, the output voltage of Si phase is 0 so that the current decreases. Then during negative half-fundamental period, when S_L=1, the current flow though switches S_2, S_{L1} and main inductance L_1 negatively, the output voltage of Si phase is 0 so that the current decreases; when S_L=0, the current flow though switches S_4, S_{L2} and main inductance L_1 negatively, the output voltage of Si phase is $-V_{in}/2$ so that the current increases. The relationship of SiC phase current flowing paths between switching function S_F and S_H is similar as Si phase. However, the switching frequency of Si IGBTs is lower (10kHz~20kHz) while SiC MOSFETs operate with relatively higher frequency (100kHz~200kHz). The current ripple of Si phase in one low-frequency period is compensated by SiC phase current to obtain smaller total current ripple. Generally speaking, the power flows from DC voltage source and flows through full-power-fundamental-frequency IGBTs every half fundamental period, then it's divided into two parts, Si and SiC phase separately processes main and auxiliary power with lower and higher switching frequency. Finally, the total current with high effective frequency and low ripple is delivered to load.

TABLE I
VARIABLE OUTPUT VOLTAGE OF EACH PHASE WITH SWITCHING STATES

S_1/S_3	S_2/S_4	S_{L1}	S_{L2}	S_{H1}	S_{H2}	U_{an}	U_{bn}
1	0	1	0	1	0	$V_{in}/2$	$V_{in}/2$
1	0	1	0	0	1	$V_{in}/2$	0
1	0	0	1	1	0	0	$V_{in}/2$
1	0	0	1	0	1	0	0
0	1	1	0	1	0	0	0
0	1	1	0	0	1	0	$-V_{in}/2$
0	1	0	1	1	0	$-V_{in}/2$	0
0	1	0	1	0	1	$-V_{in}/2$	$-V_{in}/2$

III. CURRENT RIPPLE COMPENSATION AND DYNAMIC COORDINATED OPERATION

In this section, the analysis of current ripple compensation between two phases with different operation frequency is conducted, then the dynamic coordinated operation is proposed.

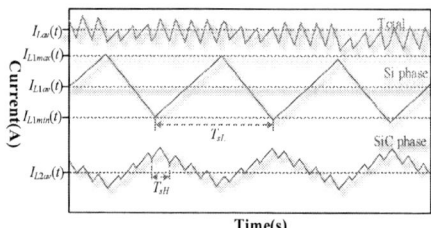

Fig. 3. Detailed Si-phase, SiC-phase and total current waveform in several low-frequency periods.

A. Current Ripple Compensation Analysis

The detailed current waveforms in several low-frequency periods when the current of Si and SiC phase compensates with each other normally is shown in Fig. 3. In one high-frequency period T_{sH}, total current variation ΔI_L is the sum of Si and SiC phase current variation ΔI_{L1H} and ΔI_{L2H}, there is:

$$\Delta I_{LH} = \Delta I_{L1H} + \Delta I_{L2H} \tag{1}$$

The current at positive and negative half-fundamental period is symmetrical, so we only analyze positive half-fundamental period. During one high-frequency period, switching state of low-frequency switches can be seen as constant, so ΔI_{L1H} is derived as:

$$\Delta I_{L1H} = \frac{0.5 S_L V_{in} - V_o}{L_1} T_{sH} \tag{2}$$

$D_{sH}^{S_L}$ denotes the high-frequency conduction ratio of SiC phase upper switch S_{H1}. Subsequently, ΔI_{L2H} is deduced as:

$$\Delta I_{L1H} = \frac{(0.5 D_{sH}^{S_L} V_{in} - V_o)}{L_2} T_{sH} \tag{3}$$

With normal current ripple compensation, the total current variation during one high-frequency period can be seen as 0, $D_{sH}^{S_L}$ is deduced as:

$$D_{sH}^{S_L} = \frac{2V_o}{V_{in}} \frac{L_1 + L_2}{L_1} - S_L \frac{L_2}{L_1} \tag{4}$$

To make SiC phase current ripple completely compensates Si phase, $D_{sH}^{S_L}$ should be more than 0 but less than 1, then the output voltage modulation ratio range for complete ripple compensation is deduced as:

$$\frac{2V_o}{V_{in}} \in [\frac{L_2}{L_1 + L_2}, \frac{L_1}{L_1 + L_2}] \tag{5}$$

In fact inverter operation, DC bus voltage is usually fixed, output voltage is sine wave. From inequation (5), the complete compensation range of output voltage can be ensured.

B. Dynamic Coordinated Working

From above analysis, the current ripple uncompensated regions are near zero crossing and peak value of output voltage. To simplify the analysis, pure resistance load is adopted so that output current is in phase with output voltage. The lower limit value near zero crossing current I_{lim} can be deduced by input DC voltage, output voltage amplitude V_m and output current amplitude I_m, there is:

$$I_{lim} = \frac{V_{in} L_2}{2 V_m (L_1 + L_2)} I_m \tag{6}$$

To address the uncompensated problem near zero crossing region, we can let SiC phase work independently for smaller current during one fundamental period, the maximum safe operation current of SiC devices I_{soa} should be larger than I_{lim}. In addition, the power loss reduction advantage of SiC MOSFETs is highlighted at smaller current because of knee voltage of Si IGBTs. Meanwhile, the independent work of SiC phase can eliminate circulating current at light load. The specific dynamic coordinated work between Si and SiC phases is shown in Fig. 4.

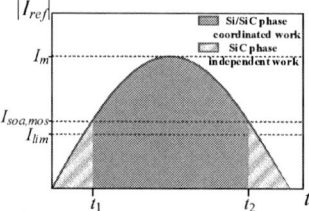

Fig. 4. Dynamic coordinated work of Si and SiC phase.

As for uncompensated problem near peak region, it's not suitable to let large-capacity Si phase work with high switching frequency independently for heavier load since the larger switching loss. The maximum output AC voltage amplitude for current ripple complete compensation is presented as (7).

$$V_{om} < \frac{V_{in} L_1}{2(L_1 + L_2)} \tag{7}$$

IV. INDUCTANCE DESIGN AND CONTROL STRATEGY

In this section, the design consideration of low-frequency Si phase and high-frequency SiC phase inductance is introduced, then the control strategy of proposed three-level ANPC inverter is presented.

A. Indutance Design

We repectively consider how to design the main and auxiliary inductance. For Si phase, the current ripple is regular, the maximum current ripple during one fundamental period can be easily deduces as:

$$\Delta I_{L1pp} = \frac{V_{in}}{8 L_1 f_{sL}} \tag{8}$$

The current ripple rate is defined as the ratio between maximum current ripple and current amplitude, there is:

$$r_{L1} = \frac{V_{in}}{8 L_1 f_{sL} I_m} \tag{9}$$

Since the relatively lower switching frequency, r_{L1} is set as a large value 0.45. Si phase switching frequency is set as 12 kHz, combining with 600 V DC bus voltage and 28 A current amplitude, the main inductance is designed as 500 μH. As for SiC phase, the total current ripple should be considered. The total current ripple presents high-frequency characteristic, the peak to peak current value in one high-frequency period can be deduced as follows:

$$\begin{cases} \Delta I_{Lpp}^{S_L=1} = \dfrac{(L_1+L_2)(0.5V_{in}-V_o)\left[2V_o(L_1+L_2)-V_{in}L_2\right]}{V_{in}L_1^2L_2f_{sH}} \\[4mm] \Delta I_{Lpp}^{S_L=0} = \dfrac{(L_1+L_2)^2 2V_o^2}{V_{in}L_1^2L_2f_{sH}} \end{cases} \quad (10)$$

The total current ripple ratio r_L is set to be 0.15 at full load, there is:

$$Max\left\{\Delta I_{Lpp}^{S_{L1}=1}, \Delta I_{Lpp}^{S_L=0}\right\} \le 0.15 I_m \quad (11)$$

Moreover, the SiC phase switching frequency is set as 120 kHz while the main inductance is 500 µH. Furthermore, the auxiliary inductance is designed as 200 µH.

B. Control Strategy

The current control block diagram of proposed three-level Si/SiC hybrid interleaved ANPC inverter is shown in Fig. 5. For fundamental moudle, the switches are directly controlled by the polarity of reference current. Power sharing ratio K ($0<K<1$) is introduced to directly adjust the current distribution between Si and SiC phase. The control loop for Si phase and control loop for SiC phase are similar except switching frequency. The modulated model predictive control(MMPC) with fixed lower and higher switching frequency is used to trace the reference current with no error. With the discretized control, the duty cycles of kth low-frequency and qth high-frequency preriod can be calculated as:

$$\begin{cases} D_{sL}^k = \dfrac{L_1\Delta I_{L1err}^k f_{sL} + V_o^k + (0.5-0.5S_F)V_{in}}{0.5V_{in}} \\[4mm] D_{sH}^q = \dfrac{L_1\Delta I_{L2err}^q f_{sL} + V_o^q + (0.5-0.5S_F)V_{in}}{0.5V_{in}} \end{cases} \quad (12)$$

To trace the Si phase and total current without error respectively, the current variation of Si and SiC phase should be predicted as follows:

$$\begin{cases} \Delta I_{L1err}^k = KI_{Lr}^k - I_{L1}^k \\[2mm] \Delta I_{L2err}^q = I_{Lr}^q - I_L^q - \Delta I_{L1H}^q \end{cases} \quad (13)$$

Because the switching frequency of SiC phase is 6~10 times than Si phase, the Si phase current variation during one high-frequency period can be deduced as:

$$\Delta I_{L1H}^q = \dfrac{0.5(S_L^q + S_F^q - 1)V_{in} - V_o^q}{L_1 f_{sH}} \quad (14)$$

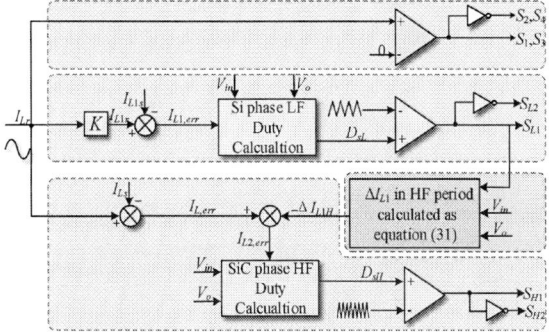

Fig. 5. The current control block diagram of proposed three-level Si/SiC hybrid interleaved ANPC inverter.

V. SIMULATION AND COMPARISON

In this section, the operation simulation result of the inverter is presented firstly, then the cost, transfer efficiency and thermal stress on devices are compared between proposed and traditional hybrid three-level topologys.

A. Operation Simulation Result

TABLE I
PARAMETERS OF THREE-LEVEL SI/SIC HYBRID INTERLEAVED INVERTER

Parameters	Value
Input Voltage	600 V
Full Power	3.3 kW
Modulation Ratio	0.55~0.78
Total Current Ripple	<0.15 @full load
Output Voltage THD	<2% @full load
SiC phase switching frequency	120 kHz

Table I lists the electrical parameters of proposed three-level Si/SiC hybrid interleaved inverter. Based on the control strategy, we use PLECS software to simualte the normal operation of proposed inverter. The simulation operation waveform at DC bus voltage of 600 V, current amplitude of 25A and output power of 2.7kW is shown in Fig. 6. Two phases dynamically coordinates with each other, the total current switching ripple displays high frequency spectrum characteristic so that the excellent output power quality is obtained.

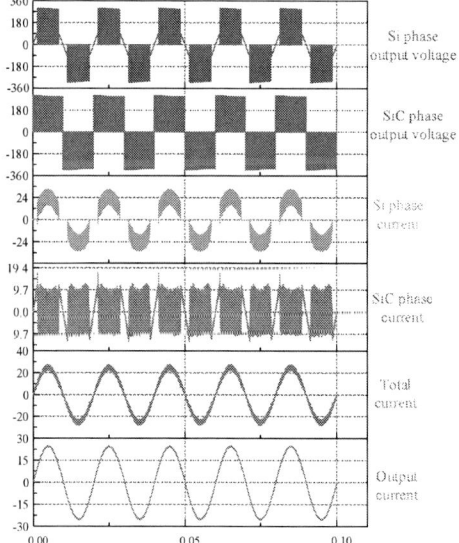

Fig. 6. The simulation operation waveform at DC bus voltage of 600 V, current amplitude of 25A and output power of 2.7kW.

B. Cost and Performance Comparison

Table II lists the specific main components and their cost of proposed and traditional three-level ANPC inverters. The unit price of power devices and inductors is considered on a per-1000 basis. In comparsion conventional 4SiC & 2Si and 4Si & 2SiC topologies, the proposed three-level hybrid interleaved inverer achieves the least total cost.

979-8-3315-1110-4/25 $31.00 © 2025 IEEE

TABLE II
MAIN COMPONENTS AND COST COMPARSION OF PROPOSED AND
CONVENTIONAL ANPC INVERTERS

Components	4SiC & 2Si	4Si & 2SiC	Proposed
Devices	IKW40N65H5×2 ($1.86×2) C3M0045065D×4 ($11.59×4)	IKW40N65H5×4 ($1.86×4) C3M0045065D×2 ($11.59×2)	IKW40N65H5×6 ($1.86×6) C3M0120065D×2 ($5.09×2)
Inductors	500μH / 30A ($3.71×1)	500μH /30A ($3.71×1)	500μH / 25A ($3.05×1) 200μH/15A ($1.90×1)
Driver ICs	Si: UCC27714×1 ($1.12×1) SiC: Si8261×4 ($1.25×4)	Si: UCC27714×2 ($1.12×2) SiC: Si8261×2 ($1.25×2)	Si: UCC27714×3 ($1.12×3) SiC: Si8261×2 ($1.25×2)
Total Cost	$59.91	$39.07	$32.15

Fig. 7 shows the simulated transfer efficiency of proposed and conventional three-level ANPC inverters using thermal simulation of PLECS software. The corresponding thermal simulated models of used devices are downloaded from official website of the manufactures, the outside thermal impedance for devices is generally set as 4.75 K/W while the ambient temperature is designed as 25°C. The power sharing ratio of proposed ANPC inverter is set as 0.7 while the switching frequency of conventional 4SiC & 2Si and 4Si & 2SiC inverters is set as 48 kHz. Moreover, the simulated peak junction temperatures of devices at different power stages are shown in Fig. 8(a), (b) and (c). At the light load ($\leq$ 500 W), SiC phase of proposed ANPC inverter works independently. Since the smaller capacity of SiC MOSFETs, the transfer efficiency is lower while the junction temperature is higher than conventional inverters; At the medium load (500~2700W), the transfer efficiency of proposed ANPC inverter exhibits the minor reduction in comparison to conventional inverters. The knee voltages of Si IGBTs at smaller current play an negative role in operation progress; At the heavier load (2700~3300W), the negative effect of Si IGBTs' knee voltages declines while the more low-frequency advantages of large-capacity IGBT demonstrates. The transfer efficiency of proposed inverter achieves near or even exceeding that of other two conventional inverters.

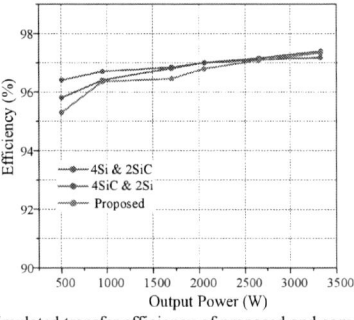

Fig. 7. The simulated transfer efficiency of proposed and conventional three-level ANPC inverter.

Fig. 9 shows the maximum peak junction temperature among devices in three types of three-level ANPC inverter. In comparison to conventional 4Si & 2SiC and 4SiC & 2Si inverter, the proposed hybrid interleaved topology achieves the more even thermal stress distribution among devices,

the overheat problem of high-frequency devices is solved at higher power stages.

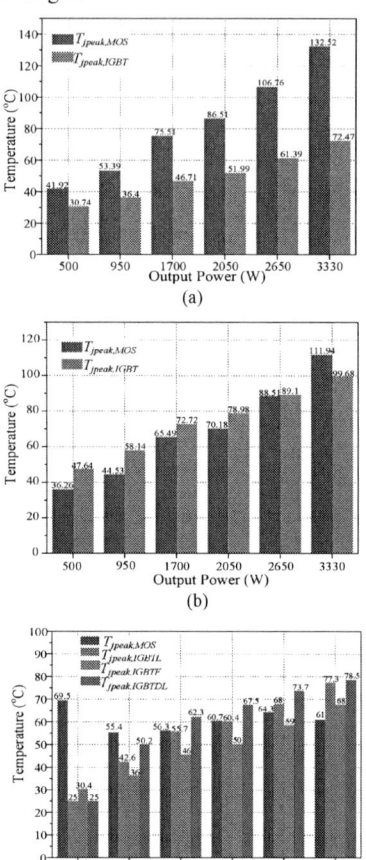

Fig. 8. Simulated peak junction temperatures of devices at different power stages. (a) 4Si & 2SiC. (b) 4SiC & 2Si. (c) Proposed.

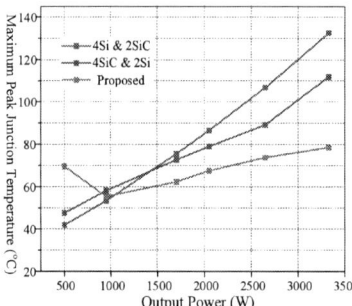

Fig. 9. The simulated transfer efficiency of proposed and conventional three-level ANPC inverter.

VI. CONCLUSION

In this article, a new Si/SiC hybrid interleaved three-level ANPC inverter based on wide fracrional power processing (WFPP) and Hybrid[2] concept is proposed for cost and performance tradeoff. Then dynamic coordinated work mode between Si/SiC phase is utilized for output power quality optimization. Finally, the cost, transfer efficiecny and thermal stress on devices are compared among proposed and conventional 4Si-&-2SiC-design and 4SiC-&-2Si-design three-level ANPC inverter through simulation.

REFERENCE

[1] Q. Guan et al., "An extremely high efficient three-level active neutralpoint-clamped converter comprising SiC and Si hybrid power stage," IEEE Trans. Power Electron., vol. 33, no. 10, pp. 8341–9352, Oct. 2018.

[2] D. Zhang, J. He, and D. Pan, "A megawatt-scale medium-voltage high efficiency high power density "SiC+Si" hybrid three-level ANPC inverter for aircraft hybrid-electric propulsion systems," IEEE Trans. Ind. Appl., vol. 55, no. 6, pp. 5971–5980, Nov./Dec. 2019.

[3] L. Zhang *et al.*, "Evaluation of Different Si/SiC Hybrid Three-Level Active NPC Inverters for High Power Density," in *IEEE Transactions on Power Electronics*, vol. 35, no. 8, pp. 8224-8236, Aug. 2020.

[4] T. Xia *et al.*, "A Hybrid Three-Level Hybrid Switch (Hybrid2) Active Neutral-Point-Clamped Converter with Optimal Switching Strategy," in *IEEE Journal of Emerging and Selected Topics in Power Electronics, Early Access.*

[5] J. Wang, Z. Li, X. Jiang, C. Zeng and Z. J. Shen, "Gate Control Optimization of Si/SiC Hybrid Switch for Junction Temperature Balance and Power Loss Reduction," in *IEEE Transactions on Power Electronics*, vol. 34, no. 2, pp. 1744-1754, Feb. 2019.

[6] C. Zhang et al., "WBG and Si Hybrid Half-Bridge Power Processing Toward Optimal Efficiency, Power Quality, and Cost Tradeoff," in IEEE Transactions on Power Electronics, vol. 37, no. 6, pp. 6844-6856, June 2022.

[7] C. Zhang et al., "A New PFC Design With Interleaved MHz-Frequency GaN Auxiliary Active Filter Phase and Low-Frequency Base Power Si Phase," in IEEE Journal of Emerging and Selected Topics in Power Electronics, vol. 8, no. 1, pp. 557-566, March 2020.

[8] C. Zhang, K. Qu, B. Hu, J. Wang, X. Yin and Z. J. Shen, "A High-Frequency Dynamically Coordinated Hybrid Si/SiC Interleaved CCM Totem-Pole Bridgeless PFC Converter," in IEEE Journal of Emerging and Selected Topics in Power Electronics, vol. 10, no. 2, pp. 2088-2100, April 2022.

[9] K. Qu, C. Zhang, W. Chen, B. Hu, J. Chen and J. Wang, "A Hybrid Si/SiC Interleaved Bidirectional DC-DC Converter to Optimal Power Quality, Efficiency, and Cost Tradeoff," 2021 IEEE Energy Conversion Congress and Exposition (ECCE), Vancouver, BC, Canada, 2021, pp. 2001-2004.

2025 IEEE Workshop on Wide Bandgap Power Devices and Applications in Asia (WiPDA Asia)

Turn-off Analysis and Modeling of Releasing Loss in Snubber Capacitor Self-Balancing Circuits for Series-Connected SiC MOSFETs Applied to High-Voltage Pulsed Power Systems

Jiaxuan Niu
School of Electrical Engineering
Xi'an Jiaotong University
Xi'an, China
niujiaxuan0810@126.com

Xu Cheng
Zhuhai Power Supply Bureau of
Guangdong Power Grid Co.,Ltd.
DC PowerDistribution and Consumption
Technology Research Center of
Guangdong Power Grid Co.,Ltd.
Zhuhai, China
664157102@qq.com

Xu Yang
School of Electrical Engineering
Xi'an Jiaotong University
Xi'an, China
yangxu@mail.xjtu.edu.cn

Yong Chen
Zhuhai Power Supply Bureau of
Guangdong Power Grid Co.,Ltd.
DC PowerDistribution and Consumption
Technology Research Center of
Guangdong Power Grid Co.,Ltd.
Zhuhai, China
35665035@qq.com

Fan Zhang
School of Electrical Engineering
Xi'an Jiaotong University
Xi'an, China
zhangfan1990@xjtu.edu.cn

Kexin Zhao
School of Electrical Engineering
Xi'an Jiaotong University
Xi'an, China
kexin@stu.xjtu.edu.cn

Abstract—**SiC MOSFETS demonstrate superior characteristics such as high switching speed and high breakdown voltage, the development of series-connected SiC MOSFETS-based pulse power switches has emerged as a primary research truth. To meet the dynamic voltage balancing demands of series-connected SiC MOSFETS under ultra-fast switching conditions, the snubber capacitor self-balancing circuit provides advantages including simplicity, high reliability, and full-speed switching capability. However, the absence of quantitative analysis on releasing loss in this topology limits circuit design and optimization. Meanwhile, the modeling of the series-connected SiC MOSFETS turn-off process under ohmic loads is understudied, which increases the difficulty of loss modeling. This paper investigates the working principles and turn-off process of the topology, simplifies the model under ohmic load conditions, and proposes a method for analyzing the drain-source voltage rise rate of series-connected SiC MOSFETS under different delay combinations while establishing a releaseing loss model. Moreover, this method is scalable for extending the number of series-connected devices. The trends demonstrated by the model is verified by comparison of multiple simulation results.**

Keywords—*SiC MOSFET, series-connection, voltage balanceing*

I. INTRODUCTION

High-voltage pulse power systems, characterized by high voltage, high current, and high repetition frequency, are widely used in multiple fields, including industrial manufacturing, medical applications, and scientific research [1]. The critical component pulse power switches require high voltage withstand capability, high repetition frequency, and fast triggering. SiC MOSFETs, owing to lightweight design, extended switching lifetime, and simplified triggering [2], have found extensive application in pulse power technology.

Current single SiC MOSFETs cannot meet the voltage withstand requirements (tens to hundreds of kV) for pulse power switches, series connection of multiple SiC MOSFETs is required. However, mismatches in parasitic capacitance and gate signal delays can lead to severe dynamic voltage imbalance, resulting in overvoltage failure issues [3]. Passive snubber circuits are simple to implement and offer advantages in voltage balancing and reliability, but the addition of passive components increases system cost and size while reducing power density. The voltage rise time with snubber capacitor exceeds 700 ns [4], significantly longer than the typical rise time of SiC MOSFETs, and switching losses increase significantly, necessitating a trade-off between losses and voltage balancing [5]. A multi-step packaging technique was proposed that redistributes parasitic capacitance [6], reducing voltage-sharing imbalance among series-connected devices. However, this approach overlooks other parameter mismatches beyond packaging layout and fails to achieve perfect voltage balancing. An active gate driver with a voltage slew rate controller is designed in [7][8], which adjusts the turn-off speed by modulating the control voltage. However, this method imposes high demands on the response speed of the sampling and control stages, and control transients occur when device voltages are unbalanced. Thus, additional clamping circuits are still required for overvoltage protection. Compared to other solutions, the self-balancing snubber capacitor circuit offers simplicity in implementation and advantages in both voltage balancing and reliability [9], enabling full-speed switching unlike RC snubbers.

Current research lacks quantitative models for releasing loss in self-balancing snubber circuits. Moreover, most modeling studies on SiC MOSFETs are based on inductive loads

---

This work was supported by the China Southern Power Grid Company, Ltd., Science and Technology Project: Research on Insulation Design and Parasitic Parameter Optimization Technology for High Voltage SiC Devices and Development of Half Bridge Device Packaging under Grant 030400KC23090015 and Grant GDKIXM20231031.

979-8-3315-1110-4/25 $31.00 © 2025 IEEE

[10], while research on resistive-load models for SiC MOSFETs is scarce and suffers from low accuracy [11]. Models for series-connected SiC MOSFETs are also limited. This paper provides a detailed analysis of the working principle of this topology and the energy accumulation in snubber capacitors during the turn-off process. By introducing appropriate simplifications, an analytical method is proposed to model the drain-source voltage rise rate and discharge loss under different delay combinations. Additionally, the relationship between delay, voltage, and releasing loss is analyzed and validated through simulations. This research facilitates the design of clamping thresholds for snubber circuits and the calculation of voltage-sharing performance for series-connected SiC MOSFETs, thereby facilitates device selection and subsequent topology improvements.

II. ANALYSIS OF SERIES-CONNECTED SiC MOSFETS SNUBBER CAPACITOR SELF-BALANCING CIRCUIT PRINCIPLE AND TURN-OFF PROCESS

As shown in Fig. 1(a), the snubber capacitor self-balancing circuits consists of three parts: snubber circuit, self-balancing circuit and releasing circuit [9]. The snubber circuit mitigates turn-off trends of earlier switches (Fig. 1(b)). Taking Q_3 turn-off fastest as an example, u_{ds3} rises first during turn-off. When u_{ds3} exceeds snubber capacitor C_3 voltage u_{C3}, snub-

ber diode D_3 conducts and reducing du_{ds3}/dt. The self-balancing circuits comprises diodes D_n and current-limiting resistors R_{an} ($n = 1,2,3$). If u_C of higher-position switch exceeds u_C of lower-position switch, the C_n, self-balancing circuits, and switches form a circuit when series-connected SiC MOSFET conducts, achieving voltage balance, as shown in Fig. 1(c).

To prevent u_C rise per switching cycle and exceeding safe limits caused by continuous switching speed differences, excess energy must be released. The self-balancing loop transfers energy from top to bottom capacitors. All u_{dsn} can be maintained within safe ranges by paralleling a release circuit with the snubber capacitor of bottom switch and setting clamp voltage U_{CLP}

The simplified u_{ds} and current waveforms at turn-off instant are shown in Fig. 2. Unlike the double-pulse test, the rise of u_{ds} and the decrease of load current i_d are performed at the same time during turn-off under resistive load. Assuming sequential turn-off from Q_3 to Q_1 with $u u_{Cn} = U_{CLP}$ before turn-off. After u_{ds3} and u_{ds2} reaches U_{CLP} respectively, the snubber circuit significantly reduces du_{dsn}/dt while i_d charges C_3 and C_2. Although C_n far exceeds the junction capacitance of SiC MOSFET, the rapid turn-off causes capacitor charging current i_{Cn} to decline immediately after reaching i_d. The switches is completely turn-off and i_d decreases to 0, when $u_{ds1} + u_{ds2} + u_{ds3} = U_{dc}$. Snubber capacitor voltages remain constant at U_{C3} and U_{C2}, while each u_{ds} shows gradual equalization due to leakage currents. The green shadow and the gray line shadow in Fig. 2 indicate the energy absorbed by C_3 and C_2 during turn-off respectively, which needs to be consumed in the release circuit.

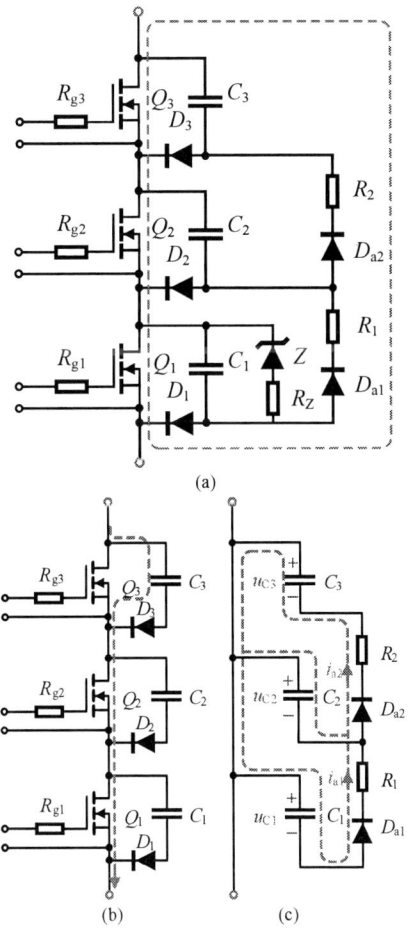

(a)

(b)   (c)

Fig.1. (a) Snubber capacitor self-balancing circuit. (b) Principle of snubber circuit. (c) Principle of self-balancing circuit.

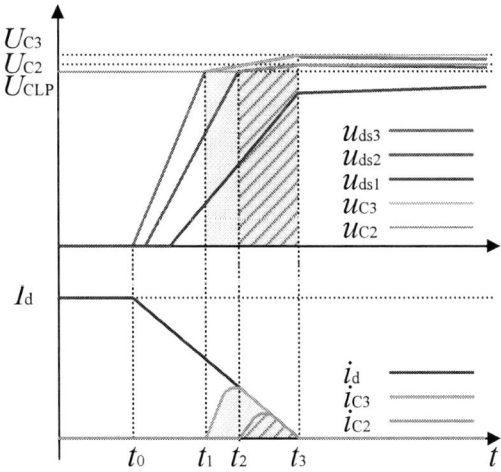

Fig. 2. Simplified voltage and current waveform at turn-off instant.

III. DELAY-BASED SLEW RATE AND RELEASING LOSS MODEL

The charging duration and i_{Cn} directly affect the energy accumulation of snubber capacitors. The duration is related to the du_{dsn}/dt of Q_n, and du_{ds}/dt is related to the turn-off delay. Assuming identical driving conditions and parasitic parameters for all switches, using the SiC MOSFET model under resistive load [12] (Fig. 3), assuming simultaneous turn-off in saturation region, then:

979-8-3315-1110-4/25 $31.00 © 2025 IEEE   475

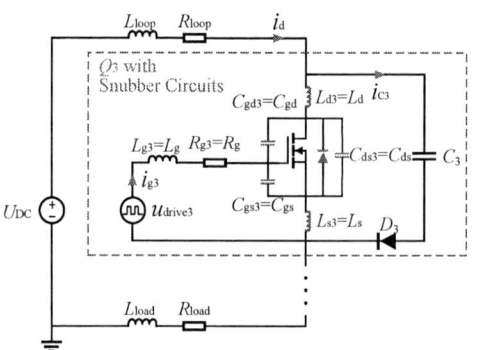

Fig. 3. Equivalent circuit model of series-connected SiC MOSFETs and snubber circuits.

$$u_{ds1} + u_{ds2} + u_{ds3} = U_{DC} - (R_{loop} + R_{load})i_d$$
$$-(L_{loop} + L_{load} + 3L_d + 3L_s)\frac{di_d}{dt} \quad (1)$$

Assuming Q_3 turns off earliest, while Q_1 and Q_2 turn off equally slowest with sufficient driving delay, when Q_3 is in the saturation region, Q_1 and Q_2 remain conducting. Since $u_{ds1} = u_{ds2} = i_d \cdot R_{ds(on)}$, it follows that:

$$u_{ds3} = U_{DC} - (R_{loop} + R_{load} + 2R_{ds(on)})i_d$$
$$-(L_{loop} + L_{load} + 3L_d + 3L_s)\frac{di_d}{dt} \quad (2)$$

Compared to the load resistance of the pulse generator, the on-resistance ($R_{ds(on)}$) of SiC MOSFETs is negligible. As shown above, under the second operating condition, du_{ds3}/dt is approximately three times that of the simultaneous turn-off scenario. Due to delay effects, turn-off states exhibit three patterns: identical, overlapping, and completely staggered. Considering the number of SiC MOSFETs in series-connection, the entire turn-off process becomes significantly complex. To simplify the analysis, the turn-off process of SiC MOSFETs is divided into three linearized stages [11][13], as illustrated in Fig. 4(a). The major voltage rise occurs during t_0-t_1. Although u_{ds} exhibits minor variations before t_0, these are negligible compared to the u_{ds} when completely turn-off. Thus, the analysis of charging/discharging dynamics of parasitic capacitances in this stage are omitted, and the system is simplified to the model in Fig. 4(b), with focus on the t_0-t_1 stage. This simplification is also reflected in the schematic diagrams of Fig. 5 and Fig. 6.

Fig. 4(b)-(d) present the equivalent circuits of individual MOSFETs and their snubber networks across the above three stages. Although u_{ds} of the earlier turn-off switch will exceed U_{CLP}, reaching U_{Cn}, LTspice simulations show $(U_{Cn}-U_{CLP})/U_{CLP} < 0.048$ with $\pm25\ ns$ driving delays. In simplified analysis, justifying $U_{Cn} \approx U_{CLP}$ for earlier turn-off switches.

Combined with the turn-off condition, three SiC MOSFETs connected in series can be divided into 9 delay conditions. Take $Q_F \leftrightarrow Q_S \cap Q_L$ as an example to analyze the du_{ds}/dt of each switch (F, S and L in the subscript indicate the first, second and last turn-off switch respectively, the symbol $\leftrightarrow$ denotes completely staggered turn-off, $\cap$ indicates overlap turn-off, and = in conditions represents identical turn-off delays). The schematic diagram of $Q_F \leftrightarrow Q_S \cap Q_L$ is shown in Fig. 5 (a), the turn-off process can be divided into four stages. Stage ①: only Q_F is turning-off, it can be obtained:

$$u_{dsF1} = U_{DC} - (R_{loop} + R_{load} + 2R_{ds(on)})i_d$$
$$-(L_{loop} + L_{load} + 3L_d + 3L_s)\frac{di_d}{dt} \quad (3)$$

Stage ②: Q_F turn-off completely, only Q_S is turning-off, it can be obtained:

$$u_{dsS2} = U_{DC} - U_{CLP} - (R_{loop} + R_{load} + R_{ds(on)})i_d$$
$$-(L_{loop} + L_{load} + 2L_d + 2L_s)\frac{di_d}{dt} \quad (4)$$

Stage ③: Q_S and Q_L are turning off, it can be obtain:

$$u_{dsS3} + u_{dsL3} = U_{DC} - U_{CLP} - (R_{loop} + R_{load})i_d$$
$$-(L_{loop} + L_{load} + 2L_d + 2L_s)\frac{di_d}{dt} \quad (5)$$

Stage ④: Q_S turn-off completely, only Q_L is turning off, it can be obtained:

$$u_{dsL4} = U_{DC} - 2U_{CLP} - (R_{loop} + R_{load})i_d$$
$$-(L_{loop} + L_{load} + L_d + L_s)\frac{di_d}{dt} \quad (6)$$

Differentiating (3)-(6) with respect to time, di''_d/dt can be neglected since i_d approximates a multi-stage first-order function. Setting $D(t) = -(R_{loop} + R_{load})di_d/dt$ yields $du_{dsF1}/dt = D(t)$, $du_{dsS2}/dt \approx D(t)$, $du_{dsS3}/dt \approx du_{dsL3}/dt = D(t)/2$, $du_{dsL4}/dt = D(t)$, as shown in Fig. 5. (b). Where $t_S - t_0 = T_S$ and $t_L - t_0 = T_L$ represent turn-off delays of Q_S and Q_L respectively. The turn-off duration of Q_n

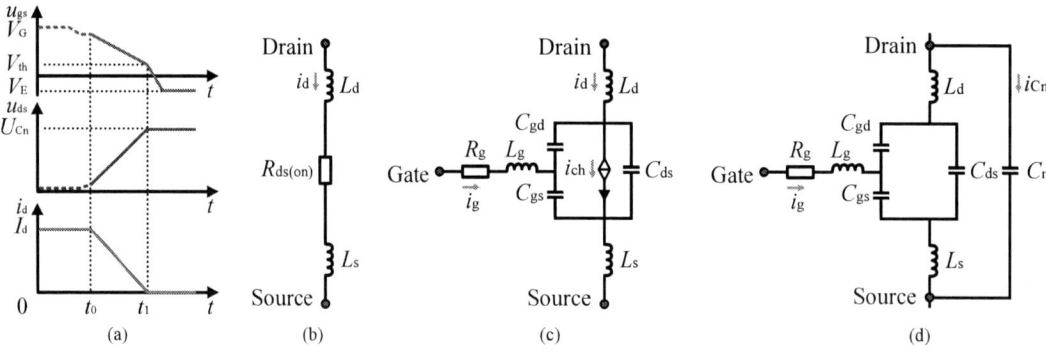

Fig. 4. (a) Simplified SiC MOSFET turn-off process. (b) Equivalent circuit before t_0. (c) Equivalent circuit during $t_0 - t_1$. (d)Equivalent circuit after t_1.

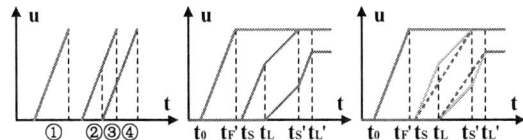

Fig.5. (a) Schematic diagram for $Q_F \leftrightarrow Q_S \cap Q_L$ (delay condition H). (b) Schematic diagram of du_{ds}/dt piecewise analysis. c) Schematic diagram of du_{ds}/dt simplified analysis.

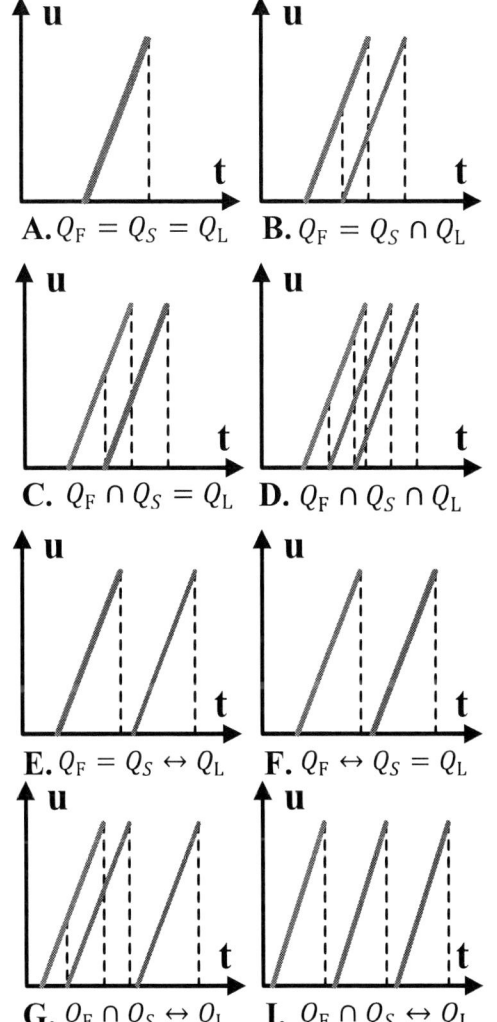

Fig.6. Schematic diagram of other delay conditions of three SiC MOSFETs connected in series.

can be obtained by solving the listed equations. For computational simplification, u_{ds} during turn-off is approximated as a first-order function as shown in Fig. 5(c), neglecting insignificant terms and it can be obtained as (7). The other 8 delay conditions for 3 SiC MOSFETs connected in series are illustrated in Fig. 6. Similarly, du_{ds}/dt and turn-off duration for each MOSFET under these conditions can be derived. For simplification, negligible terms have been omitted, as shown in TABLE I and II. This methodology can

be extended to analyze larger numbers of SiC MOSFETs connected in series.

$$\frac{du_{dsF}}{dt} = D(t)$$
$$\frac{du_{dsS}}{dt} = \frac{D(t)}{2} \qquad (7)$$
$$\frac{du_{dsL}}{dt} = \frac{(U_{DC} - 2U_{CLP})D(t)}{U_{DC} - U_{CLP}}$$

Linearizing u_{gs}, the model yields:

$$R_g i_g + (L_g + L_s)\frac{di_g}{dt} + L_s\frac{di_d}{dt} + u_{gs} = V_E \qquad (8)$$

$$i_g = C_{gs}\frac{du_{gs}}{dt} + C_{gd}\left(\frac{du_{gs}}{dt} - \frac{du_{ds}}{dt}\right) \qquad (9)$$

$$\frac{di_d}{dt} = \frac{V_E - u_{gs} - R_g i_g}{L_s} \qquad (10)$$

Combined with the turn-off waveform, the E_T can be obtained as:

$$E_T = E_{CF} + E_{CS} = \int_{t_F'}^{t_L'} u_{CF} \cdot i_{CF}dt + \int_{t_S'}^{t_L'} u_{CS} \cdot i_{CS}dt$$

$$= U_{CLP}g_F \int_{t_F'}^{t_L'}\left(\frac{U_{DC}}{R_{load}} + \frac{di_d}{dt}(t - t_0)\right)dt \qquad (11)$$

$$+ U_{CLP}g_S \int_{t_S'}^{t_L'}\left(\frac{U_{DC}}{R_{load}} + \frac{di_d}{dt}(t - t_0)\right)dt$$

Where $g_F(T_S, T_L)$ and $g_S(T_S, T_L)$ represent the ratios of time-integrated i_{CF} and i_{CS} to integrated i_d, respectively.

TABLE I. du_{DSN}/dt UNDER EACH DELAY CONDITION

Condition	du_{dsF}/dt	du_{dsS}/dt	du_{dsL}/dt
A			$D(t)/3$
B	$D(t)/3$		$(U_{DC} - 2U_{CLP})D(t)/U_{DC}$
C	$D(t)/3$		$(U_{DC} - U_{CLP})D(t)/2U_{DC}$
D		$D(t)/3$	$(U_{DC} - 2U_{CLP})D(t)/U_{DC}$
E		$D(t)/2$	$D(t)$
F	$D(t)$		$D(t)/2$
G		$D(t)/2$	$D(t)$
H	$D(t)$	$D(t)/2$	$\dfrac{(U_{DC} - 2U_{CLP})D(t)}{U_{DC} - U_{CLP}}$
I			$D(t)$

IV. SIMULATION AND ANALYSIS

The voltage and current of three SiC MOSFETs connected in series and snubber circuits are simulated in LTspice, where U_{DC} is 1800V, R_{load} is 60Ω, and MOSFET is chosen as C3M0016120D. The switching period is 0.2ms and the duty cycle is 50%. In addition, C_n is designed to be 30nF, self-balancing loop R_{an} is 3Ω, and U_{CLP} is 630V. Fig.7 shows the delay-voltage-E_T relationship for all combinations at a interval of 25ns with a turn-off delay of ±25ns, where the numbers in the combination represent turn-off delays from Q_3 to Q_1 in ns.

Results indicate that 1) switch with earlier turn-off exhibits higher u_{ds} and u_C at turn-off instant. Notably, when Q_1 functions as Q_F, u_{ds1} and u_{C1} rapidly rise then immediately drop to approach U_{CLP},2) larger delay discrepancies induce greater bleeder losses, 3) when both switches turn-off equally slowly,

TABLE II. TURN-OFF TIME t'_N OF EACH SWITCH AND TOTAL TURN-OFF DURATION $t_T(t, T_S, T_L)$ UNDER EACH DELAY CONDITION

Condition	$t'_F - t_0$	$t'_S - t_S$	$t'_L - t_L$	$t_T(t, t_S, t_L)$
A	$U_{DC}/D(t)$			$U_{DC}/D(t)$
B	$3U_{CLP}/D(t) - (t_L - t_0)/2$		$U_{DC}/D(t) - (t_L - t_0)$	$U_{DC}/D(t)$
C	$3U_{CLP}/D(t) - 2(t_S - t_0)$	$U_{DC}/D(t) - (t_S - t_0)$		$U_{DC}/D(t)$
D	$\dfrac{3U_{CLP}}{D(t)} - \dfrac{3(t_S - t_0)}{2} - \dfrac{t_L - t_0}{2}$	$\dfrac{3U_{CLP}}{D(t)} - \dfrac{t_L - t_S}{2}$	$\dfrac{U_{DC}}{D(t)} - (t_L - t_0)$	$\dfrac{U_{DC}}{D(t)}$
E	$\dfrac{2U_{CLP}}{D(t)}$		$\dfrac{U_{DC} - 2U_{CLP}}{D(t)}$	$\dfrac{U_{DC} - 2U_{CLP}}{D(t)} + t_L - t_0$
F	$\dfrac{U_{CLP}}{D(t)}$	$\dfrac{U_{DC} - U_{CLP}}{D(t)}$		$\dfrac{U_{DC} - U_{CLP}}{D(t)} + t_S - t_0$
G	$\dfrac{2U_{CLP}}{D(t)} - t_S + t_0$		$\dfrac{U_{DC} - 2U_{CLP}}{D(t)}$	$\dfrac{U_{DC} - 2U_{CLP}}{D(t)} + t_L - t_0$
H	$\dfrac{U_{CLP}}{D(t)}$	$\dfrac{2U_{CLP}}{D(t)} - t_L + t_S$	$\dfrac{U_{DC} - U_{CLP}}{D(t)} - t_L + t_S$	$\dfrac{U_{DC} - U_{CLP}}{D(t)} + t_S - t_0$
I	$\dfrac{U_{CLP}}{D(t)}$		$\dfrac{U_{DC} - 2U_{CLP}}{D(t)}$	$\dfrac{U_{DC} - 2U_{CLP}}{D(t)} + t_L - t_0$

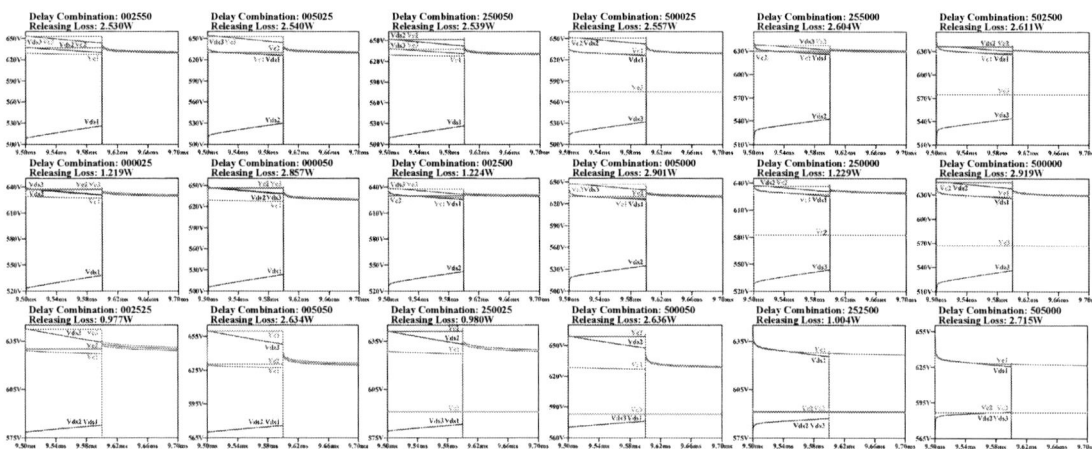

Fig.7. Voltage performance and releasing loss for each delay combination.

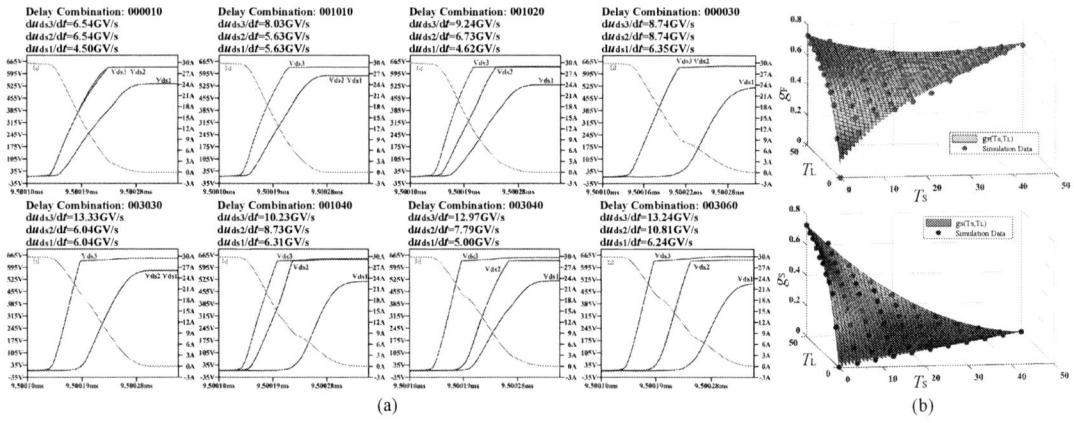

Fig. 8. (a) du_{ds}/dt of delay combinations. b) $g_F(T_S, T_L)$ and $g_S(T_S, T_L)$.

only C_F charges, resulting in smaller E_T compared to other conditions where both C_F and C_S charge, 4) While E_T shows minimal variation with same delays at different locations, however delayed switch higher/earlier switch lower configurations exhibit lower E_T due to extended releasing paths consuming energy in the self-balancing loop.

Fig. 8 (a) shows the u_{ds} and i_d waveforms and the du_{ds}/dt of each MOSFETs during the turn-off under different delay conditions. It can be seen that larger delay discrepancies result in increased du_{ds}/dt in first turn-off switches, consistent with prior analysis. di_d/dt is also related to the delay, larger delays increase t_T, leading to reduced $D(t)$ and slower du_{dsL}/dt during turn-off of Q_L. Fig. 8 (b) shows $g_F(T_S, T_L)$ and $g_S(T_S, T_L)$ fitted according to simulation data points, which demonstrate that increased T_L elevates both g_F and g_S, while larger T_S reduces g_S, validating the principle that minimized losses occur when T_S approaches T_L.

V. Conclusions

In this paper, the topology and principle of the series-connected SiC MOSFETs snubber capacitor self-balancing circuit are analyzed, and the turn-off model under resistive load is simplified. For three SiC MOSFETs connected in series, the du_{ds}/dt and turn-off time under 9 different delay conditions are calculated, and establishes a discharge loss modeling method for this topology under ohmic load for varying switching delays. The relationship between the topological voltage, the bleed loss and the turn-off delay is summarized by the simulation in LTspice according to different delay combinations, and the law of the model is verified. The above research is conducive to the design of U_{CLP}, the selection of devices and the further improvement of voltage balancing ability for this topology.

References

[1] Y. He et al., "A Polarity-Adjustable Nanosecond Pulse Generator Suitable for High Impedance Load," *IEEE Trans. Plasma Sci.*, vol. 48, no. 10, pp. 3409–3417, Oct. 2020.

[2] X. She, A. Q. Huang, O. Lucia, and B. Ozpineci, "Review of Silicon Carbide Power Devices and Their Applications," *IEEE Trans. Ind. Electron.*, vol. 64, no. 10, pp. 8193–8205, Oct. 2017.

[3] L. F. S. Alves, P. Lefranc, P.-O. Jeannin, and B. Sarrazin, "Advanced voltage balancing techniques for series-connected SiC-MOSFET devices: A comprehensive survey," *Power Electronic Devices and Components*, vol. 7, p. 100055, Apr. 2024.

[4] I. Lee and X. Yao, "Active Voltage Balancing of Series Connected SiC MOSFET Submodules Using Pulsewidth Modulation," *IEEE Open J. Power Electron.*, vol. 2, pp. 43–55, 2021.

[5] X. Lin, L. Ravi, R. Burgos, and D. Dong, "Hybrid Voltage Balancing Approach for Series-Connected SiC MOSFETs for DC–AC Medium-Voltage Power Conversion Applications," *IEEE Trans. Power Electron.*, vol. 37, no. 7, pp. 8104–8117, Jul. 2022.

[6] L. F. S. Alves, P. Lefranc, P.-O. Jeannin, B. Sarrazin, and J.-C. Crebier, "Multi-Step Packaging Concept for Series-Connected SiC MOSFETs," in *2019 21st European Conference on Power Electronics and Applications (EPE '19 ECCE Europe)*, Genova, Italy: IEEE, Sep. 2019, p. P.1-P.10.

[7] A. Marzoughi, R. Burgos, and D. Boroyevich, "Active Gate-Driver With dv/dt Controller for Dynamic Voltage Balancing in Series-Connected SiC MOSFETs," *IEEE Trans. Ind. Electron.*, vol. 66, no. 4, pp. 2488–2498, Apr. 2019.

[8] K. Sun et al., "Modeling, Design, and Evaluation of Active dv/dt Balancing for Series-Connected SiC MOSFETs," *IEEE Trans. Power Electron.*, vol. 37, no. 1, pp. 534–546, Jan. 2022.

[9] F. Zhang, X. Yang, W. Chen, and L. Wang, "Voltage Balancing Control of Series-Connected SiC MOSFETs by Using Energy Recovery Snubber Circuits," *IEEE Trans. Power Electron.*, vol. 35, no. 10, pp. 10200–10212, Oct. 2020.

[10] X. Wang, Z. Zhao, K. Li, Y. Zhu, and K. Chen, "Analytical Methodology for Loss Calculation of SiC MOSFETs," *IEEE J. Emerg. Sel. Topics Power Electron.*, vol. 7, no. 1, pp. 71–83, Mar. 2019.

[11] R. Risch and J. Biela, "Nanosecond switching of ohmic loads using SiC MOSFETs in ultra-low inductive PCB-packages," in *2019 21st European Conference on Power Electronics and Applications (EPE '19 ECCE Europe)*, Genova, Italy: IEEE, Sep. 2019, p. P.1-P.10.

[12] Z. Ma, Y. Pei, L. Wang, Q. Yang, X. Lu and F. Yang, "Research on switching characteristics of SiC MOSFET in pulsed power supply with analytical model", Proc. IEEE 12th Energy Convers. Congr. Expo.-Asia, pp. 325-330, 2021.

[13] X. Li, Y. Luo, R. Wang, Z. Shi, and F. Xiao, "A Passive Voltage-Balancing Method for Series-Connected SiC MOSFETs in Pulse Generator Based on Snubber Circuit," *IEEE Trans. Ind. Electron.*, vol. 71, no. 7, pp. 7030–7041, Jul. 2024.

Optimization and Compensation of Leakage-Induced Deviation of CTTC Magnetic Integrated Structure in CLLC Resonant Converter

1st Liwen JIA
College of Electrical Engineering
Zhejiang University
Hangzhou, China
22310084@zju.edu.cn

2nd Bodong LI
College of Electrical Engineering
Zhejiang University
Hangzhou, China
bodong_li@zju.edu.cn

3rd Yahong YANG
State Key Laboratory of Space Power Sources
Shanghai Institute of Space Power Sources
Shanghai, China
vivian19791224@163.com

4th Jianyu LAN
State Key Laboratory of Space Power Sources
Shanghai Institute of Space Power Sources
Shanghai, China
2320483340@qq.com

5th Jiarui ZHANG
College of Electrical Engineering
Zhejiang University
Hangzhou, China
3200102822@zju.edu.cn

6th Kelin CHEN
College of Electrical Engineering
Zhejiang University
Hangzhou, China
3210103075@zju.edu.cn

7th Feng JIANG
College of Electrical Engineering
Zhejiang University
Hangzhou, China
jiangfeng@zju.edu.cn

8th Min CHEN
College of Electrical Engineering
Zhejiang University
Hangzhou, China
calim@zju.edu.cn

Abstract—This paper addresses the high power density requirements of CLLC resonant converters driven by wide bandgap devices in wide voltage application scenarios, focusing on the key challenges in magnetic integration optimization—the sensitivity of leakage inductance parameters and the mismatch of equivalent relationships. This study establishes a CTTC two-port network equivalent model that includes leakage inductance effects, quantitatively reveals the modulation mechanism of leakage inductance deviation on core parameters such as voltage gain, and proposes a multi-objective optimization framework based on the NSGA-II algorithm. This framework dynamically balances magnetic network parameter sensitivity and power density indicators through the Pareto solution set. Experimental validation shows that the optimized 3kW/110kHz SiC prototype achieves an efficiency of 97% under full load conditions, with a 33% reduction in magnetic component volume. The parameter compensation mechanism effectively solves the characteristic parameter drift caused by magnetic integration, providing theoretical support for the high power density design of high-frequency and high-efficiency power supplies.

Keywords—*CLLC resonant converter, magnetic integration, parameter deviation and optimization.*

I. INTRODUCTION

In the ongoing development of contemporary power electronics technology, enhancing power density and efficiency are key objectives. As power demands escalate across diverse applications, wide voltage range operation and the application of wide bandgap devices have become crucial pathways to meet these requirements. The CLLC resonant converter, with its high efficiency, high power density, and soft-switching characteristics, exhibits significant advantages in high-power, high-voltage applications, and is widely utilized in areas such as renewable energy generation, energy storage systems, and electric vehicle charging[1][2][3][4]. However, the inherent magnetic component requirements of the CLLC topology itself result in a larger magnetic component volume, which constrains the improvement of power density. Therefore, magnetic integration technology for CLLC converters has garnered widespread attention to meet the demands of high power density applications.

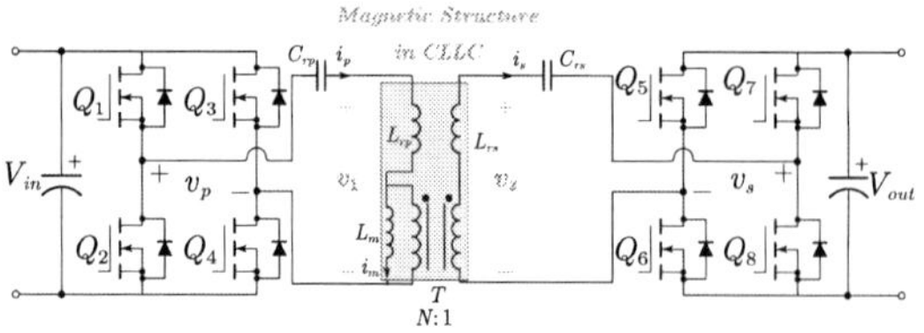

In the magnetic integration design of CLLC resonant converters, a common strategy involves reusing the transformer's leakage inductance as the resonant inductance[5][6][7]. However, the leakage inductance of the transformer is relatively small and its inductance value is difficult to adjust, which fails to meet the parameter requirements in high-power applications. Alternatively, integrating the three magnetic components into a single magnetic core by sharing the core between the resonant inductor and the transformer. But it necessitates addressing the coupling issues among different magnetic elements. One approach involves calculating the coupling relationships of each part to derive their equivalent parameters in the circuit[8]. Another method is to construct a low-reluctance common path or design windings that generate symmetrical magnetic flux, thereby canceling coupled magnetic flux in multiple windings and achieving decoupling[9][10]. Nevertheless, it is evident that magnetic integration schemes employing a shared core require customized designs for the core shape and coil spatial distribution, leading to complexity and a lack of versatility. Furthermore, the uneven distribution of magnetic flux density poses challenges in optimizing loss design.

Reference [11] proposed a CTTC structure, which simplifies the three magnetic components in a CLLC converter into two by using a two-port network equivalence method, as shown in Fig. 2. This approach avoids complex

Fig. 1 Topology of CLLC Bidirectional Resonant Converter.

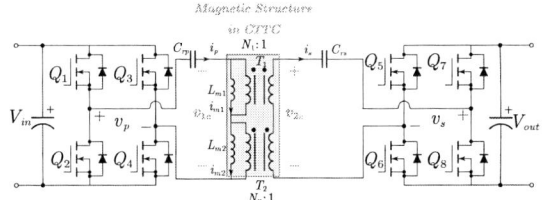

Fig. 2 Topology of CTTC Bidirectional Resonant Converter.

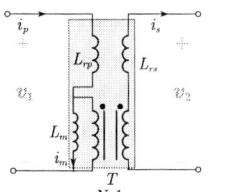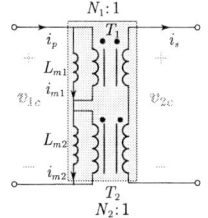

Fig. 3 Magnetic network (a) Discrete CLLC structure. (b) CTTC structure.

magnetic circuit designs and allows for the use of standard magnetic components, effectively enhancing power density and versatility. However, circuit-equivalence-derived magnetic integration strategies exhibit critical dependence on parametric accuracy. Deviations in circuit parameters would cause mismatches between actual and designed circuit performance.

In this paper, for the CLLC converter with magnetic integration using the CTTC structure, the influence of leakage inductance on its parameters is studied, and parameter compensation is performed. In Section II, a CTTC equivalent model considering the effects of leakage inductance is proposed, and the gain deviation of the magnetically integrated converter under the influence of leakage inductance is analyzed. Based on the new equivalent relationship of the CTTC structure considering leakage inductance effects, parameter compensation of the magnetic components is achieved through a multi-objective optimization method, thereby effectively reducing the parameter deviation of the converter. In Section III, a 3kW prototype is designed for experimental verification. Finally, the main conclusion is reported in Section IV.

II. OPTIMIZATION AND COMPENSATION FOR CTTC MAGNETIC INTEGRATED STRUCTURE

A. Magnetic Integration of CLLC

[11] proposed a magnetic integration method adopting the CTTC structure for the CLLC converter. Fig. 3(a) shows the magnetic network of discrete CLLC structure, and the magnetic structure of the CTTC conforms to the form of Fig. 3(b). If all parameters of T, T_1 and T_2 are regarded as unknown quantities, the port voltages v_1, v_2, v_{1c} and v_{2c} of the two networks can be expressed as (1) and (2) respectively:

$$\begin{cases} v_1 = j\omega\left(L_{rp}+L_m\right)i_p - j\omega\dfrac{L_m}{N}i_s \\ v_2 = j\omega\dfrac{L_m}{N}i_p - j\omega\left(L_{rs}+\dfrac{L_m}{N^2}\right)i_s \end{cases} \quad (1)$$

$$\begin{cases} v_{1c} = j\omega\left(L_{m1}+L_{m2}\right)i_p - j\omega\left(\dfrac{L_{m1}}{N_1}+\dfrac{L_{m2}}{N_2}\right)i_s \\ v_{2c} = j\omega\left(\dfrac{L_{m1}}{N_1}+\dfrac{L_{m2}}{N_2}\right)i_p - j\omega\left(\dfrac{L_{m1}}{N_1^2}+\dfrac{L_{m2}}{N_2^2}\right)i_s \end{cases} \quad (2)$$

Equivalent substitution conditions can be obtained by making the coefficients of the corresponding currents in v_1 and v_{1c}, v_2 and v_{2c} equal. The parameter equivalence relationship of the two structures is:

TABLE I. THE PARAMETERS OF CONVERTER

Parameters	Values
Input voltage, V_{in}	390V
Nominal output voltage, V_{out}	260V
Resonant capacitance, C_{rp}, C_{rs}	100nF, 225nF
Resonant frequency, f_r	110kHz

TABLE II. THE PARAMETERS OF MAGNETIC COMPONENT

	Parameters	Values	Magnetic core
Discrete Structure	Magnetizing inductance of T, L_m	79.53μH	PQ5050
	Transformer turn ratio of T, N	21/14	
	Primary Resonant inductance, L_{rp}	20.93μH	PQ3230
	Secondary Resonant inductance, L_{rs}	9.3μH	PQ3230
Integrated Structure	Magnetizing inductance of T_1, L_{m1}	36.18μH	PQ4040
	Transformer turn ratio of T_1, N_1	14/15	
	Magnetizing inductance of T_2, L_{m2}	64.28μH	PQ4040
	Transformer turn ratio of T_2, N_2	27/6	

$$\begin{cases} N - \sqrt{\left(L_{m1}+L_{m2}\right)\Big/\left(\dfrac{L_{m1}}{N_1^2}+\dfrac{L_{m2}}{N_2^2}\right)} \\ L_m = N\left(\dfrac{L_{m1}}{N_1}+\dfrac{L_{m2}}{N_2}\right) \\ L_{rp} = L_{m1}+L_{m2}-L_m \end{cases} \quad (3)$$

According to the method proposed in [11], the parameters of the magnetic components in both discrete structure and integrated structure are shown in TABLE II. Compared to traditional discrete CLLC structure, the volume of the magnetic network in integrated CTTC structure has been reduced by 33%[11].

B. Analysis of Leakage Inductance Influence

In transformer design, the generation of leakage inductance is an inherent challenge. Particularly in the CTTC structure, transformer leakage inductance significantly impacts key parameters such as the equivalent turns ratio,

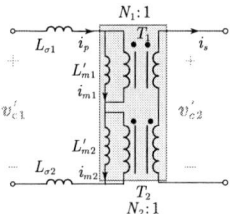

Fig. 4 Magnetic network of CTTC structure considering leakage inductance.

resonant inductance, and magnetizing inductance. Therefore, analysis of its specific effects and the implementation of effective compensation and optimization measures are crucial.

Based on Fig. 4, the port voltages v'_{1c} and v'_{2c} considering leakage inductance are:

$$
\begin{cases}
v'_{1c} = j\omega\left(L_{\sigma1} + L_{\sigma2} + L'_{m1} + L'_{m2}\right)i_p - j\omega\left(\dfrac{L'_{m1}}{N_1} + \dfrac{L'_{m2}}{N_2}\right)i_s \\[2mm]
v'_{2c} = j\omega\left(\dfrac{L'_{m1}}{N_1} + \dfrac{L'_{m2}}{N_2}\right)i_p - j\omega\left(\dfrac{L'_{m1}}{N_1^2} + \dfrac{L'_{m2}}{N_2^2}\right)i_s
\end{cases}
\tag{4}
$$

where $L_{\sigma1}$ and $L_{\sigma2}$ are the leakage inductance of T_1 and T_2. The parameter equivalence relationship of the two structures is modified to:

$$
\begin{cases}
N' = \sqrt{\left(L'_{m1} + L'_{m2} + L_{\sigma1} + L_{\sigma2}\right)\Big/\left(\dfrac{L'_{m1}}{N_1^2} + \dfrac{L'_{m2}}{N_2^2}\right)} \\[3mm]
L'_m = N'\left(\dfrac{L'_{m1}}{N_1} + \dfrac{L'_{m2}}{N_2}\right) \\[3mm]
L'_{rp} = L'_{m1} + L'_{m2} + L_{\sigma1} + L_{\sigma2} - L'_m
\end{cases}
\tag{5}
$$

For magnetically integrated solutions employing port network equivalence, consistency in parameter equivalence is crucial for ensuring the reliability of the equivalent model. The presence of leakage inductance directly affects key parameters k and Q, and these deviations directly manifest as deviations in voltage gain. The ratio of leakage inductance to inductance in high-frequency transformers can reach up to 10%. Taking a 10% leakage inductance as an example, the converter gain curves before and after the leakage inductance impact are shown in Fig. 5. The voltage gain of the converter can be calculated using (6). If optimization and compensation are not implemented during the design and manufacturing

process, the converter may fail to meet the external characteristic requirements. Deviations in key parameters such as k can also lead to variations in converter efficiency.

$$
\begin{cases}
G = \dfrac{1}{N\sqrt{\left[1 - a^2/k\right]^2 + Q^2 b^2/k^2}} \\[3mm]
a = \dfrac{1}{\omega_n^2} - 1, \ b = (2k+1)\omega_n - \dfrac{(2k+2)}{\omega_n^2} + \dfrac{1}{\omega_n^3}
\end{cases}
\tag{6}
$$

C. Parameter Optimization and Compensation

The presence of leakage inductance simultaneously causes deviations in multiple parameters. Traditional optimization algorithms, due to their single-objective optimization nature, struggle to effectively address the multi-objective problem presented in this study. Therefore, this paper employs the Non-dominated Sorting Genetic Algorithm II (NSGA-II). This algorithm is advantageous due to its high computational efficiency, good solution set distribution, and excellent optimization performance [12]. Based on the equivalent relationships proposed in (5) and (6), the specific optimization steps are illustrated in Fig. 6 .

For parameter optimization of the CTTC magnetically integrated structure, this paper directly utilizes the deviations in equivalent turns ratio N, equivalent inductance ratio k, and inductance L_{rp} under the influence of leakage inductance as optimization metrics, setting corresponding objective functions error(N), error(k), and error(L_{rp}). Implementing

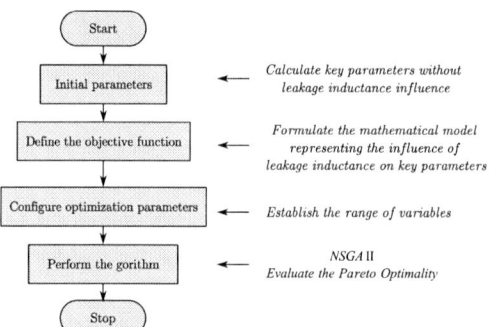

Fig. 6 Process of optimization by using NSGA-II.

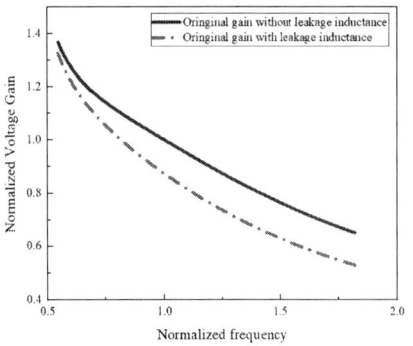

Fig. 5 Comparison of converter gain curves demonstrating the effects of leakage inductance.

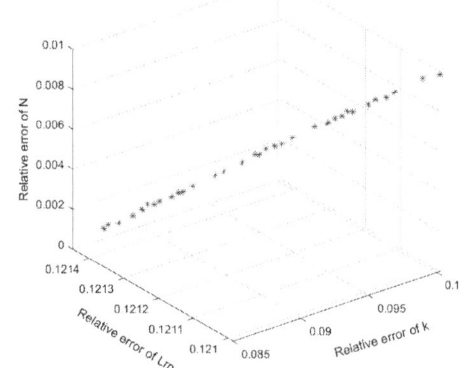

Fig. 7 The pareto optimality of system.

979-8-3315-1110-4/25 $31.00 © 2025 IEEE

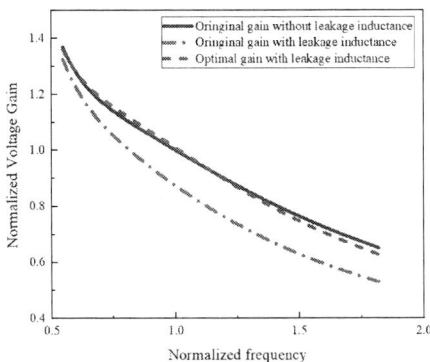

Fig. 8 Comparison of converter gain curves before and after optimization.

TABLE III. THE PARAMETERS OF TRANSFORMER

Parameters	Original Values	Optimal Values
L_{m1}, L_{m2}	36.18μH, 64.28μH	37.99μH, 57.85μH
$L_{\sigma2}, L_{\sigma2}$	3.62μH, 6.43μH	0.82μH, 4.93μH

optimization using the NSGA-II algorithm requires the pre-determination of variables and their parameter ranges. This paper selects the magnetizing inductance values and leakage inductance of the two transformers in the CTTC structure as variables. The relevant constraints remain based on a 10% leakage inductance deviation. Considering the significant difference in the turns ratio between the primary and secondary sides of T_2, the lower limit of its leakage inductance value is correspondingly increased. After determining the objective functions and parameter optimization ranges, NSGA-II-based CTTC leakage inductance parameter optimization and compensation are performed. The pareto optimality of this system is shown in Fig. 7. TABLE III presents the optimized parameters of the magnetic components, while Fig. 8 illustrates the voltage gain characteristics of the optimized converter.

III. PROTOTYPE DESIGN AND EXPERIMENTAL VERIFICATION

To validate the effectiveness of the proposed leakage inductance parameter optimization strategy for the magnetically integrated CTTC structure, an experimental prototype with a rated power of 3kW was designed and built, as shown in Fig. 9, with its operating characteristics and resonant parameters listed in TABLE I, II and III.

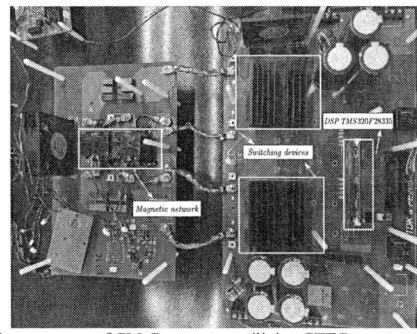

Fig. 9 The prototype of CLLC converter utilizing CTTC structure.

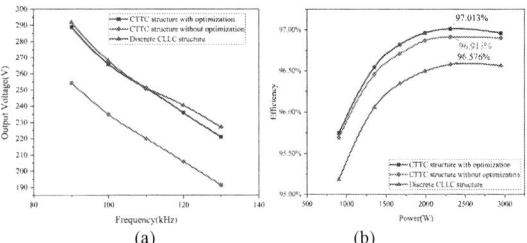

(a)   (b)

Fig. 10 Comparison of the converter before and after optimization: (a) Voltage gain (b) Efficiency.

The experimental results in Fig. 11 demonstrate that the CTTC structure with optimized parameters can achieve the same functions as the traditional discrete CLLC converter in the below-resonant-frequency mode (BRFM), resonant-frequency mode (RFM), and above-resonant-frequency mode (ARFM). The voltage gain characteristics of the converter were tested, as shown in Fig. 10(a), and the experimental results verified the effectiveness of the leakage inductance parameter optimization of the CTTC structure. Furthermore, the CTTC structure effectively improves the working efficiency of the converter by reducing the circulating current on the magnetic components [11]. However, if the influence of leakage inductance parameters is ignored, the parameter deviation caused by k will weaken the efficiency advantage, as shown in Fig. 10(b). Through the optimization of the transformer leakage inductance, the experimental prototype achieved a further improvement in efficiency.

IV. CONCLUSIONS

To mitigate the influence of leakage inductance in CTTC-structured, magnetically integrated CLLC converters, this study introduces a multi-objective optimization-based parameter deviation compensation method. By developing an

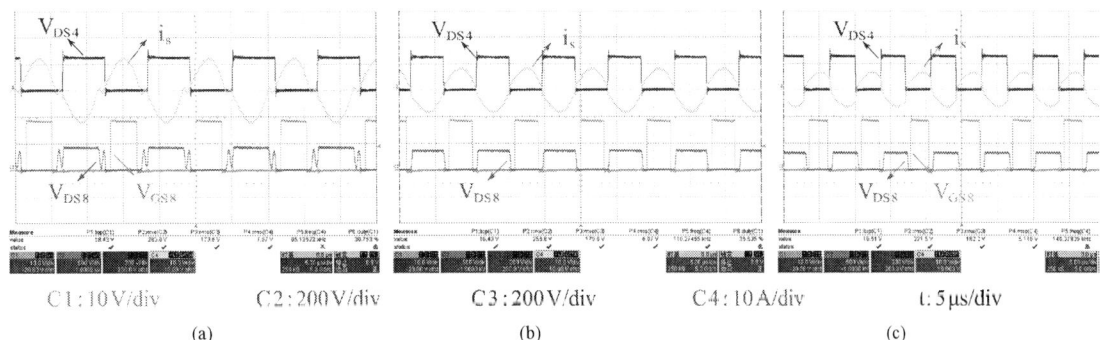

C1: 10 V/div   C2: 200 V/div   C3: 200 V/div   C4: 10 A/div   t: 5 μs/div

(a)   (b)   (c)

Fig. 11 Waveforms in CTTC structure. (a) BRFM ($f_s < f_r$). (b) RFM ($f_s = f_r$). (c) ARFM ($f_s > f_r$).

equivalent magnetic network model, the nonlinear influence of leakage inductance distribution on key parameters, including voltage gain characteristics, is analyzed. Utilizing the NSGA-II algorithm, a collaborative optimization scheme for inductance values and leakage inductance values is derived from the Pareto front within a multi-dimensional parameter space defined by equivalent turns ratio, equivalent inductance ratio, and equivalent inductance. This scheme effectively satisfies the voltage gain requirements while slightly enhancing conversion efficiency. Experimental results demonstrate that the CTTC structure with optimized parameters exhibits voltage gain characteristics more closely resembling those of a traditional discrete CLLC converter, validating the proposed method.

REFERENCES

[1] B. Li, M. Chen, X. Wang, N. Chen, X. Sun and D. Zhang, "An Optimized Digital Synchronous Rectification Scheme Based on Time-Domain Model of Resonant CLLC Circuit", *IEEE Trans. on Power Electron.*, vol. 36, no. 9, pp. 10933-10948, Sept. 2021.

[2] N. Chen et al., "Synchronous Rectification Based on Resonant Inductor Voltage for CLLC Bidirectional Converter," *IEEE Trans. on Power Electron.*, vol. 37, no. 1, pp. 547-561, Jan. 2022.

[3] B. Li, M. Chen, X. Sun, J. Wang and F. Jiang, "A Hybrid Control for Smooth Power Direction Transition of Bidirectional Resonant CLLC Converter with Wide Voltage Gain," *IEEE Trans. on Power Electron.*, early access, Aug. 23, 2024, doi: 10.1109/TPEL.2024.3448375.

[4] M. Chen et al., "A Coupled Inductor Scheme for CLLC Bidirectional Converter and Optimized Current Detection Method," *IEEE Trans. on Power Electron.*, vol. 37, no. 10, pp. 11546-11551, Oct. 2022.

[5] A. Chandwani and A. Mallik, "Parametric modeling and characterization of leakage-integrated planar transformer for CLLC DC–DC converter", *IEEE Trans. Magn.*, vol. 58, no. 6, Jun. 2022.

[6] P. He, A. Mallik, A. Sankar and A. Khaligh, "Design of a 1-MHz high-efficiency high-power-density bidirectional GaN-based CLLC converter for electric vehicles", *IEEE Trans. Veh. Technol.*, vol. 68, no. 1, pp. 213-223, Jan. 2019.

[7] S. Zou, J. Lu, A. Mallik and A. Khaligh, "Modeling and optimization of an integrated transformer for electric vehicle On-Board charger applications", *IEEE Trans. Transp. Electrific.*, vol. 4, no. 2, pp. 355-363, Jun. 2018.

[8] B. Li, Q. Li and F. C. Lee, "High-frequency PCB winding transformer with integrated inductors for bi-directional resonant converter", *IEEE Trans. Power Electron.*, vol. 34, no. 7, pp. 6123-6135, Jul. 2019.

[9] M. Noah et al., "A current sharing method utilizing single balancing transformer for a multiphase LLC resonant converter with integrated magnetics", *IEEE J. Emerg. Sel. Topics Power Electron.*, vol. 6, no. 2, pp. 977-992, Jun. 2018.

[10] S. Gao and Z. Zhao, "Magnetic integrated LLC resonant converter based on independent inductance winding", *IEEE Access*, vol. 9, pp. 660-672, 2021.

[11] M. Chen, L. Jia, B. Li, D. Zhang and F. Jiang, "A Novel CTTC Structure and Optimization Design Method for CLLC Bidirectional Resonant Converter," *IEEE Trans. on Power Electron.*, early access, Jan. 31, 2025, doi: 10.1109/TPEL.2025.3536010.

[12] D. Jiang et al., "Multiobjective optimization considering PET's vibration suppression of dual active bridge converter based on BP-NSGA-II", *IEEE Trans. Power Electron.*, vol. 39, no. 2, pp. 2226-2236, Feb. 2024.

Sustained Oscillation Characterization of GaN HEMT at Cryogenic Temperature

1st Zilong Chen
School of Electrical Engineering
Xi'an Jiaotong University
Xi'an, China.
czl0928@stu.xjtu.edu.cn

2nd Yuqi Wei
School of Electrical Engineering
Xi'an Jiaotong University
Xi'an, China.
yuqiwei@xjtu.edu.cn

3rd Yanjie He
School of Electrical Engineering
Xi'an Jiaotong University
Xi'an, China.
18970201013@stu.xjtu.edu.cn

4th Yukun Zhang
School of Electrical Engineering
Xi'an Jiaotong University
Xi'an, China.
yukunzhang@stu.xjtu.edu.cn

5th Chong Dou
Sungrow Power Supply Co.,Ltd.
Hefei, China.
douchong@sungrowpower.com

6th Qian Cui
Sungrow Power Supply Co.,Ltd.
Hefei, China.
cuiqian@sungrowpower.com

Abstract—**Cryogenic power electronics has emerged as a transformative technique to achieve ultra-high efficiency and power density. Many projects have been conducted to investigate the application of cryogenic power electronics. Compared with other types of semiconductors, gallium nitride (GaN) high electron mobility transistor (HEMT) is proved to be the best candidate for cryogenic application with large reduction of on-state resistance and switching loss. However, the unique stability issue of the GaN HEMT is not explored under cryogenic. Therefore, in this paper, the unique sustained oscillation characteristic of GaN HEMT at cryogenic temperature is investigated and analyzed. After simplifying the equivalent circuit, the small-signal model is established to conduct the stability analysis. A cryogenic clamped inductive switching test platform, operating at [133 K–298 K], is built to evaluate the sustained oscillation performance. Experimental results reveal a 75% increase in sustained oscillation voltage range span at 133 K compared to ambient conditions. which indicates that GaN HMET is more prone to sustained oscillation at cryogenic temperature.**

Index Terms—**Cryogenic power electronics, GaN HEMT, sustained oscillation**

I. INTRODUCTION

To meet the requirements like the energy sustainability and technology development, the application scenarios of power electronic converter (PEC) has extended to the cryogenic temperature (CT). For example, the deep space exploration, deep underground neutrino experiment and so on [1–3]. They needs the PEC to operate at the CT with a high reliability almost for the whole life time [1, 3]. Fortunately, not only a high reliability is found at CT, but also an improved efficiency and a higher power density were reported [4]. By leveraging the advantages of CT, the cryogenic power electronic technology provides a novel thought to inspire applications demanding extreme efficiency, miniaturization, and

environmental resilience such as cryo-electric hydrogen-powered aviation , power generation system and Magnetic Resonance Imaging [1, 5–7].

Power semiconductor devices serve as the cornerstone of PECs, with extensive studies focused on their cryogenic behavior. While conventional silicon-based components, such as Si Insulate-Gate Bipolar Transistors (IGBTs) and Metal-Oxide-Semiconductor Field-Effect TransistorS (MOSFETs), exhibit reduced on-state resistance and enhanced switching characteristics under cryogenic conditions compared to room-temperature operation [1, 3, 8]. However, they suffer from critical limitations , including substantial degradation of breakdown voltage and pronounced carrier freeze-out effects below 100 K, which paradoxically elevate conduction losses despite initial improvements [9]. Although silicon carbide (SiC) MOSFETs demonstrate superior thermal conductivity and carrier mobility over silicon counterparts, their practical adoption in cryogenic environments is hindered by anomalous increases in on-state resistance and the switching loss [10, 11]. In contrast, gallium nitride (GaN) high electron mobility transistors (HEMTs) reveal a unique advantage spectrum under cryogenic operation, achieving simultaneous reductions in both conduction and switching losses without carrier freeze-out phenomenon which makes it an outstanding candidate for CT applications[11, 12].

Although there are many advantages of the GaN based PEC operating at CT, the change of the physical parameters will cause device failure[13]. Therefore, it is of great importance to conduct the research to discover all potential factors leading to reliability problems. The sustained oscillation is a unique phenomenon for GaN-based half bridge under some certain conditions [14, 15].

It arises between the upper GaN device and the circuitry of both the gate loop and the power loop while the lower GaN device is in off state. A positive feedback which leading to sustained oscillation is determined by the backward transconductance (g_m) and the parasitic network of the circuit. However, seldom literature investigates sustained oscillation characteristic under the cryogenic condition. It is reported that the g_m increases with the decreasing temperature which might enlarge the boundary of sustained oscillation theoretically [16]. However, there is still no practical experiment data to evaluate the variation of sustained oscillation region under CT. Therefore, in this paper, a sustained oscillation experiment at CT is conducted to fill the gap.

The remainder of this paper is organized as follows. Section II provides a brief introduction to the sustained oscillation mechanism. Section III presents a detailed description of the experimental setup. The results of the sustained oscillation experiment are discussed in Section IV, and the final section draws conclusions based on these findings.

II. MECHANISM ANALYSIS OF SUSTAINED OSCILLATION FOR GaN HEMT AT CT

The clamped inductive switching (CIS) circuit is a common method to evaluate the sustained oscillation characteristic. Figure. 1 shows the topology of the CIS circuit. Due to the unique physical characteristic, GaN HEMT can conduct reversely without body diode. When V_{gd} is greater than the threshold voltage, electrons is collected on the heterostructure interface of aluminum gallium nitride and GaN, as a result, the two-dimensional electron gas (2DEG) is re-constructed which makes the GaN HEMT conductive reversely. The sustained oscillation occurs while S_1 is turned off and S_2 conducts reversely carrying the commutation current.

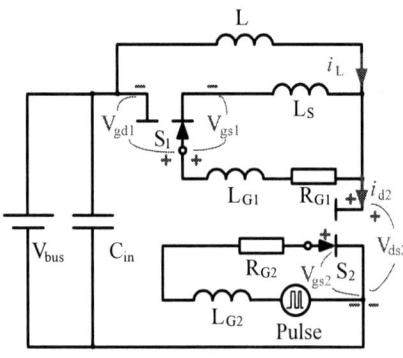

Fig. 1. Topology of the inductor clamped switching circuit.

To analyze the stability criteria, the linearized small-signal model is established for the CIS circuit in Fig.

2(a). By opening the inductor circuit and simplifying the capacitor circuit, a small-signal model is derived from Fig. 1. Because S_1 operates in saturation region, the current flows through it is controlled by V_{gd}. Therefore, the GaN HEMT can be replaced by a controlled voltage source. S_2 is switched off during the current commutation period, thus, replaced by the capacitor C_{OSS2}.

Fig. 2(b) is obtained by simplify the small-signal circuit mode via Eq. 1, Eq. 2, Eq. 3.

$$Z_4 = \frac{Z_1 Z_2 + Z_2 Z_3 + Z_3 Z_1}{Z_2}, Z_7 = Z_4 \parallel \frac{1}{sC_{ds}} \quad (1)$$

$$Z_5 = \frac{Z_1 Z_2 + Z_2 Z_3 + Z_3 Z_1}{Z_1}, Z_8 = Z_5 \parallel \frac{1}{sC_{gd}} \quad (2)$$

$$Z_6 = \frac{Z_1 Z_2 + Z_2 Z_3 + Z_3 Z_1}{Z_3}, Z_9 = Z_6 \parallel \frac{1}{sC_{gs}} \quad (3)$$

where $Z_1 = sL_s$, $Z_2 = R_{G1} + sL_{G1}$, $Z_3 = R_{loop} + sL_{loop} + 1/(sC_{OSS2})$. Z_1 to Z_9 is indicated in Fig. 2(a) and Fig. 2(b). The detailed explanation of these impedance is mentioned in [14] and will not be demonstrated here, considering the brevity.

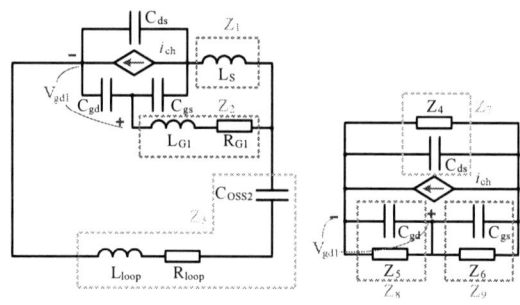

(a) original diagram   (b) simplified diagram

Fig. 2. Small-signal model of Fig. 1.

The feedback system in Fig. 3 is built up with Eq. 4, Eq. 5.

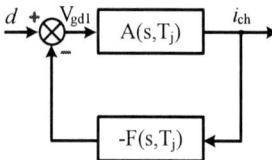

Fig. 3. Block diagram of the feedback system with the disturbance d.

$$A(s, T_j) = \frac{i_{ch}(s, T_j)}{v_{gd}(s, T_j)} = g_m(T_j) \quad (4)$$

$$F(s, T_j) = \frac{v_{gd}(s, T_j)}{i_{ch}(s, T_j)} = -\frac{Z_7 Z_8}{Z_7 + Z_8 + Z_9} \quad (5)$$

where $A(s, T_j)$ is the amplified gain of the GaN device, $-F(s, T_j)$ is the feedback transfer function of the equivalent circuit. To better analyze the loop, a disturbance signal d is introduced in the system. Then, the close-loop function is derived as

$$T(s, T_j) = \frac{V_{ch}(s, T_j)}{d(s)} = \frac{A(s, T_j)}{1 - A_{(s, T_j)} \cdot F_{(s, T_j)}} \quad (6)$$

The approach to determining whether this feedback system is experiencing a stability issue involves identifying the pole points. If the pole point close to imaginary axis, the system is more likely to oscillate sustainably. As a measurement indicator, the damping ratio is used to determine whether or not sustained oscillation occurs under the corresponding condition.

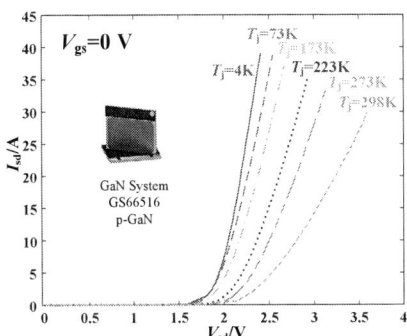

Fig. 4. Reverse conduction characteristics at different temperature [16].

As Fig. 4 shows, one of the most important changes of the device at CT is the g_m. A higher transconductance increases the possibility of obtaining a negative damping ratio. Another factor that effect the sustained oscillation is the L_{loop} resistance variation. Since the conductivity of the copper foil has a positive temperature coefficient, the resistance of the power loop decreases significantly, which also makes the system prone to sustained oscillation.

III. EXPERIMENT SETUP FOR SUSTAINED OSCILLATION AT CT

To conduct a cryogenic sustained oscillation characterization experiment, a CIS circuit for CT need to be designed. Compared with room temperature (RT), the CT environment affects the performance of many components. In order to obtain the sustained oscillation characterization of GaN HEMT, other external factors must be decoupled. For example, there is a considerable output voltage drop for isolated power supply at CT, which introduce a variation on gate driving voltage for different temperature. As Figure. 5 shown, the isolated power supply operates outside the cryogenic chamber to maintain a stable gate-drive voltage.

Fig. 5. Picture of the designed experiment setup for the sustained oscillation characterization.

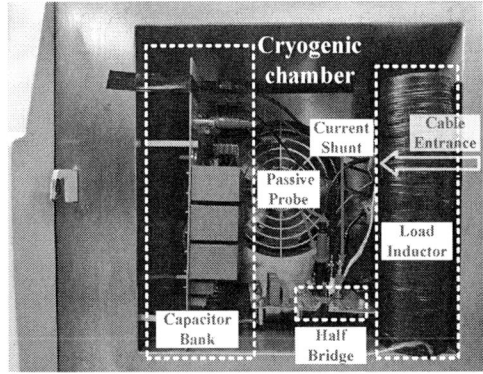

Fig. 6. Picture of the CIS circuit.

In addition to the power supply, other on-board components may also varies with temperature. The components listed in Table. I are carefully selected with a great temperature-independent performance. The whole CIS circuit and the measurement are put in the cryogenic chamber as Fig. 6 shows, except for the isolated power supply. It is worth noting that although the driver integrated circuit (IC) does not directly participate in the sustained oscillation process, its drive capability should be maintained nearly constant as the temperature decreases. Another challenge is the measurement. It is known that CT environment might cause the performance deterioration or even malfunction of the active

TABLE I: CIS Circuit Component Selection

Component	Capacitor	Resistor	Driver
Producer	FARATRONIC PSA, EPCOS	VISHAY	SILLICON LABS
Name /Series	C3D, FV B32653	TNPW	SI8271GB MCU
Category	PP film, C0G	Thin film	CMOS

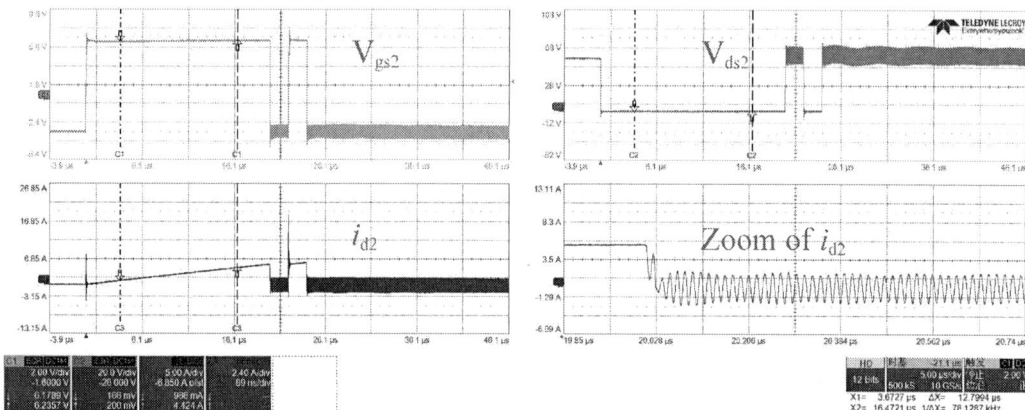

Fig. 7. The experimental sustained oscillation waveforms of S_2.

probe[17]. Instead, the passive probes always operates well at CT. After careful consideration, the high-voltage passive probe PPE6KV-A from Teledyne Lecroy, the passive probe PP026 from Teledyne Lecroy and the high band-width current shunt from T&M is used to measure V_{ds2}, V_{gs2} and i_{d2}, respectively.

The experiment is carried out with the programmable variable temperature chamber. Both the signal cable and the power supply cable is connected to the CIS circuit through the hole on the side of the chamber. After checking all connections, the temperature of CIS circuit is reduced to the objective value. When all of the components are cooled down, CIS circuit is triggered via double pulse signal with a fixed pulse width generated by the signal generator. The OS waveform can be obtained by oscilloscope with the V_{ds} changing. If the oscillation keeps for over 10 μs without a visible attenuation, the oscillation sustained phenomenon is confirmed in this experiment.

IV. RESULTS AND DISCUSSIONS

The experiment is conducted with the temperature ranging from 133 K to 298 K, voltage from several tens of volts to hundreds of volts. The sustained oscillation waveforms are recorded for every scenarios. Fig. 7 displays the experimental oscillation waveforms when $V_{ds} = 55$ V.

Fig. 8 and 9 show the waveforms when $V_{ds} = 50$ V and $V_{ds} = 100$ V at 133 K and 298 K, respectively. The two figures prove that the sustained oscillation characteristic is different with different temperature and voltage. It can be observed that the sustained oscillation is formed when the lower switch is turned off both between the two pulse and after the second pulse. Moreover, sustained oscillation phenomenon appears in the waveforms of V_{gs}, V_{ds} and i_d in the same time, which might cause great unreliability for the GaN HEMT in practice.

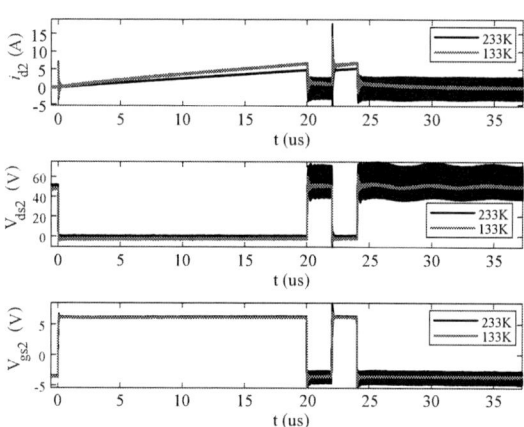

Fig. 8. Waveforms when $V_{ds} = 50$ V.

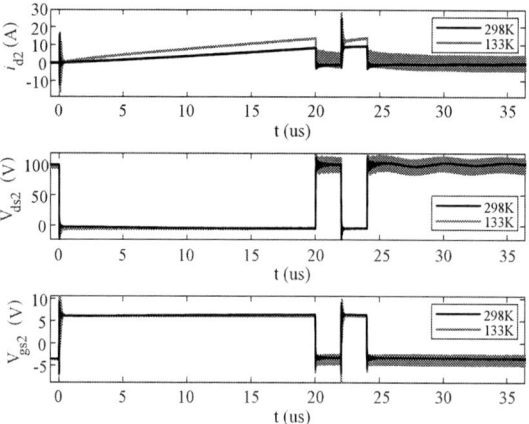

Fig. 9. Waveforms when $V_{ds} = 100$ V.

According to the theoretical analysis at CT, the sustained oscillation appears when V_{ds} locates between the lower limit and the upper limit, which is consistent with the experiment. Fig. 10 exhibit the limit with the dash line graph and the oscillation voltage range span with the green histogram. The upper limit increases rapidly (256 mV/K) with the decrease of temperature. In contrast, the lower limit increases at a lower speed (73 mV/K) under the same condition. Obviously, the upper limit dominant the sustained oscillation voltage range span. As a consequence, a 75% increase of the sustained oscillation range span takes place from RT to CT.

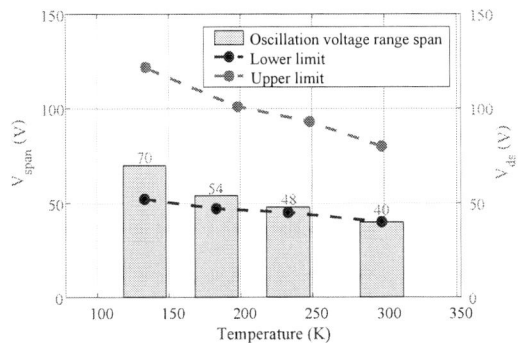

Fig. 10. The sustained oscillation voltage limits and range span.

V. CONCLUSION

This paper investigates the sustained oscillation characteristic for GaN HEMT at cryogenic temperature. Initially, the sustained oscillation mechanism of the GaN HEMT is analyzed briefly. Then, based on the clamped inductive switching circuit, the cryogenic test platform is built. Finally, oscillation waveforms are captured and analyzes from 133 K to 298 K with different voltage. A 75% increase of the sustained oscillation range is observed at CT, which indicates that the GaN HEMT based half bridge circuit is prone to sustained oscillation under low temperatures.

ACKNOWLEDGMENT

This work is supported the Aeronautical Science Fund (ASF) (Grant No. ASFC-2022Z072070001) and Sungrow Power Supply.

REFERENCES

[1] R. Singh and B. Baliga, *Cryogenic Operation of Silicon Power Devices*, ser. Power Electronics and Power Systems. Springer US, 2012.

[2] D. Santoro, N. Gallice, M. Bassani, P. Cova, N. Delmonte, M. Lazzaroni, V. Trabattoni, and A. Zani, "Dc-dc boost converter design with analog feedback control for cryogenic applications," *IEEE Access*, vol. 13, pp. 23 220–23 233, 2025.

[3] M. Elbuluk, A. Hammoud, and R. Patterson, "Power electronic components, circuits and systems for deep space missions," in *2005 IEEE 36th Power Electronics Specialists Conference*, 2005, pp. 1156–1162.

[4] K. Rajashekara and B. Akin, "A review of cryogenic power electronics - status and applications," in *2013 International Electric Machines & Drives Conference*, 2013, pp. 899–904.

[5] S. Farrukh, D. Wu, R. Al-Dadah, W. Gao, and Z. Wang, "A review of integrated cryogenic energy assisted power generation systems and desalination technologies," *Applied Thermal Engineering*, vol. 221, 2023.

[6] D. H. Johansen, J. D. Sanchez-Heredia, J. R. Petersen, T. K. Johansen, V. Zhurbenko, and J. H. Ardenkjaer-Larsen, "Cryogenic preamplifiers for magnetic resonance imaging," *IEEE Transactions on Biomedical Circuits and Systems*, vol. 12, no. 1, pp. 202–210, 2018.

[7] J. K. Noland, R. Mellerud, and C. Hartmann, "Next-generation cryo-electric hydrogen-powered aviation: A disruptive superconducting propulsion system cooled by onboard cryogenic fuels," *IEEE Industrial Electronics Magazine*, vol. 16, no. 4, pp. 6–15, 2022.

[8] A. Caiafa, *Cryogenic behavior of insulated gate bipolar transistors (IGBTs)*. PLENUM PRESS, 2004.

[9] Z. Zhang, C. Timms, J. Tang, R. Chen, J. Sangid, F. Wang, L. M. Tolbert, B. J. Blalock, and D. J. Costinett, "Characterization of high-voltage high-speed switching power semiconductors for high frequency cryogenically-cooled application," in *2017 IEEE Applied Power Electronics Conference and Exposition (APEC)*, 2017, pp. 1964–1969.

[10] H. Gui, R. Ren, Z. Zhang, R. Chen, J. Niu, F. Wang, L. M. Tolbert, B. J. Blalock, D. J. Costinett, and B. B. Choi, "Characterization of 1.2 kv sic power mosfets at cryogenic temperatures," in *2018 IEEE Energy Conversion Congress and Exposition (ECCE)*, 2018, pp. 7010–7015.

[11] Z. Zhang, H. Gui, R. Ren, F. Wang, L. M. Tolbert, D. J. Costinett, and B. J. Blalock, "Characterization of wide bandgap semiconductor devices for cryogenically-cooled power electronics in aircraft applications," in *2018 AIAA/IEEE Electric Aircraft Technologies Symposium (EATS)*, 2018, pp. 1–8.

[12] L. Ching-Hui, W. Wen-Kai, L. Po-Chen, L. Cheng-Kuo, C. Yu-Jung, and C. Yi-Jen, "Transient pulsed analysis on gan hemts at cryogenic temperatures," *IEEE Electron Device Letters*, vol. 26, no. 10, pp. 710–712, 2005.

[13] R. Ren, H. Gui, Z. Zhang, R. Chen, J. Niu, F. Wang, L. M. Tolbert, D. Costinett, B. J. Blalock, and B. B. Choi, "Characterization and failure analysis of 650-v enhancement-mode gan hemt for cryogenically cooled power electronics," *IEEE Journal of Emerging and Selected Topics in Power Electronics*, vol. 8, no. 1, pp. 66–76, 2020.

[14] K. Wang, X. Yang, L. Wang, and P. Jain, "Instability analysis and oscillation suppression of enhancement-mode gan devices in half-bridge circuits," *IEEE Transactions on Power Electronics*, vol. 33, no. 2, pp. 1585–1596, 2018.

[15] J. Chen, Q. Luo, Y. Wei, X. Zhang, and X. Du, "The sustained oscillation modeling and its quantitative suppression methodology for gan devices," *IEEE Transactions on Power Electronics*, vol. 36, no. 7, pp. 7927–7941, 2021.

[16] Z. Li, Y. Wei, M. M. Hossain, J. Liu, and H. A. Mantooth, "Is the gan hemt more prone to sustained oscillations under cryogenic conditions?" in *2023 IEEE Energy Conversion Congress and Exposition (ECCE)*, 2023, pp. 5385–5391.

[17] Y. Wei, M. M. Hossain, and H. A. Mantooth, "Comparisons and evaluations of silicon and wide band gap devices at cryogenic temperature," *IEEE Transactions on Industry Applications*, vol. 59, no. 2, pp. 1982–1994, 2023.

Design of an All-SiC On-Board Auxiliary Inverter for Urban Rail Vehicles

Xuefei Li
National Engineering Research Center of Railway Vehicles
CRRC Changchun Railway Vehicles CO., LTD
Changchun, China
lxf790917@163.com

Yongang Chen
Key Lab. of Vehicular Multi-Energy Drive Systems (VMEDS), Ministry of Education
School of Electrical Engineering, Beijing Jiaotong University
Beijing, China
24121262@bjtu.edu.cn

Zixiao Li
Key Lab. of Vehicular Multi-Energy Drive Systems (VMEDS), Ministry of Education
School of Electrical Engineering, Beijing Jiaotong University
Beijing, China
24110432@bjtu.edu.cn

Shuiyuan He
Key Lab. of Vehicular Multi-Energy Drive Systems (VMEDS), Ministry of Education
School of Electrical Engineering, Beijing Jiaotong University
Beijing, China
21126128@bjtu.edu.cn

Yuwen Qi
National Engineering Research Center of Railway Vehicles
CRRC Changchun Railway Vehicles CO., LTD
Changchun, China
q00y00w00@126.com

Lijun Diao
Key Lab. of Vehicular Multi-Energy Drive Systems (VMEDS), Ministry of Education
School of Electrical Engineering, Beijing Jiaotong University
Beijing, China
ljdiao@bjtu.edu.cn

Abstract—In order to further expand the development of urban rail transit and reduce energy consumption, an on-board auxiliary inverter for urban rail vehicles based on high-power and high-voltage SiC devices is proposed. This design is based on three-phase inverter circuit, parameter design, simulation modeland loss analysis. The two-level SVPWM control mode is adopted, and after reasonable parameter design and efficient control scheme, it runs stably in simulation. A 130kVA high-power experimental platform is established by using all SiC power devices. When the input voltage is changed, the output voltage of the three-phase inverter is still stable at the expected value. When the load changes abruptly, the three-phase inverter still has fast response and output stability. The operation reliability of three-phase inverter using SiC power device is further verified, and the output voltage is more stable, the distortion rate is lower, and the efficiency is higher than that of Si power device.

Keywords—silicon carbide devices, auxiliary inverter, low distortion

I. INTRODUCTION

With the continuous expansion of urban rail transit in China, research on energy efficiency and lightweighting of urban rail vehicle power systems has become increasingly important. Traditional Si IGBT devices suffer from high switching losses and long switching times, which limit the operating frequency and lightweighting potential of auxiliary converters. As the manufacturing process of SiC power devices matures, SiC power modules with higher voltage ratings and superior switching performance have gained attention in the urban rail transit sector[1]. SiC MOSFET devices offer faster switching speeds, significantly reducing the switching losses of converters and shortening dead-time. The reverse recovery loss of the body diode is almost zero, greatly improving the efficiency of the converter. The use of SiC power devices is of great significance for enhancing the lightweighting of auxiliary converters[4]. This paper presents the design of an on-board auxiliary inverter for urban rail vehicles using high-power, high-voltage SiC devices. The switching frequency is several times higher than that of traditional Si-based auxiliary converters, and the switching losses are reduced by 90%. The two-level SVPWM control strategy is adopted, and through theoretical calculations, simulation analysis, and prototype testing, the feasibility and stability of SiC devices in high-voltage, high-power applications for rail transit auxiliary converters are demonstrated.

II. CIRCUIT TOPOLOGY AND PARAMETER DESIGN

The topology of the three-phase inverter is shown in Fig. 1. The input voltage is 600V DC, and the inverter outputs a stable 380 V AC voltage for the vehicle's AC loads.

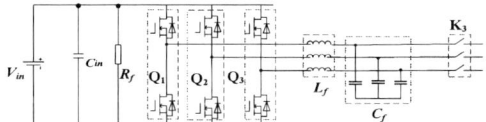

Fig. 1. Three-phase inverter circuit

In Fig. 1, V_{in} is the DC input voltage; C_{in} is the input support capacitor, which also serves as an input voltage filter and energy storage; L_f and C_f form an LC low-pass filter. The output of the three-phase inverter passes through the low-pass filter to eliminate high-frequency harmonics near the switching frequency, resulting in a clean sinusoidal voltage and current waveform.

A. Input Support Capacitor Parameter Calculation

The capacitance of the support capacitor should be selected based on the inverter's output power. The formula for selecting the support capacitor is as follows:

$$C_{in} = \frac{P_O}{4\eta f_s V_{in}\Delta V_{in}} \tag{1}$$

Where P_o is the system output power, $\triangle V_{in}$ is the DC bus ripple voltage (typically 0.5% of the DC bus voltage), V_{in} is the DC bus voltage, η is the inverter efficiency, and f_s is the inverter switching frequency. Substituting the parameters into the formula yields a support capacitor value of 614 µF.

B. Output Filter Parameter Calculation

When designing the filter, the filter inductor is determined first, and then the filter capacitor is selected based on the cutoff frequency. The formula for the filter inductor is as follows:

$$L_f = \sqrt{\frac{\left(\dfrac{\omega_1 V_{o1N}^2}{\omega_L^2} + \dfrac{\omega_1^3 V_{o1N}^2}{\omega_L^4}\right)}{\omega_1 I_{o1}^2}} \qquad (2)$$

Where ω_1 is the output voltage angular frequency, ω_L is the cutoff angular frequency, V_{o1N} is the output phase voltage, and I_{o1} is the output phase current. The cutoff frequency is initially set to 1/30 of the switching frequency, i.e., 1kHz. Substituting the parameters yields a filter inductor value of 0.178 mH.

The current THD decreases as the inductor size increases, but the voltage drop across the inductor also increases. To ensure that the voltage drop does not exceed 5% of the output voltage, the filter inductor size is set to 0.1 mH.

Considering the impact of different cutoff frequencies on the filtering effect, the filter capacitor is selected to be around 50 μF to meet practical requirements.

III. TWO-LEVEL SVPWM CONTROL STRATEGY

The six switches in the three-phase inverter form three bridge arms, and their switching states are represented by symbols S_u, S_v, and S_w. The switching state where the upper switch is on and the lower switch is off is represented by 1, and the opposite state is represented by 0. The states 000 and 111 do not produce output current (under ideal conditions) and are called zero vectors. The other six switching states are non-zero basic vectors, and their basic vector relationship is shown in Fig. 2.

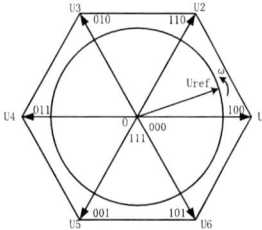

Fig. 2. Basic vector diagram

In Fig. 2, U_1 and U_2 represent the two adjacent basic voltage space vectors in the sector where the current voltage vector is located. The reference phase voltage vector is represented by U_{ref}, which can be synthesized by the linear combination of U_1, U_2 and the zero vector. The times t_1 and t_2 represent the durations of the basic voltage vectors U_1 and U_2, respectively, and the zero vector duration is t_0. The durations of the three basic vectors are as follows:

$$\begin{cases} t_1 = \dfrac{\sqrt{3}U_{amp\_ref}}{U_{in}} T_S \sin\left(\dfrac{\pi}{3} - \theta\right) \\[2ex] t_2 = \dfrac{\sqrt{3}U_{amp\_ref}}{U_{in}} T_S \sin\theta \\[2ex] t_0 = T_S - t_1 - t_2 \end{cases} \qquad (3)$$

The vector duration formulas for the other five sectors can be calculated using a normalization algorithm, where θ is replaced by θ-π/3, θ-2π/3, θ-π, θ-4π/3, and θ-5π/3, respectively.

IV. SIMULATION VERIFICATION

A. Input Voltage Transient Simulation

A simulation model was built in MATLAB/Simulink to verify the system's output regulation capability when the input voltage changes abruptly between 500V and 1000V DC under full load conditions. The simulation times 0 s, 0.15 s, 0.2 s, and 0.3 s correspond to input voltages of 500 V, 600 V, 1000 V, and 600 V, respectively.

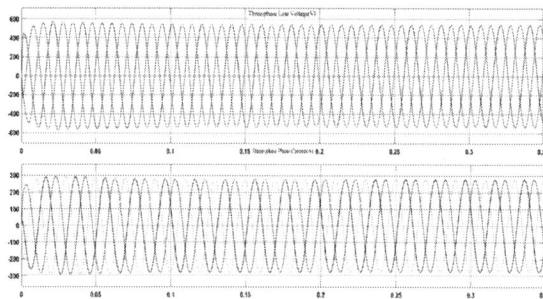

Fig. 3. Three-phase inverter output voltage and current waveform when the input voltage changes abruptly

The simulation results show that when the input voltage changes abruptly, the output of the three-phase inverter exhibits slight fluctuations but quickly stabilizes at the desired output.

B. Load Transient Simulation

The output regulation capability of the three-phase inverter under load transients was verified. The input voltage was 600V, and the full load power of the inverter was 130 kVA. The simulation times 0s, 0.15s, 0.2s, 0.25s, 0.3s, 0.35s, and 0.4s correspond to load levels of 10%, 30%, 70%, 100%, 70%, 30%, and 10%, respectively.

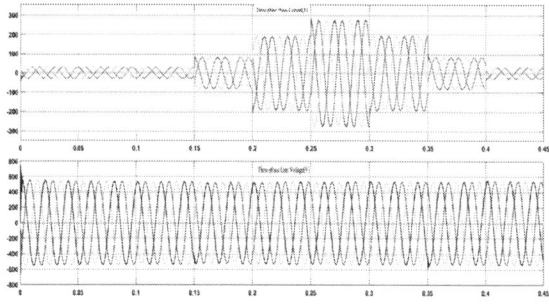

979-8-3315-1110-4/25 $31.00 © 2025 IEEE

Fig. 4. Three-phase inverter output voltage and current waveform when the load changes

The simulation results show that when the load changes from 10% to full load, the output voltage of the three-phase inverter changes slightly, with the peak line voltage varying within ±15 V. The output stabilizes at the desired 380 V within 0.03 s, demonstrating the inverter's fast response and output stability under load transients.

C. Loss Calculation Simulation

A simulation model of the three-phase inverter circuit was built in PLECS, and a SiC loss model was added to calculate the losses. The results are shown in Table 1.

TABLE I.   THREE-PHASE INVERTER LOSS SIMULATION RESULTS

Parameter	Value
Input Power	130 kW
Switching Frequency	30 kHz
SiC Module Loss	2×167.3 W
Total SiC Module Loss	1003.8 W
Total Inverter Loss	1180 W
Converter Efficiency	99.06%

V.   EXPERIMENTAL VERIFICATION AND COMPARISON

A. System Setup

A prototype of the three-phase inverter was built according to the designed parameters and component selection, as shown in Fig. 5. The prototype uses high-power, high-voltage SiC devices, and the control system's core control board consists of a DSP and FPGA to control the auxiliary converter.

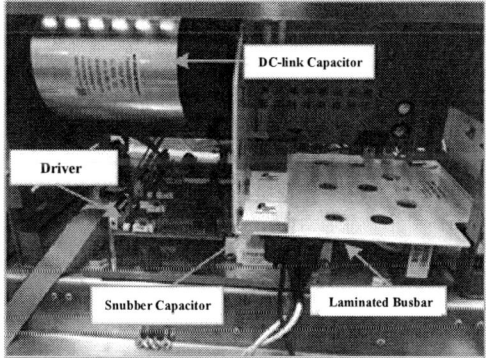

Fig. 5. Three-phase inverter prototype physical picture

B. Performance Experiment of All-SiC Inverter

Fig. 6 shows the waveform at a certain moment when the load changes abruptly under a DC input voltage of 600 V. The full load power is 130 kVA, and the load changes from 100% to 70%. It can be seen that when the load changes abruptly, the output voltage of the three-phase inverter exhibits only slight fluctuations but quickly stabilizes.

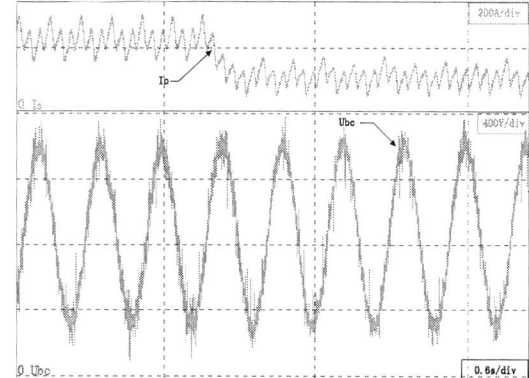

Fig. 6. Waveform with sudden load change

Fig. 7(a) shows the output voltage waveform of the three-phase inverter at an input voltage of 500 V, and Fig. 7(b) shows the output voltage waveform at an input voltage of 800 V. It can be seen that under different input voltage levels, the output voltage of the three-phase inverter remains stable at the desired value, demonstrating the stability and reliability of the all-SiC high-voltage devices.

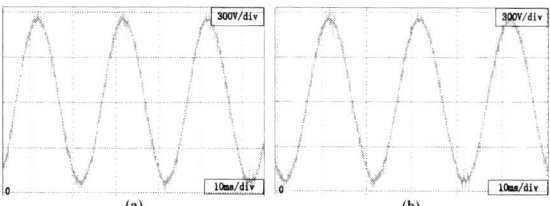

Fig. 7. Waveform of output voltage of three-phase inverter with different input voltage levels

C. Comparison of SiC and Si Inverters

As shown in Table 2, under the same input voltage, control strategy, and power level, and with identical parameters and component selection except for the power devices, the efficiency, output voltage, and output voltage distortion rate of the three-phase inverter using SiC devices are compared with those using Si devices. The data in the table show that under the same conditions, the efficiency of the three-phase inverter using SiC power devices is significantly higher than that using Si devices. The faster switching speed of SiC devices further reduces output voltage fluctuations and results in lower waveform distortion.

TABLE II.   COMPARISON OF SiC AND Si DEVICE PARAMETERS

Parameter	SiC Power Device	Si Power Device
Switching Frequency	30 kHz	5 kHz
Power Device Model	CAS300M12BM2	FD300R12KE3
Efficiency	99%	95%
Output Voltage	390 V	400 V
Output Voltage Distortion Rate	3.2%	5.6%

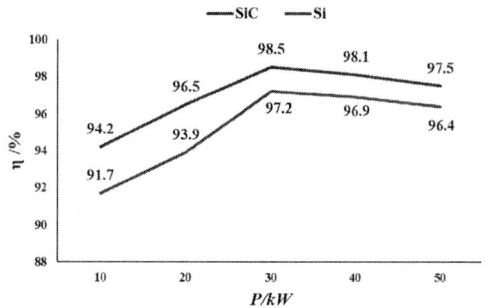

Fig. 8. Efficiency comparison of SiC devices and Si devices

Fig. 8 shows the efficiency curves of the three-phase inverter using SiC power devices compared to that using Si devices under different load levels. The experimental data show that under varying output loads, the use of SiC power devices significantly improves efficiency compared to the original Si-based inverter.

REFERENCES

[1] M. Ocklenburg, M. Döhmen, et al. Next generation DC-DC converters for Auxiliary Power Supplies with SiC MOSFETs[C]. 2018 IEEE International Conference on Electrical Systems for Aircraft, Railway, Ship Propulsion and Road Vehicles & International Transportation Electrification Conference (ESARS-ITEC). Nottingham, UK. 2018: 1-6.

[2] Yi Tao, Zhao Qingliang, Bi Bilong, et al. Design and Optimization Test Verification of High-Power High-Frequency Auxiliary Converter Cooling System for Metro Trains [J]. Urban Rail Transit Research, 2024, 27 (01).

[3] Li Yanwei, Liang Haigang, Wang Lei, et al. Design and Experimental Verification of Auxiliary Converter Based on All-Silicon Carbide Devices [J]. Urban Rail Transit Research, 2020, 23(07)

[4] M. Helsper, M. Ocklenburg. SiC MOSFET Based Auxiliary Power Supply for Rail Vehicles[C]. 2018 20th European Conference on Power Electronics and Applications (EPE'18 ECCE Europe). Riga, Latvia. 2018: 1-8.

[5] A.Maerz, A. Maerz, et al. Constraints replacing IGBTs with SiC MOSEFTs in an onboard railway power supply[C]. Proceedings of PCIM Europe 2015; International Exhibition and Conference for Power Electronics, Intelligent Motion, Renewable Energy and Energy Management. Nuremberg, Germany. 2015: 1-8.

[6] X. Liu, B. Liu, S. He, et al., A multi-objective optimization approach for modeling and analyzing the impact of drive parameters in SiC power converters. Int J Circ Theor Appl. 2024; 1-16.

A Symmetrical Double-Sided Cooled SiC Power Module for Multi-Parallel Applications

1st Guolian Guan
State Key Laboratory of Electrical Insulation and Power Equipment
Xi'an Jiaotong University
Xi'an, China
guanguolian4546@stu.xjtu.edu.cn

2nd Zhiqiang Zhao
State Key Laboratory of Electrical Insulation and Power Equipment
Xi'an Jiaotong University
Xi'an, China
zhaozq@stu.xjtu.edu.cn

3rd Mingzhi Zhao
State Key Laboratory of Electrical Insulation and Power Equipment
Xi'an Jiaotong University
Xi'an, China
1399044778@stu.xjtu.edu.cn

4th Tongyu Zhang
State Key Laboratory of Electrical Insulation and Power Equipment
Xi'an Jiaotong University
Xi'an, China
zty598979175@stu.xjtu.edu.cn

5th Laili Wang
State Key Laboratory of Electrical Insulation and Power Equipment
Xi'an Jiaotong University
Xi'an, China
llwang@mail.xjtu.edu.cn

6th Dewen Wang
China North Vehicle Research Institute
China North Vehicle Research Institute
Beijing, China
dwwangbfcl@163.com

Abstract—To address dynamic current sharing and thermal dissipation bottlenecks in multi-parallel silicon carbide (SiC) power module systems, this paper proposes a symmetrically arranged double-sided cooling (DSC) power module. The design employs dual direct bonded copper (DBC) substrate stacking and coplanar power terminal configuration, achieving 5.48 nH loop parasitic inductance through mutual inductance cancellation effects. Combined with 44.4° truncated pyramidal heat dissipation structure optimization using metallic spacers, simulations demonstrate 8.7°C reduction in maximum chip temperature and 60% thermal gradient mitigation. A 52mm×30mm×5.5mm half-bridge module was developed via triple reflow soldering and silicone gel encapsulation. Experimental results show 6A dynamic current peak deviation in six-module parallel operation under 800V/1000A conditions, with three-module static current imbalance of 9.6%, validating excellent electrical performance.

Keywords—Silicon carbide (SiC) power module, Double-sided cooling (DSC), paralleled modules.

I. INTRODUCTION

As core enablers for carbon neutrality in transportation, low-carbon vehicles are undergoing powertrain transitions from silicon to wide-bandgap semiconductors [1]. Inverter systems for hybrid/electric vehicles demand power modules with high power density, low thermal resistance, and enhanced electromagnetic compatibility, imposing unprecedented challenges on SiC packaging architectures. Although direct liquid cooling modules have become industry standards, wire-bonding-induced parasitic inductance and unidirectional heat dissipation in single-sided cooling structures severely limit system efficiency [2][3]. DSC technology theoretically reduces thermal resistance by 50% [4] and parasitic impedance by 75% [5] through 3D packaging, yet faces multiple engineering barriers.

Electromagnetic limitations of existing DSC modules primarily stem from single-ended structural designs [6]. While suitable for conventional IGBTs, such architectures conflict with SiC MOSFETs' high-speed switching characteristics. As demonstrated in [7], power loop elongation in modules integrating >4 parallel chips induces >15% dynamic current imbalance, causing uneven switching loss distribution and localized thermal runaway risks. More critically, impedance mismatch in conventional busbars exacerbates static current imbalance to 9.6% [7] in multi-module parallel scenarios,

violating stringent current balancing requirements for EV traction systems. Although active gate driving [8] and DBC layout optimization [9] have been proposed, the former suffers implementation complexity while the latter faces 3D spatial constraints.

Thermal management remains pivotal for DSC implementation. While copper clip bonding and solder ball interconnects enable dual-side cooling paths [10], existing DSC modules are constrained by 45° thermal spreading limits [11], inducing gradient accumulation at material interfaces. Typical designs with right-angle connectors create 14°C thermal gradients [12], accelerating junction temperature fluctuations and reducing module lifetime by >30%. Though copper-molybdenum alloy spacers [13] enhance thermal conduction, their cost and complexity exceed automotive industry tolerances.

Addressing these challenges, this work proposes a symmetrically configured DSC SiC module with systematic solutions for multi-parallel current sharing. Section II details the low-inductance (5.48 nH) module design. Section III elaborates manufacturing processes including reflow soldering and silicone encapsulation. Section IV validates 9.6% current imbalance via double-pulse tests and ANSYS ICEPAK thermal simulations demonstrating 8.7°C temperature reduction. Section V concludes the study.

II. NOVEL SIC DOUBLE-SIDED POWER MODULE DESIGN

This paper proposes a symmetrically configured DSC SiC power module with a three-dimensional stacked architecture, as illustrated in Fig. 1. The optimized compact packaging measures 52 mm × 30 mm × 5.5 mm, achieved through dual DBC substrate stacking.

The structural configuration leverages mutual inductance cancellation by adopting coplanar DC power terminal placement and electromagnetic path optimization, significantly reducing loop parasitic parameters. As depicted by the current paths in Fig. 1, the module incorporates a fully symmetric topology: upper and lower arm paralleled chip sets are integrated on separate DBC substrates and interconnected via metallic spacers to form interleaved 3D electrical connections. This spatial layout effectively mitigates thermal stress coupling between power devices while enhancing heat dissipation uniformity. ANSYS Q3D electromagnetic simulations confirm a total power loop parasitic inductance of

5.48 nH. The terminal arrangement further provides accessible interfaces for external thin-film decoupling capacitors.

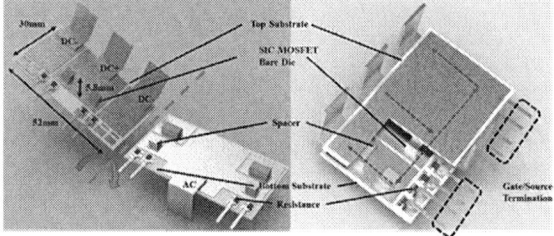

Fig. 1. DSC half-bridge power module structure with two paralleled chips.

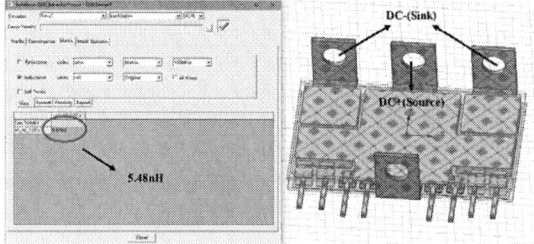

Fig. 2. Test results of parasitic parameters of sub-modules.

III. FABRICATION PROCESS

Prior to manufacturing the double-sided power module, essential components including SiC chips, ceramic substrates, and packaging materials must be prepared. The module fabrication follows a standardized process flow, which primarily includes three sequential reflow soldering stages, wire bonding, and silicone gel encapsulation.

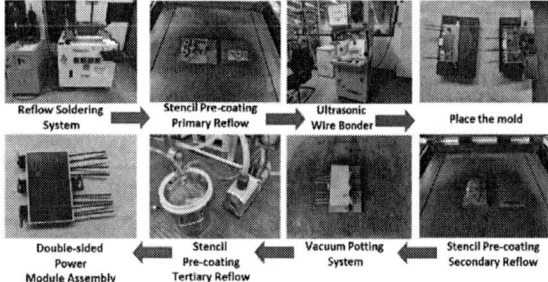

Fig. 3. Schematic diagram of the manufacturing process of self-developed double-sided silicon carbide.

A. Reflow Soldering Process

The reflow soldering process achieves multilayer interconnections through three distinct phases. The first reflow soldering establishes electrical connections between semiconductor chips, power terminals, and DBC ceramic substrates. The second phase interconnects the chips with metallic posts and integrates gate resistors onto the DBC substrates. The third reflow soldering facilitates electrical bridging between the upper and lower DBC substrates via metallic posts.

The soldering process, conducted in a vacuum environment, follows a temperature profile tailored to the solder material characteristics and is divided into four sequential stages. During the preheating stage, components are gradually heated to achieve thermal equilibrium while ensuring complete volatilization of solvents and gases from the solder. Excessive heating rates during this phase may damage sensitive components. The soaking stage maintains a stable temperature to prevent thermal stress caused by temperature lag between components and the solder. The soldering stage rapidly elevates the temperature to melt the solder, enabling full wetting of the bonding surfaces. Finally, the cooling stage ensures controlled solidification of the solder joints.

Post-soldering inspection employs ultrasonic scanning to evaluate joint quality. Chip solder layers are required to exhibit less than 2% voidage, while large-area DBC substrate joints must not exceed 5%. Excessive voidage increases thermal resistance, significantly reducing the operational lifespan of the module. This process ensures robust interlayer connections while addressing reliability challenges in multilayer power module fabrication.

B. Wire Bonding Technology

Wire bonding employs ultrasonic bonding processes to achieve electrical interconnections between chips and DBC ceramic substrates, representing the most widely adopted solution in current applications. Despite inherent limitations such as significant parasitic inductance, and although researchers have proposed various alternative methods to reduce parasitic inductance, wire bonding remains the most reliable interconnection technology due to the immaturity of competing approaches. Commonly used bonding wires include aluminum, copper, and gold wires, with aluminum wires being predominantly utilized in power modules for their cost-effectiveness and process compatibility. Ultrasonic energy effectively removes oxide layers from metal surfaces, ensuring robust wire adhesion and minimizing thermal stress in devices. The bonding process involves initially applying pressure to the bonding wire, followed by welding and subsequent cutting. Aluminum wires are available in standard diameters such as 5 mil, 10 mil, 12 mil, 15 mil, and 20 mil, with larger diameters offering higher current-carrying capacity. The selection of wire quantity and dimensions requires a balanced consideration of operational current demands, while being constrained by the resistive and thermal limitations inherent to the bonding wires themselves. This method ensures reliable electrical connectivity while addressing practical constraints in power module manufacturing.

C. Encapsulation Process

The encapsulation process primarily enhances the module's insulation performance while providing external environmental protection for DBC ceramic substrates, chips, and bonding wires. A two-component silicone gel is prepared by mixing materials in a specific ratio, with the mixed colloid curing at room temperature within 24 hours. The cured silicone gel exhibits high dielectric strength and moderate elasticity, enabling stable operation across a temperature range of −60°C to 200°C. In this study, Type 195 transparent silicone encapsulant is selected for its low shrinkage rate and dielectric strength of 25 kV/mm. To eliminate bubbles that could compromise performance, the mixed silicone gel undergoes vacuum degassing prior to application. This process ensures reliable insulation and mechanical protection under harsh operational conditions.

IV. SIMULATION AND EXPERIMENTAL VERIFICATION

A. Multi-module parallel dual-pulse simulation test

To enhance the system's power capability, this study employs self-developed symmetrical DSC power modules to construct a six-parallel power electronic system. As illustrated in Fig. 4, six DSC modules are integrated into a half-bridge topology using dual-side heat sinks. The DC side utilizes multiple symmetrically arranged busbars, while the AC side achieves electrical interconnection through optimized busbar designs. Each module incorporates decoupling capacitors at its positive/negative terminals, reducing the commutation loop's equivalent parasitic inductance to approximately 1.6 nH. The system achieves a power rating of 1700 V/1000 A.

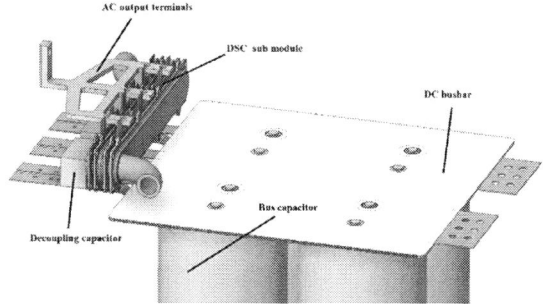

Fig. 4. The whole system with three paralleled DSC modules

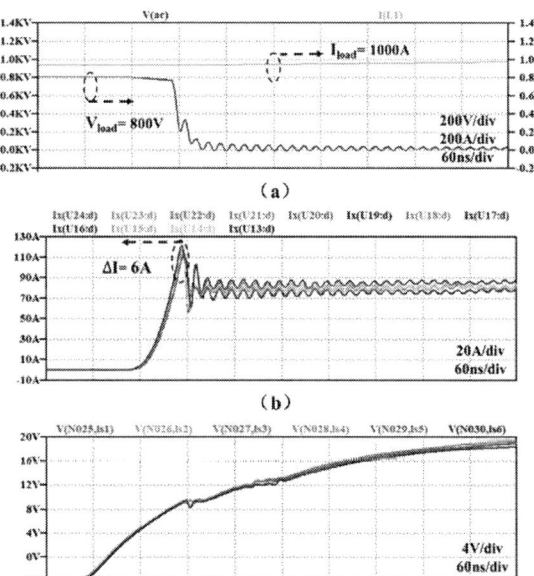

Fig. 5. 12 parallel 800V/1000A dual-pulse test: (a) Load voltage and current waveforms; (b) down-tube turn-on current waveform; (c) down-tube turn-on gate-source voltage waveform

A double-pulse test circuit was modeled in LTspice, with the lower bridge configured with twelve parallel SiC chips (six-module parallel configuration) under 800 V bus voltage and 1000 A load current. A single-channel control scheme was implemented for the gate drivers. Simulation results in Fig. 5 demonstrate a dynamic current peak deviation of 6 A among the twelve parallel chips under 800 V/1000 A operation, with a static current distribution standard deviation below 2%, meeting the parallel current-sharing design criteria.

B. The construction and testing situation of the multi-module experimental platform

To comprehensively validate the technical feasibility of the six-parallel double-sided submodule design, systematic experimental studies were conducted using the double-pulse test platform illustrated in Fig. 6. The experimental study was divided into two progressive phases: initial electrical stress testing on a single double-sided submodule, followed by dynamic current-sharing evaluation under multi-module parallel operation.

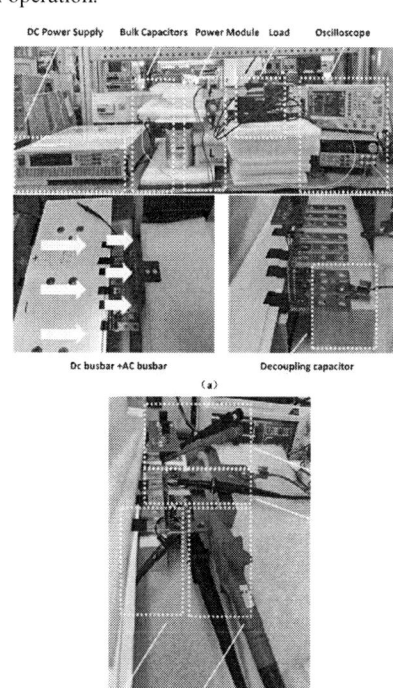

Fig. 6. Construction of the dual-pulse test platform

In the single-module test, under typical operating conditions of 600 V bus voltage and 135 A load current, the measured peak voltage stress on the lower switch was 770 V. Under extreme conditions (800 V bus voltage and 115 A load current), the peak voltage stress reached 952 V. Both datasets confirm stable voltage stress distribution and sufficient safety margins, meeting industrial-grade reliability standards for power devices.

For multi-module parallel characteristics, preliminary experiments were performed on a three-module parallel system. Under 200 V bus voltage, real-time current monitoring via Rogowski coils revealed a current imbalance of merely 9.6% among modules, significantly outperforming the 15% design threshold for power electronic parallel systems. Further 400 V/500 A double-pulse tests demonstrated excellent dynamic current-sharing performance during transient switching events, with less than 5% synchronization error in current waveforms across modules. Notably, no voltage spikes or anomalies were observed. These results validate the effectiveness of snubber circuit parameter matching and power loop layout optimization in the parallel topology, establishing a technical foundation for scaling to six-module parallel configurations to achieve the 1000 A test target.

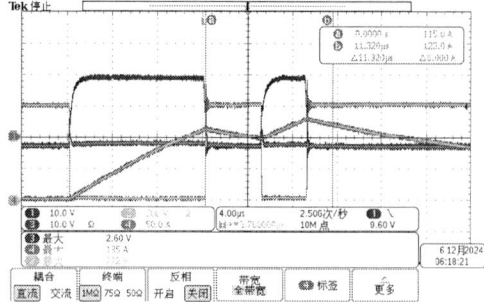

Fig. 7. Single-module bus voltage 600V dual-pulse test

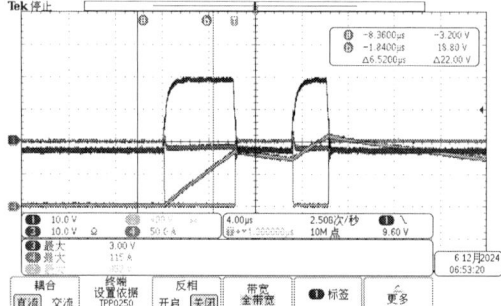

Fig. 8. Single-module bus voltage 800V dual-pulse test

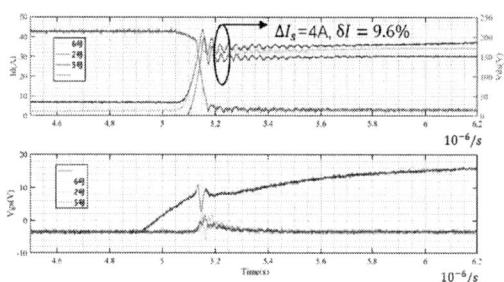

Fig. 9. The current equalization test waveform of the three modules in parallel when the bus voltage is 200V

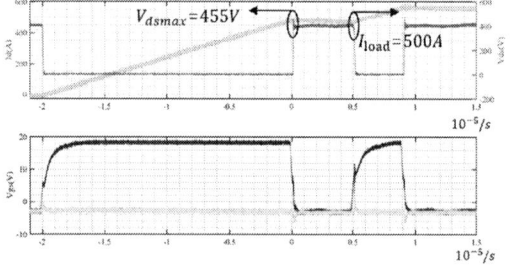

Fig. 10. Double-pulse test results when the bus voltage is 400V and the load current is 500A

It should be emphasized that the current experiments are subject to the following constraints: first, the number of parallel modules has not yet reached the design maximum; second, measurement errors of approximately 5% may be introduced due to current probe bandwidth limitations (as per the probe calibration certificates). These technical limitations will be systematically addressed in subsequent studies by increasing the number of parallel units and implementing

differential current sensors. Such optimizations aim to enhance measurement accuracy and validate the scalability of the proposed architecture under full-scale operating conditions.

C. Multi-module parallel dual-pulse simulation test

During thermal conduction in power devices, heat transfer efficiency across heterogeneous material interfaces exhibits gradient variation characteristics. As addressed in prior research [14], a comprehensive theoretical model has been established to describe this phenomenon. By leveraging the differences in thermal properties between materials, the interfacial heat spreading angle can be characterized using the following formula:

$$\alpha = \tan^{-1}\left(\frac{K_a}{K_b}\right) \qquad (1)$$

In this formula, α characterizes the actual heat conduction path angle, where K_a represents the thermal conductivity of SiC, and K_b orresponds to the thermal conductivity of the copper substrate. According to classical heat transfer theory, heat diffusion in typical electronic devices generally follows a 45° spreading pattern [15]. Building on this insight, this study proposes a truncated pyramidal metallic spacer structure for subsequent design optimization, as depicted in Fig. 11. This configuration maintains thermal conduction paths between the upper and lower DBC substrates while enabling bidirectional controllable heat dissipation angle adjustment.

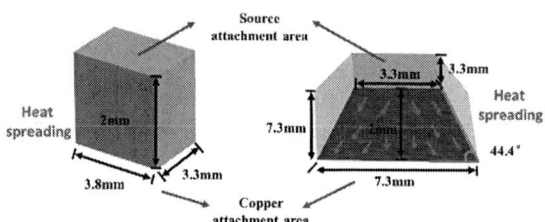

Fig. 11. 90° (left) and 44.4° (right) spacer connection blocks showing heat dissipation path and isometric view

To accommodate the standardized height of commercial interlayer heat sinks shown in Fig. 4, the spacer structure was optimized using ANSYS ICEPAK simulations. Under the constraint of maximizing chip contact area (3.3 mm × 3.3 mm), a 44.4° heat spreading angle was identified as optimal. Experimental results demonstrate that compared to conventional 90° right-angle designs, the novel structure reduces the device operating temperature from 120.1°C to 111.4°C at a standardized height of 2 mm, while the thermal gradient is significantly improved from 14°C to 5.6°C. These findings validate the technical superiority of the truncated pyramidal metallic spacer in achieving homogeneous thermal field distribution.

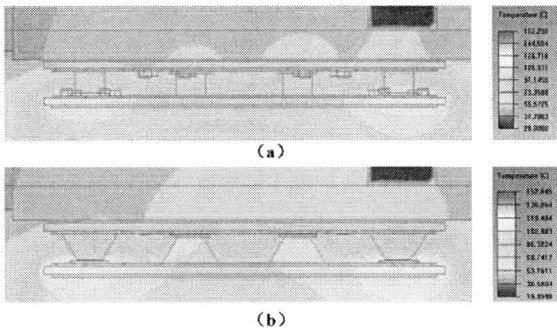

(a)

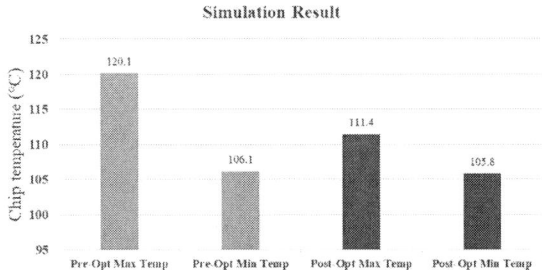

(b)

Fig. 12. Temperature distribution of power module with spacer at 90(a) degrees and 44.4(b) degrees

Simulation Result

Fig. 13. The maximum and minimum temperatures of the chip before and after optimization

The structural height parameters strictly adhere to the assembly requirements of interlayer heat sinks. The 2 mm baseline dimension of the metallic spacer not only satisfies stress-free manufacturing conditions but also ensures robust structural support for the DBC substrates. Surface treatment employs electroless nickel immersion gold plating, forming stable metallurgical bonds with the DBC substrate coatings and device sintering layers. For material selection, while copper-molybdenum alloy (coefficient of thermal expansion,CTE:7.2×10^{-6}/k) exhibits superior thermal matching with SiC (CTE:7.2×10^{-6}/k), a comprehensive evaluation of thermal conductivity, simulation results, manufacturing costs, and processing feasibility led to the selection of industrial-grade pure copper as the optimal engineering solution. This decision balances thermal performance with practical manufacturability, ensuring compatibility with high-volume production workflows.

V. CONCLUSION

This study proposes a symmetrically arranged DSC)silicon carbide power module design to address the technical requirements of multi-module parallel systems in electric vehicles. By employing dual DBC substrate stacking and coplanar power terminal configuration, a low-impedance power loop with mutual inductance cancellation effects was constructed, achieving a measured total parasitic inductance of 5.48 nH. The implementation of vertical metallic spacer interconnects enabled staggered electrical connections for upper/lower arm paralleled chip sets, coupled with optimized 44.4° truncated pyramidal heat dissipation structures,

resulting in an 8.7°C reduction in maximum chip operating temperature. Experimental results demonstrated a static current imbalance of 9.6% for a three-module parallel system, while dynamic current peak deviations in a six-module parallel simulation were controlled within 6 A under 800 V/1000 A conditions.

ACKNOWLEDGMENT

The author would also like to thank Mr. Gong for his help in writing the article.

REFERENCES

[1] Z. Liang, "Status and trend of automotive power packaging", 2012 24th International Symposium on Power Semiconductor Devices and ICs, pp. 325-331, 2012.

[2] C. Chen, F. Luo and Y. Kang, "A review of SiC power module packaging: Layout material system and integration", CPSS Trans. Power Electron. Appl., vol. 2, no. 3, pp. 170-186, Sep. 2017.

[3] G. Bower, C. Rogan, J. Kozlowski and M. Zugger, "SiC power electronics packaging prognostics", Proc. IEEE Aerosp. Conf., pp. 1-12, 2008.

[4] J. Marcinkowski, "Dual-sided cooling of power semiconductor modules", Proc. PCIM Europe; Int. Exhib. Conf. Power Electron. Intell. Motion Renewable Energy Energy Manage., pp. 1-7, 2014.

[5] Y. Li et al., "Highly Integrated Power Unit Based on Double Sided Cooling IGBT Module", PCIM Europe 2017; International Exhibition and Conference for Power Electronics Intelligent Motion Renewable Energy and Energy Management, pp. 1-6, 2017.

[6] Z. Liang, "Integrated double sided cooling packaging of planar SiC power modules", 2015 IEEE Energy Conversion Congress and Exposition (ECCE), pp. 4907-4912, 2015.

[7] J. Yang, Y. Gan, L. Wang, C. Zhao, Y. Nie and L. Ran, "Design and Current Balancing Optimization of A 1700V/1000A Multi-chip SiC Power Module," 2023 11th International Conference on Power Electronics and ECCE Asia (ICPE 2023 - ECCE Asia), Jeju Island,Korea, Republic of, 2023, pp. 1952-1958.

[8] C. Zhao, L. Wang, F. Zhang, and F. Yang, "A method to balance dynamic current of paralleled SiC MOSFETs with kelvin connection based on response surface model and nonlinear optimization," IEEE Trans. Power Electron., vol. 36, no. 2, pp. 2068–2079, Feb. 2021.

[9] Y. Wen, Y. Yang and Y. Gao, "Active Gate Driver for Improving Current Sharing Performance of Paralleled High-Power SiC MOSFET Modules," in IEEE Transactions on Power Electronics, vol. 36, no. 2, pp. 1491-1505, Feb. 2021.

[10] Q. Zhu, A. Forsyth, R. Todd and L. Mills, "Thermal characterization of a copper-clip-bonded IGBT module with double-sided cooling", Proc. 23rd Int. Workshop Thermal Investigations ICs Syst., pp. 1-6, 2017.

[11] A. B. Lostetter, F. Barlow and A. Elshabini, "An overview to integrated power module design for high power electronics packaging", Microelectron. Rel., vol. 40, no. 3, pp. 365-379, 2000

[12] J. Jeon, J. Seong, J. Lim, M. K. Kim, T. Kim and S. W. Yoon, "Finite element and experimental analysis of spacer designs for reducing the thermomechanical stress in double-sided cooling power modules", IEEE J. Emerg. Sel. Topics Power Electron., vol. 9, no. 4, pp. 3883-3891, Aug. 2021..

[13] X. Liu, Z. Wu, Y. Yan, Y. Kang and C. Chen, "A novel double-sided cooling inverter leg for high power density EV based on customized SiC power module", Proc. IEEE Energy Convers. Congr. Expo., pp. 3151-3154, 2020.

[14] N. B. Nguyen, "Properly implementing thermal spreading will cut cost while improving device reliability", Proc. Int. Symp. Microelectron., pp. 383-388, 1996.

[15] A. B. Lostetter, F. Barlow and A. Elshabini, "An overview to integrated power module design for high power electronics packaging", Microelectron. Rel., vol. 40, no. 3, pp. 365-379, 2000.

A Transistor Clamp Circuit of On-state Voltage Drop for SiC MOSFET Temperature Monitoring

Yixiang Zhao
School of Electrical Engineering
Beijing Jiaotong University
Beijing, China
23121523@bjtu.edu.cn

Hong Li*
School of Electrical Engineering
Zhejiang University
Hangzhou, China
hong_li@zju.edu.cn

Xiaofei Hu
Rundian Energy Science and Technology Co., Ltd
Zhengzhou, China
huxiaofei17@crpower.com.cn

Kuang Zhang
School of Electrical Engineering
Beijing Jiaotong University
Beijing, China
23126396@bjtu.edu.cn

Abstract—To ensure the long-term reliability of power electronic systems, accurate junction temperature monitoring of Silicon Carbide (SiC) MOSFETs is essential, as excessive thermal stress is a major cause of device degradation and failure. This paper addresses the challenge of real-time, non-intrusive temperature monitoring by proposing a fast-response measurement circuit based on the on-state voltage drop ($V_{DS(on)}$), a temperature-sensitive electrical parameter. A theoretical linear correlation between ($V_{DS(on)}$) and junction temperature is established, enabling reliable thermal state estimation during normal operation. The proposed circuit accurately extracts ($V_{DS(on)}$) within 0.6 μs during device conduction and effectively suppresses voltage spikes during the off-state through voltage clamping. This design eliminates the need for complex auxiliary circuits, reduces monitoring latency, and enhances measurement precision. Its simplicity, robustness, and high temporal resolution make it particularly suitable for predictive maintenance, fault prevention, and thermal management in high-efficiency applications such as renewable energy inverters, electric vehicle powertrains, and smart grid systems.

Keywords—*SiC MOSFET, junction temperature monitoring, on-state voltage drop, temperature-sensitive electrical parameter (TSEP), clamping circuit.*

I. INTRODUCTION

As a wide-bandgap semiconductor device, the SiC MOSFET offers superior blocking voltage, faster switching speed, and improved thermal conductivity over conventional Si MOSFETs, enabling higher efficiency, power density, and reliability in power electronic systems [1-2]. These advantages make SiC MOSFETs ideal for high-temperature, high-frequency, high-voltage, and high-power applications. With their growing deployment in smart grids, rail transportation, aerospace, and new energy vehicles, long-term reliability, condition monitoring, and health management have become critical research and industrial focuses.

As shown by the power electronic system reliability research report, power devices are the highest failure rate components in the variable current system, accounting for about 34% [3]. Among the various failure factors including temperature, humidity, vibration and dust, about 55% of power electronic system failures are mainly induced by temperature factors [4]. Therefore, real-time online monitoring of the operating junction temperature of power devices is extremely important for monitoring the reliable operation of the devices as well as the whole system.

Due to packaging limitations, direct junction temperature measurement of power semiconductor devices remains challenging. Extensive research has yielded various monitoring techniques, broadly categorized into four types based on physical principles: physical contact, optical, thermal network, and temperature-sensitive electrical parameter (TSEP) methods [6]. Physical contact methods employ embedded sensors, such as thermistors or thermocouples [7], but require intrusive package modifications, exhibit slow thermal response, and cannot capture dynamic temperature changes. Optical methods like infrared thermography [8] need chip decapsulation and surface blackening, making them destructive and unsuitable for online monitoring. Thermal network methods estimate junction temperature using transient thermal impedance models and real-time power loss data [9], but suffer from complex modeling and significant errors due to device aging. Conversely, TSEP-based methods infer junction temperature by monitoring external electrical parameters affected by internal temperature variations [10]. These enable rapid, non-intrusive, online temperature estimation at the microsecond scale, positioning them as a highly promising approach for junction temperature monitoring [11].

The temperature-sensitive electrical parameter (TSEP) method has garnered significant attention in academia and industry, prompting extensive global research. TSEPs are generally classified into static and dynamic types based on temporal characteristics [5]. Static TSEPs, such as the on-state voltage drop, the saturation current, the short-circuit current, are extracted under steady-state operation, while dynamic TSEPs, such as the turn-on/off delay times and the voltage change rates, are obtained during transient switching. Among these, the on-state voltage $V_{DS(on)}$ is a promising candidate for real-time junction temperature monitoring [12]. Existing $V_{DS(on)}$ measurement circuits are broadly divided into small-current-based [13] and load-current-based approaches [14]. Small-current-based circuits feature complex architectures and limited response speed, restricting

---

* Corresponding Author, E-mail: Hong Li, hong_li@zju.edu.cn
* Supported in part by the National Science Fund for Distinguished Young Scholars 52325704, in part by the Key Program of National Natural Science Foundation of China 52237008.

979-8-3315-1110-4/25 $31.00 © 2025 IEEE

high-frequency applications. Load-current-based circuits offer faster response but suffer from increased measurement errors, limiting industrial usability.

In junction temperature monitoring based on the on-state voltage drop $V_{DS(on)}$, numerous methods have been proposed to enhance measurement accuracy, circuit stability, and applicability under practical operating conditions. Reference [15] introduced a $V_{DS(on)}$ measurement circuit that incorporates optimized voltage spike suppression and offline calibration techniques. These enhancements improve measurement stability and reduce implementation cost, however, the circuit exhibits sensitivity to diode parameter variations, which compromises its performance, particularly in high-frequency switching applications. Reference [16] presented a novel approach based on the pass-state leakage-source voltage of SiC MOSFETs, where real-time junction temperature monitoring is achieved by effectively decoupling chip and package parasitic resistances. While this method improves temperature estimation accuracy, it suffers from a complex resistance extraction process and susceptibility to ambient environmental variations, such as temperature and humidity, limiting its robustness in industrial deployment.

Furthermore, Reference [17] proposed an advanced voltage clamp circuit specifically designed to address issues such as RC delay and voltage overshoots during high-frequency switching transitions. This solution significantly enhances the resolution and accuracy of $V_{DS(on)}$ measurements; however, it still experiences residual voltage peaks at switching edges and remains sensitive to the performance variability of discrete components, which affects long-term stability and repeatability.

To overcome these limitations—namely, circuit complexity, slow dynamic response, low measurement accuracy, and poor adaptability to high-speed switching environments—this paper proposes a novel switch-clamped-based $V_{DS(on)}$ measurement circuit tailored for SiC MOSFET applications. The proposed design simplifies the circuit topology while significantly enhancing response speed, noise immunity, and measurement precision. By enabling real-time extraction of $V_{DS(on)}$ during the conduction phase, the circuit provides a reliable and scalable solution for junction temperature monitoring using $V_{DS(on)}$ as a robust temperature-sensitive electrical parameter (TSEP), especially suited for high-frequency, high-voltage power electronic systems.

II. RELATIONSHIP BETWEEN THE ON-STATE VOLTAGE DROP AND JUNCTION TEMPERATURE

To demonstrate the feasibility of using the on-state voltage drop ($V_{DS(on)}$) of SiC MOSFETs as a critical temperature-sensitive parameter for junction temperature monitoring, this paper first establishes a theoretical model of the relationship between $V_{DS(on)}$ and junction temperature.

Fig.1 shows the typical cross-sectional structure of a SiC MOSFET. When operated in the linear conduction region, the device behaves as a resistor, and $V_{DS(on)}$ is primarily determined by the channel resistance R_{CH}, which can be expressed as (1):

$$V_{DS(on)} = I_D \cdot R_{CH} = \frac{I_D \cdot L_{CH}}{Z \mu_{ni} C_{OX} (V_{GS} - V_{TH})} \tag{1}$$

where I_D is the drain current, L_{CH} is the channel length, Z is the cell width, μ_{ni} is the inversion layer mobility, C_{OX} is the

gate oxide capacitance, V_{GS} is the gate voltage, and V_{TH} is the threshold voltage. In (1), only μ_{ni} and V_{TH} are temperature-dependent parameters related to the junction temperature T_j.

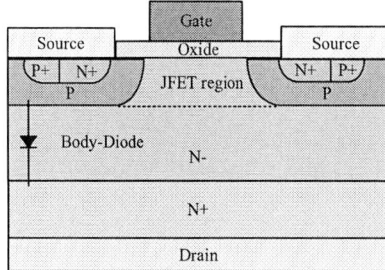

Fig. 1. The structure of SiC MOSFET

The inversion layer mobility μ_{ni} can be expressed as (2):

$$\mu_{ni} = C_1 \left(T_j \right)^{-2.7} \tag{2}$$

where C_1 is a proportionality constant of proportionality.

The gate oxygen characteristic capacitance C_{OX} can be expressed as (3):

$$C_{OX} = \frac{\varepsilon_{OX}}{t_{OX}} \tag{3}$$

where ε_{OX} is the relative dielectric constant of silica and t_{OX} is the thickness of the oxide layer.

The threshold voltage V_{TH} can be expressed as (4):

$$V_{TH} = \frac{\sqrt{4\varepsilon_S q N_A \varphi_B}}{C_{OX}} \tag{4}$$

where φ_B is the surface potential energy, ε_S is the semiconductor relative permittivity, and N_A is the doping concentration. Only φ_B in (4) is related to the temperature T_j, which can be specifically expressed as (5):

$$\varphi_B = \frac{kT_j}{q} \ln \left(\frac{N_A}{n_i} \right) \tag{5}$$

where k is the Boltzmann constant, q is the fundamental charge constant, and n_i is the concentration of intrinsic carriers, which can be expressed as (6):

$$n_i = C_2 \left(T_j \right)^{\frac{3}{2}} e^{-\frac{E_g}{2kT_j}} \tag{6}$$

where C_2 is a positive proportionality constant and E_g is the forbidden bandwidth of the silicon carbide semiconductor.

As shown in Fig. 2, the on-state voltage drop ($V_{DS(on)}$) of SiC MOSFETs exhibits a good linear correlation with junction temperature (T_j) under different drain currents (I_D). The sensitivity of this linear relationship increases with higher I_D, making $V_{DS(on)}$ a suitable temperature-sensitive parameter (TSEP) for online junction temperature monitoring of SiC MOSFET. Consequently, the problem of monitoring T_j can be transformed into real-time measurement of $V_{DS(on)}$.

979-8-3315-1110-4/25 $31.00 © 2025 IEEE

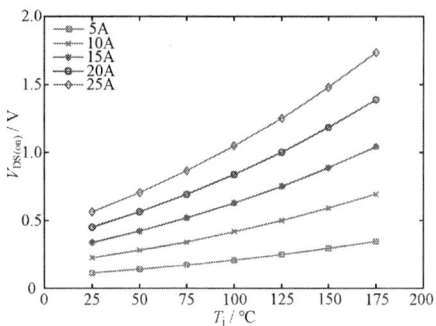

Fig. 2. The variation curves of on-state voltage drop with junction temperature at different drain currents

III. Working Principle of the Proposed SiC MOSFET On-State Voltage Drop Measurement Circuit

Existing on-state voltage drop measurement circuits often employ voltage-clamping methods based on biased current sources to isolate off-state voltages. Fig. 3 shows a conventional measurement circuit using this approach. The circuit requires designing a sub-circuit to generate a stable small current I_a, and strict matching of $I_{D1}=I_{D2}$ must be ensured during measurements to avoid significant errors. Furthermore, the slow response speed limits its applicability to high-frequency operation of SiC MOSFET. Additionally, the circuit is susceptible to turn-off voltage spikes and negative voltage overshoot in the tested device, while the use of a linear power supply for the current source not only reduces main-circuit efficiency but also causes thermal issues, adversely affecting the actual junction temperature of the device under test.

To address the issues of complex structure, slow response speed, and low reliability in conventional SiC MOSFET on-state voltage drop measurement circuits, this paper proposes a switch-clamped-based fast-response and highly reliable measurement circuit. As shown in Fig.4, this circuit enables rapid and accurate $V_{DS(on)}$ measurement for SiC MOSFET.

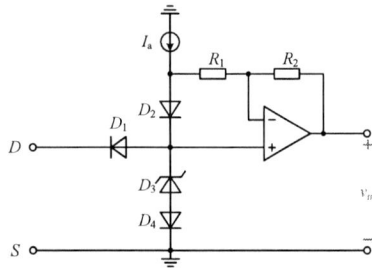

Fig. 3. Traditional on-state voltage drop measurement circuits

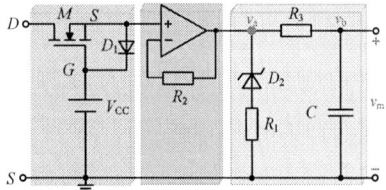

Fig. 4. On-state voltage drop measurement circuit

The conduction voltage drop measurement circuit consists of a measurement circuit with a switch transistor (M), Zener diode (D_2), and resistor (R_1), and an RC filter with resistor (R_3) and capacitor (C). The gate of transistor M is connected to a positive DC voltage (V_{CC}) greater than its threshold voltage (V_{TH}) to keep it in the on-state. Diode (D_1) between the gate and source prevents negative gate-source voltage and suppresses voltage spikes from parasitic capacitance (C_{DS}) when M turns off. The RC filter further smooths output voltage (v_a) for better measurement accuracy. The operational amplifier is added to improve the accuracy of signal detection, reduce interference, and suppress forward spikes along with the clamp diode. This design balances response speed, cost, and reliability, offering advantages in performance for applications requiring these factors.

TABLE I
MEASUREMENT CIRCUIT COMPARISON

Measurement circuit	Speed	Number of devices	Reliability
Conventional measurement circuits	2.4μs	More	Low
Proposed measurement circuit	0.25μs	Less	High

When the SiC MOSFET under test is conducting, the switch transistor M is also in the on-state. Since the resistance of the branch containing the Zener diode D_2 and resistor R_1 is much larger than the on-resistance of the SiC MOSFET, the load current will flow through the branch containing the SiC MOSFET. At this time, neglecting the conduction voltage drop of the switch transistor M, the voltage across the Zener diode D_2 and resistor R_1, denoted as v_a, is approximately equal to the conduction voltage drop of the SiC MOSFET, $V_{DS(on)}$. This voltage is then processed by the RC filter circuit and output as v_m, as shown in (7):

$$v_m \approx V_{DS(on)} \tag{7}$$

When the SiC MOSFET under test is turned off, the switch transistor M remains in the on-state at this moment, and the load current will flow through the switch transistor M, causing the voltage at the source terminal of the switch transistor M to rise. When the gate-to-source voltage of the switch transistor M drops below its threshold voltage V_{TH} (for example, V_{TH}= 4V), the switch transistor M turns off. Ultimately, a voltage balance will be established, making the gate-source voltage V_{GS} of the switch transistor M equal to V_{TH}. The voltage across the Zener diode D_2 and resistor R_1 will be clamped to $v_a=V_{CC}-V_{TH}$. This voltage also passes through the RC filter circuit and is output as v_m, as shown in (8). This process allows the range of the oscilloscope to be expanded to measure the voltage v_a, thus improving the measurement resolution.

$$v_m \approx V_{CC} - V_{TH} \tag{8}$$

IV. Simulation Results and Analysis

The simulation schematic of the double pulse circuit is shown in Fig. 5, and the simulation parameters are shown in Tab. 2, and the simulation parameters of the measurement circuit are shown in Tab. 3.

TABLE II
SIMULATION PARAMETERS OF DOUBLE PULSE

Parameters	V_{dc}	C_{dc}	L_{load}	R_{G1}	R_{G2}
Value	200V	450μF	90μH	3Ω	3Ω

The gate source voltage v_{GS}, drain source voltage v_{DS}, drain current i_D, and output v_m simulation waveforms of the SiC MOSFET Q_2 to be tested in the double-pulse circuit obtained based on the above simulation circuit, as well as the output v_m simulation waveforms of the proposed on-state voltage drop measurement circuit, are shown in Fig. 6.

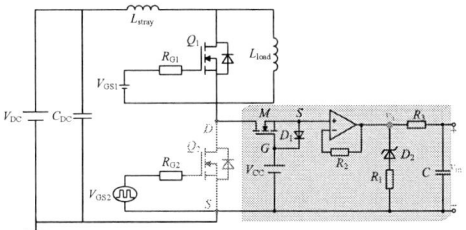

Fig. 5. The schematic of simulation circuit

TABLE III
SIMULATION PARAMETERS OF ON-STATE VOLTAGE DROP MEASUREMENT
CIRCUIT

Parameters	Value /Model	Parameters	Value /Model
M	IRFAC30	R_3	1kΩ
D_1	BAS40	R_4	1kΩ
D_2	1N4372	V_{CC}	7V
R_1	10Ω	C	200pF
R_2	230Ω	OPA	OPA350

The gate source voltage v_{GS}, drain source voltage v_{DS}, drain current i_D, and output v_m simulation waveforms of the SiC MOSFET Q_2 to be tested in the double-pulse circuit obtained based on the above simulation circuit, as well as the output v_m simulation waveforms of the proposed on-state voltage drop measurement circuit, are shown in Fig. 6.

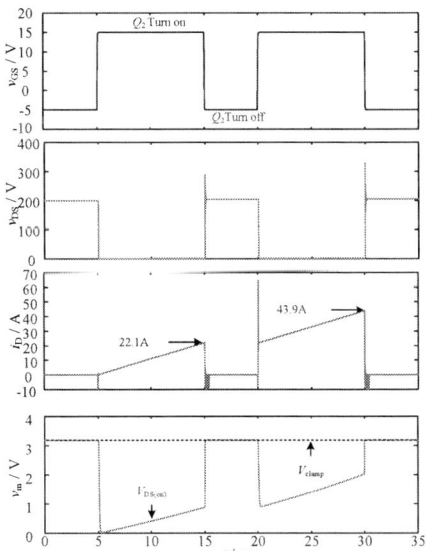

Fig. 6. Simulation waveforms for measuring SiC MOSFET on-state voltage drop based on transistor clamp circuit

A comparison of the measured output voltage v_m and the simulated waveforms of the drain-source voltage V_{DS} of the SiC MOSFET Q_2 to be tested in the double-pulse circuit obtained based on the above simulation circuit is shown in Fig.7. The simulation results demonstrate that the proposed circuit achieves accurate ($V_{DS(on)}$) measurement within 0.6 µs

during the on-state and maintains a stable low output during the off-state through effective clamping, ensuring high-speed, high-precision, and stable performance for SiC MOSFETs.

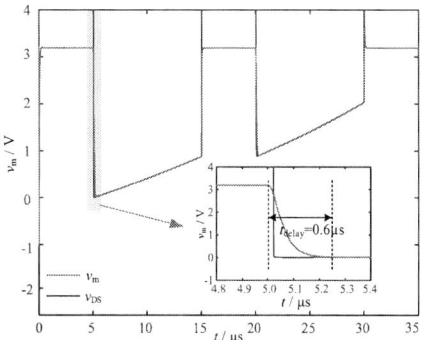

Fig. 7. Comparison of simulation waveforms between measurement output voltage v_m and V_{DS}

V. EXPERIMENTAL RESULTS AND ANALYSIS

In order to further verify the performance of the proposed on-state voltage drop measurement circuit, the experimental platform based on the double-pulse circuit and the proposed measurement circuit is shown in Fig. 8. The device to be measured in the experiment is a SiC MOSFET of Infineon Model No. IMW120R030M1H (1200V, 56A), and the driver chip is a single-channel isolated driver chip of Infineon Model No. 1EDI60H12AH single-channel isolated driver chip, the main equipment and models of the experimental platform are shown in Tab. 4.

The proposed on-state voltage drop measurement circuit was subjected to double-pulse experiments based on the experimental platform built in Fig. 8, and the experimental waveforms are shown in Fig. 9. From the double-pulse experimental waveforms, it can be seen that the proposed on-state voltage-drop measurement circuit is able to quickly and accurately measure the on-state voltage drop of the SiC MOSFET to be measured within 0.6 µs after its on-state, and during the off-state period of the SiC MOSFET to be measured, the output of the measurement circuit is clamped to a low voltage level, which ensures the accuracy of the measurement. Thus, the proposed on-state voltage drop measurement circuit is capable of realizing fast, accurate and high precision measurement of the on-state voltage drop during the on-state period of the SiC MOSFET. In addition, the output of the measurement circuit still has small voltage spikes at the turn-on and turn-off moments of the SiC MOSFET, which is due to the parasitic inductance between the measurement probe and the measurement point.

To enable the proposed on-state voltage drop $V_{DS(on)}$ measurement circuit for junction temperature monitoring, an offline calibration was conducted to establish the relationship among $V_{DS(on)}$, junction temperature T_j, and drain current I_D. The calibration was performed on a custom experimental platform, where a BK946 heating stage was employed to regulate the device temperature from 50 °C to 150 °C.

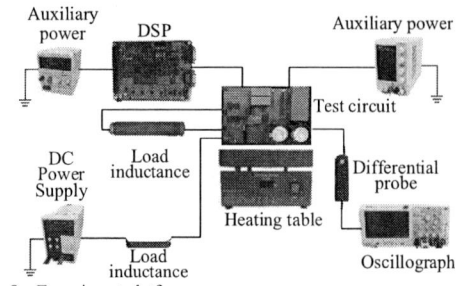

Fig. 8. Experiment platform

TABLE IV
MAIN EQUIPMENT AND MODELS

Equipment Name	Model
Oscilloscope	MSO4054
DC Power Supply	HSPY-1000-005
Auxiliary Power	UTP3315TFL-II
Heating Platform	BK946
Voltage Probe	1kΩTHDP0200
Rogowski Coil	CWTUM / 1 / B
DSP	TMS320F28335
Infrared Thermometer	FLUKE Ti32

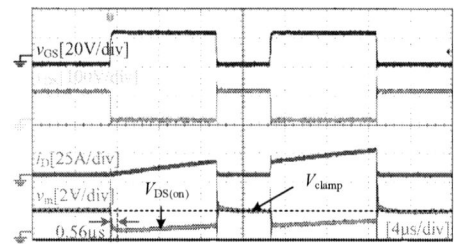

Fig. 9. Experimental waveforms

During the experiment, the SiC MOSFET under test was heated on the stage until thermal steady-state is achieved—defined as a constant temperature maintained for over five minutes. Under such conditions, due to sufficient thermal conduction, the case temperature T_c is assumed to approximate the junction temperature T_j, allowing accurate calibration.

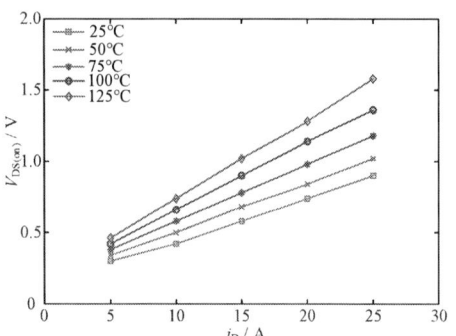

Fig. 10. The variation curves of on-state voltage drop with drain current under different junction temperatures

Measurements are taken at temperature intervals from 25 °C to 125 °C in steps of 25 °C, and for five drain current levels from 5 A to 25 A in steps of 5 A. The corresponding $V_{DS(on)}$ values are recorded under each condition. Fig. 10

further illustrates that the sensitivity of this relationship increases with higher drain current.

VI. CONCLUSION

This paper proposes a junction temperature monitoring method for SiC MOSFETs based on a switch-clamped circuit, addressing the issues of complex structure, slow response, and low measurement accuracy in conventional on-state voltage drop ($V_{DS(on)}$) measurement circuits. The key conclusions are as follows: A theoretical linear relationship between $V_{DS(on)}$ and junction temperature is established, providing a foundation for temperature monitoring. The proposed switch-clamped circuit features a simple structure, fast response speed, and enables accurate $V_{DS(on)}$ measurement within 0.6 µs after device conduction, making it suitable for high-frequency applications.

Despite these advantages, several potential limitations should be acknowledged. Firstly, the measurement accuracy may be affected by external electromagnetic interference in high-voltage or noisy environments. Secondly, the stability of the auxiliary power supply ($V_{DS(on)}$) is critical to ensure consistent clamping behavior and prevent offset errors. Thirdly, the reverse recovery characteristics of the clamp diode can introduce transient spikes during switching events, which may distort the measurement waveform. Future work will focus on improving the anti-interference of the circuit robustness, optimizing power supply regulation, and selecting ultra-fast recovery diodes to further enhance performance under complex application scenarios.

REFERENCES

[1] H. A. Mantooth, M. D. Glover, and P. Shepherd, "Wide bandgap technologies and their implications on miniaturizing power electronic systems," *IEEE J. Emerg. Sel. Topics Power Electron.*, vol. 2, no. 3, pp. 374–385, Sep. 2014.

[2] Z. J. Shen, G. Sabui, Z. Miao, et al., "Wide-bandgap solid-state circuit breakers for DC power systems: Device and circuit considerations," IEEE Trans. Electron Devices, vol. 62, no. 2, pp. 294–300, Feb. 2015.

[3] S. Yang, A. Bryant, P. Mawby, et al., "An industry-based survey of reliability in power electronic converters," IEEE Trans. Ind. Appl., vol. 47, no. 3, pp. 1441–1451, May/Jun. 2011.

[4] Choi U M, Blaabjerg F, Lee K B. Study and handling methods of power IGBT module failures in power electronic converter systems[J]. IEEE Transactions on Power Electronics, 2015, 30(5): 2517-2533．

[5] W. Li, Y. Chen, H. Luo, et al., "Review and outlook on junction temperature extraction principles for high-power power electronic devices," Proceedings of the CSEE, vol. 36, no. 13, pp. 3546–3557, 2016. (in Chinese)

[6] X. Yang, L. Zhou, X. Du, et al., "Junction temperature measurement methods for insulated gate transistors and their development," Electrical Measurement & Instrumentation, vol. 49, no. 554, pp. 7–12, 2012. (in Chinese)

[7] B. Ji, X. Song, W. Cao, *et al.*, "In situ diagnostics and prognostics of solder fatigue in IGBT modules for electric vehicle drives," *IEEE Trans. Power Electron.*, vol. 30, no. 3, pp. 1535–1543, Mar. 2015.

[8] K. Li, G. Tian, L. Cheng, et al., "State detection of bond wires in IGBT modules using eddy current pulsed thermography," IEEE Trans. Power Electron., vol. 29, no. 9, pp. 5000–5009, Sep. 2014.

[9] M. Chen and A. Hu, "Comparative study on simulation and detection methods of IGBT junction temperature," Electric Machines and Control, vol. 15, no. 12, pp. 44–49, 55, 2011. (in Chinese)

[10] Avenas Y, Dupont L, Khatir Z. Temperature measurement of power semiconductor devices by thermo-sensitive electrical parameters-a review[J]. IEEE Transactions on Power Electronics, 2012, 27(6): 3081-3092.

[11] BAKER N, LISERRE M, DUPONT L, et al. Improved reliability of power modules：a review of online junction temperature measurement methods[J]. IEEE Industrial Electronics Magazine, 2014, 8(3): 17-27.

[12] KOENIG A, PLUM T, FIDLER P, et al. On-line junction temperature measurement of CoolMOS de-vices[C]//2007 7th International

Conference on Power Electronics and Drive Systems. Bangkok: IEEE, 2007: 90-95.

[13] L. Rossetto and G. Spiazzi, "A Fast ON-State Voltage Measurement Circuit for Power Devices Characterization," in IEEE Transactions on Power Electronics, vol. 37, no. 5, pp. 4926-4930, May 2022.

[14] X. Yang, X. Ge, Y. Chai, et al., "An on-line monitoring circuit for IGBT on-state voltage drop based on reverse series Zener diode clamping," Proceedings of the CSEE, vol. 42, no. 12, pp. 4547–4561, 2022. (in Chinese)

[15] L. Rossetto and G. Spiazzi, "A Fast ON-State Voltage Measurement Circuit for Power Devices Characterization," in IEEE Transactions on Power Electronics, vol. 37, no. 5, pp. 4926-4930, May 2022.

[16] X. Chai, P. Ning, H. Cao, et al., "Research on junction temperature monitoring method for IGBT power modules based on large-current on-state voltage drop," Journal of Power Supply, vol. 18, no. 4, pp. 77–84, 2020. (in Chinese)

[17] R. Zhao, K. Yang, T. Tang, et al., "On-line junction temperature monitoring method for IGBT devices based on a novel on-state voltage drop sampling circuit," Journal of Power Supply, pp. 1–14, published online, 2024. [Online]. Available: [Accessed: Sep. 18, 2024]. (in Chinese)

2025 IEEE Workshop on Wide Bandgap Power Devices and Applications in Asia (WiPDA Asia)

A Dynamic Current Balancing Method for Multichip SiC MOSFET Modules with Separate Gate Drive Structures

1st Zicong Li
National Key Laboratory of Electromagnetic Energy,
Naval University of Engineering, Wuhan, China
Z21180807@nue.edu.cn

2nd Zenan Shi
School of Electric Engineering,
Xi'an Jiaotong University, Xi'an, China
shizenan@stu.xjtu.edu.cn

3rd Yifei Luo
National Key Laboratory of Electromagnetic Energy,
Naval University of Engineering, Wuhan, China
yfluo23@ nue.edu.cn

4th Xin Li
National Key Laboratory of Electromagnetic Energy,
Naval University of Engineering, Wuhan, China
xinlee@nue.edu.cn

Abstract—Dynamic current imbalance between parallel dies due to asymmetric layout will limit the available capacity of power modules. In high current modules such as 1000A modules, the number of parallel dies increases further and the traditional dynamic current balancing method may be difficult to achieve the ideal effect. In this paper, a separate gate drive structure is implemented to effectively suppress the unbalanced dynamic current within SiC multichip modules. By incorporating a pair of gate and kelvin terminals for each group of dies, the power and gate loops can be completely decoupled and the dynamic current balancing can be achieved. First, the mechanism of dynamic current imbalance is analyzed and based on which the separate gate drive structure is proposed. The module as well as the gate driver for two dies in parallel are fabricated and the effectiveness of the proposed method has been verified through simulations and experiments. Finally, the significance of the proposed method and future applications in industry taking into account economic factors is discussed.

Keywords—multichip SiC modules, dynamic current sharing, parasitic parameters, gate driver, package layout

I. INTRODUCTION

Compared to Si devices, SiC MOSFETs are promising devices for high-capacity converters due to their superior characteristics including a higher blocking voltage, operating temperature and switching speed [1]. However, a current imbalance between paralleled dies is inevitable due to asymmetric layout [2]. It is imperative to mitigate the dynamic current imbalance of multichip SiC power modules.

In recent years, it has been investigated by using double-ended structures [3], changing the bonding wires [4],Cu clip-bonding structures [5] or DBC layout [6] to improve the dynamic current sharing performance. So far, the main technical approach is to achieve a more symmetric layout to suppress the parasitic parameter mismatch. However, when the number of parallel dies further increases, it becomes challenging to achieve a completely symmetrical layout simply by changing the package.

This paper presents a dynamic current balancing method using a separate gate drive structure for high current modules. By incorporating a pair of gate and kelvin terminals for each group of dies, the power and gate loops can be completely decoupled and the dynamic current balancing can be achieved. To verify this method, the module as well as the gate driver

This work was supported by the Innovative Group Project of Hubei Natural Science Foundation under 2025AFA045.

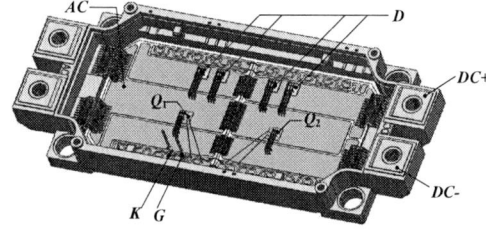

Fig. 1. The baseline SiC MOSFET module.

for two dies in parallel are fabricated and the effectiveness has been verified through simulations and experiments.

The rest of this paper is organized as follows.Section II presents the dynamic current balancing method. Section III verifies the method by simulations and experiments in the case of two dies in parallel. The application of the proposed method in industry is discussed in Section IV. The conclusions are summarized in section V.

II. PROPOSED DYNAMIC CURRENT BALANCING METHOD BASED ON SEPARATE GATE DRIVE STRUCTURES

A. The Baseline SiC Module

The layout of the baseline module is shown in Fig. 1 which is similar to the Econodual module from Rohm(1200V,600A). In Rohm's commercial module, each switching position consists of ten SiC MOSFET with Kelvin connections. In order to simplify the verification of the proposed method, only two SiC MOSFET at lower half-bridge are investigated. The upper half-bridge uses only four diodes to fulfil the freewheeling function.

There is no electrical isolation among the power-sources and Kelvin-sources of the dies, which precludes complete decoupling of the power and gate loops. This structure can not realize dynamic current balance with the mismatched parasitic parameters.

B. Mechanism of Dynamic Current Sharing

The mismatched parasitic parameters impacts the dynamic current sharing by modifying the gate-source voltage of the dies. The extent of the influence of parasitic inductance and resistance on the dynamic current sharing can be quantified by the dynamic voltage ratio coefficient β_x, which is expressed as

$$\beta_x = \frac{L_x}{IR_x}\left|\frac{di}{dt}\right| \approx \frac{L_x f}{0.35 R_x} \tag{1}$$

979-8-3315-1110-4/25 $31.00 © 2025 IEEE

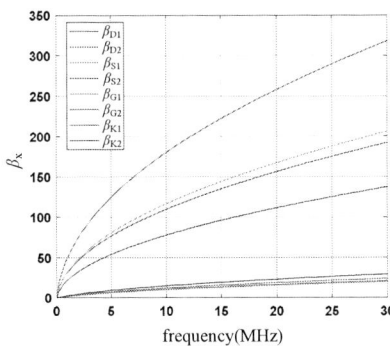

Fig. 2. Frequency-dependent dynamic voltage ratio.

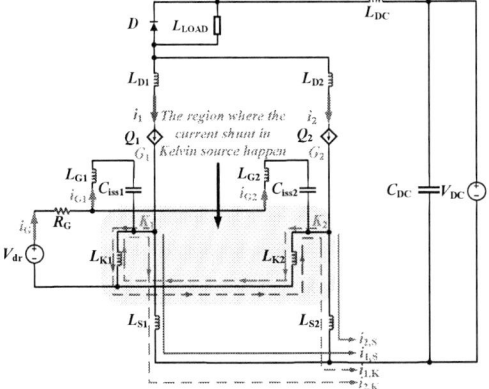

Fig. 3. Dynamic circuit model for two dies in parallel

i and I are the dynamic and static current flowing through L_x (R_x). L_x and R_x refer to partial inductance and resistance of segments inside the module. For example, L_{D1} (L_{D2}) is the partial inductance between DC+ and the drain of die Q_1 (Q_2). L_{S1} (L_{S2}) is the partial inductance between AC and the source of die Q_1 (Q_2). f is the equivalent frequency defined in [7]. Fig. 2 illustrates the sweeping results of the β_x. It can be concluded that the voltage drops on the parasitic inductance are much larger than those on the parasitic resistance. Therefore, the parasitic inductance is the primary factor to be considered in the analysis of dynamic current sharing.

The SiC dies can be equated to voltage-controlled current sources during transients. The gate-source capacitance (C_{GS}) and the Miller capacitance (C_{GD}) can be represented by the input capacitance (C_{iss}). Fig. 3. shows the dynamic circuit model. The dynamic current of each die can be expressed as

$$\begin{cases} i_1 = g\left(v_{G_1 K_1} - V_{th}\right)^2 \left(v_{G_1 K_1} > V_{th}\right) \\ i_2 = g\left(v_{G_2 K_2} - V_{th}\right)^2 \left(v_{G_2 K_2} > V_{th}\right) \end{cases} \quad (2)$$

Where g and V_{th} are the transconductance and threshold voltage of dies. According to Kirchhoff Voltage Law, there is

$$\begin{aligned} V_{dr} &= v_{G_1 K_1} + i_G R_G + (L_{G1} + L_{K1})\frac{di_{G1}}{dt} + L_{K1}\frac{d\left(i_{1,K} - i_{2,K}\right)}{dt} \\ &= v_{G_2 K_2} + i_G R_G + (L_{G2} + L_{K2})\frac{di_{G2}}{dt} + L_{K2}\frac{d\left(i_{2,K} - i_{1,K}\right)}{dt} \end{aligned} \quad (3)$$

The Kelvin-source shunt current $i_{1,K}$, $i_{2,K}$ can be calculated as

$$\begin{cases} \left(L_{S2} + L_{K1} + L_{K2}\right)\dfrac{di_{1,K}}{dt} = L_{S1}\left(\dfrac{di_1}{dt} - \dfrac{di_{1,K}}{dt}\right) \\ \left(L_{S1} + L_{K1} + L_{K2}\right)\dfrac{di_{2,K}}{dt} = L_{S2}\left(\dfrac{di_2}{dt} - \dfrac{di_{2,K}}{dt}\right) \end{cases} \quad (4)$$

The gate current(i_{G1}, i_{G2}) is relatively insignificant in comparison to the drain current(i_1, i_2) and the effect of the voltage drops of gate current on the parasitic inductance can be ignored. Combining (2) to (4), the current discrepancy can be expressed as

$$i_1 - i_2 \approx a\left(L_{S2}\frac{di_2}{dt} - L_{S1}\frac{di_1}{dt}\right) \quad (5)$$

a is a variable related to parasitic inductance and gate source voltage of each die, which can be given as

$$a = g\left(\frac{L_{K1} + L_{K2}}{L_{S1} + L_{S2} + L_{K1} + L_{K2}}\right)\left(v_{G_1 K_1} + v_{G_2 K_2} - 2V_{th}\right) \quad (6)$$

According to (5), the current imbalance can be attributed to discrepancy in the power source inductance of each die. When the power source inductance reaches equilibrium, the dynamic current difference will no longer exist.

Based on (3) and (4), it can be seen that the gate-source voltage of paralleled dies contains the drain current (i_1, i_2) due to the Kelvin-source shunt phenomenon. The Kelvin-source fail to fully decouple the power and gate loops under a traditional structure which allows for the influence of a mismatched power source inductance on dynamic current sharing.

C. Proposed Dynamic Current Balancing Method

As previously discussed, a method of balancing dynamic current between parallel dies is to match the power source inductance by changing the package layout. However, this method may result in a complex structure. Furthermore, as the number of parallel dies increases, it may become challenging to achieve satisfactory effect merely by adjusting the package layout.

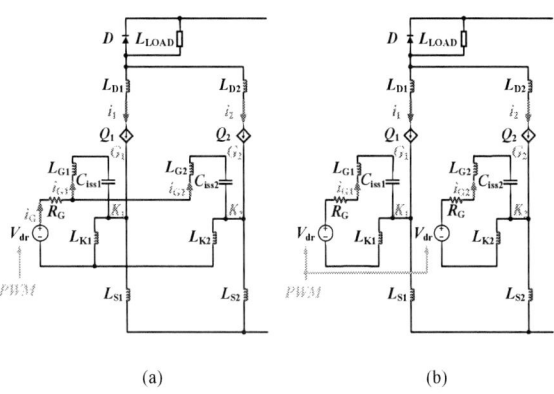

Fig. 4. Dynamic circuit models of (a) the traditional structure and (b) the proposed structure.

Another way to balance dynamic current is to decouple the coupling relationship between the power and gate loops by the implementation of a separate gate drive structure, as proposed in this article. Fig. 4. compares the dynamic circuit models of the traditional and the proposed structure. As shown in Fig. 4.(b), the proposed structure continues to utilize a single PWM signal for control while separates the gate drive circuits of each die. In this case, the gate-source voltage of each die can be calculated as

$$
\begin{aligned}
V_{dr} &= v_{G_1K_1} + i_{G1}R_G + (L_{G1}+L_{K1})\frac{di_{G1}}{dt} \\
&= v_{G_2K_2} + i_{G2}R_G + (L_{G2}+L_{K2})\frac{di_{G2}}{dt}
\end{aligned}
\tag{7}
$$

It can be seen that the drain current is no longer observable in the gate-source voltage calculation as a consequence of the elimination of the Kelvin shunt paths. The proposed structure decouple the power and gate loops of parallel dies.The gate current can be calculated as

$$
\begin{cases}
(L_{G1}+L_{K1})\dfrac{d^2i_{G1}}{dt^2} + R_G\dfrac{di_{G1}}{dt} + \dfrac{i_{G1}}{C_{iss1}} = \dfrac{dV_{dr}}{dt} \\[3mm]
(L_{G2}+L_{K2})\dfrac{d^2i_{G2}}{dt^2} + R_G\dfrac{di_{G2}}{dt} + \dfrac{i_{G2}}{C_{iss2}} = \dfrac{dV_{dr}}{dt}
\end{cases}
\tag{8}
$$

The correlation between gate current and drain current can be deduced as

$$
\begin{cases}
i_{G1} = \dfrac{C_{iss1}}{2\sqrt{g}}\left(i_1\right)^{-\frac{1}{2}}\dfrac{di_1}{dt} \\[3mm]
i_{G2} = \dfrac{C_{iss2}}{2\sqrt{g}}\left(i_2\right)^{-\frac{1}{2}}\dfrac{di_2}{dt}
\end{cases}
\tag{9}
$$

Based on (8) and (9), it can be concluded that the power source inductance mismatch will not impact dynamic current sharing with a separate gate drive structure. The dynamic current of paralleled dies can be matched by an approximately symmetric gate loop and chip preselections.

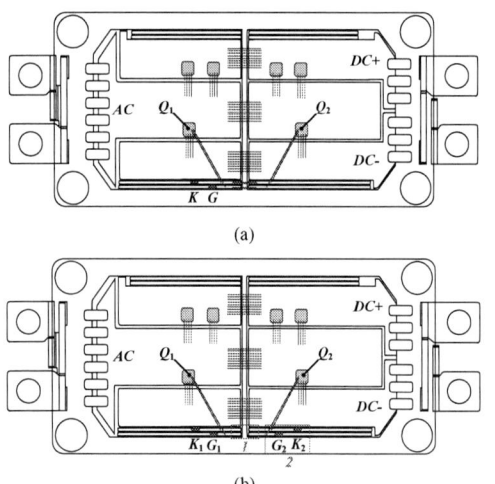

(a)

(b)

Fig. 5. Configurations with (a) a traditional single gate drive structure and (b) a proposed separate gate drive structure.

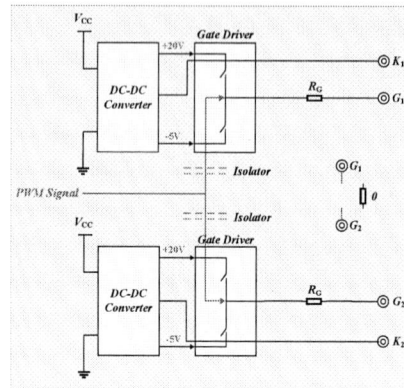

Fig. 6. Configuration of the driver board

III. SIMULATION AND EXPERIMENTAL VERIFICATION FOR TWO DIES IN PARALLEL

A. The Module and Gate Driver for Two Dies in Parallel

Fig. 5. illustrates the configurations of the baseline and the optimized modules for two dies in parallel. As shown in Fig.5.(a), the gates and Kelvin sources of paralleled dies are connected in the baseline module. In the optimized module, the respective gate and Kelvin loops of paralleled dies are disconnected. Moreover, each die is connected to a pair of gate and Kelvin terminals separately. Consequently, there is no Kelvin-source shunt current.

The gate driver for the the baseline and optimized modules for two dies in parallel is depicted in Fig.6. The PWM signal controls the switching behaviours and the DC-DC converters provide the gate voltage. Two push-pull gate drivers are incorporated to drive the SiC power die.A pad is located between the G_1 and G_2 terminals on the driver board. When the proposed structure is employed, two gate drive circuits drive the paralleled dies separately. When the single gate drive is implemented, a 0Ω resistor is connected on the pad and the G_1 and K_1 are used to connect the baseline modules,thereby enabling comparative validation.

B. Simulation Verification

To verify the effect of the proposed method, the parasitic parameters of the baseline module and the optimized module are extracted respectively and the circuit with a single and separate gate drive structure are simulated in LTspice under double pulse conditions. The extraction results at 5 MHz is substituted into the simulated model.The schematic is similar to that depicted in Fig. 4.

The SiC MOSFET dies with a voltage rating of 1200V and a current rating of 118A at 25°C and the 1200V/50A SiC SBD dies were employed in the experiments.Since it is not feasible to obtain the SPICE models of the utilized dies directly from the manufacturer, it is necessary to create the models of the dies based on the data provided in the datasheet.

Considering the efficacy and precision,this article employs the behavioral model to construct the SPICE models of the SiC MOSFET dies by fitting the curves present within the datasheet.The simplified circuit is illustrated in Fig. 7. R_{gint} and D_B are the internal gate resistance and body diode, respectively. M is the core utilized to simulate the output and transfer characteristics of dies.

The simulated turn-on and turn-off waveforms at 200V

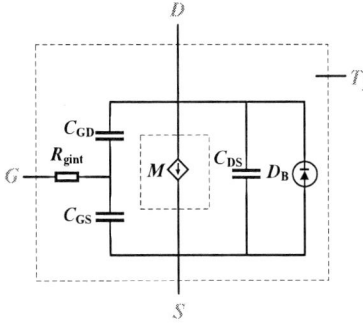

Fig. 7. Simplified circuit of SiC MOSFET SPICE model

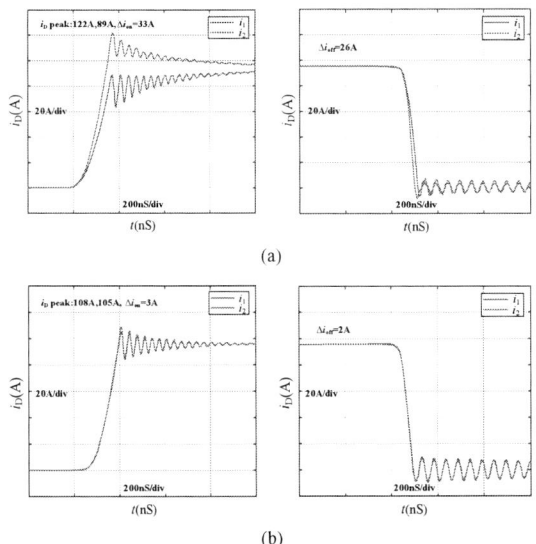

Fig. 8. Simulated turn-on and turn-off current waveforms with (a) a single gate drive structure and (b)a separate gate drive structure.

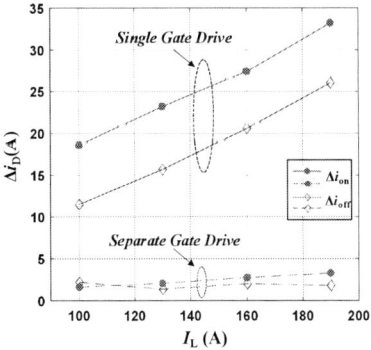

Fig. 9. Simulated turn-on and turn-off current discrepancy at 200V

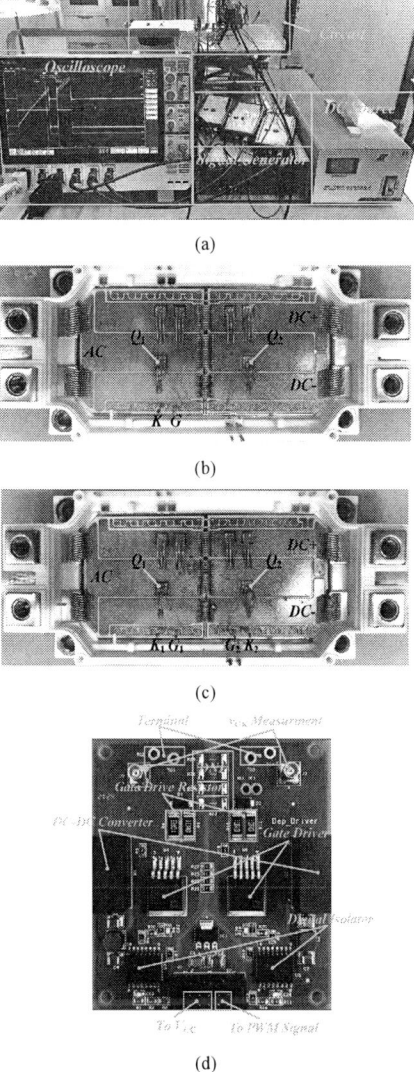

Fig. 10. Experimental double-pulse test platform.(a)Overview of the test bench. (b)Baseline module. (c)Optimized module.(d) Gate drive board

turn-off current discrepancy at 200V are shown in Fig. 9. which indicates that the separate gate drive structure can improve the current sharing performance of paralleled SiC dies.

C. Experimental Verification

To verify the effect of the proposed method, the double-pulse test platform was bulit,as shown in Fig. 10. The basline and optimiazed modules are shown in Fig. 10. (b)-(c).The silica gel has been removed from the modules to facilitate current measurement.

Waveforms are observed by oscilloscope Tektronix MSO56. The drain-source voltage is measured by the high voltage differential probe (1000V, 200 MHz). The gate-source voltage is measured by Optical-fiber isolated probes Micsig MOIP200P (±25V, 200 MHz).The current going through each die and the total current are measured by Rogowski coils CWTMini50HF (300A,50MHz) from PEM. The Rogowski coils will have a measurement delay depending on the transducer bandwidth and the length of connecting cables.The

and 190A are presented in Fig. 8. The maximum current discrepancy at transients between two dies is employed to quantify the extent of dynamic current imbalance. The original structure which incapable of decoupling the power and gate loops, has permitted the Δi_{on} to reach 33A and Δi_{off} to reach 26A as a consequence of the mismatched power source inductance. When utilized the proposed structure, Δi_{on} and Δi_{off} are reduced to 3A and 2A. The simulated turn-on and

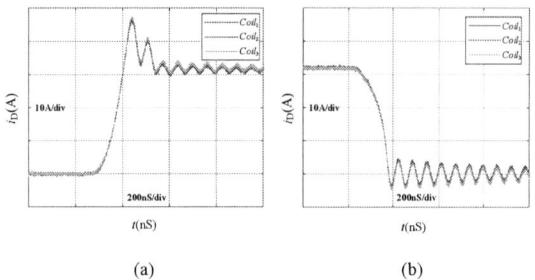

Fig. 11. Verification of the consistency of three Rogowski coils.(a)Turn-on.(b)Turn-off

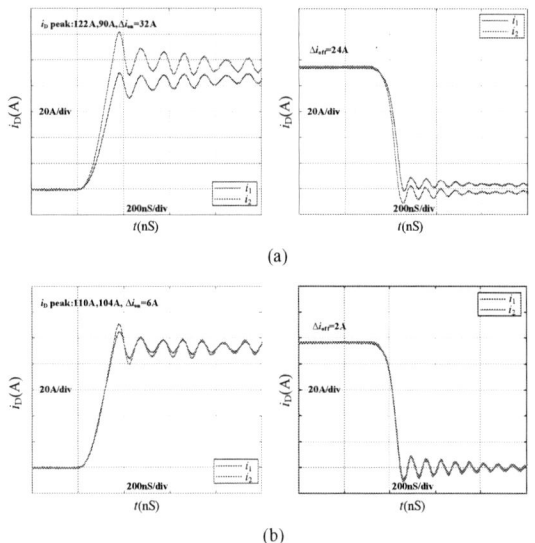

Fig. 12. Experimental turn-on and turn-off current waveforms with (a) a single gate drive structure and (b)a separate gate drive structure.

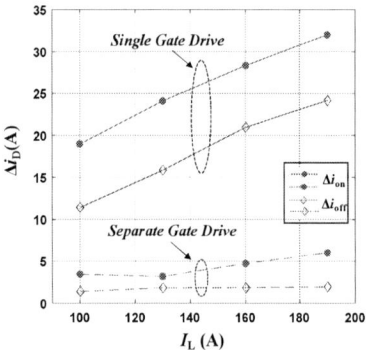

Fig. 13. Experimental turn-on and turn-off current discrepancy at 200V

consistency of three Rogowski coils, Coil1, Coil2 and Coil3, is verified, as shown in Fig.11. Before the experiment, the dies need to be preselected and the V_{th} and R_{on} difference are controlled below 0.2%.During the experiment, the modules are immersed in insulating oil to prevent partial discharge.

The experimental turn-on and turn-off waveforms at 200V and 190A are presented in Fig. 11. The result shows that the original structure which incapable of decoupling the power

and gate loops, has permitted the Δi_{on} to reach 32A and Δi_{off} to reach 24A as a consequence of the mismatched power source inductance. When utilized the proposed structure, Δi_{on} and Δi_{off} are reduced to 6A and 2A. The experimental turn-on and turn-off current discrepancy at 200V are shown in Fig. 12. which indicates that the separate gate drive structure can improve the current sharing performance of paralleled SiC dies.The result demonstrates the efficacy of the simulations.

IV. DISCUSSION FOR INDUSTRIAL APPLICATION

The traditional current balancing methods are designed to optimize the current sharing performance by matching the equivalent power source inductance. Once the matching has been achieved, the Kelvin source shunt current can be offset to decouple the power and gate loops. However, as the number of parallel dies increases, it becomes challenging to achieve satisfactory effect merely by adjusting the package layout. Compared to modifying the package layout, the separate gate drive structure proposed in this article will not make the structure complex . By decoupling the power and gate loops, it is possible to achieve dynamic current balance under an asymmetric layout.

There are still some potential issues that require further investigation before the proposed method can be considered for industrial applications. The main challenges are increased cost and potential desynchronisation between gate drive signals.Verified through practice, it can be stated that the cost increase with the proposed method is predominantly attributed to the fabrication of the driver board.As illustrated in Fig. 6. it can be seen that if a separate gate driver is used for each die in a multichip module, the number of DC-DC converters and gate drivers must be further increased. For the multichip applications, a reduction in the required power level of each individual component compensates for the cost increase to some extent.Furthermore, the delay between two gate voltages in the practice of two parallel dies is less than 0.5 nS. It can be seen that the gate signals synchronisation can be achieved by designing the gate loops to be approximately equal in length.

However, taking into account the cost increase and signal desynchronisation, a separate gate driver for each parallel die is not an economical or reliable option in future industrial applications. Separate gate drive structures can be considered to be applied in high current modules like 600A module as shown in Fig. 14. The number of dies in parallel reaches 10, in this case, traditional methods will not be able to balance the dynamic current between the dies farther apart.The separate gate drive structures can be employed to balance the dynamic current between the two groups of dies located in different coppertrace.The dynamic current sharing between dies in each group can be modified by changing package to take full advantage of different methods.

Similarly, for higher current modules such as 1000A , the

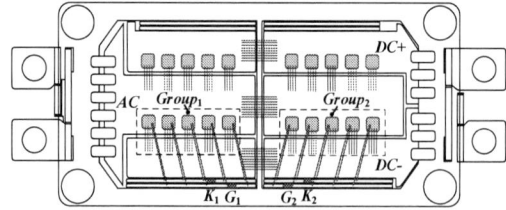

Fig. 14. Configuration of the 600A module with a proposed separate gate drive structure.

number of parallel dies will reach 13 to 20, in which the parallel dies are generally divided into multiple groups separately inside the module. The separate gate drive structure will effectively optimise the dynamic current sharing between multiple groups of dies. In future work, the relationship between the number of separate gate drivers and the degree of dynamic current imbalance will be analysed , and the design will be optimised by weighing the economy, reliability and the dynamic current balancing effect.

V. CONCLUSIONS

This article presents a separate gate drive structure to balance dynamic current between paralleled SiC MOSFET dies with Kelvin connections. The proposed structure can effectively suppress the dynamic current imbalance under an asymmetric layout by completely decoupling the power and gate loops. The method requires no additional efforts other than some modifications to the modules and the driver boards. In this article, the separate gate drive prototypes of two dies in parallel is completed and the effect was verified by some simulations and experiments.The feasibility of applying this technology to multichip modules was discussed considering economy, reliability and the current balancing effect.

REFERENCES

[1] J. Millán, P. Godignon, X. Perpiñà, A. Pérez-Tomás, and J. Rebollo, "A Survey of Wide Bandgap Power Semiconductor Devices," *IEEE Transactions on Power Electronics,* vol. 29, no. 5, pp. 2155-2163, 2014.

[2] C. Zhao, L. Wang, and F. J. I. T. o. P. E. Zhang, "Effect of Asymmetric Layout and Unequal Junction Temperature on Current Sharing of Paralleled SiC MOSFETs with Kevin-Source Connection," *IEEE Transactions on Power Electronics,* vol. PP, no. 99, pp. 1-1, 2019.

[3] W. Miao, L. Fang, L. J. I. J. o. E. Xu, and S. T. i. P. Electronics, "A Double-End Sourced Wire-Bonded Multi-Chip SiC MOSFET Power Module with Improved Dynamic Current Sharing," *IEEE Journal of Emerging and Selected Topics in Power Electronics,* vol. PP, no. 4, pp. 1-1, 2017.

[4] C. Zhao, L. Wang, F. Zhang, and F. Yang, "A Method to Balance Dynamic Current of Paralleled SiC MOSFETs with Kelvin Connection by Response Surface Model and Nonlinear Optimization," *IEEE Transactions on Power Electronics,* vol. PP, no. 99, pp. 1-1, 2020.

[5] L. Wang *et al.*, "Cu Clip-Bonding Method With Optimized Source Inductance for Current Balancing in Multichip SiC MOSFET Power Module," *IEEE Transactions on Power Electronics,* vol. 37, no. 7, pp. 7952-7964, 2022.

[6] H. Li, S. Munk-Nielsen, S. Bęczkowski, and X. Wang, "A Novel DBC Layout for Current Imbalance Mitigation in SiC MOSFET Multichip Power Modules," *IEEE Transactions on Power Electronics,* vol. 31, no. 12, pp. 8042-8045, 2016.

[7] S. H. Hall and G. W. Hall, *High-Speed Digital System Design : A Handbook of Interconnect Theory and Design Practices.* 2000.

2025 IEEE Workshop on Wide Bandgap Power Devices and Applications in Asia (WiPDA Asia)

A Novel Analytical Physical Model of Gate-drain Capacitance and Output Characteristics for SiC-MOSFET

Ze Tao
National Key Laboratory of Electromagnetic Energy
Naval University of Engineering
Wuhan, China
35112019007@nue.edu.cn

Zenan Shi
State Key Laboratory of Electrical Insulation and Power Equipment
Xi'an Jiaotong University
Xi'an, China
shizenan@stu.xjtu.edu.cn

Yifei Luo
National Key Laboratory of Electromagnetic Energy
Naval University of Engineering
Wuhan, China
yfluo23@nue.edu.cn

Xin Li
National Key Laboratory of Electromagnetic Energy
Naval University of Engineering
Wuhan, China
xinlee@nue.edu.cn

Abstract—**In this paper, a new physical model for gate-drain capacitance and output characteristics of SiC-MOSFET is developed. The proposed model takes the secondary effects of gate-drain capacitance and output characteristics into account. Methods for adjusting parameters with deviations in the simulation main loop and datasheet are also proposed. By the comparison of experiment and simulation under 600V/160A, 400V/108A, and 200V/54A, the maximum relative errors of the switching transient are about 6%. The proposed model has sufficient accuracy in predicting the dynamic behavior of SiC MOSFETs.**

Keywords—SiC-MOSFET, output characteristics, physical model, gate-drain capacitance

I. INTRODUCTION

In recent years, a new generation of wide-bandgap semiconductor devices, represented by silicon carbide metal-oxide semiconductor field-effect transistors (SiC-MOSFETs), has been widely used in the electrical field [1]. However, due to the faster switching speed and shorter switching transients compared with Si-MOSFETs, accurate and detailed characterization of their switching transients becomes more complex.

Domestic and foreign research has been done on the modeling of SiC-MOSFETs. It is quite common to study the physical and behavioral modeling of gate-drain capacitance C_{gd} [1~6]. The waveform of C_{gd} modeled in Ref. [4] is approximated as inverse proportional function, and this is a deviation from the quasi-S-shaped curve in the datasheet. The convergence of the segmentation function of the output characteristics used in Ref. [3] affects the success of the simulation.

The basic idea of this paper is to investigate a problem of accuracy in solving C_{gd} and output characteristic with a physical model. Analytical expressions are proposed to solve the segmented function problem.

II. PROPOSED MODEL FUNCTION

In this section, the capacitance C_{gd} as well as the output characteristics are modeled.

A. Modeling of gate-drain Capacitance

The gate drain capacitance C_{gd} of SiC MOSFET is determined by the gate oxide capacitance C_{ox} and the depletion region capacitance C_{dep}. The expression for these three capacitances is shown in the following equation:

$$C_{gd} = \begin{cases} C_{ox} & V_{DG} < 0 \\ C_{ox} \cdot C_{dep}/(C_{dep} + C_{ox}) & V_{DG} \geqslant 0 \end{cases} \quad (1)$$

$$C_{ox} = A_{gd}\varepsilon_{ox}/\omega_{ox} \quad (2)$$

$$C_{dep} = A_{gd}^{*}\varepsilon_{sic}/\omega_{dep} \quad (3)$$

A_{gd} and ω_{ox} represent the area and thickness of C_{ox}, respectively. ε_{ox} and ε_{sic} are the dielectric constants of the oxide layer and depletion region. A_{gd}^{*} and ω_{dep} are shown as the area of the JFET region where the gate electrode overlaps the N-drift region and the thickness of the depletion region, respectively. C_{ox} can be treated as a constant in the switching transient. C_{dep} will vary with drain gate voltage V_{DG}. This is because the area of the actual depletion layer A_{gd}^{*} is not constant at low gate drain voltages, as shown in Fig.1. At the beginning of the drain bias voltage increase, the depletion region below the gate area has an expanding process, during which the junction depth ω_{dep} increases with the increase of V_{DG}. The area of the depletion region A_{gd}^{*} obviously has a process of increasing, then decreasing, and finally stays unchanged with the increase of ω_{dep}, which corresponds to the first to third, the fourth to sixth, and the seventh to ninth point respectively, in Fig.1. Since the change trend of A_{gd}^{*} is a quasi-S-shaped curve, the fitting function can be constructed of the hyperbolic tangent function, which can be derived as follows:

$$\omega_{dep} = \sqrt{2 \cdot \varepsilon_{sic} \cdot V_{dg} \cdot N_{N-}/q} \quad (4)$$

$$C_{dep} = $$
$$\left(1 + \left(k_a + k_1 \cdot e^{\frac{-(V_{DG}-k_2)^2}{k_3}}\right) \cdot \tanh(k_b V_{DG} - k_c)\right)^{k_d} \quad (5)$$
$$\cdot \varepsilon_{sic}/\omega_{dep} \ , \ V_{DG} \geqslant 0$$

This work was supported by the Innovative Group Project of Hubei Natural Science Foundation under 2025AFA045.

979-8-3315-1110-4/25 $31.00 © 2025 IEEE

The parameters k_a, k_b, k_c, k_d, k_1, k_2 and k_3 in (5) are determined by A_{gd}^*. Compared with the traditional model, the proposed model fitted the datasheet better in low V_{DS} area (Fig.1).

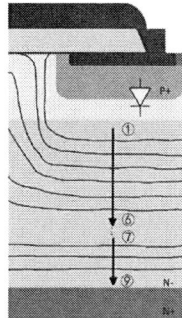

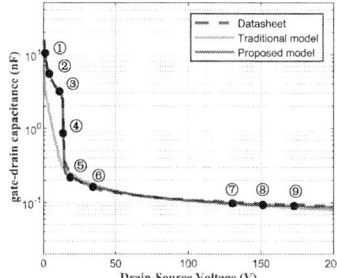

Fig. 1. Variation of the Depletion Region and Comparison of gate-drain capacitance Model Curves

B. Modeling of Output Characteristics

In traditional models, the output characteristics are usually expressed as:

$$I_D = \begin{cases} \dfrac{Z\mu_n C_{ox}}{2L_{CH}}(V_{GS} - V_{th})^2, & V_{GS} - V_{th} \geqq 0 \\ 0, & V_{GS} - V_{th} < 0 \end{cases} \quad (6)$$

μ_n, Z, L_{CH} represent the electron mobility, channel thickness and channel length, respectively. However, when $V_{GS} < V_{th}$, there is still a weakly inverted layer, and the drain current I_D is not zero, but shows an exponential relationship with V_{GS}, which can be expressed as :

$$I_D = I_0 e^{V_{GS}/\zeta\phi_t} \quad (7)$$

I_0 is the drain current corresponding to the zero-bias gate source voltage proportional to the structural parameter Z/L_{CH}. The physical significance of ζ and ϕ_t is the subthreshold slope correlation parameter and the thermal potential. In order to form a continuous expression for the drain current, the constructive form functions are as follows:

$$I_D = 2\frac{Z\mu_n C_{ox}}{L_{CH}} \cdot \zeta^2\phi_t^2 \left[\ln\left(1 + e^{\frac{V_{GS} - V_{th}}{2\zeta\phi_t}}\right)\right]^2 \quad (8)$$

The expression of this form function as it tends to positive and negative infinity in the domain of definition can be expressed as:

$$I_D = \begin{cases} \dfrac{Z\mu_n C_{ox}}{2L_{CH}} \cdot (V_{GS} - V_{th})^2, & \dfrac{V_{gs} - V_{th}}{2\zeta\phi_t} \to +\infty \\ 2\zeta^2\phi_t^2\dfrac{Z\mu_n C_{ox}}{L_{CH}}e^{\frac{V_{gs} - V_{th}}{2\zeta\phi_t}}, & \dfrac{V_{gs} - V_{th}}{2\zeta\phi_t} \to -\infty \end{cases} \quad (9)$$

At normal temperature, $\zeta \cdot \phi_t$ is about hundreds of millivolts, so the voltage range studied satisfies this condition for $V_{GS} - V_{th} \gg 2\zeta\phi_t$. Note that when $V_{GS} - V_{th} = 2\zeta\phi_t$, it is the junction corresponding to the traditional physical model, and the expression for I_D is:

$$I_D = 2Z\mu_n C_{ox}/L_{CH} \cdot \zeta^2\phi_t^2[\ln(1 + e)]^2 \quad (10)$$

Typical values of I_D at this point are in the hundred milliampere range, and the shape function satisfies the need to fit the output characteristic curve.

Further, considering the disordered square law caused by the defects present in short channel transistors, the square can be replaced by the k power. And the channel length modulation effect cause the change of channel length rate of variation, which is proportional to the V_{DS} in the strong inversion type with ratio λ, (8) can be further rewritten as:

$$I_D = 2\frac{Z\mu_n C_{ox}}{L_{CH}} \cdot \zeta^2\phi_t^2 \cdot \left[\ln\left(1 + e^{\frac{V_{GS} - V_{th}}{2\zeta\phi_t}}\right)\right]^k (1 + \lambda V_{DS}) \quad (11)$$

The continuous equation not only has the physical significance which is consistent with the segmented function constructed by the traditional physical model, but also meets the physical significance.

III. SIMULATION AND EXPERIMENT

In order to verify the correctness of the above theory, the simulations and experiments described below are carried out.

A. Setup of Experiments and Simulations

The experimental setup is shown in Fig. 2. After the spice model is initially constructed according to the datasheet, the double pulse test circuit is built by LTspice simulation software.

B. Adjustment of Model Parameters

Because of different work environments and individual differences, it is necessary to adjust the parameter data in the model. The process is shown in Fig.3.

Parameter adjustment needs to be started with the waveforms of V_{GS}. The consistency of the waveform of V_{GS} is the basis for subsequent parameter tuning. The waveform can be achieved by adjusting the input capacitance C_{iss} (C_{gs} and C_{gd}).

Fig. 2. Experiment Circuit

Then the I_D waveform is adjusted. The threshold voltage V_{th} of the model can first be modified by recording the value of the current I_D in the experimental waveform when it reaches (10). Second, the current rise rate dI_D/dt can only be adjusted by g_m, since the V_{GS} waveform has already been adjusted. Third, the current peak that occurs after I_D first reaches circuit current depend on a diode model D_{ap}. The value of peak is determined by the junction capacitance and voltage of the diode. The amplitude of the subsequent oscillations can be adjusted by the soft reverse recovery parameter V_P (vary from 0 to 1) of the diode.

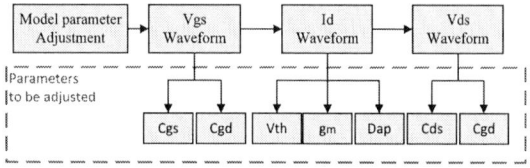

Fig. 3.   The Process of Parameter Adjustment

Finally, the V_{DS} waveform is adjusted. The voltage drop rate dV_{DS}/dt of this process is determined by the output capacitance C_{oss} (C_{gd} and C_{gs}). Since V_{GS} is at the Miller plateau during the voltage drop phase, the drive current charges C_{gd} almost exclusively, hence $I_G \approx I_{GD}$. Since I_G can be regarded as a constant value, the larger C_{gd} is, the longer its charging time is, and the smaller dV_{DS}/dt is.

The parameter adjustments for the turn-on process apply to the turn-off process. Therefore, the parameters of the model do not need to be changed.

C.   Comparison of Experiments and Simulations

The comparison of simulated and experimental switching transient waveforms is shown in Fig.4. The data and relative errors are exhibited in Table I and Table II. The simulation results achieve high accuracy in all working conditions, especially V_{GS}. However, there are still some discrepancies in the oscillatory part of the switching transient follow-up, which is due to the lack of accuracy of the capacitance data at large voltages in the datasheet, and the subsequent focus will be on adjusting the model for this part. The average relative errors of the simulation to the experiment are about 4%.

TABLE I.   TESTED AND MODELED VALUE WITH RELATIVE ERROR OF TURN-ON TRANSIENT UNDER DIFFERENT WORKING CONDITIONS

working condition		Turn-on trajectory			
		Delay	di/dt	I_{peak}	dv/dt
600V/160A	Tested value	495ns	0.447A/ns	208.2A	2.01V/ns
	Modeled value	513ns	0.475A/ns	205.5A	1.90V/ns
	Relative error	3.60%	6.20%	1.30%	5.50%
400V/107A	Tested value	511ns	0.387A/ns	142.1A	1.49V/ns
	Modeled value	528ns	0.408A/ns	143.5A	1.42V/ns
	Relative error	3.30%	5.40%	0.99%	4.70%
200V/54A	Tested value	510ns	0.255A/ns	76.53A	0.87V/ns
	Modeled value	523ns	0.267A/ns	75.76A	0.90V/ns
	Relative error	2.50%	4.70%	1.01%	4.63%

TABLE II.   TESTED AND MODELED VALUE WITH RELATIVE ERROR OF TURN-OFF TRANSIENT UNDER DIFFERENT WORKING CONDITIONS

working condition		Turn-off trajectory			
		Delay	dv/dt	V_{peak}	di/dt
600V/160A	Tested value	975ns	2.27V/ns	779.5V	2.03A/ns
	Modeled value	989ns	2.34V/ns	778.3V	1.99A/ns
	Relative error	1.40%	3.10%	0.20%	1.97%
400V/107A	Tested value	964ns	1.56V/ns	537.3V	1.55A/ns
	Modeled value	975ns	1.64V/ns	545.8V	1.64A/ns
	Relative error	1.15%	5.13%	1.60%	5.81%
200V/54A	Tested value	937ns	0.52V/ns	276.2V	0.74A/ns
	Modeled value	951ns	0.54V/ns	277.7V	0.73A/ns
	Relative error	1.50%	3.80%	0.54%	1.62%

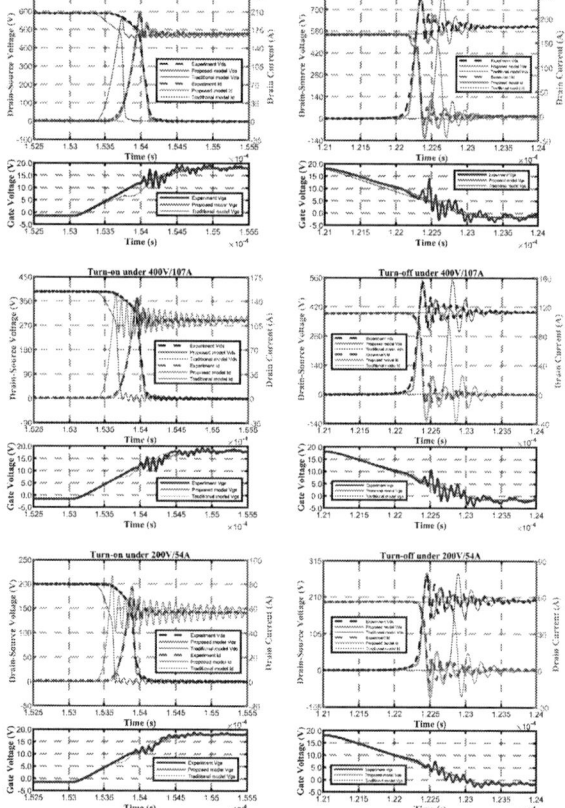

Fig. 4.   The Turn-on (up) and Turn-off (down) Waveforms of Experiments and Simulations in the Condition of 600V/160A, 400V/107A, 200V/54A

IV.   CONCLUSIONS AND FUTURE WORK

In this paper, the gate-drain capacitance and output characteristics of SiC-MOSFETs are analyzed. The modeling process takes the secondary effects in the depletion region into account and considers the two patterns in the subthreshold region in connection with the strong and weak inverted layer to form a continuous expression. A method for adjusting physical quantities with deviations in the datasheet based on the differences between simulation and experiment is proposed. The average relative errors of the simulation to the experiment are about 4%, which can be reduced by an average of 3.5% compared to Ref. [1]. Follow-up will be more fine-tuning of the model and fitting function, and modeling studies of the subsequent oscillatory part of the switching transient.

ACKNOWLEDGMENT

This work was supported by the National Key Laboratory of Electromagnetic Energy.

REFERENCES

[1] N. Wang and J. Zhang, "Nonlinear Capacitance Model of SiC MOSFET Considering Envelope of Switching Trajectory," in IEEE Transactions on Power Electronics, vol. 37, no. 7, pp. 7977-7988, July 2022.

[2] K. Chen, Z. Zhao, L. Yuan, T. Lu and F. He, "The Impact of Nonlinear Junction Capacitance on Switching Transient and Its Modeling for SiC MOSFET," in IEEE Transactions on Electron Devices, vol. 62, no. 2, pp. 333-338, Feb. 2015.

[3] Z. Duan, T. Fan, X. Wen and D. Zhang, "Improved SiC Power MOSFET Model Considering Nonlinear Junction Capacitances," in IEEE Transactions on Power Electronics, vol. 33, no. 3, pp. 2509-2517, March 2018.

[4] X. Li, F. Xiao, Y. Luo, R. Wang and Z. Shi, "An Improved Equivalent Circuit Model of SiC MOSFET and Its Switching Behavior Predicting Method," in IEEE Transactions on Industrial Electronics, vol. 69, no. 9, pp. 9462-9471, Sept. 2022.

[5] Y. Wu, Z. Xue, J. Xie, J. Wang, L. Wang and K. Gao, "A Novel Compact Model of Gate Capacitance for SiC MOSFET with Easy Parameter Extraction Method," 2023 IEEE 2nd International Power Electronics and Application Symposium (PEAS), Guangzhou, China, 2023, pp. 115-119.

[6] R. Stark, A. Tsibizov, N. Nain, U. Grossner and I. Kovacevic-Badstuebner, "Accuracy of Three Interterminal Capacitance Models for SiC Power MOSFETs Under Fast Switching," in IEEE Transactions on Power Electronics, vol. 36, no. 8, pp. 9398-9410, Aug. 2021.

Reliability Analysis for 1200V SiC MOSFETs under Repetitive Surge Current Operation with Negative Gate-source Bias

Xinbin Zhan
Faculty of Integrated Circuit,
Xidian University,
Xi'an, China,
xbzhan@stu.xidian.edu.cn

Yanjing He
Faculty of Integrated Circuit, Xidian
University, Xi'an, China,
Guangzhou Institute of Technology,
Xidian University, Guangzhou, China
hyj@xidian.edu.cn

Jiankun Lai
Guangzhou Institute of Technology,
Xidian University,
Guangzhou, China
ljk1997xdu@163.com

Xi Jiang
Guangzhou Institute of Technology,
Xidian University,
Guangzhou, 510555, China
xjiang@xidian.edu.cn

Song Yuan
Guangzhou Institute of Technology,
Xidian University,
Guangzhou, 510555, China
syuan@xidian.edu.cn

Hao Yuan
Faculty of Integrated Circuit, Xidian
University, Xi'an, China,
haoyuan@xidian.edu.cn

Qingwen Song
Faculty of Integrated Circuit, Xidian
University, Xi'an, China,
qwsong@xidian.edu.cn

Xiaoyan Tang
Faculty of Integrated Circuit, Xidian
University, Xi'an, China,
xytang@mail.xidian.edu.cn

XIAOWU GONG
Faculty of Integrated Circuit, Xidian
University, Xi'an, China,
Guangzhou Institute of Technology,
Xidian University,
Guangzhou, China
xwgong@xidian.edu.cn

Yuming Zhang
Faculty of Integrated Circuit,
Xidian University,
Xi'an, China,
zhangym@xidian.edu.cn

Abstract—**In new energy vehicle electric drive inverters and DC/DC power conversion systems, the body diode of silicon carbide MOSFETs (SiC MOSFETs) is increasingly utilized as a freewheeling element due to its inherently low reverse recovery charge. While prior research has extensively characterized the single-pulse surge current capability of commercial SiC MOSFET body diodes, critical gaps remain in understanding their long-term reliability under repetitive surge stress. In this study, the failure mechanisms of SiC MOSFETs with different structures are systematically elucidated based on the quantitative coupling of the evolution characteristics of static electrical parameters and the degree of device degradation, combined with the chip morphology damage patterns revealed by microstructure characterization techniques. The experimental results show that device degradation is mainly concentrated in three key regions: the increase of interface state density in the gate-oxide layer, the accumulation of lattice defects in the drift region, and the concentration of thermo-mechanical stresses in the encapsulation layer.**

Keywords—*SiC MOSFET, repetitive surge current, reliability, gate-oxide, bipolar degradation.*

I. INTRODUCTION

Power-switching devices typically require antiparallel diodes to ensure reverse current continuity during dead-time intervals in high-power converter topologies. However, the body diode of conventional silicon (Si) MOSFETs proves unsuitable for this role due to material limitations: the lower critical avalanche field strength of silicon necessitates thicker drift regions, which, combined with inherent carrier lifetime constraints, induces pronounced transient current spikes and switching losses during reverse recovery[1-5]. In contrast,

SiC exhibits a critical electric field strength tenfold higher than silicon, enabling a tenfold reduction in drift region thickness for equivalent breakdown voltage ratings. This structural optimization markedly diminishes stored charge during the forward conduction of SiC MOSFET body diodes while leveraging SiC's shorter minority carrier lifetimes to enhance recombination efficiency, thereby achieving superior reverse recovery performance[1]. While these attributes expand the utility of SiC MOSFETs in third-quadrant operation, the long-term reliability of devices under sustained reverse-bias conditions remains insufficiently characterized, necessitating further investigation into degradation mechanisms and operational limits.

Under actual operating conditions, SiC MOSFET body diodes are subjected to periodic surge current impacts. Although the intensity of a single impact is usually lower than the rated inrush tolerance limit of the device, the micro-structural degradation and electrical parameter drift induced by the repetitive stress will lead to the degradation of the converter's performance, which will threaten the operational stability and service life of the power electronic system[6]. Consequently, establishing a long-term surge reliability assessment method and a system-level stability verification mechanism for SiC MOSFET devices is of great engineering significance.

In this study, the failure mechanisms of SiC MOSFETs with different structures are systematically elucidated based on the quantitative coupling of the evolution characteristics of static electrical parameters and the degree of device degradation, combined with the chip morphology damage patterns revealed by microstructure characterization techniques. The

979-8-3315-1110-4/25 $31.00 © 2025 IEEE

experimental results show that device degradation is mainly concentrated in three key regions: the increase of interface state density in the gate-oxide layer, the accumulation of lattice defects in the drift region, and the concentration of thermo-mechanical stresses in the encapsulation layer.

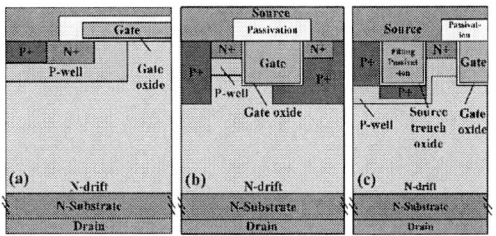

Fig. 1. Cell structure of (a) P-MOSFET, (b) A-MOSFET, and (c) D-MOSFET.

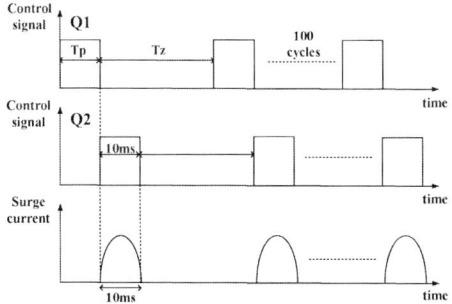

Fig. 2. Control signals in repetitive surge tests.

II. REPETITIVE SURGE TEST

Currently, the market primarily offers two types of gate structures for devices: planar-type SiC MOSFETs (P-MOSFETs), asymmetric trench-type SiC MOSFETs (A-MOSFETs), and dual-trench SiC MOSFETs (D-MOSFETs), as shown in Fig. 1. In Fig. 2, this study's repeated surge current I_{surge} is set to three times the rated current (120 A) under the $V_{GS} = -5V$, the pulse width T_P remains at 10 ms, and the interval between two surge pulses T_Z is 30 s. These parameters are selected based on the thermal performance of the SiC MOSFET and the average junction temperature during the test. This configuration accelerates device degradation and experimental progress while effectively preventing internal heat buildup.

III. DEGRADATION PHENOMENON

As shown in Fig. 3, under the continuous impact of surge current, the V_{SD} waveforms of all DUTs exhibit significant shifts during the test. After the test, the A-MOSFET and D-MOSFET do not exhibit any failure phenomena. However, after being subjected to repeated surge current impacts, the P-MOSFET shows distortion in the V_{SD} waveform. After 1,200 surge cycles, an abnormal spike in the V_{SD} waveform is observed around 3 ms. After the device failed, the source-drain resistance was 3.9 Ω, indicating that the body diode had lost its blocking capability. All DUTs maintain their basic electrical performance after repeated surge current stress, confirming the feasibility of the intrinsic body diode for

practical applications. However, the reliability risk of progressive degradation leading to a functional failure caused by periodic thermal-mechanical stress should be paid attention to. To assess and compare the degradation mechanisms of SiC MOSFETs prior to final failure under repeated surge conditions, static and dynamic parameters of the DUTs are measured during surge cycles.

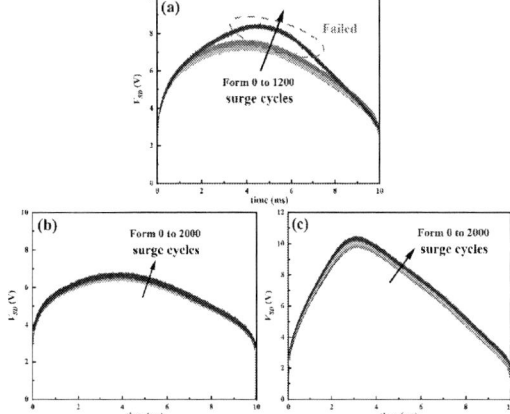

Fig. 3. Source-drain voltage waveform bias under repetitive surge stress shift . (a) P-MOSFET, (a) A-MOSFET, (a) D-MOSFET.

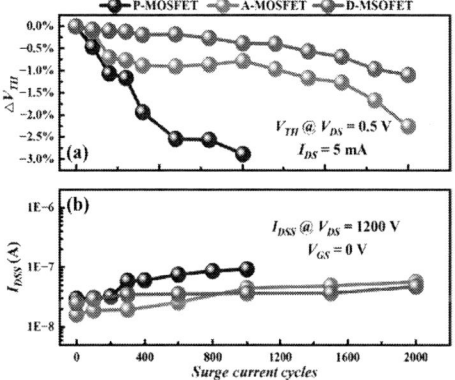

Fig. 4. Evolution of V_{TH} and I_{DSS} with the increase in surge current cycles under surge current stress.

Fig. 4 shows the variation of threshold voltage (V_{TH}) and reverse breakdown leakage current (I_{DSS}). It is found that the V_{TH} of the DUTs decreased continuously with repeated surge current impacts, while Idss increased continuously. Among them, the degradation of the P-MOSFET is the most significant, with Vth decreasing by 2.66% before final failure. The degradation trends of the A-MOSFET and D-MOSFET are relatively minor, and the devices do not fail even after 2000 surge cycles. Combining the subthreshold curves in Fig.5 and the internal electric field distribution analysis of SiC MOSFETs under 120A surge current in Fig. 6, the mechanisms behind the decrease in V_{TH} and the increase in Idss are analyzed.

To investigate the degradation of the channel region in SiC MOSFETs, it is necessary to monitor the reliability of the gate oxide layer. Fig. 5 shows the subthreshold characteristics of the MOSFET. As shown in the figure, when V_{DS} is fixed, the device operates in the subthreshold region where $V_{GS} < V_{TH}$. In the subthreshold region, the inverse of the slope of the I_{DS}

versus V_{DS} plot on the logarithmic axis is defined as the subthreshold swing, also known as the S-factor[6]:

$$S = \frac{dV_{GS}}{d \log I_{DS}} \quad (1)$$

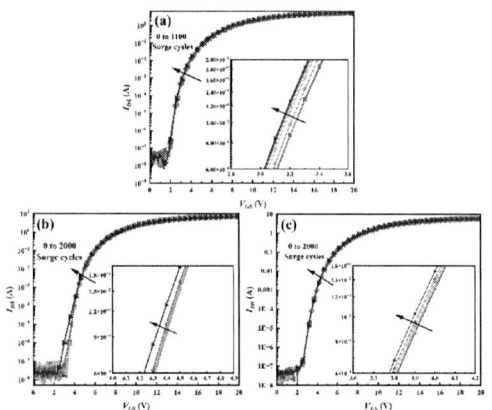

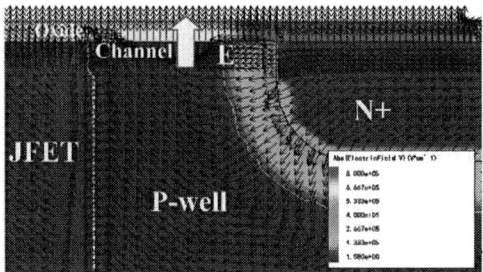

Fig. 5. Variation of the transfer characteri-stics (subthreshold region) with the cycles of surge current. (a) P-MOSFET, (b) A-MOSFET, and (c) D-MOSFET.

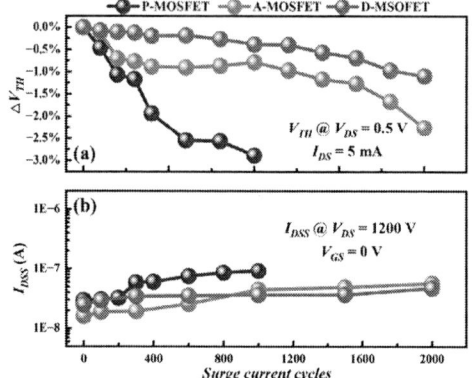

Fig. 6. Electric field distribution inside SiC MOSFET at V_{GS_OFF}= -5V.

S represents the change in gate-source voltage required for a change in leakage current by one order of magnitude, which is related to the gate interface state density D_{it}. Dit can be expressed as:

$$D_{it} = \frac{C_{OX}}{q^2} \left(\frac{qS}{\ln(10)kT} - 1 \right) - \frac{C_b}{q^2} \quad (2)$$

C_b is the parasitic capacitance of the device. Equation (2) shows that changes in the subthreshold swing can be used to determine changes in the interface state density of the device. It is observed that the transfer characteristic curves of the DUTs shifted overall to the left as the number of surge current impacts increased, indicating that the threshold voltage is continuously decreasing. Combining the changes in threshold voltage and subthreshold swing, it can be inferred that the gate oxide layer is positively charged, affecting the gate-oxide potential. When V_{GS} is -5V, the electric field direction is from SiC toward the gate oxide layer, and holes tunnel into the gate-oxide traps. In SiC MOSFETs, the gate oxide layer becomes positively charged due to the capture of holes by gate oxide traps and the excitation of positively charged ions in the oxide layer under repeated surge stresses[7]. Simultaneously, new conductive pathways are provided in the gate oxide layer, increasing Idss.

Fig. 7 shows the trend of on-resistance ($R_{DS(ON)}$) and source-drain on-voltage drop (V_{SD}). $R_{DS(ON)}$ degradation patterns for different SiC MOSFETs vary significantly. For P-

MOSFETs, gate oxide degradation is the primary degradation factor during the initial stage of repeated surges, leading to decreased channel resistance and reduced $R_{DS(ON)}$. Subsequently, degradation of the packaging or drift region begins to dominate, causing $R_{DS(ON)}$ to increase. Nevertheless,

Fig. 7. Evolution of $R_{DS(ON)}$ and V_{SD} with the increase in surge current cycles under surge current

during the growth process, the V_{TH}'s drift also influences the channel resistance. The degradation trends of A-MOSFET and D-MOSFET $R_{DS(ON)}$ are similar to those of P-MOSFET during the later stages of repeated surges, indicating that the drift region resistance or packaging dominates the changes in $R_{DS(ON)}$. However, the degradation magnitude of A-MOSFET $R_{DS(ON)}$ is greater than that of P-MOSFET and D-MOSFET. Combined with the V_{SD} change trend, bipolar degradation is suspected to have occurred, which was later confirmed through reverse recovery characteristics. The current path of the PiN diode lacks the channel portion compared to the MOSFET channel current path. In most cases, drift zone degradation and packaging degradation are the causes of the positive drift of V_{SD}[8, 9].

To better verify the occurrence of bipolar degradation in the device, it is necessary to monitor the changes in the reverse recovery characteristics of the DUTs during the surge cycle, as shown in Fig. 7. It can be observed that the reverse recovery current I_{RR} of the P-MOSFET and D-MOSFET remains essentially unchanged. However, the reverse recovery current I_{RR} of the A-MOSFET shows a decreasing trend. The reverse recovery charge Q_{RR} of the A-MOSFET body diode also continues to decrease. Combined with the similar growth trends in $R_{DS(ON)}$ of the A-MOSFET, it can be inferred that bipolar degradation has occurred in the drift region. When a large current flows through the body diode, the device operates in bipolar mode, causing holes and electrons to be injected into the N-drift, resulting in increased hole-electron recombination[2, 10]. The energy released during recombination causes stacking faults propagating at the basal plane dislocations. Bipolar degradation reduces the active region of the SiC device, leading to an increase in $R_{DS(ON)}$ and V_{SD} of the SiC MOSFET.

To validate the conclusions regarding packaging degradation, the DUTs are decapsulated to examine chip damage in V_{DS}=600V, I_{SD}=12A, as shown in Fig. 9. Except for the A-MOSFET, all other devices exhibit noticeable damage. The active region metal of the P-MOSFET shows slight melting, with no other apparent damage observed, indicating that the packaging-induced decrease in resistivity

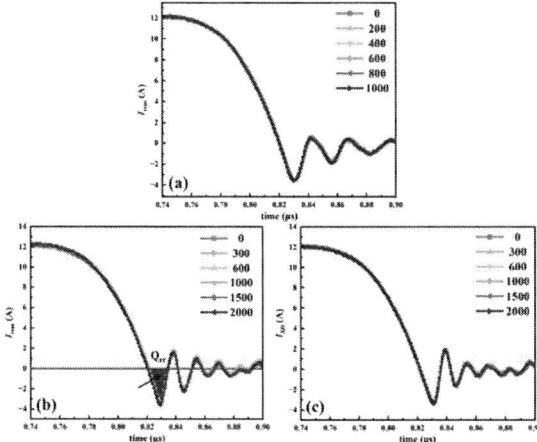

Fig. 8. Degradation of the reverse recovery chara-cteristic under surge current cycles. (a) P-MOSFET, (b) A-MOSFET, and (c) D-MOSFET.

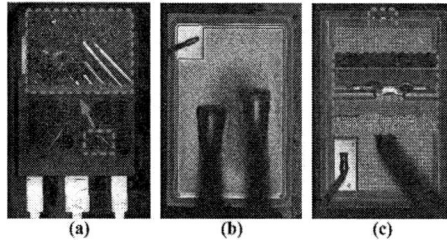

Fig. 9. Package degradation (a) P-MOSFET, (b) A-MOSFET, and (c) D-MOSFET.

was not significant. No signs of gate-source short circuits are observed at the packaging level, confirming the hypothesis that gate oxide breakdown is the cause of device failure. Burned areas are observed in the terminal regions adjacent to the gate bus at the top of the D-MOSFET chip. When operating in the third quadrant, the D-MOSFET exhibits a new current path in addition to the body diode and trench channel, namely conduction in the terminal region, where current concentration effects may occur near the terminal region.

IV. CONCLUSIONS

Significant drifts in static parameters such as threshold voltage (V_{TH}), on-resistance ($R_{DS(ON)}$), body diode forward voltage drop (V_{SD}), and gate-source leakage current (I_{DSS}) are observed under repetitive surge stress. The microanalysis shows that the thermo-mechanical failure of the package layer and the accumulation of lattice defects in the drift region are the dominant factors leading to the elevation of $R_{DS(ON)}$ and V_{SD}. In contrast, the increase in the interfacial density of states in the gate oxide layer is mainly attributed to the negative drift of V_{TH} and the anomalous increase of I_{DSS}. The degradation of each parameter exhibits significant coupling characteristics: the channel carrier transport process establishes the correlation between the $R_{DS(ON)}$-V_{SD} degradation and V_{TH} drift, and the gate oxide trap charge accumulation triggers the synergistic degradation mechanism of I_{DSS} and V_{TH}. Therefore, enhancing the gate oxide dielectric reliability and main-taining the threshold voltage stability are the core optimization paths to enhance the surge tolerance of MOSFET devices.

It is noteworthy that, thanks to the cont-inuous optimization of the fabrication process, the bipolar degradation phenomenon of SiC devices has been effectively suppressed under DC bias stress conditions. However, it is found in this study that asymmetric structure SiC MOSFETs still exhibit typical bipolar degrad-ation characteristics in repeated surge stress experiments. Further analysis shows that encap-sulation technology is still the key technical bottleneck to limiting the reliability improvement of SiC MOSFETs: the delamination of the encapsulation interface and bonding line fatigue lead to significant parameter drifts

in $R_{DS(ON)}$ and V_{SD}. The intrinsic operating temperature of the SiC chip can reach more than 600 °C. However, it is limited by the current encapsulation mate-rial's thermal conductivity (typical value<200 W/mK) and the coefficient of thermal expansion mismatch (CTE). The coefficient of thermal expa-nsion mismatch problem(CTE difference>5ppm/K) of curr-ent packaging materials, as well as commercial devices, is generally limited to a rated junction temperature of 175°C. In addition, insufficient optimization of Gate Bars and metallurgical reactions at bonding wire inter-faces exacerbate package reliability degradation under high-frequency surge stress.

REFERENCES

[1] X. Jiang *et al.*, "Comparison Study of Surge Current Capability of Body Diode of SiC MOSFET and SiC Schottky Diode," in *Proc. IEEE Energy Convers. Congr. Expo.* , 2018, pp. 845-849, doi: 10.1109/ECCE.2018.8558388.

[2] S. Palanisamy, T. Basler, J. Lutz, C. Künzel, L. Wehrhahn-Kilian, and R. Elpelt, "Investigation of the bipolar degradation of SiC MOSFET body diodes and the influence of current density," in *2021 IEEE Int. Rel. Phys. Symp. (IRPS)*, 21-25 March 2021, pp. 1-6, doi: 10.1109/IRPS46558.2021.9405183.

[3] J. Wei *et al.*, "Review on the Reliability Mechanisms of SiC Power MOSFETs: A Comparison Between Planar-Gate and Trench-Gate Structures," *IEEE Trans. Power Electron.*, vol. 38, no. 7, pp. 8990-9005, 2023, doi: 10.1109/TPEL.2023.3265864.

[4] V. Pala *et al.*, "Physics of bipolar, unipolar and intermediate conduction modes in Silicon Carbide MOSFET body diodes," in *Proc. 28th Int. Symp. Power Semiconductor Devices ICs (ISPSD)*, June 2016, pp. 227-230, doi: 10.1109/ISPSD.2016.7520819.

[5] K. Han and B. J. Baliga, "Comprehensive Physics of Third Quadrant Characteristics for Accumulation- and Inversion-Channel 1.2-kV 4H-SiC MOSFETs," *IEEE Trans. Electron Devices*, vol. 66, no. 9, pp. 3916-3921, 2019, doi: 10.1109/TED.2019.2929733.

[6] H. Xu *et al.*, "Methodology for Enhanced Surge Robustness of 1.2-kV SiC MOSFET Body Diode," *IEEE J. Emerg. Sel. Topics Power Electron.*, vol. 10, no. 5, pp. 5039-5047, 2022, doi: 10.1109/JESTPE.2021.3106742.

[7] X. Jiang *et al.*, "Investigation on Degradation of SiC MOSFET Under Surge Current Stress of Body Diode," *IEEE J. Emerg. Sel. Topics Power Electron.*, vol. 8, no. 1, pp. 77-89, 2020, doi: 10.1109/JESTPE.2019.2952214.

[8] H. Li, W. J, R. N, X. H, and S. K, "Investigation of 1200 V SiC MOSFETs' Surge Reliability," *Micromachines*, vol. 10, no. 7, p. 485, Jul. 2019, doi: 10.3390/mi10070485.

[9] D. Ma *et al.*, "Degradation Evaluation and Defects Analysis for 1.2-kV Planar-Gate SiC MOSFETs Under Repetitive Surge Current Stress," *IEEE Trans. Electron Devices*, vol. 70, no. 12, pp. 6473-6479, 2023, doi: 10.1109/TED.2023.3323912.

[10] Z. Zhu *et al.*, "Degradation of 4H-SiC MOSFET body diode under repetitive surge current stress," in *Proc. IEEE 32th Int. Symp. Power Semiconductor Devices IC's (ISPSD)*, 13-18 Sept. 2020, pp. 182-185, doi: 10.1109/ISPSD46842.2020.9170166.

AUTHOR INDEX

Afanasenko, Valentyna 147
Al-Haddad, Kamal 246, 393, 644
Alhosaini, Waleed 853
An, Wenbo 609
An, Ziyang 949
Bai, Jinyang 398
Bai, Song 462
Bai, Zhitong 849
Bao, Dingyuan 26
Bao, Guojun 430
Barón, Kevin Muñoz 147
Ben, Hongqi 133
Cai, Xuchong 885, 968
Cailin, Wang 1006
Cao, Hanlin 11
Cao, Jiaying 21, 164
Cao, Junhou 462, 659
Cao, Liqiang 52
Cao, Pei 692
Cao, Pingyu 11, 232
Cao, Ruibo 52
Cao, Wenping... 71, 355
Cao, Xiaobo 857
Cao, Ying 410
Cao, Yufeng 902
Chan, Ian Yj 582
Chang, Given Shucheng 813
Chang, Liuchen... 328, 468
Chang, Yifei 21, 164
Chang, Yongqi 717
Chao, Zhang 1006
Chen, Bingxin 450
Chen, Cai 118, 177, 280, 520
Chen, Cen 577, 630
Chen, Cong 944
Chen, Dawei 430
Chen, Fuao 963
Chen, Gang 867
Chen, Haobin 221
Chen, Jian 32, 524
Chen, Jiawei 990
Chen, Jia-Xiang 562
Chen, Jiaxiang 704
Chen, Jiwen 190
Chen, Junyang 762
Chen, Kelin 480, 839
Chen, Min 47, 62, 211, 480, 839
Chen, Pengyu 544

Chen, Tiwei 713
Chen, Wei 920, 1033
Chen, Xiao 109, 201
Chen, Xingye 122
Chen, Yang 963
Chen, Yenan 104
Chen, Yidan 867
Chen, Yifeng 151
Chen, Yong 314, 324, 416, 474, 753, 817, 916, 944
Chen, Yongang 491
Chen, Yongxin 355
Chen, Yu 867
Chen, Yue 933
Chen, Zihao 404
Chen, Zilong 343, 485
Chen, ... 688
Cheng, Haifeng 138
Cheng, Haoyuan 67
Cheng, Huaihao 753
Cheng, Ji 221
Cheng, Jinpeng 339
Cheng, Xinhong 550, 954
Cheng, Xu 275, 324, 416, 474, 753, 817, 916, 944
Cheng, Yujie 138
Cheng, Zhijie 920, 1033
Cheng, Zizhen 747
Chi, Jialiang 304
Chi, Qingguo 696
Chiu, Yi 762
Chu, Hao 118
Cui, Miao 11, 232, 911
Cui, Qian 343, 485
Cui, Wentao 206
Cui, Xinchun 572
Cui, Yingxin 731
Dai, Haohao 295
Dai, Jianxun 169
Dai, Siqi 177
Dai, Xinyue 898
Dai, Yuxing 246, 328, 393, 468, 644
Deng, Chaofan 726
Deng, Gaoqiang 644
Deng, Xiaochuan 295, 737
Diao, Lijun 491
Ding, Desheng 462
Ding, Lijian 445, 667, 788, 853
Ding, Xiang-Jin 562
Ding, Xiangjin 704

Dong, Jintong	128
Dong, Ruixiao	328, 468
Dong, Xiaojun	924
Dong, Xiaonan	186
Dou, Chong	343, 485
Dou, Wenzheng	16
Dou, Yu	949
Du, Yifei	47, 211
Duan, Yuhan	21, 164
Fahlbusch, Sebastian	638
Fan, Meng-Qi	562
Fan, Mengqi	898
Fan, Minfan	582
Fan, Shanzhen	700
Fan, Wendi	104
Fang, Liu	445
Fang, Mingzhu	788
Fang, Yin	260
Feng, Chao	873, 990
Feng, Hao	339
Feng, Siyuan	280
Feng, Yu	445
Fu, Hao	462, 659
Fu, Minfan	544
Fu, Xichen	566
Fu, Yehan	833
Fu, Yu	308, 334
Gao, Mingyang	83
Gao, Peng	767, 822
Gao, Shanshan	133, 376
Gao, Tianyu	717
Ge, Xinglai	741
Geng, Guifeng	678, 798
Geng, Yuqing	849
Gong, Jiakun	933
Gong, Jie	924
Gong, Kaixiang	314, 416
Gong, Shoulai	731
Gong, Xiaowu	366, 516, 709, 726
Gu, Caixin	867
Gu, Yitian	873
Gu, Yu	264
Gu, Zihang	382, 902
Guan, Guolian	495
Guan, Hao	21, 164
Guan, Jiajia	118, 280, 520
Guan, Lei	877
Guan, Renfeng	974
Guan, Yueshi	528
Guang, Yang	160
Guangzhou, Jiahang Wang	366
Guangzhou, Xi Jiang	366

Gui, Qingzhong	424
Guo, Fei	920, 1033
Guo, Gaofu	713
Guo, Suxia	71
Guo, Yanao	731
Hai, Dong	26
Han, Jisheng	731
Han, Shouhui	684
Han, Yiting	767, 822
Han, Yu	304
Han, Zhiyun	398, 450, 684
Hao, Xiamin	757
Hao, Yue	655
He, Daozhen	673
He, Feng	757
He, Hailong	290
He, Jinliang	867
He, Quanbo	371
He, Shuiyuan	491
He, Wenzhi	450
He, Xiaomin	91, 440
He, Yanjie	485
He, Yanjing	516, 709
He, Yingfeng	813
He, Yuanheng	673
He, Yuying	606, 609, 833
He, Zhixing	974
Ho, Carl Ngai Man	260
Hou, Chunyao	181
Hou, Fengze	52
Hu, Cungang	71, 355
Hu, Haipeng	788
Hu, Haolin	572, 704, 873
Hu, Jichao	91, 890
Hu, Jingyang	939
Hu, Pei	87
Hu, Qiang	32
Hu, Qin	867
Hu, Qingmao	424
Hu, Qingrong	366
Hu, Tingwen	524
Hu, Xiaofei	500
Hu, Yaoyu	614
Hu, Yifan	290
Hu, Yirui	269
Hu, Zhuangzhuang	138
Hu, Zijian	929
Hua, Mengyuan	704
Huang, Baoying	349, 588
Huang, Hao	26, 939
Huang, Huolin	169
Huang, Jun	211

Huang, Kai 83
Huang, Lei 659
Huang, Qian 295
Huang, Yongle 1017, 1022
Huang, Zhi-Ying 762
Hui, Xiaoshuang 142, 156
Huo, Yiting 588
Iannuzzo, Francesco 26
Jayamaha, Shan..260
Ji, Kai 128
Ji, Mingming..109
Ji, Runyang 792
Ji, Yanfei 190
Jia, Liwen 480, 839
Jia, Pengyu 190, 195, 777, 783
Jia, Ziqi 290
Jiang, Dong 420
Jiang, Dongjun...933
Jiang, Feng 47, 211, 480, 839
Jiang, Haonan 655
Jiang, Junsong 71
Jiang, Runquan 678, 798
Jiang, Shaoyan 566
Jiang, Xi 516, 655, 709, 726
Jiang, Yu 290
Jiang, Yuteng 450
Jiang, Zepeng 741
Jiang, Zuoheng 990
Jiao, Teng........................... 562, 572, 704
Jiao, Zhiwei 382, 902
Jiarui, Guo 1006
Jie, Huamin 717
Jin, Dongxin..924
Jin, Rui 77, 757, 813
Jin, Yuting 26, 939
Jingfei, Wang 160
Jun, Yuan..1033
Kallfass, Ingmar.......................................147
Kang, Qian..849
Kang, Yong 118, 177, 280, 520
Kang, Yuhui 156
Kefan, Yu 160
Kong, Hang 100
Kong, Jie 827
Kong, Liudan 21, 164
Kong, Weixuan 577
Lai, Jia-Jun..62
Lai, Jiankun 516
Lai, Wei...236
Lam, Sang 11, 911
Lan, Jianyu..480
Lan, Xin 16, 151

Lei, Guangyin 924
Lei, Yun 169
Lei, Zhengzi 609
Li, Binbin 456
Li, Bingru 57
Li, Bodong 480, 839
Li, Chao 77
Li, Chi 863
Li, Chuangye 667
Li, Chushan 949
Li, Dawei 299
Li, Fang 600
Li, Haiyang 898
Li, Hao 410
Li, Helong 122, 445, 667, 788
Li, Hong 500, 594, 624, 673, 772, 1011
Li, Honghong 215, 226, 808, 1000
Li, Jianbiao....................753, 817, 916, 944
Li, Jie 47, 211
Li, Longnv....................................57, 319, 959
Li, Mingfu 62
Li, Ningbo 382
Li, Qingmin 688, 700
Li, Shouxiang 308, 334
Li, Shuanglong 349
Li, Tianxi 280
Li, Wuhua 26, 939
Li, Xianfeng 688
Li, Xiangdong 655, 726
Li, Xiao 269, 619
Li, Xin 151, 436, 506, 512, 808, 1022
Li, Xing 38
Li, Xu 295, 737
Li, Xuan 295, 737
Li, Xuebao 77, 410
Li, Xuefei 491
Li, Yanjun 624
Li, Yanzuo 709, 726
Li, Yaqi 849
Li, Yi 817
Li, Yitong 614
Li, Zeyu 692
Li, Zhen 398
Li, Zheyang 813
Li, Zhihui 700
Li, Zhucheng 713
Li, Zicong 506
Li, Zikang 980
Li, Zixiao 491
Li, Zongjian 974
Liang, Haiyan 844
Liang, Lin 1022

Liang, Mei	190
Liang, Senhao	933
Liang, Shiwei	393, 644
Liang, Yidi	594, 772
Liang, Zhiqing	974
Liao, Aojie	624
Liao, Hui	186
Liao, Kaiju	867
Liao, Yiyang	614
Lin, Junru	985
Lin, Xinpeng	704
Lin, Xiyuan	544
Lin, Xuanyu	630
Lin, Yen-Liang	762
Liping, Jia	1006
Liu, Baihan	177
Liu, Bin	349, 588
Liu, Boyang	304
Liu, Chang	71, 731
Liu, Chao	388
Liu, Chaohui	221
Liu, Chuang	388
Liu, Chuyuan	319
Liu, Decai	813
Liu, Guoyou	524
Liu, Hangzhi	907
Liu, Hongchen	959
Liu, Hui	264, 420
Liu, Jiahong	827
Liu, Jianwei	863
Liu, Jing	857
Liu, Jingrui	994
Liu, Ming	556
Liu, Mingjun	195, 777, 783
Liu, Pan	21, 164
Liu, Qifan	709
Liu, Qingchang	582
Liu, Qingsong	398, 684
Liu, Shuyu	339
Liu, Sinuo	264
Liu, Siyang	462, 659
Liu, Wei	95, 556
Liu, Yan	562, 572, 704
Liu, Yang	974
Liu, Yi	802
Liu, Yipeng	177
Liu, Yitao	844
Liu, Yu	236
Liu, Yuan	1028
Liu, Yushan	269, 619
Liu, Zhaocheng	410
Liu, Zheng	684

Liu, Zhicheng	802, 833
Liu, Zhiqiang	667
Liu, Zhongyue	692
Liu, Zixuan	361, 649
Lou, An	181
Lu, Bohang	630
Lu, Jie	71
Lu, Zijian	844
Luan, Aozu	181
Luo, Anjing	713
Luo, Cheng	924
Luo, Haoze	26, 939
Luo, Jian	83
Luo, Shengjie	939
Luo, Xixi	867
Luo, Yifei	506, 512, 1017, 1022
Lv, Jiahui	1
Lv, Jianwei	177
Ma, Hao	349, 588
Ma, Hongbo	114
Ma, Jun	704
Ma, Mingcheng	534, 885, 968
Mai, Ruikun	963
Man, Lichang	133
Mao, Danfeng	873, 898, 990
Mao, Weina	215, 226, 1000
Mao, Yanfang	792
Mei, Yun-Hui	57, 319
Mei, Yunhui	802, 959
Mi, Tianhe	572
Ming, Jie	949
Ming, Lei	382, 902
Mo, Guorui	709
Mo, Zhili	550
Ngo, Bac-Bien	994
Ning, Jianping	241, 250
Ning, Puqi	142, 156
Ning, Wenjie	614
Niu, Chunping	290
Niu, Jiaxuan	474
Niu, Yukun	275, 314, 416
Nuerdebieke, Yeerzhati	299
Ou, Zhenyuan	974
Ou, Zhujian	792
Ouyang, Runze	366
Pan, Siyu	349, 588
Pan, Xiang	215
Pan, Yanchen	885, 968
Pan, Yunbin	62
Pang, Zhuo	762, 767
Pei, Xingyu	753, 817, 916, 944
Pei, Xuejun	678, 798

Peng, Bo	91
Peng, Wensong	737
Pu, Hongbin	890
Pu, Xuhui	890
Qi, Bin	87
Qi, Jingjing	284
Qi, Xianbin	122, 445, 788
Qi, Yuwen	491
Qi, Zhiyuan	404
Qin, Peng	445, 788
Qiu, Kai	195
Qiu, Maohang	114
Qu, Vickie	582
Ran, Li	339
Ren, Hanwen	688, 692, 700
Ren, Na	929
Ruan, Jiabin	566
Ruan, Xinbo	255
Sang, Xikun	376
Sang, Zihan	410
Shangguan, Miaomiao	236
Shangguan, Xu	594, 772
Shao, Shuai	181, 206
Shao, Tiancong	87, 849
Shao, Zhe	456
Shen, Lingyan	954
Sheng, Haoyi	974
Sheng, Kuang	6, 67, 173, 881, 929
Sheng, Tsung-Huan	762
Shi, Mingxin	624, 673
Shi, Qianru	582
Shi, Zenan	506, 512
Shi, Ziliang	62
Shu, Peng	410
Shu, Zhou	717
Sihao, Shen	1006
Solomakha, Oleksandr	147
Song, Jian	52
Song, Lei	867
Song, Qingwen	404, 516
Song, Ruya	853
Song, Wensheng	32, 524
Song, Xuanting	246, 393, 644
Song, Xuhui	275, 314, 416
Song, Yanlin	1011
Su, Yan	122
Sun, Hao	62
Sun, Jiahua	539
Sun, Peng	594, 772, 817
Sun, Weifeng	462, 659
Sun, Xinnan	47, 211, 839
Sun, Xizhi	534

Sun, Xuchen	118, 280
Sun, Yuhan	87
Sun, Zhen	241, 250
Tan, Jingyang	980
Tan, Kun	71, 355
Tang, Jiuyang	21, 164
Tang, Tao	524
Tang, Xi	71, 355
Tang, Xiaoyan	516
Tang, Yi	994
Tang, Yihui	122
Tao, Ze	512
Taylor, Stephen	911
Thomas, Rony	638
Tian, Bowen	817
Tian, Jiamin	867
Tian, Ye	430, 949
Tian, Yufei	550
Udrea, Florin	371
Wan, Fayu	663
Wan, Yu-Xi	562
Wan, Yuxi	572, 704, 873, 898, 990
Wang, Cailin	877
Wang, Ce	6
Wang, Changdong	717
Wang, Chaojun	206
Wang, Chen	361, 649
Wang, Chenghai	696
Wang, Chenlu	659
Wang, Chenyi	577
Wang, Denggui	138
Wang, Dewen	495, 747
Wang, Feng	700
Wang, Hao	236
Wang, Haodong	577
Wang, Haoyun	963
Wang, Hengyu	6, 67, 173, 371, 881
Wang, Huai	827
Wang, Jiahui	839
Wang, Jian	688, 700
Wang, Jianing	38, 215, 226, 436, 808, 1000
Wang, Jun	246, 328, 393, 468, 634, 644, 974
Wang, Junbo	655
Wang, Kai	87
Wang, Kuan	920, 1033
Wang, Laili	100, 495, 747
Wang, Lei	77
Wang, Lu	319
Wang, Lurenhang	534, 885
Wang, Meiyu	762, 767, 822
Wang, Nianzheng	100
Wang, Peng	572

Wang, Pengfei 582
Wang, Qian 83, 634
Wang, Qidong 52
Wang, Runze 38, 226, 436, 808
Wang, Ruoyin 721
Wang, Shuyu 959
Wang, Tiefu 67
Wang, Wei 688, 700
Wang, Xi 890
Wang, Xiahao 38, 215, 226, 436
Wang, Xiao-Ping 562
Wang, Xiaoping 898
Wang, Xulong 361, 649
Wang, Yan 898
Wang, Yao 241, 250
Wang, Yifan 881
Wang, Yijie 128, 133, 376, 528, 980
Wang, Ying 366, 709
Wang, Youzheng 57, 319, 959
Wang, Yuanfeng 539
Wang, Yuchen 924
Wang, Yudong 343
Wang, Yulin 328, 468
Wang, Yuwei 634
Wang, Zhao 911
Wang, Zhi 450
Wang, Zhiqiang 304, 920, 1033
Wang, Zhiyuan 456
Wang, Zhongjie 109, 186, 201
Wang, Zicheng 630
Wang, Ziyang 32
Wang, Zuoxing 624
Wangliang, ... 43
Wei, Jiaxing 462, 659
Wei, Jinxiao 445, 853
Wei, Mingbo 594
Wei, Suhang 177
Wei, Yuqi 343, 485
Wei, Zhaoxiang 462, 659
Wei, Zheng 614
Wen, Yi 737
Wu, Enyou 280
Wu, Hongfei 138
Wu, Hongyuan 753, 817, 916, 944
Wu, Junye 873, 990
Wu, Keping 990
Wu, Min 314
Wu, Shunqing 388
Wu, Tuanzhuang 462, 659
Wu, Yangyang 920, 1033
Wu, Yanlin 873, 990
Wu, Yi 290

Wu, Yifan 863
Wu, Yucheng 47, 211
Wu, Yue 181
Wu, Yunjie 236
Wu, Yuzhen 255
Xi, Wenyu 877
Xia, Runze 6
Xia, Zhenchun 38, 215, 226, 436, 1000
Xiao, Chuanwei 566
Xiao, Xiangan 118
Xiao, Ziheng 994
Xiaoguang, Wei 43, 160
Xie, Dong 741
Xie, Lihong 255
Xie, Minglei 450
Xin, Guoqing 920, 1033
Xin, Sixiao 450
Xin, Zhen 382, 902
Xing, Wenbin 95
Xing, Yimei 195, 777, 783
Xinling, Tang 43, 160
Xiong, Zhuofan 619
Xu, An 709
Xu, Aoxue 83
Xu, Binbo 968
Xu, Dianguo 133, 376, 528, 534, 885, 968, 980
Xu, Hongyi 929
Xu, Jun 600
Xu, Ke 57
Xu, Lei 355
Xu, Shengxiu 959
Xu, Shikang 737
Xu, Tao 382, 902
Xu, Wenjie 747
Xu, Xiaohui 128
Xu, Xiaoyi 792
Xu, Xingque 566
Xu, Xunjin 62
Xu, Yihao 232
Xu, Zhiliang 741
Xue, Fei 232
Xue, Lingxiao 109, 186, 201
Xue, Tangman 696
Xue, Yao 600
Xue, Zedong 404
Yan, Baolong 38, 436, 808, 1000
Yan, Haidong 221, 767, 822
Yan, Na 52
Yan, Yishun 534, 885, 968
Yan, Yuxin 334
Yan, Zhangzhe 138
Yan, Zhaoheng 709, 726

Yang, Dongsheng	606
Yang, Fengming	1022
Yang, Fengtao	747
Yang, Jiajun	142, 156
Yang, Jingli	717
Yang, Lei	539
Yang, Mao-Jin	562
Yang, Maojin	704
Yang, Wuhua	1006
Yang, Xiaodong	91
Yang, Xiaofeng	849
Yang, Xiaolei	462
Yang, Xin	424, 894
Yang, Xiong	62
Yang, Xu	474
Yang, Yahong	480
Yang, Ying	361, 649
Yang, Yingkun	767
Yang, Yuan	1, 95, 890
Yang, Yun	250
Yang, Zeqi	388, 1028
Yang, Zhichang	1011
Yang, Zhihao	91
Yang, Zhiqing	122, 445, 788
Yany, Yateng	700
Yao, Bo	827
Yao, Jianguang	792
Yao, Rui Ray	911
Yao, Sankun	994
Yao, Siyi	544
Yao, Zhigang	994
Yin, Shuangyan	717
Yin, Xunran	717
Yin, Zhonggang	857
You, Shuzhen	655
You, Xiangan	52
Yu, Jinfeng	264
Yu, Jinhui	91
Yu, Kanghua	634
Yu, Shaolin	38, 215, 226, 436, 808, 1000
Yu, Zhe	638
Yu, Zheyuan	314
Yuan, Hao	404, 516
Yuan, Jun	920
Yuan, Song	366, 516, 709, 726
Yue, Hao	32, 524
Yujie, Du	43
Zalinge, Harm Van	232
Zeng, Chunhong	713
Zeng, Wei	873
Zeng, Zheng	933
Zeng, Zhijie	430

Zeng, Zhongming	713
Zha, Xian-Hu	562
Zha, Xianhu	704
Zhai, Fan	201
Zhan, Xinbin	516, 709
Zhang, Baoshun	713
Zhang, Bo	737
Zhang, Changhai	696
Zhang, Chao	468
Zhang, Chi	67
Zhang, Chuanqi	393
Zhang, Congcong	544
Zhang, Dao-Hua	562, 704
Zhang, Daohua	572
Zhang, Dezheng	833
Zhang, Donglei	1000
Zhang, Fan	275, 314, 416, 474
Zhang, Haining	284
Zhang, Haitao	440
Zhang, Haochen	528
Zhang, Haoyu	304
Zhang, Huanyu	713
Zhang, Jiarui	480, 839
Zhang, Jiayu	169, 741
Zhang, Jincheng	655
Zhang, Junming	206, 985
Zhang, Junzhao	420
Zhang, Kai	284
Zhang, Kuang	500
Zhang, Li	606, 609, 713, 833
Zhang, Maosheng	221
Zhang, Minmin	1
Zhang, Peng	77
Zhang, Ping	232
Zhang, Qian	857
Zhang, Qingchun	21, 164
Zhang, Rong	920, 1033
Zhang, Ruihao	663
Zhang, Shaowei	857
Zhang, Taohui	26
Zhang, Tiandong	696
Zhang, Tongyu	495
Zhang, Xiangqian	186
Zhang, Xiaodong	713
Zhang, Xiaojun	963
Zhang, Xiaolu	275, 314, 416
Zhang, Xiaotian	324
Zhang, Xingye	704
Zhang, Xinlc	696
Zhang, Xinyu	308
Zhang, Xuelun	62
Zhang, Yajing	349, 588, 600

Zhang, Yakun 221
Zhang, Yaodong 324
Zhang, Yi 827
Zhang, Yichi 827
Zhang, Yifan 177
Zhang, Yihang 190
Zhang, Yiheng 777
Zhang, Yujie 1017
Zhang, Yukun 343, 485
Zhang, Yulei 890
Zhang, Yuming 404, 516
Zhang, Zewei 762
Zhang, Zhaofu 71
Zhang, Zhe 87, 388, 1028
Zhang, Zhihao 109, 201
Zhang, Ziyang 151
Zhao, Bin 817, 916
Zhao, Dengrui 713
Zhao, Dingkun 894
Zhao, Fangwei 87, 600
Zhao, Haiqiang 762
Zhao, Hang 236
Zhao, Hanqing 737
Zhao, Jianyu 169
Zhao, Kepeng 232
Zhao, Kexin 474
Zhao, Mingzhi 495
Zhao, Ning 16
Zhao, Shuang 241, 853
Zhao, Yao 304
Zhao, Yixiang 500
Zhao, Yuanzhi 366
Zhao, Yucheng 308, 334
Zhao, Yunxuan 709, 726
Zhao, Zhenyu 717
Zhao, Zhibin 817
Zhao, Zhiqiang 495, 747
Zheng, Hong 721
Zheng, Hongjun 544
Zheng, Li 550, 954
Zheng, Trillion 849
Zheng, Yuze 275, 314, 416
Zheng, Zedong 863
Zheng, Zexiang 177
Zheng, Zheyi 963
Zheng, Zhongshu 609
Zheng, Zijie 246, 644
Zhi, Shuaiqing 534
Zhong, Linhai 709
Zhongkang, Lin..160
Zhou, Chengyuan 173
Zhou, David 873, 898, 990

Zhou, Han 47, 211
Zhou, Jianjun 138
Zhou, Liang 83
Zhou, Peng 678, 798
Zhou, Qiang 619
Zhou, Shaoze 614
Zhou, Xiang 299
Zhou, Xinyu 659
Zhou, Xuelei 881
Zhou, Xuetong 550, 954
Zhou, Yuming.. 907
Zhou, Zekun 737
Zhou, Zhenning 577
Zhu, Gaojia 57, 319, 959
Zhu, Hexin 295
Zhu, Lixuan 430
Zhu, Shuangxi 520
Zhu, Tao 813
Zhu, Xiaoyong 721
Zhu, Ye 62
Zhu, Zhanshan 667
Zhu, Zhengyun 83
Zhuang, Huizhu 594, 772
Zhuang, Qiongyang 867
Zong, Yujian 206
Zou, Liang 398, 450, 684
Zou, Liwen 186
Zou, Mingrui 933
Zou, Wen 606
Zou, Yang 173
Zou, Yongzhou 246, 644
Zou, Zhili 713
Zou, Zhixiang 430
Zuo, Qingyuan 169

IEEE
445 Hoes Lane
Piscataway, NJ 08854-4141

ISBN 979-8-3315-1110-4

2025 IEEE Workshop on Wide Bandgap Power Devices and Applications in Asia (WiPDA Asia 2025)

Beijing, China
15-17 August 2025

Pages 520-1036

IEEE Catalog Number: CFP25O09-POD
ISBN: 979-8-3315-1110-4